ANIMAL

ANIMAL

EDITOR-IN-CHIEF

David Burnie

LONDON, NEW YORK, SYDNEY, DELHI,
PARIS, MUNICH, & JOHANNESBURG

DORLING KINDERSLEY, LONDON

SENIOR ART EDITORS Ina Stradins, Vanessa Hamilton
ART EDITORS Kirsten Cashman, Philip Ormerod, Sara Freeman
DESIGNER Amir Reuveni
DTP DESIGNERS Louise Waller, Rajen Shah
PICTURE RESEARCHERS Cheryl Dubyk-Yates, Kate Duncan, Sean Hunter
ILLUSTRATORS Richard Tibbets, Evi Antoniou, Paul Banville

ART DIRECTOR Bryn Walls

SENIOR EDITORS Angeles Gavira, Peter Frances
PROJECT EDITORS David Summers, Sean O'Connor
EDITORS Alison Copland, Lesley Riley, Polly Boyd
ZOOLOGICAL EDITORS Kim Dennis-Bryan, Stephen Parker
INDEXER Jane Parker
PRODUCTION CONTROLLERS Elizabeth Cherry, Michelle Thomas

CATEGORY PUBLISHER Jonathan Metcalf

DORLING KINDERSLEY, DELHI

MANAGING ART EDITOR Shuka Jain
PROJECT DESIGNER Shefali Upadhyay
DESIGNER Pallavi Narain
DESIGN ASSISTANCE Elizabeth Thomas, Suresh Kumar
DTP DESIGNER Pankaj Sharma

MANAGING EDITORS Ira Pande, Prita Maitra
PROJECT EDITORS Ranjana Saklani, Atanu Raychaudhuri
EDITORS Rimli Borooah, Kajori Aikat
DTP COORDINATORS Jacob Joshua, Rajesh Bisht

studiocactus ©
13 SOUTHGATE STREET WINCHESTER HAMPSHIRE SO23 9DZ

SENIOR DESIGNER Sharon Moore
PROJECT EDITOR Lizzie Mallard-Shaw

SCHERMULY DESIGN COMPANY
THE CHURCH HALL, YORK RISE, DARTMOUTH PARK, LONDON NW5
SENIOR DESIGNER Hugh Schermuly
DESIGNERS Phil Gamble, Nick Buzzard, Sally Geeve, Paul Stork
SENIOR EDITOR Cathy Meeus
EDITORS Joanna Chisholm, Jackie Jackson, Mary Pickles

First published in Great Britain in 2001 by
Dorling Kindersley Limited, 80 Strand, London WC2R 0RL

Copyright © 2001 Dorling Kindersley Limited, London

2 4 6 8 10 9 7 5 3

A CIP catalogue record for this book is available from the British Library

ISBN 0-7513-3427-8

Colour reproduction by Colourscan, Singapore
Printed and bound in Italy by Mondadori

see our complete catalogue at
www.dk.com

Editor-in-Chief DAVID BURNIE

Main Consultants

MAMMALS

Dr Juliet Clutton-Brock

Managing Editor of the Zoological Society of London's *Journal of Zoology* and Research Associate of the Natural History Museum, London

BIRDS

Dr François Vuilleumier

Curator at the Department of Ornithology, American Museum of Natural History

REPTILES

Chris Mattison

Zoologist and writer specializing in herpetology. Fellow of the Royal Photographic Society

AMPHIBIANS

Professor Tim Halliday

International Director of the Declining Amphibian Population Task Force, Open University (UK)

FISH

Professor Richard Rosenblatt

Professor of Marine Biology and Curator of Marine Vertebrates, Scripps Institution of Oceanography, University of California

INVERTEBRATES

Dr George C. McGavin

Assistant Curator of the Hope Entomological Collections, Oxford University Museum of Natural History. Lecturer in Biological and Human Sciences, Jesus College, Oxford

Dr Richard Barnes

Lecturer in the Department of Zoology, University of Cambridge

Dr Frances Dipper

Marine biologist and writer

Contributors and Consultants

Dr Richard Barnes
Dr Paul Bates
Dr Simon K. Bearder
Deborah Behler
John Behler
Keith Betton
Dr Michael de L. Brooke
Dr Charles R. Brown
Dr Donald Bruning
George H. Burgess
Dr Kent E. Carpenter
Norma G. Chapman
Ben Clemens

Dr Richard Cloutier
Malcolm C. Coulter
Dominic M. Couzens
Dr Timothy M. Crowe
Dr Kim Dennis-Bryan
Dr Christopher Dickman
Joseph A. DiCostanzo
Dr Philip Donoghue
Dr Nigel Dunstone
Dr S. Keith Eltringham
Prof. Brock Fenton
Joseph Forshaw
Susan D. Gardieff

Dr Anthony Gill
Dr Joshua Ginsberg
Dr Colin Groves
Dr Jurgen H. Haffer
Prof. Tim Halliday
Gavin Hanke
Dr Cindy Hull
Dr Barry J. Hutchins
Dr Paul A. Johnsgard
Dr Angela Kepler
Dr Jiro Kikkawa
Dr Nigel Leader-Williams
Dr Douglas Long

Dr Manuel Marin
Chris Mattison
Dr George C. McGavin
Dr Jeremy McNeil
Dr Rodrigo A. Medellin
Dr Fridtjof Mehlum
Chris Morgan
Rick Morris
Dr Bryan G. Nelson
Dr Gary L. Nuechterlein
Jemima Parry-Jones
Malcolm Pearch
Prof. Christopher Perrins

Dr Ted Pietsch
Dr Tony Prater
Dr Galen B. Rathbun
Dr Ian Redmond
Dr James D. Rising
Robert H. Robins
Jeff Sailer
Dr Scott A. Schaeffer
Dr Karl Schuchmann
Prof. John D. Skinner
Dr Andrew Smith
Dr Ronald L. Smith
Dr David D. Stone

Dr Mark Taylor
Dr David H. Thomas
Dr Dominic Tollit
Dr Jane Wheeler
Dr Ben Wilson
Dr David B. Wingate
Dr Hans Winkler
Dr Kevin Zippel

CONTENTS

PREFACE

THIS BOOK IS A CELEBRATION of the richness of the animal kingdom, produced at a critical moment in its history. Compiled by an international team of over 70 zoologists, biologists, and naturalists, it forms a complete survey of the animal world, from familiar species that are encountered almost every day, to extreme rarities that only a few scientists have ever seen and fewer still have had the chance to study. It includes not only vertebrates, but also a broad selection of invertebrate life – creatures that form the immense but often overlooked majority of animal species on earth.

Altogether, over 2,000 species are profiled in *Animal*. Nearly all are illustrated with either a colour photograph or artwork. These images represent the combined work of hundreds of wildlife photographers and professional zoologists, and they show not only the animals themselves, but also key aspects of their biology and ecology. Because of the difficulties of locating and photographing some species in the wild, a number of the photographs show specimens in captivity, but the majority are of animals living in their native habitat. The pictures include award-winning portraits taken by some of the world's leading nature photographers, as well as rare images that have been tracked down after months of research, with

the patient help of experts in the field. The result is an out-standing collection of photographs that we believe to be unparalleled in a book of this kind.

The species profiles themselves include a range of new additions to the catalogue of the world's animal life. Among them are newly recognized kinds of monkeys, a third species of elephant, and some of the world's most elusive whales. At the other end of the spectrum of size, new thinking is reflected in the coverage of invertebrates, where classification and species totals draw on the latest research in this rapidly moving field. In addition, some long-accepted aspects of behaviour are brought up to date as new studies reveal that, even with well-known animals, our knowledge is often far from complete.

Animal follows today's general consensus on how different animal groups should be classified, but it also points out places where opinions differ, or where – often as a result of cladistic analysis – views have recently changed. Throughout the book, colour-coded panels give a summary of the classification levels for each group of animals, making it easy to see how they fit into the broad sweep of the animal world. The book is organized with mammals at the start, and invertebrates at the end, and, within each group, the sequence of animals follows the order generally adopted by experts working in each field.

At a time when animal life faces a great variety of threats, the remorseless decline in species numbers forms a sobering backdrop to modern life. Statistics suggest that, during the production of this book, as many as 5,000 animal species may have disappeared. The vast bulk of these will have been small invertebrates, unnoticed even by experts, but some more visible casualties also feature on the list. Among them was the last wild Spix's macaw, which vanished in December 2000, after a decade as the last free-ranging member of its kind.

It is a poignant fact that, in this book's lifetime, some other animals it features are likely to share the same fate. At the end of each section of this book, lists summarize species that are the most likely victims – animals that are currently classed as critically endangered by the International Union for the Conservation of Nature (IUCN). These include a host of well-known species, such as the tiger, the black rhinoceros, and the orang-utan, as well as a collection of amphibians, fish, insects, and other invertebrates, whose general status is often poorly known. Some, like the Spix's macaw, may be preserved in captivity. Others are likely to vanish forever.

While most people would mourn the loss of animals such as tigers, bears, and rhinoceroses, which have become very familiar to most of us through zoos and nature programmes on television, the loss of smaller and less appealing animals rarely elicits the same concern. Yet many biologists believe that the current steep decline in the earth's biodiversity – triggered mainly by habitat change, but also by pollution and unsustainable exploitation – is one of the greatest threats that affects our planet today. When species disappear, ecological links become disrupted, and the checks and balances that operate in nature risk being stretched until they finally break. If this happens, the result is ecological instability – a situation that can have grave implications for animal and plant life, and therefore for human welfare as well. In the distant past, life has climbed back from even greater crises, although the process has taken immense lengths of time. But there is no precedent for a sustained assault on biodiversity triggered by a single species – one that now dominates life on earth.

Unlike previous generations, we live in an age in which there is growing recognition of this danger, and of the potentially fragile state of the living world. Wildlife can no longer be taken for granted, and the wish to conserve rather than exploit it is something that unites people across the globe and has given rise to a number of high-profile conservation organizations. Knowledge and awareness are essential first steps: this book aims to provide both in a way that conveys the variety, beauty, and richness of animal life as a whole.

DAVID BURNIE
Editor-in-Chief

ABOUT THIS BOOK

Animal is organized into three main sections: a general INTRODUCTION to animals and their lives; a section on the world's main HABITATS; and the main part of the book, THE ANIMAL KINGDOM, which is dedicated to the description of animal groups and species. At the end of the book, a detailed glossary defines all zoological and technical terms used, while a full index lists all the groups and species featured, by both Latin name and common name, including alternative common names.

INTRODUCTION

Animal begins with an overview of all aspects of animal life. This includes an account of what an animal is – and how it differs from other living things. It also examines animal anatomy, life cycles, evolution, behaviour, and conservation. In addition, there is a comprehensive presentation of the classification scheme that underpins the species profiles in THE ANIMAL KINGDOM.

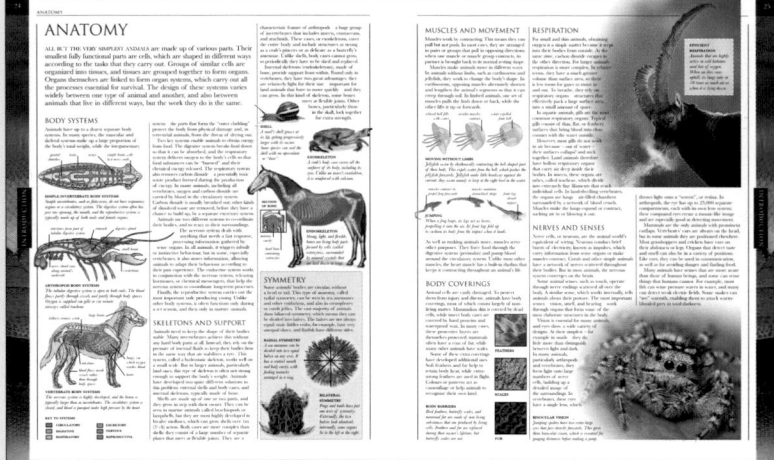

HABITATS

This section looks at the world's main animal habitats. Coverage of each habitat is divided into two parts. The first (illustrated below) describes the habitat itself, including its climate and plant life, and the types of animals found there. The pages that immediately follow describe how those animals have adapted their anatomy and behaviour to suit the conditions in their habitat.

general introduction to habitat type   *description of more specific habitat sub-type*   *dotted lines identify distinct zones within habitat*

map showing global distribution of habitat   *photographs show representative animals found within each zone*   *feature on conservation issues*

THE ANIMAL KINGDOM

This section of *Animal* is divided into six chapters: mammals, birds, reptiles, amphibians, fish, and invertebrates. Each chapter begins with an introduction to the animal group. Lower-ranked groups such as orders and families are then introduced, followed by profiles of the species classified within them. The invertebrates are organized slightly differently (see right), while passerine birds are profiled at family level, with only representative species shown. A list of critically endangered species ends each chapter.

HABITAT SYMBOLS

Symbols in profiles are listed as shown below, not in order of main occurrence.

 Temperate forest, including woodland

 Coniferous forest, including woodland

Tropical forest and rainforest

Mountains, highlands, scree slopes

Desert and semi-desert

Open habitats including grassland, moor, heath, savanna, fields, scrub

Wetlands and all still bodies of water, including lakes, ponds, pools, marshes, bogs, and swamps

Rivers, streams, and all flowing water

Mangrove swamps, above or below the waterline

Coastal areas, including beaches and cliffs, areas just above high tide, in the intertidal zone, and in shallow, offshore waters

Seas and oceans

Coral reefs and waters immediately around them

Polar regions, including tundra and icebergs

Urban areas, including buildings, parks, and gardens

Parasitic; living on or inside another animal

DATA FIELD

Summary information is given at the start of each profile. Measurements are for adult males of the species and may be a typical range, single-figure average, or maximum, depending on available records. † indicates an estimate based on available data.

LENGTH (all groups, except Invertebrates) MAMMALS: head and body. BIRDS: tip of bill to tip of tail. REPTILES: length of carapace for tortoises and turtles; head and body, including tail, for all other species. FISH & AMPHIBIANS: head and body, including tail.

TAIL (Mammals only) Length.

WEIGHT (Mammals, Birds, and Fish only) Body weight.

SOCIAL UNIT (Mammals only) Whether a species lives mainly alone (*Individual*), in a *Group*, in a *Pair*, or varies between these units (*Variable*).

PLUMAGE (Birds only) Whether sexes are *alike* or *differ*.

MIGRATION (Birds only) Whether a bird is a *Migrant, Partial migrant, Non-migrant,* or *Nomadic*.

BREEDING (Reptiles and Fish only) Whether the species is *Viviparous, Oviparous,* or *Ovoviviparous*.

HABIT (Reptiles and Amphibians only) Whether the species is partly or wholly *Terrestrial, Aquatic, Burrowing,* or *Arboreal*.

BREEDING SEASON (Amphibians only) The time of year in which breeding occurs.

SEX (Fish only) Whether the species has separate males and females (*Male/Female*), is a *Hermaphrodite,* or a *Sequential hermaphrodite*.

OCCURRENCE (Invertebrates only) Number of species in the family, class, or phylum; their distribution and the micro-habitats they can be found in.

STATUS (all groups) *Animal* uses the IUCN (see p.31) and other threat categories, as follows:

EXTINCT IN THE WILD (IUCN) Known only to survive in captivity or as a naturalized population well outside its natural range.
CRITICALLY ENDANGERED (IUCN) Facing an extremely high risk of extinction in the wild in the immediate future.
ENDANGERED (IUCN) Facing a very high risk of extinction in the wild in the near future.
VULNERABLE (IUCN) Facing a high risk of extinction in the wild in the medium-term future.
LOWER RISK (IUCN) Dependent on conservation measures in order not to qualify for one of the above threat categories; or close to qualifying for vulnerable; or of least concern.
COMMON (non-IUCN) fairly widely distributed and found at relatively high densities throughout the range.
LOCALLY COMMON (non-IUCN) found at relatively high densities within a restricted area or areas.

For fuller details see the IUCN web site (www.iucn.org)

◁ MAJOR GROUPS
Each of the 6 major animal groups is introduced with an overview of its defining characteristics. These pages also cover key aspects of evolution, life cycle, and behaviour.

◁ PHYLA, CLASSES, AND ORDERS
Within each major group, lower-ranked groups such as orders are described in separate introductions. These identify the animals found in the group, describe their defining properties, and explain many of the terms used in the species profiles.

INVERTEBRATE ANIMALS
There are so many invertebrate species that in order to provide representative coverage of the group as a whole, this book focuses on profiling the larger taxonomic groups – phylum, class, order, and family – rather than on individual species. Groups are presented in 3 ways: a brief introduction to an order or class, supported by a description of one or more constituent families, with each family illustrated by representative species (e.g. insects); a profile of a class, illustrated by species that represent distinct groups or types within it (e.g. crustaceans); a description of a phylum illustrated by a representative species of that phylum (used for minor phyla only).

CLASSIFICATION KEY
In each animal group introduction, a colour-coded panel shows the position in the animal kingdom of the group being described. The group is identified by a white outline and an arrow. The taxonomic ranks above the group are named in the upper part of the table, while the number of lower ranks that it contains are shown below it. In some introductions, an extra Classification Note is included, to list subgroups or identify areas of debate.

◁ LOWER-RANKED GROUPS
In many chapters, animal groups are divided into smaller groups, such as families. An introduction to each of these smaller sections describes common anatomical features. Relevant aspects of reproduction and behaviour are also introduced.

▽ SPECIES PROFILES
Over 2,000 wild animals are profiled in THE ANIMAL KINGDOM. Every profile contains a text summary and, in most cases, a colour illustration and a distribution map.

LATIN NAME
The species' Latin name appears in a coloured band at the beginning of each entry

DATA FIELD
Core information is provided in summary form at the beginning of each species entry. The categories vary between groups (see left)

Felis yagouaroundi
Jaguarundi

Length	55 – 77 cm (22 – 30 in)
Tail	33 – 60 cm (13 – 23 1/2 in)
Weight	4.5 – 9 kg (10 – 20 lb)

Location S. USA to South America

Social unit Individual

Status Lower risk

COMMON NAME
The species' common name appears below the Latin name, with alternative common names given in the text profile below

LOCATION
Colour maps indicate the world distribution for each species with further details given in the caption below

HABITAT SYMBOLS
These define the environment in which the animal is found (see left for key). Additional information may also be given in the text profile below

More mustelid than felid in overall proportions, with a pointed snout, long body, and shortish legs, the jaguarundi has several colour forms of unpatterned fur, from black – mainly in forests – to pale grey-brown or red – in dry shrubland. This cat hunts by day, often on the ground, in habitats ranging from semi-arid scrub to rainforest and swamp. Its main prey are birds, rodents, rabbits, reptiles, and invertebrates.

TEXT PROFILE
Each entry has a description of the most characteristic and noteworthy features of the species

ILLUSTRATIONS
Most entries include a colour photograph or artwork of the species. The animals pictured are normally adult males

FEATURE BOXES
Profiles may include features on anatomy or species variation (blue box), behaviour (yellow box), or conservation (green box)

FEATURE PROFILES ▷
Species of particular interest are given a double-page entry. These consist of a profile with feature boxes, a spectacular, close-up photograph, and, in some cases, an action sequence.

INTRODUCTION

Animals form the largest of the natural world's five kingdoms. Although they evolved after other living things, they are now the dominant form of life on earth. Among the attributes that have made them so successful are the abilities to adapt their behaviour and to move. As well as the features that set animals apart, this section looks at how they have evolved, their often complex responses to the world around them, and the way biologists classify them. It also examines the many threats they face.

WHAT ARE ANIMALS?

WITH ALMOST TWO MILLION SPECIES identified to date, and even more than that awaiting discovery, animals are the most varied living things on the planet. For over a billion years, they have adapted to the changing world around them, developing a vast array of different lifestyles in the struggle to survive. At one extreme, animals include fast-moving predators, such as sharks, big cats, and birds of prey; at the other, there are the inconspicuous sorters and sifters of the animal world's leftovers, living unseen in the soil or on the deep sea bed. Together, they make up the animal kingdom – a vast collection of living things that are linked by a shared biology and that occupy a dominant place in life on earth.

CHARACTERISTICS OF ANIMALS

Animals are usually easy to distinguish from other forms of life because most of them can move. However, while this rule works for most animals that live on land, it does not apply to many that live in water. Here, many animals spend their adult lives in one place, and some have trailing arms or tentacles that make them look very much like plants. A more reliable way of identifying animals is by their basic biological features: their bodies are composed of many cells, and they have nerves and muscles that enable them to respond to the world around them. Most important of all, they get the energy they need by taking in food.

Animals are highly complex, and remarkably responsive, compared with other forms of life. Even the simplest animals react quickly to changes around them, shrinking away from potential danger or reaching out for food. Animals with well-developed nervous systems can go much further: they can learn from experience – an ability that is unique to the animal world.

STATIC LIFE
Sea squirts are typical "sessile" animals: they spend their adult lives fixed to a solid surface. The young, which resemble tadpoles, can move freely, enabling sea squirts to spread.

THE SCALE OF ANIMAL LIFE

The world's largest living animals, baleen whales, can be up to 25m (82ft) long and weigh 120 tonnes. At the other end of the spectrum are microscopic organisms – rotifers and tardigrades (only 0.05mm long) – and sub-microscopic flies and beetles (about 0.2mm long). These animals are so tiny that their weight is negligible. Even so, they possess all the body systems needed for survival.

Different body sizes allow animals to live in different ways. Whales have few natural predators; and the same is true of elephants, the largest land animals. Their massive bodies are highly energy-efficient because they process food on such a large scale. However, they take a long time to reach maturity, which means they are

MINIATURE ANIMALS
Microscopic tardigrades (also called water bears) move about on tiny legs. Their behaviour is as complex as that of animals thousands of times their size.

slow to reproduce. Insects, on the other hand, are easy prey for many animals, and their small size means that their bodies are not as energy-efficient as those of large animals. But, since they can breed rapidly when conditions are favourable, their numbers can climb at a prodigious rate.

VERTEBRATES AND INVERTEBRATES

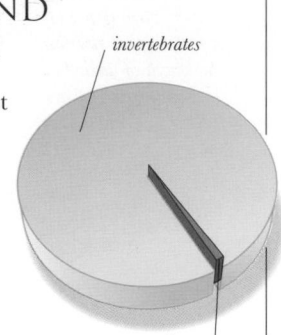
invertebrates

vertebrates

Almost all the world's largest and most familiar animals are vertebrates – animals that have backbones. They include the fastest animals on land, in water, and in the air, and also the world's most intelligent species, one of which – *Homo sapiens* – is quite outstanding in this latter respect.

Vertebrates are all related to each other, sharing a common ancestry that stretches back millions of years. However, despite leading the animal kingdom in so many fields, vertebrates make up a tiny minority of the animal species known today. The vast majority of species are invertebrates – animals without backbones.

Unlike vertebrates, invertebrates often have very little in common with each other, apart from their lack of a backbone. The giant squid, the largest invertebrate, can measure over 16m (52ft) long, but

INVERTEBRATE MAJORITY
Vertebrates make up less than 3 per cent of the world's animals. The remaining 97 per cent are made up by invertebrates.

OCEAN GIANTS
A humpback whale bursts out of the sea. Animals this large depend on aquatic habitats because water can support most of their weight.

KINGDOMS OF LIFE

Biologists classify all living things into overall groups, called kingdoms. The members of each kingdom are alike in fundamental ways, such as in the nature of their cells or in the way they obtain energy. In the most widely used system of classification there are five kingdoms, of which the animal kingdom is the largest. In recent years, a new classification system has been proposed. In this, there are three "superkingdoms": Archaebacteria, the Eubacteria, and Eukaryota. The first and second reflect chemical and physical differences within bacteria. The third contains the living things that, unlike bacteria, have complex cells: protists, fungi, plants, and animals.

ANIMALS	PLANTS	FUNGI	PROTISTS	MONERANS
Animals are multicellular organisms that obtain energy by ingesting food. All animals are capable of moving at least some parts of themselves, and many can move from place to place.	Plants are multicellular organisms that grow by harnessing the energy in light. Through a process called photosynthesis, they use this energy to build up organic matter from simple materials, creating most of the food on which animals rely.	Most fungi are multicellular. They collect energy from organic matter, which they do not ingest but break down externally using microscopic threads that spread throughout their food. Many fungi are too small to be seen, but some form large fruiting bodies.	Protists are single-celled organisms that typically live in water, or in permanently moist habitats. Their cells are larger and more complex than those of bacteria. Some protists behave like plants, collecting energy from sunlight; others, known as protozoa, are more like animals, acquiring energy by ingesting food.	Monerans, or bacteria, are the simplest fully independent living things. Their cells are prokaryotic, meaning they lack organelles – the specialized structures more complex cells use for carrying out different tasks. Bacteria gather energy from various sources, including organic and inorganic matter, and sunlight.

MAMMAL

FLOWERING PLANT

TERRESTRIAL FUNGUS

PARAMECIUM

BACTERIUM

SUPPORT SYSTEMS

Many invertebrates – such as leeches – have no hard body parts; to keep their shape they rely on the pressure of their body fluids. Simple chordates, which include lancelets, have a strengthening rod or notochord that runs along the length of their bodies. Vertebrates, which are more advanced chordates, are the only animals that have internal skeletons made of bone.

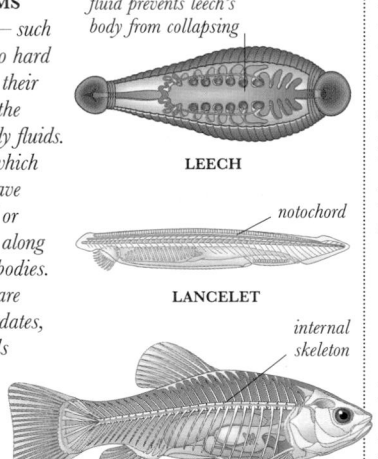

fluid prevents leech's body from collapsing

LEECH

notochord

LANCELET

internal skeleton

BONY FISH

it is very much an exception. Most invertebrates are tiny, and many live in inaccessible habitats, which explains why they are still poorly known compared with vertebrates.

WARM- AND COLD-BLOODED ANIMALS

Most animals are cold-blooded (or ectothermic), which means that their body temperature is determined by that of their surroundings. Birds and mammals are warm-blooded (or endothermic), which means that they generate their own heat and maintain a constant internal temperature regardless of the conditions outside.

This difference in body temperature has some far-reaching effects on the way animals live; this is because animal bodies work best when they are warm. Cold-blooded animals, such as reptiles, amphibians, and insects, operate very effectively

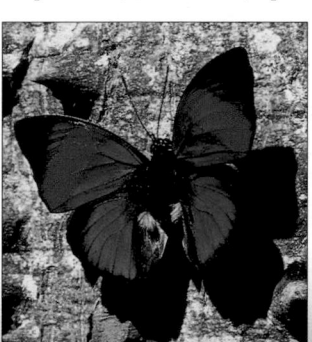

HEAT CONTROL

A basking butterfly soaks up the sunshine. By basking, or by hiding in the shade, butterflies and other cold-blooded animals can adjust their body temperature. Even so, they have difficulty coping with extreme temperatures, especially severe cold.

in warm conditions, but they slow down if the temperature drops. They can absorb some heat by basking in sunshine, but if the temperature falls below about 10°C (50°F) their muscles work so slowly that they find it difficult to move. Birds and mammals, on the other hand, are barely affected by this kind of temperature change. Their internal heat and good insulation enable them to remain active even when temperatures fall below freezing.

INDIVIDUALS AND COLONIES

Physically, vertebrates function as separate units, even though they may live together in families or larger groups. In the invertebrate world it is not unusual for animals to be permanently linked together, forming clusters called colonies. Colonies often look and behave like single animals. Most are static, but some – particularly ones that live in the sea – can move about.

LIVING TOGETHER

This branching coral is covered with a living "skin" that connects its individual animals, or polyps. The polyps are in constant contact with each other but otherwise lead separate lives. Each has a set of stinging tentacles and catches its own food.

Colonial species include some of the world's most remarkable invertebrates. Pyrosomes, for example, form colonies shaped like test tubes that are large enough for a diver to enter. But, in ecological terms, the most important colonial animals are reef-building corals, which create complex structures that provide havens for a range of other animals. In reef-building corals, the members of each colony are usually identical. But in some colonial species, the members have different shapes designed for different tasks. For example, the Portuguese man o'war – an oceanic drifter that has a highly potent sting – looks like a jellyfish but consists of

separate animals, called polyps, that capture food, digest it, or reproduce. They dangle beneath a giant gas-filled polyp that acts as the colony's float.

FUEL FOR LIFE

Animals obtain their energy from organic matter, or food. They break food up by digesting it, and then they absorb the substances that are released. These substances are carried into the animal's cells, where they are combined with oxygen to release energy. This process – called cellular respiration – is like a highly controlled form of burning, with food acting as the fuel.

The majority of animals are either herbivores, which eat plants, or carnivores, which eat other animals. Carnivores include predators, which hunt and kill prey, and parasites, which feed in or on the living bodies of other animals. There are also omnivores, which eat both animal and plant food, and scavengers, which feed on dead matter – from decaying leaves and corpses to fur and even bones.

All animals, regardless of lifestyle, ultimately provide food for other animals. All are connected by food chains, which pass food – and its energy – from one species to another. However, individual food chains are rarely more than five or six links long. This is because up to 90 per cent of an animal's energy cannot be passed on: it is used up in making the animal's own body work.

CARNIVORE 3

The food chain ends with a "top predator" – in this case an osprey. When it dies, the energy in its body is used by scavengers, such as insects and bacteria.

CARNIVORE 2

The perch lives almost exclusively on other animals. A perch feeding on dragonfly nymphs is a second-level carnivore, receiving food that has already been through two other animals.

CARNIVORE 1

Dragonfly nymphs are typical first-level carnivores, using a mixture of speed and stealth to hunt small prey. Tadpoles are a good food source and often feature in their diet.

HERBIVORE

During their early lives, tadpoles use their jaws to feed on water plants. By digesting the plant food, and therefore turning it into animal tissue, they change plant food into a form that carnivores can use.

PLANT

By synthesizing sunlight, plants gather the energy that drives most of life on earth. In this food chain, waterweed is the first link, creating food that can then be passed on.

FOOD CHAIN

This is a typical five-stage food chain from a freshwater habitat. Food and energy move upwards through the chain, being passed on every time one organism eats another. The chain ends when it reaches an animal that has no natural predators. The energy in the final animal's body is ultimately passed on to scavengers and to recyclers, known as decomposers, many of which live in soil.

INSULATION

In subzero conditions, superb insulation keeps the body temperature of these young emperor penguins at an almost constant 40°C (104°F).

EVOLUTION

LIKE ALL LIVING THINGS, animals undergo changes as each new generation succeeds the one before. These changes are usually so slight that they are very difficult to see, but over thousands or millions of years they can completely alter the way animals look and also the way they behave. This process of change is called evolution. It allows animals to exploit new opportunities and to adapt to changes that take place in the world around them. Evolution works by modifying existing characteristics, usually through a series of extremely small mutations. The result of this is that every animal is a living store of evolutionary history – one that helps to show how different species are related.

ANIMAL ADAPTATION

Evolution is made possible by the variations that exist within animals, and it occurs because animals compete with each other for limited resources, such as space and food. In this competition, some characteristics prove to be more useful than others, which means that their owners are more likely to

BLENDING IN
Most owls have brown plumage, which helps them to hide in trees. However, the snowy owl, from the treeless Arctic tundra, is mainly white – an adaptation that increases its chances of catching food and raising young.

TAWNY OWL

SNOWY OWL

thrive, and to produce the most young. Animals with less useful features face more difficulties and find it harder to breed. The least successful animals are therefore gradually weeded out, while those with "winning" characteristics become widespread.

A HISTORY OF LIFE
Geologists divide earth's history into periods characterized by major physical changes, such as bursts of volcanic activity, collisions between continents, or alterations in climate. Many of these periods have ended in worldwide mass extinctions, both on land and in the sea. For each period described below, the date shown indicates the period's end.

This weeding-out process is called natural selection. Natural selection operates all the time, invisibly screening all the subtle variations that come about when animals reproduce. For example, for many animals, camouflage is a valuable aid to survival. Natural selection ensures that any improvements in an animal's camouflage – a slight change in colour, pattern, or behaviour – are passed on to the next generation, increasing its chances of survival and therefore its chances of producing young.

Adaptations such as camouflage last only as long as they are useful: if an animal's lifestyle changes, the path that evolution follows changes too. This has happened many times with birds: some lineages have evolved the power of flight only to lose it when they take up life on land.

SPECIES AND SPECIATION

A species is any group of animals that has the potential to interbreed and that, under normal circumstances, does not breed with any other group. Speciation is the evolutionary process that brings about new species. It usually occurs when an existing species becomes split up into two or

GERMAN FORM

JAPANESE FORM

SPECIATION
The Apollo butterfly is a highly variable species. Many forms are restricted to specific parts of the northern hemisphere. Since the Apollo often lives on mountains, its different forms tend to remain apart. In time, these forms could evolve into new species.

more isolated groups, which are kept apart either by physical barriers, such as seas and mountain ranges, or by changes in behaviour. If these groups remain separate for long enough, they evolve their own characteristic adaptations and become so different that they can no longer interbreed.

Speciation is difficult to observe because it occurs so slowly, but evidence of it is not hard to find. Many animals – from butterflies to freshwater fish – show distinct regional differences. In time, these local forms or subspecies can become species in their own right.

EXTINCTION

Extinction is a natural feature of evolution because for some species to succeed, others must fail. Since life began, about 99 per cent of the earth's species have disappeared and, on at least five occasions, huge numbers have died out in a relatively short time. The most recent of these mass extinctions, about 65 million years ago, swept away the dinosaurs and many other forms of life. However, despite such catastrophes, the total number of living species has, until recently, followed a generally upward trend. Today, the extinction rate is increasing rapidly as a result of human interference in natural ecosystems.

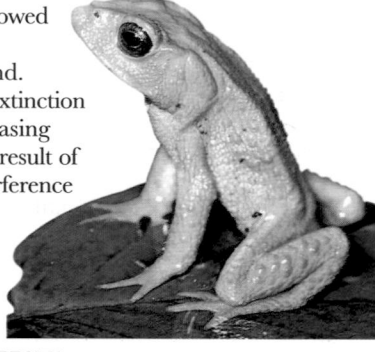

VANISHING SPECIES
The golden toad, from the highlands of Central America, is one of the few species whose apparent extinction has been closely monitored. It normally breeds in forest-floor pools, but since the early 1990s the species has vanished without trace.

Primates, tropical birds, and many amphibians are particularly threatened. For the foreseeable future, this decline is set to continue because evolution generates new species far more slowly than the current rate of extinction.

CONVERGENCE

Unrelated animals often develop very noticeable similarities. For example, sharks and dolphins are fundamentally very different, but both have streamlined bodies with an upright dorsal fin – a shape that gives them speed and stability underwater – while moles and marsupial moles

PRECAMBRIAN	CAMBRIAN	ORDOVICIAN	SILURIAN	DEVONIAN	CARBONIFEROUS
This vast expanse of geological time stretches from when the continents first formed to when animals with hard body parts first appeared in the fossil record. Life itself emerged near the beginning of Precambrian time, about 3.8 billion years ago. The date for the first animals is less certain: being soft-bodied, they left few traces, although their existence is evident from fossils of burrows and tracks (about one billion years old).	The start of the Cambrian Period was marked by an extraordinary explosion of animal life in the seas. Cambrian animals included molluscs, echinoderms, and arthropods. These were among the first creatures to have hard, easily fossilized body parts. By the end of this period, all the major divisions of animal life, or phyla, that exist today had been established.	During this period, life was still confined to the sea. Animals included the earliest crustaceans and some of the earliest jawless fish. Trilobites (arthropods with a three-lobed body) were numerous, as were long-shelled nautiloids (predatory molluscs with sucker-bearing arms). Like the Cambrian Period, the Ordovician ended in a mass extinction, probably caused by climatic changes.	Silurian times saw the evolution of the first fish with jaws and of giant sea scorpions – relatives of today's arachnids. The first land plants appeared.	In Devonian times, jawed and jawless fish diversified rapidly, which is why the period is known as the "Age of Fishes". Many fish lived in fresh water, where warm conditions and falling water levels encouraged the evolution of primitive lungs. As a result, amphibians evolved, becoming the first vertebrates to live on land. On land, insects became widespread, and the first true forests began to form. The Devonian ended with the third mass extinction, which killed up to 70 per cent of animal species.	During this period, which is also called the "Coal Age", a warm global climate encouraged the growth of forests on swampy ground – home to amphibians and flying insects, including dragonflies with wingspans of up to 60cm (23½ in). In the sea, ammonoids (molluscs related to today's nautiluses), were common.

YEARS AGO (MILLIONS): 545   495   443   417   354   290

share a range of adaptations for life under ground. Amphisbaenians and caecilians also look very similar, although the former are reptiles and the latter are highly unusual amphibians. Such similarities are the result of convergent evolution, a process in which natural selection comes up with the same set of adaptations to a particular way of life, reshaping

EURASIAN MOLE

MARSUPIAL MOLE

SIMILAR SHAPES
Apart from their difference in colour, Eurasian moles and marsupial moles are alike in many ways. However, they are not at all closely related; their similarities are a result of convergence.

body parts or whole animals until they outwardly look the same. Convergence is responsible for a whole series of striking similarities in the animal kingdom. It can make the task of tracing evolution extremely difficult, which is why the details of animal classification often change.

ANIMAL PARTNERSHIPS

Over millions of years, animals have evolved complex partnerships with each other and with other forms of life. In one common form of partnership, called mutualism or symbiosis, both species benefit from the arrangement. Examples

MIXED BLESSING
Oxpeckers eat ticks and other parasites that live on the skin of large animals, but they also feed on blood from wounds – a habit that is less helpful to their hosts.

are oxpeckers and large mammals, and corals and microscopic algae. Many partnerships are loose ones but, as with pollinating insects and plants, some are so highly evolved that the two partners cannot survive without each other. By contrast, a commensal partnership – such as that between remoras and their host fish (see p.514) – is one in which one species gains but the other neither gains nor loses.

Partnerships may appear to be mutually obliging, but each partner is driven purely by self-interest. If one partner can tip the balance in its favour, natural selection will lead it to do so. The ultimate outcome is parasitism, in which one animal, the parasite, lives on or inside another, entirely at the host's expense.

BIOGEOGRAPHY

The present-day distribution of animals is the combined result of many factors. Among them are continental drift and volcanic activity, which constantly reshape the surface of the earth. By splitting up groups of animals, and creating completely new habitats, these geological processes have had a profound impact on animal life. One

of the most important effects can be seen on remote islands, such as Australia and Madagascar, which have been isolated from the rest of the world for millions of years. Until humans arrived, their land-based animals lived in total seclusion, unaffected by competitors from outside. The result is a whole range of indigenous species, such as kangaroos and lemurs, which are found nowhere outside their native homes.

Animals are separated when continents drift apart, and they are brought together when they collide. The distribution of animals is evidence of such events long after they occur. For example, Australasia and Southeast Asia became close neighbours long ago, but their wildlife remains entirely different: it is divided by "Wallace's line", an invisible boundary that indicates where the continents came together.

BRAZILIAN TAPIR

■ BRAZILIAN TAPIR DISTRIBUTION

■ MALAYSIAN TAPIR DISTRIBUTION

SPLIT BY CONTINENTAL DRIFT
One species of tapir lives in Southeast Asia; another three live in Central and South America. This indicates that these two landmasses were once joined.

MALAYSIAN TAPIR

ARTIFICIAL SELECTION

The variations that natural selection works on are often difficult to see. One herring, for example, looks very much like another, while starlings in a flock are almost impossible to tell apart. This is because natural selection operates on a huge range of features among many individuals in a species. However, when animals are bred in controlled conditions, their hidden variations are very easy to bring out. Animal breeders rigorously concentrate on the reproduction of particular features, such as a specific size or colour, and by selecting only those animals with the desired features they can exaggerate those features with remarkable speed. This process, called artificial selection, is responsible for all the world's domesticated animals and all cultivated plants.

FAMILY LIKENESS
All domesticated dogs are descended from the grey wolf. Through artificial selection, individual breeds can be established in a very short time.

CHIHUAHUA　　**GREY WOLF**

PERMIAN	TRIASSIC	JURASSIC	CRETACEOUS	TERTIARY	QUARTERNARY
Reptiles became the dominant land animals. The continents formed a single landmass. The period ended with the mass extinction – probably from climate change and volcanic activity – of over 75 per cent of land species and over 90 per cent of marine species.	During the Triassic Period, reptiles dominated life on earth. They included flying pterosaurs, swimming forms such as nothosaurs and ichthyosaurs, and the first true dinosaurs. Early mammals also existed, but with reptiles in the ascendant they made up only a minor part of earth's land-based fauna.	During this period, known as the "Age of Dinosaurs", reptiles strengthened their position as the dominant form of animal life. They included a wide range of plant- and meat-eating species. There was a rich variety of vegetation and, towards the end of the Jurassic, flowering plants appeared, creating new opportunities for animals – particularly pollinating insects. Birds gradually evolved from dinosaurs. The first known bird, *Archaeopteryx*, probably took to the air over 150 million years ago.	The Cretaceous Period saw rapid evolution in flowering plants and in animals that could use them as food. But the most striking feature of this period was the rise of the most gigantic land animals that have ever lived. They were all dinosaurs, and they included immense plant-eating sauropods, perhaps weighing up to 80 tonnes, as well as colossal two-legged carnivores, such as *Tyrannosaurus*, and its even bigger relative *Carcharodontosaurus*. Towards the end of the Cretaceous, dinosaurs were already in decline, but they were wiped out by a mass extinction that also erased many other forms of life.	Mammals existed long before this period began, but the disappearance of the dinosaurs enabled mammals to evolve rapidly, becoming larger and much more diverse. By this time in the earth's history, the continents had split apart, allowing different groups of mammals to evolve in isolation from each other.	This period has seen abrupt changes in climate. Mammals have become supreme, and humans have become the dominant form of life.
248	205	142	65	1.8	PRESENT

MESOZOIC ERA　　　　　CENOZOIC ERA

CLASSIFICATION

AT LEAST 1.5 MILLION DIFFERENT KINDS OF ANIMAL have been identified by scientists, and more are added to the list every year. The total number of animals in existence may be as high as 30 million, which means that the task of locating and identifying all the world's fauna might never be complete. To make sense of this bewildering diversity, biologists use formal classification systems, in which all the past and present life on the planet is named and organized into groups (known as taxa). Each type of animal is assigned a unique species name and placed in the group containing the species to which its evolutionary development suggests it is most closely related.

PRINCIPLES OF CLASSIFICATION

Modern scientific classification dates back to the eighteenth century, to the work of Karl Linnaeus, a Swedish botanist and explorer. Linnaeus devised a system in which all organisms are identified by two-part scientific names, which are written in Latin. The first part of the name indicates the genus to which the animal belongs; the second part denotes its species. This system remains in use today. At first sight, these names may seem cumbersome, but they have two immense advantages: unlike common names, they can be understood by scientists all over the world, and, as well as labelling animals (and all other living things), they act like signposts, showing exactly where a species fits into the living world.

In scientific classification, the species is the basic unit. Species are organized into groups, called genera, which in turn are organized into families. This grouping process continues working upwards through orders, classes, phyla, and, finally, into kingdoms, which are the largest groups of all and the highest level in the hierarchy.

Each group in the hierarchy has a common ancestor, which becomes increasingly remote from a species with each higher

level of the hierarchy. Identifying the groups is a key part of the classification process, and often involves complicated research to tease apart the confusing clues that evolution leaves behind.

CLASSIFICATION LEVELS

The diagram below shows a classification hierarchy for the tiger. However, for many other animals – for example insects or snakes – the hierarchy would look different in that it would have one or more extra levels. This is because animals do not always fit neatly into the categories that scientists devise. Intermediate levels, such as superclasses or suborders, have to be created to accommodate certain groups.

These additional levels do not mean that the classification system does not work; they simply reflect the fact that classification levels are labels rather than things that actually exist in the natural world. The only category that really exists is the species, and even at this level an animal can be very difficult to pin down.

Traditionally, species are defined as groups of living things that share a set of characteristics and that are able to interbreed in the wild. However, deciding whether a group fits this definition is sometimes difficult, and various researchers may come to different conclusions. As a result, many of today's scientists prefer to determine species by examining genetic data, which provides a more detailed record of the similarities and differences between animals. This approach has resulted in the "discovery" of new species among animals that are already known – the forest elephant (see p.221) is one. However, genetic data do not solve all classification puzzles because researchers still have to decide if two animals' genes are similar enough to warrant their being classed as one species.

GATHERING EVIDENCE

Nearly all of today's animal groups were identified before genetic analysis became available. The groups were established on the basis of anatomical evidence both in living species and in those preserved as fossils. Anatomical studies can be very useful in tracing the path of evolution because they often show how certain body parts – such as limbs, jaws, or teeth – have been modified, sometimes drastically, for different uses.

PHYLUM	Chordata
CLASS	Mammalia
ORDER	Carnivora
FAMILY	Felidae
SPECIES	*Panthera tigris*

TIGER CLASSIFICATION
In this book, panels such as the one above are used to identify the position of animal groups in the taxonomic hierarchy. The larger panel on the right defines the various taxonomic ranks, starting with the kingdom (the highest rank) at the top and ending with the species. Taking the tiger as an example, it also shows how a particular animal's physical characteristics are used to determine its place in the classification scheme.

KINGDOM A kingdom is an overall division containing organisms that work in fundamentally similar ways.	**ANIMALIA** *The kingdom Animalia contains multicellular organisms that obtain energy by eating food. Most have nerves and muscles, and are mobile.*
PHYLUM A phylum is a major subdivision of a kingdom, and it contains one or more classes and their subgroups.	**CHORDATA** *The phylum Chordata contains animals with a strengthening rod or notochord running the length of their bodies, for all or part of their lives.*
CLASS A class is a major subdivision of a phylum, and it contains one or more orders and their subgroups.	**MAMMALIA** *The class Mammalia contains chordates that are warm-blooded, have hair, and suckle their young. The majority of them give birth to live young.*
ORDER An order is a major subdivision of a class, and it contains one or more families and their subgroups.	**CARNIVORA** *The order Carnivora contains mammals that have teeth specialized for biting and shearing. Many of them, including the tiger, live primarily on meat.*
FAMILY A family is a subdivision of an order, and it contains one or more genera and their subgroups.	**FELIDAE** *The family Felidae contains carnivores with short skulls and well-developed claws. In most cases, the claws are retractable.*
GENUS A genus is a subdivision of a family, and it contains one or more species and their subgroups.	***PANTHERA*** *The genus Panthera contains large cats that have a specialized larynx with elastic ligaments. Unlike other cats, they can roar as well as purr.*
SPECIES A species is a group of similar individuals that are able to interbreed in the wild.	***PANTHERA TIGRIS*** 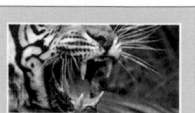 *The tiger is the only member of the genus Panthera that has a striped coat when adult. There are several varieties, or subspecies.*

NEW DISCOVERIES

Two centuries after scientific classification began, zoologists believe that relatively few terrestrial vertebrates remain completely unknown to science. Marine life is less well known, so it is quite likely that many types of fish have yet to be found – the Indonesian coelacanth (see p.480), discovered in 1997, is an example of a recent find. But in coming decades, the vast majority of "new" animals will be invertebrates. According to some estimates, less than one in 20 invertebrate species have so far been identified and named.

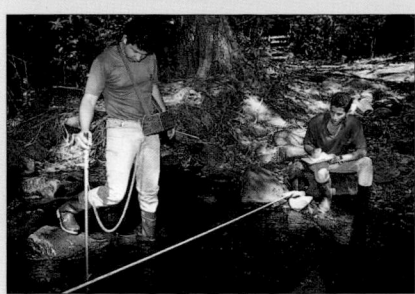

SAMPLING FOREST LIFE
Surveys of some habitats – particularly tropical forests and the shallow seabed – frequently turn up new species of invertebrates. The vast majority of these are fairly easy to identify because it is clear that they belong to groups that are already known and well documented.

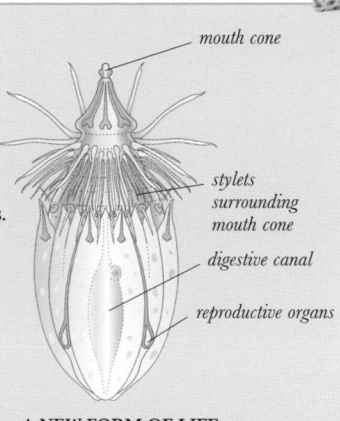

mouth cone

stylets surrounding mouth cone

digestive canal

reproductive organs

A NEW FORM OF LIFE
Very occasionally, a completely new kind of animal comes to light. This one is a loriciferan – a type of marine invertebrate first seen in 1983. Its unique features warranted the creation of a phylum of its own.

CARTILAGINOUS FISH
Although they look superficially like bony fish, cartilaginous fish are different in many ways. The most notable of these differences is their skeleton, which is made of rubbery cartilage instead of bone. They also lack a swim bladder.

BONY FISH
Unlike cartilaginous fish, bony fish have a single flap covering their gills, a gas-filled swim bladder, and a skeleton made of bone. These key features show that cartilaginous and bony fish are not close relatives.

The limbs of four-legged vertebrates, or tetrapods, are an excellent example of this kind of evolutionary modification. The basic limb pattern, dating back over 300 million years, is built around three main sets of bones: a single bone at the point where the limb meets the body, two bones further away, and five sets of smaller bones at the limb's outermost point. As tetrapods spread to various habitats – on land, in water, and in air – the limbs of the different groups became specialized in shape and size, and in the way they worked. However, since evolution always modifies what already exists – rather than starting from scratch – the underlying pattern of bones was preserved in all the different groups. For classification specialists, this is one of many pieces of evidence showing that all tetrapods evolved from a shared ancestor.

In classification, tetrapods are said to form a natural group, or clade. A clade is made up of an ancestral species and all its descendants. Together, they form a complete and self-contained branch from the evolutionary tree of life.

bodies, adaptations to a highly active life in water. Judged by these two characteristics alone, these two fish would appear to make up a single group of animals. However, when they are examined in more detail, these similarities turn out to be skin-deep: the many underlying differences between them show that cartilaginous and bony fish belong to separate branches of the evolutionary tree.

A specialist would be unlikely to confuse these two kinds of fish but, with animals that are already related, convergence can make relationships extremely difficult to unravel. One example of this can be seen with walruses, seals, and sea lions. These animals are traditionally classified in a

single, separate order of mammals – Pinnepedia – but a detailed study of their anatomy suggests that pinnipeds may actually consist of two or perhaps three separate lines of mammals that have evolved independently from different kinds of carnivores. If this is true, the pinnipeds are an artificial grouping, rather than one that actually indicates the path of evolution, and they could therefore be divided into separate clades.

As today's scientists apply cladistics to the animal world, many well-known and long-established groups are being shown to have been built on shaky foundations. Re-examination of animal groups is part of a constant process in classification science in the same way that reinterpretation of past events is constantly applied to human history. This reassessment helps to produce an ever more precise picture of how animals are related, and how the animal kingdom came to look as it does today.

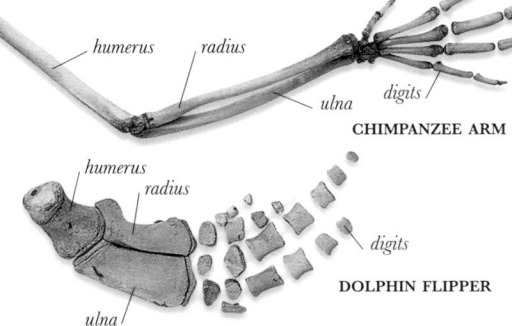

humerus   radius

ulna   *digits*

CHIMPANZEE ARM

humerus
radius

digits

DOLPHIN FLIPPER

ulna

ARMS AND FLIPPERS
A human arm and a dolphin's flipper look very different in life, but they both contain the same arrangement of bones. This similarity is strong evidence that chimpanzees (and humans) and dolphins evolved from a common ancestor.

CLADES

The clade concept has had a major impact on the science of classification (see panel, right), because it helps to cut through some of the confusing clues that evolution leaves behind. The greatest source of this confusion is convergence – a process that makes unrelated species develop in particular ways to suit similar ways of life (see p.16). Cartilaginous and bony fish are a typical example of this: both have fins and streamlined

CLADISTICS

Pioneered in the 1950s, cladistics is a classification technique that involves comparing large numbers of characteristics to see how closely different species are related. The resulting data are used to produce a cladogram – a diagram that shows branch points in evolution, together with the sets of nested groups, or clades, that the branch points create.

Cladistics is concerned with new, or derived, features rather than with primitive ones, so the cladogram does not indicate when different groups evolved. Instead, it shows the most likely order in which derived features arose, and the number of these features that different groups share. In the cladogram below, gibbons are shown to have many derived features in common with great apes

and humans, and relatively few with tarsiers. So, it is reasonable to deduce that gibbons are among the human being's close relatives, while tarsiers are not.

When cladistic analysis first became widely used it generated heated debate because it divides many of the traditionally recognized groups and it combines some that traditional classification keeps apart. For example, birds and dinosaurs probably form a clade so, cladistically, a bird is simply a feathered dinosaur that flies. Today, cladistics is generally seen as a very useful means of tracing evolution.

PRIMATE CLADOGRAM
This diagram shows five nested clades containing all the "advanced" (non-prosimian) primates. For comparison, the groups used in conventional classification are shown at the top. Each clade contains all the species, living or extinct, that share a set of derived features – some of these are listed on each clade's baseline. Each clade also shares all the features of the clades within which it is nested.

PRIMATES

TARSIERS   MONKEYS   APES

TARSIERS   NEW WORLD MONKEYS   OLD WORLD MONKEYS   GIBBONS   GREAT APES AND HUMANS

orbits higher than broad; four or five lumbar vertebrae
CLADE **Hominidae**

frontal bone wide; modified vertebral column; no tail
CLADE **Hominoidae**

nasal aperture high; premaxilla short; upper incisors of similar size
CLADE **Catarhini**

tubular ear canal; dental formula 2/2-1/1-2/2-3/3; characteristic ankle and thumb bones
CLADE **Anthropoidea**

short cranium; skull contains postorbital plate
CLADE **Haplorhini**

ANIMAL NAMES AND GROUPS

THE FOLLOWING FOUR PAGES summarize the classification scheme that is used in this book. Like all classification schemes, it represents current thinking, which is always liable to change. Groups are "nested" to show how they are related; informal groups, which have no distinct biological identity, are bounded by dotted lines. Some of the species totals, particularly for invertebrates, are based on estimates; these figures are often the subject of much debate. For ease of reference, some groups – marsupials, for example – are presented here in their consolidated form, although some biologists now split them apart.

VERTEBRATES

MAMMALS

CLASS Mammalia

GREVY'S ZEBRA

Mammals are vertebrates belonging to the phylum Chordata. Biologists differ on how many orders make up the class Mammalia. In this book, mammals are classified into 21 orders. Marsupials are traditionally classed as a single order – putting them on an equal footing with egg-laying mammals – but some mammologists feel that they are so diverse that they should be classified in as many as seven different orders. There is also disagreement about seals and sea lions: many mammalogists think that they are simply aquatic carnivores rather than animals deserving their traditional place in an order of their own.

EGG-LAYING MAMMALS		
ORDER Monotremata	**FAMILIES** 2	**SPECIES** 5

MARSUPIALS		
ORDER Marsupialia	**FAMILIES** 22	**SPECIES** 292

INSECTIVORES		
ORDER Insectivora	**FAMILIES** 6	**SPECIES** 365

BATS		
ORDER Chiroptera	**FAMILIES** 18	**SPECIES** 977

FLYING LEMURS		
ORDER Dermoptera	**FAMILIES** 1	**SPECIES** 2

ELEPHANT-SHREWS		
ORDER Macroscelidea	**FAMILIES** 1	**SPECIES** 15

TREE SHREWS		
ORDER Scandentia	**FAMILIES** 1	**SPECIES** 19

PRIMATES		
ORDER Primates	**FAMILIES** 11	**SPECIES** 356
PROSIMIANS		
SUBORDER Strepsirhini	**FAMILIES** 6	**SPECIES** 85
MONKEYS AND APES		
SUBORDER Haplorhini		
MONKEYS		
FAMILIES 3	**SPECIES** 242	
APES		
FAMILIES 2	**SPECIES** 21	

ANTEATERS AND RELATIVES		
ORDER Xenarthra	**FAMILIES** 4	**SPECIES** 29

PANGOLINS		
ORDER Pholidota	**FAMILIES** 1	**SPECIES** 7

RABBITS, HARES, AND PIKAS		
ORDER Lagomorpha	**FAMILIES** 2	**SPECIES** 80

RODENTS		
ORDER Rodentia	**FAMILIES** 30	**SPECIES** 1,702
SQUIRREL-LIKE RODENTS		
SUBORDER Sciuromorpha	**FAMILIES** 7	**SPECIES** 377
MOUSE-LIKE RODENTS		
SUBORDER Myomorpha	**FAMILIES** 5	**SPECIES** 1,137
CAVY-LIKE RODENTS		
SUBORDER Caviomorpha	**FAMILIES** 18	**SPECIES** 188

CETACEANS		
ORDER Cetacea	**FAMILIES** 13	**SPECIES** 83
BALEEN WHALES		
SUBORDER Mysticeti	**FAMILIES** 4	**SPECIES** 12
TOOTHED WHALES		
SUBORDER Odontoceti	**FAMILIES** 9	**SPECIES** 71

CARNIVORES		
ORDER Carnivora	**FAMILIES** 7	**SPECIES** 249
DOGS AND RELATIVES		
FAMILY Canidae	**SPECIES** 36	
BEARS		
FAMILY Ursidae	**SPECIES** 8	
RACCOONS AND RELATIVES		
FAMILY Procyonidae	**SPECIES** 20	
MUSTELIDS		
FAMILY Mustelidae	**SPECIES** 67	
CIVETS AND RELATIVES		
FAMILY Viverridae	**SPECIES** 76	
HYENAS AND AARDWOLF		
FAMILY Hyaenidae	**SPECIES** 4	
CATS		
FAMILY Felidae	**SPECIES** 38	

SEALS AND SEA LIONS		
ORDER Pinnipedia	**FAMILIES** 3	**SPECIES** 34

ELEPHANTS		
ORDER Proboscidea	**FAMILIES** 1	**SPECIES** 3

AARDVARK		
ORDER Tubulidentata	**FAMILIES** 1	**SPECIES** 1

HYRAXES		
ORDER Hyracoidea	**FAMILIES** 1	**SPECIES** 8

DUGONG AND MANATEES		
ORDER Sirenia	**FAMILIES** 2	**SPECIES** 4

HOOFED MAMMALS

ODD-TOED HOOFED MAMMALS
ORDER Perissodactyla   **FAMILIES** 3   **SPECIES** 19

HORSES AND RELATIVES
FAMILY Equidae   **SPECIES** 10
RHINOCEROSES
FAMILY Rhinocerotidae   **SPECIES** 5
TAPIRS
FAMILY Tapiridae   **SPECIES** 4

EVEN-TOED HOOFED MAMMALS
ORDER Artiodactyla   **FAMILIES** 10   **SPECIES** 225

PIGS
FAMILY Suidae   **SPECIES** 14
PECCARIES
FAMILY Tayassuidae   **SPECIES** 3
HIPPOPOTAMUSES
FAMILY Hippopotamidae   **SPECIES** 2
CAMELS AND RELATIVES
FAMILY Camelidae   **SPECIES** 7
DEER
FAMILY Cervidae   **SPECIES** 45
MUSK DEER
FAMILY Moschidae   **SPECIES** 7
CHEVROTAINS
FAMILY Tragulidae   **SPECIES** 4
PRONGHORN
FAMILY Antilocapridae   **SPECIES** 1
GIRAFFE AND OKAPI
FAMILY Giraffidae   **SPECIES** 2
CATTLE AND RELATIVES
FAMILY Bovidae   **SPECIES** 140

BIRDS

CLASS Aves

NORTHERN ORIOLE

In the classification scheme that is used in this book, birds are separated into 29 orders. One of these orders – the passeriformes, or perching, birds – is vast, and contains as many species as all the other orders combined. By comparison, several orders of birds contain just a single family, consisting of less than 10 species. In bird classification, some alternative schemes combine the large flightless birds, or ratites, into a single order while others combine flamingos with herons and their relatives. At a more detailed level, there is considerable disagreement about how many families of passerine birds there are. Some ornithologists consider that there are 60 or fewer, while others put the total at more than 80.

OSTRICH
ORDER Struthioniformes   **FAMILIES** 1   **SPECIES** 1

RHEAS
ORDER Rheiformes   **FAMILIES** 1   **SPECIES** 2

CASSOWARIES AND EMUS
ORDER Casuariiformes   **FAMILIES** 2   **SPECIES** 4

KIWIS
ORDER Apterygiformes   **FAMILIES** 1   **SPECIES** 3

TINAMOUS
ORDER Tinamiformes   **FAMILIES** 1   **SPECIES** 45

PENGUINS
ORDER Sphenisciformes   **FAMILIES** 1   **SPECIES** 17

DIVERS
ORDER Gaviiformes   **FAMILIES** 1   **SPECIES** 5

GREBES
ORDER Podicipediformes   **FAMILIES** 1   **SPECIES** 22

ALBATROSSES AND PETRELS
ORDER Procellariiformes   **FAMILIES** 4   **SPECIES** 108

PELICANS AND RELATIVES
ORDER Pelecaniformes   **FAMILIES** 6   **SPECIES** 65

HERONS AND RELATIVES
ORDER Ciconiiformes   **FAMILIES** 6   **SPECIES** 119

FLAMINGOS
ORDER Phoenicopteriformes   **FAMILIES** 1   **SPECIES** 5

WATERFOWL
ORDER Anseriformes   **FAMILIES** 2   **SPECIES** 149

BIRDS OF PREY
ORDER Falconiformes   **FAMILIES** 5   **SPECIES** 307

GAMEBIRDS
ORDER Galliformes   **FAMILIES** 6   **SPECIES** 281

CRANES AND RELATIVES
ORDER Gruiformes   **FAMILIES** 12   **SPECIES** 204

WADERS, GULLS, AND AUKS
ORDER Charadriiformes   **FAMILIES** 18   **SPECIES** 343

PIGEONS
ORDER Columbiformes   **FAMILIES** 1   **SPECIES** 309

SANDGROUSE
ORDER Pteroclidiformes   **FAMILIES** 1   **SPECIES** 16

PARROTS
ORDER Psittaciformes   **FAMILIES** 2   **SPECIES** 353

CUCKOOS AND TURACOS
ORDER Cuculiformes   **FAMILIES** 3   **SPECIES** 160

OWLS
ORDER Strigiformes   **FAMILIES** 2   **SPECIES** 205

NIGHTJARS AND FROGMOUTHS
ORDER Caprimulgiformes   **FAMILIES** 5   **SPECIES** 118

HUMMINGBIRDS AND SWIFTS
ORDER Apodiformes   **FAMILIES** 3   **SPECIES** 424

MOUSEBIRDS
ORDER Coliiformes   **FAMILIES** 1   **SPECIES** 6

TROGONS
ORDER Trogoniformes   **FAMILIES** 1   **SPECIES** 35

KINGFISHERS AND RELATIVES
ORDER Coraciiformes   **FAMILIES** 10   **SPECIES** 191

WOODPECKERS AND TOUCANS
ORDER Piciformes   **FAMILIES** 6   **SPECIES** 380

PASSERINES
ORDER Passeriformes   **FAMILIES** c.80   **SPECIES** 5,200–5,500

REPTILES

CLASS Reptilia

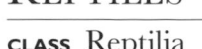

AFRICAN STRIPED SKINK

Snakes, lizards, and amphisbaenians, which make up the largest reptile order, Squamata, account for over 95 per cent of living reptile species. While reptiles share many features, their evolutionary history is very varied. As a result, many biologists believe that they form an informal group rather than a natural group or clade.

TORTOISES AND TURTLES		
ORDER Chelonia	FAMILIES 11	SPECIES 294

TUATARAS		
ORDER Rhyncocephalia	FAMILIES 1	SPECIES 2

SQUAMATES		
ORDER Squamata	FAMILIES 40	SPECIES c.7,558

SNAKES		
SUBORDER Serpentes	FAMILIES 18	SPECIES c.2,900

BOAS, PYTHONS, AND RELATIVES		
SUPERFAMILY Henophidia	FAMILIES 11	SPECIES 149

COLUBRIDS AND RELATIVES		
SUPERFAMILY Caenophidia	FAMILIES 4	SPECIES 2,439

COLUBRIDS	
FAMILY Colubridae	SPECIES 1,858
BURROWING ASPS	
FAMILY Atractaspidae	SPECIES 62
VIPERS	
FAMILY Viperidae	SPECIES 228
ELAPIDS	
FAMILY Elapidae	SPECIES 291

BLIND AND THREAD SNAKES		
SUPERFAMILY Scolecophidia	FAMILIES 3	SPECIES 319

LIZARDS		
SUBORDER Lacertilia	FAMILIES 19	SPECIES c.4,500

IGUANAS AND RELATIVES		
SUPERFAMILY Iguania	FAMILIES 3	SPECIES 1,412

GECKOS AND SNAKE LIZARDS		
SUPERFAMILY Gekkota	FAMILIES 4	SPECIES 1,054

SKINKS AND RELATIVES		
SUPERFAMILY Scincomorpha	FAMILIES 7	SPECIES 1,890

ANGUIMORPH LIZARDS		
SUPERFAMILY Anguimorpha	FAMILIES 6 or 7	SPECIES 173

AMPHISBAENIANS		
SUBORDER Amphisbaenia	FAMILIES 3	SPECIES 158

CROCODILIANS		
ORDER Crocodilia	FAMILIES 3	SPECIES 23

AMPHIBIANS

CLASS Amphibia

PACIFIC GIANT SALAMANDER

Frogs and toads make up the largest order of amphibians, and show the widest range of adaptations for terrestrial life. Newts and salamanders most closely resemble ancestral amphibians; caecilians are an aberrant and relatively little known group.

FROGS AND TOADS		
ORDER Anura	FAMILIES 29	SPECIES c.4,380

NEWTS AND SALAMANDERS		
ORDER Caudata	FAMILIES 10	SPECIES c.470

CAECILIANS		
ORDER Gymnophiona	FAMILIES 5	SPECIES c.170

FISH

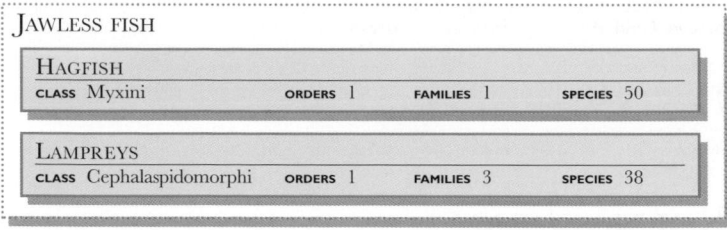

AMERICAN PADDLEFISH

Despite superficial similarities, fish are only an informal collection of relatively diverse animals. Of the four classes in which living fish species are generally classified, two are jawless – a characteristic shared by the earliest vertebrates (from over 500 million years ago). Today, bony fish make up the largest class (about 96 per cent of species). The major subclass of this group (ray-finned fish) contains so many orders and families that, in this book, it is dealt with at superorder level.

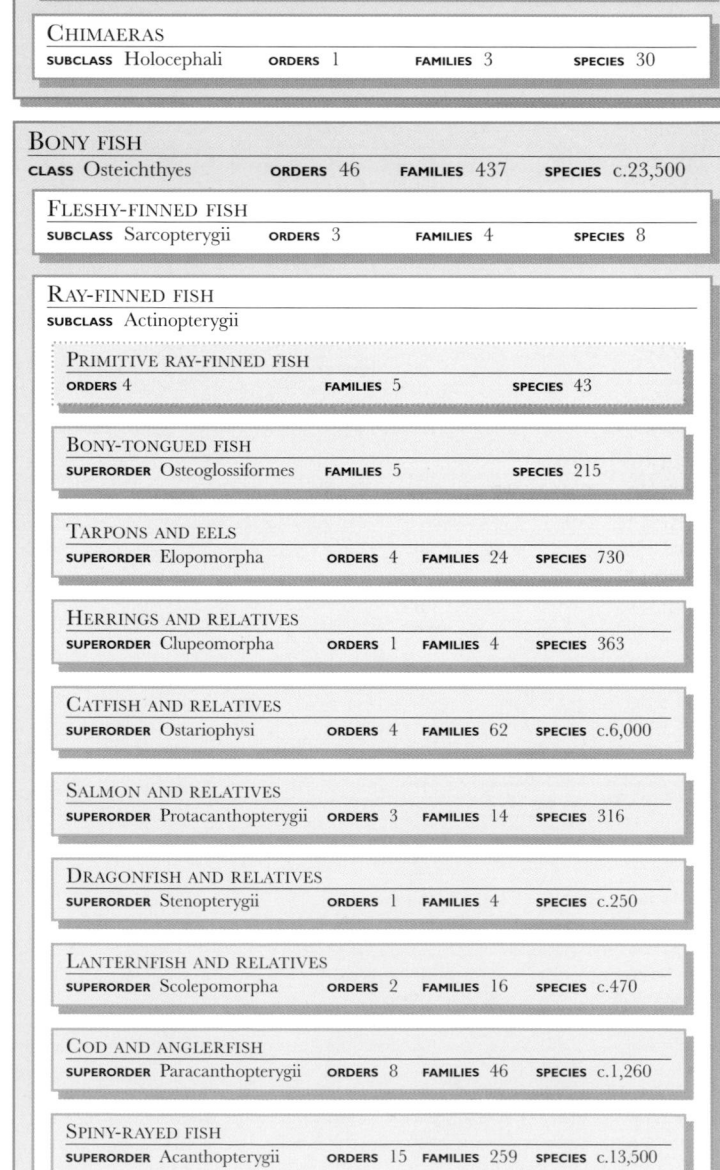

JAWLESS FISH			

HAGFISH			
CLASS Myxini	ORDERS 1	FAMILIES 1	SPECIES 50

LAMPREYS			
CLASS Cephalaspidomorphi	ORDERS 1	FAMILIES 3	SPECIES 38

CARTILAGINOUS FISH			
CLASS Chondrichthyes	ORDERS 14	FAMILIES 50	SPECIES c.860

SHARKS AND RAYS			
SUBCLASS Elasmobranchi			

SHARKS			
ORDERS 9		FAMILIES 33	SPECIES c.330

SKATES AND RAYS			
ORDERS 4		FAMILIES 14	SPECIES c.450

CHIMAERAS			
SUBCLASS Holocephali	ORDERS 1	FAMILIES 3	SPECIES 30

BONY FISH			
CLASS Osteichthyes	ORDERS 46	FAMILIES 437	SPECIES c.23,500

FLESHY-FINNED FISH			
SUBCLASS Sarcopterygii	ORDERS 3	FAMILIES 4	SPECIES 8

RAY-FINNED FISH			
SUBCLASS Actinopterygii			

PRIMITIVE RAY-FINNED FISH			
ORDERS 4		FAMILIES 5	SPECIES 43

BONY-TONGUED FISH			
SUPERORDER Osteoglossiformes		FAMILIES 5	SPECIES 215

TARPONS AND EELS			
SUPERORDER Elopomorpha	ORDERS 4	FAMILIES 24	SPECIES 730

HERRINGS AND RELATIVES			
SUPERORDER Clupeomorpha	ORDERS 1	FAMILIES 4	SPECIES 363

CATFISH AND RELATIVES			
SUPERORDER Ostariophysi	ORDERS 4	FAMILIES 62	SPECIES c.6,000

SALMON AND RELATIVES			
SUPERORDER Protacanthopterygii	ORDERS 3	FAMILIES 14	SPECIES 316

DRAGONFISH AND RELATIVES			
SUPERORDER Stenopterygii	ORDERS 1	FAMILIES 4	SPECIES c.250

LANTERNFISH AND RELATIVES			
SUPERORDER Scolepomorpha	ORDERS 2	FAMILIES 16	SPECIES c.470

COD AND ANGLERFISH			
SUPERORDER Paracanthopterygii	ORDERS 8	FAMILIES 46	SPECIES c.1,260

SPINY-RAYED FISH			
SUPERORDER Acanthopterygii	ORDERS 15	FAMILIES 259	SPECIES c.13,500

INVERTEBRATES

SPONGES

PHYLUM Porifera	**CLASSES** 4	**ORDERS** 18	**FAMILIES** 80	**SPECIES** c.10,000

CNIDARIANS

PHYLUM Cnidaria	**CLASSES** 4	**ORDERS** 27	**FAMILIES** 236	**SPECIES** 8,000–9,000

FLATWORMS

PHYLUM Platyhelminthes	**CLASSES** 4	**ORDERS** 35	**FAMILIES** 360	**SPECIES** c.17,500

ROUNDWORMS

PHYLUM Nematoda	**CLASSES** 4	**ORDERS** 20	**FAMILIES** 185	**SPECIES** c.20,000

SEGMENTED WORMS

PHYLUM Annelida	**CLASSES** 3	**ORDERS** 31	**FAMILIES** 130	**SPECIES** c.12,000

MOLLUSCS

PHYLUM Mollusca	**CLASSES** 8	**ORDERS** 35	**FAMILIES** 232	**SPECIES** c.100,000

MINOR PHYLA

ROTIFER

Invertebrates are classified in about 30 phyla, which vary considerably in size. In this book, major phyla are treated separately, but a selection of minor phyla also appears on pages 536–7. Almost all these minor phyla contain marine animals or ones that live in damp habitats.

BRACHIOPODS		WATER BEARS	
PHYLUM Brachiopoda	**SPECIES** c.350	**PHYLUM** Tardigrada	**SPECIES** c.600
ARROW WORMS		RIBBON WORMS	
PHYLUM Chaetognatha	**SPECIES** c.90	**PHYLUM** Nemertea	**SPECIES** c.1,200
COMB JELLIES		VELVET WORMS	
PHYLUM Ctenophora	**SPECIES** c.100	**PHYLUM** Onychophora	**SPECIES** c.70
BRYOZOANS		SPOONWORMS	
PHYLUM Bryozoa	**SPECIES** c.4,300	**PHYLUM** Echiura	**SPECIES** c.150
HEMICHORDATES		HORSESHOE WORMS	
PHYLUM Hemichordata	**SPECIES** c.85	**PHYLUM** Phorona	**SPECIES** c.20
PEANUT WORMS		10 OTHER MINOR INVERTEBRATE GROUPS	
PHYLUM Sipuncula	**SPECIES** c.350		
ROTIFERS			
PHYLUM Rotifera	**SPECIES** c.1,800		

ECHINODERMS

PHYLUM Echinodermata	**CLASSES** 6	**ORDERS** 36	**FAMILIES** 145	**SPECIES** c.6,000

INVERTEBRATE CHORDATES

TUNICATE

Invertebrate chordates are animals that share some characteristics with vertebrates but lack a bony skeleton. They are divided into two subphyla – the tunicates (the majority of species) and the lancelets. Tunicates are immobile, and they often have bag-like bodies as adults; lancelets are mobile and bear strong resemblances to vertebrates, despite their lack of bone.

TUNICATES			
SUBPHYLUM Urochordata			
CLASSES 3	**ORDERS** 8	**FAMILIES** 45	**SPECIES** c.2,100

LANCELETS			
SUBPHYLUM Cephalochordata			
CLASSES 1	**ORDERS** 1	**FAMILIES** 2	**SPECIES** 25

ARTHROPODS

PHYLUM Arthropoda

SPIDER-HUNTING WASP

Arthropods form the largest phylum in the animal kingdom. Insects make up the biggest subgroup, but the phylum also contains two other giant classes – crustaceans and arachnids – which dwarf many phyla in the invertebrate world.

MANDIBULATES

SUBPHYLUM Mandibulata	**CLASSES** 6	**ORDERS** 85	**FAMILIES** c.1660	**SPECIES** c.1 million

HEXAPODS

SUPERCLASS Hexapoda	**CLASSES** 4	**ORDERS** 32	**FAMILIES** 980	**SPECIES** c.1 million

INSECTS

CLASS Insecta	**ORDERS** 29	**FAMILIES** 949	**SPECIES** c.1 million

BRISTLETAILS		PARASITIC LICE	
ORDER Archaeognatha	**SPECIES** c.350	**ORDER** Phthiraptera	**SPECIES** c.6,000
SILVERFISH		BUGS	
ORDER Thysanura	**SPECIES** c.370	**ORDER** Hemiptera	**SPECIES** c.82,000
MAYFLIES		THRIPS	
ORDER Ephemeroptera	**SPECIES** c.2,500	**ORDER** Thysanoptera	**SPECIES** c.5,000
DAMSELFLIES AND DRAGONFLIES		DOBSONFLIES AND ALDERFLIES	
ORDER Odonata	**SPECIES** c.5,500	**ORDER** Megaloptera	**SPECIES** c.300
CRICKETS AND GRASSHOPPERS		SNAKEFLIES	
ORDER Orthoptera	**SPECIES** c.20,000	**ORDER** Rapdhidioptera	**SPECIES** c.150
STONEFLIES		ANTLIONS, LACEWINGS, AND RELATIVES	
ORDER Plecoptera	**SPECIES** c.2,000	**ORDER** Neuroptera	**SPECIES** c.4,000
ROCK CRAWLERS		BEETLES	
ORDER Grylloblattodea	**SPECIES** 25	**ORDER** Coleoptera	**SPECIES** c.370,000
STICK AND LEAF INSECTS		STREPSIPTERANS	
ORDER Phasmatodea	**SPECIES** c.2,500	**ORDER** Strepsiptera	**SPECIES** c.560
EARWIGS		SCORPIONFLIES	
ORDER Dermaptera	**SPECIES** c.1,900	**ORDER** Mecoptera	**SPECIES** c.550
MANTIDS		FLEAS	
ORDER Mantodea	**SPECIES** c.2,000	**ORDER** Siphonaptera	**SPECIES** c.2,000
COCKROACHES		FLIES	
ORDER Blattodea	**SPECIES** c.4,000	**ORDER** Diptera	**SPECIES** c.122,000
TERMITES		CADDISFLIES	
ORDER Isoptera	**SPECIES** c.2,750	**ORDER** Trichoptera	**SPECIES** c.8,000
WEB-SPINNERS		MOTHS AND BUTTERFLIES	
ORDER Embioptera	**SPECIES** c.300	**ORDER** Lepidoptera	**SPECIES** c.165,000
ANGEL INSECTS		BEES, WASPS, ANTS, AND SAWFLIES	
ORDER Zoraptera	**SPECIES** c.30	**ORDER** Hymenoptera	**SPECIES** c.198,000
BARKLICE AND BOOKLICE			
ORDER Psocoptera	**SPECIES** c.3,500		

SPRINGTAILS

CLASS Collembola	**ORDERS** 1	**FAMILIES** 18	**SPECIES** c.6,500

PROTURANS

CLASS Protura	**ORDERS** 1	**FAMILIES** 4	**SPECIES** c.400

DIPLURANS

CLASS Diplura	**ORDERS** 1	**FAMILIES** 9	**SPECIES** c.800

MYRIAPODS

SUPERCLASS Myriapoda	**CLASSES** 4	**ORDERS** 16	**FAMILIES** 144	**SPECIES** c.13,700

CRUSTACEANS

SUPERCLASS Crustacea	**CLASSES** 6	**ORDERS** 37	**FAMILIES** 540	**SPECIES** c.40,000

CHELICERATES

SUBPHYLUM Chelicerata	**CLASSES** 3	**ORDERS** 14	**FAMILIES** 480	**SPECIES** c.77,500

ARACHNIDS

CLASS Arachnida	**ORDERS** 12	**FAMILIES** 450	**SPECIES** c.75,500

SEA SPIDERS

CLASS Pycnogonida	**ORDERS** 1	**FAMILIES** 9	**SPECIES** c.1,000

HORSESHOE CRABS

CLASS Merostomata	**ORDERS** 1	**FAMILIES** 1	**SPECIES** 4

INTRODUCTION

ANATOMY

ALL BUT THE VERY SIMPLEST ANIMALS are made up of various parts. Their smallest fully functional parts are cells, which are shaped in different ways according to the tasks that they carry out. Groups of similar cells are organized into tissues, and tissues are grouped together to form organs. Organs themselves are linked to form organ systems, which carry out all the processes essential for survival. The design of these systems varies widely between one type of animal and another, and also between animals that live in different ways, but the work they do is the same.

BODY SYSTEMS

Animals have up to a dozen separate body systems. In many species, the muscular and skeletal systems make up a large proportion of the body's total weight, while the integumentary

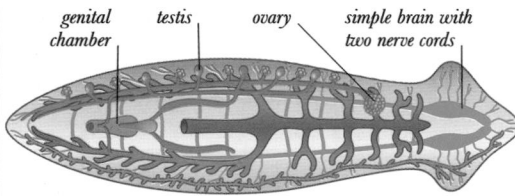

genital chamber   testis   ovary   simple brain with two nerve cords

SIMPLE INVERTEBRATE BODY SYSTEMS
Simple invertebrates, such as flatworms, do not have respiratory organs or a circulatory system. The digestive system often has just one opening, the mouth, and the reproductive system is typically made up of both male and female organs.

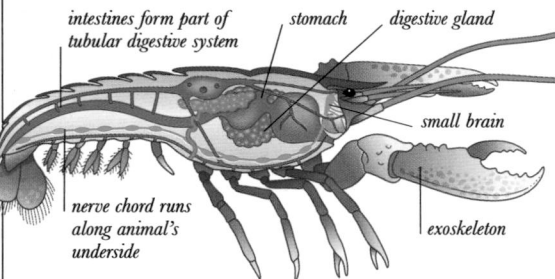

intestines form part of tubular digestive system   stomach   digestive gland

small brain

nerve chord runs along animal's underside   exoskeleton

ARTHROPOD BODY SYSTEMS
The tubular digestive system is open at both ends. The blood flows partly through vessels and partly through body spaces. Oxygen is supplied via gills or via minute airways called tracheae.

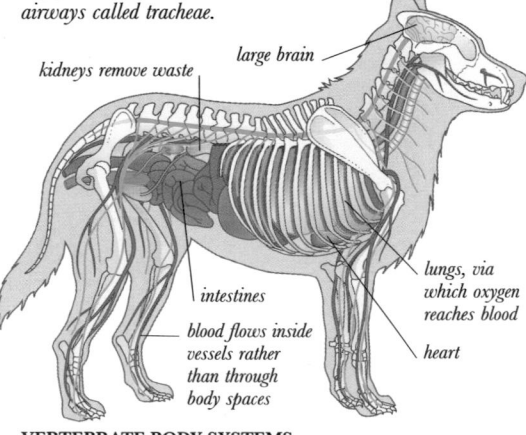

large brain

kidneys remove waste

intestines

blood flows inside vessels rather than through body spaces

lungs, via which oxygen reaches blood

heart

VERTEBRATE BODY SYSTEMS
The nervous system is highly developed, and the brain is typically larger than in invertebrates. The circulatory system is closed, and blood is pumped under high pressure by the heart.

KEY TO SYSTEMS

■	CIRCULATORY	■	EXCRETORY
■	DIGESTIVE	■	NERVOUS
■	RESPIRATORY	■	REPRODUCTIVE

system – the parts that form the "outer cladding" – protect the body from physical damage and, in terrestrial animals, from the threat of drying out.

Two key systems enable animals to obtain energy from food. The digestive system breaks food down so that it can be absorbed; and the respiratory system delivers oxygen to the body's cells so that food substances can be "burned" and their chemical energy released. The respiratory system also removes carbon dioxide – a potentially toxic waste product formed during the production of energy. In many animals, including all vertebrates, oxygen and carbon dioxide are carried by blood in the circulatory system. Carbon dioxide is usually breathed out; other kinds of dissolved waste are removed, before they have a chance to build up, by a separate excretory system.

Animals use two different systems to co-ordinate their bodies, and to react to their surroundings. The nervous system deals with anything that needs a fast response, processing information gathered by sense organs. In all animals, it triggers inbuilt or instinctive behaviour, but in some, especially vertebrates, it also stores information, allowing animals to adapt their behaviour according to their past experience. The endocrine system works in conjunction with the nervous system, releasing hormones, or chemical messengers, that help the nervous system to co-ordinate long-term processes.

Finally, the reproductive system carries out the most important task: producing young. Unlike other body systems, it often functions only during a set season, and then only in mature animals.

SKELETONS AND SUPPORT

Animals need to keep the shape of their bodies stable. Many invertebrates achieve this without any hard body parts at all. Instead, they rely on the pressure of internal fluids to keep their bodies firm in the same way that air stabilizes a tyre. This system, called a hydrostatic skeleton, works well on a small scale. But in larger animals, particularly land ones, this type of skeleton is often not strong enough to support the body's weight. Animals have developed two quite different solutions to this problem: external shells and body cases; and internal skeletons, typically made of bone.

Shells are made up of one or two parts, and they grow in step with their owner. They can be seen in marine animals called brachiopods or lampshells, but they are most highly developed in bivalve molluscs, which can grow shells over 1m (3¼ ft) across. Body cases are more complex than shells: they consist of a large number of separate plates that meet at flexible joints. They are a

characteristic feature of arthropods – a huge group of invertebrates that includes insects, crustaceans, and arachnids. These cases, or exoskeletons, cover the entire body and include structures as strong as a crab's pincers or as delicate as a butterfly's antennae. Unlike shells, body cases cannot grow, so periodically they have to be shed and replaced.

Internal skeletons (endoskeletons), made of bone, provide support from within. Found only in vertebrates, they have two great advantages: they are relatively light for their size – important for land animals that have to move quickly – and they can grow. In this kind of skeleton, some bones meet at flexible joints. Other bones, particularly those in the skull, lock together for extra strength.

SHELL
A snail's shell grows at its lip, getting progressively larger with its owner. Some species can seal the shell with an operculum or "door".

EXOSKELETON
A crab's body case covers all the surfaces of its body, including its eyes. Unlike an insect's exoskeleton, it is reinforced with calcium.

SECTION OF BONE

marrow cavity

hard bone containing osteocytes

ENDOSKELETON
Strong, light, and flexible, bones are living body parts formed by cells (called osteocytes), surrounded by mineral crystals that are laid down in rings.

SYMMETRY

Some animals' bodies are circular, without a head or tail. This type of anatomy, called radial symmetry, can be seen in sea anemones and other cnidarians, and also in ctenophores or comb jellies. The vast majority of animals show bilateral symmetry, which means they can be divided into halves. The halves are not always equal: male fiddler crabs, for example, have very unequal claws, and flatfish have different sides.

RADIAL SYMMETRY
A sea anemone can be divided into two equal halves on any axis. It has a central mouth and body cavity, with feeding tentacles arranged in a ring.

BILATERAL SYMMETRY
Frogs and toads have just one axis of symmetry. Externally, the two halves look identical; internally, some organs lie to the left or right.

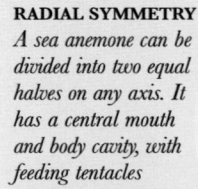

MUSCLES AND MOVEMENT

Muscles work by contracting. This means they can pull but not push. In most cases, they are arranged in pairs or groups that pull in opposing directions: when one muscle or muscle group contracts, its partner is brought back to its normal resting shape.

Muscles make animals move in different ways. In animals without limbs, such as earthworms and jellyfish, they work to change the body's shape. In earthworms, opposing muscles alternately shorten and lengthen the animal's segments so that it can creep through soil. In limbed animals, one set of muscles pulls the limb down or back, while the other lifts it up or forwards.

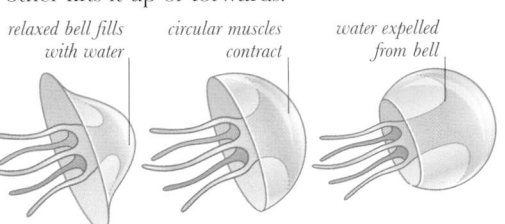

relaxed bell fills with water   *circular muscles contract*   *water expelled from bell*

MOVING WITHOUT LIMBS
Jellyfish swim by rhythmically contracting the bell-shaped part of their body. This expels water from the bell, which pushes the jellyfish forwards. Jellyfish make little headway against the current: they swim mainly to keep at the right level in the water.

muscles contract to propel frog forwards   *muscles maintain streamlined shape*   *front legs absorb impact*

JUMPING
When a frog leaps, its legs act as levers, propelling it into the air. Its front legs fold up to cushion its body from the impact when it lands.

As well as making animals move, muscles serve other purposes. They force food through the digestive system (peristalsis) and pump blood around the circulatory system. Unlike most other muscles, the heart muscle has a built-in rhythm that keeps it contracting throughout an animal's life.

BODY COVERINGS

Animal cells are easily damaged. To protect them from injury and disease, animals have body coverings, most of which consist largely of non-living matter. Mammalian skin is covered by dead cells, while insect body cases are covered by hard proteins and waterproof wax. In many cases, these protective layers are themselves protected: mammals often have a coat of fur, while many other animals have scales.

Some of these extra coverings have developed additional uses. Soft feathers and fur help to retain body heat, while extra-strong feathers are used in flight. Colours or patterns act as camouflage or help animals to recognize their own kind.

FEATHERS

SCALES

FUR

BODY BARRIERS
Bird feathers, butterfly scales, and mammal fur are made of non-living substances that are produced by living cells. Feathers and fur are replaced during their owner's lifetime, but butterfly scales are not.

RESPIRATION

For small and thin animals, obtaining oxygen is a simple matter because it seeps into their bodies from outside. At the same time, carbon dioxide escapes in the other direction. For larger animals, respiration is more complex. In relative terms, they have a much greater volume than surface area, so there is less room for gases to move in and out. To breathe, they rely on respiratory organs – structures that effectively pack a large surface area into a small amount of space.

In aquatic animals, gills are the most common respiratory organs. Typical gills consist of thin, flat, or feathery surfaces that bring blood into close contact with the water outside.

However, most gills do not work in air because – out of water – their surfaces collapse and stick together. Land animals therefore have hollow respiratory organs that carry air deep inside their bodies. In insects, these organs are tubes, called tracheae, which divide into extremely fine filaments that reach individual cells. In land-dwelling vertebrates, the organs are lungs – air-filled chambers surrounded by a network of blood vessels. Muscles make the lungs expand or contract, sucking air in or blowing it out.

NERVES AND SENSES

Nerve cells, or neurons, are the animal world's equivalent of wiring. Neurons conduct brief bursts of electricity, known as impulses, which carry information from sense organs or make muscles contract. Corals and other simple animals have a network of nerves scattered throughout their bodies. But in most animals, the nervous system converges on the brain.

Some animal senses, such as touch, operate through nerve endings scattered all over the body. A similar sense, which works internally, tells animals about their posture. The most important senses – vision, smell, and hearing – work through organs that form some of the most elaborate structures in the body.

Vision is essential for many animals, and eyes show a wide variety of designs. At their simplest – for example in snails – they do little more than distinguish between light and dark. In many animals, particularly arthropods and vertebrates, they focus light onto large numbers of nerve cells, building up a detailed image of the surroundings. In vertebrates, these eyes have a single lens, which throws light onto a "screen", or retina. In arthropods, the eye has up to 25,000 separate compartments, each with its own lens system; these compound eyes create a mosaic-like image and are especially good at detecting movement.

Mammals are the only animals with prominent earflaps. Vertebrates' ears are always on the head, but in some animals they are positioned elsewhere. Most grasshoppers and crickets have ears on their abdomen or legs. Organs that detect taste and smell can also be in a variety of positions. Like ears, they can be used in communication, as well as for avoiding danger and finding food.

Many animals have senses that are more acute than those of human beings, and some can sense things that humans cannot. For example, most fish can sense pressure waves in water, and many can detect weak electric fields. Some snakes can "see" warmth, enabling them to attack warm-blooded prey in total darkness.

EFFICIENT RESPIRATION
Animals that are highly active in cold habitats need lots of oxygen. When an ibex runs uphill, its lungs take in 10 times as much air as when it is lying down.

BINOCULAR VISION
Jumping spiders have two extra-large eyes that face directly forwards. This gives them binocular vision, which is essential for gauging distances before making a jump.

BEHAVIOUR

AN ANIMAL'S BEHAVIOUR encompasses all the things that it does as well as the way that it does them. Behaviour ranges from simple actions, such as eating or keeping clean, to highly elaborate activities, such as hunting in a pack, courting a mate, or building a nest. In some animals, behaviour is almost entirely predictable; in others it develops with experience, so the more that an animal does something the more skilled it becomes. As with all aspects of animal biology, behaviour is the product of evolution, which means that it gradually changes as time goes by. These changes enable species to react in the most effective way to the opportunities and dangers that they encounter in daily life.

INSTINCT AND LEARNING

In simple animals, behaviour is "hard wired" – governed by inherited instincts – which means that it consists of fixed sequences of actions prompted by triggers. For example, day-old birds instinctively beg for food when their parents appear at the nest. At this stage, they are usually blind: their behaviour is triggered by noise and movement rather than by the sight of food itself.

Instinctive behaviour may seem basic, but it can produce quite remarkable results. The structures that animals build – from nests to dams – are the results of inborn behavioural impulses. When beavers set out to make a dam, they do so without any knowledge of engineering principles. Yet the structure they make is shaped to withstand water pressure as if it had been scientifically designed.

Beavers do not have to think how to build, just as spiders do not need to work out how to weave webs. Even so, the results of instinctive behaviour

can change. As animals repeat certain tasks, such as making a nest, their performance often improves. This is particularly important for some animals – such as male weaverbirds – which use their nestbuilding skills to attract a mate.

Apart from octopuses and their relatives, most invertebrates have narrow limits when it comes to learning. For vertebrates, on the other hand, learned behaviour is often extremely important. Frogs and toads quickly learn to avoid animals that taste unpleasant, while mammals acquire a wide range of skills from their parents, including how to hunt. Among primates, individuals very occasionally "invent" new behaviour, which is then copied by their neighbours. This copying process produces culture – patterns of behaviour that are handed on down the generations. Culture is something that humans, as a species, have developed to a unique degree.

INSTINCTIVE WEAVING
Spiders often build highly complex webs, but they always produce them to one particular design. As a result, it is often possible to identify a spider from its web alone. This garden spider will use its web for just one day; then it will eat it before starting on a replacement.

LEARNING TO FEED
Eurasian oystercatchers learn how to feed by watching their parents. Some birds hammer at shells to break them, while others stab at the shells' hinges to force them to open. Once a bird has learnt one technique, it uses it for life.

ANIMAL INTELLIGENCE

Intelligence is the ability to make decisions based on insight and understanding. This ability is rare in the animal world and difficult to identify and measure. For example, some herons use bait to catch fish. They look as if they understand their actions, but their behaviour is instinctive – it involves no insight at all. Chimpanzees, however, will probe for insects with twigs that they shape to increase their effectiveness. To shape the twig, they must first form a mental picture of how it will work – a process that humans also use. They also have a complex vocabulary of sounds that allows them to exchange information.

THE SIGNS OF INTELLIGENCE
Chimpanzees can solve problems, and they are able to learn sign language to communicate with humans. They can learn symbols for objects and actions, and they occasionally combine the symbols in ways that resemble spoken phrases.

COMMUNICATION

For most animals, keeping in touch with their own kind is essential to their survival. Animals communicate with each other for a range of reasons, including finding food, attracting a mate, and bringing up their young. Different methods of communication have their own advantages and drawbacks. Body language – which includes facial expressions and physical displays – works well at close quarters but is ineffective at a distance and in habitats where dense vegetation gets in the way. In such cases, communication by sound is much more practical. Whales call to each other over immense distances, while some small animals produce remarkably loud sounds for their size. Treefrogs, cicadas, and mole crickets, for example, can often be heard at a distance of over 2km (1 mile). Each species uses its own distinctive "call sign", and many behave like ventriloquists, pitching their calls in a way that throws predators off their track.

Animals that are capable of producing light also use identifying call signs. These can consist of specific sequences of flashes or – in many deep-sea fish – illuminated body patterns. But, like body language and sound, this form of communication works only when the signaller is actively signalling. Scent communication is quite different because

NIGHT LIGHT
Female glow worms use light to signal their presence to the males, which fly overhead. Signalling with light can be dangerous because it can attract predators as well as potential mates. If a glow worm senses danger, she quickly "switches off" her light.

JOINING THE CHORUS
By howling, wolves advertise their ownership of a hunting territory to any other wolves that may be in the area. Wolves often howl at night, after they have made a successful kill.

the signal lingers long after the animal that made it has moved on. Animal scents are specific, allowing animals to lay trails and to advertise their presence to potential mates. Some male insects are able to respond to individual molecules of airborne scent, allowing them to track down females far upwind.

LIVING IN GROUPS

Some animals spend all their lives alone and never encounter another member of their species. But, for many, getting together is an important part of life. Animal groups vary in size as well as in how long they last: mayflies, for example, form mating swarms that last just a few hours, while migrating

GROUP FORMATION
By forming a V (or skein), geese can reduce the amount of energy needed to migrate by flying in the leading birds' slipstream. They take turns to lead.

SAFETY IN NUMBERS
In open habitats, such as grassland, where predators tend to be fast and hiding places scarce, prey mammals typically live in large groups. This group of zebras has further increased its security by grazing near a herd of wildebeest.

birds often assemble for several weeks. Many other animals, including fish and grazing mammals, form groups that are maintained for life.

Groups of animals may seem to be easy targets for predators, but the opposite is usually true. Predators find it difficult to single out individuals from a group, so living together gives animals a better chance of survival. Groups are also more difficult to catch by surprise because there is always more than one animal alert for signs of danger.

In most animal groups, the members belong to a single species but do not necessarily share the same parents. However, in the most tight-knit groups,

the members are all closely related. Examples of such extended families include wolf packs and kookaburra "clans", where the young remain with their parents instead of setting up independently. This kind of group-living reaches its extreme in social insects, such as termites and ants, which cannot survive alone.

DEFENCE AND ATTACK

Both predatory and prey animals use specialized behaviour to help them survive. For example, while many prey animals simply try to escape, others keep perfectly still, relying on camouflage to protect them. A wide range of species, from moths to lizards, try to make themselves appear dangerous by

RESPONDING TO THREAT
When threatened, puffer fish enlarge themselves by gulping water. Once distended, they can barely move, but their spines make them practically impossible to attack.

exaggerating their size or by revealing coloured spots that look like eyes. Sometimes such threats are real: for example, the brilliant colours of poison-arrow frogs indicate that they contain some of the animal kingdom's most potent poisons.

Predatory animals use one of two techniques to catch prey: they either wait for it to come their way or they track it down. "Sit-and-wait" predators are often camouflaged, and some actively entice their victims within range. In anglerfish, for example, the snout has a long, luminous protuberance, called a lure, which the fish dangle appetizingly in front of their mouths: anything swimming close to inspect this lure is snapped up whole. For active hunting, nature puts a premium on speed and keen senses, which is why animals such as cheetahs, peregrine falcons, and blue marlins are among the fastest in the world. Some active predators operate in groups. By working together, grey wolves, African wild dogs, and lions can tackle prey much larger than themselves.

PACK HUNTING
A pack of African wild dogs pulls down a wildebeest that they have run to exhaustion. Once their victim is dead, the dogs will share out the remains and take some back to any pups that have been left behind at their den.

BEHAVIOURAL CYCLES

Some kinds of behaviour, including self-defence, can be provoked at any time. Others are cyclical, triggered by cues that keep animals in step with changes around them. One of the most important cycles is the alternation between night and day. Others include the rise and fall of the tide and the annual sequence of changing seasons.

Cyclical behaviours are all instinctive. They may be stimulated by external changes, by inbuilt "biological clocks", or by a combination of the two. Birds, for example, often gather to roost late in the day, a form of cyclical behaviour that is triggered by falling light levels as the sun nears the horizon. On a much longer time scale, ground squirrels show an annual cycle in body weight, getting heavier before they enter hibernation. However, ground squirrels maintain their cycle even if kept in conditions of constant temperature and day length, which shows that the rhythm is controlled biologically. Biological clocks often involve hormones, but the way they work is not yet fully understood.

FEEDING TIME
Fiddler crabs emerge from their burrows to pick over the nutrient-rich sediment for small particles of food. They cannot feed underwater, so their feeding behaviour is governed by the daily rise and fall of the tide.

LIFE CYCLES

AN ANIMAL'S LIFE CYCLE CONSISTS OF ALL THE STAGES between the beginning of one generation and the beginning of the next. In some species – especially insects and other small invertebrates – the entire cycle is completed within a few weeks; in much larger animals it often takes many years. Regardless of how long it takes to complete, an animal's life cycle always involves two main steps: a period of growth and development, followed by reproduction. Some animals reproduce once and then die: for them, reproduction marks the end of life as well as the completion of the life cycle. For many, reproduction continues throughout adulthood, giving animals more than one chance to produce young.

REPRODUCTION

The ability to reproduce is the cornerstone of life because it allows living things to multiply, exploit new opportunities around them, and evolve. Animals reproduce in one of two ways: asexually (without sex) or sexually.

In asexual reproduction, a single parent partitions off part of itself to form a new animal. The partitioning process can happen in

SEXUAL REPRODUCTION
As a female common frog lays her eggs, her mate sprays them with sperm to fertilize them. Each tadpole will be genetically unique.

ASEXUAL REPRODUCTION
This sea anemone is in the final stages of reproducing, which it does by tearing itself in half. The result will be two individuals that have exactly the same genetic make-up.

a variety of ways. Hydras, for example, produce small buds that grow into complete new animals, while sea anemones literally tear themselves in two. Some animals produce eggs that develop without fertilization – a process called parthenogenesis. This is common in aphids and other sap-sucking insects, but it is rare in vertebrates (whiptail lizards are one of the few examples). Asexual reproduction is relatively quick and simple, but it has one important disadvantage: since only one parent is involved, the offspring are either genetically identical to that parent or very similar to it. As a result, parent and offspring are equally vulnerable to threats such as disease: if one animal dies, the rest will often follow suit. Sexual reproduction gets around this problem because the involvement of two parents means that the offspring are genetically varied. Each one has a unique combination of characteristics, which means the fittest survive and the species slowly evolves. However, the disadvantage of sexual reproduction is that it is much more complicated: the parents must be of the right species and correct sex, and in most cases they must co-operate to breed. In addition, only one parent – the female – actually produces young, so some reproductive potential is lost. Despite these difficulties, sexual

reproduction is widespread throughout the animal world, which demonstrates its long-term value. Even species that normally reproduce without sex periodically include a sexual phase in their life cycle, thereby getting the best of both worlds.

ANIMAL SEXES

All animals that reproduce sexually with a partner of the opposite sex show dimorphism – that is, the males and females are anatomically different and behave in different ways. In some species, the differences are not obvious, but in others they are quite distinct. Dimorphism exists because the sexes have developed different roles in reproduction and need different body forms to carry them out.

However, not all sexually reproducing animals are of opposite sexes. Some – earthworms and terrestrial snails, for example – are hermaphrodite (they have both male and female sex organs).

DIMORPHIC PARTNERS
In some animals, sex differences are extreme. Here, a wingless female vapourer moth (seen on the right) has attracted a winged male. After mating, the female will crawl away to lay her eggs.

This simplifies sexual reproduction because any adult of a species is potentially a suitable mate for another. A further variation is that some species have separate sexes but individuals can change sex during adult life. Parrotfish, for example, often live in small shoals dominated by a single male. If the male dies, a female changes sex and takes his place.

COURTSHIP

Before an animal can mate, it has to find a partner. This is easy enough for species that live in groups, but for animals that live alone it poses problems. Solitary animals locate potential mates by sending out signals, such as sounds or airborne scents. Each species has its own characteristic "call sign", ensuring that it finds others of its own kind.

Once the sexes are in contact, one partner – usually the male – has to overcome the other's wariness and demonstrate his suitability as a mate. This process is known as courtship. It often takes the form of ritualized behaviour that displays the male's physical fitness or his ownership of a good provision of food. If the female is sufficiently impressed, she will accept him as her mate.

While some species form lifelong partnerships, many go their separate ways after mating. In the latter case, the males typically mate with several females but take no part in raising the young. More rarely, things work in the opposite way, with one female mating with several males. Where this happens – for example in phalaropes (see p.306) – the female is more brightly coloured than the male and often takes the lead in courtship. In general, these females take little or no part in rearing young.

LIFESPANS

In general, animal lifespans are directly related to adult body size: the larger an animal is, the longer it is likely to live, although there are exceptions to this rule. One important factor affecting lifespan is metabolic rate – the rate at which an animal uses energy to make its body work. "Cold-blooded" animals, such as amphibians and reptiles, have a relatively low metabolic rate and tend to be long-lived, while "warm-blooded" animals, such as birds and mammals, have a high metabolic rate and tend to be relatively short-lived. This is especially true of small species because a low body mass means that heat escapes quickly from the body and has to be constantly replaced by food. Environmental factors, such as temperature and humidity, also have an effect. Houseflies, for example, often die within six weeks in warm conditions but may survive for many months if it is cool, and some microscopic animals survive for decades if they remain in a dormant state. In general, animals rarely live for long after their reproductive life has come to an end.

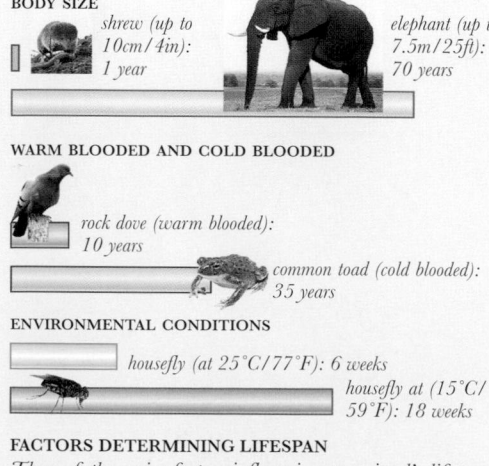

BODY SIZE
shrew (up to 10cm/4in): 1 year
elephant (up to 7.5m/25ft): 70 years

WARM BLOODED AND COLD BLOODED
rock dove (warm blooded): 10 years
common toad (cold blooded): 35 years

ENVIRONMENTAL CONDITIONS
housefly (at 25°C/77°F): 6 weeks
housefly at (15°C/59°F): 18 weeks

FACTORS DETERMINING LIFESPAN
Three of the major factors influencing an animal's lifespan are illustrated above. The life expectancy of each animal is represented by a coloured bar, indicating the degree to which a factor can affect its lifespan.

COURTSHIP RITUAL
A male frigatebird inflates his throat pouch to attract the attention of a female. Birds have good colour vision, which explains why males often have striking plumage.

FERTILIZATION

Marking the start of a new life, fertilization occurs when a male sperm and female egg cell fuse. In animals, it takes place in one of two ways: either outside or inside the female's body.

External fertilization is used by many animals that either live permanently in water or return to it to breed. In the simplest version of this process – seen in static invertebrates such as corals – vast numbers of the male and female sex cells are shed into the water, where they mingle so that fertilization can occur. A more advanced version of this, shown by animals that can move about, such as frogs, involves two partners pairing up. Although they appear to mate, fertilization nevertheless takes place in water rather than inside the female's body.

External fertilization does not work on land because sex cells soon dry out and die when exposed to air. Most terrestrial animals therefore use internal fertilization. In general, this involves the male injecting sperm into the female. However, some terrestrial animals, such as salamanders and newts, do not copulate. Instead, the male deposits a package of sperm (a spermatophore) near the female; she then collects it with her reproductive organ so that internal fertilization can take place.

INTERNAL FERTILIZATION
Like all insects, flat-footed bugs have to pair up so that the female's eggs can be fertilized. Mating takes several hours.

EXTERNAL FERTILIZATION
Corals release their sex cells into the water. They are triggered to do this by the changing phases of the moon – a natural clock that many animals use to synchronize their breeding behaviour.

STARTING LIFE

Most animals – apart from the ones that use asexual reproduction – start life as a single fertilized egg cell. If the egg has been fertilized externally, it will already be outside the mother's body, perhaps drifting in the sea or glued to sea-bed plants or sand. If the egg has been fertilized internally, it will either be laid, to hatch afterwards, or it will be retained inside the mother while it begins to develop into a young animal.

The degree of development that takes place at any one stage varies from one type of animal to another. Oviparous species, such as birds, lay their

LIVE BIRTH
Animals that give birth have relatively few offspring, but they often show a high level of parental care. This female seal will look after her cub for many weeks.

fertilized eggs before foetal development begins. In birds, development is often deferred for several more days until the clutch is complete; it begins as soon as the parent starts to incubate the eggs. Ovoviviparous species, which include many reptiles and sharks, incubate their eggs internally, "giving birth" at the moment when the eggs are about to hatch. Viviparous species – which include virtually all mammals, as well as some reptiles, amphibians, and fish – give birth to live young.

METAMORPHOSIS

All animals change shape as they grow and develop. In some the changes are gradual and relatively minor, but in others they are so far-reaching that the animal is completely transformed. This transformation is called

LIVING TOGETHER

Many animals associate with their own kind but continue to lead independent lives. However, eusocial species, such as termites, ants, and bees, form permanent groups or colonies in which just one member – the queen – produces all the colony's young. As a result, the colony's members are closely related and in many ways behave like a single organism. The great success of the system is apparent in the fact that these insects are among the most numerous on earth.

QUEEN TERMITE
Hidden deep inside a termite nest, where she is attended by her workers, a queen termite lays up to 30,000 eggs a day. In her complete dependence on the workers for food, she represents the ultimate form of reproductive specialization.

metamorphosis. It allows animals to live in different ways – and often in different habitats – during their young and adult lives.

Although metamorphosis is most common in invertebrates, it does occur in amphibians and some fish. Animals that undergo the metamorphic process spend the early part of their lives as larvae. In the sea, larvae often drift near the surface as part of the plankton, and because they are carried far and wide they play an important role in helping their species to spread.

In the insect world, metamorphosis occurs in two ways. Incomplete metamorphosis, shown by grasshoppers and bugs, involves a series of gradual changes that are made as the young insect, or nymph, matures. Complete metamorphosis – shown by butterflies, beetles, and flies – involves more drastic changes, which occur during a resting stage, called pupation, when the body is broken down and rearranged.

CRAB EGGS

ZOEA LARVA

ADULT CRAB

FROM LARVA TO ADULT
Like most crustaceans, shore crabs start life as eggs (top). The eggs hatch to produce the first larval stage, called a zoea (centre), which floats in plankton. This changes into a megalopa larva, then sinks to the sea bed and finally changes into an adult crab.

ANIMALS IN DANGER

UNTIL A CENTURY AGO, the earth still contained large areas of wilderness where animals had little or no contact with human beings. Since then, the human population has grown rapidly, use of energy and raw materials has escalated with it, and there are now very few parts of the planet that are truly beyond human reach. Natural habitats are disappearing, and earth's biodiversity – the sum total of all living species – is in rapid decline. This is potentially dangerous for human beings because it makes the world a less stable place and wipes out resources that may one day be needed. For animals, the results can be disastrous. The worst affected species have now reached the point where, ironically, they rely on human help to survive.

HABITAT CHANGE

Humans first started to alter habitats when they discovered fire, but it was with the start of farming, about 10,000 years ago, that habitat destruction became rapid. Agriculture has been the main driving force behind deforestation, which swept the northern hemisphere in previous centuries and continues in the tropics today. It has also been responsible for the destruction of some of the world's major natural grasslands, as well as of marshes and other freshwater wetlands.

In recent years, mangrove forests have been destroyed to make way for shrimp farms, while coral reefs have been "mined" for building materials. Both these activities have a far-reaching effect on marine animals because they remove the nursery areas that many species use to breed. Coasts and rivers are affected by dam-building: the silt that normally accumulates in a river's delta, helping to protect the shoreline, is trapped behind the dam. This results in the erosion of the shoreline as well as allowing salt

NATURAL PRAIRIE

CEREAL FIELD

FROM PRAIRIE TO FARMLAND
North America's forests and prairies underwent an explosive burst of habitat change with the arrival of European settlers. Both were originally home to a wide variety of animals, such as bears and bison, which were hunted and then displaced as agriculture expanded. Much of the forest has grown back, but the prairies are now almost entirely devoted to cereal farming: the original grassland habitat has largely disappeared.

water to flow into freshwater habitats. Fast-growing cities have created further pressure on the natural habitats that remain.

The pattern that the process of habitat change takes is often as important as its scale. For example, if large areas of a habitat are preserved in their original state then the habitat can often function as it did previously, albeit on a reduced scale. But if the same amount of habitat is fragmented – divided up into smaller isolated pieces – the effect on animals is much more severe. This is because many species, particularly predators at the top of food chains, need extensive territories if they are to

find enough to eat. Fragmented habitats are also exposed to more intrusion and disturbance from humans and domestic animals, making it much harder for wild animals to feed and breed.

POLLUTION

Pollution occurs when chemicals or other agents infiltrate and disrupt natural ecosystems. Sometimes pollution has a natural origin, but in most cases it is the result of human activity. It can affect animals physically – for example, by entangling them in waste or clogging them with oil – but its chemical effects inside the body or in an animal's environment are often more serious, as well as being harder to identify and predict.

The most problematic chemical pollutants are synthetic-organic (carbon-containing) substances, such as solvents, pesticides, and herbicides. Hundreds of thousands of these chemicals now exist, and new kinds are produced every year. Their chemical structure means that they are often absorbed by living tissue, where they cause the most damage. Some of these substances are toxic to all life forms, but others are more selective. They are passed on when predators eat their prey, and so they accumulate in species at the top of food chains, such as whales, polar bears, and birds of prey. The low breeding success of some predators – beluga whales in the Gulf of St Lawrence, for example – is thought to be a direct result of this kind of pollution.

Animals are also affected by air pollution, which occurs mainly when fuels are burned, causing poisonous gases to be released into the atmosphere. This creates localized problems such as acid rain, which has a highly damaging effect on freshwater fish. On a much broader scale, it is also responsible for global warming – an environmental change that could ultimately affect almost every animal species on earth.

WARNING SIGN
Fish are particularly sensitive to chemical pollutants and to changes that reduce the water's oxygen content. Mass fish deaths are often the first sign that a stretch of water has been seriously polluted.

HUNTING, FISHING, AND COLLECTING

Unlike many of the world's resources, animals can reproduce. This means that – in theory – useful species can be harvested without their ever running out. However, many species have been exploited to excess, with the result that a considerable number of them are now in serious danger while others have died out altogether.

The list of casualties of hunting includes the African blue buck, which died out in about 1800, and the North American passenger pigeon – once the most numerous bird in the world – which became extinct in 1914. These animals were killed primarily for food, but animals are now

AWAITING EXPORT
Packed in tiny wire cages, these parakeets are destined to supply the pet-bird trade. The trade is driven by money from dealers overseas. However, in the birds' native countries, collecting birds may sometimes be the only way of earning a living.

often hunted to meet less pressing needs. Elephants are in demand for their tusks, and rhinoceroses are killed for their horns. Tigers are hunted for their fur and body parts, which fetch increasingly inflated prices as the number of surviving animals falls. At sea, fish have become victim to the kind of excessive exploitation once reserved for animals on land: plummeting stocks of once-common species, such as the Atlantic cod, are a warning sign that many commercially important fish are experiencing major difficulties.

A range of animals are collected for the pet trade. Apes and tropical parrots are particularly at risk. Some are trapped as adults, but many are taken when they are young. Deprived of parental care, they often fail to survive.

INTRODUCED SPECIES

Even before Christopher Columbus first travelled to the Americas, explorers and colonists had introduced animals to new parts of the world. The process increased rapidly during the sixteenth and seventeenth centuries, and the result – hundreds of years later – is that the wildlife of some isolated regions has been overwhelmed by a host of over-assertive intruders, from rats and cats to sheep, deer, and goats. Some of these introduced species cause problems by preying on local wildlife. Others harm native animals indirectly by competing with them for territory or food.

In Australia, introduced species have disrupted the ecology of an entire continent. Kangaroos still thrive, but many small marsupials now live in a tiny fraction of their original range, in marginal habitats that introduced species find difficult to reach. Similar problems affect New

UNWELCOME ARRIVALS
Rabbits were introduced to Australia in the mid-nineteenth century for food and their fur. They quickly spread inland, displacing native animals and destroying vegetation. In dry areas, soil erosion set in, permanently changing the landscape.

Zealand and Madagascar. On much smaller oceanic islands the situation can be even more severe: their native birds are often wiped out by cats and rats – tenacious newcomers that are extremely difficult to eradicate. In this age of rapid travel and expanding tourism, the threat from introduced species increases unabated.

CLIMATE CHANGE

The earth's climate is so complex that changes are not easy to explain and even more difficult to predict. However, over recent decades one trend has become clear: the planet's surface is quite rapidly warming up. This is almost certainly the result of increasing carbon dioxide levels caused by burning fossil fuels.

Animals are well equipped to deal with gradual changes in climate because they are able to adapt and evolve. But the changes expected during the coming century are likely to be much faster

SHRINKING ICE
Polar bears use winter sea ice as a platform for catching seals. With global warming, the Arctic Ocean's ice cover is diminishing, making it more difficult for polar bears to stock up on food during this crucial time of year.

than those that have occurred in the past, and evolution does not work at such speed. Extra warmth – in itself – will not necessarily be the problem: in many cases, the difficulty will come from changing vegetation zones on land, and changing water levels and current patterns at sea.

There are already some signs that animals are feeling the effects. Studies have shown that some mountain animals are shifting to higher altitudes, and some birds are beginning their breeding cycle earlier in spring. In the sea, coral bleaching (see pp.74 and 529) is widely believed to be linked to climate change. These changes are happening now: whatever agreements are reached to tackle climate change, the problems are bound to multiply in the decades to come.

ANIMALS ON THE BRINK

In the 2000 IUCN Red List of Threatened Species (see below), over 5,400 species of animal are reported to face a high risk of extinction in the near future. In almost every case, the threat is the result of human activities. The number of threatened species represents less than a tenth of one per cent of all the animals that have been identified to date, but IUCN recognizes that the figure is seriously underestimated. This is because many of the world's animals are poorly known, which makes it very difficult to assess changes in their population size. For birds and mammals, the statistics are more reliable, and they make disturbing reading: nearly one in four mammals are included in the list, with 180 species being critically endangered. For birds, the total is one in eight, of which 182 species are at critical risk. Nearly 6,000 species of plant are also listed – a figure of crucial importance to the animals that depend on them for survival.

Being "on the brink" means different things for different species. Some, particularly invertebrates, can reproduce rapidly when conditions are good, which means that they have the potential to make a fast comeback. But many species are slow breeders and take a long time to recover if their numbers fall. Albatrosses are a typical example: they take up to seven years to become mature, they lay just one egg, and they often breed only in alternate years. To

complicate matters, animals cannot necessarily breed even if they find a suitable habitat and a partner. This is because many species breed in groups in which the presence of numerous others is essential to trigger reproductive behaviour such as courtship and nestbuilding. The passenger pigeon was a classic example of a communal breeder, nesting in colonies that extended for many square kilometres. Even when several thousand remained, the species had already stepped over the threshold into oblivion.

EXTINCT IN THE WILD: PRZEWALSKI'S HORSE

CRITICALLY ENDANGERED: BLACK RHINOCEROS

ENDANGERED: QUEEN ALEXANDRA'S BIRDWING

THREAT CATEGORIES
The IUCN Red List of Threatened Species places animals in one of eight categories according to the degree of risk they face: the most threatened species (such as the tiger and the black rhinoceros) are "critically endangered"; the next category (including the orang-utan and Queen Alexandra's birdwing) are "endangered", and so on. Details of the categories – which are also used in this book – can be found on page 10.

VULNERABLE: WANDERING ALBATROSS

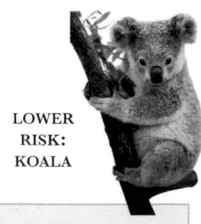

LOWER RISK: KOALA

THE RED LIST

The Red List of Threatened Species is published by the International Union for the Conservation of Nature and Natural Resources (IUCN). The IUCN, founded in 1948 by the United Nations, carries out a range of activities aimed at safeguarding the natural world. Part of its work is the regular compilation of the Red List, which draws together information provided by over 10,000 scientists from all over the world; this list has become a global directory to the state of living things on our planet.

The current Red List shows that threatened species are often grouped in particular parts of the world. Today's "hot spots" include East Africa, Southeast Asia, and the American tropics. One of the reasons for this is that these regions have a much greater diversity of species than regions further north or south: the American tropics, for example, are particularly rich in bird species. In recent years, these areas have seen rapid habitat change – particularly deforestation – which has come about partly because an expanding human population needs more land on which to grow food.

CONSERVATION

THE NEGATIVE IMPACT THAT HUMAN BEINGS have on wildlife grows day by day, but so, too, does the impact of conservation. Across the world, organizations large and small are engaged in a concerted effort to preserve nature in its original state or to ensure that it is used in a sustainable way. It is a huge task, and one that raises some difficult practical questions. Which is the best way of safeguarding species? How do you go about saving an animal that is on the verge of extinction? And, if resources are limited, are some animals more "important" than others? Experts do not always agree on the answers, but there is no doubt that conservation is an urgent priority if today's threatened species are to survive.

HABITAT PROTECTION

By far the most effective way of safeguarding animals is to protect their natural habitats. An animal's habitat provides everything necessary for its survival, and in its natural state it can continue to do this indefinitely as food and energy is passed from one species to another.

This is the thinking behind the establishment of national parks and wildlife reserves. Even small parks can be effective – particularly when they protect breeding grounds – but, as a general rule, the larger the area that is protected, the greater the number of species that benefit and the more likely it is that the habitat will become genuinely self-sustaining. Canada's Wood Buffalo National Park – which straddles northern Alberta and the Northwest Territories – is the world's largest, with an area of 44,807 square km (17,300 square miles). It protects North America's largest surviving buffalo herds, and its remote location ensures that human intrusion and disturbance are kept to a minimum.

WATERSIDE VANTAGE POINT
Specially constructed hides allow visitors to watch birds in a wetland reserve. After centuries of drainage for agriculture, reserves like these are vitally important to many wetland species.

In more accessible parts of the world, national parks and reserves can suffer as a result of their own popularity as well as from the pressure for resources. In the Galapagos Islands, for example, conservationists are engaged in an often difficult struggle to balance the needs of wildlife against the demands of an expanding human population and increasing numbers of visitors.

CAPTIVE BREEDING

When an animal species is in immediate danger of extinction, captive breeding can be a highly effective way of bringing it back from the brink. In 1982, this was the situation with the California condor, when only about 24 birds were left in the wild. During the 1980s, a breeding programme was initiated. As the wild population continued to decline, all the remaining birds were caught – a drastic measure that caused considerable controversy at the time. Two decades later, the intervention seems to have been vindicated: the total population has reached about 170, and about 40 of these birds have been released into the wild. Even better results have been achieved with the Hawaiian goose, or nene, and the Arabian oryx. With new developments in genetic engineering, there is even the possibility that recently extinct species could be "brought back to life".

READY FOR RELEASE
Raised in captivity, this California condor may one day help to swell the population in the wild. However, in comparison with life in captivity, life in the wild can be difficult and even hazardous.

ANIMAL APPEAL

One problem with animal conservation is that human reactions to individual species differ. Some endangered species arouse immense interest and sympathy, while others elicit indifference or even distaste. A good example of the former is the giant panda, which has become an international symbol for nature conservation. The latter includes thousands of animals, such as beetles and other insects, that are generally dismissed as "creepy crawlies". To be effective, conservation must protect all animals in a habitat, from the so-called "charismatic megafauna" – big animals with star appeal – to invertebrates that few people see and even fewer can name.

POLES APART
The giant panda and the European longhorn beetle are at opposite ends of the spectrum of public interest and concern. While the panda attracts funds and media attention, the beetle and its like rarely arouse comment.

However, most conservationists believe that captive breeding is not a long-term route to improving survival rates. There are two reasons for this: first, it requires a huge commitment of time, money, and space; and, second, if the reason for a species' decline is habitat loss then captive breeding will not help because there will be nowhere for the animals to live when they are released into the wild.

CONTROLLING INCOMERS

In isolated parts of the world, introduced – or "alien" – species can make life extremely difficult for native animals, not just by preying on them but by competing with them for food and territory.

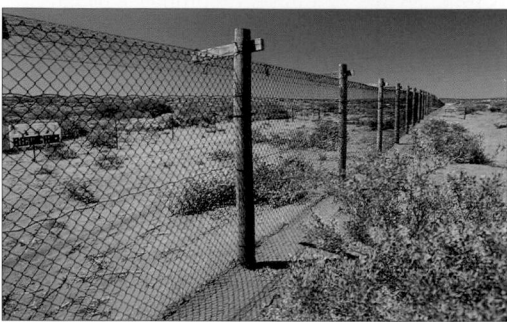

EXCLUDING INTRUDERS
In Western Australia, this electric fence protects the Peron Peninsula from introduced mammals, such as cats. The entire peninsula – covering 1,000 square km (390 square miles) – is to become an "alien-free" haven for endangered marsupials.

Cats, foxes, and rats head the list of these troublesome incomers, although plant-eating mammals can also cause immense damage. In some of the worst affected regions, such as Australia and New Zealand, conservation programmes are now underway to reduce the threat to indigenous animals. In an island that is as vast as Australia, eradicating feral cats and foxes is not a feasible goal. However, in some parts of the country, large areas have been fenced off to protect bandicoots, bilbies, and other vulnerable marsupials from immigrant predators. In these giant enclosures, alien species are either trapped

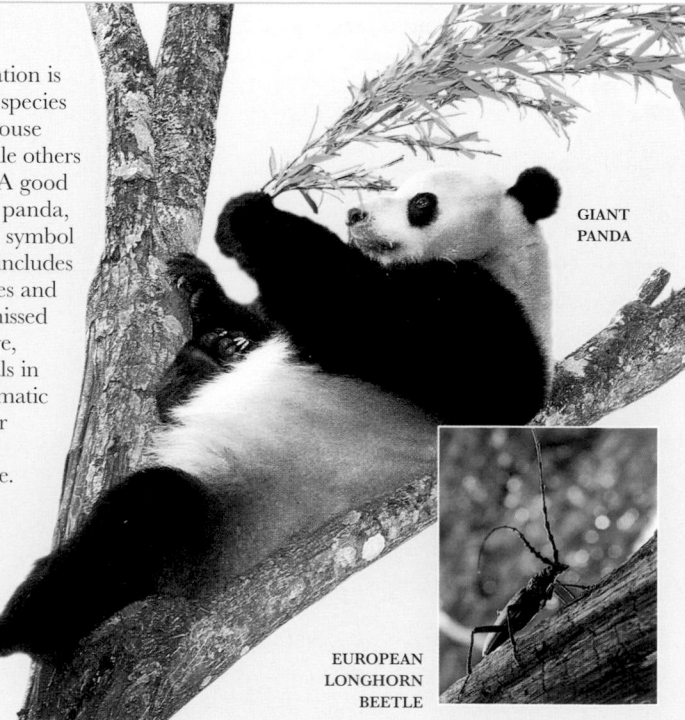

GIANT PANDA

EUROPEAN LONGHORN BEETLE

or controlled by poison bait. The poisons used are substances taken from native plants: these are ideal because they affect only the alien species, leaving native ones unharmed. Killing for conservation is a difficult and divisive issue, particularly when the victims are cats that have run wild. However, there is no doubt – as far as Australian marsupials are concerned – that it is highly effective.

Introduced species are even more of a problem on offshore islands, where they can devastate land animals and colonies of nesting birds. Many of the world's remotest islands, such as Kerguelen in the Southern Indian Ocean, have been overrun by rats, which arrived aboard ships several centuries ago. Rats can be extremely difficult to control. On Kerguelen, eradication programmes have failed, although they have succeeded on several islands off the coast of New Zealand where rats have been removed to create havens for tuataras – one of the world's most endangered reptiles. The small size of these islands makes them ideal "arks" because they are relatively easy to keep mammal-free.

TOURIST ATTRACTION
Watched by a group of tourists, a cheetah relaxes in the evening sunshine. Its tameness is unusual and is a sign that tourism's intrusion into its habitat is affecting its natural behaviour.

LEGAL PROTECTION

After centuries of indiscriminate exploitation, endangered animals are now protected by a host of international agreements and national laws. One of the most important of these is the Convention on International Trade in Endangered Species, or CITES (see panel, right). Other agreements protect specific habitats, such as wetlands or – as with the Antarctic Treaty – entire continents. Some international bodies promote conservation as a way of managing wildlife resources. For example, the International Whaling Commission, which was set up in 1946, originally supervised the "sustainable harvesting" of whales. When it became clear that numbers of most great whales were plummeting, whaling limits were gradually tightened until, in 1986, a complete ban on commercial whaling was introduced.

Such legal protection is an essential part of wildlife conservation. In the case of whales, for example, it has almost certainly saved some species from extinction. However, its effectiveness is sometimes undermined by loopholes or illegal activity. Two notorious examples of the latter are the poaching of black rhinoceroses, whose numbers have collapsed from about 100,000 in the early

1960s to about 2,000 today, and tigers, which are reduced to a total population of about 7,000. Both these animals are killed for their body parts, which fetch extremely high prices in Asia.

COMMERCIAL EXPLOITATION

Few people would condone the sale of rhinoceros horn or tiger bones, but some conservationists do believe that – where possible – wild animals should be made to "pay their way". They take the view that animals are best conserved if they generate income, for this provides an incentive to protect them. There are two main ways in which this can happen: wildlife tourism can be encouraged, with some of the revenue it generates being used for conservation work; or the animals themselves can be managed as a resource.

Wildlife tourism is a booming business, although it has undeniable drawbacks, such as habitat disturbance. On the use of animals as a resource, however, wildlife experts are more sharply divided. The African elephant is a case in point: different conservation bodies are at odds about the exploitation of ivory. One side believes that the legal, albeit controlled, sale of ivory is bound to have a damaging effect on elephant numbers (and inevitably help to sustain poaching). The other believes that, if it is carefully regulated, the sale of ivory could actually safeguard the species by generating money to protect it.

At present, no one knows for certain whether commerce has a real and useful place in wildlife conservation. If it does, one thing is certain: the income generated by wild animals will have to benefit local people, for their co-operation is essential to making conservation work.

CITES

The Convention on International Trade in Endangered Species (CITES) was instituted to control the movement of live wild animals, animal products, and wild plants across international borders. Signed in 1973 by 81 countries, it came into force in 1975 and today has over 120 signatories. CITES completely prohibits trade in over 400 species, including many turtles and birds, and requires special permits for trade in others. It has been successful in some areas but, despite increasing vigilance at ports and airports, smuggling is still a problem.

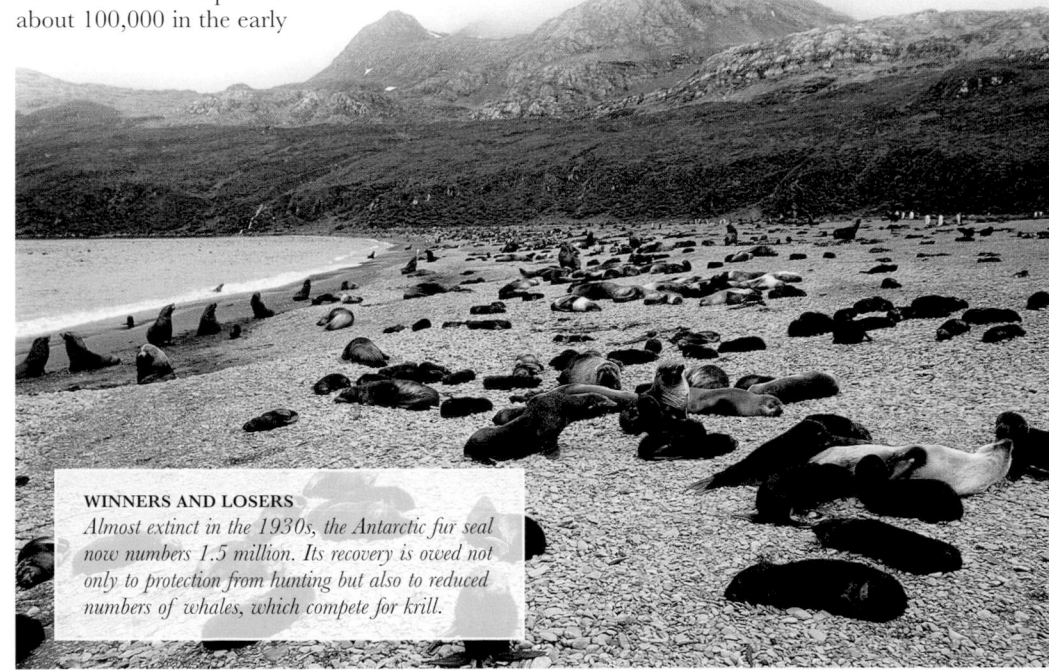

WINNERS AND LOSERS
Almost extinct in the 1930s, the Antarctic fur seal now numbers 1.5 million. Its recovery is owed not only to protection from hunting but also to reduced numbers of whales, which compete for krill.

BANNED GOODS
All the items shown in this photograph are made of turtle shell – an animal product that cannot be exported under CITES regulations. In some countries, national laws make it illegal to own animals or objects listed by CITES.

HABITATS

The earth is the only place in the universe known to support life. Like all other living things, animals occupy a zone between the lower part of the earth's atmosphere and the floors of its oceans. Although animal life is richest in the warm, wet conditions of the tropics, some animals have adapted to life in the hostile conditions found in arid deserts, high mountains, and the polar ice-caps. Wherever they live, animals interact with each other, with other living things, and with their non-living surroundings to produce complex, constantly changing environments known as habitats. This section looks at the habitats of the world and the animals that live in them.

WORLD HABITATS

SEEN FROM SPACE, THE MOST STRIKING FEATURE of the earth is its sheer physical variety. Not only is there dry land and open sea, but there are also mountains, plains, rivers, coastal shelves, and deep oceanic trenches. The earth also varies in its climate: in some parts of the world, weeks or months pass by under almost cloudless skies, while in others, the ground is scoured by icy winds or soaked by intense tropical storms. Differences like these create a complex jigsaw of varied habitats, enabling the earth to support a rich diversity of animal life. Some species are highly adaptable, and can survive in a wide range of conditions, but the vast majority are found in one kind of habitat and nowhere else.

WHAT ARE HABITATS?

In its narrowest sense, a habitat is the environment in which something lives. For some animals, a habitat might be as restricted as a temporary pool in a desert or as small as a piece of decaying wood. In a broader sense – the one used in this book – a habitat can mean a characteristic grouping of living things, together with the setting in which they are found. In ecology, a habitat defined in this way is known as a biome.

Habitats contain both living and non-living matter. In some – for example, true desert – living things are thinly scattered, so the non-living part of the environment is dominant. In others, such as forest and coral reefs, living things are so abundant that they fill all the available space and create habitats for each other. In these habitats, huge numbers of species exist side by side, forming extraordinarily complex webs of life.

FACTORS THAT SHAPE HABITATS

Geology plays a part in shaping habitats, but by far the most important factor is climate. As a result, differences in climate – which sometimes occur over remarkably small distances – can have a huge effect on plant and animal life. A classic example of this occurs where mountain ranges intercept

EFFECTS OF CLIMATE
These two habitats – in Argentina (left) and Chile (right) – are at the same latitude, but they have very different climates, and therefore very different plant and animal life. The contrast is created by the Andes, a mountain barrier that blocks rain-bearing winds. Chile is on the windward side of the Andes.

rain-bearing winds. On the windward side of the mountains, heavy rainfall often creates lush forests teeming with all kinds of animal life. But in the "rainshadow", to the lee of the mountains, low rainfall can produce desert or scrub, where only drought-tolerant animals can survive.

Temperature is another climatic factor that has an important effect both on land and in the sea. For example, in the far north, coniferous forest eventually peters out in the face of biting winter frosts. This northern treeline, which runs like a ragged ring around the Arctic, marks the outer range of crossbills, wood wasps, and many other animals that depend on conifers for survival.

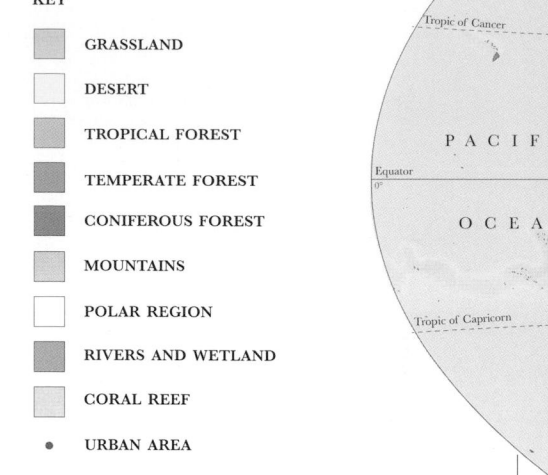

On coasts and at sea, temperature changes are usually more gradual than they are inland. However, warmth – or lack of it – still determines where some habitats are found. For example, reef-building corals do not thrive at temperatures of less than about 20°C (68°F), so most reefs are found in the tropics. However, on the west coast of Africa and the Americas, reefs are rare because, although the climate is warm, cold currents pass close to the shore. Mangrove swamps present a similar pattern: in the southern hemisphere they reach as far as South Australia; in the northern hemisphere they extend only just out of the tropics.

BIODIVERSITY

From the earliest days of scientific exploration, naturalists noticed great variations in biodiversity, or species richness. In the far north and south, species totals are low compared with the numbers found near the equator. Arctic tundra, for example, is inhabited by just a few hundred species of insect, while in tropical forests the total is probably at least a million. A similar picture – albeit on a smaller scale – is true for mammals and birds. However, high-latitude habitats make up for their lack of biodiversity by having some phenomenally large species populations. The seas around Antarctica, for example, harbour perhaps 40 million crabeater seals – the most numerous large wild mammals on earth.

The reasons for such variation in biodiversity are still not fully understood, although climate almost certainly plays a part. However, in an age

CHEMICAL CYCLES

In all habitats, living things take part in cycles that shuttle chemical elements between living and non-living matter. About 25 elements are essential to life and, of these, just four make up the bulk of living things. These are hydrogen, oxygen, nitrogen, and – the key element – carbon. In the non-living world, carbon can be found in the atmosphere (as a gas), in water (in dissolved form), and in the ground (in rocks and fossil fuels). Plants absorb carbon dioxide from the atmosphere, and most other forms of life give off carbon dioxide when they break down carbon-containing substances to release energy. Carbon is also released by burning fossil fuels.

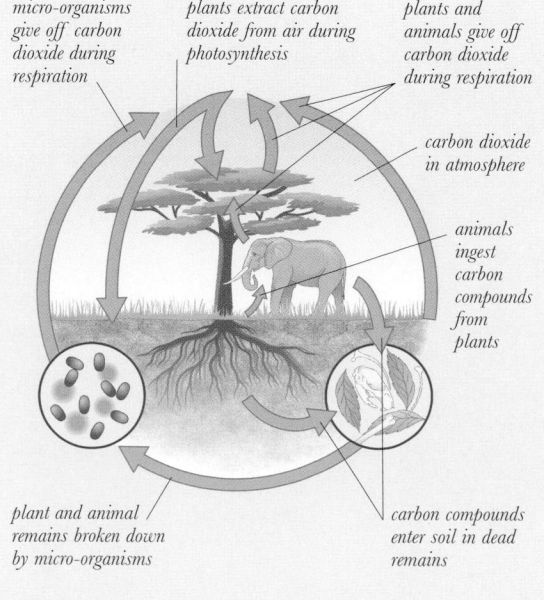

micro-organisms give off carbon dioxide during respiration

plants extract carbon dioxide from air during photosynthesis

plants and animals give off carbon dioxide during respiration

carbon dioxide in atmosphere

animals ingest carbon compounds from plants

plant and animal remains broken down by micro-organisms

carbon compounds enter soil in dead remains

THE CARBON CYCLE
This diagram shows some of the main pathways in the carbon cycle. The time taken for each part of the cycle to be completed varies greatly. Carbon may stay in living things for only a few days, but it can remain locked up underground for thousands of years.

NUMBER OF SPECIES	
3	165
10	110
55	60
155	20
	1

TROPICAL TREASURY
The hummingbird family vividly demonstrates the high level of species diversity that is to be found in the tropics. Only a handful of hummingbird species live at high latitudes – and most of these are migrants – but on the equator, the number of species rises to over 150.

to form scattered "refugia" – islands of forest surrounded by grassland. Even today, these areas of forest still contain a wider variety of birds than the relatively new forest that has grown back. Ice ages also affect sea levels, by locking up water as ice. When sea levels fall, land habitats expand; when they rise, the land is drowned again and plants and animals are forced to retreat.

Since the last ice age ended, natural changes have not been the only ones that have affected the world's habitats. On a local and global scale, human activity has had an increasing impact and, as a result, the pattern we now see is partly man-made. This is especially true of forests, which have been cut back to make space for agriculture, but it is also true of some grasslands, wetlands, and even deserts. In some remote regions – particularly in the far north – the original pattern still remains, but in populated regions it has been transformed, creating a world where wild animals can have difficulty finding a home.

in which many animal species are endangered, biodiversity – and ways of maintaining it – has become an important topic. Tropical forests and coral reefs are especially rich in species, which is why so much attention is currently focused on preserving them and their animal life.

ANIMAL DISTRIBUTIONS

A glance at the map on this page shows that various types of habitat are spread across large expanses of the world. However, with a few exceptions, most of their animals are not. Instead, each species has a characteristic distribution, which comes about partly through its evolutionary history (see p.17), and partly as a result of its way of life.

In many cases, an animal's lifestyle shapes its distribution in unexpectedly subtle ways. For example, in the Americas, the brown pelican is found all along the western coast, apart from the far north and south; in the east, it does not reach south of the Caribbean. The reason for this is that, unlike its relatives, the brown pelican feeds by diving for fish, and needs clear water to spot its prey. The Caribbean is clear, but further south lies the River Amazon, which pours vast amounts of mud-laden water into the sea. For the pelican, this muddy water is a barrier that cannot be crossed.

Many animals' distributions are linked to those of particular plants. Extreme examples include the yucca moth, which depends on yucca plants; the fig wasp, which

develops inside figs; and countless bees that depend on particular flowers. Not all plant-dependent animals are insects, however. The robber crab – the largest and heaviest terrestrial crustacean – feeds predominantly on rotting coconuts, which it scavenges along the shore (see below). As a result, it is found only where coconuts grow. Mammals can be just as particular. The giant panda – one of the most famous examples – depends on about two dozen species of bamboo, which are found only at mid-altitudes in the mountains of central China.

CHANGING HABITATS

In nature, habitats change all the time. Forest and grassland catch fire, rivers burst their banks, and storms batter coral reefs and coasts. Such unpredictable occurrences are facts of life, and animals – along with other living things – have evolved ways of surviving them. Habitats can also change in much more profound ways, over much longer periods of time. Here the driving force is usually climate change, a natural process that is triggered by a host of factors, including continental drift. On several occasions in the distant past – most recently about 12,000 years ago – the polar ice caps have expanded, destroying existing habitats and evicting their animals. On each occasion, when the ice has eventually melted, plants have moved back into the empty landscape, and animals have followed suit.

The world's climates are inter-related, which means that changes in one area can have long-term effects all over the globe. For example, during the last ice age, the climate in the tropics became drier, and the Amazon rainforest shrank

COCONUT-EATERS
Grasping a fallen coconut, two robber crabs tussle over their favourite food. The robber crab is one of many species whose distribution is determined by the presence of a single plant – in this case, the coconut palm.

LEVELS OF LIFE

Even in the remotest places on earth, animals very rarely live entirely on their own. Instead, they interact with other individuals, and with other species. Taken together, these interactions create a range of different ecological levels, from local populations, communities, and ecosystems to the whole biosphere – the sum total of all the places where living things can be found. Because microorganisms are so widespread, the biosphere extends high into the atmosphere, and probably several kilometres underground.

INDIVIDUAL
An individual animal is normally an independent unit that finds its own food. It often lives within a set home range.

POPULATION
A population is a group of individuals that belong to the same species, live in one area, and interbreed.

COMMUNITY
A community is a collection of populations. Although they belong to different species, they depend on each other for survival.

ECOSYSTEM
An ecosystem is made up of a community and its physical surroundings. Characteristic ecosystem types are known as biomes (or habitats).

BIOSPHERE
The biosphere consists of all the ecosystems on earth, and therefore all the places – from the earth's crust to the atmosphere – that living things inhabit.

GRASSLAND

IN PARTS OF THE WORLD where it is too dry for trees to grow, yet moist enough to prevent the land from becoming desert, grasses are the dominant plants. Grasses are unusual in that their stems grow from a point near the ground. This means that unlike most other plants, which grow from their tips, grasses are unharmed by grazing. In fact, grazing animals help grasses to maintain their dominance by stunting the growth of competing plants. This creates a vast, open habitat in which there is plenty of plant food – for those that can digest it – but little shelter from the elements.

TEMPERATE GRASSLAND

Before the advent of farming, grassland covered large parts of the temperate world, notably in the northern hemisphere. These vast grasslands – which include the prairies of North America, and the steppes of Europe and central Asia – are nearly all in the centre of large landmasses, far away from coasts and their moisture-laden winds. Summers are often warm, but winters can be long and cold, with biting winds.

An unusual feature of this kind of habitat is that the majority of the plant matter is hidden away below ground – the exact opposite of the situation elsewhere on land. This is because grass plants direct much more energy into growing roots than into producing leaves, and their roots form a continuous mat, which protects the surface of the ground by holding the soil in place. If grassland is burned, or hit by drought, it can soon recover because the grass can draw on its buried reserves in order to start growing again. The root mat makes a useful source of food for insects and other small animals. It is also a perfect medium for burrowers, because it is easy to dig through and, unlike loose soil, rarely caves in.

Above ground, the food supply is closely tied to the seasons. In temperate grassland, most of the year's water usually comes in the form of spring rain or melting snow. This creates a flush of growth during spring and early summer, which is the time that most grazing animals breed. By late summer, the grass is brown and dry, although for a while grass seeds make a valuable autumn harvest. Winter is a difficult time for all grassland animals, but particularly for grazers because they often have to survive on low-grade food that is hidden under snow.

GRASSLAND SONGBIRDS, *such as the skylark, have no trees to perch on, so they broadcast their songs and calls from the air.*

SAIGAS *form large herds. They look and live like true antelopes (of which there are few in temperate grassland).*

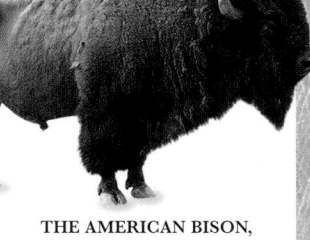

THE AMERICAN BISON, *now relatively rare, is a large, heavy grazer that once had a profound effect on the ecology of the prairies.*

GRASSLAND REPTILES, *such as this sand lizard, spend the winter underground and are active in summer.*

SOME BIRDS, *such as this burrowing owl, lay their eggs in underground burrows because there are no trees to provide nest holes.*

THE LARGEST AREAS *of temperate grassland are found in North America, South America, eastern Europe, and central and eastern Asia.*

GROUND

AERIAL SCAVENGERS, *such as this African white-backed vulture, exploit the savanna's strong thermals, uninterrupted views, and abundance of wildlife.*

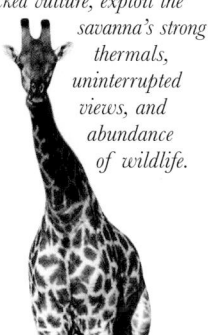

THE GIRAFFE'S *extraordinary reach is responsible for the umbrella-like shape of many African grassland trees.*

FLIGHTLESS *birds, such as the ostrich, thrive in savanna, where running is an effective means of escape.*

THE CHEETAH *attacks in the open, so it relies on speed, rather than stalking, to enable it to make a kill.*

TERMITES *carry plant matter below ground, helping to recycle vital nutrients.*

SAVANNA

Savanna is tropical or subtropical grassland that contains scattered shrubs and trees. The grasslands of east Africa are a familiar example, with their diverse wildlife and distinctive vegetation (particularly the flat-topped acacia trees). Compared with temperate grassland, savanna is very variable: in some savanna habitats, trees are few and far between; in others, they form scattered thickets, merging into open woodland.

Trees have a major impact on the savanna's animal life. This is because they produce a wide variety of food, including wood, leaves, flowers, and seeds; and they also create shelter and breeding sites for animals that live off the ground. The balance between trees and grass is a delicate one that is sometimes changed by the

SAVANNA IS FOUND *mainly in Central and South America, tropical Africa, southern Asia, and northern Australia.*

animals themselves. For example, elephants destroy trees by pushing them over so that they can reach their leaves. However, elephants also help trees to reproduce because they ingest the trees' seeds, which are then passed in their dung – an ideal medium for promoting seed growth. Browsing mammals often keep trees in check by nibbling saplings before they have had a chance to become established. Fire also helps to hold back trees, and its effect is most apparent in places where trees grow close together.

Unlike temperate grassland habitats, savanna is usually warm all year round. There is often a long dry season, when most trees lose their leaves, followed by a wet, or "rainy", season, which produces a rapid burst of growth that turns the landscape green. During this wet season, plant-eating animals rarely have to contend with a shortage of food; in the dry season, the threat of starvation is never far away, and many animals travel long distances to find water and food.

CONSERVATION

At one time, natural grassland covered about two-fifths of the earth's land surface. With the spread of agriculture, much of this was adapted for growing crops or raising livestock, and now there are only small remnants of grassland containing its original vegetation and wildlife. However, farming has also created grassland: the hill pastures of Europe and New Zealand, for example, are the result of deforestation that occurred several centuries ago. Old pasture can be a rich resource for wildlife; modern pasture has a much narrower range of plants and, consequently, less varied animal life.

CHANGING HARVESTS *In the space of less than 150 years, over 90 per cent of North America's prairie grassland has been converted to agricultural land.*

LIFE IN GRASSLAND

DESPITE CENTURIES OF HUMAN DISTURBANCE, grassland supports some of the largest concentrations of animal life on earth. Survival in grassland habitats is far from easy, however: aside from the lack of shelter and plant diversity, there are hazards such as drought and fire to contend with. Added to this is the ever-present risk of attack by some of the world's fastest and most powerful predators.

HERDING

Life in open grassland is often dangerous because there are few places to hide. To increase their chances of survival, many large plant-eaters live in herds. This makes it more difficult for predators to attack because while most members of the herd are eating, some are always on the lookout for danger.

Today's largest herds are found on Africa's plains. Here, migrating wildebeest can form herds over a quarter of a million strong, and 40km (25 miles) long, although even these herds are small compared with some that existed in the past. During the nineteenth century, springbok herds in southern Africa sometimes contained more than 10 million animals. In North America, bison herds probably reached similar sizes before hunting brought the species to the edge of extinction.

Life in herds does have its problems, one of which is the risk that an animal might wander off and become lost. Most herding species have scent glands on their hooves so that if an animal becomes isolated it can follow the scent tracks to rejoin the herd. Another problem is giving birth. To prevent their young from being trampled, or attacked, many grazers give birth in cover, and rejoin the herd a few weeks later. Some, however, are born in the open and have to be able to keep up with the herd when they are just a few hours old.

MOVEMENT

In grassland and savanna, there is a premium on speed. It is no accident that the world's fastest land animals, such as the cheetah and the pronghorn, are found in this habitat. Natural selection favours predators that are fast enough to catch food, and prey animals that are fast enough to escape.

Most of the fast runners are mammals, however grassland also has nature's fastest-running birds, including ostriches, rheas, and emus – giant species that have lost the ability to fly. These birds can reach speeds of up to 70kph (44mph). More importantly, they are able to maintain such speeds for up to 30 minutes – long enough to outrun most of their enemies unless a predator launches an attack from a very close range.

Despite the many fast runners, grassland life often appears tranquil. This is because running is extremely energy-intensive, and animals run only when they absolutely have to. Prey animals have invisible "security thresholds" that vary according to the threat they face. For example, gazelles often let lions approach to within about 200m (650ft) because they are instinctively aware that lions that are visible at this distance are unlikely to be stalking prey. A solitary cheetah, on the other hand, will send a gazelle herd sprinting, even if it is seen to be four times further away.

FLIGHTLESS BIRDS
Standing guard over its chicks, a greater rhea watches for danger. Like other flightless birds that live in grassland, it uses its height and large eyes to spot potential predators at a distance so that it has time to run for safety.

LIVING UNDERGROUND

Some grassland animals find safety not by running away but in retreating into burrows below ground. There they can stay out of reach of most predators and find some protection from the worst of the elements. Subterranean animals include a wide variety of species, ranging from mammals to insects. Some animals, particularly snakes, do not excavate their own burrows; instead they adopt existing ones. The largest burrows, made by African aardvarks, are big enough to accommodate a person, and are a serious

MAKING A HOME BELOW GROUND
A black-tailed prairie dog collects grass to line a nesting chamber. The colony, or town, consists of tunnels up to 5m (16ft) deep.

entrance | entrance mound | nest chamber | passing place

PRAIRIE-DOG BURROWS

MIGRATING WILDEBEEST
In Tanzania's Serengeti National Park, wildebeest migrate in a cyclical path to take advantage of fresh plant growth at different times of the year. They spend the wet season in open grassland, and the dry season in wooded savanna.

FIRE

Fires, ignited by lightning, are a natural feature of grassland life, clearing away dead growth, and allowing fresh grass to sprout. In the long run, fires help grassland wildlife, but while they are burning they can be lethal. As the flame front advances, most animals react by running or flying for safety, often abandoning their usual caution in an urgent bid to escape. Some animals, such as bustards and storks, have learned to make the most of this frenzied exodus from the flames. They gather close to fires and snap up insects and other small animals as they scurry away; and once the fire has moved on, they pick over the charred ground for casualties.

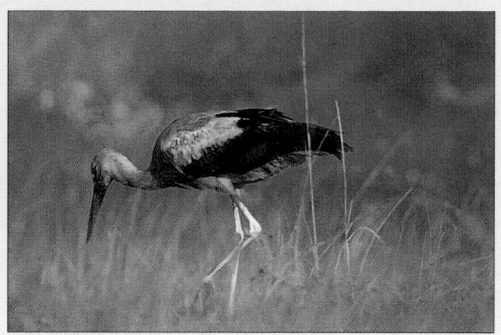

FEEDING BY THE FLAMES
A European white stork searches for small animals fleeing before an advancing fire. Soaring high over the plains, the stork will have been attracted by the fire's smoke.

hazard to vehicles; the most extensive are made by prairie dogs and other rodents. Before farming became widespread in North America's prairies, some prairie-dog burrow systems covered several thousand square kilometres and housed millions of animals.

Termites are also accomplished builders, constructing giant, elaborate, subterranean nests that extend high above ground level. These nests house large, co-operative communities that can contain over 30 million inhabitants. Along with ants, they make up a very large part of the habitat's animal life, and provide food for the large insect-eaters.

INSECT-EATER
The giant anteater uses its powerful claws to break open termite mounds and reach the insects inside.

FEEDING

Although grass is rich in nutrients and easy to find, it is difficult to digest. Many mammals, including humans, cannot break it down at all because it contains large amounts of cellulose – a carbohydrate that most animals cannot digest. Grazing mammals, however, have special micro-organisms in the gut that break down cellulose so that the body is able to use it. Some non-mammal species also use microbes to digest plant material. In tropical savanna, for example, termites rely on them to break down dead leaves and wood.

The animals that are most efficient at using cellulose are ruminant mammals – antelopes, buffaloes, and giraffes, for example – which helps to explain why these animals dominate grasslands. The ruminant's complex stomach acts like a fermentation tank, working to extract the maximum amount of nutrients from food. The animal assists in the process by regurgitating its food and chewing it a second time, making it even easier to break down. Non-ruminant plant-eaters, such as zebras, have less efficient digestive systems and must therefore eat more to survive.

In pure grassland, plant-eaters compete for the same food, although each may have a preference for a different type of grass. In savanna, the presence of trees and shrubs makes for a wider range of food, and browsing mammals have minimized competition further by evolving specific ways of feeding. This means that a remarkable number of species can live side by side. For

COLLECTING DUNG
Dung beetles make use of the large amount of dung produced by grazing mammals. They gather the dung into balls, which they then roll away and bury below ground where it is used as food for the beetle's grubs.

example, the small Kirk's dik dik antelope feeds on shoots and fruit, and rarely touches grass, while the much larger eland will eat almost anything from fruit and seeds to roots scraped up from the ground.

Scavengers, which also play an important part in the habitat's ecology, include birds as well as coyotes, jackals, and hyenas. Most airborne scavengers are vultures, but there are several species of stork, one of which, the marabou stork, rivals the Andean condor for the title of the world's largest flying animal.

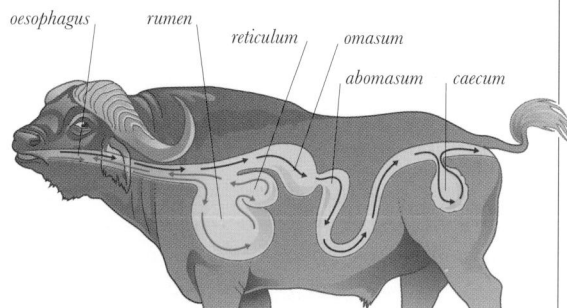

DIGESTING GRASS
The largest of the four chambers in a buffalo's stomach, the rumen, contains millions of bacteria and protozoans. These simple organisms produce cellulose-splitting enzymes that turn cellulose into simple nutrients that can then be absorbed.

oesophagus   rumen   reticulum   omasum   abomasum   caecum

REACHING FOR FOOD
Unlike most antelopes, the gerenuk can stand on its back legs. This means that although it measures only about 1m (39in) at the shoulder, it can browse leaves over 2m (6½ft) up – much higher than the reach of other antelopes of similar size.

DESERT

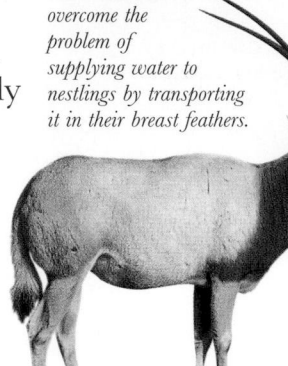

SANDGROUSE
overcome the problem of supplying water to nestlings by transporting it in their breast feathers.

DESERTS ARE PLACES OF EXTREMES. Besides being dry, they experience intense sunshine, and a greater daily temperature range than any other land habitat. Rain – when it comes – typically falls in brief but torrential downpours, while strong winds pick up sand and grit, carrying it almost horizontally through the air. Although no two deserts are identical, true desert is usually defined as having less than 15cm (6in) of rainfall a year. Semi-desert has more rainfall – up to 40cm (16in) a year – which typically falls during a relatively short spring or wet season followed by months of drought.

THE SCIMITAR-HORNED
oryx, like other non-drinkers, can get all the water it needs from its food.

TRUE DESERT

Most of the world's true desert is found in two belts, one straddling each of the tropics. Here, zones of high atmospheric pressure persist for months at a time, preventing low-pressure air from bringing in rain. Desert also forms where mountains block rain-bearing winds, and where cold coastal currents chill the air so that it carries very little moisture inland.

In true desert, the amount of rain is so meagre, and so unpredictable, that very few plants can survive. The ones that do – such as cacti and other succulents – are highly effective at collecting and conserving what little water nature provides: they have large networks of shallow roots, which drain the surrounding ground so thoroughly that, often, nothing else can grow near by.

THE GOLDEN JACKAL
is able to live nocturnally, which means that it can avoid the worst of the desert's daytime heat.

For animals, this arid environment creates some interesting effects. With so few plants, there is very little soil, which severely limits invertebrate life. Most small animals, such as insects, are found either on the plants themselves or in the debris that accumulates immediately beneath them. Larger animals, such as reptiles and rodents, venture away from these pockets of greenery, but even they have to be careful to avoid the worst of the daytime heat.

Lack of vegetation means that most of the ground is exposed. Bare ground absorbs warmth very quickly when the sun rises, and re-radiates it once the sun has set. The dry air accentuates this effect, allowing daytime surface temperatures to soar to over 70°C (158°F). As a result, most animals living in true desert are active after dark. During the day they hide away, leaving little sign of themselves apart from their tracks.

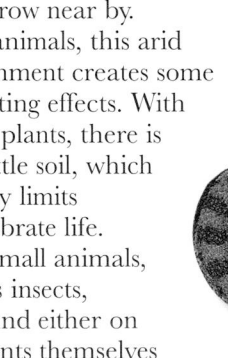

TRUE DESERT *occurs at mid-latitudes in the northern and southern hemispheres, and it can have less than 5cm (2in) of rain a year.*

THE THORNY DEVIL *has protective spines in addition to the good camouflage that other slow-moving ground-dwellers often depend on for survival.*

DESERT SNAKES, *such as the desert horned viper, will pursue their prey into their burrows.*

COUCH'S SPADEFOOT
toad is one of the desert frogs and toads that spend most of their lives underground.

GROUND

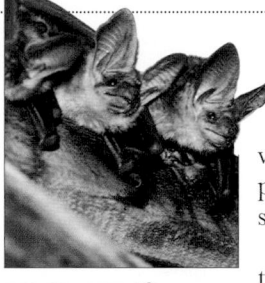

DESERT BATS *play a vital ecological role because they feed on insects and pollinate flowers.*

SEMI-DESERT BIRDS, *such as the roadrunner, often nest among spiny plants to protect their young from predators.*

THE MEERKAT *lives in large colonies. Its varied diet and co-operative foraging technique help it to survive when food is scarce.*

THE LOCUST'S *migratory lifestyle is an adaptation to a habitat where food supplies are erratic and unpredictable.*

TARANTULAS *hunt mainly by touch, enabling them to find their prey after dark.*

SEMI-DESERT

Compared with true desert, semi-desert is more widespread, and it is also much more biologically productive. It is found in every continent, including some regions far outside the tropics.

The modest but nevertheless reliable rainfall that semi-desert receives has a dramatic effect on the landscape and the types of animal that it can support. Plants often grow in profusion, creating tangled thickets of vegetation that provide plenty of cover. There are woody species, which store most of their water in underground roots, and fleshy succulents, which store it above ground in their stems and leaves. Most of these desert species are well protected from plant-eaters – cacti, for example, have extremely sharp spines, while spurges exude a poisonous milky sap when they are damaged – but, for animals that can overcome these defences, they are an important source of food. Semi-desert also has plants known as desert ephemerals, which spring up rapidly after rain, flower, set seed, and then die. This short life cycle produces extra fresh food for animals, and adds to the stock of seeds scattered over the desert floor.

While some semi-deserts are warm or hot throughout the year, others are surprisingly cold in winter. In the deserts of central Asia, and in the northern parts of America's Great Basin – the desert region between the Rocky Mountains and the coastal ranges further west – temperatures can fall to –30°C (–22°F). In these areas, animals need protection against winter cold as well as against summer heat: small animals, such as insects, usually become dormant in winter, and many burrowing mammals hibernate until the spring.

SEMI-DESERT *is most widespread in the world's major landmasses, and it extends well into the temperate zone in both hemispheres.*

CONSERVATION

Compared to many other habitats in the world, desert has experienced relatively little human interference. Oil extraction can create environmental problems in deserts, but a more serious difficulty is caused by the exploitation of the habitat by livestock. Goats and other browsing animals can stunt or kill shrubs and other plants; and when these plants die, the sparse soil is often eroded, making it difficult for vegetation to recover. The result is desertification, or the spread of desert into areas where they did not previously exist. These newly created deserts are of little value to wildlife.

ANIMAL IMPACT
Owing to their exceptional climbing skills, goats have a major impact on desert shrubs. In areas where large numbers of goats are raised, desert vegetation is often badly affected.

LIFE IN DESERT

IN A HABITAT WHERE MOISTURE IS SCARCE, obtaining and conserving water are every animal's top priorities. Desert animals practise a tight "water economy", which means collecting water wherever they can, and minimizing water loss wherever possible. However, being economical with water is not in itself enough to guarantee survival: desert species have had to evolve various other adaptations to enable them to cope with a wide range of temperatures and the ever-present threat of food shortage. As a result, these animals are able to live in some of the driest places on earth.

CONSERVING WATER

Most deserts have a scattering of oases, where animals gather to drink. Some species need to drink daily, which restricts how far they can roam from an oasis. Others can survive on their on-board reserves for days or even weeks, depending on the temperature. A remarkable feature of desert life is that some animals can manage without drinking at all. Instead, they get all their water from their food. Some extract it from the moisture contained in food; but most use the food to manufacture metabolic water, which is created by chemical reactions when the energy in food is released. Seed-eating rodents are expert at this: although their food looks dry, they are able to metabolize all the water they need.

moisture in food (10%)   metabolic water released by digesting food (90%)

WATER IN

urine (23%)   moisture in droppings (4%)   moisture lost from skin and in breath (73%)

WATER OUT

WATER BALANCE
This diagram illustrates how a kangaroo rat survives entirely on the water in its food. The water taken in has to balance that which is lost to prevent the animal from becoming dehydrated.

WATER-STORING FROG
The Australian water-holding frog stores water in its bladder and beneath its skin. To prevent this water from drying out, the frog then seals itself in a semi-permeable cocoon underground.

For drinkers and non-drinkers alike, water has to be eked out to make sure that it lasts. Compared with animals from other habitats, desert species lose very little moisture in their urine and droppings, and only a small amount is released from their skin and in their breath. Desert species are also good at withstanding dehydration. The dromedary, or one-humped camel, can lose nearly one half of its body water and survive. For humans, losing just a fifth can be fatal.

STORING FOOD

To enable them to cope with erratic food supplies, many animals keep their own food stores. Some do this by hiding food away. The North American kangaroo rat, for example, constructs underground granaries that

RAPID REFILL
After going without water for several days, a camel can drink over 50 litres (11 gallons) in just a few minutes. It also metabolizes water from surplus food, laid down as fat in its hump. Its salt tolerance is high, which is useful in a habitat where water is often brackish.

contain up to 5kg (11lb) of seeds. But for predators, and for animals that browse on shrubs, creating such larders is not possible. Their food is difficult to collect and to transport, and even if it could be hoarded it would be unlikely to remain usable for more than just a few days. The answer is to store food inside the body. The classic example of this is the camel, which stores surplus food, in the form of fat, in its hump. Several other species, such as the Gila monster and fat-tailed dunnart, store food in their tails.

COPING WITH HEAT AND COLD

In desert, the temperature rarely stays steady for more than a few hours, and it can reach extremes of both heat and cold very quickly. Humans lose excess heat by sweating, but at very high temperatures this cooling system can use as much as 1 litre (35 fl. oz) of water an hour – far more than any desert animal could afford.

Desert animals tackle the heat problem in two ways: by reducing the heat they absorb, and by increasing the heat they give out. Light-coloured skin or fur reflects some of the sun's rays, minimizing heat absorption; but a much more effective method – used by many desert animals – is to avoid the most intense heat by being nocturnal, spending the day sheltering underground. Burrows do not have to be very deep to make a difference: while the desert surface may be too hot to touch, the ground just a few centimetres below it will be relatively cool.

Getting rid of excess heat is more difficult, particularly when an animal's body temperature is dangerously high. Lizards

COLD DESERT
This graph shows the average annual temperature on the western edge of the Gobi Desert. From November through to March, temperatures fall below freezing.

CHANGING COAT
In the deserts of central Asia, freezing winters are followed by soaring temperatures in spring. The Bactrian camel copes by growing a thick coat that falls off as soon as the spring warm-up begins. This camel is about halfway through this process.

and snakes are often described as "cold blooded", but this actually means that their body temperature rises and falls with that of their surroundings. Although they thrive in warm habitats, and can survive with a body temperature of up to 44°C (111°F), they often have to sit out the hottest part of the day in shade. Some desert birds cool down by panting, which involves fluttering the flap of skin over their throats. Desert kangaroos and wallabies lick their front legs, covering them with saliva. As the saliva evaporates, the animal's blood cools down.

In high-latitude desert regions, such as the Gobi Desert of central Asia and the Great Basin Desert of North America, winter can be extremely cold. Animals have various ways of coping with this. Most reptiles

NOCTURNAL ACTIVITY
Desert jerboas are typical of the small rodents that live in dry habitats. They are entirely nocturnal, and feed mainly on seeds. Jerboas can travel over 10km (6 miles) in their search for food, hopping on their long back legs and balancing with their tails.

hibernate, while birds often fly to warmer climates. Mammals keep warm by growing thick fur, or by sheltering underground.

EXPLOSIVE BREEDING

Desert animals often have highly variable breeding seasons. Instead of reproducing at a fixed time of year, many produce young when there is the best chance of finding food. Female kangaroos, for example, give birth extremely regularly when food is plentiful, but when food is scarce they stop breeding altogether. This flexible system is an efficient way of using resources because it prevents parents having to tend hungry youngsters when they are hungry themselves.

Some desert species carry irregular breeding to extremes. Desert wildlife includes a number of animals that, paradoxically, live or breed in water and, for those species, reproducing is a highly unpredictable and time-sensitive business. Such animals include burrowing frogs and toads, and also freshwater shrimps that live in temporary pools. For months or even years at a time, they are an invisible part of desert wildlife, with the amphibian species lying hidden underground, and the shrimps present only as eggs in dried-up ground. But immediately after a heavy storm, the frogs and toads dig their way to the surface, and the shrimp eggs hatch. Once

LIFE IN BRIEF
Trapped in a rapidly shrinking desert pool, these adult tadpole shrimps have only a few days to live, but the eggs they leave behind can survive in a dried-out state for several years – long enough to last until the next heavy storm, when they will hatch.

NOMADIC ANIMALS

Where food and water supply is patchy, some animals adopt a wandering lifestyle. This is common in desert habitats, especially in animals that can fly. Desert locusts are famous for their huge nomadic swarms, and some desert birds, particularly seed-eaters, form large nomadic flocks. Unlike migrants, nomadic animals do not follow fixed routes – the weather often dictates their course – and they breed erratically, wherever they find a good food supply.

WILD BUDGERIGARS
Australia's nomadic birds include the budgerigar, the cockatiel, and several pigeon species. Budgerigars can breed when just one month old; and, since parents can raise several families in quick succession, flocks can build to prodigious numbers.

active, these animals immediately set about finding mates because they have to complete their life cycles before the pools dry up again.

MOVEMENT

Desert sand makes life difficult for animals on the move. Large animals sink into it, while small ones struggle to climb up and down

MOVING ON SAND
Flaps of skin between the web-footed gecko's toes make for snowshoe-like feet that allow it to run across dunes.

ATTACK FROM BELOW
Guided by vibrations overhead, this desert golden mole has emerged from its burrow to ambush a gecko.

slopes of shifting grains. To combat the problem, some animals, such as golden and marsupial moles, move through the sand rather than above it. Others, such as camels and geckos, have extra large feet, which help to spread their body weight over the surface of the sand and so increase stability. Sidewinding snakes have a different solution: they throw themselves forwards in a succession of sideways jumps, leaving a characteristic pattern of J-shaped tracks. In addition to saving energy, this method helps to minimize contact with hot ground.

Some insects and lizards have learnt to tolerate hot ground by alternating the feet that are in contact with the ground at any one time. Having long legs also helps as they hold the animal's body away from the sand's surface, where the heat is fiercest.

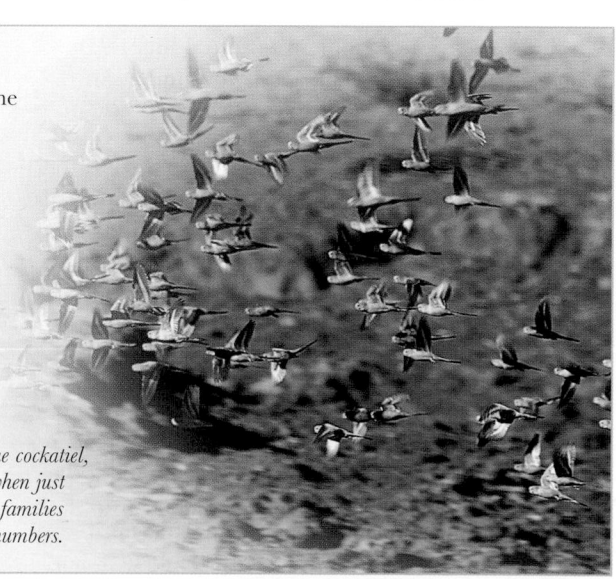

TROPICAL FOREST

FORESTS HAVE FLOURISHED in the tropics for longer than they have existed anywhere else on earth, which helps to explain why the animal species that live there outnumber those of all other land habitats combined. Most large tropical forest animals have been identified and classified, but the invertebrate life is so diverse that the task of cataloguing it will never be complete. There are two main types of tropical forest: rainforest, which is closest to the equator; and seasonal or monsoon forest, which grows towards the edges of the tropical belt.

TROPICAL RAINFOREST

Near the equator, the climate is warm and moist all year round, creating ideal conditions for plant growth. As a result, trees and other forest plants grow almost incessantly in an endless competition for light. Some plants put all their resources into growing towering trunks, while others are adapted for survival in partial shade. As a result of these different growth patterns, the forest is divided into clearly defined layers, each with its own characteristic animal species.

The highest layer, at about 75m (245ft), consists of giant, isolated trees called emergents. These provide nest sites for predatory birds and feeding platforms for monkeys. Beneath this level is the canopy, where copious light, combined with some protection provided by the emergents, results in a continuous layer of branches and lush foliage up to 20m (65ft) deep. This layer feeds or harbours most of the forest's animal life. Below the canopy is the understorey – a more open layer made up of shade-tolerant trees. On the forest floor, leaf litter is food for some very small animals as well as support for plants and saplings that grow where sufficient light filters through from above.

This zonal pattern is characteristic of lowland rainforest (the most common rainforest type). At higher altitudes, the trees are lower and the layers are more compressed – an effect that is exacerbated as altitude increases until eventually the trees form elfin forests little more than head high. Soil is also an important factor in shaping the forest. In some parts of the tropics, such as the Rio Negro region of South America, infertile sand results in the growth of stunted trees with leathery leaves.

FOREST EAGLES *have broad wings that enable them to glide through the canopy in their search for prey.*

SLOTHS *spend most of their lives suspended from branches in the forest canopy, relying on camouflage to avoid attack.*

TROPICAL RAINFOREST *harbours a greater variety of treefrogs than any other habitat.*

HOWLER MONKEYS *of South America's rainforest are among the few primates that survive on leaves.*

LEAFCUTTER ANTS *are found at all levels in rainforest, from the ground to the highest treetops.*

TAPIRS, *and other hoofed mammals, gather fallen fruit from the tropical forest floor.*

TROPICAL RAINFOREST *is found near the equator, where annual rainfall exceeds 2.5m (8¼ft) and is evenly spread throughout the year.*

HORNBILLS *use their long bill to collect fruit growing on the high branches in seasonal forest. In tropical rainforest, toucans feed in the same way.*

FOR STICK INSECTS *and leaf insects, highly developed camouflage is the key to survival in seasonal forest.*

FRUIT BATS, *the world's largest bats, are almost all found in tropical forest habitats.*

THE ATLAS MOTH *is found in seasonal forest, as are many of the world's other large, winged insects.*

FOREST CATS *make up more than half of the world's cat species. The tiger is the largest and one of the most endangered.*

SNAKES *in seasonal forest include species that hunt on the ground, and those that climb trees to search for prey.*

SEASONAL (MONSOON) FOREST

Unlike rainforest, where the climate is very stable, seasonal forest grows where rainfall is concentrated into a wet, or rainy, season, which is known as a monsoon. Up to 2.5m (8¼ft) of rain can fall in just three months – as much as some tropical rainforests receive throughout the whole year. As a consequence, seasonal forest is not as tall as tropical rainforest and, typically, the canopy is more open and extends further towards the forest floor. Immediately after the monsoon, seasonal forest is lush and green; but in the long dry season that follows, many of the trees shed their leaves, and the piercing sunlight is able to reach through the bare branches to the ground. Some seasonal forest trees are unusual in that they flower and fruit after losing their leaves. Where this happens, birds, insects, and mammals congregate in large numbers to feed. In the rainy season, the forest's animals are well hidden by the foliage; once the leaves have fallen, they become much easier to find.

Despite the yearly cycle of deluge and drought, the animal life of seasonal forest is some of the most numerous and varied in the world. In southern Asia, which has the largest area of this type of forest, the habitat supports elephants, monkeys, leopards, and also tigers. In Asia's seasonal forests there are some spectacular birds, including giant hornbills, and some of the world's largest snakes. In Africa, seasonal forests abound with browsing antelopes, while, in Central America, they are inhabited by pumas, coatis, and white-tailed deer. Most of these animals breed during the wet season, when they can take advantage of the abundant supply of fresh leaves.

SEASONAL FOREST *grows on either side of the equator. The dry season lasts longer the further the region is from the equator.*

CONSERVATION

Deforestation is currently threatening wildlife in forests throughout the tropics. Nearly half of the world's original tropical rainforest cover has been destroyed during the last 40 years, and large areas of seasonal forest have been cleared either for timber or for use as farmland.

International efforts are being made to preserve forests – in Costa Rica, for example, almost 30 per cent of the country has been designated as national park – but in many countries throughout the tropics, forest clearance is continuing at a rapid rate.

FORMER FOREST
Tropical forest clearance follows the pattern set in forests in other parts of the world, but its impact on plants and animals is greater because there are so many species at stake.

LIFE IN TROPICAL FOREST

SOME TROPICAL FOREST ANIMALS spend all their time on the ground. For most, however, daily life involves getting about among trees. The canopy holds most of the forest's food, so an animal that is good at moving around in the treetops has the greatest chance of thriving. Some animals are so well adapted to life in the trees – breeding as well as feeding there – that they very rarely have to visit the forest floor.

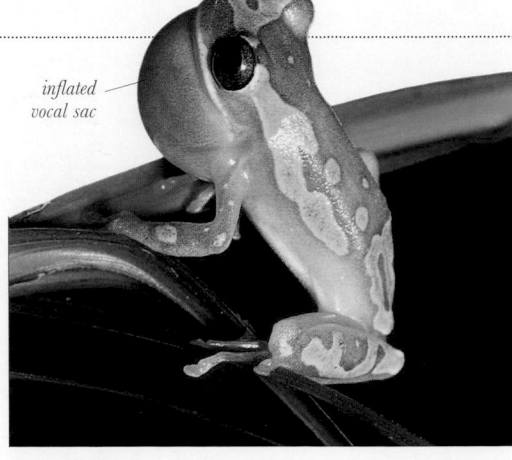

PRIVATE CALL
With its vocal sac inflated, a male treefrog calls to deter rival males and to attract potential mates. Female frogs react only to the call sign of their own species; and the louder and longer a male calls, the more likely females are to respond.

inflated vocal sac

MOVING IN TREES

Small animals need relatively few special adaptations for moving about in trees. Ants, for example, are so light that it makes little difference to them whether they are travelling up and down trees or across the forest floor. But for larger animals, such as apes, monkeys, and other primates, climbing is a dangerous occupation: if they lose their grip – as occasionally happens – they risk a fatal fall.

Most primates climb by running or leaping along the tops of branches, often using familiar routes that act like highways through the trees. Monkeys follow these routes mainly by sight,

PREHENSILE TAILS
Climbing snakes – such as this emerald tree boa – use their tails to clamp themselves to branches. The front of the snake's body folds up, ready to straighten out as it strikes anything that comes within reach.

but many of the more primitive primates, such as bushbabies, move about after dark, identifying their pathways partly by smell. Gibbons are different again: they travel underneath the branches by swinging hand-over-hand in a breathtakingly acrobatic manner. This unusual but highly effective form of movement is called brachiation.

Tropical forest harbours a huge variety of flying animals – birds, bats, and flying insects – that swoop or hover among the treetops. However, during the course of evolution, many unrelated animals, including mammals, frogs, and even snakes, have developed wing-like flaps of skin that enable them to glide. Some of these gliders can travel over 100m (330ft) from tree to tree and, remarkably, many of them are most active after dark.

COMMUNICATION

In any kind of forest, animals face problems keeping in touch. In the canopy, leaves and branches make it difficult to

see for more than a few metres, while tree trunks get in the way on the ground. As a result, many forest animals rely on sound and scent, rather than visual signals, to claim territories and attract partners.

Some of the loudest animals in the world live in tropical forest. They include howler monkeys, bellbirds, parrots, cicadas, and an enormous variety of treefrogs. Like mammals and birds, each species of treefrog has its own characteristic call: some produce a short metallic "tink", others generate a sustained trilling that sounds like machinery.

Signalling with sound can be dangerous because it can attract predators as well as potential mates. Treefrogs and cicadas minimize the problem by pitching their calls so that the source is very difficult to locate. Other animals, including many mammals and flying insects, avoid the problem by using scent to stay in touch. One great advantage of scent is that it lingers: for example, in marking its territory, a jaguar or okapi leaves a signal that will last for several days.

KEEPING OUT OF DANGER

Tropical forest abounds with camouflaged animals as well as species that mimic others. Animals that use camouflage – chiefly insects and spiders, but also snakes, lizards, frogs, and toads – resemble a huge variety of inanimate objects, from bark, thorns, and bird droppings to branches and fallen leaves. Many animals use camouflage to avoid being spotted and eaten, but some predators also use it to enable them to ambush their prey.

Mimicry, in which one species "pretends" to be another, is a subtler means of avoiding attack. It involves a relatively harmless species evolving to look like one that is dangerous, and it is most common in invertebrates. Some tropical forest spiders, for example, closely resemble stinging ants and even move like them as well. Matters are complicated where several species come to look alike. Some groups of unrelated butterflies, which contain poisons that are distasteful to birds, imitate each other; thus they have evolved the insect equivalent of a shared warning trademark.

ARM OVER ARM
Swinging or "brachiation" is a highly efficient way of moving about in trees. This lar gibbon can leap over 5m (16ft) between hand-holds and can easily overtake someone running on the ground.

DOUBLE PROTECTION
South American thorn bugs not only look like thorns but feel like them. The thorn-like extension on the thorax both disguises them and makes them difficult to swallow.

DECEPTIVE "EYES"
Peanut bugs rely primarily on camouflage for protection but, if threatened, they display a pair of alarming-looking eye-spots. This is a bluff because these bugs are harmless.

Warning signals are most developed in extremely toxic animals. For example, unlike other frogs, tiny poison-arrow frogs hop nonchalantly about the forest floor, relying on their extraordinarily vivid colours to warn other animals that they are not merely unpalatable but highly dangerous to eat.

FEEDING

Near the equator, trees grow, flower, and set seed all year round, generating a non-stop supply of food. Many forest animals – including bats, birds, and insects – live almost exclusively on the abundant nectar and fruit. Some of these animals help trees to spread their pollen and seeds. Quetzals, for example, swallow fruit whole, and then regurgitate the stones onto the forest floor where they can germinate.

NECTAR FEEDERS
A bat laps up flower nectar. Flowers pollinated by bats and birds have to be robust if they are not to be damaged by their visitors. Birds are attracted to certain flowers mainly by their bright colours, bats by their pungent scent.

RODENT OPPORTUNISTS
Agoutis and other small, ground-dwelling mammals often follow the progress of parrots and monkeys through the trees, gathering the pieces of food that accidentally fall to the ground.

Compared with flowers and fruit, tropical forest leaves are difficult to digest. Animals that feed on these leaves generally pick them while they are still young – before protective toxins have had a chance to build up inside them. Insects are the most prolific leaf-eaters, but some of the forest's larger animals also rely on this difficult diet. They include several kinds of monkeys and sloths, and the hoatzin – a highly unusual bird from South America. The hoatzin processes its food much as a grazing mammal does; after eating, it is often so heavy that it can barely fly.

Tropical forest predators range from some of the world's smallest insects to the largest cats. In an environment that provides lots of cover, most of them stalk their prey rather than running it down. Army and driver ants are the most remarkable exceptions: they hunt in "packs" over 50,000 strong, overpowering and eating anything that cannot escape.

REPRODUCTION

For animals that live in trees, breeding can sometimes involve unusual adaptations. Some treefrogs come down to the forest floor to lay their eggs, but many lay them high up in the canopy, either in water-filled tree-holes or in the pools of water that gather in plants. Some frog species are more creative, laying their eggs in nests of foam that keep their eggs moist until the tadpoles are ready to hatch.

Many tropical birds start life in the safety of tree-holes, but climbing mammals rarely build nests, and many of their young start life in the open. Young monkeys often cling to their mothers' chests, keeping a tight grip as the parent runs along branches or leaps through the air. For young murine (mouse-like) opossums, which live in the American tropics, early life is even more precarious because their mothers do not have well-developed pouches. This means that until their feet have developed the ability to cling, these tiny marsupials hang from their mothers' teats by their mouths, their legs dangling in the air.

PROTECTIVE NEST
These African treefrogs have grouped together to make a giant foam nest hanging from a branch. The nest's exterior hardens to protect the eggs and to keep the interior foam moist. When the tadpoles hatch, they break out and drop into water below.

ANT FOLLOWERS

In Central and South American forests, a single column of marauding army ants attracts up to 30 different kinds of bird, all swooping down to catch the tiny animals that burst out of the leaf litter to escape the ants. Some birds flutter ahead of the ant column; others dart among the ants themselves. Some are full-time ant followers, while others follow the throng only for as long as the ants are marching through their territories. A few small animals, including wasps and rove beetles, travel with the army on the ground. Although this sounds dangerous, these animals mimic the ants very closely – an adaptation that prevents them being attacked while gaining them protection and a supply of free food. Army ants also attract lizards and frogs, as well as parasitic flies that lay their eggs on the fleeing animals.

ARMY ANTS

AERIAL ATTACK
The blue-crowned motmot, a part-time ant follower, is distantly related to kingfishers. Like them, it swoops down on its prey, carries it back to a perch, and hits it against a branch before eating it.

TEMPERATE FOREST

TEMPERATE FOREST grows in regions that have a wide range of climates. In some, winters are cold and summers are cool; in others, the winter is relatively mild, and the summer heat rivals that in the tropics. Where winters are cold, temperate forest trees are usually deciduous, shedding their leaves in winter and growing a new set in spring; in warmer regions, many trees keep their leaves all year. Although temperate forest does not have as many animal species as tropical forest, it is still among the richest wildlife habitats on land.

DECIDUOUS FOREST

In the depths of winter, deciduous forest can seem gaunt and empty, and largely devoid of animal life. But as the days lengthen in spring, and buds begin to burst, the habitat becomes alive with birdsong and animals on the move. This transformation is triggered by a sudden abundance of plant food – one that nourishes large numbers of plant-eating insects as well as the animals that feed on them. Many of these forest animals are permanent residents, but they also include migratory birds that fly in from distant parts of the world.

Compared with tropical forest, temperate deciduous forest has relatively few tree species: the maximum number – found in some of the forests of eastern North America – is several hundred, while tropical forest might contain several thousand. Nevertheless, temperate forest trees are powerhouses of life. Large oak trees, for example, can produce over a quarter of a million leaves a year – enough to sustain the army of weevils, gall wasps, and moth caterpillars that feed rapidly in spring and early summer while the leaf crop is at its freshest and most nutritious.

LARGE AREAS *of temperate deciduous forest are found in the northern hemisphere; most temperate forest in the southern hemisphere is evergreen.*

Like tropical forest, deciduous forest has a clear, vertically layered structure, but there are some important differences. The trees are rarely more than 30m (100ft) tall, and the canopy layer is usually deep but open, allowing light to reach the understorey and encourage plant growth. Fallen leaves rot slowly in cool conditions, so deciduous forest has an unusually deep layer of leaf litter that insects, woodlice, and millipedes use as food and cover. This means that while many small animals live in the cracks and crevices in bark, the place that is richest in invertebrate life is not the trees but the ground.

MOTHS *lay their eggs on leaves, buds, or bark. Their caterpillars are often the most numerous leaf-eaters in the forest.*

TREECREEPERS *patrol tree trunks in search of small insects hidden in bark crevices.*

SQUIRRELS *spend the autumn collecting food, which they then store away for use in the winter.*

DEER *feed on leaves in summer; in winter, they often strip bark from shrubs and young understorey trees.*

WILD BOAR *dig up the leaf litter with their snouts, feeding on animals, nuts, and roots.*

IN THE MOIST *conditions of the forest floor, lungless salamanders can absorb oxygen through their skin.*

CANOPY

THE RINGTAILED
possum has a prehensile tail, which helps it to climb along high branches to reach flowers and fruit.

HOOPOES *swoop down from the canopy to catch animals on the ground.*

KOALAS
live in the forest canopy, but also walk across the ground to reach isolated clumps of trees.

UNDERSTOREY

WHITE'S TREEFROG *has exceptionally thick skin, which minimizes moisture loss and enables it to cope with the dry conditions in evergreen forest.*

GRASSHOPPERS
and other insects, such as beetles, provide food for reptiles and mammals that live on the ground.

GROUND

SUN-SEEKING *lizards and snakes often bask in the warm sunlight that bathes the forest floor.*

EVERGREEN FOREST

In warm parts of the temperate world, many broadleaved trees are evergreen. Unlike trees of deciduous forest, which grow in spring and summer, evergreens grow in winter and spring, when temperatures are low but not cold, and when water is readily available. Described by botanists as sclerophyllous (meaning hard-leaved) forest, this habitat is found in several widely scattered regions of the world, including parts of California and western South America, the Mediterranean region in Europe, and large areas of eastern and southwestern Australia. In some of these places, the forest is low-growing, but in Australia, where eucalyptus is the dominant species, it includes the tallest broadleaved trees in the world.

TEMPERATE EVERGREEN *forest is found in regions with a Mediterranean-type climate: mild, damp winters, and warm, dry summers.*

Temperate evergreen trees usually have open crowns, which means that the vertical layers are usually less pronounced than they are in forests in cooler regions, and plenty of light is able to reach the forest floor. As a result, these forests are rich in ground-based wildlife, and warmth-loving animals – such as lizards and butterflies, which are usually associated with higher levels – can often be seen sunbathing on the floor. The open structure also makes it easy for birds, such as kookaburras and other forest kingfishers, rollers, and hoopoes, to swoop down on animals moving about on the ground.

The air in evergreen forest often smells pleasantly aromatic because most of the leaves are filled with pungent oils. These oils help to stop the leaves from drying out, and they also protect them from animals. They are a highly effective deterrent, for relatively few animals – apart from specialists such as the koala – include these leaves in their diet.

HABITATS

CONSERVATION

Deciduous forest once covered huge swathes of the northern hemisphere. Now, after centuries of clearance for agriculture, only small pockets of the original forest remain. In general, the forest cover is no longer declining, and in some countries it is actually on the increase. However, evergreen forests continue to be threatened both by logging and by clearance for industrial and residential development. In Australia, conservationists have been engaged in a protracted struggle to safeguard areas of untouched forest from the logging industry and from farming.

SALINIZATION
These Australian eucalyptuses have been killed by salinization – rising salt levels in the ground. Salinization is caused by forest clearance, which changes the water balance of the ground. Salinized land is useless for agriculture.

LIFE IN TEMPERATE FOREST

THE FACTOR THAT MOST AFFECTS LIFE in temperate forest is the variable food supply. At all levels – from the treetops to the forest floor – the life cycles of forest animals living in temperate forest move in step with the seasons so that they produce their young when food is easiest to find. Life is relatively easy in spring and summer; but in winter, when the food supply is at its lowest, special adaptations are needed for survival.

FEEDING

As a habitat, deciduous temperate forest – the kind of forest found across much of the northern hemisphere – has one very useful feature. The trees that grow in it, such as oaks and beeches, produce leaves that are designed to last for just one growing season. As a result, these leaves are usually thin and easy to eat, which is why vast numbers of insects feast on them from the moment they begin to appear in spring.

This sudden explosion of insect life attracts an army of highly specialized avian predators. In Europe, northern Asia, and North America, dozens of warbler species migrate north as the buds open. These birds have extremely acute eyesight, enabling them to scour leaves for the tiniest grubs and caterpillars, which they then pick up with their tweezer-like beaks. Other birds, including treecreepers, woodpeckers, and nuthatches, concentrate on the bark, seeking out and pecking at the tiny animals hidden among the crevices.

FOREST-FLOOR FORAGERS
Watching over their piglets, adult wild boars root in the leaf litter for food. Wild boars feed on acorns and other nuts, and they also use their spade-like snouts to dig up roots, fungi, and small animals hidden in soil or among fallen leaves.

acorns in crevices

WINTER GRANARY
In western North America, acorn woodpeckers store acorns in trees, cramming them into holes that they have previously drilled in the bark. These storage trees are called granaries. A single granary may hold up to 50,000 acorns and be stocked by up to a dozen woodpeckers working and nesting together. The woodpeckers will also store acorns in fence posts and telephone poles.

By midsummer, leaves stop growing, and animal feeding behaviour changes. Most temperate trees are pollinated by wind, which means that they do not produce enticing nectar-rich flowers. However, they do produce large crops of nuts and other seeds, which are extremely important foods for animals because, unlike leaves, they can be stored away and used when other food is scarce.

TREE GALLS

Common in broadleaved forest, galls are growths whose development is triggered by insects and other organisms. Galls often look like berries or buttons on leaves and twigs. They act as both home and larder, or nursery, allowing their occupants to feed unseen. In broadleaved forest, most galls are produced by tiny wasps and flies. Each species of insect affects a particular tree, and each produces galls of a characteristic shape, making them easy to identify.

OAK GALLS
Oak marble galls are produced by Andricus kollari, *a wasp belonging to the family* Cynipidae. *The wasp's larvae develop inside the galls. They then chew an exit hole through the outer surface to emerge as adults.*

wasp larvae inside gall

Food storage, or "caching", is practised by many forest birds and mammals. Jays bury acorns in the ground, while acorn woodpeckers store them in trees. Squirrels bury seeds of all kinds, and red foxes bury anything that is even faintly edible, from half-eaten remains to food wrappers and discarded shoes. Some animals locate their stores by scent, but most are very good at pinpointing them by memory alone, finding and digging up their food even when it is covered by snow.

Seed-caching has an important impact on forest ecology. Although animals that bury seeds have good memories, some of what they hide is always forgotten about. This means that provided the seeds are not discovered by rodents or other animals, they remain effectively planted and ready to germinate, helping the forest's trees to reproduce.

HIBERNATION

In autumn, many insect-eating birds migrate to warmer climes, leaving the forest's remaining animals to face the winter cold. Animals that store food can remain active throughout this difficult time of year, but others use a very different survival strategy: they hibernate, living on the fat reserves they have built up during the summer months.

How long and how deeply an animal hibernates depends on where it lives. In the forests of northwest Europe, hedgehogs may hibernate for up to six months, whereas further south their winter sleep is much shorter. In eastern North America, woodchucks – or groundhogs – typically hibernate from October to February; their wanderings early in the year are a traditional sign that spring is not far off.

Some hibernating animals, such as the common dormouse, hardly ever interrupt their winter break, even if they are picked

SLEEPING THROUGH THE COLD
Common dormice use leaves and moss to make well-insulated winter nests among brambles and other plants. Here, part of the nest has been temporarily removed to reveal the hibernating animal, fast asleep with its tail wrapped tightly around its body.

up. However, many hibernators behave in a different way. If the weather turns warm, they briefly rouse themselves: bats, for example, take to the air for feeding flights, while hedgehogs often move out of one hibernation nest and into another. But forest hibernators have to be careful not to do this

NIGHT FLIGHT
Unlike most squirrels, North American flying squirrels feed at night. The elastic skin flaps between their legs enable them to glide for over 50m (165ft). They steer with their front legs, and use their large eyes to navigate safely in the dark.

too often: activity uses up their bodily food reserves, and it therefore puts them at risk of running out before the winter is truly over.

Many insects also hibernate, often hidden under bark; but, in some species, the adults die out, leaving behind tough, overwintering eggs that will hatch in spring.

MOVEMENT

While monkeys and gibbons are the most impressive climbers in tropical forest, squirrels are the experts in temperate forest. Unlike many climbing mammals, they can run head-first down tree trunks, as well as up them, by hooking their long, curved hind claws into bark. Squirrels have excellent eyesight, and they instinctively scuttle to the back of a tree if they spot a potential predator – a simple behaviour that makes them difficult to catch.

Temperate forest is inhabited by gliding rodents, and also – in Australasia – by gliding marsupials. But for precise manoeuvring among trees, owls and birds of prey are unrivalled. Unlike their relatives in open habitats, most of these aerial hunters have relatively short, broad wings that enable them to twist and turn effectively. A prime example of this adaptation to woodland life is the Eurasian sparrowhawk: rather than soaring and then swooping, it speeds among trees and along hedgerows – sometimes only a metre or so above ground – ambushing small birds in mid-air and carrying them away in its talons.

Temperate forest provides ground-dwelling animals with lots of cover. As a result, small mammals, such as voles and shrews, abound on the forest floor. To avoid being seen as they move about, these animals often use runways partly covered by grass or fallen leaves. Voles use a combination of vision, smell, and touch to find their way along these runs, but shrews, which have very poor eyesight, navigate partly

by emitting high-pitched pulses of sound. These signals bounce back from nearby objects in the same way as those sent out by a bat's sonar system.

LIVING IN LEAF LITTER

The leaf litter in temperate forest is one of the world's richest animal micro-habitats. This deep layer of decomposing matter harbours vertebrates, including mammals and salamanders, but its principal inhabitants are invertebrates that feed on leaf fragments, on fungi and bacteria, or on each other. Some of these animals – such as centipedes and woodlice – are large enough to be easily seen, but many others are microscopic.

HUNTER IN THE LEAVES
Lithobiid centipedes, such as this one, have flattened bodies that allow them to crawl under leaves and fallen logs as well as on the ground surface. Geophilid centipedes live permanently underground, and therefore have narrow, almost worm-like bodies.

Animals that live deep in leaf litter exist in total darkness, so most of them rely on their sense of touch to find food. This is especially true of predators: centipedes locate their prey with long antennae, while tiny pseudoscorpions use the sensory hairs that cover their pincers. Like true scorpions, pseudoscorpions are venomous; but they are so small that they pose no threat to anything much bigger than themselves. This is fortunate because in just a few square metres of leaf litter their numbers can run into millions.

Dead leaves are a useful screen, hiding leaf-litter dwellers from other animals foraging on the forest floor. However, it is not totally secure. Some temperate forest birds, particularly thrushes, pick up leaves and toss them aside, snapping up leaf-litter animals as they try to rush away from the light.

INSECT ENGRAVERS

Female bark beetles tunnel through the sapwood beneath living bark, laying eggs at intervals along the way. The hatched larvae eat their way out at right angles to the original tunnel, creating distinctive "galleries" that can be seen when dead bark falls away. The side-tunnels end in exit holes, from which the developed beetle emerges and flies away. Common in deciduous forest, bark beetles can be highly destructive because they often infect trees with fungi. One species – the elm bark beetle – spreads Dutch elm disease, a fungus that has wiped out elms in parts of Europe and North America.

BARK GALLERIES
The adult bark beetle has a cylindrical body and a round-fronted thorax hiding most of the head. The gallery patterns vary from one species to another.

CONIFEROUS FOREST

CONIFERS ARE THE WORLD'S toughest trees. Their small, needle-shaped leaves can withstand extreme cold and are impervious to strong sunshine and wind, and their relatively narrow, upright habit enables them to grow closely together to form dense, sheltered forest. As a result, conifers thrive where few broadleaved trees can survive, such as the far north and in mountain ranges. They also flourish in places that have very heavy rainfall. In such areas they form temperate rainforest, home to some of the largest trees in the world.

BOREAL FOREST

Named after Boreas, the ancient Greek god of the north wind, boreal forest, or taiga, is the largest continuous expanse of forest on earth. It covers about 15 million square km (6 million square miles) and stretches in an almost unbroken belt across the far north, often reaching deep into the Arctic. In some places, the belt is over 1,600km (1,000 miles) wide. Across the boreal forest belt as a whole, winter temperatures routinely drop below – 25ºC (–13ºF), but in some of the coldest regions, such as northeast Siberia, temperatures can fall below – 45ºC (– 49ºF). Summers in boreal forest are brief but can be warm.

Compared with the types of forest that occur at lower latitudes, boreal forest has only a handful of tree species and therefore provides only a limited variety of food for herbivores. Plant diversity is also restricted both by the amount of light that can reach the forest floor and by the high acidity of pine needles. Even in summer, the interior of the forest is often dark, with a thick layer of dead needles carpeting the floor. Fungi thrive in these conditions, but the only forest-floor plants that live here are the ones that can tolerate low light levels and acidic soil conditions.

BOREAL FOREST *stretches across much of the far north. There is no equivalent habitat in the southern hemisphere.*

Apart from insects, few animals can digest conifer leaves or wood, so most plant-eaters concentrate on seeds, buds, and bark, or on berries from low-growing shrubs. However, what this habitat lacks in variety, it more than makes up for in quantity, especially as there is relatively little competition for food. This is one of the reasons why many boreal forest animals, from birds to bears, have far more extensive ranges than species that live in warmer parts of the world.

THE ANIMAL LIFE *in coniferous forest provides birds of prey, such as the northern goshawk, with a year-round food supply.*

TREE TRUNKS *provide the growing larvae of horntails and other wood wasps with protection from the worst of the winter cold.*

THE BROWN BEAR *can climb trees, but it finds most of its food on the ground.*

GREY WOLVES *hunt in packs, a strategy that enables them to kill animals larger than themselves. This is particularly important during winter, when food is scarce.*

FOR ELK *and other hoofed mammals, boreal forest offers protection from the cold winds and blizzard conditions of winter.*

CANOPY

UNDERSTOREY

GROUND

THE GOLDEN-CROWNED *kinglet is one of many small insect-eaters that feed and nest high above the ground.*

OWLS, *such as this great horned owl, fly at night, hunting the many small mammals and birds that live in temperate rainforest.*

THE NORTH AMERICAN PORCUPINE *is a slow and awkward tree-climber, but it manages to reach the treetops to eat buds and bark.*

THE FISHER *belies its name in that it rarely catches fish. It is one of the few animals that includes porcupines among its prey.*

THE STOAT'S *slim body enables it to pursue animals into their burrows, leaving them little chance of escape.*

SLUGS *in temperate rainforest include plant-eating species and some that prey on other slugs.*

TEMPERATE RAINFOREST

The world's largest areas of rainforest are found in the tropics, but rainforest also exists in parts of the temperate world. It grows where west-facing mountains intercept moist air blowing in from the sea and, unlike boreal forest, it experiences relatively mild temperatures all year round.

Compared with boreal forest, temperate rainforest is a rare habitat, occurring in a few widely separated areas. In the southern hemisphere, it is found in the South Island of New Zealand, and in parts of southern Chile. In both of these places, most rainforest trees are broadleaved species, but in America's Pacific northwest – where the largest temperate rainforest in the world can be found – the trees are almost entirely conifers. Some are over 75m (nearly 250ft) high, and more than 500 years old. This kind of coniferous forest looks unlike any other. On the ground, and in the understorey, every surface is draped with ferns or waterlogged moss. Densely packed trunks, some over 3m (9¾ ft) across, rise up to the canopy high overhead, where the sky is always laden with rain.

Temperate rainforest supports many animals that are found in coniferous forest all over the world, but it has some additional features that set it apart: the mild, damp conditions, which make it a haven for slugs and salamanders, and the immense amount of fallen timber, which creates opportunities for insects that feed on dead wood. In its natural state, the forest teems with mammals, as well as with owls and other birds that need large, old trees as nest sites. Unfortunately, these trees are in great demand by the timber industry and, as a result, untouched temperate rainforest is increasingly rare.

TEMPERATE RAINFOREST *is found on west-facing coasts, where heavy rain falls throughout much of the year.*

CONSERVATION

Of all the world's forests, boreal forest has been least altered by human activity. However, this is changing as demand for softwood and paper pulp increases. Rainforest, on the other hand, has been subjected to immense change. In North America, much of the Pacific coast rainforest has been cut down, leaving only small remnants of original, "old-growth" forest, separated by planted forest. Of the many remnants that have been threatened by logging, some have been saved.

INDISCRIMINATE FELLING
Clear-felling removes all the trees in large areas of forest, forcing the animal inhabitants to move out.

LIFE IN CONIFEROUS FOREST

COUP DE GRACE
At the end of a high-speed chase, a Canadian lynx catches a snowshoe hare. Population cycles of both species are closely linked.

FOR ANIMALS AND TREES ALIKE, life in boreal forest is dominated by the need to survive long and extremely cold winters. Animals that remain active in winter, such as wolves, need a constant supply of food simply to avoid freezing. Conifers are difficult to exploit for food, which means that animals that rely on them have developed some highly specialized physical and behavioural characteristics.

FEEDING

Compared with many broadleaved trees, conifers are well protected against attack. In addition to tough leaves, they often have oily resins that make both their leaves and their wood difficult to digest. Furthermore, if the sapwood is injured, this resin oozes out and traps insects and spiders as effectively as glue.

Despite these defences, some animals manage to live entirely on coniferous trees. Among the most successful are sawfly larvae. These caterpillar-like grubs bore deep inside the trunks, leaving cylindrical tunnels in the wood. They do not eat the wood itself; instead they feed on a fungus that grows on the walls of the tunnels they have built. Female sawflies carry small amounts of this fungus with them when they emerge as adults, and they infect new trees when they lay their eggs. This kind of symbiotic partnership is vital to sawflies, but it is not entirely unique: in other habitats, particularly in the tropics, ants and termites also "cultivate" fungi as food.

Although wood-boring grubs are safe from most predators, they are not entirely immune from attack. Coniferous forest is the habitat of some of the world's largest woodpeckers, which hammer their way into tree trunks to reach the grubs inside. Sawflies also face a threat from the ichneumon wasp, which drills through the wood with its long ovipositor to lay its eggs on the sawfly larvae. It is thought that the wasp locates the larvae by

EXTRACTING SEEDS
A crossbill's beak overlaps at its tips, making it an ideal tool for extracting cone seeds. For short periods, crossbill nestlings can survive temperatures as low as −35°C (−31°F).

smell. When the eggs hatch, the ichneumon grubs eat their host larvae alive.

Some animals, such as the capercaillie and North American porcupine, eat large quantities of conifer needles, but moth caterpillars are the leading leaf-eaters: as always in coniferous forest, the number of species involved is small, but the damage they inflict can be vast. This is especially true of species such as the gypsy moth, which has been accidentally introduced to many parts of the world.

FINDING WINTER FOOD

Conifers do not have true flowers, but they nevertheless produce seeds. For birds and small mammals, this seed crop is a valuable winter fuel. However, accessing conifer seeds is not easy: they develop inside woody cones, and the cones stay tightly closed until the seeds inside are mature.

Coniferous forest animals have developed a variety of ways of removing these seeds before the trees scatter them on the forest floor. Squirrels gnaw through the soft, unripe cone while it is still attached to the branch, eating the seeds and dropping the remains of the cone on the ground below. Woodpeckers often take fallen cones and wedge them into tree-holes or broken stumps, using these to hold the cone firm while they peck out all the seeds. Crossbills are even more proficient: their beaks are uniquely adapted for dealing with cones, enabling them to extract the seeds with surgical precision.

MOTH ATTACK
Although the gypsy moth itself does no damage, its caterpillars feed voraciously on conifers as well as on a range of deciduous trees.

DRILLING THROUGH WOOD
Bracing itself with its feet, a female ichneumon wasp bores into a branch to lay eggs on a sawfly grub.

BROWSING DEER
Deer have a major impact on coniferous forest by eating young tree shoots and bark. Where deer are numerous, such browsing can kill small saplings (inset). However, this creates open glades, which are useful to other animals because they let in light, allowing fruit-forming plants – a good food source – to grow.

Compared with seeds, bark is a low-quality food, but in winter it is vital to some species' survival. Deer strip it away from the base of young saplings, while bank voles and porcupines often climb trees to attack the bark higher up. Bark stripping often stunts a tree's growth, and a severe attack can kill it.

COPING WITH COLD

Coniferous forest has its share of seasonal visitors, principally insect-eating birds that arrive in spring and then leave for the south once they have raised their young. But for the rest of its animals, long winters are an inescapable fact of forest life. Some species

POPULATION GRAPH
Fur trappers' records vividly show how snowshoe-hare numbers in Canada's northern forest rise and fall. The peaks, around ten years apart, are separated by periods of population depletion.

POPULATION CYCLES

Since there are relatively few animal species in the northern coniferous forests, the lives of predators and prey are very closely linked. During mild years, strong tree growth can trigger a population explosion among small animals; as a result, the predators that feed on them begin to increase in number. These conditions never last for long, though: as the plant-eaters begin to outstrip the food supply, their numbers start to fall again. And as the rate of the fall accelerates, the predators soon follow suit.

Despite the unpredictability of the northern climate, these ups and downs occur with surprising regularity. In North America, fur trappers' records dating back over a century provide some long-term evidence of population swings. For example, they show that the snowshoe hare population roughly follows a 10-year cycle, with two or three good years, followed by a lengthy slump. The Canadian lynx – one of the snowshoe hare's main predators – follows the same pattern, but with a one- or two-year time-lag. Similar cycles involving lemmings and other small mammals take place in tundra (see p.65).

While there is little that they can do to prevent this boom-and-bust pattern from occurring, animals like the snowshoe hare are able to make a fairly fast recovery from a population slump by breeding quickly when conditions are favourable.

hibernate, but many remain active even during the coldest months, relying on their insulation for survival.

One group of forest animals, the mustelids, have coats that are exceptionally well insulating. This group includes pine martens, wolverines, minks, and sables, all of them agile hunters renowned for their thick and luxurious fur. As with most mammals, their coats contain two different kinds of hair: long, outer guard hairs form the coat's water-repellent surface, while shorter, much denser hairs – the underfur – trap a layer of air close to the body, keeping the animal warm. Northern species all grow an extra-thick coat after their late-summer

moult; and some species, such as the stoat, use this moult to change colour, developing a white coat that provides better camouflage for the winter.

Keeping warm is relatively easy for large mammals because their bodies contain a large store of heat. But, for the smallest warm-blooded inhabitants of coniferous forest, winter conditions test their cold tolerance to its limits. Voles and other rodents can hide in burrows, but birds spend most of their lives in the open. For wrens and tits, which often weigh less than 10g (⅜oz), winter nights are a particularly dangerous time. With such minute bodies, their fuel reserves are tiny, and so they must make special provision if they are to stay alive until dawn when the search for food can resume. Some of them make the most of what body heat they have by huddling together in tree-holes, but a few, such as the Siberian tit, bed down in the snow, using it as an insulating material.

RUNNING ON SNOW
Cold winters actually make it easier for the wolverine to catch food: its ability to run at high speed over frozen snow enables it to catch large animals, such as reindeer, which can outpace it during the summer months.

IRRUPTIONS

In the northern coniferous forests, food supplies are affected by the weather and the degree of competition. In winters that follow cool summers, the supply of seeds and berries can be thin, leaving seed- and fruit-eating birds with little to survive on. Rather than starve, these birds fly south in waves, called irruptions, which may involve travelling beyond their normal winter range by as much as 1,500km (930 miles). Species that frequently irrupt include crossbills, waxwings, and tits, as well as nutcrackers and other seed-eating members of the crow family.

FLYING SOUTH
A flock of Bohemian waxwings pause during their southward flight. Like other irrupting species, waxwings usually spread south when their population reaches a high level. Shortage of winter food can trigger a mass exodus.

MOUNTAINS

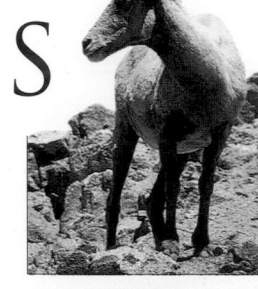

IN MANY LAND HABITATS, climatic
conditions vary only slightly within
a region. In mountains, however, the
average air temperature drops by about
1°C (1.8°F) for every 200m (650ft)
gained in height, oxygen becomes
scarcer, and the air becomes less effective
at screening out ultraviolet light. As a
result, mountains can be divided into
distinct zones, each supporting plant
and animal life that is very different
from that of the zone above and below.
A wide range of animals live in the low-
altitude foothills, but only the hardiest
survive year-round in the harsh
environment above the tree-line.

HOOFED MAMMALS, *such
as the bighorn sheep, survive
at high altitude. They can
often be found as high as the
snowline.*

TEMPERATE MOUNTAINS

In temperate regions, a mountain's climate is
relatively cool throughout the year. However,
seasonal changes are much more marked than they
are in the tropics. At high altitude, above the tree-
line, there is a sudden burst of plant growth in
spring and summer. Some animals
migrate upwards to make use of
this brief abundance of food, but
others, such as the marmot, are permanent residents
between mid- and high altitude, surviving the winter
cold by living in burrows, and by hibernating for up
to eight months a year. High-altitude insects spend
many months in a dormant state, coming to life
when warm weather arrives. For many, this dormant
period is spent inside the egg, which hatches when
the days lengthen and the temperatures rise.

MARMOTS *hibernate in
burrows, while some other
mountain rodents, such as
pikas, stockpile food to
survive the winter.*

THE PUMA *is an
adaptable predator that
ranges from sea level to
over 3,000m
(9,900ft).*

At lower altitudes, the
climate is warmer, and
generally more like that
of the surrounding land.
However, because the
sloping, rocky ground
is difficult to farm, the
mountainsides often
retain more of their
tree cover than does
flatter ground. In
undisturbed conditions,
these montane forests are
the natural habitat of large
mammals, such as mountain
lions, bears, and deer, and also of a wide range of
seed- and insect-eating birds.

**THE PEREGRINE
FALCON** *uses the ledges
of steep cliffs as vantage
points and nest sites.*

THE NORTHERN
*temperate zone's mountain
ranges include the highest
peaks on earth. In the
southern hemisphere,
mountains are smaller
and more isolated.*

Temperate mountains abound in birds of prey.
Some, such as the peregrine falcon, pursue their prey
on fast-flapping wings, while eagles and buzzards
soar high up, riding on updraughts. One characteristic
mountain species, a vulture called the lammergeier,
turns the mountain landscape to its advantage by
carrying carrion bones aloft and dropping them
onto the rocks to expose the edible marrow inside.

LIKE MANY DEER *species,
the white-tailed deer migrates
vertically, feeding above the
tree-line in summer.*

MOUNTAIN *hummingbirds survive low night-time temperatures in the Andes by becoming torpid – a form of overnight hibernation.*

ALPINE ZONE

THE VICUNA'S *high levels of oxygen in its blood enable it to range at high altitude.*

FOREST ZONE

THE EASTERN GORILLA *uses the dense vegetation of the forest zone for both food and cover.*

LOW ALTITUDE

MONKEYS *are common in tropical mountain forests. Africa's black and white colobus is a striking example.*

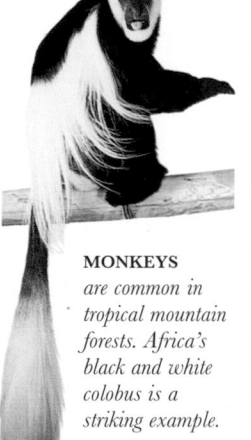

TROPICAL MOUNTAIN *tree and ground frogs live at all altitudes, from foothill forests to high, fast-flowing streams.*

TROPICAL MOUNTAINS

In the tropics, the generally warm climate means that mountain vegetation zones extend much higher than they do in montane habitats elsewhere in the world. For example, near the equator, trees often grow at altitudes of up to 4,000m (13,200ft), which is why many tropical mountains are forested to their summits. Above this altitude is the tropical alpine zone, an open landscape dominated by grass and some highly specialized plants. This zone is often above the clouds, which means that nights are cold and frosty and yet the sunshine is fierce.

Many tropical animals have successfully adapted to life at high altitude. They include the vicuna, which can be found up to 5,500m (18,100ft) in the South American Andes, and the yak, which reaches a record 6,000m (19,800ft) just north of the tropics in the Himalayas. Birds also live at great heights: in South America, for example, mountain hummingbirds called Andean hillstars often feed at over 4,000m (13,200ft). The Andean hillstar's minute size means it has difficulty in storing enough energy to enable it to survive the cold nights. To combat the problem, its nocturnal heartbeat slows down and its temperature plummets, conserving energy.

The cloud-covered forest below the alpine zone is the habitat of some of the world's most endangered animals. They include the eastern gorilla – a species restricted to the mountains of central Africa – and the resplendent quetzal, a bird that lives in the cloud forests of Central America. The abundant moisture means that the forest zone also teems with many different species of frog, living both on the ground and in trees.

TROPICAL MOUNTAINS *include a major chain, the Andes, and isolated ranges in East Africa and Southeast Asia.*

HABITATS

CONSERVATION

Among the current threats to mountain wildlife are deforestation, quarrying, and development for hydroelectricity and skiing resorts. However, in the long term, global warming threatens even greater change by altering the pattern of vegetation zones. Some mountain animals will benefit from this, but others will be faced with a steadily shrinking habitat. This is particularly problematic for those species that live at specific altitudes on isolated ranges or peaks because many of them are unable to descend to lower ground in order to find other areas of suitable habitat.

ROCK REMOVAL *In some parts of the world, quarrying has destroyed important wildlife habitats. As road building and construction projects continue to increase, the demand for rock continues to grow.*

LIFE IN MOUNTAINS

LIFE AT HIGH ALTITUDE CAN BE HARSH. Food is often scarce, the weather can be treacherous, and the thin air can make it difficult to breathe. However, there is also more space, relatively little interference from humans, and fewer predators than there are lower down. Many animals are "incomers", using high ground as an extension of their normal range, but there are also some that live only on high ground. In large mountain chains, many animals have a wide distribution; but isolated peaks are often inhabited by animals that are found nowhere else.

BREATHING THIN AIR

At 6,000m (19,800ft), air is half as dense as it is at sea level. As a result, it contains only half the normal amount of oxygen – so little that anyone trying to breathe at this height would have difficulty remaining conscious. Yet some mountain animals live even higher than this because they have evolved specialized body systems that enable them to get the maximum amount of oxygen into their blood.

In the vertebrate world, birds are the unrivalled experts at high-altitude living. This is because air passes through their lungs in only one direction, not in and out, which ensures that a high proportion of the air's oxygen enters the blood – far more than enters a mammal's bloodstream in the same conditions. This fact is apparent from the height at which birds are capable of flying. In the Himalayas, choughs have been seen fluttering around campsites at over 8,000m (26,400ft), and there are records of vultures colliding with planes at over 11,000m (36,300ft) – far higher than Mount Everest.

Birds are unusual in being able to cope with rapid changes in altitude without experiencing any ill effects. For mammals, moving from one altitude to another necessitates special adjustment by the body, which is achieved by acclimatization – a process that can take several weeks to complete. During acclimatization, the number of red cells in the blood slowly increases, boosting its oxygen-carrying capacity. This physical adjustment, which is shown by a broad range of mammals – including humans – is temporary. If an animal moves back to lower ground, the process is reversed.

However, for mountain mammals, such as the vicuna and ibex, adaptation to life high up is a permanent state, not something that can be switched on and off. When measured as a proportion of volume, vicunas have three times more red cells in their blood than most other mammals, and the haemoglobin in their red cells is unusually good at collecting oxygen. As a result, vicunas can run almost effortlessly on the altiplano – the high-altitude plateau that runs the length of the Andes.

Compared with mammals, cold-blooded animals, such as reptiles, have fewer problems with thin air because they use oxygen more slowly. For them, the main problem with mountain life is cold: if the temperature is too low, their body processes slow down, and their muscles have difficulty working.

MOVEMENT

Mountains seem almost purpose-made for soaring birds because strong air currents make it easy for them to gain height. For animals on or near the ground, moving about is not so easy. Many insects are wingless, and species that do fly usually keep close to the rock to reduce the risk of being blown away by the wind.

For larger animals, the situation is even more hazardous, for a single misjudged move can lead to a fatal fall. Many rock-dwelling mammals therefore have feet designed to prevent slipping. In Australia, rock wallabies use their large back feet to take leaps of up to 4m (13ft), with their long tails helping them to balance.

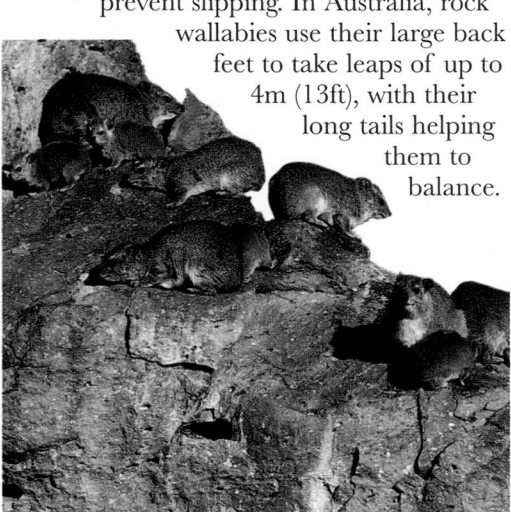

NON-SLIP FEET
The hyrax's small, adhesive feet are perfect for negotiating rock, although some species also use them for climbing trees. As well as living high up, rock hyraxes inhabit kopjes – miniature mountains of eroded boulders scattered on the African savanna.

In Africa and the Middle East, hyraxes run over rocks and boulders with the help of specialized soles, which work like suction cups. But in most of the world's mountains, the most adept climbers are hoofed mammals.

Hooves may seem far from ideal for climbing, and it is true that some hoofed mammals, such as horses, have great difficulty moving about on rocky slopes. But in climbing mammals, such as mountain goats and klipspringers, hooves have evolved into perfect aids for moving about in mountains: they are small and compact, allowing them to fit onto narrow ledges, and they have hard edges surrounding rough non-slip pads. These combined characteristics make for good grip in all conditions, including rain and snow.

Getting a firm grip is essential for moving on rock, but equally important is a strong sense of balance and head for heights. Most terrestrial mammals are instinctively afraid of steep drops, but, from an early age, mountain-dwellers show what appears to be a reckless

SOARING
Updraughts provide an almost effortless means of travel for birds like the Andean condor, which can cover hundreds of kilometres in a day by riding waves of rising air.

spiralling flight path

wind direction

HABITATS

exposed mountain top to the forests lower down, but some mountain forest species also migrate. Among these latter migrants are birds such as nutcrackers, which feed on conifer seeds. If the seed crop fails, they fly downhill in a form of sporadic migration called an irruption (see p.57). Clark's nutcracker, from the Rocky Mountains, is a typical example: normally found at up to 2,500m (8,200ft), it descends as low as sea level when food becomes hard to find.

FINDING FOOD

As in most land habitats, a mountain's animal life depends ultimately on plants, for plants provide food for herbivorous animals, which are, in turn, eaten by a wide range of predators. However, some mountain animals make use of a very different food source – the cargo of small animals, mainly insects, that are carried uphill by the wind to be stranded among the rocks, snow, and ice.

Most of these wind-blown animals are so tiny that they are practically invisible; yet they provide useful nourishment for scavengers that live above the snowline. They consist almost entirely of invertebrates, such as springtails and snow fleas, which can survive the very low temperatures of high-altitude winters. During the depths of winter, they hide among rocks and moss, but when the weather warms up they can often be seen hopping across banks of snow, feasting on the debris that the wind has brought up from lower ground.

disregard for their own safety. Adult chamois take 6m (20ft) leaps, and can run down near-vertical slopes as easily as they can run up them; and their young are able to keep up with them when just a few weeks old.

COPING WITH WINTER

In tropical mountains, conditions are often much the same all year round, which means that animals can stay at one altitude all their lives. But in temperate mountains, seasonal changes affect the food supply. Winter is the critical time: anything that cannot survive the cold weather conditions and the shortage of food has to move to lower ground or hibernate until the return of spring.

Animals that are resident at high altitude have a variety of ways of coping with the changes. Insects often enter a dormant state, called diapause, which puts their development on hold. Many small mammals, such as marmots, survive mountain winters by hibernating, while many of those that remain active live on food stores accumulated earlier in the year. Pikas, for example, gather up leaves and grass and build them into "haystacks" among the broken rocks around their homes. Before adding fresh supplies to a stack, they sometimes spread them out to dry in the sunshine, which reduces the chances of the food rotting before it is eaten.

VERTICAL MIGRANTS
In mountainous regions, red deer spend the summer high up, where food is plentiful and there are relatively few biting flies. Their downhill migration in autumn often coincides with the start of the rutting season, when males grow a mane of hair on the neck and compete with each other for the right to mate.

For other animals, the first autumn snows are the signal to move downhill. These vertical migrations are a common feature of mountain life in temperate regions, and they are demonstrated by a wide range of mammals and birds, from mountain sheep and deer to choughs and grouse. In many cases, the migration involves moving from the

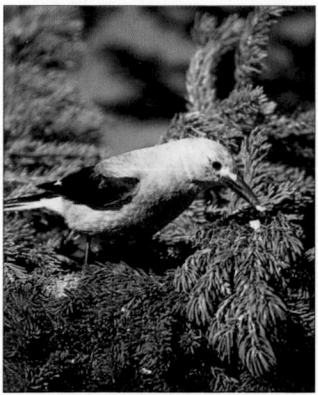

FLEXIBLE MIGRANT
Like many temperate mountain animals, the nutcracker lives at various altitudes, depending on the conditions: severe cold drives it downhill, but because there is more competition for food on low ground it returns uphill as soon as the weather improves.

LIFE IN CAVES

Many animals use caves temporarily, but some have adapted to spend their entire lives in them. These permanent cave-dwellers, called troglodytes, feed either on each other or on the droppings deposited by roosting bats and birds.

As a habitat, deep caves have the advantage that temperatures remain fairly constant throughout the year. However, they are also completely dark, which means that eyes are useless. Bats, oilbirds, and swiftlets use echolocation to navigate while underground, but permanent cave-dwellers sense their surroundings largely by touch, often using smell to track down food. Cave crickets detect food using antennae, while spiders and harvestmen use their feet. In subterranean streams and pools, cave salamanders, such as the olm, sense vibrations in the water. The cave fish has a row of pressure

sensors along each side of its body that enables it to detect other animals several metres away.

Although food is scarce, animals exist many kilometres underground and have even been found in pothole systems that have no direct contact with the surface except via water trickling its way underground.

hard, waterproof exoskeleton

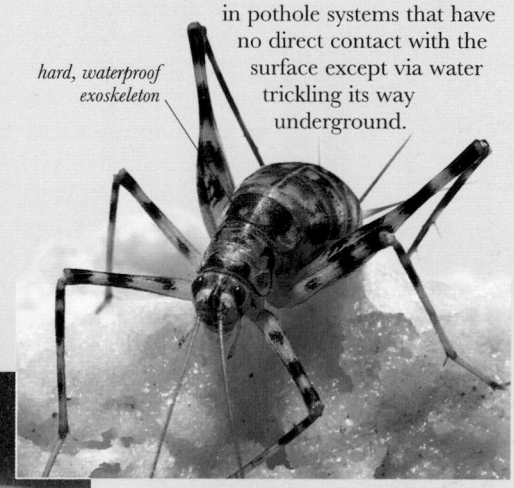

GIANT ANTENNAE
Cave crickets' antennae, which can be two or three times the body length, locate the dead remains and bat droppings that make up the animal's diet.

UNDERWATER LIFE
Like most aquatic vertebrates in caves, the olm has very little pigment in its skin. Its eyes are vestigial and covered with skin. Olms breed in water and remain in it throughout adult life.

POLAR REGIONS

ARCTIC OWLS *are forced to hunt by daylight during the weeks of high summer.*

THE ARCTIC AND ANTARCTIC are the coldest places on earth. The Arctic is a partly frozen ocean, hemmed in by large expanses of windswept tundra; the Antarctic is an ice-covered continent, surrounded by the world's stormiest seas. They are similar to each other – and unlike any other habitat – in that they have 24-hour daylight in summer and perpetual darkness in winter, but they are physically different in ways that have important effects on animal life. In the Arctic, many animals live on land; in the Antarctic, animal life is based almost entirely in the ocean.

WILLOW GROUSE, *like many other animals, turn white in winter to blend in with the snow.*

ARCTIC AND TUNDRA

Covering about 12 million square km (4.6 million square miles), the Arctic Ocean is both the smallest and the shallowest ocean in the world. For several months in summer, permanent daylight produces a constant supply of energy, which is harnessed by vast quantities of planktonic algae. These form the first link in the Arctic Ocean's food chain, which ultimately nourishes animals as large as whales and polar bears.

Sea ice – or the lack of it – is a major factor in determining where large mammals live, especially during winter when the surface area of the ice is at its greatest. Polar bears and Arctic foxes can traverse the ice to find food, but seals and some other marine mammals must maintain breathing holes to survive.

MUSK OXEN *and other tundra grazers reach food by scraping away snow with their hooves.*

LEMMINGS *provide food for many of the Arctic's predators, including foxes.*

MOSQUITOES, *which emerge in early summer, make life uncomfortable for mammals and birds.*

TRUE TUNDRA *is found north of the Arctic Circle, but tundra-like conditions exist on some mountains elsewhere.*

Despite the icy conditions, sea life is plentiful in the Arctic because cold water is rich in oxygen and the sea-bed sediment is rich in nutrients. On land, though, intense winter cold means that trees cannot survive. The result is tundra – an open, often featureless, landscape, scraped smooth by glaciers during the last ice age. Today, Arctic glaciers are restricted mainly to mountains and to the ice-cap that covers Greenland, but large areas of tundra remain permanently frozen underground. This frozen zone – the permafrost layer – prevents spring melt-water from draining away, creating waterlogged landscapes in a region where rainfall, or snow, is paradoxically very low.

In late spring and early summer, tundra plants grow, and flower, very rapidly. Geese and other migratory birds arrive to breed, and vast numbers of mosquitoes emerge from tundra pools. The migrants' departure, when the short summer draws to a close, marks the end of another biological year.

THE POLAR BEAR *is a superb swimmer, and it is as much at home on shifting pack-ice as it is on tundra.*

ANTARCTIC

Unlike the Arctic, mainland Antarctica is isolated from the rest of the world. It is covered with ice, up to 4,000m (13,200ft) thick, which continues out to sea forming large ice shelves. On the Antarctic Peninsula – a finger of land pointing towards South America – summer temperatures rise to a few degrees above freezing point, but in the rest of the continent average temperatures are below freezing all year round.

Algae and lichens grow on bare rocks in many parts of the Antarctic coastline, but the Antarctic Peninsula is the only part of the continent where terrestrial plants can survive. This is also the only place that has a significant range of terrestrial animals, although these are chiefly springtails, mites, and nematode worms – few of which are over 5mm (⅕in) long. The rest of Antarctica's land-based animal life consists of species that feed in the sea and come ashore to breed, such as penguins, or those that scavenge food at these animals' breeding grounds, such as skuas. With the exception of emperor penguins, vertebrates desert the ice at the end of summer to spend the winter at sea.

The Southern Ocean, which surrounds Antarctica, is one of the most biologically productive seas in the world. Although species numbers are relatively low, population sizes are often enormous because the non-stop summer daylight generates a vast supply of food. Krill – small crustaceans that form the diet of seals and whales – are especially prolific: some of their swarms are estimated to weigh in excess of 10 million tonnes and are large enough to be seen by satellites in space. Although the Southern Ocean is always cold, it maintains a minimum temperature of about –1.8°C (28.8°F); below this, seawater freezes. As a result, the ocean is quite warm compared with Antarctica itself.

MOST OF ANTARCTICA, *excluding the relatively mild Antarctic Peninsula, lies south of the Antarctic Circle.*

SKUAS, *found in both the Arctic and the Antarctic, feed on carrion and on other birds' eggs and chicks.*

NESTING COLONIES *of some birds, such as Adelie penguins, can contain over a million members.*

SPRINGTAILS *eat decaying plant remains – unlike most Antarctic animals, which rely on food from the sea.*

KRILL *harvest plankton, forming the basis of many Antarctic food chains.*

LEOPARD SEALS *feed on krill, fish, and seabirds. This predatory lifestyle has no direct equivalent in the Arctic.*

BALEEN WHALES *play a major part in the Southern Ocean's ecology by eating vast amounts of plankton.*

CONSERVATION

Under the terms of the Antarctic Treaty, which came into effect in 1961, Antarctica is currently protected against commercial exploitation. In the Arctic, the situation is quite different, with oil production, mineral extraction, and hunting taking place in many areas. In both polar regions – but particularly in the Antarctic – wildlife faces habitat change brought about by atmospheric increases in carbon dioxide (causing global warming) and by thinning of the atmosphere's ozone layer (causing increased levels of ultraviolet radiation).

GAPS IN THE ICE *In the year 2000, oceanographers reported finding large stretches of open water at the North Pole – a region that is normally covered by floating ice. This unprecedented melting of the polar ice-sheet was almost certainly triggered by global warming.*

LIFE IN POLAR REGIONS

ALTHOUGH THEY LIVE AT OPPOSITE ENDS OF THE EARTH, the animals that inhabit the Arctic and Antarctic share many adaptations. Resilience to extreme cold is first and foremost among these, but almost as important is the ability to cope with a highly seasonal food supply. For some animals, winter is a good time for catching food, but, for most, hunger and cold make the long winter months a critical time of year. Such testing conditions mean that in comparison with other parts of the world the poles are inhabited by very few animal species. However, those that do thrive can be extraordinarily numerous.

COPING WITH COLD

Warm-blooded animals have to maintain a constant body temperature, which means that combating heat loss is a major priority in the polar environment. Cold-blooded animals can function with a fluctuating body temperature, but even they have limits – in sub-zero conditions they can freeze solid. Fish are particularly at risk of freezing, for while their body fluids normally freeze at about −0.8°C (30.6°F) polar sea water is often slightly colder still. To help prevent freezing, the blood of many cold-blooded species contains proteins that lower its normal freezing point. Some insects can survive at −45°C (−49°F) without any ice forming in their bodies.

Since mammals and birds cannot afford to let their internal temperature fall even slightly, they need insulation to keep warm. Fur and

THE ANTARCTIC'S ICE FISH
In the Antarctic, fish live in a world that is often roofed by ice. The high level of oxygen in cold water allows some Antarctic fish to survive without any haemoglobin in their blood. As a result, their blood is almost clear rather than red.

feathers are among the finest insulating materials that nature has devised, but many polar animals, such as whales, seals, and penguins, have additional insulation in the form of blubber – a layer of yellowish fat that is laid down under the skin. Blubber, which can be up to 30cm (12in) thick, is such an effective insulator that all these animals feel cold to the touch even when their internal body temperature is above 38°C (100.5°F). Blubber is particularly useful in the sea because water conducts heat away from the body 25 times faster than air. It also has another valuable function: because fat contains lots of energy, blubber can be used as a food reserve when supplies are low.

SLEEPING ON ICE
A polar bear cub rests at its mother's side. Polar bears have long body fur, and their feet have furry soles. This keeps them warm on the ice, while a thick layer of blubber enables them to retain body heat while swimming in the sea.

SURVIVING UNDER ICE

Whales and seals face problems during the long polar winter because sea ice restricts their access to air. They dive under the ice to feed, but they must then surface to breathe. Some species avoid the problem by moving to lower latitudes. Those that remain behind survive either by maintaining breathing holes or by congregating in polynia – areas where the wind and currents keep the water ice-free.

Seals start making breathing holes when the ice is thin, rasping away at it with their teeth. As the depth of the ice increases with the progress of winter, they continue to visit and work on their holes to ensure they remain clear. The Weddell seal, which lives further south than any other species, spends so much time keeping its breathing holes open that its teeth develop distinctive patterns of wear, and by late winter its breathing holes can be 2m (6½ft) deep. The seals have to find their holes in almost complete darkness because during the Antarctic winter the sun stays below the horizon for weeks.

PAUSE FOR BREATH
A Weddell seal heads up towards a breathing hole in the ice. Weddell seals can stay underwater for over an hour, making return journeys of up to 10km (6 miles) before they have to come up for air.

Whales rarely make breathing holes. Instead they head for polynia, where they can come up for air in open water. This less laborious strategy means they are not tied to one place, but it does have its dangers: groups of whales can become ensnared in shrinking polynia, unable to reach the next stretch of open water. There are records of narwhals – the world's most northerly whales – being trapped in their hundreds, making them easy targets for hunters.

CHANGING COLOUR

In the treeless Arctic tundra, camouflage is one of the most effective ways both of avoiding attack and of making an attack unseen. The summer and winter landscapes look so different that many tundra animals change their camouflage twice a year. The Arctic fox is a classic example: its summer coat is usually brownish grey, but in early autumn it turns white; in spring, the process is reversed so that the fox blends with the rapidly thawing tundra. In some parts of the far north – particularly western Alaska and northern Greenland – Arctic foxes develop a bluish winter coat instead. Some researchers have suggested that this blue coat is an adaptation to coastal landscapes, where there is less winter snow, but as these foxes have been widely introduced by fur farmers the theory is difficult to prove.

The least weasel, the world's smallest carnivore, changes its coat in a similar way, as do ptarmigans and many other tundra birds. Some, such as the snowy owl, keep their white plumage all year round, which suggests that good camouflage is most important in winter and less so in summer when food is easier to find.

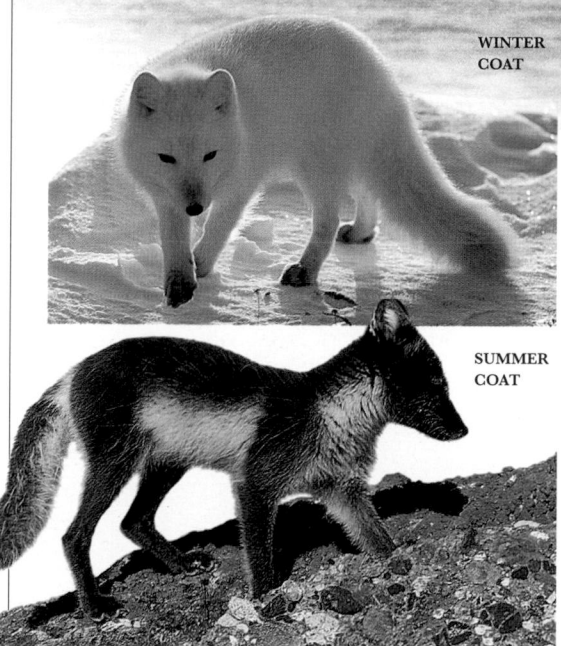

LONG-DISTANCE MIGRANTS
In Alaska and northern Canada, vast herds of reindeer migrate between their summer grounds on open tundra and their winter grounds in coniferous forest. Some travel over 1,000km (620 miles) each way, swimming across rivers and sea inlets en route. Pregnant females lead the herd during the spring leg of the journey.

WINTER
COAT

SUMMER
COAT

SEASONAL COAT
When the Arctic fox changes colour, the consistency of its coat changes as well. Its white winter coat has long guard hairs and thick underfur, making for superb insulation. Its brown summer coat is shorter, with thinner underfur, to help prevent overheating.

SUMMER MIGRANTS

Near the poles, 24-hour daylight in summer creates the ideal conditions for rapid plant growth. This short-lived but profuse supply of food has a dramatic effect on tundra life, attracting vast numbers of migrants. Geese come to crop the plants with their beaks, waders arrive to feed on worms and insects that live in swampy ground, while terns find food both on the tundra and in water close to the shore.

This annual influx of visitors is mirrored in the sea. Many of the world's baleen whales head towards polar waters during summer to make use of the annual upsurge in planktonic life. However, unlike migratory birds, these huge mammals do not breed at high latitudes. Instead, they put on weight and then return to warmer waters to give birth. During the breeding season, they often do not feed at all.

WINTER FOOD

Although the food supply in polar seas slowly falls in autumn, there is still a reasonable amount for animals to eat. On land, life is not so easy. The growth of tundra plants comes to a complete halt and, to make matters more difficult, the plants themselves are often covered by deep snow. For herbivores, this lack of accessibility is a major problem at a critical time of year. In the Arctic tundra, plant-eating animals reach plants in one of two different ways. Reindeer (caribou) and musk oxen use their hooves to clear away the snow to reveal the lichens and dwarf willows underneath. Lemmings turn the snow to their advantage by burrowing in it. The snow protects them from predators and from the weather outside: no matter how cold or windy it is on the surface, the lemmings enjoy a benign microclimate that allows them to feed all year.

In Antarctica, there are very few terrestrial plants, which means that almost no animals stay active in winter on food gathered from land. With so much ice, even food from the sea can be difficult to reach. Male emperor penguins, guarding their eggs, do not even attempt to find it: huddling on the ice through the long night of winter, they go without food until spring.

RUNWAYS UNDER THE SNOW
Lemmings need cover to survive. In winter, they develop enlarged front claws that enable them to tunnel in snow. The tunnels keep them secure, although they must still contend with sparse winter food supplies. When the snow melts, they move underground.

TUNDRA INSECTS

Unlike Antarctica, the Arctic tundra teems with insect life, including aphids, bumble bees, damselflies, and unimaginable numbers of mosquitoes. The mosquitoes spend their larval and pupal stages in tundra pools, emerging as adults in early summer. For warm-blooded animals, tundra mosquitoes are a serious problem. The females need blood before they can breed, and they are relentless in their attempts to get it. When the mosquito season is at its height, some animals head for high ground, but most have no alternative but to sit it out. Mosquitoes do bring some benefits: the larvae and pupae are a useful source of protein for waterfowl and waders.

FROM WATER TO AIR
A cloud of mosquitoes, caught in the summer sunlight. As adults, tundra mosquitoes live for only a few weeks — just long enough to feed, mate, and lay eggs, before autumn frosts glaze their breeding pools with ice.

FRESH WATER

EVERY YEAR, ABOUT 100,000 cubic km (24,000 cubic miles) of water evaporates from the world's oceans, condenses, and then falls as rain or snow. Most of this water disappears back into the atmosphere to continue this cycle, but about a third returns to the oceans by flowing either over ground or beneath the land surface. This steady supply of fresh water sustains all the world's land-based life, as well as creating highly diverse habitats – from streams, rivers, and lakes, to reedbeds, marshes, and swamps – in which a wide range of different animal and plant life can thrive.

HERONS HUNT *by stealth, waiting along river banks and lake shores for fish to come within range.*

LAKES AND RIVERS

For permanent water dwellers, life in lakes and rivers is shaped by many different factors. One of these is the water's chemical make-up, which is often dictated by the type of rock that forms the bed of a river or lake. Hard water, for example, is good for animals that grow shells because it contains calcium that can be used as shell-building material, while water that is especially rich in oxygen is important for highly active predatory fish, such as salmon and trout. Water that is very deficient in oxygen, on the other hand, provides a poor environment for animal life as a whole because relatively few aquatic species, apart from specialized worms, can survive in it.

In lakes and rivers, aquatic animals usually occupy clearly defined zones. The brightly lit

water close to the surface often teems with water-fleas, copepods, and other microscopic forms of animal life. They live here in order to feed on phytoplankton – the microscopic algae that become extremely abundant during the summer months. Feeding in the middle zone and near the water's surface are larger animals, such as

IN GEOLOGICAL TERMS, *lakes and rivers are highly changeable. Lakes gradually fill with sediment, while rivers often change course.*

fish, that are able to hold their own against the strength of the current. Weak swimmers live near the banks, where the current is slower, or among stones and sediment on the bottom. In still and slow-flowing water, surface tension supports insects that hunt by walking or running over the water.

Some animals that are associated with freshwater habitats are not necessarily permanent water dwellers; instead they divide their time between water and the adjacent land, entering the lake or river to hunt or breed, or using it as a nursery for their young.

DRAGONFLIES, *like many other insects, spend the early part of their lives in fresh water, leaving it only when they are ready to breed.*

FROGS *vary in their affinity for water. The American bullfrog rarely leaves it, even as an adult.*

MANY AMPHIBIANS, *such as the great crested newt, use still or slow-flowing water containing plenty of plants as a repository for their eggs.*

SMALL CRUSTACEANS *are part of the animal plankton living at all levels in ponds, and at the surface in lakes.*

PIKE *are sit-and-wait predators, using dense waterside vegetation as cover from which to ambush their prey.*

SURFACE

MID-WATER

BOTTOM

FISH EAGLES *use their long talons to snatch fish from the surface of the water.*

THE JACANA, *or lilytrotter, has long toes that spread the bird's weight so that it can stand on floating leaves.*

THE CAPYBARA, *the world's largest rodent, feeds on land, but uses water as a refuge from attack by predators.*

CAIMANS, *and many other crocodilians, can survive for many weeks when wetlands dry out by wallowing in moist mud.*

THE SNAKE-NECKED *turtle's slow metabolism enables it to remain underwater for long periods.*

LUNGFISH *survive times of drought by lying dormant in mud, protected by their waterproof cocoons.*

WETLAND

A wetland is any waterlogged or flooded area with a covering of water plants. In some wetlands – reedbeds and bogs, for example – the plants hide the water completely. However, in most wetland, areas of open water and dense vegetation are mixed, creating a rich and complex habitat that can be exploited by animals of almost every kind.

Biologically, wetland is among the most productive inland habitats, sometimes surpassing even rainforest in the amount of food that it generates for animals. In temperate parts of the world, this productivity reaches a peak during spring and summer, but in the tropics and subtropics it is more affected by the water supply. Some tropical wetlands – South America's Pantanal, for example – largely dry out during the dry season, but then look like vast lakes once it has rained.

FRESHWATER WETLANDS *include the Pantanal, South America, and the Okavango, an inland delta in southern Africa.*

In many wetlands, the water is no more than a metre (3¼ ft) or so deep, which means that bottom-living animals and surface dwellers are rarely far apart. This kind of environment is ideal for air-breathing swimmers, such as snakes and turtles, and also for land-based animals that use water as a temporary refuge from danger. Unlike large lakes, wetlands have an extra dimension in the form of emergent plants, which grow up through the water's surface and into the air. These plants range in size from small grasses and rushes just tall enough to keep insects clear of the water, to water-loving trees that grow to over 35m (115ft) high. Trees act as important roosting and breeding sites for waterbirds, providing them with shelter and relative safety from predators as well as keeping them close to the source of their food.

CONSERVATION

Freshwater habitats are affected by human activities in many ways. Water pollution is a general problem that has profound consequences for a wide range of animal life. However, in addition, large areas of wetland have been drained for agricultural use, often because wetland conditions produce highly fertile soil. Water abstraction, both for agriculture and for domestic use, is also an increasing threat to freshwater habitats. Efforts to protect wetlands have produced several international agreements. One of these, the Ramsar Convention, adopted in 1971, focuses on wetlands used by migrating waterfowl.

HIGH AND DRY *In central Asia, the Aral Sea – a giant inland lake – has shrunk dramatically because incoming river water has been diverted to irrigate fields.*

LIFE IN FRESH WATER

FRESH WATER IS AN ESSENTIAL RESOURCE for all land-dwelling animals, but it is also an important habitat in its own right. Freshwater habitats vary immensely, from temporary pools to giant lakes, and from tiny streams to rivers thousands of kilometres long. As a result, the problems that animals must overcome in order to survive in their environment can be very different from one freshwater habitat to another. Strong currents, periodic drought, and intense competition for food are some of the difficulties for which animals must find solutions.

STAYING IN PLACE

Animals living in fast-flowing rivers face a constant battle with the current, and they cope with it in one of two main ways. The first is to avoid the problem by staying close to the river-bed, where the current is relatively slow. Many invertebrates, such as stonefly and mayfly larvae, never venture into open water; and to improve their staying power they often have a flat profile, so that if they position themselves with their heads facing upstream the current presses down on their backs, helping them to stay in place. Remarkably, the same technique is used by dippers – the only songbirds that feed underwater: they are naturally buoyant, but as they walk upstream over the river-bed, the force of the current keeps them submerged.

The second solution is to compete with the current by swimming against it. Species that cannot avoid the current match their swimming speed to the water flow, enabling them to stay in place – and they keep swimming even when they are asleep.

COPING WITH THE CURRENT
Brown trout live in cool, fast-flowing streams, where it takes a lot of energy simply to stay in place. Trout can hold their own because fast-moving water is usually rich in oxygen – exactly what the fish need to keep their muscles working.

MIGRATION

As well as supporting freshwater residents, rivers and lakes receive visitors from the sea. These are migratory fish, which divide their time between fresh- and saltwater habitats. Anadromous species, such as salmon, breed in rivers but spend most of their adult lives offshore. Catadromous species, such as some eels, do exactly the opposite: they live in fresh water, but swim out to sea to breed.

For most fish, fresh water makes a much safer nursery than the ocean, and this justifies their long journey upriver to lay their eggs. However, because the food supply in fresh water is limited, the young eventually make for the much more dangerous, but also more fruitful, environment of the open sea. The advantages of this are apparent where some, but not all, members of a species migrate: those that swim out to sea usually grow much bigger than those that stay behind.

Like migrating birds, migrating fish are often remarkably accurate in pinpointing the place of their birth, returning there to spawn even though it might mean an inland journey of over 2,500km (1,500 miles). Each river has its own characteristic chemical fingerprint and, with their acute sense of smell, migratory fish are able to identify the estuary that they left as young fish. By monitoring the scent of the water as they progress upriver, they home in on their own spawning ground.

Migrating fish often meet barriers to their progress. Salmon are famous for jumping waterfalls and rapids, while eels tackle obstacles by slithering around them over land. Eels usually do this after dark, in damp conditions, when they can survive out of water by breathing through their skin.

LIVING IN AND OUT OF WATER

Most amphibians, such as frogs and toads, develop in freshwater habitats and then, as they approach adulthood, take up life on land, returning to water to breed. However, insects evolved amphibious lifestyles long before the first true amphibians appeared. Today, almost every patch of fresh water, from the smallest puddle to the largest lake, is inhabited by insects. Mosquito larvae feed on microscopic freshwater life, whereas the larvae of dragonflies and damselflies stalk larger prey, catching it with a set of extensible jaws called a mask. These insects leave the water when they become adult, but water beetles and bugs remain in it for life, although their ability to fly makes it easy for them to spread from one pool to another.

Some larger animals have developed lifestyles that straddle water and land. Snakes are good swimmers, and a number

A DAYTIME REFUGE
Hippos lounge in rivers and lakes by day and emerge at night, sometimes travelling over 15km (9 miles) to find good grazing. Their skin is thin and almost hairless, but it exudes a special secretion that helps to protect them from the daytime sun.

JOURNEY'S END
Most migratory fish travel upriver to breed several times during their lives. Sockeye salmon, however, almost always die once the adults have spawned. Sockeyes that are ready to breed typically develop hooked jaws and humped backs.

BREATHING AT THE SURFACE
Mosquito larvae hang upside down at the water's surface, breathing through snorkel-like tubes. The tubes are tipped with water-repellent hairs that break through the surface film.

of species, such as the grass snake, specialize in catching aquatic animals. The anaconda, the heaviest snake in the world, uses water for cover and for support: despite its size, water buoys it up, reducing its effective weight to almost nil and enabling it to swim at considerable speed.

As well as providing food, fresh water is a valuable resource for some land-based animals seeking refuge from predators or daytime heat. At dusk, birds often roost on lakes and reservoirs while, during the day,

ROOSTING AMONG THE REEDS
Swallows, starlings, and other flock-forming songbirds often use reeds as overnight roosts. The sleeping birds are relatively safe from predators because reeds are usually surrounded by water.

hippos, capybaras, and beavers spend a lot of time in the water, emerging under the cover of darkness to feed on land.

SURVIVING DROUGHT

In warm parts of the world – particularly in the tropics – rivers and lakes often dry out completely for several months each year. In these conditions, freshwater animals need some unusual adaptations to survive.

When a lake's water level drops, its temperature increases. Warm, stagnant water often contains so little oxygen that many fish suffocate, but lungfish – characteristic inhabitants of tropical lakes and wetlands – are experts at coping with drought. They gulp air at the surface and, when their home starts to dry out, they burrow into the mud, sealing themselves inside mucous cocoons. Later in the year, when heavy rain falls and soaks the mud, the cocoon breaks down, and the fish wriggle away.

DRY TIMES
Two staring yellow eyes identify a common caiman in the mud of a dried-out lake. During drought, caimans can survive being entombed for many months as long as the mud stays moist.

Caimans and turtles also hide away in this manner, although their scaly, waterproof skins mean that cocoons are not needed. Being "cold blooded", or ectothermic, they need relatively little energy to stay alive, so they can survive drought-induced food shortages for months.

Animals that cannot survive drought often leave behind drought-resistant eggs, which hatch when water returns – an effective way of bridging the gap between one wet period and the next. This survival strategy is used by water fleas, rotifers, tardigrades, and many other "micro-animals" that live in temporary pools or in the film of fresh water that covers mosses and other plants.

COMING UP FOR AIR

Freshwater animals need to breathe oxygen. Fish use gills to extract it from the water, but many invertebrates collect it from the air. These air-breathers include water snails, many insects, and the water spider – the only spider to have evolved a fully aquatic lifestyle.

For animals living at the water's surface, air is easy to reach. For fully submerged ones, breathing requires periodic trips to the surface to replenish air reserves, which are stored in or on the insect's body, often forming a film-like bubble that gives the animal a silvery sheen. The water spider has an exceptionally elaborate storage system. It constructs a "diving bell" from strands of silk, trapping a large bubble of air inside it. The bell acts as both lair and nursery – a unique example of an animal creating a submerged habitat that resembles dry land.

LIVING IN SALT LAKES

Salt and soda lakes are the most saline habitats on earth, with up to 10 times as much dissolved salt as the sea. They form where rainfall is low and evaporation rates are high because temperatures often climb to over 40°C (104°F). Relatively few animals can cope with these extreme conditions, but the ones that do can be extremely numerous because they face very little competition. Salt-lake food chains are based on cyanobacteria – plant-like micro-organisms that harness the energy in sunlight, turning it into food.

LIVING IN BRINE
Brine shrimps are small crustaceans that thrive in salt-rich desert lakes. They breed rapidly, producing eggs that can withstand several years of drought.

FILTER-FEEDERS
Lesser flamingos use their beaks to filter micro-organisms out of the water. A large colony can consume more than one tonne of cyanobacteria a day.

OCEANS

THE OCEANS FORM by far the largest continuous habitat on earth, and they were almost certainly the environment in which life first evolved. The underwater landscape is made up of mountains and volcanoes, cliffs, deep valleys, and vast, flat plains, many of them far larger than any found on land. The oceans are so immense – they cover more than three quarters of the earth's surface – and difficult to explore that scientific knowledge of ocean wildlife lags behind that of life on land. However, research has shown that life is found at all levels, from the sunlit surface to the deepest trenches over 11km (7 miles) down.

INSHORE WATERS

Some inshore waters are so shallow that if the world's oceans were lowered by just 75m (245ft), huge areas of sea bed would be exposed. Off western Europe, for example, the coast would be extended by about 200km (125 miles), and in parts of Siberia by more than 700km (435 miles). These shallow waters owe their existence to continental shelves – the gently sloping plateaux that flank many of the deep ocean basins. Continental shelves are a key habitat for sea life, supporting large shoals of fish and a diverse collection of other animals, from lobsters and crabs to molluscs and burrowing worms. This wealth of life is possible because, in shallow water, sunlight can reach the sea bed, promoting the growth of algae, seagrasses, and countless other organisms that need energy from light to survive. Just like plants on land, these provide animals with a year-round supply of food, as well as with plenty of cover and places to breed.

Some inshore animals, such as lobsters and flatfish, spend most of their lives on the sea bed, while a few live entirely in mid-water or at the surface. Others live in quite different habitats at different stages of their lives. For example, some larvae develop as part of the surface plankton, then move to mid-water or the sea bed as adults.

Inshore waters are also visited by animals from the open sea – most often by passive drifters, such as jellyfish, but also by powerful swimmers, such as whales and sharks. Some come to breed, while others arrive by accident and then return to deeper water. Occasionally, the latter get into difficulties: jellyfish often end up on the shore, and whales can become stranded when their navigation systems guide them into shallow water instead of safely out to sea.

CONTINENTAL SHELVES *vary in width from a few kilometres – in coasts close to deep-sea trenches – to over 1,000km (620 miles).*

SURFACE

MID-WATER

SEA BED

PLANKTONIC LIFE *near the surface of inshore waters consists partly of sea animals starting their lives as larvae.*

JELLYFISH, *the largest individual planktonic animals, can swim as well as drift near the surface. Some are over 2m (6½ft) across.*

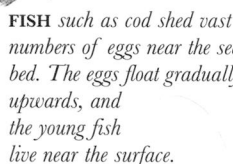

MOST SHARKS *live in the relatively shallow water of the continental shelves rather than in the deep sea.*

FISH *such as cod shed vast numbers of eggs near the sea bed. The eggs float gradually upwards, and the young fish live near the surface.*

LOBSTERS *are common sea-bed scavengers, crawling over the bottom in search of live prey and dead remains.*

LIKE MANY FLATFISH, *the European plaice lives on the sea bed. It is camouflaged to match the sediment.*

SURFACE

MID-WATER

ABYSSAL ZONE

SEA BED

OCEAN WANDERERS, *such as tropicbirds, spend most of their lives far out at sea, returning to land only to breed.*

THE HUGE SUNFISH *is a typical pelagic species: it spends its life in open water, often hundreds of kilometres from land.*

MARINE TURTLES *often migrate long distances between their feeding grounds and the beaches where they breed.*

DOLPHINS *have streamlined bodies that give them exceptional swimming ability. They often roam far out of sight of land.*

THE VIPERFISH *is one of many deep-sea species that produce light in order to attract their prey.*

DEEP-SEA SCAVENGERS, *such as brittlestars, feed on organic matter that drifts down to the sea bed.*

OPEN SEA

Even in the clearest seawater, light penetrates no further than about 250m (825ft) below the surface. More than any other factor, this has a crucial effect on sea life because it determines what there is to eat. In the brightly lit surface zone, microscopic algae grow by harnessing the energy in sunlight, creating an invisible harvest for planktonic animals. This food is passed on when the plankton are themselves eaten in food chains that eventually end with sharks and other large predators. Below the point where the light fades and finally disappears, there is no home-grown supply of food. At this depth, and further below, animals feed either on each other or on the supply of dead remains that constantly drifts down from the surface.

IF THE CONTINENTAL *shelves are excluded, the average depth of the world's ocean basins is about 4,000m (13,200ft).*

Despite the unimaginable volume of the oceans, few of the world's animal species – perhaps 5 per cent – live in open water. Since most of these stay near the surface, where they can take advantage of the relatively plentiful food supply, animal life in the huge mid-water zone and the deeper abyssal zone is comparatively sparse.

In contrast with these central zones, much of the sea floor abounds with animals. Deep-sea creatures, which are known collectively as benthic animals, include species that swim or crawl over the sea bed as well as those that burrow through it, mining the soft sediment for food. Many of these animals appear to have changed little over millions of years, for although the water is extremely cold, and the pressure intense, the deep-sea bed is not subjected to the changeable conditions that can affect the surface. It is therefore one of the most stable habitats on earth.

CONSERVATION

The main threats to marine wildlife are pollution and overfishing. Dissolved pollutants and floating waste are now found in oceanic water all over the world, but the greater problem for inshore marine wildlife is fishing. Improvements in technology, and a steady increase in the size of the world's fishing fleets, have brought about the collapse of some fish stocks and may have permanently altered the ecology of the continental shelf. International agreements now restrict fishing in some environmentally sensitive areas, but commercial

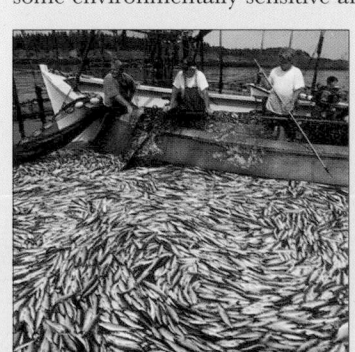

pressures mean that many fish are still being harvested at an unsustainable rate.

LONG REACH *While traditional fishing methods – as seen here – sometimes produce good catches, modern trawlers can track down shoals with pinpoint accuracy, ensuring repeated success.*

LIFE IN OCEANS

IN THE BILLION OR MORE YEARS since the first animal species evolved, competition for survival in the sea has become intense. Today, the oceans are home to the largest predators on the planet, as well as to vast numbers of microscopic animals that drift by unseen. As on land, ocean wildlife is affected by local conditions, the most important being the supply of food. For many, survival also depends on being able to defend themselves against attack. In some regions, life is thinly spread, but in others animals are found in greater numbers than they are anywhere else on earth.

FEEDING AT SEA

The oceans are so huge that, although they contain plenty of nutritious food, marine animals face a challenge in finding enough to eat without expending too much energy in the process.

Some concentrate on large prey. The sperm whale, for example, hunts giant squid at depths of over 1,000m (3,300ft), although most pursuit-hunters search for food near the surface rather than in the depths. Other large sea animals eat smaller fare, scooping it up in huge amounts, often sieving it out with their gills. This technique, known as filter feeding, is used by baleen whales and some of the largest sharks and rays. Most filter feeders live on plankton, which is so abundant that it allows them to reach a gigantic size. Drifting animals also feed on plankton, although on a much smaller scale. Comb jellies or sea gooseberries, for example, haul it in with stinging tentacles that work like fishing nets.

On the sea bed, animal life depends almost entirely on the dead organic matter that drifts steadily down from above. Brittlestars are typical of these scavengers, collecting food particles with their arms. However, there are also predators – bizarre fish, for example – that hunt in the total darkness on or near the sea bed. Animal life at the bottom of the sea can be sparse, so these predators cannot

afford to miss any opportunity to feed. Many of them therefore have gigantic mouths and elastic stomachs that enable them to swallow prey that is almost as large as themselves.

CONTROLLING BUOYANCY

Very few marine animals – with the exception of deep-diving mammals – are found at all levels in the sea. Instead, most are adapted for life at a particular depth, and have buoyancy devices that help to keep them there. Surface drifters, such as the violet sea-snail, have simple floats: filled with material that is lighter than water, they ensure that the animal stays at the surface, even in a heavy swell.

For animals that spend their lives fully immersed, remaining at one level requires more complex apparatus. They have to be neutrally buoyant at their optimal depth and yet able to rise or sink as the need arises. To

ANIMALS ADRIFT
The violet sea-snail hangs from a raft made of hardened bubbles of mucus. To improve buoyancy, its shell is unusually thin. It drifts with the currents, feeding on other surface animals.

FILTER FEEDING
Using large flaps on either side of its head, a manta ray channels plankton into its mouth. Filter feeders like the manta are indiscriminate eaters, swallowing anything that becomes trapped by their sieve-like gills.

BIOLUMINESCENCE

A variety of marine animals produce light – some to maintain contact with their own kind, others to lure prey. In some species, notably planktonic invertebrates, the value of light is less easy to explain. Bioluminescence is most common in bathypelagic species (those that live in very deep, open water). Typically, the light is produced by skin organs called photophores. Light production is not always confined to the animal itself: some fish can eject luminous clouds to distract predators while they make an escape.

LIGHTS IN THE DARK
Bristlemouths have photophores that give out a yellowish green light. As in most luminous fish, the photophores are arranged in characteristic patterns that enable members of a species to recognize each other in the dark. Photophores may shine for extended periods or they may be flashed on and off.

do this, some use adjustable buoyancy aids hidden inside their bodies. Bony fish, for example, have a gas-filled chamber, called the swim bladder, just below the backbone. If the fish needs to sink, it removes some of the gas from the swim bladder by pumping it into the bloodstream; if it needs to rise, it pumps it back into the bladder. Cartilaginous fish, such as sharks, do not have swim bladders; instead they rely on their large, oily liver to keep them afloat. Many sharks are actually slightly heavier than sea water: swimming provides the lift that allows them to control their depth.

AVOIDING PREDATORS

In the open sea, there is nowhere to hide, which leaves animals highly vulnerable to predators. To survive, some rely on camouflage or disguise; others behave in ways that make them difficult to attack.

For slow-moving invertebrates, such as those that make up plankton, one of the most effective disguises is transparency. Planktonic animals are often as clear as glass, which makes them difficult to see, even at close quarters. Most of these animals are only a few millimetres long, although some tunicates form translucent, tube-shaped colonies that can be over 3m (9¾ ft) in length.

Some fish are transparent when they are very small, but then use camouflage of a different kind as they get older. Almost all species that live in brightly lit, open water have dark backs but much paler undersides.

SYNCHRONIZED SWIMMING
A shoal of silversides scatters from a predator. Their extraordinary degree of co-ordination makes it seem as though they are obeying a central command. In reality, each fish simply responds to the others.

HYDROTHERMAL VENTS

First seen in 1977, off the Galapagos Islands, hydrothermal vents are remarkable ecological oases on the deep-sea bed. Vents are created by volcanic activity releasing streams of intensely hot, mineral-laden water into the ocean. They are among the few habitats where life does not rely on energy from the sun – specialized bacteria use the minerals to produce energy, and vent animals either consume food created by the bacteria or eat each other. Over 300 animal species have been discovered around vents; most of these are not found elsewhere.

GHOSTLY SHELLFISH
Like most vent animals, these crabs and clams – caught in the powerful lights of a submersible – have very little pigmentation, which explains their white appearance.

This pattern, known as countershading, protects fish in two ways: it hides them from predators deeper down by disguising their silhouette against the bright sky, and it conceals them from surface hunters, including seabirds, by making them blend in with the dark water beneath.

TRANSPARENT LIFE
Barrel-shaped salps have clear, jelly-like bodies that are difficult for predators to see. Up to 10cm (4in) long, they live in swarms that can extend hundreds of kilometres across the ocean's surface.

LIVING IN GROUPS

On land, animal groups can be very large, but none rivals the size of those that can occur at sea. Fish often live in shoals thousands or even millions strong, while some planktonic animals form swarms that can be over 100km (60 miles) long.

These giant aggregations of underwater life often seem easy targets for predators. Whales gorge on krill with almost nonchalant ease, while other predators make huge inroads into shoals of fish. But, in general, animals living in groups are safer than they would be alone: they are more difficult to single out and much more difficult to take by surprise.

A few animals, such as dolphins, live in sophisticated social groups. Dolphins use echolocation to locate prey, to warn each other of danger, and to organize themselves during hunting.

MIGRATION

Many marine animals – including most of the largest whales – migrate between breeding grounds in the tropics and feeding grounds at higher latitudes. The grey whale probably travels the greatest distances: its lifetime annual migrations total up to 800,000km (500,000 miles) – twice the distance to the moon. Tuna are also known to undertake immense journeys: some shoals travel the length of the Mediterranean; in the Pacific, one fish tagged in Mexico was recovered off Japan. Turtles show very precise migration patterns, returning to the same stretch of beach year after year to lay their eggs. This is all the more remarkable because it takes two

or three decades for the animal to mature: during this time a turtle remembers precisely where it hatched so that when it is ready to breed it can make the long journey back.

On a smaller scale, many fish swim inshore to spawn, while other slow-moving animals migrate across the sea bed. Caribbean spiny lobsters travel between shallow reefs, where they breed, and deeper water, where they overwinter; they set off in single file, each one following the tail of the animal in front.

Journeys like these are usually annual events. But some animals migrate daily to feed. Planktonic animals often rise to the surface at night, sinking back into the depths by day. Some of the plankton's predators copy this pattern, creating a 24-hour cycle involving many animals. These vertical migrations are clearly revealed by shipboard sonar, which shows a reflective layer rising at sunset and sinking at dawn.

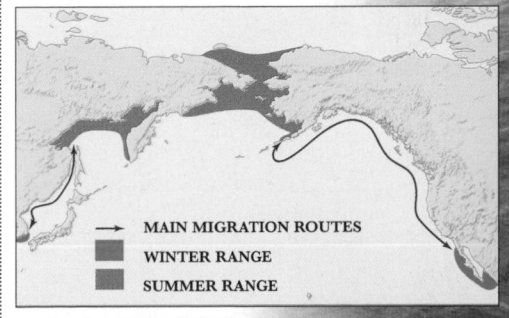

→ **MAIN MIGRATION ROUTES**
■ **WINTER RANGE**
■ **SUMMER RANGE**

WHALES ON THE MOVE
North American grey whales make an annual return trip between their winter breeding grounds off northwest Mexico and their summer feeding grounds around the Bering Sea. A smaller number migrate between Korea and eastern Siberia.

COASTS AND CORAL REEFS

IN THE NATURAL WORLD, the boundaries between different environments are often rich habitats for wildlife. The world's coasts are the ultimate example of this meeting of habitats because they bring together animals that live on the land and those that live in the sea. Coastal wildlife varies according to local conditions but, in rocky coasts as well as sandy ones, shore animals are associated with clear-cut zones that are usually determined by the tides. Coral reefs are a special kind of coastal habitat. Famed for their spectacular shape and colour, they can grow to vast proportions and are unrivalled in the immense variety of aquatic life that they support. Reef-building corals require specific conditions, the most important being warmth and bright sunlight all year round. As a result, coral reefs are largely restricted to the tropics.

CONSERVATION

The chief threat to coastal wildlife is pollution. Some kinds, such as oil spills, have very visible effects, but coastal life is also affected by contaminants that often go unnoticed, such as sewage and industrial waste. Tourism can aggravate the problem: in heavily developed areas, pollution levels rise, and animals that normally breed on the coast, such as turtles and birds, are often scared away. Coral reefs face problems from overfishing, pollution, and quarrying for coral stone but, worldwide, the most serious threat to coral is global warming. When sea temperatures rise above certain levels, corals expel their symbiotic algae, cutting off one of their main sources of food. Such coral is called "bleached". If the temperature remains high, the coral may die. Mass bleaching has occurred in nearly all the world's major reefs and, in some, the coral has failed to recover.

VICTIM OF OIL
The trade in oil is a major source of pollution around the world's coasts. Oil kills seabirds by clogging their feathers, which prevents them from flying and hunting. It also kills fish as well as limpets, mussels, and other invertebrates that live in the zone between the tides.

COASTS

Some habitats look much the same from one year to the next, but the seashore is always changing. Waves pound away at rocks, undermining them and breaking them up, while coastal currents reshape the shore in a less dramatic way by moving shingle and sand. Superimposed on this is the rhythmic movement of the tides – a twice-daily cycle that has a profound impact on seashore wildlife.

Tides vary enormously in different parts of the world. Around islands in mid-ocean, the total rise and fall is often less than 30cm (12in), while in deep bays and inlets on continental coasts it can be over 10m (33ft). Whatever its height range, the tide divides the shore into three different zones, each with its own distinctive animal life.

The highest of these zones is the supralittoral, which is the part of the shore just beyond the reach of the highest tides. Although this zone is never actually submerged, it is affected by salt spray, which means that animals that are sensitive to salt – and also salt-intolerant plants – are rarely found in this zone. Below this is the littoral, a zone that is regularly covered and then exposed as the tide floods and recedes. The animals found here, such as mussels and limpets, lead a double life in that they have to be able to survive both in water and in air. The next zone, or infralittoral, is always submerged, even during the lowest tides. Most of the animals that live in this zone are fully marine, although a few leave the water to breed.

Coastal wildlife is also affected by the geology of the shore. Many animals live on rocky coasts, while others specialize in living in sand or coastal mud. Compared with these, shingle is a difficult habitat for animals, although some shoreline waders use it as a place to nest.

CORAL REEFS

There are two main types of coral: hard and soft. Coral reefs are formed by hard corals. The individual coral animals, called polyps, secrete external skeletons that persist after they die. Soft corals are found all over the world, but hard, reef-building corals grow only in clear, nutrient-poor water, where their symbiotic algae can harness the energy in sunlight. Such is the richness of the habitat created by coral reefs that a huge number of animals are able to live side by side without competing for the same food.

There are three main types of reef: fringing reefs, which grow close to the shore; barrier reefs, which are separated from the coast by deep channels, sometimes over 100km (62 miles) wide; and atolls, which are ring- or horseshoe-shaped reefs that grow around oceanic islands, often where volcanoes have subsided into the sea.

CORAL REEFS *are found in parts of the world where the sea temperature never drops below about 18°C (64°F). The world's largest reefs are in the Indo-Pacific region.*

Although every reef is unique, reefs share a common structure. Depth and exposure are two important factors that influence this because corals vary in their need for light and in their ability to withstand the force of the waves. The fastest-growing corals, which need lots of light and relatively calm water, typically form the central zone of the reef, projecting just above the surface at the lowest tides. On its inshore edge, the central zone is often backed by lagoons – large pools of open water lying over coral sand. By contrast, the seaward edge of the reef often forms a submarine cliff that drops steeply into the depths. The corals in this zone are solid and resilient because they have to withstand heavy breakers rolling in from the open sea.

CLIFFS *are important nesting sites for many seabirds, although some, such as terns, nest on the ground above the high-tide line.*

SUPRALITTORAL

LAGOON

RAYS, *and many other bottom-dwelling fish, feed animals buried in the coral sand of reef lagoons. Most are well camouflaged.*

THE LUMPFISH *stays in place by sticking itself to submerged rocks with its sucker-like pelvic fins.*

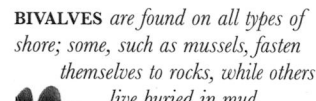

STARFISH *are common predators in the littoral zone, particularly on rocky shores with plenty of bivalves to eat.*

BIVALVES *are found on all types of shore; some, such as mussels, fasten themselves to rocks, while others live buried in mud.*

SEALS *spend most of their time in the water but have to come ashore to breed. Some breeding colonies contain thousands of animals.*

SEASHORE WORMS, *including ragworms, hunt and scavenge on the sea bed. Many species burrow, filtering food from the water.*

LITTORAL   INFRALITTORAL

REEF   SEAWARD SHELF

TURTLES *use their toothless beaks to feed on algae and the various small animals that live in all actively growing parts of reefs.*

MANY SEA SNAKES *live in inshore waters. Some hunt among coral reefs, preying upon small fish and eels.*

PARROTFISH *use beak-like jaws to feed on corals. The chewed-up skeletons are excreted as sand, which builds up on the reef.*

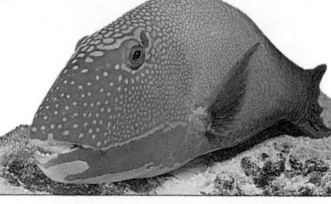

BRANCHING CORALS, *mainly of the genus Acropora, are characteristic of the central zone of most reefs. Some can gain 15cm (6in) in height each year.*

SPONGES *filter food from water around reefs. Some are minute, but deep-water forms can reach 1m (39in) in height.*

LIFE ON COASTS AND CORAL REEFS

COMPARED WITH THE OPEN SEA, coasts and coral reefs are disproportionately rich in marine wildlife. Rocky coasts and mudflats abound with seabirds and invertebrates, while coral reefs probably contain at least a third of the world's fish species. Conditions in coasts and reefs are variable, so the animals that use them are often extremely specialized. Some roam over large areas of the shore, but for most a distance of just a few metres makes the difference between an ideal habitat and one in which it is impossible to survive.

ADAPTING TO TIDES

For coastal animals, life is governed by the rhythm of the tides. As well as adapting to the twice-daily ebb and flow, they have to adapt to the drawn-out rhythm of extra-large spring tides, which are normally 14 days apart. Knowing where the tide stands is important because animals caught unprepared run the risk of either drying out or drowning.

Some shore animals, such as barnacles, adjust their behaviour according to whether or not they are submerged. Most aquatic animals, however, are much more sophisticated, reacting to their internal

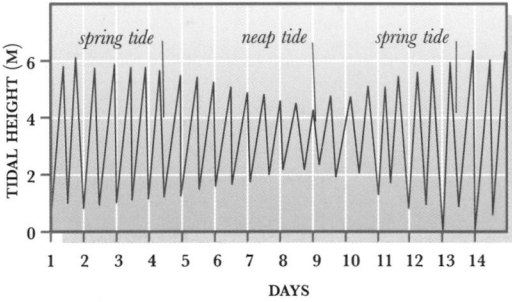

TIDAL RHYTHMS
This chart shows how the sea's height varies over 14 days. Spring and neap tides occur when the sun, moon, and earth form a right angle so that their gravitational pull is combined.

CLOSED ANEMONE   OPEN ANEMONE

PREVENTING WATER LOSS
To prevent themselves from drying out at low tide, many sea anemones withdraw their tentacles, turning into jelly-like blobs that can be seen on coastal rocks. They often live under overhangs, which keeps them out of direct sunlight.

biological clocks, which march in step with the tides. Even if they are taken away from the shore and placed in a tank, their tidal clocks continue to tick.

This in-built ability to keep time enables animals to anticipate events. For example, submerged limpets crawl over rocks to graze on microscopic algae. But since they are vulnerable away from their niche on the rock, they have to return home before the sea ebbs away. Their biological clocks tell them when to head back, ensuring that they are securely in place, and airtight, before the tide goes out. Conversely, the fiddler crab's biological clock prompts it to come out and feed at low tide and then return to its underground burrow before the incoming tide engulfs it.

FEEDING ON THE SHORE

Coasts provide a much greater variety of food than the open sea. High up the shore, an abundance of animal and plant remains left in strandlines by the retreating tide are consumed by beach-hoppers, springtails, and other scavengers, which are then preyed upon by gulls and waders.

In the intertidal zone, filter-feeders are common. Unlike the giant filter-feeders of the open sea (see p.72), coastal ones are generally small and often spend their entire adult lives fixed in one place. Mussels and other bivalves, which filter particles of food using modified gills, are examples of these. Barnacles have a different filtering technique: despite their resemblance to molluscs, barnacles are crustaceans, with a set of feathery legs. At high tide, the legs protrude from the barnacle's case to collect food particles.

Rocky shores are home to some fast-moving swimmers as well as to many other animals that take a more leisurely approach

FEEDING AT THE SHORE
Sandpipers wade into the sea in their search for food. Like many coastal waders, they follow each wave as it retreats, and then they scuttle back up the shore as the next one breaks.

to finding food. Starfish and sea urchins are among the slowest, crawling over rocks on hundreds of fluid-filled feet.

LIVING IN MUD AND SAND

Compared with rocky shores, coastal mud and sand seem to harbour only a limited amount of animal life. But appearances are deceptive. Mud in particular often teems with hidden life feeding on organic matter brought in by the tide.

One of the advantages of living below ground is that despite the attacks of curlews and other long-billed birds, it provides good protection from predators. The chief disadvantage is that the surface is constantly shifting, cutting off buried animals from the water above – and from oxygen and food.

Some of these buried animals have specialized body parts that enable them to connect with the surface. Clams, for example, have leathery tubes or siphons. In many species the siphons can be retracted inside the shell, but in some they are too long to be stowed away. Species that do not have such accessories often live in burrows. The

MANGROVE SWAMPS

Mangrove swamps are formed by trees that are adapted for life in salty intertidal mud. Found only in the tropics and subtropics, they play an important ecological role by stabilizing the coast and providing inshore nurseries for fish and other marine animals. Most mangroves develop arching prop roots, which are exposed when the tide is out and submerged when it is in. These provide anchorage for molluscs, and act as convenient perches for mudskippers – the air-breathing fish that thrive in this habitat. Mangrove foliage is tough and leathery, and eaten mainly by insects; but the dense canopy attracts large numbers of birds, which use mangroves as breeding sites and overnight roosts. The mud in mangrove swamps is deep and often foul-smelling, but it is rich in organic matter that is replenished each day by the tide.

Until the late twentieth century, mangroves escaped much of the large-scale deforestation in the tropics. But, with the increase in shrimp farming, large areas of mangroves have been cut down, raising concerns about the long-term impact on coastal wildlife.

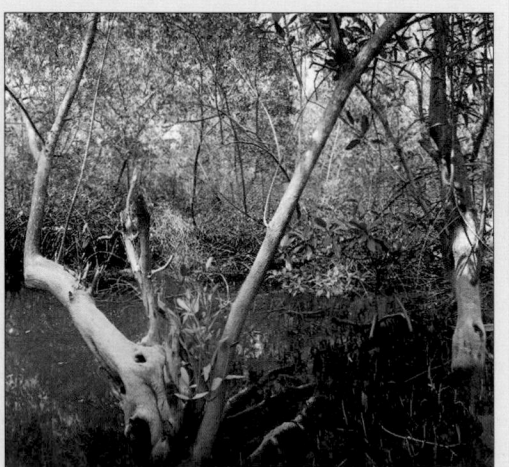

SWAMP VEGETATION
Mangroves are the only trees that can grow in mud periodically flooded by the tide. Few animals eat the trees' leaves, partly because they are often rich in salt, but these coastal forests are important breeding grounds for birds and marine animals.

LUGWORM CASTS
Lugworm burrows have two holes – one for incoming water and another for ejected waste. The waste forms a pile or "cast" that looks like muddy toothpaste squirted out of a tube.

lugworm is an example: it makes a U-shaped burrow, lining it with mucus to keep it intact so that seawater does not make it collapse.

SYMBIOTIC PARTNERSHIPS

Symbiosis is a feature of life in all habitats, but it is particularly apparent on coasts. In some cases, the symbiotic partners are animals; but, in many others, one is an animal and the other is microscopic algae that live inside its body. These algae, known as zooxanthellae, can be found in thousands of coastal animals, including corals, jellyfish, and giant clams.

Zooxanthellae live by photosynthesis – the same process by which plants grow. Through a complex series of chemical reactions, they harness the energy in sunlight and use it to build up organic matter. The host animal provides the zooxanthellae with protection from the outside and, in return, the algae surrender some of their manufactured food.

These partnerships are very important to reef-building corals because they allow them to live in places where the supply of food is

MUTUAL BENEFIT
Rather than harming clownfish, the sea anemone's stinging tentacles protect them from predators, while the fish probably help to keep the anemone clean. However, the partnership is not entirely equal for while anemones can live alone, clownfish cannot.

NEW GENERATION
A giant clam expels eggs into the sea. The larvae will drift away to settle on distant parts of the reef. Giant clams live on the food generated by the symbiotic algae harbouring in their fleshy lips.

otherwise low. For the partnership between coral and algae to work, the corals have to encourage algae growth, which means they must live in bright sunshine near the water surface. However, this limits their upward growth because few corals can survive more than an hour of exposure at low tide.

REPRODUCTION

Many land animals have adapted to life at sea. Some are now fully marine, but others – such as turtles, some sea snakes, and seals and their relatives – must come ashore to breed. These animals are often scattered over a wide area, so they tend to form colonies

A LIFE IN THE SEA
The sea otter is the only otter to live entirely in water: it gives birth in the sea, and floats upside down to suckle its young. This female is swimming with her cub on her chest.

during the breeding season, congregating in the same place each year to maximize their chances of finding a mate. Many coastal invertebrates, on the other hand, spend their entire adult lives in one place. For them, reproduction is an opportunity not only to multiply but also to disperse. Their eggs hatch into planktonic larvae, which may then drift long distances in coastal currents before eventually settling down. For some species, such as barnacles, choosing a home is an irrevocable decision because a larva cannot detach itself once it has settled. Chemical cues help it to "make up its mind" before it takes this momentous step.

Some fish come inshore to breed because the shoreline offers plenty of hiding places for their eggs. An extreme example of such a fish is the California grunion, which lays its eggs not in water but in damp sand on beaches. Grunions stage mass spawnings during high spring tides at night. At the next spring tide, the eggs hatch and the young are washed into the sea.

URBAN AREAS

TWO HUNDRED YEARS AGO, only about 3 per cent of the world's population lived in cities. Today, the figure is rapidly approaching 50 per cent, and the human population has increased nearly sixfold. This phenomenal growth in urban living has transformed large areas of the planet. It has created a wide range of artificial habitats – both in and out of doors – that animals can use as their homes, as well as vast amounts of waste that form the basis of animal food chains. As a result, there is a wealth of wildlife living with and among us.

CITIES *that have a population of more than one million people are spread throughout the world. There are more than 250 of these cities, and the total is rapidly increasing.*

OUTDOORS

For animals that can cope with disturbance, cities and towns can be good places to live. They have plenty of suitable places for sheltering or raising young, from trees and window ledges to underground passageways, and for omnivorous species they provide a constant supply of leftover food. In winter, the artificial heat that escapes from buildings offers additional benefit. Even better, cities are relatively safe: cats and dogs aside, they are free of many of the predators that animals would face in their natural homes.

Animals have adapted to urban expansion with different degrees of success. Some species are never found in cities, always retreating as the concrete advances. Others, such as migrating birds and insects, are occasional visitors, touching down briefly before moving on. More adaptable animals – raccoons and red foxes, for example – are equally at home in town or country, and treat built-up areas as extensions of their natural habitat. True urban specialists, such as the ubiquitous feral pigeon and the house sparrow, are now so fully adapted to city living that they are rarely seen anywhere else.

While feral pigeons can survive in the busiest city centres, many urban animals are found chiefly in parks and gardens – the small-scale versions of their habitats in the wild. These animals vary from one part of the world to another, but they include tree-dwelling mammals, such as squirrels and opossums, and a wide range of birds. The spread of suburbia is normally a threat to wildlife, but for these species it can actually be a help because it creates a patchwork of suitable habitats, sometimes with the bonus of food hand-outs.

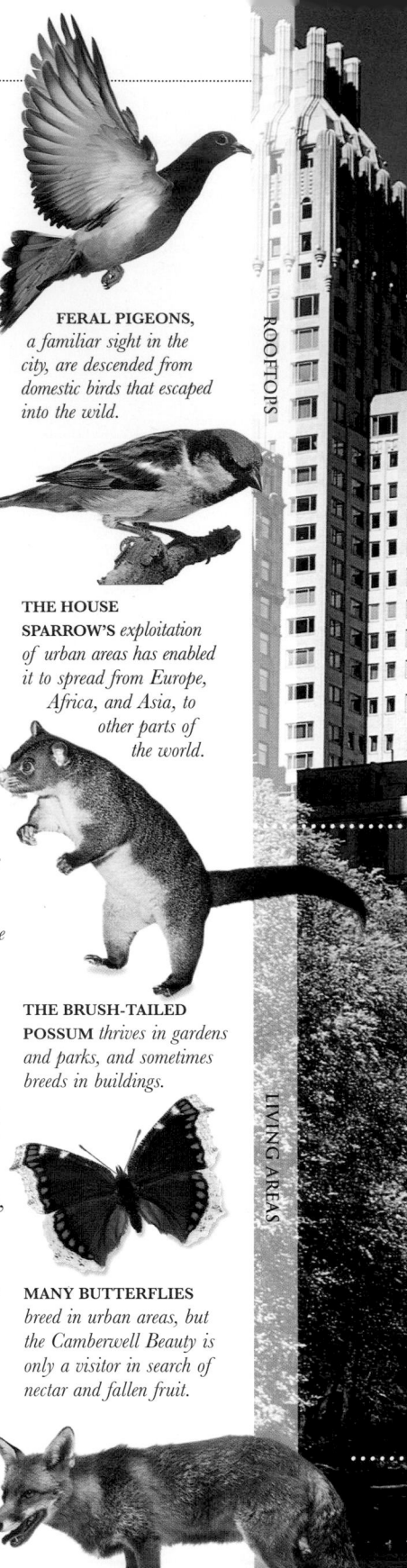

FERAL PIGEONS, *a familiar sight in the city, are descended from domestic birds that escaped into the wild.*

THE HOUSE SPARROW'S *exploitation of urban areas has enabled it to spread from Europe, Africa, and Asia, to other parts of the world.*

THE BRUSH-TAILED POSSUM *thrives in gardens and parks, and sometimes breeds in buildings.*

MANY BUTTERFLIES *breed in urban areas, but the Camberwell Beauty is only a visitor in search of nectar and fallen fruit.*

THE RED FOX *is active at night, reducing the risk of its being disturbed while foraging.*

THE BROWN RAT'S *burrowing lifestyle enables it to live close to people without being easily seen.*

HABITATS

ROOFTOPS

LIVING AREAS

GROUND

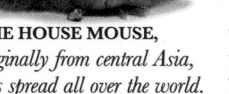

THE HOUSE MOUSE, *originally from central Asia, has spread all over the world. Other household rodents include rats and some species of dormice.*

ROOF SPACES

FOR GECKOS, *house walls make good nocturnal hunting grounds – especially if there are lights nearby to attract flying insects.*

HOUSEFLIES *enter houses to find food, but some other species – such as the cluster fly – use attics and undisturbed empty rooms for hibernation.*

LIVING AREAS

COCKROACHES *can become a serious pest in warm parts of the world and in houses with central heating.*

SILVERFISH *are widespread but largely harmless. They feed nocturnally on all kinds of starchy substances, from flour to wallpaper paste.*

BASEMENTS

THE DADDY-LONG-LEGS *spider is found in caves in the wild. Cellars and dark rooms make perfect artificial alternatives.*

INDOORS

In the natural world, many animals inadvertently create habitats for other species when they build their nests. Humans do exactly the same. However, because our "nests" are so extensive and complex, they can host an exceptionally wide range of animal life. Much of it is harmless, but some can cause problems or at least inconvenience.

Most indoor animals are small and nocturnal, which helps them to avoid being noticed by their human hosts. This is especially true of species that share daily living areas and that scavenge leftover food. Silverfish, for example, emerge after dark to search for flour and other starchy produce, scuttling away if cupboards or drawers are suddenly opened, exposing them to the light. Cockroaches behave in a similar way, but they are more of a nuisance because they spread disease. At dawn, nocturnal animals hide away, leaving the day shift to take over. Houseflies, for example, are most active during the day because they navigate by sight. In basements and attics, wildlife is less affected by the cycle of light and dark, and it is less frequently disturbed by human comings and goings. For wild animals, attics resemble extra-large tree-holes, while basements resemble caves. Wasps, birds, and house mice will all nest in attics – if they can get in – and they sometimes share this habitat with roosting bats. Basements and cellars provide a haven for spiders, which can survive for long periods without food and, in many cases, catch their prey in total darkness.

The advent of central heating has been an important factor in the increase in the number of animals that choose to share our homes. For example, cockroaches, which were originally found mainly in warm parts of the world, are now widespread in cooler regions. Soft furnishings and carpeting also play a part: as well as helping to keep the home warm, they provide hiding places and nesting material for various animals.

HABITATS

CONSERVATION

Urban animals face a form of "unnatural selection" in that any species that are not suited to city life are ruthlessly weeded out, regardless of how successful they would be in the wild. Shortage of space, noise, and pollution prevent many animals from establishing themselves in cities, while other environmental factors, such as bright street lighting, deter many more. Such are the hazards of city life that even experienced, and otherwise successful, urban animals sometimes succumb to the dangers. Traffic, of course, is a major hazard that claims the lives of thousands of animals (including humans) every day.

ROAD HAZARD
Urban traffic, which has increased massively over the past 50 years, is one of the worst threats faced by animals that live in cities. It is particularly dangerous for nocturnal animals, such as foxes.

LIFE IN URBAN AREAS

ANIMALS HAVE HAD MILLIONS OF YEARS to adapt to earth's natural habitats, but only a fraction of that time to adjust to life in cities. Despite this, animals are never far away in built-up places. Their success is owed mainly to "preadaptation" – characteristics that evolved to suit one way of life, or habitat, but that accidentally turn out to be useful for another. Thus, some animals thrive in man-made habitats that resemble the ones they would use in nature. Others succeed because they are highly adaptable and can exploit the opportunities that we inadvertently provide.

INTRODUCED SPECIES

Species that have set up home in unfamiliar parts of the world are plentiful in urban areas. Some have been introduced deliberately, while others arrive with imported food. Pets, also, are sometimes released, or manage to escape, into the wild. Some of these animals remain urban, but a few – such as the starling in North America – have gone on to colonize entire continents.

FAR FROM HOME
The rose-ringed parakeet from tropical Africa and southern Asia is good at coping with cold, so it can thrive in parts of North America and Europe.

FEEDING

Some outdoor urban animals live on the same foods that they eat in the wild, but for scavenging species – such as raccoons, foxes, and pigeons – the daily fare is often very different to that of their natural homes. These versatile creatures will try any kind of leftover food, however unfamiliar it looks and smells, and this highly opportunistic streak is the secret of their great success. Modern food packaging can sometimes present problems, but they quickly learn how to tear or peck away at plastic and paper to get at the edible contents within.

Indoor animals get their food from one of three sources: the things we eat, the animals that eat those things, and the fabric of our homes. The first category contains a wide range of household pests, such as rats, mice, houseflies, and cockroaches; the second category consists chiefly of spiders, but also centipedes and geckos in warm parts of the world. Spiders are almost perfectly adapted to indoor life, and although widely disliked, they make a positive contribution by keeping indoor insect numbers in check. Animals in the third category are the least welcome of these uninvited guests. They include wood-eaters such as termites and beetles, as well

as insects that attack other organic materials, such as wool. In many parts of the world these animals are serious pests.

LIGHT AND WARMTH

In cities, street lamps light up the night sky, while heat from buildings and traffic makes them far warmer than nearby countryside. Artificial lights confuse insect navigation systems and interfere with birds' biological clocks. As a result, songbirds sometimes sing late at night, and some species start building nests in winter, convinced by the bright light

that it is spring. Extra warmth is appreciated by a range of animals, from butterflies to birds. In some regions, starlings commute into cities on winter afternoons to roost on buildings where they are relatively warm.

ROOFTOP ANIMALS

For birds and bats, the tops of buildings can make ideal homes. High above the ground, and relatively undisturbed by people, animals feed and breed unmolested. Some species nest in attics or under eaves, while others favour the tops of chimneys. Swifts, swallows, and martins are foremost among rooftop-dwellers and are prime examples of preadaptation at work: they naturally nest on cliffs or in crevices, but the rapid spread of towns and cities has provided alternatives that have enabled them to extend their range to places where they would otherwise be rare.

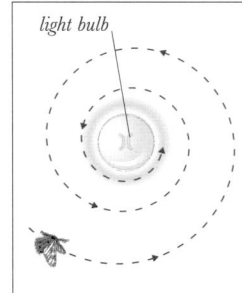

light bulb

moon

DRAWN TO THE LIGHT
Moths navigate by maintaining a set angle to the moon, which, because it is so far away, works like a compass, enabling the moth to follow a straight line. With closer lights, such as street lamps, the "compass" appears to the moth to drift. The moth adjusts its path accordingly, and ends up spiralling inwards.

DAWN RAID
Alert for signs of danger, and using its keen sense of smell and highly dexterous front paws, a raccoon sorts through rubbish from a bin that it has overturned.

LIFE ON THE ROOFTOPS
A pair of European white storks look down on a town from their chimney-top nest. Although they often nest on buildings, they feed in fields. They were once common urban visitors in northern Europe, but changing agricultural practices have reduced their numbers.

STRUCTURAL PESTS

Since humans first started to build timber houses, wood-eating insects have been a problem. Wood-eating beetles are common in temperate regions, while termites attack timber in the tropics. At one time, little could be done to control severe attacks; today, insecticides are effective at keeping them in check. Even so, many of these animals, especially wood-boring beetles, are far more widespread than they were because they have been exported in the timber shipped all over the world.

EATEN FROM INSIDE
Deathwatch beetle larvae chew tunnels through old timber. During the five years they take to turn into adults, they can weaken timber so much that it eventually collapses. A related species attacks furniture wood.

URBAN HUNTER

The peregrine falcon is one of the most adaptable of all birds of prey. It feeds mainly on other birds, which it usually catches in mid-air. City rooftops and ledges provide an excellent vantage point for the falcon to pick out its quarry before launching into a high-speed downwards dive, known as a stoop.

THE ANIMAL KINGDOM

Almost two million species of animals have so far been identified – and it is thought that the true number may be several times greater than this. Members of the animal kingdom range from invertebrates that are too small to be seen with the naked eye to the blue whale, which is more than 30m (100ft) long. The variety of lifestyles is equally immense, from animals that filter food from water while permanently fixed to the sea floor to agile predators that can travel across the ground at over 100 kph (60 mph). This section reflects the great diversity of the animal kingdom, profiling over 2,000 species, from the largest and best known to the tiniest and most obscure.

MAMMALS

MAMMALS

PHYLUM	Chordata
CLASS	Mammalia
ORDERS	21
FAMILIES	140
SPECIES	4,475

MAMMALS, THE MOST FAMILIAR group of vertebrates, all nourish their young on milk produced by the female's mammary glands (the unique skin structures after which the class is named). Most also give birth to live young and, with only a few exceptions, have a covering of hair on their body. Mammals are most widespread and diverse on land, but they have also colonized air and water. Their success is largely due to their ability to maintain a constant internal body temperature, regardless of changing external conditions. They are also highly adaptable and often modify their behaviour to suit changing circumstances. Some mammals, especially primates (the group that includes humans), form complex societies.

EVOLUTION

The ancestors of mammals were a group of reptiles known as the therapsids. These small, active carnivores lived during the Triassic Period (which began 225 million years ago). The various features that now separate reptiles and mammals in fact developed over a long period of time and at different rates. As well as several important differences in skull shape (see panel, below), the therapsids evolved a lighter, more flexible skeleton. Another significant change was the alignment of the limbs beneath rather than at the sides of the body, which helped the early mammals to become faster and more agile.

The transition from reptiles to mammals was completed towards the end of the Triassic Period (195 million years ago). This evolutionary change coincided with the rise of the dinosaurs, the reptiles that became the dominant animals on earth during the Mesozoic Era (225–65 million years ago) and threatened the mammals with extinction. The mammals' ability to regulate their internal temperature may explain why they outlasted the dinosaurs. During the Mesozoic Era, the climate became cooler and daily temperatures became more variable. Like other reptiles, the bodies of dinosaurs would have tended

YOUNG MAMMALS
One of the key features that sets mammals apart from other animals is the way that females suckle their young on milk from their mammary glands. Some mammals look after their young until they are several years old.

to shut down in low temperatures. Mammals, however, were unaffected by these fluctuations in temperature and so were able to survive the change in climate.

ANATOMY

Several differences between the skeletons of mammals and other vertebrates can be seen in the skull. Mammals are unique in having a lower jaw that is hinged directly to the rest of the skull – in other vertebrates the connection is indirect, with at least one other bone between the 2 parts of the jaw. This direct articulation, and the fact that the lower jaw is made of a single bone (see panel, left), makes the jaw a powerful tool for cutting and dismembering food. Unlike other vertebrates, mammals also have a set of specialized teeth, which reflects and facilitates their varied diet. All mammals have 3 types of teeth: incisors, which are used for biting, canines for gripping and tearing, and cheek teeth (molars and premolars), which are used for grinding. The shape and size of each type varies according to diet.

Most mammals, unlike other animals, have mobile external ears (pinnae) to locate sounds and then funnel them to the eardrum, where they are transmitted to the inner ear

EVOLVING SKULL

DENTARY BONE

hinge at back of skull

uniform teeth

EARLY REPTILE

zygomatic arch

hinge further forward than in early reptile

TRIASSIC MAMMAL

large zygomatic arch

specialized teeth

MODERN MAMMAL

Mammals' reptilian ancestors had uniform teeth and a jaw that was made of several bones and limited to an up-and-down movement. By the Triassic, mammal skulls had both a zygomatic arch to house powerful jaw muscles and only one jawbone (the dentary). Modern skulls have specialized teeth and a single jawbone often capable of complex movements. These features, in combination with strong jaw muscles, allow mammals to chew their food – a unique feature of the group.

JAWS
Mammals have a unique jaw, in which the lower jaw is hinged directly to the skull. The hippopotamus has an unusually large mouth and a wide jaw extension.

molars   premolars   canine   incisors

TEETH OF A RACCOON

sharp cutting edge

broad, grinding surface

multi-cusped chewing surface

CARNIVORE CHEEK TOOTH

HERBIVORE CHEEK TOOTH

OMNIVORE CHEEK TOOTH

TEETH IN MAMMALS
Many mammals, such as a raccoon (top), have incisors, canines, and premolars and molars (cheek teeth) that reflect their diet. Carnivores have some sharp cheek teeth for cutting bones, while herbivores have broad cheek teeth to grind up vegetation. Omnivore cheek teeth are broad and multi-cusped for chewing a variety of foods.

HAIR

Mammals are the only animals with a covering of hair on the body. A hair consists of a rod of cells strengthened by the protein keratin. Hair can take several forms, including whiskers, spines, prickles, and even horns (as in rhinoceroses). Most common is fur, which usually consists of an insulating layer of underfur and a projecting guard coat that protects the skin and gives hair its colour (which may aid camouflage). Hairs such as whiskers may also have a sensory function.

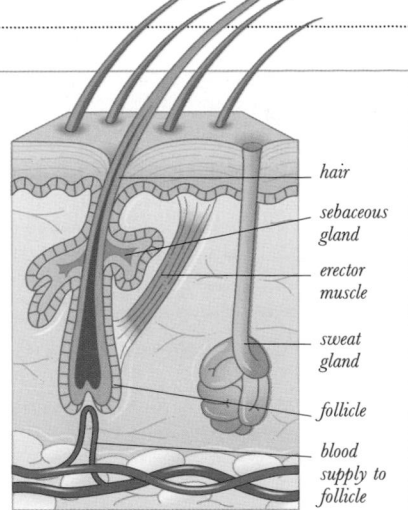

hair
sebaceous gland
erector muscle
sweat gland
follicle
blood supply to follicle

SKIN SECTION
Each hair arises from a pocket in the skin called the follicle. Next to the follicle is an erector muscle that raises or lowers the hair, changing the insulating properties of the coat.

MODIFIED HAIR
Porcupines have whiskers, long body hairs, and defensive spines (which are modified hairs).

by 3 tiny bones, and so to the brain. The fennec fox has enormous, sensitive pinnae; by contrast, true seals have lost their pinnae.

One of the most distinctive parts of a mammal's body is its skin. This consists of 2 layers: a protective outer layer of dead cells (the epidermis) and an inner layer (the dermis) that contains blood vessels, nerve-endings, and glands. It is the glands in the dermis that are particularly unusual: the sebaceous (or scent) glands secrete chemicals that mammals use to communicate with one another, the mammary glands produce the milk used to nourish newborn young, and the sweat glands – together with the hair, which also arises from the dermis (see panel, above) – play an important part in regulating temperature.

REPRODUCTION

Depending on the way they reproduce, mammals are usually divided into 3 groups. In all 3,

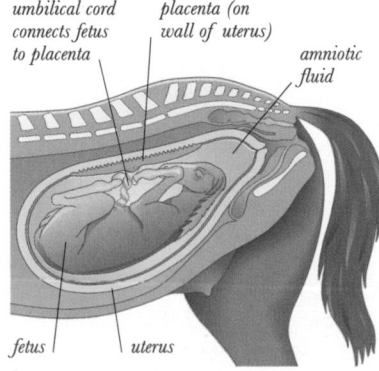

umbilical cord connects fetus to placenta
placenta (on wall of uterus)
amniotic fluid
fetus
uterus

PLACENTAL MAMMALS
The fertilized egg of a placental mammal, such as a horse, divides many times and eventually becomes a fetus. As the fetus grows, the uterus expands in size and weight. In the horse, pregnancy lasts about 11 months.

fertilization is internal (see p.29). The first group, the monotremes (which comprises the duck-billed platypus and the echidnas), lay eggs. Members of the other 2 groups give birth to live young. Of these 2, however, the marsupials have no true placenta (see below). When they are born, marsupials are usually poorly developed and are nourished by their mother's milk. In some species the young are kept in a pouch on the outside of the mother's body until they are more fully grown. The largest reproductive group is referred to as the placental mammals, in which the unborn young develop in the mother's uterus. During pregnancy, food and oxygen pass from mother to fetus through an organ known as the placenta, while waste substances move in the opposite direction. When born, infant placental mammals are more highly developed than those of marsupials.

The juveniles of all mammals are fed on milk secreted by the mother's mammary glands, which become active after the young are

born. Except in monotremes, the milk is delivered through teats. As well as providing nourishment (it is rich in proteins and fats), the milk contains antibodies, which help to establish resistance to infection. Being nourished on milk during the early weeks of life also means that young mammals do not have to forage for their own food, which greatly increases their chances of survival.

Litter sizes vary from 20 (Virginia opossum) to only 1 (orang-utan); gestation periods range from 12 days (short-nosed bandicoot) to 22 months (African elephant).

TEMPERATURE CONTROL

Mammals, like birds, are endothermic, meaning that they maintain a constant body temperature and can therefore remain active at extremely high or low external temperatures. This is why mammals are able to occupy every major habitat and are more widespread than any other vertebrates (except birds). Many species, such as the seals and whales of the Antarctic, live in regions where the temperature is well below freezing for much or all of the year. An area of the brain known as the hypothalamus monitors body temperature and adjusts it if necessary. Body temperature can be altered by increasing or decreasing the metabolic rate, widening or constricting blood

vessels that carry heat to the skin's surface, raising or lowering body hair to trap or release an insulating layer of air, and causing shivering (to gain heat) or evaporation via sweating or panting (to lose heat). Mammals can also control their body temperature by adopting special body postures: a monkey, for example, will hunch up in the cold (and many mammals huddle together in small groups to keep warm); lemurs warm up in the early morning sun by sitting up and spreading their arms to expose their thinly haired undersurface.

Behavioural patterns also help to regulate body temperature. For example, in the desert and in tropical grasslands, in the heat of the day

STAYING WARM
Large mammals, such as red deer, that do not hibernate in winter, survive and keep warm by using fat reserves built up during the summer. Over time, this leads to weight loss.

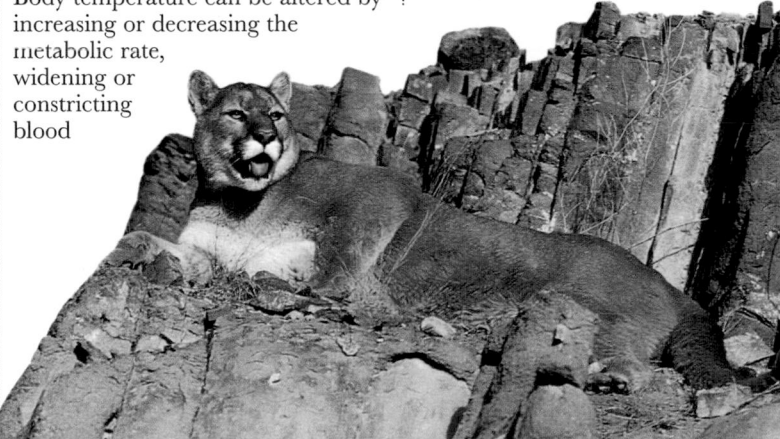

KEEPING COOL
In hot climates, mammals avoid overheating by resting in the hottest part of the day. They may also keep cool by panting, as this puma is doing. Panting helps to lower the body temperature through the evaporation of water from internal surfaces, such as the tongue.

HIBERNATION

Some mammals, especially small species, conserve energy during the cold months by hibernating, just as some reptiles (for example the garter snake) do. Their temperature falls, their breathing slows, their metabolism drops to almost imperceptible levels, and they fast, drawing on stored fat reserves. When hibernating, the animal is torpid and difficult to rouse. A West European hedgehog, for example, begins hibernation when the outside temperature falls below 15°C (59°F), and in midwinter its body temperature lowers to about 6°C (43°F). In some bats, rectal temperatures of 0°C (32°F) have been recorded during hibernation. Larger mammals, such as the American black bear, do not truly hibernate – they sleep and become cold but are roused more easily. Related to hibernation is aestivation, which is torpidity during the summer. Like hibernation, aestivation saves energy when food is short.

COLD SLEEP
Many bats that live in temperate regions hibernate over the winter. While sleeping, the temperature of their body falls to that of the roost site, as shown by the dew on this Daubenton's bat.

WINTER HIDEAWAY
For hedgehogs, effective camouflage is an important part of winter hibernation.

AQUATIC MAMMALS

Three groups of mammals have adapted to aquatic life, developing a streamlined body and the ability to stay underwater for long periods (although all of them return to the surface to breathe). The largest group, the whales and dolphins, are the most specialized. They have lost their body hair and spend their entire life in water. Like whales, seals and sea lions rely on subcutaneous fat to keep warm, but they have retained their fur, which is kept waterproof by an oily secretion from the sebaceous glands. Sirenians (the manatees and dugong) live in warm coastal waters and estuaries and are the only herbivorous aquatic mammals. The sea otter is the only other mammal that spends most of its life in water. It lacks subcutaneous fat and instead keeps itself warm by trapping air in its dense fur.

LIVING IN THE SEA
Aquatic mammals, such as this humpback whale, evolved from land-living mammals. Although the back limbs have been lost, some whales retain a vestigial pelvic girdle.

GRACEFUL MOVERS
Seals, such as this harbour seal, are highly acrobatic swimmers, with front and back limbs that have been modified into webbed flippers. Ungainly on land, they rarely venture far from the water, even to give birth to their young.

rodents retreat into cool, moist burrows, while larger mammals rest in the shade in cool depressions in the ground. Body colouring is another important factor: dark colours absorb heat; light colours reflect it. Desert mammals are therefore often light fawn in colour, while those living in cool climates are dark. This arrangement may conflict with the need for

camouflage – where there is winter snow, mammals may turn white (stoats) or have a permanently white coloration (polar bears), and possess extra-thick fur to compensate.

FEEDING

To maintain a high, constant body temperature is "energy expensive" – mammals therefore need a nutritious and plentiful diet. While the earliest mammals were probably predators, different species have since adapted to meet their dietary requirements in

FEEDING HIGH UP
Mammals avoid competing with one another by eating different foods or by getting the same food in different ways. For example, the long neck of a giraffe allows it to feed at a height beyond the reach of other herbivores.

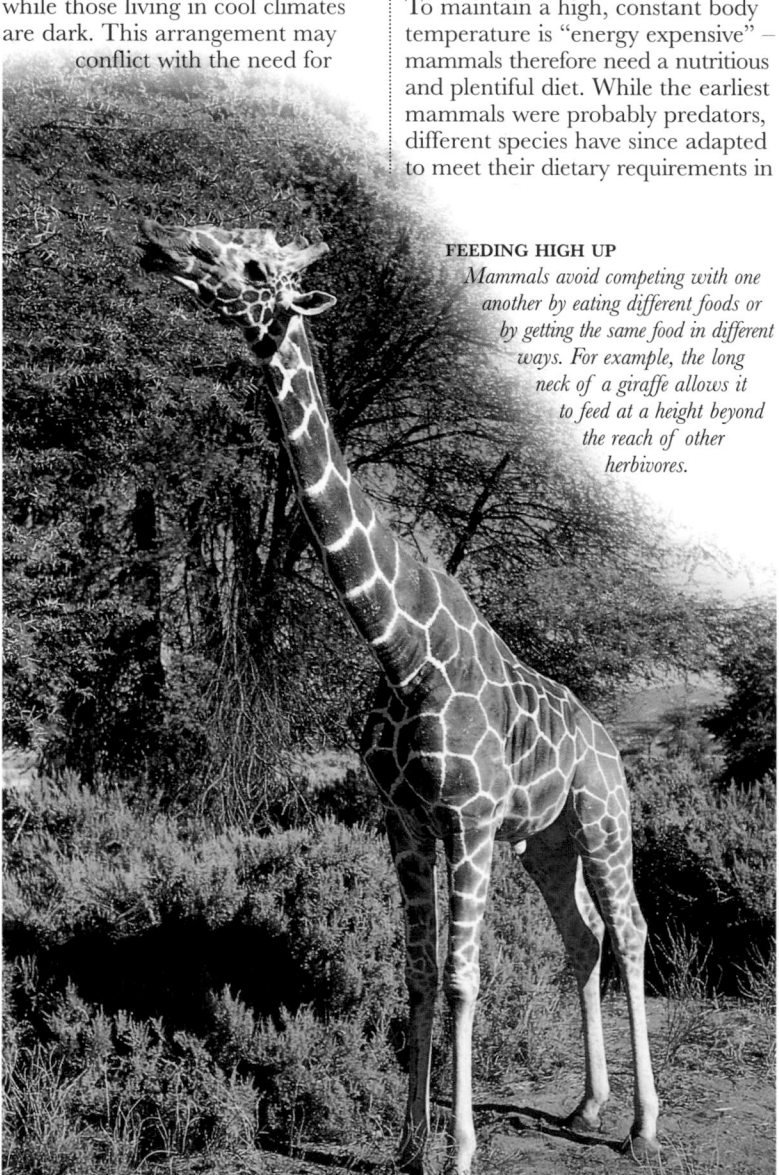

a variety of ways. Some eat animal prey – this is a carnivorous diet (and includes insect-eating). Other mammals, called herbivores, eat plants. A herbivorous diet includes sub-types such as fruit-eating and grass-eating. An omnivore eats both animal prey and plants.

Carnivorous mammals have a simple digestive tract, because the proteins, lipids, and minerals found in meat require little in the way of specialized digestion. Plants, on the other hand, contain complex carbohydrates, such as cellulose. The digestive tract of a herbivore is therefore host to bacteria that ferment these substances, and make them available for digestion. The bacteria are housed either in a multi-chambered stomach or in a large caecum (a blind-ending sac in the large intestine).

The size of an animal is a factor in determining diet type. Small mammals have a high ratio of heat-losing surface area to heat-generating volume, which means that they tend to have high energy requirements and a high metabolic rate. As they cannot tolerate the slow, complex digestive process of a herbivore, mammals that weigh less than about 500g (18oz) are mostly insectivorous. Larger mammals on the other hand generate more heat and less of this heat is lost. They can therefore tolerate either a slower collection process (those that prey on large vertebrates) or a slower digestive process (herbivores). Furthermore, mammals that weigh more than about 500g (18oz) usually cannot collect enough insects during their waking hours to sustain themselves. The only large insectivorous mammals are those that feed on huge quantities of colonial insects (ants or termites).

SOCIAL STRUCTURES

Mammals communicate socially by scent, either from glands (which may be located in the face, in the feet, or in the groin) or in their urine or faeces (which contain sexual hormones). They also communicate by body posture or facial gesture,

| **FEAR** | **SUBMISSION** | **EXCITEMENT** |

FACIAL EXPRESSION
Some mammals communicate using facial expressions. This ability is well developed in primates, such as the chimpanzee. Fear is shown by baring the teeth, submission by a pouting smile, and excitement by exposed teeth and an open mouth.

PLAY

A great deal of a young mammal's learning occurs through play, when infants experiment with adult behaviours, such as fighting and hunting. Play can also take the form of exploring and displaying to one another. Hoofed mammals, such as deer, establish dominance rankings when young to avoid conflict as adults. This reduces the risk of serious injury that would make them vulnerable to predators. Predators must learn to stalk and kill prey to survive as adults.

SURVIVAL SKILLS
Potential predators, such as these lion cubs, use play to practise pouncing, biting, and raking with the back feet, stopping before real damage is done.

FINDING THEIR PLACE
Like many herbivores, these elephant calves use play to establish their rank in the herd. Young elephants also use their trunks to examine novel objects.

INTELLIGENCE

In mammals thought to be intelligent – such as primates – the "thinking" part of the brain, the cerebrum, is larger relative to the rest of the brain. Intelligence itself is not easy to define, but indications of intelligence include the ability to learn, matched with behavioural flexibility. Rats, for example, are considered to be highly intelligent as they can learn to perform new tasks, an ability that may be important when they first colonize a fresh habitat. In some mammals, food-gathering appears to be related to intelligence: a deer feeding on plants has a brain relatively smaller than a cat that must "think" to outwit its prey.

RAT IN A MAZE
Animal intelligence is often assessed by a creature's ability to complete simple tasks, usually using a food incentive. This albino brown rat is finding its way through a maze.

FOOD FOR THOUGHT
Primates are rare among mammals in that some can use a variety of objects as tools, while others clean their food prior to eating it. This Japanese macaque has learnt to separate wheat grains from sand by immersing the food in water. The sand sinks; the wheat grains float on the surface.

by touch, and by sound, which may develop into complex messages.

Socialization begins at birth, when signals between a mother and her infant facilitate suckling. The process continues throughout the juvenile period, when the young interact with each other through play (see panel opposite, below) and gradually learn to interpret the behaviour of adults.

Some mammals only interact to mate or to nurture young, but many form temporary or permanent social groupings beyond these minimal needs. The hoofed mammals, for example, form herds for safety as predators are more likely to be spotted. Interwoven with social behaviour is the use of space: most mammals

have a home range (the area within which an animal, or group, performs most of its activities), and in some of these areas are developed into territories (an area that the animal, or group, defends against others of the same species).

Solitary mammals, such as the aardvark and most cats, usually occupy a territory that they defend against members of the same sex. A tigress, for example, will not permit other females to enter her territory, whereas the territory of a male tiger overlaps those of several females.

Some mammals, such as gibbons, live in monogamous pairs (with their immature offspring), and each pair occupies and defends a territory. However, recent studies indicate that the assumption of "lifelong fidelity" is wrong: surreptitious copulations outside the pair are frequent, and many pairs break up and find new mates after some years.

Other mammals form larger social groups, of varied composition. In some species, such as seals and elephants, the sexes live separately for most of the year. Related females form social units, and the males either live alone or form small bachelor groups. In mammals living under this system there is ferocious competition among the males for the right to mate. The most successful males are usually the largest and strongest individuals that also have the best weapons (antlers,

horns, or tusks). In elephant seals and sea lions, as the females become sexually receptive the males bellow and display to establish dominance.

Other species, such as Burchell's zebras, form small, permanent harem groups, in which one male leads and mates with a group of females. The subordinate males are relegated to bachelor bands unless one can successfully challenge and expel a harem male.

The final type of larger social group exhibited by mammals is one in which the collection of individuals consists of several males and females. Such societies are restricted to primates and the social carnivores (although some other species, such as migrating wildebeests, form temporary mixed-sex groups). Among baboons, for example, several adult males join a troop of females related to each other (but not to the males) and compete for dominance and the right to mate with the females. These power struggles result in a continual change in the dominance hierarchy. In lion prides, the males are related to each other and cooperate in the defence of the females rather than competing for them. In wolf packs, only the dominant pair breeds; the other adults are the offspring from previous years who, instead of leaving to form their own packs, stay with their parents to help rear their younger siblings.

LOCOMOTION

Mammals dwell in a huge variety of habitats and have accordingly developed a number of different ways of moving around. While a quadrupedal (4-footed) gait is the commonest form of mammal locomotion, some species, such as kangaroos, are bipedal (2-footed gait). As mammals move, the way in which the foot comes into

contact with the ground varies in one of 3 different ways (see below).

Locomotion is also related to the lifestyle of the animal. Predators, which require short, explosive bursts of speed to catch their prey, have a flexible spine. For example, in the cheetah – the fastest terrestrial animal – the backbone coils and uncoils with every stride, propelling the animal forwards. It has been estimated that these movements increase the cheetah's speed by about 30kph (19mph). In contrast, prey species, such as gazelles, rely on endurance to outrun predators. They have a rigid back, and the "energy cost" of their movement is reduced as a result of a lengthening of the lower segments of the limbs and a concentration of the muscles close to the body.

metapodials
digits
calcaneum
calcaneum
claws
PLANTIGRADE
metapodial
single digit
calcaneum
hoof
UNGULIGRADE
digits
metapodial
claws
DIGITIGRADE

WALKING GAITS
When moving, plantigrade mammals, such as bears, keep the calcaneum (heel bone), metapodials, and digits (toes) of each foot on the ground. Digitigrade mammals, such as dogs, move with only the digits touching the ground. Unguligrade mammals, such as horses, walk and run on the tip of each digit.

AERIAL MAMMAL: LONG-EARED BAT

ARBOREAL MAMMAL: BLACK LEMUR

TERRESTRIAL MAMMAL: SNOW LEOPARD

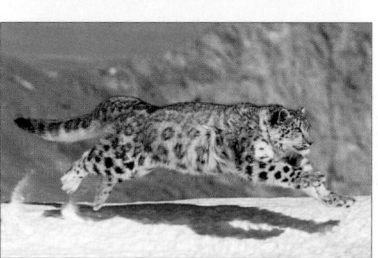

AQUATIC MAMMAL: DOLPHIN

ADAPTABLE MAMMALS
Over time, the limbs of some mammals have evolved into either wings, flippers, or grasping hands and feet, enabling bats to take to the air, whales and dolphins to adopt an aquatic existence, and primates to move freely through the trees. Predatory land mammals, such as cats, have developed a flexible spine for speed and manoeuvrability, and a long tail for balance.

BENEFITS OF COMMUNAL LIVING
Living in a group has several advantages for small mammals, such as these dwarf mongooses. For example, there is more chance of spotting predators, the burden of rearing young can be shared, and the territory can be defended collectively.

MAMMALS

MAMMALS

EGG-LAYING MAMMALS

PHYLUM	Chordata
CLASS	Mammalia
ORDER	Monotremata
FAMILIES	2
SPECIES	5

ALSO KNOWN AS monotremes, this group comprises the duck-billed platypus and the echidnas. Monotremes are the only mammals that lay eggs. The duck-billed platypus has a duck-like bill, a beaver-like tail, and webbed feet. Echidnas have a tubular beak, huge claws for digging, and spines. The duck-billed platypus lives in freshwater in eastern Australia; echidnas exist in varied habitats in Australia and New Guinea.

AQUATIC LIFE
The duck-billed platypus is a strong swimmer. Its coat is waterproof and has a dense underfur for insulation. When diving, the eyes and ears are closed.

ANATOMY

Monotremes are short-legged animals with a small head and tiny eyes. In all species, the digestive, urinary, and reproductive tracts empty into a common chamber (the cloaca), which terminates in a single exit ("monotreme" means "one hole").

PLATYPUS

ECHIDNA

webbing between digits

DIFFERENT SNOUTS
The platypus (top) has a flattened bill, covered with sensitive skin. Echidnas, however, have a cylindrical beak. In both, the snout is used to probe for and locate food.

WEBBED FEET
In water, the duck-billed platypus is propelled by powerful thrusts of its fully webbed front feet.

STICKY CATCH
The short-nosed echidna uses its long, slender tongue to catch termites and ants. A sticky coating of saliva ensures that the insects do not escape.

FEEDING

The platypus uses its sensitive bill to probe the beds of rivers or lakes for crustaceans and insect larvae. The short-nosed echidna digs with its powerful claws for ants, termites, or earthworms; long-nosed echidnas mainly eat earthworms. Adult monotremes lack teeth: food is ground between plates or spines in the mouth.

Tachyglossus aculeatus

Short-nosed echidna

Length	30–45 cm (12–18 in)
Tail	1 cm (3/8 in)
Weight	2.5–7 kg (5 1/2–15 lb)

Location Australia (including Tasmania), New Guinea

Social unit Individual

Status Lower risk

The spines of the short-nosed echidna – also called the spiny anteater – are longer than the fur between them. Active both day and night, this echidna is solitary and can become torpid in very cold or hot weather, when its temperature falls from the normal 31–33°C (88–92°F) to as low as 4°C (39°F). It eats a variety of ants, termites, grubs, and worms. These are detected by smell and perhaps by sensors on the long snout that detect electric signals. The small head joins the shoulders with no external neck.

Zaglossus bartoni

Long-nosed echidna

Length	60–100 cm (23 1/2–39 in)
Tail	None
Weight	5–10 kg (11–22 lb)

Location New Guinea

Social unit Individual

Status Endangered

The downcurved snout of this species may exceed 20cm (8in) in length, with a tiny mouth at the tip and small, close-set eyes at its base. As in other echidnas, the female long-nosed echidna digs a burrow for her egg, but in this species she carries and suckles the hatched infant in her pouch. The largest monotreme, it is slow moving, rolling into a spiny ball for defence. The spines are often only just visible through its long black fur.

Ornithorhynchus anatinus

Duck-billed platypus

Length	40–60 cm (16–23 1/2 in)
Tail	8.5–15 cm (3 1/4–6 in)
Weight	0.8–2.5 kg (1 3/4–5 1/2 lb)

Location E. Australia, Tasmania

Social unit Individual

Status Vulnerable†

The duck-like, beak-shaped mouth, sprawling, reptile-type gait, and flattened, almost scaly, beaver-like tail make the duck-billed platypus an unmistakable animal. Its waterproof plum-coloured body fur has a plush texture that is reminiscent of that of a mole. The probing bill is very sensitive to both touch and water-borne electrical signals from the muscles of its small aquatic prey. It has an unusually high density of red blood cells to enable it to make deep dives in pursuit of its quarry. Platypuses are solitary, yet occupy overlapping home ranges. However, males in the breeding season defend these territories. The male has a poisonous spur on each hind foot, with which he attempts to wound his rivals. Home is a bankside burrow, usually about 5m (16ft) long but may reach up to 30m (98ft). After a gestation period of one month, the female incubates her clutch of 1–2 soft, leathery-shelled eggs here for 10 days. After the young hatch, she suckles them for 3–4 months in the burrow, leaving them walled in for periods of up to 38 hours while she forages.

partially webbed back feet

fully webbed front feet

MARSUPIALS

PHYLUM	Chordata
CLASS	Mammalia
ORDER	Marsupialia
FAMILIES	22
SPECIES	292

CLASSIFICATION NOTE

Many scientific authorities now divide marsupials into 7 distinct orders: American opossums; "shrew" opossums; the colocolo; Australasian carnivorous marsupials; bandicoots; the marsupial "mole"; and wombats, possums, wallabies, kangaroos, and the koala. For ease of reference, and because of their shared reproductive biology, this book treats all the marsupials as one group (order Marsupialia).

LIKE OTHER MAMMALS (apart from monotremes), marsupials bear live offspring and have nipples to supply their young with milk, yet they are distinct from all other live-bearers (together described as placental mammals), in that they give birth at a very early stage of the embryo's development and nourish the newborn by milk instead of by a placenta. They are amazingly diverse, including animals such as kangaroos, possums, and bandicoots. The Australasian marsupials, through a lack of competing species, have diversified and become specialized insect-eaters, carnivores, and herbivores. In South America, the marsupials are small and mostly arboreal; only one species, the Virginia opossum, has spread to North America. Marsupials inhabit a huge range of habitats, from desert to rainforest.

ANATOMY

Externally, marsupials are highly varied, although many have long back legs and feet (for example kangaroos) and elongated snouts; almost all have large eyes and ears. Female marsupials have a unique "doubled" reproductive tract (see below), and in some males the penis is forked. The testes are held in a pendulous scrotum with a long, thin stalk, which swings in front of the penis. Apart from their specialized reproductive system, marsupials also differ from placental mammals in that their brain is relatively smaller and lacks a corpus callosum (the nerve tract connecting the two cerebral hemispheres). The group that contains the kangaroos, possums, wombats, and the koala, and the group containing the bandicoots, have an arrangement called syndactyly: the second and third toes of the back foot are combined to form a single digit with 2 claws.

EARLY LIFE

Marsupial offspring are born in an almost embryonic state after a very short gestation (only 12 days in some bandicoot species). The newborn makes its way to one of the mother's nipples, where it remains attached for several weeks. Larger species have single births, but the small quolls and dunnarts have litters of up to 8. In many kangaroo species the female mates again while pregnant, but the new embryo remains dormant until the previous young leaves the pouch.

OVERCROWDING
A mother's pouch may be too small to carry several offspring. This mouse opossum has no option but to carry her babies externally.

AMAZING JOURNEY
This tiny tammar wallaby baby is hauling itself over its mother's fur to reach a nipple, to which it will firmly attach itself. As in most marsupials, the nipples are protected by a pouch. The joey will not relinquish the nipple until, months later, it begins to explore the outside world.

SPECIAL POUCHES
Most marsupials have a pouch to carry their young. This western grey kangaroo has a forward-facing pouch, from which her joey's head is protruding. Some pouches face backwards.

MAMMALS

FEMALE REPRODUCTIVE SYSTEM
Unlike placental mammals, which have a single uterus and vagina (far right), female marsupials have a double system with 2 uteri, each with its own lateral vagina (right). The young is born through a separate, central birth canal. In some marsupials this canal forms before each birth. In others it remains after the first birth (but sometimes fills with connective tissue).

ovary
uterus
uterus
ovary
lateral vagina
lateral vagina
birth canal
uterus
birth canal
vagina

MARSUPIAL

PLACENTAL MAMMAL

MOVEMENT

Although most marsupials run or scurry, there are several variations. For example, wombats waddle, while koalas and possums climb. Kangaroos and wallabies hop on their long back legs, using the extended middle toe as an extra limb segment. Although hopping at low speeds uses more energy than running on all fours, above approximately 1.8m per second (6ft per second) the larger species begin to conserve energy. This is because energy is stored in the tendons of the foot, and the heavy tail swings up and down like a pendulum, providing momentum.

HOPPING
The faster a large kangaroo (such as this red kangaroo) moves, the more energy efficient it becomes.

GLIDING
Some possums, known as gliders, have a membrane between their front and back legs, which they use as a parachute.

Marmosa murina
Common mouse opossum

Length 11–14.5 cm
(4¼–5¾ in)

Tail 13.5–21 cm
(5¼–4¾ in)

Weight 15–45 g
(9/16–1⅝ oz)

Location N. and C. South America

Social unit Individual

Status Lower risk†

Found near forest streams and human habitation, this opossum is pale buff to grey on the upperparts, creamy white below, with a black face mask. Its diet includes small invertebrates such as insects and spiders, small vertebrates such as lizards, birds' eggs and chicks, and some fruit. It is a fast, agile climber and rests by day in a tree-hole, old bird's nest, or tangle of twigs among the branches. The 5–8 young are weaned at 60–80 days.

strongly prehensile tail

short, fine, velvety fur

prominent eyes

Chironectes minimus
Water opossum

Length 26–40 cm
(10–16 in)

Tail 31–43 cm
(12–17 in)

Weight 550–800 g
(20–29 oz)

Location S. Mexico to C. South America

Social unit Individual

Status Lower risk

Also called the yapok, this is the only aquatic marsupial. It has fine, dense,

water-repellent fur, long, webbed toes on its rear feet, and – in both male and female – a pouch with a muscular opening that can close tight underwater. The diet includes fish, frogs, and similar freshwater prey, detected and grabbed by the dextrous, clawless front toes. It rests by day in a leaf-lined river-bank den.

white-tipped tail

Caluromysiops irrupta
Black-shouldered opossum

Length 21–26 cm
(8½–10 in)

Tail 22–31 cm
(9–12 in)

Weight 200–600 g
(7–21 oz)

Location W. South America

Social unit Variable

Status Vulnerable

A broad black stripe from shoulders to front legs identifies this woolly furred opossum. It forages by night in trees for a wide diet including fruits and grubs. Its densely furred tail is grey, turning white at the end, with a hairless underside. Gestation is 13–14 days and litter size 2.

Neophascogale lorentzii
Long-clawed marsupial mouse

Length 16–23 cm
(6½–9 in)

Tail 17–22 cm
(6½–9 in)

Weight 200–250 g
(7–9 oz)

Location New Guinea

Social unit Individual

Status Lower risk†

Also called the speckled dasyure, from the sprinkling of long white hairs in the dark grey upper fur, this species has short, powerful limbs with very long claws on all toes. It digs by day for grubs, worms, and similar prey. Information on nesting and breeding is very sparse; 4 young have been recorded in the female's pouch.

Didelphis virginiana
Virginia opossum

Length 33–50 cm
(13–20 in)

Tail 25–54 cm
(10–21½ in)

Weight 2–5.5 kg
(4½–12 lb)

Location W., C., and E. USA, Mexico, Central America

Social unit Individual

Status Lower risk†

The highly adaptable Virginia opossum is actively expanding its range in North America. It benefits from human habitation, both for shelter as it nests in piles of debris or outbuildings, and for food as it scavenges for scraps. A true omnivore, its diet ranges from grubs and eggs to flowers, fruits, and carrion; it may raid farm poultry and damage garden plants. The largest American marsupial, the Virginia opossum is nocturnal and usually terrestrial, but also climbs well and swims strongly. Although not territorial, individuals avoid each other

except when breeding. The female bears up to 18 per litter, yet has only enough teats for 13. The survivors attach to the teats for 50 days, leaving the pouch at 70 days.

"PLAYING POSSUM"

When threatened, the Virginia opossum feigns death in an attempt to make predators lose interest. It may lie immobile, partly curled on its side, for up to 6 hours. The mouth and eyes are open, and it does not respond to stimuli, including touch. It may void a foul-smelling fluid from its anus for added effect.

pale grey-white face

5 long-clawed toes per foot

hairless, partly prehensile tail

SCRUFFY APPEARANCE
The Virginia opossum's fur varies from grey to red, brown, and black. Long, white-tipped guard hairs with thick underfur make it look unkempt.

Ningauri ridei
Inland ningaui

Length 5–7.5 cm
(2–3 in)

Tail 5–7 cm
(2–2¾ in)

Weight 6–12 g
(7/32–7/16 oz)

Location C. Australia

Social unit Individual

Status Lower risk†

Formally described for science only in 1975, the inland or wongai ningaui is a small, fierce, solitary, nocturnal, shrew-like predator of invertebrates such as beetles, crickets, and spiders, mostly less than 1cm (3/8 in) long.

It hunts mainly by smell and hearing in clumps of spinifex (hummock grass), resting by day in thick undergrowth or an old hole dug by a lizard, rodent, or large spider. After an incubation of 13–21 days, 6–7 newborn ningaui attach to the female's teats in her relatively open pouch area for 6 weeks. They are weaned by 13 weeks.

Sminthopsis crassicaudata
Fat-tailed dunnart

Length 6–9 cm
(2 1/4 –3 1/2 in)
Tail 4–7 cm
(1 1/2 –2 3/4 in)
Weight 10–20 g
(3/8 –11/16 oz)

Location Australia

Social unit Individual

Status Lower risk†

Widespread in wood, scrub, and dry grassland, the fat-tailed dunnart has benefited from the spread of grazing farmland. It has fawn or brown upper fur and white underparts. It actively searches by night across open soil and leaf litter for grubs, worms, and similar small invertebrates. It is solitary during the breeding season of August to March, but at other times small groups, of less than 10, may huddle together in sheltered crevices, logs, or burrows during cold periods.

wide tail base to store fat

Antechinomys laniger
Kultarr

Length 7–10 cm
(2 3/4 –4 in)
Tail 10–15 cm
(4–6 in)
Weight 25–35 g
(7/8 –11/4 oz)

Location C. and S. Australia

Social unit Individual

Status Vulnerable†

Well adapted to a range of habitats, from scattered woodland to scrubby semi-desert, the kultarr preys at night on grubs and

Parantechinus apicalis
Dibbler

Length 10–16 cm
(4–6 1/2 in)
Tail 7.5–12 cm
(3–4 3/4 in)
Weight 40–125 g
(1 7/16 –4 oz)

Location S.W. Australia

Social unit Individual

Status Endangered

After a gap of some 80 years, this species was rediscovered in 1967 in extreme southwest Australia and in 1985 on 2 offshore islands. Distinctive

similar small invertebrates. It bounds on its relatively large, 4-toed rear feet, but lands on its front legs (like a rabbit). It has large ears and dark-ringed eyes, a thin, pointed face, and a compact body. Its long tail has a dark-tufted tip. The fur is fawn or brown above and white underneath.

features include a white eye-ring, white-flecked "grizzled" grey-brown fur, a well-haired, short, tapering tail, and large canine teeth for killing prey. This includes small vertebrates such as mice, birds, and lizards, as well as insects and other invertebrates.

Pseudantechinus macdonnellensis
Red-eared antechinus

Length 9.5–10.5 cm
(3 3/4 –4 1/4 in)
Tail 7.5–8.5 cm
(3–3 1/4 in)
Weight 20–45 g
(11/16 –1 5/8 oz)

Location W. and C. Australia

Social unit Individual

Status Lower risk†

When well fed, this insectivorous, nocturnal species, also called the fat-tailed pseudantechinus, has a carrot-shaped tail swollen with fat (as a food store). Its fur is grey-brown, with red patches behind the ears and grey-white underparts. It prefers dry, rocky areas with clumped bushes and grass. After a gestation of 45–55 days, the female carries 6 young, one on each teat in her pouch, for 40 days. She suckles them in a nest until 14 weeks.

Dasycercus byrnei
Kowari

Length 13.5–18 cm
(5 1/4 –7 in)
Tail 11–14 cm
(4 1/4 –5 1/2 in)
Weight 70–150 g
(2 1/2 –5 oz)

Location C. Australia

Social unit Individual

Status Vulnerable

Despite a squirrel-like appearance and bushy, black-tufted tail, the kowari is a burrow-dwelling carnivore. It kills large prey such as lizards, birds, and rodents by holding them in its front paws and biting repeatedly and rapidly. It also eats insects and similar creatures, mostly more than 1cm

Dasycercus cristicauda
Mulgara

Length 12–20 cm
(4 3/4 –8 in)
Tail 7.5–11 cm
(3–4 1/4 in)
Weight 50–150 g
(1 3/4 –5 oz)

Location W. and C. Australia

Social unit Individual

Status Vulnerable

(3/8 in) long. The kowari is solitary and extends its burrow after rain, which softens the stony soil in its harsh, sparsely vegetated desert habitat. It marks its burrow and home range with urine, faeces, and chin- and chest-gland scents, and communicates with neighbours through chatters, hisses, and flicks of its tail.

This inhabitant of sandy desert with scattered spinifex (hummock grass) resembles the kowari (above) in body form and, like several similar species, has a fat-swollen tail base. Its fur is sandy brown on the upperparts, off-white below, with a crest of black hairs on the tail's upper surface towards the tip. Its burrow varies from single, simple, and short to a system of tunnels with 5–6 entrances. At night, the mulgara hunts small rodents, lizards, birds, and large insects, and may also tackle large centipedes.

Dasyurus viverrinus
Eastern quoll

Length 28–45 cm
(11–18 in)
Tail 17–28 cm
(6 1/2 –11 in)
Weight 0.7–2 kg
(1 1/2 –4 1/2 lb)

Location Tasmania

Social unit Individual

Status Lower risk

Last seen on the Australian mainland in the 1960s, in Sydney's suburbs, this cat-like marsupial (sometimes called the spotted native cat) now survives only in Tasmania. It hunts on the ground by night for small mammals, birds, and other vertebrates, and large insects; it also eats fruits, grass, and carrion. Quolls occupy varied woody, shrubby, and grassy habitats, as provided by mixed agriculture. The male is longer and 50 per cent heavier than the female.

MATERNAL CARE

Like many female marsupials, the quoll has a large litter, up to 24, but only 6 nipples in her pouch, so only this many can survive. They attach for 8 weeks, then remain in the den as the mother goes off to feed and returns to suckle. She regularly takes them to new dens, often by piggyback.

SPOTTED COAT
Slim and agile, the eastern quoll has white spots on its brown or black body, but not on the tail. The ears are large and erect.

Sarcophilus harrisii
Tasmanian devil

Length 52–80 cm (20 1/2 –32 in)	
Tail 23–30 cm (9–12 in)	
Weight 4–12 kg (8 3/4 –26 lb)	

Location Tasmania	Social unit Individual
	Status Lower risk

The largest marsupial carnivore, the "devil" occurs in all Tasmania's major habitats and hunts prey of varied sizes, from insects to possums and wallabies, as well as scavenging. Nocturnal, it screeches or barks and rears up if alarmed. It makes its den in a burrow or among rocks or roots. Born after a gestation of 30–31 days, the offspring attach to the female's 4 teats in her shallow, rear-opening pouch.

white patch on rump

COMPETING AT A KILL

The Tasmanian devil is normally solitary, but its excellent sense of smell means many can converge on a large carcass. Its powerful jaws and sharp, sturdy teeth rip up the hide and crush gristle and bone. The devils growl and snarl at each other for the prime parts, but stop short of physical clashes.

WHITE ON BLACK
The devil has long white patches on its chest and rump.

Myrmecobius fasciatus
Numbat

Length 20–28 cm (8–11 in)	
Tail 16–21 cm (6 1/2 –8 1/2 in)	
Weight 300–725 g (11–26 oz)	

Location S.W. Australia	Social unit Individual
	Status Vulnerable

The numbat, or banded anteater, has a long (10 cm/4 in) tongue to lick up termites, or very rarely ants, after the nest has been ripped open using the large-clawed, powerful forefeet. It also has more teeth – 52 – than any other land mammal, although these are very small. Day-active and solitary, the numbat chases same-sex intruders from its territory. Offspring attach to the female's 4 teats for 4 months, then are suckled in the nest for a further 2–3 months.

bushy tail

Perameles gunnii
Eastern barred bandicoot

Length 27–35 cm (10 1/2 –14 in)	
Tail 7–11 cm (2 3/4 –4 1/4 in)	
Weight 0.5–1.5 kg (1–3 1/4 lb)	

Location S.E. Australia, Tasmania	Social unit Individual
	Status Vulnerable

A few hundred of these bandicoots survive in mainland Australia, in Victoria, the majority being found in Tasmania. This omnivorous opportunist forages alone at night and shelters by day in a simple nest of grasses, leaves, and twigs. The eastern barred bandicoot has rabbit-like ears, 3–4 whitish back stripes, and a white tail. Its life cycle is rapid: 12-day gestation, weaning by 60 days, and sexual maturity at 3 months.

Echymipera kalubu
New Guinean spiny bandicoot

Length 20–50 cm (8–20 in)	
Tail 5–12.5 cm (2–5 in)	
Weight 0.5–1.5 kg (1–3 1/4 lb)	

Location New Guinea and surrounding islands	Social unit Individual
	Status Lower risk

Bandicoots are mostly insectivorous, rat- to rabbit-sized marsupials. The New Guinean spiny bandicoot has a long, mobile snout, an outer coat of stiff, stout hairs in shades of brown, copper, yellow, and black, buff underparts, and a hairless tail. It forages alone at night for prey, takes fruits, berries, and other plant matter, and aggressively repels others of its kind. It shelters by day in a hollow log, leaf pile, or self-dug burrow.

Macrotis lagotis
Greater bilby

Length 30–55 cm (12–22 in)	
Tail 20–29 cm (8–11 1/2 in)	
Weight 1–2.5 kg (2 1/4 –5 1/2 lb)	

Location W. and C. Australia	Social unit Pair
	Status Vulnerable

Also known as the rabbit-eared bandicoot, this omnivorous species is recognized by its huge ears, long back feet, and tricoloured tail: blue-grey at the base (as on the body), then black, with the last half white and feathery. It digs powerfully to shelter by day in a burrow 3m (10 ft) long and some 2m (7ft) deep. Pairs associate for the breeding season and the 2 offspring, born after 13–16 days' gestation, leave the pouch after 80 days.

Notoryctes typhlops
Marsupial mole

Length 12–18 cm (4 3/4 –7 in)	
Tail 2–2.5 cm (3/4 –1 in)	
Weight 40–70 g (1 7/16 –2 1/2 oz)	

Location S. Australia	Social unit Individual
	Status Endangered

The marsupial mole tunnels down to 2.5m (8ft) in sandy deserts, loose-soiled grassland, and scrubby bush. It "swims" through light sand, which collapses behind leaving no permanent tunnel. Food includes fungi and tubers as well as animal prey, which is eaten whenever encountered. The silky, off-white to cinnamon fur is rubbed shiny by burrowing, and may be stained deep red by iron minerals in the soil. The female's pouch, in which she carries 1–2 young, opens rearwards so it does not fill with soil.

BURROWING MACHINE
Sand and soil are probed by the horny nose pad, scooped aside by the front feet, and kicked up and back by the 3 large claws on each rear foot.

MOLE FOOD

The marsupial mole eats soil-dwelling worms, grubs, centipedes, and even small lizards, such as the gecko held here by the huge front claws. Using smell and touch, it follows small tunnels made by potential prey in order to catch its victims. It may forage on the surface after rain.

tiny eyes and small ears hidden in fur

twin, spade-like front claws

Phascolarctos cinereus
Koala

	Length 65–82 cm (26–32 in)
	Tail 1–2 cm (3/8 –3/4 in)
	Weight 4–15 kg (8 3/4 –33 lb)
Location E. Australia	Social unit Individual
	Status Lower risk

The koala lives nearly all its life in eucalyptus trees. It feeds for about 4 hours at night, eating some 500g (1lb) of leaves, and dozes for the remaining time, wedged securely in a branch fork. Occasionally, it descends to change trees or promote digestion by eating soil, bark, and gravel. Individuals bellow to each other during the breeding season and dominant males mate with more females than do junior males. Copulation is brief and often accompanied by bites and scratches. After a gestation of 35 days the single offspring crawls into the pouch, where it is suckled for 6 months. It then clings to the mother's back. Among its early solid food is the mother's droppings, which carry helpful food-digesting and disease-fighting microbes. Despite a placid appearance, the koala readily bites and scratches.

CONSERVATION

Adult koalas have few natural enemies apart from large birds of prey. However, domesticated dogs attack koalas when on the ground and so they sometimes need veterinary first aid (shown left). Disease caused by the *Chlamydia* bacterium, which can be treated by antibiotics, has also reduced koala numbers in some areas. Land clearance for farming, forestry, or building also poses a considerable threat.

large, rounded, white-tufted ears

BEAR-LIKE MARSUPIAL
The koala's face is large and wide, like that of a bear, with a smooth black muzzle; the body is compact and thick set. The fur is soft, long, and mainly grey to grey-brown, with slightly paler underparts, and mottling on the rump.

HOLDING ON

The koala's short, powerful limbs have sharp claws (except on the first rear toe) and granular paw pads, to grip bark and branches. Toes 1 and 2 on the front foot can oppose the other 3, allowing a pincer-like grasp of thin boughs.

Vombatus ursinus
Common wombat

	Length 70–120 cm (28–47 in)
	Tail 2–3 cm (3/4 –1 1/4 in)
	Weight 25–40 kg (55–88 lb)
Location E. Australia, Tasmania	Social unit Individual
	Status Lower risk†

The common wombat has a hairless nose and remarkably bear-like form. A prolific burrower, its tunnel system generally has only one entrance but many underground branches, with a total length of up to 200m (650ft). Preferred sites are slopes above creeks and valleys, where the wombat grazes, mainly at night, on grasses, sedges, roots, and tubers. In winter it may graze or sunbathe by day. Although wombats are generally solitary, and adult males tend to chase intruders from their home areas, they also seem to visit each other's burrows on occasion. After the offspring has left its mother's pouch, it stays at first in a nest chamber lined with dry grass and leaves, inside the burrow. The common wombat is not dangerous to humans, but its holes may trip horses, harbour dingoes, undermine banks, and damage rabbit-proof fences.

coarse, dense, grey-tinged, brown fur

BROAD BURROWER
The common wombat's broad angular head, compact body, and strong limbs with wide, large-clawed feet make it an accomplished digger.

OUT OF THE POUCH

The common wombat usually produces a single offspring, which remains in the mother's pouch for 6–7 months. It returns there occasionally over the following 3 months to suckle or seek shelter. Weaning occurs when it is 15 months old.

Lasiorhinus latifrons
Southern hairy-nosed wombat

	Length 77–95 cm (30–37 in)
	Tail 3–6 cm (1 1/4 –2 1/4 in)
	Weight 19–32 kg (42–71 lb)
Location S. Australia	Social unit Group
	Status Lower risk†

Even stockier and shorter-limbed than the common wombat (see left), this species has longer ears, long, silky fur mottled brown and grey, and grey-white hairs on the snout. It is colonial, groups of 5–10 occupying a burrow system (warren) covering several hundred square metres. This wombat grazes at night on grass and other low vegetation. Senior males repel strangers of their kind unless these are receptive females at breeding time.

Common brush-tailed possum

Trichosurus vulpecula

Length 35–58 cm (14–23 in)	
Tail 25–40 cm (10–16 in)	
Weight 1.5–4.5 kg (3 1/4–10 lb)	

Location Australia (including Tasmania)

Social unit Individual

Status Lower risk

sharp, curved claws

opposable first toe on rear foot

partly prehensile tail

Familiar in many habitats, including parks and gardens, the common brush-tailed possum has mainly silver-grey body fur, which is short and tinged red in the north of its range, and longer and darker grey in the south. It bounds and climbs with ease as it forages for eucalypt, acacia, and other leaves, as well as flowers, fruits, and, occasionally, birds' eggs and chicks. This vocal possum, with its range of hisses, chitters, grunts, and growls, is solitary apart from the brief mating season. It lives in a crevice or hollow in a tree, log, rock, or roof. The usually single young remains in the pouch for 5 months.

Squirrel glider

Petaurus norfolcensis

Length 18–23 cm (7–9 in)	
Tail 22–30 cm (9–12 in)	
Weight 200–300 g (7–11 oz)	

Location E. Australia

Social unit Group

Status Lower risk

The squirrel glider's furry gliding membrane extends from each fifth front toe to the back foot. The long, bushy, soft-furred, squirrel-like tail acts as a rudder when parachuting as far as 50m (165ft) to other trees or to escape enemies. This possum lives in small groups of one adult male, 1–3 females, and their offspring of the season. It feeds on insects and similar small tree creatures, as well as sap, gum, pollen, and seeds. The 1–2 young remain in the pouch for up to 3 months and are weaned by 4 months.

Common ringtail

Pseudocheirus peregrinus

Length 30–35 cm (12–14 in)	
Tail 30–35 cm (12–14 in)	
Weight 700–1,100g (25–39 oz)	

Location E. Australia, Tasmania

Social unit Group

Status Lower risk

The common ringtail's reddish or grey-brown fur extends to the base of the tapering, strongly prehensile tail, becomes darker along the first half of the tail, then changes to white. Found in many habitats, including parks and gardens, the main food of this group-dwelling mammal is leaves, especially eucalypts and acacia. In the north of its range, it nests in a tree-hole; in the south, in a squirrel-like drey.

Common cuscus

Phalanger orientalis

Length 38–48 cm (15–19 in)	
Tail 28–43 cm (11–17 in)	
Weight 1.5–3.5 kg (3 1/4–7 3/4 lb)	

Location New Guinea, Solomon Islands

Social unit Individual

Status Lower risk

Also known as the northern common cuscus, the colouring of this species varies across its range of many islands, from almost white to black. There is generally, however, a stripe along the back, and part of the tail is always hairless and white. Resembling a combination of sloth and monkey, it has deliberate but agile climbing habits as it forages for leaves, fruits, and other plant material. Docile by nature, it is a common household pet.

Brushy-tailed ringtail

Hemibelideus lemuroides

Length 31–40 cm (12–15 1/2 in)	
Tail 32–37 cm (12 1/2–14 1/2 in)	
Weight 800–1,000 g (29–36 oz)	

Location N.E. Australia

Social unit Variable

Status Lower risk

Lemur-like in the face and body proportions (see pp.120–1), the long, charcoal-coloured tail of the brushy-tailed ringtail has a hairless tip. Strongly prehensile, the tail is used to grip boughs and to steer as this nocturnal possum makes "free-fall" leaps of some 3m (9ft) through the rainforest canopy. The main diet is leaves, with also some flowers and fruits. Social life is flexible, with male–female pairs, mother–young duos, or larger groups. The single offspring leaves the pouch after 6–7 weeks.

Leadbeater's possum

Gymnobelideus leadbeateri

Length 15–17 cm (6–7 in)	
Tail 14.5–20 cm (6–8 in)	
Weight 100–175 g (3 5/8–6 oz)	

Location S.E. Australia (Central Highlands, Victoria)

Social unit Group

Status Endangered

Notable for its rediscovery in 1961, after 52 years of presumed extinction, Leadbeater's possum has a grey back with a dark central stripe from head to tail, and off-white underside. Speedy and elusive in trees, it feeds at night on small invertebrates and tree juices such as gum, sap, and nectar. Colonies of up to 8 comprise a senior breeding pair and their offspring, mainly males. Females defend the group's territory.

Striped possum

Dactylopsila trivirgata

Length 24–28 cm (9 1/2–11 in)	
Tail 31–39 cm (12–15 1/2 in)	
Weight 250–525 g (9–19 oz)	

Location N.E. Australia, New Guinea

Social unit Variable

Status Lower risk

Skunk-like black and white stripes and a bushy, black, white-tipped tail are the distinguishing features of the striped possum. Like the skunk (see panel, p.200), it can emit a foul, penetrating odour from glands in the genital region. At night, it forages alone through tree branches, probing for wood-boring grubs, ants, and termites with the extra-long claw on its front fourth toe. It also eats fruits, birds, and small mammals.

Greater glider

Petauroides volans

Length	35–48 cm (14–19 in)
Tail	45–60 cm (18–23 1/2 in)
Weight	0.9–1.5 kg (2–3 1/2 lb)
Location E. Australia	Social unit Pair
	Status Lower risk

Groups of scratch marks on tree trunks betray regular landing sites of the greater glider, the largest gliding marsupial. Like other tree-dwelling marsupials, it has sharp claws to grip bark, and 2 toes on the front foot oppose the other 3, giving a pincer-like grip. This nocturnal forager feeds in eucalypt woodland (not rainforest), preferring the tender young leaves of just a few eucalypt species. It forms female–male pairs, which share the same tree-hollow den for most of the year. The single offspring stays in the pouch for 5 months, then remains in the den or rides on the mother's back for another 1–2 months. By 10 months, young males are driven away by the father.

furred gliding membrane between elbow and rear foot

underparts off-white in both colour phases

non-prehensile tail used for steering during glides

FACING THE FRONT

The greater glider's large eyes and huge ears face the front, so it can judge distances accurately by stereoscopic vision and stereophonic sound. This enables the glider to parachute horizontal distances of more than 100m (330ft) at night and yet land precisely on a tree trunk.

COLOUR PHASES
The greater glider occurs in 2 colour forms or phases, often in the same area. One is charcoal-grey to black, tinged with brown (as seen here); the other is very pale grey or mottled cream.

Feather-tailed possum

Distoechurus pennatus

Length	10.5–13.5 cm (4 1/2–5 1/2 in)
Tail	12.5–15.5 cm (5–6 1/2 in)
Weight	40–60 g (1 7/16–2 1/8 oz)
Location New Guinea	Social unit Group
	Status Lower risk

The feather-tailed possum is identified by its white face with 4 black stripes and its feather- or quill-like tail with a prehensile tip. It moves through the tree branches by darting leaps, gripping with the sharp claws on all toes, and consumes insects (especially cicadas), as well as flowers, nectar, and fruits. These possums appear to live in groups of 2–3, and the female carries 1–2 young in her pouch, then riding on her back near the nest. Details of the social and breeding habits are, however, lacking.

black stripes over eye and behind ear

sharp, curved claws

Little pygmy-possum

Cercartetus lepidus

Length	5–6.5 cm (2–2 3/4 in)
Tail	6–7.5 cm (2 1/2–3 in)
Weight	5–9 g (3/16–11/32 oz)
Location S.E. Australia, Tasmania	Social unit Individual
	Status Lower risk†

The smallest possum species, with a thumb-sized body, this is the only pygmy-possum with grey fur on the underside; the upperparts are fawn or brown. It has a short, blunt face and large, erect ears. The prehensile tail can support the animal's entire weight, and expands at its base to store excess food as fat. Nocturnal and usually solitary, this marsupial feeds in low bushes and shrubs, or on the ground, on a variety of small animals from insects to lizards.

Musky rat-kangaroo

Hypsiprymnodon moschatus

Length	16–28 cm (6 1/2–11 in)
Tail	12–17 cm (4 3/4–7 in)
Weight	375–675 g (13–24 oz)
Location N.E. Australia	Social unit Individual
	Status Lower risk

Neither rat nor kangaroo, this is a potoroid marsupial (see right). Its preferred habitat is thick rainforest. It forages alone by day, mainly on the ground for fallen fruits such as figs, as well as palm nuts, seeds, and fungi. It also hoards food at scattered sites – unusual behaviour among marsupials – and bounds on all fours, gripping with the opposable big toe on each rear foot. Both sexes produce a musky odour, particularly during breeding.

Pygmy glider

Acrobates pygmaeus

Length	6.5–8 cm (2 3/4–3 in)
Tail	7–8 cm (2 3/4–3 in)
Weight	9–15 g (11/32–9/16 oz)
Location E. Australia	Social unit Group
	Status Lower risk

Also called the feather-tailed glider, this tiny, agile marsupial has a long tail with a row of stiff hairs on either side. The gliding membrane extends between the front and rear limbs and the "dual-purpose" toes have sharp claws to dig into bark. Expanded, pad-like tips on the toes grip smooth, shiny surfaces such as leaves – even glass windows. The tongue is long and brush tipped, to gather nectar, pollen, and small insects from flowers.

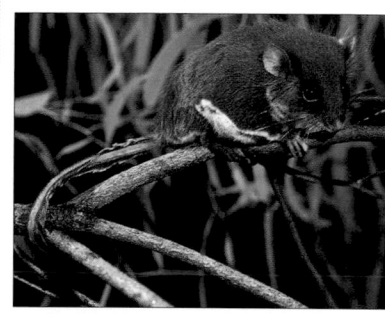

Long-footed potoroo

Potorous longipes

Length	38–42 cm (15–16 1/2 in)
Tail	31–33 cm (12–13 in)
Weight	1.5–2 kg (3 1/4–4 1/2 lb)
Location S.E. Australia	Social unit Individual
	Status Endangered

This potoroo ("rat-kangaroo") is an active, solitary, nocturnal fungus-eater. It bounds at speed, kangaroo-like, on its large back feet, and scrabbles for food with the shorter but strong front limbs. Some 30 fungus species make up four-fifths of its diet. It also eats insects and green plant matter. After an incubation period of 38 days, the single young suckles in the pouch for up to 5 months, and remains with its mother for a further 2–3 months.

Macropus rufus
Red kangaroo

Length	1–1.6 m (3¹⁄₄–5¹⁄₄ ft)
Tail	75–120 cm (2¹⁄₂–4 ft)
Weight	25–90 kg (55–200 lb)

Location Australia

Social unit Group

Status Common

The largest living marsupial, the red kangaroo is found over much of Australia, with the highest numbers living in open savanna woodland. Its population varies greatly from year to year: when rainfall is high, numbers may reach as many as 12 million, falling to 5 million in times of drought. Females will conceive only if there has been enough rain to produce plenty of green vegetation. In a prolonged drought the males no longer produce sperm. Red kangaroos use a highly developed sense of smell to detect water, and if it is scarce will migrate up to 200km (125 miles) from their usual grazing grounds to find it. They forage mainly at night, eating succulent grass shoots, herbs, and leaves. They live in groups of 2 to 10 animals, although 1,500 or so may gather at a water hole during a drought. The dominant male of the group mates with several females. The red kangaroo is regarded as a pest in its native Australia, and is hunted for its meat and skin. Apart from humans and occasionally the wedge-tailed eagle, its only predator is the dingo.

ALERT TO DANGER
Red kangaroos usually browse and graze head down, but remain alert, looking and listening for predators. Their sharp eyes can spot a dingo 350m (1,150ft) away, and their large ears are very sensitive.

DEFENSIVE MEASURES
The dominant male of a group of red kangaroos will fight off any challenge for supremacy from other males. If one of a group spies a potential predator, it will warn the rest by stamping its foot or thumping its tail on the ground. The group will then flee, taking refuge in water if it can.

PLAY FIGHTING
When playing or fighting among themselves, kangaroos may stand up and spar, but their normal defence is to deliver a powerful kick.

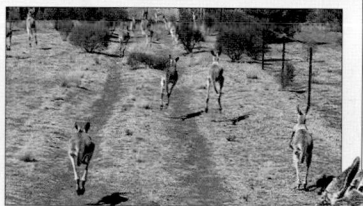

IN FULL FLIGHT
Fleeing from danger, the red kangaroo bounds on its hindlegs. It can reach 50kph (30mph) for short periods.

very long, strong tail

short forelegs

large hindlimbs

RUFOUS MALE
Males, which may be more than double the weight of females, are usually orange-red, while females are blue-grey – although coloration varies.

IN MOTHER'S POUCH
The infant kangaroo spends 190 days entirely in its mother's pouch, before making its first foray outside. It becomes independent when a year old.

Bettongia penicillata

Brush-tailed bettong

Length	30–38 cm (12–15 in)
Tail	29–36 cm (11 1/2 –14 in)
Weight	1–1.5 kg (2 1/4 – 3 1/4 lb)
Location S.W. Australia	Social unit Individual
	Status Lower risk

Like the potoroo family (see p.97), the brush-tailed bettong – also called the woylie – is a fungivore. It forages in woodland soil by night, scraping earth to find the fungi, which form 90 per cent of its diet. The remainder of its diet comprises roots, bulbs, tubers, and worms. This bettong has a tail as long as its head and body, with a crest of black fur along the upper side. By day it shelters in a large, domed nest of bark, leaves, and grass.

Aepyprymnus rufescens

Rufous bettong

Length	37–52 cm (14 1/2 – 20 1/2 in)
Tail	34–40 cm (13 1/2 –16 in)
Weight	2.5–3.5 kg (5 1/2 – 7 3/4 lb)
Location E. Australia	Social unit Variable
	Status Lower risk

white underparts

Grizzled red-brown fur over the body gives this rat-kangaroo its name of rufous bettong, although some fur is white. It uses its grasping tail to gather grasses and stems for its tall, cone-shaped nest, which it builds on the woodland floor, often against a log or tree; one individual may have 5 such refuges. The diet is fungi, grass, roots, leaves, flowers, seeds, and small invertebrates. The single young leaves the pouch after 16 weeks.

Lagorchestes conspicillatus

Spectacled hare wallaby

Length	40–48 cm (16–19 in)
Tail	37–50 cm (14 1/2 –20 in)
Weight	1.5–4.5 kg (3 1/4 –10 lb)
Location N. Australia	Social unit Individual
	Status Lower risk

A conspicuous orange eye patch in the generally shaggy, white-grizzled, grey-brown coat gives this small member of

the kangaroo family the name of spectacled hare wallaby. It grazes by night, usually alone, on grasses and herbs, and hides in a burrow or thicket by day. After 29–31 days' gestation, the single young stays in the pouch for 5 months, and is weaned by 7 months.

Setonix brachyurus

Quokka

Length	40–54 cm (16–21 1/2 in)
Tail	25–35 cm (10–14 in)
Weight	2.5–5 kg (5 1/2 –11 lb)
Location S.W. Australia (Rottnest and Bald Islands)	Social unit Group
	Status Vulnerable

Very rare on mainland Australia, the quokka survives on 2 islands off the southwest coast, mainly because introduced predators such as foxes are absent. This small wallaby frequents thick forest, open woodland, low scrub, and swamp edges or river banks where available. After resting by day in

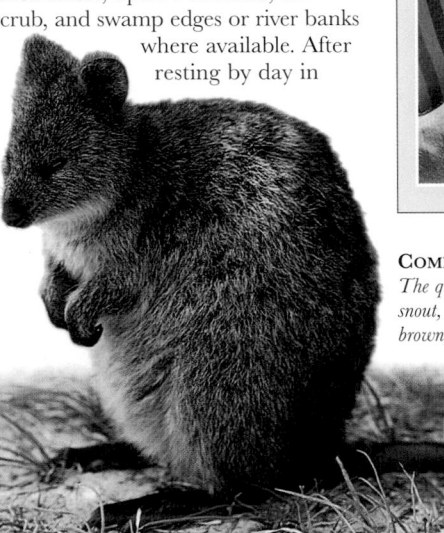

dense vegetation, it feeds at night on leaves, grasses, and fruits. Small family groups associate as larger gatherings, which maintain a group territory. Gestation is 27 days; the single young leaves the pouch after 6 months.

CONSERVATION

Quokkas roam freely around their stronghold, Rottnest Island, and are a popular tourist attraction. Well-meaning visitors may feed these appealing wallabies, but modern foods upset their sensitive digestive system.

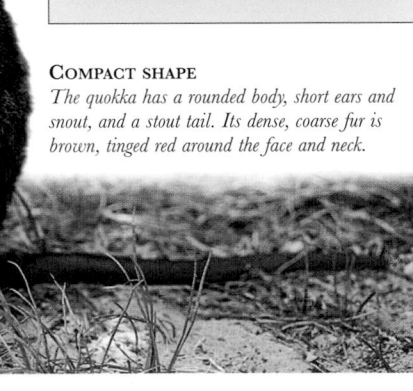

COMPACT SHAPE
The quokka has a rounded body, short ears and snout, and a stout tail. Its dense, coarse fur is brown, tinged red around the face and neck.

Thylogale stigmatica

Red-legged pademelon

Length	38–58 cm (15 –23 in)
Tail	30–47 cm (12 –18 1/2 in)
Weight	2.5–7 kg (5 1/2 –15 lb)
Location N. and E. Australia, New Guinea	Social unit Individual
	Status Lower risk

This slender-headed, stout-bodied, thick-tailed pademelon (a type of small wallaby) tends to be brown-grey in rainforests but paler fawn in open woodland. Active day and night, it is usually solitary but may gather in groups at a fruiting tree to feed. Other foods include leaves and seeds.

Petrogale penicillata

Brush-tailed rock wallaby

Length	50–60 cm (20 –23 1/2 in)
Tail	50–70 cm (20 –28 in)
Weight	5 –11 kg (11 –24 lb)
Location S.E. Australia	Social unit Group
	Status Vulnerable

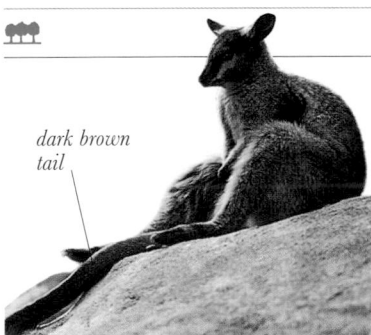

dark brown tail

The rock wallaby is specialized for leaping and scrambling over boulders, cliffs, ledges, and outcrops. It can make single bounds of 4 m (13 ft), and the soles of the rear feet are enlarged, padded, and roughened for excellent grip. By day this wallaby rests in a cool rock crevice or cave, occasionally sunbathing. At night it feeds on grasses, ferns, bush leaves, and fruits. Colonies may exceed 50 in number.

Petrogale concinna

Little rock wallaby

Length	29–35 cm (11 1/2 –14 in)
Tail	22–31 cm (9–12 in)
Weight	1–1.5 kg (2 1/4 –3 1/4 lb)
Location N. Australia	Social unit Individual
	Status Lower risk

Similar to the brush-tailed rock wallaby (see left) in form and habits, this species – also called the nabarlek – has short, silky, reddish brown fur with a dark shoulder stripe and brush-tipped tail. It grazes alone by night on grass and sedge, but its main food in the dry season is nardoo (a tough fern). To cope with this abrasive food, the wallaby's molar teeth slowly move forward in the jaw as they wear, and are then replaced by another set – a process of repeated tooth replacement unique among marsupials.

Dendrolagus dorianus
Doria's tree-kangaroo

Length 51–78 cm
(20–31 in)

Tail 44–66 cm
(17 1/2–26 in)

Weight 6.5–14.5 kg
(14–32 lb)

Location New Guinea | Social unit Individual

Status Vulnerable

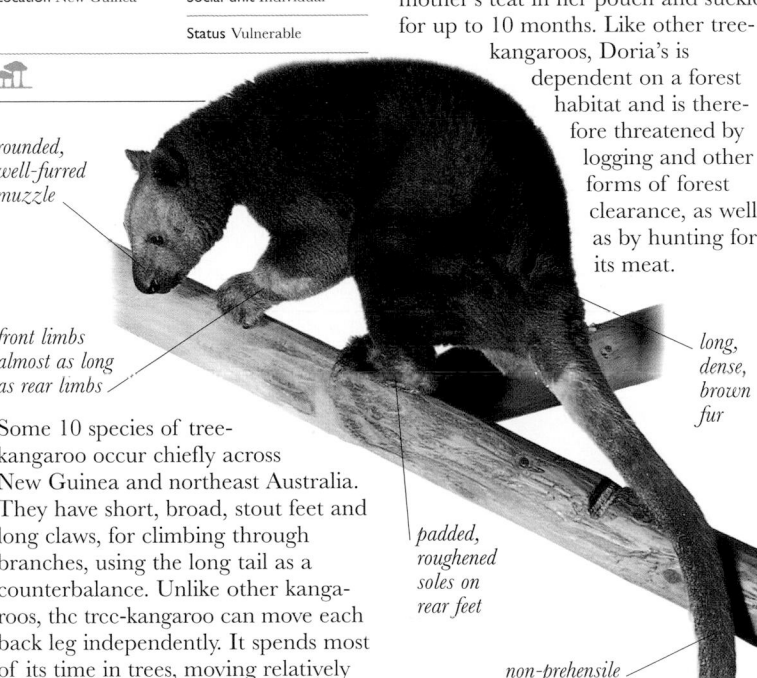

rounded,
well-furred
muzzle

front limbs
almost as long
as rear limbs

padded,
roughened
soles on
rear feet

long,
dense,
brown
fur

non-prehensile
tail for balance

Some 10 species of tree-kangaroo occur chiefly across New Guinea and northeast Australia. They have short, broad, stout feet and long claws, for climbing through branches, using the long tail as a counterbalance. Unlike other kanga-roos, the tree-kangaroo can move each back leg independently. It spends most of its time in trees, moving relatively slowly but with great precision,

although it can walk and bound quickly on the ground. Doria's tree-kangaroo is one of the largest and, like most other species in its genus, is mostly solitary and nocturnal, with a diet of various leaves, buds, flowers, and fruits. It has black ears, a whorled fur pattern in the middle of the back, and a pale brown or cream tail. After a gestation period of about 30 days, the single young attaches to the mother's teat in her pouch and suckles for up to 10 months. Like other tree-kangaroos, Doria's is dependent on a forest habitat and is there-fore threatened by logging and other forms of forest clearance, as well as by hunting for its meat.

Macropus fulginosus
Western grey kangaroo

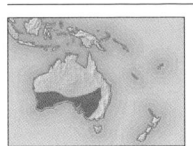

Length 0.9–1.4 m
(3–4 1/2 ft)

Tail 75–100 cm
(30–39 in)

Weight 15–54 kg
(33–120 lb)

Location S. Australia | Social unit Group

Status Lower risk

One of the largest, most abundant kangaroos, the western grey kangaroo has thick, coarse fur that varies from pale grey-brown to chocolate-brown, with a paler chest and belly. It lopes like a rabbit when moving slowly, using all four limbs, with the tail as a brace, but bounds on its back legs at high speed. Males are up to twice the size of females and can cover 10m (33ft) in one leap. This species grazes primarily at night, mainly on grasses, but it also browses on leafy shrubs and low trees. It lives in stable groups of up to 15. The dominant male is usually the only male to breed. The incubation period is 30–31 days.

"BOXING" KANGAROOS

Male kangaroos battle for females during the mating season, and also for food or resting sites if these are limited. The antagonists lock arms and attempt to push each other over. They may also lean back on their tails and kick with the rear feet. As in most such contests, serious injury is rare.

JOEY ON BOARD
The young or joey remains attached to the teat in the pouch for 130–150 days. From about 250 days it begins to leave the pouch for short periods, but quickly returns if danger threatens.

powerful tail

MAMMALS

Macropus parma
Parma wallaby

Length 45–53 cm
(18–21 in)

Tail 41–54 cm
(16–21 1/2 in)

Weight 3.5–6 kg
(7 3/4–13 lb)

Location E. Australia | Social unit Individual

Status Lower risk

The parma wallaby's distinguishing marks are a black stripe in the centre of the red- or grey-brown fur on the back, from the neck to the mid-back, and a white stripe on the side of the muzzle and cheek. Solitary, shy, and well camouflaged in dense vegetation, for a century it was believed extinct on the Australian mainland, but was rediscovered in 1967. This night-time grazer and browser feeds on a very wide range of plants.

Macropus robustus
Wallaroo

Length 0.8–1.4 m
(2 1/2–4 1/2 ft)

Tail 60–90 cm
(23 1/2–35 in)

Weight 15–47 kg
(33–105 lb)

Location Australia | Social unit Individual

Status Lower risk

Also called the euro or hill kangaroo, the wallaroo is found in a range of habitats, but usually in and around rocky outcrops, cliffs, and boulder

piles. It shelters there by day, thereby surviving very hot, dry conditions, and by late afternoon moves out to forage on grasses, sedges, and other leafy foods. The wallaroo resembles other brown wallabies but may adopt a distinctive pose with shoulders back, elbows together, and wrists raised.

Wallabia bicolor
Swamp wallaby

Length 66–85 cm
(26–34 in)

Tail 65–86 cm
(26–34 in)

Weight 10.5–20.5 kg
(23–45 lb)

Location E. Australia | Social unit Individual

Status Lower risk

Unlike other wallabies, this species moves with its head low and tail held out straight behind. It has coarse, brown-black fur and a much darker

orange-tinged under-parts

face, snout, front and back feet, and tail. Also called the stinker or black wallaby, it feeds at night on a wide diet of plant material, including toxic types such as hemlock.

Tarsipes rostratus
Honey possum

Length 6.5–9 cm
(2 1/2–3 1/2 in)

Tail 7–10.5 cm
(2 3/4–4 1/4 in)

Weight 7–16 g
(1/4–9/16 oz)

Location S.W. Australia | Social unit Group

Status Lower risk

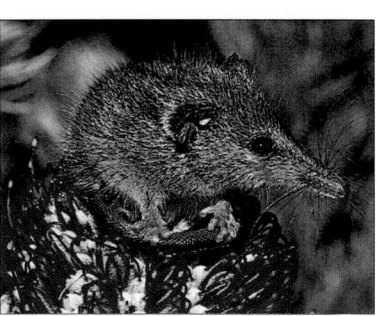

One of the smallest possums, this tiny, agile, nocturnal species has a long, pointed, well-whiskered snout and a long, prehensile tail. It lives in small groups and its toes have soft, pad-like tips and small, sharp claws, to grip both bark and glossy leaves. The 2.5-cm (1-in), bristle-tipped tongue gathers pollen and nectar ("honey") from flowers, which are then scraped onto the much-reduced teeth (upper canines and lower incisors only).

INSECTIVORES

PHYLUM	Chordata
CLASS	Mammalia
ORDER	Insectivora
FAMILIES	6
SPECIES	365

THESE SMALL, MOSTLY nocturnal mammals have a diet that consists mainly of invertebrates, such as insects, spiders, and worms. The anatomy of insectivores differs according to species and lifestyle, but generally their eyes and ears are small, and their snouts well developed. Some species are terrestrial, such as hedgehogs; some, such as moles, are adapted for burrowing; others are semi-aquatic, such as the web-footed tenrec. Three of the families – hedgehogs and moonrats; moles, shrew-moles, and desmans; and shrews – are widespread in most parts of the world. The remaining families – solenodons, tenrecs and otter-shrews, and golden moles – are much more localized.

ANATOMY

Many species of insectivores are characterized by a long, slender, mobile snout, 5 clawed toes on each foot, and very primitive dentition. They have many different body shapes, the 3 main ones being elongated (otter-shrews and moonrats), cylindrical (moles), and squat (hedgehogs and tenrecs). Insectivores are considered to be among the most primitive mammals – they retain many features of their ancient ancestors, such as a small brain with few folds and testes that do not descend into a scrotum.

Insectivores usually move by placing the sole and heel of the foot on the ground at the same time – this is known as plantigrade locomotion (see p.89).

EUROPEAN MOLE

COAT TYPES
Some species, such as the streaked tenrec (below left), have defensive spines interspersed among their body hairs. In hedgehogs, this mechanism is even further developed. In contrast, the European mole (above left) has short, close-set fur that lies equally well in any direction. This helps it move freely either backwards or forwards in its underground tunnels.

STREAKED TENREC

SENSES

Although insectivores usually have small ears and minute eyes, they can locate prey easily using their acute sense of smell. Moles, for example, live underground, where vision and hearing are not as important as the ability to smell and feel. As a result, they have no external ear flaps, and their eyes are hidden in the fur. Instead they have a highly tactile snout, a strong capacity for detecting smells, and many sensory hairs. Hedgehogs, however, rely most on their sense of hearing. It is believed that some species of shrews are able to echolocate, which means that they can orientate themselves or hunt by bouncing sound off objects or other animals in their path.

UNUSUAL NOSE
The fleshy pink tentacles that surround the snout of the star-nosed mole enhance the mole's sense of touch. They also help with navigation and prey manipulation.

COMPETENT CLIMBER
Like most insectivores, this Eurasian shrew has small eyes and a well-developed snout. Despite poor vision, this shrew is an adept climber. It also has acute hearing, even though there is no external ear flap.

FEEDING

Despite the name "insectivore", the diet of these mammals is not entirely restricted to insects (and insectivores are not the only mammals that eat insects) – a wide range of plants and animals is consumed by different species. Water shrews, for example, feed on small fish, frogs, and crabs, while desert hedgehogs consume birds' eggs, small vertebrates, and scorpions. Many insectivores are adapted for dealing with a specialized diet. The highly mobile, jointed nose of a solenodon, and the slender body of a moonrat, for example, allow both species to forage for food in restricted spaces.

FEEDING TECHNIQUE
The west European hedgehog locates food, such as earthworms, using its long, flexible snout and by digging with its clawed front limbs. As its mouth closes, the prey is impaled on the 2 front lower incisors.

MAMMALS

Hylomys suillus

Lesser moonrat

Length 10–15 cm
(4–6 in)
Tail 1–3 cm
(3/8–1 1/4 in)
Weight 12–80 g
(7/16–2 7/8 oz)

Location S.E. Asia

Social unit Individual

Status Lower risk†

The lesser moonrat (lesser gymnure) resembles a large mouse but with a long, flexible snout and a distinctively short, hairless tail. Widespread in lowland and mountain forests, it can climb in bushes, but feeds mainly on the ground by day or night on insects, worms, and other small animals, as well as seasonal fruits. Its soft, dense fur is brown above with paler underparts. The lesser moonrat shelters in nests of dead leaves under rocks or logs. Its main self-defence is to run away from danger.

Erinaceus europaeus

West European hedgehog

Length 22–27 cm
(9–10 1/2 in)
Tail None
Weight 0.9–1 kg
(2–2 1/4 lb)

Location Europe

Social unit Individual

Status Common

The densely spined West European hedgehog roams in urban parks and gardens as well as in hedgerows, fields, and woods at night, nosing – with pig-like snuffling (hence "-hog") – for small animals such as worms, insects, and spiders. It also takes birds' eggs and carrion. Its day shelter is a nest of grass and leaves under a bush, log, or outbuilding or in an old burrow. During hibernation it may wake on mild nights to feed. Mating takes place from May to October and gestation takes 31–35 days. The spines of the 4–5 young appear within hours of birth.

SELF-ANOINTING

The purpose of "self-anointing", when a hedgehog twists around to lick and smear its spines and skin with its own frothy saliva, is not clear. It may be a form of scent-marking that helps these normally solitary animals to recognize neighbours on their occasional nightly encounters.

AGILE DEFENCE
The hedgehog runs and climbs with surprising agility. In self-defence it tucks its non-spiny head and legs onto its belly and rolls into a prickly ball.

Paraechinus micropus

Indian hedgehog

Length 14–23 cm
(5 1/2–9 in)
Tail 1–4 cm
(3/8–1 1/2 in)
Weight 300–450 g
(11–16 oz)

Location S. Asia

Social unit Individual

Status Lower risk†

Similar to the long-eared desert hedgehog (see right) in appearance and habits, although slightly spinier, this species has a bare area of skin on the head. It is adapted to desert, dry scrub and arid grassland, remaining inactive at times of harsh conditions (usually drought). Natural shelters such as rock crevices serve for nests, but the Indian hedgehog may dig a short burrow for this purpose. It hunts insects, scorpions, and other small creatures, and also eats birds' eggs and scavenges for carrion. This hedgehog may cache (store) food, carrying it back to the nest for later consumption. Coloration is extremely variable with very dark (melanic) and almost white (albino) hues relatively common in the wild, and also brown and yellow banding patterns. The litter size is small, usually 1–2.

Echinosorex gymnura

Moonrat

Length 26–46 cm
(10–18 in)
Tail 16–30 cm
(6 1/2–12 in)
Weight 0.5–2 kg
(1–4 1/2 lb)

Location S.E. Asia

Social unit Individual

Status Lower risk†

Like other moonrats (gymnures), this species makes a territory-marking scent likened to rotting onions. It resembles a combination of hedgehog and small pig with harsh, rough, spiky outer fur, streaked with black and grey-white, and a long, scaly, almost hairless tail. Solitary, the moonrat rests in a burrow or crevice by day; at night it forages for small creatures such as insects, and also swims after fish and other aquatic prey.

Hemiechinus auritus

Long-eared desert hedgehog

Length 15–27 cm
(6–10 1/2 in)
Tail 1–5 cm
(3/8–2 in)
Weight 250–275 g
(9/16–5/8 oz)

Location W., C., and E. Asia, N. Africa

Social unit Individual

Status Locally common

Similar in appearance to the larger West European hedgehog, the long-eared desert hedgehog has coarse fur on the face, limbs, and belly, but spines elsewhere. Banding on the spines varies from black through brown to yellow and white. It uses natural daytime resting places under rocks and shrubs. It eats various small animals and can become inactive when food or water are scarce.

Podogymnura truei

Mindanao moonrat

Length 13–15 cm
(5–6 in)
Tail 4–7 cm
(1 1/2–2 3/4 in)
Weight 150–175 g
(5–6 oz)

Location S.E. Asia (Mindanao)

Social unit Individual

Status Endangered

Inhabiting only one island in the Philippines, this poorly known and endangered species probably forages on the forest floor by day or night, especially around marshes and streams, for varied small animal prey. It lives alone, sheltering in a simple nest of leaves under a rock or log, or in an abandoned burrow. The long, soft fur is mainly grey-red shading to grey-white on the underside. The distinctively pointed lower snout extends beyond the bottom lip, and the tail is short and coarse furred.

Solenodon paradoxus

Hispaniolan solenodon

Length 28–32 cm
(11–12 1/2 in)
Tail 17–26 cm
(6 1/2–10 in)
Weight 1 kg
(2 1/4 lb)

Location Caribbean (Hispaniola)

Social unit Individual

Status Endangered

The two species of solenodon – Cuban and Hispaniolan – are large, long-tailed, shrew-like, nocturnal insectivores. Both are under threat. The snout of the Hispaniolan solenodon is long and mobile. Its fur varies from black to red-brown, and its feet, tail, and upper ears are almost hairless. It is fast and agile. It noses and scrabbles on the forest floor with its sharp claws for insects, worms, small lizards, fruits, and other plant matter. Its poisonous bite is used for defence and to stun prey.

Potamogale velox
Giant otter-shrew

Length	29–35 cm (11 1/2 –14 in)
Tail	24–90 cm (9 1/2 –35 in)
Weight	350 g (13 oz)

Location W. and C. Africa	Social unit Individual
	Status Endangered

Easily mistaken at a glance for a small otter (which is a mustelid rather than an insectivore), the giant otter-shrew has a rounded muzzle, a long, flexible body, and a long, muscular tail flattened from side to side, which it uses to propel itself through water. This species inhabits a variety of freshwater habitats, from still pools to mountain torrents at altitudes of 1,800m (6,000ft). The eyes and ears are small and high-set, for swimming low in the water, yet, unlike an otter, the toes lack webs. Mainly nocturnal and solitary outside the breeding season, the giant otter-shrew hunts primarily at night for fish, frogs, shellfish, and other freshwater animals. Its bankside burrow, where it shelters by day, has an underwater entrance.

Limnogale mergulus
Web-footed tenrec

Length	12–17 cm (4 3/4 –6 1/2 in)
Tail	12–16 cm (4 3/4 –6 1/2 in)
Weight	40–60 g (1 7/16 –2 1/8 oz)

Location E. Madagascar	Social unit Individual
	Status Endangered

Once believed extinct, the only aquatic species of tenrec has been relocated in the wild but information is very scarce. It has an otter-like form with a broad, round-muzzled head, long whiskers, small high-set eyes and ears, and webbed rear feet with sharp claws, used to grip slippery rocks and prey. The short, dense, water-repellent fur is reddish brown. The web-footed tenrec noses among stones and weeds for water insects, crabs, and crayfish, propelled mainly by means of its long, scantily haired tail.

ears hardly visible in fur

Tenrec ecaudatus
Common tenrec

Length	26–39 cm (10–15 1/2 in)
Tail	1–1.5 cm (3/8 –1/2 in)
Weight	1.5–2.5 kg (3 1/4 –5 1/2 lb)

Location Madagascar	Social unit Individual
	Status Common

The 25 species of tenrecs are mainly from central Africa and Madagascar. Most resemble a combination of shrew and hedgehog. The nocturnal, common tenrec is the largest land-dwelling species. It has coarse, grey to reddish grey fur and sharp spines. Using its long, mobile snout, it grubs among leaves for worms and other small creatures. It also scavenges and hunts frogs and mice. In defence, a tenrec squeals, erects the spiny hairs on its neck into a crest, jumps and bucks, and readily bites. It shelters by day in a nest of grass and leaves under a log, rock, or bush. After a gestation of 50–60 days, a litter of 10–12 is born. When young, they are striped black and white.

Micropotamogale lamottei
Nimba otter-shrew

Length	12–20 cm (4 3/4 –8 in)
Tail	10–15 cm (4–6 in)
Weight	125 g (4 oz)

Location W. Africa (Mount Nimba area)	Social unit Individual
	Status Endangered

This otter-shrew is largely restricted to upland forest streams in an area of about 1,500 square km (580 square miles) around Mount Nimba in West Africa. It has a fleshy nose, rounded head, compact body, and long tail. Long, grey or dark brown fur usually hides the eyes and most of the ears. It is mostly nocturnal, catching small fish, crabs, water insects, and similar prey on short dives or along river banks, and eating them on land. It digs a short nesting burrow in soft soil.

Setifer setosus
Greater hedgehog-tenrec

Length	15–22 cm (6–9 in)
Tail	1.5 cm (1/2 in)
Weight	175–275 g (6–10 oz)

Location Madagascar	Social unit Individual
	Status Locally common

This hedgehog-like tenrec has short, pointed, white-tipped spines on its long body and coarse hair, varying from grey to black, on the head and legs. Unlike true hedgehogs, it is active by day but, like them, it rolls into a prickly ball when threatened. It climbs well and eats a wide variety of worms, amphibians, reptiles, insects, carrion, fruits, and berries. It becomes torpid for several weeks in adverse conditions.

Hemicentetes semispinosus
Streaked tenrec

Length	16–19 m (6 1/2 –7 1/2 in)
Tail	None
Weight	80–275 g (2 7/8 –10 oz)

Location Madagascar	Social unit Group
	Status Locally common

This species is distinguished from other tenrecs by its two-tone colour of background black with variable stripes of white, yellow, or brown. The fur is coarse with scattered spines and a patch of spiky hairs on the crown, which erect as a crest. The main foods are worms and grubs. The streaked tenrec lives in groups of 15 or more, and all help to protect each female's litter of 2–4 young.

Chrysochloris asiatica
Cape golden mole

Length	9–14 cm (3 1/2 –5 1/2 in)
Tail	None
Weight	Not recorded

Location Southern Africa	Social unit Individual
	Status Locally common

African golden moles (chrysochlorids) form a separate family to true moles (talpids). The Cape golden mole has the soft, dense fur typical of moles, which may appear olive, brown, or grey depending on the direction of the light. Its snout has a hairless nose pad, its eyes and ears are tiny, and each front leg has two large digging claws – all adaptations for a tunnelling lifestyle. It is solitary and eats worms, grubs and other soil creatures found when burrowing or that have fallen into its more permanent tunnels.

dense, shiny fur

Grant's golden mole

Eremitalpa granti

Length	7–8 cm
	(2³/₄ – 3¹/₄ in)
Tail	None
Weight	15–30 g
	(⁹/₁₆ – 1¹/₁₆ lb)

Location Southern Africa	Social unit Individual
	Status Vulnerable

Long, soft, silky fur covers almost the entire body of Grant's golden mole, and varies from steel-grey to buff or white. This mole has tiny, almost

grey or buff upperparts

invisible eyes and ears, a hard, hairless nose pad, and three long, broad claws on each foot. It pushes through loose sand as though "swimming", making more permanent tunnels deeper down in sand or near the surface in harder, more compacted soil. The main components of its diet are various small desert animals, from ants, termites, and beetles to lizards and snakes. Grant's golden mole is solitary, and probably active for short periods through the day and night. It hardly, if ever, comes to the surface except to locate a mate. The specialized desert habitat of this species is under increasing threat from mining and other human activities.

Eurasian shrew

Sorex araneus

Length	5–8 cm
	(2–3¹/₄ in)
Tail	2.5–4.5 cm
	(1–1³/₄ in)
Weight	5–14 g
	(³/₁₆ – ¹/₂ oz)

Location Europe to N. Asia	Social unit Individual
	Status Common

One of the smallest mammals, the Eurasian shrew is adaptable, aggressive, and voracious. It must eat 80–90 per cent of its body weight every 24 hours, and it hunts in up to 10 bursts of activity, according to season and conditions. Food includes insects, worms, and carrion. Adults are solitary except for a brief courtship in spring or early

autumn. After a gestation period of 24–25 days, 6–7 young are born in a special breeding nest made from woven grass and dry leaves. Larger than the usual resting nest, this is similarly sited under a log, root, rock, or in an old burrow. The Eurasian shrew has a pointed, flexible snout, and short legs. Its fur is dark brown to black on the back, paler brown on the flanks, and grey-white on the under-side. It is territorial, making ultrasonic squeaks, especially when a female gathers her offspring. If cornered, this shrew readily bites.

long whiskers

Giant Mexican shrew

Megasorex gigas

Length	8–9 cm
	(3¹/₄ – 3¹/₂ in)
Tail	4–5 cm
	(1¹/₂ – 2 in)
Weight	10–12 g
	(³/₈ – ⁷/₁₆ oz)

Location S.W. Mexico	Social unit Individual
	Status Unconfirmed

The sole species in the genus *Megasorex*, this large, compact-bodied, but short-tailed shrew prods with its prominent, pointed snout among leaves and loose soil for worms, grubs, millipedes, spiders, and other small prey. The upperparts are dark brown or greyish brown, becoming paler on the underside. This shrew prefers areas of damp soil and moist undergrowth in grassland and forest, ranging from the lowlands to altitudes of 1,700m (5,600ft). Only some 20 giant Mexican shrews have been studied and their nesting and breeding habits are not yet known.

North American short-tailed shrew

Blarina brevicauda

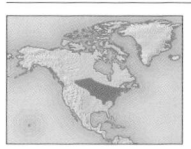

Length	12–14 cm
	(4³/₄ – 5¹/₂ in)
Tail	3 cm
	(1¹/₄ in)
Weight	20 g
	(¹¹/₁₆ oz)

Location S. Canada to N. and E. USA	Social unit Variable
	Status Common

Like most shrews, this large, robust species has poor sight but excellent sense of smell, and its poisonous bite (due to toxic saliva) helps disable prey. It uses mainly scent and touch to hunt its

main prey of soil-living animals and – unusually for the shrew group – smaller mammals such as voles and mice, and even some plant matter. This shrew rests and feeds largely underground in runways and old mole or vole tunnels, usually 10–50cm (4–20in) deep, and stores items in cold weather. The eyes and ears are tiny, and the snout stouter and less pointed than in other shrews.

ears concealed by fur

greyish black fur

Eurasian water shrew

Neomys fodiens

Length	6.5–9.5 cm
	(2¹/₂ – 3³/₄ in)
Tail	4.5–8 cm
	(1³/₄ – 3¹/₄ in)
Weight	8–25 g
	(³/₁₆ – ⁷/₈ oz)

Location Europe to N. Asia	Social unit Individual
	Status Common

This shrew hunts aquatic insects, small fish, and frogs. It also feeds on land, on worms, beetles, and grubs, and so can survive in damp woods. The small eyes and ears, and the long, pointed snout are typically shrew-like. Solitary but less aggressive than other shrews, the Eurasian water shrew establishes a series of runways and burrows and has a nest of dry grass and old leaves. Here, after 14–21 days' gestation, the female suckles her litter of 4–7 young for approximately 6 weeks.

UNDERWATER ACTION

The water shrew swims by kicking hard with its back feet, which have fringing hairs to increase thrust. A row of hairs forms a keel along the tail to help steering. Trapped air forms silvery streaks in the shrew's water-repellent fur.

COLORATION

This water shrew has distinctive demarcation between its shiny black back and sides and its grey-white underside.

Nectogale elegans
Tibetan water shrew

Length 9–13 cm
(3½–5 in)

Tail 8–11 cm
(3–4¼ in)

Weight Not recorded

Location S. Asia

Social unit Individual

Status Unconfirmed

A small, wary, secretive inhabitant of cold, fast mountain streams in the Himalayas and nearby mountains, this tubby, semi-aquatic shrew is slate-grey on the upperparts and silvery white beneath. Its snout is blunt, its eyes and ears tiny, and its black tail is fringed by rows of hairs along each side. It probably eats water insects, fish fry, and other small prey, carrying them to a bank or a mid-stream rock for consumption. Little is known of its nesting or breeding habits.

Crocidura leucodon
Bicoloured white-toothed shrew

Length 4–18 cm
(1½–7 in)

Tail 4–11 cm
(1½–4¼ in)

Weight 6–13 g
(7/32–7/16 oz)

Location Europe to W. Asia

Social unit Individual

Status Lower risk†

Sharp demarcation between its grey-white upperparts and whitish yellow underside give this

Chimarrogale hantu
Malayan water shrew

Length 8–12 cm
(3–4¾ in)

Tail 6–10 cm
(2½–4 in)

Weight 30 g
(1 1/16 oz)

Location S.E. Asia
(Selangor)

Social unit Individual

Status Critically endangered

Known only from the Ulu Langat Forest Reserve in Malaysia, this relatively large but exceptionally rare shrew has a long tail relative to body size, streamlined head and body, tiny eyes and ears, and feet fringed with stiff hairs to aid kick-propulsion in water. Like many water shrews, it spreads skin oils through its fur by regular grooming, to maintain its coat's water-repellent properties. Threats include fish traps, pollution, and deforestation.

shrew its common name. It also has long, thick whiskers (vibrissae) on its sharp-pointed snout, and a bicoloured tail that is less than half the length of the head and body. This adaptable forager in grassland, scrub, wood edge, parks, and gardens eats worms, grubs and other small invertebrates, hunting mainly by night but also for short periods by daylight. Like similar shrew species, it builds a nest of dry grass in a hole or thick undergrowth. The male produces strong scents from glands on his flanks during the breeding season, generally March to October. Gestation is 31 days, average litter size is 4, and weaning occurs after 26 days.

Suncus etruscus
Pygmy white-toothed shrew

Length 4–5 cm
(1½–2 in)

Tail 2–3 cm
(¾–1¼ in)

Weight 2–3 g
(1/16–1/8 oz)

Location S. Europe, S. to S.E. Asia, Sri Lanka, N. to E. Africa, W. Africa

Social unit Individual

Status Unconfirmed

Also called the Etruscan or Savi's shrew, this species actively hunts for small prey such as insects, worms, snails, and spiders, then rests for a few hours, through day and night. It nests in a small hole or crevice and is solitary most of the year, forming pairs only in the breeding season. The gestation period is 27–28 days and litter size 2–5, with up to 6 litters per year.

The pygmy white-toothed shrew relies on its high-speed reactions and its ability to launch a sudden, ferocious, biting attack to over-come cold-blooded creatures much larger than itself. It also scavenges for recently deceased prey such as beetles and grasshoppers (above).

CLASSIC SHREW
This minute mammal has a pointed snout and large ears. Its fur is mainly greyish brown.

Diplomesodon pulchellum
Piebald shrew

Length 5–7 cm
(2–2¾ in)

Tail 2–3 cm
(¾–1¼ in)

Weight 7–13 g
(¼–7/16 oz)

Location C. Asia

Social unit Individual

Status Unconfirmed

This species derives its name from its coloration: grey upperparts with a distinct oval white patch in the middle of the back, and

white underparts, feet, and tail. It has a very pointed snout and long whiskers, even for a shrew. Active at night when the desert habitat is cooler and prey such as insects and small lizards become more energetic, it hunts mainly on the surface but may also dig in loose sand for grubs and worms. The average litter size is 5, with several litters in a good year.

Sylvisorex megalura
Forest musk shrew

Length 5–7 cm
(2–2¾ in)

Tail 8–9 cm
(3–3½ in)

Weight 5 g
(3/16 oz)

Location W., E., C., and southern Africa

Social unit Individual

Status Unconfirmed

This species occurs mainly in forests, at both low and high altitudes, and especially in moist conditions, but it is also found in drier grasslands. It is

typically shrew-like in most respects, although it has a relatively long, streamlined body, large and prominent ears, and a tail longer than its head and body. Its soft, velvety fur is brown on the upperparts and almost white on the belly. Both terrestrial and arboreal, the forest musk shrew hunts in soil, leaf litter, and among branches (using its tail to balance), for worms, beetles, grubs, caterpillars, and similar prey. Like the other 10 or so *Sylvisorex* species, it is probably active in bursts through the day and night, and must eat almost its own body weight in food every 24 hours. Details of its ecology and breeding are not known.

Scutisorex somereni
Armoured shrew

Length 10–15 cm
(4–6 in)

Tail 6.5–9.5 cm
(2¾–3¾ in)

Weight 70–125 g
(2½–4 oz)

Location C. to E. Africa

Social unit Individual

Status Unconfirmed

This large, woolly coated shrew, also known as the hero shrew, has a distinctively arched and tremendously strong back. This is

because its vertebrae have interlocking flanges or spines not only along their sides, as in other mammals, but also above and below. The armoured shrew is solitary, and is a skilful climber. Its diet consists of worms, insects, spiders, and carrion.

grey fur

Desmana moschata
Russian desman

Length 18–21 cm (7–8¹/₂ in)	
Tail 17–21 cm (7–8¹/₂ in)	
Weight 450 g (16 oz)	
Location E. Europe to C. Asia	Social unit Group
	Status Vulnerable

Desmans belong to the mole family, but resemble water shrews. The tail is as long as the head and body, and flattened from side to side for use as both a paddle and a rudder. The rear feet are fully webbed to the toe tips, the front ones partly so. Using its long, sensitive nose, it probes by night for prey in river-bed mud and stones. Unusually for an insectivore, the desman lives in groups and several may share a bank burrow. After 40–50 days' gestation, the 3–5 young are cared for by the female, and weaned by 4 weeks.

CONSERVATION

After centuries of being hunted, desmans are now subject to legal protection and reintroduction programmes. However, reliant on a diet of aquatic insects, frogs, crayfish, and similar water creatures, they are still at risk through the loss of their fresh-water habitat. This continues, especially through land clearance, water diversion, and agricultural and industrial contamination.

OUTER AND INNER COATS
The soft, dense underfur of the Russian desman is covered by long, coarse guard hairs, the coat being rich brown on the head and body, fading to ash-grey on the underside.

Galemys pyrenaicus
Pyrenean desman

Length 12.5 cm (5 in)	
Tail 14 cm (5¹/₂ in)	
Weight 40–50 g (1⁷/₁₆–1³/₄ oz)	
Location W. Europe	Social unit Individual/Pair
	Status Vulnerable

The Pyrenean desman generally resembles the only other species of desman (see left), although it is smaller, takes lesser prey, such as the aquatic larvae of mayflies and stoneflies, and is more suited to fast-flowing streams. Its long black snout is almost hairless, its thick fur brown above and silvery below, and its tail slightly flattened from side to side like a rudder. A male and a female may form a loose pair bond, with the male chasing away rivals and the female nesting in a bank burrow, but little more is known of these mammals.

Euroscaptor micrura
Asian mole

Length 7–10 cm (2³/₄–4 in)	
Tail 1–2 cm (³/₈–³/₄ in)	
Weight 50–70 g (1³/₄–2¹/₂ oz)	
Location S. to S.E. Asia	Social unit Individual
	Status Lower risk†

Similar to the European mole (see below left), this species is slightly shorter and slimmer, and coloured dark brown with a silvery sheen. Also known as the Himalayan mole, its massive front legs and sharp-clawed feet, dense fur, tiny eyes, small ears, and naked, sensitive nose are all adaptations for tunnelling. Like most other mole species, the Asian mole is probably active in sessions of 3–4 hours, patrolling tunnels and digging new ones, with rests in between. Its average litter size is 3–4.

Talpa europaea
European mole

Length 1–16 cm (4¹/₄–6¹/₂ in)	
Tail 2 cm (³/₄ in)	
Weight 65–125 g (2³/₈–5 oz)	
Location Europe to N. Asia	Social unit Individual
	Status Common

Virtually blind, this mole lives mainly underground in tunnels radiating from a central chamber, feeding on worms and other soil animals. When plentiful, the worms are bitten to paralyse them for future use, but, if uneaten, they recover and escape. The female digs a large nest chamber, and after 4 weeks' gestation, gives birth to 3–4 young.

TUNNELLING TOOLS

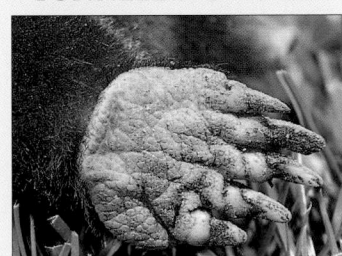

This mole's large front legs have powerful shoulder muscles and broad, outward-facing paws (above) with a strong, spade-like claw on each toe. Anchoring itself with its back feet, it uses its front legs to scoop soil sideways and back, pushing it up as molehills.

REVERSIBLE FUR
Short, dense, black fur that can lie at any angle allows this mole to go forwards or backwards in its tunnels.

Condylura cristata
Star-nosed mole

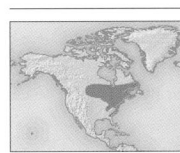

Length 18–19 cm (7–7¹/₂ in)	
Tail 6–8 cm (2¹/₂–3 in)	
Weight 45 g (1⁵/₈ oz)	
Location E. Canada, N.E. USA	Social unit Variable
	Status Locally common

This mole is an expert swimmer. It has a long, sparse-haired, scaly tail, which enlarges in winter with fatty food reserves. In lifestyle and habits it resembles other moles, yet it is less solitary and tolerates meeting others of its kind. Its tunnels are about 4cm (1¹/₂ in) in diameter and 5–60cm (2–24in) in depth.

VARIED DIET
The star-nosed mole eats leeches, snails, small fish, and other aquatic prey, as well as soil animals.

STAR-SHAPED NOSE

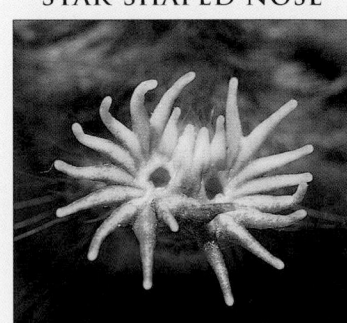

An unmistakable snout, with 22 pale, fleshy rays (tentacles) around the nostrils, allows the star-nosed mole to sniff and feel prey in water. It also forages for food among reeds, mosses, and other vegetation. As it hunts, the fleshy rays around its nose wriggle and flex in constant motion.

dense, near-black fur

BATS

PHYLUM	**Chordata**
CLASS	**Mammalia**
ORDER	**Chiroptera**
FAMILIES	**18**
SPECIES	**977**

CLASSIFICATION NOTE

Chiroptera has 2 suborders: Megachiroptera, comprising one family (Pteropodidae), and Microchiroptera (all other bats). Megachiropterans have a fox-like face that lacks features for echolocation (see below).

BATS ARE THE ONLY mammals that possess true, flapping wings and the ability to fly (as opposed to flying lemurs, for example, which glide). Bats' wing membrane (the patagium), an extension of the skin of the back and belly, provides a high degree of manoeuvrability in flight. Wingspans vary widely, ranging from over 1.5m (5ft) in the large flying fox to as little as 15cm (6in) in the hog-nosed bat. Over half the species echolocate to capture prey and to navigate at night. Chiroptera is a huge order that comprises nearly a quarter of all mammal species and is exceeded only by rodents in terms of species numbers. Bats are common in tropical and temperate habitats worldwide but are not found in environments that are too cold to support a source of food, such as the polar regions.

ANATOMY

Perhaps the most distinctive feature of bats is their wings, which are formed from a double layer of skin stretched between the side of the body and the 4 elongated fingers on each hand. Blood vessels and nerves run between these 2 layers. Extra support is required for the arms to be used as wings, and this is provided by features such as fused vertebrae, flattened ribs, and a strong collarbone. The sternum (breastbone) has a central ridge to which the large muscles used in the downward stroke of the wing are attached. A short, clawed thumb is present in most species at the point where the fingers join, and a cartilaginous spur (the calcar) on the inside of the ankle joint assists in spreading the tail membrane.

Labels: thumb, second finger, third finger, fourth finger, fifth finger, upper leg, foot, collarbone, fused vertebrae, forearm, upper arm, flattened ribs, knee, elbow, lower leg

SKELETAL FEATURES
A bat's arms, legs, and greatly elongated fingers provide the framework for the wings. Bats' legs have been rotated 180 degrees so that the knee and the foot bend in the opposite direction to the knee and back foot of other mammals.

Labels: large external ears, tragus, nose leaf, large nostrils, small eyes

EQUIPPED FOR ECHOLOCATION
Although bats' eyes are well developed, hearing and the sense of smell are more important than sight. Many microchiropterans, such as this neotropical fruit bat, also have a large nose-leaf, which assists echolocation. The function of the tragus (a lobe in the front of the ear) is uncertain: it may improve the accuracy of echolocation. Megachiropteran bats generally do not echolocate and have larger eyes (for detecting prey) and smaller ears.

ECHOLOCATION

All microchiropterans echolocate. When used in flight, this navigation system makes them formidable hunters. Sounds ("clicks") are produced in the larynx, emitted through the nose or mouth, and directed or focused by the nose-leaf (if present). Once the clicks have reflected off an object, the returning echo is picked up via the bat's sensitive ears. The time it takes to receive the echo reveals the size and location of anything in the bat's path.

seek | approach | seize

PHASES OF ECHOLOCATION AS A BAT HUNTS AN INSECT

USING SOUND TO FIND PREY
All of the insect-eating bats use echolocation to find airborne prey. When searching for food, such as mosquitoes and moths, the bat emits a series of clicks, represented by red bars on this diagram. As the bat approaches its prey, the time between clicks shortens. This helps the bat to pinpoint its target.

CATCHING PREY FROM THE AIR

① CLOSING IN ON THE PREY
Using echolocation, this fisherman bat has pinpointed a small surface-swimming fish.

② CAPTURE
In one swift movement, the bat catches the fish by raking the water with its long, sharp claws.

③ HOLDING ON
The bat quickly transfers the fish from the claws to the mouth, so that the fish cannot escape.

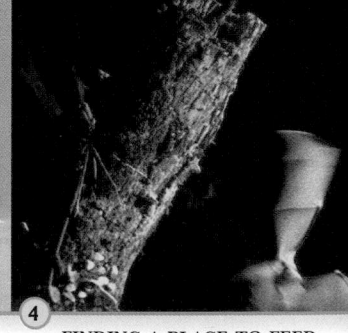

④ FINDING A PLACE TO FEED
In order to eat the fish, the bat must first locate a tree on which to land.

FORAGING AND DIET

Many bats eat insects: some species forage for them among shrubs and trees, while others skim the surface of the forest canopy to catch higher-flying insects. A single bat may eat hundreds of mosquitoes in one night (thereby lessening the incidence of malaria in other animals). Several bats eat fruits, and some use their long tongue to feed on pollen and nectar. Vampire bats use their sharp teeth to make a small incision in the skin of an animal while it sleeps and then drink the blood. Carnivorous bats prey on lizards and frogs; fish-eating bats use the hooked claws on their powerful feet to capture fish (see below).

FEEDING ON FRUITS
Fruit-eating bats use their senses of sight and smell to locate food and so lack facial ornaments used in echolocation. This Wahlberg's epauletted fruit bat is eating a mango: a group of these bats in a plantation can cause considerable damage. As fruit bats require a constant supply of ripe fruits, they are found mostly in tropical areas. Fruit bats often feed in groups and fly long distances in search of food.

FEEDING ON BLOOD
Vampire bats are well adapted to feed on blood. They have sharp incisors to cut into flesh, and they produce saliva that prevents the blood from clotting. This white-winged vampire bat commonly feeds on the blood of chickens.

CONSERVATION

Despite having very few natural predators, bat populations have declined in recent years. Globally, the main cause is loss of suitable habitats as, for example, woodlands are destroyed and old buildings are converted. To learn more about bats and to aid conservation efforts, scientists can fit bats, such as this fruit bat (below), with radio-transmitter collars and monitor their movements. Bats are now protected by law throughout Europe and in some parts of North America (for example), yet in many countries worldwide they still receive no protection at all.

FLYING MAMMAL
Bats, the only mammals that can fly, have structural adaptations that allow them to make up-and-down movements of their wings (in the same way as birds). The open wings of this New World leaf-nosed bat reveal the extent of the wing membranes.

ROOSTING

Bats often gather in great numbers at a single site, which may be a cave, the roof of an old building, or a hollow tree. All roosting sites must provide a resting place that offers protection from predators, the heat of the sun, the low temperatures of winter (hibernating roosts), and rain. Bamboo bats are small enough to roost in the hollow stems of a plant, while some species of leaf-nosed bats bite into leaf stems so that the leaf droops downwards, forming a tent around them. Why bats gather in such large numbers is not fully understood; however, at the end of hibernation bats living in colonies often weigh more than species that do not.

A PLACE TO REST
Bats, such as these fruit bats, commonly roost in caves during the day, emerging at dusk to feed. Some species use the same roost for many years and gather in groups of many thousands.

MAMMALS

LANDING
...e bat attaches itself to a tree (upside down), ...tinues to hold the fish firmly in its mouth.

6

DEVOURING THE PREY
The fish is eaten head first. The bat may use its wings to manipulate the food.

7

ROOM FOR PLENTY
The bat has highly elastic cheeks, which can be extended during feeding.

8

ALMOST GONE
With its meal nearly finished, the fisherman bat will soon begin the hunt for more food.

Rousettus egyptiacus
Egyptian fruit bat

	Length 14–16 cm (5 1/2 – 6 1/2 in)
	Tail 1.5–2 cm (1/2 – 3/4 in)
	Weight 80–100 g (2 7/8 – 3 5/8 oz)
Location W. Asia, N. Africa (Egypt), W., E., and southern Africa	Social unit Group
	Status Common

Widespread and adaptable, these fruit- and leaf-eating bats are sometimes so common that they reach pest status and damage farm crops. They are also the only fruit bat species to use the high-pitched clicks of echolocation (like smaller insectivorous bats). This means the Egyptian fruit bat, also known as the Egyptian rousette, can find its way around and roost in dark caves, rather than sleeping in trees like other fruit bats.

smoky grey underparts

FUR-COVERED FOREARMS
Egyptian fruit bats vary from dark brown to slate grey on the back, with lighter, smoky grey underparts. Unusually among bats, fur extends about halfway along each forearm.

collar of yellow or buff fur most obvious in males

ROOSTING IN CAVES
In the daytime Egyptian fruit bats congregate in the safety of dark caves. Conditions here are more humid and have a more constant temperature than outside. The bats compensate for the coolness by crowding together, to retain body warmth more effectively.

Epomops franqueti
Franquet's epauletted bat

	Length 11–15 cm (4 1/4 – 6 in)
	Tail None
	Weight 85–100 g (3 – 3 5/8 oz)
Location W. and C. Africa	Social unit Group
	Status Unconfirmed

At night, male Franquet's epauletted bats make monotonous high-pitched whistling calls to attract females for mating. The male is slightly heavier than the female and has shoulder patches of long pale hairs. This species breeds at any time, and twice yearly if its main diet of figs, guavas, bananas, other fruits, and soft, young leaves is abundant.

pale patch of fur at base of ear

Epomophorus wahlbergi
Wahlberg's epauletted fruit bat

	Length 12–15.5 cm (4 3/4 – 6 1/4 in)
	Tail None
	Weight 65–125 g (2 3/8 – 4 oz)
Location E., C., and southern Africa	Social unit Variable
	Status Common

In addition to white "epaulette" fur patches on the male's shoulders, both sexes have two white patches at the base of each ear. These pale tufts may be disruptive camouflage, breaking up the bat's outline when seen from below against dappled leaves. In the breeding season, the male's distinctive call to attract females resembles a squeaky bicycle pump.

Pteropus rodricensis
Rodriguez flying fox

	Length 35 cm (14 in)
	Tail None
	Weight 250–275 g (9 – 10 oz)
Location Indian Ocean (Rodriguez Island)	Social unit Group
	Status Critically endangered

Formerly the daytime roosts or "camps" of this flying fox contained more than 500 individuals. Due to habitat loss, through storm damage and human intervention, and also local hunting for food, the species currently numbers just a few hundred in the wild. Several centres, however, have established successful captive breeding programmes. At night the bats forage in dry woodland for fruits of various trees such as tamarinds, rose-apples, mangoes, palms, and figs. Like many other fruit bats they squeeze out the juices and soft pulp, rarely swallowing the harder parts. Observations in captivity show that each dominant male gathers a harem of up to 10 females, with which he roosts and mates. Subordinate and immature males tend to roost in another part of the camp.

hook-like foot claws permit roosting without muscle tension

brown wing membranes

brown fur

Rhynchonycteris naso
Proboscis bat

	Length 3.5–5 cm (1 1/2 – 2 in)
	Tail 1–1.5 cm (3/8 – 1/2 in)
	Weight 3–6 g (1/8 – 7/32 oz)
Location Mexico to C. South America	Social unit Group
	Status Common

Also called the sharp-nosed bat, this species has a long, pointed nose and streamlined appearance. It is a typical small, insectivorous bat, but shows unique roosting behaviour as groups of 5–10 (rarely more than 40) rest by day in a line, nose-to-tail on a branch or wooden beam. One adult male may dominate the group and may defend their feeding area – a nearby patch of water where they catch small insects.

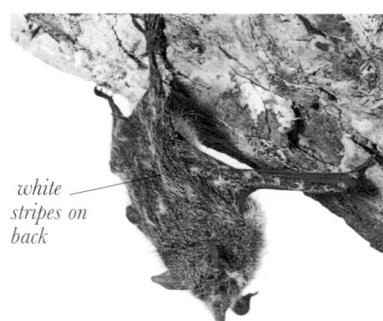

white stripes on back

Taphozous mauritianus
Mauritian tomb bat

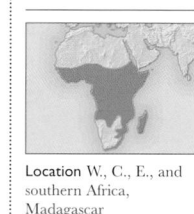

	Length 7.5–9.5 cm (2 3/4 – 3 1/2 in)
	Tail 2–3 cm (3/4 – 1 1/4 in)
	Weight 15–30 g (9/16 – 1 1/16 oz)
Location W., C., E., and southern Africa, Madagascar	Social unit Group
	Status Common

A member of the sheath-tailed group, the Mauritian tomb bat is known across Africa for its wide range of clicks, squeaks, and other noises just audible to some humans. It also makes ultrasonic sounds for echolocation. This active bat is watchful as it roosts by day in the open on tree trunks and walls, including town buildings. It hunts in clearings, especially over water, for flying insects. The back is grizzled brown-black; the underparts and wings are white.

Macroderma gigas
Australian false vampire bat

Length	10–12 cm (4–4³/₄ in)
Tail	None
Weight	75–150 g (2⁵/₈–5 oz)

Location W. and N. Australia

Social unit Group

Status Vulnerable

Also known as the Australian ghost bat, this bat derives its name from the mistaken belief that it feeds on blood. One of the largest microchiropterans, it preys on insects, birds, frogs, and lizards. Its decline may be partly due to the increasing use of its rocky roosting sites for mines and quarries.

Rhinopoma hardwickei
Lesser mouse-tailed bat

Length	5.5–7 cm (2¹/₂–2³/₄ in)
Tail	4.5–7.5 cm (1³/₄–3 in)
Weight	10–15 g (³/₈–⁹/₁₆ oz)

Location W. to S. Asia, N. and E. Africa

Social unit Group

Status Common

Also called long-tailed bats, the four *Rhinopoma* species are the world's only small, insectivorous bats with thin, trailing tails. The tail may be as long as both head and body. This species lives in scrub, semi-desert, and tropical forest. When food is plentiful, it may double its body weight, storing fat for several weeks of dry-season inactivity.

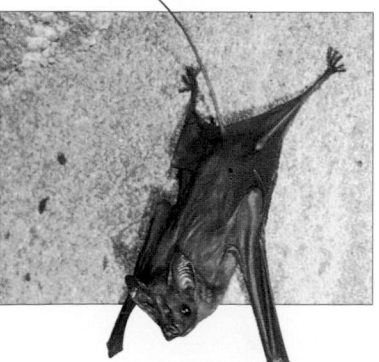

Anoura geoffroyi
Geoffroy's tail-less bat

Length	6–7.5 cm (2¹/₂–2³/₄ in)
Tail	Up to 7 mm (⁷/₃₂ in)
Weight	13–18 g (⁷/₁₆–⁵/₈ oz)

Location Mexico to N. South America

Social unit Group

Status Common

The small, fur-covered tail membrane of this bat gives it the appearance of hairy legs. It also has a small, triangular, upright nose leaf and a long muzzle with protruding lower jaw. These features give it its alternative names of Geoffroy's hairy-legged or long-nosed bat. It hovers in front of night-blooming flowers to sip nectar and gather pollen with its unusual brush-tipped tongue, which is the length of its head. Geoffroy's tail-less bat also eats insects, such as beetles or moths. It roosts in caves and tunnels.

Pteronotus davyi
Davy's naked-backed bat

Length	4–5.5 cm (1¹/₂–2¹/₂ in)
Tail	2–2.5 cm (³/₄–1 in)
Weight	5–10 g (³/₁₆–³/₈ oz)

Location Mexico to N. and E. South America

Social unit Group

Status Common

This bat is a common sight at night near towns as it feeds on flies, moths, and other insects attracted to street lights. By day it roosts in large colonies in caves and old mines, often some distance away from its feeding areas. Davy's naked-backed bat has wings that join along the centre of the back, obscuring the fur beneath. Bats within this genus are also known as moustached or leaf-lipped bats.

Rhinolophus hipposideros
Lesser horseshoe bat

Length	4 cm (1¹/₂ in)
Tail	2–3.5 cm (³/₄–1¹/₂ in)
Weight	4–10 g (⁵/₃₂–³/₈ oz)

Location Europe, N. Africa to W. Asia

Social unit Group

Status Vulnerable

This diminutive bat, with its head and body smaller than a human thumb, is widespread in woods and scrub – yet the species is at risk. Its underground, winter hibernation sites, such as deep caverns, have been disturbed, as have summer day roosts in tree holes, caves, chimneys, and mine shafts. Domestic cats take a heavy toll and destruction of woods and hedges has reduced the availability of its prey of small flying insects. The lesser horseshoe is one of the smallest of the 62 species of horseshoe bats.

broad wings allow slow, hovering flight

horseshoe-shaped nose leaf

relatively large head

Hipposideros speoris
Schneider's leaf-nosed bat

Length	4.5–6 cm (1³/₄–2¹/₂ in)
Tail	2–3 cm (³/₄–1¹/₄ in)
Weight	9–12 g (¹¹/₃₂–⁷/₁₆ oz)

Location S. Asia

Social unit Group

Status Common

A medium-sized member of this genus of about 60 species, this bat is a typical, small insect-eater. It has a flap-like "leaf" on the upper muzzle around the nostrils, with a U-shaped part below. By day, thousands roost in caves, tunnels, and buildings.

Noctilio leporinus
Greater bulldog bat

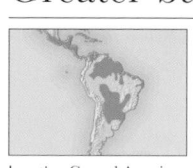

Length	6–8 cm (2¹/₂–3 in)
Tail	1.5–2 cm (¹/₂–³/₄ in)
Weight	15–35 g (⁹/₁₆–1¹/₄ oz)

Location Central America, N., E., and C. South America

Social unit Group

Status Lower risk†

Also called fisherman bats, the two species of bulldog bats have large nose pads, drooping upper lips, and ridged chins. The greater bulldog bat has velvety fur – orange, brown, or grey – with a distinctive pale stripe

Nycteris grandis
Large slit-faced bat

Length	7–9.5 cm (2³/₄–3³/₄ in)
Tail	6.5–7.5 cm (2¹/₂–3 in)
Weight	25–40 g (⁷/₈–1⁷/₁₆ oz)

Location W., C., E., and southern Africa

Social unit Group

Status Unconfirmed

This slit-faced bat has a furrow down the face. This may be partly covered by nose "leaves", so it looks like two slits running from nostrils to eyes. A powerful species, it swoops onto other bats, birds, scorpions, sun-spiders, frogs, and even fish near the surface. By day it roosts in groups of up to 60 in trees, caves, and buildings.

along the middle of the back. It roosts by day in hollow trees or caves. At night it hunts over water or sandy beaches for fish, crabs, and other prey, which it snatches from the ground or water using its large and powerful, sharp-clawed back feet.

Trachops cirrhosus

Fringe-lipped bat

Length 6.5–9 cm (2 1/2–3 1/2 in)	
Tail 1–2 cm (3/8–3/4 in)	
Weight 25–35 g (7/8–1 1/4 oz)	

Location Mexico to N. South America

Social unit Group

Status Unconfirmed

Why this broad-winged, strong-flying bat's lips are studded with papillae (small, wart-like bumps) is not clear. Also known as the frog-eating bat, it hunts along streams, ditches, and similar waterways, killing prey such as insects, frogs, and lizards with its powerful bite. It locates its victims by hearing their sounds, such as the croaking of male frogs, rather than by its own echolocation. In other respects it is a typical bat, roosting in tree holes, hollow logs, and caves.

Pipistrellus pipistrellus

Common pipistrelle

Length 3.5–4.5 cm (1 1/2–2 in)	
Tail 3–3.5 cm (1 1/4–1 1/2 in)	
Weight 3–8 g (1/8–5/16 oz)	

Location Europe to N. Africa, W., and C. Asia

Social unit Group

Status Common

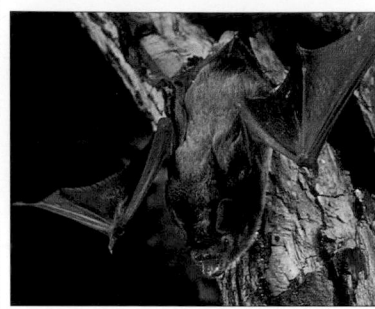

The genus *Pipistrellus* includes almost 70 similar species, of which this species is one of the smallest and most widespread, found in habitats from forest to city park. It is among the first bats to emerge each evening in pursuit of small flying insects. It roosts by day in crevices, buildings, and bat boxes, and hibernates through winter in similar sheltered places. Nursery colonies may contain up to 1,000 mothers, each with a single young.

Desmodus rotundus

Vampire bat

Length 7–9.5 cm (2 3/4–3 3/4 in)	
Tail 6.5 cm (2 1/2 in)	
Weight 19–45 g (11/16–1 5/8 oz)	

Location Mexico to South America

Social unit Group

Status Common

The vampire bat is a strong flier, yet it can also scuttle over the ground with amazing speed and agility, propped up on its forearms and back legs. From dusk it searches for a warm-blooded victim, such as a bird, tapir, or farm animal – even a seal or a human. The bat lands nearby, crawls closer, bites away any fur or feathers, and laps some 25ml (1fl oz) of blood

strong forearms and legs for hopping on ground

long thumb

RAZOR FANGS

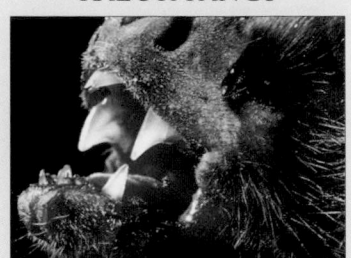

The vampire bat's thin, pointed, blade-like upper incisors are so sharp that its victim rarely notices as they slice away a piece of flesh about 5mm (3/16 in) across.

over 30 minutes, its saliva preventing clotting. This bat has a communal roost in a hollow tree, cave, mine, or old building, which it shares with hundreds of others.

COLORATION
The fur is dark brownish grey; the underparts are paler, with a buff tinge.

Thyroptera tricolor

Spix's disc-winged bat

Length 4 cm (1 1/2 in)	
Tail 2.5–3.5 cm (1–1 1/2 in)	
Weight 3–5 g (1/8–3/16 oz)	

Location Mexico to Central America, South America, Trinidad

Social unit Group

Status Common

The disc-winged or sucker-footed bat has a rounded, sucker-like structure near each thumb claw in the middle front edge of the wing, and a smaller one on the sole of each foot. Like tiny suction cups, these grip smooth, glossy leaf surfaces so the bat can shelter within part-furled leaves – roosting head-up, unlike nearly all other bats. Alternatively, it rests with others of its kind in small groups of up to 10, among unfurled leaves. Spix's is the smallest of the three New World disc-winged species, and feeds on small creatures, many of them non-flying such as jumping spiders. It is slim and delicate, with a dark or reddish brown back and whitish brown or yellow underparts.

Nyctalus noctula

Noctule

Length 7–8 cm (2 3/4–3 in)	
Tail 5–5.5 cm (2–2 1/4 in)	
Weight 15–50 g (9/16–1 3/4 oz)	

Location Europe to W., E., and S. Asia

Social unit Group

Status Lower risk†

The most widespread of the 6 noctule species, this bat flies high and powerfully before diving steeply to grab flying insects as large as crickets and chafers. By day, it roosts, usually alone, in any available small hollow – for example, in a tree, building, or among rocks. The noctule migrates 2,000km (1,200 miles) or more between its winter and summer sites. In spring, the female may produce 3 young, in contrast to the single offspring of most small bats.

short, broad ears

reddish yellow or golden fur

Uroderma bilobatum

Tent-building bat

Length 6–6.5 cm (2 1/4–2 1/2 in)	
Tail 4–4.5 cm (1 1/2–1 3/4 in)	
Weight 13–20 g (7/16–11/16 oz)	

Location Mexico to C. South America, Trinidad

Social unit Group

Status Common

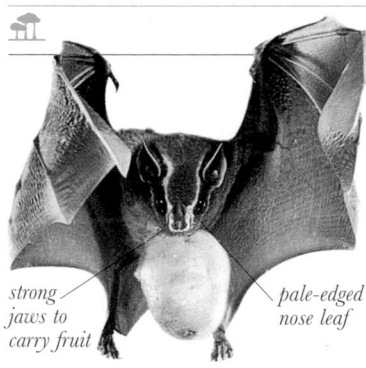

strong jaws to carry fruit

pale-edged nose leaf

About 15 bat species from the leaf-nosed group roost in "tents" – shelters shaped like umbrellas, cylinders, cones, or flasks, and made by biting leaves such as palm or banana so they droop or fold over. Each tent protects 2–50 or more bats from sunlight, rain, and predators, and lasts up to 3 months. The tent-building bat feeds on a variety of leaves and fruits, chewing them to a pulp and sucking out the juices. It is grey-brown with white stripes on its face and back.

Vampyrum spectrum

False vampire bat

Length 13.5–15 cm (5 1/4–6 in)	
Tail None	
Weight 150–200 g (6–7 oz)	

Location Mexico to N. South America, Trinidad

Social unit Variable

Status Lower risk

Also called Linnaeus' false or spectral vampire, the 1m (3 1/2 ft) wingspan makes this by far the largest bat in the Americas. As the name suggests, it is not a blood-sucker, but it is a powerful predator near the top of the food web and so at particular risk from habitat loss. It hunts other bats, small rodents such as mice and rats, and birds such as wrens, orioles, and parakeets. By day it roosts in hollow trees in groups of up to 5 individuals.

Molossus ater
Red mastiff bat

Length 7–10 cm (2 3/4–4 in)	
Tail 4–5 cm (1 1/2–2 in)	
Weight 30–40 g (1 1/16–1 7/16 oz)	
Location Mexico to C. South America, Trinidad	Social unit Group
	Status Common

Medium in size, mastiff bats are also called free-tailed velvety bats for their short, soft fur. This species often roosts in buildings and feeds on insects

attracted to street lights. It is more active around dawn and dusk than most bats, and roosts in the middle of the night as well as by day. Huge numbers of insects are stored in its cheek pouches and only chewed and swallowed on return to the roost.

Plecotus auritus
Brown long-eared bat

Length 4–5 cm (1 1/2–2 in)	
Tail 4–5 cm (1 1/2–2 in)	
Weight 7–14 g (1/4–1/2 oz)	
Location Europe, C. Asia	Social unit Group
	Status Lower risk†

A compact face and relatively large ears identify the four *Plecotus* species of Old World bats. The fur is usually brownish grey, the face darker. This species takes a variety of insects, including moths and beetles, carrying the meal to a perch to eat. Its other habits are typical of small, insect-eating bats, with winter hibernation in caves, mines, and cellars.

ears joined at base above forehead

Antrozous pallidus
Pallid bat

Length 5.5–8 cm (2 1/2–3 in)	
Tail 3.5–5.5 cm (1 1/2–2 1/4 in)	
Weight 14–30 g (1/2–1 1/16 oz)	
Location W. North America to Mexico, Cuba	Social unit Group
	Status Vulnerable†

This medium-sized, pig-nosed, pale bat tolerates a range of dry habitats, from grassland to scrubby desert and even the intense heat of California's Death Valley. It detects victims by the sounds they make, consuming beetles, crickets, spiders, centipedes, scorpions, lizards, and pocket mice. The pallid bat utters piercing directive cries audible to humans as it "rallies", flying in groups to locate its roost in rocky outcrops, trees, or attics.

Vespertilio murinus
Parti-coloured bat

Length 5–6.5 cm (2–2 1/2 in)	
Tail 3.5–4.5 cm (1 1/2–2 in)	
Weight 10–25 g (3/8–7/8 oz)	
Location Europe to W., C., and E. Asia	Social unit Group
	Status Lower risk†

 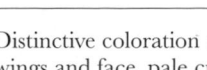

Distinctive coloration of almost black wings and face, pale cream fur below, and brown back hairs tipped with white give the parti-coloured bat its

various common names, including frosted bat. It roosts by day in small crevices in cliffs or buildings. In late autumn, males fly high near steep rock faces and tall buildings, their courting calls resembling the shrill whine of a high-speed metal-grinder.

Natalus stramineus
Mexican funnel-eared bat

Length 4–4.5 cm (1 1/2–1 3/4 in)	
Tail 4.7–5.2 cm (1 3/4–2 in)	
Weight 3–5 g (1/8–3/16 oz)	
Location W. USA to N. South America	Social unit Group
	Status Unconfirmed

The 5 species of tropical American funnel-eared bats are tiny and delicate with rounded ears, soft woolly fur, and a tail joined by flight membranes to the legs. The Mexican funnel-eared bat has a rapid, agile flight, almost like a butterfly. It eats small flying insects and roosts by day in caves.

tail longer than head and body

orange- or yellow-brown fur on back

Mops condylurus
Angolan free-tailed bat

Length 7–8.5 cm (2 3/4–3 in)	
Tail 4 cm (1 1/2 in)	
Weight 18–35 g (5/8–1 1/4 oz)	
Location W., C., E. and southern Africa, Madagascar	Social unit Group
	Status Common

Common and widespread from deserts to rainforests, this bat has a long, mouse-like, "free" tail, not enclosed by the tail membranes. It displays the typical bat habit of emerging from its daytime roost in noisy, flapping groups, thereby lessening each individual's risk of being caught by a predator such as an owl, hawk, or snake. It hunts flying insects, eating them in the air and dropping hard parts such as the legs.

Otonycteris hemprichi
Hemprich's long-eared bat

Length 6–7 cm (2 1/2–2 3/4 in)	
Tail 4.7–4.9 cm (1 7/8–2 in)	
Weight 20–30 g (11/16–1 1/16 oz)	
Location N. Africa to W. Asia	Social unit Group
	Status Unconfirmed

Also known as the desert long-eared bat, this is one of the few bat species content in dry, barren habitats. During periods of harsh weather, such as drought, it probably enters a period of hibernation-like inactivity. Its spectacular ears are 4 cm (1 1/2 in) long and often held almost horizontally in flight, such as when swooping onto the ground for insects and spiders. This bat roosts by day in groups of at least 20 in a crevice, cave, or building.

Myotis daubentonii
Daubenton's bat

Length 4–6 cm (1 1/2–2 1/2 in)	
Tail 2.5–5 cm (1–2 in)	
Weight 5–15 g (3/16–9/16 oz)	
Location Europe to N. and E. Asia	Social unit Group
	Status Common

Daubenton's is one of about 87 species in the widespread bat genus *Myotis* – little brown or mouse-eared bats. It flutters 1–2 m (3 1/4–6 1/2 ft) above water to catch flying insects by mouth or in the pouch of its curled wing or tail membrane – hence its other name of water bat. It also skims the surface and grabs small fish in its large back feet. By day, this bat roosts in trees, buildings, old walls, and bridges. It flies up to 300 km (180 miles) to its winter hibernation site in a cave or mine.

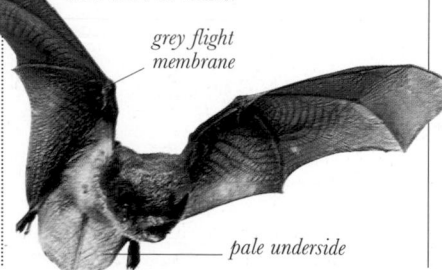

grey flight membrane

pale underside

FLYING LEMURS

PHYLUM	**Chordata**
CLASS	**Mammalia**
ORDER	**Dermoptera**
FAMILIES	**1 (Cynocephalidae)**
SPECIES	**2**

BY EXTENDING a strong membrane that surrounds their body (the patagium), flying lemurs (also known as colugos) are able to glide more than 100m (330ft) between trees, with very little loss in height. The term "flying lemur" is misleading, however, as these mammals are not lemurs and are not capable of true flight. About the size of a domestic cat, they have large eyes, a blunt muzzle, strong claws for climbing, and mottled fur (for camouflage). Flying lemurs, which live in the rainforests of Southeast Asia, strain food (fruits and flowers) through peculiar comb-shaped lower teeth, which are also used to groom the fur.

GLIDER
A kite-shaped membrane called the patagium allows flying lemurs to glide from tree to tree. It is attached to the neck and to the tips of the fingers, toes, and tail.

Cynocephalus variegatus
Malayan flying lemur

Length	33–42 cm (13–16 ¹/₂ in)
Tail	17.5–27 cm (7–10 ¹/₂ in)
Weight	0.9–2 kg (2–4 ¹/₂ lb)

Location S.E. Asia

Social unit Variable

Status Common

Although still common in some areas, numbers of this species have declined rapidly in logged and farmed regions. It has a small head relative to its body with large eyes, little rounded ears and a blunt muzzle. The fine, short fur is brownish grey with red or grey on the back, often with lighter flecks to mimic lichen-covered branches, and paler on the underside. The Malayan flying lemur is active in twilight and at night, eating soft plant parts such as flowers, fruits, buds and young leaves, and scraping up nectar and sap with its comb-like lower incisor teeth. It lives alone or in small, loose groups, inhabiting tree-holes or resting among dense foliage high in the treetops. After a gestation period of about 2 months, the single offspring clings to its mother as she leaps and glides among the trees and is weaned by 6 months.

large, forward-facing eyes

offspring enclosed within mother's gliding membrane

ELEPHANT-SHREWS

PHYLUM	**Chordata**
CLASS	**Mammalia**
ORDER	**Macroscelidea**
FAMILIES	**1 (Macroscelididae)**
SPECIES	**15**

THE LONG, POINTED SNOUT, which is flexible and highly sensitive, gives elephant-shrews their name. All species have keen senses of hearing and vision, and long, powerful back legs for running swiftly around their territory. These secretive animals are found only in Africa, where their habitat ranges from stony ground and grassland to forest undergrowth. They forage mostly by day, feeding on invertebrates.

BURROWS AND TRAILS
Elephant-shrews are strictly terrestrial and live in a variety of habitats. The rock elephant-shrew (right) favours rocky areas where it will either dig its own burrow or occupy one that has been abandoned. It maintains a system of trails that lead from the burrow to feeding areas. This behavioural pattern is seen in several elephant-shrew species.

Elephantulus rufescens
Rufous elephant-shrew

Length	12–12.5 cm (4³/₄–5 in)
Tail	13–13.5 cm (5–5 ¹/₂ in)
Weight	50–60 g (1³/₄–2 ¹/₈ oz)

Location E. Africa

Social unit Individual/Pair

Status Lower risk†

Grey to brown with white underparts, this species has a white eye-ring with a dark patch on the outer edge. In addition to small creatures, it eats soft fruits, seeds, and buds. The 1–2 young are born after a gestation period of 60 days. A male–female pair defend their territory by drumming their back feet and chasing away the intruder. The male chases out other males, and the female other females.

Rhynchocyon chrysopygus
Golden-rumped elephant-shrew

Length	27–29 cm (10 ¹/₂–11 ¹/₂ in)
Tail	23–26 cm (9–10 in)
Weight	525–550 g (1 ¹/₄ lb)

Location E. Africa

Social unit Individual/Group

Status Endangered

This colourful species has hairless black feet, legs, and ears, a mainly black, white-tipped, sparsely furred tail, russet-coloured head and body, and a golden patch on the rump. Its diet consists of small invertebrates such as worms, insects, and centipedes. When threatened, this elephant-shrew slaps its tail on the leaves to warn others, then bounds away at great speed, often with a stiff-legged leap to demonstrate its vigour.

TREE SHREWS

PHYLUM	Chordata
CLASS	Mammalia
ORDER	Scandentia
FAMILIES	1 (Tupaiidae)
SPECIES	19

THESE SMALL, squirrel-like mammals are not wholly arboreal (they spend much of their time on the ground) and are not true shrews. In fact, they have some features that are associated with primates, such as a large braincase and, in males, testicles that descend into a scrotum. Tree shrews lack whiskers and seek prey, such as insects, using well-developed senses of hearing, smell, and vision. Most species have a long, thickly furred tail. Tree shrews inhabit the tropical forests of Southeast Asia.

SEEKING FOOD
Tree shrews are skilful climbers and agile runners. They forage with their hands and pointed snout.

Ptilocercus lowi

Pen-tailed tree shrew

Length	10–14cm (4–5 1/2 in)
Tail	13–19cm (5–7 1/2 in)
Weight	25–60g (7/8–2 1/8 oz)

Location S.E. Asia	Social unit Variable
	Status Unconfirmed

Named after its mainly naked tail with bushy white hairs at the tip, like a bottle-brush or old-fashioned quill pen, this species is greyish brown above and greyish yellow on the underside. It climbs skilfully with its strong, sharp-clawed limbs, using its tail for balance, and rarely descends to the ground. It makes simple nests in hollow trees or on branches and lives in pairs or small groups. Its varied diet includes worms, insects, mice, small birds, lizards, and fruits.

Dendrogale melanura

Bornean smooth-tailed tree shrew

Length	10–15cm (4–6 in)
Tail	9–14cm (3 1/2–5 1/2 in)
Weight	35–60g (1 1/4–2 1/8 oz)

Location S.E. Asia	Social unit Individual/Pair
	Status Vulnerable

This species lives in mountain forests at altitudes of 900–1,500m (3,000–5,000ft). Its long limbs and long claws equip it to grasp branches and it spends more time in trees than other tree shrews. The coloration of its upperparts is a mixture of black and buff or cinnamon; the underparts and legs are more red or ochre. It lives alone, feeding by day and sleeping in a leaf-lined tree nest at night. After a gestation period of about 50 days, the female produces 3–4 young.

Anathana ellioti

Indian tree shrew

Length	17–20cm (6 1/2–8in)
Tail	16–19cm (6 1/2–7 1/2 in)
Weight	150g (5 oz)

Location S. Asia	Social unit Individual
	Status Lower risk

Also known as the Madras tree shrew, this mammal resembles a small, slender grey squirrel. It is speckled with yellow and brown on the upperparts and has a distinctive cream-coloured shoulder stripe. The Indian tree shrew has a prominent head with a pointed muzzle, large eyes, and furred ears. It forages actively by day on the ground and among low bushes, as well as in trees, for small edible items, mainly worms and insects but also some fruits. At night it sleeps in a rock crevice or tree-hole. The species is probably solitary outside the breeding season and does not, as far as is known, defend a specific territory, but information is lacking. The female cares for the young.

long, bushy tail helps balance when climbing

Tupaia minor

Lesser tree shrew

Length	11.5–13.5cm (4 3/4–5 1/2 in)
Tail	13–17cm (5–6 1/2 in)
Weight	30–70g (1 1/16–2 1/2 oz)

Location S.E. Asia	Social unit Individual
	Status Lower risk†

This species is more tolerant than most other tree shrews of the loss of natural forest habitat to plantations, parks, and gardens. It is a skilled climber and has a wide diet. It forages by day along branches, in bushes, among fallen logs, and under rocks for small animals, fruits, leaves, seeds, and carrion. Usually 2–3 young are born after a gestation period of 46–50 days. The female leaves the offspring in a leafy nest among the foliage while she feeds, returning sporadically to suckle. Tree shrews fall victim to snakes, mongooses, tree-dwelling cats, and diurnal birds of prey. They often hold food by their front paws while sitting up on their haunches, in the manner of a squirrel, to watch for danger.

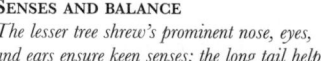

GRIPPING FEET
The splayed toes, sharp claws, and pimple-like protrusions on the foot pads of the lesser tree shrew all give an excellent grip on bark and rock. It uses its front legs to pull itself along branches on its belly. This spreads scent produced by abdominal glands. This scent is likely to be recognizable to others of its species, and is probably a means of marking its home range, which it defends against others of the same sex.

SENSES AND BALANCE
The lesser tree shrew's prominent nose, eyes, and ears ensure keen senses; the long tail helps with balance while climbing at speed through the branches.

speckled olive-brown or red fur on upperparts

MAMMALS

PRIMATES

PHYLUM	Chordata
CLASS	Mammalia
ORDER	Primates
FAMILIES	11
SPECIES	356

CLASSIFICATION NOTE

Recent research suggests that primates may comprise 2 suborders: Strepsirhini (lemurs, galagos, lorises, and pottos) and Haplorhini (tarsiers, apes, and monkeys). Tarsiers have features linking them with both suborders; in this book, they are grouped with Strepsirhini (traditionally known as the prosimians). To allow adequate coverage, monkeys and apes are presented here as separate groups. Humans do not appear in this book but are classified, with the apes, as hominoids.

Prosimians *see pp.118–21*
Monkeys *see pp.122–31*
Apes *see pp.132–7*

MEMBERS OF THE ORDER PRIMATES – prosimians, monkeys, and apes (but see the Classification note, left) – are a diverse group that form highly complex social units. They are found in South and Central America, Africa and Madagascar, and Southeast and eastern Asia. Primates mostly inhabit tropical rainforests, and their dextrous prehensile (grasping) hands and feet are an adaptation to a largely arboreal lifestyle (some species also have a prehensile tail). Traditionally, primate authorities have recognized about 180–200 species. However, in recent years there has been a thorough reconsideration of primate species, partly because of the need to identify significant populations for conservation purposes. As a result, many new species have been described since 1990.

SOCIAL ORGANIZATION

Only orang-utans and a few of the lemurs and galagos lead solitary lives; all other primates exist in social groups. Many species – including most monkeys – live in troops that consist of several females and either one or several adult males. In some species, such as the mandrill, drill, and gelada, huge troops of several hundred individuals periodically split into subgroups of bachelor bands and one male and his harem, both of fixed composition. Chimpanzees and spider monkeys live in large communities of 20–100 individuals that divide into groups of varying composition. Some species, mostly New World monkeys, live in monogamous pairs.

MUTUAL GROOMING
The bonds between members of primate groups are partly maintained by mutual grooming, as shown by these olive baboons. Low-ranking members groom those of higher rank to gain favour and support (in disputes).

ANATOMY

Primates form a highly varied group, with members as diverse as mouse lemurs, which may weigh only 35g (1¼oz), and gorillas, which can reach over 200kg (440lb). Most species have flat nails on the fingers and toes – only a few have claws – and the big toe always has a nail. Most have a tail, except the apes. Apart from some toothed whale species, the higher primates have the largest brain relative to body size of all the mammals, which helps explain their high intelligence. The cerebral hemispheres (which process sensory information and coordinate responses) are highly developed, allowing sight keen enough for accurate tree-to-tree leaps.

large braincase

large, forward-facing eye sockets

flattened facial profile

SKULL STRUCTURE
Primates have a large, domed braincase and forward-facing eye sockets. This monkey skull also shows the flattened facial profile that is characteristic of most monkeys and apes.

elongated finger

opposable big toe

clawed fingers

AYE-AYE HAND **AYE-AYE FOOT**

fingers with rounded pads

grooming claw

grooming claw

TARSIER HAND **TARSIER FOOT**

mobile fingers

larger gap between big toe and other toes

CHIMPANZEE HAND **CHIMPANZEE FOOT**

HANDS AND FEET
Hand and foot structure varies according to lifestyle. Claws (aye-aye) and rounded pads (tarsier) improve grip in arboreal species; highly mobile hands and feet (chimpanzee) are vital for species that live on the ground and in trees.

USING A SIMPLE TOOL

1 SEEKING TERMITES
This common chimpanzee has made a stick tool to obtain food that would otherwise be inaccessible.

2 DEXTROUS HANDS
The chimpanzee uses its grasping hands to hold and insert the stick into the termite mound.

3 REMOVING THE TOOL
The termites on guard, provoked by the intrusion, bite the stick with their pincer-like jaws.

4 SUCCESSFUL ENDEAVOUR
As the chimpanzee removes the stick, the termites remain firmly attached.

FEEDING

As a general rule, small primates tend to eat insects, whereas larger species mostly eat leaves and fruits (a large primate cannot sustain itself on insects alone). Small primates have a high metabolic rate and cannot afford the long digestion times needed to process vegetable matter. Some leaf-eating species, such as colobus monkeys and langurs, have a complex stomach containing bacteria to ferment cellulose; other species have bacteria in the caecum or in the colon. A few species, including chimpanzees and baboons, hunt vertebrate prey as well as eating vegetable matter. Only tarsiers are entirely carnivorous.

EATING VEGETATION
Of all primates, gorillas probably eat the most plant material. In order to break down cellulose (thereby releasing the cell nutrients), gorillas have large molar teeth and strong jaw muscles for chewing. Their pot belly houses a long digestive tract.

EATING MEAT
Given the opportunity, some primates will hunt (sometimes cooperatively) and kill other animals. This common chimpanzee is eating part of a duiker.

CONSERVATION

Primate numbers are declining severely in the wild due to habitat loss and, more recently, through the illegal hunting of protected species (such as the gorilla) for meat. As a result, many species are now endangered. Primates are also widely used in medical and space research. Although a few species are being re-introduced to the wild from zoos with captive breeding programmes, and sanctuaries, such as this chimpanzee orphanage in Zambia (below), provide some refuge, the situation remains bleak.

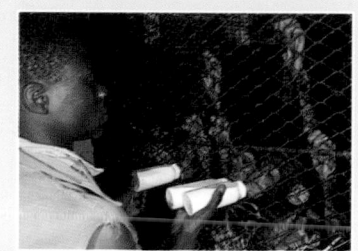

MOVEMENT

Most primates spend at least part of their life in trees and have adapted accordingly. To provide a strong hold on branches, the big toe is separated from the other toes in all species except humans, and the thumb is always separated from the fingers, although it is fully opposable (that is, it can turn, face, and touch the other digits in the same hand) only in apes and in some Old World monkeys. The arm and wrist bones are not fused, which increases dexterity. Primates also have "free" limbs – the upper part of each limb is outside the body wall, which allows great freedom of movement. (In other mammals, such as horses, the upper part is inside the body wall – the "armpit" is in fact the elbow joint, which means movement is more restricted.) Some species have a long, prehensile tail, used as a "fifth limb".

MAMMALS

LIFE IN THE TREES
These douc langurs are typical primates in that they are arboreal, have prehensile hands and feet, live in groups, and eat leaves and fruits. Their colourful markings help provide camouflage. Like almost all primates, loss of habitat is a major threat to their survival.

SWINGING
Spider monkeys and gibbons, such as this lar gibbon, use their long arms to swing from branch to branch. This is called brachiation.

STANDING
Primates, such as chimpanzees, that are capable of standing and walking on 2 legs (known as bipedalism), tend to have long legs.

CLIMBING
The most common way of moving is on all fours. Quadrupeds, such as this woolly spider monkey, usually have limbs of about the same length.

CLINGING
Vertical clingers and leapers, such as these indri, move with the back held vertically. They have well-developed back limbs for long leaps between trees.

BALANCING ACT
chimpanzee stabilizes the termite-covered by resting it on the opposite forearm.

6

EATING QUICKLY
Most of the termites are swiftly gobbled before they can escape, including any on the forearm.

7

MORE OF THE SAME
Once all the termites have either been eaten or escaped, the chimpanzee will repeat the process.

8

LEARNED BEHAVIOUR
Making and using tools is rare in mammals. Juvenile chimpanzees acquire these skills by copying adults.

PROSIMIANS

PHYLUM	Chordata
CLASS	Mammalia
ORDER	Primates
SUBORDER	Strepsirhini
FAMILIES	6
SPECIES	85

MORE PRIMITIVE than monkeys and apes, prosimians comprise the lemurs of Madagascar, the galagos and pottos of Africa, and the lorises of Asia. Lemurs (which include sifakas, the indri, and the aye-aye) have large ears, an elongated body, long limbs, and most have a long, bushy tail. Lorises, pottos, and galagos are generally smaller than lemurs and tend to have larger eyes. Prosimians mostly inhabit forests and are usually nocturnal (some lemurs are diurnal). Deforestation has endangered many prosimians, especially the lemurs.

ANATOMY

Prosimians have a black, dog-like snout and a sense of smell that is more highly developed than in other primates. They have large eye sockets and a crystalline layer behind the retina of the eye that reflects light. This increases the amount of light falling on the visual cells and improves night vision. As most prosimians are arboreal, their hands and feet are adapted for grasping (although they are still less dextrous than monkeys and apes). They have flat nails on all digits except the second toe, which instead has a long claw (the "toilet claw") used for grooming. All species except the aye-aye have a "dental comb" – 4 to 6 of the lower front teeth that are pressed together and grow slightly forwards. This is used for mutual grooming.

MOVEMENT

Most lemurs are quadrupedal and usually run or leap from branch to branch. Weasel lemurs, the avahi, sifakas, the indri, and galagos are vertical clingers and leapers (see below). On the ground, sifakas and indri move on 2 feet, employing sideways hops with their arms in the air for balance. Lorises and pottos usually clamber slowly along branches, clinging tightly at every step (although when startled they can move swiftly).

COMMUNICATION

Lemurs produce various calls, both to signal alarm (there are often different calls to distinguish aerial and ground predators) and to communicate within or between troops. Pairs of indri occupy treetop territories, and these are marked with loud wailing calls; small sifaka troops define their areas with "shi-fak" calls that sound like hiccups. Male ring-tailed lemurs and bamboo lemurs have a wrist gland with a spur, which they use to mark territory by drawing the wrist sharply across a sapling. This produces a click, creates a scar, and leaves a scent – a gesture that is auditory, visual, and olfactory. Galagos leave scent trails around their territory by placing urine on their feet.

TARSIERS

Not strictly prosimians (see the Classification note on p.116), tarsiers are unusual in that they have many primitive features similar to prosimians, yet they also exhibit characteristics linking them to monkeys and apes, such as a dry, hairy nose. The most striking feature of tarsiers is their enormous eyes: each is slightly heavier than the brain. Other characteristics include a large head and ears, long digits with disc-like pads at the tips, very long legs with elongated ankles, and a long tail. The 8 species of tarsiers are found in the forests of Southeast Asia, where they spend much of their time clinging to upright tree stems, scanning the forest floor for prey.

VERTICAL LEAP
When leaping from tree to tree, this Verreaux's sifaka uses its long tail for balance, its muscular legs for propulsion, and its large, grasping hands and feet for a secure landing. These features typify prosimians that keep their bodies vertical when climbing. In mid-air, a semi-upright posture is maintained.

MAMMALS

Arctocebus calabarensis
Angwantibo

Length 22–26 cm (9–10 in)	
Tail 1 cm (3/8 in)	
Weight 225–475 g (8–17 oz)	
Location W. Africa	**Social unit** Group
	Status Lower risk

One of only 2 *Arctocebus* species, this one is orange to yellow on its upper-parts and buff beneath. It climbs carefully and deliberately using all 4 equal-length limbs. The second toe is tiny and the first widely separated from the other 3, giving a clamp-like grip. Solitary and nocturnal, the angwantibo eats small creatures such as caterpillars, plus a few fruits. The other species, the golden angwantibo, lives farther south.

Loris tardigradus
Slender loris

Length 17.5–26 cm (7–10 in)	
Tail None	
Weight 85–350 g (3–13 oz)	
Location S. Asia, Sri Lanka	**Social unit** Individual/Pair
	Status Vulnerable

large, forward facing eyes for precise depth judgment

dark face mask with central pale stripe

FOUR-WAY GRIP
On each foot the big toe opposes the other 4 toes for a pincer-like grip. The slender loris can even sleep holding onto branches.

This small, slim primate moves with great deliberation, gripping twigs with all 4 feet – until it suddenly smells, sees, and with its front legs snatches a small creature such as an insect or lizard. It also eats soft leaves and buds, fruits, and birds' eggs,. By day, the slender loris curls up in a tree hole, dense leaf nest, or similar secure place. Its fur colour ranges from yellow-grey to dark brown on the back, with silver-grey underparts.

REARING A FAMILY
The female slender loris reaches sexual maturity at 10 months and is receptive to the male twice a year thereafter. After a gestation period of 166–169 days, she bears 1–2 young, which she suckles for 6–7 months.

Nycticebus coucang
Slow loris

Length 26–38 cm (10–15 in)	
Tail 1–2 cm (3/8–3/4 in)	
Weight 225–650 g (8–23 oz)	
Location S.E. Asia	**Social unit** Variable
	Status Unconfirmed

Aptly named because of its lack of speed, the slow loris has gripping hands like the slender loris (see left) and is also a nocturnal tree-dweller. It creeps carefully towards its prey, then lunges with its front limbs. The loris lives alone or in pairs or groups, and the adult male chases other males from its territory, which is marked by urine. The dense, soft fur is brown with a white face and underparts, and dark eye rings and ears.

Perodicticus potto
Potto

Length 30–40 cm (12–16 in)	
Tail 3.5–15 cm (1 1/2–6 in)	
Weight 0.85–1.5 kg (1 3/4–3 1/4 lb)	
Location W. and C. Africa	**Social unit** Variable
	Status Locally common

A careful, nocturnal climber, the potto has very mobile limb joints and reaches out at any angle to bridge gaps between branches. Its fur may be grey, brown, or red, the eyes and ears are small, and the diet consists mainly of fruits, sap, gum, and small animals. A potto can remain immobile for hours to escape attention. If attacked, it will tuck down its head and batter the enemy with the "shield" of horny skin covering spiny bones on the upper back. The gestation period is about 200 days, and the potential life span is over 25 years.

Euoticus elegantulus
Western needle-clawed galago

Length 10.5–27 cm (4 1/4–10 1/2 in)	
Tail 19.5–34 cm (7 1/2–13 1/2 in)	
Weight 275–350 g (10–13 oz)	
Location W. Africa	**Social unit** Individual/Pair
	Status Locally common

Resembling other galagos in form, this species has an orange back, grey underside, pink hands and feet, and a long, grey, white-tipped tail. The oval eyes are rimmed by pale fur, and thin, sharp claws grip well for climbing. Like all galagos and lorises, the second toe on the rear foot has an upward-facing, "toilet" claw, for scratching and combing fur. The specially enlarged front teeth scrape wood and bark to obtain gum and sap, which make up three-quarters of the diet, along with a variety of fruits and insects. It may make 1,000 "gum scrapes" per night.

Galago crassicaudatus
Thick-tailed galago

Length 25–40 cm (10–16 in)	
Tail 34–49 cm (13 1/2–19 1/2 in)	
Weight 1–2 kg (2 1/4–4 1/2 lb)	
Location C., E. and southern Africa	**Social unit** Variable
	Status Locally common

The largest galago, this species locates insects at night using its huge eyes and ears, and snatches them by hand in a split second. It also scrapes gum and sap with its comb-like, protruding lower incisor and canine teeth. Deriving its alternative name of bushbaby from its child-like wails, it lives in small family groups of a male–female pair or 2 females, with young. Compared to closely related species, it tends to run on all fours more frequently and to leap in an upright position less often.

main colour varies from silver to grey, brown, or black

strong feet with roughened friction pads on soles

Galago moholi
South African galago

Length 15–17 cm (6–6 1/2 in)	
Tail 12–27 cm (4 3/4–10 1/2 in)	
Weight 150–250 g (5–9 oz)	
Location E., C., and southern Africa	Social unit Variable
	Status Common

The South African galago is also called the lesser bushbaby. It leaps vertically, in enormous jumps of 5m (16ft), and its grasping hands and feet are moistened regularly with urine, to maintain grip. It snatches insects from mid-air by hand or scrapes gum from trees with its comb-like lower front teeth. This galago lives in small family groups, which sleep huddled together by day.

diamond-shaped black eye rings

large rear legs

Cheirogaleus medius
Fat-tailed dwarf lemur

Length 17–26 cm (7–10 in)	
Tail 19–30 cm (7 1/2–12 in)	
Weight 175 g (6 oz)	
Location W. and S. Madagascar	Social unit Individual/Group
	Status Lower risk

grey, red, or buff fur on upper parts

The fat-tailed dwarf lemur stores food as fat in its body and tail, to survive the 8-month dry season, during which it remains torpid, huddled with others of its kind. When active again, it becomes solitary and clambers in trees and bushes at night, seeking flowers, fruits, and insects. It rests by day in a leaf-and-twig nest in a tree hole or fork.

Lemur fulvus
Brown lemur

Length 38–50 cm (15–20 in)	
Tail 46–60 cm (18–23 1/2 in)	
Weight 2–4 kg (4 1/2–8 3/4 lb)	
Location N. and W. Madagascar	Social unit Group
	Status Lower risk†

Extremely variable in colour, from brown to yellow or grey, according to subspecies, the brown lemur usually has a dark face with pale eyebrow patches. Adaptable to different forest habitats, it feeds on a wide range of fruits, flowers, leaves, and sap. Living in often-changing groups numbering from 3 to 12 individuals, each member smears urine on itself for scent recognition. Group size and composition change frequently.

Lemur macaco
Black lemur

Length 30–45 cm (12–18 in)	
Tail 40–60 cm (16–23 1/2 in)	
Weight 2–3 kg (4 1/2–6 1/2 lb)	
Location N. Madagascar	Social unit Group
	Status Lower risk†

A medium-sized species, only the male has the long, soft black fur for which it is named. Females are red, brown, or grey. Unusually for lemurs,

ruff of long fur around neck and ears

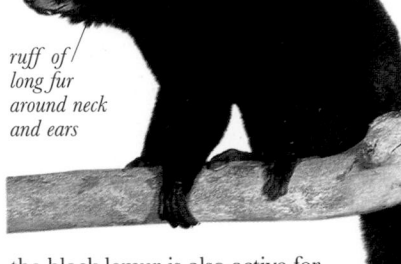

the black lemur is also active for part of the night. This behaviour may result from hunting and other human disturbance. Led by one female, groups of 5–15 black lemurs forage mainly in trees for fruits, flowers, leaves, and soft bark.

Hapalemur griseus
Grey gentle lemur

Length 40 cm (16 in)	
Tail 40 cm (16 in)	
Weight 1 kg (2 1/4 lb)	
Location N. and E. Madagascar	Social unit Group
	Status Unconfirmed

This is the only lemur – indeed the only primate – adapted to living in lake-fringing reeds and rushes. All-over grey with a distinctive blunt snout, it jumps between reed stalks to eat their leaves, shoots, buds, and pith. Groups of 3–5 may expand up to 40, led by one dominant male. Disturbance of its restricted habitat threatens this specialized lemur.

Lemur catta
Ring-tailed lemur

Length 39–46 cm (15 1/2–18 in)	
Tail 56–62 cm (22–24 in)	
Weight 2.5–3.5 kg (5 1/2–7 3/4 lb)	
Location S. and S.W. Madagascar	Social unit Group
	Status Vulnerable

This lemur spends more time than its relatives on the ground, even though it is a skilled climber. Very sociable, it forms groups of 5–25, with a core of adult females showing a well-defined hierarchy among themselves and over any males. Young females remain with their mothers and sisters; juvenile males move to other groups. Using their hands, they gather flowers, fruits, leaves, bark, and sap. After 134–138 days' gestation, the female bears one (sometimes 2) offspring. It first clings to the mother's underside and then rides on her back. Like most lemurs, this lemur faces many threats, especially habitat loss.

CAT-LIKE APPEARANCE
This lemur is cat-like both in body proportions and graceful movements. The white face has a dark nose and eye patches; the underparts are whitish grey. The distinctive tail is used for visual and scent signalling.

black and white tail rings

dark, triangular eye patch

SUNBATHING

Unlike most other lemurs, the ring-tailed lemur enjoys basking in sunshine, whether on the ground or in trees. It sits upright, with hands on knees in a "sun-worship" posture. Loud alarm calls from others in the group alert it, when necessary, to possible danger – one of the benefits of a highly social lifestyle.

upper fur brown-grey to rosy brown

Varecia variegata
Ruffed lemur

	Length 55 cm (22 in)
	Tail 1.1–1.2 m (3 1/2 –4 ft)
	Weight 3.5–4.5 kg (7 3/4 –10 lb)
Location E. Madagascar	Social unit Group
	Status Endangered

This large species is white or reddish white except for black on the face, shoulders, chest, flanks, feet, and tail. The ruffed lemur eats a higher

proportion of fruit than any other lemur. It builds a leafy nest in a tree hole or fork for its 2–3 young. The female gives birth after a gestation period of 90–102 days. The offspring remain in the nest for several weeks after which they cling to her. Group size is 2–20, with several dominant females defending the territory.

Propithecus verreauxi
Verreaux's sifaka

	Length 43–45 cm (17–18 in)
	Tail 56–60 cm (22–23 1/2 in)
	Weight 3–5 kg (6 1/2 –11 lb)
Location W. and S. Madagascar	Social unit Group
	Status Critically endangered†

This large, mostly white lemur has brown-black areas on the face, crown, and undersides of the limbs. It uses its powerful legs to move among cactus-like trees with massive, spring-loaded leaps. It eats a wide range of leaves, fruits, flowers, and bark. Living in variable social groupings, it makes its "sifaka" call when two groups dispute territorial boundaries.

Lepilemur mustelinus
Weasel lemur

	Length 30–35 cm (12–14 in)
	Tail 25–30 cm (10–12 in)
	Weight 1 kg (2 1/4 lb)
Location N.E. Madagascar	Social unit Individual
	Status Lower risk

This nocturnal prosimian feeds mainly on fruits and leaves. It has grasping feet and long, soft, brown fur, which darkens towards the tail tip. As with other leaping primates, both eyes face forwards for good stereoscopic vision that enables it to judge distances precisely. The weasel lemur occupies a territory of 0.15–0.5 hectares (1/2 –1 1/4 acres).

Indri indri
Indri

	Length 60 cm (23 1/2 in)
	Tail 5 cm (2 in)
	Weight 6–7 kg (13–15 lb)
Location E. Madagascar	Social unit Pair
	Status Endangered

Largest of the lemur group, the indri or babakoto ("little father") has very long rear legs, for enormous leaps, but a very short tail. Although diurnal, it is inactive for long periods in the day. The diet is chiefly young leaves, but

also fruits, flowers, and seeds. Indris live in pairs with their offspring. The male defends their territory, and the female has first access to food. Indris are mainly black, with white patches.

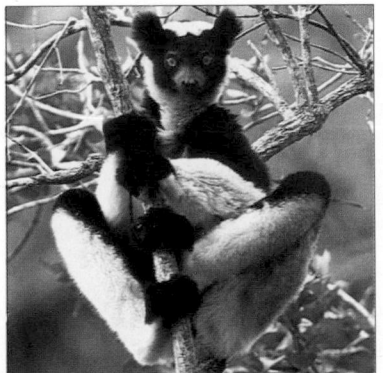

Daubentonia madagascariensis
Aye-aye

	Length 40 cm (16 in)
	Tail 40 cm (16 in)
	Weight 2.5–3 kg (5 1/2 –6 1/2 lb)
Location N.W. and E. Madagascar	Social unit Variable
	Status Endangered

The aye-aye has coarse, shaggy black fur with a mantle of white guard hairs. It is specialized as a night-time primate "woodpecker". It taps trees with its long middle finger, listens intently with its huge ears for wood-boring grubs under the bark, exposes them by gnawing with its rodent-like, ever-growing front teeth, and extracts them with the middle finger. It also eats fruits, including coconut flesh, seeds, and fungi. Aye-ayes share large, stick-made nests but use them in succession, not together. The offspring stays with its mother for 2 years.

LONG FINGERS
The aye-aye uses its elongated middle finger, which has a double-jointed tip, to extract grubs from under tree bark.

CONSERVATION
Several small captive breeding colonies of aye-ayes have been established, for study and possible re-introduction of individuals into suitable habitats. Below, a one-day-old aye-aye, born after a gestation period of about 170 days, is being weighed to improve our scant knowledge of this secretive species – believed extinct until rediscovered in 1957.

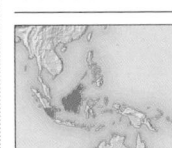

Tarsius bancanus
Western tarsier

	Length 12–15 cm (4 3/4 –6 in)
	Tail 18–23 cm (7–9 in)
	Weight 85–135 g (3–4 oz)
Location S.E. Asia	Social unit Individual
	Status Unconfirmed

This long-tailed, nocturnal Asian prosimian is small and compact-bodied. Mainly tree-dwelling, the western tarsier has slender fingers and its toes have pads and sharp claws, for gripping branches. It can turn its head to look backwards and detect possible predators or prey with its huge eyes and keen ears. Its diet consists mainly of insects. Having spotted its victim, it creeps nearer and then leaps on it and seizes it with its

front paws. The female gives birth to one offspring after a gestation period of about 180 days. At first the young is carried by the mother, but it soon learns to cling to her fur.

MAMMALS

MONKEYS

PHYLUM	Chordata
CLASS	Mammalia
ORDER	Primates
SUBORDER	Haplorhini (part)
FAMILIES	3
SPECIES	242

THIS LARGE, diverse group is split into 2 broad, geographically separate subgroups: the Old World monkeys (larger species such as baboons, as well as colobus monkeys and langurs) and the New World monkeys (such as marmosets and spider monkeys), which are distinguished mainly by nose shape. Monkeys are normally found in forests throughout the tropics. Most have short, flat, human-like faces, although baboons and mandrills have a dog-like snout. Many species are endangered by loss of habitat, and the rhesus macaque is one example of a monkey used widely in laboratory research.

ANATOMY

Monkeys are characterized by a flattened chest, a hairy nose, a relatively large brain, a deep lower jaw, and sharp canine teeth. Although monkeys are quadrupedal, they are able to sit upright (and will occasionally stand erect), so that the dextrous hands are freed for manipulative tasks (such as picking apart fruits). They have grasping hands and feet, each with 5 digits. Their legs are slightly longer than the arms – much longer in leaping species (such as the red colobus), which also have a long, flexible spine. Monkeys also have a tail that is usually longer than the body, although in some species it is tiny and undeveloped. A few species, such as spider monkeys, have a prehensile tail, sometimes with a bare area at the end with creases and ridges that increase friction for grasping. The tail may also be used as a balancing organ and to indicate social gestures.

There are several anatomical differences between Old and New World monkeys. Old World monkeys, which are more closely related to apes than the New World monkeys, have a narrow nasal septum and the nostrils face forwards or downwards. New World monkeys have a broad nasal septum and nostrils that face sideways. Another major difference is that Old World monkeys have hard sitting pads on the rump, which are absent in New World monkeys.

SOCIAL GROUPS

New World monkeys have a great variety of social organizations. Marmosets, for example, usually live in groups consisting of a monogamous pair and subadult offspring that help rear the recent young. Squirrel monkeys, on the other hand, live in very large groups, sometimes over 100, with many females and few males. Spider monkeys live in large communities that split into small groups of varying composition when foraging.

In contrast, Old World monkeys usually exhibit just 2 types of social organization: baboons and macaques live in large, multi-male troops; the mandrill, the drill, the gelada, guenons, and most langurs live in "harems", with one adult male and several females. Within all monkey social groups, relationships are commonly very close, and grooming is a significant social glue. However, males in some species, such as baboons, will fight furiously with their long, sharp canines for dominance.

INTELLIGENCE

Monkeys are intelligent mammals. They are quick to learn, inquisitive, and have an excellent memory. These abilities have helped monkeys succeed in a range of habitats, where they must learn (for example) what they can eat and then remember when and where to find the food again.

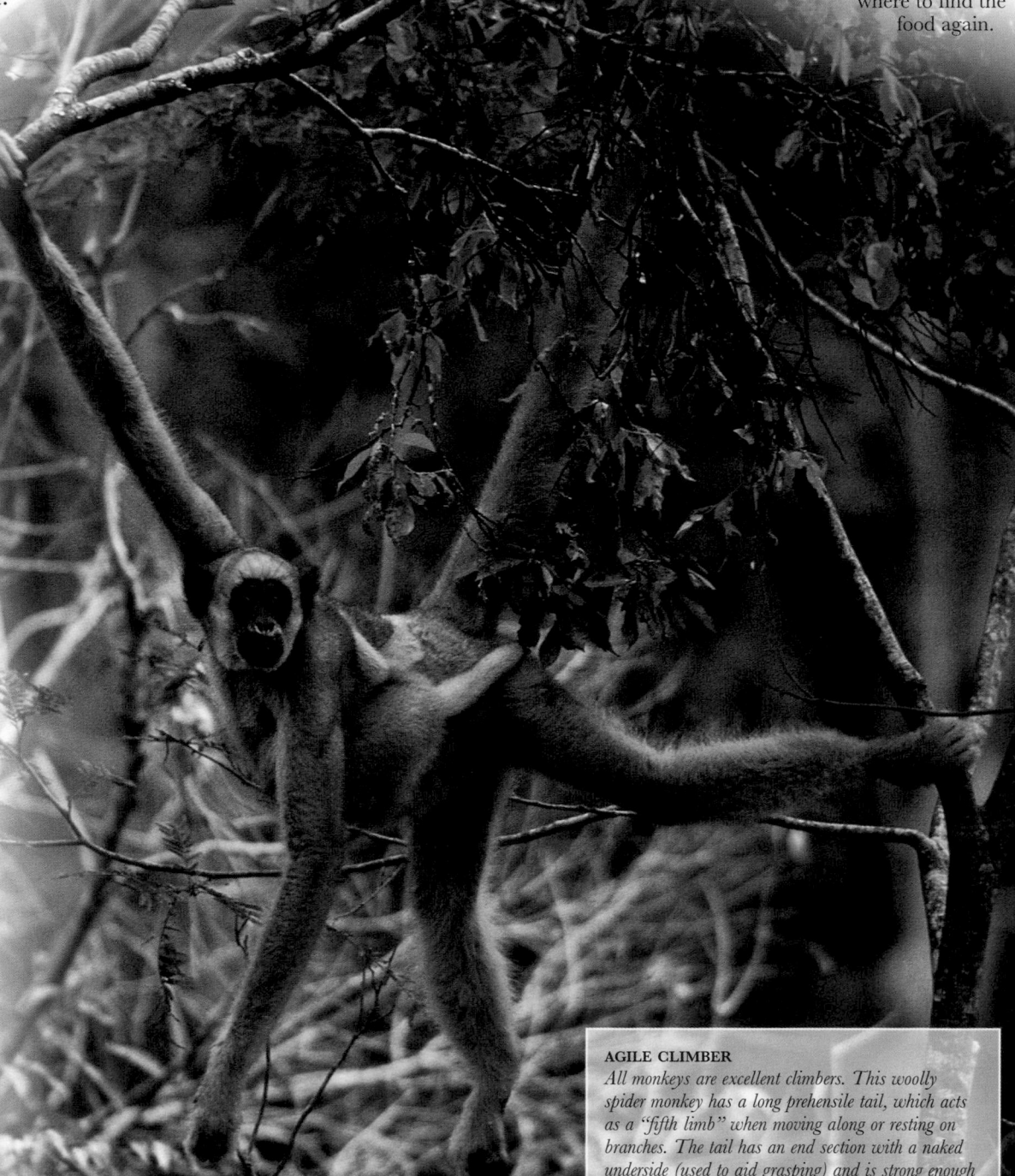

AGILE CLIMBER
All monkeys are excellent climbers. This woolly spider monkey has a long prehensile tail, which acts as a "fifth limb" when moving along or resting on branches. The tail has an end section with a naked underside (used to aid grasping) and is strong enough to support both the monkey and its offspring.

Lagothrix cana
Grey woolly monkey

Length 50–65 cm (20–26 in)	
Tail 55–77 cm (22–30 in)	
Weight 4–10 kg (8¾–22 lb)	

Location C. South America	Social unit Group
	Status Vulnerable

Woolly monkeys have thick, soft, close-curled fur, which in this species is grey with black flecks, darker grey on the head, hands, feet, and tail tip. In some individuals, there is a reddish tinge on the underparts. It lives in a mixed troop, which breaks into subgroups to forage primarily for

CONSERVATION

Monkeys such as the grey woolly need large areas in which they can roam freely. However, their forest habitat is increasingly fragmented, and their survival is additionally threatened by hunting for their meat, which is prized.

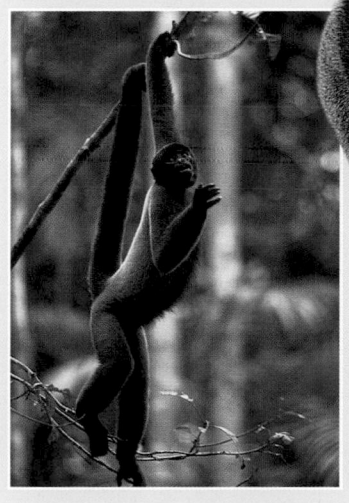

fruits but also for leaves, flowers, sap, seeds, and small creatures. This peaceful, gentle monkey often allows members of another troop into its own troop's territory. After a gestation period of 233 days, the single offspring is born and clings to its mother's underside, moving onto her back at 7 days. It is weaned after 6 months. Males grow larger than females when mature.

MUSCULAR MOVER
Woolly monkeys are stout bodied and almost pot bellied, with powerful shoulders, hips, and tail, for hanging from and swinging through trees.

naked gripping pad on underside near tail tip

large forehead and braincase

Ateles geoffroyi
Black-handed spider monkey

Length 50–63 cm (20–25 in)	
Tail 63–84 cm (25–33 in)	
Weight 7.5–9 kg (17–20 lb)	

Location S. Mexico, Central America	Social unit Group
	Status Vulnerable

Also known as Geoffroy's spider monkey, this species has black hands, head, and feet, and a cowl-like face surround. Like other spider monkeys, the thumbless hand acts as a simple hook, to swing agilely through trees or to pull fruit-laden branches to the mouth.

Ateles chamek
Black spider monkey

Length 40–52 cm (16–20½ in)	
Tail 80–88 cm (32–35 in)	
Weight 9.5 kg (21 lb)	

Location W. South America	Social unit Group
	Status Lower risk

Long black fur and black facial skin identify this otherwise typical spider monkey, which feeds mainly on fruits, berries, flowers, soft leaves, grubs, termites, and honey. They live in large territorial troops, each occupying 150–230 hectares (370–570 acres). The troop splits into variable subgroups to feed, and meets in the evening with greeting calls.

Ateles hybridus
Long-haired spider monkey

Length 42–58 cm (16½–23 in)	
Tail 68–90 cm (27–35 in)	
Weight 7.5–10.5 kg (17–23 lb)	

Location N.W. South America	Social unit Group
	Status Endangered

The long-haired spider monkey has a conspicuous triangular white patch on the forehead. It is brown above with pale underparts – giving rise to its other common name of white-bellied spider monkey. Mixed troops of about 20 split into single-sex subgroups of 3–4 to feed on fruits, juicy leaves,

and, oddly, soft, decaying wood. They whoop and whinny as they meet up again. Like all spider monkeys, it has long limbs, a slim body, thumbless hands, and a prehensile tail.

pale inner surfaces of limbs

long, prehensile tail

Brachyteles arachnoides
Woolly spider monkey

Length 55–61 cm (22–24 in)	
Tail 67–84 cm (26–33 in)	
Weight 9.5–12 kg (21–26 lb)	

Location C. South America (S.E. Brazil)	Social unit Variable
	Status Critically endangered

Also called the muriqui, the woolly spider monkey has a heavy body, long limbs, and thumbless hands with hook-like fingers. It is the largest New World monkey, and, being a leaf-eater, is critically endangered due to destruction of Atlantic coastal forests. Just a few hundred are left in 4 or 5 tiny groups on private ranches.

On reaching maturity, females leave to join another troop. After a gestation of 225 days, the single young is born. After 16 weeks it rides on its mother's back and is weaned by 18 months. Offspring of females high in the dominance hierarchy are more likely to survive to adulthood, when they move to other troops.

long, prehensile tail

long, slender, thumbless hands

Alouatta pigra
Mexican howler monkey

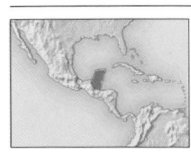

	Length 52–64 cm (20 1/2–25 in)
	Tail 59–69 cm (23–27 in)
	Weight 6.5–11.5 kg (14–25 lb)
Location Mexico, Central America	Social unit Group
	Status Vulnerable

Once considered a subspecies of the mantled howler (*Alouatta palliata*), this monkey is completely black, apart from the male's white scrotal sac. Loud whoops and howls at dawn and dusk proclaim a troop's territory, which may be up to 25 hectares (62 acres). Most troops comprise about 7 members with just one male, who is up to twice as heavy as a female. This monkey eats a plentiful, low-nutrient, leafy diet.

Alouatta seniculus
Red howler monkey

	Length 51–63 cm (20–25 in)
	Tail 55–68 cm (22–27 in)
	Weight 7–9 kg (15–20 lb)
Location N.W. South America	Social unit Group
	Status Lower risk

Red howlers are the largest of the nine howler species, whose loud howls, whoops, and other calls carry more than 2 km (1 1/4 miles) through the forest, informing others of a troop's presence or conveying alarm signals. They live mainly in groups of one male and 3–4 females. The male is much heavier than the female. When a new male ousts an existing male from a troop, he may kill the latter's offspring, so the females become ready to breed with him sooner.

GOLDEN SADDLE
The northern red howler (shown here) has a reddish gold "saddle" on the body, which contrasts with the maroon head, shoulders, and limbs. Other populations are a uniform red colour.

strong, prehensile tail that can support body weight

TREETOP LIFE

Like all howlers, the red howler is wonderfully adapted to life among trees. Its prehensile tail lacks fur on the underside near the tip, to help it grip and work as a fifth grasping limb. Its climbing ability provides access to a wide variety of leaves and fruit. Howlers need to consume up to 1 kg (2 1/4 lb) daily of this plentiful, but low-energy, diet. They spend up to three-quarters of the day resting to conserve energy.

long face

Pithecia monachus
Monk saki

	Length 37–48 cm (14 1/2–19 in)
	Tail 40–50 cm (16–20 in)
	Weight 1.5–3 kg (3 1/4–6 1/2 lb)
Location N. and W. South America	Social unit Group
	Status Lower risk

All 7 saki species have broad noses (especially the septum, the central portion between the nostrils), lank fur that falls to the sides from the back and neck, a bushy tail, and extra-long fur around the face,

grey-black body coloration

which in the monk, or red-bearded, saki forms a hood or cowl. Compared to many other New World monkeys, this saki is shy and quiet, preferring to stay high in the trees, keep still, and remain unnoticed. It can make a loud alarm call, but under threat its main defence is to bare its teeth. Its diet is fruits and seeds, and it lives in close-knit troops of 4–5, who spend much time grooming each other.

white stripe down side of nose

Pithecia pithecia
White-faced saki

	Length 34–35 cm (13 1/2–14 in)
	Tail 34–44 cm (13 1/2–17 1/2 in)
	Weight 2 kg (4 1/2 lb)
Location N. South America	Social unit Group/Pair
	Status Lower risk

No other New World monkey has such differences between the sexes: the male is black with a white or pale gold face and a black nose, while the

female is grey-brown with pale tipped hairs and a dark face. Although it is vegetarian, its teeth have a predatory appearance, with sharp incisors for impaling fruit and long canines to crack seeds and nuts. A typical group is a female, male, and 1–3 young.

pale face (male only)

white nose stripe (female only)

Chiropotes satanas
Black-bearded saki

	Length 33–46 cm (13–18 in)
	Tail 30–46 cm (12–18 in)
	Weight 2–4 kg (4 1/2–8 3/4 lb)
Location N. South America	Social unit Group
	Status Endangered

grip powerfully and this saki can hang by one limb as it feeds on seeds, hard fruits, and small animals, crushing them with well-developed molar teeth. When excited, it "switches" its tail like a cat and emits a piercing whistle.

In this species, the typical saki's long chin fur forms a bushy beard, and the long, dense head fur makes a thick forehead fringe. Both feet and hands

MAMMALS

Cacajao calvus
Bald uakari

Length 38–57 cm
(15–22 1/2 in)

Tail 14–18.5 cm
(5 1/2–7 1/2 in)

Weight 3–3.5 kg
(6 1/2–7 3/4 lb)

Location N.W. South America

Social unit Group

Status Endangered

Bald uakaris prefer "blackwater", part-flooded forests along small rivers, lakes, and swamps, rather than forests fringing large rivers. They forage by day in trees for seeds, fruits, flowers, and small animals, in a large troop of males and females, usually 10–20 but sometimes up to 100. The troop may mix with similar primates such as squirrel monkeys to feed.

COLOUR FORMS

Bald uakari subspecies have a variety of fur colours, leading to different names: white uakari (above) in northwestern Brazil, golden uakari on the Brazil–Peru border, red uakari on the Colombia–Brazil border, and pale-backed red uakari further east.

HAIRLESS FACE
The bald uakari has a hairless face and forehead, the skin colour varying from pink to deep red.

short tail in relation to body

Callicebus moloch
Dusky titi monkey

Length 27–43 cm
(10 1/2–17 in)

Tail 35–55 cm
(14–22 in)

Weight 0.7–1 kg
(1 1/2–2 1/4 lb)

Location N. South America

Social unit Pair

Status Lower risk

The 20 or more titi species all have thick, soft fur, chunky bodies, short limbs, and – distinctively among New World monkeys of their size – ears almost hidden in fur. They eat mainly fruits, leaves, seeds, and grubs. The dusky titi monkey's back is speckled brown, the underparts mainly orange. It relies on its drab coloration and slow movements for camouflage in trees near swamps and pools. Female and male form a close pair-bond and defend a territory of 6–12 hectares (15–30 acres). They intertwine tails and sing a "duet" just after dawn to maintain their family and pair-bonds, and to proclaim their territory.

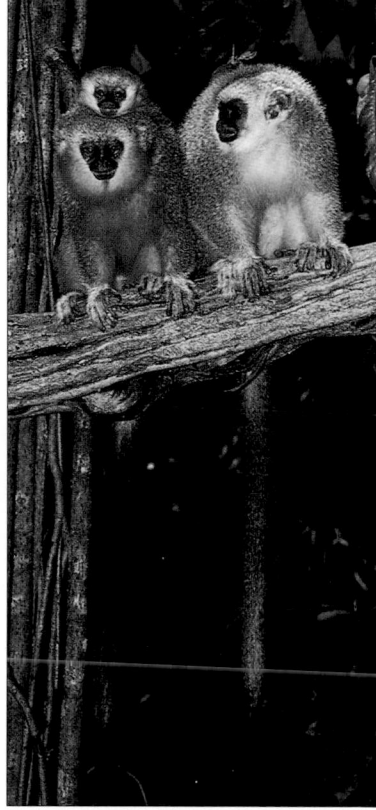

Aotus lemurinus
Night monkey

Length 30–42 cm
(12–16 1/2 in)

Tail 29–44 cm
(11 1/2–17 1/2 in)

Weight 900–950 g
(32–34 oz)

Location Central America to N.W. South America

Social unit Pair

Status Vulnerable

Formerly regarded as one species, genetic studies show there may be 10 species of night monkeys, also called douroucoulis, or owl monkeys from their hooting calls in the darkness. They are the only nocturnal monkeys, eating a mixed diet of fruits, leaves, and insects as they clamber cautiously through the branches. They live in male–female pairs and communicate by scents in urine and chest gland secretions. The single young is born after a gestation of 120 days. Weaning takes 8 months and, since juveniles may also stay with parents, close-knit family groups of 4–5 develop.

speckled grey fur on back

dark, bushy tail tip

yellow or grey underparts

Callicebus torquatus
Yellow-handed titi monkey

Length 30–46 cm
(12–18 in)

Tail 37–49 cm
(14 1/2–19 1/2 in)

Weight 1–1.5 kg
(2 1/4–3 1/4 lb)

Location N.W. South America

Social unit Group/Pair

Status Lower risk

Mainly dark brown in colour, except for its distinctive white neck collar, and yellow hands, this titi monkey has the typical *Callicebus* features of thick fur that obscures the ears, stout body, relatively short limbs, and bushy, non-prehensile tail. Its main foods are fruits, especially of *Jessenia* palms, and seeds, leaves, and insects. The yellow-handed titi monkey is less vocal than other titi monkey species. However, the male and female remain close at all times, groom each other, and intertwine tails. Family groups develop as youngsters stay with their parents for up to 3 years.

Cebus apella
Brown capuchin

Length 33–42 cm
(13–17 in)

Tail 41–49 cm
(16–19 1/2 in)

Weight 3–4.5 kg
(6 1/2–10 lb)

Location N., C., and E. South America

Social unit Group

Status Lower risk

Capuchins are often regarded as the most adept New World monkeys, and the brown capuchin uses a variety of tools, such as stones to crack hard nuts. It also eats fruits, insects, and some vertebrates such as frogs, lizards, and even small bats. This species, also known as the tufted capuchin from the furry "horns" above each ear, has the widest range of any New World monkey. Mixed groups of 8–14 are usual as members are not sexually mature until 7 years old – later than most monkeys of similar size.

prehensile tail often curled at tip

Cebus olivaceus

Weeper capuchin

Length	37–46 cm (14 1/2 –18 in)
Tail	40–55 cm (16–22 in)
Weight	2.5–3.5 kg (5 1/2 – 7 3/4 lb)

Location N.E. South America

Social unit Group

Status Lower risk†

Like the other 5 or so capuchin species, the weeper has a robust build with relatively short limbs, and a prehensile tail. There is no bare skin on the tail's underside, as in many other American monkeys. The main colour is brown, becoming paler on the arms, and even grey or yellow on the face. Seeds, fruits, and small creatures, especially snails and insects, comprise the main diet. The weeper capuchin forms bands of 30 or more, containing mostly females and young, with several males, but only one dominant male breeds. Mothers may look after each other's offspring (known as "allomothering"). Troop members keep in contact with a plaintive, "weeping" cry.

tail habitually held in curled position

paler fur on limbs

small face

Saguinus imperator

Emperor tamarin

Length	23–26 cm (9–10 in)
Tail	39–42 cm (15 1/2 –16 1/2 in)
Weight	450 g (16 oz)

Location W. South America

Social unit Group

Status Vulnerable

Marmosets and tamarins form a distinct group of about 35 American primate species. They are similar to other New World monkeys but differ in body chemistry, have claws rather than nails, and bear 2 rather than one offspring. Identified by its flowing white moustache, this species feeds on fruits in the wet season, flower nectar and tree sap in the dry season, and insects, especially crickets, all year round. It often forms a mixed troop with related species such as the saddleback tamarin. Each species responds to the other's alarm calls if a predator is detected nearby.

MOUSTACHE
The white, curly-ended moustache of both males and females contrasts with the black face, speckled red- or grey-brown body, and red-orange tail fur.

TAMARIN BABYCARE

Tamarins and marmosets have relatively long gestation periods for such small mammals, the emperor tamarin's being 140–145 days. There are almost always 2 offspring, and the father carries these, as here, except when they are being suckled by their mother.

Saimiri boliviensis

Bolivian squirrel monkey

Length	27–32 cm (10 1/2 –12 1/2 in)
Tail	38–42 cm (15–16 1/2 in)
Weight	950 g (34 oz)

Location W. to C. South America

Social unit Group

Status Lower risk†

No other New World monkeys form such large, active troops as the 5 species of squirrel monkey. They regularly number 40–50, occasionally 200 or more, moving noisily by day with twitters and clucks to disturb their food of varied, small creatures. They may follow in the path of other monkeys to obtain insects in their wake. They also consume fruits and seeds. The mature male becomes "fatted" around the shoulders in the breeding season and competes aggressively – the winner mating with most females.

FAMILY MEAL

Within a squirrel monkey troop, members coalesce into peer sub-groups: adult males, pregnant females, females with young, and juveniles. If an individual locates food, subgroup members quickly gather around to exploit the find.

tail longer than head and body

SQUIRREL FEATURES
The Bolivian squirrel monkey has a small, white face, a large forehead, dark head crown above, black nose tip and muzzle, and well-furred ears.

Callimico goeldii

Goeldi's marmoset

Length	22–23 cm (9 in)
Tail	26–32 cm (10–12 1/2 in)
Weight	575 g (21 oz)

Location N.W. South America

Social unit Group

Status Vulnerable

Goeldi's marmoset is larger than most marmosets and tamarins. Its long fur is black, and it has a "cape" of longer hair on its head and neck. Unlike similar species, it has wisdom teeth. It eats fruits, sap and gum (digging its incisor teeth into bark to make these flow), insects, and small vertebrates such as lizards. The species forms stable, close-knit groups of up to 10, mainly male–female pairs and their young, and keeps to dense vegetation such as creeper-tangled bamboo.

slim tail expands to black, bushy tip

Cotton-top tamarin

Saguinus oedipus

Length 20–25 cm (8–10 in)	
Tail 33–40 cm (13–16 in)	
Weight 400–450 g (14–16 oz)	

Location N.W. South America

Social unit Group/Pair

Status Endangered

The long, white, flowing fur on the head crown distinguishes this tamarin, found in an extremely restricted range in Colombia. Like many marmosets and tamarins, this species has a varied diet and a "helper" system of rearing offspring, in which males and older siblings carry the very young. However, in each troop of 10–12 only one male–female pair breeds; 4 out of 5 births are twins. The cotton-top tamarin has been used in medical research, and currently there are more of these monkeys in captivity than in the wild.

Silvery marmoset

Callithrix argentata

Length 20–23 cm (8–9 in)	
Tail 30–34 cm (12–13 1/2 in)	
Weight 325–350 g (12–13 oz)	

Location C. South America

Social unit Group

Status Lower risk†

Colour is the main distinguishing feature among 10–15 very similar, closely related marmosets from south of the Amazon. This species has pale silver-grey fur on the back, creamy yellow on the underside, and black on the tail; the face has pink skin and ears. In each small troop of silvery marmosets, only one male and female breed. The rest are siblings, cousins, and other "helpers", who assist in carrying and protecting the infants.

Pygmy marmoset

Callithrix pygmaea

Length 12–15 cm (4 3/4–6 in)	
Tail 17–23 cm (7–9 in)	
Weight 100–125 g (3 5/8–4 oz)	

Location W. South America

Social unit Group/Pair

Status Lower risk†

The smallest monkey in the world, a curled-up pygmy marmoset fits into a human palm. However, it is long-lived for such a diminutive mammal and may reach 12 years of age. The pygmy marmoset differs from other marmosets in the way it eats gum: it gouges out 10 or more new holes in bark each day, scent-marks them, and returns to these and older holes at intervals to scrape up the sticky, oozing liquid with its long, lower incisor teeth. It also takes flower nectar, fruits, and small creatures such as grubs and spiders. The pygmy marmoset follows the reproductive pattern of other marmosets, with one breeding pair per troop of 5–10. Other troop members, usually older offspring, are childcare "helpers", who carry the two newest youngsters after the breeding male has cared for them during the first few weeks.

long cape or hood of hair on head

speckled tawny fur

clawed fingers and toes

indistinct black and tawny tail rings

Geoffroy's marmoset

Callithrix geoffroyi

Length 20 cm (8 in)	
Tail 29 cm (11 1/2 in)	
Weight 350 g (13 oz)	

Location E. South America

Social unit Group/Pair

Status Vulnerable

ringed tail

This marmoset prefers secondary forest, that is, regrown after logging or other disturbance, rather than primary areas. It digs holes in tree bark with its long incisors, marks "ownership" with scent from its perineal gland, and then returns at intervals to scrape up the oozed gum. Like other marmosets, this species also eats fruits and insects. There is only one breeding pair in each troop, but the other members also carry and protect the young.

Golden lion tamarin

Leontopithecus rosalia

Length 20–25 cm (8–10 in)	
Tail 32–37 cm (12 1/2–14 1/2 in)	
Weight 400–800 g (14–29 oz)	

Location E. South America (Rio São João basin)

Social unit Group/Pair

Status Critically endangered

This tamarin weighs twice the average for the marmoset and tamarin group. The face is dark grey and the hands, fingers, and claws are long and thin, to probe into bark and crevices for grubs. However some four-fifths of the diet is fruit, supplemented by gum and nectar. It forages by day and sleeps at night in tangled vegetation or, more often, a hole in a tree. There is only one male–female breeding pair in the troop of 4–11. In its "helper" system of rearing young – 2 offspring being typical for the group – sexual activity of the helpers is not suppressed, although they do not produce young until they form the dominant partnership.

CONSERVATION

The plight of the golden lion tamarin became symbolic of the consequences of forest clearance and has been the subject of conservation efforts since the 1960s. It bred well in captivity (here, a golden lion tamarin is seen using a nesting box in a zoo) and since the mid-1980s it has been re-introduced into southeast Brazil. Although over 150 captive-bred tamarins and their offspring are now living in the wild, its situation is still precarious.

LION'S MANE
The long, silky, red-gold head hair of this species cascades over the shoulders in the manner of a lion's mane. Its attractive appearance is a further threat, making it vulnerable to the illegal pet trade.

Mandrillus sphinx

Mandrill

Length	63–81cm (25–32in)
Tail	7–9cm (2¾–3½in)
Weight	11–37kg (24–82lb)

Location W. Central Africa | **Social unit** Group

Status Vulnerable

A scarlet nose with prominent, bony blue flanges on either side are the unmistakable characteristics of the male mandrill. The female's facial colour is much more subdued. She is about a third the size of the male, which, at 60cm (23½in) high to the shoulder, is the largest of all monkeys. Mandrills live in groups in dense African rainforest, spending most of their day on the ground looking for fruits and seeds, as well as eggs and small animals. When night falls, they take to the trees for safety. Troops move over a range of up to 50 square km (20 square miles); they mark their territory with scent and defend it against rivals. Much of what is known about this species comes from studies of captive animals: in the wild, hunted for meat, and with its habitat being destroyed by logging, the mandrill is becoming increasingly scarce.

A WARNING YAWN
When threatened by a predator or approached by rivals, the male will yawn widely to reveal his fearsome teeth, which can be 6.5cm (2½in) long.

speckled, olive-grey fur

stumpy tail

facial ridges

4 limbs of similar length

BOLD ADVERTISEMENT
With his spectacularly coloured face – scarlet nose with blue flanges, and yellow beard – the mandrill boldly declares his identity to other animals in the forest. Together with his mauve-blue rump, these colours also announce the male's sex and display his virility to females.

LIFE ON THE GROUND
The mandrill walks and runs on all fours – a form of locomotion made all the more efficient by having fore- and hindlimbs of nearly equal length.

MANDRILL SOCIETY

Mandrills usually live in mixed groups of about 20, which come together to form troops of as many as 250 individuals. There is a strict hierarchy within the group: a dominant male heads each group, mating with fertile females and fathering almost all the infants. Non-breeding males make up the lower ranks.

MALE AND FEMALE
The male is much larger than the female, with a far more colourful face. Some mandrills may live in harems, with one mature male to 20 females.

MOTHER AND CHILD
The mandrill gives birth to one offspring every 18 months or so. At first, the mother will carry the infant on her belly, but as it grows heavier, it will ride through the forest on her back.

FORAGING PARTY
Mandrills often forage in small parties, grunting constantly to stay in touch. When it is time to move on, the dominant male will round up his group with a 2-phase grunt or roar.

Erythrocebus patas
Patas monkey

	Length 60–88 cm (23 1/2 – 35 in)
	Tail 43–72 cm (17–28 in)
	Weight 10–13 kg (22–29 lb)
Location W. to E. Africa	Social unit Group
	Status Lower risk†

This is one of the fastest monkeys when running. It has a long, slim body, lengthy limbs, hands, and feet, and short digits. The white moustache and beard contrast with the darker face; the red back contrasts with the white limbs and underside. It lives in troops of up to 10 members, and the single male stays on the periphery, to act as decoy, drawing predators while the females and young hide.

Macaca fascicularis
Crab-eating macaque

	Length 37–63 cm (14 1/2 – 23 in)
	Tail 36–72 cm (14–28 in)
	Weight 3.5–12 kg (7 3/4 – 26 lb)
Location S.E. Asia	Social unit Group
	Status Lower risk

The crab-eating macaque is southeast Asia's commonest monkey, occurring mainly in forests and mangroves along rivers, coasts, and offshore islands. It climbs and swims well, spends time on the ground, and exploits human habitation. The 20 or so macaque species are mostly adaptable, with varied diets. This species eats fruits and seeds as well as aquatic animals, and it forms quarrelsome troops.

Cercopithecus neglectus
De Brazza's monkey

	Length 50–59 cm (20–23 in)
	Tail 59–78 cm (23–31 in)
	Weight 7–8 kg (15–18 lb)
Location C. to E. Africa	Social unit Pair
	Status Lower risk†

One of the most terrestrial of the 20 or so *Cercopithecus* species, De Brazza's monkey is also the only one of the genus that forms male–female pair bonds. Widespread yet inconspicuous over its large range, it marks its territory with saliva and scent, yet avoids rather than challenges any intruders. It has blue-white upper-lip and chin fur, and a thin, white thigh stripe. It uses deep, booming calls to communicate, and eats mainly seeds and fruits.

Macaca nigra
Celebes macaque

	Length 52–57 cm (20 1/2 – 22 1/2 in)
	Tail 2.5 cm (1 1/4 in)
	Weight 10 kg (22 lb)
Location S.E. Asia (northern Sulawesi)	Social unit Group
	Status Endangered

The all-black fur, very short tail, and homeland of Celebes (now Sulawesi) give this species the alternative name of Celebes black ape. A crest runs from the forehead back over the crown. Usually flat, it rises when the animal is aroused. This species forms huge, mixed-sex troops of more than 100, yet is generally an inconspicuous, quiet, fruit-eating forest dweller.

mid-line head crest

Cercocebus torquatus
White-collared mangabey

	Length 50–60 cm (20–23 1/2 in)
	Tail 60–75 cm (23 1/2 – 30 in)
	Weight 10 kg (22 lb)
Location W. Africa	Social unit Group
	Status Critically endangered

Troops of up to 90 white-collared mangabeys forage by day for fruits and seeds in their home range, usually along a river. Mainly terrestrial, this species has sooty fur, a pink-grey face,

Papio anubis
Olive baboon

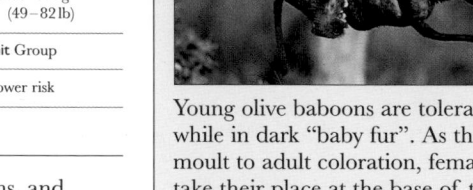

	Length 60–86 cm (20–34 in)
	Tail 41–58 cm (16–23 in)
	Weight 22–37 kg (49–82 lb)
Location W. to E. Africa	Social unit Group
	Status Lower risk

One of the largest baboons, and common across west and northeast Africa, the male may be twice the female's weight. Both sexes have a thick grey ruff around the cheeks. Average troop size is 20–50, occasionally over 100. They eat fruit, leaves, insects, lizards, and sometimes larger prey such as gazelle fawns.

Papio papio
Guinea baboon

	Length 69 cm (27 in)
	Tail 56 cm (22 in)
	Weight 17.5 kg (39 lb)
Location W. Africa	Social unit Group
	Status Lower risk

The male of the smallest baboon species has proportionally the longest mane, reaching almost to his rump. Even for a baboon, the diet is

and fossae (pits) below the cheeks. Its powerful jaws and teeth easily crack most nuts, which are held by hand. Troops are mixed males, females, and young. Males establish a hierarchy yet subordinates also mate – sometimes more often than their seniors.

GROUP ORDER

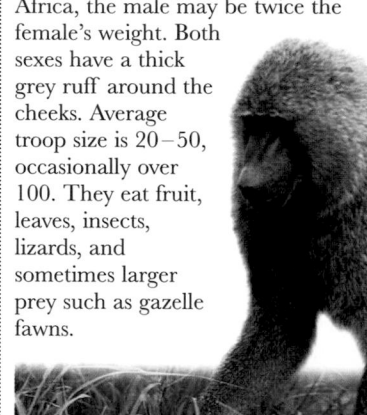

Young olive baboons are tolerated while in dark "baby fur". As they moult to adult coloration, females take their place at the base of the troop hierarchy. Males are driven off, and must battle their way into another troop.

PRIMATE "DOG"
Typically powerful and dog-like, the olive baboon is speckled olive-green, with a black face and rump.

varied: from tough roots to juicy grubs and eggs, and sometimes farm crops. Troops number up to 200, although about 40 is more usual, with several males gathering their own harem of females.

Papio ursinus
Chacma baboon

Length 60–82 cm (23 1/2 – 32 in)	
Tail 53–84 cm (21–33 in)	
Weight 15–30 kg (33–66 lb)	
Location Southern Africa	Social unit Group
	Status Lower risk†

The largest baboon, the chacma has a drooping snout and protruding nostrils. Fur colour varies from yellow-grey to black, paler on the muzzle. This intelligent, adaptable primate enjoys very fluid troop composition in some parts of its range. Its wide diet ranges from roots and seeds to insects and young gazelles. In the wild, it is known to use tools such as sticks.

Theropithecus gelada
Gelada

Length 70–74 cm (28–29 in)	
Tail 46–50 cm (18–20 in)	
Weight 19 kg (42 lb)	
Location E. Africa	Social unit Group
	Status Lower risk

A close cousin of baboons, the pink-chested gelada is restricted to the windy, grassy Ethiopian highlands. Its limited diet of grass blades, stems, and seeds is picked by rapid, dextrous hand movements as it sits and shuffles along. Small groups of females and young led by a male may band into huge but loose troops.

Nasalis larvatus
Proboscis monkey

Length 73–76 cm (29–30 in)	
Tail 66–67 cm (26–26 1/2 in)	
Weight 21 kg (46 lb)	
Location S.E. Asia	Social unit Group
	Status Endangered

Among the most specialized and distinctive of mammals, the proboscis monkey occupies very restricted habitats near water in lowland rainforest, mangrove swamps, and coasts on the island of Borneo. The average troop is one male with 6–10 females and their young. The male defends his group, honks loudly, bares his teeth, and waves his erect penis at an aggressor.

partly webbed feet

CHANGING FACE

At birth, the young have a blue face, dark fur, and a "normal" monkey nose. The coloration changes and the nose grows with age. In females the nose is much smaller than in the male, but it is still large compared to related species.

LARGE NOSE
The proboscis monkey's defining feature is its large nose. In mature males it is long and pendulous, and may play a role in attracting a mate.

Semnopithecus entellus
Hanuman langur

Length 51–78 cm (20–31 in)	
Tail 69–102 cm (27–40 in)	
Weight 10.5–20 kg (23–44 lb)	
Location S. Asia, Sri Lanka	Social unit Group
	Status Lower risk

The Hanuman langur is found throughout south Asia, except for rainforest areas. Coloration varies from dark brown in individuals from the Himalayas to pale fawn in those from Sri Lanka. Groups of females and young may be led by one or several males, while other males form bachelor troops.

MIXED DIET

The Hanuman langur feeds mostly on leaves, fruits, buds, and shoots, which it digests easily in its compartmentalized stomach. Like many wild creatures, these monkeys supplement their diet by licking salt or eating mineral-rich soil. Troops that live around villages also often benefit from leftovers and offerings from local people, who hold the species sacred, identifying it with the Hindu deity Hanuman.

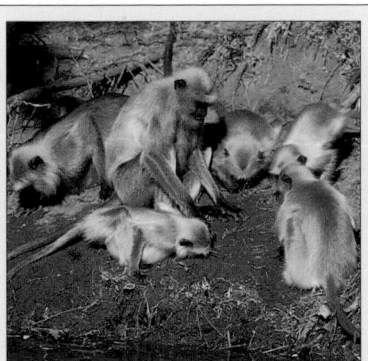

Colobus guereza
Eastern black and white colobus

Length 52–57 cm (20 1/2 – 22 1/2 in)	
Tail 53–83 cm (21–33 in)	
Weight 8–13.5 kg (18–30 lb)	
Location C. and E. Africa	Social unit Group
	Status Lower risk†

Also known as the mantled guereza, this colobus has a white face border, "veils" down its flanks and rump, and a bushy tail tip, but is otherwise black. Each male leads a small troop of 4–5 females and young, defending his territory with roars and tremendous leaps. The 3-part stomach houses gut microbes that break down cellulose-based plant food, enabling this species to gain twice as much nutrition than other monkeys from such a leaf-rich diet.

Rhinopithecus roxellana
Snub-nosed monkey

Length 54–71 cm (21 1/2 – 28 in)	
Tail 52–76 cm (20 1/2 – 30 in)	
Weight 12.5–21 kg (28–46 lb)	
Location E. Asia	Social unit Group
	Status Vulnerable

In cold mountain forests this monkey endures average winter temperatures of -5°C (23°F). Its long fur and bushy tail give insulation, and it moves in trees or on the ground on its strong, stout limbs. Stubby fingers pluck leaves, fruits, seeds, and lichens. Large troops of several hundred split into bands of one male and many females for foraging and breeding.

APES

PHYLUM	Chordata
CLASS	Mammalia
ORDER	Primates
SUBORDER	Haplorhini (part)
FAMILIES	2
SPECIES	21

APES ARE THE CLOSEST relatives to humans. They are similar to people not only in appearance but also in that they are highly intelligent and form complex social groups. Apes are divided into the lesser apes (the gibbons) and the larger, more human-like great apes – the orang-utan, gorillas, and chimpanzees. They are found in western and central Africa and throughout southern and Southeast Asia, mainly in tropical rainforests. Apes are essentially vegetarian and mostly eat fruits, although some are omnivorous. They are threatened by loss of their forest habitat, hunting and poaching (often for their skins or skulls), and capture for zoos and the pet trade. Chimpanzees were once widely used in medical research.

ANATOMY

A shortened spine and a relatively short, broad pelvis lower the centre of gravity in apes, thereby facilitating a more upright posture. Apes have a broad chest, with the shoulder blades at the back, which allows an exceptionally wide range of movement in the shoulder joint. A gorilla, for example, can sit on the ground and reach out in any direction to pull in vegetation. Apes also have a flattened face, well-developed jaws, grasping hands and feet, and downward-directed, close-set nostrils. The great apes are very large: the orang-utan is the largest arboreal mammal and the gorilla may weigh over 200kg (440lb).

SOCIAL ORGANIZATION

The lesser apes form monogamous pairs. They mark their treetop territories with loud, musical songs in which the male and female sing different parts.

A maturing young will attempt to establish its own territory by singing alone, until it finds a mate.

The orang-utan is the only great ape that is solitary. A mature male controls a large territory with deep, resonant calling. He has access to all the females that enter his domain. The other great apes have well-defined social groups. Gorillas live in troops of 5–10 (occasionally up to 30), consisting of several females, one dominant mature ("silverback") male, and possibly one or 2 other silverbacks (the sons or younger brothers of the dominant male). Chimpanzees live in communities of 40–100. Although there is a dominant male and a social hierarchy, individuals have almost complete freedom to come and go. Foraging occurs in small groups, the composition of which changes daily. Chimpanzees found in west Africa are particularly fond of hunting, and the males cooperate closely to catch several monkeys (for example) at once. Chimpanzees apparently pass on customs and technologies socially, by example, rather than genetically.

INTELLIGENCE

Apes are extremely intelligent – even more so than monkeys. They appear to work through problems in the same way that humans do. Chimpanzees, for example, use and sometimes make simple tools, as does at least one population of orang-utans in northern Sumatra. The orang-utan is one example of an ape that has performed several complex tasks – such as solving puzzles, using sign language, and learning to recognize symbols – in research centres.

MUTUAL GROOMING
Grooming is important in ape societies because it strengthens and maintains bonds between individuals, such as these 2 chimpanzees. However, it is also a means of gaining favour and is used, for example, by males to ensure support from friends in the event of a challenge to their supremacy.

Hylobates lar
Lar gibbon

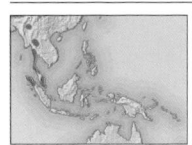

Length 42–59cm (16½–23in)	
Tail None	
Weight 4.5–7.5kg (10–17lb)	
Location S.E. Asia	**Social unit** Pair
	Status Endangered

The lar gibbon becomes active shortly after dawn, when a female and male "duet" to reinforce their pair-bond. She begins with a series of loud, long hoots that rise to a crescendo; the male responds as these fade, with simpler, more tremulous hoots. Each duet, repeated numerous times, lasts 15–20 seconds according to region. Most of this gibbon's day is spent finding food and eating. Half the diet is fruits, the rest being leaves, insects, and flowers. Some 15 minutes each day is spent in mutual grooming between partners. The lar gibbon rarely moves at night, resting among tree branches or forks. Also called the common or white-handed gibbon, the lar gibbon was considered to make lifelong pair-bonds, but recent studies show some serial monogamy with occasional partner changes, and even non-monogamous groupings. Each pair or group defends a territory comprising about three-quarters of its total home range. Gestation is 7–8 months; the single offspring is suckled for 18 months, is adult size by 6 years, and fully mature by 9 years. Lifespan in the wild averages 25–30 years. Deforestation and hunting by humans are the major threats.

young clings to mother's chest

arms approximately 40 per cent longer than legs

WHITE FRINGES
This gibbon's skin is black and the fur around its face, hands, and feet is white. The rest of the fur is uniformly coloured for each individual, but varies from cream to red, brown, or almost black.

ischial callosities (hard-skinned sitting pads)

BRACHIATION

The gibbon's arm-hanging, hand-swinging method of movement is called brachiation. It saves energy by maintaining momentum, using the body as a pendulum. The lar gibbon releases its grip with one hand at the height of its arc of swing, as its forward-facing eyes allow stereoscopic, distance-judging vision that helps to determine the next hand-hold, which may be 3m (9¾ft) away. The small thumb is set far back near the wrist and the fingers work like a hook.

HANDY FEET

The feet of the lar and other gibbons, like the palms of their hands, have bare, leathery skinned soles for effective grip. The big toe is able to grasp in opposition to the toes, so that they can walk upright along branches.

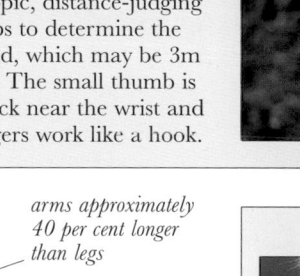

MAMMALS

Hylobates syndactylus
Siamang

Length 90cm (35in)	
Tail None	
Weight 10–15kg (22–33lb)	
Location S.E. Asia	**Social unit** Group/Pair
	Status Lower risk

The largest gibbon, with a "standing" height of 1.5m (5ft), the siamang also has the loudest calls of the group, and the closest-knit families. The female (who is dominant), male, and 1–2 offspring rarely stray more than 30m (100ft) from each other, and are usually less than 10m (33ft) apart. About three-fifths of their food intake is leaves, and one-third fruits, with a few blossoms and small creatures such as grubs. The family occupies a home range of about 47 hectares (116 acres), but defends only some 60 per cent of this as their territory, chiefly using their powerful calls.

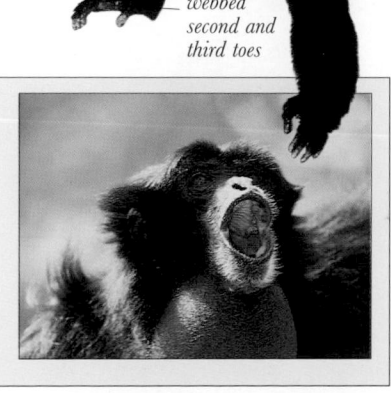

arm spread up to 1.5m (5ft)

ALL BLACK
The siamang is covered with black, shaggy fur. The male is slightly larger than the female and has a tuft of hair on his genital region that, at a glance, looks like a tail.

webbed second and third toes

BARKS AND SCREAMS

The siamang's dark grey, elastic throat skin inflates to the size of a grapefruit, to act as a resonator and amplify its amazingly loud calls. The male's screams are thought to discourage other males, while the female's longer, more distinctive series of bark-like sounds is associated with territory defence.

Hylobates concolor
Crested gibbon

Length 45–64cm (18–25in)	
Tail None	
Weight 4.5–9kg (10–20lb)	
Location S.E. Asia	**Social unit** Group/Pair
	Status Critically endangered

Unlike most other "flat-headed" gibbons, the crested gibbon has long fur on the head crown. The offspring is born with yellow fur, becoming black as it matures; females continue to change to brown or grey. This gibbon eats mainly leaf buds, shoots, and fruits, but rarely animals. It lives in groups, usually a female, a male, and their young.

crest of fur on crown

white cheek patch

Hylobates leucogenys
White-cheeked gibbon

Length 45–64cm (18–25in)	
Tail None	
Weight 4.5–9kg (10–20lb)	
Location S.E. Asia	**Social unit** Group/Pair
	Status Unconfirmed

Considered a subspecies of the crested gibbon until 1989, the infants, adult females, and males of this species have similar coloration (see left), and also the "crest" of longer fur on the head. The 2 species are distinguished mainly by geography: the crested gibbon occurs northeast of the Song Ma and Song Bo rivers in Vietnam, while the white-cheeked is found to the southwest. The single young is born after a gestation period of 7–8 months and is dependent on its mother for 18 months.

arms longer than legs

Hylobates moloch

Silvery gibbon

	Length 45–64 cm (18–25 in)
	Tail None
	Weight 5.5 kg (12 lb)

Location S.E. Asia	Social unit Pair
	Status Critically endangered

Pale eyebrows, cheeks, and beard merge with the mainly silvery fur of this gibbon. Its diet is fruits, some leaves, and sometimes nectar and grubs. Like most other gibbons, the family group (female, male, offspring) use calls to defend their territory. Unlike the lar gibbon (p.133), the male and female do not "duet" together.

dark blue-grey head cap

Pan paniscus

Bonobo

	Length 70–83 cm (28–33 in)
	Tail None
	Weight Up to 39 kg (86 lb)

Location C. Africa	Social unit Group
	Status Endangered

The bonobo or pygmy chimpanzee is only slightly smaller than the Pan troglodytes chimpanzee (see below), but has a slimmer body and relatively long, more slender limbs. It feeds chiefly on the ground mainly on fruits and seeds, but also on leaves, flowers, fungi, eggs, and small animals. This ape can live in groups of up to 80 but is usually found in smaller groups, as it forages and grooms. Sexual relations are common between males, females, and young in various combinations, and may be used to ease social tensions. Females are dominant and leave their family groups when mature; males tend to stay.

FACE AND FUR
The skin of the pygmy chimpanzee is mostly black, even on a juvenile's face. The fur on the crown has a central "parting".

LONG INFANCY

The bonobo is born after 8 months' gestation, and suckled by its mother for up to 3 years. The female continues to protect, groom, and share nests with her offspring for another year or 2. The interval between births is about 5 years.

Pan troglodytes

Chimpanzee

	Length 63–90 cm (25–35 in)
	Tail None
	Weight 30–60 kg (66–130 lb)

Location W. to C. Africa	Social unit Group
	Status Critically endangered

Chimpanzees live in communities of 15–120. Subgroup composition varies almost hourly for activities such as grooming, feeding, travelling, and defending the territory. This last task is usually carried out by adult male parties, who may attack and kill stray chimps from other communities. Most daylight hours are spent eating – mainly fruits and leaves, but also flowers and seeds. Raiding parties sometimes cooperate to kill and eat animal prey such as monkeys, birds, and small antelopes. Social bonds may last years, but there are no long-term male–female bonds for reproduction. The single young (rarely twins), born after a gestation period of 8 months, is fed, carried, and groomed by its mother for 3–4 years. It also learns her feeding techniques. Chimpanzees not only use tools but also make them – for example, stripping side branches from a twig, which it uses to scoop out termites from their nest. The 2 chimpanzee species are our closest living relatives, and their intelligence, range of emotions, and communication and learning skills have made them valuable to animal trainers, collectors, and researchers. They are also killed for the bushmeat trade.

bare skin on face darkens with age

sparse black hair over most of body

arms longer than legs

knuckles used for walking

big toe opposes other toes

VOCAL CHIMPANZEES

The chimpanzee makes more than 30 distinct calls including the pant-hoot (shown right), which consists of shrieks and roars that can be heard up to 2km (1¼ miles) away. It is used in many situations and is the most common adult sound. It is thought to identify the caller within its community and solicit information from other members.

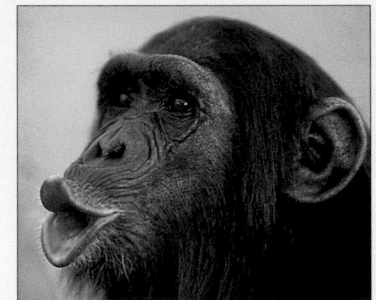

FACIAL FEATURES
The chimpanzee uses a wide variety of expressions, clearly visible on the hairless facial area. In particular it uses its flexible, protrusile lips to make grimace-like "smiles" that actually signify fear.

NIGHT NESTS

Each adult chimpanzee builds a new, individual tree nest each night, for sleeping. (Rarely an old nest is renovated or reused.) The chimpanzee bends over and inter-twines many branches to make a firm, leafy platform, usually 3–10m (10–33ft) above the forest floor, away from ground-based predators. Infants generally sleep in their mothers' nests until they are about 5–6 years of age.

Gorilla gorilla
Western gorilla

Height	1.3–1.9m (4 1/4 – 6 1/4 ft)
Tail	None
Weight	68–200 kg (150–430 lb)

Location C. Africa	Social unit Group
	Status Endangered

"silverback" of mature male

opposable big toe

mainly black fur

The largest living primates, gorillas are day-active forest dwellers that feed on fruits, leaves, stems, and seeds, as well as a few small creatures such as termites. At night they bend tree twigs and branches to form a sleeping nest. Living in small, stable groups of 3–20 with strong bonds between the silverback (dominant male) and the females with offspring, the western gorilla uses a wide range of facial expressions, body gestures, and sounds to communicate, including whimpers, grunts, rumbles of contentment, and alarm barks. Its home range of 800–1,800 hectares (2,000– 4,450 acres) may overlap with neighbouring groups but is not actively defended. The single young is born after a gestation of 8 1/2 months. It clings to the mother's belly for 4–6 months, then rides on her shoulders or back. It first chews vegetation at 4 months and is weaned by 3 years.

SILVERBACK
The adult male gorilla is almost twice the weight of the female, and has a taller bony head crest, longer canine teeth, and a "saddle" of silvery fur.

NERVOUS BEHAVIOUR

When nervous, gorillas may "yawn" (above). They usually avoid danger by walking quietly in single file into thick forest – termed "silent flight". In active defence the silverback barks and stares. If the threat persists, he may charge (see panel below).

CONSERVATION

The main threats to western gorillas are slash-and-burn clearance of their forest habitat, illegal hunting for the commercial bushmeat trade, and trophy poaching. Almost all gorillas in zoos and parks belong to this species, but reintroductions of captive-bred individuals into the wild are rarely attempted, partly due to the gorilla's complex, close-knit social life. Habitat conservation remains the long-term priority.

MAMMALS

Gorilla beringei
Eastern gorilla

Height	1.3–1.9m (4 1/4 – 6 1/4 ft)
Tail	None
Weight	68–210 kg (150–460 lb)

Location C. and E. Africa	Social unit Group
	Status Critically endangered

Previously considered a subspecies of the western gorilla, this species includes eastern lowland gorillas (*Gorilla beringei graueri*), and one or more subspecies of mountain gorillas (such as *Gorilla beringei beringei*). Each group roams a home range of 400–800 hectares (1,000–2,000 acres), which, apart from a central core area, may overlap with neighbouring groups. Main foods are leaves, shoots, and stems, especially bamboo; also fruits, roots, soft bark, and fungi. Occasionally ants are scooped up and swallowed quickly before they bite. At dusk the group settles to rest – the adult males on the ground, females and young sometimes in nests in trees. Each female shares a nest with her current offspring. The dominant silverback fathers most or all young in the group. He gains the attention of receptive females with mock feeding, hoots, chest-beating, thumping plants, or jump kicking. A female who has bred, is unlikely to transfer to another male unless her original mate is killed. The offspring remains with its mother until her next birth, after about 4 years. Some eastern gorillas attract tourist income and receive protection, but others are at continued risk from poaching.

GORILLA GROUP
Long-term groups of up to 40 usually comprise one silverback, females, and young. Sometimes brothers, or father and son, stay together, resulting in a multi-male group.

WARM FUR

The mountain subspecies of eastern gorilla has long, shaggy fur to retain body warmth at altitudes of up to 4,000m (13,200ft). Only the face, hands, and feet (and the male's chest) lack hair.

DEFENCE DISPLAY

If barking does not deter an attacker (see panel above), the silverback may begin to hoot. He then stands upright, beats his chest with cupped hands (the bare skin amplifies the sound), and throws vegetation. If this fails, he may charge with a huge roar, and hand-swipe or knock over and bite the invader.

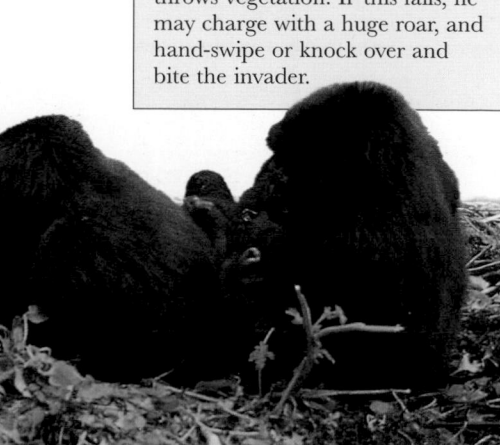

Pongo pygmaeus

Bornean orang-utan

Height 1.1–1.4 m	(3½–4½ft)
Weight 40–80 kg	(88–175 lb)

Location S.E. Asia (Borneo)

Social unit Individual

Status Endangered

The orang-utan is very much a tree-dwelling animal, feeding, sleeping, and breeding in the forest canopy, with only males occasionally coming to ground. It spends most of the day looking for and eating fruits and other food, and at night it builds a sleeping platform by weaving branches together. The female gives birth in a treetop nest, and the tiny infant clings to its mother as she clambers about the canopy. The pair will stay together until the youngster is about 8 years old. Orang-utans live in widely scattered communities – probably determined by the availability of food. They are mainly solitary, but may meet up with others at fruit trees, and adolescent females may travel together for 2 or 3 days. All will be aware of neighbouring males from their roaring "long-calls". Until fairly recently orang-utans were considered to be a single species – *Pongo pygmaeus*. Genetic research has

CONSERVATION

Although the orang-utan is protected by law, infants are still captured and sold illegally as pets. Projects to rehabilitate rescued orang-utans – adults as well as juveniles – have a good success rate, but some animals find it difficult to readjust to life in their natural habitat – and in any case this is rapidly being destroyed.

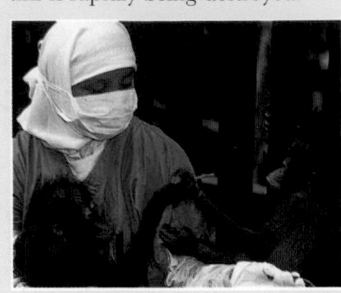

now led to 2 distinct species being recognized: the Bornean and the Sumatran (*Pongo abelii*). Loss of habitat presents the main threat to both species as their forest homelands are destroyed by logging and fire. It is estimated that Bornean orang-utans now number 12–15,000, while as few as 3–5,000 Sumatran orang-utans survive.

"PERSON OF THE FOREST"
The male orang-utan – which means "person of the forest" in Malay – looks strikingly different from the female, with large cheek pads (which grow bigger as the animals age), a long beard and moustache, and a hanging throat pouch. He also has long arm hair, which hangs like a cape when the arms are outstretched.

strong, grasping hands

very long arms compared to body

hand-like feet

ARBOREAL APE
With its long arms, spanning up to 2.2m (7¼ft), and feet that can grasp branches like hands, this male orang-utan is well adapted to life in the treetops. Its limbs are also extremely flexible, with wrist, hip, and shoulder joints allowing a greater range of movement than in the other great apes.

FOOD FROM THE FOREST

Fruits are the favourite food of orang-utans, but they will also eat other parts of plants, as well as honey, small animals such as lizards, termites, nestling birds, and eggs.

FAMILY GROUP
A mother and her offspring forage together in the forest, plucking fruits and leaves from the trees.

EATING HABITS
Orang-utans use their hands and teeth to prepare their food, stripping plants and peeling fruits to expose the succulent flesh.

ANTEATERS AND RELATIVES

PHYLUM	Chordata
CLASS	Mammalia
ORDER	Xenarthra
FAMILIES	4
SPECIES	29

ANTEATERS, SLOTHS, AND ARMADILLOS – now collectively known as xenarthrans; previously called edentates – are among the strangest of mammals in terms of appearance. Although the 3 families bear no physical resemblance to each other, they all have unusual joints in the backbone, a small brain, and reduced dentition (anteaters have no teeth at all). Armadillos and anteaters are mainly insectivorous; sloths eat leaves, shoots, and fruits. Xenarthrans are found in a variety of habitats from South America to the southern USA.

TERMITE MEAL
After breaking into a termite mound using the curved claws on its front feet, the giant anteater gathers termites with its long, sticky tongue.

ANATOMY

Xenarthrans are unique in that they have special articulations (called xenarthrales) on the vertebrae of the lower back. These provide extra support, which is especially useful to armadillos when burrowing. Additionally, anteaters are characterized by a long, tubular snout; armadillos have a protective, armour-like covering of hardened skin on their head, back, sides, and limbs; sloths have a short, rounded head and long, coarse fur.

ANTEATER — elongated snout

ARMADILLO — armoured head

SLOTH — shortened face

HEAD SHAPES
The elongated head of anteaters (far left) contrasts with the short, rounded head of sloths (left). The armadillo's head (centre) is intermediate in length and is protected by bony plates.

SPECIAL CLAWS
Sloths have highly modified hands and feet with long, curved claws, which are ideal for their totally arboreal life.

Choloepus didactylus

Lime's two-toed sloth

Length	46–86 cm (18–34 in)
Tail	1.5–3.5 cm (1/2–1 1/2 in)
Weight	4–8.5 kg (8 3/4 –19 lb)

Location N. South America

Social unit Individual

Status Unconfirmed

foreleg longer than hindleg

Lime's two-toed sloth has 2 hook-clawed toes on each front foot but 3 on each rear one. Its coarse fur is grey-brown, paler on the face, but, like other sloths, it may be tinged green by algae growing on the hairs. It eats a typical sloth's diet of leaves and fruits and, in common with all sloths, is solitary, moves extremely slowly, and descends to the ground only to defecate (about once a week).

Bradypus torquatus

Maned three-toed sloth

Length	45–50 cm (18–20 in)
Tail	4–5 cm (1 1/2 –2 in)
Weight	3.5–4 kg (7 3/4 –8 3/4 lb)

Location E. South America

Social unit Individual

Status Endangered

This species has the typical sloth's small head, tiny eyes and ears, and small tail hidden in the fur, which contrasts with the large body and powerful limbs. Algae, mites, ticks, beetles, and even moths live in the coarse outer coat, which is longer, darker, and mane-like around the head, neck, and shoulders. The underfur is fine, dense, and pale. The maned three-toed sloth eats the leaves, buds, and soft twigs of a few forest trees, notably *Cecropia*. It comes to the ground only to defecate, or to move to another tree if it cannot travel through the branches. On the ground the sloth drags itself along by its longer, stronger front legs and claws. Surprisingly, it can swim well. In addition to its physical slowness, the sloth's muscles are small and weak for its overall body size, and even its metabolism is less rapid than most

other mammals, giving a low body temperature of just above 30°C (86°F). Its main defence is to stay still and unnoticed or to lash out with its formidable claws. After 5–6 months' gestation, one offspring is born, weighing about 250g (9oz). It clings to its mother's abdomen with its well-formed, hook-shaped claws. The young suckles for up to 4 weeks and after weaning stays with her, being carried and learning feeding patterns, for a further 6 months.

dark mane

WRONG-WAY FUR
The outer hairs are angled "up" in the opposite direction to most mammals, so they hang down when the sloth is in its normal inverted position.

CONSERVATION

Sloths are entirely tree-dependent and cannot race away when loggers appear, so in some areas they are carefully captured before the forest is felled. The sloths are marked for future reference, measured, weighed, and sexed. Other useful data is collected, including blood samples for genetic studies, to increase our knowledge of these still little-understood mammals. Then the sloths are released at safer sites.

Giant anteater

Myrmecophaga tridactyla

Length	1–2 m (3¼–6½ ft)
Tail	64–90 cm (25–35 in)
Weight	22–39 kg (49–86 lb)

Location Central to South America

Social unit Individual

Status Vulnerable

The giant anteater has a long and tubular snout that widens to a small face with tiny ears and eyes. With its massive front legs and smaller rear legs, it walks with an ambling gait, protecting its large front claws from wear by walking on its knuckles. Active day

black, white-centred stripe along each side

ANTS' EYE VIEW

This anteater rips open ant nests and termite mounds with its big-clawed front feet, and then uses its tongue, which can protrude more than 60 cm (2 ft), to take its prey. The tongue is covered with minute, backward-pointing spines and sticky saliva, to which tiny prey adhere. The remaining termites then repair the nest.

and night, this species wanders its home range, which may cover up to 25 square km (10 square miles), depending on the availability of food.

It sleeps in a hollow or in the shelter of bush, its tail draped over its head and body. This species is threatened by hunting, and habitat destruction.

NARROW BODY AND BUSHY TAIL
The giant anteater has coarse, long fur on its narrow body. Its coloration is mainly grey with black and white markings. It has a very bushy, brown tail.

Nine-banded armadillo

Dasypus novemcinctus

Length	35–57 cm (14–22½ in)
Tail	24–45 cm (9½–18 in)
Weight	2.5–6.5 kg (5½–14 lb)

Location S. USA, Mexico, Caribbean, Central America, South America

Social unit Individual

Status Locally common

The most commonly seen armadillo, this species has 8–10 bony bands around its middle, which allows some flexibility. The bony armour and leathery skin account for one-sixth of the total weight. Like most armadillos, it digs an extensive burrow system. It takes most foods, from ants and birds to fruits and roots, and is solitary yet may share a burrow with others of its kind. Offspring are nearly always quadruplets – 4 of the same sex.

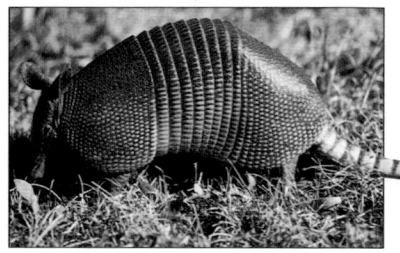

Silky anteater

Cyclopes didactylus

Length	16–21 cm (6½–8½ in)
Tail	16–23 cm (6½–9 in)
Weight	150–275 g (5–10 oz)

Location Central America to N. South America

Social unit Individual

Status Unconfirmed

underside of tail bare at tip

brown body stripe

Solitary and nocturnal, this anteater has long, dense, fine fur, usually smoky grey with a silver sheen and a variable brown stripe on the side of the body. Specialized for arboreal life, it grips strongly with its feet and hook-like claws, supported by its prehensile tail, which is bare on the underside near the tip. It breaks open tree-ant nests and licks out the ants with its long, saliva-coated tongue.

Southern tamandua

Tamandua tetradactyla

Length	53–88 cm (21–35 in)
Tail	40 cm (16 in)
Weight	3.5–8.5 kg (7¾–19 lb)

Location N. and E. South America

Social unit Individual

Status Vulnerable†

Also called the lesser anteater, the southern tamandua has a long, narrow head and sparsely haired, prehensile tail. It climbs well and feeds both in the branches and on the ground, breaking into the nests of ants, termites, and bees. It is active for 8-hour periods, day or night. A single offspring is born after 4–5 months' gestation and rides on the mother.

BLACK VEST
The tamandua is pale yellow with a black "vest" over the shoulders, chest, sides, and lower back.

SELF-DEFENCE

Under threat, a tamandua backs against a trunk or rock, rears up onto its hindlegs, props itself up by its tail, and holds its power-ful front legs outstretched. In this position it can slash out at its attacker with the long, sharp claws on its front feet.

Hairy armadillo

Chaetophractus villosus

Length	22–40 cm (9–16 in)
Tail	9–17 cm (3½–6½ in)
Weight	1–3 kg (2½–6½ lb)

Location S. South America

Social unit Individual

Status Unconfirmed

This armadillo of arid habitats has long, coarse hairs projecting between the 18 or so distinct bands of bony, skin-covered armour on its body. Some 7 or 8 bands are hinged, allowing the armadillo to roll into a ball and protect its softer, white- or brown-furred underside. In summer, it is mainly nocturnal and eats varied small prey, from grubs to rodents. In winter, activity is mainly by day and the diet includes more plant matter.

Zaedyus pichiy
Pichi

Length	26–34 cm (10–13 1/2 in)
Tail	10–12 cm (4–4 3/4 in)
Weight	1–2 kg (2 1/4–4 1/2 lb)

Location S. South America

Social unit Individual

Status Common

When threatened, the pichi squats down and draws its sharp-clawed feet under its body, relying on its wide, low, domed body armour for protection; or it wedges itself into a burrow with its armour facing outwards. This small armadillo digs a short tunnel for shelter and eats various small insects, worms, other invertebrates, and sometimes also carrion.

Cabassous centralis
Northern naked-toed armadillo

Length	30–40 cm (12–16 in)
Tail	13–18 cm (5–7 in)
Weight	2–3.5 kg (4 1/2–7 3/4 lb)

Location Central America and N. South America

Social unit Individual

Status Unconfirmed

Occupying a wide range of habitats, this big-eared armadillo has large claws, especially the middle forefoot claw, for digging up prey and making a burrow to shelter in by day. It also has a long, sticky tongue to lick up termites and ants in the manner of an anteater. In common with other armadillos, it is mainly silent, perhaps growling when threatened, as it tries to dig itself into the ground so only the armoured upper body is exposed.

Priodontes maximus
Giant armadillo

Length	75–100 cm (30–39 in)
Tail	50 cm (20 in)
Weight	30 kg (66 lb)

Location N. and C. South America

Social unit Individual

Status Endangered

By far the largest armadillo, this species has 11–13 slightly flexible, hinged plates over its body, and 3–4 over its neck. The long, tapering tail is likewise armoured. The main body colour is brown, with a pale yellow-white head, tail, and band along the lower edges of the plates. The especially large third front claw is used to rip up soil for small food items – mainly termites, ants, worms, spiders, small snakes, and lizards. The front claws also dig a burrow in which the giant armadillo shelters by day. It feeds in an area for 2–3 weeks, then moves on. Like most other armadillos, it shows little social or territorial behaviour. The gestation period is 4 months, and the 1–2 offspring are weaned by 6 weeks and are sexually mature by 12 months.

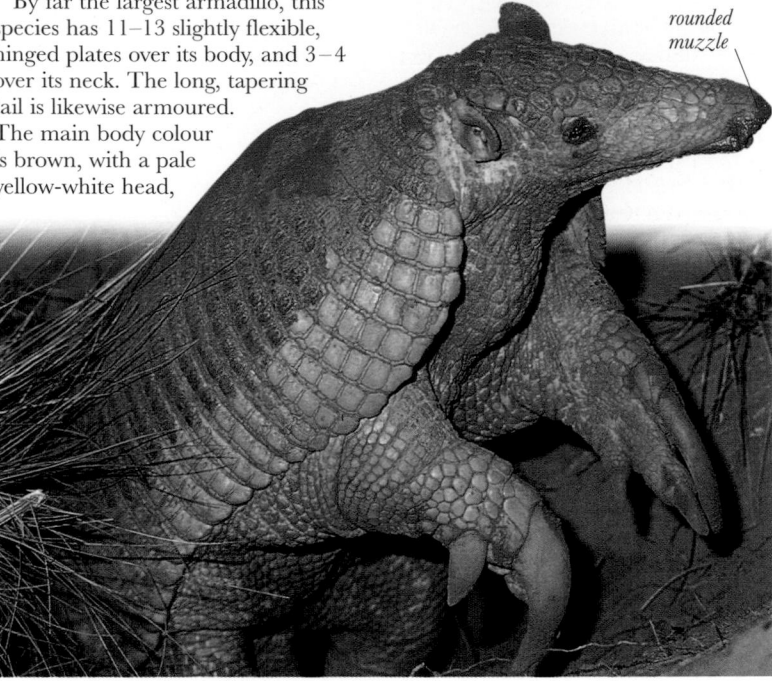

rounded muzzle

PANGOLINS

PHYLUM	Chordata
CLASS	Mammalia
ORDER	Pholidota
FAMILIES	1 (Manidae)
SPECIES	7

SIMILAR IN SHAPE to armadillos and anteaters, pangolins are covered in overlapping scales, which act as armour and camouflage. Pangolins lack teeth: prey (ants and termites) is collected with the tongue, and powerful muscles in the stomach "chew" the food. Pangolins are found in southern Asia and Africa, in habitats ranging from forest to savanna.

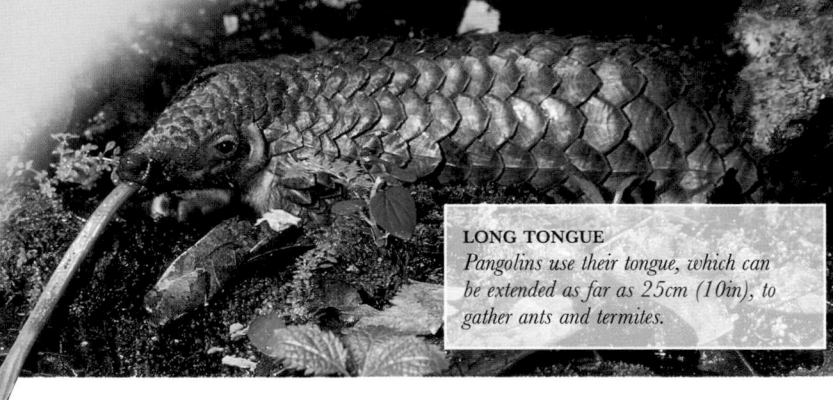

LONG TONGUE
Pangolins use their tongue, which can be extended as far as 25cm (10in), to gather ants and termites.

Manis pentadactyla
Chinese pangolin

Length	54–80 cm (21 1/2–32 in)
Tail	26–34 cm (10–13 1/2 in)
Weight	2–7 kg (4 1/2–15 lb)

Location E. to S.E. Asia

Social unit Individual

Status Lower risk

Bony, pale or yellow-brown scales, up to 5cm (2in) across, cover all parts of the Chinese pangolin except for its snout, cheeks, throat, inner limbs, and belly. When rolled into a ball, no soft areas are exposed. The thin tongue, as long as 40cm (16in), scoops up ants and termites. The strongly prehensile tail and long claws make this pangolin surprisingly agile in trees and a powerful burrower.

Manis temmincki
Temminck's pangolin

Length	50–60 cm (20–23 1/2 in)
Tail	40–50 cm (16–20 in)
Weight	15–18 kg (33–40 lb)

Location E. to southern Africa

Social unit Individual

Status Lower risk

Similar in most respects to the Chinese pangolin (see left), this species rips open termite mounds and ants' nests, both in trees and on the ground, with its large claws, and licks up the occupants. This pangolin shows little territorial behaviour. The 1–2 young are born after a gestation of about 120 days.

dark or yellow-brown scales

RABBITS, HARES, AND PIKAS

PHYLUM	Chordata
CLASS	Mammalia
ORDER	Lagomorpha
FAMILIES	2
SPECIES	80

ALTHOUGH THESE SMALL to medium-sized gnawing animals are similar to rodents in many ways (for example, both groups have large incisors), they differ in several respects, including the presence of a second set of upper incisors and a lighter skull structure. Rabbits, hares, and pikas (also termed "calling hares" on account of their varied vocalizations) – the lagomorphs – are among the most hunted of all animals. Their natural predators are carnivores and birds, and they are also hunted by humans for sport, for food, and for their fur. All species are terrestrial and are found worldwide (except the West Indies, the southern parts of South America, Madagascar, and several islands of Southeast Asia) in habitats ranging from arctic tundra to semi-desert.

ANATOMY

The physical features of rabbits and hares reflect their need to perceive danger and elude predators. Large ears provide excellent hearing, eyes positioned high on each side of the head give almost 360-degree vision, and elongated back legs enable impressive running speeds – hares can reach up to 56kph (35mph). Unlike rodents, rabbits and hares have tails that are small and spherical, and they have well-furred feet with hair on the soles, which provides grip when running. Pikas tend to hide in crevices or burrows when threatened, and are more mouse-like in appearance, having legs of approximately the same length (they cannot run as swiftly). They also have shorter, rounded ears and no visible tail. All species have slit-like nostrils that can be closed completely. Unusually for mammals, the females of some species are larger than the males.

eye socket   *molars*
premolars
upper incisor
peg tooth
diastema   *lower incisor*

RABBIT SKULL
Lagomorphs have well-developed, continually growing incisor teeth. Behind the upper incisors is a second pair of smaller incisors ("peg teeth"). There is a large gap between the incisors and premolars in both jaws, called the diastema.

FLEXIBLE NECK
In the lagomorphs, self-grooming is important: mutual grooming between individuals is rare. Great flexibility in the neck – this European rabbit is able to rotate its head through 180 degrees – allows them to reach the fur on the back.

FEEDING

Lagomorphs are herbivorous and generally eat grass and other succulent plants. Matter that cannot be digested initially is expelled in the form of a moist pellet and eaten, usually straight from the anus. It is then held in the stomach to be mixed with other food for second digestion before being excreted as a dry pellet. In this way, most food travels through the digestive system twice, enabling the animal to derive maximum nutrition from its diet. This process is called refection.

PLANT-EATERS
All lagomorphs, such as this North American pika, spend much of their time feeding. During the summer, the pika also gathers and stores food for winter, creating a "haystack" of dried foliage outside its rocky shelter.

AERIAL BOXING
During the mating season, fights between hares occur frequently. The front feet are used for boxing, and the powerful back feet are used for kicking. Males battle for access to females, and a female will drive off a male if she is not ready to mate (as shown here).

REPRODUCTION

Although lagomorphs are hunted intensely by many predators, they are able to maintain healthy population levels through a high reproductive rate. Because ovulation is not cyclical but is instead triggered in response to copulation, females can become pregant directly after giving birth. Some species may even conceive a second litter before giving birth to the first. Rabbits, the most prolific breeders of all the lagomorphs, can produce litters of up to 12 young 6 times annually. Furthermore, rabbits are sexually mature at a young age (the European rabbit is able to conceive when only 3 months old), and the gestation period may be very short (the Florida cottontail rabbit gestates for as little as 26 days).

FEEDING TIME
Although some rabbits give birth in their burrows, all hares are born above ground. These young brown hares remain hidden during the day but are collected together at sunset when the mother visits to nurse them.

Ochotona princeps
North American pika

Length 16–22 cm (6½–9 in)	
Tail None	
Weight 125–175 g (4–6 oz)	
Location S.W. Canada, W. USA	**Social unit** Individual
	Status Locally common

Found at high altitudes, this pika relies on a talus – an area of piled, broken rocks fringed by alpine meadows and similar low, grassy vegetation. At each talus, solitary pikas use whistling calls to defend territories that alternate across the area by gender, giving a female–male patchwork. A typical territory is 600 square m (6,400 square ft), and has a foraging area and a den in a burrow or rock crevice.

EGG-SHAPED MAMMAL
A crouched pika has a rounded outline resembling an egg. The fur is varying shades of brown.

READY FOR WINTER

During late summer each pika gathers grasses, herbs, and other greenery in a "hay pile" near its burrow. This serves as a readily accessible winter food store when snow covers the area. Pikas are careful to choose and store those plants that decompose most slowly.

dark ears, furred on both surfaces

Brachylagus idahoensis
Pygmy rabbit

Length 22–29 cm (9–11½ in)	
Tail 1.5–2.5 cm (½–1 in)	
Weight 350–450 g (13–16 oz)	
Location W. USA	**Social unit** Individual
	Status Lower risk

The world's smallest rabbit, this arid-adapted species digs a large burrow system and feeds on big sagebrush and closely related species. The long, silky fur on its back is grey in winter and brown in summer; the abdomen is whitish. Unusually for a rabbit, it climbs well into bushes to feed, and, although solitary, it also makes pika-like whistling calls to warn neighbours of approaching predators. Breeding details are poorly known: the gestation period may be 26–28 days, and litter size 4–8, with up to 3 litters per year.

short ears, furred on inside edge only

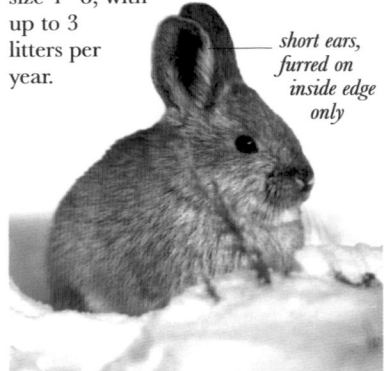

Sylvilagus aquaticus
Swamp rabbit

Length 45–55 cm (18–22 in)	
Tail 4 cm (1½ in)	
Weight 1.5–2.5 kg (3¼–5½ lb)	
Location S.E. USA	**Social unit** Group
	Status Locally common

cinnamon eye-rings

Always associated with water, in marshes, creeks, and pools, the swamp rabbit has black to rusty brown fur. It swims well and readily, especially if under threat, and feeds by day or night on sedges, rushes, and other aquatic plants, including swamp bamboos (*Arundinaria*). The swamp rabbit lives in small groups, usually controlled by a dominant territorial male, and builds an above-ground nest of plant stalks and stems. The female lines the breeding nest with her fur, in the typical rabbit manner. The average litter size is 3.

Ochotona curzoniae
Black-lipped pika

Length 14–18.5 cm (5½–7½ in)	
Tail None	
Weight 125–175 g (4–6 oz)	
Location E. Asia	**Social unit** Group
	Status Common

This pika is sandy brown above and dull yellow-white on the underside, with a rust-hued patch behind the ear and a dark nose and lips. An extended family occupies each burrow system and members are sociable. In some areas this pika is so numerous as to be considered a pest. Females can have up to 5 litters of 8 young per year, which are cared for by both parents.

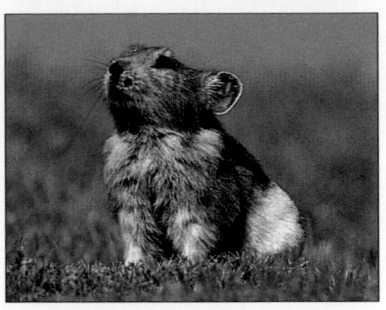

Romerolagus diazi
Volcano rabbit

Length 23–32 cm (9–12½ in)	
Tail 1–3 cm (⅜–1¼ in)	
Weight 375–600 g (13–21 oz)	
Location C. Mexico	**Social unit** Group
	Status Endangered

Restricted to open pine forests on volcanic peaks near Mexico City, the volcano rabbit lives in groups of 2–5. It has very short, rounded ears for a rabbit, relatively small back legs and feet, and communicates by means of penetrating whistles. Its diet is mainly the tall, dense grass, in which it also makes its burrow.

coat of yellow and black guard hairs

Pentalagus furnessi
Amami rabbit

Length 42–51 cm (16½–20 in)	
Tail 1–3.5 cm (⅜–1½ in)	
Weight 2–3 kg (4½–6½ lb)	
Location Amami and Tokuno Islands	**Social unit** Individual/Group
	Status Endangered

Found only on 2 small Japanese islands, this rabbit's many distinctive features include an all-black coat, pointed snout, small eyes and ears, and short, long-nailed limbs for digging nest-holes. Nocturnal in habit, it eats forest plants such as pampas grass leaves, sweet potato runners, bamboo sprouts, nuts, and bark. Social and breeding habits are little known, although it communicates by means of clicking sounds. The female has 2 litters of 2–3 young each year.

Sylvilagus floridanus
Eastern cottontail

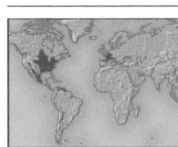

Length 38–49 cm (15–19½ in)	
Tail 2.5–7 cm (1–2¾ in)	
Weight 1–1.5 kg (2¼–3¼ lb)	
Location S.E. Canada to Mexico, Central America, N. South America, Europe	**Social unit** Group
	Status Common

Widely distributed in many habitats over its natural range, and introduced to new areas of North America and Europe, the cottontail has the typical rabbit body form with a reddish-topped white tail. In summer it feeds on lush green vegetation; bark and twigs predominate in winter. Groups have established dominance hierarchies. Average litter size ranges from 5 in North America to 2 in South America.

Oryctolagus cuniculus

European rabbit

Length 34–50 cm (13 1/2–20 in)	
Tail 4–8 cm (1 1/2–3 1/4 in)	
Weight 1–2.5 kg (2 1/4–5 1/2 lb)	
Location Europe, N.W. Africa, Australia, New Zealand, S. South America	**Social unit** Group
	Status Common

Originally from southwest Europe and perhaps northwest Africa, this species has been introduced to many other regions and has reached severe pest status in some, devastating farmland and local wildlife. It is the ancestor of all breeds of domestic rabbits. Nocturnal and very sociable, the European rabbit lives in colonies and digs complex tunnel systems (warrens) with many entrances and emergency exits. It eats mainly grass, herbs, twigs, and some bark. Senior females nest in the main warren but lower-ranking mothers may dig separate short burrows (stops) for their young.

long, black-tipped ears

BUFF PATCHES
The European rabbit is buff coloured between the shoulders, and has a pale buff eye-ring, inner limb surfaces, and underside.

FAST BREEDER

The rabbit's legendary powers of reproduction include a gestation period of 28–33 days, and, in good conditions, a litter size of up to 8 (average 5), with as many as 6 litters a year. The newborn are helpless with eyes closed, and for warmth the mother lines the nursery chamber with dry grass, moss, and fur plucked from her own belly. She visits to suckle them for only a few minutes daily.

mixed black and brown hairs in upper fur

Lepus californicus

Black-tailed jackrabbit

Length 47–63 cm (18 1/2–25 in)	
Tail 5–11 cm (2–4 1/4 in)	
Weight 1.5–3.5 kg (3 1/4–7 3/4 lb)	
Location W., C., and S. USA to N. Mexico	**Social unit** Individual
	Status Common

This lean, long-legged hare's huge ears (up to 15 cm/6 in) detect the faint sounds of predators, and also rid the body of excess warmth in the hot summers of its generally arid habitat. It prefers succulent grass and herbs, but can survive by nibbling woody twigs during winter or drought. One of the speediest lagomorphs, it can run at 56kph (35mph). The complex courtship involves the pair jumping, chasing, and fighting.

Caprolagus hispidus

Hispid hare

Length 38–50 cm (15–20 in)	
Tail 2.5–5.5 cm (1–2 1/4 in)	
Weight 2–2.5 kg (4 1/2–5 1/2 lb)	
Location S. Asia	**Social unit** Individual/Pair
	Status Endangered

Also called the bristly rabbit after its coarse, dark brown fur, the hispid hare lives in tall "elephant grass" country, feeding by night on the soft shoots and roots of the grass. It has short ears, and its back legs are not much larger than the front legs. It does not burrow but shelters in surface vegetation, living alone or as a female–male pair. Reproductive information is scant but the suspected litter size is small for a lagomorph, perhaps 2–5, with 2 or possibly 3 litters produced each year.

Lepus europaeus

Brown hare

Length 48–70 cm (19–28 in)	
Tail 7–13 cm (2 3/4–5 in)	
Weight 2.5–7 kg (5 1/2–15 lb)	
Location Europe, Australia, New Zealand, North and South America	**Social unit** Individual
	Status Common

This hare has a conspicuous tail, black on top and white below. The grey ears have a black patch near the outer tip. This species has been introduced from Europe and West Asia to many other regions, and adapts to open woods, bush, mixed farmland, and even scrubby semi-desert. Its diet is grass, herbs, bark, and rarely carrion. Nocturnal and solitary, hares gather in courting pairs or groups in late winter and spring. At this time, "boxing" may occur between rival males or unreceptive females and rejected males. The young (litter size 1–10) are reared in a shallow depression (form) in grass or bushes.

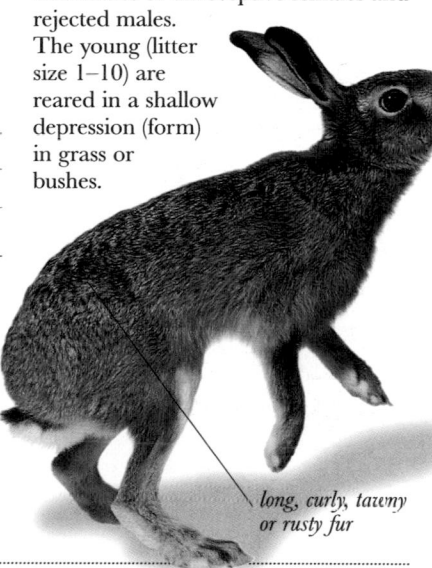

long, curly, tawny or rusty fur

Lepus arcticus

Arctic hare

Length 43–66 cm (17–26 in)	
Tail 4.5–10 cm (1 3/4–4 in)	
Weight 3–7 kg (6 1/2–15 lb)	
Location N. Canada, Greenland	**Social unit** Variable
	Status Common

The Arctic hare (sometimes confused with a similar white-in-winter species, the snowshoe hare, *Lepus americanus*) is a true tundra species. It can survive in an open, treeless habitat through the long and bitter cold season. Preferred sites are rocky outcrops or hillsides with crevices and crannies for shelter. This hare may be solitary but, especially in winter, and uniquely among lagomorphs, it shows "flocking" behaviour in which large groups of up to 300 gather, move, run, and change direction almost as one. The diet is a variety of low-growing grasses, herbs, and shrubs, including lichens, mosses, and most parts of the arctic dwarf willow. However, these opportunistic hares may also eat small animals or larger carrion. During the aggressive spring courtship, the male follows the female

WINTER COAT

The Arctic hare's thick winter coat is almost pure white with black ear tips. It provides both warmth and camouflage in snow and ice. In most regions the spring moult produces the grey-brown summer coat, although in some areas this is also white. The timing of the moult depends on the number of daylight hours, detected by the eyes and then controlled via the body's hormonal (endocrine) system.

and may bite her neck so hard that bleeding occurs. Litter size is 1–8 with 1–3 litters per year. The young hares (leverets) stay in their nest (form), a hollow near rocks, lined with grass, moss, and fur. The mother visits to suckle them for only 2 minutes every 18 hours.

large feet spread body weight on soft snow

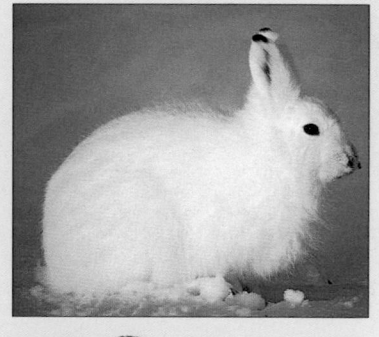

COMPACT BODY
The relatively large, compact body, with short ears and other extremities, helps to reduce heat loss in cold conditions. The ears are darker in front than at the rear.

RODENTS

PHYLUM	**Chordata**
CLASS	**Mammalia**
ORDER	**Rodentia**
FAMILIES	**30**
SPECIES	**1,702**

CLASSIFICATION NOTE

Some mammalogists prefer to divide the order Rodentia into 2 suborders (Sciurognathi and Hystricognathi); while others advocate a division into 3 suborders (defined by jaw musculature): Sciuromorpha (squirrel-like rodents), Myomorpha (mouse-like rodents), and Caviomorpha (cavy-like rodents). For greater ease of reference, the latter division is used in this book.

Squirrel-like rodents
see pp.146–9
Mouse-like rodents
see pp.150–6
Cavy-like rodents
see pp.157–9

REPRESENTING OVER 40 per cent of all mammal species, rodents form a successful and highly adaptable order. They are found worldwide (except Antarctica) in almost every habitat: lemmings, for example, favour the cold climate of the arctic tundra, while gundis prefer the heat of African desert regions. Despite the variety of lifestyles and habitats exhibited by members of this order, there are many common characteristics: most rodents are small quadrupeds with a long tail, clawed feet, long whiskers, and teeth (especially the long incisors) and jaws specialized for gnawing. Although generally terrestrial, some species are arboreal (such as tree squirrels), burrowing (mole-rats, for instance, live almost wholly underground), or semi-aquatic (such as beavers). Although some species, such as the woodchuck, are solitary, most are highly social and form large communities.

ANATOMY

While the anatomy of rodents is more uniform than that of most other mammals, some characteristics, such as a compact body and a long tail and whiskers, are shared by many species. The front foot usually has 5 digits (although the thumb may be vestigial or absent), the back foot has 3–5 digits, and the method of locomotion is generally plantigrade. Different species use their tail to perform distinct functions: beavers have a flat, wide tail for propulsion when swimming; the Eurasian harvest mouse uses its prehensile tail when climbing in long grass. In some species, part of the tail skin, or the tail itself, will break off if caught, enabling the animal to escape. Because rodent anatomy is more generalized than that of other mammals, they can adapt easily and are able to thrive in many different habitats.

SENSES

Most rodents enjoy acute senses of smell and hearing, which, in combination with their long and numerous touch-sensitive whiskers, provide them with a heightened awareness of their surroundings. Nocturnal species have larger eyes than diurnal species to maximize the amount of light received by the retina (the greater the amount of light the brighter and clearer the image). Rodents communicate by smell (odours are secreted from scent glands on the body) and by an extensive range of vocalizations.

large, tubular ears

large eyes

elongated snout

long whiskers

HIGHLY TUNED
Well-developed sense organs are present in most rodents and may contribute to the adaptability of species such as this brown rat. The large ears and eyes, elongated snout, and long whiskers are typical of many rodents.

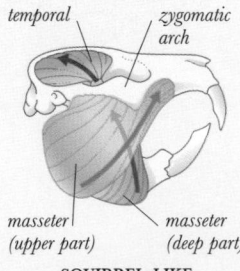

JAW MUSCLES
Rodents have an enlarged chewing muscle (the masseter), which permits both a vertical and a forwards and backwards motion of the lower jaw. In squirrel-like rodents, the upper part of the masseter reaches the back of the skull, the deep part extends to the zygomatic arch, and the temporal muscle is small. This system allows a strong forward motion when biting. In mouse-like rodents, the deep part of the masseter extends on to the upper jaw, the upper part is located forwards, and the temporal is large. This permits a versatile chewing action. In cavy-like rodents, the deep part of the masseter extends in front of the eye, and the temporal is small. This gives a strong forward bite.

temporal — zygomatic arch

masseter (upper part) — masseter (deep part)

SQUIRREL-LIKE

temporal — eye socket

masseter (upper part) — masseter (deep part)

CAVY-LIKE

temporal

masseter (upper part) — masseter (deep part)

MOUSE-LIKE

LARGE INCISORS
The 4 huge incisors (seen here in a marmot) distinguish rodents from most other mammal orders. These teeth are long, curved, and grow continually. Only the front surface of these teeth has enamel, however — the back surface consists of softer dentine, which is eroded by constant gnawing. This ensures that the teeth remain sharp.

FEEDING

Most rodents have a plant-based diet that may include leaves, fruits, seeds, and roots. However, many species have alternative diets: water rats and the wood mouse eat snails; rice rats take young turtles; muskrats eat clams and crayfish; the southern grasshopper mouse eats ants and scorpions; while the black rat scavenges in human food supplies. To assist digestion, rodents have a large caecum, a blind-ending sac in the large intestine. This contains bacteria that split cellulose, the main component of plant cell walls, into digestible carbohydrates. In some rodents, after food is processed in the caecum, it progresses to the anus and is eaten again. Once in the stomach, the carbohydrates (amounting to 80 per cent of the energy contained in the food) are absorbed. This highly efficient process, known as refection, leaves only a dry faecal pellet to be excreted.

OMNIVOROUS DIET
Some rodents are omnivorous: they feed on both plants and animals, depending on availability. This spiny mouse eats mainly vegetation but will also feed on insects.

HERBIVOROUS DIET
Most rodents are herbivorous, eating only plants. The European water vole shown here feeds on aquatic and land plants. Food may be stored for consumption during winter shortages.

REPRODUCTION

The high birth rate among rodents enables them to maintain stable population levels in adverse conditions. This means that predation and human controls (such as poisoning) have little effect on the survival of a species, and in favourable conditions numbers may increase rapidly. A brown rat, for example, is able to breed at only 2 months of age and may yield litters of more than 10 young every month or so. Voles are also prolific breeders: some species may produce more than 13 litters annually. Smaller rodents tend to produce more young than larger species (such as the capybara) – as a result, small rodents form the staple diet of a wider range of animals. In rodents, the complete cycle of reproduction, from sexual attraction right through to raising young, is influenced by the emission of pungent glandular secretions. Female rats, for example, produce a pheromone about 8 days after giving birth. This scented chemical is secreted into the mother's faeces and helps prevent the offspring from becoming separated from her.

CAPYBARA FAMILY
Not all rodent species breed as prolifically as mice and rats do. The capybara, for example, usually produces only one litter a year unless conditions are particularly favourable. Litter size varies between one and 8 but is usually 5. Capybara offspring are well developed at birth and are soon able to follow their mother and eat solid food.

SOCIAL ANIMALS
Among rodents, many species live in organized communities, although some are solitary. These black-tailed prairie dogs, like most ground squirrels, are highly sociable. They live in a system of burrows called "towns", each of which may cover an area up to 1 square km (⅔ square mile). The interconnected burrows in a town provide a refuge from predators and a safe place to rear young. Within the town, prairie dogs form subgroups known as coteries. Members of a coterie act cooperatively, for example to defend their territory.

RODENTS AND PEOPLE

Some rodents, mainly rats and mice, are considered pests by people because they are often in direct competition with humankind (they occupy the same habitats and eat the same food) and are highly adaptable. Rodents consume over 40 million tonnes (39 million tons) of human food every year, contaminate stored food with their urine and faecal pellets, and are known to transmit more than 20 disease-causing organisms. Although some control of rodent populations is brought about by the use of traps and poisons, many species are sufficiently intelligent to learn to avoid such measures. Only a few of the 1,702 species of rodents, however, are genuine pests: many benefit people – for example, by destroying insects and weeds or by maintaining the health of forests by spreading fungi. Beavers and chinchillas are farmed for their fur, while rats, mice, and guinea pigs are kept as pets and are used extensively in medical research.

GREAT OPPORTUNISTS
Rodents are highly gregarious animals and have successfully colonized many habitats, especially those created by human settlement, such as refuse tips and sewers. These brown rats are scavenging for food among the rubbish.

GNAWING DAMAGE
All rodents use their well-developed incisor teeth to gnaw. A beaver can fell trees, such as this birch, by gnawing through the trunk. Branches and smaller trunks from the tree are used to build a lodge or to dam a river.

SQUIRREL-LIKE RODENTS

PHYLUM	Chordata
CLASS	Mammalia
ORDER	Rodentia
SUBORDER	Sciuromorpha
FAMILIES	7
SPECIES	377

THIS GROUP, which is defined by the arrangement of the jaw muscles (see p.144), embraces a variety of rodents, such as beavers and squirrels. Members of the squirrel family all have long whiskers, a cylindrical body, and a well-haired tail. Species from other families vary anatomically according to lifestyle. Beavers, for example, are semi-aquatic, and have webbed back feet. Squirrel-like rodents are distributed worldwide, in a variety of habitats.

BALANCING ACT
The Eurasian red squirrel uses its tail for balance when running along branches.

Aplodontia rufa
Mountain beaver

Length 30–46 cm (12–18 in)	
Tail 2–4 cm (³/₄–1¹/₂ in)	
Weight 0.8–1.5 kg (1³/₄–3¹/₄ lb)	
Location S.W. Canada to S.W. USA	**Social unit** Individual
	Status Lower risk

This rodent digs its tunnels and home (lodge) under felled trees, and so is increasing where commercial logging occurs. It lives alone and its tunnel openings lead directly to food such as bark, twigs, shoots, and soft plants, which it brings back to eat or store. Also known as the sewellel, it climbs well and destroys many small trees. The long fur is black to red-brown above, yellow-brown beneath, with a white spot below each ear.

Marmota monax
Woodchuck

Length 32–52 cm (12¹/₂–20¹/₂ in)	
Tail 7.5–11.5 cm (3–4¹/₂ in)	
Weight 3–5 kg (6¹/₂–11 lb)	
Location Alaska and W. Canada to E. Canada and E. USA	**Social unit** Variable
	Status Common

Also called the groundhog, this is one of the largest, strongest ground squirrels. It feeds mainly in the afternoon, often in a loose group with others of its kind, on a variety of seeds, grasses, clovers, fruits, and small animals, such as grasshoppers and snails. Despite its size, it is an able climber and swims well. In autumn it excavates a deeper burrow for its long hibernation. In North America, 2 February is Groundhog Day, when the woodchuck supposedly peers from its winter burrow to assess the weather. The woodchuck regularly shows aggressive behaviour to its own kind, especially as males fight for dominance in the spring mating season. It also vigorously defends its burrow by arching its back, jumping, flicking its stiffened tail, and chattering its bared teeth.

white area around nose

white-tipped, "grizzled" hairs on upperparts

POWERFUL BUILD
The woodchuck has a stout body, small ears, short legs, and a bushy tail.

Tamias striatus
Eastern chipmunk

Length 15.5–16.5 cm (6–6¹/₂ in)	
Tail 7–10 cm (2³/₄–4 in)	
Weight 80–125 g (2⁷/₈–4 oz)	
Location S.E Canada to C. and E. USA	**Social unit** Individual
	Status Locally common

The eastern chipmunk, occasionally kept as a pet, is also familiar in the wild, as a bold visitor to picnic sites. It frequents mainly deciduous woods, especially birch, as well as woody areas containing abundant rock crevices. It can climb but forages mainly on the ground for seeds and nuts, especially during mid-morning and mid-afternoon, carrying food items in its cheek pouches. The basic colouring of the eastern chipmunk is greyish or reddish brown, becoming paler red on the rump. It lives alone in its burrow system and hibernates during winter, although it may emerge on mild days to feed. The noisy "chip" and "cuk" calls act as alarm signals for fellow chipmunks and other small animals living nearby.

pale eye and ear borders

pale-bordered body stripes

Marmota flaviventris
Yellow-bellied marmot

Length 34–50 cm (13¹/₂–20 in)	
Tail 13–22 cm (5–9 in)	
Weight 1.5–5 kg (3¹/₄–11 lb)	
Location S.W. Canada to W. USA	**Social unit** Group
	Status Locally common

Adaptable in habitat, the yellow-bellied marmot takes a wide diet of grasses, flowers, herbs, and seeds. It feeds mainly in the morning and late afternoon, then grooms with others in its colony – usually one male and several females. Its long hibernation lasts up to 8 months. The 3–8 young are weaned after 20–30 days.

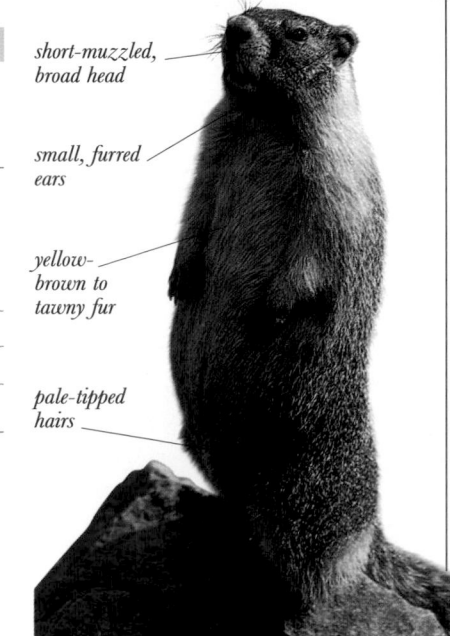

short-muzzled, broad head

small, furred ears

yellow-brown to tawny fur

pale-tipped hairs

Spermophilus columbianus
Columbian ground squirrel

Length 25–29 cm
(10–11 1/2 in)
Tail 8–11.5 cm
(3 1/4–4 1/2 in)
Weight 850–1,000 g
(30–36 oz)

Location W. Canada to N.W. USA

Social unit Group

Status Locally common

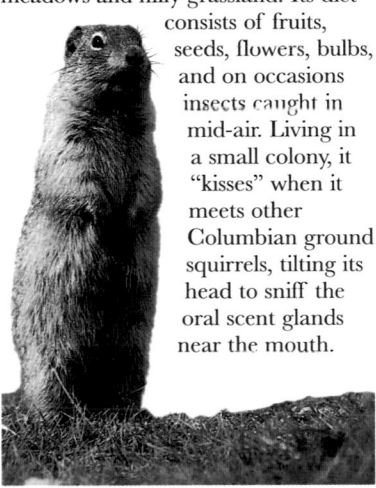

The Columbian ground squirrel ("Columbia" being British Columbia, Canada) comes from mountain meadows and hilly grassland. Its diet consists of fruits, seeds, flowers, bulbs, and on occasions insects caught in mid-air. Living in a small colony, it "kisses" when it meets other Columbian ground squirrels, tilting its head to sniff the oral scent glands near the mouth.

Cynomys ludovicianus
Black-tailed prairie dog

Length 28–30 cm
(11–12 in)
Tail 7–11.5 cm
(2 3/4–4 1/2 in)
Weight 0.7–1.5 kg
(1 1/2–3 1/4 lb)

Location S.W. Canada to N. Mexico

Social unit Group

Status Lower risk

Preference for a grassy habitat and a bark-like, doggy "yip" give 5 species of ground squirrels their common name of prairie dogs. The black-tailed or plains prairie dog dwells at altitudes of 1,300–2,000 m (4,300–6,600 ft) across the Great Plains of North America and south into the more arid, extreme north of Mexico. The body hairs of the black-tailed prairie dog are tipped black in winter but white in summer, and the whiskers and end third of the tail are black. This rodent eats a range of seasonal plants such as wheat grass, buffalo grass, globe mallow, and rabbit brush in summer, and thistles, cacti such as prickly pear, and underground roots and bulbs in winter. Prairie dogs breed rapidly, with up to 8 young born after a gestation period of 33–38 days.

In the past, their feeding habits led to massive destruction of wheat and other cereal crops, and their burrows tripped horses and farm stock. So prairie dogs were subject to extermination campaigns, which were very successful. In recent times, the rodents have been restricted mainly to parks and reserves. Their drastic reduction in numbers has seriously threatened the black-footed ferret (see p.197), for which they were virtually sole prey.

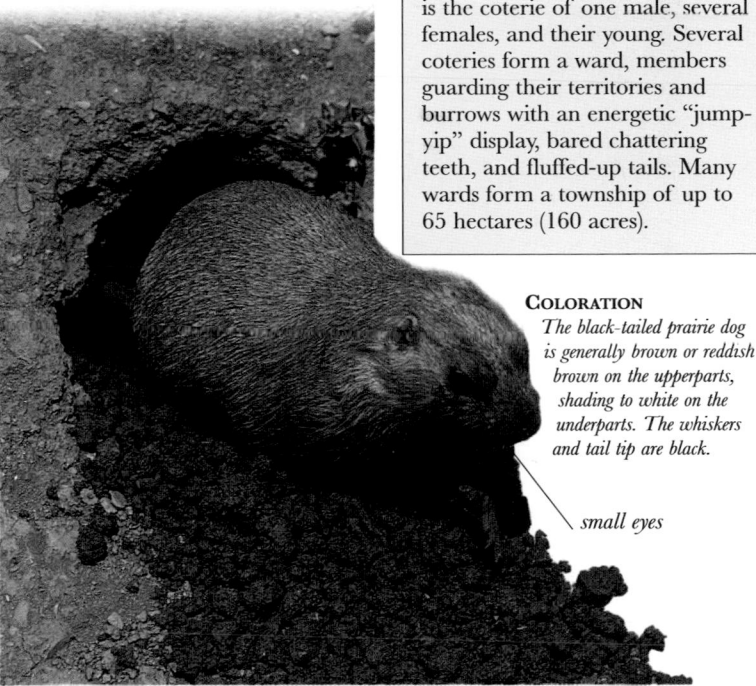

COLORATION
The black-tailed prairie dog is generally brown or reddish brown on the upperparts, shading to white on the underparts. The whiskers and tail tip are black.

small eyes

COTERIES AND WARDS

The basic prairie dog social unit is the coterie of one male, several females, and their young. Several coteries form a ward, members guarding their territories and burrows with an energetic "jump-yip" display, bared chattering teeth, and fluffed-up tails. Many wards form a township of up to 65 hectares (160 acres).

Xerus inauris
Cape ground squirrel

Length 20–30 cm
(8–12 in)
Tail 18–26 cm
(7–10 in)
Weight 575 g
(21 oz)

Location Southern Africa

Social unit Group

Status Locally common†

The large claws of the Cape ground squirrel can burrow in hard, dry, stony soil. The upper fur is brownish pink with a white flank stripe and belly. The prominent eyes are circled with white and the muzzle and feet are also white. There are black bands near the base and tip of the tail. Its diet is opportunistic, ranging from seeds, bulbs, and roots to insects and birds' eggs. This ground squirrel lives in colonies of 6–10, and in some cases up to 30.

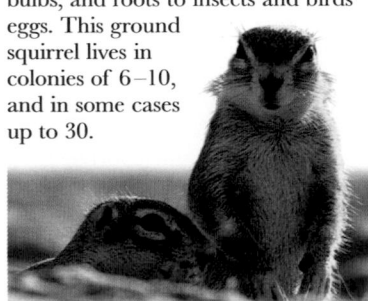

Sciurus vulgaris
Eurasian red squirrel

Length 20–25 cm
(8–10 in)
Tail 15–20 cm
(6–8 in)
Weight 200–475 g
(7–17 oz)

Location W. Europe to E. Asia

Social unit Individual

Status Lower risk

The "red" squirrel varies from red to brown, grey, or black on its back, and may turn grey-brown in winter. The underparts are always pale or white. An excellent climber and leaper, it feeds on the ground and in branches on seeds (especially those of conifers), nuts, mushrooms and other fungi, shoots, fruits, soft bark, and sap. It lives alone except when a female is nursing young.

BUSHY TAIL, TUFTED EARS
The red squirrel's bushy tail is about the same length as the body, densely furred with long hairs. The ears are always tufted, particularly in winter.

food held in forepaws

BREEDING NEST

The female red squirrel gives birth to 2–5 blind, naked babies. She nurses them for 12 weeks in the breeding nest, which may be a larger version of the ball-shaped, twig drey in a branch fork, or in a tree hole. The nest is lined with soft, fine material, to keep the young warm in her absence.

Sciurus carolinensis
Eastern grey squirrel

Length 23–28 cm
(9–11 in)
Tail 15–25 cm
(6–10 in)
Weight 300–700 g
(11–25 oz)

Location S. and S.E. Canada to S. USA, Europe

Social unit Variable

Status Common

Introduced from North America into parts of Europe, the grey squirrel has a grey back and white or pale underparts. The face, back, and forelegs are brown tinged. This opportunistic feeder takes nuts, seeds, flowers, fruits, buds, and fungi. It may emerge from its grass- and bark-lined twig nest (drey) in winter to forage.

no ear tufts

MAMMALS

Petaurista elegans
Giant flying squirrel

Length	30–45 cm (12–18 in)
Tail	32–61 cm (12 1/2–24 in)
Weight	0.5–2 kg (1–4 1/2 lb)
Location S. and S.E. Asia	Social unit Pair
	Status Common

This squirrel stretches out its limbs to extend large patches of thin, loose, furred skin between front and rear legs, in a parachute-like shape. A glider rather than a true flier, it can travel more than 400m (1,300ft) at a 3-in-1 glide ratio (covering 3 times the distance that it loses in height). It changes the glide angle using its front legs. Glides are generally from high in one tree to lower in another, rather than to the ground, and are often to escape danger. Unlike most squirrels, this species is nocturnal, leaving its tree-hole at night to find conifer seeds, nuts, fruits, leaves, shoots, and buds. Fur colour varies but is usually tawny to reddish brown above, with black rings around the large, "night-vision" eyes and hairless ears.

pale brown or buff underparts

black-tipped tail

large gliding membrane

Callosciurus prevosti
Prevost's squirrel

Length	13–28 cm (5–11 in)
Tail	8–26 cm (3 1/4–10 in)
Weight	150–500 g (5–18 oz)
Location S.E. Asia	Social unit Variable
	Status Lower risk†

One of the most brightly coloured mammals, this tree-dwelling, twilight-active squirrel is black on its upperparts, vivid chestnut-red on the underparts, with a broad, gleaming white band along each side from nose to thigh. It lives alone or in small family groups, communicating by bird-like calls and visual displays of the bushy tail. The diet includes fruits, soft seeds, oily nuts, and buds, as well as termites, ants, grubs, and birds' eggs. After 46–48 days' gestation, the 2–3 young are born in a large nest of twigs and leaves, in a tree-hole or among branches.

Ratufa indica
Indian giant squirrel

Length	35–40 cm (14–16 in)
Tail	35–60 cm (14–23 1/2 in)
Weight	1.5–2 kg (3 1/4–4 1/2 lb)
Location S. Asia	Social unit Individual/Pair
	Status Vulnerable

The huge, bushy tail of the Indian giant squirrel is usually longer than the head and body combined. The upperparts are dark, the head and limbs red-brown, and the underside whitish. This squirrel, alert and wary, makes massive leaps of 6m (20ft) among branches as it forages for fruits, nuts, bark, insects, and eggs. Its characteristic feeding posture is not upright but leaning forwards or down, perched on its back legs with the tail as a counterbalance. It builds the typical squirrel-type nest (drey) for resting and for rearing young.

brown, black, or dark red upperparts

broad front paws

Heliosciurus gambianus
Gambian sun squirrel

Length	15.5–21 cm (6–8 1/2 in)
Tail	15.5–31 cm (6–12 in)
Weight	250–350 g (9–13 oz)
Location Africa	Social unit Individual/Pair
	Status Lower risk†

Banded hairs in yellows, browns, and greys give the sun squirrel a speckled olive-brown appearance. The tail has 14 rings along its length and the eyes are white ringed. In habit, this squirrel is a typical ground-and-tree species with a diet ranging from seeds to birds' eggs. Its distinctive behaviours include "basking" on sunny branches, as its name suggests, and relining its nest each night with freshly plucked leaves.

Thomomys bottae
Botta's pocket gopher

Length	11.5–30 cm (4 1/2–12 in)
Tail	4–9.5 cm (1 1/2–3 3/4 in)
Weight	45–55 g (1 5/8–2 oz)
Location W. USA to N. Mexico	Social unit Individual
	Status Common

Botta's pocket gopher lives mainly alone. It digs an extensive burrow system (lodge) in loose soil, using its strong forelegs equipped with large claws. This gopher stays underground most of its life, consuming roots, tubers, bulbs, and other subsurface plant parts. It is greyish brown above and browny orange below, and has adaptations typical of burrowers, such as a flat head, long whiskers, small eyes, and small ears closed by flaps.

Crategeomys merriami
Merriam's pocket gopher

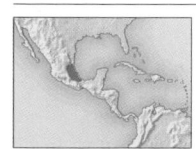

Length	14–26 cm (5 1/2–10 in)
Tail	6–13 cm (2 1/4–5 in)
Weight	225–900 g (1/2–2 lb)
Location E. Mexico	Social unit Individual
	Status Unconfirmed

The pocket gophers have a pocket-like pouch of furred skin on each cheek, for carrying food back to the nest. This species occupies a variety of habitats from sea level up to almost 4,000m (13,200ft). Its front incisor teeth are very long and can cope with many kinds of plant foods, from spiky cacti and farm crops (gophers sometimes becomes a pest) to fir tree needles and seeds. The upperparts are yellow, brown, or almost black; the underside is paler.

Orthogeomys grandis
Large pocket gopher

Length	10–35 cm (4–14 in)
Tail	4–14 cm (1 1/2–5 1/2 in)
Weight	300–900 g (11–32 oz)
Location Mexico to Central America	Social unit Individual
	Status Common

Like other pocket gophers, the large pocket gopher digs a burrow system (lodge) using its strong, large-clawed forefeet. It feeds on roots, bulbs, and other underground plant parts and also comes above ground at night to forage for stems and shoots, which it carries to its lodge in its fur-lined cheek pouches. Normally solitary, during the breeding season the large pocket gopher forms groups of one male and 4 females. Two or more young are born to each female in a grass-lined nesting chamber at the lowest level of the lodge.

Pedetes capensis
Springhare

Length 27–40 cm (10 1/2–16 in)	
Tail 30–47 cm (12–18 1/2 in)	
Weight 3–4 kg (6 1/2–8 3/4 lb)	
Location C. and E. Africa to southern Africa	**Social unit** Individual
	Status Vulnerable

The only member in its rodent family, the springhare resembles a rabbit-kangaroo, with a long, bushy tail and large, narrow, upright ears. It makes huge leaps, easily covering 2m (6 1/2 ft) per bound. The diet consists mainly of seeds, stems, bulbs, and other plant parts, but also includes locusts, beetles, and other invertebrates. When feeding, it tends to bend forwards and lope rabbit-like on all fours. The 5 large front claws are specialized for digging. When bounding at full speed, its tail curves up; when sitting up, its tail is used as a support. The springhare sleeps resting on its haunches, head tucked between its hindlegs, and the tail wrapped around the whole body. Nocturnal, the springhare lives alone or as a male–female pair, and digs several extensive burrows. There appears to be no peak breeding season; the single young is born at any time of year and is suckled in the main breeding burrow for approximately 7 weeks.

grey to brownish pink upperparts

black-tipped tail

very large hindfeet

Dipodomys merriami
Merriam's kangaroo rat

Length 8–14 cm (3 1/4–5 1/2 in)	
Tail 14–16 cm (5 1/2–6 1/2 in)	
Weight 40–45 g (7/16–1 5/8 oz)	
Location S.W. USA to N. Mexico	**Social unit** Group
	Status Common

Relatively large back legs and feet, for kangaroo-like hopping, and a very long, slender, tufted tail, for balance, allow Merriam's kangaroo rat to move at speed over sandy soil in its desert habitat. The silky fur is grey above, white on the underparts, with narrow dark grey and white stripes along the flanks. This kangaroo rat regularly has energetic dust-baths to keep its fur and skin clean. It digs, burrows, or searches for food, especially cockle- and sand-burr seeds in winter, and cactus seeds in summer.

Castor fiber
Eurasian beaver

Length 83–100 cm (33–39 in)	
Tail 30–38 cm (12–15 in)	
Weight 23–35 kg (51–77 lb)	
Location Europe	**Social unit** Group
	Status Lower risk

Similar to the American beaver (see right) in appearance, habits, and lifestyle, the Eurasian beaver is usually heavier. As in its close cousin, glands at the base of the tail produce an oily, waterproofing secretion that is spread through the fur when grooming. In areas with many natural waterways, the Eurasian beaver does not build a sticks-and-mud lodge but digs tunnels in the bank, with underwater entrances. It eats bark, leaves, and plants, and can stay submerged for 20 minutes.

Microdipodops megacephalus
Dark kangaroo mouse

Length 6.5–7.5 cm (2 1/2–3 in)	
Tail 6.5–10.5 cm (2 1/2–4 1/4 in)	
Weight 10–17 g (3/8–5/8 oz)	
Location S.W. USA	**Social unit** Individual
	Status Locally common

The dark kangaroo mouse is named after its dark-furred large back legs, which it uses for hopping and leaping over sand-dunes and soft, dry soil. The upperparts are grey-brown, the underparts pure white. The prominent eyes, large ears, long snout, and bushy whiskers indicate adaptation for a nocturnal lifestyle. Like many small desert rodents, this species is a seasonal opportunist. In summer it consumes mainly insects; in winter it switches to being a seed-hoarder in its burrow nest, carrying food items in its external cheek pouches. It also stores food as body fat in its dumpy tail. Each male dark kangaroo mouse aggressively defends a territory of up to 6,600 square m (7,900 square yd) against others of its species. Female territories tend to be much smaller, only about 400 square m (480 square yd). Each female produces a litter of 2–7 offspring.

Castor canadensis
American beaver

Length 74–88 cm (29–35 in)	
Tail 26–33 cm (10–13 in)	
Weight 11–26 kg (24–57 lb)	
Location North America	**Social unit** Group
	Status Locally common

The American beaver is well adapted to aquatic life. Its feet are webbed for swimming and the flat, scaly tail slaps the surface as a loud alarm signal. Underwater, the ears and nose shut with valve-like flaps and the lips close behind the incisor teeth, which can then be used for nibbling and gnawing. The eyes have a third, transparent eyelid (nictitating membrane) to see below the surface. Long whiskers feel the way in the dark. The American beaver feeds on the leaves, twigs, and bark of bankside trees and water plants. It also gnaws and fells small trees to eat the tender shoots and leaves. It uses the fallen branches and small trunks for building its lodge (see panel) and for dam construction, dragging them to the dam site in their strong jaws. Following a gestation period of 107 days, the 3–4 kits (young beavers) are born fully furred and are weaned within 2 months.

flat, scaly tail

LAKESIDE LODGE

Beavers rest by day in their lodge – a pile of mud and sticks built in a pool or lake. The lodge's underwater entrances keep out land-based predators. The beaver digs channels and builds dams of mud, stones, and branches to maintain a system of waterways. These activities are believed by some to harm crops and trees, and affect local wildlife. Another view is that beavers reduce local floods, and help to return the habitat to its natural state.

TWO COATS
The American beaver's long guard coat (outer fur) varies from yellowish brown to black, although it is usually reddish brown. The dense underfur is dark grey and retains body warmth even in freezing water.

MOUSE-LIKE RODENTS

PHYLUM	Chordata
CLASS	Mammalia
ORDER	Rodentia
SUBORDER	Myomorpha
FAMILIES	5
SPECIES	1,137

THIS SUBORDER – which is distinguished from the other 2 rodent suborders by the way the jaw muscles are arranged (see p.144) – constitutes over a quarter of all mammal species. Within the group are rats and mice (including voles, lemmings, hamsters, and gerbils), and jerboas. Mouse-like rodents often have a pointed face and long whiskers, and are usually small, nocturnal seed-eaters. They are found worldwide (except Antarctica), in almost all terrestrial habitats. Some species, such as the naked mole-rat, live underground; others, such as the water vole, occupy aquatic habitats. Species that live in open areas may have longer legs and feet (for quick escapes), and larger ears to detect danger from a distance.

LARGE LITTERS
Mouse-like rodents, such as this edible dormouse, are usually prolific breeders.

Oxymycterus nasutus

Burrowing mouse

Length 9.5–17 cm
(3³/₄–6¹/₂ in)

Tail 7–14.5 cm
(2³/₄–5³/₄ in)

Weight 90 g
(3¹/₄ oz)

Location E. South America

Social unit Individual

Status Locally common

This large mouse rarely comes out into the open. It forages by day under leaves, logs, and stones for creatures such as grubs and worms, sniffing with its long, flexible, shrew-like snout and scrabbling up prey with its large-clawed forefeet. Its back is black, tinged with red or yellow, and is darker along the mid-line; its flanks are yellow-brown; the underside is yellow-orange mixed with grey; and the short, scaly tail is sparsely haired. It rarely excavates its own tunnels or pathways, using those of other rodents.

Baiomys taylori

Northern pygmy mouse

Length 5–6.5 cm
(2–2¹/₂ in)

Tail 3.5–4.5 cm
(1¹/₂–1³/₄ in)

Weight 7–9 g
(¹/₄–¹¹/₃₂ oz)

Location S. USA to C. Mexico

Social unit Individual

Status Locally common

The female of this species, North America's smallest rodent, can become pregnant at the youngest age of any New World mouse – just 4 weeks. The fur is mid-brown on the back and grey underneath. This mouse occupies a territory up to 30m (100ft) across and feeds at twilight on plants and seeds. Its nest is in a burrow under logs or plants.

Nyctomys sumichrasti

Sumichrast's vesper rat

Length 11–13 cm
(4¹/₄–5 in)

Tail 8.5–15.5 cm
(3¹/₄–6 in)

Weight Not recorded

Location S. Mexico to S. Central America

Social unit Group

Status Locally common

This large mouse rarely comes out into the open. It forages by day under

Similar in appearance to the house mouse (see p.156), the American harvest mouse has large ears and a long, slender, scaly, well-furred tail. It builds a strong summer nest of grass above ground in a bush or undergrowth, and feeds on seeds, shoots, and insects. In winter, it moves into a tunnel or burrow dug, but then deserted, by another rodent.

Reithrodontomys raviventris

American harvest mouse

Length 7–7.5 cm
(2³/₄–3 in)

Tail 4.5–11.5 cm
(1³/₄–4¹/₂ in)

Weight 6–20 g
(2¹/₄–²/₃ oz)

Location W. USA (San Francisco Bay area)

Social unit Individual

Status Vulnerable

One of the more arboreal and brightly coloured rats, this species has tawny or pinkish brown upperparts with darker hairs along the centre of the back. It has pale flanks, white underparts, a dark ring around the eye, and a brown, scaly, hairy, tail. The ears are short and finely furred. On each foot the first toe is almost thumb-like, developed for gripping twigs. Sumichrast's vesper rat lives in a colony, builds squirrel-like nests of twigs, leaves, and creepers, and rarely descends to the ground. It is active mainly at night, eating a variety of plants including figs and avocados.

Peromyscus leucopus

White-footed mouse

Length 9–10.5 cm
(3¹/₂–4¹/₄ in)

Tail 6–10 cm
(2¹/₄–4 in)

Weight 14–30 g
(¹/₂–1¹/₁₆ oz)

Location S.E. Canada to Mexico

Social unit Individual

Status Common

A widespread and common species, this mouse resembles others in the deer mouse group, with its white feet and underparts, and brown fur on the back. White-footed mice usually live in pairs, which occupy a small

---

den in a sheltered place under tree roots, below a log or stone, or in a thicket; or they may dig a den in soil, or take over an abandoned burrow. The nest is made from soft, dry plant matter, including shredded stems. The pair forage principally at night for fruits, berries, seeds, and insects, staying mainly on the ground even though they climb well. Food is stored in the den near the nest, covered with soil. In cold weather the white-footed mouse may hibernate for a few hours each day. The female is pregnant for 22–23 days, and the average litter size is 4–5.

white feet and underfur

Kunsia tomentosus

Giant South American water rat

Length 29cm (11 1/2 in)	
Tail 15cm (6 in)	
Weight Not recorded	
Location C. South America	**Social unit** Individual
	Status Vulnerable

A little-known species, this large, small-eared, short-tailed rat spends much time burrowing, using its strong feet equipped with long, curved claws. Its upper fur is dark brown with stiff hairs, and the underneath hairs are grey with white tips. As the giant South American water rat tunnels, it consumes roots, tubers, and other underground plant parts. During the flood season, however, water fills its tunnels, so this rat stays mainly above ground, where its diet changes to mostly grasses and shoots.

Phodopus roborovskii

Desert hamster

Length 5.5–10cm (2 1/4 –4 in)	
Tail 7–11cm (2 3/4 –4 1/4 in)	
Weight 25–50g (7/8 –1 3/4 oz)	
Location E. Asia	**Social unit** Individual
	Status Locally common

Also called the dwarf hamster, this small, short-tailed, prominent-eared rodent is pale brown on its upperparts with pure white underparts. Its rear feet are short and broad, with dense fur on the underside for jumping across hot, loose desert sand. Like other hamsters, the desert hamster crams seeds into its internal cheek pouches and takes them back to its burrow for storage. It also eats insects such as beetles, locusts, and earwigs. The nesting burrow is dug in firm sand and is lined with hair shed by camels and sheep.

Sigmodon hispidus

Hispid cotton rat

Length 13–20cm (5–8 in)	
Tail 8–16.5cm (3 1/4 –6 1/2 in)	
Weight 100–225g (3 5/8 –8 oz)	
Location S. USA to N. South America	**Social unit** Individual
	Status Lower risk†

This is one of 10 cotton rat species in the Americas. Some are extremely rare, others locally numerous, even increasing to pest status when food is plentiful. The hispid cotton rat feeds on many foods including plants (sometimes ruining crops such as sweet potato and sugar cane), insects, and grubs. Being a good swimmer, it also takes freshwater crabs, crayfish, and frogs, and climbs reeds to eat birds' eggs and chicks. It is active day and night, usually living alone in a grassy nest in a sheltered depression in the ground or in a burrow up to 75cm (30in) deep. The hispid cotton rat digs a shallow burrow as it feeds, and establishes well-worn foraging runways. Its stiff fur is brown to brownish grey on the back, grey-white underneath. The female becomes sexually mature in 6–8 weeks, and, after 27 days' gestation, she gives birth to up to 12 young.

Mesocricetus auratus

Golden hamster

Length 13–13.5cm (5–5 1/4 in)	
Tail 1.5cm (1/2 in)	
Weight 100–125g (3 5/8 –4 oz)	
Location W. Asia	**Social unit** Individual
	Status Endangered

The golden hamster, familiar around the world as a pet, is restricted in the wild to a small area of western Asia. Its golden coat may show a darker patch on the forehead and a black stripe from each cheek to the upper neck. The fur pales to grey-white on the underside. Grooming is important to keep the coat in good condition, and both front teeth and claws are used for this purpose. This hamster excavates a burrow down to 2m (6½ ft), which it rarely leaves except to feed on a diet of seeds, nuts, and small creatures such as ants, flies, cockroaches, bugs, and even wasps. It is aggressive towards other hamsters.

rich golden orange fur

cheek pouches filled with food

SOLITARY GROOMING

Like many other hamsters, the golden hamster lives alone, and so has to rid its own coat of dirt, old fur, tangles, and pests such as fleas, rather than relying on mutual grooming like many social rodents.

ROBUST RODENT
The golden hamster has a blunt muzzle, broad face, small eyes, prominent ears, and tiny tail.

Calomys laucha

Vesper mouse

Length 7cm (2 3/4 in)	
Tail 5.5cm (2 1/4 in)	
Weight 13g (7/16 oz)	
Location C. and E. South America	**Social unit** Variable
	Status Common

The vesper mouse resembles the house mouse (see p.156) not only in appearance but also in living near human dwellings and in its occasional population surges, when it may become a pest. It is grey-white above and pale to dark brown on the underside, with a white patch behind each medium-sized ear, and a long, sparsely furred tail. It builds a grassy nest in any crevice such as under a log, rock, floorboards, or even – being a skilful climber – in a tree fork. The main diet is plants of all kinds, supplemented by a few insects such as beetles and caterpillars.

Cricetus cricetus

Common hamster

Length 20–34cm (8–13 1/2 in)	
Tail 4–6cm (1 1/2 –2 1/4 in)	
Weight 100–900g (3 5/8 –32 oz)	
Location Europe to C. Asia	**Social unit** Individual
	Status Lower risk†

The largest hamster, this species has distinctively thick fur, which is red-brown on the back and mainly black on the underside, with white patches on the nose, cheeks, throat, flanks, and paws – darker underside color-ation compared to the upperparts is very unusual among mammals. In autumn the common hamster hoards seeds, roots, and other plant matter, carrying items in its large, external cheek pouches back to its burrow. It then hibernates until spring, waking every 5–7 days to feed. In summer it also eats grubs, worms, and other creatures. After 18–20 days' gestation, the female produces a litter of up to 12, which are weaned by 3 weeks and fully grown by 8 weeks.

Hypogeomys antimena

Malagasy giant rat

Length	30–35 cm (12–14 in)
Tail	21–25 cm (8 1/2–10 in)
Weight	1–1.5 kg (2 1/4–3 1/4 lb)
Location W. Madagascar	Social unit Group
	Status Endangered

Known locally as the votsotsa, this rat has tall, rabbit-like ears and large rear feet. Similar to a rabbit in behaviour, too, it hops rather than runs and lives in a family group of male, female, and offspring of the past 2–3 years. The Malagasy giant rat digs a burrow system with up to 6 entrances in the sandy soil of coastal forests. It eats fruits, shoots, and soft bark, holding them in the forefeet. It is threatened by habitat loss and competition from introduced black rats.

rabbit-like ears

well-developed, digging claws

Parotomys brantsii

Brant's whistling rat

Length	12.5–16.5 cm (5–6 1/2 in)
Tail	7.5–10.5 cm (3–4 1/4 in)
Weight	85–125 g (3–4 oz)
Location Southern Africa	Social unit Group
	Status Locally common†

Also known as the karroo rat from its arid, rocky, saltbush habitat, this wary species feeds near its extensive burrow system dug in hard earth. Its whistle warns others in its colony of danger. The fur is patchy yellow and brown-black, with yellow ears, grey-white underparts, and reddish orange nose.

Rhizomys sinensis

Bamboo rat

Length	22–40 cm (9–16 in)
Tail	5–9.5 cm (2–3 3/4 in)
Weight	1–3 kg (2 1/4–6 1/2 lb)
Location E. Asia	Social unit Individual
	Status Locally common†

This large, beaver-like rat has thick, grey-brown fur, a squat snout, stocky body, and short, rounded, scaly tail. It lives in bamboo thickets, digging tunnels and nest chambers between the roots. An excellent climber, it feeds on bamboo roots and shoots as well as various seeds and fruits.

Tachyoryctes macrocephalus

Giant African mole rat

Length	31 cm (12 in)
Tail	9–10 cm (3 1/2–4 in)
Weight	350–1,000 g (13–36 oz)
Location E. Africa	Social unit Variable
	Status Lower risk†

The giant African mole rat lives like a mole, spending a great deal of time in its burrow. This may exceed 50m (165ft) in length and houses one rat.

Burrowing adaptations include a blunt and rounded head, robust body, short limbs, small eyes and ears, and thick fur. This mole rat gnaws roots and other plant parts with its large, projecting, orange-yellow incisors, and digs its extensive tunnel system in the same way.

Pachyuromys duprasi

Fat-tailed gerbil

Length	9.5–13 cm (3 3/4–5 in)
Tail	5.5–16.5 cm (2 1/4–6 1/2 in)
Weight	20–50 g (11/16–1 3/4 oz)
Location N. Africa	Social unit Variable
	Status Locally common

A Sahara dweller, the fat-tailed gerbil has long, soft fur, a pointed snout, and long rear feet. Coloration is chestnut-cinnamon on the back and sides, with black tips to the hairs, shading to white underneath. The club-shaped tail contains a store of body fat for nourishment and water. This gerbil emerges from its burrow at dusk to search for insects such as crickets. It also eats leaves, seeds, and other plant matter.

Clethrionomys glareolus

Bank vole

Length	7–13.5 cm (2 3/4–5 1/4 in)
Tail	3.5–6.5 cm (1 1/2–2 1/2 in)
Weight	12–35 g (7/16–1 1/4 oz)
Location W. Europe to N. Asia	Social unit Group
	Status Common

Typically blunt headed, the bank vole's upperparts vary from yellowish to reddish or brown, with grey flanks, grey-white rump, white feet, and a slightly bushy-tipped tail. The species also varies greatly in size, being twice as long and 3 times as heavy in some regions compared to others. The bank vole is very adaptable, nesting in burrows, thickets, and tree stumps, and eating a huge range of foods, from fungi and mosses to seeds, buds, insects, and birds' eggs.

Meriones unguiculatus

Mongolian gerbil

Length	10–12.5 cm (4–5 in)
Tail	9.5–11 cm (3 3/4–4 1/4 in)
Weight	50–60 g (2–2 1/8 oz)
Location E. Asia	Social unit Group
	Status Locally common

long hind legs for leaping

Familiar as a pet, this is one of some 13 gerbil species native to the Middle East and Asia. It burrows in dry steppes and is active by day and night, summer and winter. It eats mainly seeds, storing excess in its elaborate burrow, which it may share with its mate and up to 12 young. Family members groom each other but are quick to attack a stranger. Mainly brown haired with black tips, the underparts are grey or white.

Arvicola terrestris

European water vole

Length	12–23 cm (4 3/4–9 in)
Tail	7–11 cm (2 3/4–4 1/4 in)
Weight	60–300 g (2 1/8–11 oz)
Location W. Europe to W. and N. Asia	Social unit Individual/Pair
	Status Locally common

Water voles that mainly burrow in meadows and woods are almost half the size of those that live near rivers, lakes, and marshes. Both types eat plant foods and have thick fur, which is grey, brown, or black on the upperparts and dark grey to white below. The rounded tail is half the body length. This vole is threatened by pollution, loss of habitat, and an introduced predator, mink.

Microtus arvalis
Common vole

Length 9–12 cm (3 1/2 – 4 3/4 in)	
Tail 3–4.5 cm (1 1/4 – 1 3/4 in)	
Weight 20–45 g (1 1/16 – 1 5/8 oz)	
Location W. Europe to W. and C. Asia	**Social unit** Group
	Status Common

This medium-sized vole is one of the most numerous rodents in grassy and farmland habitats. It has short fur, grey-brown or sandy on the back changing to grey underneath, a blunt snout, small eyes and ears, stocky body, and short, furred tail. It digs burrows to make nest chambers and food stores, and eats chiefly green plant parts such as grass blades. In winter it may take refuge in a barn or haystack, and gnaw soft bark.

Ondatra zibethicus
Muskrat

Length 25–35 cm (10–14 in)	
Tail 20–25 cm (8–10 in)	
Weight 0.6–2 kg (1 1/4 – 4 1/2 lb)	
Location North America, W. Europe to N. and E. Asia	**Social unit** Group
	Status Common

The muskrat usually lives in a group of up to 10, digging tunnels in the bank or building a beaver-like home (lodge) from mud, plant stems, and twigs. It eats reeds and other water plants, and occasionally hunts for crayfish, frogs, fish, and molluscs. The female builds a nest in a dry tunnel chamber or lodge platform, where the litter of 1–3 young are born. Musk secretions are used, with droppings and urine, to mark out territories.

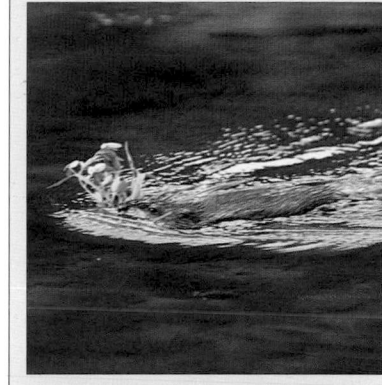

EXPERT SWIMMER
The largest species of burrowing vole, the muskrat is well adapted to swimming with large back feet that have small webs between the toes and a row of stiff hairs along one edge, forming a "swimming fringe". Its long, furless tail is flattened from side to side, which enables it to be used as a rudder. The nostrils and small ears are closed by flaps during dives, which may last 20 minutes. It can swim up to 100m (330ft) underwater without surfacing.

fine underfur with long, coarse guard hairs

MUSKY SMELL
The muskrat is named after the musky-smelling secretions from glands around its genital and anal region. The glands are especially prominent in males and enlarge at breeding time.

Apodemus flavicollis
Yellow-necked field mouse

Length 8.5–13 cm (3 1/4 – 5 in)	
Tail 9–13.5 cm (3 1/2 – 5 1/4 in)	
Weight 18–50 g (5/8 – 1 3/4 oz)	
Location W. Europe to W. and C. Asia	**Social unit** Individual
	Status Common

The yellow throat of this large, long-tailed mouse contrasts with the brown back and yellowish white underparts. The large, prominent eyes and big ears indicate twilight and nocturnal habits, and the long rear feet allow prodigious jumps. The yellow-necked field mouse climbs trees to 20m (66ft), searching for seeds, berries, and small creatures such as caterpillars, spiders, and millipedes. It nests in any suitable hole, among roots, or high in a tree trunk, and aggressively chases away other mice including similar wood mice (see right).

Apodemus sylvaticus
Wood mouse

Length 8–11 cm (3–4 1/4 in)	
Tail 7–11 cm (2 3/4 – 4 1/4 in)	
Weight 15–30 g (9/16 – 1 1/16 oz)	
Location W. Europe to N. Asia, N. Africa	**Social unit** Individual
	Status Common

Sometimes mistaken for a small yellow-necked field mouse (see left), the wood mouse may have not only a yellow throat but also an orange-brown chest patch. Its upperparts are grey-brown, the underparts grey-white. Fast and agile, the wood mouse consumes many foods including mushrooms, berries, fruits, worms, and insects. It nests in a burrow or tree hole, marks its territory with urine, and fights intruders violently.

Lemmus sibericus
Brown lemming

Length 12–15 cm (4 3/4 – 6 in)	
Tail 1–1.5 cm (3/8 – 1/2 in)	
Weight 45–150 g (1 5/8 – 5 1/2 oz)	
Location N.E. Europe to N. Asia, Alaska to N.W. Canada	**Social unit** Group
	Status Common

The brown lemming lives in large colonies, breeds prolifically, and makes small-scale seasonal migrations between high, shrubby grassland and moors, and sheltered lowlands for winter. Its migrations are much less spectacular than those of the Norway lemming (*Lemmus lemmus*), which is sometimes driven by instinct to try to swim rivers or scramble down cliffs. The brown lemming eats mosses, sedges, herbs, and soft twigs, and sometimes birds' eggs. The female builds a nest from grass and her own fur and, after a gestation period of 18 days, produces up to 12 young.

WELL-ROUNDED SHAPE
The brown lemming has a blunt snout, thick fur, and short tail. The black stripe on the back is absent in individuals from parts of North America.

AUTUMN RETREAT

Lively and energetic, the brown lemming is an active, noisily squeaking burrower and feeder. In autumn it descends to lower tundra areas, lakes, and rivers, to dig sheltered tunnels under peaty mounds or make nests in clumps of vegetation. It feeds along runways under the snow.

small, well-furred ears

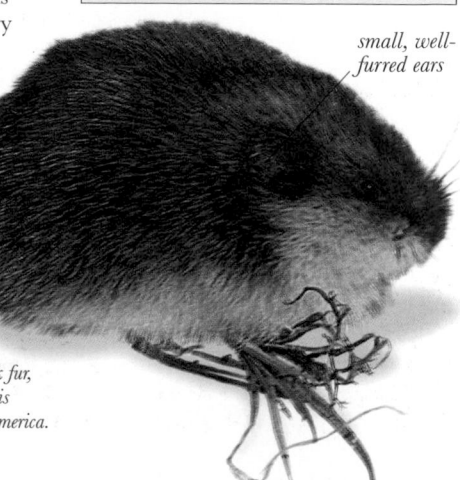

MAMMALS

Lagurus lagurus
Steppe lemming

Length 8–12 cm (3 1/4 – 4 3/4 in)	
Tail 0.7–2 cm (7/3 – 3/4 in)	
Weight 25–35 g (7/8 – 1 1/4 oz)	

Location E. Europe to E. Asia	Social unit Variable
	Status Lower risk

Long, waterproof fur, even covering feet and ears, keeps the steppe lemming warm in the cold north Asian steppes. It has a black stripe along the middle of its pale grey or cinnamon back, and pale underparts. Burrows up to 30cm (12in) deep give temporary shelter, while more permanent ones – 3 times as deep – house grass-lined nests. There are 5 litters per year of up to 12 young.

Cricetomys gambianus
Giant pouched rat

Length 35–40cm (14–16in)	
Tail 37–45cm (14 1/2 –18in)	
Weight 1–1.5 kg (2 1/4 –3 1/4 lb)	

Location W., C., E., and southern Africa	Social unit Variable
	Status Common

This rat eats a great variety of moist or fleshy foods, from termites to avocados, and also ground nuts and maize. Its huge cheek pouches carry items back to its burrow, which has extensive chambers for food, resting, breeding, and defecation. Its bristly hair is buff-brown on the back, fading to white on the throat and underside, with a dark brown eye-ring. This large, big-eared, docile rat is kept both as a pet and for meat.

Micromys minutus
Eurasian harvest mouse

Length 5–8cm (2–3in)	
Tail 4.5–7.5 cm (1 3/4 –3in)	
Weight 5–7 g (3/16 –1/4 oz)	

Location W. Europe to E. Asia	Social unit Individual
	Status Lower risk

This tiny mouse is the only Old World rodent with a prehensile tail. It feeds on seeds, including the heads of wheat and other farmed cereals, berries, and small animals such as insects and spiders. In the breeding season, the female gives birth to about 2–6 young, occasionally 12 or more, after a pregnancy lasting 21 days. However, if food becomes scarce, she may eat them – a self-survival strategy that occurs in various rodents.

BALL-LIKE NEST
The harvest mouse constructs a spherical nest of finely shredded grass blades and stems, perhaps using an old bird's nest as a base, in a thicket or grassy clump. It is about 8–12cm (3–4 3/4in) across, and is located some 50–130cm (20–51in) above ground level. The female's breeding nest is more substantial than the regular nest.

COLORATION
The Eurasian harvest mouse has yellowish or reddish brown upperparts and a mainly white underside. The face is rounded but the snout is pointed.

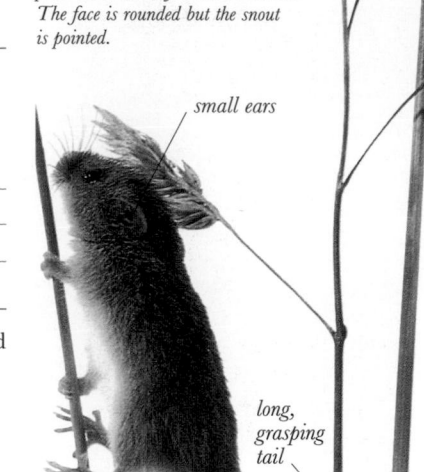

small ears

long, grasping tail

broad, gripping feet

Lemniscomys striatus
Striped grass mouse

Length 10–14cm (4–5 1/2 in)	
Tail 10–15.5 cm (4–6in)	
Weight 20–70g (11/16 –2 1/2 oz)	

Location W. Africa, southern Africa, to E. Africa	Social unit Individual
	Status Common

The striped grass mouse is paler in the west of its range compared to the east, with light stripes along the buff or reddish orange back; underparts are brown-tinged white. It lives mainly on the ground with runways leading to feeding areas of grass stems, leaves, farm crops, and the occasional insect. This nervous, jumpy mouse lives alone and may feign death – "play dead" – or shed the skin from its tail when caught by a predator.

dark central stripe down the back

Rattus muelleri
Müller's rat

Length 18.5–24cm (7 1/2 –9 1/2 in)	
Tail 22–28cm (9–11 in)	
Weight 150–300g (5–11 oz)	

Location S.E. Asia	Social unit Individual
	Status Locally common

Also called the giant Sunda rat, due to its size and geographical area of origin, this large, coarse-haired rodent has grizzled buff-black upperparts with pale grey or brown underparts. It has a typical, rat-like face, small ears, almost hairless feet, and a brown tail. Müller's rat eats many plants and also small animals such as snails, insects, and lizards. It is familiar across southeast Asia in various habitats, primarily on and near the ground in tropical forests, and also frequents human dwellings and outhouses.

Rattus rattus
Black rat

Length 16–24cm (6 1/2 –9 1/2 in)	
Tail 18–26cm (7–10 in)	
Weight 150–250g (5–9 oz)	

Location Worldwide (except polar regions)	Social unit Variable
	Status Common

Also called the ship rat, roof rat, or house rat, in early Roman times the black rat spread around the world from Asia in ships or crates of cargo. It prefers plant matter such as seeds and fruits, but it can survive on insects, dead animals, faeces, and refuse. Black rats gather in "packs" of 20–60 and may intimidate larger animals such as dogs. The basic coloration is black, but shades of brown also occur, with grey to white on the belly, and whitish or pink feet. The female produces 4–10 young after a gestation period of 20–24 days. This rat can run, climb, and swim very well, and makes a nest of twigs and grass, but often nests in a roof cavity and uses almost any material. The fleas carried by black rats spread diseases among humans, including bubonic plague, which has killed hundreds of millions of people through the centuries.

long, hairless tail, for balance

Rattus norvegicus

Brown rat

Length 20–28 cm (8–11 in)	
Tail 17–23 cm (7–9 in)	
Weight 275–575 g (10–21 oz)	

Location Worldwide (except polar regions)

Social unit Group

Status Common

A hugely varied, opportunistic diet, sharp senses, and great agility have enabled the brown rat, also called the Norway or common rat, to spread all around the world. "Packs" of up to 200, dominated by large males, will attack rabbits, large birds, and even fish. After a gestation of 22–24 days, the female gives birth to 6–9 young in a nest of grass, leaves, paper, rags, or almost any other material. This species is the ancestor of rats bred as pets and for scientific research.

mainly hairless tail

tail held out for balance

SWIMMING RAT
The brown rat is an exceptional swimmer and catches small fish and crayfish, and crunches up water snails and aquatic insects.

small eyes

COLORATION
The brown rat varies from brown to grey-brown or black on its back. It is paler on the underside, and has a long, sparsely haired tail.

MAMMALS

Mallomys rothschildi

Smooth-tailed giant rat

Length 34–38 cm (13½–15 in)	
Tail 36–42 cm (14–16½ in)	
Weight 0.95–1.5 kg (2–3¼ lb)	

Location New Guinea

Social unit Individual

Status Locally common

This large rat has an almost black back, reddish brown upperparts, and a pale underside, perhaps with a white band from the belly up each side. It scampers through trees in search of shoots, leaves, and other plant foods, gripping with its sharp-clawed feet.

Hydromys chrysogaster

Australian water rat

Length 29–39 cm (11½–15½ in)	
Tail 23–33 cm (9–13 in)	
Weight 0.65–1.25 kg (1½–3 lb)	

Location New Guinea, Australia (including Tasmania)

Social unit Individual

Status Locally common

Australia's heaviest native rodent has broad back feet and webbed toes, for swimming. The upperparts vary from brown to grey, the underparts brown

---

Leporillus conditor

Greater stick-nest rat

Length 17–26 cm (7–10 in)	
Tail 14.5–24 cm (5¾–9½ in)	
Weight 150–450 g (5½–16 oz)	

Location S. Australia (Franklin Island)

Social unit Group

Status Endangered

Now confined to one island off Australia, the greater stick-nest rat is almost rabbit sized, with long ears, a rounded nose, and slim tail. It is grey-brown above and white underneath, and builds a strong surface nest of sticks and twigs up to 1.5 m (5 ft) high.

to golden-yellow, cream, or even white, and the tail is thick and white tipped. This water rat is active at dusk and dawn as a powerful predator of shellfish, water snails, fish, frogs, turtles, birds, mice, and even bats.

Acomys minous

Spiny mouse

Length 9–12 cm (3½–4¾ in)	
Tail 9–12 cm (3½–4¾ in)	
Weight 11–90 g (⅜–3¼ oz)	

Location Europe (Crete)

Social unit Variable

Status Vulnerable

Coarse, stiff hairs on the back and tail give this mouse its common name. Its fur varies from yellow to red, grey, or brown above, and white below. A nocturnal forager, it takes anything edible – mainly grass blades and seeds – and builds only a rudimentary nest. A gregarious species, gestation is 5–6 weeks (long for a mouse). Other females clean and assist the mother at the birth, when the young are already well developed, with open eyes.

Notomys alexis

Spinifex hopping mouse

Length 9–18 cm (3½–7 in)	
Tail 12.5–23 cm (5–9 in)	
Weight 20–50 g (¹¹⁄₁₆–1¾ oz)	

Location W. and C. Australia

Social unit Group

Status Common

Also called the dargawarra, this large mouse is named after the bushy, spiky desert grass called spinifex, where it often lives. It obtains all its moisture from leaves, seeds, berries, and other plant foods, never needs to drink, and produces some of the most concentrated urine of any rodent. Very sociable, this mouse lives in mixed-sex groups of up to 10, shares nests, and breeds soon after any rainfall.

Mus musculus
House mouse

Length	7–10.5 cm (2¾–4¼ in)
Tail	5–10 cm (2–4 in)
Weight	10–35 g (³⁄₈–1 oz)

Location Worldwide (except polar regions)

Social unit Group

Status Common

The second most widely distributed mammal, after humans, this mouse survives on a huge range of foods and lives in a family group of dominant male and several females. They communicate by high-pitched squeaks and mark their territory with scent and urine. Maturity is at 8–10 weeks, gestation 18–24 days, and litter size 3–8, with 10 litters in favourable years. The species has been widely bred for pets and scientific research.

upperparts vary from grey-black to red-brown

mostly hairless tail

Muscardinus avellanarius
Hazel dormouse

Length	6.5–8.5 cm (2½–3¼ in)
Tail	5.5–8 cm (2¼–3 in)
Weight	15–35 g (⁹⁄₁₆–1¼ oz)

Location Europe

Social unit Individual

Status Lower risk

About the size of a house mouse (see above), the hazel dormouse is an excellent climber and jumper. Feeding mainly in trees, it changes its diet with the season from flowers, grubs, and birds' eggs in spring and summer to seeds, berries, fruits, and nuts in autumn. Dormice are the only rodents that do not have a caecum (a part of the large intestine), which may indicate that their diet is low in cellulose. The nest is made of grass in a thick bush or tree hole. Several individuals may live nearby and share the feeding area. They communicate using a wide range of whistles and growling noises. The female is pregnant for 22–24 days and has a litter of 2–7 young, with up to 2 litters per year. The tail is bushy and its skin can detach if seized by a predator.

DEEP WINTER SLEEP

Most dormice hibernate deeply in winter ("dor" meaning "sleeping"). The hazel dormouse rests for about 7 months in a nest about 12 cm (4¾ in) in diameter – larger than its summer quarters. This nest may be in a burrow or under moss or leaves. It stores food both in its nest and in its body as fat, to enable it to survive through the cold season.

VARIABLE COLORATION
The hazel or common dormouse has yellow, red, orange, or brown upperparts and a white underside.

Glis glis
Edible dormouse

Length	13–20 cm (5–8 in)
Tail	10–18 cm (4–7 in)
Weight	70–150 g (2½–5 oz)

Location C. and S. Europe to W. Asia

Social unit Group

Status Lower risk

This species inhabits woods and out-buildings, nesting in tree holes or crevices in roofs and under floors. It lays down fat in autumn to provide reserves of food for a 7-month hibernation in a large, strong nest deep in a burrow. The edible dormouse eats leaves, seeds, fruits, nuts, bark, mushrooms, small creatures such as insects, and birds' eggs and chicks. Like other dormice, it forms loose social groups and communicates by squeaks and twitters. While pregnant, the female is solitary.

ONCE "EDIBLE"
This mouse was kept in ancient Rome and fattened with extra food in autumn, to be served at the dinner table – hence its name.

SQUIRREL-LIKE LEAPER

The edible dormouse resembles a squirrel with its long, bushy tail, large back legs for leaping in branches, and semi-upright posture. Its fine, dense fur is brown or silver-grey with dark eye patches and white underparts.

bushy tail

Jaculus jaculus
Desert jerboa

Length	10–12 cm (4–4¾ in)
Tail	16–20 cm (6½–8 in)
Weight	45–75 g (1⅝–2⅝ oz)

Location N. Africa to W. Asia

Social unit Individual

Status Common

Well adapted for desert sand, this jerboa has very long hind legs, each with 3 toes on a pad of hairs. It hops at speed, balancing with its very long tail, which has a black band near the fluffy white tip. The fur is brown-orange on the back, grey-orange along the sides, and white below with a whitish hip band. The desert jerboa feeds at night on seeds, roots, and leaves. By day, it plugs the entrances to its burrow to keep out the heat, predators, and other animals.

Allactaga tetradactyla
Four-toed jerboa

Length	10–12 cm (4–4¾ in)
Tail	15.5–18 cm (6–7 in)
Weight	50–55 g (1¾–2 oz)

Location N. Africa

Social unit Individual

Status Endangered

Each back foot of this jerboa has an extra, fourth toe, which is small compared to the 3 functional toes. In other respects it is a typical jerboa, with its huge, hopping back feet and tall, rabbit-like ears. The upperparts are speckled black and orange, the rump orange, the sides grey, and the underparts white. The long, balancing tail has a black band near the white, feathery tip. Emerging at night, the jerboa eats grass, leaves, and soft seeds.

CAVY-LIKE RODENTS

INCLUDED IN THIS "umbrella" group are species as diverse as the semi-aquatic capybara, the largest living rodent; the New World porcupines, which are arboreal and possess distinctive spines and a prehensile tail; and African mole-rats, which live underground. The defining feature of the suborder is the organization of the jaw muscles (see p.144), and most species are characterized by a relatively large head, a sturdy body, a short tail, and slender legs. Cavy-like rodents are found throughout Africa, the Americas, and Asia.

PHYLUM	Chordata
CLASS	Mammalia
ORDER	Rodentia
SUBORDER	Caviomorpha
FAMILIES	18
SPECIES	188

SAFETY IN NUMBERS
By living in a family group, these capybaras greatly increase the chances of spotting a predator before it can attack.

North American porcupine

Erethizon dorsatum

Length 65–80 cm (26–32 in)
Tail 15–30 cm (6–12 in)
Weight 3.5–7 kg (7¾–15 lb)

Location Canada, USA

Social unit Individual

Status Common

The large crest of long quills, up to 8cm (3in) long on the head, identifies this stocky, short-limbed porcupine. The strong, broad feet have sharp claws and naked soles, to aid grip. The North American porcupine is very vocal, especially during courtship in early winter, when it whines, screeches, grunts, mews, and hoots.

crest of long quills

QUILLS AND FUR
The main quills are yellow-white, with black or brown tips. The rest of the body is furred and spined in shades of brown.

CLUMSY CLIMBER

The North American porcupine seems clumsy in trees. Yet it climbs extensively for buds, blossom, shoots, leaves, berries, and nuts. It also eats grass and farm crops in summer, and soft bark and conifer needles in winter. Solitary most of the year, it does not hibernate and may share a den in cold spells.

Cape porcupine

Hystrix africaeaustralis

Length 63–80 cm (25–32 in)
Tail 10.5–13 cm (4¼–5 in)
Weight 20 kg (44 lb)

Location C. to southern Africa

Social unit Variable

Status Common

These rodents sniff and forage for distances of up to 15km (9 miles) at night, alone, in pairs, or in small groups, seeking roots, bulbs, berries, and fruits. By day they rest in caves or rocky crevices. After 6–8 weeks' gestation, the female produces 1–4 young, and the male helps to care for them. The cape (or crested) porcupine communicates by means of quill-rattles, piping squeaks, and grunts.

spines intermingled with ordinary hairs

short bristle-covered legs

eyes set far back in rounded head

SPINES AND WHISKERS
The back is covered by banded brown-black and white spines with white tips, intermingled with hairs. The nose has long, stout whiskers.

BARBED DEFENCE

The Cape porcupine cannot shoot out its quills; but, if alarmed, it raises them and charges backwards at the enemy. The quills detach easily and their barbed tips work into the aggressor's flesh. These defences effectively deter attacks by other animals, but these porcupines are killed in large numbers by humans because they damage crops, and for their prized meat.

Prehensile-tailed porcupine

Coendou prehensilis

Length 52 cm (20½ in)
Tail 52 cm (20½ in)
Weight 5 kg (11 lb)

Location N. and E. South America, Trinidad

Social unit Individual

Status Locally common†

This large, muscular porcupine climbs slowly but surely with curved claws, naked-soled feet, and prehensile tail, which is almost as long as the head and body and lacks fur towards the tip. By day it sleeps in a hole, in a tree trunk, or in the ground. At dusk the prehensile-tailed porcupine forages for leaves, bark, fruits, and shoots, and occasionally small animals.

MAMMALS

Cavia aperea

Guinea pig

Length 20–30 cm (8–12 in)	
Tail None	
Weight 500–600 g (18–21 oz)	
Location N.W. to E. South America	Social unit Group
	Status Common

A close relative and perhaps ancestor of the domestic (pet) guinea pig (*Cavia porcellus*), this is one of the smallest cavy-like rodents. It has a large head, blunt snout, tail-less body, and short legs, with 4-toed front feet and 3-toed rear feet. The long, coarse fur is dark grey-brown to black. It eats leaves, grasses, seeds, flowers, and bark. Guinea pigs live as close neighbours in shrubby grassland, with communal feeding runways but separate nests.

Hydrochaerus hydrochaeris

Capybara

Length 1.1–1.3 m (3 1/2–4 1/4 ft)	
Tail Vestigial	
Weight 35–66 kg (77–145 lb)	
Location N. and E. South America	Social unit Variable
	Status Common

The world's heaviest rodent, the capybara has partially webbed toes, and its nostrils, eyes, and ears are set on top of its head, so that it can smell, see, and hear when swimming.

STOUT SWIMMER
The capybara is heavy bodied, with short but sturdy limbs, hoof-like claws, and almost no tail.

It is an excellent swimmer and diver in rivers, lakes, and swamps. It forms various groupings including male–female pairs, families with young, and larger, mixed herds dominated by one male who mates with all females in his group. They roam a home range, marked with scent, and chase away intruders. After a gestation period of 150 days, 1–8 offspring (usually 5) are born fully furred and can run, swim, and dive within hours of birth. The capybara's numerous predators include humans, who value the meat and hide.

coarse fur

pale to dark brown fur, tinged with yellow or grey

Dolichotis patagonum

Mara

Length 43–78 cm (17–31 in)	
Tail 2.5 cm (1 in)	
Weight 2 kg (4 1/2 lb)	
Location S. South America	Social unit Pair
	Status Lower risk

The mara or Patagonian cavy is a large, long-legged rodent that resembles a small deer in looks and behaviour. It has a white, collar-like neck patch and whitish fringe to the short tail. The muzzle is long; the eyes and ears large. The mara runs and jumps well, and is mainly a grazer on grass and low shrubs. Male–female pairs stay together for life, digging a large burrow for their 1–3 offspring.

Dasyprocta azarae

Azara's agouti

Length 50 cm (20 in)	
Tail 2.5 cm (1 in)	
Weight 3 kg (6 1/2 lb)	
Location E. South America	Social unit Variable
	Status Vulnerable

This large, social, day-active species has prominent ears, short legs, and a tiny tail. Its fur is speckled pale to mid-brown, perhaps tinged yellowish around the underside. The distinctive feet have 5 toes on the front feet, but 3 on the hindfeet. Azara's agouti eats a variety of seeds, fruits, and other plant material, and barks when alarmed. People hunt it for meat across much of its range.

Agouti paca

Paca

Length 60–80 cm (23 1/2–32 in)	
Tail 1.5–3.5 cm (1/2–1 1/2 in)	
Weight 6–12 kg (13–26 lb)	
Location S. Mexico to E. South America	Social unit Individual
	Status Common

Pacas live alone, apart from the mother and single young, which she suckles for 6 weeks. This expert swimmer has a big head, squared-off muzzle, robust body, short, sturdy legs, and tiny tail. The upper body is brown, red, or grey with 4 pale "dotted-line" stripes along each side; the underparts are white or buff. The paca rests in a burrow or hollow tree by day and emerges to eat fruits, leaves, buds, and flowers at night.

Lagostomus maximus

Plains viscacha

Length 4.5–6.5 cm (1 3/4–2 1/2 in)	
Tail 1.5–2 cm (1/2–3/4 in)	
Weight 7–9 kg (15–20 lb)	
Location C. and S. South America	Social unit Group
	Status Endangered†

The largest member of the chinchilla family, the plains viscacha lives in noisy colonies of 20–50. It digs tunnel systems extending more than 300 m (980 ft). At night it eats grasses, seeds, and roots, and collects sticks, stones, and bones to pile at tunnel entrances. The face is black and white striped, the back is grey-brown fading to white underneath, and the tail is brown tipped with black. The male is twice as heavy as the female.

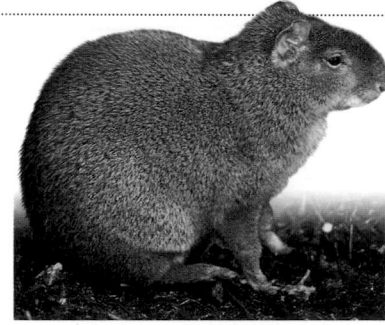

Chinchilla lanigera
Chinchilla

Length 22–23 cm (8–9 in)	
Tail 13–15.5 cm (5–6 in)	
Weight 400–500 g (14–18 oz)	

Location S.W. South America	Social unit Group
	Status Vulnerable

This rodent's thick, soft, silky fur, which keeps out the bitter cold of its mountain habitat, has long been prized by humans and, although now protected, has been hunted and farmed in many regions.

thickly furred, bushy tail

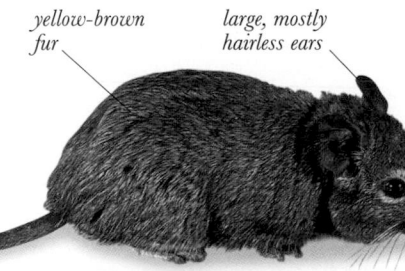

long whiskers

large rear legs for jumping

Endangered in the wild, the chinchilla is common as a pet, with its appealing appearance and usually friendly nature. The fur is silvery grey-blue above and cream or yellowish on the underside, with long grey and black hairs on the tail's upper surface. In the wild, the chinchilla forms colonies of 100 or more in rocky areas, sheltering in caves and crevices. It eats most plant foods, especially grass and leaves, sitting up to hold items in the front feet while watching for danger. If threatened, it rears up and spits hard at the aggressor. During the winter breeding season, the female, which is larger than the male, becomes more aggressive towards other females. Up to 4 young, but usually 2–3, are born after a gestation period of 111 days and are suckled for 6–8 weeks.

Octodon degus
Degu

Length 25–31 cm (10–12 in)	
Tail 7.5–13 cm (3–5 in)	
Weight 175–300 g (6–11 oz)	

Location S.W. South America	Social unit Variable
	Status Locally common

yellow-brown fur

large, mostly hairless ears

Resembling a very large mouse, this mountain-dwelling rodent has a stout body. Mainly brown, it has pale underparts, and there may also be yellow furry "lids" above and below each eye, and a yellow neck ring. The long tail, which has a tufted black tip, breaks off easily if grabbed by a predator. The degu lives in colonies and excavates extensive burrow systems. It feeds by day on a varied diet of plant matter and, in the dry season, cattle droppings. It stores excess food in the burrow for winter.

Myocastor coypus
Coypu

Length 47–58 cm (18 1/2–23 in)	
Tail 34–41 cm (13 1/2–16 in)	
Weight 6.5 kg (14 lb)	

Location S. South America	Social unit Group
	Status Common

The coypu or nutria was farmed for its dense brown fur, but escapees established colonies in many areas outside its native region. It has a large head, small, high-set eyes and ears, a robust body, arched hindquarters, long, rounded tail, and webbed rear feet for rapid swimming. The coypu eats most vegetation, especially water plants, and lives in bankside tunnels in family groups.

Capromys pilorides
Desmarest's hutia

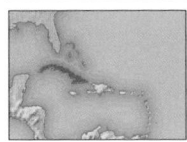

Length 55–60 cm (22–23 1/2 in)	
Tail 15–26 cm (6–10 in)	
Weight 4.5–7 kg (10–15 lb)	

Location Caribbean	Social unit Individual/Pair
	Status Endangered†

Desmarest's hutia resembles a huge vole with its typically blunt nose, large head, short neck, small ears, stocky body, and short limbs. It has a white nose, red-brown to black upperparts, and grey or yellow-brown underparts.

The strong, tapering, hairy tail and sharp, curved claws are adapted for support and gripping when tree climbing and when foraging for its varied diet of fruits, leaves, soft bark, and, occasionally, lizards. Litter size is 1–4. Hutias live only in the Caribbean, and most species are either severely threatened or already extinct.

Heterocephalus glaber
Naked mole-rat

Length 8–9 cm (3 1/4–3 1/2 in)	
Tail 3–4.5 cm (1 1/4–1 3/4 in)	
Weight 30–80 g (1 1/16–2 7/8 oz)	

Location E. Africa	Social unit Group
	Status Locally common

This mole-rat's social system is unique among mammals. Only one dominant female, the "queen", breeds. She may have more than 20 pups per litter and is tended by several non-workers. The colony's workers form head-to-tail "digging chains" to tunnel and gather food. The naked mole-rat is not truly naked but has pale, sparse hairs over its pinkish grey skin. As in other mole-rats, the massive incisor teeth are for digging and eating, and the eyes and ears are minute. The tail is rounded and the limbs strong, with 5 thick-clawed toes, for digging. Colonies comprise 70–80 individuals that inhabit elaborate tunnel systems. Food-gathering galleries 15–50cm (6–20in) deep radiate up to 40m (130ft) from the central chamber. New galleries are dug regularly for roots, bulbs, tubers, and other underground plant parts. This rodent only surfaces to travel to another colony.

pale, sparse hairs over body

Cryptomys hottentotus
Hottentot mole-rat

Length 10–18.5 cm (4–7 1/4 in)	
Tail 1–3 cm (3/8–1 1/4 in)	
Weight 100–150 g (3 5/8–5 1/2 oz)	

Location Southern Africa	Social unit Group
	Status Locally common

This mole-rat has a wide head, small eyes and ears, short, sturdy limbs, stocky body, and thick, dense fur. Coloration is pink-brown or grey above, with paler underparts. Extra-large incisor teeth bite through soil as well as bulbs, tubers, roots, and other subterranean plant parts. Colonies of 5–15 individuals dig large, complex tunnel systems and store excess food. Usually only the senior pair breeds.

Bathyergus janetta
Namaqua dune mole-rat

Length 17–24 cm (7–9 1/2 in)	
Tail 4–5 cm (1 1/2–2 in)	
Weight Not recorded	

Location S.W. Africa	Social unit Variable
	Status Lower risk

This species tunnels in shifting sands, either on the coast or inland. It lives in small colonies in a tunnel system up to 200m (655ft) long. The large head has massive protruding incisor teeth, the body is cylindrical, the legs are short but strong, and the tail is tiny. The back is dark brown, the sides grey, and the head and belly dark grey.

CETACEANS

PHYLUM	Chordata
CLASS	Mammalia
ORDER	Cetacea
FAMILIES	13
SPECIES	83

CLASSIFICATION NOTE

This book adopts the traditional method of classifying cetaceans, whereby the order is divided into 2 suborders: the baleen whales and the toothed whales. Although early molecular evidence seemed to indicate that baleen whales originated from the toothed whales, this relationship is now in doubt.

Baleen whales
see pp.162–5
Toothed whales
see pp.166–77

CETACEANS – WHALES, DOLPHINS, and porpoises – are perhaps the most specialized of all mammals, with their fish-shaped, hairless body, a flipper-like front limb, and vestigial back limbs (located within the body wall). However, they are true mammals: they breathe air with lungs, and they have mammary glands with which they suckle their young. Cetaceans, which can be divided into baleen whales (such as the humpback whale and the right whale), and toothed whales (such as dolphins and porpoises), are found throughout the world's seas, and some species live in tropical and subtropical rivers. Many species, including the blue whale, have been hunted in such numbers that they are in danger of extinction.

ANATOMY

Cetaceans have a hairless, streamlined body to reduce water turbulence. External projections are reduced to the essentials: flippers for steering, a tail with 2 boneless, horizontal flukes (fish have vertical flukes), and usually a dorsal fin for stability. Even the genitals are concealed within folds. Other adaptations to underwater life include a thick layer of blubber (fat and oil) beneath the skin that conserves body heat, and light, spongy, oil-filled bones. Cetaceans breathe through one (toothed whales) or 2 (baleen whales) blowholes – muscular nostrils usually situated on the top of the head. Toothed whales have a brain that is relatively as large as that of primates, and they are known for their intelligence. Baleen whales have a relatively smaller brain.

BALEEN WHALE
Instead of teeth, a baleen whale has 130–400 horny (baleen) plates on each side of the upper jaw. The inner edge of each plate has bristles used to sieve food.

baleen plates
long jaws

TOOTHED WHALE
The teeth of the toothed whale group are simple, conical, and in most species numerous. The jaws may be extended to form a beak, as seen in dolphins.

conical teeth
beak

POWERFUL TAIL
In cetaceans, the main propulsive force is the up-and-down movement of the tail. This motion is powerful enough for a whale, such as this humpback, to push two-thirds of its body out of the water when breaching.

SENSES

Cetaceans have extraordinarily sensitive hearing. Vision is reasonable – underwater it is excellent up to about 1m (3¼ft), and in air up to 2.5m (8¼ft), but colour vision is very limited. Some species can focus both eyes ahead, above, or behind them, and some can move their eyes independently. Some freshwater dolphins, however, are nearly or entirely blind. Members of the toothed whale group produce high-frequency clicks for echolocation and can also communicate using a wide range of sounds audible to humans. Other cetaceans employ a variety of vocalizations, but these are less well studied. Cetaceans have no sense of smell.

blowhole
outgoing clicks (to prey)
melon
incoming (reflected) clicks
incoming (reflected) clicks

USING ECHOLOCATION
Toothed whales avoid obstacles and catch prey by emitting high-frequency clicks that are reflected off objects in their path. The melon (a fluid-filled swelling) changes shape to focus the clicks. Incoming clicks pass through the jaw.

BREACHING

① THE ASCENT
As this southern right whale begins to breach, it surfaces vertically, creating little water turbulence.

② THE FLIPPERS APPEAR
The front flippers clear the surface as the whale continues to propel itself upwards.

③ STARTING TO GO DOWN
As the upward surge, which is powered by the tail, is completed, nearly half of the whale is exposed.

④ ACROBATIC MANOEUVRES
As the whale crashes back into the water, it also turns on to its side.

OCEANIC GIANT
This sperm whale has a typically large, elongated head that is continuous with the long, torpedo-shaped body. Like many cetaceans, sperm whales grow to enormous proportions – in males, up to 18m (59ft) in length and over 50 tonnes (49 tons) in weight.

REPRODUCTION

Some whales, including rorquals (such as the humpback whale), breed during the winter. They migrate from their summer feeding grounds in polar seas to tropical waters (usually around island groups or close inshore), where they give birth and then immediately become pregnant again. In spring they return to colder waters to feed. Other whales may breed seasonally, but do not migrate to do so. In all cetaceans, copulation is very brief. The male's penis, which is S-shaped, is held inside the body wall. It becomes erect not by filling with blood but by straightening as a result of muscular action. After giving birth, the mother (and in some dolphins other members of the pod) assists the newborn to the surface to take its first breath.

CALF SUCKLING
Young cetaceans are suckled on milk until they are old enough to eat solid food. The nipples of this Atlantic spotted dolphin are housed in a pouch in the body wall, as is typical of cetaceans.

CONSERVATION

Whales have been hunted by humans throughout history for their meat, bones, and blubber. In the 20th century, with the emergence of intensive hunting and new technology, such as factory ships (above), species numbers declined dramatically. However, the International Whaling Commission banned hunting of certain species in 1966, and these have recovered somewhat. Commercial whaling was completely banned in 1986. The most serious threat to smaller cetaceans today is drowning when they are caught in fishing nets.

SURVIVING UNDERWATER

Even though all cetaceans breathe air, they are able to remain underwater for extended periods, returning to the surface only to exhale ("blow"). In order to remain underwater, the heart rate slows by a half. Also, the water pressure squeezes blood out of the vessels near the skin – the blood can then nourish the vital organs. The water pressure compresses the lungs, which forces air into the trachea and nasal passages where some of the air is absorbed by foamy secretions along the respiratory tract wall. Some toothed whales dive to great depths in search of prey.

SEARCHING THE DEEP
Sperm whales in search of squid dive deeper than other cetaceans. They are known to descend to depths of up to 1,000m (3,300ft) and can remain underwater for 90 minutes.

BLOWING
Cetaceans open their blowholes as they surface and explosively release air and a spray of oil droplets before taking another breath.

RE-ENTERING THE WATER
...noise made by a whale striking the surface ...be heard up to 1,000m (3,300ft) away.

6

GOING UNDER
As the head disappears, only one front flipper and part of the body is visible.

7

MAKING A SPLASH
The whale's re-entry creates a huge splash compared with the initial breach of the water's surface.

8

THE WHALE DISAPPEARS
With the sequence complete, the whale may breach again. Why whales breach is not clearly understood.

BALEEN WHALES

PHYLUM	Chordata
CLASS	Mammalia
ORDER	Cetacea
SUBORDER	Mysticeti
FAMILIES	4
SPECIES	12

THE MOST STRIKING feature of baleen whales is their size, which ranges from up to 6.5m (21ft) in the pygmy right whale to 33m (110ft) in the blue whale. Also characteristic are the baleen plates, which filter prey from the water. The group consists of the gray whale, the rorquals (which include the humpback whale and the blue whale), and right whales. Although commonly found in Antarctic and Arctic regions, they are distributed throughout all the world's oceans, usually in deep water.

ANATOMY

All whales in this group have 2 rows of baleen plates (see p.160) that are anchored to either side of the upper jaw. To support these structures, the jaws are elongated. This means that the head, which has 2 blowholes, is large in relation to the body. In right whales, the head comprises up to half the body length, and the jaws are deep (to accommodate long baleen plates). The body is relatively short and stout. Rorquals, on the other hand, have a long, slender body, and shorter baleens.

FEEDING

While the gray whale feeds on small crustaceans called amphipods (which are found on the sea bottom), right whales and rorquals eat planktonic crustaceans, which live near the surface. Right whales feed by swimming slowly through dense shoals of prey with their mouths open, skimming the plankton out of the water. Rorquals are more active hunters and usually surround their prey. As they feed, the grooves in the throat are relaxed, and the lower jaw becomes a vast sac into which water is gulped. When the water is expelled, the prey remains on the baleen plates. The larger species of rorquals have finer bristles on their baleen plates, which trap crustaceans such as krill and sometimes small fish. Smaller species have coarser bristles to catch larger crustaceans and small fish. The gray whale feeds by stirring up sediments with its relatively short, firm snout then gulping down its prey, often together with sand, silt, and pebbles.

MIGRATION

Most baleen whale species are migratory. Rorquals, which are known for travelling long distances, feed in the Arctic and Antarctic (where krill is abundant) during the summer. In autumn and winter they migrate to tropical waters. Here, although they feed little, they give birth and mate again immediately. In spring they then slowly move back towards higher latitudes, with their suckling calves. Although humpback whales usually migrate along coastlines, other rorquals prefer deeper water. Gray whales migrate further than any other mammal: approximately 20,000km (12,450 miles) each year.

COMMUNICATION

Baleen whales communicate using a variety of sounds, from squeals to rumbles. The most famous whale sounds are the "songs" of male humpbacks. These are produced during winter breeding and consist of a repeated series of high and low notes that are gradually varied as the season progresses. The songs are vital for communication in the vast oceans. Other species employ different sounds: the fin whale, for example, produces a call that is below the range of human hearing and travels huge distances through the ocean.

UNUSUAL FEEDING TECHNIQUE
Baleen whales feed by taking large quantities of small prey. This humpback whale, having circled upwards through a shoal of small fish, is gulping a large mouthful of water and fish. As the mouth closes, the water presses against the rigid upper jaws and flows out over the baleen plates, leaving the fish behind.

MAMMALS

Eschrichtius robustus
Gray whale

Length	13–15 m (43–49 ft)
Weight	14–35 tonnes (14–34 1/2 tons)

Location North Pacific | **Social unit** Group

Status Endangered

Gray whales filter-feed like other baleen whales. Unusually, however, they also rely on diving to the shallow sea bed, scooping up huge mouthfuls of mud, and filtering worms, starfish, shrimps, and other small creatures with their short, coarse baleen. This whale's sounds include grunts, wails, moans, and clacking knocks, but the function of these sounds is little understood.

CONSERVATION

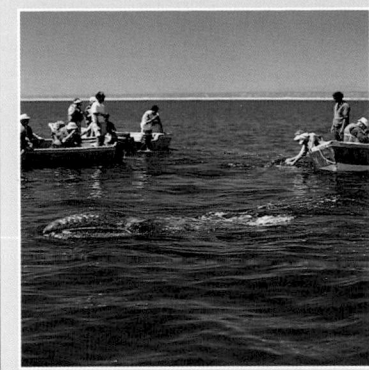

Gray whales feed near to shore and so are accessible to watchers, especially in the East Pacific. These whales migrate in groups of up to 10, north to the Arctic for summer feeding, and south to warm-water lagoons to rest and produce calves in winter. They make the longest migrations of any mammal, up to 20,000 km (12,500 miles) yearly. Populations in the East Pacific have risen since legal protection was introduced in 1946. However, West Pacific numbers are still low.

SPY-HOPPING

Many baleen whales, including the gray whale, "spy-hop", becoming vertical in the water with the head well out. They may be detecting other whales, viewing landmarks, or checking water currents for migration. Gray whales swim in a coordinated way, staying in line or arching out of the water together.

series of 8–9 bumps replace dorsal fin

mottled grey skin encrusted with barnacles, whale lice, and other growths

notched flukes

long, slender head

COMPARATIVELY SLEEK
The gray whale is sleeker than right whales but stockier than the rorquals. It has short baleen, the plates being only about 40cm (16in) long. There are 130–180 plates on each side of the jaw.

Eubalaena glacialis
Northern right whale

Length	13–17 m (43–56 ft)
Weight	40–80 tonnes (39 1/2–79 tons)

Location Temperate and subpolar waters worldwide | **Social unit** Variable

Status Endangered

The most endangered of the large whales, this massive whale is at risk from collisions with ships and from fishing equipment because it swims slowly, dives for only a few minutes, and feeds near the surface. It migrates to the far north or south in summer and returns to warmer, mid-latitude waters in winter, when the female produces one calf, 4–6m (13–20ft) long. Twin blowholes give a bushy, V-shaped blow. Sounds include flipper-slaps and breaching.

HUGE HEAD
The head takes up about one-quarter of the total length and is encrusted with barnacles and whale lice.

scattered white patches

FILTER FEEDING

The northern right whale feeds alone or in small groups, swimming with mouth open to filter plankton as water rams into its mouth. The narrow baleen plates, mainly blue-black but sometimes white, are up to 3m (9 3/4 ft) long. They number 200–270 on each side of the characteristically down-curved upper jaw.

fibrous growths

Balaena mysticetus
Bowhead whale

Length	14–18 m (45–59 ft)
Weight	50–60 tonnes (49 1/2–59 tons)

Location Arctic and sub-arctic waters | **Social unit** Group

Status Endangered

The bowhead whale has a massive head, around one-third of the total weight. The bowhead stays near spreading polar ice all year, enduring weeks of darkness when it may echo-sound to navigate between and under floes. It surface-skims with its mouth slightly agape. Dives of 5–15 minutes also allow it to feed in mid-water or even grub at the bottom. The newborn calf is 4–4.5m (13–15ft) long and is suckled for 5–6 months.

white around lower jaw, chin, and base of tail

SMOOTH SURFACE
The mainly black body is clear of whale lice, barnacles, skin callosities, and similar growths.

BIGGEST BALEEN

The bowhead has the longest baleen of any whale. The brown- or blue-black baleen plates reach 4.6m (15ft) in length, with 240–340 plates on each side of the strongly curved or "bowed" upper jaws (hence the name bowhead). This whale has no throat grooves and no dorsal fin either, giving its body surface a remarkably uncluttered appearance.

Balaenoptera physalus
Fin whale

Length 19–22 m (62–72 ft)	
Weight 45–75 tonnes (44 1/2–74 tons)	
Location Worldwide (except E. Mediterranean, Baltic, Red Sea, Arabian Gulf)	**Social unit** Variable
	Status Endangered

The fin whale is the second-largest whale, and one of the fastest. In addition to some hums and squeals, it produces an immensely loud, deep moan that can be heard hundreds of kilometres away. Like other great whales, the fin whale undertakes long migrations from high latitudes in summer, where it feeds on fish and krill, towards tropical regions for winter and breeding. The gestation period is 11 months, the newborn calf 6.4m (21ft) long, and weaning takes place at 9–10 months. This means each female has only one offspring every 2 years, so low populations take decades to recover.

FIN, FLIPPERS, AND FLUKES
The fin whale's back, flippers, and flukes are grey. The dorsal fin, set two-thirds of the way along the back, has a concave rear edge.

RIGHT-SIDED FEEDING
Fin whales carry out high-speed lunge-feeding, on krill and fish such as capelin or herring. The whales synchronize attacks and take in huge volumes of water, close the mouth, and force out the water to trap fish on the baleen. The whale swims on its right side, which may be why the left part of the mouth is black but the right is white. Such asymmetry in colour is very unusual in mammals.

55–100 throat pleats

white underside

Balaenoptera musculus
Blue whale

Length 20–30 m (66–98 ft)	
Weight 100–160 tonnes (98 1/2–157 1/2 tons)	
Location Worldwide (except Mediterranean, Baltic, Red Sea, Arabian Gulf)	**Social unit** Individual
	Status Endangered

The biggest animal on the planet, the blue whale can consume more than 6 tonnes of Euphausiidae (small crustaceans), of which its diet almost exclusively consists. It lunges into a shoal of prey and its throat swells to four times the normal width. It closes its mouth, expels the water, and swallows the thousands of food items retained by the baleen. Feeding occurs mainly in summer, in and near rich polar waters. The blue whale is thought to migrate to warmer, lower latitudes for winter, when the females give birth. The calf is 7m (23ft) long, 2.5 tonnes in weight, and is suckled for 6–8 months. Blue whales are usually solitary or in mother-calf pairs, although they may gather as loose groups to feed. They make grunts, hums, and moans, which at volumes greater than 180 decibels are the loudest of any creature sounds.

VENTING THE BLOWHOLE
On return to the surface, the whale releases a "blow", about 9m (30ft) high, comprising warm, humid air from the lungs, mucus, and ocean water.

ANATOMY OF A DIVE

When making a deep dive, the whale "headstands", exposing its distinctively wide tail flukes, then descends steeply to a maximum depth of some 200m (655ft). The long, narrow flippers play no part in propulsion, the thrust coming from the powerful back muscles which swish the rear body and flukes up and down.

STREAMLINED GIANT
The blue whale has a slim outline, especially in winter, although it fattens in summer. The tiny dorsal fin is set well to the rear, near the tail. Its coloration is mainly pale blue-grey.

55–68 skin grooves or pleats run along half the body length

Balaenoptera edeni
Pygmy Bryde's whale

Length 9–12 m (30–39 ft)	
Weight 16–25 tonnes (16–24 1/2 tons)	
Location Tropical and warm temperate waters worldwide	**Social unit** Variable
	Status Locally common

One of two similar species with this name (pronounced "Broodah's"), the pygmy Bryde's whale is mainly coastal in the Eastern Indian and West Pacific Oceans. The larger, more offshore Bryde's whale (*Balaenoptera brydei*) is found in the Atlantic and Indian Oceans, and parts of the North, East and South Pacific. Apart from their size difference, both species have blue-grey coloration that is paler on the underside, 40–70 throat grooves, coarse-bristled baleen plates, and a small dorsal fin set two-thirds of the way along the stocky body. These whales dive for up to 20 minutes to feed mainly on shoaling fish and krill. They usually live alone or in small, loose herds, but may gather in dozens where prey is plentiful. Both species of Bryde's whale are rapid swimmers, changing speed and direction frequently. They surface steeply to reveal the head first, arch over, and expose the rear of the body but not the tail flukes as they dive. The female reaches sexual maturity at 8–11 years and produces a single calf after a gestation of 12 months.

small, crescent-shaped dorsal fin

3 ridges on snout

40–70 skin grooves or pleats

pale underside

Balaenoptera borealis
Sei whale

Length	14–16 m (43–52 ft)
Weight	20–25 tonnes (20–24½ tons)

Location Worldwide (except Mediterranean, Baltic, Red Sea, Arabian Gulf)

Social unit Group

Status Endangered

Sei whales frequent more temperate waters (8–25°C/46–77°F) than their more widespread close cousins, the blue and fin whales. The sei whale is slender with a long, slim, pointed head and slightly downcurved jaw line. Its upper surfaces are dark grey, with a relatively sharp transition to the white or pale grey underside.

There are 320–340 close-spaced, delicate-fringed baleen plates hanging from each side of the upper jaw. These enable sei whales to take a variety of plankton, from copepods (water-flea-like crustaceans) less than 1cm (²/₅in) across to schooling fish and squid 30cm (12in) long. Sei whales are usually found in

schools of 2–5 individuals. They rarely dive deeper than 300m (1,000ft), staying underwater for up to 20 minutes. Most births among sei whales are single, although twins occur occasionally.

distinctively tall, pointed dorsal fin

small flukes

approximately 50 throat grooves

Balaenoptera acutorostrata
Minke whale

Length	8–10 m (26–33 ft)
Weight	8–13 tonnes (7½–13 tons)

Location Worldwide (except E. Mediterranean)

Social unit Individual

Status Lower risk

The smallest of the rorqual whales, the minke has small baleen plates, 30cm (12in) long, with 230–360 in each side of the upper jaw. It has no universal migration pattern and is found in a variety of waters, including open ocean and near ice sheets, coasts, fjords, and estuaries. It feeds alone by gulping in tight-knit swarms of krill or fish. Yet it is not shy, and will approach stationary boats. The 3m (9¾ ft) calf

is born in midwinter and weaned after about 4 months. A smaller minke whale was recently recognized as a separate species – the Antarctic minke (*Balaenoptera bonaerensis*).

SMOKY PATTERNS
The minke has grey, "smoky" patches where the black back meets the white belly. The white band on the flipper may extend to the chest.

smoky patches

DOLPHIN-LIKE WHALE
The minke has a dolphin-like shape, with a sharp snout, pointed head, and a ridge from the snout up to the paired blowholes, which are set well back on the forehead. It is a speedy, agile swimmer, able to "surf" huge ocean breakers, and makes an occasional spectacular surface lunge.

Megaptera novaeangliae
Humpback whale

Length	13–14 m (43–46 ft)
Weight	25–30 tonnes (24⅝–29½ tons)

Worldwide (except Mediterranean, Baltic, Red Sea, Arabian Gulf)

Social unit Variable

Status Vulnerable

the knobbly leading edges to the longest flippers of any animal. In spite of its size, it is graceful and athletic, being able to leap out of the water (see below). The humpback whale migrates from cold, food-rich summer waters near the poles to warmer, coastal shallows at lower latitudes for winter calving. Pregnant females spend the longest in the feeding areas. It has a wide array of food-gathering methods and small groups cooperate in blowing "curtains" or

"cylinders" of underwater bubbles to herd fish. An extremely vocal whale, it also emits numerous sounds, perhaps partly to coordinate its feeding. In winter calving areas, solitary males produce a long, complex series of sounds of amazing variety (see below right).

BODY AND FINS
The humpback whale has a dumpy body and a fatty pad at the base of the dorsal fin, which is variable in shape, from almost flat to tall and triangular. The span of the tail flukes is about one-third the length of the head and body.

The humpback whale is dark blue-black above, with paler or white patches below. Colour variations, particularly on the underside of the tail, can be used to identify individual humpbacks, as can

fatty pad

12–36 throat grooves or pleats

slight ridge from dorsal fin to tail

SINGING

BREACHING
The humpback whale generates sufficient upward force with its tail to lift almost all of its 20-plus tonnes out of the water. It then twists in the air and falls on its back with a tremendous splash. It is not clear why whales "breach", as such a movement out of the water is called. It may be to create massive sound waves or to ease irritation from skin parasites.

flippers up to one-third of total body length

leading edge to flippers (pectoral fins) has knobs or tubercles

The song of the male humpback whale develops from year to year. Lasting up to 30 minutes, it may attract females, warn off males, or be a form of sonar to detect other whales. Above, a male humpback moves into a singing posture, hanging vertically with head downward some 10–40m (33–130ft) below the surface.

MAMMALS

TOOTHED WHALES

PHYLUM	Chordata
CLASS	Mammalia
ORDER	Cetacea
SUBORDER	Odontoceti
FAMILIES	9
SPECIES	71

A MUCH MORE diverse group than the baleen whales, toothed whales make up almost 90 per cent of all cetaceans. The group consists of dolphins (which include the killer whale), river dolphins, porpoises, white whales, sperm whales, and beaked whales. Most are medium-sized – although the sperm whale grows to up to 18m (59ft) – and all possess teeth instead of baleen plates. On the forehead, there is a fluid-filled swelling called the melon, in front of which there is usually a beak. Most species are found the world over in deep water and coastal shallows, although a few live in fresh water. Some toothed whales migrate, but only the sperm whale travels long distances.

ANATOMY

Toothed whales have simple, conical, pointed teeth that are not divided into incisors, canines, premolars, and molars (as in most other mammals). Each tooth has a single root and is either straight or slightly curved. One set of teeth lasts the whale's lifetime. The number of teeth present varies from more than 40 pairs in each jaw in some dolphins, to a single pair (in the lower jaw) in beaked whales. Unlike baleen whales, toothed whales have only one blowhole, which means that the skull is asymmetrical. The single blowhole usually opens at the top of the head (except in the sperm whale, where the opening is at the front of the skull). Another feature of many toothed whales is the streamlined head and the long, narrow beak seen in most species.

FEEDING

While baleen whales trap their prey en masse, toothed whales capture victims individually. The even, conical teeth found in this group are perfect for catching slippery fish, which form the diet of most species. Sperm whales, however, mostly eat squid (but also other prey, such as octopuses), and beaked whales have a mixed diet of fish and squid. Killer whales eat other whales (which they attack in packs), fish, and seals, which they sometimes seize from land, using waves to slide onto the beach. They will also upset sheets of pack ice to knock prey into the water.

Fish-eating species (including killer whales) usually have numerous teeth to hold wriggling prey, while those that eat squid or octopuses, for example, have fewer teeth. Sperm whales have functional teeth only in the lower jaw (which is very narrow), and prey is held firmly between these teeth and the roughened palate in the upper jaw.

All toothed whales use echolocation (see p.160) to help find prey (and to avoid objects in their path).

SOCIAL BEHAVIOUR

Most toothed whales live in pods, which vary in size from less than 10 to over 1,000 (as seen in some dolphin species). The exact organization of these associations is poorly understood, although it is thought that subgroups form to perform independent tasks, such as feeding. This suggests the presence of complex social structures. Some species, especially killer whales, appear to practise co-operative hunting, whereby prey is "herded" until trapped or surrounded. When swimming in formation, dolphins often leap out of the water, which may simply be playful behaviour, or it may function as a type of communication.

STREAMLINED FOR SPEED
Atlantic spotted dolphins are fast, energetic swimmers, and, like several other toothed whale species, they often travel in large pods. The torpedo-shaped body is propelled through the water by powerful thrusts of the tail. The mottled coloration camouflages them from predators and prey in shallow, sunlit waters.

Platanista gangetica
Ganges river dolphin

Length	2.1–2.5 m (7–8 1/4 ft)
Weight	85 kg (185 lb)

Location S. Asia (Indus and Ganges-Brahmaputra river systems)	Social unit Variable
	Status Endangered

Also known by the local names susu and bhulan, this exclusively freshwater species has distinctively broad flippers and a long, narrow beak armed with 26–39 pairs of upper, and 26–35 pairs of lower, sharp teeth. The front teeth extend outside the beak's tip, to form a catching cage for fish, shellfish, and similar prey on or near the bottom. The flexible neck allows the head to bend at right angles as the dolphin grubs in the mud or "scans" the area with echolocating sound pulses. The Ganges river dolphin lives in small groups, usually 4–6 but occasionally up to 30. However, its social life and breeding habits are largely unknown. There are 2 subspecies: *Platanista gangetica minor* in the Indus river and its tributaries, and *Platanista gangetica gangetica* in the Ganges–Brahmaputra river system. Both are extremely rare and face numerous threats from humans.

grey back

tiny eyes

long, narrow beak

pinkish underside

Lipotes vexillifer
Chinese river dolphin

Length	2.2–2.5 m (7 1/4–8 1/4 ft)
Weight	125–160 kg (275–350 lb)

Location E. Asia (Chang Jiang)	Social unit Variable
	Status Critically endangered

One of the least-known dolphins, this species is also called the baiji. It inhabits the lower and middle Chang Jiang (Yangtse) River and associated lakes and waterways. Like other river dolphins, the Chinese river dolphin has a long, thin beak and flexible neck, to grub in the mud for fish and other prey. There are 30–35 pairs of peg-like teeth in both upper and lower jaws. Its tiny eyes are not fully formed, so this dolphin probably navigates by touch and echolocation. Shy and secretive, it lives in small groups of 2–6. It faces a multitude of threats: depletion of its prey by over-fishing, blockage of its migratory travel routes by dams, collision with boats, pollution by chemicals and waste, and engine noises that interfere with its echolocation. It has also been hunted illegally for its meat and for its body parts, used in traditional medicines.

blue-grey or brown back

Inia geoffrensis
Amazon river dolphin

Length	2–2.6 m (6 1/2–8 1/2 ft)
Weight	100–160 kg (220–350 lb)

Location South America (Amazon and Orinoco basins)	Social unit Variable
	Status Vulnerable

Also called the boto or boutu, this species ranges through the Amazon and Orinoco river systems. It possesses a long, slim beak and flexible neck, for poking in mud for prey during short dives of 1–2 minutes. Being small-eyed and living in murky water, it probably finds its way and food mainly by echolocating sound pulses. The Amazon river dolphin has 25–35 pairs of teeth in both upper and lower jaws. At the front, the teeth are peg-like, for seizing prey, but at the rear they are flatter with peaks or cusps, and thus suited to crushing freshwater crabs, river turtles, and armoured catfish. The single calf, 80cm (32in) long, is born between May and July.

CAUTIOUS APPROACH
Known for its slow, apparently lethargic lifestyle, the Amazon river dolphin usually lives alone or in twos, occasionally in groups of up to 20. In some areas it may approach boats or swimmers out of curiosity, but in regions where it is hunted it has become more cautious.

BACK HUMP
Unlike other river dolphins, this species lacks a proper dorsal fin and in its place has a low back hump.

touch sensitive bristles on beak

variable coloration with pink and grey blotches

Delphinapterus leucas
Beluga

Length	4–5.5 m (13–18 ft)
Weight	1–1.5 tonnes (1–1 1/2 tons)

Location Arctic Ocean	Social unit Group
	Status Vulnerable

The beluga, or white whale, keeps mainly to the edges of the Arctic ice-fields and has been radio-tracked diving to 300m (985ft), presumably navigating by echolocation to find prey and breathing holes. This vocal whale emits varied calls including squeaks, whistles, mews, clicks, and hums. These are audible through the hulls of boats, giving the beluga its nickname of sea canary. The outgoing echolocation sounds are focused by the bulging melon. The beluga feeds on fish, molluscs, crustaceans, and other prey, which it crushes with its 8–11 pairs of teeth in the upper jaw and 8–9 pairs below. The calf is dark grey and 1.5m (5ft) long at birth. It is paler by 2 years old, and blue-tinged white by the age of 5 years.

ALL-WHITE WHALE
The only all-white cetacean, the adult beluga blends with the Arctic ice floes and icebergs. Its skin may be tinged yellow before the summer moult.

CONSERVATION
There is growing risk posed to this species by pollution of its native waters. Moreover, hunting of the beluga in small numbers is still permitted, and may add to the threat to populations of this whale in some areas of its Arctic range.

fibrous ridge along back (no dorsal fin)

Monodon monoceros
Narwhal

Length 4–4.5 m (13–15 ft)	
Weight 0.8–1.6 tonnes (³/₄–1⁵/₈ tons)	
Location Arctic Ocean	**Social unit** Group
	Status Locally common

The narwhal has the most northerly range of any mammal, among ice-fields and floes of Arctic waters. It has only one growing tooth, the upper left incisor, which forms a long tusk (see right). Fish, molluscs, crustaceans, and other prey are probably sucked into the mouth by the narwhal's powerful lips and tongue. Like belugas – indeed, often with them – narwhals can sometimes form large schools of thousands, which may be segregated by age and sex. They communicate with a wide variety of sounds, including hum-like tones which may be for individual recognition.

COLORATION
The narwhal (female pictured here) is speckled grey and black on a pale background, with more patches on the back merging into larger areas of dark grey.

MALE NARWHAL'S TUSK
The narwhal's tusk grows with age, through the upper lip, spiralling clockwise, to reach 3m (9³/₄ ft) long. It may be a sword-like weapon to "fence" rival males at breeding time. Any use in feeding seems unlikely, since most females lack tusks, but survive well. An occasional male has 2 tusks, or a female has one.

C-shaped flukes

small flippers with upturned tip

Phocoena phocoena
Harbour porpoise

Length 1.4–2 m (4¹/₂–6¹/₂ ft)	
Weight 50–90 kg (110–200 lb)	
Location North Pacific, North Atlantic, Black Sea	**Social unit** Variable
	Status Vulnerable

The most numerous cetacean over much of its range, the harbour or common porpoise has nevertheless suffered, like other inshore sea creatures, as a result of human activity. A major danger is being snared in underwater fishing nets when, as an air-breather, it asphyxiates. Apart from humans, its main predators are killer whales, bottlenose dolphins, and large sharks. The harbour porpoise generally forages alone, in waters down to 200m (655ft), for sea-bed dwellers such as fish and shellfish. It uses very high-pitched echolocation and grips prey with its spade-shaped teeth, numbering 22–28 pairs in the upper jaw and 21–25 in the lower. Groups usually form because prey such as shoaling fish are particularly rich in small areas. The single calf is born in early summer and cared for by the mother for up to 12 months.

black or chocolate-brown back

triangular, slightly shark-like dorsal fin

stripe from chin to flipper

cream underside

Phocoena sinus
Vaquita

Length 1.5 m (5 ft)	
Weight 48 kg (105 lb)	
Location Gulf of California	**Social unit** Variable
	Status Critically endangered

This small porpoise's very restricted range, in shallow waters less than 40m (130ft) deep at the northern end of the Gulf of California, makes it one of the most vulnerable of all sea mammals. Also called the Gulf of California porpoise, its habits are little known. It is often solitary but may also form small groups of up to 7 individuals. It takes a mixed diet of small fish, squid, and other prey on or near the sea bed, and uses high-frequency clicks for echolocation. The main body colour is grey, darker above than below and also around the eyes and mouth. The calf is only 70–80cm (28–32in) long at birth and is suckled by the mother for several months. Such threats as entanglement in fishing nets (leading to over 30 deaths a year), pollution, boat noise, and oil exploration mean that the vaquita's future is bleak.

dark grey back

pale grey to white underside

stripe from chin to flipper base

Neophocaena phocaenoides
Finless porpoise

Length 1.5–2 m (5–6¹/₂ ft)	
Weight 72 kg (160 lb)	
Location Indian Ocean, W. Pacific	**Social unit** Variable
	Status Locally common

Lack of a dorsal fin makes this medium-sized porpoise difficult to identify, or even see, as it surfaces briefly and rolls gently while taking breaths. The vaguely dolphin-like bulging forehead and a slightly beaked snout make it distinctive among the 6 porpoise species. The finless porpoise frequents coastal waters around the Indian and West Pacific oceans, including estuaries and up rivers. Like other porpoises (and unlike dolphins) it leaps from the water only rarely, yet it may "spyhop" in the manner of whales, holding its body vertically, half out of the water, as though standing up to look around. It feeds alone or in small groups of 3–5, occasionally 10 or more. Small fish, molluscs, and crustaceans on or near the sea bed are caught using the 13–22 pairs of spade-shaped teeth in both upper and lower jaws. They sometimes follow prey on small-scale, seasonal migrations.

slightly pointed snout

pale underside

low, warty back ridge from above flippers to tail

Phocoenoides dalli

Dall's porpoise

Length 2.2–2.4 m	
(7¼–7¾ ft)	
Weight 170–200 kg	
(380–440 lb)	

Location North Pacific   **Social unit** Group

Status Lower risk

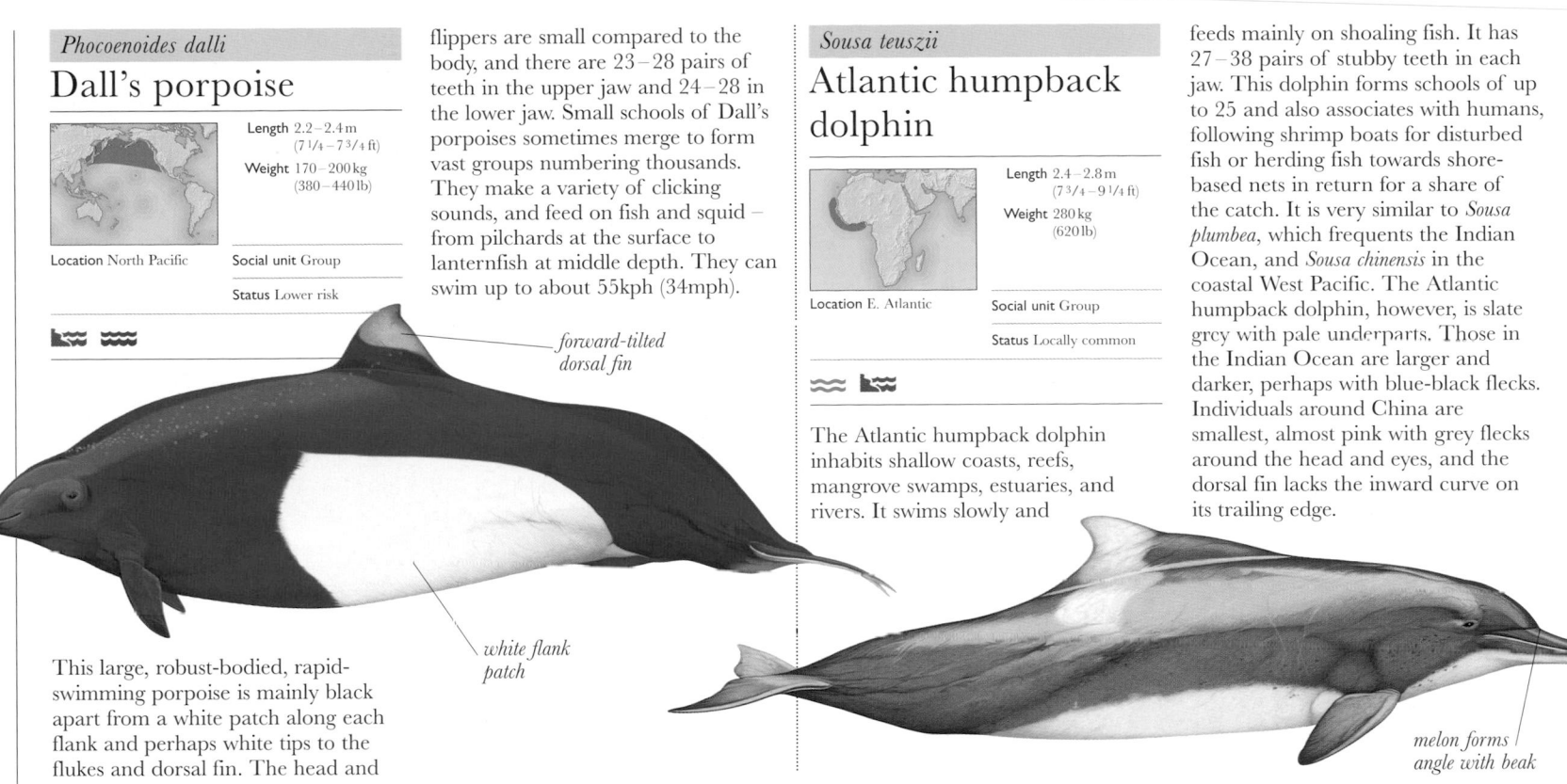

forward-tilted dorsal fin

white flank patch

flippers are small compared to the body, and there are 23–28 pairs of teeth in the upper jaw and 24–28 in the lower jaw. Small schools of Dall's porpoises sometimes merge to form vast groups numbering thousands. They make a variety of clicking sounds, and feed on fish and squid – from pilchards at the surface to lanternfish at middle depth. They can swim up to about 55kph (34mph).

This large, robust-bodied, rapid-swimming porpoise is mainly black apart from a white patch along each flank and perhaps white tips to the flukes and dorsal fin. The head and

Sousa teuszii

Atlantic humpback dolphin

Length 2.4–2.8 m	
(7¾–9¼ ft)	
Weight 280 kg	
(620 lb)	

Location E. Atlantic   **Social unit** Group

Status Locally common

The Atlantic humpback dolphin inhabits shallow coasts, reefs, mangrove swamps, estuaries, and rivers. It swims slowly and

melon forms angle with beak

feeds mainly on shoaling fish. It has 27–38 pairs of stubby teeth in each jaw. This dolphin forms schools of up to 25 and also associates with humans, following shrimp boats for disturbed fish or herding fish towards shore-based nets in return for a share of the catch. It is very similar to *Sousa plumbea*, which frequents the Indian Ocean, and *Sousa chinensis* in the coastal West Pacific. The Atlantic humpback dolphin, however, is slate grey with pale underparts. Those in the Indian Ocean are larger and darker, perhaps with blue-black flecks. Individuals around China are smallest, almost pink with grey flecks around the head and eyes, and the dorsal fin lacks the inward curve on its trailing edge.

Sotalia fluviatilis

Tucuxi

Length 1.3–2 m	
(4¼–6½ ft)	
Weight 35–40 kg	
(77–88 lb)	

Location Central America, N.E. South America   **Social unit** Variable

Status Locally common

The tucuxi (pronounced "too-koo-shee") is one of the smallest dolphins, stocky with large flippers, flukes, and dorsal fin. It occurs as a marine form

around the coasts and river estuaries of northeast South America, and as a river form in the lower reaches and lakes of the Amazon system. It is a different species to the solely fresh-water Amazon river dolphin (see p.167). The tucuxi lives alone or in twos although larger schools occur, up to 10 in rivers and 30 around the

coast. It may leap, somersault, and surf waves – but the reason for this is not clear. Prey up to 35cm (14in) long is swallowed whole, including fish such as anchovies and catfish, and

large dorsal fin

large, broad tail flukes

pale underside

prominent beak

line from eye to flipper base

squid. In some places, local tradition forbids hunting these dolphins. However, many are accidentally asphyxiated in fishing nets, and some are deliberately killed for meat or fishing bait; the eyes and other parts have been valued as love charms.

Lagenorhynchus obscurus

Dusky dolphin

Length 1.7–2.1 m	
(5½–7 ft)	
Weight 70–85 kg	
(155–185 lb)	

Location S. Atlantic, Indian Ocean, S. Pacific   **Social unit** Variable

Status Locally common

The dusky dolphin is mainly dark grey to blue-black on its upperside, and pale grey or white on the underside. These areas are separated by a tapering grey stripe from the face along the flank to near the tail base. The head has a smooth profile that widens gradually from beak to blowhole. The species occurs

as 3 subspecies based respectively around South America (*Lagenorhynchus obscurus fitzroyi*), southern Africa (*Lagenorhynchus obscurus obscurus*), and New Zealand (unnamed). It prefers waters of 10–18°C (50–64°F) and less than 200m (655ft) deep. Off South America it feeds by day on

shoaling fish, such as anchovies, and squid; off New Zealand feeding is mainly by night at middle depths. Schools of dusky dolphins change rapidly in size and composition, varying in number from 2 up to 1,000. Groups often engage in much leaping, chasing, and rubbing.

COMPLICATED PATTERNS
The dusky dolphin has complex markings in shades of blue-black, dark grey, pale grey, and white. The black "lips" and beak tip are a distinctive feature.

tall, crescent-like dorsal fin

pale grey flank stripe

forked pale patch on upper flank

CONSERVATION

Like many cetaceans, the dusky dolphin is at risk because it chases its shoaling prey into nets that people have set to catch the same fish. Tangled and unable to swim or breathe properly, it soon asphyxiates. This species is also hunted off the coast of Peru for its flesh, used as bait for fishing, and as meat for direct human consumption.

MAMMALS

Lagenorhynchus obliquidens
Pacific white-sided dolphin

Length	2.1–2.5 m (7–8¼ ft)
Weight	75–90 kg (165–200 lb)

Location North Pacific	**Social unit** Group
	Status Common

Distinguished by its tapering head, coloration, and tall dorsal fin, this dolphin often rides ship bow waves in the North Pacific. It has a dark back, a grey-white belly, and a lower flank patch from

side stripe extends from tail to above eye

dark leading edge to flippers

the beak to level with the dorsal fin. A pale stripe from the tail base extends forwards to level with the dorsal fin, then it may fork, arching over the shoulder region to near the eye. This dolphin eats a variety of fish and squid, using 23–36 pairs of small teeth in both the upper and lower jaw. It forms fast-changing schools numbering tens to thousands, and often associates with other dolphins and whales. After a gestation period of 10–12 months, the female gives birth to a single calf, which is 90cm (3ft) long. This dolphin makes up a sizeable part of the catch in some fishing industries of the northwest Pacific Ocean.

Lagenorhynchus albirostris
White-beaked dolphin

Length	2.8 m (9¼ ft)
Weight	350 kg (770 lb)

Location North Atlantic	**Social unit** Group
	Status Common

This large, stocky dolphin is an acrobatic swimmer. Its stubby beak meets the bulging melon of the forehead at a

tall, black, sickle-shaped dorsal fin

distinct angle. Coloration is mainly dark grey or black with variable grey or white patches extending along the upper and lower flanks; the underside, including the beak, tends to be white. It has 22–27 pairs of robust, cone-shaped teeth in both upper and lower jaws, for feeding on open-water shoaling fish such as herring, although seabed-dwelling flatfish and squid are also taken. The single calf, 110–120cm (3½–4ft) long, is born, like many other dolphins, in summer. White-beaked dolphins congregate in rapidly changing schools numbering from 5 to more than 1,000. They produce machinegun-like bursts of clicks and a range of squeals to communicate with each other and also to navigate and find prey.

Grampus griseus
Risso's dolphin

Length	3.8 m (12½ ft)
Weight	400 kg (880 lb)

Location Pacific, Atlantic, Mediterranean, Indian Ocean	**Social unit** Group
	Status Common

This distinctive dolphin is readily identified by its large size, blunt and "beakless" head, central crease down the bulging melon, tall dorsal fin, and overall grey, white-scarred

tall, sickle-shaped dorsal fin

coloration. It forms schools of 10–15, but these may aggregate into groups of several hundreds mixed with other cetaceans, especially smaller dolphins and pilot whales. In common with

many other dolphins, Risso's dolphin suffers from human activities. It is asphyxiated in fishing nets, accumulates pollutants in the body through the food chain, and swallows pieces of plastic and other refuse.

WHITE WITH AGE
Wounds and scratches accumulate with age. These heal as pale, scarred patches. Older individuals, perhaps over 30 years, may appear almost white.

long, sickle-shaped flippers

COMMUNICATION

Risso's dolphins dive deeply and eat mainly squid. Water darkens with depth, so sounds of echolocation are more important than sight for locating prey. Parties of dolphins may monitor each other's clicks and echoes, to find prey more efficiently.

Tursiops truncatus
Bottlenose dolphin

Length	1.9–4 m (6¼–13 ft)
Weight	500 kg (1,100 lb)

Location Worldwide (except polar regions)	**Social unit** Variable
	Status Common

This largest of the beaked dolphins is the "performing" species of marine life centres. In fact there are probably two species: *Tursiops truncatus*, which is found worldwide, and the smaller *Tursiops aduncus*, which has a more limited range off the coast of India and in

the West Pacific. There may even be more species, considering the wide variation across regions and habitats. Around tropical coasts, the bottlenose dolphin is an average of 2m (6½ ft) long, and has relatively large flippers, flukes, and dorsal

large, sickle-shaped dorsal fin

fin. In colder, open oceans they are almost twice as long, with proportionately smaller extremities. Social groupings and feeding methods are similarly varied over its range, as are the leaps, water-slaps, sounds, and calls of the bottlenose dolphin.

VARIABLE COLORATION
The basic coloration of dark grey or black back, fading to cream underneath, varies in hue and pattern between individuals.

cream underside

pointed flippers

ADAPTABILITY

The bottlenose dolphin, named after its short and robust beak, is an adaptable, successful generalist. It eats many kinds of fish, molluscs, and crustaceans, which it grips with 18–27 pairs of small, conical teeth in both jaws.

Stenella longirostris
Spinner dolphin

Length 1.3–2 m
(4 1/4–6 1/2 ft)

Weight 45–75 kg
(99–165 lb)

Location Tropical waters worldwide

Social unit Group

Status Lower risk

No other dolphin, or possibly cetacean, varies so greatly in body proportions, colour, and pattern across its range. There are at least three spinner subspecies: one worldwide and two in the eastern Pacific, with possibly a fourth, dwarf subspecies in shallow coral reefs in the Gulf of Thailand. There are 45–65 pairs of sharp teeth in each of the upper and lower jaws. Coloration is black or dark grey fading to a pale or white underside, with black edging to the eyes and lips, although this is very variable. The spinner dolphin dives deep to eat mid-water fish and squid. It forms massive schools of hundreds or thousands, which often associate with other cetaceans, and even with predatory fish such as tuna – although the reason for this is not clear.

BODY FORM
The spinner is slender but muscular, with a long, slim beak, tall, crescent-shaped or triangular dorsal fin, and pointed flippers and flukes.

grey flanks

The acrobatic spinner dolphin performs high leaps when it twists in the air or spins on its vertical axis several times. The purpose of this behaviour is unclear. It may be to demonstrate fitness and/or attract a mate, or to dislodge parasites.

Stenella attenuata
Pantropical spotted dolphin

Length 1.6–2.6 m
(5 1/4–8 1/2 ft)

Weight Up to 120 kg
(260 lb)

Location Tropical and temperate waters worldwide

Social unit Group

Status Lower risk

One of the commonest cetaceans, this dolphin is usually found in waters warmer than 22°C (72°F). It has a slender, streamlined body and slim beak with 40 pairs of teeth in both upper and lower jaws. An elongated, oval, dark grey "cape" extends from the forehead to just behind the dorsal fin. The flanks are lighter grey and the underside is pale. Large schools of thousands are often segregated into mothers with young, older juveniles, and other subgroups. These all associate with other cetaceans, especially spinner dolphins and tuna fish. Pantropical spotted dolphins eat mainly mackerel, flying fish, squid, and other near-surface prey.

spotting increases with age and varies with habitat

adult may have white lips

Stenella frontalis
Atlantic spotted dolphin

Length 1.7–2.3 m
(5 1/2–7 1/2 ft)

Weight 140 kg
(310 lb)

Location Atlantic

Social unit Variable

Status Locally common

The Atlantic spotted dolphin digs with its beak in the sandy sea bed, poking so deep that the whole head is immersed. It also uses various techniques to catch fish, squid, and other prey from surface and mid-waters. It has 32–42 pairs of teeth in the upper jaw and the same in the lower jaw. The Atlantic spotted dolphin forms schools of up to 15 individuals near coasts, but these may gather into larger groups to follow seasonal food. Like its pantropical relative (above), the newborn calf lacks spots. These develop with age, starting on the belly and extending to the sides and back over several years.

STOUTER SPECIES
The Atlantic spotted dolphin is distinguished from its pantropical cousin mainly by its stouter body and beak.

SPOT THE HABITAT

In several spotted dolphin species, including the Atlantic and pantropical, the dark patches and flecks increase in extent with age but may also vary with habitat. These marks are generally less prominent in oceanic groups. In populations that stay near coasts the spots may be so extensive as to obscure the background colour.

Stenella coeruleoalba
Striped dolphin

Length 1.8–2.5 m
(6–8 1/4 ft)

Weight 110–165 kg
(240–360 lb)

Location Tropical and temperate waters worldwide

Social unit Group

Status Lower risk

This species lives in areas of changeable water temperature and may dive as deep as 200m (655ft) to seize small fish and squid, using the 40–55 sharp-pointed pairs of teeth present in each jaw. It derives its name from the complex pattern of black and grey stripes along its back and flanks. This active dolphin is capable of a wide variety of acrobatic leaps and spins. Although it is relatively common, its numbers have declined in recent years. In particular, infection by a virus (morbilli) drastically reduced Mediterranean populations in the early 1990s.

SURFING SCHOOL

Schools of 50–500 striped dolphins race across the ocean and may gather in thousands, leaping high and whistling to keep in contact. They ride pressure waves just in front of migrating great whales, and the man-made equivalent – ship bow waves, especially in Mediterranean and Atlantic waters.

wide, pale grey stripe

COLORATION
The thin black and wider grey stripes branch or fork and overlay a dark, blue-grey back, blue-grey sides, and pale cream or pink underside. It has a black beak and eye patches.

Delphinus delphis
Common dolphin

	Length 2.3–2.6 m (7½–8½ ft)
	Weight 80 kg (175 lb)
Location Temperate and tropical waters worldwide	Social unit Group
	Status Common

HOURGLASS PATTERN
The flanks have yellow and creamy grey areas that form a distinctive hourglass shape.

This species has a yellow to buff blaze along the flank, from the face to below the dorsal fin. It tapers to a point, then widens again towards the tail, but as pale grey. There are also dark, narrow stripes from mouth corner to eye and chin to flipper. The common dolphin has 40–55 pairs of small, sharp teeth in the upper jaw, and the same in the lower jaw. It lives offshore in the deep ocean; the inshore form is sometimes regarded as a separate species, *Delphinus capensis*. Both species hunt shoaling fish and squid to a depth of 300m (985ft).

SCHOOLING

Common dolphins are very social, forming fast-swimming schools of thousands. They leap and tumble, ride waves from ships and great whales, and make many sounds such as clicks, squeaks, and creaks. Their whistles are loud enough to be heard from nearby boats.

Lissodelphis borealis
Northern right-whale dolphin

	Length Up to 3m (9¾ft)
	Weight Up to 115 kg (250 lb)
Location North Pacific	Social unit Group
	Status Common

This sociable dolphin forms schools of 100–200, which merge into gatherings of thousands. It makes varied sounds and associates with other cetaceans, leaping high and riding ship bow waves. Its main foods are mid-water fish and squid, down to 200m (655ft). A similar species, *Lissodelphis peronii*, is found in southern oceans.

CONSERVATION

Due in part to its schooling habits, the northern right-whale dolphin is at high risk of asphyxiation in vast drift nets set for squid in the central Pacific. In the 1980s some 20,000 died yearly, reducing its numbers by perhaps 70 per cent.

SLIM AND FAST
The body is slender, with relatively small flippers and tail flukes, indicating that this dolphin is a speedy swimmer. It has no dorsal fin.

narrow white band along underside

Orcaella brevirostris
Irrawaddy dolphin

	Length 2.1–2.8 m (7–9¼ ft)
	Weight 90–150 kg (200–330 lb)
Location S.E. Asia and N. Australia	Social unit Variable
	Status Locally common

This river dolphin's distinctive features include a bulging forehead, no proper beak but ridge-like lips, and "creases" demarcating a neck region between head and body. The head muscles allow a wide range of facial expressions, giving this dolphin an animated appearance to human observers; relevance of such expressions to its own species is extremely conjectural. Some Irrawaddy dolphins live solely in fresh water, roaming almost 1,500km (930 miles) inland in the major waterways of the Irrawaddy (Burma) and Mekong (Vietnam). Overall, however, this is an estuarine and coastal species frequenting muddy, silt-laden river mouths and deltas. It swims slowly in schools of 15 or fewer, feeding on fish, squid, octopus, prawns, and similar prey on or near the sea bed, using its 15–20 pairs of teeth in each of the upper and lower jaws. Gestation is estimated at 14 months and the single calf is about 100cm (39in) long and 12kg (26lb) at birth. In some regions, these dolphins traditionally work with people to herd fish into nets, receiving rewards of food for their cooperation, and even being revered as sacred. However, in other places Irrawaddy dolphins are viewed as pests at river fisheries and/or killed for their meat.

bulging forehead

small dorsal fin set just behind midpoint of back

slightly paler underside

Cephalorhynchus commersonii
Commerson's dolphin

	Length 1.4–1.7 m (4½–5½ ft)
	Weight Up to 86 kg (190 lb)
Location S. South America, Falkland Islands, Indian Ocean (Kerguelen Islands)	Social unit Group
	Status Locally common

Commerson's dolphin has similar coloration to that of the killer whale. Its forehead slopes smoothly from the snout, merging with its stocky body. The newborn calf, 65–75cm (26–30in) long, is grey and becomes two-tone with age. The species forms schools of less than 10 that sometimes expand up to 100. It feeds on sea-bed dwellers such as fish, crabs, starfish, and squid. There are 2 populations (probably subspecies), separated by more than 8,000km (4,970 miles) – those around South America being 25–30cm (10–12in) shorter than those in the Indian Ocean.

rounded dorsal fin

Cephalorhynchus hectori
Hector's dolphin

	Length 1.2–1.5 m (4–5 ft)
	Weight Up to 57 kg (125 lb)
Location New Zealand	Social unit Group
	Status Endangered

One of the smallest dolphins, this species is similar to a porpoise in outline, with a smoothly tapering snout and no distinct beak or melon bulge. It is grey with black flippers, dorsal fin, and tail. The white underside extends a projection up each flank towards the tail. Active and sociable, it forms small schools of up to 5, and spends much time chasing, touching, flipper-slapping, and generally interacting with others. It feeds at various depths, mainly on fish and squid. As an inshore species, Hector's dolphin is at particular risk from entanglement in fishing nets and from pollution.

finger-like white flank markings

Pseudorca crassidens
False killer whale

Length 5–6 m (16–20 ft)	
Weight 1.3–1.4 tonnes (1 1/4 –1 3/8 tons)	
Location Temperate and tropical waters worldwide	**Social unit** Group
	Status Locally common

One of the largest dolphins, this rapid swimmer has a long, slim body and a tall, sickle-like dorsal fin. Coloration is uniform black or slate-grey, with a paler underside patch between the flippers, and perhaps pale patches on the sides of the head. It prefers deep oceans but appears occasionally off oceanic islands, in schools of 10–20, rarely up to 300. Equipped with 8–11 pairs of large, conical teeth in both the upper and lower jaws, this formidable hunter pursues large oceanic fish such as salmon, tuna, and barracuda. It also takes squid and even smaller dolphins.

CONSERVATION

The false killer whale makes a wide variety of echolocating and communicating sounds, such as clicks and whistles. It also leaps with agility and skilfully surfs breakers and ship bow waves. However, this confident navigator is often stranded on beaches in vast groups of up to 1,000. Why this happens is not known, but, with help to return to the water, some survive.

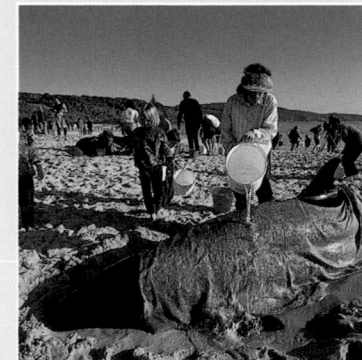

BULL-NOSED WHALE
The pronounced, rounded melon combined with the absence of a beak gives the false killer whale a bull-nosed profile.

angled flippers

Globicephala macrorhynchus
Short-finned pilot whale

Length 5–7 m (16–23 ft)	
Weight 1–1.8 tonnes (1–1 3/4 tons)	
Location Temperate and tropical waters worldwide	**Social unit** Group
	Status Lower risk

Nocturnally active, the pilot whale feeds mainly on deep-water squid and octopus, diving below 500m (1,600ft) for more than 15 minutes. There are 2 very similar species: the short-finned (*Globicephala macrorhynchus*) has smaller flippers (pectoral fins) than the long-finned pilot whale (*Globicephala melas*). In both there is an anchor-shaped pale patch on the throat and chest, and white streaks behind the dorsal fin and each eye. Males are almost twice the weight of females, who live 15 years longer, up to 60 years. Along with with skin scarring, these features suggest competition between males with battles for females. After a gestation of almost 15 months, the female gives birth to a single calf 1.4–1.8m (4 1/2 –6ft) long. Pilot whales are still hunted, by driving them into the shallows for slaughter.

BONDING

Pilot whales form schools of tens to hundreds and also associate with other cetaceans such as bottlenose and common dolphins, and minke whales. Within a pilot whale school, adults form long-term bonds and probably recognize each other by individual "signature" whistles. However, many offspring are not closely related genetically to the school's males, suggesting that mating occurs between schools. Females past reproductive age may suckle calves that are not their own.

STOCKY BODY
The pilot whale has a stocky body, a markedly bulbous forehead, and a dorsal fin set about one-third of the way along the body.

large, rounded dorsal fin

overall grey, brown, or black coloration

Hyperoodon ampullatus
Northern bottlenose whale

Length 6–10 m (20–33 ft)	
Weight Not recorded	
Location North Atlantic	**Social unit** Individual/Group
	Status Lower risk

The northern bottlenose whale is one of about 19 species of beaked whales – mostly medium-sized, open-ocean cetaceans that make long, deep dives for squid, starfish, fish, crab, and other food. This species has a bulging forehead and dolphin-like beak. Males, which grow larger than females, have 2 tusk-like teeth growing at the tip of the beak. Other teeth are present but the degree of growth above the surface of the gums (eruption) is very variable. Feeding is probably by suction, using the tongue as a piston to draw in water, sea-bed mud, and prey. The body of the northern bottlenose whale is long and slim, the flippers small, and the erect dorsal fin is set about two-thirds of the way to the tail. Like most other beaked whales, this species congregates in small schools of up to 10, usually containing individuals of the same sex and of similar ages.

scars from combat with other males of the species

orange- or grey-brown back

streamlined flukes

slim body

pale brown underside

MAMMALS

Orcinus orca

Killer whale

Length	Up to 9m (30ft)
Weight	Up to 10 tonnes (10 tons)

Location Worldwide

Social unit Group

Status Lower risk

Distinctive black and white markings make the killer whale the most easily recognized of the toothed whales and dolphins. It is a highly social whale, living in long-lasting family groups called pods, which consist of adult males and females, and calves of various ages. Pods typically number up to 30 individuals, but groups of up to 150 whales occur when pods come together to form superpods. Pods are matriarchal and both male and female calves tend to stay with their mother for life. When the young reproduce, their offspring remain to build up multi-generational groupings around the original mother.

Generally, the killer whale's diet is as diverse as its hunting techniques. Fish, ranging from herring to great white sharks, and marine mammals, including whales and seals, are taken, as well as turtles and birds. In some regions such as the northeast Pacific, however, there appear to be 2 forms of killer whale: transients, who feed on mammals, turtles, and birds; and residents, who eat only fish. Despite its name, the killer whale is approachable and very inquisitive. It has a variety of elaborate surface habits, including spy hopping (rising slowly vertically, until its head is above the water), tail and flipper slapping, and breaching.

PARENTAL CARE
The whole pod may provide parental care, but a newborn calf stays closest to its mother – a bond that will remain strong for the rest of its life.

POD FORMATIONS
Pods may travel in tight formations, with the females and calves at the centre, and the males on the fringes, or spread across distances of up to 1km (⅔ mile). They communicate using sets of highly distinctive cries and screams, which also act as social signals that reinforce group identity.

INGENIOUS HUNTER
Killer whales are versatile hunters, using a number of different techniques that give them one of the most varied diets in the oceans. They often hunt in pods, chasing down prey or herding fish together before attacking from different angles. Those living off southern South America have perfected the technique of catching sea lions by intentionally beaching themselves to pursue the sea lions in the shallows (shown right), while other methods include tipping over ice floes to unbalance seals and penguins, and breaching next to rocks to wash birds into the sea.

taller, less curved dorsal fin in male

rounded, tapering head

conspicuous eye patches

grey saddle patch

wide tail flukes

large, paddle-shaped flippers

white underside

BUILT FOR HUNTING
The killer whale has a powerfully built, stocky body ideally suited to hunting. The broad tail flukes help to propel the whale at high speeds, and the dorsal fin – which reaches up to 1.8m (6ft) in males – and paddle-like flippers provide stability. Underwater its markings have the effect of camouflaging the whale from above and below.

Ziphius cavirostris

Cuvier's beaked whale

| Length 7–7.5 m (23–25 ft) |
| Weight 3–4 tonnes (3–3⅞ tons) |

| Location Temperate and tropical waters worldwide | Social unit Variable |
| | Status Locally common |

This species has the long, slim body and small dorsal fin set well to the rear typical of the beaked whale family. It feeds on deep-living squid and other creatures using suction (see northern bottlenose whale, p.173). The jaw-line curves up at the tip of the snout and then down. Along with the relatively smooth forehead, this leads to the alternative name of goose-beak whale. The small flippers fit into indentations in the body so that the whale is streamlined for fast swimming, using only its tail flukes, down to great depths. The pale brown to blue-grey body is scarred by parasites and also, in males, by bites from males of the same species. These wounds probably occur during dominance battles at breeding time.

Older male Cuvier's beaked whales tend to live alone. Most males have 2 cone-shaped teeth, which project like tusks from the lower jaw. Females and young are toothless. Younger males, females, and offspring, which are 2.7m (nearly 9ft) long at birth, form schools of rarely more than 10.

"goose-beak" snout

2 throat grooves

tan or pale brown to grey-blue coloration

scars from skin parasites and inter-male battles

Kogia simus

Dwarf sperm whale

| Length 2.7 m (8¾ ft) |
| Weight 135–270 kg (300–600 lb) |

| Location Temperate and tropical waters worldwide | Social unit Variable |
| | Status Locally common |

Smallest of the 3 species of sperm whale, this whale dives to 300m (nearly 1,000ft) for fish, squid, crustaceans, and molluscs. The lower jaw has 7–13 pairs of sharp teeth, and is slung almost shark-like under the large, bulbous head; the upper jaw has only 3 pairs of teeth. Back, fin, flippers, and flukes are blue-grey, shading to cream below. Just behind the mouth and eye is a contrasting, pale crescent which, in size and position, resembles the gill slit of a fish. A shy creature, the dwarf sperm whale lives alone or in small schools (fewer than 10), and releases a cloud of faeces to repel predators. Little is known about its breeding habits: the gestation period appears to vary between 9 and 11 months, and the single calf, about 1m (3ft) long, is usually born in autumn. These whales seem prone to group strandings.

tall, dolphin-like dorsal fin

pale, crescent-shaped markings

Physeter catodon

Sperm whale

| Length 11–20 m (36–65 ft) |
| Weight 20–57 tonnes (19½–56 tons) |

| Location Deep waters worldwide | Social unit Individual/Group |
| | Status Vulnerable |

The world's largest carnivore, this massive cetacean makes extremely deep dives for food, mainly squid and octopus, but also fish and, sometimes, giant squid. It can stay submerged for almost 2 hours and has been tracked by sonar to 1,200m (nearly 4,000ft) deep, with indirect evidence, such as types of bottom-dwelling fish found in its stomach, of dives below 3,100m (10,170ft). The ability to make such deep dives may be in part due to its spermaceti organ (see panel, right). Between dives, the whale lies log-like at the surface, a plume of misty air emerging with each exhalation at 45 degrees from the single blowhole. The long, narrow lower jaw is slung under the huge head and bears 50 pairs of conical, round-tipped teeth; there are no visible upper teeth. The main colour is dark grey or brown, paler on the underside, with white or cream around the lower jaw. Bull (male) sperm whales are twice the weight of females and tend to migrate farther north and south, into colder waters, for summer feeding. They form loose, bachelor pods when young but then become more solitary. Cows (females) stay nearer the tropics and form mixed groups with their young and juveniles to about 10 years old. In summer or autumn, after a gestation period of 14–15 months, the single calf is born, 4m (13ft) long, and suckles for 4 years or more.

KEEPING IN TOUCH
Sperm whales often swim near, touch, and caress others in their school. They also produce loud, rhythmic clicking and banging sounds which may aid individual recognition.

SPERMACETI ORGAN

The sperm whale's enormous head contains a spermaceti organ – a large mass of waxy oil. This may work as a buoyancy aid, becoming denser with changing temperature and pressure, to assist very deep dives. It may also function as a sound-lens, like a dolphin's melon, to focus the whale's clicks and other noises.

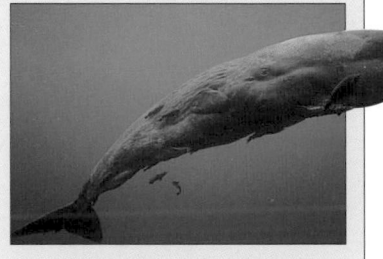

LARGE HEAD
The sperm whale's tall, narrow, box-like head comprises up to one-third of its total length. The low dorsal fin is set well to the rear.

wrinkled skin on rear parts of body

row of knobs along back between dorsal fin and tail

pale underparts

DIVING DEEP

A sperm whale leaves the sunlit upper levels of the ocean and heads into the dark ocean depths in search of food. Diving to greater depths than any other mammal, it descends at a rate of up to 3m (10ft) per second. Away from the light, its eyes are of little use and the whale relies on echolocation.

CARNIVORES

PHYLUM	Chordata
CLASS	Mammalia
ORDER	Carnivora
FAMILIES	7
SPECIES	249

CLASSIFICATION NOTE

In this book, carnivores are
arranged in 7 families. Some
zoologists recognize 8 families,
with mongooses in a separate
family from civets and relatives.

ALTHOUGH THE TERM carnivore is commonly used
to describe an animal that eats meat, it also refers
specifically to members of the order Carnivora. While
most members of the group eat meat, some have
a mixed diet or are entirely herbivorous. Meat-
eating carnivores are the dominant predators on
land in all habitats: their bodies and lifestyles
are highly adapted for hunting. However, there
is great variety within the group, which
includes species as diverse as the giant
panda and the spotted hyena. Uniquely
among mammals, carnivores have 4
carnassial teeth. They also have a penis
bone (baculum). Indigenous to most parts
of the world (except Antarctica), carnivores
have also been introduced to Australasia.

HUNTING

Carnivores include some of nature's most skilful
and efficient predators. Most use keen senses of
sight, hearing, or smell to locate prey, which they
catch either by pouncing from a concealed place
or by stalking and then running down their
quarry in a lengthy chase or swift rush.
Many can kill animals larger than themselves. Weasels kill by biting the back of
the head and cracking the skull, while
cats bite into the neck, damaging the
spinal cord, or into the
throat, causing suffocation.
Dogs shake prey vigorously
to dislocate the neck.

SOLITARY HUNTER
*The bobcat, which feeds mainly on
small prey such as the snowshoe
hare, hunts alone.*

ANATOMY

Although carnivores vary considerably in size and shape, most
share several features that make them well suited to a hunting
lifestyle. A typical carnivore is a fast and agile runner with
sharp teeth and claws, acute hearing and eyesight, and a well-
developed sense of smell. Carnassial teeth (see below) are present
in predacious living carnivores but are less well developed
among the omnivorous and herbivorous species. Carnivores
have either 4 or 5 digits on each limb. Members of the cat
family (except cheetahs) have sharp, retractable claws used to
rake prey, defend themselves, and climb. Most other carnivores
have non-retractable claws, often used for digging.

HUNTING IN PACKS
*Lions generally hunt in groups to capture large
animals, such as wildebeests. The females (adult
males rarely join the hunt) usually stalk to within
30m (98ft), fanning out to encircle the prey.
After a short charge, the animal is brought
down with a grab to the flank, then killed
by suffocation with a bite to the throat.*

JAWS AND TEETH

*Most carnivores have sharp teeth
and powerful jaws for killing and
disembowelling prey. The temporalis
muscles, which are most effective when
the jaws are open, are used to deliver
a powerful stab from the sharp
canines. The carnassial teeth are
sharpened cheek teeth in the upper
and lower jaws that mesh together
perfectly. In combination with the
masseter muscles, which can be used
when the jaw is almost completely
closed, they form a powerful shearing
tool for tearing flesh.*

upper
carnassial
tooth

temporalis
muscle

upper
canine

lower carnassial
tooth

masseter
muscle

lower
canine

HYENA SKULL

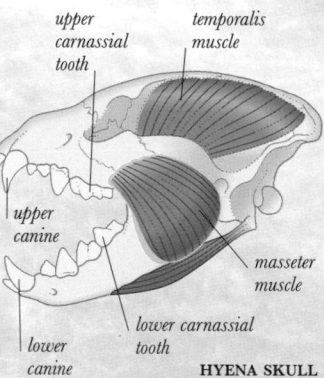

**TIGER
SKELETON**

*separate radius and ulna
maximize flexibility*

*fused "wrist"
bones*

*flexible spine
enables back to
bend while running*

SKELETON AND MOVEMENT

*Predatory carnivores have several physical
adaptations that enable them to move
quickly over the ground in pursuit of
prey. The spine is generally flexible, the
limbs are relatively long, and the collarbone
is reduced, maximizing the mobility of
their shoulders. To increase the length of
their stride and to add to their speed, all
carnivores have fused wrist bones, and
dogs and cats walk on their toes (rather
than the soles of the feet).*

SIBERIAN TIGER

SOCIAL GROUPS

Although many carnivores
live alone or associate only
in pairs, others form groups.
These take many different
forms, and their structure is
often complex. Lion prides,
for example, consist of
several related families,
although most males leave the
pride into which they are born. Lions spend most
of their time together, hunt cooperatively, and tend
each other's young. In most other groups, the ties
between individuals are looser. Red and Arctic foxes,
for example, live in groups of one adult male and
several vixens, but each adult hunts alone in a
different part of the group's territory.

PLAY
*Young carnivores develop their fighting skills
through play. By playing together, these red
foxes learn to test another animal's strength
without suffering painful consequences.*

SHARED PARENTHOOD
*These young slender-tailed meerkats are
not necessarily the offspring of the adult
watching over them. Sharing parental duties
is common in many carnivore societies.*

FEEDING

Most carnivores live on a diet of freshly killed animal prey, ranging in size from insects, other invertebrates, and small vertebrates to animals as large as buffaloes and reindeer. Carnivores are generally adaptable feeders, seldom restricting themselves to a single food-type. Some – such as bears, badgers, and foxes – eat a mixed diet of meat and plants while a few, notably the giant panda, are almost entirely herbivorous.

CARRION-FEEDERS
Hyenas feed on live prey and the remains of other animals' kills. Their particularly sharp teeth and strong jaws enable them to break bones and tendons that are too tough for other carnivores.

PLANT-EATERS
The diet of the giant panda consists mainly of bamboo shoots and roots, on which it feeds for up to 12 hours a day. Pandas are slow-moving compared with other carnivores, and their flat cheek teeth are better suited for grinding than for cutting.

COMMUNICATION

Carnivores communicate with each other with scent-markings, visual signals, and vocalizations. Scent messages, which have the advantage of being persistent, are used to define territory or to find potential sexual partners. They are left by spraying urine or leaving piles of faeces, although some animals also rub scent onto objects from glands on their face, between their claws, or at the base of their tail. When animals meet face-to-face, posture, facial expression, and sound are used to pass on a wealth of information, including threats, submissions, advances to partners, and warnings of approaching danger.

GREETING POSTURES
Body language is an important form of communication for African wild dogs, which live in large packs. In this greeting ritual, adult dogs push their muzzles into each other's faces.

MARKING TERRITORY
Bears use trees to leave both scent marks and visual signs. Here scent is being transferred in saliva and from glands in the bear's feet, while the sharp teeth and claws are being used to rip the bark.

DOGS AND RELATIVES

PHYLUM	Chordata
CLASS	Mammalia
ORDER	Carnivora
FAMILY	Canidae
SPECIES	36

MEMBERS OF THE dog family – dogs, wolves, coyotes, jackals, and foxes – are collectively described as canids. They are known for great endurance (rather than sudden bursts of speed) and for opportunistic and adaptable behaviour. Dogs are characterized by a slender build, long legs, and a long, bushy tail. Wild canids generally inhabit open grassland habitats the world over, and are absent only from isolated areas such as Madagascar and New Zealand (where the domestic dog is present, however).

ANATOMY

Canids have a muscular, deep-chested body covered with a fur coat that is usually uniformly coloured or speckled. The lower limbs are developed for strength and stamina: some of the wrist bones are fused, and the front legs cannot be rotated (the bones at the front of the leg are locked). There are 4 digits on the back feet and 5 on the front feet, and each digit has a hard pad. The claws are short, non-retractable, and blunt (other carnivores have sharp claws). Canids also have long jaws, long, fang-like canines (for stabbing prey), and well-developed carnassials (the slashing teeth at the back of the jaws). Canids track their prey by scent, and the long, pointed muzzle houses large olfactory organs. Hearing is also acute, and the ears are large, erect, and usually pointed. Sight is less important, but is still well developed.

SOCIAL GROUPS

Smaller species, which usually feed mainly on small rodents and insects, tend to have a flexible social organization but often live either in pairs (for example jackals) or alone (for example foxes). However, larger species, such as the wolf and the African wild dog, live in social groups called packs. These packs, which consist of a dominant pair and their offspring, occupy and defend territories, which they mark with urine. The young in a pack are of different ages because older offspring remain in the group for some years, and may help rear new young. Only the dominant pair breed, and the female digs a den in which to give birth. The pack will often perform a bonding ritual, which involves mutual licking, whining, and tail wagging. When hunting, the usual tactic is to track a herd of deer or antelopes (for example) and then cooperatively manoeuvre so as to separate one animal. This individual is then run down and slashed at by pack members until it falls, exhausted. On returning to the den, the hunters regurgitate meat for the cubs to eat.

CANIDS AND PEOPLE

Throughout history, canids have proved useful to humankind in many ways. Wild canids are, for example, important controllers of rodent populations, which can spiral quickly if left unchecked. Furthermore, the domestic dog – which descended from the wolf over 10,000 years ago – has always played an important role in a number of human activities. From the tiny chihuahua to the huge St Bernard (the domestic dog displays more variation between types than any other domestic animal), there are breeds specialized for hunting, herding, guarding, performing, carrying or dragging loads, and companionship.

However, despite these factors, many canid species are considered pests. The wolf, for example, has been hunted and persecuted as a killer of valuable livestock. As a result, this species is now rare throughout its vast range and is extinct in many regions. Other species have fared even worse: the bush dog and the maned wolf, for example, are on the brink of extinction; the red wolf now survives only in zoos. On the other hand, the coyote and the red fox – both great opportunists – have benefited from the spread of urban environments and are more abundant than ever before.

STRENGTHENING BONDS
Establishing and maintaining bonds between pack members is essential for the survival of dogs that live in social groups. In African wild dogs, bonding behaviour, such as licking and whining, frequently occurs before a hunt. It is only by cooperating and hunting as a team that the dogs are able to bring down and kill prey larger than themselves.

MAMMALS

Vulpes vulpes
Red fox

Length	58–90 cm (23–35 in)
Tail	32–49 cm (12 1/2–19 1/2 in)
Weight	3–11 kg (6 1/2–24 lb)

Location Arctic, North America, Europe, Asia, N. Africa, Australia

Social unit Pair

Status Common

Active by day and night, the red fox is exceptionally widespread, and adaptable both in habitat, from Arctic tundra to city centre, and in diet. Home is an earth (den) in a sheltered place – for example, an enlarged rabbit burrow, a crevice among rocks or roots, or a space under an out-building. The basic social unit is a female (vixen) and male (dog), who mark their territory of 1–10 square km (0.4–4 square miles) with urine, droppings, and scent from anal and other glands. Single-male–multi-female groups also occur but only senior females breed. Mating is in late winter or early spring when females make eerie shrieks. Gestation is 49–55 days and litter size up to 12; averages vary from 4–5 in Europe, to 6–8 in North America. Both parents and "helper" non-breeding females care for the cubs and feed them after weaning at 6–12 weeks.

THE UNFUSSY FOX
In grassy or farmed areas, a large part of the red fox's diet comprises lagomorphs, especially rabbits and young hares. The fox stealthily stalks its prey, then makes a dash to catch the victim before it reaches its burrow (rabbit) or accelerates away (hare). The prey is carried by the neck to a secluded spot where the fox can eat at leisure. The red fox also consumes beetles, worms, frogs, birds, eggs, mice, voles, fruits, carrion, and refuse – in fact, almost anything edible.

NOT ALWAYS RED
Coat colour varies from greyish or rusty red to almost orange, usually with black on the backs of the ears, sometimes on the lower limbs and feet, and an often black-tinged but pale-tipped tail. All-black and silver-white forms also occur.

bushy tail (brush)

Vulpes cana
Blanford's fox

Length	42 cm (16 1/2 in)
Tail	30 cm (12 in)
Weight	0.9–1.5 kg (2–3 1/4 lb)

Location W. and S. Asia

Social unit Individual

Status Locally common

This small fox has relatively large ears and tail, patchy body colouring in black, grey, and white, white under-parts, a dark stripe along the middle of the back, and a stealthy, feline gait. It is a solitary, nocturnal hunter of small creatures, including insects, mainly in barren, rocky hills and grassy uplands. It also consumes appreciable amounts of fruits and is found near orchards and groves. Litter size is 1–3.

Vulpes macrotis
Kit fox

Length	38–52 cm (12 1/2–20 1/2 in)
Tail	22–32 cm (8 1/2–12 1/2 in)
Weight	1.5–3 kg (3 1/4–6 1/2 lb)

Location W. USA

Social unit Pair

Status Vulnerable†

Similar to the swift fox (see right) in appearance and habits, this species has a more westerly range, but with overlap and perhaps interbreeding in Texas. The kit fox has longer, closer-set ears, a more angular head, and is more heavily built overall. There are 3 colour forms: pale grey-brown, dark grey-brown, and intermediate grey. Its habitats vary from grassland to desert, and its diet is omnivorous. Both parents raise the 3–6 young.

Vulpes velox
Swift fox

Length	38–53 cm (12 1/2–21 in)
Tail	18–26 cm (7–10 in)
Weight	1.5–3 kg (3 1/4–6 1/2 lb)

Location C. USA

Social unit Pair

Status Unconfirmed

Recently recognized as a separate species from the kit fox (see left), the swift fox has a more easterly distribution. Its coloration is similar to that of the kit fox, but it is greyer on the upperparts and buff-orange under-neath. It has a bushy, black-tipped tail. Both species dig dens about 1m (3ft) deep, with 4m (13ft) of tunnels, and mate from December to January – later in northern areas. The gestation period is 50–60 days.

Vulpes zerda
Fennec fox

Length	24–41 cm (9 1/2–16 in)
Tail	18–31 cm (7–12 in)
Weight	1–1.5 g (2 1/4–3 1/4 lb)

Location N. Africa, W. Asia

Social unit Group

Status Unconfirmed

The smallest fox, the fennec has relatively large ears and a black-tinged tail tip. Its furred soles are adapted for walking on

soft, hot sand. Mostly nocturnal, the diet of this fox ranges from fruits and seeds to eggs, termites, and lizards. Unusual among foxes, it associates in groups of up to 10, but relationships are not clear. Each member digs a den several metres into soft earth. Mating occurs in January–February and the 2–5 cubs remain in the den, protected by the female, for 2 months. They are fully mature by 11 months. Hunted for its fur, the fennec fox is also trapped as a pet.

cream to yellowish cream fur

white underparts

Vulpes rueppelli
Rueppell's fox

Length 40–52 cm (16–20 1/2 in)	
Tail 25–39 cm (10–15 1/2 in)	
Weight 1–3.5 kg (2 1/4–6 1/2 lb)	
Location N. Africa, W. Asia	Social unit Group/Pair
	Status Unconfirmed

white chin, bib, and belly

Rueppell's fox (also called the sand fox) is similar to but slighter in build than the red fox. It has soft, dense, sandy or silver-grey fur to match its arid habitat, black patches on the sides of the muzzle, and a white tail tip. In some regions this species forms monogamous pairs, but in others it gathers in groups of up to 15. It rests by day in a sheltered crevice or burrow, and changes its den every few days. Average litter size is 2–3, born in early spring. It eats a wide variety of foods, from grass to insects, reptiles, and mammals.

Alopex lagopus
Arctic fox

Length 53–55 cm (21–22 in)	
Tail 30 cm (12 in)	
Weight 4 kg (8 3/4 lb)	
Location N. Canada, Alaska, Greenland, N. Europe, N. Asia	Social unit Group
	Status Common

The Arctic fox has 2 colour types or "phases". Foxes that are "white" phase are almost pure white in winter for camouflage in snow and ice. This phase is associated with the true tundra of open, treeless plains and grassy hillocks. Those of the "blue" phase are more prevalent in mixed coastal and shrubby habitats and are pale grey-brown tinged with blue in winter. The Arctic fox eats a huge variety of foods – mainly lemmings, but also birds, eggs, crabs, fish, insects, seal and whale carcasses, fruits, seeds, and human refuse. Its social group is likewise flexible, with male–female pairs, larger groups of non-breeders, or a breeding pair plus "helper" females. The den site is extensive, with complex burrow systems for shelter and breeding. Reproduction is closely tied to available food, with more than 15 cubs per litter when lemmings are plentiful, and 6–10 in an average year.

stout, rounded body under thick fur

The Arctic fox's summer coat is half as thick as its winter one, with less than half of the underfur. In summer, white-phase animals are grey-brown to grey above and grey below; those of the blue phase are browner and darker.

FURRED BUNDLE
The Arctic fox has small ears, a blunt muzzle, and short legs and tail, since these areas lose heat fastest. Every part of its body except its nose is thickly furred.

Urocyon cinereoargenteus
Grey fox

Length 53–81 cm (21–32 in)	
Tail 27–44 cm (10 1/2–17 1/2 in)	
Weight 3–7 kg (6 1/2–15 lb)	
Location S. Canada to N. South America	Social unit Pair
	Status Common

Also called the tree fox, this long-bodied species prefers woodland. It climbs skilfully, leaping up tree trunks and between branches with almost cat-like agility. Active at night, it consumes various insects and small mammals, but may rely more on fruits and seeds in certain seasons. The grey fox has a small, dark grey neck mane and central back stripe, and a red tinge to the neck, flanks, and legs, with a buff or white chin and belly. Its den may be in an old burrow or log, but more often in a tree-hole up to 9m (30ft) above ground, or on a building ledge or in a roof space. Most grey foxes live as breeding pairs.

GRIZZLED GREY
The speckled or grizzled coat is due to individual hairs banded in white, grey, and black.

The average litter size for the grey fox is 4 (range 1–10). Each new-born cub is black furred and, like most foxes at birth, helpless, with eyes closed. Its eyes open at 9–12 days and by 4 weeks it ventures from the den and begins to climb, guarded by a parent. It starts to take solid food 2 weeks later.

Atelocynus microtis
Small-eared dog

Length 72–100 cm (28–39 in)	
Tail 25–35 cm (10–14 in)	
Weight 9–10 kg (20–22 lb)	
Location South America	Social unit Individual
	Status Locally common

With its rounded, short ears, the small-eared dog resembles a raccoon dog (see p.183) but its fur is much shorter and more velvety, grey to black on the back and varying shades of grey tinged with red-brown on the underside. The black tail is more bushy and fox-like. Also known as the small-eared zorro, this mainly nocturnal and solitary dog is a secretive, little-known inhabitant of tropical forests. It moves with cat-like stealth and probably eats mainly small rodents, with some plant matter.

Pseudalopex culpaeus
Culpeo fox

Length 60–120 cm (23 1/2–47 in)	
Tail 30–45 cm (12–18 in)	
Weight 5–13.5 g (11–30 lb)	
Location W. South America	Social unit Pair
	Status Locally common

A species of open upland and pampas grassland, this large, powerful fox is extensively hunted for its fur and to prevent predation of livestock such as lambs and poultry. Its diet also includes rodents, rabbits, birds and their eggs, and seasonal berries and fruits. Like many foxes, it stores food during times of plenty, burying the excess or wedging it under logs and rocks, for later consumption. The culpeo fox's coat is grizzled grey on the back and shoulders, more tawny on the head, neck, ears, and legs, with a fluffy, black-tipped tail.

Cerdocyon thous
Crab-eating fox

Length 64 cm
(25 in)

Tail 29 cm
(11 1/2 in)

Weight 5–8 kg
(11–18 lb)

Location N. and E. South America

Social unit Group/Pair

Status Common

As well as crabs – both coastal and freshwater – this medium-sized fox eats much else, including fish, reptiles, birds, mammals, grubs, and fruits. Widespread in many habitats, it shows much variation across its range, although the body is generally grey-brown, with reddish brown face, ears, and front legs, a white underside, and black on the tips of the ears, tail end, and the backs of the legs. Active at night, it lives in loose social groups of an adult pair and their offspring.

Nyctereutes procyonoides
Raccoon dog

Length 50–60 cm
(20–23 1/2 in)

Tail 18 cm
(7 in)

Weight 7.5 kg
(17 lb)

Location Europe, C., N., and E. Asia

Social unit Group/Pair

Status Locally common

This canid resembles a combination of raccoon and dog, with its black face "mask" and variable black fur on the shoulders and upperside of the tail. It is nocturnal and has a huge dietary range, from fruits to birds, mice, crabs, and fish. It also forages

along river banks, lakesides, and the seashore. It lives in pairs or loose family groups; litters average 4–6. The raccoon dog is abundant in Japan, extinct in parts of China, yet spreading rapidly in areas of Europe, where it has been introduced.

READY FOR WINTER

The raccoon dog is an unusual member of the dog family – even youngsters can climb well. It is also unique (for a canid) in that it hibernates in winter. Autumn feasting increases body weight by up to 50 per cent. A deep sleep follows, usually sheltered in an old fox or badger den.

COLORATION
This raccoon-like dog has long, yellow-tinged, brown-black body fur (especially in winter), black facial patches below the eyes, a white muzzle, short-furred legs, and a bushy tail.

Speothos venaticus
Bush dog

Length 57–75 cm
(22 1/2–30 in)

Tail 12.5–15 cm
(5–6 in)

Weight 5–7 kg
(11–15 lb)

Location Central America to N. and C. South America

Social unit Group

Status Vulnerable

Long-bodied and short-legged, this day-active predator lives in family-based packs of up to 10. It is a powerful and persistent hunter of ground birds and rodents up to the size of Azara's agouti (see p.158). The pack, however, tackles larger prey such as rheas and capybaras, swimming efficiently after victims. By night the group members sleep in dens, in deserted burrows, hollow logs, or under rocks. Average litter size is 4, born after a gestation of 67 days. The male brings food to the suckling female in the den.

Chrysocyon brachyurus
Maned wolf

Length 1.2–1.3 m
(4–4 1/4 ft)

Tail 28–45 cm
(11–18 in)

Weight 20–23 kg
(44–51 lb)

Location C. and E. South America

Social unit Individual

Status Lower risk

Similar to a red fox (see p.181), but with very long legs, this wolf has long, thick, reddish yellow fur, a black neck crest, central back stripe, and black muzzle. It prefers open, grassy, or low-scrub habitats where it can peer over vegetation for prey and danger. Female and male share a territory, and mate each year, usually in May or June, but otherwise rarely associate. Active at twilight and night, the maned wolf takes a varied diet, including rabbits, birds, and mice, as well as smaller creatures such as grubs and ants, and also appreciable amounts of plant material such as

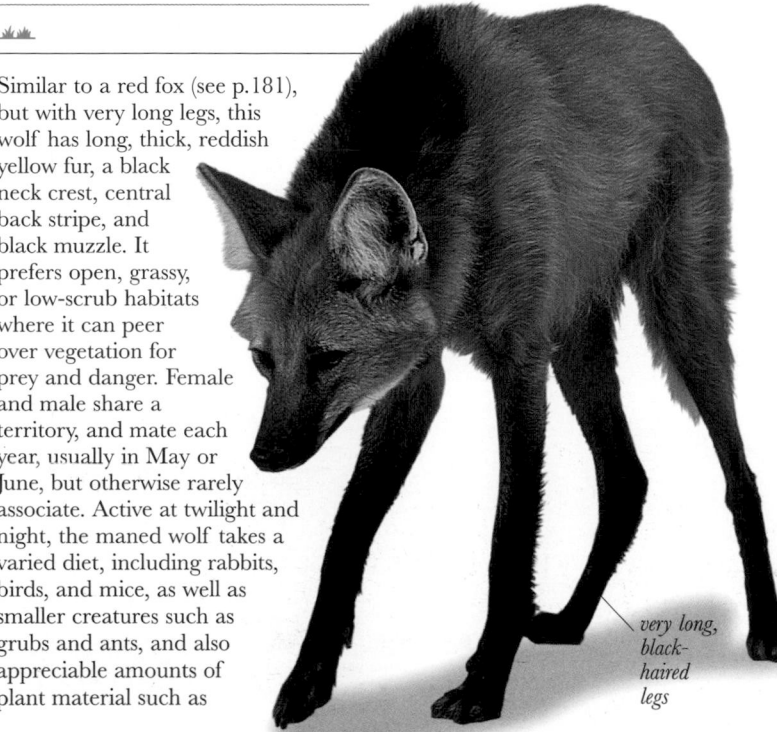

very long, black-haired legs

fruits and berries. It is said to kill livestock, especially poultry, and so is hunted as a pest in some areas – yet it is kept as a pet in others. Disease is another major threat. The gestation period is 62–66 days and the 1–5 pups (average 2) are born in an above-ground den in thick grass or bushes. The mother cares for them alone, suckling them for up to 15 weeks.

Canis adustus
Side-striped jackal

Length 65–81 cm
(26–32 in)

Tail 30–41 cm
(12–16 in)

Weight 6.5–14 kg
(14–31 lb)

Location C., E., and southern Africa

Social unit Pair

Status Unconfirmed

often indistinct white and black side stripes

grey-yellow coat paler on underside

white tail tip

Sometimes sighted foraging at night near city centres, the side-striped jackal is also found in grass, along forest edges, and in mixed farmland. More omnivorous than other jackals, it takes rodents, birds, eggs, lizards, insects and other invertebrates, refuse, carrion, and plant material such as fruits and berries. The basic social group is a female–male pair with their young, which can number up to 6 (average 5 per litter). Offspring are born after 57–70 days' gestation, in a secure den such as an old termite mound or aardvark burrow. Weaned by 10 weeks, they become independent at about 8 months.

Golden jackal

Canis aureus

Length 60–110 cm	(23 ½ – 43 in)
Tail 20–30 cm	(8–12 in)
Weight 7–15 kg	(15–33 lb)

Location S.E. Europe, N. and E. Africa, W. to S.E. Asia

Social unit Pair

Status Common

These omnivorous, opportunistic jackals usually live as breeding pairs, but in areas with plentiful food, such as refuse dumps near human habitation, they form packs of up to 20. The gestation period is 60–63 days. Average litter size is 5–6 pups (range 1–9), which are cared for in a secure den.

SHADES OF GOLD
The coat is mainly pale yellow, gold, or light brown, greyer on the back and gingery on the belly.

MEAL TIME

Golden jackal pups transfer from milk to solid food at about 8–10 weeks. At this age they are too young to hunt, so parents, older siblings, and other young adults regurgitate meals for them on return from a successful outing.

ginger-coloured nose and ears

Coyote

Canis latrans

Length 70–97 cm	(28–38 in)
Tail 30–38 cm	(12–15 in)
Weight 9–16 kg	(20–35 lb)

Location North America to N. Central America

Social unit Variable

Status Common

The coyote, like many canids, is highly adaptable in habitat and opportunistic in diet. Once believed to be always solitary, it may also form a breeding pair or, when larger prey is common, gather as a small hunting pack. Food varies from pronghorns, deer, and mountain sheep to fish, carrion, and refuse. The coyote is a rapid sprinter (65kph/40mph) and often runs down jackrabbits. Its well-known nocturnal howl usually announces an individual's territory or location to neighbours. Mating occurs from January to March, gestation takes 63 days, and the litter size is 6–18 (average 6). The pups are born in a secure den.

COLORATION
The grizzled buff coat is yellowish on the outer ears, legs, and feet. Underparts are grey or white. The shoulders, back, and tail may be tinged black.

STALK AND POUNCE

The coyote, like many similar species, uses a characteristic pounce to catch small prey such as mice in snow or grass. It moves forward slowly, watching and listening intently. Having located the prey, it leaps almost vertically into the air and brings its front feet down onto the animal, pinning it to the ground before killing it with a bite.

Black-backed jackal

Canis mesomelas

Length 45–90 cm	(18–35 in)
Tail 26–40 cm	(10–16 in)
Weight 6–12 kg	(13–26 lb)

Location E. and southern Africa

Social unit Pair

Status Common

This jackal's range extends from city suburbs to the deserts of southern Africa. The main coloration is ginger to red-brown with a distinctive black saddle over its shoulders and back, and a black, bushy tail. Female and male mate for life and hunt together as adaptable omnivores. Their prey includes livestock such as sheep or young cattle. Their breeding habits resemble those of other jackals.

Ethiopian wolf

Canis simensis

Length 1 m	(3 ¼ ft)
Tail 33 cm	(13 in)
Weight 15–18 kg	(33–40 lb)

Location E. Africa

Social unit Group

Status Critically endangered

Formerly known as the Simien jackal, this species' 3 remnant populations (in the Ethiopian highlands) are at risk from habitat loss, competition and diseases from domestic dogs, and overgrazing, which has reduced their prey of hares, rodents, and giant mole rats. Groups of up to 12 wolves congregate noisily at morning, noon, and evening; most hunting is around dawn and dusk. Both parents and young adult "helpers" protect and regurgitate food for the cubs.

Red wolf

Canis rufus

Length 1–1.2 m	(3 ¼ – 4 ft)
Tail 25–35 cm	(10–14 in)
Weight 18–41 kg	(40–90 lb)

Location Reintroduced to E. USA (North Carolina)

Social unit Group

Status Critically endangered†

By the 1970s red wolves were believed to be extinct in the wild as a pure species, due chiefly to persecution and interbreeding with coyotes. Reintroduced from 1987 in North Carolina, they have established a population of more than 50. They hunt mammals such as rabbits, coypu, and raccoons, and live in packs with a social organization comparable to that of the grey wolf (see p.186). The coat is tawny-cinnamon mixed with grey and black, and is darkest on the back.

Dingo
Canis dingo

Length	72–110 cm (28–43 in)
Tail	21–36 cm (8 1/2–14 in)
Weight	9–21.5 kg (20–47 lb)

Location Australia

Social unit Group

Status Locally common

This dog has variously been regarded as a subspecies of the domestic dog; as a subspecies of the domestic dog's ancestor, the grey wolf (see p.186); or as a full and separate species in its own right. It is likely that the dingo descended from the domestic dog within the past 10,000 years, and is now able to survive in the wild in many habitats. Dingoes are found throughout Australia, except for the southwest and southeast, where dingo fences exclude them from livestock; they are classed as pests both to farm animals and for rabies control. Dingo-like dogs also occur wild, semi-feral, or semi-domesticated, on the mainland and many islands of South and Southeast Asia. Dingoes interbreed readily with domestic dogs, and in parts of Australia one-third of individuals are such hybrids. Prey includes rabbits, rodents, wallabies, small kangaroos, and birds. However, the opportunistic dingo can survive on fruits, plant matter, and carrion. In social behaviour and pack system it resembles the grey wolf.

irregular white patches on muzzle, chest, belly, and feet

bushy tail may be white-tipped

PACK HIERARCHY

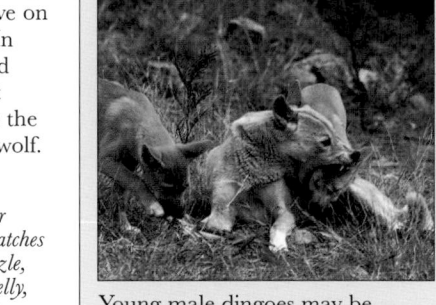

Young male dingoes may be solitary and nomadic. Breeding adults usually form settled packs, unless the population is widely spaced, when pairs are likely. About 5 pups (range 1–10) are born after a gestation of 63 days. Senior pack members teach them their place in the hierarchy by nips and other rebuffs.

DINGO OR DOG?
The dingo's coat varies from light sandy to deep red-ginger. Dingo-domestic dog hybrids can look very similar, but may be distinguished by their canine and carnassial tooth shape.

Bat-eared fox
Otocyon megalotis

Length	46–66 cm (18–26 in)
Tail	23–34 cm (9–13 1/2 in)
Weight	2–4.5 kg (4 1/2–10 lb)

Location E. and southern Africa

Social unit Variable

Status Locally common

Huge ears and a small face with a pointed muzzle are the bat-eared fox's main external features, but its teeth, too, are very unusual. They are much smaller than those of a typical canid, and with up to 8 extra molars may number 48 – more than any other non-marsupial mammal. Its main diet is insects, especially termites and dung beetles. However, the breeding and social habits of this species are more typically fox-like.

Dhole
Cuon alpinus

Length	90 cm (35 in)
Tail	40–45 cm (16–18 in)
Weight	15–20 kg (33–44 lb)

Location S., E., and S.E. Asia

Social unit Group

Status Vulnerable

Also called the Asian red dog, this species is widespread, but with a shrinking overall range and declining numbers. It lives in territorial, day-active packs of 5–12, usually based on an extended family. The fur colour is evenly tawny or dark red, with a darker tail and lighter underparts; the legs are relatively short. The main prey is medium-sized hoofed mammals, supplemented by smaller creatures, fruits, and other plant food.

African wild dog
Lycaon pictus

Length	76–110 cm (30–43 in)
Tail	30–41 cm (12–16 in)
Weight	17–36 kg (37–79 lb)

Location Africa

Social unit Group

Status Endangered

Probably the most social canid, the African wild or hunting dog lives in packs of 30 or more adults and young. Only the dominant pair breed, producing a litter of 10–12 (range 2–19) after a gestation of 69–73 days. However, the whole pack cares for and protects the pups, regurgitating food for them until they develop hunting skills by about 12 months. The pack also cooperates to hunt very large prey such as wildebeest, zebra, and impala. This dog has long legs, and a lean build, with a relatively small head, large ears, and a short, broad muzzle. Unusually for a canid, it has only 4 toes on each foot. Its coat pattern is exceptionally variable, but the muzzle is usually black and the tail tip is white.

CONSERVATION

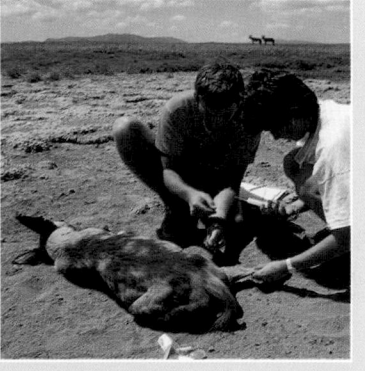

Once widespread across Africa, in many habitats, this wild dog is now reduced to scattered, fragmented populations. It is still persecuted, trapped, shot, and snared, and is also killed accidentally by road vehicles. It suffers both habitat loss and diseases (rabies, distemper) from domestic dogs. Survival depends on active conservation, including tracking pack movements by fitting radio collars (shown here).

large, rounded ears

dark muzzle and forehead stripe

PAINTED WOLF
This species' scientific name means "painted wolf" and aptly describes the coat pattern of variable patches and swirls in black, grey, yellow, and white.

MAMMALS

Canis lupus
Grey wolf

Length	1–1.5 m (3¼–5 ft)
Tail	30–51 cm (1–1¾ ft)
Weight	16–60 kg (35–130 lb)

Location North America, Greenland, Europe, Asia

Social unit Group

Status Vulnerable

The grey wolf is the largest wild member of the canid family and the ancestor of the domestic dog. Once the world's most widely ranging carnivore, its distribution has since been restricted by widespread human persecution

and habitat destruction. An intelligent and social animal, its survival and great success as a predator is dependent on its organization into packs – family groups that commonly consist of 8 to 12 wolves. Packs patrol territories, covering very wide areas, which they maintain by scent markings. The clearly defined hierarchy within a pack centres around a dominant breeding pair that usually mates for life. By hunting in packs the grey wolf is able to take a wide range of prey, including moose and caribou, that may be up to 10 times a wolf's weight.

FEEDING PACK
Having captured their prey, the pack members wait behind the dominant pair for access to the kill.

CARE OF THE YOUNG

During the breeding season, which lasts from January to April, the dominant female gives birth to between 4 and 7 pups. After about a month of suckling, pups emerge from the den to receive scraps of food regurgitated by their parents and other pack members. If their food supply has been plentiful, pups will have developed enough to travel with the pack after 3 to 5 months, and by the next breeding season some juveniles will have chosen to leave the pack altogether.

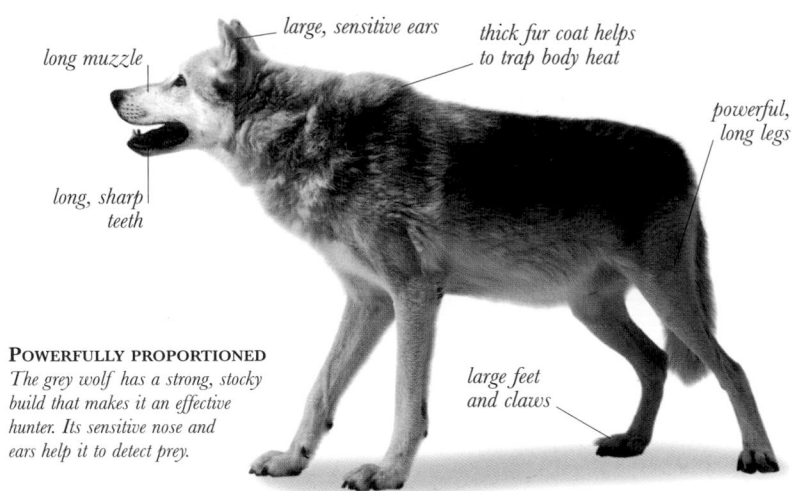

large, sensitive ears

long muzzle

thick fur coat helps
to trap body heat

powerful,
long legs

long, sharp
teeth

large feet
and claws

POWERFULLY PROPORTIONED
The grey wolf has a strong, stocky build that makes it an effective hunter. Its sensitive nose and ears help it to detect prey.

PREDATOR AND VICTIM
Although grey wolves can occur close to human settlements, their mythical reputation for ferocity has led to their near extermination. Today, most grey wolves live in remote areas, where they hunt herds of large deer or musk oxen.

CALL OF THE WILD
Grey wolves howl to announce their presence and to define and defend their territories. Heard at distances of up to 10km (6 miles), howling allows rival packs to stay well separated, avoiding confrontation.

MAMMALS

BEARS

PHYLUM	Chordata
CLASS	Mammalia
ORDER	Carnivora
FAMILY	Ursidae
SPECIES	8

THE BEAR FAMILY includes the world's largest terrestrial carnivore, the brown bear, which can stand up to 3.5m (11ft) tall. Bears have a heavy build, a large skull, thick legs, and a short tail. They are found throughout Eurasia and North America, and in parts of north Africa and South America, mainly in forests. Unlike most carnivores, bears rely heavily on vegetation as a food source.

ANATOMY

Bears are either large or medium-sized, and males are up to 20 per cent larger than females. Although the giant panda is one notable exception, most bears have a black, brown, or white coat, and many feature a white or yellow mark on the chest. Despite the fact that they have a keen sense of smell, bears' sight and hearing are less well developed, and this is reflected in their large snout and small eyes and ears. Most bears have lost the carnassial (shearing) function of the molar teeth. Instead, the molars are flat with rounded cusps, making them effective tools for grinding vegetation. Bears have large, strong paws – a single blow can often kill another animal – and long, non-retractable claws.

MOVEMENT

Compared with other carnivores, bears walk slowly and deliberately, with all 5 toes as well as their heels touching the ground (plantigrade gait). They can, however, move quickly if the need arises. When threatened or defending their territory, many bears stand on their back legs to increase their already considerable size. The majority of bears are agile climbers.

FEEDING

The diet of most bears consists of a mixture of meat (including insects and fish) and plant material (from roots and shoots to fruits and nuts). Only the polar bear lives exclusively on meat, while the giant panda (which is sometimes classified in a family with the lesser panda and is occasionally grouped with the raccoon family) is almost entirely herbivorous. Because bears depend on plants more than other carnivores, they spend more of their time feeding. Most forage during the day.

DENS AND DORMANCY

Many bears, especially those in cold regions, become dormant in winter. During this time, they retreat to a prepared den and live on reserves of body fat. This state differs from true hibernation (see p.87), which involves a drop in body temperature. Cubs are often born during dormancy. Since they have no fur, the newborn cubs are highly vulnerable and benefit from the snug environment created by their mother's body heat.

CONSERVATION

Only 2 of the 8 species of bear, the American black bear and the polar bear, are not endangered – and this is due to considerable conservation efforts. The main threats to bears come from hunting (mainly for their gall bladders, which are used in traditional medicine, especially in China) and destruction of habitat to clear land for logging or settlement.

PHYSICAL INTIMIDATION
Bears can be aggressive animals, particularly when competing with each other during the breeding season. When male brown bears come into conflict, they will often try to intimidate one another by making themselves look as large as possible, growling and displaying their teeth. Smaller individuals will usually give way to larger ones, but if a warning is ignored actual fighting will break out, often leading to serious injury or death.

Ursus maritimus

Polar bear

Length	2.1–3.4 m (7–11 ft)
Tail	8–13 cm (3 1/4–5 in)
Weight	400–680 kg (880–1,500 lb)

Location Arctic, N. Canada

Social unit Individual

Status Lower risk

Vying with the brown bear as the largest land-based carnivore, the male polar bear can weigh twice as much as a female. Its favoured habitat is a mix of pack ice, shoreline, and open water where seals are found. Some bears move 100 km (60 miles) inland in summer and vary their diet with birds' eggs, lemmings, lichens, mosses, and carrion such as caribou and musk-oxen. Mating occurs on sea ice in April–May. The pregnant female digs a den in snow or earth and gives birth to 2 cubs (range 1–4) from November to January. The cubs take solid food at 5 months but are not weaned for a further 2–3 years.

straight profile

longer neck than other bears

STALK OR STILL HUNT

The polar bear's chief prey of seals and an occasional walrus, are caught by 2 main hunting methods. In the stalk, the bear moves slowly nearer its prey, relying on its camou-flaging white coat and "freezing" if the seal looks up. It charges the last 15–30 m (50–100 ft) at up to 55 kph (34 mph). In the still hunt, the bear waits motionless next to a seal's breathing hole, and grabs the prey as it surfaces. The bear bites the seal's head and drags it a short distance for consumption.

PAW PADDLES

Polar bears swim readily across open water at up to 10 kph (6 mph). They paddle with the massive fore-paws – the rear legs trailing as rudders. The coat's hollow, air-filled guard hairs (see also below) aid buoyancy. When diving, the eyes remain open but the nostrils close as the bear holds its breath for up to 2 minutes, coming up stealthily beneath prey such as seabirds or surface-basking seals.

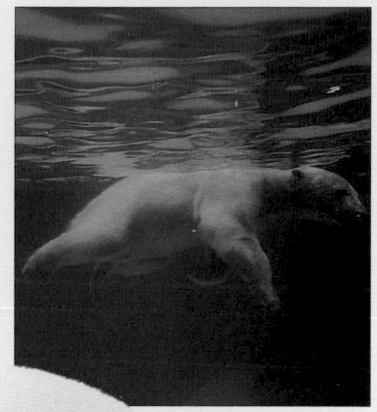

NOT QUITE WHITE
The polar bear's guard (outer) fur is creamy rather than pure white. Hollow and translucent, the guard hairs transmit the sun's heat internally down to their bases, where it is absorbed by the black skin. The dense underfur and thick blubber (fatty layer) under the skin aid insulation.

partially furred paw pads retain heat

Ursus americanus

American black bear

Length	1.3–1.9 cm (4 1/4–6 1/4 ft)
Tail	7–15 cm (2 3/4–6 in)
Weight	55–300 kg (120–660 lb)

Location North America, Mexico

Social unit Individual

Status Lower risk†

The American black bear is adaptable in habitat, but generally prefers forested country. Its powerful limbs and short claws tear open old logs to search for worms and grubs, and are also excellent for tree climbing, when this bear plucks fruits with its prehensile lips. The ears are larger and more erect than those of the brown bear, and it lacks a prominent shoulder hump. It sleeps in winter, which lasts up to 6 months in the north of its range. This bear may break into outbuildings or vehicles to obtain food left by humans, but it usually flees on confrontation.

BLACK TO BLUE BEARS
In the east black is the main fur colour, but in the west it may be cinnamon or yellow-brown, and on the Pacific coast grey-blue.

MAINLY VEGETARIAN

Some 95 per cent of this bear's diet is plant-based, including, according to season, roots, buds, shoots, fruits, berries, and nuts, which are often obtained by climbing. They may also become adept at hunting deer fawns and at catching fish.

Ursus thibetanus

Asiatic black bear

Length	1.3–1.9 m (4 1/4–6 1/4 ft)
Tail	Not recorded
Weight	100–200 kg (220–440 lb)

Location E., S., and S.E. Asia

Social unit Individual

Status Vulnerable

Similar to the American black bear (see left) in appearance and habits, the Asiatic black bear may spend up to half its time in trees. Its main foods include acorns, beech, and other nuts, fruits such as cherries, bamboo shoots and leaves, grasses, herbs, grubs, and insects such as ants. Where its natural forest habitat has been farmed and become fragmented, this bear may raid maize and other crops and, on occasions, has caused human fatalities. Eight months after mating the female gives birth in her winter den to 1–3 (usually 2) cubs. The Asiatic black bear is hunted for its body parts (especially the gall bladder), which are used in Asian cuisine and medicines.

whitish yellow chest patch gives alternative name of moon bear

strong legs, adept at bipedal walking

MAMMALS

Helarctos malayanus
Sun bear

Length 1.1–1.4 m	
	(3½–4½ ft)
Tail Not recorded	
Weight 50–65 kg	
	(110–145 lb)

Location S.E. Asia	Social unit Individual
	Status Endangered

The only truly tropical bear, the sun bear is an elusive, nocturnal, little-known omnivore of hardwood lowland forest. Its sleek, smooth fur varies from black to grey or rusty. It is paler on the muzzle, which is comparatively short. Its stocky, dog-like body proportions and small size have led to the local name of dog bear. The sun bear spends much time in trees, even sleeping in a rough nest of bent-over or broken branches. It eats a range of fruits, shoots, eggs, small mammals, grubs, honey (another name is honey bear), and varied plant food. Habitat loss, as forests are logged and converted to agriculture, is the major threat to the species, and it may raid crops, notably palm-tree plantations for the shoots, leading to persecution by farmers.

white to reddish "sun" chest patch varies from a U shape to a circle or irregular spot

SMALLEST BEAR
The smallest bear species, the sun bear also has the shortest fur. If seized by a tiger or other predator, the loose skin around its neck allows it to turn and fight.

CLIMBING CLAWS

The sun bear has extremely long, curved claws, an adaptation for tree climbing. It also hugs the trunk with its front limbs and grips with its teeth, to haul itself up. The claws are also used to dig for worms and insects, and to tear up bark or old logs to expose and extract termites and remove honey from wild bees' nests.

LONG TONGUE

The sun bear's tongue can protrude 25cm (10in) to extract grubs, honey, and similar food from holes and crevices. It may also place each front paw alternately in a termite nest; the occupants crawl onto the paw, and the bear then licks them off.

Melursus ursinus
Sloth bear

Length 1.4–1.8 m	
	(4½–6 ft)
Tail 7–12 cm	
	(2¾–4¾ in)
Weight 55–190 kg	
	(120–420 lb)

Location S. Asia	Social unit Variable
	Status Endangered

This small to medium-sized bear has a stocky body and short, powerful limbs. It can survive in a variety of habitats, including thorn scrub, grassland, and forest, providing its three major foods – ants, termites, and fruits – are present (it also eats honey and eggs). Its long, non-

white chest mark varies from a U shape to a Y or O

SHAGGY BEAR
The sloth bear is distinguished by its long, rough fur, especially around the ears, rear neck, and shoulders. Colour varies from black to brown or reddish; the muzzle is usually much paler or even white.

ON THE SCENT

Bears locate food mainly by smell, sniffing with their long, mobile snout. The sloth bear specializes in ants and termites, tearing open their nests in soil, old logs, or trees with its foreclaws, which are 8cm (3in) long. Closing its nostrils and pursing its lips, it sucks the insects through the gap formed by its missing upper incisor teeth. The sucking noises it makes can be heard up to 100m (330ft) away.

retractile foreclaws (see panel above) do not, in this bear, permit effective tree climbing. Like other bears, the adult sloth bear is mainly solitary except during the mating season (June–July). However, brief groupings of 5–7 bears have been observed, even without local concentrations of food to attract them. The female occupies a natural hollow or digs one and usually has 2 cubs (in November–January). The cubs stay in the den for 2–3 months, ride on the mother's back clinging to her long fur for another 6 months, and become independent after about 2 years. Threats include habitat loss, poaching for body parts (used in traditional medicines), persecution, and cub capture for training and performance.

STANDING SURVEY

Like other bears, the sloth bear can stand just on its back legs. Once believed to be an aggressive posture, in fact the bear is gaining a clearer view of its surroundings and, more importantly, scenting the air to assess possible food or danger. Injuries to humans, mainly clawings, are usually the result of surprise encounters.

Spectacled bear

Tremarctos ornatus

Length	1.5–2 m (5–6 1/2 ft)
Tail	7–12 cm (2 3/4–4 3/4 in)
Weight	140–175 kg (310–390 lb)
Location W. South America	
Social unit Individual	
Status Vulnerable	

South America's only bear, and its largest land mammal after the tapirs, the spectacled or Andean bear is also the most arboreal of the family. Playing a prominent role in the folklore and mythology of many Andean peoples, its multitude of local names include oso achupayero (bromeliad-eating bear), yura mateo (white-fronted bear), yanapuma (black puma), and ucucu (after one of its rare vocalizations). The species once occupied habitats from coastal desert to high-elevation grassland, but human presence increasingly limits it to cloud forests at 1,000–2,700m (3,300–8,900ft). Mainly vegetarian, its massive jaw muscles and cheek teeth grind the toughest plants. For example, low-growing plants are simply torn up and chewed, despite cactus spines or leaf barbs on puya bromeliads. A common tree-feeding technique is to edge along a branch,

CUBS IN THEIR DEN
Newborn spectacled bear cubs, like other baby bears, are tiny, each weighing only 325g (12oz). Most are born between December and February.

bending other branches to bring fruits within reach. These bears may also create a simple tree platform, 5m (16ft) or more across, for feeding and resting. Most matings occur from April to June, but may happen at almost any time so that the birth corresponds with greatest food availability. The pair stay together for 1–2 weeks. The cubs' eyes open at 42 days and they may leave the den, in a hollow among rocks or tree roots, by 3 months. The cubs stay with the mother for 2 years, learning about feeding methods, food types, and threats. As with other bears, the male plays no part in rearing the cubs and, if he encounters them by chance, may attack and kill them.

CROPS AND CONFLICT

The diet of the spectacled bear, one of the most herbivorous of bears, includes a huge range of fruits, as well as bromeliads, bulbs of wild orchids and similar flowers, palm shoots and leaf stalks, and, in drier areas, grass stems and cacti. Animal foods include insects, birds, eggs, small mammals, and carrion. Raids on crops, especially maize, and occasional attacks on livestock provoke revenge killings by farmers.

SPECTACLES
The creamy white eye markings vary from complete circles to "eyebrows" above or "teardrops" below, and allow identification of individual bears.

usually black, occasionally red-brown, fur

Giant panda

Ailuropoda melanoleuca

Length	1.6–1.9 m (5 1/4–6 3/4 ft)
Tail	10–15 cm (4–6 in)
Weight	70–125 kg (155–280 lb)
Location E. Asia	
Social unit Individual	
Status Endangered	

Instantly recognizable as the world-wide symbol of conservation, the giant panda's own survival is still far from secure. It has a highly restricted diet – 99 per cent bamboo, using different parts of the 30 or more bamboo species, taking new shoots in spring, leaves in summer, and stems in winter. Carrion, grubs, and eggs are eaten when available. Normally solitary, the panda feeds mainly at dawn and dusk, and sleeps in a bamboo thicket. It marks its home ranges with anal gland scents, urine, and claw scratches, avoiding confrontation by using overlapping areas at different times. The female, four-fifths the male's size, indicates readiness to mate by moans, bleats, and barks (11 distinct panda calls have been identified). Males gather, chase, and fight each other for the female. One to 2 cubs are born in

a den in a hollow tree or rocky cave after 45 days' gestation. Only 15cm (6in) long, 100g (3 5/8 oz) in weight, the newborn cubs are nearly naked.

coarse, oily guard (outer) hairs up to 10cm (4in) long

front limbs more muscled than rear limbs, for climbing

BLACK ON WHITE
The giant panda is white with black ears, oval-shaped eye patches, nose, shoulder "saddle", and limbs. It has erect ears, a broad face, and small eyes.

CONSERVATION

More than 100 pandas are kept in captivity but only 30 per cent of cubs born there survive more than 6 months. In the wild, the panda cub stays in the den for 4–6 weeks, then rides on its mother's back, trots with her by 6 months, and is independent by 18 months. Most wild populations are so small and fragmented as to be genetically non-viable and are still at risk from poaching. The future may lie in habitat corridors and local support.

SIXTH "FINGER"

The giant panda handles bamboo with great dexterity due to an extension of the sesamoid bone in the wrist, which projects as a pad-like "false thumb". This can flex and oppose the true thumb (first digit) to grip stems and leaves.

MAMMALS

Ursus arctos

Brown bear

Length 2–3 m	
(6¹/₂–9³/₄ ft)	
Tail 5–20 cm	
(2 in–8 in)	
Weight 100–1,000 kg	
(220–2,200 lb)	

| Location N. and N.W North America, N. Europe, Asia | Social unit Individual |
| | Status Lower risk† |

The brown bear enjoys the widest distribution of all bear species and varies widely in size depending on its food and habitat. Large areas of open wilderness are important to its survival, which explains why populations can be found in isolated areas such as parts of Alaska and the Yukon, while habitat destruction in the rest of North America and Europe has seen a drastic reduction in its numbers there. The brown bear's distinctive features are its shoulder hump of muscle, and long claws that help it to dig for roots and bulbs. It can stand upright on its hind paws in order to identify a threat or a food source. Although mainly herbivorous, brown bears will readily eat meat when it is available. To avoid winter food shortage, they can den up in dug-outs in hillsides, or in brush, for as long as 6 months, during which period the female also gives birth to her cubs. The lifespan of the brown bear is about 25 years in the wild and longer in captivity.

BROWN BEAR SUBSPECIES

■ KODIAK ■ GRIZZLY ■ EURASIAN

Several races of brown bear are commonly recognized: grizzly bear, Kodiak bear, Alaskan bear, Eurasian brown bear, Syrian bear, Siberian bear, Manchurian bear, and Hokkaido bear. However, their taxonomy is disputed, and rather than being "true" subspecies, they may simply represent size variations due to disparities in food supply.

KODIAK BEAR
Weighing up to 780kg (1,720lb), the impressive Kodiak bear, Ursus arctos middendorfi, *is the largest of the subspecies.*

EURASIAN BROWN BEAR
This bear, Ursus arctos arctos, *is the smallest member of the species. Rapid loss of habitat has now restricted it to small pockets of mountain woodland.*

GRIZZLY BEAR
This subspecies, Ursus arctos horribilis, *gets its common name from its "grizzled" coat, the hair being lighter at the tips than at the base.*

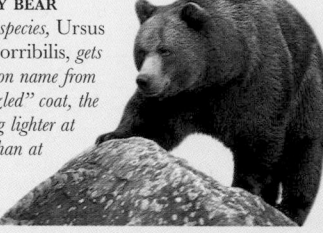

prominent shoulder hump

concave profile

powerful limbs

coat usually dark brown, but varies from blonde to black

long, non-retractable front claws

dense coat

SALMON HARVEST

Brown bears are comfortable in water and may wait for hours at waterfalls or in the shallows of a stream ready to dive on their prey. As spawning salmon swim upstream the brown bear pounces, delivering a crushing bite with its powerful jaws or a stunning blow from one of its large, clawed paws. Salmon is a vital source of protein for coastal populations of brown bear, which are usually the largest of the species; once caught, the fish rarely escape the bear's grasp.

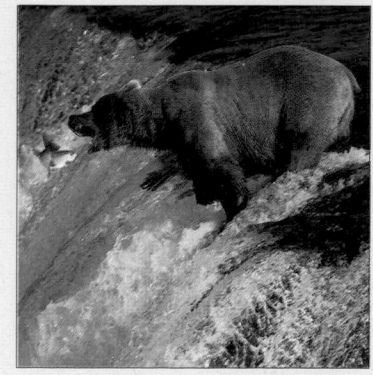

POWERFUL BUILD
Brown bears are large, powerfully built animals. Although there is little difference in body length between the sexes, males can be up to twice as heavy as females, which have a smaller, lighter frame. Both sexes feed intensively from spring to autumn in order to put on weight in preparation for winter sleep.

AGGRESSIVE BEHAVIOUR
Due to its power, size, and unpredictable behaviour, the brown bear has long been considered a threat to humans and livestock. Grizzly bears evolved in open habitat and may act aggressively when defending themselves as they have little opportunity to find the safety of any cover. A mother with young cubs is a particularly dangerous animal to encounter. Nevertheless, bears will usually avoid coming into contact with humans.

RACCOONS AND RELATIVES

PHYLUM	Chordata
CLASS	Mammalia
ORDER	Carnivora
FAMILY	Procyonidae
SPECIES	20

CLASSIFICATION NOTE

Many authorities still group the lesser panda with raccoons and relatives (family Procyonidae), but recent studies suggest that it is quite distinct and should be classified in the family Ailuridae, with the giant panda. It is treated here as a member of the informal "raccoon group".

THIS FAMILY is typified by the raccoons, which are well known for their mischievousness and dexterity. They will boldly approach humans for food and can use their highly mobile hands for opening doors and the like. In addition to raccoons, the family also includes coatis, ringtails, kinkajous, olingos, and the lesser (or red) panda (but see the Classification note, left). Common characteristics include distinctive tail rings and a dark mask pattern on the face. The animals in this family are typically arboreal, and are found in the forests of the Americas. The only exception is the lesser panda, which lives in western China and the Himalayas.

ANATOMY

Members of the raccoon family are all medium sized and short legged, and have a flat-footed (plantigrade), bear-like gait. They commonly have a pointed snout, a relatively long body, a broad face, round or pointed ears, and brown or grey fur (except the lesser panda, which has a short face and red fur). All species have short claws, and raccoons feature front paws developed into sensitive, mobile hands. Highly arboreal species, such as ringtails and kinkajous, have the ability to rotate their ankle joints and hang by their feet when feeding or descending tree trunks. Kinkajous are able to hang by their tail alone.

FEEDING

Most species in this family are omnivorous, and their diet varies with location, season, and the availability of food sources. They will eat fruits, roots, shoots, and nuts, as well as insects and small vertebrates, such as birds, amphibians, and reptiles. Kinkajous mostly eat fruits and are important seed-dispersers in their rainforest habitat. Raccoons use their dextrous hands to reach into streams to feel for crustaceans, fish, and other prey. Their unfussy palate also brings them into urban areas, where they raid garbage bins and fearlessly solicit food from house-holders. The lesser panda mainly eats bamboo leaves, thereby avoiding competition with the giant panda, which is found in some of the same regions and eats mostly bamboo stems.

OPPORTUNISTIC FEEDING
Raccoons and relatives eat a great variety of food. This common raccoon has used its mobile hands and finely tuned hunting skills to capture a fish.

Ailurus fulgens

Lesser panda

Length	50–64 cm (20–25 in)
Tail	28–50 cm (11–20 in)
Weight	3–6 kg (6 1/2–13 lb)

Location S. to S.E. Asia

Social unit Individual

Status Endangered

In addition to bamboo leaves and shoots, the lesser panda eats other grasses, roots, fruits, and also grubs, small vertebrates such as mice and lizards, and birds' eggs and chicks. It is mainly nocturnal and solitary, but forms pairs during the mating season, and offspring stay with their mother for up to a year. The panda scent-marks its territory with droppings, urine, and powerful musk-like secretions from the anal glands. It communicates by short whistles and squeaks. Studies in captivity show the gestation period is probably 90 days plus a variable time of delayed implantation. Prime habitats are dense temperate mountain forests, at 1,800–4,000m (6,000–13,200ft).

TREE CLIMBER

The lesser panda has partly retractable claws and climbs well. It uses trees not only for feeding but also to escape ground-based predators, and to sunbathe high in the canopy during winter. The female's nest may be in a tree-hole, lined with leaves, moss, and other soft plant material, where she rears her 1–5 (usually 2) offspring. Other nesting sites are branch forks, tree roots, and bamboo thickets.

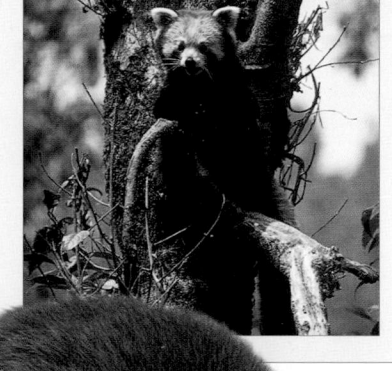

alternating light and dark rings on tail

CHESTNUT COLOURING
The lesser or "red" panda is red-brown or chestnut with almost white ear rims, cheeks, muzzle, and spots above the eyes. There are also brown, facial "teardrop" stripes.

Bassariscus astutus

Ringtail

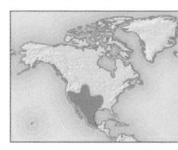

Length	30–42 cm (12–16 1/2 in)
Tail	31–44 cm (12–17 1/2 in)
Weight	0.8–1.5 kg (1 3/4–3 1/4 lb)

Location C. and W. USA to S. Mexico

Social unit Individual

Status Unconfirmed

Also called the ringtailed cat or cacomistle, this slender, agile predator has a tail boldly ringed in black and white. The upper body is grey-brown or buff, with black eye-rings and white muzzle and "eyebrows". Nocturnal and solitary, this procyonid hunts small birds, mammals, and reptiles, and also forages for grubs, raids birds' nests, and eats fruits and nuts. It marks its territory with droppings and urine and defends it against others of the same sex. Gestation is 51–60 days, with an average litter size of 2–3.

MUSTELIDS

PHYLUM	Chordata
CLASS	Mammalia
ORDER	Carnivora
FAMILY	Mustelidae
SPECIES	67

OF ALL THE CARNIVORES the mustelid family is the most diverse and contains the highest number of species. The group includes terrestrial forms (such as skunks), arboreal species (such as martens), burrowing species (such as badgers), semi-aquatic species (such as minks), and fully aquatic species (such as otters). With such a range of life-styles, the main physical link between species is short legs and an elongated body. Mustelids are found throughout Eurasia, Africa, and the Americas. Although mostly occurring in forest or bush, they have adapted to populate almost every habitat type.

ANATOMY

All mustelids have short ears and 5 toes on each foot (most carnivores have only 4 on each back foot); most have a short snout, a long braincase, a long tail, and long, non-retractable, curved claws. Body form tends to be either slender (as in weasels) or heavy and squat (as in badgers). Slimmer forms have a flexible backbone and employ a scampering, bounding gait; stocky forms move with a rolling shuffle. Mustelids have a fur coat that consists of warm underfur and longer, sparser guard hairs. Coat colour varies from dark brown or black to spotted or striped. In some species, contrasting patterns – for example the black-and-white stripes of skunks – are thought to convey a warning to predators.

Most mustelids have an excellent sense of smell for tracking prey and for communication. All have scent glands in the anus, which generally produce an oily, strong-smelling liquid known as musk. This is secreted into the faeces, which are used to mark territory.

FEEDING

Reflecting their diverse lifestyles, mustelids have a varied diet. Weasels and stoats, for example, are agile and aggressive and are capable of killing prey larger than themselves, such as rabbits. Some otter species actively hunt fish; others eat mainly shellfish collected by feeling along river beds with their sensitive paws. The sea otter cracks open abalone shells by floating on its back on the surface and then hitting the shell against a rock balanced on its chest. Martens, which are arboreal, catch and eat squirrels and birds, while the zorilla, a burrowing species, catches small rodents, lizards, and insects.

REPRODUCTION

Female mustelids do not ovulate automatically. Instead, ovulation is stimulated by copulation, which may last up to 2 hours. This lengthy procedure does expose pairs to predators, but fertilization is almost guaranteed. In many species, the fertilized egg remains dormant and does not implant until conditions are favourable. Therefore, although the gestation period is only 1–2 months, pregnancy may last more than 12 months. Most mustelids are solitary, except during the breeding season.

FUR TRADE

Mustelid fur is much sought after because it is soft, dense, and water repellent. Of all the mammal families, the mustelid family has more species with fur valued by humans than any other. Most highly prized is the fur of minks, the sable, and the stoat (when it has its white winter coat). The American mink is farmed for its fur, and in Europe these have escaped in a number of locations and established wild populations, often at the expense of the indigenous European mink. The quest for sea otter fur has brought this species to the brink of extinction, although recently numbers have increased in North America.

AQUATIC HUNTER
The American mink is a voracious hunter, as is typical of mustelids. This species, like all minks and otters, is an excellent swimmer. It can remain submerged for distances of up to 30m (98ft) in pursuit of fish, which forms an important part of its diet. Adaptations to a semi-aquatic lifestyle include partially webbed feet and a thick, waterproof coat.

Mustela erminea
Stoat

Length 17–24cm (6¹/₂–9¹/₂in)	
Tail 9–12cm (3¹/₂–4³/₄in)	
Weight 60–200g (2¹/₈–7oz)	
Location North America, Greenland, Europe to N. and E. Asia	**Social unit** Individual
	Status Locally common

This extremely widespread mustelid adapts to many habitats, and hunts varied prey from mice, voles, and small birds to rats and, often, rabbits. It has the typical mustelid's long, slim, flexible body, pointed muzzle, small eyes and ears, and short legs. In the north of its range, the summer coat of russet to ginger-brown above, demarcated from cream or white below, turns all-white for winter, when the stoat is often called the ermine. However, the tail tip is always black.

Mustela nivalis
Weasel

Length 16.5–24cm (6¹/₂–9¹/₂in)	
Tail 3–9cm (1¹/₄–3¹/₂in)	
Weight 35–250g (1¹/₄–9oz)	
Location North America, Europe to N., C., and E. Asia	**Social unit** Individual
	Status Locally common

One of the smallest and most widespread mustelids, this weasel has a small, flattened head hardly wider than the neck, allowing it to enter mouse burrows. It also eats voles, other small rodents, and birds. It is active day and night and must consume one-third of its body weight each day to survive. Overall size varies greatly across the huge range and between sexes, with the male being a quarter as long and up to twice the weight of the female. A small

stoat and large weasel are distinguished by the former's black-tipped tail. Like stoats, weasels in northern lands turn white in winter, for camouflage in snow. Also like the stoat, this species lives alone, making several nests lined with grass and prey fur or feathers, in a crevice, tree root, or abandoned burrow. After a gestation period of 34–37 days, the female produces a litter of 1–7 (average 5). She cares for them for 9–12 weeks.

russet or chocolate-brown upperparts, legs, and tail

Mustela lutreola
European mink

Length 30–40cm (12–16in)	
Tail 12–19cm (4³/₄–7¹/₂in)	
Weight 500–800g (18–29oz)	
Location Europe	**Social unit** Individual
	Status Endangered

The European mink is similar in habits and appearance to the American mink (below), although slightly smaller. It hunts small prey such as birds, mammals, frogs, fish, and crayfish on land and in water. The fur is dark brown to almost black with a narrow white edging to the lips. This endangered mink is being bred in captivity for release.

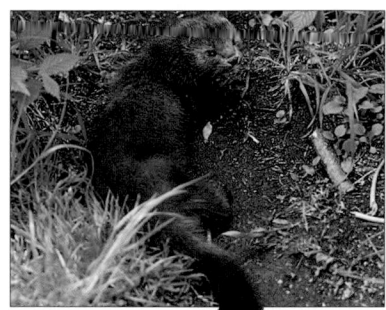

Mustela putorious
European polecat

Length 35–51cm (14–20in)	
Tail 12–19cm (4³/₄–7¹/₂in)	
Weight 0.7–1.5kg (1¹/₂–3¹/₄lb)	
Location Europe	**Social unit** Individual
	Status Locally common

Regarded as the domestic ferret's ancestor, the polecat has long buff to black hairs with cream or yellow underfur visible between them, and a "mask" across the face. It runs, climbs, and swims well. If threatened, it releases very pungent-smelling anal gland secretions. As in many similar mustelids, male neighbours defend separate territories, as do females, but male territories usually overlap female ones.

long, sinuous body

Mustela vison
American mink

Length 30–54cm (12–21¹/₂in)	
Tail 14–21cm (5¹/₂–8¹/₂in)	
Weight 0.7–2kg (1¹/₄–4¹/₂lb)	
Location North America, S. South America, N. and W. Europe, N. and E. Asia	**Social unit** Individual
	Status Locally common

This opportunistic predator uses its partly webbed feet to hunt on land and in water for a variety of small animals including rats, rabbits, birds, frogs, fish, and crayfish. As its eyesight is not well adapted for underwater vision, prey is located at the surface and then pursued. In her nesting den among tree roots or rocks, the female suckles 3–6 young for 5–6 weeks. In its natural range the American mink was once trapped by the thousand; in about 1900 it was taken to South America, Europe, and Russia for fur farming. Escapes established wild populations and are regarded as threats to local wildlife – not only to prey but also to rival predators.

COLOUR VARIATION
Most American minks are dark brown to almost black, but approximately 1 in 10 is grey-blue.

Mustela nigripes
Black-footed ferret

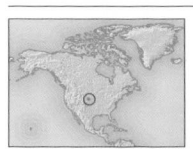

Length 38–41cm (15–16in)	
Tail 11–13cm (4¹/₄–5in)	
Weight 900–1,000g (32–36oz)	
Location Reintroduced to C. USA	**Social unit** Individual/Pair
	Status Extinct in wild

Exceptionally rare, partly due to extermination of its almost exclusive prey of prairie dogs (see p.147), this species was considered extinct in the wild. Captive breeding and release have re-established a few in Wyoming, USA. It chases its prey down burrows and also sets up a nest there, the female giving birth to 3–6 kits.

black tail tip

MINK TERRITORIES

American mink are territorial, with typical territory sizes of 1–3km (¹/₂–1³/₄ miles) across for the female and 2–5km (1¹/₄–3 miles) for the male. Each marks its territory with urine, droppings, and scents from the anal glands. From February to April, males try to mate with females from adjacent territories.

Martes foina
Beech marten

Length	42–48 cm (16 1/2–19 in)
Tail	26 cm (10 in)
Weight	1.5–2.5 kg (3 1/4–5 1/2 lb)

Location Europe	Social unit Individual
	Status Common

This marten has adapted to human habitation, and hunts around farms and other buildings for small animals

of all kinds. It also scavenges in refuse and eats fruits. It has a relatively short body (for a marten), long legs, and a wide, wedge-shaped head. The fur is brown, with a paler "bow tie" throat patch. Typically solitary, the den is a rocky crevice, tree-hole, or old rodent burrow. The territory of up to 80 hectares (200 acres) is marked by droppings. However, near towns, the marten may nest in an outbuilding and be less territorial, foraging with others of its kind. As in many mustelids, there is a period of delayed implantation after mating (230–275 days) followed by 30 days' gestation. Litter size averages 3–4.

bushy tail

large, rounded ears

Martes martes
European pine marten

Length	40–55 cm (16–22 in)
Tail	20–28 cm (8–11 in)
Weight	0.9–2 kg (2–4 1/2 lb)

Location Europe to W. and N. Asia	Social unit Individual
	Status Lower risk†

Almost cat-like in its movements, this sharp-clawed mustelid climbs well and nests in a tree-hole or old squirrel drey (nest). It has a long, slender body and the fur is chestnut to dark brown, with a cream to orange throat "bib". Its bushy tail is used for balance in

branches. Although extremely agile in trees and capable of great leaps, the pine marten takes most prey on the ground, feeding on small rodents, birds, insects, and fruits.

large, rounded ears

cream to orange fur on throat and chest

Martes zibellina
Sable

Length	32–46 cm (12 1/2–18 in)
Tail	14–18 cm (5 1/2–7 in)
Weight	0.9–2 kg (2–4 1/2 lb)

Location N. and E. Asia	Social unit Individual
	Status Endangered†

Few mustelids suffered more hunting for fur than the sable, which is now protected in some regions. It has a brown-black coat, with an indistinct, paler brown throat patch, and the head is wide, with rounded ears.

Compared to other species, its legs are longer, its tail bushier, and its sharp claws partly retractable. Fast and agile on the ground, the sable climbs well but rarely. It has the typical marten diet of small animals and fruits, and takes over an old burrow for its main nest; it also has various temporary dens.

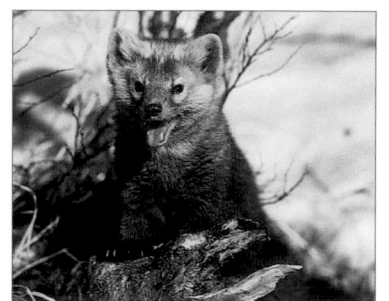

Martes pennanti
Fisher

Length	47–75 cm (18 1/2–30 in)
Tail	30–42 cm (12–16 1/2 in)
Weight	2–5 kg (4 1/2–11 lb)

Location Canada to N. USA	Social unit Individual
	Status Lower risk†

Despite its name, the fisher hunts ground prey, from mice to porcupines, as well as scavenging on carcasses. It makes dens in rocks, roots, bushes, and stumps, or high in trees, where it prefers to raise young. Hunted for its long, dense fur, fisher numbers have recovered in some areas but new threats include disturbance by logging.

Vormela peregusna
Marbled polecat

Length	33–35 cm (13–14 in)
Tail	12–22 cm (4 3/4–9 in)
Weight	700 g (25 oz)

Location S.E. Europe to W., C., and E. Asia	Social unit Individual
	Status Vulnerable

This polecat is black with variable white or yellow spots and stripes, and the typical black-and-white "face mask". A species of steppes and other dry, open regions, it hunts at twilight and at night for a variety of small animals, especially hamsters. The den is an old, enlarged rodent burrow where the female gives birth to a litter of 4–8 young. When threatened, the marbled polecat arches its head and curls its tail over its body, and may release its pungent anal gland odour.

Martes flavigula
Yellow-throated marten

Length	48–70 cm (19–28 in)
Tail	35–45 cm (14–18 in)
Weight	1–5 kg (2 1/4–11 lb)

Location S., S.E., and E. Asia	Social unit Individual
	Status Endangered

Similar to the European pine marten (see left) in its climbing agility and long leaps, the yellow-throated marten is larger and has long, dense fur and a bushy tail. Its colour varies from dark orange-yellow to brown, with a yellow or white throat patch. It feeds on small rodents, birds, insects, and fruits, on the ground and in trees. Pairs or family groups may catch young deer.

Poecilogale albinucha
African striped weasel

Length	25–35 cm (10–14 in)
Tail	15–23 cm (6–9 in)
Weight	225–350 g (8–13 oz)

Location C. to southern Africa	Social unit Individual
	Status Lower risk†

This exceptionally long, slim mustelid is black apart from a white patch running from the forehead over the head to the neck, where it splits into 2 white stripes, which each divide again into 2, along the back and sides of the body. All 4 stripes unite at the tail, which is white and bushy. The African striped weasel digs well with its long-clawed front feet, and eats almost exclusively mice and similar small rodents, as well as occasional birds and eggs. In defence, it can squirt a pungent spray from its anal glands more than 1m (3 1/4 ft).

Galictis vittata

Greater grison

Length 47–55 cm (18 1/2–22 in)	
Tail 14–20 cm (5 1/2–8 in)	
Weight 1.5–2 kg (3 1/4–4 1/2 lb)	

Location S. Mexico, Central and South America

Social unit Individual

Status Lower risk†

Even for a mustelid, the greater grison is long and sinuous, with a slim, pointed head and flexible neck; the tail, however, is relatively short. Coloration is all grey, apart from a white, U-shaped stripe running across the forehead, passing just above each eye and over each ear, tapering towards the shoulder. There is also black below on the muzzle, throat, chest, and front legs. The grison lives alone or in a male–female pair, is an agile runner, swimmer, and climber, and eats various small animals (even worms) and fruits. Its sounds include snorts, growls, screams, and barks.

black underparts

Ictonyx striatus

Zorilla

Length 28–38 cm (11–15 in)	
Tail 20–30 cm (8–12 in)	
Weight 1.5 kg (3 1/4 lb)	

Location W. to E. and southern Africa

Social unit Individual

Status Locally common

Resembling a small skunk (see p.200), the zorilla (African or striped polecat) is jet black, apart from 4 pure white stripes that fan out from the head, along the back and sides, to the tail base. The fluffy tail is mottled white and grey. If threatened, the zorilla hisses, screams, and raises its tail to spray noxious fluid from its anal glands. It digs with its long-clawed, front feet for grubs, mice, and other small creatures.

Gulo gulo

Wolverine

Length 65–105 cm (26–41 in)	
Tail 17–26 cm (6 1/2–10 in)	
Weight 8–14 kg (18–31 lb)	

Location Canada, N.W. USA, N. Europe to N. and E. Asia

Social unit Individual

Status Vulnerable

The second biggest mustelid after the giant otter, the male wolverine is a third larger than the female. The species is stocky, strong, and bear-like, with extremely powerful jaws that can crunch frozen meat and bone of caribou or similar carcasses in winter; parts of the carcass may be buried for later use. As well as scavenging, the wolverine runs down victims even on soft snow, with its broad feet and muscular limbs. Also called the glutton, it takes various prey, from deer and hares to mice, as well as birds and eggs, and seasonal fruits. Active all year, it covers up to 50km (31 miles) daily. The long, dense fur is blackish brown, with a pale brown band along each side, from the shoulder and flank over the rump to the base of the tail. There may be a white chest patch. The wolverine lives in a den among roots or rocks or dug 2m (6 1/2 ft) into a snowdrift. The average litter is 2–3, and the female suckles her young for 8–10 weeks.

pale bands on sides and rump

stocky, bear-like build

variable white chest patch

Mellivora capensis

Honey badger

Length 60–77 cm (23 1/2–30 in)	
Tail 20–30 cm (8–12 in)	
Weight 7–13 kg (15–29 lb)	

Location W., C., E., and southern Africa, W. and S. Asia

Social unit Variable

Status Lower risk†

The heavily built honey badger or ratel is silver-grey on the upper head, back, and tail, and black or dark brown elsewhere. Its long front claws are well adapted for digging. The prey of this mustelid includes worms, termites, scorpions, porcupines, and hares. It also cooperates with the black-throated honeyguide bird (see p.334), which leads it to bees' nests. The badger opens the nest to provide honey and grubs for both.

Meles meles

Eurasian badger

Length 56–90 cm (22–35 in)	
Tail 12–20 cm (4 3/4–8 in)	
Weight 10–12 kg (22–26 lb)	

Location Europe to E. Asia

Social unit Group

Status Locally common

One of the few group-dwelling mustelids, the badger has a small, pointed head and short neck, widening to a powerful body, with short, strong limbs and a small tail. The underparts are black, the main body and tail grey, and the face and neck white, with a black stripe on each side, from the nose over the small eye to the ear. The sight of the Eurasian badger is poor, its hearing better, but it has a keen sense of smell. Nocturnal and omnivorous, this badger varies its diet with season and availability. Earthworms are a staple, supplemented by insects and grubs, as well as frogs, lizards, small mammals, birds and their eggs, carrion, fruits, and other plant matter. Inside the group's sett (see panel), the nests are lined with bedding of dry grass, leaves, and moss, which is changed regularly. After up to 10 months of delayed implantation (when the eggs are fertilized but do not immediately implant into the womb lining), and a gestation period of 7 weeks, the female gives birth to up to 6 cubs. She suckles them for 10 weeks.

COMMUNAL LIVING

Each badger clan averages 6 members: usually a dominant boar (male) with one or more sows (females) and cubs. Their sett – an extensive system of tunnels and chambers – may have 10 or more entrances. Setts are kept clean, and are maintained and enlarged over many generations. Badgers range over a territory of 50–150 hectares (125–370 acres), which they defend against other clans.

VARIABLE STRIPE
The badger's distinctive striped face varies slightly between individuals. It may allow clan members to recognize each other or act as camouflage.

short tail

grey upper body

black underparts

MAMMALS

Arctonyx collaris
Hog-badger

Length 55–70 cm (22–28 in)	
Tail 12–17 cm (4³/₄–6¹/₂ in)	
Weight 7–14 kg (15–31 lb)	

Location S.E. and E. Asia | Social unit Group
| Status Lower risk†

Named after its looks and its tendency to root and snuffle in the ground like a pig, the hog-badger is an excellent digger, using the very long claws on its front feet. In appearance and habits it resembles the Eurasian badger (see p.199), being a seasonal omnivore with a very wide range of foods, from fruits and honey to worms and mice. The whitish face has 2 black stripes on either side. The upper body is grey tinged with yellow, and the underparts are black.

Taxidea taxus
American badger

Length 42–72 cm (16¹/₂–28 in)	
Tail 10–16 cm (4–6¹/₂ in)	
Weight 4–12 kg (8³/₄–26 lb)	

Location S.W. Canada to USA, N. Mexico | Social unit Individual
| Status Common

Smaller but similar in appearance to the Eurasian badger (see p.199), the American species is, however, solitary. It prefers open country with loose soil, and feeds mainly on small mammals such as prairie dogs and other ground squirrels, which it digs from their burrows. The shaggy fur is grizzled grey on the upperparts, yellow-white below, and black on the limbs.

Melogale personata
Burmese ferret-badger

Length 33–43 cm (13–17 in)	
Tail 15–23 cm (6–9 in)	
Weight 1–3 kg (2¹/₄–6¹/₂ lb)	

Location S.E. and E. Asia | Social unit Individual
| Status Lower risk†

Smaller than other badgers, the bushy-tailed Burmese ferret-badger has a long, flexible body resembling a ferret's. It feeds at night on insects,

snails, small mammals, birds, and plant matter such as fruits, and sometimes climbs trees. This badger is dark grey or brown, with white or yellow patches on the cheeks and between the eyes. It lives in a burrow and – like most badgers – is fearless if threatened, releasing an offensive anal gland odour and biting hard.

white or yellow stripe on top of the head and neck

Mydaus marchei
Palawan stink badger

Length 32–46 cm (12¹/₂–18 in)	
Tail 1–4 cm (3/8–1¹/₂ in)	
Weight 3 kg (6¹/₂ lb)	

Location Palawan and Busuanga islands | Social unit Individual
| Status Vulnerable

Slow and ponderous, when attacked the Palawan stink badger accurately squirts an extremely noxious fluid from its anal glands over a distance of 1m (3¹/₄ft). It has a short tail, small ears and eyes, a typically badger-like, stocky body, and a long, flexible, almost hairless snout for sniffing and rooting out small soil dwellers such as worms, grubs, and slugs. The fur is dark brown with a yellow head cap that tapers to a stripe between the shoulders. This badger lives alone in a rocky den or old porcupine burrow.

Conepatus humboldti
Patagonian hog-nosed skunk

Length 25–37 cm (10–14¹/₂ in)	
Tail 30–57 cm (12–22¹/₂ in)	
Weight 1.5–3 kg (3¹/₄–6¹/₂ lb)	

Location S. South America | Social unit Individual
| Status Vulnerable†

Along with the typical small skunk head, stocky body, and fluffy tail, this hog-nosed skunk also has a broad nose pad for rooting up food. Its fur is black or reddish brown with a white stripe along each side; the stripes meet on the head and extend onto the tail. It will feed on virtually anything edible, but mainly takes insects. Like other skunks, it occupies a secure den under a rock, in a burrow, or among bushes, and it can spray enemies with foul-smelling fluid.

Spilogale putorius
Eastern spotted skunk

Length 30–34 cm (12–13¹/₂ in)	
Tail 17–21 cm (6¹/₂–8¹/₂ in)	
Weight 500–1,000 g (1–2¹/₄ lb)	

Location E. to C. USA, N.E. Mexico | Social unit Individual
| Status Lower risk†

Heavy bodied and short legged, this mustelid relies on striking coloration to warn its enemies of the noxious fluid sprayed from its anal glands. The white markings on the body differ in every individual; there is usually, however, a white patch on the forehead and a white tail tip. Food includes a variety of small animals, fruits, and vegetable matter. Generally solitary, up to 8 eastern spotted skunks may share a den in winter.

Mephitis mephitis
Striped skunk

Length 55–75 cm (22–30 in)	
Tail 17.5–25 m (7–10 in)	
Weight 2.5–6.5 kg (5¹/₂–14 lb)	

Location C. Canada to N. Mexico | Social unit Individual
| Status Common

In common with other skunks, the striped skunk has black-and-white warning coloration. Its diet includes insects, small mammals, birds and their eggs, fish, molluscs, fruits, seeds, and human leftover food. It is solitary but may gather in communal winter dens among rocks, in old burrows, or under outbuildings. The 5–6 young may stay with their mother for more than a year.

NOXIOUS DEFENCE

If threatened, the striped skunk fluffs its fur, arches its back, and lifts its tail. Should the aggressor remain, it stands on its front feet, rear feet in the air, and twists its body. It then ejects a foul-smelling liquid over its head for up to 3m (9³/₄ft), from 2 nozzle-like ducts protruding from its anus.

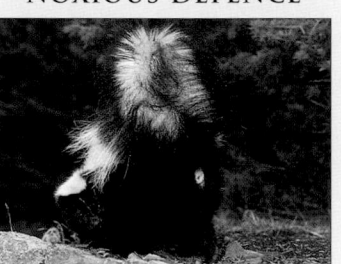

COLORATION
This skunk is black with a thin white stripe on the muzzle, and wider, upper-back stripes from head to tail.

Lontra canadensis

North American river otter

Length	66–110 cm (26–43 in)
Tail	32–46 cm (12 1/2–18 in)
Weight	6–9 kg (13–20 lb)
Location Canada, USA	Social unit Individual
	Status Lower risk†

Probably the most numerous otter, this species is similar to the European otter (see below) and is solitary except at mating time, when it makes squeaks, chitters, and whistles. It dwells along river banks, lake shores, and coasts, maintaining territories of 5–25km (3–15 miles). Home is a den in a riverside burrow, under a pile of rocks or a thicket near water, or in a beaver home (lodge).

pale throat

muscular tail to aid swimming

HUNTER BY DAY OR NIGHT

The river otter's main prey is fish, which it catches by day, except in areas disturbed by people, where the otter becomes more nocturnal. Crayfish, frogs, snakes, lizards, and insects in the water are also devoured. Occasionally this otter may also take an aquatic mammal such as a water vole, or a small water bird such as a duckling.

SINUOUS AND SILVERY
This long-bodied otter has red- or grey-brown to black velvety fur on the back, paler silvery or grey-brown fur on the underparts, with the cheeks and throat tinged silver or yellow-grey.

Lutra lutra

European otter

Length	57–70 cm (22 1/2–28 in)
Tail	35–40 cm (10–16 in)
Weight	7–10 kg (15–22 lb)
Location Europe, Asia	Social unit Individual
	Status Vulnerable

The European otter has suffered through hunting for fur, fishery protection, and sport, as well as from water pollution and loss of river habitat caused by bank clearance, irrigation, leisure, and water

sports. With its waterproof coat, webbed paws, and stiff whiskers (to feel currents from prey movement), this otter is well adapted to its aquatic habitat. It hunts mainly fish, as well as frogs and other aquatic or amphibious prey. Its coloration is mainly brown, with a paler throat, and the muscular tail is flattened from top to bottom. Inland, the European otter hunts chiefly in twilight or darkness, while along coasts, it is more active in daylight. The burrow (holt) is in a bankside territory, 4–20km (2 1/2–12 miles) long, marked by scent and droppings. The European otter is mostly solitary, pairing for 2–3 months in early spring. After 60–70 days' gestation, the 2–3 cubs are suckled for 3 months, and stay with the mother for more than a year.

coat of strong, outer guard hairs and dense underfur

Aonyx capensis

African clawless otter

Length	73–95 cm (29–37 in)
Tail	41–67 cm (16–26 in)
Weight	10–16 kg (22–35 lb)
Location W., E., C., and southern Africa	Social unit Pair
	Status Lower risk†

In common with other otters, the African clawless otter's long, sinuous body, muscular tail, and short limbs make it well adapted to swim and dive. It catches crabs, frogs, and fish, as well as lobster and octopus along coasts, crushing them with its large

teeth. The rear feet are webbed with small claws on toes 3 and 4. The clawless front toes resemble fingers, and are able to feel and hold prey. The clans of paired adults and 2–3 young are exceptionally playful, enjoying mock fighting, mud-sliding, and noisy, yapping chases.

Aonyx cinerea

Oriental short-clawed otter

Length	45–61 cm (18–24 in)
Tail	25–35 cm (10–14 in)
Weight	1–5 kg (2 1/4–11 lb)
Location S., E., and S.E. Asia	Social unit Individual/Group
	Status Lower risk†

Smallest of the otters, the oriental short-clawed otter has very short claws that do not extend beyond the fleshy end pads of the partly webbed toes. The cheek teeth are broad, for crushing hard-cased food such as mussels and other shellfish, crabs, and frogs; unusually for otters, fish are relatively unimportant in the diet. The upperparts are brown; the underside is paler, with variable white areas on the lower face, throat, and chest. These otters form loose social groups

of about 12, members keeping in touch with noises and scents. Male–female pair bonds are especially strong. As in many similar species, territories are marked with scent from paired glands at the base of the tail, urine, and droppings (spraints). Litter size varies from one to 6, average 2; both parents care for the young.

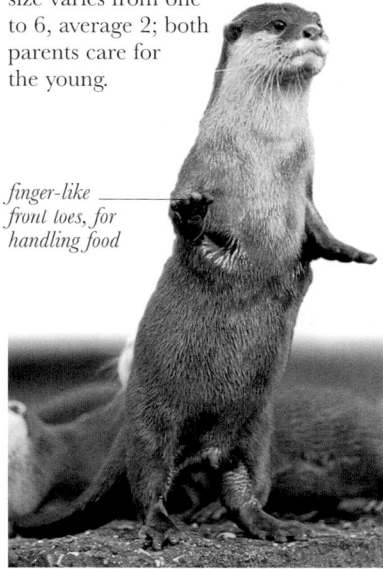

finger-like front toes, for handling food

Pteroneura brasiliensis

Giant otter

Length	1–1.4 m (3 1/4–4 1/2 ft)
Tail	45–65 cm (18–26 in)
Weight	22–32 kg (49–71 lb)
Location N. and C. South America	Social unit Group
	Status Endangered

The biggest mustelid, this species is similar to a very large river otter. It has short legs, well-webbed toes, and a flattened, wide-based tail, for swimming and diving. Its stout whiskers and sensitive

eyes effectively detect prey movements in water. Being so aquatic, it is ungainly on land. The short, dense fur is dark brown, appearing black when wet, with cream spots and patches on the chin, throat, and chest, which may merge into a "bib". The giant otter forms groups of 5–10: usually 2 parents, their offspring, and various younger adults. They live in a communal bank den or burrow and hunt by day for fish, crabs, and other aquatic prey.

EXPERT DIVER
The sea otter spends much of its time diving for food on the sea bed, and is well adapted to an aquatic existence. Its lungs are two and a half times the size of those found in land mammals of similar size, and, when underwater, it can survive at a depth of 30m (98ft) for up to 4 minutes.

Enhydra lutris

Sea otter

Length 55–130 cm	(1³/4–4¹/4ft)
Tail 13–33 cm	(5–13 in)
Weight 21–28 kg	(46–62 lb)

Location North Pacific   **Social unit** Group

Status Endangered

The smallest marine mammal, the sea otter lives and feeds in the ocean. It comes ashore only rarely, but is commonly seen close to the coast, particularly near marine kelp forests, floating on its back with its paws out of the water. Superbly adapted to its aquatic way of life, the sea otter has luxuriantly thick fur that keeps it warm in the cold waters it inhabits. It has a strong, flat tail that acts as a rudder, and large, flipper-like hindfeet that propel it through the water; the smaller forepaws have retractable claws, like a cat's, which the sea otter uses for holding food and grooming its fur. Excellent eyesight – both underwater and at the surface – a good sense of smell, and sensitive whiskers help it to find food. It forages mainly on the sea bed, looking for crabs, clams, sea urchins, and abalone, and has immensely strong teeth for crushing the shells. Before the sea otter sleeps, it may anchor itself by wrapping its body in kelp. Sea otters are social animals, usually found in groups (rafts), with males forming separate rafts from females. In Alaska hundreds of animals may be found together. Although hunted in the past for its fur – almost to extinction in some areas – the sea otter is now a protected species.

CONSERVATION

Protected from hunting, sea otters are now increasing in number, but only in part of their original range. Efforts to relocate them to other areas are continuing. Promoting the growth of kelp beds – one of the sea otter's favourite habitats (shown below) – may also help the animal to thrive once again.

long, flat tail   *long brown body fur*   *flipper-like hindfeet*   *straw-coloured fur on head*

GROOMING TO STAY WARM
The sea otter has the densest fur of all animals – up to 150,000 hairs per square cm (1 million per square in). Grooming the fur, to keep it clean and waterproof, is essential.

SPECIALIST BEHAVIOUR

The sea otter is a resourceful and adept animal, adapting its behaviour in a variety of ways in order to make the most of its environment. Perhaps most remarkably, it has learnt to use stones as tools for cracking open shellfish.

TOOL USER
An abalone shell can be hard to crack, but the sea otter has devised a reliable method. Lying on its back, it breaks the shell open by hitting it against a rock collected from the sea bed.

WATER BABY
The sea otter will sometimes give birth on the rocks, but soon takes her pup into the water. Until the pup learns to groom its fur, the mother keeps it dry by carrying it on her belly, while she floats on her back.

CIVETS AND RELATIVES

PHYLUM	Chordata
CLASS	Mammalia
ORDER	Carnivora
FAMILY	Viverridae
SPECIES	76

ANIMALS IN THIS FAMILY – which includes civets, genets, binturong, mongooses (often treated as separate from the viverrids), and fossa – usually have a slender body and a long tail. They are related to cats and hyenas but are more primitive, with a longer snout and additional teeth. Viverrids are distributed throughout Africa, Madagascar, southern Asia, and southeastern Europe, and are found in forest, desert, and savanna. Most are terrestrial, but there are arboreal species (for example, the binturong) and semi-aquatic ones (for example, otter genets).

ANATOMY

Viverrids typically have a long body and tail, short legs, an elongated neck and head, and a tapered snout. Most civets and genets have spots in longitudinal rows along the body; mongooses have a plain or striped coat. All species have scent glands in the anal region, and in civets these glands produce a substance used to make perfume.

FEEDING

Some viverrids are carnivorous, while others, such as the two-spotted palm civet, eat only fruits. Most, however, are omnivorous, and eat small mammals and birds, lizards, or invertebrates. Species that hunt prey usually stalk their victims in a cat-like manner.

SOCIAL GROUPS

Most species in this family are solitary, but dwarf mongooses live in large social groups that forage over a considerable area. Individuals communicate with high-pitched "beeps". The meerkats of southern Africa live in burrows, each with several entrances and each occupied by a social group. While some members are foraging, others stand on their back legs, on small vantage points, watching out for predators.

ARBOREAL HUNTERS
Many civets and genets hunt for food in trees as well as on the ground. This large-spotted genet is searching for nesting or roosting birds.

Viverra tangalunga
Malayan civet

Length	62–66 cm (24–26 in)
Tail	28–35 cm (11–14 in)
Weight	3.5–4.5 kg (7 3/4–10 lb)

Location S.E. Asia | Social unit Individual

Status Common

In addition to the typical civet coat of many dark spots forming lines along the body, this species has a distinctive black-and-white neck collar, white underside, black legs and feet, and about 15 bands along the tail. It climbs only occasionally and feeds mainly on forest-floor creatures including millipedes, giant centipedes, scorpions, and small mammals such as mice. Widespread and common throughout Southeast Asia, the Malayan civet is nocturnal and solitary, and lives for up to 11 years.

Genetta genetta
Small-spotted genet

Length	40–55 cm (16–22 in)
Tail	40–51 cm (16–20 in)
Weight	1.5–2.5 kg (3 1/4–5 1/2 lb)

Location W., E., and southern Africa, W. Europe | Social unit Individual

Status Common

This very cat-like species, also known as the common genet, has semi-retractile claws and is an excellent climber. It takes a variety of small mammals, birds, eggs, grubs, and fruits. In some areas it raids farms for poultry and is considered a pest. The den is a sheltered area under roots or in thick bushes. After the gestation period of 70 days, typically 2–3 young are born with eyes closed.

Prionodon pardicolor
Oriental linsang

Length	37–43 cm (14 1/2–17 in)
Tail	30–36 cm (12–14 in)
Weight	0.6–1.2 kg (1 1/4–3 lb)

Location S., E., and S.E. Asia | Social unit Individual

Status Unconfirmed

This small, slender, sinuous linsang seems to "flow" through the branches with amazing grace, using its tail to balance and brake. Both sexes are generally solitary and nocturnal, with large ears and big eyes adapted for night vision. The male is almost twice the size of the female. Its diet consists of small animals such as frogs, snakes, rats, and mice, as well as carrion. The average litter size is 2–3, with peak births in February and August.

Paradoxurus hermaphroditus
Palm civet

Length	43–71 cm (17–28 in)
Tail	40–66 cm (16–26 in)
Weight	1.5–4.5 kg (3 1/4–10 lb)

Location S., E., and S.E. Asia | Social unit Individual

Status Common

This adaptable, bushy-tailed civet is brownish grey with black stripes on its back, dark flank spots, and a polecat-like "face mask". It stays mainly in trees but may rest by day in a house or on an outbuilding roof. The diet includes many fruits, especially figs, as well as buds, grasses, small animals such as insects and mice, and sometimes poultry. Fermented palm tree juice is a favourite, giving rise to a local name of "toddy cat".

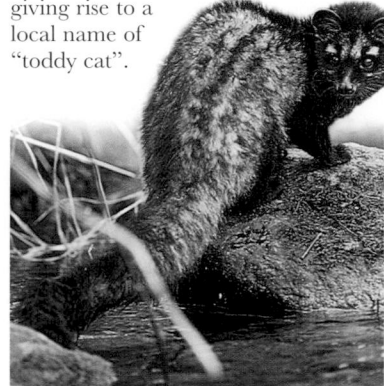

Arctictis binturong
Binturong

Length	61–96 cm (24–38 in)
Tail	56–89 cm (22–35 in)
Weight	9–14 kg (20–31 lb)

Location S. and S.E. Asia

Social unit Individual

Status Common

The binturong has shaggy black fur and tufted ears. Its tail is long-haired and prehensile at the tip. It moves slowly and cautiously among branches in search of fruits, shoots, and small animals such as insects, birds, and rodents. By day it curls up on a secluded branch to rest, but may continue to feed. Individuals mark their territories with scent. The 1–3 young, born after a gestation of 92 days, reach adult size in one year.

Eupleres goudotii
Falanouc

Length	48–56 cm (19–22 in)
Tail	22–25 cm (9–10 in)
Weight	1.5–4.5 kg (3 1/4 – 10 lb)

Location E. and N. Madagascar

Social unit Individual

Status Endangered

Brown on the upper parts, whitish grey on the underside, with a long, slender snout and a short, bushy tail, the falanouc inhabits the Madagascan rainforests and marshes. It uses its long front claws to dig in soil for worms, grubs, insects, slugs, snails, and rodents. The single young, born with eyes open, can follow its mother after only 2 days, and is weaned by 9 weeks. Falanoucs are threatened by habitat loss, humans, dogs, and an introduced competitor, the small Indian civet (*Viverricula indica*).

Cynictis penicillata
Yellow mongoose

Length	23–33 cm (9–13 in)
Tail	18–25 cm (7–10 in)
Weight	450–800 g (16–29 oz)

Location Southern Africa

Social unit Group

Status Common

This mongoose is a yellowish buff colour in the south of its range but greyer in the north. The family group (a breeding pair, their offspring, and non-breeding young adults) occupies and extends a tunnel system taken over from meerkats or ground squirrels. Occasionally these species all co-exist in a large burrow. The main diet is insects such as termites, ants, beetles, and locusts, as well as birds, eggs, frogs, lizards, and small rodents.

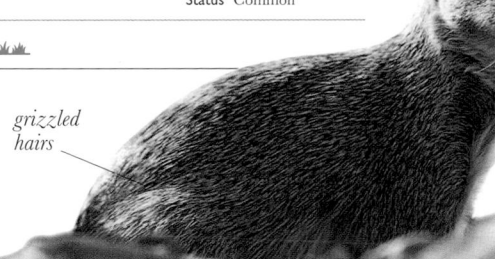

grizzled hairs

Mungos mungo
Banded mongoose

Length	30–45 cm (12–18 in)
Tail	15–30 cm (6–12 in)
Weight	1.5–2.5 kg (3 1/4 – 5 1/2 lb)

Location Africa

Social unit Variable

Status Common

Common, lively, and opportunistic, this stocky mongoose has distinctive crosswise bands over the rear of the body. The fur is coarse and grizzled, and populations from moist habitats are darker brown than drier-region individuals. Often kept as a pet, the banded mongoose eats varied small items, from termites to birds' eggs. It is often found in packs of 15–20 that include one dominant male.

about 12 bands over rump

Helogale parvula
Dwarf mongoose

Length	18–28 cm (7–11 in)
Tail	14–19 cm (5 1/2 – 7 1/2 in)
Weight	200–350 g (7–13 oz)

Location E. and southern Africa

Social unit Group

Status Common

The smallest mongoose, this species has thick fur, brown but fine-grizzled in red or black, very small eyes and ears, and long-clawed front feet. It forms packs of 2–20, which "rotate" around the termite mounds of their range. They use the mound as shelter for a few days and feed on insects, lizards, snakes, birds, eggs, and mice. All members of the pack help to care for offspring, which may number up to 6 per female.

Suricata suricatta
Meerkat

Length	25–35 cm (10–14 in)
Tail	17–25 cm (6 1/2 – 10 in)
Weight	600–975 g (21–35 oz)

Location Southern Africa

Social unit Group

Status Common

EXPERT DIGGERS
The meerkat's long front claws are used to dig its burrow and to find food, mainly insects, spiders, and other small animals, as well as roots and bulbs.

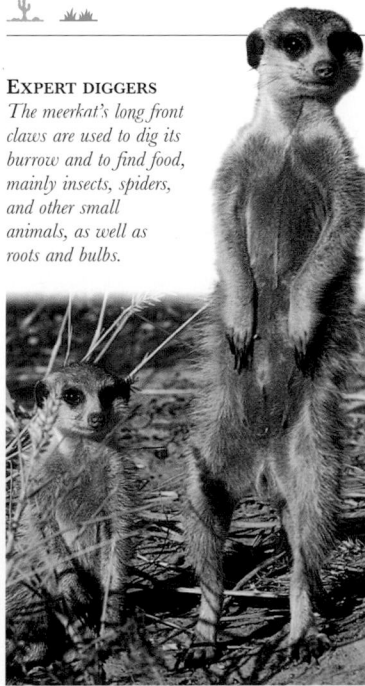

Day-active and sociable, the meerkat forms colonies of up to 30, which enlarge the former burrow systems of ground squirrels. In early morning it emerges to sit up and sunbathe, then forages for small prey. Coloration is pale brown on the underside and face, silver-brown on the upper parts, with 8 darker bands on the rear back, dark eye-rings, and a dark tip to the slender tail. After a gestation of 11 weeks, the 2–5 young are born in a grass-lined nursery chamber in the burrow.

ON GUARD

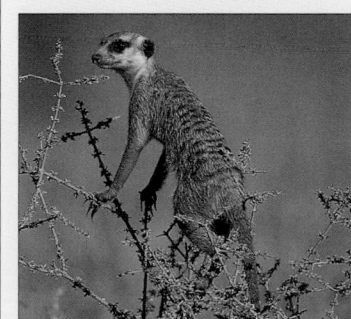

While most pack members forage, some act as lookout sentries, especially for hawks and other aerial predators. Sentries stand at vantage points such as on mounds and in bushes, and cheep or cluck warnings. Sharp barks or growls denote more urgent threats and the meerkats dive for cover.

Cryptoprocta ferox
Fossa

Length	60–76 cm (23 1/2 – 30 in)
Tail	55–70 cm (22–28 in)
Weight	9.5–14 kg (21–31 lb)

Location Madagascar

Social unit Individual

Status Endangered

Resembling a diminutive brown "big cat", the fossa is lithe, agile, and an excellent leaper and climber. Madagascar's largest carnivore, this muscular, powerful predator hunts by day or night, using the stalk-and-pounce method. It originally specialized in hunting lemurs but now also takes pigs, poultry, and other domesticated animals. A top carnivore, the solitary fossa naturally has a large territory, more than 4 square km (1 1/2 square miles), and thus a low population density. It is threatened by loss of its habitat and persecuted for its attacks on livestock.

HYENAS AND AARDWOLF

PHYLUM	Chordata
CLASS	Mammalia
ORDER	Carnivora
FAMILY	Hyaenidae
SPECIES	4

ALTHOUGH MEMBERS of this family superficially resemble dogs, they are in fact more closely related to cats and civets and genets. They all feature a distinctive back line that slopes downwards from the shoulders to the tail. Hyenas and aardwolves are primarily found in Africa (although the striped hyena extends to southern parts of Asia), in savanna, scrub, and semi-arid habitats. They are primarily nocturnal and dig dens that are used to shelter adults and cubs (except in the spotted hyena, where only the cubs seek refuge in dens).

ANATOMY

Physical characteristics common to the species in this family include a large head and ears, long front legs and short back legs, a mane on the nape that (except in the spotted hyena) extends down the back, a bushy tail, and short, blunt, non-retractable claws. Hyenas have 4 toes on both the front and back feet; the aardwolf has 5 toes on the front feet and 4 on the back feet. The coat is spotted or striped (the brown hyena has stripes on the limbs only).

FEEDING

Hyenas have a broad muzzle and immensely strong jaws (indeed the jaws are the most powerful of any mammal of comparable size) and teeth for crushing bones. Of the 3 hyena species, the spotted hyena is the most voracious hunter. When hunting cooperatively, spotted hyenas are able to take down large prey, such as zebras. This species is also a highly efficient scavenger – groups of spotted hyenas are capable of driving a lion off its prey. Striped and brown hyenas, on the other hand, scavenge most of the time, although they may also capture small prey. All hyenas are able to digest parts of a kill that other mammals cannot process, such as skin and bone. This means that they occupy a niche not filled by most other mammals. Portions of the meal that they cannot digest, such as ligaments, hair, and hooves, are regurgitated in the form of pellets.

Given these digestive adaptations, it is surprising that one member of this family – the aardwolf – does not eat large prey. Instead it uses its smaller teeth and sticky tongue to feed on termites. There is little competition for this food source, it requires minimal effort to obtain, and the nutritional value is surprisingly high. An aardwolf may eat as many as 200,000 termites in a single night.

SOCIAL GROUPS

While the aardwolf is solitary, and striped and brown hyenas live in pairs or small groups, spotted hyenas live in larger groups called clans. These clans may consist of up to 80 individuals (males and females and their cubs). When cubs are 2–3 months old, they are transferred to a communal den, where all are suckled by any lactating female (brown hyenas employ this system, too). The cubs remain in the den until they are weaned and ready to accompany adults hunting and foraging, which may be at as little as 7 months of age. All hyenas are territorial, and their territories are marked using an anal scent gland (which can be turned inside out). Spotted hyena clans patrol and defend their territory communally.

THE ULTIMATE MEAT-EATER
The kill on which this spotted hyena is feeding may have been hunted down by the clan or scavenged from another predator. Because hyenas have powerful, bone-crushing jaws, and eat almost anything, little will be left of this hartebeest carcass when the hyenas leave.

MAMMALS

Proteles cristatus
Aardwolf

Length	67 cm (26 in)
Tail	24 cm (9¹/₂ in)
Weight	9 kg (20 lb)

Location E. and southern Africa

Social unit Individual

Status Lower risk†

A small relative of the hyenas, the aardwolf's specialized diet is termites, particularly surface-foraging nasute (snouted) harvester termites. It also licks up maggots, grubs, and other small, soft-bodied creatures. Its longer front legs and down-sloping body are accentuated by the crest-like back mane, most prominent on the neck and shoulders. This erects under stress so the animal appears larger. The fur is pale buff or yellow-white with 3 vertical stripes on each side, and diagonal stripes across the fore- and hindquarters. The front teeth are hyena-like but the molars are small pegs, the food being ground up by the muscular stomach. The aardwolf is solitary and nocturnal, resting in a burrow by day. It marks its territory with urine, dung, and anal gland secretions. The 2–4 cubs are born after a gestation period of 90 days, emerge from the den at 4 weeks, forage with the mother from 9–11 weeks, and are weaned by 16 weeks.

Parahyaena brunnea
Brown hyena

Length	1.3 m (4¹/₄ ft)
Tail	21 cm (8¹/₂ in)
Weight	38–47 kg (84–105 lb)

Location Southern Africa

Social unit Variable

Status Lower risk†

This species ranges farther into deserts than other hyenas, and can scent carrion from 14km (8¹/₂ miles). It has the typical hyena's powerful jaws and shearing teeth for scavenging on any carcass, including seal pups along the Namib Desert coast. It also catches prey such as springhares. The brown hyena forms loose clans that defend their territory. Its coat is shaggy, dark brown to black with a pale tawny neck mantle, a grey-patched face, and barred legs.

Hyaena hyaena
Striped hyena

Length	1.1 m (3¹/₄ ft)
Tail	20 cm (8 in)
Weight	35–40 kg (77–88 lb)

Location W., N., and E. Africa, W. to S. Asia

Social unit Individual

Status Lower risk

Preferring savanna and open woodland, this hyena avoids extreme habitats such as deserts. It is grey or pale brown with 5–6 vertical flank bars. The neck mane lessens on the back and merges with the bushy black-and-white tail. Generally solitary, this hyena may form a family group when breeding. It scavenges, hunts prey from insects to hares, and eats fruits and other plant matter.

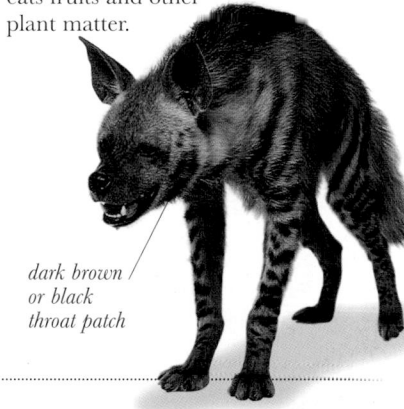
dark brown or black throat patch

MAMMALS

Crocuta crocuta
Spotted hyena

Length	1.3 m (4¹/₄ ft)
Tail	25 cm (10 in)
Weight	62–70 kg (135–155 lb)

Location W. to E. and southern Africa

Social unit Group

Status Lower risk

The spotted hyena is the largest hyena species. The female is some 10 per cent larger than the male, and her external sexual organs are enlarged so that they are difficult to distinguish from the male's. The social system is female-dominated and based on the clan, which varies from 5 or fewer adults and young in deserts to 50 or more in prey-rich savanna. The clan occupies a communal den, uses communal latrines, and jointly defends its territory of 40–1,000 square km (15–390 square miles), delineated by calls, scent-marking, and boundary patrols. The spotted hyena makes many sounds, including whoops, to rally its group or locate cubs, and the famous hyena's "laugh", to signify submission to a senior clan member.

HUNTING HYENA
The spotted hyena is a powerful hunter. Several clan members may form a pack to bring down large prey such as an adult zebra or wildebeest. When hunting alone, it catches hares, ground birds, and fish in shallows and swamps. It gorges on food, and is able to consume up to one-third of its body weight at one meal.

short, rounded ears

REVERSED MANE
The spotted hyena's neck and back mane is reversed and erectile – the hairs slope forwards rather than back, and stand erect when the hyena is excited.

sandy to grey-brown coat with dark spots that fade with age

relatively long front legs

SIBLING RIVALRY
The spotted hyena mother is solely responsible for cub rearing; the male plays no part. The average litter is 2 (range 1–3), born after 100 days' gestation and weaned at 14–18 months, when nearly full grown. The dominant cub controls access to the mother when suckling, and, in times of milk shortage, may kill its sister or brother to improve its own chance of survival.

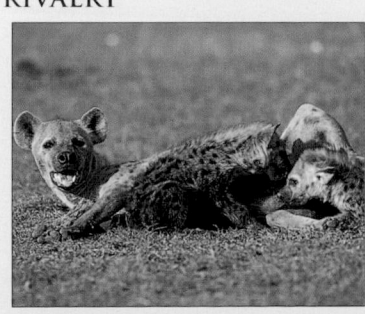

CATS

PHYLUM	Chordata
CLASS	Mammalia
ORDER	Carnivora
FAMILY	Felidae
SPECIES	38

WITH A LITHE, muscular body, acute senses, highly evolved teeth and claws, lightning reflexes, and camouflage coloration, cats are model hunters.

In fact cats are the most specialized of the mammalian flesh-eaters. They are unusual in that all species appear remarkably similar: the differences between tigers and the domestic cat, for example, are surprisingly small. As a result, classifying smaller cats remains problematic: they can be either grouped together in the genus *Felis* or subdivided into several genera. Cats are found throughout Eurasia, Africa, and the Americas (the domestic cat is found worldwide), from alpine heights to deserts. Many species live in forests. All except the largest cats are expert climbers, and several are excellent swimmers. Most cats are solitary.

ANATOMY

Cats have a rounded face and a relatively short muzzle (but a wide gape). The heavy lower jaw helps deliver a powerful bite, and the long canines are used for stabbing and gripping. The carnassials, modified cheek teeth that slice bones and tendons, are highly developed. Cats are covered with soft fur, which is often striped or spotted, and have a tail that is haired, flexible, and usually long. There are 5 digits on the front feet and 4 on the back feet, and each digit has a curved, retractable claw for holding prey. The claws are normally retracted, which helps keep them sharp. However, when required (during climbing, for example), they spring forward via a mechanism similar to a jackknife. The naked pads on the soles of the feet are surrounded by hair, which assists with silent stalking.

SENSES

All cats have keen senses. Large, forward-facing eyes enable them to judge distances accurately. The pupils can contract to a slit or to a pinhole (depending on species) in bright light and can dilate widely for excellent night vision. The ears are large, mobile, and funnel-shaped to draw in sounds made by prey, while long, stiff, highly sensitive whiskers aid navigation and night hunting. The sense of smell is also well developed, and in the roof of the mouth is a "smell-taste" organ, called the Jacobson's organ, which detects sexual odours. Secretions from scent glands on the cheeks and forehead, under the tail, and between the claws communicate information such as age and sex.

HUNTING TECHNIQUES

While some cats actively search for prey, others conceal themselves and await passing victims. Many employ a combination of these 2 methods. In either case, the cat's fur usually provides camouflage: a tiger's stripes, for example, blend in with tall grasses, while many forest-living species are spotted, to mimic the effect of sunlight through leaves.

The distance a cat will chase its quarry varies between species. Heavier-built cats, such as tigers, prefer to stalk and pounce; the cheetah uses explosive speed – up to 110kph (68mph). Some small cats, such as the serval, hunt in long grass, and use "jack-in-the-box" leaps to surprise and flush out their prey.

Cats hunt any animal they can catch and overpower. The big cats specialize in prey larger than themselves, and are capable of dragging a carcass some distance to a safe feeding spot. Smaller cats seek out rodents and birds – some, such as the fishing cat, wade into streams and scoop out fish.

CONSERVATION

There is a conservation crisis for cats: the entire family is listed by CITES, and some species are critically endangered. Even the leopard – one of the great survivors – is in danger. The main problem is hunting – either for fur (there was a huge increase in demand for fur during the 20th century) or for use in traditional medicines (each year huge quantities of tiger bones are smuggled into China to be made into tiger bone wine). Habitat destruction has also reduced numbers, as cats require large areas to maintain their food supply. In a few areas, the trend of declining populations has been reversed: puma, tiger, and wild cat numbers have all increased slightly after the introduction of protective measures.

TERRITORIAL CONFLICT
All cats are territorial and will fight if their scent marks and vocal warnings are ignored. These male jaguars are threatening one another, and a conflict will ensue if neither of them gives way. The flattened ears show fear, and dilated pupils and bared teeth indicate aggression.

Felis sylvestris

Wild cat

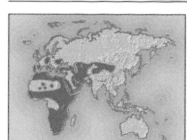

Length 50–75 cm (20–30 in)	
Tail 21–35 cm (8 1/2–14 in)	
Weight 3–8 kg (6 1/2–18 lb)	
Location Europe, W. and C. Asia, Africa	**Social unit** Individual
Status Lower risk†	

The wild cat resembles a slightly larger, heavier-built, usually longer-furred (especially in winter) version of the domestic tabby cat. The species interbreeds with domestic cats, of which the African subspecies *Felis sylvestris libyca* is presumed to be the ancestor. The wild cat's preferred habitat is mixed broadleaved woodland, but habitat loss and amenity use of woods has driven it to marginal habitats such as conifer forest, rocky upland, moor, scrub, swamp, and coast. It feeds on rabbits and small rodents such as rats, mice, voles, and lemmings. This cat climbs well and catches young squirrels or birds in the branches; carrion is also eaten. It mates between January and March and the gestation period is 63–68 days. The female gives birth to an average of 3–4 (range 1–8) cubs in its den in a tree-hole, among rocks or tree roots, or in an old rabbit or badger burrow.

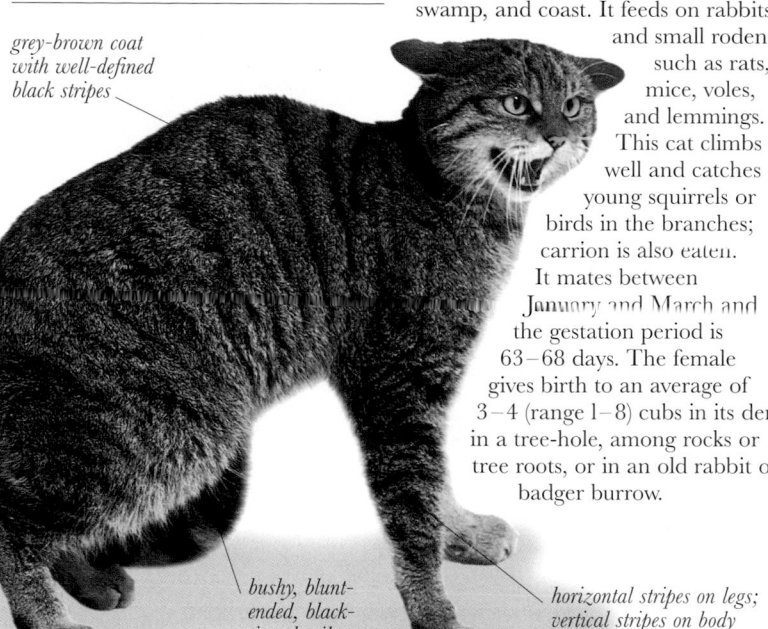

grey-brown coat with well-defined black stripes

bushy, blunt-ended, black-tipped tail

horizontal stripes on legs; vertical stripes on body

Felis chaus

Jungle cat

Length 50–94 cm (20–37 in)	
Tail 23–31 cm (9–12 in)	
Weight 4–16 kg (8 3/4–35 lb)	
Location W., C., S., and S.E. Asia, N.E. Africa	**Social unit** Individual
Status Lower risk†	

Despite its main common name, this slender, long-legged cat is also known more aptly as the swamp or reed cat. It hunts along marshes, river banks, shores, and also ditches and ponds around human settlements, taking mammals up to the size of coypu, birds (including poultry), reptiles, and, being a strong swimmer, fish and amphibians. Female and male may stay together, and both protect the cubs.

unpatterned body

Felis margarita

Sand cat

Length 45–57 cm (18–22 1/2 in)	
Tail 28–35 cm (11–14 in)	
Weight 1.5–3.5 kg (3 1/4–7 3/4 lb)	
Location N. Africa, W., C., and S. Asia	**Social unit** Individual
Status Endangered†	

Surviving on fluids in its food and very little additional water, the blunt-clawed sand cat digs well for its main prey of gerbils and similar rodents, as well as an occasional lizard or snake. It also excavates a den for daytime shelter. The average litter of 3 grows quickly and may be independent in just 6 months.

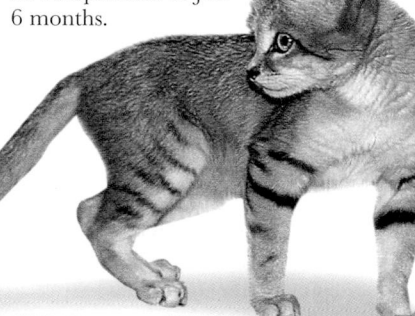

Felis pardina

Iberian lynx

Length 85–110 cm (34–43 in)	
Tail 13 cm (5 in)	
Weight 10–13 kg (22–29 lb)	
Location S.W. Europe	**Social unit** Individual
Status Endangered	

Very rare, critically restricted in distribution, and fully protected by law, the Iberian lynx is about half the size of the Eurasian species (see right). Now mainly confined to remote wetlands and uplands, it uses thickets for shelter and open areas that favour its main food of rabbit – with deer fawns, ducks, and other prey in winter.

tufted ears

distinctive spotted coat

Felis lynx

Eurasian lynx

Length 0.8–1.3 m (2 1/2–4 1/4 ft)	
Tail 11–25 cm (4 1/4–10 in)	
Weight 8–38 kg (18–84 lb)	
Location N. Europe to E. Asia	**Social unit** Individual
Status Lower risk†	

Primarily a cat of mixed forest, this lynx has been driven by human presence and persecution to more open woods and rocky mountain slopes. However, it still has one of the widest ranges of all cat species. Its major prey is deer, goats, sheep, and similar hoofed mammals up to 4 times its own size, but, if these are lacking, it hunts hares and pikas.

COAT PATTERNS
The Eurasian lynx has 3 predominant coat patterns: mainly striped, mostly spotted (as here), and plain.

CONSERVATION

Despite conservation efforts and reintroductions in western Europe, lynx remain rare because of revenge hunting by livestock farmers, road kills, and mysterious loss of male cubs – possibly caused by a genetic problem.

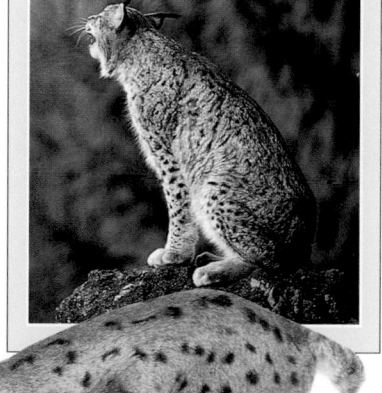

Felis rufus

Bobcat

Length 65–110 cm (26–43 in)	
Tail 11–19 cm (4 1/4–7 1/2 in)	
Weight 4–15.5 kg (8 3/4–34 lb)	
Location S. Canada, USA, Mexico	**Social unit** Individual
Status Lower risk†	

Named after its short "bobbed" tail, this medium-sized cat has a ruff-like facial border. Mainly tawny in colour, it always has some spots; they may be prominent all over or only on the underside. It hunts lagomorphs, such as cottontail rabbits in the south of its range and snowshoe hares farther north, but can survive on rodents, deer, and carrion. Its habitat varies from desert to mixed woodland and conifer forests.

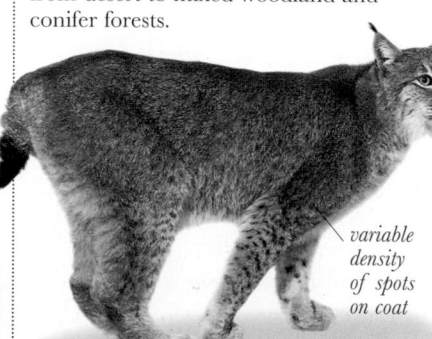

rusty or yellowish grey background colour

variable density of spots on coat

Felis caracal
Caracal

	Length 60–91 cm (23 1/2 – 36 in)
	Tail 23–31 cm (9–12 in)
	Weight 6–19 kg (13 – 42 lb)
Location Africa, W., C., and S. Asia	Social unit Individual
	Status Lower risk†

Also called the desert lynx, because of its arid scrubby habitat, the caracal is tawny or reddish, although – as with many cats – occasional black (melanic) individuals occur (see Black panther, p.214). It is famed for its ability to spring 3m (10ft) vertically and to "bat" flying birds with its paw. Its main foods are rodents, hyraxes, hares, small antelopes, poultry, and other livestock.

narrow, tufted ears

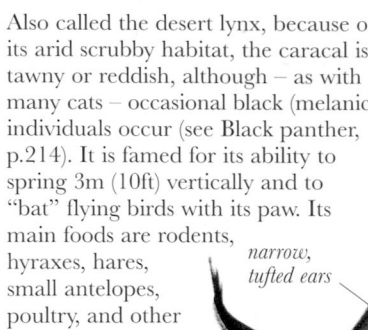

Felis serval
Serval

	Length 60–100 cm (23 1/2 – 39 in)
	Tail 24–45 cm (9 1/2 – 18 in)
	Weight 9–18 kg (20 – 40 lb)
Location Africa	Social unit Individual
	Status Lower risk†

Resembling a small cheetah, with its lean body and long limbs, the serval has yellowish fur with dark spots. It prefers to live among reeds and rushes fringing wetlands. Here it hunts rats and similar prey, helping to control rodents and thereby aiding farmers (servals rarely attack livestock). The average litter of 2 young is born after 73 days' gestation.

CAT POUNCE

Having located prey, usually at dusk, and mainly by hearing, the serval excels at the cat pounce. It leaps up to 4m (13ft) horizontally and more than 1m (3 1/4 ft) high, to strike the victim with its forepaws. This cat eats rats and similar-sized rodents, birds, fish, and large insects such as locusts. Frogs are a favourite with wetland-dwelling servals.

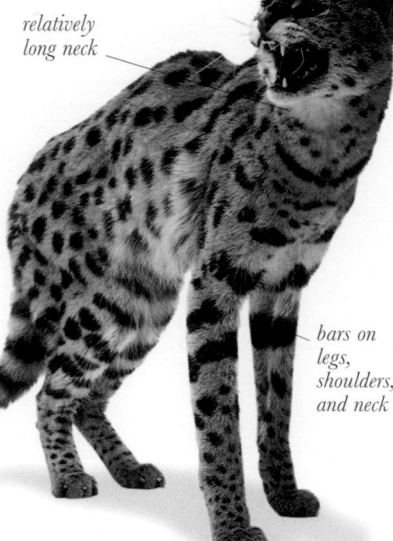

relatively long neck

bars on legs, shoulders, and neck

HEAD UP
The serval's relatively long legs and neck elevate its head to 75cm (30in) above ground, enabling it to see and hear clearly in long grass.

Felis marmorata
Marbled cat

	Length 45–53 cm (18–21 in)
	Tail 47–55 cm (18 1/2 – 22 in)
	Weight 2–5 kg (4 1/2 – 11 lb)
Location S. to S.E. Asia	Social unit Individual
	Status Vulnerable†

A little-known species, found mainly in moist, lowland tropical forest, the marbled cat resembles a small clouded leopard (see p.214). It is long tailed, a proficient climber, and probably a nocturnal predator of squirrels, birds, and similar tree dwellers. Little is known about this cat, but data being collected in Thailand should help to shed some light on this species. Gestation is 81 days, litter size 1–4, and sexual maturity is at 21 months.

Felis aurata
African golden cat

	Length 61–100 cm (24 –39 in)
	Tail 16–46 cm (6 1/2 –18 in)
	Weight 5.5–16 kg (12 –35 lb)
Location W. and C. Africa	Social unit Individual
	Status Lower risk†

This little-studied, medium-sized cat may vary from grey to red-brown, and it may be faintly spotted or plain. It occurs primarily in tropical rainforest and other forest habitats, especially near rivers. Its prey is mainly rats and other rodents, hyraxes, small forest antelopes, monkeys (perhaps already injured), and similar small mammals. It may also catch birds, mainly on the ground but also in trees.

Felis nigripes
Black-footed cat

	Length 34–50 cm (13 1/2 – 20 in)
	Tail 15–20 cm (6 – 8 in)
	Weight 1.5–3 kg (3 1/4 – 6 1/2 lb)
Location Southern Africa	Social unit Individual
	Status Lower risk†

One of the smallest cats, the size of a very small pet cat, this species is pale brown with bold stripes, which thicken on the legs and merge into the all-black undersides of the feet. Prey is likewise small: for example mice, insects from termites to locusts, spiders, small lizards, and birds. Well adapted to the Karoo, the Kalahari, and other arid regions of southern Africa, it seldom needs to drink water.

Felis viverrinus
Fishing cat

	Length 75–86 cm (30 –34 in)
	Tail 25–33 cm (10 – 13 in)
	Weight 8–14 kg (18 – 31 lb)
Location S. to S.E. Asia	Social unit Individual
	Status Lower risk

Olive-grey with black markings and a short tail, the fishing cat is mostly confined to rivers, lakes, marshes, and coastal mangrove swamps. However, its adaptations to water are largely behavioural: its toes are only slightly webbed, and its teeth are not especially suited to grasping slippery prey. Locally common, its dependence on water-edge habitats means its population may suffer as wetlands are affected by drainage, intensive agriculture, human habitation, and pollution.

CIVET-LIKE CAT
This cat's scientific name, viverrinus, *reflects its viverrid- or civet-like proportions: long, stocky body and relatively short legs.*

FISH DIET

As its name implies, the fishing cat is a semi-aquatic hunter of fish, frogs, snakes, water insects, crabs, crayfish, and shellfish. It scoops prey from the water with its paws, dives in pursuit, sometimes surfacing under a water bird. It also hunts small land mammals such as mice.

Felis planiceps
Flat-headed cat

Length 41–50 cm (16–20 in)	
Tail 13–15 cm (5–6 in)	
Weight 1½–2 kg (3¼–4½ lb)	
Location S.E. Asia	Social unit Individual
	Status Vulnerable

Unusually small and low-set ears accentuate the flattened forehead of this semi-aquatic fish predator, which also takes shrimps, frogs, rodents, and small birds. Its toes are partly webbed and claws not fully retractable. The upper premolar teeth are relatively large and sharp to grip slippery food. Slightly smaller than a typical pet cat, the flat-headed cat is usually sighted around rivers, lakes, and swamps, and along irrigation ditches and canals.

Felis guigna
Kodkod

Length 42–51 cm (16½–20 in)	
Tail 19.5–25 cm (7½–10 in)	
Weight 2–2.5 kg (4½–5½ lb)	
Location W. South America	Social unit Individual
	Status Vulnerable

Felis geoffroyi
Geoffroy's cat

Length 42–66 cm (16½–26 in)	
Tail 24–36 cm (9½–14 in)	
Weight 2–6 kg (4½–13 lb)	
Location C. to S. South America	Social unit Individual
	Status Vulnerable†

Geoffroy's cat, which is sometimes called Geoffroy's ocelot, prefers scrub and shrub to forest and open grassland. It hunts in the branches, on the ground, and in water, for frogs and fish as well as the usual small-cat fare of rodents, lizards, birds, and the (introduced) brown hare. Now

Felis jacobita
Andean cat

Length 58–64 cm (23–25 in)	
Tail 41–48 cm (16–19 in)	
Weight 4 kg (8¾ lb)	
Location W. South America	Social unit Individual
	Status Endangered†

Little is known about the habits of the rare Andean, or mountain, cat. Small and sturdy, with a long, bushy tail, it has thick, warm, grey-brown fur that is marked with vertical stripes along the upper back, rosette-type spots on the flanks, and bands around the legs and tail. This feline inhabits dry, rocky slopes above the tree-line – generally 3,000m (9,900ft) – and preys mainly on rodents such as viscachas and formerly chinchillas. Unlike many other cats, this species is not directly at risk from hunting or habitat loss, but hunting threatens some of the prey species on which it depends.

The smallest cat in the Americas, the kodkod or güiña closely resembles Geoffroy's cat (see below) but with a thicker tail and smaller head. Its coat is black-spotted grey to ochre, with a ringed tail and a dark throat stripe. Found in moist, cool forests in the Andean foothills of Argentina and Chile, it makes its den in bamboo thickets in the understorey. Its main prey are rodents such as mice and rats, and lizards, captured on the ground as well as in trees. It probably hunts by day as well as at night.

protected, this species was hunted for its yellow-brown to silver-grey fur after the trade in ocelot fur (see above, right) declined in the 1980s.

regularly sized and spaced black spots

Felis pardalis
Ocelot

Length 50–100 cm (20–39 in)	
Tail 30–45 cm (12–18 in)	
Weight 11.5–16 kg (25–35 lb)	
Location S. USA to Central and South America	Social unit Individual
	Status Lower risk†

The chain-like rosettes and spots on this cat's body are highly distinctive. Typically cat-like in its nocturnal, solitary, tree-climbing lifestyle, the ocelot is wide-ranging and adaptable, living in a variety of habitats from grassland to swamp, as well as most types of forest. It takes a huge variety of prey: chiefly small rodents but also birds, lizards, fish, bats, and larger animals such as monkeys, turtles, young deer, armadillos, and anteaters. The ocelot has a gestation period of 79–85 days and an average litter size of 3 (range 1–2). Females breed from 2 years of age; males from about 2½ years.

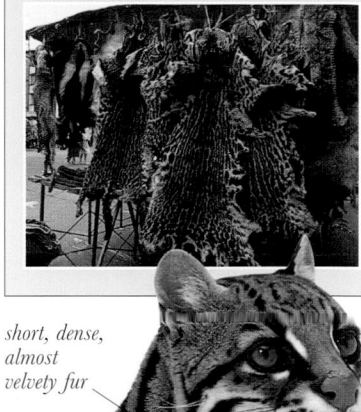

short, dense, almost velvety fur

pale background colour on underparts

SPOTS AND STRIPES
Similar in overall pattern to the much smaller margay (see below), the ocelot has a variable tawny background colour. The black rosettes on the back and sides grade into spots on the limbs and stripes on the head.

Felis wiedi
Margay

Length 46–79 cm (18–31 in)	
Tail 33–51 cm (13–20 in)	
Weight 2.5–4 kg (5½–8¾ lb)	
Location S. North America to Central and South America	Social unit Individual
	Status Vulnerable†

The margay has exceptional climbing abilities because of its almost "reversible" rear feet, being able to run head-first down a trunk or hang from a bough by one paw. Its mainly tree-dwelling prey includes rats, mice, squirrels, possums, young sloths, small birds, and invertebrates such as

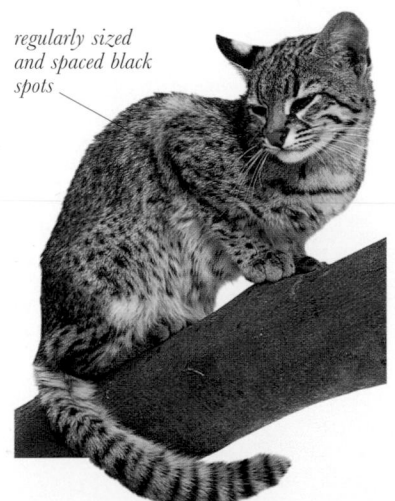

CONSERVATION

Ocelots were extensively hunted for fur in the 1960s and 1970s, trading at some 200,000 skins per year. The species is now protected over much of its range by national legislation, and its numbers in certain areas have since increased. However, the general threat of habitat loss due to deforestation poses new problems for this and many other forest-dwelling species.

grubs and spiders. It also occasionally eats fruits. The margay is nocturnal, resting in the safety of a tree fork by day. The gestation period is 76–85 days and litter size is one, rarely 2. Following the decline in the availability of ocelot fur (see above), margays became one of the most sought-after small cats for the fur trade. Hunting may still continue illegally in a few areas, but the primary threat to the margay is now forest clearance.

YOUNG LIFE

A young tiger is dependent on its mother for food for the first year or so of its life. By the time it is 2 years old, it will have enough power, strength, and experience to be able to catch prey for itself. It may start breeding in its fourth or fifth year, and lives on average until the age of 8–10.

TIGER IN ACTION

Tigers in the Ranthambhore reserve in northwest India have been known to charge into lakes after samba deer and become fully submerged with their prey. The tigers in this area have also been reported to have killed and eaten crocodiles.

Panthera tigris

Tiger

Length 1.4–2.8 m (4¹⁄₂–9¹⁄₄ ft)	
Tail 60cm–1.1m (2–3¹⁄₂ ft)	
Weight 100–300 kg (220–660 lb)	

Location S. and E. Asia   Social unit Individual

Status Endangered

The tiger is the largest member of the cat family, and its orange coat with black stripes and white markings is instantly recognizable. Its size, coat colour, and markings vary according to subspecies. Although 8 subspecies are recognized, 3 have become extinct since the 1950s, and the 5 remaining are endangered. The geographical distribution of the tiger once extended as far west as eastern Turkey, but it is now restricted to pockets of southern and eastern Asia. The tiger's habitat varies widely, from the tropical forests of Southeast Asia to the coniferous woodlands of Siberia, but its basic requirements are dense cover, access to water, and sufficient large prey. Hunting mainly by night, it takes mostly deer and wild pigs, and cattle in some regions, but it also eats smaller animals including monkeys, birds, reptiles, and fish, and readily feeds on carrion. Tigers will also attack young rhinoceroses and elephants. They may eat up to 40kg (88lb) of meat at a time and return to a large kill for 3–6 days. Tigers are usually solitary, but are not necessarily anti-social. A male is occasionally seen resting or feeding with a female and cubs, and tigers may also travel in groups.

CONSERVATION

Poaching, loss of habitat, and lack of prey because of over-hunting, pose the biggest threats to the world's tigers, which survive only in scattered populations in eastern Russia, China, and Sumatra, and from Vietnam to India. Although protected in most areas, tigers are still killed illegally, mainly for their skins and for use in traditional Chinese medicine. Various projects exist – such as at Chitwan National Park in Nepal, below – to monitor tiger populations, and to identify the conservation measures needed. Anti-poaching task forces have been set up, and there is an international campaign to educate users of traditional Chinese remedies about the role they are playing in the tiger's decline.

long, sensitive whiskers

white underside

sharp, retractable claws

DISTINCTIVE COAT

The Bengal tiger, Panther tigris tigris, *the most common of the subspecies, exhibits the classic tiger coat: deep orange with white undersides, cheeks, and eye areas, and distinctive black markings. Tiger stripes, which range from brown to jet-black, vary in number, width, and tendency to split. No 2 tigers have the same markings.*

LIVING TIGER SUBSPECIES

The 5 surviving subspecies of tiger differ markedly from one another. In general, those animals from southern areas are smaller and more deeply coloured than their larger and paler northern kin. Tigers that live in cold climates also tend to have thicker fur.

SUMATRAN TIGER

The smallest and darkest of the tiger subspecies is the Sumatran tiger, Panthera tigris sumatrae. *Only about 600 are thought to exist today.*

SIBERIAN YOUNGSTERS

The Siberian tiger, Panthera tigris altaica, *is the largest subspecies, and the lightest in colour, with the longest coat. Its numbers may be as low as 150–200.*

Felis yagouaroundi
Jaguarundi

	Length 55–77 cm (22–30 in)
	Tail 33–60 cm (13–23½ in)
	Weight 4.5–9 kg (10–20 lb)

Location S. USA to South America | Social unit Individual

| | Status Lower risk |

More mustelid than felid in overall proportions, with a pointed snout, long body, and shortish legs, the jaguarundi has several colour forms of unpatterned fur, from black – mainly in forests – to pale grey-brown or red – in dry shrubland. This cat hunts by day, often on the ground, in habitats ranging from semi-arid scrub to rainforest and swamp. Its main prey are birds, rodents, rabbits, reptiles, and invertebrates.

Felis concolor
Puma

	Length 1.1–2 m (3½–6½ ft)
	Tail 66–78 cm (26–31 in)
	Weight 67–105 kg (150–230 lb)

Location W. and S. North America, Central America, South America | Social unit Individual

| | Status Lower risk |

Larger than some "big" cats, the puma – also called panther, cougar, or mountain lion – is probably related more closely to smaller cats. Most of its fur is uniformly buff coloured. It makes many sounds, including an eerily human-like scream when courting, but it cannot roar. Amazingly adaptable, the puma lives in habitats ranging from tropical rainforest, high mountains, and conifer forest to desert. Small mammals such as mice, rats, rabbits, and hares, form the staple diet in many areas, as well as occasional sheep, young cattle, moose, and other livestock; the puma rarely scavenges. Births peak from February to September. The 2–3 (range 1–6) spotted cubs are born after an average gestation of 92 days, in a den among rocks or in a thicket. They take solid food from 6–7 weeks.

long, muscular rear legs, for powerful leaping

very large paws relative to overall size

Neofelis nebulosa
Clouded leopard

	Length 60–110 cm (23½–43 in)
	Tail 55–91 cm (22–36 in)
	Weight 16–23 kg (35–51 lb)

Location S., S.E., and E. Asia | Social unit Individual

| | Status Vulnerable |

Black-edged, darker-centred, irregular elliptical "clouds", on a tawny, grey, or silver background, give this smallest of the big cats its name. Information on this extremely elusive species is scarce, but its climbing skills rival many small cats, and it stalks, avoids danger, and rests mainly in trees. Prey includes monkeys, gibbons, birds, porcupines, and young wild boar and deer on the ground. Its principal threats are loss of its forest habitat and hunting for fur and meat.

Panthera pardus
Leopard

	Length 0.9–1.9 m (3–6¼ ft)
	Tail 60–110 cm (24–43 in)
	Weight 37–90 kg (82–200 lb)

Location W., C., S., E., and S.E. Asia, Africa | Social unit Individual

| | Status Lower risk |

Few other wild cats are as varied in appearance or in prey preference as the leopard, or have a wider geographical range. Its varied diet includes small creatures such as dung beetles and large mammals many times its own weight, such as antelopes. A large victim may provide enough food for 2 weeks, although such kills are usually made about every 3 days, twice as often for a female with cubs. The average litter of 2 is born after 90–105 days' gestation, and is cared for by the mother. Weaned by 3 months, they stay with her for a year or more, and siblings may associate for longer. Adaptable to human presence, leopards hunt to within a few kilometres of big cities. They have survived well, despite numerous threats.

BLACK PANTHER

Like many species of cats and other mammals, leopards may exhibit melanism. As a result of this genetic change (mutation), the skin and fur contain large amounts of the dark pigment, melanin. Most common in moist, dense forests, melanic leopards, known as "black panthers", were once viewed as a separate species. In deserts, leopards are pale yellow; in grass, they are deeper yellow.

rosettes and spots may be faintly visible

CACHING IN TREES

The leopard is an adept climber and uses its immense strength to drag its prey up into trees, for immediate consumption or for caching (hiding for future use). In the branches, it can eat undisturbed, and the meat is safe from scavenging hyenas and jackals.

HEAD AND SHOULDERS
The leopard's large head houses powerful jaw muscles to bite, kill, and dismember prey. The shoulders and forelimbs are also heavily muscled, to hold down victims and drag or haul prey into trees.

solid black patches and spots on and limbs and head

pale-centred rosettes on body

pale background colour on underparts

ringed tail

Panthera uncia
Snow leopard

Length	1–1.3 m
	(3 1/4 – 4 1/4 ft)
Tail	80–100 cm
	(32–39 in)
Weight	25–75 kg
	(55–165 lb)

Location C., S., and E. Asia

Social unit Individual

Status Endangered

Resembling the leopard (see p.214) in its wide range of prey, this woolly-furred big cat prefers crags and ridges in steppe, rocky shrub, and open conifer forest to altitudes of 5,000m (16,500ft). It can hunt yak or asses, but most prey are smaller – wild sheep, goats, marmots, pikas, hares, and birds. Breeding habits resemble similar-sized felids, although 4–5 cubs may be reared.

short, stocky limbs, for climbing

Panthera onca
Jaguar

Length	1.1–1.9 m
	(3 1/2 – 6 1/4 ft)
Tail	45–75 cm
	(18–30 in)
Weight	36–160 kg
	(79–350 lb)

Location Central America to N. South America

Social unit Individual

Status Lower risk

The New World's only "big" cat, the jaguar resembles the leopard (see p.214) but has rosettes with dark centres, and is more squat and powerful, with a large, broad head and heavily muscled quarters. It prefers a watery environment, such as permanent swamps and seasonally flooded forest, where its main prey are medium-sized mammals such as deer, peccaries, and tapirs. Despite legal protection and reduced hunting for fur, jaguars are increasingly at risk from habitat destruction and their elimination from cattle ranches.

Acinonyx jubatus
Cheetah

Length	1.1–1.5 m
	(3 1/2 – 5 ft)
Tail	60–80 cm
	(23 1/2 – 32 in)
Weight	21–72 kg
	(46–160 lb)

Location Africa, W. Asia

Social unit Individual/Pair

Status Vulnerable

Renowned as the world's fastest land animal, the cheetah can sprint at over 100kph (62mph) for 10–20 seconds, before it begins to overheat. If its prey can stay ahead for longer than this, it invariably escapes. The cheetah eats medium-sized ungulates such as Thomson's gazelle, as well as larger antelopes and smaller animals such as hares. It is more sociable than any other big cat except lions. Siblings leave their mother at 13–20 months, but may stay together for several more months – indeed, brothers may stay together for years.

CONSERVATION

In Namibia, in order to monitor the effect of wildlife management and livestock protection measures, some cheetahs are caught and fitted with radio collars. Following their release (below), the animal's movements are tracked.

COLORATION
The cheetah is yellowish with small black spots. Desert animals tend to be paler with smaller spots, and the "king cheetah", from southeast Africa, has the largest spots.

ringed tail

Panthera leo
Lion

Length	1.7–2.5 m
	(5 1/2 – 8 1/4 ft)
Tail	0.9–1.1 m
	(3 – 3 1/2 ft)
Weight	150–250 kg
	(330–550 lb)

Location Africa, S. Asia (Gir Forest)

Social unit Group

Status Vulnerable

Unique among felids, lions form close-knit, long-term social groups. In females these are called prides, and average 4–6 related adults and their cubs. Females tend to give birth at the same time and suckle each other's young. Prides occupy home ranges and members cooperate to hunt large prey such as zebra, wildebeest, impala, and buffalo. Individuals also forage alone for small rodents, hares, and reptiles. Adult males live alone, or in coalitions of usually 2–3 unrelated members or 4–5 relatives (originating from the same pride). A coalition defends a large area against other male coalitions, and holds mating rights over prides within it, but this tenure generally lasts only 2–3 years.

MALE AND FEMALE
The lion averages 180kg (400lb) against the lioness's 125kg (280lb). The male's skull is also significantly larger than that of the female.

ROLE-PLAYING

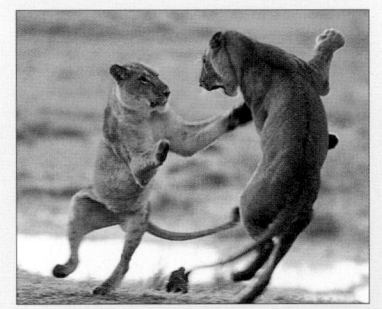

Retracted claws and non-exposed teeth show that these 2-year-old lionesses are "play-fighting" to develop skills for the hunt. The tussle helps to determine whether a lioness will be one of those who chase and direct prey – or one who carries out the ambush and kill. Play also helps to establish relative social status within the pride.

uniform tawny coat

thick mane for protection

ASIATIC LION

The Asiatic lion (*Panthera leo persica*) survives only in the Gir Forest region of northwest India, with a probable population of 200–300. It tends to be smaller than the African lion, with a fold of skin along the central underside of the belly; males have shorter manes. Prides are also smaller, usually 2 related females and young.

SEALS AND SEA LIONS

MAMMALS

PHYLUM	Chordata
CLASS	Mammalia
ORDER	Pinnipedia
FAMILIES	3
SPECIES	34

CLASSIFICATION NOTE
Seals and their allies were
traditionally classified in the
order Pinnipedia – the system
used here. Today, however,
most zoologists believe that
these mammals belong to
the order Carnivora. Seals
and their allies are broken
down into 3 families. It is
possible that sea lions (family
Otariidae) and walruses (family
Odobenidae) originated from
bear-like ancestors, while true
seals (family Phocidae) are
more closely related to otters.

ALTHOUGH CLUMSY ON LAND, pinnipeds (seals,
sea lions, and the walrus) are supremely agile
underwater. They have a streamlined body
and powerful flippers and can dive to depths
of over 100m (330ft). Some species can remain
underwater for over an hour. The order consists
of 3 groups: true seals, which have no external
ears and back flippers that point backwards;
eared seals (sea lions and fur seals), which have
small external ears and back flippers that can
be rotated forward for movement on land;
and the walrus, with its distinctive tusks. Only
eared seals and the walrus can support themselves in a semi-upright position on land.
Pinnipeds are found worldwide, mostly
in temperate and polar seas.

TEMPERATURE CONTROL

Pinnipeds have several heat-
regulating adaptations. In cold water
the blubber insulates the internal
organs, and blood flow to the flippers
is restricted. In warm conditions,
some species wave their flippers to
expel excess heat. In addition, true
seals and walruses can either contract
the blood vessels near the skin's
surface (to reduce heat loss in icy
water) or they can dilate these vessels
to gain heat when basking in the
sun. Eared seals, however, will enter
the water to avoid overheating.

COLOUR CHANGE
*The blood vessels in the skin
of these walruses are dilated
to maximize the amount of
heat they can absorb by lying
in the sun. As a result, their
bodies turn pink.*

UNDERWATER ACROBATS
*In water, pinnipeds such as these South
American sea lions are graceful, athletic, and
capable of swimming at high speed. While
underwater, they can communicate by sounds
produced using air retained in the lungs.*

ANATOMY

Most pinnipeds have a short face, a thick neck, and a torpedo-
shaped, flexible body. A layer of blubber beneath the skin
provides insulation, aids buoyancy, acts as an energy store, and
protects the organs. All species are covered with hair, except
the walrus, which is nearly hairless. Pinnipeds have large eyes
for good deep-water vision, excellent hearing, ear passages and
nostrils that can be closed underwater, and long whiskers that
enhance the sense of touch. Many species
display marked sexual dimorphism:
elephant seal males weigh 4 times
more than females.

*flattened
head aids
underwater
movement*

*flexible
backbone*

*smaller front
flippers*

*more powerful
back flippers*

SKELETAL FEATURES
*Pinniped limbs have been
modified to form flippers:
the arm and leg bones are
short, stout, and strong, and
the digits are elongated and
flattened. Also, as the backbone's
vertebrae have fewer interlocking
projections than most other mammals, the
spine is highly flexible. In true seals (shown
here), the back limbs are directed backwards.*

LIFE CYCLE

Unlike the other marine mammals
(cetaceans and manatees and the
dugong), pinnipeds have not abandoned
land entirely. In most species, during
the annual breeding season males
attempt to set up territories on suitable
beaches, fighting savagely for space
and excluding weaker males. Females
move onto the beaches, sometimes
several weeks after the males, and give
birth. A few days after a pup is born
(usually only one young is produced),
the female mates with the male in
whose territory she has settled. For
the majority of the gestation period,
which lasts approximately 8–15 months,
pinnipeds are mostly at sea and return
to land only when it is time to repeat
the breeding process.

MALE AGGRESSION
*In the breeding season,
there is fierce competition
between males, such as
these 2 elephant seals,
for mating rights. Only
the strongest males
are able to secure a
breeding territory.*

LARGE COLONY
*Breeding beaches
are always crowded.
Colonies of Cape fur
seals number several
thousand, with males
controlling harems of
7–9 females.*

Callorhinus ursinus
Northern fur seal

Length Up to 2.1 m
(7 ft)

Weight 180–270 kg
(400–600 lb)

Location North Pacific

Social unit Variable

Status Vulnerable

The male northern fur seal is brown-grey, while females and juveniles are silver-grey above, red-brown below, with a grey-white chest patch. The front flippers are long and appear "cut off" at the wrist. The diet includes many fish, and also birds such as loons and petrels. Most populations are migratory, with adult males heading south in August. The young stay on land for 4 months and by November follow with their mothers.

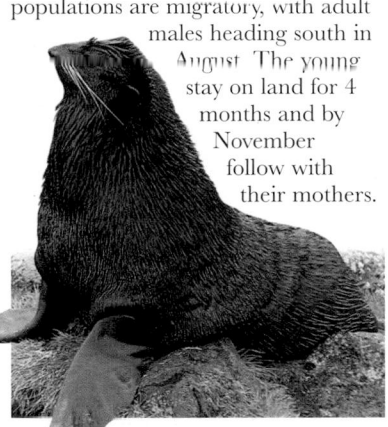

Arctocephalus pusillus
Cape fur seal

Length 1.8–2.3 m
(6–7 1/2 ft)

Weight 200–360 kg
(440–790 lb)

Location Southern Africa, S.E. Australia, Tasmania

Social unit Variable

Status Locally common

Cape fur seals from off South Africa tend to be a darker grey-brown than Australian individuals, and they dive twice as deep (to 400m/1,300ft). The young are about 70cm (28in) long and weigh 6kg (13lb) when born, in November–December. They play in "nursery" tidal pools, while the mothers feed at sea for several days at a time.

long, conspicuous ear flaps

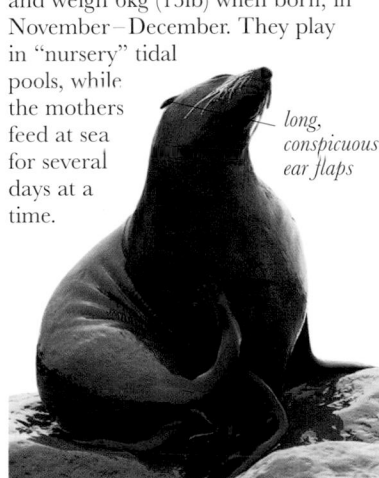

Zalophus californianus
Californian sea lion

Length Up to 2.4 m
(7 3/4 ft)

Weight 275–390 kg
(610–860 lb)

Location W. USA, Galapagos Islands

Social unit Variable

Status Vulnerable†

The "performing seal" of marine parks and aquaria, this species rarely strays more than 16km (10 miles) out to sea, and often enters harbours and estuaries for food and shelter. The male is dark brown and females and juveniles are a uniform tan colour. Males have a peak-like head crest. The main prey is shoaling fish such as herring and squid, and, around the Galapagos Islands, sardines, caught on short (2-minute) dives down to about 75m (245ft). In "El Niño years" the diet switches to whiting, salmon, and also birds such as guillemots. During the breeding season (May–July), males fight for small territories on the beaches and rock pools. However, after 2 weeks they must swim off to feed, and on return battle to regain a territory. The mother cares for her single pup (rarely 2) for 8 days, then enters a cycle of 2–4 days feeding at sea and 1–3 days suckling on land. This usually lasts for 8 months until the next birth.

Phocarctos hookeri
New Zealand sea lion

Length 2–3.3 m
(6 1/2–11 ft)

Weight 300–450 kg
(660–990 lb)

Location Subantarctic islands S. of New Zealand

Social unit Variable

Status Vulnerable

Also called Hooker's sea lion, this species is restricted to a few islands south of New Zealand. It forages up to 150km (95 miles) from land and then retires perhaps 1km (2/3 mile) inland, to rest among cliffs or trees. Males are dark brown with silver-grey hindquarters and a shoulder mane; females and juveniles are silvery or brown-grey above, yellowish or tan underneath. The diet includes fish, crabs, penguins, and seal pups.

Otaria byronia
South American sea lion

Length 2.3–2.8 m
(7 1/2–9 1/4 ft)

Weight 300–350 kg
(660–770 lb)

Location W., S., and E. South America, Falkland Islands

Social unit Variable

Status Vulnerable†

An enormous, heavy head and brown fur that is paler or yellow on the underside identify this powerful sea lion. The male of this non-migratory species has a copious shoulder and chest mane, and is twice the weight of the female (as in many sea lions). Its breeding (rookery) areas are used year-round for resting. The mothers coax their pups into the water after 1–2 months – a relatively early age for a sea lion.

Eumetopias jubata
Steller's sea lion

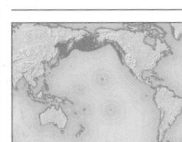

Length 3–3.3 m
(9 3/4–11 ft)

Weight 585–1,120 kg
(1,290–2,500 lb)

Location North Pacific rim

Social unit Group

Status Endangered

The largest sea lion, a male Steller's sea lion has a wide muzzle, a huge head, and a thick neck, and may be three times the weight of the female. Both are tawny or buff coloured, with black flippers furred only on the upper side. Breeding habits resemble those of other sea lions – colonies numbering over 1,000. Steller's sea lions dive deeply for fish, seals, and otters.

dark brown or black pup

Arctocephalus gazella
Antarctic fur seal

Length 1.6–2 m
(5 1/4–6 1/2 ft)

Weight 90–210 kg
(200–460 lb)

Location Antarctic and subantarctic waters

Social unit Variable

Status Locally common

The mane of the male Antarctic fur seal is accentuated by extra muscle and fat, deposited under the skin. He is dark grey-brown while the female is mid-grey. Males arrive at their breeding islands in November and compete for territories in which they can mate with about 5 females. Almost exterminated by fur hunting in the 19th century, this species is again threatened – by increased fishing of krill, a major component of its diet.

Odobenus rosmarus
Walrus

Length	3–3.6 m (9¾–12 ft)
Weight	1.2–2 tonnes (1⅛–2 tons)

Location Arctic waters	Social unit Group
	Status Vulnerable†

The male walrus is twice the weight of the female. The walrus's front flippers resemble those of sea lions, and the rear flippers those of seals. Its blunt, thickly whiskered muzzle widens rapidly to the head, neck, and chest, then tapers markedly to the tail, which is embedded in a web of skin. The walrus feeds mainly on sea-bed dwelling worms, shellfish, sea-snails, shrimps, and slow-

moving fish. It dives more than 100m (330ft) deep, for 25 minutes or more, to find prey using the touch of its whiskers and snout. It then excavates the food with its nose, aided by jets of water squirted from the mouth. Items are eaten mainly by suction with the mouth and tongue, rather than using the teeth. Walruses are social and huddle on land or ice floes in large, mixed groups of hundreds, which split at sea into smaller bands of less than 10; bachelor males tend to form their

own gatherings. Courting males make underwater pulses and bellows to attract partners, and mating occurs between January and March. The pup, born after 15 months (which includes 4–5 months' delayed implantation), is up to 1.2m (4ft) long and 75kg (165lb) in weight. It suckles for 6 months and is weaned over the following 18 months. Mothers are extremely protective of their young, and other females may "adopt" pups that have been orphaned.

CHANGEABLE COLOUR
The walrus's skin shows through its short, coarse hair, and changes colour with activity. It is usually greyish or cinnamon-brown, but, when sunbathing, the skin flushes rose-red, as though sunburned.

rough, heavily creased skin

tusks grow to 1m (3¼ft) in mature males

SPARRING RIVALS

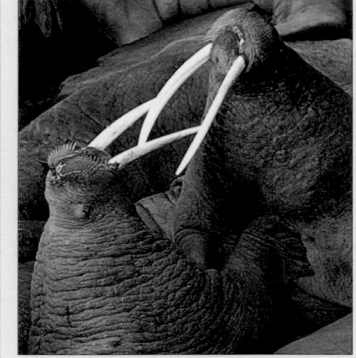

Most male walruses begin to breed at about 10 years. Males display and spar with their tusks (extra-long upper canine teeth), for a favoured position at the breeding site and, with it, access to females. Stab wounds may occur but are usually non-fatal; older males have many scars.

Monachus monachus
Mediterranean monk seal

Length	2.4–2.8 m (7¾–9¼ ft)
Weight	250–400 kg (550–880 lb)

Location Mediterranean, Black Sea, Atlantic (N.W. Africa)	Social unit Individua/Group
	Status Critically endangered

This species has smooth, dark brown fur, paler beneath, and eats mainly fish, such as eels, sardines, and tuna, as well as lobster and octopus. On land it is less social than most other seals, with the mother–pup pairs or small groups being widely spaced. This extremely rare species is sensitive to human disturbance such as tourism, so hides in sea caves. Cave collapses, pollution, overfishing, and viral infection are also serious threats.

Lobodon carcinophagus
Crabeater seal

Length	2.2–2.6 m (7¼–8½ ft)
Weight	220 kg (490 lb)

Location Antarctic and subantarctic waters	Social unit Variable
	Status Common

The long, lithe crabeater seal has oar-shaped, pointed front flippers and a silver-grey to yellow-brown coat with irregular darker spots and rings. One of the most abundant and fastest seals, it swims at 25kph (16mph). A typical feeding dive is down to 40m (130ft) for 5 minutes. Despite its name, the crabeater seal sieves krill using its unusual lobed teeth. Breeding habits are typical for seals, although the male stays with female and pup until weaning takes place at 3 weeks.

Hydrurga leptonyx
Leopard seal

Length	2.5–3.2 m (8¼–10 ft)
Weight	200–455 kg (440–1,000 lb)

Location Antarctic and subantarctic waters	Social unit Individual
	Status Locally common

The solitary, sinuous leopard seal is widest at the shoulders and, unusually for a true seal (phocid), swims with its front flippers, which have claws on the fingertips. The head is reptile-like in proportions, with no forehead and a wide, deep lower jaw. The 2.5-cm (1-in) canine teeth are adapted for seizing smaller seals, penguins, and other birds; the diet also includes squid and krill.

silver or grey coat with variable dark spots

Leptonychotes weddelli
Weddell seal

Length	2.5–2.9 m (8¼–9½ ft)
Weight	400–600 kg (880–1,320 lb)

Location Antarctic and subantarctic waters	Social unit Variable
	Status Locally common

Bulky, but small-headed and short-flippered, the Weddell seal has a short, blunt muzzle and few, short whiskers. To find fish, squid, and

other prey, it dives to depths of 500m (1,600ft) for one hour. This seal bites breathing holes in sheet ice with its long, upper incisor teeth.

silver-grey back

Ross seal

Ommatophoca rossi

Length 1.7–3 m (5 1/2 – 9 3/4 ft)	
Weight 130–215 kg (290–470 lb)	
Location Antarctic waters	**Social unit** Individual
	Status Vulnerable

The Ross seal has a distinctively blunt muzzle, a wide head, and long rear flippers. Its fur is the shortest of any seal and is dark grey to chestnut-brown with a buff underside and broad, dark bands along the body in both adults and pups. Less social than most seals, on ice it lives alone or as a mother–pup pair. The main food is squid, krill, and fish, caught at depths of several hundred metres. In November and December, males battle for territories around breathing holes in the ice used by females.

Southern elephant-seal

Mirounga leonina

Length 4.2–6 m (14–20 ft)	
Weight 2.2–5 tonnes (2 1/8 – 4 7/8 tons)	
Location Antarctic and subantarctic waters	**Social unit** Variable
	Status Locally common

The largest pinniped, the male southern elephant-seal is 4–5 times the weight of the female. His huge nose resembles an elephant's trunk, which he inflates when roaring at rivals during the 2-month breeding season. To establish dominance, he also rears up, slaps, and butts. The single pup is born after a gestation of 11 months (including 4 months' delayed implantation) and is suckled for 19–23 days by the ever-present mother, who loses one-third of her body weight during this time. After breeding and moulting, these elephant-seals migrate south, feeding on fish and squid and diving to 600m (2,000ft) for 20 minutes on average.

silvery grey fur, with scars and wounds in the male

fleshy, inflatable nose

Hooded seal

Cystophora cristata

Length 2.5–2.7 m (8 1/4 – 8 3/4 ft)	
Weight 300–410 kg (660–900 lb)	
Location North Atlantic to Arctic Ocean	**Social unit** Variable
	Status Locally common

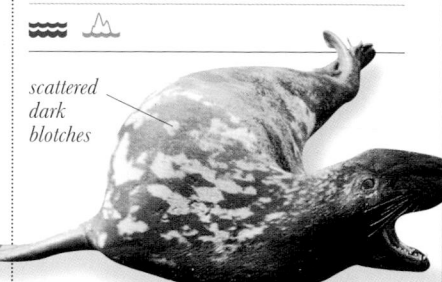

scattered dark blotches

The hooded seal has a wide, fleshy muzzle that droops over the mouth. The male intimidates his rivals at breeding time by inflating his nasal chamber to form a "hood", which doubles his head size, and also by extruding an internal membrane from his left nostril, which also inflates like a brown-red balloon. Mainly solitary, this seal migrates when not breeding or moulting, to follow the pack ice. The pup, born on an ice floe, is up to 1.1m (3 1/2 ft) long and 30kg (66lb) in weight, and is weaned in 4–5 days – the shortest time of any mammal.

Grey seal

Halichoerus grypus

Length 2–2.5 m (6 1/2 – 8 1/4 in)	
Weight 170–310 kg (370–680 lb)	
Location North Atlantic, Baltic Sea	**Social unit** Variable
	Status Common

tan. The face has small eyes, widely separated nostrils, and an angular nose. There are 3 populations: coastal northwest Atlantic, coastal northeast Atlantic, and the Baltic Sea. The first group are 20 per cent heavier and breed from December to February; Baltic grey seals breed until April, and those from the northeast July to December. The male grey seal does not defend a set territory.

The male grey seal is grey-brown with a few pale patches; the female is paler grey-

Harp seal

Pagophilus groenlandicus

Length 1.7 m (5 1/2 ft)	
Weight 130 kg (290 lb)	
Location North Atlantic to Arctic Ocean	**Social unit** Variable
	Status Common

The harp seal has close-set eyes in a wide face, black fingertip claws, and silver-white fur with curved dark marks on the back that form a harp shape. It eats cod, capelin, and similar fish, migrating with the edges of pack ice. Social both on ice and in water, harp seals travel in dense, noisy groups. Pups, born February–March on ice, have yellow fur that whitens for 2 weeks before the first moult.

Baikal seal

Phoca sibirica

Length 1.2–1.4 m (4 – 4 1/2 ft)	
Weight 80–90 kg (175 –200 lb)	
Location E. Asia (Lake Baikal)	**Social unit** Individual
	Status Lower risk

One of the smaller seals, and the only solely freshwater pinniped species, the Baikal seal resembles its marine cousins in most respects. However, it is mainly solitary; and females tend to mate with the same male over years (serial monogamy). The single pup is born in an ice lair, moults its woolly white coat to the silvery grey adult fur after 6–8 weeks, and may suckle for 10 weeks. At 50–55 years, its lifespan is longer than many other seals.

Common seal

Phoca vitulina

Length 1.4–1.9 m (4 1/2 – 6 1/4 ft)	
Weight 55–170 kg (120 –370 lb)	
Location North Atlantic, North Pacific	**Social unit** Variable
	Status Common

Also known as the harbour seal, this is the most widespread pinniped, with at least 5 subspecies – one of them, the Ungava seal, lives in fresh water in northern Quebec, Canada. Below the large, set-back eyes, the angled, close-set nostrils form a V shape. Colour is extremely variable, mainly dark to pale grey-brown with small rings and blotches. A wide-ranging opportunist, the common seal may cause problems in fisheries. Its main prey are herring, sand-eels, gobies, hake, and whiting, caught on dives of 3–5 minutes.

ELEPHANTS

PHYLUM	**Chordata**
CLASS	**Mammalia**
ORDER	**Proboscidea**
FAMILIES	**1 (Elephantidae)**
SPECIES	**3**

THE LARGEST LIVING ANIMALS on land – the male African elephant may be as tall as 4m (13ft) and weigh nearly 10 tonnes ($9^7/_8$'tons) – elephants are characterized by pillar-like legs, a thick-set body with a convexly curved spine, large ears (relatively smaller in Asian elephants), and a heavy head with a long, mobile trunk. African and Asian elephants live in savanna and light forest; African forest elephants (recently given species status) mainly live deep in the African rainforest (they occasionally venture on to the savanna). Elephants live for about 60 years – longer than any other mammals except humans. Males (and females to a lesser degree) grow throughout life: 50-year-old males are noticeably larger than 20-year-olds.

ANATOMY

Perhaps the most distinctive feature is the trunk, a flexible elongation of the upper lip and nose that consists of thousands of muscle pairs. It is used like a "5th limb" to pluck grass, pull down branches, lift logs, or squirt water or dust. Also immediately noticeable are the tusks (upper incisors), which are large, thick, and curved in most bull elephants; cows have smaller tusks (in female Asian elephants they do not protrude beyond the lip). The skeleton consists of thick, heavy bones, which are able to support the animal's great weight. The large, fan-shaped ears, which contain a network of blood vessels, are constantly in motion to aid heat loss. In aggressive displays, the ears are spread sideways. The skin is thick, finely wrinkled, and sparsely haired.

LIGHTENED SKULL
air cells
The skull is filled with air cells to lighten the weight of the bone. The long incisors (tusks) have deep, downward-pointing sockets. The lower jaw has a spout-like chin that, unlike in most mammals, moves horizontally during chewing.
molar *long chin*
AFRICAN ELEPHANT SKULL *incisor (tusk)*

FEEDING

Elephants have large, ridged cheek teeth (molars and premolars) to deal with their coarse diet of bark, leaves, branches, and grass (African forest elephants also eat fruits). In eating these foods elephants cause enormous damage: grass is pulled up in tufts, branches are broken off, bark is stripped, and small trees are sometimes uprooted. Some areas have alternated between closed woodland and open savanna, depending on the number of elephants living there.

STRIPPING FOLIAGE
Elephants use their mobile trunk to pull down branches. An adult needs to eat about 160kg (350lb) of food daily.

DIGGING FOR SALT
Elephants often need to supplement their diet with extra salt. This juvenile African elephant is loosening fragments of salt-rich soil with its long tusks. Juveniles learn from older members of the herd where to find salt.

upper process
single process
lower process
AFRICAN
ASIAN

TRUNK SHAPE
African elephants have 2 opposing, finger-like outgrowths (processes) at the tip of the trunk; Asian elephants have one. In both, the processes are used to pick up small objects.

TAKING A DUST-BATH

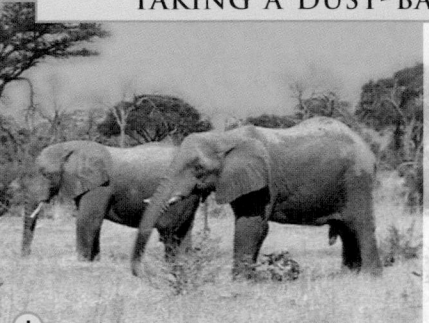

DAILY ROUTINE
To keep the skin healthy, African elephants take a daily dust-bath.

SUCKING UP THE DUST
Dust is sucked up into the trunk, which is a tubular extension of the upper lip.

PROTECTION
The dust acts as a sunscreen, protecting the elephant's skin from the direct rays of the sun.

REPELLING INSECTS
Dust is also a good insect repellent, deterring insects from biting the sensitive skin.

SOCIAL GROUPS

Elephants live in family groups that consist of the oldest, most experienced female (the matriarch) and other females of various ages (and their young). For protection, or when feeding in lush areas, small herds of African elephants may join together to form groups made up of several hundred individuals. African forest elephants and Asian elephants live in small family groups only. Males, however, only join the herd when a female is sexually receptive and are otherwise either solitary (older bulls) or live in bachelor groups (young bulls). Adult bull Asian elephants have annual periods of sexual excitement, called "musth" (bull African elephants have an equivalent condition, about which less is known).

PROTECTING YOUNG

Elephant calves are protected from predators and other dangers by all members of the herd, which are usually blood relatives. This Asian elephant calf is only a few weeks old and remains close to its mother. Two smaller females are close by, ready to assist the mother if necessary.

CONSERVATION

The Asian elephant is endangered due to competition with people for land. Although this is becoming a problem for African elephants, the real damage to African elephant populations was caused by hunting – elephant tusks are the major source of ivory. As the gestation period of elephants is about 22 months, numbers are slow to recover. In 1989 a government-approved burning of stock-piled ivory in Kenya (below) sent a clear message that the ivory trade would no longer be tolerated. In 1990, the international sale of ivory was banned, but demand remains. Despite concerns about poaching, controlled sales resumed in 1997 in 3 African countries, and some of the revenue is used for conservation purposes.

LIVING TOGETHER

This is a typical African elephant family group. Communication within a herd takes many forms, including vocalizations (some of which are below the range of human hearing), touch, foot stamping, and body postures. Cooperative behaviour – such as employing a system of lookouts while bathing – is common.

DEPOSITING THE DUST
The elephant blows out the dust through the trunk, depositing it on the back and head.

KEEPING THE SKIN HEALTHY
To maintain good skin condition, regular dust-baths are as important as water-baths.

Loxodonta africana

African elephant

Length 4–5 m
(13–16 ft)

Tail 1–1.5 m
(3 1/4 –5 ft)

Weight 4–7 tonnes
(3 7/8 –6 7/8 tons)

Location Africa

Social unit Group

Status Endangered

Although it is also called the African bush or savanna elephant, this species – the largest of the 3 elephant species – lives in varied habitats from desert to high rainforest. It has larger ears than the Asian elephant, a concave curve to its back, and 2 processes on the tip of the trunk rather than one (see left). Both male and female African elephants have forward-curving tusks (incisor teeth), which are sometimes used as tools to loosen mineral rich soil that is then eaten. Requiring substantial amounts of food and a large area in which to forage for it, a herd of African elephants may cause dramatic changes to the environment, especially during prolonged periods of drought.

forward-curving tusk

Loxodonta cyclotis

African forest elephant

Length 3–4 m
(9 3/4 –13 ft)

Tail 50–120 cm
(2 1/2 –4 ft)

Weight 0.9–3 tonnes
(7/8 –3 tons)

Location W. and C. Africa

Social unit Group

Status Endangered

Formerly regarded as a subspecies of the African elephant, this species is smaller, has darker skin, more rounded ears, and a hairier trunk. The yellow or brownish tusks are parallel and point downwards, adaptations that allow the African forest elephant to move freely through dense vegetation.

Elephas maximus

Asian elephant

Length Up to 3.5 m
(11 ft)

Tail 1–1.5 m
(3 1/4 –5 ft)

Weight 2–5 tonnes
(2 –4 7/8 tons)

Location S. and S.E. Asia

Social unit Group

Status Endangered

Asian elephants have smaller ears than the African species, and a unique tip to their trunk (see opposite). The tusks are small and may be absent in females (above). The molar teeth are very like those of the extinct mammoth, which suggests a close relationship between them. The Asian elephant has a long association with humans, and animals from all 4 subspecies – Malaysian, Sumatran, mainland Indian, and Sri Lankan – have been domesticated.

AARDVARK

PHYLUM	Chordata
CLASS	Mammalia
ORDER	Tubulidentata
FAMILIES	1 (Orycteropodidae)
SPECIES	1

THE ONLY SURVIVING MEMBER of the order Tubulidentata is the aardvark, a solitary animal found in Africa. It is characterized by non-functional, columnar cheek teeth, a long snout, large ears, a pig-like body, and powerful limbs and shovel-shaped claws for digging. Ears that can be folded back (and a profusion of nostril hairs) help keep out dirt when burrowing. The aardvark has a primitive brain and poor eyesight. It does, however, have an excellent sense of smell. This is used to locate termites and ants, which are then captured, using the animal's long, sticky tongue.

MASTER BURROWER
The aardvark is a fast and prolific burrower, using its strong, clawed front limbs to dig and its back feet to push away excavated soil. Some aardvark burrows consist of an extensive tunnel network; others are shorter and provide temporary refuge. The aardvark always leaves its burrow head first, as shown here.

Orycteropus afer
Aardvark

Length	1.6 m (5 1/4 ft)
Tail	55 cm (22 in)
Weight	38–64 kg (84–140 lb)

Location Africa (south of the Sahara)

Social unit Individual

Status Unconfirmed

Also known as ant-bear, the aardvark is one of the most powerful mammal diggers, excavating burrows up to 10 m (33 ft) long around its home range of 2–5 square km (3/4 – 2 square miles). The single young is born after a gestation period of 243 days and weighs 1.7 kg (3 3/4 lb). Aardvarks chew one species of ant with their molar teeth, but other species of ant and termites are swallowed whole and probably ground up in their muscular stomach.

large ears for acute hearing

SEASONAL DIET

The aardvark favours ants as its food, which are more abundant in summer, but it also eats termites at times when ants are not available. It breaks into a nest or mound using its front feet. The dense mat of hairs that surrounds the aardvark's nostrils effectively filters dust as it digs.

HUNCHED BACK
The aardvark has a distinctive curved back, and its snout, ears, and tail are long and tapering. The bristly, scant, brown fur is tinged with yellows and greys.

HYRAXES

PHYLUM	Chordata
CLASS	Mammalia
ORDER	Hyracoidea
FAMILIES	1 (Procaviidae)
SPECIES	8

ALTHOUGH HYRAXES resemble rabbits in size and shape, genetic evidence suggests that they are more closely related to primitive hoofed mammals. The pads of their feet are moistened by glandular secretions, which make the soles more adhesive. This, along with the opposable toe on each back foot, enables hyraxes to climb steep rock faces. They are found in Africa and parts of the Middle East. Some species inhabit rocky outcrops; others are arboreal. They can survive food shortages as they eat almost every type of vegetation and require very little water (their kidneys are highly efficient).

KEEPING WARM
Hyraxes have poor temperature control. These rock hyraxes are huddling together for warmth.

Procavia capensis
Rock hyrax

Length	30–58 cm (12–23 in)
Tail	20–31 cm (8–12 in)
Weight	3–5 kg (6 1/2 –11 lb)

Location S. and E. Africa, W. Asia

Social unit Group

Status Locally common

The rock hyrax has a plump body and short, dense, grey or grey-brown fur, which is paler below. Also called the rock dassie, it lives in colonies of 4–40, usually of one dominant male, other males, and females and young. Although found in a wide variety of habitats, it usually lives in rocky outcrops and crags, among boulders, where it makes a grass-lined nest.

small, rounded ears

Dendrohyrax arboreus
Tree hyrax

Length	40–70 cm (16–28 in)
Tail	1–3 cm (3/8 –1 1/4 in)
Weight	1.5–4.5 kg (3 1/4 –10 lb)

Location E. and southern Africa

Social unit Variable

Status Vulnerable

The tree hyrax or tree dassie has a yellowish patch on the back near the rump, but is otherwise grey-brown with buff underparts. The head, legs, and tail seem almost too small for the stocky body. True to its name, the tree hyrax lives among trees, shrubs, and creepers, and nests in a tree-hole. It feeds on the ground only occasionally. Like all hyraxes, this species has poor internal control of body temperature, so groups often sunbathe to warm up, or rest in shade to cool down. Litters of 1–3 are born after a gestation period of 7–8 months.

DUGONG AND MANATEES

PHYLUM	Chordata
CLASS	Mammalia
ORDER	Sirenia
FAMILIES	2
SPECIES	4

SIRENIANS – THE DUGONG and manatees – are large, slow-moving creatures with a streamlined body. They are the only marine mammals that feed purely on plants. Sirenians must rise to the surface to breathe, but they can remain submerged for up to 20 minutes. Even though they have no enemies apart from humans, sirenians number only around 130,000 – this is less than any other mammal order (see panel, right).

ANATOMY

Sirenians have paddle-like front limbs and a flat tail to aid propulsion. Their skin is thick and tough, and they have a relatively small brain. Due to the large volume of gas given off during the digestion of plant matter, sirenians are highly buoyant. To compensate, their bones are heavy and dense. Under water, sirenians can close their nostrils and contract their eyelids.

FEEDING

Sirenians strip vegetation using their large, mobile upper lip. Food is then crushed between horny plates on the front part of the palate and on the lower jaw, and finally ground between the teeth.

FORAGING ON THE SEA BED
Sirenians, such as this dugong, frequently search the sea bed for the rhizomes (underground stems) of sea grasses, which have high concentrations of carbohydrates.

CONSERVATION

In the past sirenians were hunted extensively for their meat, hide, and oil. Today they are vulnerable to injury or death caused by boat propellers (below), fishing nets, and the pollution of coastal waters.

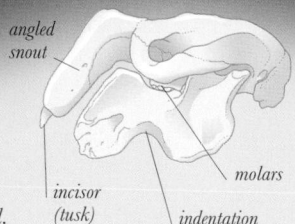

DUGONG SKULL
The steeply angled snout bearing a single pair of tusks and a marked indentation in the lower jaw are characteristic features of a dugong skull.

angled snout

incisor (tusk)

molars

indentation

MOTHER AND CALF
Sirenians are slow breeders: usually only one calf is produced every 2 years. The "mouthing" contact between this manatee mother and calf helps to preserve the family bond.

Dugong dugon

Dugong

Length	2.5–4 m (8 1/4–13 ft)
Weight	250–900 kg (550–1,985 lb)

Location E. Africa, W. Asia, S. Asia, S.E. Asia, Australia, Pacific islands

Social unit Group

Status Vulnerable

The dugong has a crescent-shaped tail, and short foreflippers. Primarily diurnal, it moves regularly each day between on- and offshore areas depending on the tides and food supply; in some areas, it undertakes longer seasonal migrations – perhaps hundreds of kilometres –

to follow marine plant growth and avoid cold-water currents. Some dugongs are solitary but most form loose groups averaging 10–20, exceptionally 100 or more, with little social structure. Members may gather to intimidate and butt predators such as sharks. Males compete for females by sounds and pushes. Courtship and mating (which is monogamous) are similarly auditory and tactile. One offspring, up to 1.2m (4ft) long and 35kg (77lb) in weight, is born after a gestation of 13–14 months. It is cared for by the mother, with help from older siblings and female relatives, and is weaned by 18 months. A dugong may live for up to 70 years.

crescent-shaped tail

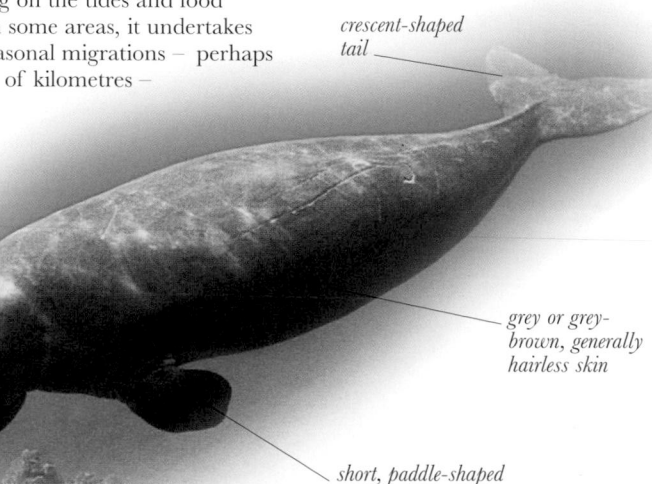

grey or grey-brown, generally hairless skin

short, paddle-shaped foreflipper

Trichechus manatus

West Indian manatee

Length	2.5–4.5 m (8 1/4–15 ft)
Weight	200–600 kg (440–1,320 lb)

Location S.E. USA to N.E. South America, Caribbean

Social unit Variable

Status Vulnerable

Best known of the 3 manatee species (the other 2 being the Amazonian and West African), the West Indian manatee lives along shallow shores and estuaries, and in nearby rivers and freshwater lagoons. Groups of up to 20 swell to 100 or more where food is plentiful; however, there is little cohesion, and individuals come and go, ranging widely. Reproduction is similar to the dugong's (see left), although the manatee is polygynous.

FOOD INTAKE

Feeding occurs from the surface down to about 4m (13ft). The manatee holds food with its flippers, and directs it into the mouth using its flexible lips. Daily food intake is up to one-quarter of its body weight, and may include a few fish (for protein).

BLUNT FRONT
Like other sirenians, this species has tiny eyes, and no external ear flaps. Its grey or grey-brown skin, paler below, may harbour growths of algae.

HOOFED MAMMALS

PHYLUM	Chordata
CLASS	Mammalia
ORDERS	Perissodactyla, Artiodactyla
FAMILIES	13
SPECIES	244

CLASSIFICATION NOTE

Hoofed mammals are classified in 2 orders: Perissodactyla (odd-toed hoofed mammals) and Artiodactyla (even-toed hoofed mammals). Although superficially similar, the 2 orders are not closely related. However, they are often grouped together (as here) because they share a range of common traits. Elephants, hyraxes, and the aardvark are also sometimes regarded as having hooves and may be included in the group, which is also known as the "ungulates".

HOOFED MAMMALS ARE a highly successful group. Their position as the dominant terrestrial herbivores can be attributed largely to their speed and endurance (they are able to outrun most predators), and to the fact that they are well equipped to break down the cellulose in their plant diet. Despite a variety of body forms, most species have a long muzzle, a complex battery of grinding teeth, and a barrel-shaped body. The group consists of odd-toed hoofed mammals (such as tapirs) and even-toed hoofed mammals (such as deer). Wild odd-toed hoofed mammals are found in Africa, Asia, and South and Central America. Wild even-toed hoofed mammals are distributed worldwide (except the West Indies, Australasia, and Antarctica). Hoofed mammals are mostly found in open habitats, such as savanna. Domestic hoofed mammals are found almost anywhere there are humans.

ANATOMY

The ability of hoofed mammals to run swiftly for long distances is largely due to the structure of the limbs, which are adapted for simple but powerful forwards and backwards movement. Each limb is embedded in the body wall as far down as the elbow or knee joint. Below this joint are the radius and ulna (front limbs) or tibia and fibula (back limbs), then the greatly elongated metapodials (the palm and foot bones in humans). This longer lower limb (and increased movement in the shoulder joint) give a longer stride length and hence more speed. These animals also have a reduced number of toes, which means fewer muscles and tendons and therefore lower energy demands (which aids endurance). They run on their toes (unguligrade gait), which are encased in hooves.

single (third) toe
third toe
RHINOCEROS

metapodial
reduced fifth toe
HORSE

reduced second toe
third toe
PIG

metapodials mostly fused
third toe
CAMEL

ODD AND EVEN TOES

In odd-toed hoofed mammals (rhinoceros and horse), the leg's weight rests on the central (third) toe. In even-toed hoofed mammals (pig and camel) the weight is borne by the third and fourth toes; the second and fifth toes are greatly reduced (pig) or lost (camel).

complex, branched antlers

simple, unbranched horn

REINDEER

RHINOCEROS

HORNS AND ANTLERS

Most hoofed mammals have antlers or horns. Antlers are bony outgrowths of the skull, which in deer are shed each year. Horns are permanent and have a keratinous sheath over a bony core. The horn of the rhinoceros is entirely keratinous.

ESCAPING DANGER

Hoofed mammals must be able to detect danger quickly to survive. To do this they have mobile, tubular ears and acute hearing, an excellent sense of smell, and eyes on the side of the head, giving all-round vision. When frightened, they flee at great speed. Antelopes, such as these impala, often make spectacular leaps as they escape.

BIRTH ON THE HOOF

① THE BIRTH BEGINS
During labour, this female wildebeest is vulnerable and restless, alternately lying down and standing up.

② THE HEAD APPEARS
The calf is born head first, but the long front legs are first into the outside world.

③ A NEW LIFE
The newborn wildebeest emerges, partially covered by the amniotic membrane.

④ SYNCHRONIZED CALVING
Most wildebeests calve in the same 3-wee... so that, despite some losses, many calves su...

MAMMALS

FEEDING

Almost all hoofed mammals are herbivores. The plants they eat contain indigestible cellulose (the major component of plant cell walls), which is split into digestible carbohydrates via bacterial fermentation. In ruminants (see below), food passes slowly through their system to maximize the nutrition gained. These animals thrive where food is limited but of high quality. In hindgut fermenters (see below), food is not retained in the stomach and passes through the system more quickly. These animals live where food is plentiful but of poor quality. As a result, a larger volume of food must be eaten to obtain enough nutrients.

GRAZING
Grazers feed almost exclusively on grass. To find fresh pasture, some hoofed mammals migrate long distances. The hippopotamus (above) usually feeds at night, using its horny lips to crop the grass.

abomasum   omasum
colon
caecum
passage of food (first time)
passage of food (second time)
small intestine   reticulum   rumen   **RUMINANT**

caecum
colon
passage of food
small intestine
stomach
HINDGUT FERMENTER

DIGESTIVE SYSTEMS
Ruminants, such as buffaloes, have a complex stomach. In the first chamber, the rumen, bacterial fermentation occurs; food is then regurgitated, re-chewed ("chewing the cud") and swallowed, this time passing through the digestive system. Hindgut fermenters, such as zebras, have a simple stomach: the fermenting bacteria are in the caecum and at the start of the colon.

BROWSING
Browsers eat almost any plant material. These mountain goats spend much of their time in mountainous areas where grass is scarce – they supplement their diet by eating mosses, lichens, herbs, and woody plants.

DEFENDING RESOURCES
Hoofed mammals living in arid areas, such as these onagers, often form mixed-sex herds. Males tend to defend resources rather than guard a harem of females.

DEFENDING A HAREM
During the breeding season, some male hoofed mammals, such as this red deer, gather a harem, which they defend fiercely. The loud roaring and large antlers of the stag help deter rivals.

SOCIAL SYSTEMS

Hoofed mammals have a variety of social systems, depending on factors such as habitat, body size, whether breeding is seasonal, and whether they migrate. The following examples are typical hoofed mammal social organizations. Tapirs, rhinoceroses, and some forest antelopes are mostly solitary, and a male's territory covers that of several females. Dik-diks live in pairs that occupy small territories. In gazelles, males set up small territories, and female herds wander in and out of these areas. Hartebeests and most zebras live in harems consisting of a male and a number of females. Red deer, on the other hand, form separate-sex herds except during the breeding season (which is known as the "rut"). Males fight with each other, and the most successful will collect a large harem of females.

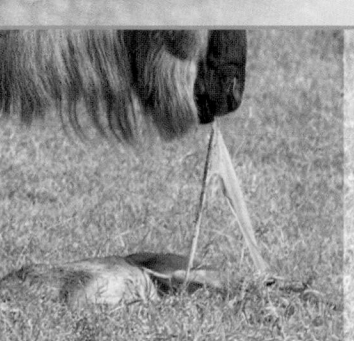

NUTRITIOUS MEAL
mother eats the afterbirth. Once this is she will lick the calf to stimulate it.

6

PROTECTIVE INSTINCTS
A newborn calf is vulnerable, and the mother is reluctant to let even another wildebeest approach.

7

STANDING UP
The calf attempts to stand. In less than 45 minutes it will be able to run, reducing the risk of predation.

8

REJOINING THE HERD
Mother and calf quickly rejoin the herd: in a group the young calf is less conspicuous.

HORSES AND RELATIVES

PHYLUM	Chordata
CLASS	Mammalia
ORDER	Perissodactyla
FAMILY	Equidae
SPECIES	10

AN ENDURING SYMBOL of grace and freedom, horses and their relatives (equids) are the ultimate odd-toed hoofed mammals – they have just a single toe on each foot. Equids – horses, asses (including the onager and the kiang), and zebras – have a long neck and head and long, slender legs. They possess great stamina and can run at speed: the swiftest wild equid, the onager, can attain 70kph (43mph) for short periods. They are found in grasslands and deserts of Africa and Asia, and various species have been widely introduced across the world.

ANATOMY

Equids are characterized by a deep chest, a mane on the neck, a tufted or long-haired tail, a solid hoof on each foot, areas of hard, thickened skin (called chestnuts) on the inside of the front legs above the knee, and mobile lips and nostrils. The eyes, which have oblong pupils, are at the sides of the head for good all-round sight (to help detect predators). Day and night vision is excellent. The ears are long and can twist to locate the sources of sounds, without having to move the body. Hearing is acute.

All species have a heavily haired coat, which is usually uniform in colour in horses and asses. Zebras have striking black-and-white stripes; the function of these is thought to be either social recognition, temperature regulation, or to create a "dazzle" effect to confuse predators.

FEEDING

Horses and relatives eat mainly grass (they have a battery of hard-wearing cheek teeth for shearing grass), although they will also feed on desert vegetation

and may browse on bark, leaves, buds, and fruits. Unlike cattle (for example), they do not ruminate but instead employ a hindgut fermentation system (see p.225). This allows them to take in large amounts of food, which passes rapidly through the digestive tract. Quality of food is therefore less important than quantity, which means equids can survive in arid habitats. They usually rest during the heat of the day and forage in the morning, evening, and night.

SOCIAL GROUPS

Wild horses, Burchell's zebras, and mountain zebras live in groups consisting of mares and their young, led by a "harem" stallion, who protects and herds them. This stallion also defends the group's territory and attempts to prevent other stallions mating with his mares. Young females may remain in the same group as their mothers

or they may join a different group; young males leave at maturity and try to collect their own harem. Wild asses and Grevy's zebras, on the other hand, have a different social organization, without long-term associations. Breeding stallions defend large territories – up to 15 square km (6 square miles) – and these are marked by dung piles. The stallion mates with mares that range through his territory.

Equids communicate with each other by whinnying or braying, and these vocalizations vary depending on species. To assess the sexual condition of mares, stallions sniff mares' urine. To analyse the scent in detail, they roll back their upper lip to induce the inhaled air into the Jacobson's organ, a special pouch in the roof of the mouth. This is called the "flehmen" response.

HORSES AND PEOPLE

The donkey, a descendant of wild asses, was domesticated in the Middle East before 3,000BC. The domestic horse, derived from the wild horse, appeared later (about 2,500BC) – probably from central Asia. It displaced the donkey for a number of purposes – including transport, agriculture, warfare, and recreation – although the donkey remained a popular beast of burden. Domesticated equids continue to serve people, and most horses alive today are domestic.

Most of the wild equids are highly endangered due to habitat loss and hunting: only Burchell's zebra and the kiang are relatively numerous.

FIGHTING FOR DOMINANCE
Conflicts between equids are common throughout the mating season. During this fight, the Burchell's zebra stallions may bite, rear up and strike out with their front feet, and kick with their back feet. The loser is driven off, leaving the victorious stallion with control of the harem, usually between one and 6 mares.

Equus africanus

African wild ass

Length	2–2.3 m
	(6 1/2–7 1/2 ft)
Tail	45 cm
	(18 in)
Weight	200–230 kg
	(440–510 lb)
Location E. Africa	Social unit Group
	Status Critically endangered

The African wild ass lives in rocky deserts where the ground temperature exceeds 50°C (122°F). It eats almost any plant food, from grasses to thorny acacia bushes, and goes without water for several days. The upperparts are buff-grey in summer and iron-grey in winter; the mane is sparse but erect. Females mate only with mature males that hold territories.

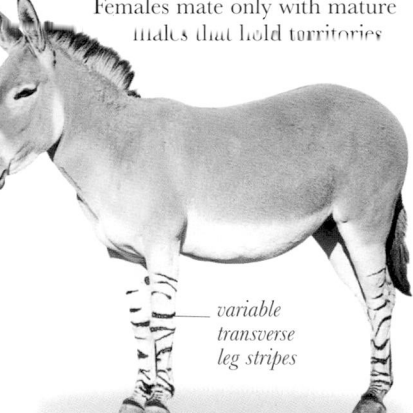

variable transverse leg stripes

Equus hemionus

Onager

Length	2–2.5 m
	(6 1/2–8 1/4 ft)
Tail	30–49 cm
	(12–19 1/2 in)
Weight	200–260 kg
	(440–570 lb)
Location W., C., and S. Asia	Social unit Group
	Status Vulnerable

This Asian wild ass eats a variety of vegetation, including grasses and succulent plants. Females and young onagers form loose, wandering herds while immature males gather in bachelor groups. Solitary mature males kick and bite rivals to occupy the territory they need for breeding. Mainly buff, tawny, or grey in coloration, the onager is white underneath, with a dark mane, back stripe, ear tips, and feathered tail tip.

Equus przewalskii

Przewalski's wild horse

Length	2.2–2.6 m
	(7 1/4–8 1/2 ft)
Tail	80–110 cm
	(32–43 in)
Weight	200–300 kg
	(440–660 lb)
Location Reintroduced to E. Asia	Social unit Group
	Status Extinct in wild

This horse now survives in zoos, parks, and field stations, although there have been several attempts to reintroduce it in Mongolia. It lives in cohesive, long-term herds that wander great distances for grass, leaves, and buds. A typical herd is led by a senior mare and has 2–4 other mares, their offspring, and one stallion who stays on the periphery. A single foal is born after a gestation of 333–345 days.

SMALL BUT STOCKY
Przewalski's wild horse is heavily built, with a thick neck, a large head, and short legs compared to domesticated horses.

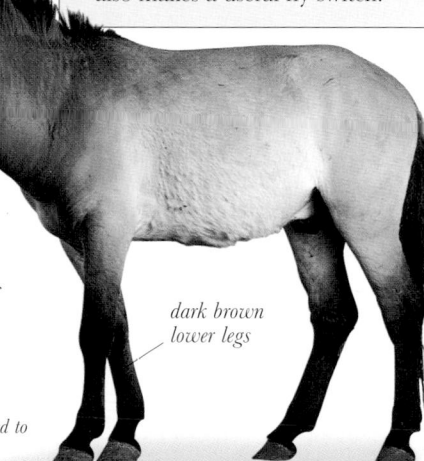

dark brown lower legs

MUTUAL GROOMING

Social grooming is important for this species, as for most wild horses, serving to reinforce herd bonds. Usually two animals stand nose to tail so that they can look out for danger in both directions, and each nibbles the other's shoulder and withers. The tail also makes a useful fly switch.

Equus grevyi

Grevy's zebra

Length	2.5–3 m
	(8 1/4–9 3/4 ft)
Tail	38–60 cm
	(15–23 1/2 in)
Weight	350–450 kg
	(770–990 lb)
Location E. Africa	Social unit Group
	Status Endangered

The largest zebra and biggest wild equid, Grevy's zebra has dense, narrow stripes that remain distinct all the way down to the hooves; the belly and tail base are white. Males occupy huge territories, up to 10 square km (4 square miles). Females and foals roam freely, perhaps gathering in small, loose herds to graze, but there are no long-term herd bonds.

Equus burchelli

Burchell's zebra

Length	2.2–2.5 m
	(7 1/4–8 1/4 ft)
Tail	47–56 cm
	(18 1/2–22 in)
Weight	175–385 kg
	(390–850 lb)
Location E. and southern Africa	Social unit Group
	Status Lower risk†

This successful and widespread zebra has a diet of nine-tenths grass; the remainder is leaves and buds. The main, long-term social unit is a stallion, his harem of one or several mares, and their offspring. Maturing stallions form loose bachelor herds and may challenge for the harem with fierce fights of biting and kicking. The single foal can stand within a few minutes of birth and is grazing after a week.

broad stripes on upper body

stripes extend under belly

FLEHMEN RESPONSE

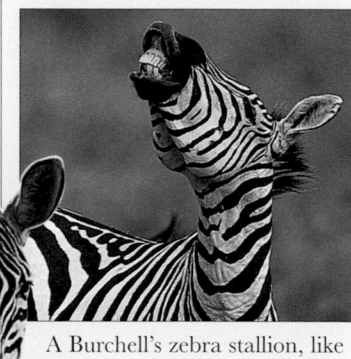

A Burchell's zebra stallion, like other male equids, curls up his top lip in order to heighten his sense of smell. Known as the flehmen response or, more colloquially, the "horse laugh", this behaviour is usually seen when the male (stallion) assesses the readiness of a female (mare) for breeding. By this means, he is able to detect certain scents in the mare's urine to determine if she is receptive for mating.

DISTINCTIVE STRIPES
Burchell's zebra has a different stripe pattern from Grevy's (see left). Some Burchell's zebras have faint "shadow" stripes between the large flank stripes.

RHINOCEROSES

PHYLUM	Chordata
CLASS	Mammalia
ORDER	Perissodactyla
FAMILY	Rhinocerotidae
SPECIES	5

WITH THEIR huge size, bare, sometimes pleated skin, relatively short limbs, and horned snout, rhinoceroses are almost dinosaur-like in appearance. They live in the savannas of Africa, and in the tropical and subtropical forests and swampy grasslands of Asia. Although rhinoceroses are often considered aggressive – they will charge to scare off an intruder – they are generally timid. All 5 species are endangered, 3 of them critically so. All species require a large daily intake of food (either grass or stems, branches, and leaves) to support the massive body.

ANATOMY

Rhinoceroses are large, heavily built animals – the white rhino may weigh up to 2.3 tonnes (2⅜ tons). The species found in Asia support their bulky frame with thick legs, but African species have surprisingly slim legs and are capable of running at speeds of up to 45kph (28mph). Each foot has 3 toes, each with a hoof. The large head features one or 2 horns (depending on species) on the snout. Instead of a bony core (as in the horns of cattle and their relatives, for example), the horns are composed entirely of keratin – a tough protein also found in hair and nails – and the horn perches on a roughened area of the skull (rather than being "rooted" in the skull). Rhinoceroses also have skin up to 2cm (¾in) thick, and body hair is usually inconspicuous (although all species have a tail tuft and ear fringes). Asian rhinoceroses have heavily folded skin, giving the appearance of plates of armour.

Of all the senses, smell is the strongest, although mobile, tubular ears provide good hearing. The eyes, however, are small, and rhinoceroses have poor vision.

CONSERVATION

The entire rhinoceros family is listed by CITES, mainly because rhinoceros horn is, literally, worth its weight in gold. In China it is powdered and used as a drug thought to reduce fever, and in Yemen horns are carved to make traditional dagger handles. These 2 factors are mainly responsible for the dramatic drop in rhinoceros populations (it is a myth that rhinoceros horn is widely used as an aphrodisiac): rhinoceroses now number only about 12,000. All rhinoceros species are protected, but this has largely proved ineffective due to habitat loss and poaching.

SOCIAL SYSTEMS

Although rhinoceroses are mostly solitary, subadults may travel in pairs, female white rhinoceroses sometimes form groups, and Indian rhinoceroses will share a bathing pond without aggression. Adult males of all species tend to be territorial; however, females are not. Stronger male Indian rhinoceroses have a home range that overlaps with the ranges of several females, and is marked by dung piles up to 1m (3⅓ft) high. Weaker males share the ranges of stronger males but do not attempt to mate. When 2 strong Indian rhinoceros males meet, they may fight using their tusk-like lower incisors. Many such conflicts end in the death of one of the combatants. The white rhinoceros, like the Indian species, also employs a "strong" male/"weak" male system. The strong males actively herd females into their territories, and then prevent them leaving. Black rhinoceroses, however, have less well-defined territories. Little is known about the social behaviour of either the Javan or the Sumatran rhinoceros.

HIGH-SPEED CHARGE
Rhinoceroses have a heavy, awkward appearance but are immensely strong and well muscled. If disturbed, an individual such as this black rhinoceros is capable of charging at speeds of 45kph (28mph). Even at high speed, it can make rapid changes in direction.

Sumatran rhinoceros

Dicerorhinus sumatrensis

Length	2.5–3.2 m (8¼–10 ft)
Tail	Not recorded
Weight	Under 800 kg (1,760 lb)

Location S. and S.E. Asia

Social unit Individual

Status Critically endangered

Solitary and secretive, this is the smallest and hairiest rhinoceros. It rests in wallows by day and browses at night on twigs, leaves, and fruits, also felling saplings for tender shoots. Its high-altitude habitat, once relatively safe, is now being lost to

loggers; horn poachers are also a major threat (the front horn may reach 40cm/16in in length). A single calf is born after 7–8 months' gestation and stays with its mother for 18 months, until she next gives birth.

3 toes for grip on slippery ground

WALLOWING

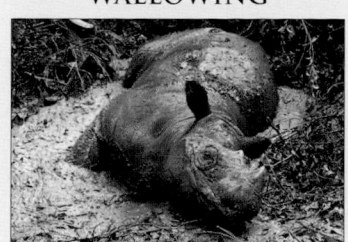

The Sumatran rhinoceros, like all rhinoceroses, hippopotamuses, and similar sparsely haired mammals, wallows in mud, which dries onto the skin. This is cooling and also protects its delicate surface areas from flies and other biting insects.

FEW WRINKLES
This small rhinoceros has relatively few skin wrinkles except around the neck. Sparse hairs cover the skin surface.

Javan rhinoceros

Rhinoceros sondaicus

Length	3–3.5 m (9¾–11 ft)
Tail	Not recorded
Weight	Up to 1.4 tonnes (1½ tons)

Location S.E. Asia

Social unit Individual

Status Critically endangered

One of the rarest large mammals, this species is hairless apart from its ears and tail tip. Its thick grey skin is divided by deep folds to make a "saddle" over the neck with lumps or nodules giving an armour-plated effect. The single horn rarely exceeds 25cm (10in) long and is lacking in some females. A solitary, nocturnal browser, it eats a wide range of plants. The Javan – or lesser one-horned – rhinoceros was decimated by lowland forest removal. Two remnant populations survive, taking advantage of coastal mangrove and bamboo marshes. The male is probably territorial, marking his area with dung piles and urine pools; he encounters potential mates at suitable muddy wallows. After 16 months' gestation, one calf is born and stays with its mother for 2 years, possibly longer.

Indian rhinoceros

Rhinoceros unicornis

Length	Up to 3.8 m (12 ft)
Tail	70–80 cm (28–32 ft)
Weight	Up to 2.2 tonnes (2⅛ tons)

Location S. Asia (Brahmaputra valley)

Social unit Individual

Status Endangered

Like other rhinoceros species, the Indian rhinoceros is generally solitary except for temporary male–female associations when mating, and a mother with her calf. Both male and female have a single horn, up to 60cm (24in) long. Males have larger, sharp, tusk-like incisors for fighting rivals at breeding time. Otherwise they usually tolerate intruders into their ranges, which vary from 2 to 8 square km (¾–3 square miles) depending on the quality of habitat. The gestation period is 16 months and the calf may remain with its mother until her next offspring is born, which may be 3 years later. Despite protection schemes and an encouraging recovery in numbers, remaining populations of the Indian rhinoceros are scattered and fragmented, and therefore still at some risk. These animals are also still subject to poaching for their horns and other body parts.

grey-brown coloration

deep skin folds

PLATES AND RIVETS

The Indian rhinoceros's skin has deep folds, its armour-plated appearance accentuated by tubercles (lumps), especially on the sides and rear, which resemble rivets. The pink skin within the folds is prone to parasites that are removed by egrets and tick birds, who also act as sentinels.

IN THE TALL GRASS

Tall grasses make up the bulk of the Indian rhinoceros's diet (they grow to 8m/26ft tall during the wet season, giving excellent cover). This species feeds mainly by twilight and darkness, curling its upper lip around the stems to bend and bite off the tender tips. It is also the most aquatic rhinoceros, wading and swimming with ease.

MAMMALS

Diceros bicornis
Black rhinoceros

	Length 2.9–3.1 m (9 1/2 –10 ft)
	Tail 60 cm (23 1/2 in)
	Weight 0.9–1.3 tonnes (1–1 1/4 tons)
Location E. and southern Africa	Social unit Individual
	Status Critically endangered

Browsing on a variety of bushes and low trees, the black rhinoceros occupies a variety of habitats but mainly wooded savanna with mosaics of grass and trees. As in other rhinoceros species, its sight is poor but hearing and smell excellent. It feeds mainly by twilight and at night; days are spent dozing in shade or wallowing in mud. The black rhinoceros is solitary, and marks its home range with piles of dung and copious squirts of urine. It may tolerate intruders (of its own species, or human), but this unpredictable rhinoceros may suddenly charge or jab with its horns. Two black rhino-ceroses together are usually a mating pair, associating for just a few days, or a mother and calf. She gives birth after 15 months' gestation. The new-born weighs 40kg (88lb), begins to take solid food after a few weeks, and is weaned at around 2 years.

PREHENSILE LIP

The black rhinoceros, also known as the hook-lipped rhinoceros, has a pointed, prehensile upper lip. This curls around new twigs and shoots to draw them into the mouth, where they are bitten off by the molar teeth.

CONSERVATION

Intense demand for rhinoceros horn for traditional Chinese medicines and for dagger handles and similar items in the Middle East led to a massive decline in black rhinoceros numbers from 65,000 in 1970 to just 2,500 in the mid-1990s. Despite legal protection and wildlife patrols, the risk from poachers remains so great that some black rhinoceroses have 24-hour armed guards.

front horn up to 1.4m (4 1/2 ft) long

GREY SKIN, BLACK APPEARANCE
The black rhinoceros's skin is grey with hairs only on the eyelashes, ear tips, and tail end. Its dark appearance is the result of mud dried on the skin.

Ceratotherium simum
White rhinoceros

	Length 3.7–4 m (12–13 ft)
	Tail 70 cm (28 in)
	Weight Up to 2.3 tonnes (2 1/4 tons)
Location W., E., and southern Africa	Social unit Group
	Status Lower risk

The largest and most numerous rhinoceros, the white rhinoceros rivals the hippopotamus as the biggest land animal after the elephants. Males weigh up to half a tonne more than females, and have longer horns and a more pronounced nuchal crest. The front horn may reach 1.3m (4 1/4 ft) in length, the rear one 40cm (16in). This is also the most sociable rhinoceros, generally placid, with mother–calf pairs staying together for long periods and up to 7 juveniles forming small herds. Mature males, however, tend to be solitary. An almost exclusive grazer, the wide, straight upper lip – giving the alternative name of square-lipped rhinoceros – and hard lip pads crop grass extremely close. Populations of southern white rhinoceros (*Cerato-therium simum simum*) are reasonably secure, numbering more than 8,500, although still conservation-dependent. There may be fewer than 30 of the critically endangered northern white rhinoceros (*Ceratotherium simum cottoni*).

URINE-MARKING

A male rhinoceros's penis faces to the rear, so urine sprays out between the back legs. Male white rhinoceroses mark home ranges in this way. Each needs a territory of about 1 square km (1/2 square mile) to be selected for breeding.

HUMPED SHOULDER
The prominent nuchal crest behind the ears is formed by the bones, muscles, and ligaments to support the huge head.

slate-grey to yellow-brown coloration

pronounced nuchal crest

skin folds only on flanks and elbows

elongated head reaches to ground for grazing

CONSERVATION

A rhinoceros's horns can be quickly cut off under anaesthetic, thereby removing the main target of poachers, who have been largely responsible for the catastrophic decline in white and black rhinoceros numbers over the past 200 years. Since the horn is made of a hair-like material, the procedure is painless, and has little effect on the rhinoceros's social life.

TAPIRS

PHYLUM	**Chordata**
CLASS	**Mammalia**
ORDER	**Perissodactyla**
FAMILY	**Tapiridae**
SPECIES	**4**

THESE MAMMALS can truly be called "living fossils" because as a group they have changed little over the past 35 million years. They are medium-sized animals with a pig-like body on relatively high, slender legs, and a short, extensible trunk. Tapirs live in forest areas (never far from water) in Southeast Asia and South and Central America. Much of their time is spent in water with just their trunk exposed (like a snorkel), to escape predators and to keep cool. Malayan, mountain, and Baird's tapirs are all endangered by habitat destruction and hunting.

ANATOMY

Tapirs have a streamlined shape, which allows them to move more easily through dense undergrowth. They have a very deep face because their nasal passages are greatly enlarged, with nostrils positioned at the tip of the snout. The sense of smell is acute and is vital in finding food and smelling danger and other tapirs. These animals also have large, erect ears (providing good hearing) and small eyes that are deep in the socket, protected from thorns and sharp branches. Their 3 toes spread out on soft ground, which helps support their weight and prevent sinking. Tapirs have a short, broad tail and hard skin (except on the soles of the feet, which are soft and sensitive). The body hair is usually sparse, except in the mountain tapir, which has a thick coat. Baird's tapir and the South American tapir have a short, bristly mane (which provides protection if a jaguar bites). Most species are entirely brown, grey, or blackish, except for white ear rims in some. The mountain tapir has striking white lips, while South American and Baird's tapirs have light patches on the cheeks, throat, and chin. Only the Malayan tapir has extensive white body markings. Newborn tapirs have white spots and stripes in rows along the flanks and limbs, which provide good camouflage.

JUVENILE COAT
A red-brown coat with horizontal white stripes and spots is typical of young tapirs. The adult coloration begins to appear at about 6 months.

Tapirus terrestris
South American tapir

Length	1.7–2 m (5 1/2 – 6 1/2 ft)
Tail	46–100 cm (18 – 39 in)
Weight	225–250 kg (500 – 550 lb)
Location N. and C. South America	Social unit Individual
	Status Lower risk

This bristly coated tapir has white-tipped ears and a short, narrow mane. It favours waterside habitats, swims well, and dives to escape predators such as pumas and jaguars. Browsing selectively by night, it feeds on a wide range of grasses, reeds, fruits, and other vegetation.

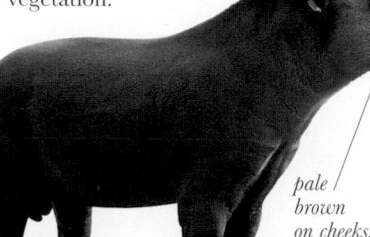

pale brown on cheeks, throat, and chest

Tapirus pinchaque
Mountain tapir

Length	1.8 m (6 ft)
Tail	50 cm (1 1/2 ft)
Weight	150 kg (330 lb)
Location N.W. South America	
Social unit Individual	
Status Endangered	

The furriest of the 4 tapir species, the mountain tapir has thick, dark brown to black fur that keeps out the cold of its high-altitude habitat. The lips and ears are usually white fringed. It browses on a variety of dwarf trees and shrubs, mainly at dawn and dusk. Like other tapirs, the mountain tapir hides in thickets by day. Its droppings contain many intact seeds, thereby assisting plant dispersal and consequently forest regeneration.

Tapirus bairdii
Baird's tapir

Length	2 m (6 1/2 ft)
Tail	7–13 cm (2 3/4 – 5 in)
Weight	240–400 kg (530 – 880 lb)
Location S. Mexico to N. South America	Social unit Individual
	Status Vulnerable

The largest American tapir, Baird's is dark brown, with pale grey-yellow cheeks and throat, and white-edged ears. It eats many different plant

Tapirus indicus
Malayan tapir

Length	1.8–2.5 m (6 – 8 1/4 ft)
Tail	5–10 cm (2 – 4 in)
Weight	250–540 kg (550 –1,190 lb)
Location S.E. Asia	Social unit Individual
	Status Vulnerable

A striking, two-tone pattern distinguishes the largest and only Old World tapir. It is black with a white "saddle" over the back and rump,

parts, from buds and leaves to fallen fruits. The gestation period is about 390–400 days and the female produces one young (very rarely 2), which weighs 5–8 kg (11–18 lb). Baird's tapir uses shrill whistles to communicate with its young or warn other adults away from its territory.

which helps break up its body outline in shady forests. The Malayan tapir feeds on soft twigs and young leaves of bushes and saplings, and also on fallen fruits. The male's average range of 13 square km (5 square miles) overlaps the ranges of several females.

PIGS

PHYLUM	Chordata
CLASS	Mammalia
ORDER	Artiodactyla
FAMILY	Suidae
SPECIES	14

DESPITE A REPUTATION for gluttony, wild pigs rarely overeat and are intelligent, adaptable animals. Members of the pig family – which also includes hogs, boars, and the babirusa – are omnivorous (rather than purely herbivorous) and are characterized by a barrel-shaped body, curiously slender legs, a short neck, and a large head. They are found in forests and grasslands throughout Africa and Eurasia and have also been successfully introduced to Australia, New Zealand, and the Americas. The babirusa and the pygmy hog are both endangered because of habitat destruction. Almost all domestic pigs are descended from the wild boar.

ANATOMY

One of the most interesting features of a pig's anatomy is its snout, which has a cartilaginous disk at its tip, enclosing the nostrils. The disk is supported by a small bone (the prenasal), not present in other mammals, and is used like a bulldozer when foraging for food. Most species also have upper and lower canines that grow outwards and upwards to form tusks. The tips of the lower canines wear against the underside of the upper canines. In the babirusa, however, the male's upper tusks grow up through the skin of the face and then curve backwards. Female pigs have smaller canines (female babirusas have no canines).

Pigs also have cloven feet. Two large, flattened hooves bear the animal's weight, but on soft ground the 2 shorter, lateral hooves may touch the ground and help spread the weight. Pigs have thick skin, with hair that is either long and bristly (as in the wild boar) or sparse (as in the babirusa). Most species have a mane down the back of the neck. The tail is thin, twisted, mobile, and usually sparsely tufted.

FIGHTING

The long tusks of males are used for defence against predators and for fighting other males for social status or mates. Pigs demonstrate 2 distinct fighting styles: lateral and head-to-head. Lateral conflict involves the combatants slashing at each other's shoulders. This fighting style is practised by pigs, such as the wild boar, that have a long, narrow face, no facial warts, and small tusks. Pigs with a broad head, thick skull, long tusks, and facial warts (for protection against wounds) – such as warthogs and giant forest hogs – tend to fight head-to-head.

FAMILY GROUPS

Pigs live in sounders (families of a sow and her offspring), which communicate by squeals and grunts. Boars join sounders during the mating season. Pigs are the only hoofed mammals to have litters rather than one or 2 young (only the babirusa has twins).

PECCARIES

Although the 3 species of peccary together form a separate family (Tayassuidae) from pigs, the 2 groups are closely related. Peccaries differ from pigs by having short canines that point downwards instead of upwards, only one lateral hoof (instead of 2) on each back foot, an extremely short tail, a special gland on the upper surface of the rump, and a more complex stomach (3 chambers). Peccaries live in very large, mixed-sex sounders. These units are quite fearless and will attack even big cats. Peccaries are found mostly in forest areas from southwestern USA through Central America to northern Argentina. The Chacoan peccary is heavily hunted and is in danger of extinction.

HEAD-TO-HEAD FIGHTING
When warthogs clash, head-to-head, they are in fact only trying to push each other off balance. Lateral fighting is far more damaging.

Hylochoerus meinertzhageni
Giant forest hog

Length	1.3 – 2.1 m (4¼ – 7 ft)
Tail	30 – 45 cm (12 – 18 in)
Weight	130 – 275 kg (290 – 610 lb)

Location W., C., and E. Africa

Social unit Group

Status Endangered†

Largest of the pig family, this massive-headed hog has 2 large, wart-like skin growths (excrescences) below and behind each eye, and canines that grow horizontally from the jaw. The straw-coloured piglets turn brown and then black as they mature, and the long, coarse hair becomes sparser. Unlike its relatives, this pig does not root; instead it grazes and low-browses on grasses, sedges, and shrubby vegetation.

Sus salvanius
Pygmy hog

Length	50 – 71 cm (20 – 28 in)
Tail	3 cm (1¼ in)
Weight	6.5 – 9.5 kg (14 – 21 lb)

Location S. Asia

Social unit Group

Status Critically endangered

This chunky, short-legged pig, the smallest species in the pig family, has a tapering snout and head to push through dense undergrowth. Its overall colour is dark brown; the male's upper canines poke slightly out of the sides of the mouth. Both sexes dig large troughs and line them with grassy layers to form nests. Although legally protected, this species is still at risk from poaching and continued removal of its grassy, riverside habitat.

Sus scrofa
Wild boar

Length 0.9–1.8 m
(3–6 ft)

Tail 30 cm
(12 in)

Weight Up to 200 kg
(440 lb)

Location Europe, Asia, N. Africa

Social unit Individual/Group

Status Locally common

The wild boar, or Eurasian wild pig, is one of the most widely distributed terrestrial mammals, and is also the main ancestor of domestic breeds. It occupies a wide variety of habitats, eats almost any food, runs fast, and swims well. Males live alone except for the mating season, when they join with females and fight rival males for harems. Females are very protective of their young and may band together into groups (sounders) of 20 or more.

FADING STRIPES
Wild boar piglets are typically pale brown with paler stripes along the back and sides of the body. This provides camouflage in their nest of grass, moss, and leaves in a dense thicket. The mother rarely leaves her litter (usually 4–6 piglets), for the first 1–2 weeks. Gradually she and the young venture from the nest to forage. From 2 to 6 months, as the piglets become less vulnerable and camouflage is no longer so important, their stripes fade. They become independent at 7 months.

long, tufted tail

BRISTLY PIG
The wild boar has thick, coarse hair and a narrow mane of longer hair along its spine. Compared with some other wild pigs, it has small eyes and tusks. It has no facial warts.

Potamochoerus porcus
Bush pig

Length 1–1.5 m
(3 1/2–5 ft)

Tail 30–43 cm
(12–17 in)

Weight 46–130 kg
(100–290 lb)

Location W. to C. Africa

Social unit Group

Status Locally common

By far the reddest pig, this species – also known as the red river hog – has long, pointed ears with prominent tufts, a narrow white stripe along the back, and white facial stripes. It is omnivorous and nocturnal, like other pigs. Highly sociable, the male stays with his harem of females and offspring, and helps to defend them. Sometimes these family parties of 4–6 congregate into wandering bands of 50 or more.

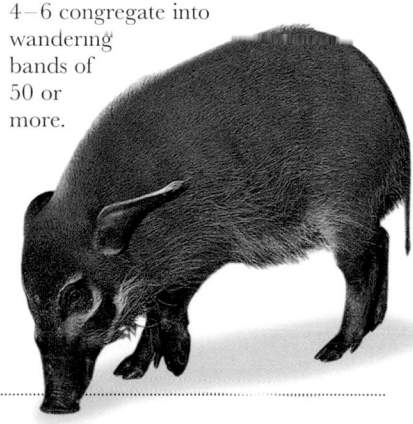

Babirousa babyrussa
Babirusa

Length 0.9–1.1 m
(6 ft 3 in)

Tail 27–32 cm
(10 1/2–12 1/2 in)

Weight Up to 100 kg
(220 lb)

Location S.E. Asia (Sulawesi, Togian, and Mangole islands)

Social unit Individual/Group

Status Vulnerable

The distinctive upper canines of the male babirusa grow through the muzzle and curve back towards the face. Up to 30cm (12in) long, they are also loose socketed and brittle. The almost hairless hide varies in colour from brown to grey. Males are primarily solitary, while females and their young travel in groups of about 8. The gestation period of 155–158 days is fairly typical for the pig family; the litter size is only 1–2.

Phacochoerus africanus
Warthog

Length 0.9–1.5 m
(3–5 ft)

Tail 25–50 cm
(10–20 in)

Weight 50–150 kg
(110–330 lb)

Location Africa (south of the Sahara)

Social unit Group

Status Locally common

The warthog is a long-legged pig, with a large head. When running, its tail is held straight and upright. Generally active by day, it lives in mixed groups of 4–16 young males or females with young. They shelter and raise their young in grass-lined burrows, dug by themselves or by aardvarks.

GRAZING

The warthog is the only pig adapted for grazing in grassland. Typically, it kneels on its padded "wrists" to nip off the growing tips of grass, using its lips or its incisor teeth. In the dry season, it feeds on underground stems (rhizomes), rooting for them with its toughened snout.

facial "wart"

MANED PIG
The warthog's long, dark mane extends from the nape of the neck to the middle of the back, where there is a gap; it then continues on the rump.

Pecari tajacu
Collared peccary

Length 75–100 cm
(30–39 in)

Tail 1.5–5.5 cm
(1/2–2 1/4 in)

Weight 14–30 kg
(29–66 lb)

Location S.W. USA to S. South America

Social unit Group

Status Locally common

Also called the javelina, this is the smallest of the 3 peccary species. It is generally dark grey with a whitish, often indistinct, neck collar; the young are reddish with a narrow black stripe along the back. Found in a variety of habitats, the collared peccary eats mainly plant matter such as berries, shoots, tubers, and bulbs, but also grubs, worms, and small vertebrates such as snakes and lizards. Peccaries are notably gregarious and form bands of up to 15, of mixed age and sex, which cooperate to repel enemies. Members stand side by side and rub faces in mutual grooming.

MAMMALS

HIPPOPOTAMUSES

PHYLUM	Chordata
CLASS	Mammalia
ORDER	Artiodactyla
FAMILY	Hippopotamidae
SPECIES	2

MEMBERS OF THIS FAMILY have a semi-aquatic lifestyle and are thought to be more closely related to whales than to other even-toed hoofed mammals. They can float and swim, and may remain submerged for over 5 minutes. The hippopotamus lives along rivers and lakes in Africa, while the smaller, less aquatic pygmy hippopotamus is restricted to swampy forest areas in west Africa. The hippopotamus is abundant, but the pygmy hippopotamus is endangered due to habitat destruction and hunting.

ANATOMY

Hippopotamuses have a long, heavy body with short, surprisingly insubstantial-looking legs. The enormous head features jaws that allow a huge gape (up to 150 degrees) and carry long, tusk-like canine and incisor teeth. The nose is wide and covered with sensitive bristles. The tail is short, tufted, and flattened.

Adaptations for life in the water include webbed toes; eyes, ears, and nostrils located at the top of the head (these are often the only part of the animal protruding above water); and the ability to close the nostrils underwater.

The skin is grey with a pinkish tinge on the underside and around the eyes and skin-folds. It is almost hairless in most parts, and extremely thick and fatty. Hippopotamus skin is unusual in that there are no sebaceous glands. Instead, there are mucous glands (which are modified sweat glands) that produce a viscous fluid to keep the skin moist when exposed to air. This fluid, which is pink due to the presence of a red pigment, may also protect against infection and prevent sunburn.

While the hippopotamus weighs about 1.4 tonnes (1³⁄₈ tons), the pygmy hippopotamus, which has relatively longer legs, a much smaller head, darker skin, and eyes at the side of the head, averages only about 250kg (550lb).

FEEDING

The hippopotamus moves inland at night to graze, generally following trails – marked by dung piles – that lead to its feeding grounds. Each night, an individual will eat about 40kg (88lb) of grass. The food is digested in a compartmentalized stomach (the fore-stomach contains bacteria that ferments cellulose). Although this system is slow, the hippopotamus requires less food than animals of a similar size because much of its life is spent supported in water. The pygmy hippopotamus eats roots, grasses, shoots, and fallen fruits, although little is known about its feeding habits.

SOCIAL GROUPS

The pygmy hippopotamus is usually found in small groups of up to 3. Female hippopotamuses and their young, however, form groups of usually 10–20 (but sometimes of up to 100) during the day (night foraging is a solitary affair). They communicate via staccato grunts and deep rumbles that carry some distance in the water. Each group occupies a home range along a section of river bank or lake shore, within the territory of a dominant male. This male marks his territory with heaps of dung that he scatters by furiously wagging his tail. Other males enter the territory, but they are tolerated only if they behave submissively and do not attempt to mate. Mating occurs in the water, but the calves are usually born on land.

TERRITORIAL BATTLE
Territorial conflicts between male hippopotamuses are not uncommon, particularly where population densities are high. If, after a period of roaring and ritualized displaying, neither male gives way, a fight will ensue. Using their lower canines as weapons, a battle may last for hours and result in serious injury.

Hippopotamus amphibus

Hippopotamus

Length	2.7 m (9 ft)
Tail	56 cm (22 in)
Weight	1.4–1.5 tonnes (1 3/8–1 1/2 tons)

Location Africa

Social unit Individual

Status Common

Despite its massive bulk, the hippopotamus swims and walks underwater with grace, and trots with surprising rapidity on land on its short legs. A truly amphibious mammal, its skin has a thin outer layer (epidermis), which dries out easily and is sensitive to the bites of pests such as flies. Despite specialized mucus-producing skin glands, the hide soon cracks unless moistened regularly in water or mud. However, the skin's inner layer (dermis) is up to 3.5cm (1 1/2 in) thick and formed of a dense mat of fibres that provides great strength. The hippopotamus's main diet is grass, grazed at night, although they have been observed eating small animals or scavenging. The dominant male mates with females in his territory, and the usually single calf is born, generally underwater, after a gestation of 210 days – short for such a large mammal. The mother is fiercely protective and the calf has few natural predators apart from big cats and hyenas. Hippopotamuses have been known to attack humans, if they feel threatened.

BUOYANCY
The density of the hippopotamus's body is slightly greater than that of water, so it sinks gently and can walk light-footed along the bottom. However, if it keeps its lungs well inflated when breathing at the surface, the extra air reduces this density and it can stay afloat with minimal effort.

SENSES UP TOP

The hippopotamus's nostrils, eyes, and ears are all on top of its head, so it can be almost totally submerged yet breathe easily and remain receptive to its surroundings. The nostrils and ears are closed to water entry when diving.

TEMPORARY HERDS

During the dry season hippopotamuses must wander to find grazing. Instead of each animal returning to its home area by day, some use a nearby pool as a short-term "stop-over" wallow, thereby extending its grazing range. This leads to large gatherings at certain pools, but these lack long-term social or territorial structures.

thin outer skin (epidermis)

lips pluck grass when grazing

MOTHER AND CALVES
Calves remain with their mothers after weaning (at 6–8 months) until about 5 years. In this way, family groups develop.

Hexaprotodon liberiensis

Pygmy hippopotamus

Length	1.4–1.6 m (4 1/2–5 ft)
Tail	15 cm (6 in)
Weight	245–275 kg (540–610 lb)

Location W. Africa

Social unit Individual

Status Vulnerable

Only one-fifth the weight of its huge cousin (above), the pygmy hippopotamus has a relatively small, less angular head and narrower feet with fewer webbed toes, as adaptations for spending more time on land. It also feeds on a wider variety of plant material including shrubs, ferns, and fruits. Pygmy hippopotamuses are usually solitary; although their home ranges overlap, there seems to be little fighting for territory or other form of interaction. They forage at night, following well-worn trails, and spend the day hidden in swamps or sometimes in a river-bank den enlarged from the burrow of some other animal. After a gestation period of 196–201 days, the single calf is born, in water or in the den. The calf risks falling prey to a crocodile or python, but adults have few predators except for leopards and humans. Captive pygmy hippopotamuses have lived longer than many hippopotamuses: 55 compared to about 45 years.

LAND-BASED FORAGER
A pygmy hippopotamus's squat, narrow-fronted form is suited to pushing head-down through dense forest vegetation during night foraging on land.

mostly black hide

CONSERVATION

Evidence suggests that the pygmy hippopotamus has always been rare. It survives in dense forests and marshes in Liberia and neighbouring West African countries. Despite legal protection, these areas are difficult to patrol and subject to uncontrolled logging and widespread hunting for the bushmeat trade. The flesh is said to resemble pork, yet genetic studies indicate that hippopotamuses are more closely related to whales than to pigs.

CAMELS AND RELATIVES

PHYLUM	Chordata
CLASS	Mammalia
ORDER	Artiodactyla
FAMILY	Camelidae
SPECIES	7

CAMELIDS – CAMELS AND THEIR RELATIVES – have long, slender legs and a distinctive gait known as pacing, whereby the front and back legs on the same side move forward together in a rocking motion. Of the Old World camelids – the camels – only one species (the Bactrian camel, from the border area between western China and Mongolia) now survives in the wild. A camel is able to drink up to a quarter of its weight at a time, and can store the water for several days. The New World members of this family, the guanaco and vicuña, are found wild in South America; their domestic descendants, the llama and alpaca, have been bred in the Andes since the time of the Inca civilization. All domesticated camelids are vital to human survival: they provide people with hair, milk, and transportation.

ANATOMY

Camelids have a relatively small head, a long, thin neck, and a split upper lip. Camels have either one hump (the dromedary) or 2 humps (the Bactrian camel): these store fat that can be drawn on during lean times. All camelids have a thick coat that provides insulation against daytime heat and warmth during the cooler nights or at altitude. Unlike other hoofed mammals, camelids rest their weight not on their hooves but on the undersides of the 2 digits on each foot, which are cushioned by a fatty pad. This is an adaptation to walking on sandy soil. Camelids are unique among mammals in that they have oval red blood cells, possibly so that these cells can be transported around the body easily, even if the blood is thickened due to dehydration.

SOCIAL INTERACTION

In the wild, camelids form groups that consist of one dominant male and a harem of females. "Surplus" males form bachelor bands. While the social systems of the South American camelids have not been studied in great detail, more is known about the way camels interact with each other. Both species go through an elaborate and dramatic ritual when a dominant male is faced with a challenger. First, the harem leader grinds his teeth, rubs a gland on the back of his head against his hump (or front hump in the case of the Bactrian camel), smacks his tail loudly against his rump, and urinates on his back legs, rump, and tail. The 2 then pace side by side, display their tall, humped profiles, and extrude a red bladder-like sac (the dulaa) from the corner of the mouth.

WILD AND DOMESTIC

The only living Old World wild camelid is the Bactrian camel, although its numbers are now reduced to only 1,000–2,000 individuals. It is taller and slimmer than its domestic counterpart and has more compact, pointed humps. Domestic Bactrian camels are used for transport in cold regions from northern China to Turkey. The dromedary is extinct in the wild, although feral herds now live in central Australia. Domestic dromedaries are found in hot regions in north and northeastern Africa, the Middle East, and northern India through to Kazakhstan. As with other domestic animals there are different breeds, one of which is kept for its speed and is used in camel racing.

The llama is a domestic animal bred from the wild guanaco and is the traditional pack animal in the Andes. The alpaca is also a domestic species, bred for its fine wool. In the past, the alpaca was thought to be a descendant of the guanaco, but current genetic evidence indicates that it is in fact a descendant of the vicuña, the fine-fleeced, wild camelid of the high-altitude Andes.

DESERT SPECIALISTS
Camels, such as these dromedaries, are well adapted to life in hot climates. Their broad feet provide stability on desert sand, and they have long eyelashes and slit-like, closable nostrils that afford protection during dust storms.

Lama guanicoe
Guanaco

Length 0.9–2.1 m	
(3–7 ft)	
Tail 24–27 cm	
(9¹/₂–10¹/₂ in)	
Weight 96–130 kg	
(210–290 lb)	

Location W. to S. South America

Social unit Group

Status Vulnerable

The guanaco prefers cold habitats, mainly grasslands, but also scrub and forest, at heights up to 4,000m (13,000ft). It browses and grazes on many grasses and shrubs, as well as lichens and fungi. Typical family groups consist of one male and 4–7 females with young. In the north of its range, the offspring leave their group at about one year of age, compared to nearer 2 years in more southerly populations. Young males form bachelor bands; old males are mostly solitary in their lifestyle.

COLORATION
A typical guanaco is pale to dark brown, with whitish chest, belly, and inner legs, and grey to black head with white-edged eyes, lips, and ears.

LLAMA

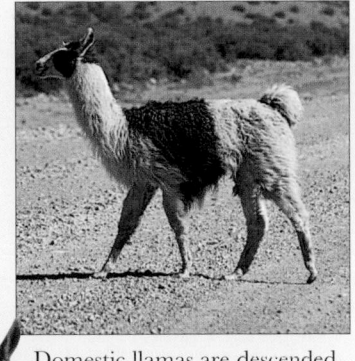

Domestic llamas are descended from wild guanacos domesticated 6,000–7,000 years ago. They have been raised by Andean peoples for their wool (fibre), meat, and skins, and have also been used as pack animals.

Vicugna vicugna
Vicuña

Length 1.5–1.6 m	
(5–5¹/₄ ft)	
Tail 20–25 cm	
(8–10 in)	
Weight 40–55 kg	
(88–120 lb)	

Location W. South America

Social unit Group

Status Endangered

Found in tundra-like grasslands at 3,600–4,800m (12,000–16,000ft), the vicuña is a strict and selective grazer, grasping perennial grasses with its prehensile, cleft upper lip and snipping off the tips against the tough upper pad. It needs to drink daily. Family groups of one male, 5–10 females, and their young occupy territories, delineated mainly by dung; bachelor males form roving bands.

ALPACA

Alpacas, bred mainly for their thick fibre, were once thought to be domesticated from the guanaco, in the high Andes of central Peru, 6,000–7,000 years ago. New evidence now suggests a vicuña ancestry.

WHITE BIB
The vicuña is pale to dark cinnamon, with a variable whitish chest "bib".

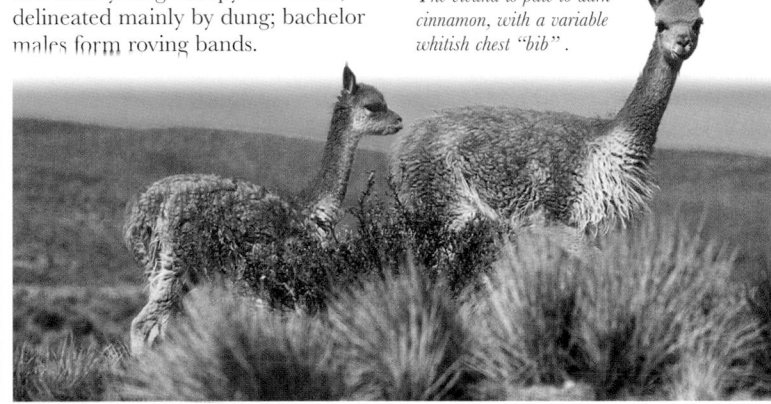

Camelus bactrianus
Bactrian camel

Length 2.5–3 m	
(8¹/₄–9³/₄ in)	
Tail 53 cm	
(21 in)	
Weight 450–690 kg	
(990–1,520 lb)	

Location E. Asia

Social unit Group

Status Endangered

Endangered in the wild (a domesticated animal is shown), this 2-humped camel can withstand temperatures from -29°C (-20°F) to 38°C (100°F). After a drought, it can drink 110 litres (24 gallons) of water in 10 minutes. It eats grasses, leaves, and shrubs. During the rut, males puff out their cheeks, toss their heads back, and grind their teeth. The winner gains 6–30 females and offspring. One young (rarely twins) is born after 406 days' gestation and is suckled for 1–2 years. Females attain sexual maturity at 3–4 years; males at 5–6 years.

relatively small ears

erect humps indicate a well fed animal

shaggy, pale beige to dark brown winter coat

long, almost U-shaped neck

broad foot pad for stability in sand and snow

Camelus dromedarius
Dromedary

Length 2.2–3.4 m	
(7¹/₄–11 ft)	
Tail 50 cm	
(20 in)	
Weight 450–550 kg	
(990–1,210 lb)	

Location N. and E. Africa, W. and S. Asia

Social unit Group

Status Common

Extinct in the wild, this one-humped domestic camel shows many adaptations to desert life, losing up to 40 per cent of its body weight when food and water are scarce. Allowing its temperature to rise in hot conditions, it reduces sweating to conserve moisture. It feeds on a huge variety of plants, including salty and thorny species, and also scavenges on bones and dried-out carcasses. Dromedaries form small herds of several females and young, and one male, who defends them by spitting, biting, and leaning on opponents.

single hump for fat storage

cream to brown or black coloration

double-row of eyelashes and brows to keep out sand

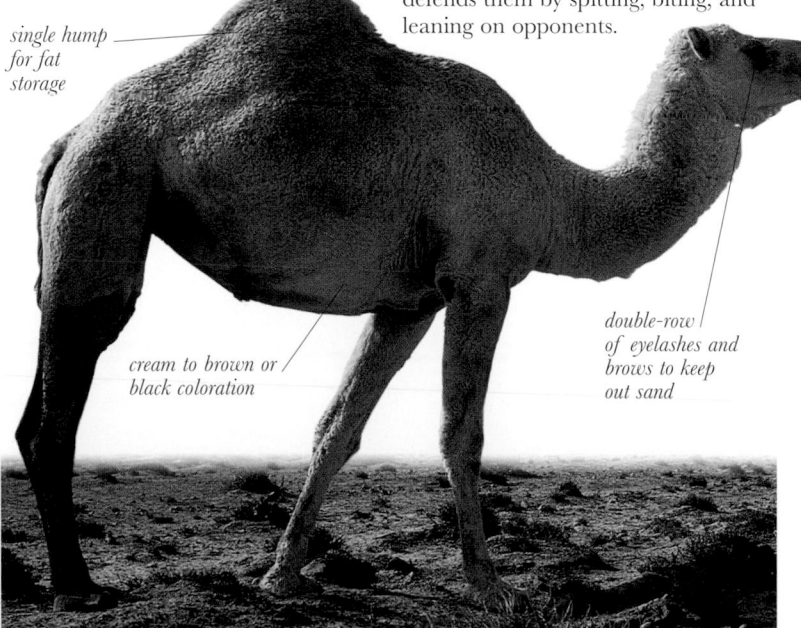

DEER

PHYLUM	Chordata
CLASS	Mammalia
ORDER	Artiodactyla
FAMILY	Cervidae
SPECIES	45

ALTHOUGH SIMILAR in appearance to antelopes, deer are distinguished by their antlers, which are solid, usually branched, and are shed and regrown each year. Deer are mainly woodland and forest dwellers but can be found in a range of other habitats, from arctic tundra to grassland. They live in northwest Africa, Eurasia, and the Americas. Some species have also been introduced beyond their natural range, for example to New Zealand and to Britain and mainland Europe.

(autumn). After breeding (also known as the "rut"), the antlers are shed. It is not certain why antlers are shed and regrown each year, as annual regrowth is metabolically costly – it could be a chance to renew antlers damaged during the rut. Antlers are first grown at 1–2 years of age and are initially simple spikes. As the years progress, the antlers generally become larger and more branched, but regress again in old age. Antler size is an indication of general body condition. Those species that do not have antlers, such as the Chinese water deer, instead possess canine teeth modified into tusks.

MUSK DEER AND CHEVROTAINS

Musk deer (family Moschidae) and chevrotains (family Tragulidae) – also known as mouse deer – have several features that separate them from other deer (family Cervidae). For example, both have long upper canine teeth instead of antlers. In males, these teeth project below the level of the lower jaw and are used for fighting. The 7 species of musk deer (found throughout Asia) are endangered due to demand for musk, which is secreted from the groin of males during the breeding season and is used as an ingredient in many perfumes. Chevrotains (4 species) live in the rainforests of Africa and Asia.

ANATOMY

Most deer have an elongated body, a long neck, large eyes situated at the side of the head, high-set ears, and a small tail. The well-developed third and fourth digits bear the weight, while the second and fifth toes are smaller and usually do not touch the ground. Each year the coat is moulted at least once. In many species the young have spots on the coat, for camouflage.

The most striking feature of deer, however, is their antlers, which are present only in males (except in reindeer, where both sexes have them). In spring each year, the antlers begin to develop. They grow directly from the skull and are initially covered by finely haired skin (called velvet). As they develop, the velvet dries and is rubbed off so that the antlers are ready to be used in fighting during the breeding season

SOCIAL GROUPS

Social organization depends largely on diet. Smaller species are usually browsers and generally live singly or in small groups. This is because their food occurs in small pockets, which produces competition. Larger species tend to graze more open habitats and therefore compete less for food. These deer often live in

herds for protection from predators. Such groups are usually single-sex, except during the rut, when males fight for possession of a harem, using their antlers as weapons and sexual ornaments (to attract mates).

SURVIVING HARD TIMES
Those deer that inhabit regions where food becomes scarce during winter browse on a variety of plants to survive. These red deer stags will usually feed for longer periods than hinds, possibly because during the autumn rut they must fight for and defend their harem and can therefore spend less time building fat reserves.

MAMMALS

Hyemoschus aquaticus
Water chevrotain

Length 70–80 cm
(28–32 in)

Tail 10–14 cm
(4–5 1/2 in)

Weight 10–12 kg
(22–26 lb)

Location W. to C. Africa

Social unit Individual/Group

Status Lower risk†

Tragulus meminna
Indian spotted chevrotain

Length 50–58 cm
(20–23 in)

Tail 3 cm
(1 1/4 in)

Weight 3 kg
(6 1/2 lb)

Location S. Asia

Social unit Individual

Status Unconfirmed

Like other chevrotains (mouse deer), this species has 4 fully developed toes on each foot (true deer have 2). It skulks in cover, preferring rocky patches within tropical rainforest. Nocturnal and solitary, it has a spotted

Generally found within 250m (820ft) of water, this chevrotain is dark olive-brown, with white markings that include spots on the back, 1–3 stripes along each flank, and chin, throat, and chest bands. It has short legs and ears, a stout body, and swims well, which enables it to escape land-based predators successfully most of the time, but incurs the risk of crocodile attack. Water chevrotains forage for leaves and fallen fruits. Males live singly, while females and their young form small groups.

back, and striped flanks and throat; however, these pale markings are less distinct than those of the water chevrotain (above) and the main coat colour is brown with tiny yellow speckles. Males compete using their sharp, tusk-like upper canine teeth. The single young is born after a gestation of 5 months.

Moschus chrysogaster
Alpine musk deer

Length 70–100 cm
(28–39 in)

Tail 2–6 cm
(3/4–2 1/4 in)

Weight 7–18 kg
(15–40 lb)

Location S. Asia

Social unit Individual

Status Lower risk

An inhabitant of rocky, forested slopes at an altitude of 2,600–3,600m (8,600–11,900ft), the musk deer has well-developed side toes to climb rocks and even trees, and to move through soft snow. Its coat is rich, dark brown mottled with grey, paler underneath, with a whitish chin and ear edgings. Its musk secretions are valued in the perfume industry. This has led to hunting and a decline in numbers in the wild.

Muntiacus reevesi
Reeves' muntjac

Length 75–95 cm
(30–37 in)

Tail 17 cm
(6 1/2 in)

Weight 10–18 kg
(22–40 lb)

Location E. Asia

Social unit Individual/Group

Status Locally common

The muntjac's diet varies from succulent shoots, herbs, and blossom to tough grasses and nuts. The male is in breeding condition all year, and he uses his short – 10cm (4in) or less – pointed antlers to push rivals off balance, so that he can inflict wounds with his tusk-like upper canines. Also called barking deer, these small deer have been successfully introduced into England and the Netherlands. There are some 8 other species of muntjac, all found in Asia.

Dama dama
Fallow deer

Length 1.4–1.9 m
(4 1/2–6 1/4 ft)

Tail 14–25 cm
(5 1/2–10 in)

Weight 35–150 kg
(77–330 lb)

Location Europe

Social unit Group

Status Locally common

Fallow deer have long been kept semi-domesticated for their beauty and meat, and have been introduced to the Americas, Africa, and Australia. They are active at twilight, consuming many plant foods, from grasses to acorns. Herds may exceed 100. Bucks rut to establish a small patch of land, where they mate.

broad antlers

EARTHY HUES
Commonly brown with white spots as here, the fallow deer can also be pale brown, black, or white.

LYING UP

The fallow deer fawn, like most young deer, "lies up" in thick vegetation or among leaf litter, as the mother feeds. The fawn's instincts are to stay still and silent, camouflaged from predators by its spotted coat. The mother returns at intervals to allow the fawn to suckle, and occasionally leads it to a new lying-up site.

Axis axis
Axis deer

Length 1–1.5 m
(3 1/4–5 ft)

Tail 10–25 cm
(4–10 in)

Weight 70–79 kg
(155–175 lb)

Location S. Asia

Social unit Group

Status Locally common

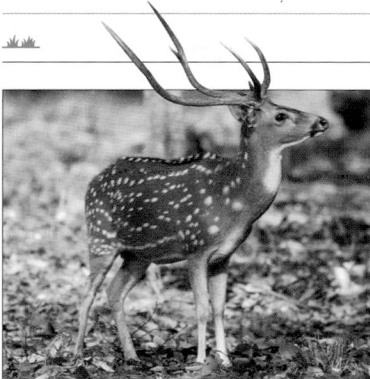

Also called the chital, the axis deer lives in large herds of 100 or more, comprised of mixed males, females, and young. They graze in grassland and browse in open woodland, often below troops of langurs (see p.131), who knock fruits to the ground and emit warning calls. Axis deer dash for thick cover at 65kph (40mph) when disturbed. The male's antler has one brow tine (prong), and a rear-directed main beam that forks into two points.

Cervus unicolor
Sambar

Length 2–2.5 m
(6 1/2–8 1/4 ft)

Tail 15–20 cm
(6–8 in)

Weight 230–350 kg
(510–770 lb)

Location S. and S.E. Asia

Social unit Individual

Status Locally common

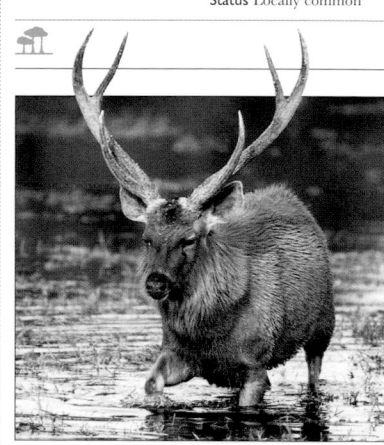

The sambar is dark brown except for rusty hues on the chin, inner legs, and tail underside (which has a black tip). The male's 3-point antlers grow to 1.2m (4ft) long. Both sexes have a neck mane of thicker fur; this is more prominent in rutting males. Solitary, except for a female with a fawn, and perhaps a yearling, too, these nocturnal deer eat a variety of vegetation.

Cervus elaphus
Red deer

Length	1.5–2 m (5–6 1/2 ft)
Tail	12 cm (4 3/4 in)
Weight	65–190 kg (145–420 lb)

Location Europe to E. Asia, North America

Social unit Group

Status Locally common

antlers with multiple points

Highly adaptable in habitat and diet, the red deer has been introduced to most continents and is widely farmed for meat, hides, and antler velvet. There is great variation among the 28 or so subspecies, which include the wapitis of China and North America. Females (hinds) form herds led by a dominant hind, with separate male bands except during the autumn rut.

MORE BROWN THAN RED
Red deer are red-brown in summer, perhaps with a dark line along the neck and back, and vague flank spots, and then turn dull brown in winter.

RUTTING

Red deer males (stags), like most other male deer, battle during the rutting season. The contest is part display and part physical tussle. Stags roar and bellow, thrash their antlers against bushes and trees, and walk parallel to each other, as they assess whether to fight. If so, they lock antlers, push, twist, and shove. The winner gains a harem.

Cervus nippon
Sika

Length	1.5–2 m (5–6 1/2 ft)
Tail	12–20 cm (4 3/4–8 in)
Weight	35–55 kg (77–120 lb)

Location E. and S.E. Asia

Social unit Group

Status Critically endangered

The sika has been kept in parks and farmed for centuries, and has been introduced into many regions. Appearance varies among the 14 subspecies, some of which are endangered, but is generally rich red-brown with white spots in summer, and almost black in winter, with perhaps vague spotting on females, The white rump hairs can be flared. The sika eats mainly grasses, including bamboo, and twigs and buds.

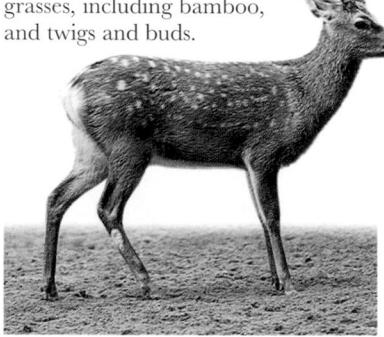

Elaphurus davidianus
Père David's deer

Length	2.2 m (7 1/4 ft)
Tail	66 cm (26 in)
Weight	150–215 kg (330–470 lb)

Location E. Asia

Social unit Group

Status Critically endangered

Distinct from other deer in its body form, this species has a long, horse-like face, wide hooves, and a long tail. Also unusual are the male's (stag's) "back-to-front" antlers. The coat is dark grey-fawn in winter, bright red-brown in summer, with a dark central back stripe, and a whorled hair pattern on the rump. Extinct in the wild, the species was saved by captive breeding in England from about 1900. Since the 1980s, it has been reintroduced into the wild in China.

Odocoileus hemionus
Mule deer

Length	0.85–2.1 m (2 3/4–7 ft)
Tail	10–35 cm (4–14 in)
Weight	55–210 kg (120–460 lb)

Location W. North America

Social unit Group

Status Lower risk

The mule deer is widely distributed in many habitats, and is recorded as eating hundreds of plant species. The main colour is grey-brown in winter, and rusty brown in summer. Despite its other name of black-tailed deer, the tail is black on the upper surface only; the rest is white. The face and throat also have variable white areas, with black chin and forehead bands. Rutting is in September–November; 1–2 young are usually born in June.

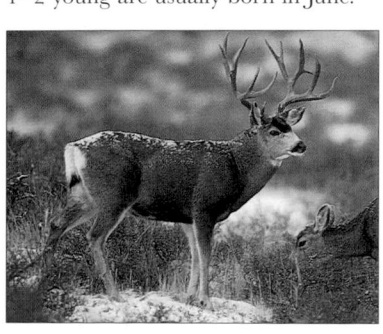

Odocoileus virginianus
White-tailed deer

Length	1.8–2.4 m (6–7 3/4 ft)
Tail	15–30 cm (6–12 in)
Weight	52–140 kg (115–310 lb)

Location S. Canada to N. South America

Social unit Group

Status Locally common

This species is extremely similar in appearance and habits to the mule deer (see left), and in zoos and parks the two may interbreed. This rarely happens in the wild, however, even though their ranges overlap. The numerous subspecies become smaller towards the south of the range, shoulder height ranging from 1.1m (3 1/2ft) in Canada to 60cm (2ft) in Venezuela, where it is known as the venado.

WHITE WARNING
When danger threatens, this deer raises its tail and flashes the bright white underside, as a warning to others in its herd.

YEARLY ANTLER CYCLE

By February, the male white-tailed deer will have cast its antlers. In April–May, they start to grow again, protected by fur-covered skin (velvet), but in September all the velvet will have been rubbed off against trees, leaving the clean bone, before the rut.

Blastocerus dichotomus
Marsh deer

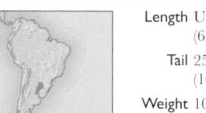

Length	Up to 2m (6 1/2 ft)
Tail	25 cm (10 in)
Weight	100–140 kg (220–310 lb)

Location S. South America

Social unit Individual/Group

Status Vulnerable

Long legs and wide hooves allow the marsh deer to move easily in swamps and floodplains. The largest South American deer, it is reddish brown in summer and darker in winter, with black lower legs, a pale face, and black around the lips and nose. It eats grasses, reeds, water plants, and bushes, and lives alone or in groups of 2–3. It is threatened by habitat loss due to irrigation and conversion to pasture or crops, water pollution, and competition from livestock.

Pudu puda
Southern pudu

Length 85 cm (34 in)	
Tail 5 cm (2 in)	
Weight Up to 15 kg (33 lb)	

Location S.W. South America	Social unit Individual
	Status Vulnerable

The southern pudu is one of 2 small, stocky species of pudu. It is buff to red-brown with rounded ears. The male's antlers are simple spikes, 8cm (3¼ in) long. Solitary and diurnal, it dwells in moist forests, hiding in understorey thickets, where it feeds on bark, buds, fruits, and flowers – but seldom eats grass. It is sexually mature at 6 months.

Rangifer tarandus
Reindeer

Length 1.2 – 2.2 m (4 – 7 ¼ ft)	
Tail 10 – 25 cm (4 – 10 in)	
Weight 120 – 300 kg (260 – 660 lb)	

Location N. North America, Greenland, N. Europe to E. Asia	Social unit Group
	Status Endangered†

Known as caribou in North America, the reindeer has long antlers with a distinctive shovel-like brow tine on one side. It grazes grasses, sedges, and herbs in summer, and mosses, lichens, and fungi in the long winter. One calf is born in May–June after a gestation of 210–240 days.

COLOUR VARIATION
American forms have mainly brown coats with darker legs; European and Asian reindeer (shown) are greyer.

MIGRATION

Some reindeer travel 15–65km (9–40 miles) daily within the same region; others migrate up to 1,200km (750 miles) twice yearly. In some populations, females and young move to the calving grounds in spring, males following later.

Capreolus capreolus
Roe deer

Length 1–1.3 m (3¼ – 4¼ ft)	
Tail 5 cm (2 in)	
Weight 20 – 30 kg (44 – 66 lb)	

Location Europe, W. Asia	Social unit Variable
	Status Locally common

This deer has a black muzzle band and variable white chin and throat patches. The white rump patch, which can be fluffed out when the deer is alarmed, is heart shaped in the female, kidney shaped in the male (shown). The male has rough-surfaced, 3-point antlers. The sleek, bright red-brown summer coat moults to a longer, denser grey coat in winter.

Alces alces
Elk

Length 2.5 – 3.5 m (8¼ – 11 ft)	
Tail 10 cm (4 in)	
Weight 500 – 700 kg (1,100 – 1,550 lb)	

Location Alaska, Canada, N. Europe to N. and E. Asia	Social unit Variable
	Status Locally common

Called the moose in North America, the elk is the largest deer, with males up to twice as heavy as females. Found in woods close to swamps, lakes, and other water, it may submerge in summer to reach roots of lilies and other aquatic plants. The winter diet is mainly twigs of trees such as willow and poplar. Elk live alone or in small family groups. Males rut in September–October. The 1–2 young are born after 242–250 days' gestation and weaned by 6 months.

COLORATION
The elk is brownish grey in summer, greyer in winter. Its paler-hued, long legs have wide hooves for wading in mud and walking on soft snow.

furred throat flap (dewlap)

ELK HEAD

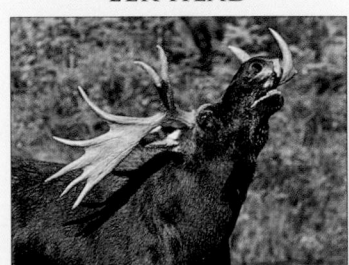

The elk's very broad muzzle and flexible lips help to grasp water plants and to strip leaves from twigs. The male's massive antlers may span 2m (6½ft); each has up to 20 points, mostly growing from the palm-shaped "beam".

PRONGHORN

PHYLUM	**Chordata**
CLASS	**Mammalia**
ORDER	**Artiodactyla**
FAMILY	**Antilocapridae**
SPECIES	**1**

NAMED AFTER THE "prong" on its horns, the pronghorn is the only species in its family. The horns are unusual in that they consist of a horny sheath on a bony core (as in antelopes), but are forked and shed yearly (as are deer's antlers).

SWIFT RUNNERS
The pronghorn is one of the fastest mammals – it can attain speeds of over 65kph (40mph). In winter, herds of over 1,000 animals may gather.

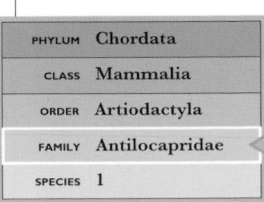

Antilocapra americana
Pronghorn

Length 1.0 – 1.5 m (3¼ – 5 ft)	
Tail 7.5 – 18 cm (3 – 7 in)	
Weight 36 – 70 kg (79 – 155 lb)	

Location W. and C. North America	Social unit Variable
	Status Locally common

The pronghorn is red-brown to tan, with white underside, face, rump, and neck bands. The male has a black neck patch, and horns that are longer

than the ears, with a forward-facing prong halfway up. The female's horns are shorter than the ears. Herds of over 1,000 form in winter, splitting into smaller groups in summer. Many plants feature in the pronghorn's diet.

MAMMALS

GIRAFFE AND OKAPI

PHYLUM	Chordata
CLASS	Mammalia
ORDER	Artiodactyla
FAMILY	Giraffidae
SPECIES	2

THE GIRAFFE and the okapi are the last surviving representatives of a once highly diverse family. They have long legs, a long, narrow head with small horns, and peculiar, lobed lower canines. The giraffe, with its distinctive long neck, is the tallest living animal – males can reach 5.5m (18ft). The 2 species, together known as giraffids, differ in their behaviour and ecology because the giraffe lives in woodland savanna (in Africa, south of the Sahara), while the okapi inhabits rainforest (in northeastern Zaire).

ANATOMY

With front legs longer than the back legs, the front parts of the giraffe and the okapi are raised (to facilitate easier browsing). The giraffe has especially long front legs and these are surprisingly sturdy. They are sometimes used in defence: one kick can kill a lion. Both species have a long neck, but what is unexpected about the giraffe's hugely lengthened neck is that it contains only 7 vertebrae, as do almost all other mammals. Each vertebra is greatly elongated, however.

In the giraffe, both sexes have horns, while only the male okapi has horns. The horns are different from those found in other mammals in that they form as cartilage, turn to bone from the tips down, and are covered with skin.

Giraffids have a thick hide to help ward off predators. The okapi has a velvety, dark brown coat with white stripes on the haunches and upper legs (the legs are white below the knees). Giraffes have spotted coats, the pattern of which varies according to location. Both species have a long, copiously tufted tail.

FEEDING

Giraffids are browsers with unique 2- or 3-lobed canine teeth, which can be used like a comb to strip leaves from small branches. They also use their thin, mobile lips and long black tongue – which can be extended more than 45cm (18in) in giraffes – to gather leaves and shoots. They have a 4-chambered, ruminating stomach (see p.225). Male giraffes tend to be taller than females and are therefore able to feed at higher levels.

SOCIAL SYSTEMS

The giraffe and the okapi have contrasting social systems. Giraffes have home ranges of, on average, approximately 160 square km (62 square miles). When these overlap (which is often), loose associations of up to 25 individuals form. The composition of these herds changes daily. Males are non-territorial, but a dominance hierarchy is determined in a ritualized fight called "necking" – 2 adult males stand side by side, alternately swinging their heads and hitting each other on the neck. Male giraffes have extra bone all over the skull, which provides reinforcement. Only males of high social standing have the right to mate.

Okapis, on the other hand, are mainly solitary animals and are never found in herds. They have much smaller home ranges, and only the dominant males maintain a territory (females move freely from one territory to the next). Male okapis mate with females that wander through their territory.

SWIFT RUNNERS
When danger threatens, a giraffe will take flight (without giving a vocal warning) and can reach speeds of over 50kph (31mph). This action leads others nearby to follow suit. Giraffes move directly from a walk to a gallop because their long legs and short body make trotting impossible without tripping.

Okapi

Okapia johnstoni

Length	2–2.2 m (6½–7¼ ft)
Tail	30–42 cm (12–16½ in)
Weight	200–350 kg (440–770 lb)

Location C. Africa

Social unit Individual/Pair

Status Lower risk

An elusive browser of thick tropical rainforest, the okapi feeds by day on leaves, soft twigs, shoots, fruits and other plant parts. It relies mainly on hearing in the dense forest and makes a "chuff" sound on meeting another okapi. Rival males "neck-fight" like giraffes in the presence of a receptive female, and emit soft moaning sounds during courtship; the female indicates her readiness with similar calls and territorial scent-marking. She is slightly taller than the male and 25–50 kg (55–110 lb) heavier. Both sexes look similar: long head and neck, dark muzzle and body (which slopes down from the shoulders), large, rear-set ears, and zebra-like stripes on the rump and upper parts of the legs. The female bears a single calf in August–October, after a gestation of 425–491 days. She defends her offspring against predators, but the bond between mother and young is not as strong as in many hoofed mammals.

horn-like structures (ossicones) in male

white or cream, mid-facial markings

SLEEK COAT
The okapi's coat is short and sleek. The darker parts appear deep red, purple, maroon, brown, or black, according to the angle of the light.

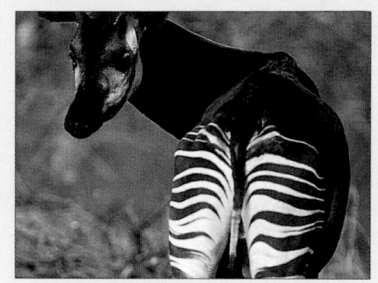

"FOREST ZEBRA"
The okapi was not identified as a distinct species until 1900–1901. Before that time it had been sighted occasionally, but mainly from the rear. Being shy and secretive, it had dashed away from humans into the dense forest. The impression of a forest-dwelling zebra was reinforced by the few specimens of old skin.

LONG TONGUE

The okapi curls its long, black, prehensile tongue around leaves, buds, and small branches, to draw them into the mouth. The tongue is also used for self grooming and, in the female, for cleaning her calf.

Giraffe

Giraffa camelopardalis

Length	3.8–4.7 m (9¾–15 ft)
Tail	78–100 cm (31–39 in)
Weight	0.6–1.9 tonnes (½–2 tons)

Location Africa

Social unit Variable

Status Lower risk

A native of dry savannas and open woodland, the giraffe browses higher than any other mammal, mainly for leaves of acacia and wild apricot, but also shoots, fruits, and other vegetation. The combination of greatly elongated tongue, skull, neck, shoulder region (pectoral girdle), and front legs provides the giraffe's great reach when browsing. Usually a small branch is drawn into the mouth with the long, flexible tongue; then the head is pulled away to rake off leaves between the lobe edged teeth. Among the giraffe's many distinctive features are large eyes and ears; a back that slopes steeply from shoulder to rump; stilt-like legs with large, heavy feet; and a thin tail with a long black tuft for whisking away flies. The 2–4 specialized horns, called ossicones, are more developed in males than females. Feeding, drinking, and other activities occur in the morning and evening, occupying about 12 hours, with rest (standing up, as in most hoofed mammals) taking place at night, and cud-chewing in the hot midday. Cows mate with local dominant bulls, who have competed with each other by swaying and intercurling necks, and even clashing heads. This activity, "necking", is more a slow-motion ritual than a forceful encounter. It occurs mainly among young bulls, and when a new male arrives in the area. The winner reinforces success by sexually mounting his defeated rival. Cows, after a gestation period of 457 days, give birth to one calf (in rare cases 2), usually in the dry season. The newborn weighs up to 70 kg (155 lb) and stands 2 m (6½ ft) tall. For 10–30 days, the mother keeps it away from the herd; weaning takes place by 13 months. The giraffe's main predators are lions, leopards, and hyenas.

TALLEST ANIMAL
A mature female measures 4.5 m (14¾ ft) to her horn-tips; the male may be 1 m (3 ft) taller. This difference helps the sexes to avoid feeding competition as they utilize different levels. The maned neck is the longest of any animal, yet still has only 7 neck bones, or cervical vertebrae, as in most other mammals.

patches fade to white towards feet

PATCHY DISTRIBUTION
Some 9 subspecies of giraffe are recognized by skin pattern. The reticulated giraffe (*G. c. reticulata*), pictured here, has very large, sharp-edged, deep chestnut patches separated by fine white lines. Other subspecies have smaller, irregular, fuzzy-edged patches, which vary from yellow to almost black, with much more white between. Each giraffe's pattern is constant through life, and may help other giraffes to recognize it, although the colours tend to vary with season and health.

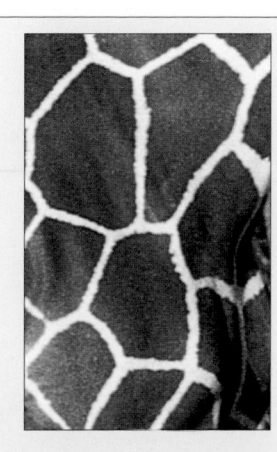

DRINKING PROBLEM
The giraffe's great height means that, to drink water, it must splay its front legs, and even bend at the knees. When upright, its heart has to pump blood upwards at enormous pressure to reach the brain, but when the head lowers to drink, a series of one-way valves regulate the blood's force and flow to prevent damage to the brain.

CATTLE AND RELATIVES

PHYLUM	Chordata
CLASS	Mammalia
ORDER	Artiodactyla
FAMILY	Bovidae
SPECIES	140

THE SPECIES THAT make up this family – the bovids – form a highly diverse group. Members include cattle (wild and domesticated) and their immediate allies (such as bison); sheep and goats and their relatives (such as the chamois); and antelopes (such as the impala), which is a catch-all term for bovids with long, slender limbs. The highest diversity of bovids occurs in Africa, where each species occupies a slightly different niche. They are also found in Eurasia and North America, and a number of species have been introduced to Australasia. Bovids mostly prefer grassland, desert, scrub, and forest habitats.

ANATOMY

Although bovids exhibit a wide range of body forms, from sleek, graceful gazelles to the massive, stocky buffalo, all species have unbranched horns consisting of a keratin sheath surrounding a bony core. Unlike the antlers in deer, horns are never shed, and in most species they are present in both sexes. Horns may be straight, curved, or spiralled; keeled, ridged, or smooth; short or long. All have pointed tips.

Bovids have divided ("cloven") hooves: the weight of the animal rests on the 2 central toes of each foot (a pair of small lateral toes is also usually present). The tail varies from small and triangular to long and tufted, and the coat may be smooth and sleek or long and shaggy. As bovids are often hunted by large predators, they have large, sideways-facing eyes (for good all-round vision); long, mobile ears; and an acute sense of smell. Most species have scent glands located on the face, between the hoofs, and/or in the groin. The glands between the hooves release a scent onto the ground that an isolated animal can follow back to the herd.

Bovids also have a 4-chambered ruminating stomach (see p.225). Food (usually grass or leaves) is pulled in using the dextrous tongue and is then crushed between the lower incisors and a toothless pad in the upper jaw.

SOCIAL SYSTEMS

There is a huge variety of social and breeding systems among bovids. Duikers, for example, are solitary, and dik-diks live in pairs. The impala, however, lives in groups of a male with several females (younger and weaker males form bachelor herds). Male gazelles are territorial: they mate with females that move, in groups, through their territories. Wild cattle and buffalo, on the other hand, live in less structured groups, although most of the mating is performed by the dominant males.

BOVIDS AND PEOPLE

Domestic sheep, goats, and cattle are farmed on large and small scales in most countries around the globe and are therefore of huge economic importance to people. Sheep and goats were probably domesticated 8,000–9,000 years ago, in southwest Asia, and their wild ancestors still live in the same region. Cattle, however, were domesticated about 2,000 years after sheep and goats (also in southwest Asia), but the ancestor of most domestic cattle, the aurochs, is now extinct.

Although many wild species are abundant – the wildebeest, for example, numbers in the millions – some, including several species of gazelle, are close to extinction due to hunting.

ESCAPING DANGER
Like most hoofed mammals, the common eland has keen senses, and flees when frightened or chased by predators. Despite their size and massive build, they can gallop at speeds of up to 70kph (43mph) and are capable of jumping 1.5m (5ft) in the air to clear obstacles in their path.

Tragelaphus spekei
Sitatunga

Length	1.2–1.7 m (4–5 1/2 ft)
Tail	20–26 cm (8–10 in)
Weight	50–125 kg (110–280 lb)

Location W. and C. Africa	**Social unit** Variable
	Status Lower risk

Amphibious in habit, the sitatunga occurs in permanent swamps, marshes, and similar watery habitats. Its long, pointed, widely splayed hooves and extremely flexible foot joints are specialized for soft, muddy ground. When in danger from a land-based predator, this antelope retreats to water, and may submerge with only its eyes and nose

exposed. At night, males bark warnings or avoidance calls to other males; if they meet, they posture and "horn" the ground. The sitatunga eats many kinds of aquatic and terrestrial plants, including reeds, grasses, and shrubby foliage. While feeding, it may stand in water up to its shoulders. Only the male has ridged, spiralling horns. It also has a greyish tinge to its coat, while the female's is brown to chestnut; both sexes have white around the eyes, and on the cheeks and body. Females are solitary or live in groups of up to 3. The single calf (rarely twins) is born after 247 days' gestation.

grey-brown coloration in male

Tragelaphus euryceros
Bongo

Length	1.7–2.5 m (5 1/2–8 1/4 ft)
Tail	45–65 cm (18–26 in)
Weight	210–405 kg (460–890 lb)

Location W. and C. Africa	**Social unit** Individual/Group
	Status Lower risk†

The largest and most distinctive forest antelope, the bongo has vertical white stripes along the body, and a white chest crescent, cheek spots, nose chevron, and leg bands. The coat is chestnut above, darker below, and it darkens in older males. This selective browser has lyre-shaped horns, longer (up to 95cm/37in) in the male, who is solitary. Females form herds of up to 50, perhaps coalescing as their calves associate into nursery bands.

vertical body stripes

Tragelaphus angasi
Nyala

Length	1.4–1.6 m (4 1/2– 5 1/4 ft)
Tail	40–55 cm (16 –22 in)
Weight	55–125 kg (120–280 lb)

Location Southern Africa	**Social unit** Group
	Status Lower risk

dark, bushy tail

The male nyala is larger and heavier than the female, with a charcoal-grey head and body, indistinct body stripes, tan lower legs, and horns up to 70cm (28in) long. Females have no horns and, like juveniles, are red-brown, with a white V between the eyes, and vertical white body stripes. Nyalas prefer dense bush near water, and both graze and browse, rearing up to reach higher leaves.

Tragelaphus scriptus
Bushbuck

Length	1.1–1.5 m (3 1/2–5 ft)
Tail	20–27 cm (8–10 1/2 in)
Weight	25–80 kg (55–175 lb)

Location W., C., E., and southern Africa	**Social unit** Individual
	Status Locally common

The bushbuck resembles a small sitatunga (see above), with less twisted horns. Females from bush-dwelling populations are pale tawny; those in forests are redder. Males in both habitats are dark brown to black. The white marks on the throat, neck, and body vary with the subspecies. This antelope is a browser and solitary, except when courting or when a mother is with her young.

Tragelaphus strepsiceros
Greater kudu

Length	2–2.5 m (6 1/2 –8 1/4 ft)
Tail	37–48 cm (14 1/2 –19 in)
Weight	120–315 kg (260–690 lb)

Location E. to southern Africa	**Social unit** Group
	Status Lower risk†

The male greater kudu is one of the tallest and longest-horned (average 1.7m/5 1/2ft) antelopes, and has a long throat fringe. Coloration in both sexes is grey-tinged red or brown, with 6–10 white body stripes, and white nose and cheek marks. This kudu eats leaves, flowers, fruits, herbs, and tubers. Females form groups of up to 6, as do males, except when they compete during the breeding season.

crest along neck and back in male

Taurotragus oryx
Common eland

Length	2.1–3.5 m (7 –11 ft)
Tail	60–90 cm (23 1/2 –35 in)
Weight	300–1,000 kg (660 –2,210 lb)

Location C. E., and southern Africa	**Social unit** Variable
	Status Lower risk

The two species of eland are the biggest, most cow-like antelopes. Male common elands can weigh up to 1,000kg (2,210lb), and have spiralling horns up to 1.2m (4ft) long, and a brown-black "topknot" of matted hair on the head. Females are up to 600kg (1,320lb), with horns half the length of those of the male. Overall body proportions resemble cattle more than antelopes. Also in the manner of cattle, females band together to defend young and repel predators such as lions. The common eland browses in bush, plains and open woodland. It gathers food with its lips, and digs up bulbs and tubers with its hooves. Like the camel, it allows its body temperature to rise by up to 7°C (13°F) to avoid losing water as sweat, and thereby survives drought. Herds usually consist mainly of females with calves and juveniles; older males may be solitary. These antelopes are relatively docile and are kept for meat, milk, and hides in Africa. They are also exported to regions such as Asia.

WHITE STRIPES
Common eland are mainly greyish fawn, with a black stripe along the centre of the back, and up to 15 whitish cream vertical body stripes.

short mane

shoulder hump

MAMMALS

Boselaphus tragocamelus
Nilgai

	Length 1.8–2.1 m (6–7 ft)
	Tail 45–53 cm (18–21 in)
	Weight Up to 300 kg (660 lb)
Location S. Asia	Social unit Group
	Status Lower risk

Also called bluebuck or blue bull, the nilgai is a small-headed bovid with longer front than rear legs, and, in the male, stout, tapering horns 20 cm (8 in)

coarse, iron-grey or "blue" coat in male

long. The male's coat is grey or bluish grey; the female's is tawny. The nilgai prefers open woods to thick forest, is very wary, with sharp senses, and flees rapidly from predators such as tigers. It grazes and browses on a wide variety of grasses, leaves, and fruits, from early to mid morning and in the early evening. Males compete for territories – and thus access to groups of 2–10 females – by kneeling in front of each other and lunging with their horns. Breeding occurs during much of the year but most calves, either one or 2, are born in June–October, after a gestation of 243–247 days.

throat tuft in male

darker lower limbs

white ring just above fetlocks

Bubalus depressicornis
Lowland anoa

	Length 1.6–1.7 m (5¼–5½ ft)
	Tail 18–31 cm (7–12 in)
	Weight 150–300 kg (330–660 lb)
Location Sulawesi	Social unit Individual
	Status Endangered

One of the smallest wild cattle species, this anoa is dark brown to black, with a pale throat bib and facial and leg patches. It has a thick neck, plump body, and short legs, and horns that sweep diagonally backwards – adaptations for pushing through dense, swampy forest. Solitary except when breeding, it feeds mainly in the morning on leaves, fruits, ferns, saplings, and twigs. The single calf is born after 9–10 months' gestation.

Bubalus quarlesi
Mountain anoa

	Length 1.5 m (5 ft)
	Tail 24 cm (9½ in)
	Weight 150–300 kg (330–660 lb)
Location Sulawesi and Buton Island	Social unit Individual/Pair
	Status Endangered

Resembling its lowland cousin (see left) in size and overall form, the mountain anoa has a woollier coat, even when fully grown, and fewer white markings, especially on the lower legs and throat. Its horns are smooth surfaced and short, about 15–20 cm (6–8 in) long (those of the lowland anoa being 18–38 cm/7–15 in long). In general, the male is larger and darker than the female, with bigger horns. The mountain anoa is one of the few wild cattle species in Southeast Asia that relies on undisturbed forest habitat, but such terrain is remote and difficult for field studies, and so the detailed habits and population of the species remain unclear. Most sightings are of solitary adults or females with a single young of the year, born after a gestation of 9–10 months. The diet is probably a variety of leaves, and also plentiful mosses, but relatively few grasses.

Tetracerus quadricornis
Chousingha

	Length 80–100 cm (32–39 in)
	Tail 12 cm (4¾ in)
	Weight 17–21 kg (37–46 lb)
Location S. Asia	Social unit Individual
	Status Vulnerable

Also called the four-horned antelope, the male chousingha has 2 pairs of horns – a feature unique among bovids. The front pair, at only 3–4 cm (1¼–1½ in) long, are half the length of the rear pair. The habits of this shy,

fast-moving antelope are little known. It grazes on grasses, sedges, and other plants, usually near water in wooded hills, and communicates by a low whistle for identification or barks for alarm. The brownish coat has a dark stripe on the front of each leg, and a black muzzle and outer ear surfaces.

Bubalus bubalis
Asian water buffalo

	Length 2.4–3 m (7¾–9¾ ft)
	Tail 60–100 cm (23½–39 in)
	Weight Up to 1.2 tonnes (1 ton)
Location S. Asia	Social unit Group
	Status Endangered

At more than 1 tonne in weight, the Asian water buffalo – or arni as it is also known – is a massive, powerful animal, with the widest horn span of any bovid – more than 2 m (6½ ft). Its large, splayed feet and flexible fetlock joints are suited to the muddy, marshy ground on which this water buffalo grazes in the morning and evening, and occasionally at night, on lush grass and leafy aquatic vegetation. A stable clan of females with young is led by a dominant matriarch (as in elephants), while males form bachelor groups of about 10. Young males spar to assert dominance but avoid serious fighting, and then mix with females at mating time. The Asian water buffalo has been domesticated for thousands of years and spread around the world into various, mostly smaller breeds of less than 500 kg (1,100 lb). The remaining wild populations are scattered and scant, restricted mainly to India, Nepal, and perhaps Thailand.

WALLOWING

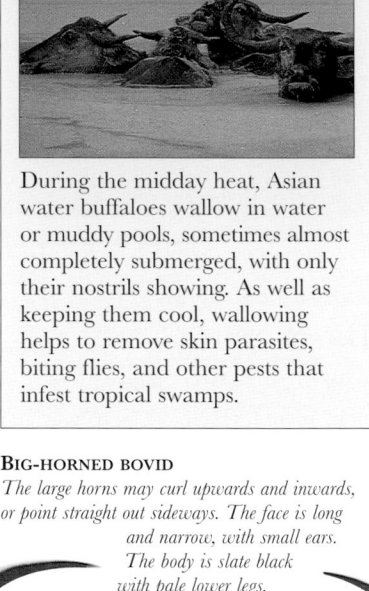

During the midday heat, Asian water buffaloes wallow in water or muddy pools, sometimes almost completely submerged, with only their nostrils showing. As well as keeping them cool, wallowing helps to remove skin parasites, biting flies, and other pests that infest tropical swamps.

BIG-HORNED BOVID
The large horns may curl upwards and inwards, or point straight out sideways. The face is long and narrow, with small ears. The body is slate black with pale lower legs.

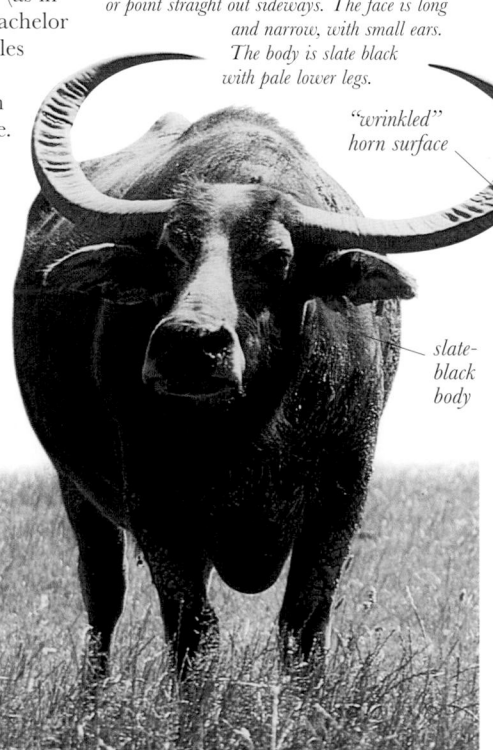

"wrinkled" horn surface

slate-black body

Syncerus caffer

African buffalo

Length 2.1–3.4 m
(7–11 ft)

Tail 75–110 cm
(30–43 in)

Weight Up to 685 kg
(1,510 lb)

Location W., C., E., and southern Africa

Social unit Group

Status Locally common†

Africa's only cow-like mammal frequents varied habitats at altitudes of up to 4,000m (13,200ft). However, the African buffalo needs a daily drink and is never farther than about 15km (9 miles) from water. The male may approach twice the weight of the female and has more robust horns that meet in a "boss" on the forehead, a thicker neck, a shoulder hump, and a small hanging fringe of hairs on the throat (dewlap). This buffalo feeds at night and at cool periods of the day, on various grasses, leaves, and other herbage. Males posture for females and dominance, and may fight by pressing or ramming heads. Herd members mutually groom and use mainly sound signals for coordinated actions such as moving on, flight, or warning. They may also cooperate to mob a predator such as a lion. The single calf (rarely twins) is born after 340 days' gestation, and is fiercely protected by the mother and often by other members of the herd.

HERD BEHAVIOUR

African buffaloes are very gregarious and gather at times of plentiful food in herds of 2,000 or more. In the dry season they split into smaller groups of females and young (including males up to 3 years), or bachelor bands of mature males; older males are more solitary. In any herd, large males dominate smaller ones and also any females.

DARK COLORATION
The African buffalo has a sparse, dark brown coat, big, drooping, hair-fringed ears, a naked muzzle, a long tail, and large feet with rounded hooves.

tapering horn curves in C shape

MIXED BENEFIT

Buffaloes may be infested with lice, ticks, fleas, and similar skin parasites, which birds such as oxpeckers nip off, to eat. The bird is fed, and the buffalo cleaned. However, the birds are also known to keep wounds open to feed on the blood, indicating a somewhat parasitic relationship.

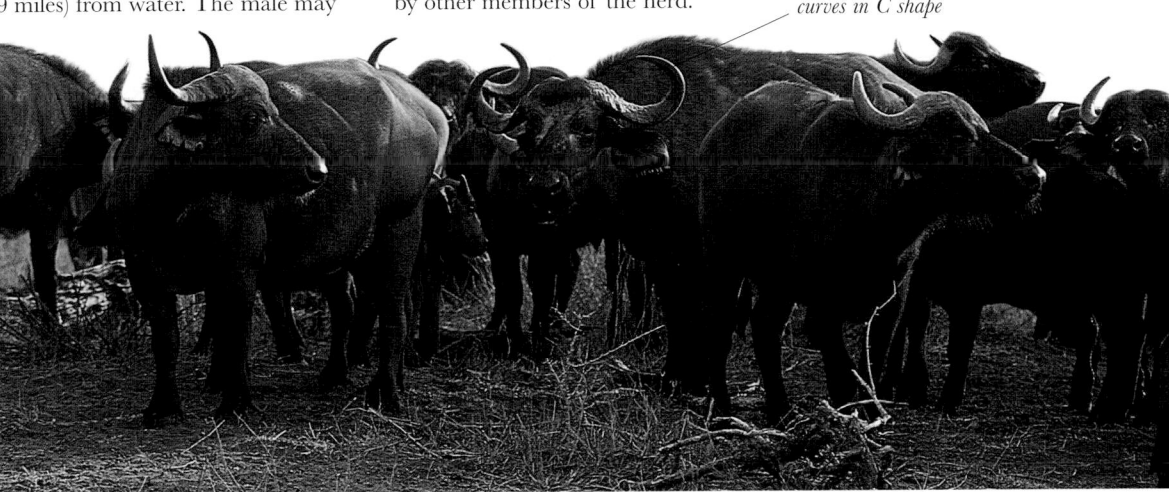

Bos javanicus

Banteng

Length 1.8–2.3 m
(6–7½ ft)

Tail 65–70 cm
(26–28 in)

Weight 400–900 kg
(880–1,990 lb)

Location S.E. Asia

Social unit Group

Status Endangered

Ancestor of the domestic banteng and resembling domestic cattle in overall form, the male (bull) banteng is black-brown to dark chestnut; the female (cow) and young are red-brown. All have white undersides, legs, and rump patches. The male's horns are up to 75cm (30in) long, angled outward, and then upwards; the female's are smaller and crescent-shaped. Banteng live in female–young herds of 2–40 with one adult male, or in bachelor herds. During the monsoon, they move to the hills, returning to the lowlands for the dry season. The wild populations of this bovid are scarce, and their habitat is fast diminishing.

Bos sauveli

Kouprey

Length 2.1–2.2 m
(7–7¼ ft)

Tail 1–1.1 m
(3¼–3½ ft)

Weight 700–900 kg
(1,544–1,985 lb)

Location S.E. Asia

Social unit Variable

Status Critically endangered

The kouprey (also called Cambodian forest ox) is similar to domestic cattle in general proportions. The male may be black or dark brown with a pendulous dewlap, and L-shaped horns that split at the tips after 3 years' growth. Females may be greyer. Both sexes have off-white legs and pale underparts. They probably form small, loose herds. One of the world's rarest species, the kouprey was identified only in 1937 and is at risk from habitat loss, political instability, poaching, and many other dangers.

Bos grunniens

Yak

Length Up to 3.3 m
(11 ft)

Tail 60 cm
(23½ in)

Weight Up to 525 kg
(1,160 lb)

Location S. and E. Asia

Social unit Individual/Group

Status Vulnerable

Domesticated yak are important to many peoples in South and Southeast Asia, for milk, meat, wool, and transport. The wild yak is larger and extremely rare, and is restricted to windy, desolate, bitterly cold steppe at an altitude of up to 6,000m (19,800ft), mainly from Kashmir (India) east to Tibet and Qinghai (China). Its dense undercoat of soft, close-matted hair is covered by generally dark brown to black outer hair. The yak grazes grasses, herbs, mosses, and lichens, and crunches ice or snow as a source of water. Females and young form herds, joined by males in the breeding season; otherwise males form bachelor groups or live alone. A single calf is born every other year, after a gestation period of 258 days.

high, humped shoulders

extremely long outer hairs

Bison bison

American bison

Length	2.1–3.5 m (7–11 ft)
Tail	30–60 cm (1–2 ft)
Weight	350–1,000 kg (770–2,200 lb)

Location W. and N. North America

Population 200,000

Status Lower risk

Also known as the American buffalo, the American bison is massively built, but also deceptively tall, with shoulder heights reaching 2m (6½ft). Despite its huge bulk, the bison can run at speeds of up to 60kph (37mph). Its sense of hearing and smell are excellent, and essential for detecting danger. Bison spend much of their time grazing and browsing, in loose groups, with rest periods for ruminating. Adult females and their young live in hierarchical herds, which are led by a dominant female. Males usually live apart in bachelor groups and only join females during the mating season. Bison are not territorial, and migration is dictated by season change and adequate food supply. A subspecies, *B. bison athabascae*, is recognized, but the validity of its taxonomy is debated. Recent genetic evidence suggests that the European bison, *Bison bonasus*, is more similar to its American kin than previously thought (see p.250).

MALES BATTLE FOR SUPREMACY

During the mating season male bison will fight fiercely for the possession of females, usually by head-to-head ramming. Females wanting to mate with the dominant bull gallop about to incite competition between rival males.

CONSERVATION

Despite once numbering around 50 million, the American bison is now virtually extinct in the wild, largely due to widespread commercial hunting since the arrival of European settlers. Although subsequent conservation efforts have led to a significant increase in numbers, most American bison are either captive or have come from captive stock. Yellowstone National Park (USA) and Wood Buffalo National Park (Canada) are the only places in which wild herds survive.

WINTER COATS
Although bison are associated with the warm, dry plains of the American West, they are also found in mountainous regions, where they may experience extreme temperatures. In fact, winters offer bison few problems as their heavy coats and thick manes protect them from the ice and cold.

GRAZING MIGRANT
In the days of the free-roaming herds, bison would make annual migrations of hundreds of kilometres along traditional routes. However, of a current total population of 200,000 bison only a very small fraction of these animals roam free in herds, and then only within the boundaries of the national parks.

short, upturned horns

broad forehead

pronounced shoulder hump in male

light brown, short hair

shaggy coat

straggly beard

MASSIVE FRAME
The bison's massive build is characterized by the towering shoulder hump. The brownish, black hair on the neck, head, shoulders, and forelegs is long and shaggy, but the rest of the body is covered in shorter and lighter-coloured hair. The large and heavy head sits on a short, thick neck, and features a wide forehead and straggly bearded chin.

Bos gaurus
Gaur

Length	2.5–3.3 m (8 1/4 –11 ft)
Tail	70–100 cm (28–39 in)
Weight	650–1,000 kg (1,430–2,210 lb)

Location S. to S.E. Asia   Social unit Group

Status Vulnerable†

Among the biggest cattle, the gaur (seladang or Indian bison) has a huge head and deep body in shades of red, brown, and black, and sturdy, whitish limbs. The S-shaped horns are up to 1.1m (3 1/2 ft) long, and the shoulder hump dips in the lower back before rising to the rump. Rutting males "sing" with bellows that deepen in pitch and carry for 1.5km (1 mile). The social system and mixed herbage diet are typical of other wild cattle.

Bison bonasus
European bison

Length	2.1–3.4 m (7–11 ft)
Tail	30–60 cm (12–23 1/2 in)
Weight	300–920 kg (660–2,030 lb)

Location E. Europe   Social unit Group

Status Endangered

Various lines of evidence, including some genetic studies, suggest that the European bison, or wisent, may be the same species as the American bison (see page 248). Formerly extinct in the wild, the European bison has been park-bred and reintroduced into the coniferous Bialoweiza Forest (which straddles the Polish and Belarussian border), where original populations survived until the 1910s. It has a slightly lighter and shorter coat than the American bison, but its habits and herding behaviour are similar. The horns, larger in the male, are short and upturned. This bison browses on leaves, twigs, and bark, and grazes on low vegetation. The single calf is born after a gestation period of 260–270 days. It can run within 3 hours and is weaned by one year.

well-developed shoulder hump

Cephalophus natalensis
Red forest duiker

Length	70–100 cm (28–39 in)
Tail	9–14 cm (3 1/2 –5 1/2 in)
Weight	13 kg (29 lb)

Location S.E. Africa   Social unit Pair

Status Lower risk

The red forest duiker is one of 18 or so species of duiker – a small, arch-backed antelope with longer hindlegs than forelegs. This species has a red-orange to dark brown coat, with long, coarse neck hairs, and a tail that is red at the base and tipped with black and white. It lives mainly in woods and forests, and its habits are similar to most other duikers (see below left).

Sylvicapra grimmia
Common duiker

Length	0.7–1.2 m (2 1/4 –4 ft)
Tail	7–19 cm (2 3/4 –7 1/2 in)
Weight	12–25 kg (26–55 lb)

Location W., C., E., and southern Africa   Social unit Individual/Pair

Status Locally common

The common duiker has a tufted forehead, large, pointed ears, and, usually only in males, sharp-pointed horns about 11cm (4 1/4 in) long. It is grey to red-yellow haired above, with white underparts, and has a dark nose stripe. Adaptable in habitat, it is a nocturnal browser and also takes small animals and carrion. It lives alone or in pairs, and males defend their territories against rivals.

Kobus ellipsiprymnus
Waterbuck

Length	1.3–2.4 m (4 1/4 –7 3/4 ft)
Tail	10–45 cm (4 –18 in)
Weight	50–300 kg (110 –660 lb)

Location W., C., and E. Africa   Social unit Individual/Group

Status Lower risk

One of the heaviest antelopes, the waterbuck has coarse, long, oily hair that ranges in colour from grey to red-brown and darkens with age. There are white markings on the rump, throat, and muzzle, and white "eyebrows", rings above the hooves, and underparts. The horns, normally present only in the male, are up to 1m (3 1/4 ft) long. Some 90 per cent of the diet is grass, the rest browsed leaves. When threatened, the waterbuck usually dashes to water, where it swims fast or submerges except for the nose. Herds of younger males, usually 2–5 but rarely 50 or more, have a hierarchy based on visual displays, horn length, and frequent fights. Older (6–10 years) breeding males occupy territories. Females are solitary or form loose groups of up to 10.

prominently ringed horns

Kobus leche
Lechwe

Length	1.3–2.4 m (4 1/4 –7 3/4 ft)
Tail	10–45 cm (4 –18 in)
Weight	79–103 kg (175 –230 lb)

Location C. to southern Africa   Social unit Group

Status Lower risk

The lechwe, or marsh antelope, eats grasses and aquatic plants exposed by seasonal variations in the weather, responding to water levels across floodplains and swamps. It wades and swims well, forms large herds, and has the lek breeding system (see kob, p.251). The chestnut to black coat contrasts with the white underparts and black leg stripes. Only the males are horned.

Kobus kob
Kob

Length 1.3–2.4 m
(4 1/4 – 7 3/4 ft)

Tail 10–45 cm
(4 –18 in)

Weight 50 –300 kg
(110 – 660 lb)

Location W. to E. Africa

Social unit Group

Status Lower risk

The kob is a graceful but strong grazing antelope, pale cinnamon to brown-black, with white facial and throat markings, and black leg stripes and feet. The male has ringed, lyre-shaped horns. Kobs live at very high densities, and males compete for a patch of ground (known as the lek), perhaps only 15m (50ft) across. The winner of this contest gains mating rights over many females.

Kobus vardonii
Puku

Length 1.3–1.8 m
(4 1/4 – 6 ft)

Tail 18 –30 cm
(7–12 in)

Weight 66 –77 kg
(145 –170 lb)

Location W. to E. Africa

Social unit Individual/Group

Status Lower risk

The puku resembles the kob (see left) in its breeding system: it has leks at high population densities; territories at lower ones. The long coat is a

Redunca redunca
Bohor reedbuck

Length 1.1–1.6 m
(3 1/2 – 5 1/4 ft)

Tail 15 –45 cm
(6 –18 in)

Weight 19 –95 kg
(42 – 210 lb)

Location W. to E. Africa

Social unit Variable

Status Lower risk

This small, fawn, lightweight savanna antelope has white underparts, throat, and eye-rings. A conspicuous grey patch under the ear marks a scent

uniform golden yellow, and the horns are about 50cm (20in) in length. The puku grazes morning and evening, and, like other plains antelopes, flees rapidly from danger.

gland. The bohor reedbuck feeds on grasses and tender reed shoots. Small groups, of a female and young, or horned males, merge into dry-season herds.

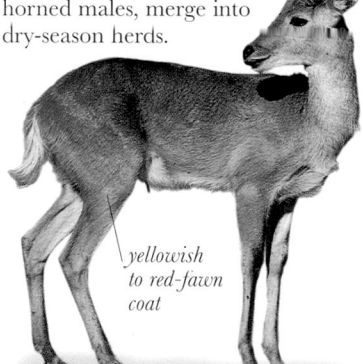

yellowish to red-fawn coat

Hippotragus equinus
Roan antelope

Length 1.9–2.7 m
(6 1/4 – 8 3/4 ft)

Tail 37–76 cm
(14 1/2 –30 in)

Weight 150 –300 kg
(330 – 660 lb)

Location W., C., and E. Africa

Social unit Individual/Group

Status Lower risk

The roan antelope is red to brown coated, with white underneath and black and white facial markings. Both sexes are horned and maned. They survive on poor grass growth and need to drink 2–3 times daily. Herds consist of 12–15 females and young with one dominant male, or younger bachelor males.

back-curved horns

Hippotragus niger
Sable antelope

Length 1.9–2.7 m
(6 1/4 – 8 3/4 ft)

Tail 37–76 cm
(14 1/2 –30 in)

Weight 150 –300 kg
(330 – 660 lb)

Location E. to S.E. Africa

Social unit Group

Status Lower risk

Similar to the roan antelope (see above right) in many respects, the sable antelope gathers in herds of 100 or more during the dry season, when

ADULT MALE
Mature males are black but with the same facial pattern as females: white with a central dark blaze and cheek stripes. Both sexes have stout, heavily ringed horns.

browsing replaces the usual grazing. In the wet season, herds split into bachelor groups of 2–12 males, while dominant males occupy territories and mate with the females there. After an initial concealment, calves form nursery groups, only joining their mothers to suckle.

facial markings resemble adult's

SABLE CALF
Calf coloration is similar to that of adult females – chestnut or sorrel. It is born after 240 – 280 days' gestation, and initially is kept away from the herd.

Oryx dammah
Scimitar-horned oryx

Length 1.5–2.4 m
(5 1/4 – 7 3/4 ft)

Tail 45 –90 cm
(18 –35 in)

Weight 100 –210 kg
(220 – 460 lb)

Location N. Africa

Social unit Group

Status Critically endangered†

Specialized for deserts, arid plains, and rocky hills, the scimitar-horned oryx has many physiological adaptations to conserve body water. Its kidneys are very efficient, and it sweats only when its body temperature exceeds 46°C (116°F). The large hooves spread the stocky body's weight on soft sand. This oryx feeds on a wide variety of plants in the early morning and evening, and on moonlit nights,,

resting by day in any available shade. It forms nomadic mixed herds of 20–40, males displaying and tussling to mate with females. After 222–253 days' gestation, the mother leaves the herd to calve, but returns within hours. Young are weaned by 14 weeks and are sexually mature by 2 years. This species was hunted almost to extinction, surviving only on a reserve in north-central Chad. Captive-bred animals were reintroduced in Tunisia in 1991; further releases are planned.

smooth horns

faint stripe along lower flank

ruddy brown neck and chest

MAMMALS

Oryx gazella
Gemsbok

Length	1.6–2.4 m (5¼–7¾ ft)
Tail	45–90 cm (18–35 in)
Weight	100–210 kg (220–460 lb)

Location S.W. Africa | Social unit Group

Status Lower risk

The gemsbok, or southern oryx, is a large, distinctively coloured antelope of arid grassy scrub and true desert. Its many adaptations for moisture conservation include not panting or sweating until the body temperature exceeds 45°C (113°F), kidneys that make very concentrated urine, and very dry droppings. The main diet is grasses and low shrubs, with wild cucumbers, melons, and similar plants

supplying water. Nomadic herds of up to 25 contain females, young, and a few males. Like many desert species, gemsbok breed opportunistically, year-round whenever food is available. The single calf (rarely twins) is born after 260–300 days' gestation, and stays concealed but near the main herd, with occasional visits from the mother to suckle, for up to 6 weeks.

ringed horns

black lower side stripe

PEACE IN THE SHADE

Gemsbok cope in their hot, dry habitat by grazing at cool times, mainly twilight and by night. They gather in available shade from about 10am to 2–3pm. In most other antelope species, individuals grouped so closely would begin dominance or mating disputes, but gemsbok suspend social interaction for the greater need of survival.

BLACK, WHITE, AND SHADES OF GREY
The gemsbok has a fawn-tinged grey body and contrasting black and white face, ears, belly, and legs. The broad muzzle and wide row of incisor teeth are adapted to crop coarse grasses.

Addax nasomaculatus
Addax

Length	1.5–1.7 m (5–5½ ft)
Tail	25–35 cm (10–14 in)
Weight	60–125 kg (130–280 lb)

Location N.W. Africa | Social unit Variable

Status Critically endangered

Rare and remote, the addax has similar adaptations to the gemsbok (see left) for desert survival. It wanders in search of almost any vegetation, following rains, and rests in shade at midday. Formerly more numerous, a mixed herd of up to 20 was led by an older male; individuals now live alone or in small bands of 2–4. The coat is grey-brown in winter, sandy to white in summer, with a white facial patch topped by a chestnut forehead tuft, and spiral horns (1½–3 turns).

Damaliscus dorcas
Bontebok

Length	1.2–2.1 m (4–7 ft)
Tail	10–60 cm (4–23½ in)
Weight	68–155 kg (150–340 lb)

Location Southern Africa | Social unit Group

Status Vulnerable

Also known as the blesbok, the bontebok has a white blaze on its long muzzle, which narrows at the eyes, and continues to the

flattened, lyre-shaped horns, which are ringed for most of their 70-cm (28-in) length. Adult males posture and spar with their horns – although rarely fight – to gain a territory. This allows them to dominate a herd of females and young. The male keeps its members together and initiates the herd's travels. Females give birth at traditional calving grounds, after a gestation period of about 8 months; unlike many similar antelopes, they do not isolate or conceal their young. The single newborn can walk within 5 minutes, and soon follows its mother; it is weaned by 6 months. The bontebok grazes mixed grasses and herbs, early and late in the day. It was almost exterminated in the wild by the 1830s, but herds were preserved in parks and reserves. It is still South Africa's rarest antelope.

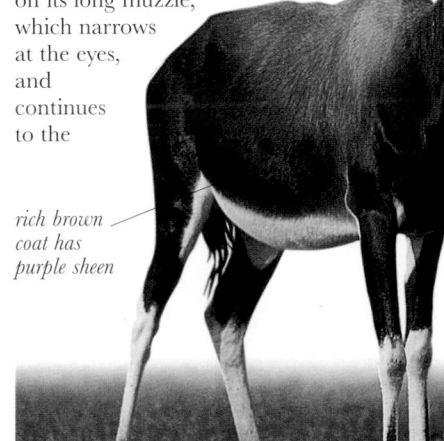

rich brown coat has purple sheen

Damaliscus lunatus
Topi

Length	1.2–2.1 m (4–7 ft)
Tail	10–60 cm (4–23½ in)
Weight	68–155 kg (150–340 lb)

Location W., C., E., and southern Africa | Social unit Group

Status Lower risk

The topi, or tsessebe, has a long head, a shoulder hump, and a downward-sloping back. Its glossy, red-brown coat is purplish on the upper limbs and darker on the upper muzzle, belly, and lower limbs. It has L-shaped, ringed horns. The topi lives in seasonally flooded grasslands, and has 2 breeding systems according to conditions: leks (see kob, p.251) in migratory populations; and male territories with harems when resident.

Alcephalus buselaphus
Hartebeest

Length	1.5–2.5 m (5–8¼ ft)
Tail	30–70 cm (12–28 in)
Weight	100–225 kg (220–500 lb)

Location W., E., and southern Africa | Social unit Group

Status Lower risk

Resembling the topi (see left) in diet and sloping profile, the hartebeest has heavily ringed horns, which are spiral or lyre shaped depending on the subspecies. Colour varies from chestnut or fawn to brown-grey, with paler hip patches, and black on the forehead, muzzle, shoulders, and thighs. This antelope forms large, highly organized herds of up to 300, dominated by territorial males but also containing sub-herds of non-territorial males and of females with young.

Connochaetes taurinus
Wildebeest

Length	1.5–2.4 m (5–7 3/4 ft)
Tail	35–56 cm (14–22 in)
Weight	120–275 kg (260–610 lb)

Location E. and southern Africa

Social unit Group

Status Lower risk

Also called the brindled gnu or blue wildebeest, this antelope has an unmistakable, large, long-muzzled head, cow-like horns, and high shoulders. The single calf is born after 8–9 months' gestation. It bleats like a lamb, and its fiercely protective mother lows like a domestic cow in reply. Males form bachelor herds at 1–4 years of age, then try to establish solitary territories, with "ge-nuu" calls, ritual posturing, and pushing. Only winning males can mate.

horns up to 80cm (32in) long in male

LONG MANE
The wildebeest's copious black mane extends from neck to shoulders, and hairs spill over the forehead. The main coat is silver-grey with brown hues fading rearwards, and the tail is long and black.

DANGEROUS JOURNEY

Although some wildebeest remain in a home range, most band into large herds and migrate hundreds of kilometres on an annual journey to find seasonal grazing. The route may be altered by unusual rains stimulating grass elsewhere. At river crossings (above), they are vulnerable to crocodile attack.

Raphicerus campestris
Steenbok

Length	61–95 cm (24–37 in)
Tail	4–8 cm (1 1/2–3 1/4 in)
Weight	7–16 kg (15–35 lb)

Location E. and southern Africa

Social unit Individual/Pair

Status Locally common

The steenbok, or steinbuck, lives alone or as a pair with largely separate lives, within a territory marked by scents and dung. It both browses and grazes, and digs up

roots and tubers with its feet, which lack the 2 lateral toes. Its colour is bright rufous-fawn, sometimes tinged silver-grey, pale beneath. There is also a white eye-stripe or ring and black "finger lines" in the ear, patches on the nose, and between the horns, which are present only in the male.

Oreotragus oreotragus
Klipspringer

Length	0.8–1.2 m (2 1/2–4 ft)
Tail	5–13 cm (2–5 in)
Weight	8–18 kg (18–40 lb)

Location E., C., and southern Africa

Social unit Pair

Status Lower risk

This small, short-muzzled, tiny-hoofed antelope leaps skilfully over the steep, rocky terrain of its native mountains and river gorges. It has a short tail and a dense, glossy olive-yellow coat speckled with yellow and brown, fading to white on the underparts and legs. The male has small, spiky horns. The klipspringer browses on evergreen and other shrubs, and lives in pairs with 1–2 offspring.

tiny hooves

Madoqua kirkii
Kirk's dik-dik

Length	52–72 cm (20 1/2–28 in)
Tail	35–56 cm (14–22 in)
Weight	3–7 kg (6 1/2–15 lb)

Location E. and S.W. Africa

Social unit Pair

Status Locally common

The 4 dik-dik species are named after their alarm call. Kirk's dik-dik has a soft, lank coat, grizzled grey to brown, and more reddish brown on the head, with a forehead crest. The rubbery-based hooves grip effectively on rocks. It eats a wide range of plants and lives as a close-bonded male–female pair and a single offspring, born after 169–174 days' gestation. The newborn remains hidden for 2–3 weeks, and suckles for 3–4 months.

Ourebia ourebi
Oribi

Length	0.9–1.4 m (3–4 1/2 ft)
Tail	6–15 cm (2 1/4–6 in)
Weight	14–21 kg (31–46 lb)

Location W. to E. and southern Africa

Social unit Variable

Status Lower risk

Small, slender, and long necked, the oribi has fine, silky fur, sandy to rufous above, and white below and on the chin and rump, with longer tufts on the knees. The male has 2 small, spiky, ringed horns. The diet consists of grasses and, in the dry season, bushy leaves. Oribi are socially flexible, with male–female pairs or small herds of 7–8 with 2–3 adult males. The male may help to clean and guard the calf, which is weaned at 2 months.

Antilope cervicapra
Blackbuck

Length	1.2 m (4 ft)
Tail	18 cm (7 in)
Weight	32–43 kg (71–95 lb)

Location S. Asia

Social unit Group

Status Vulnerable

The blackbuck grazes on grasses, including cereal crops. Females are fawn to yellow, dominant males become black with age, and other males are brown. All have a white underside, rump, muzzle, and eye-ring. The male's horns, up to 68cm (27in) long, are ringed at the base and spiral up to 5 turns. When breeding, the male defends his territory and harem.

Aepyceros melampus

Impala

Length	1.1–1.5 m (3 1/2 – 5 ft)
Tail	25–40 cm (10–16 in)
Weight	40–65 kg (88–145 lb)

Location E. and southern Africa

Social unit Group

Status Lower risk

Impala are noisy antelopes. Males (which are horned) make loud, hoarse grunts when rutting; calves bleat; and all emit loud warning snorts as they race off with high leaps, kicking the hindlegs out straight and landing on the forelegs. Adaptable grazer-browsers, impala form mixed herds in the dry season. At breeding time bachelor males compete for territories and females. The coat is reddish fawn with black streaks on the hips and tail.

Litocranius walleri

Gerenuk

Length	1.4–1.6 m (4 1/2 – 5 1/4 ft)
Tail	22–35 cm (9–14 in)
Weight	28–52 kg (62–115 lb)

Location E. Africa

Social unit Individual/Group

Status Lower risk

Also called the giraffe-gazelle, the gerenuk has a very long, slender neck and similar legs. It can curve its spine into an S shape, balancing its weight over its rear legs, in order to stand vertically for long periods. This allows it to browse higher than similar-sized herbivores in open woodland and scattered bush. Seen from the front, its neck, head, and long, wedge-shaped muzzle are extremely narrow, for probing into acacias and other thorny foliage. The gerenuk uses its long, pointed tongue, mobile lips, and sharp-edged incisors to pluck and nip the smallest leaves. It is mainly reddish fawn, with a broad, dark band along the back and upper sides, and white on the underparts, neck, chin, lips, and around the eyes; the tail is black tufted. Only the males have horns, which are 35cm (14in) long, relatively thick and curved. Social units are male–female pairs, or small groups

of one male and 2–4 females, with offspring. Only territory-holding males breed, from about 3 years old. Younger males form bachelor herds, although one male may attach to the periphery of a female band, and the occasional female may be solitary.

Antidorcas marsupialis

Springbok

Length	1.2–1.4 m (4 – 4 1/2 ft)
Tail	15–30 cm (6 – 12 in)
Weight	30–48 kg (66 – 105 lb)

Location Southern Africa

Social unit Group

Status Lower risk

The springbok is among several bovids that "stott" or "pronk" – leap stiff-legged, high and repeatedly, as if bouncing. This behaviour may serve to deter predators. This adaptable herbivore is highly gregarious, but migratory herds once millions-strong now number only 1,500. Breeding habits are as for other gazelle species.

reddish brown band on face

white underparts

Gazella thomsonii

Thomson's gazelle

Length	0.9–1.2 m (3 – 4 ft)
Tail	15–20 cm (6 – 8 in)
Weight	15–30 kg (33 – 66 lb)

Location E. Africa

Social unit Group

Status Lower risk

This small gazelle is graceful and speedy. It "stotts" (see springbok, above right) energetically when threatened by, for example, many of the big cats, hyenas, jackals, and similar carnivores, for which it is a staple part of their diet. The most common gazelle in its region, it some-times forms mixed herds with impala and other gazelle species. It mainly grazes short grasses, yet also browses. Small herds of 10–30 females and young join male

RAPID BREEDING

Thomson's gazelle is one of the few bovids that can breed twice yearly. The first calf is born in January or February, after the rains; the second in July. A newborn is quickly on its feet, but spends the first few weeks lying hidden, until it can keep up with the herd, at 3–4 weeks. It is weaned by 4 months.

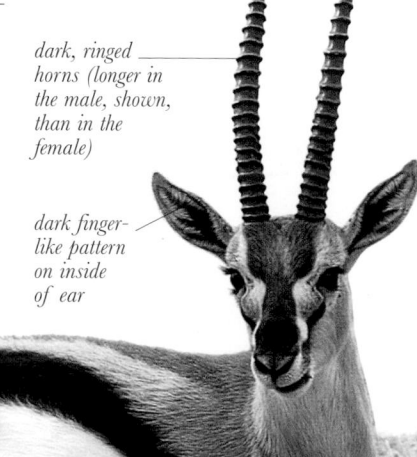

dark, ringed horns (longer in the male, shown, than in the female)

dark finger-like pattern on inside of ear

bachelor bands, and even lone males, to migrate between grasslands (in the rainy season) and bush (when it is drier). Thomson's gazelle usually produces a single offspring, which is born after 160–180 days' gestation. Although initially mottled darker than the parent, the coat of the young gazelle lightens in 1–2 weeks.

DISTINCTIVE MARKINGS

A black flank band separates the sandy fawn back from the white underside. The rufous head has a darker blaze, and white eye-rings that extend along the muzzle, above the black cheek stripes.

Saiga tatarica

Saiga

Length	1–1.4 m (3 1/4 – 4 1/2 ft)
Tail	6–12 cm (2 1/4 – 4 3/4 in)
Weight	26–69 kg (57 – 150 lb)

Location C. Asia

Social unit Group

Status Vulnerable

This medium-sized Asian "goat-antelope" has an enlarged nose with down-pointing nostrils, perhaps to control body temperature and/or give a keen sense of smell. The thick, woolly coat is cinnamon-buff above, paler on the underparts, and thickens greatly for winter. The saiga lives on dry steppe and eats varied plants. Smaller breeding groups join to form huge herds for migration. Only males of the species have horns.

Oreamnos americanus
Mountain goat

Length	1.2–1.6 m (4–5¼ ft)
Tail	10–20 cm (4–8 in)
Weight	46–140 kg (100–310 lb)

Location W. Canada, N. and W. USA

Social unit Variable

Status Locally common

This woolly haired goat survives among ice, snow, rocks, and glaciers. Its large, sharp hooves have hard rims and soft inner pads, to grip slippery surfaces, and its long, yellow-white outer coat and thick, dense underfur conserve body heat. The sharp, back-curved horns are 25cm (10in) long, and are slightly thicker in males, which may be 30 per cent larger than females. These goats feed on grass, moss, lichens, and twigs. It forms groups of up to 4 in summer, which join to form larger herds in winter.

Rupicapra rupicapra
Chamois

Length	0.9–1.3 m (3–4¼ ft)
Tail	3–4 cm (1¼–1½ in)
Weight	24–50 kg (53–110 lb)

Location Europe, W. Asia

Social unit Individual/Group

Status Critically endangered

The chamois is an agile climber, able to leap 2m (6½ft) high, spring 6m (20ft) along, and run at 50kph (31mph) – its flexible hoof pads giving sure grip on uneven, slippery terrain. Both sexes have slender, black, close-set horns, which are up to 20cm (8in) long, and curve back at the tips. It feeds on herbs and flowers in alpine pastures during summer, and moves lower in winter for mosses, lichens, and shoots, as groups disperse.

tawny fur, darker in winter

white patch on throat

Capra ibex
Ibex

Length	1.2–1.7 m (4–5½ ft)
Tail	10–20 cm (4–8 in)
Weight	35–150 kg (77–330 lb)

Location S. Europe, W. and S. Asia, N. Africa

Social unit Individual/Group

Status Endangered†

The ibex dwells at or above the treeline, up to 6,700m (22,000ft). The female has a tan coat in summer and the male a rich brown one with yellow-white patches on the back and rump. Both sexes grow a thicker winter coat of more variable colour. In spring, they migrate up to alpine pasture, descending in autumn to browse buds and shoots. Females and young form stable groups of 10–20; males form single-sex herds.

woolly beard

CLASH OF HORNS

Like many wild (and domestic) goats, male ibex compete for herd dominance and females by postures, head tosses, and fights. Opponents rear up on their back legs, then lunge forwards and clash heads and horns with skull-jarring force.

scimitar-shaped horns

HUGE HORNS
The ibex has thick, curved horns, which are up to 1.4m (4½ft) in males, but one-quarter this length in females.

Hemitragus jemlahicus
Tahr

Length	0.9–1.4 m (3–4½ ft)
Tail	9–12 cm (3½–4¾ in)
Weight	50–100 kg (110–220 lb)

Location S. Asia

Social unit Group

Status Vulnerable

The sure-footed tahr has a shaggy, conspicuous mane on its neck and shoulders, extending to its knees; its face and head fur is contrastingly short. The horns, flattened from side to side, reach 40cm (16in) in males – twice as long as in females. Like many mountain mammals, the tahr migrates high into the Himalayas in spring, to browse and graze in mixed forests at altitudes of 5,000m (16,500ft). It returns to temperate forests as low as 2,500m (8,250ft) in autumn, when herds of 2–23 females are joined by rutting males, who lock horns and try to topple each other off balance. Young are born the following May or June.

reddish brown coat

very short tail

Ovibos moschatus
Muskox

Length	1.9–2.3 m (6¼–7½ ft)
Tail	9–10 cm (3½–4 in)
Weight	200–410 kg (440–900 lb)

Location N. North America, Greenland

Social unit Group

Status Locally common

The muskox is named after the strong odour of rutting males, who charge and ram each other for females. Both sexes have broad horns, which nearly meet at a central boss, curving down and then up at the tips. The body is massive; the neck, legs, and tail are short. In summer, the muskox grazes valley sedges and grasses. In winter, it browses on higher ground, where wind keeps the ground clear of snow. Almost exterminated by humans, the muskox has recovered through wildlife management and reintroductions.

DOUBLE COAT
The muskox's outer coat of dark brown guard hairs reaches almost to the ground, to shed rain and snow effectively. The undercoat of fine, soft, pale brown hair provides excellent insulation.

paler mid back

shoulder hump

pale legs

DEFENSIVE CIRCLE

When under threat – for example, from wolves or a polar bear – adult male and female muskoxen gather in a circle, facing outwards. The vulnerable youngsters are protected in the middle. Larger herd members may break from the circle to charge or otherwise intimidate the enemy.

MAMMALS

MAMMALS

Capra aegagrus
Wild goat

	Length 1.2–1.6 m (4–5 1/4 ft)
	Tail 15–20 cm (6–8 in)
	Weight 25–95 kg (55–210 lb)

Location W. Asia | Social unit Group

Status Vulnerable

The wild or Bezoar goat, probably the ancestor of domestic goats, grazes and browses in varied habitats, from arid scrub to alpine pasture, at elevations up to 4,200m (13,800ft). Females are red-grey to yellow-brown; adult males (a young male is pictured) are bearded and silver-grey with dark markings. Both are horned. Males fight to dominate their bachelor herd and for females.

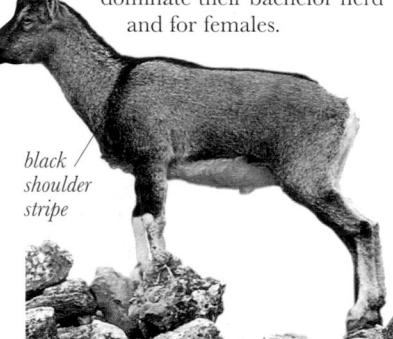

black shoulder stripe

Capra falconeri
Markhor

	Length 1.4–1.8 m (4 1/2–6 ft)
	Tail 8–14 cm (3 1/4–5 1/2 in)
	Weight 32–110 kg (70–245 lb)

Location C. and S. Asia | Social unit Variable

Status Endangered

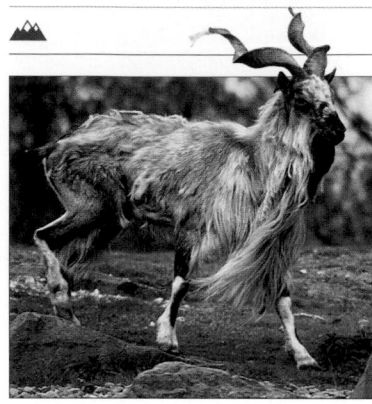

The markhor occupies various habitats at 700–4,000m (2,300–13,200ft), eating tussock grass in summer, and shrubby leaves and twigs on lower slopes in winter, when its short, red-grey coat becomes longer and greyer. The spectacular spiral horns reach 1.6m (5 1/4ft) in the shaggy-throated males, yet only 25cm (10in) in the females. These are one reason why the markhor is threatened by hunting.

Pseudois nayaur
Blue sheep

	Length 1.2–1.7 m (4–5 1/2 ft)
	Tail 10–20 cm (4–8 in)
	Weight 25–80 kg (55–175 lb)

Location S. to E. Asia | Social unit Group

Status Lower risk

Also called the bharal, the blue sheep is camouflaged to survive in rocky, icy alpine zones between tree- and snow-lines. Males are brown-grey, tinged with slate-blue, white below, with a white eyebrow strip, and black flanks and leg stripes. The smooth, 80-cm (32-in) horns splay outwards. Females are smaller with shorter horns, and lack most of the black markings. Breeding habits resemble other sheep, with rams competing for harems.

Ammotragus lervia
Barbary sheep

	Length 1.3–1.7 m (4 1/4–5 1/2 ft)
	Tail 15–25 cm (6–10 in)
	Weight 40–145 kg (88–320 lb)

Location N. Africa | Social unit Individual/Group

Status Vulnerable

The rufous-tawny coat of the Barbary sheep, or aoudad, has a short, upright mane on the neck and shoulders, and a much longer one on the throat, chest, and upper forelegs. Both manes and the crescent-shaped horns, to 84cm (33in) long, are more developed in males. These sheep eat a wide variety of plant matter. Males charge each other, heads lowered, for dominance and access to female groups for breeding.

Ovis ammon
Argali

	Length 1.2–1.8 m (4–6 ft)
	Tail 7–15 cm (2 3/4–6 in)
	Weight Up to 200 kg (440 lb)

Location C., S., and E. Asia | Social unit Individual/Group

Status Endangered

Male argali not only charge and clash heads when rutting but also run in parallel and butt the opponent's flank and chest. Also called arkhar, this is the largest wild sheep, with elaborate, ridged horns – up to 1.5m (5ft) in males – that corkscrew sideways, twisting 360 degrees or more with age. The coat of this very gregarious bovid is variably pale brown with white legs and rump patch. Like many wild sheep, it is threatened by human hunting, and habitat loss to livestock. Twin young are relatively common.

Ovis orientalis
Asiatic mouflon

	Length 1.1–1.3 m (3 1/2–4 1/4 ft)
	Tail 7–12 cm (2 3/4–4 3/4 in)
	Weight 25–55 kg (55–120 lb)

Location W. Asia | Social unit Individual/Group

Status Endangered

The smallest wild sheep, and probable ancestor of all domestic breeds, the mouflon frequents uplands and shrubby, grassy plains. The coat is red-brown with a dark central back stripe flanked by a paler "saddle" patch, a short, broad, dark tail, and paler underparts. The curved horns

are longer, up to 65cm (26in), in males. As in many other wild sheep, females live in small groups with their young, while lone or bachelor-band males compete for dominance and access to females. Success depends on a male's strength in pushing, butting, and ramming, and so most do not begin breeding until 6–7 years old, rutting in late autumn.

Ovis canadensis
Bighorn sheep

	Length 1.5–1.8 m (5–6 ft)
	Tail 7–15 cm (2 3/4–6 in)
	Weight 55–125 kg (120–280 lb)

Location S.W. Canada, W. and C. USA, N. Mexico | Social unit Individual/Group

Status Lower risk

The bighorn sheep's glossy brown summer coat of brittle guard hairs over crimped grey underfur fades in winter. Before the rut, males display, walk away from each other, turn, advance with a threat jump, and then lunge to head butt with enormous force. This may continue for hours until one gives up. The 1–3 young are born after a gestation period of 150–180 days.

pale patch on rump

OUT OF REACH

Like many wild sheep, when under threat, bighorn sheep use their gripping hooves and climbing ability to retreat to near-vertical rocky bluffs and cliffs, where few predators can follow. The young bighorns learn about seasonal pathways and suitable habitats from adults in their group.

BIGGER HORNS
The male's horns curl almost in a circle and may weigh as much as the rest of the skeleton – up to 14kg (31lb). The female's horns are smaller and only slightly curved.

MAMMALS IN DANGER

The list below includes all those mammals classed as critically endangered by the IUCN (see p.31). Gravely threatened species occur in all major mammal groups, but some are particularly badly affected. In relative terms, primates fare worst of all, with over 40 species or sub-species in imminent danger of extinction – largely due to the destruction of forest habitats. Habitat change is also the main threat facing insectivores, bats, lagomorphs and rodents, but with cetaceans, carnivores, and hoofed mammals, hunting has played a major part in cutting numbers to the precarious levels seen today. While the number of marsupials listed is relatively short, many other species in this group have already become extinct, particularly in Australia, where small mammals have faced an onslaught from aggressive introduced species.

MARSUPIALS

Gracilinanus aceramarcae (Bolivian gracile opossum)
Lagorchestes hirsutus (1 subsp: rufous hare-wallaby)
Lasiorhinus krefftii (northern hairy-nosed wombat)
Marmosa andersoni (Anderson's mouse opossum)
Marmosops handleyi (Handley's slender mouse opossum)
Perameles gunnii (1 subsp: eastern barred bandicoot)
Sminthopsis griseoventer (1 subsp: Boullanger Island dunnart)

INSECTIVORES & BATS

Amblysomus julianae (Juliana's golden mole)
Aproteles bulmerae (Bulmer's fruit bat)
Biswamoyopterus biswasi (Namdapha flying squirrel)
Chaerephon gallagheri (Gallagher's free-tailed bat)
Chimarrogale hantu (Malayan water shrew);
   C. sumatrana (Sumatra water shrew)
Chlorotalpa tytonis (Somali golden mole)
Chrysochloris visagiei (Visagie's golden mole)
Coleura seychellensis (Seychelles sheath-tailed bat)
Congosorex polli (Congo shrew)
Crocidura ansellorum (Ansell's shrew); *C. caliginea*
   (African foggy shrew); *C. desperata* (desperate
   shrew); *C. dhofarensis* (Dhofar shrew);
   C. eisentrauti (Eisentraut's shrew); *C. gracilipes*
   (Peters' musk shrew); *C. harenna* (Harenna
   shrew); *C. jenkinsi* (white-toothed shrew
   species); *C. macmillani* (Macmillan's shrew);
   C. macowi (Macow's shrew); *C. negrina* (Negros
   shrew); *C. phaeura* (Guramba shrew); *C. picea*
   (pitch shrew); *C. polia* (fuscous shrew); *C. raineyi*
   (Rainey shrew); *C. telfordi* (Telford's shrew);
   C. ultima (ultimate shrew)
Cryptochloris zyli (Van Zyl's golden mole)
Euroscaptor parvidens (small-toothed mole)
Hipposideros nequam (Malayan roundleaf bat)
Hylomys parvus (dwarf gymnure)
Latidens salimalii (Indian fruit bat)
Mops niangarae (Niangara free-tailed bat)
Microgale dryas (tree shrew tenrec)
Murina tenebrosa (gloomy tube-nosed bat)
Myosorex rumpii (Rumpi mouse shrew); *M. schalleri*
   (Schaller's mouse shrew)
Myotis cobanensis (Guatemalan myotis); *M. planiceps*
   (flat-headed myotis)
Nyctimene rabori (Philippines tube-nosed fruit bat)
Otomops wroughtoni (Wroughton's free-tailed bat)
Paracoelops megalotis (Vietnam leaf-nosed bat)
Paracrocidura graueri (Grauer's shrew)
Pharotis imogene (New Guinea big-eared bat)
Pipistrellus anthonyi (Anthony's pipistrelle); *P. joffrei*
   (Joffre's pipistrelle)
Pteralopex acrodonta (Fijian monkey-faced bat);
   P. anceps (Bougainville monkey-faced bat);
   P. atrata (cusp-toothed flying-fox); *P. pulchra*
   (montane monkey-faced bat)
Pteropus insularis (Chuuk flying-fox); *P. livingstonei*
   (Comoro black flying-fox); *P. molossinus*
   (Pohnpei flying-fox); *P. phaeocephalus* (Mortlock
   flying-fox); *P. pselaphon* (Bonin fruit bat); *P.
   rodricensis* (Rodriguez flying-fox); *P. voeltzkowi*
   (Pemba flying-fox)
Rhinolophus convexus (horseshoe bat species)
Scotophilus borbonicus (lesser yellow bat)
Sorex cansulus (Gansu shrew); *S. kozlovi* (Kozlov's
   shrew)
Soriculus salenskii (Salenski's shrew)

Suncus ater (black shrew); *S. mertensi* (Flores shrew);
   S. remyi (Gabon dwarf shrew)
Talpa streeti (Persian mole)
Taphozous troughtoni (sharp-nosed tomb bat)

PRIMATES

Alouatta belzebul (1 subsp: red-handed howling
   monkey); *A. coibensis* (1 subsp: Azuero howling
   monkey); *A. guariba* (1 subsp: northern brown
   howling monkey)
Ateles geoffroyi (2 subsp: Azuero spider monkey,
   brown-headed spider monkey)
Brachyteles arachnoides (woolly spider monkey);
   B. hypoxanthus (northern muriqui)
Callicebus barbarabrownae (northern Bahian blond
   titi); *C. coimbrai* (Coimbra's titi)
Cebus albifrons (1 subsp: Trinidad white-fronted
   capuchin); *C. apella* (1 subsp: Margarita Island
   capuchin); *C. xanthosternos* (yellow-breasted
   capuchin)
Cercocebus atys (1 subsp: white-collared mangabey);
   C. galeritus (1 subsp: Tana River mangabey)
Cercopithecus diana (1 subsp: Roloway monkey);
   C. nictitans (1 subsp: Stampf's greater spot-
   nosed guenon)
Eulemur fulvus (1 subsp: white-collared brown
   lemur); *E. macaco* (1 subsp: Sclater's black
   lemur)
Gorilla beringei (2 subsp: Bwindi gorilla, mountain
   gorilla); *G. gorilla* (1 subsp: Cross River gorilla)
Hapalemur aureus (golden lemur); *H. griseus*
   (1 subsp: Alaotran gentle lemur); *H. simus*
   (greater bamboo lemur)
Hylobates lar (1 subsp: white-handed gibbon);
   H. moloch (3 subsp: silvery gibbon, western
   Javan gibbon, central Javan gibbon)
Leontopithecus caissara (black-faced lion tamarin);
   L. chrysopygus (black lion tamarin); *L. rosalia*
   (golden lion tamarin)
Macaca pagensis (3 subsp: Mentawai macaque,
   Pagai macaque, Siberut macaque)
Nomascus concolor (1 subsp: West Yunnan black
   crested gibbon); *N. concolor* (1 subsp: Central
   Yunnan black crested gibbon)
Oreonax flavicauda (Yellow-tailed woolly monkey)
Pongo abelii (Sumatran orang-utan)
Procolobus badius (1 subsp: Miss Waldron's Bay
   colobus); *P. pennantii* (1 subsp: Bouvier's red
   colobus); *P. rufomitratus* (Tana River red
   colobus)
Propithecus diadema (3 subsp: silky sifaka,
   diademed sifaka, Perrier's sifaka); *P. tattersalli*
   (golden-crowned sifaka); *P. verreauxi* (1 subsp:
   crowned sifaka)
Rhinopithecus avunculus (Tonkin snub-nosed
   monkey)
Saimiri oerstedii (1 subsp: grey-crowned Central
   American squirrel monkey)
Trachypithecus delacouri (Dealacour's langur);
   T. poliocephalus (François's leaf monkey)
Varecia variegata (1 subsp: red-ruffed lemur)

LAGOMORPHS & RODENTS

Acomys cilicicus (African spiny mouse)
Allactaga firouzi (Iranian jerboa)
Chinchilla brevicaudata (short-tailed chinchilla)
Chrotomys gonzalesi (Mt Isarog striped rat)
Crateromys paulus (Ilin Island cloud rat)

Crunomys fallax (Northern Luzon shrew rat)
Dendromus vernayi (African climbing mouse)
Dicrostonyx vinogradovi (collared lemming)
Dipodomys heermanni (1 subsp: Morro Bay
   kangaroo rat); *D. ingens* (giant kangaroo rat);
   D. insularis (San Jose Island kangaroo rat);
   D. margaritae (Margarita Island kangaroo rat);
   D. nitratoides (2 subsp: Fresno kangaroo rat,
   Tipton kangaroo rat)
Eliurus penicillatus (white-tipped tufted-tailed rat)
Gerbillus bilensis (Bilen gerbil); *G. burtoni* (Burton's
   gerbil); *G. cosensis* (Cosen's gerbil); *G. dalloni*
   (Dallon's gerbil); *G. floweri* (Flower's gerbil);
   G. grobbeni (Grobben's gerbil); *G. hoogstraali*
   (Hoogstraal's gerbil); *G. lowei* (Lowe's gerbil);
   G. mauritaniae (Mauritian gerbil); *G. occiduus*
   (occidental gerbil); *G. principulus* (principal
   gerbil); *G. quadrimaculatus* (four-spotted gerbil);
   G. syrticus (sand gerbil)
Heteromys nelsoni (Nelson's spiny pocket mouse)
Hylopetes winstoni (arrow-tailed flying squirrel)
Isolobodon portoricensis (Puerto Rican hutia)
Leimacomys buettneri (groove-toothed forest mouse)
Leptomys elegans (long-footed water rat); *L. signatus*
   (Fly River water rat)
Macrotarsomys ingens (greater bigfooted mouse)
Macruromys elegans (lesser small-toothed rat)
Makalata occasius (armored spiny rat)
Mullomys gunung (alpine woolly rat)
Melomys rubicola (Bramble Cay mosaic-tailed rat)
Meriones chengi (Cheng's jird)
Mesocapromys angelcabrerai (Cabrera's hutia);
   M. auritus (large-eared hutia); *M. nanus* (dwarf
   hutia); *M. sanfelipensis* (little earth hutia)
Microtus evoronensis (Evorsk vole); *M. mujanensis*
   (Muisk vole)
Mus kasaicus (Kasai mouse)
Mysateles garridoi (Garrido's hutia)
Neotoma fuscipes (1 subsp: San Joaquin Valley
   woodrat)
Nectomys parvipes (small-footed water rat)
Nesolagus netscheri (Sumatran rabbit)
Ochotona helanshanensis (Helan shan pika);
   O. pallasi (1 subsp: Pallas pika); *O. thibetana*
   (1 subsp: Moupin pika)
Orthogeomys cuniculus (Oaxacan pocket gopher)
Oryzomys galapagoensis (Galapagos rice rat);
   O. gorgasi (Gorgas's rice rat)
Pappogeomys neglectus (Querétaro pocket gopher)
Perognathus alticola (1 subsp: white-eared pocket
   mouse); *P. longimembris* (1 subsp: Pacific pocket
   mouse)
Peromyscus polionotus (1 subsp: Perdido Key beach
   mouse); *P. pseudocrinitus* (false canyon mouse);
   P. slevini (Slevin's mouse)
Pogonomelomys bruijni (lowland brush mouse)
Potorous gilbertii (Gilbert's potoroo)
Pseudohydromys murinus (eastern shrew mouse)
Pseudomys fieldi (Alice Springs mouse); *P. glaucus*
   (blue-gray mouse)
Rattus enganus (Enggano rat); *R. montanus*
   (Nillu rat)
Rhagomys rufescens (Brazilian arboreal mouse)
Sicista armenica (Armenian birch mouse)
Sigmodontomys aphrastus (Harris's rice water rat)
Spermophilus brunneus (1 subsp: northern Idaho
   ground squirrel)
Sylvilagus insonus (Omiltemi rabbit)
Tamias minimus (1 subsp: New Mexico least
   chipmunk); *T. umbrinus* (1 subsp: hidden forest
   chipmunk)
Tamiasciurus hudsonicus (1 subsp: Mount Graham
   red squirrel)
Thomomys mazama (1 subsp: Cathlamet pocket
   gopher)
Tokudaia muenninki (Muennik's spiny rat)
Tylomys bullaris (Chiapan climbing rat);
   T. tumbalensis (Tumbala climbing rat)
Typhlomys chapensis (Chapa pygmy dormouse)
Zyzomys palatilis (carpentarian rock-rat);
   Z. pedunculatus (central rock-rat)

CETACEANS

Balaena mysticetus (bowhead whale)
Cephalorhynchus hectori (Hector's dolphin)
Eschrichtius robustus (Gray whale)
Lipotes vexillifer (Chinese river dolphin)
Orcaella brevirostris (Irrawaddy dolphin –
   Mahakam subpopulation)
Phocoena sinus (vaquita)

CARNIVORES

Acinonyx jubatus (1 subsp: Asiatic cheetah)
Canis rufus (Red wolf); *C. simensis* (Ethiopian wolf)
Panthera leo (1 subsp: Asiatic lion); *P. pardus*
   (4 subsp: South Arabian leopard, Amur
   leopard, North African leopard, Anatolian
   leopard); *P. tigris* (3 subsp: Amur tiger, South
   China tiger, Sumatran tiger)
Puma concolor (2 subsp: Florida cougar, eastern
   cougar)
Ursus thibetanus (1 subsp: Baluchistan bear)
Viverra civettina (Malabar large-spotted civet)

SEALS AND SEA LIONS

Monachus monachus (Mediterranean monk seal)

HOOFED MAMMALS

Addax nasomaculatus (addax)
Antilocapra americana (1 subsp: Baja California
   pronghorn)
Bos sauveli (kouprey)
Bubalus mindorensis (tamaraw)
Capra aegagrus (1 subsp: Chiltan goat); *C. falconeri*
   (1 subsp: Tadjik markhor); *C. walie* (Walia
   ibex)
Ceratotherium simum (1 subsp: northern white
   rhinoceros)
Cervus duvaucelii (1 subsp: swamp deer); *C. eldii*
   (1 subsp: Manipur brow-antlered deer);
   C. nippon (5 subsp: Shansi sika, Ryukyu sika,
   North China sika, South China sika,
   Formosan sika)
Damaliscus hunteri (Hunter's hartebeest)
Dicerorhinus sumatrensis (Sumatran rhinoceros)
Diceros bicornis (black rhinoceros)
Elaphurus davidianus (Père David's deer)
Equus africanus (African wild ass)
Gazella gazella (2 subsp: Acacia gazelle, Muscat
   gazelle)
Hexaprotodon liberiensis (1 subsp: pygmy
   hippopotamus)
Hippotragus niger (1 subsp: giant sable antelope)
Ovis ammon (2 subsp: Northern Chinese argali,
   Kara Tau argali); *O. canadensis* (1 subsp:
   Weems' bighorn sheep)
Procapra przewalskii (Przewalski's gazelle)
Rhinoceros sondaicus (Javan rhinoceros)
Rupicapra rupicapra (2 subsp: Chartreuse chamois,
   Tatra chamois)
Sus cebifrons (Visayan warty pig);
   S. salvanius (pygmy hog)

SNOW LEOPARD
*Prized for its fur, and persecuted for
attacking livestock, the snow leopard is
currently listed as endangered, but its
remote habitat – in the mountains
of Central Asia – makes its population
levels difficult to assess.*

MAMMALS

BIRDS

BIRDS

PHYLUM	Chordata
CLASS	Aves
ORDER	29
FAMILIES	About 180
SPECIES	About 9,000

BIRDS ARE THE MOST accomplished of all flying animals. Their ability to fly has allowed them to spread throughout the world, often to places, such as remote islands and Antarctica, that are beyond the reach of many other animals. Like mammals, birds are endothermic (warm-blooded) vertebrates. However, unlike most mammals, they reproduce by laying eggs. Birds have several adaptations for flight, including wings, feathers, a light but strong skeleton, and a highly efficient respiratory system.

EVOLUTION

Birds evolved from reptile-like ancestors, possibly from tree-dwelling dinosaurs that fed on insects. A lifestyle of arboreal hunting would have promoted the development of such bird-like characteristics as large eyes, grasping feet, and a long snout, later to evolve into a bill. It might also have resulted in the transition from being cold- to warm-blooded –

impressions made by feathers

ARCHAEOPTERYX FOSSIL
Archaeopteryx lithographica is thought to represent a link between reptiles and birds: the jaws, snout, and tail (which is supported by vertebrae) are reptilian, but the wings and feathers are bird-like.

an advantage for animals that feed on insects, which become inactive and slow-moving in cold conditions. Feathers, which were derived from reptilian scales, probably first evolved to provide insulation, although they were no doubt put to the purpose of flight very early on.

The earliest-known fossil of a bird is about 150 million years old, dating from the Jurassic Period (205–142 million years ago). Known as *Archaeopteryx lithographica*, this animal was about the size of a crow and showed a combination of reptilian and avian features: it had wings and feathers like a bird but also had a snout, rather than a bill, and the toothed jaws of a reptile. There is some doubt about whether

Archaeopteryx could fly or just glide, since it lacked the keeled breastbone that provides attachment sides for the muscles needed for powered flight.

During the Cretaceous Period (142–65 million years ago), birds diversified and their anatomy evolved to make possible increasingly efficient flight. It was in this period that the ancestors of living birds appeared. Towards the end of the Cretaceous, a wave of mass extinctions saw the end of the age of the dinosaurs. It is not clear why birds survived – perhaps being warm-blooded helped them to withstand a prevailing climatic catastrophe. But having survived, birds as a group flourished into the diversity of forms alive today.

ANATOMY

A bird has several physical adaptations for flight. Its body is short, strong, and compact, with powerful muscles for moving the wings and strong legs to launch it into the air and cushion the impact of landing. Its feathers form the flight surface (see

TAKING OFF
The power of flight has enabled birds to exploit aerial niches. In many habitats, birds of prey (such as this tawny eagle), with their powerful wings, acute vision, and sharp bill and talons, are the top predators.

panel below and p.262); they also provide protection and insulation.

The skeleton of a bird combines remarkable lightness with strength, attributes that are essential for powerful flight. To restrict the bird's weight, a number of bones are

CROSS-SECTION OF BONE
Many bones in a bird are hollow, including the major limb bones, parts of the skull, and pelvis. This reduces the bird's weight, to conserve energy for flight. The bones are strengthened by internal struts.

FEATHERS

Feathers are highly complex structures that are unique to birds. However, they are formed from the same material, keratin, that is found in the hair of mammals and the scales of reptiles. Feathers are subject to considerable wear and tear. To keep them in good condition, birds regularly clean, oil, and reshape their plumage using their bill, an activity known as preening. Other forms of maintenance include scratching, bathing, and sunning. Feathers are shed (moulted) and replaced at least once a year. Flying birds have 4 different kinds of feathers, each of which is modified to serve a distinct function: there are 2 types of flight feathers (wing and tail feathers), as well as down feathers and contour feathers.

inner vane (leeward edge)

outer vane (windward edge)

quill

WING FEATHER
The feathers along the edge of the wing are long and rigid with a clearly defined shape, providing the lift for flight and manoeuvring. Unlike the tail feathers, which are often symmetrical, wing feathers are unevenly shaped.

FEATHER TYPES
Down feathers form an insulating underlayer, while small contour feathers provide a stream-lined covering over the body. The long tail feathers are used for flying and steering.

DOWN FEATHER

CONTOUR FEATHER

TAIL FEATHER

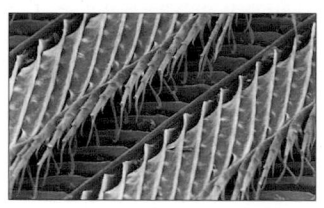

FEATHER STRUCTURE
The structure of feathers is extremely complex. Most feathers have a central shaft (rachis), from which closely spaced branches (barbs) project outwards to form a continuous, flat surface. Minute side branches (barbules) lock the barbs together.

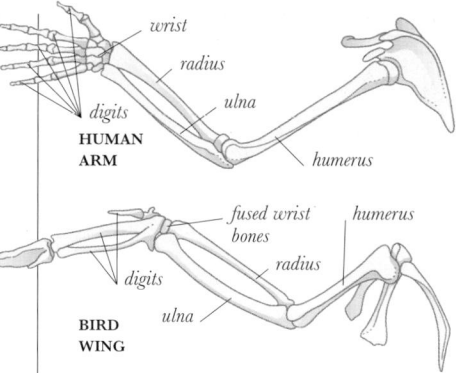

HUMAN ARM

BIRD WING

ARM BONES AND WING BONES
In comparison with a human arm, a bird's forelimb (or wing) has undergone a dramatic reduction in the number of bones, especially around the wrist and hand.

greatly reduced in size, and many have fused together, making a rigid frame without the need for large muscles and ligaments to hold the bones together. Most birds' bones are also hollow, lacking marrow. To compensate for their lightness, they are supported by a network of internal struts (trabeculae) at points of stress, giving them great strength. Several bones, including the humerus, contain air sacs, which are connected to the respiratory system.

A bird's wing is the anatomical counterpart of the human arm, although birds have relatively few digits, and some of the "hand" bones are fused together, which contributes to the general rigidity of the structure. Similarly, the "elbow" and "wrist" joints of the wing bones are inflexible in the vertical plane. As a result, when the bird is flying, the wings are held straight out and flapped only at the shoulder, providing rigidity and saving energy by preventing unnecessary movement. The powerful wing muscles are attached to a keel, a large projection that extends at right angles from the breastbone.

Another feature unique to birds is their light, flexible bill (or beak). When the bill opens, both the upper and lower jaws (or mandibles) move, giving a wide gape – in mammals, only the lower jaw moves. The bill itself has an external covering of keratin, which is light and strong and allows for great variety in form. The bill shape is always adapted to a bird's particular method of feeding,

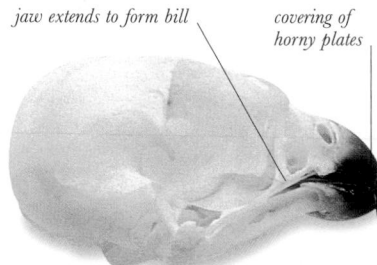

jaw extends to form bill

covering of horny plates

SKULL
The skull of a bird is very light, with many of the bones fused together. The bill consists of an upper and lower mandible, with a highly variable horny sheath on both.

DIGESTIVE SYSTEM

Since they lack teeth, birds are forced to break up particles of food further down the alimentary canal, in the stomach. The lower part of the stomach (gizzard) grinds down material, often aided by an abrasive paste containing ingested grit or stones, while the upper part (proventriculus) secretes gastric juices. The food is often stored in the oesophagus for later assimilation. In some birds, the oesophagus expands near its base to make a sac-like crop, a further storage vessel. Food can be transferred there very quickly, allowing for large quantities to be ingested in a relatively short time. This is valuable for birds that must risk exposure to predators while feeding.

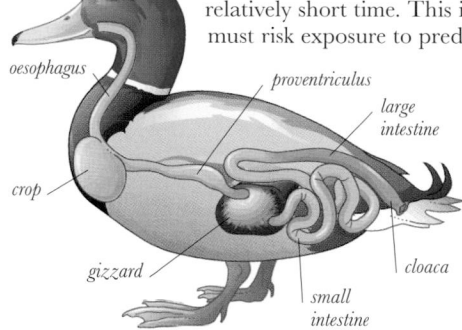

oesophagus
proventriculus
large intestine
crop
gizzard
small intestine
cloaca

FEEDING AND STORING FOOD
A bird's digestive system is adapted for a dynamic lifestyle. Food is ingested without chewing, and can be stored in the oesophagus or the crop. This allows the bird to feed quickly, but to digest in a safer place.

CROP FEEDING
Adult pigeons are unusual in producing a milky secretion from the crop to feed to their young. Here, a turtle dove gives crop milk to a nestling.

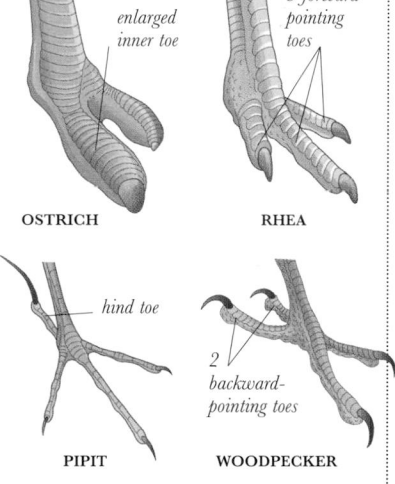

enlarged inner toe

3 forward-pointing toes

OSTRICH

RHEA

hind toe

2 backward-pointing toes

PIPIT

WOODPECKER

FEET AND TOES
Ostriches and rheas have fewer toes than most birds, enabling them to run quickly. The pipit has the perching foot typical of passerines, with 3 toes pointing forwards and 1 back. The arrangement of a woodpecker's toes allows it to climb easily in all directions.

and reflects its diet – for example, finches have a strong, conical bill for cracking seeds, whereas herons have a pointed, dagger-like bill for seizing fish. Birds that feed on insects, such as warblers, have a small, slender bill, while in flesh-eaters, such as hawks, the bill is sharp and hooked for tearing open prey.

Birds have between 2 and 4 toes on their feet, their arrangement and shape varying according to the bird's way of life. Most have 4 toes, of which 3 point forwards and one back, a good system for perching (see p.336). Many swimming birds have webs between their toes.

Like humans, birds rely primarily on their sense of sight, followed by hearing. Indeed, many birds have

exceptional vision, particularly owls and birds of prey. With their high-speed aerial lifestyle, birds rarely (and in some cases never) use their sense of smell.

RESPIRATION AND CIRCULATION

Birds are active animals that have a high metabolic rate. They have an efficient respiratory system that extracts large amounts of oxygen from air, and a circulation system that can move the oxygen rapidly around the body. Efficient oxygen extraction is also necessary so that birds can remain active at high altitude, where oxygen is less abundant. Although the lungs are small, they are connected to a series of air sacs found throughout the body, which help to inflate and

RESTRICTING HEAT LOSS
To control the rate of heat loss from the body, birds (here, a fieldfare) adjust the volume of insulating air trapped between the skin and the outermost feathers by erecting, or fluffing, the feathers. The greater the volume of air, the less heat is lost.

deflate the lungs. Instead of air flowing alternately in and out, as it does in mammals, it flows in a single direction. Working in tandem with this efficient oxygen-extraction system, birds have a large heart that pumps at a relatively rapid rate.

Birds need to maintain a constant internal temperature of about 40°C (104°F) to remain active. To achieve this, they control their rate of heat loss (see below). In some species, when there is not enough energy to sustain the metabolism, the internal temperature falls and the bird enters an inactive state called torpor. Some birds that live at high altitude, such as certain hummingbirds, become torpid overnight; others, such as some swifts and nightjars, can stay in this state for days or even weeks.

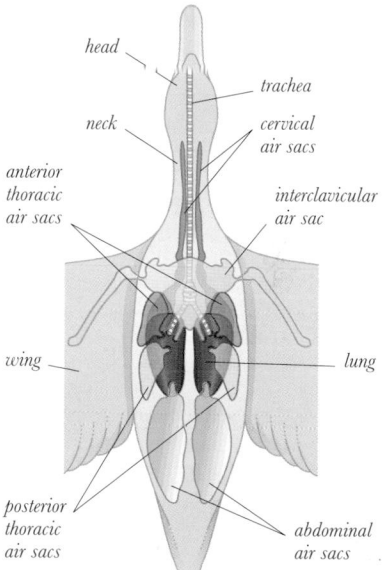

head
trachea
neck
cervical air sacs
anterior thoracic air sacs
interclavicular air sac
wing
lung
posterior thoracic air sacs
abdominal air sacs

AIR SACS
A bird's lungs are only part of its air-intake system. Connected to the lungs is a system of air sacs, which have thin, non-muscular walls and perform a similar function to the diaphragm in humans. In all, there are 9 air sacs throughout the body, 8 of which are arranged in pairs.

BIRDS

FAST FLYER
A blue and yellow macaw in flight displays its highly colourful plumage and long, elegant tail. The relatively narrow, tapering wings of macaws enable these birds to fly remarkably quickly, despite their large size.

FLIGHT

Birds are not the only animals capable of powered flight (bats and insects are similarly able), but they include among their number the largest and most powerful of all airborne animals.

A bird's wing achieves lift in the same way as does the wing of an aircraft. As the bird moves forwards, air flows more quickly over the upper surface of the wing than the lower one, creating a pressure differential between the

surfaces. Since it is the slower-moving air that exerts the greater pressure, the resulting force is upwards. The strength of this upward force (called lift) increases with the size of the bird's wing and its forward speed (see panel, below).

There are 2 main modes of flight: flapping and gliding (or soaring). Flapping does not create lift directly; instead, it generates the horizontal

motion needed to increase airflow over the surface of the wing and produce lift. This is why most birds have to flap their wings to take off. When gliding, birds hold their wings outstretched and steady to the wind. A gliding bird consumes little energy, but it loses height as its speed, and hence the amount of lift, drop (unless it is riding a current of rising air).

Birds exhibit a great variety of wing shapes, reflecting the way they fly and their general lifestyle. Those that rely on a rapid take-off and a short burst of speed to escape from predators tend to have broad, rounded wings, since these provide the required acceleration. Species that fly for long periods and need to save energy have long wings. Fast, powerful fliers, such as swifts and falcons, have long, curved wings with pointed tips to reduce drag. Tail shape is also important. Birds that make sudden changes of direction in mid-air, such as swallows and kites, often have forked tails.

Although most birds can fly, some are flightless – these include the ostrich, rheas, cassowaries, kiwis, the emu, and penguins. It is generally thought that these birds evolved from ancestors that once had the power of flight, since non-flying birds have many adaptations for flight, such as hollow bones and wings (although the wings of many flightless birds have since become reduced in size). Flightlessness seems to have come about in one of several ways. For some birds in environments with relatively few predators, flight seems to have become unnecessary – many birds of isolated islands, such as the weka and takahe of New Zealand (which are closely related to cranes), are in this category. Alternatively, flightlessness may have arisen in situations where size and strength on the ground were more important for

AERIAL MANOEUVRING
Frigate birds are highly distinctive in outline, soaring and gliding above the open sea in search of food. Their narrow, angular wings enable them to fly with great speed and agility, and the characteristic long, forked tail helps them to steer.

survival than flight. This is the case for the ostrich, which can run quickly, using its wings for balance; its long legs can also deliver a powerful kick to fend off predators. In penguins, the wings have become similar to flippers and are used by the bird to propel itself through water.

REPRODUCTION

Most birds are monogamous, breeding in pairs. In general, courtship consists of a male attracting a female with a visual display or by singing. Some birds perform spectacular displays: they may show off physical attributes, evident in the dazzling plumage displays of peacocks and birds of paradise; engage in dancing (as, for example, in cranes, see below); or make aerobatic flights, such as the dramatic swoops and dives of eagles. Many birds also draw attention to themselves using songs. The song, which is unique to the species, serves to repel male rivals and establish a breeding territory; it may also be used to attract a mate.

All birds reproduce by laying eggs, and they go to great lengths to ensure the survival of their clutch. Most do this by placing their egg (or eggs) in a specially built nest, which is either hidden or placed out of reach of predators. The eggs are

FLAPPING FLIGHT
In the upstroke of flapping flight (at the left in this sequence), a bird bends its wings so that the tips are close to its body. On the downstroke, during which most power is generated, the wings are fully extended.

HOW BIRDS FLY

Birds have a large, keeled breastbone, to which massive flight muscles are attached. When the muscles contract, they bring about the powerful downstroke of the wing, which produces forward propulsion. When the muscles relax, the wing is pulled back up. Feathers (see p.260) also enable birds to fly efficiently. Those on the trailing edge of the wings and on the tail are designed to provide lift and aid manoeuvrability, while the visible body feathers (contour feathers) streamline the body in flight. In most birds, flying consists of wing-flapping or gliding. Several birds are capable of hovering in mid-air; hummingbirds can also fly backwards.

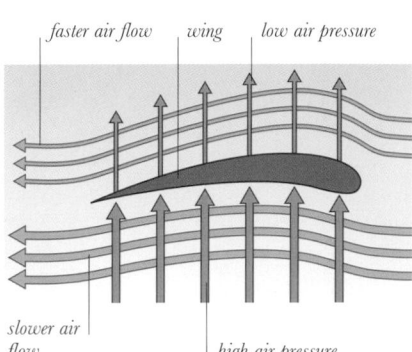

faster air flow   wing   low air pressure

AERODYNAMICS OF THE WING
A bird's wing is curved outwards along the upper surface. As a result, air passing over it has to travel a slightly longer distance at a faster pace than the air travelling along the lower surface. The slow-moving air passing under the wing exerts greater pressure, producing an upward force known as lift.

slower air flow

high air pressure

COURTSHIP DANCE
A male and female Japanese crane take part in their elaborate dance at the onset of the breeding season. One or both birds bow or bob their heads and leap into the air.

EGGS AND NESTING

Although many birds lay their eggs directly onto the ground or some other surface, most build a structure, known as a nest, in which to put them. Nests provide safety, insulation, and a fixed point for the adults to concentrate on nurturing their eggs and young. An important part of nest-building is choosing a suitable site – one that will provide concealment or inaccessibility from predators. Nests are made from various materials, depending on what is available – usually vegetation, but also animal hair, feathers, or even shed snake skin or human artefacts.

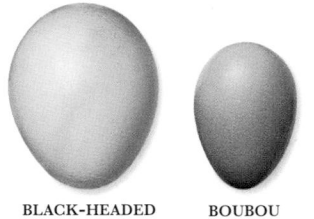

PROTECTIVE NEST
The most common type of nest is an open cup. Most nests are made of more than one material: this chaffinch's nest is bound by spiders' webs and built up with moss, lichen, grass, and feathers.

BLACK-HEADED TROGON

BOUBOU SHRIKE

NATAL CHAT

CETTI'S WARBLER

BLUE SHORTWING

EGG COLOUR
Birds' eggs have a wide range of colours. However, all shell colours are the product of just 2 pigments, one derived from haemoglobin and the other from bile. These are added as the egg moves down the female's genital tract.

incubated by one or more adults, almost always including the female. The number of eggs laid in a clutch varies greatly between species, and some birds produce more than a single clutch in a year.

A bird's egg is contained by a light but strong shell that protects the developing embryo and acts as a barrier against bacteria. The shell is made of calcium carbonate, which the female absorbs from her food. Despite appearing to be hard, the shell is porous, allowing for the free exchange of oxygen and carbon dioxide across its surface. Inside, the embryo is nourished by a large reservoir of nutritive material that promotes its growth. Even so, when young birds hatch out, they are often poorly developed. In most species, the hatchlings are blind and without feathers; they also lack the ability to regulate their internal temperature, so they must be brooded. These dependent chicks are referred to as nidicolous young and are completely reliant on their parents for warmth and food. In contrast, some groups of birds, such as waterfowl and gamebirds, hatch nidifugous young, which are covered with down and are able to feed themselves within a few hours of hatching.

BLUE TIT HATCHLINGS

DUCK HATCHLING
THE FIRST DAYS OF LIFE
Most young birds, including passerines (such as the blue tit) depend on adults for food and warmth. Their eggs are laid in well-protected nests. Waders, waterfowl, and gamebirds produce more independent young that can walk and fend for themselves within hours.

SOCIAL GROUPS

Birds vary greatly in the way that they relate to their own kind. Some are solitary or form pairs within a defined territory. Many meet up with others for specific activities, such as roosting, feeding, or breeding. Others live in groups throughout their lives.

A sociable lifestyle carries both advantages and disadvantages. Birds in groups benefit from the collective effort put into searching for food, particularly where supplies are sparsely distributed. Birds in flocks are also less at risk from predators: to be one among many reduces the chances of being singled out and caught, and individuals are more likely to have early warning of a predator's approach. This state of collective vigilance means that birds in a group can spend more time feeding or sleeping than they would if they were alone. In addition, some birds huddle

COMMUNAL FEEDING
Flamingos are among the most sociable of all birds, living in vast groups (sometimes flocks of a million or more) throughout their lives and performing all living functions in company with others. Even their courtship displays are communal, synchronized affairs.

together while roosting to share body heat and reduce heat loss. However, there are also disadvantages in being part of a flock. If food is scarce, competition between group members can cause weaker individuals to become marginalized, unable to feed properly and vulnerable to predators; such birds would be better off on their own. Another disadvantage of social living is the threat of diseases, which spread more rapidly through a group than they do among a population of solitary individuals.

FLOCKING TOGETHER
Common starlings feed and roost together, and most also breed in small colonies. Starling roosts are famous for their spectacular aerial manoeuvres prior to settling down. Up to 2 million birds may share the same roost site.

MIGRATION

Many birds undertake seasonal migrations. Most migratory species breed in spring and summer in high latitudes, taking advantage of the relatively long days, but move in winter to lower latitudes. Many species show complex patterns of movement, with only certain populations – or in some cases just females – leaving the breeding areas while others remain behind. Birds that behave in this way are referred to as partial migrants. Not all birds migrate, partly because migration consumes a lot of energy; those that do not migrate at all (including many tropical species) are referred to as sedentary.

The urge to migrate is triggered by a combination of internal physiological cycles (such as hormone levels) and changes in day length. As the time to migrate approaches, birds lay down reserves of fat to sustain them on their journey and show signs of restlessness. Some birds have an impressive ability to navigate, travelling thousands of kilometres to arrive at a destination with pinpoint accuracy (see panel, below).

MIGRATION ROUTES

Birds have various ways of navigating. They orient themselves mainly using their efficient internal body clock, which enables them to measure changes in day length, as well as the position of the sun and, at night, of the moon and stars. Many species, if not all, can also detect variations in the earth's magnetic field, which they use as a compass. Birds that have already made several migratory journeys recall landmarks and may use clues such as smells or ultrasounds.

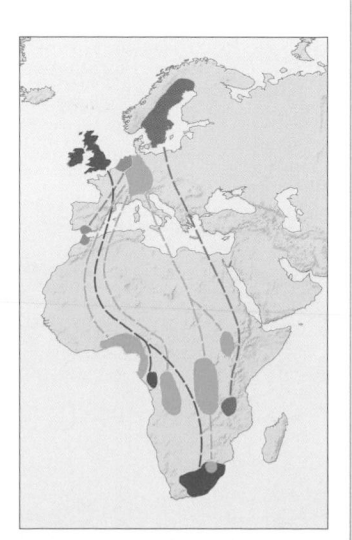

LONG-DISTANCE MIGRANTS
Swallow populations that winter in different parts of northern Europe follow a range of migration routes. Some travel the relatively short distance to southern Spain or North Africa. Others cover the length of the African continent.

OSTRICH

PHYLUM	**Chordata**
CLASS	**Aves**
ORDER	**Struthioniformes**
FAMILIES	**1 (Struthionidae)**
SPECIES	**1**

THE LARGEST OF ALL BIRDS, the ostrich is unmistakable in appearance. It has a long, bare neck and small head, a massive body, and long, muscular legs. The wings are small and covered with loosely packed feathers. The ostrich is unique among birds in having only 2 toes on each foot. Although too heavy to fly, it is capable of running with remarkable speed and stamina – it can travel at up to 70kph (45mph) for as long as 30 minutes. The single species of ostrich is placed in a separate order from other birds.

FIGHTING MALES
Male ostriches compete for territory and social status with aggressive displays and occasionally by fighting. The winner of a contest acquires a territory and several females, although only one female, the "major hen", remains during incubation and chick-rearing.

Ostrich

Height	2.1–2.8 m (7–9¼ ft)
Weight	100–160 kg (220–350 lb)
Plumage	Sexes differ

Location W. to E. Africa (south of Sahara), southern Africa	Migration Non-migrant
	Status Locally common

Once widespread in Africa and West Asia, ostriches are now restricted largely to eastern and southern Africa, although they are also farmed in other parts of the world. In the wild, they are semi-nomadic, travelling long distances to find grass and other plant food. They typically form mixed-sex herds, and are rarely found on their own. During the breeding season, males make loud, booming calls and perform elaborate displays. Several females often lay in a single nest, producing a joint clutch of up to 30 eggs. The male takes part in incubation; once the eggs hatch – after about 40 days – he is usually in sole charge of the young.

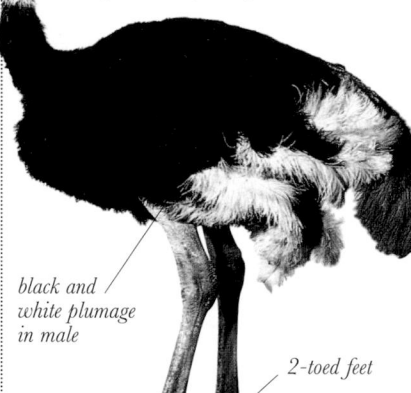

black and white plumage in male

2-toed feet

RHEAS

PHYLUM	**Chordata**
CLASS	**Aves**
ORDER	**Rheiformes**
FAMILIES	**1 (Rheidae)**
SPECIES	**2**

ALTHOUGH RHEAS LOOK similar to the ostrich, they are much smaller and have 3 (rather than 2) toes on each foot. The head, neck, and thighs are all covered with feathers, and each wing ends in a claw, which is used to fight predators. When running, rheas use their wings for balance. They occur on upland and lowland plains in most parts of South America.

CARE OF THE YOUNG
Only the male bird incubates the eggs and watches over the young (lesser rheas are shown here). The female goes off in search of other males.

Common rhea

Height	0.9–1.5 m (3–5 ft)
Weight	15–30 kg (33–66 lb)
Plumage	Sexes differ

Location E. and S.E. South America	Migration Non-migrant
	Status Lower risk

This long-legged, flightless bird lives in grassland and semi-arid scrubland. It is generally grey to brown, and white, the plumage offering effective camouflage against tall grass and scrubby vegetation. The male has a dark collar at the base of its neck during the breeding season. A sociable bird, it lives in groups, although breeding males are solitary for part of the year. In the mating season, males produce a booming call to attract females, then court them with an

impressive wing display. The male mates with up to 12 females and then scrapes out a nest, in which the females lay up to 60 eggs between them. After the eggs are laid, the male takes sole charge of incubation and protecting the young chicks. This rhea has suffered through hunting and habitat loss, although it is still relatively abundant.

dark collar in male

shaggy wing feathers

3-toed feet

CASSOWARIES AND EMUS

PHYLUM	Chordata
CLASS	Aves
ORDER	Casuariiformes
FAMILIES	2
SPECIES	4

THESE LARGE, FLIGHTLESS BIRDS, from Australia and New Guinea, have a long neck, long legs, and small wings hidden under loose, hair-like plumage. As well as being able to run quickly, they can also swim. Cassowaries and emus have 3 toes on each foot; in cassowaries, the innermost toe bears a sharp claw, up to 10cm (4in) long, which can inflict lethal wounds. Cassowaries have a protective casque on their head, and neck wattles that can change colour according to the bird's mood.

PATERNAL CARE
The male emu (shown here) does not eat or drink while incubating. After hatching, he cares for the young for up to 8 months.

Casuarius casuarius
Southern cassowary

Height	1.3–1.7 m (4¼–5½ ft)
Weight	17–70 kg (37–155 lb)
Plumage	Sexes alike

Location New Guinea, N.E. Australia

Migration Non-migrant

Status Vulnerable

All cassowaries inhabit dense tropical forest, and as a result are rarely seen. The southern cassowary – the largest of 3 cassowary species – is the only

one found in Australia as well as in New Guinea. During the breeding season, males make a low booming sound to attract females, but when not breeding the adults are solitary. Their diet consists chiefly of fallen fruits.

powerful legs

Dromaius novaehollandiae
Emu

Height	1.5–1.9 m (5–6¼ ft)
Weight	30–60 kg (66–130 lb)
Plumage	Sexes alike

Location Australia

Migration Non-migrant

Status Locally common

Australia's largest native bird, the emu has shaggy, drooping, grey-brown feathers, large legs, but tiny wings. It is highly gregarious, and lives in loose

flocks that can contain dozens of birds. It feeds mainly on seeds and berries, and will travel long distances when food is hard to find. Emus are now extinct in Tasmania, but on the Australian mainland they have benefited from cereal farming, and are now a serious pest in some areas.

fur-like feathers

3-toed feet

BIRDS

KIWIS

PHYLUM	Chordata
CLASS	Aves
ORDER	Apterygiformes
FAMILIES	1 (Apterygidae)
SPECIES	3

THE STRANGE, flightless kiwis have a stout body covered in soft, hair-like plumage, a slender, curved bill, and no tail. Unlike larger flightless birds, kiwis have 4 toes on each foot. These nocturnal birds are native to New Zealand.

SENSES
Although kiwis have poor eyesight, they have acute senses of hearing and smell and a highly touch-sensitive bill.

Apteryx australis
Brown kiwi

Height	50–65 cm (20–26 in)
Weight	1.5–4 kg (3¼–8¾ lb)
Plumage	Sexes alike

Location New Zealand

Migration Non-migrant

Status Vulnerable

Once widespread throughout New Zealand, this short, stout brown bird has been severely affected by deforestation and introduced

predators such as pigs, dogs, cats, and stoats. When feeding, it walks slowly, tapping the ground with its bill and sniffing. It may push its entire bill – up to 15cm (6in) long – into the ground to get at food, which consists of earthworms, cicadas, beetle larvae, centipedes, and fallen fruits. Females lay one to 2 eggs that are very large in proportion to the bird's size.

long bill

short, stubby legs

TINAMOUS

PHYLUM	Chordata
CLASS	Aves
ORDER	Tinamiformes
FAMILIES	1 (Tinamidae)
SPECIES	45

CAMOUFLAGE
When threatened, this Andean tinamou remains motionless, protected by its excellent camouflage.

THESE GROUND-LIVING birds, which look similar to grouse and partridges, have a plump body and small wings. When running or flying, they tire easily because they have a small heart. Tinamous are common in woodland, scrub, and grassland throughout Central and South America.

Eudromia elegans
Elegant crested-tinamou

Height	37–41 cm (14½–16 in)
Weight	400–800 g (14–29 oz)
Plumage	Sexes alike

Location S. South America

Migration Non-migrant

Status Locally common

The long, usually forward-curving crest is the distinguishing feature of this tinamou. Light to dark brown, the

elegant crested-tinamou has wings that are spotted white. One light stripe runs behind the eye, and another runs below it. This relatively shy bird usually moves in small to moderately large groups through woodland, grassland, and brush.

PENGUINS

PHYLUM	Chordata
CLASS	Aves
ORDER	Sphenisciformes
FAMILIES	1 (Spheniscidae)
SPECIES	17

THESE DISTINCTIVE FLIGHTLESS seabirds are adapted for swimming and surviving life in extreme cold. Penguins spend most of their lives in water, propelling themselves with flipper-like wings in pursuit of fish, krill, and squid. They have a thick coat of short, stiff, overlapping feathers that streamlines the body, repels water, and conserves heat. Most species come ashore during the warmer months to breed, usually forming large colonies. On land, they do not perch but instead stand upright and move with a waddling gait. Penguins are confined to the seas of the southern hemisphere. Although most common in cold climates, several species are also found where cold currents flow north into tropical regions. Only 2 species, the emperor and Adelie penguins, spend winter in Antarctica.

ANATOMY

Penguins have a plump body with short legs and webbed feet that are set so far back that they must stand upright on land, balancing on their feet and short, stiff tail. When walking, they put their weight on the soles of their feet, hence their awkward gait. On snow or ice, they may toboggan on their bellies, using their feet and flippers for propulsion. A penguin's body is streamlined in water, and is covered with extremely short feathers that form a sleek, friction-free surface. The wings are specially flattened into flippers (see below). When underwater, they use their feet and tail as a rudder. Penguins have 3 layers that provide waterproofing and insulation: a dense mass of overlapping, oil-tipped feathers; a thick layer of fat under the skin; and, in between, a layer of air that is warmed by the body. The plumage of penguins is black or grey above and white below. Any coloration or ornamentation (such as crests and eye tufts) is confined to the head and neck.

WING STRUCTURE
The wing of a penguin is unlike that of any other bird. The bones are flattened to make a flipper, and are solid instead of hollow, increasing their density and strength. The wing as a whole forms a rigid structure since free movement is possible only at the shoulder – the joints that form the equivalents of the wrist and elbow in humans are relatively inflexible. Well-developed muscles, as large as those of flying birds, help to power the flapping of the wings.

flat, solid bones   *"elbow"*
short feathers   *"wrist"*

SWIMMING

Penguins use 3 different swimming techniques. When idling, they swim slowly at the surface, paddling with their wings, and with their head and tail raised. When hunting, they dive below the surface and effectively fly underwater, flapping their wings to provide power. Most dives last about a minute, but dives of 20 minutes have been recorded. The third form of movement is called porpoising, in which penguins swimming near the surface periodically leap out of the water to breathe.

DIVING
Penguins (here a king penguin) move much more efficiently in water than they do on land. Some species can swim at speeds of 14kph (9mph).

COMMUNICATION IN COLONIES
Apart from the yellow-eyed penguin, all penguins (such as these king penguins) form colonies. When gathered in large numbers, they use calls and visual displays to locate their mates and young.

COLONIES

Most penguins breed in colonies, which can consist of hundreds of thousands of birds. Nests are made of grass, feathers, or pebbles, and the female lays either one or 2 eggs. In some species, once an egg is laid, the female leaves the nest to feed, while the male keeps the egg warm, holding it on top of his feet and beneath folds of skin on his belly. Penguins are still able to walk when the egg is in this position. The first period of incubation can last weeks or even months, during which the male does not feed but lives on stored fat. When the female returns, the pair take turns to guard the eggs while the off-duty bird feeds at sea.

BURROWING
Not all penguins nest above ground. In open terrain, Magellanic penguins nest in shallow burrows to protect themselves from the elements and from predators, including mammals and other birds.

HUDDLING
In harsh conditions, emperor penguins and chicks huddle together for warmth. The centre of a huddle can be 10°C (about 18°F) warmer than the edges. Birds take turns to occupy the outermost positions.

Aptenodytes forsteri

Emperor penguin

Height 1.1 m (3 1/2 ft)	
Weight Up to 37 kg (82 lb)	
Plumage Sexes alike	
Location Circumpolar around Antarctica	**Migration** Non-migrant
	Status Locally common

The largest of all the penguins, this bird has highly unusual breeding habits. In winter, the female lays one egg and goes to sea, not returning until the spring. The male carries the egg on his feet and protects it with a "pouch" of feathery skin, huddling with other incubating males to keep warm. He fasts until his mate returns, just as the chick hatches, and then goes to sea, later returning to help rear the chick. This species can dive to depths of 530m (1,750ft) for as long as 20 minutes, and travel up to 1,000km (625 miles) on foraging trips.

Pygoscelis adeliae

Adelie penguin

Height 46–61 cm (18–24 in)	
Weight 4–5.5 kg (8 3/4–12 lb)	
Plumage Sexes alike	
Location Circumpolar around Antarctica	**Migration** Migrant
	Status Common

One of the few penguins that nests on mainland Antarctica, along shores that are free of ice in summer, the Adelie penguin is mainly blue-black with pure white underparts and a distinctive white ring around its eye. It breeds in summer in vast colonies of 200,000 or more synchronized pairs, each female laying 2 eggs roughly 2 days apart. Both the male and the female incubate the eggs in shifts. The Adelie penguin is relatively aggressive, and adults are often observed stealing rocks from their neighbours' nests.

white ring around eyes

white underparts

Pygoscelis antarctica

Chinstrap penguin

Height 71–76 cm (28–30 in)	
Weight 3–4.5 kg (6 1/2–10 lb)	
Plumage Sexes alike	
Location Circumpolar around Antarctica	**Migration** Non-migrant
	Status Common

Characterized by a thin black line that runs from ear to ear under the chin, the chinstrap penguin is mostly blue-black with white underparts, cheeks, chin, and throat. It prefers areas of light pack ice, and breeds in high-density, sometimes large, colonies on ice-free areas of coasts. The nests comprise a circular platform of small stones, with a shallow nest cup, and often include bones and feathers. The breeding success of chinstrap penguins is highly variable: it is lower in years when sea ice persists close to colonies, since this restricts access to the sea for foraging adults.

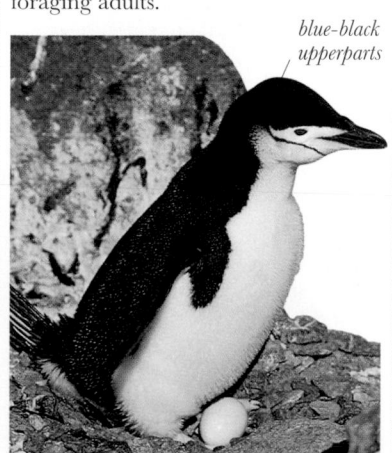

blue-black upperparts

Spheniscus humboldti

Humboldt penguin

Height 56–66 cm (22–26 in)	
Weight 4.5–5 kg (10–11 lb)	
Plumage Sexes alike	
Location W. South America	**Migration** Non-migrant
	Status Vulnerable

Small colonies of this penguin occur along the west coast of South America, in the cold but fish-rich Humboldt Current. Coloration is mostly blackish grey with white underparts, but adults have a distinctive black, horseshoe-shaped breast band and a white head stripe. The penguins usually hunt in groups in shallow water, pursuing small, shoaling fish such as anchovies

and sardines, and they nest in underground burrows, caves, or crevices between boulders. Direct hunting and over-fishing by humans have led to a decline in their population.

white head stripe

fleshy pink area at base of bill

black band from flank to thigh

black legs and feet

Eudyptula minor

Little penguin

Height 41 cm (16 in)	
Weight 1 kg (2 1/4 lb)	
Plumage Sexes alike	
Location S. Australia (including Tasmania), New Zealand	**Migration** Partial migrant
	Status Common

This mainly blue-grey bird is the smallest species of penguin, and also one of the few that remain fully active after dark. During the day it forages at sea, but after sunset in the breeding season, it returns to land, coming ashore under the cover of darkness. A highly vocal bird, it calls at sea and on land. Little penguins normally nest in underground burrows, but they also breed in caves, crevices, under vegetation among rocks, and sometimes under houses. They lay 2 eggs on a

bed of sticks or grass, and the parents take turns at incubation. Chicks are brooded for 7–10 days and guarded for a further 13–20 days. At sea, the little penguin feeds alone or in small groups, catching small fish and swallowing them beneath the surface.

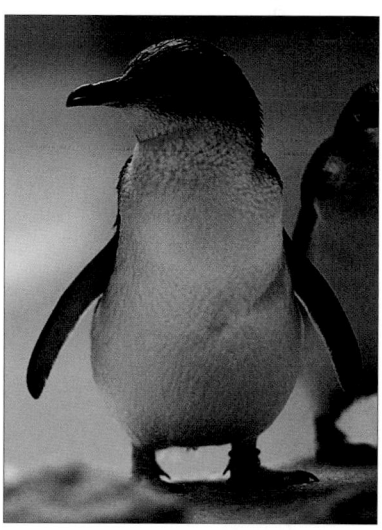

Eudyptes chrysolophus

Macaroni penguin

Height 71 cm (28 in)	
Weight 3.5–6.5 kg (7 3/4–14 lb)	
Plumage Sexes alike	
Location S. Chile, South Atlantic, S. Indian Ocean	**Migration** Non-migrant
	Status Vulnerable

Like all 6 species of crested penguin, the macaroni penguin has a conspicuous crest of golden plumes on its head. It also has a large, bulbous, orange-brown bill, often ridged in adults. Noisy, aggressive, and demonstrative, it makes raucous braying sounds in its colonies on land and gives short barks at sea. Although 2 eggs are laid, only

one chick ever survives to the fledgling stage – usually the chick from the second, larger egg. Both sexes share incubation duties but, unlike most other penguins, the female takes the first shift.

DIVERS

PHYLUM	Chordata
CLASS	Aves
ORDER	Gaviiformes
FAMILIES	1 (Gaviidae)
SPECIES	5

ALSO KNOWN AS LOONS, divers are superbly adapted for underwater swimming. With their streamlined bodies, legs set far back on the body for efficient propulsion, and strong, webbed feet, they can reach depths of 75m (250ft) and stay underwater for several minutes. Dense plumage insulates them in the Arctic and sub-Arctic waters where they live. Despite having small, pointed wings and a pin-like tail, divers fly well. But with feet set so far back, they are almost incapable of walking on land.

PLUMAGE
Divers spend a lot of time taking care of their feathers and can often be seen rolling on their side to reach their belly feathers. A varied breeding plumage (seen here in the Pacific loon) is replaced by simpler, less colourful patterns in winter.

Gavia stellata

Red-throated diver

Length 55–70 cm (22–28 in)
Weight 1–2.5 kg (2¼–5½ lb)
Plumage Sexes alike

Location North America, Greenland, Europe, Asia
Migration Migrant
Status Common

Distinguished from other divers by the reddish brown throat patch at the base of its thick, long neck, this bird has a plump, oval body that is dark greyish to black with white spots; the underparts are whitish. An inhabitant of coastal bays and inlets, it moves inland to lakes and marshes in northern forests and Arctic tundra during the breeding season, where it emits a loud yodel or wail to attract a mate or establish territory. Courtship behaviour involves splashing dives, dipping and shaking of bills, and rushing across the water in pairs. Nests are usually a simple platform of reed, rushes, and grass.

Gavia immer

Great northern diver

Length 70–90 cm (28–35 in)
Weight 3–4.5 kg (6½–10 lb)
Plumage Sexes alike

Location North America, Greenland, W. Europe
Migration Migrant
Status Common

This grey to black and white bird is usually solitary or found in pairs, although large groups of up to 300 may be seen feeding along the coast when not breeding. Like other divers, this species can dive to great depths when hunting for fish and aquatic invertebrates, which it seizes with its bill, sometimes spearing them. It may also dive to avoid predators such as mammals, hawks, and eagles. Both sexes care for the young, and the chicks stay with their parents until fledged, at about 10–11 weeks. Parents sometimes swim with the young on their backs.

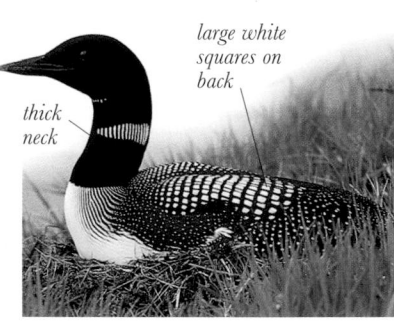

large white squares on back
thick neck

GREBES

PHYLUM	Chordata
CLASS	Aves
ORDER	Podicipediformes
FAMILIES	1 (Podicipedidae)
SPECIES	22

FOUND IN SHELTERED WATERS throughout the world, grebes are strong swimmers and accomplished divers. They are ideally suited to aquatic life: their small head and thin neck enable them to dive easily when hunting for food, and their feet – which are set well back on the body – have lobed toes with highly flexible joints that give these birds great agility when swimming. Their soft, dense plumage is very water-resistant. Grebes are remarkable for their often elaborate courtship rituals.

WEED CEREMONY
During the weed ceremony, a courting pair (here, great crested grebes) dive for vegetation, then rise out of the water to face each other, holding the plants up high.

Tachybaptus ruficollis

Little grebe

Length 25–29 cm (10–11½ in)
Weight 125–225 g (4–8 oz)
Plumage Sexes alike

Location Europe, Asia, Africa, Madagascar, New Guinea
Migration Partial migrant
Status Locally common

The smallest grebe in Europe and northern Asia, this short-billed bird is brown with a bright chestnut throat and cheeks. It stands and walks on land more easily than other grebes, and is also more likely to fly outside of migration. Both sexes build nests of aquatic vegetation, anchoring them to surface water plants or branches of bushes growing in water.

short, stout bill
dark brown upperparts

Podiceps nigricollis

Black-necked grebe

Length 30–35 cm (12–14 in)
Weight 250–600 g (9–21 oz)
Plumage Sexes alike

Location North America, Europe, Asia, N. and southern Africa
Migration Migrant
Status Locally common

This small, dark grebe (non-breeding plumage shown), with a slender, slightly upturned bill, is highly social all year. During the non-breeding season, it prefers saline waters, with thousands gathering on some saltwater lakes. However, it breeds in freshwater lakes and marshes, where it makes floating nests. Adults carry young on their backs for the first few weeks, but little parental care is required after 3 weeks.

Podiceps cristatus
Great crested grebe

Length	46–51 cm (18–20 in)
Weight	0.6–1.5 kg (1 1/4–3 1/4 lb)
Plumage	Sexes alike
Location Europe, Asia, Africa, Australia, New Zealand	**Migration** Partial migrant
	Status Locally common

The largest grebe in Europe and northern Africa, the great crested grebe is famous for its remarkably elaborate courtship displays, which involve complex, ritualized postures, dives, and head-shaking; during the displays, its crest is raised and the tippets are flared. It often peers into the water while swimming on the surface to locate fish, then dives up to 30m (100ft) deep.

greyish brown upperparts

brown and black tippets

white underparts

CREST AND TIPPETS
This elegant bird is distinguished by the ornamental black crest on its crown and its elongated, chestnut and black cheek feathers (tippets).

Podiceps gallardoi
Hooded grebe

Length	34 cm (13 1/2 in)
Weight	525 g (19 oz)
Plumage	Sexes alike
Location S. South America	**Migration** Non-migrant
	Status Lower risk

First described in 1974, this medium-sized, black and white grebe has a cinnamon and black crest and a small, pointed bill. Courtship involves a

unique "sky jabbing" display in which males and females come face to face and vigorously move their heads up and down. Populations are small, occurring on isolated lakes in remote areas of Patagonia; the species winters in sheltered bays along the Patagonian coast. It may be vulnerable to pollution, and to habitat disturbance during breeding.

Podilymbus podiceps
Pied-billed grebe

Length	31–38 cm (12–15 in)
Weight	250–575 g (9–21 oz)
Plumage	Sexes alike
Location North America, Central America, N.W. and S.E. South America	**Migration** Partial migrant
	Status Common

Small and stocky, with a short, arched bill, this brown and white bird is found in freshwater bodies in the breeding season, and winters in areas with open

water or in brackish estuaries. Strongly territorial and aggressive, it threatens and chases birds of the same species as well as other waterbirds. Mates may perform a "triumph ceremony" after chasing off an intruder; this involves the pair rising into an upright posture facing each other, and turning back and forth.

dark bar on bill

black throat patch

Parents take turns carrying the young on their backs and bringing them food. Each parent develops a preference for particular young, resulting in the adults dividing the brood in order to care exclusively for their favourites.

Populations of the great crested grebe declined in the mid-19th century, but have recovered with protection and an increase in man-made habitats, such as gravel pits and reservoirs.

Aechmophorus occidentalis
Western grebe

Length	55–75 cm (22–30 in)
Weight	1–2 kg (2 1/4–4 1/2 lb)
Plumage	Sexes alike
Location C. and S. North America	**Migration** Migrant
	Status Locally common

A large grebe with black and white plumage, this bird has a long, slender neck. Mates are in close contact from pairing to taking care of the young. Both sexes build nests on water, anchored to surface plants. Parents take turns carrying young on their backs for the first 2–4 weeks, and feed them until they are 8 weeks old. At the end of the 19th century, tens of thousands of these grebes were killed as they were hunted for their silky belly feathers, used in coats and hats. Populations have recovered, although they are still in danger from pollution, oil spills, habitat loss, and disturbance by humans.

black crown

black upperparts

COURTSHIP RUSH

This species engages in an energetic courtship, following the same rituals as the great crested grebe. The most famous display is the "rush" (shown above), in which birds run across the water side by side. If they continue to be interested in each other, they then perform the "weed dance" (see p.268), in which each bird dives for aquatic vegetation, then holds it high while turning from side to side. In established pairs, the male brings fish to his mate.

PLUMAGE FOR ALL SEASONS
Unlike its smaller North American relatives, the western grebe does not develop any special plumage in the breeding season, and keeps its black crown all year round.

short, vestigial tail tuft

Rollandia micropterum
Short-winged grebe

Length	39–45 cm (15 1/2–18 in)
Weight	625 g (22 oz)
Plumage	Sexes alike
Location W. South America (Lakes Titicaca and Poopo)	**Migration** Non-migrant
	Status Vulnerable†

From a distance, this medium-sized South American grebe looks like many of its relatives, with a crested head, brown back and flanks, and white underside. However, its wings are so small that it cannot fly. If threatened, the short-winged grebe patters across the water at speed, flapping its wings rapidly without being able to take off. This unusual bird is found primarily on 2 large lakes of the Central Andes, at an altitude of about 3,600m (11,900ft).

Poliocephalus poliocephalus
Hoary-headed grebe

Length	29–31 cm (11 1/2–12 in)
Weight	225–250 g (8–9 oz)
Plumage	Sexes alike
Location Australia (including Tasmania), S. New Zealand	**Migration** Nomadic
	Status Lower risk†

Distinguished by the "brushed back" white, hair-like plumes on its head and upper neck, this bird differs from other grebes in numerous ways. It flies more than other species of grebes, and is probably the least vocal and most social, often found in huge flocks. Courtship is relatively simple compared with that of other grebes, and the nests are typically built in shallow water. The hoary-headed grebe has a dark bill with a pale tip; the males have longer bills.

ALBATROSSES AND PETRELS

PHYLUM	Chordata
CLASS	Aves
ORDER	Procellariiformes
FAMILIES	4
SPECIES	108

ALBATROSSES AND THEIR RELATIVES are oceanic birds that occur throughout the world. They are usually encountered far from land, flying low over waves or dipping into water to feed on fish, plankton, or various other marine animals. Besides the very large albatrosses, this group includes the smaller fulmars, gadfly petrels, and shearwaters, as well as the tiny storm petrels and diving petrels. All members of the group have tubular nostrils on their upper bill – a unique feature among birds; as a result, they are often known as tubenoses.

REPRODUCTION

All birds in this group breed on land, mostly on inaccessible islands or cliffs. They return each year to established breeding sites, where they form large colonies of up to a million pairs. The female always lays a single egg, often in a burrow dug in soft soil or in rock crevices. After a long incubation, the parents feed the chick on their highly nutritious but foul-smelling stomach oil.

BREEDING COLONY
Like other birds in the group, fulmars often form dense colonies in safe, inaccessible places. Many species visit the colony only after dark, as protection against predators.

ANATOMY

Albatrosses and their relatives have a short neck, tail, and legs. The front 3 toes are joined by webbing. Most species have very long wings; the wandering albatross, with a wingspan of 3.5m (11ft), has the longest wingspan recorded for any bird. Another notable feature of birds in this group is their exceptionally acute sense of smell, which they use to detect food and locate nest sites in the dark. It is thought that each bird may exude its own particular scent.

> **SOARING**
> *The long wings of this black-browed albatross are adapted for extended soaring, often for hours on end without a wingbeat.*

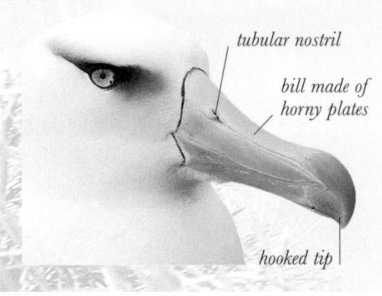

tubular nostril

bill made of horny plates

hooked tip

ALBATROSS BILL
Birds in this group have a bill with a hooked tip and sharp blades for dealing with slippery prey. Albatrosses differ from other species in having nostrils positioned on either side of the upper bill rather than being fused into one on the top (see left).

FLIGHT

These ocean-going birds are adapted to continuous flight, often travelling great distances and riding out powerful storms. Their flight method varies according to the bird's size and wingspan. The larger species, including the albatrosses, typically glide rather than fly, making use of the updrafts of wind over waves to give them lift for long periods without expending too much energy (a practice known as dynamic soaring). The tiny, lightweight storm petrels fly with a combination of wingbeats and glides.

FINDING FOOD
Storm petrels fly low over water in search of food. Some species patter on the surface with their feet to alarm their prey or draw it towards the surface.

Diomedea exulans

Wandering albatross

Length	1.1 m (3 ½ ft)
Weight	8–11.5 kg (18–25 lb)
Plumage	Sexes alike

Location Circumpolar around Antarctica	**Migration** Migrant
	Status Vulnerable

This large albatross has a wingspan of up to 3.5m (11ft), one of the longest of any bird. It feeds mainly at the surface, primarily on squid. It breeds only every other year (at best), since rearing a chick takes a year, and moults in the intervening year. The main cause of this bird's population decline is long-line fishing (for fish such as tuna), in which birds get caught on baited hooks.

VARYING PLUMAGE
As a juvenile, this albatross is chocolate-brown all over except for a white face mask and underwing, but it becomes whiter with age. The adult male (left in picture) is mostly white with black-tipped wings and tail; the female is slightly browner.

NESTING

This albatross' nest is a crude but substantial mound of grasses and moss, built on the ground. The parents take turns to incubate the solitary egg. Once it has hatched, they tend their chick for 9 months or more.

Thalassarche chlororhynchos

Atlantic yellow-nosed albatross

Length	76 cm (30 in)
Weight	2.5 kg (5 ½ lb)
Plumage	Sexes alike

Location South Atlantic	**Migration** Migrant
	Status Lower risk

The smallest albatross in the southern ocean, this black and white species takes its name from the yellow ridge and orange tip on its black bill. Squid and fish make up the bulk of its diet, but it also follows ships, feeding on offal thrown overboard. Both parents rear the chick, which is able to fly relatively quickly for an albatross – within about 4 months.

Macronectes giganteus
Southern giant petrel

Length 92 cm (36 in)	
Weight 5 kg (11 lb)	
Plumage Sexes alike	
Location Circumpolar around Antarctica	**Migration** Migrant
	Status Common

Large and aggressive, with a large bill capable of opening intact carcasses, this bird is one of a few petrels to obtain a significant amount of its food on land. Males in particular prey at seal and penguin colonies and feed on the bodies of whales washed ashore. The adult usually has mottled, greyish brown plumage, but some individuals may be white with black flecks.

Fulmarus glacialis
Northern fulmar

Length 45–50 cm (18–20 in)	
Weight 700–850 g (25–30 oz)	
Plumage Sexes alike	
Location Arctic, North Pacific, and North Atlantic oceans	**Migration** Migrant
	Status Lower risk†

The fulmar is a common bird of northern waters, flying on characteristic stiff, straight wings. Its numbers have increased dramatically in the past 200 years, particularly in temperate waters of the North Atlantic, possibly due to the increase in offal available from trawlers gutting fish at sea. As well as scavenging from ships, it eats fish, squid, and animal plankton, seizing most of its prey at the surface but sometimes plunging into the water. Most

wings held straight for gliding

petrels return to breed at the colony where they were born, but the northern fulmar rarely does so. It lays its egg in a barely lined hollow on an earthy or grassy ledge, usually on a cliff, but it will nest on flatter ground where there is no danger from predators. The adults go off in search of food once the chick is about 2 weeks old. If threatened, the youngster defends itself by vigorously spitting an unpleasant smelling oil.

grey upperparts

Pagodroma nivea
Snow petrel

Length 32 cm (12½ in)	
Weight 250–450 g (9–16 oz)	
Plumage Sexes alike	
Location Circumpolar around Antarctica	**Migration** Migrant
	Status Lower risk†

pure white plumage   *dark eyes*

One of the few birds, apart from penguins, that breed on the Antarctic continent, the dove-like snow petrel is rarely seen away from pack ice. It may nest inland, up to 300km (185 miles) from open water, but breeding is much affected by snowfall, and in some years only one in 5 nesting sites may be occupied. It fiercely defends its nest from other petrels, spitting a foul-smelling oil at them.

Pterodroma cahow
Cahow

Length 33 cm (13 in)	
Weight 250 g (9 oz)	
Plumage Sexes alike	
Location Bermuda (ocean range 1,000km/625 miles)	**Migration** Migrant
	Status Endangered

One of the world's rarest seabirds, the cahow was brought to the brink of extinction as long ago as the 17th century, as a result of predation by introduced mammals and competition for nesting sites with the white-tailed tropicbird. Conservation measures are helping this black, grey, and white petrel to stage a gradual recovery.

Pachyptila vittata
Broad-billed prion

Length 28 cm (11 in)	
Weight 150–225 g (5–8 oz)	
Plumage Sexes alike	
Location South Atlantic, South Pacific, S. Indian Ocean	**Migration** Migrant
	Status Lower risk†

This mid-grey bird has a darker "M" across both wings, and white underparts. Its bill is broad, with comb-like plates used for filtering planktonic prey from the sea. It feeds mainly by "hydroplaning" – with wings outstretched and feet paddling, pushing the sieving bill through the surface water to trap its prey.

Oceanites oceanicus
Wilson's storm petrel

Length 17 cm (6½ in)	
Weight 40 g (1 7/16 oz)	
Plumage Sexes alike	
Location Pacific, Atlantic, Indian Ocean	**Migration** Migrant
	Status Common

A small, soot-black bird with a conspicuous white rump, the storm petrel is often said to be one of the world's most numerous seabirds, with populations running into millions. It breeds around Antarctica, but during the southern winter it flies north, especially to the northern Indian and Atlantic oceans. It rarely alights on

the sea, but catches small fish and crustaceans when pattering on or hovering just above the surface. It can detect prey by smell, and it has been suggested that prey are attracted when the bird stirs its yellow-webbed feet around in the water. When faced with predators, this petrel may squeak and eject stomach oil at them.

Puffinus puffinus
Manx shearwater

Length 31–36 cm (12–14 in)	
Weight 375–500 g (13–18 oz)	
Plumage Sexes alike	
Location Atlantic	**Migration** Migrant
	Status Lower risk†

This shearwater breeds in the North Atlantic, forming colonies on offshore islands – mainly Skomer and Skokholm off the western coast of Wales, and Rhum in the Outer Hebrides – filling the air with calls that sound like screams and wails. Although it feeds at sea by day, it is active at the colonies only after dark, to avoid attack by gulls. This bird usually nests in an earth burrow, but it

HOOKED BILL
The Manx shearwater has a relatively slender and hooked bill, with which it seizes fish at the water surface. It may swim a few metres underwater to pursue its prey.

sometimes lays its single egg under rocks. Both parents incubate the egg, working in shifts of 6–7 days, and they feed the chick for up to 70 days, leaving it about a week before it starts to fly. As winter approaches, the shearwater flies south to warmer waters off Brazil.

black upperparts

COLORATION

Black above and white below, this bird produces a characteristic black then white flash as it flies over the sea, looking for small shoaling fish such as sprat.

PELICANS AND RELATIVES

PHYLUM	Chordata
CLASS	Aves
ORDER	Pelecaniformes
FAMILIES	6
SPECIES	65

THIS GROUP OF LARGE seabirds includes pelicans, cormorants, darters, tropic birds, frigatebirds, gannets, and boobies. They are strong swimmers, being the only birds with webbing between all 4 toes. Most species have broad wings, and frigatebirds and tropic birds spend much of their lives on the wing, while gannets, cormorants, and darters are also capable of sustained flight over open sea. All the birds in this group feed on fish. Some species catch their food by making spectacular plunge-dives from a considerable height, and have several adaptations to protect themselves when they hit the water at high speed. Pelicans and their relatives are found in most of the world's seas. Darters, cormorants, and pelicans also occur on bodies of water far inland.

ANATOMY

A range of physical features helps pelicans and their relatives to catch their prey underwater. The most aquatic members of the group, the cormorants and darters, have surface feathers that become soaked easily, reducing buoyancy and making diving easier. The inner feathers remain waterproof and provide insulation. To reduce their buoyancy further, these birds also swallow stones, and their bones have fewer air spaces than those of most other birds. In contrast, the other members of this group have water-repellent plumage, and their bones contain a relatively large volume of air space so that the birds can float and fly with ease. In fact, the plunge-diving species also have extra air sacs beneath their skin to cushion the force of the impact when they hit the water.

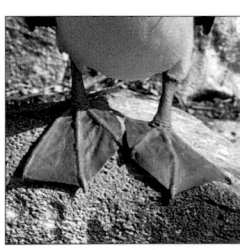

FEET
The large, webbed feet of pelicans and their relatives are used for both propulsion and steering. In some species, such as this blue-footed booby, the feet are used in displays, and to keep their eggs warm.

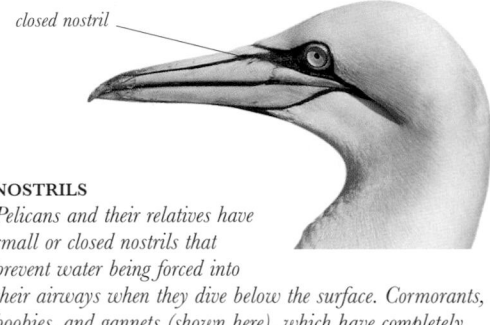

closed nostril

NOSTRILS
Pelicans and their relatives have small or closed nostrils that prevent water being forced into their airways when they dive below the surface. Cormorants, boobies, and gannets (shown here), which have completely sealed nostrils, breathe through their mouth.

FEEDING

Pelicans and their relatives use a range of methods for catching fish. Cormorants swim underwater in pursuit of prey. Darters also hunt below the surface but wait in ambush for up to a minute before spearing fish with their bill. Gannets, tropic birds, boobies, and the brown pelican dive from heights up to 30m (100ft), dropping onto fish at speeds up to 95kph (60mph). Frigatebirds steal fish from other birds, harrying them in flight until they give up their catch. In contrast, pelicans often hunt co-operatively, forming lines and herding fish into shallows.

PLUNGE-DIVING
The brown pelican is the only pelican that catches fish by plunge-diving. Just before impact, the bird pushes its wings and feet back to maximize its speed, and opens its mouth, ready to seize its prey.

SURFACE FISHING
Pelicans have a large throat pouch, which they use to scoop fish from just below the surface. The pouch is also used to collect rainwater for drinking and to dissipate heat in hot weather.

REPRODUCTION

Often choosing remote islands or cliffs that are inaccessible to predatory mammals, pelicans and their relatives nest in colonies. Some pelicans, darters, cormorants, and frigatebirds nest in trees. Males use spectacular displays to attract females – for example, boobies point their head and tail skywards and raise their wings (known as sky-pointing). Nests are often large and usually built by both members of a pair. The young, which are helpless at birth, are fed on regurgitated food. Some young, notably those of frigatebirds, are looked after for many months after fledging.

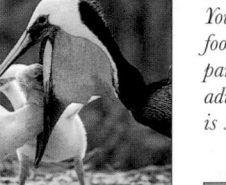

NESTING
Most of the birds in this group (such as these king cormorants) build substantial nests. Collecting enough material can require many journeys back and forth from the nest site.

FEEDING YOUNG
Young birds are fed on regurgitated food, which they remove from their parents' bills. The food provided by adult pelicans (here, a brown pelican) is sometimes known as fish soup.

Phaethon aethereus

Red-billed tropic bird

Length 78–81 cm (31–32 in)	
Weight 600–825 g (21–29 oz)	
Plumage Sexes alike	
Location E. Pacific, C. Atlantic, N. Indian Ocean	**Migration** Migrant
	Status Lower risk†

Although it is a poor swimmer, spending little time on the water, this highly aerial seabird is often found hundreds of kilometres from land. It feeds by plunge-diving, sometimes

from a considerable height, to catch squid and fish, especially flying fish. An elegant bird, with silky white, pink-flushed plumage, it has extremely long tail streamers that it switches from side to side in an elaborate, airborne courtship display. The red-billed tropic bird is one of only 3 species in its family, all restricted to the tropics.

Pelecanus crispus

Dalmatian pelican

Length 1.6–1.8 m (5¼–6 ft)	
Weight 10–13 kg (22–29 lb)	
Plumage Sexes alike	
Location S.E. Europe, S. and S.W. Asia, N.E. Africa	**Migration** Partial migrant
	Status Lower risk

This pelican is the largest found in Europe. Despite its size and weight, it is a strong flier, soaring high up during migration on its very large, broad wings. Its plumage is essentially silvery white, with black tips to the wings, and it has an orange-red pouch beneath a pale yellow, orange-tipped bill. The Dalmatian pelican feeds on a wide variety of fish, needing on average 1kg (2¼lb) a day. It swims on the surface, up-ending to seize prey. Sometimes a number of birds feed

cooperatively, forming a semi-circle and driving fish into shallow water, where they can then scoop them up. These pelicans often breed in large colonies, each pair usually rearing one young. Once chicks are 3 or 4 weeks old, they live in "pods" largely unattended by their parents although they do continue to bring them food. At about 6 weeks, the chicks begin to catch fish, and a month or so later they make their first proper flight.

pouch enables ingestion of large quantities of food

black wing margins

lead-grey legs and feet

Pelecanus occidentalis

Brown pelican

Length 1–1.5 m (3¼–5 ft)	
Weight Over 3.5 kg (7¾ lb)	
Plumage Sexes alike	
Location North, Central, and South America, Caribbean	**Migration** Partial migrant
	Status Common

The brown pelican is mainly silvery-grey and brown, with a white or white and yellow head and chestnut mane. The greenish skin of the face and throat pouch becomes much more vivid in the breeding season. The only pelican that feeds by plunge-diving instead of by swimming and fishing at the surface, it glides low over the water on outstretched wings, and on sighting a fish, flies up as high as 10m (30ft), before folding back its wings and plunging into the sea. It usually nests in trees or shrubs – often in mangroves. However, the Peruvian pelican – a race of the brown pelican – nests on the ground.

RAISING THE YOUNG

Both parents incubate the eggs in shifts of several hours. One or 2 young – very seldom 3 – may be raised. Initially, the adults regurgitate food onto the nest floor, but when the chicks are about 10 days old, they take fish directly from the parent's bill.

deep brown hind neck

greenish skin of face

GROUP LIFE
Highly gregarious, the brown pelican roosts, migrates, and often feeds in groups. It also breeds colonially, on bare, low-lying islands.

Sula nebouxii

Blue-footed booby

Length 80–85 cm (32–34 in)	
Weight 1.5 kg (3¼ lb)	
Plumage Sexes alike	
Location W. Mexico to N.W. South America, Galapagos Islands	**Migration** Partial migrant
	Status Locally common

This distinctive seabird, with vivid blue legs and feet, is one of 9 species that make up the gannet and booby family. Like its relatives, it has a cigar-shaped body adapted for plunge-diving, and narrow wings that angle back just before it hits the surface. The male is much smaller and lighter than the female and so especially

adept at diving into very shallow, inshore water – even in rock pools. Small groups of birds sometimes dive together for food – mainly fish such as flying fish, sardines, anchovy, and Pacific mackerel, or squid. The blue-footed is one of the rarer boobies, with a limited distribution. Even its dispersal is relatively restricted; for example, the Galapagos birds move only to nearby Ecuador.

densely streaked head

long, pointed tail

vivid blue feet

Sula leucogaster

Brown booby

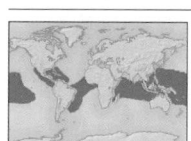

Length 64–74 cm (25–29 in)	
Weight 0.7–1.5 kg (1½–3¼ lb)	
Plumage Sexes alike	
Location Tropical Pacific, Atlantic, and Indian Oceans	**Migration** Partial migrant
	Status Locally common

Like all boobies and gannets, this species is superbly adapted to marine feeding and diving. It has a streamlined, cigar-shaped body, airsacs to cushion the impact of plunge-diving, and a long, tapering bill

for grabbing fish. Making spectacular dives from 30m (100ft) or more, it penetrates the water to great depths, bombarding shoals of squid and fish such as mullet. With its long, sturdy wings, the brown booby is also a graceful flier, and the only member of the family in which the male advertises for a mate in flight.

long, sturdy bill

chocolate-brown plumage

white belly

Morus bassanus

Northern gannet

Length	80–90 cm (32–35 in)
Weight	2.5–3 kg (5½–6½ lb)
Plumage	Sexes similar

Location N. Atlantic, Mediterranean	Migration Partial migrant
	Status Locally common

The northern gannet is a very streamlined seabird, with a torpedo-shaped body, long, narrow wings, and a dagger-like bill, perfectly adapted to plunge-diving for fish. It nests in densely packed colonies or "gannetries" on steep cliffs and raised slopes. Highly developed pair-bonding behaviour persists throughout the nesting season and, once paired, gannets will remain together for years, returning to the same nest season after season. First breeding occurs between the ages of 3 and 5 years. One bluish white egg is laid and then incubated by both parents. Young gannets typically leave the nests around 3 months after hatching and migrate without their parents. Juveniles achieve full adult plumage within 5 years.

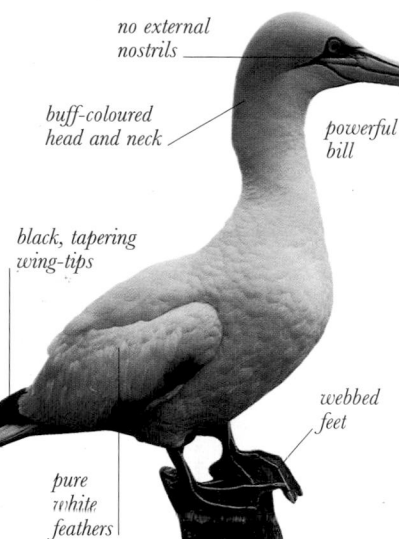

no external nostrils

buff-coloured head and neck

powerful bill

black, tapering wing-tips

webbed feet

pure white feathers

DAZZLING PLUMAGE
The northern gannet is a very distinctive seabird. Most of its plumage is dazzling white, except for its jet-black wing-tips, and the head and back of the neck which have a buff-coloured hue, contrasting with its icy blue eyes and bill.

NESTING GANNETS
Within their colonies gannets space their nests at a distance of two birds' reach. Gannetries are so densely populated that viewed from afar the high slopes and cliffs can have the appearance of being covered in snow.

SHOW OF AGGRESSION
The northern gannet can be exceptionally aggressive when defending its nest site, using its powerful bill to stab and grip when fighting.

BEHAVIOURAL PATTERNS

Few seabirds are as sophisticated in their behavioural patterns as the northern gannet. Both its plunge-diving technique and pair-bonding behaviour are highly dramatic, while the level of parental care it offers is advanced.

COURTING COUPLE
During spectacular bonding displays, partners fence with their bills, as the female aims to allay the male's aggression.

DRAMATIC PLUNGE-DIVERS
From up to 45m (150ft) above the water, the gannet uses its binocular vision to track fish before plunging at speeds of up to 100kph (60mph).

FEEDING A NESTLING
The young gannet reaches deep into its parent's throat to feed on regurgitated fish.

Phalacrocorax auritus
Double-crested cormorant

Length	76–91 cm (30–36 in)
Weight	1.5–2 kg (3 1/4–4 1/2 lb)
Plumage	Sexes alike
Location	North America
Migration	Partial migrant
Status	Common

This bird is the only cormorant in North America that is widespread inland as well as on the coast. As with other cormorants, it has a streamlined body, a serpentine neck, and large, webbed feet. Unusually, for a brief period in spring it has a double crest on the head. Its increasing population has made it a serious pest of catfish farms on the Mississippi.

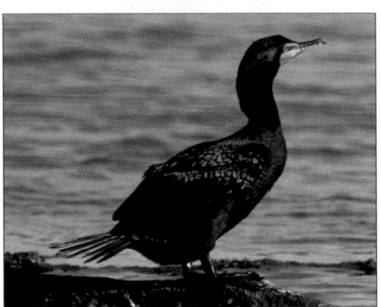

Anhinga melanogaster
Indian darter

Length	85–97 cm (34–38 in)
Weight	1–2 kg (2 1/4–4 1/2 lb)
Plumage	Sexes differ
Location	Africa (south of Sahara), S. and S.E. Asia, Australia, New Guinea
Migration	Partial migrant
Status	Lower risk

One of 4 similar species, this cormorant-like bird usually swims with only its head and neck above the waterline, the rest of the body being underwater. It has specially adapted vertebrae in its neck that form a Z-shaped kink, which straightens explosively and enables the darter to spear its prey with its pointed bill. It builds a nest of twigs sometimes as high as 5m (16ft) above water.

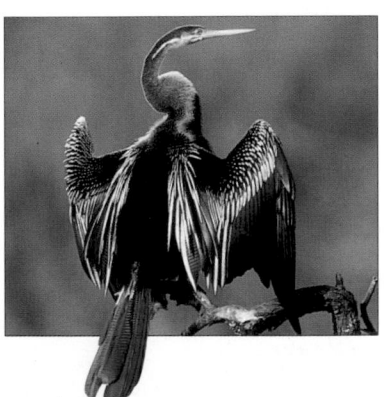

Phalacrocorax carbo
Great cormorant

Length	80–100 cm (32–39 in)
Weight	Up to 3.5 kg (7 3/4 lb)
Plumage	Sexes alike
Location	E. North America, S. Greenland, Europe, Asia, C. to southern Africa
Migration	Partial migrant
Status	Common

One of the world's most widespread coastal seabirds, the great cormorant is also common inland and can be found on almost any large area of water – fresh, brackish, or saline, natural or artificial – in Europe, Asia, and Africa. It makes its nest in a wide range of habitats, from cliffs and bare rocks to trees and reedbeds. The great cormorant breeds in colonies, in some cases returning to the same location year after year, although pairs of birds usually remain together only for a single season.

black body with bronze sheen

stout legs

LONG AND SLEEK
The great cormorant has a spare, streamlined body, a flexible, serpentine neck, and a stout, hooked bill to help grasp fish.

POWER SWIMMING

Although the great cormorant prefers to hunt in shallow water, it can dive to depths of 30m (100ft) or more. When underwater, it swims under its prey with its wings pressed closely to its body. Sturdy legs with large webs make for powerful propulsion.

Phalacrocorax atriceps
Blue-eyed cormorant

Length	68–76 cm (27–30 in)
Weight	2.5–3.5 kg (5 1/2–7 3/4 lb)
Plumage	Sexes alike
Location	Circumpolar around Antarctica, S. South America, Falkland Islands
Migration	Partial migrant
Status	Common

This distinctive, black and white cormorant, with blue eye rings, is from the southern hemisphere and found mostly on rocky coasts and islands. A highly gregarious bird, it forms dense winter flocks that forage offshore. In summer, it mainly feeds alone. There are several subspecies of the blue-eyed cormorant, differing slightly in colour and markings.

Nannopterum harrisi
Flightless cormorant

Length	89–100 cm (35–39 in)
Weight	2.5–4 kg (5 1/2–8 3/4 lb)
Plumage	Sexes differ
Location	Galapagos Islands (Isabela and Fernandina)
Migration	Non-migrant
Status	Endangered

This large cormorant has only the tiniest of wings and long since lost the ability to fly – the absence of terrestrial predators on the Galapagos Islands possibly having made flight unnecessary. It produces very little oil from its preen gland, but the soft, dense body plumage, more like hair than feathers, traps air, which prevents the bird from becoming waterlogged. The flightless cormorant has a long, strong bill that is useful for flushing octopus and fish from the sea bottom. It breeds in small groups. The male rears the young, continuing to feed them for several months, while the female deserts them to find a new mate.

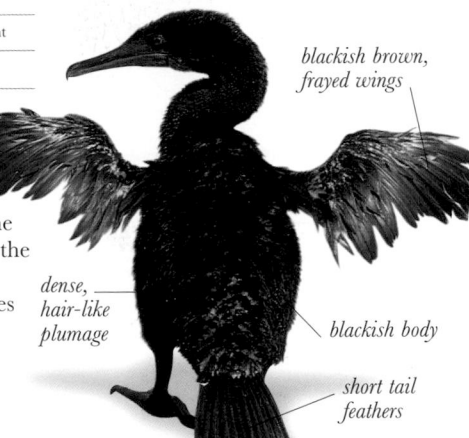

blackish brown, frayed wings

dense, hair-like plumage

blackish body

short tail feathers

Fregata minor
Great frigatebird

Length	85–105 cm (34–41 in)
Weight	1–1.5 kg (2 1/4–3 1/4 lb)
Plumage	Sexes differ
Location	Tropical Pacific, Atlantic, and Indian oceans
Migration	Partial migrant
Status	Common

With their remarkably light bodies and immense, slender wings, great frigatebirds spend their lives gliding effortlessly over the sea, taking food from the surface or from other birds. Their plumage is not waterproof so, when feeding, they dip only their bill into the water, holding the wings up to avoid getting them wet. This species – one of 5 in the family – has a typically long, hooked black bill and a long, forked tail. The male is black, with a glossy green sheen and long, pale wingbars, and when courting, exhibits a scarlet, balloon-like throat pouch. The female is black and white. The great frigatebird breeds on small, mostly uninhabited islands, forming colonies of up to several thousand pairs.

slender, blue-black bill

inflated throat pouch on male

HERONS AND RELATIVES

PHYLUM	Chordata
CLASS	Aves
ORDER	Ciconiiformes
FAMILIES	6
SPECIES	119

THIS GROUP OF wading birds includes herons, egrets, bitterns, storks, ibises, and spoonbills. All have large, plump bodies with long necks and powerful bills. Their long legs help them to wade in shallow water, where they feed on fish, amphibians, snails, and crabs. Most feed alone to avoid other birds disturbing their prey, but at night they gather in groups to roost, and many species breed in colonies. These birds are found in freshwater habitats throughout the world, mostly in warm regions not subject to winter freezing.

ANATOMY

Herons and their relatives have several adaptations for walking and feeding in shallow water. Long legs allow the bird to keep its plumage dry while it extends its flexible neck to lower its head towards the water. The feet have 4 widely spaced toes, the front 3 joined by webbing to spread the bird's weight as it walks on mud or marshy vegetation. All birds in this group have broad wings. During flight, herons, bitterns, and egrets retract their neck, whereas most storks and all ibises and spoonbills hold it straight out.

BILL SHAPES
A long bill is a useful tool for pulling animals from water or mud. Ibises have a long, thin, downcurving bill. Herons have a long, straight bill that tapers to a point. A stork's bill is similar in shape but often curves up or down at the tip, while a spoonbill's flat bill broadens at the end to form a shape similar to a spatula.

IBIS

HERON

STORK

SPOONBILL

FEEDING

Most herons and their relatives use keen eyesight to find their prey, scanning the water for signs of movement below the surface. Once sighted, a fish or frog will be grabbed (but rarely speared) with a thrust of the bill. Some birds refine the technique – for example, the black heron spreads its wings to cast a shadow on the water and lure fish into the shade. Not all of these birds feed on live prey: the marabou stork is primarily a scavenger that feeds on carcasses.

FEEDING INLAND
The cattle egret finds most of its food away from water. Flocks typically gather around cattle and other large mammals to feed on insects disturbed by their movements.

STRIKING AT PREY
When hunting, a heron bends its neck into a characteristic S-shape. On seeing its prey, the heron thrusts its head forwards with lightning speed. Once the prey is caught, it may be subdued by a stab from the bill.

WADING
Herons and their relatives wade slowly and deliberately. Spoonbills (seen here) sweep the water with their bills, using touch-sensitive cells to find small fish and crustaceans.

Ardea cinerea

Grey heron

Length 90–98 cm (35–39 in)	
Weight 1–2 kg (2¼–4½ lb)	
Plumage Sexes alike	

Location Europe, Asia, Africa

Migration Partial migrant

Status Common

The commonest and most widespread heron in Europe, this bird can be found in all kinds of shallow, freshwater habitats from reedbeds to lakes. It flies with slowly flapping, deeply bowed wings. Grey herons pair for life; the male and female build the nest together, often in a tall tree, but sometimes on the ground. The female usually lays 3–5 eggs. Like other herons, they use ritual courtship and defence displays, such as stretching upwards with an arching neck.

APPEARANCE
A large and distinctive bird, the grey heron has a long neck and long legs. Its plumage is grey, white, and black.

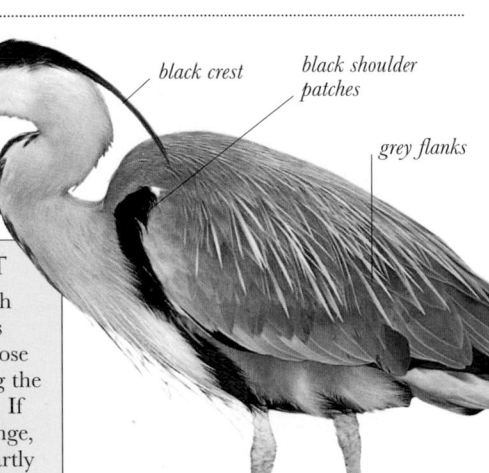

black crest

black shoulder patches

grey flanks

long, narrow toes

FISHING BY SIGHT
The grey heron relies on stealth and rapid reactions to catch its prey. When fishing, it stands close to the water's edge, monitoring the movement of any fish close by. If a fish comes within striking range, the heron tips forwards and partly extends its neck, before suddenly stabbing with its bill. It swallows small fish whole, but takes larger ones back to land.

Casmerodius albus
Great egret

Length 85–100 cm (34–39 in)	
Weight 950–1,000 g (34–36 oz)	
Plumage Sexes alike	

Location North, Central, and South America, Africa, Asia, Australia	**Migration** Partial migrant
	Status Common

In the breeding season, long plumes develop on the back, trailing over the relatively short tail. These egrets stalk their prey alone or in loose groups, but within groups individuals defend small territories. Their young in the nest are extremely aggressive, often resulting in the death of weaker chicks.

The most widespread of all herons, the great egret is found in all kinds of wetlands from the Americas to Asia and Australia. Its plumage is entirely white, and the legs and feet are black.

Butorides striatus
Green-backed heron

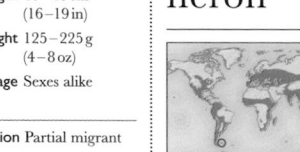

Length 40–48 cm (16–19 in)	
Weight 125–225 g (4–8 oz)	
Plumage Sexes alike	

Location North, Central, and South America, Africa, Asia, Australia	**Migration** Partial migrant
	Status Common

This highly variable, medium-sized, secretive heron is one of the few birds known to drop bait into the water to attract fish. It has about 30 subspecies, spread across the tropics and sub-tropics. One common form, from South America, is generally grey-green above. The lava heron (*Butorides striatus sundevalii*), from the Galapagos Islands, however, is entirely dark grey.

Botaurus lentiginosus
American bittern

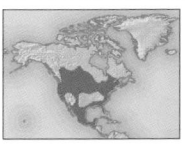

Height 60–85 cm (23 1/2–34 in)	
Weight 500–900 g (18–32 oz)	
Plumage Sexes alike	

Location North America, Central America, Caribbean	**Migration** Migrant
	Status Common

With its streaked and mottled brown plumage, the American bittern is superbly camouflaged for life in dense reedbeds and overgrown marshes. This daytime feeder preys primarily on fish and frogs, stabbing with a lightning-fast lunge of its dagger-like

Nycticorax nycticorax
Black-crowned night heron

Height 58–65 cm (23–26 in)	
Weight 500–800 g (18–29 oz)	
Plumage Sexes alike	

Location North, Central, and South America, Africa, Asia	**Migration** Partial migrant
	Status Common

Unlike most other herons, this small bird feeds at night or at dusk, using its exceptional vision to pinpoint its prey in dim light, and occasionally also responding to sounds. It grasps food with its bill while standing still or walking slowly in the water. It usually feeds singly, but nests and breeds colonially. A good climber, it is often seen clambering over roots or branches close to the water's edge.

glossy black crown

bill. Like other bitterns, when disturbed it assumes a peculiar defensive posture, with its bill upturned and its neck extended, and may sway in the wind with the surrounding grass in order to remain inconspicuous.

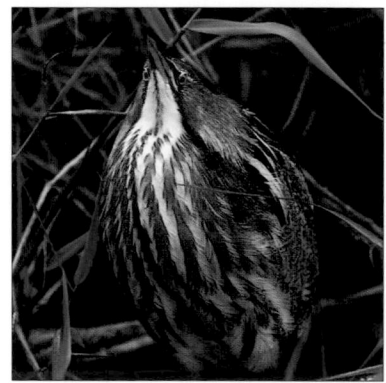

Scopus umbretta
Hamerkop

Length 40–56 cm (16–22 in)	
Weight 425 g (15 oz)	
Plumage Sexes alike	

Location Africa (south of Sahara), Madagascar, S.W. Asia	**Migration** Non-migrant
	Status Locally common

From the German "hammerkopf" meaning hammerhead, this bird's common name is a good description of its unusual shape, with its relatively short bill and heavy crest on the back of its head. Dull brown, it has darker brown primary feathers, and a paler chin and throat. While many water-birds feed primarily on fish, the hamerkop's main diet consists of amphibians. It feeds in shallow water, using its bill to rake the bottom for frogs and fish; it also flies above groups

of tadpoles and snatches them up. The hamerkop builds the largest roofed nest of any bird. Made of twigs, mud, and grass, it is oven-shaped with an entrance tunnel, up to 2m (6½ft) across and deep, and sited high up in a tree. While the hamerkop breeds solitarily, nests are often found near each other.

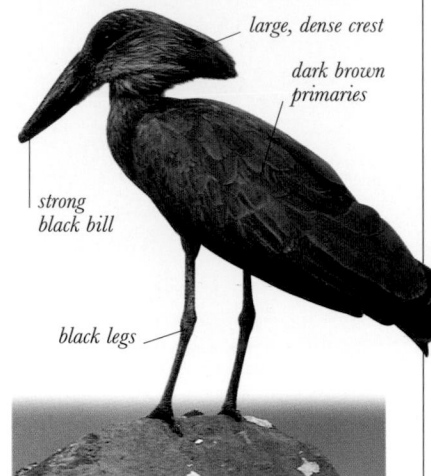

large, dense crest

dark brown primaries

strong black bill

black legs

Balaeniceps rex
Shoebill

Height 1.1–1.4 m (3 1/2–4 1/2 ft)	
Weight 4.5–6.5 kg (10–14 lb)	
Plumage Sexes alike	

Location Central Africa	**Migration** Non-migrant
	Status Lower risk†

The shoebill, or whale-headed stork, has a highly distinctive, large, broad, clog-shaped bill. It often hunts for prey in pools that are drying out, feeding mainly on lungfish, and also on frogs and small mammals. The shoebill frequently adopts an unusual method of hunting: it plunges on prey as if falling on it and then

cuts up the food with a scissor-like action before swallowing. The shoebill breeds in almost inaccessible papyrus swamps, and usually nests on floating vegetation. In hot weather, it pours beakfuls of water over the eggs to cool them.

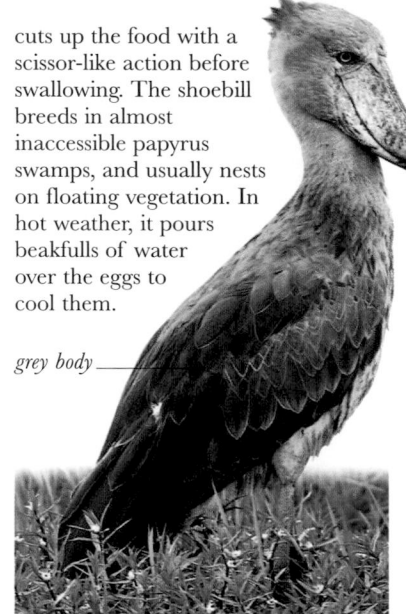

grey body

Mycteria americana
American wood stork

Height 85–110 cm (34–43 in)	
Weight 2.5 kg (5 1/2 lb)	
Plumage Sexes alike	

Location North, Central, and South America, Caribbean	**Migration** Partial migrant
	Status Common

The American wood stork locates food by both sight and touch, enabling it to feed by day or night, and in murky water. The scaly head and neck are bald, as feathers would be soiled when dipped in mud. While feeding by touch, the bird walks through the water with its long, thick bill open, moving it from side to side. When it encounters prey, its reactions

are extremely quick. White, with black wing-tips, it has a dark grey head and neck. The wood stork feeds either alone or in groups, and breeds colonially, building a nest often over water at treetop level, and very occasionally on the ground where there are no ground predators.

Ciconia ciconia
European white stork

Height 1–1.3 m (3¼–4¼ ft)	
Weight 2.5 kg (5½ lb)	
Plumage Sexes alike	
Location Europe, Africa, Asia	**Migration** Migrant
	Status Locally common

In parts of northern Europe, this elegant, black and white bird is a harbinger of spring, migrating from as far away as southern Africa, where it spends the winter. Today, largely due to habitat change, far fewer storks breed in northern Europe, but their rooftop nests are still considered good luck. The white stork feeds primarily in shallow water and grassland, but also at the edges of crop fields. It migrates mainly over land, where thermals help its soaring flight.

Leptoptilos crumeniferus
Marabou stork

Height 1.2 m (4 ft)	
Weight 5–7.5 kg (11–17 lb)	
Plumage Sexes alike	
Location Africa (south of Sahara)	**Migration** Non-migrant
	Status Common

This huge, ungainly looking stork is slaty grey, black, and white, and has a wingspan of nearly 3m (9¾ft), one of the largest of any land bird. Its elegant, soaring flight contrasts markedly with its hunched and unappealing aspect when on the ground. Because of its feeding habits (see panel below), the marabou stork has done very well, and is increasing in numbers throughout its range

large black wings

throat wattle

UNGAINLY AIR
Often described as "ugly", this large bird has a nearly featherless head and neck and a massive, wedge-shaped bill. Its pinkish to pale magenta pouch, or wattle, may extend to 35cm (14in) from the base of the bill.

SCAVENGING

Although it is a stork, the marabou often behaves more like a vulture. It soars high up in search of food, and uses its large beak to tear rotting meat from carcasses. Like a true vulture, it has an almost featherless head and neck – an adaptation that helps to prevent it from getting soiled while it feeds.

Threskiornis aethiopicus
Sacred ibis

Length 65–89 cm (26–35 in)	
Weight 1.5 kg (3¼ lb)	
Plumage Sexes alike	
Location Africa (south of Sahara), Madagascar, Aldabra Island, W. Asia	**Migration** Migrant
	Status Locally common

This medium-sized white ibis, with its strikingly black, featherless head and upper neck, was revered in ancient Egypt. A generalized feeder, it easily cohabits with people and consumes a variety of foods – refuse and offal around human habitations, insects in grassland areas, and aquatic animals in shallow pools.

bald head and neck

black legs

Eudocimus ruber
Scarlet ibis

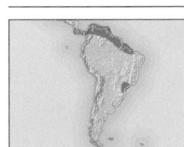

Length 56–68 cm (22–27 in)	
Weight 775–925 g (28–33 oz)	
Plumage Sexes alike	
Location S. Central America, N. and E. South America	**Migration** Non-migrant
	Status Locally common

This vividly coloured bird gathers in large flocks – sometimes of tens of thousands of birds – in the breeding season, on the coastal wetlands of northern South America, in swamps, lagoons, and mangroves, as well as on tidal rivers. Birds of a colony pair off for breeding (although they often mate with other partners too), building their nests in trees close to water. Like other ibises, the scarlet ibis finds its food primarily by touch instead of by sight, probing into soft mud with its long, gently curving bill, usually while walking along. It may also feed like a spoonbill, sweeping its bill from side to side in the water. Crabs, shellfish, and aquatic insects are the major items of prey.

long, gently curving bill

TREETOP ROOSTS

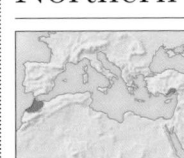

The scarlet ibis feeds on the ground during daytime, but at dusk, it flies up into waterside trees to roost. This behaviour – common among ibises, herons, and their relatives – reduces the danger of being attacked by predators at night.

black wing-tips

VIVID COLORATION
With its bright red plumage and black-tipped wings, the scarlet ibis is one of the world's most strikingly coloured birds.

Geronticus eremita
Northern bald ibis

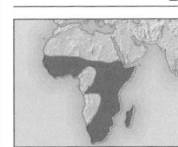

Length 80 cm (32 in)	
Weight Not recorded	
Plumage Sexes alike	
Location N.W. Africa, W. Asia	**Migration** Migrant
	Status Critically endangered

Once widespread throughout southern Europe, North Africa, and West Asia, this ibis is now extremely rare. It has a bare red head, a straggly

Platalea alba
African spoonbill

Length 75–90 cm (30–35 in)	
Weight 2 kg (4½ lb)	
Plumage Sexes alike	
Location Africa (south of Sahara), Madagascar	**Migration** Partial migrant
	Status Common

This graceful, red-legged white bird belongs to a group of 6 species that get their name from the broad, spoon-shaped tips of their bills. Like its

black crest on the nape of its neck, and brownish black iridescent plumage. It feeds near pools and dry river beds, using its long, curved bill to probe for insects and other small animals in loose ground. It nests in colonies on steep cliffs, where its young are relatively safe from attack.

relatives, it swings its bill from side to side in the water, creating currents that bring fish within range, which the spoonbill speedily snaps up. It may also feed by night, looking for fish and running after them in a sprint.

FLAMINGOS

PHYLUM	Chordata
CLASS	Aves
ORDER	Phoenicopteriformes
FAMILIES	1 (Phoenicopteridae)
SPECIES	5

WITH THEIR EXTREMELY long legs and neck, and colourful pink or red plumage, these tall wading birds are striking and instantly recognizable. They are found in the tropics and subtropics, usually in the shallows of salt or brackish water or alkaline lakes, sometimes massed in vast flocks of up to one million birds. Despite their fragile appearance, flamingos occur in many parts of the world where few other animals can survive. They are frequently found in conditions of extreme salinity or alkalinity, and are remarkably tolerant of changes in temperature. Flamingos are specialized feeders, using their unusual downturned bill to filter tiny plants and animals from water.

COLONIES

Flamingos are among the most sociable of birds, even performing their courtship displays in groups. Thousands of individuals open their wings or lift up and turn their heads in one vast, synchronized movement. These group displays seem to bring all the birds of the colony to the same readiness to mate, to ensure rapid, synchronized egg-laying as soon as conditions allow. Breeding colonies often form when the water level of a salt lake drops – the surface is suddenly covered with piles of raised mud, which form the flamingos' nests. The parents care for their young for the first week or two. Once a young flamingo is able to walk and swim, it joins a "creche" (see right).

GROUP DISPLAY
During courtship displays, flamingos may engage in "head-flagging", raising their neck and bill, and turning their head from side to side.

CRECHES
Young flamingos form large groups of up to 300,000 birds (here, lesser and greater flamingos). They are still fed by their parents but are guarded by an unrelated adult.

FLIGHT
Flying flamingos, with their thin neck and legs outstretched, are unmistakable. Large flocks often form long, curving lines. Their long wings and light body enable a relatively easy take-off.

ANATOMY

Flamingos have a slender body set on greatly elongated legs, which are longer in relation to body size than those of any other bird. The limbs are completely bare, allowing the birds to wade deep in highly saline or alkaline water without soiling their plumage. Flamingos have a long, flexible neck that, in the case of the larger species, is also remarkably thin. The head is small, and the bill has a characteristic downward bend. The way in which the upper and lower parts of the bill fit together, combined with the comb-like plates on the bill's edges, make it a useful tool for sieving food from water (see right). The flamingo's unique pink or red coloration arises from a dye extracted from food such as algae and shrimps.

STANDING IN WATER
Flamingos (here, a greater flamingo) often stand on one leg for long periods, even when asleep, with the head laid on the body and the other leg tucked under the abdomen. This posture cuts heat loss through the legs and feet.

FEET
Compared with their long legs, the feet of flamingos are relatively small. The front 3 toes are webbed, and the back one is either tiny or missing. The webbing is useful for walking over mud or other soft surfaces.

FEEDING

A flamingo usually feeds while wading in shallows, using its feet to stir up mud on the bottom. To feed, it puts its head to the surface (so that its bill is upside down and the tip points backwards) and sweeps its head from side to side. By a rapid action of the tongue, it pumps water in and out of the slightly opened bill. Along the inside edge of the bill are rows of plates, known as lamellae, some of which have tiny hairs on them. The flamingo filters out food particles by sieving them past the lamellae and their hairs. The size of particles taken differs between species. Larger flamingos tend to feed on crustaceans, molluscs, and worms, and the smaller species on algae.

BILL
By opening its bill only slightly when feeding, a flamingo filters out unwanted large items. A second filtration happens inside the bill, where rows of tiny plates, often with minute hairs attached, act as a sieve to trap the smaller particles that the flamingo needs.

tongue
moving water
lower bill
lamellae
upper bill
hook for fastening bill

CROSS-SECTION

tongue

lamellae

FEEDING

FEEDING ON ALGAE
Lesser flamingos (left) and other smaller species usually feed from the water surface. Larger flamingos immerse their heads completely to feed.

BIRDS

Phoenicopterus ruber
Greater flamingo

	Height 1.5m (5ft)
	Weight Up to 4kg (8 3/4 lb)
	Plumage Sexes alike
Location Central America, South America, Caribbean, S.W. Europe, Asia, Africa	**Migration** Partial migrant
	Status Locally common

With its exceptionally long neck and legs, the greater flamingo is the largest member of the flamingo family. It is found in a wide variety of freshwater and saline habitats, particularly salt lakes, estuaries, and lagoons. Greater flamingos outside the tropics often migrate to warmer regions for winter, and in all regions they fly long distances – sometimes up to 500km (300 miles) – to find food, travelling at night. The greater flamingo's large size enables it to wade out into relatively deep water, whereas other flamingos are restricted to the shallows; it also sometimes swims, upending like a duck in order to reach food. This flamingo usually feeds with its head fully immersed, sometimes keeping it underwater for up to 20 seconds. Unlike smaller flamingos, it rarely takes food from the surface, which reduces competition with them. Diet is varied, and includes insects, worms, microscopic algae, and pieces of vegetation. The greater flamingo feeds mainly during the day, even in the hottest conditions. Highly gregarious, it breeds in colonies of up to 200,000 monogamous pairs.

NESTS AND CHICKS

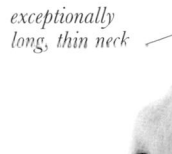

As with all flamingos, nests of greater flamingos are flattened cones of mud, often surrounded by a shallow "moat", and spaced about 1.5m (5ft) apart, just beyond pecking distance. Both parents incubate the single egg. They defend the nest when breeding, but are otherwise non-territorial. Once the chicks can walk, they gather in large creches under the supervision of a small number of adults.

MASS BREEDING

For a social species, the greater flamingo shows a remarkable range of colony size. In some parts of the world, breeding colonies contain huge numbers of birds, but in others – such as the Galapagos Islands – they may contain only a few dozen. This variability is one of the reasons why greater flamingos, unlike some of their relatives, are relatively easy to breed in captivity.

Courtship involves complex, synchronized dances – neck stretching, ritualized preening, loud honking – performed by large numbers of males and females. It has a quieter contact call while feeding.

exceptionally long, thin neck

PINK OR RED
The greater flamingo is generally pink, but 2 distinct subspecies differ in hue, with the New World subspecies being noticeably redder than its Old World counterpart.

nostril

blunt, angled bill

short, webbed toes

extremely long legs

Phoenicoparrus jamesi
Puna flamingo

	Height 1.1m (3 1/2 ft)
	Weight 2kg (4 1/2 lb)
	Plumage Sexes alike
Location W. South America	**Migration** Non-migrant
	Status Lower risk

This flamingo derives its common name from its habitat – "puna" is the local word for a high Andean plateau. Its diet consists of diatoms – microscopic algae that abound in salt lakes. It feeds by day in shallow water, walking slowly forwards with its bill dipped just under the water surface. Diatoms are filtered out of the water in the bill and swallowed. This flamingo has a variety of calls, both when feeding and on the wing. It is one of 3 South American flamingos, the other 2 being the Chilean flamingo (*Phoenicopterus chilensis*) and the Andean flamingo (*Phoenicoparrus andinus*). All 3 species can be found together in salt lakes, usually at altitudes over 3,000m (9,900ft). Most puna flamingos migrate to lower altitudes in winter, but some remain at lakes where there are hot springs. The courtship and breeding habits of this bird are similar to those of the greater flamingo (see above).

CONSERVATION

With its small population and localized distribution, the puna flamingo is vulnerable to anything that affects its reproductive success. For centuries, its eggs have been collected by Andean people for food, which may account for its current low numbers. This bird is now protected by guards at one of its most important breeding sites. Pollution is also a potential threat, as it could jeopardize the flamingo's food supply.

pink coloration

red streaks

COLORATION
The puna flamingo is pink with red streaking, especially in the breeding season.

short-hooked bill

long, slender legs

Phoeniconaias minor
Lesser flamingo

	Height Up to 1m (3 1/4 ft)
	Weight Up to 2kg (4 1/2 lb)
	Plumage Sexes alike
Location W., C., and southern Africa	**Migration** Nomadic
	Status Locally common

The smallest but most numerous flamingo, the lesser flamingo may be light to dark pink, with a relatively long, dark-coloured beak. It is non-migratory, but will readily move to a new site in search of food. Its diet consists almost entirely of microscopic blue-green algae, abundant in soda lakes. It feeds in shallow water at dusk and after dark, avoiding strong daytime winds. Some colonies are over a million strong – among the largest bird aggregations in the world. Courtship rituals often involve hundreds of birds moving and displaying in synchrony.

WATERFOWL

PHYLUM	Chordata
CLASS	Aves
ORDER	Anseriformes
FAMILIES	2
SPECIES	149

ALSO KNOWN AS wildfowl, members of this group include ducks, geese, and swans, as well as a family of 3 South American species known as screamers. Aided by waterproof plumage and webbed feet, they are among the dominant birds of freshwater wetlands. They are also found in estuaries and inshore coastal waters, while a few species are entirely marine. Waterfowl are strong swimmers. They feed mostly from the surface of water, although many ducks dive in search of food, while some species (notably geese, swans, and screamers) graze on land. Waterfowl are powerful fliers, and some species undertake annual migrations of thousands of kilometres between their wintering and breeding areas.

ANATOMY

Waterfowl have a plump, buoyant body, with a small head and, usually, a short tail. Most species have a broad, flattened bill (see below) and a long neck for reaching down to feed underwater. Many species are gaudily coloured. When moulting, most lose all their flight feathers simultaneously. To hide them from predators during the period when they are unable to fly, the males (drakes) acquire a drab colouring, known as an eclipse plumage. Waterfowl are insulated from cold by a layer of down and a layer of fat beneath the skin. Screamers look different from other waterfowl in several respects: they have longer legs, spurred wings, a chicken-like bill, and toes webbed only at the base.

BILL
Like most waterfowl, the common shelduck has a wide, flattened bill. The edges have small ridges, called lamellae, which are used to grip prey or filter edible particles out of the water. The tip of the upper bill may be hardened into a nail for tearing vegetation.

hardened tip (nail)

broad bill

PREENING
Plumage care is important for waterfowl such as this greylag goose, because it is essential that their feathers remain waterproof. A gland on the bird's rump secretes a water-repellent oil. The bird stimulates the gland with its bill and then spreads the oil around its body by rubbing and preening its feathers.

FEET
Waterfowl use their webbed feet to help propel themselves through water. Only the front 3 toes are webbed, the smaller hind toe being raised. This enables them to walk on land, although they do so with a waddling gait.

REPRODUCTION

The reproductive cycle of waterfowl is suited to a life spent on water, which often involves exposure to cold and predators. Unusually among birds, the males have an organ similar to a penis, which is inserted inside the female's cloaca, making it possible for them to mate on the surface of water. Waterfowl usually nest near water, most often among ground vegetation, although some species use holes in trees or rock crevices. The females of some species, such as eiders, pluck their own down to make an insulating cover for their eggs.

YOUNG
Waterfowl hatch in a well-developed state, with their eyes open and a covering of down that dries quickly. Hours after hatching, the chicks (such as these Canada geese) can walk or swim. Leaving the nest, they follow the adult to the relative safety of water, and soon begin to search for their own food.

TAKING OFF
Although some waterfowl can spring almost vertically from water, the heavier species have to resort to a running take-off. Swans, in particular, need a long runway, gaining speed by pattering with their feet across the surface of the water.

MOVEMENT

Several adaptations help waterfowl to move easily through water. The smooth outline of their body reduces water resistance, while their webbed feet (see left) act as powerful paddles. Thick plumage adds to their buoyancy, although many duck species that dive to find food sleek their plumage to expel air and reduce their buoyancy before submerging. With their huge wings and exceptionally light bones, screamers can fly for hours on end, often soaring. Other species of waterfowl have smaller wings that must be flapped constantly when in flight. Once airborne, however, they are fast and powerful fliers, with some species capable of exceeding 100kph (60mph) in level flight.

FLYING FORMATION
Flocks of waterfowl (here, snow geese) fly in a characteristic V-formation. By staying in the slipstream of the leading bird, those further back meet with less turbulence and thereby save energy. The position of hard-working leader is changed regularly.

Anhima cornuta
Horned screamer

	Length 84 cm (33 in)
	Weight 2–3 kg (4 1/2 – 6 1/2 lb)
	Plumage Sexes alike
Location N. South America	Migration Non-migrant
	Status Locally common

Although related to other waterfowl, screamers have many distinctive features, including turkey-like bodies, narrow bills, and long legs ending in partly webbed feet. The largest of the 3 species in the screamer family, the horned screamer is black and white, with a remarkable slender spike, up to 10 cm (4 in) long, that curves upwards and forwards from its forehead. Like its relatives, it often feeds on land rather than in water. Females lay 4–6 eggs in a nest on the ground, and these take 6 weeks to hatch.

Dendrocygna eytoni
Plumed whistling duck

	Length 40–60 cm (16–23 1/2 in)
	Weight 0.5–1.5 kg (1–3 1/4 lb)
	Plumage Sexes alike
Location N. and E. Australia	Migration Partial migrant
	Status Locally common

Like all 8 species of whistling ducks, this species has long legs and large, webbed feet, with a net-like pattern on the ankle and heel, a characteristic that is more typical of a goose than a duck. The plumed whistling duck also has some goose-like behavioural traits: it mostly feeds on land, pulling and clipping grasses. It is believed to remain monogamous for life. The male also cares for the young.

straw-coloured, pointed flank plumes

long legs

Cygnus olor
Mute swan

	Length 1.5 m (5 ft)
	Weight 12 kg (26 lb)
	Plumage Sexes alike
Location North America, Europe, Africa, Asia, Australia	Migration Partial migrant
	Status Locally common

Originally from Europe and Central Asia, this extremely elegant bird has been introduced as an ornamental species in many other parts of the world. When young, the mute swan is greyish brown, but the adult has pure white plumage, an orange-red bill, and black legs and feet. One of the world's heaviest flying birds, it runs or paddles across the water to take off, but once airborne, is a powerful flier, making a distinctive pulsating sound with its wings. Mute swans feed mainly on water, often up-ending to reach plants and small animals in underwater mud. They mate for life, and nest by the water's edge or on small islands, making a mound of vegetation often more than 1 m (3 1/4 ft) across. Females lay up to 8 eggs in a clutch, and take sole charge of incubation. Once the young have hatched, both parents look after them, until they become

independent at about 5 months. Young swans often remain with their parents for longer than this, but they are driven away by the male at the onset of the following breeding season. It takes 3–4 years for the young to become fully mature and able to raise young of their own.

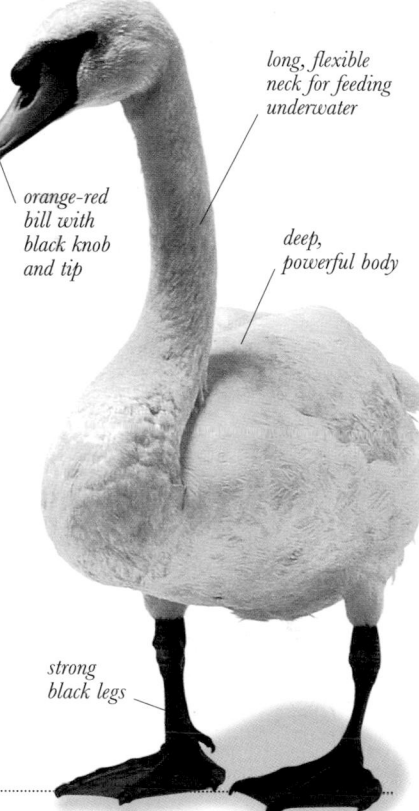

long, flexible neck for feeding underwater

orange-red bill with black knob and tip

deep, powerful body

strong black legs

Cygnus atratus
Black swan

	Length 1.1–1.4 m (3 1/2 – 4 1/2 ft)
	Weight 6 kg (13 lb)
	Plumage Sexes alike
Location Australia (including Tasmania), New Zealand	Migration Nomadic
	Status Locally common

This is the only swan that is almost entirely black. The innermost flight feathers are curiously twisted, and when threatened, the swan raises these feathers and exposes the white primaries. Exclusively vegetarian, it feeds on aquatic plants; occasionally, it may graze on land. Compared to

LONG-NECKED SWAN
The black swan has the longest neck of all swans, making up much more than half the total length of the bird in flight.

FAMILY TIES

Like other swans, this species is strictly monogamous. The male and female have a strong bond and both participate in building the nest and caring for the young.

most other swans, black swans are highly sociable, and also semi-nomadic. After breeding, they sometimes gather in flocks thousands strong.

bright red bill

white primaries

Anseranas semipalmata
Magpie goose

	Length 75–90 cm (30–35 in)
	Weight 3 kg (6 1/2 lb)
	Plumage Sexes alike
Location S. New Guinea, N. Australia	Migration Partial migrant
	Status Locally common

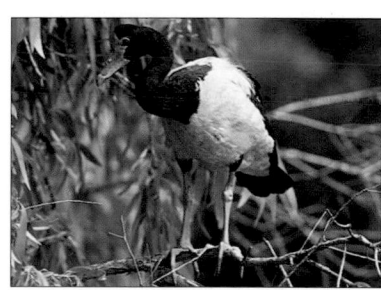

The only species in its family, this goose-like bird is unique among waterfowl in many ways: its toes are only slightly webbed; the hind toes are unusually long, enabling it to perch easily on small branches; and it is the only waterfowl with legs so long that the ends of the feet extend beyond the tail during flight. It is also the only waterbird to form breeding groups of one male and 2 females. The male is larger, with a more distinctive bony "bump" on his crown.

Anser anser
Greylag goose

	Length 75–90 cm (30–35 in)
	Weight 3–4 kg (6 1/2 – 8 3/4 lb)
	Plumage Sexes alike
Location Europe (including Iceland), Asia	Migration Migrant
	Status Common

The greylag goose is the wild ancestor of barnyard geese, originally bred in central Europe, and the basis for almost all European goose folklore. The name "greylag" derives from its grey coloration and the fact that it undertakes late migrations, or "lags" behind other geese. Males swim in a haughty posture to attract the female, and mating is often preceded by mutual head-dipping.

pink bill

Branta canadensis
Canada goose

Length 55–100 cm
(22–39 in)

Weight 3–6 kg
(6 1/2–13 lb)

Plumage Sexes alike

Location North America, N. Europe, N.E. Asia, New Zealand

Migration Migrant

Status Common

Originally from North America, the highly sociable Canada goose has been introduced widely to northern Europe and New Zealand. It is becoming increasingly common, a result of its ability to adapt to a wide range of climates, tolerate ecological change, and feed on grass as well as other low vegetation. One of the most variable of all geese in size, subspecies range from the small Arctic forms, up to 2kg (4½lb), to the large, most southerly populations, which may occasionally weigh as much as 8kg (18lb).

COLORATION
This goose has a black head and neck, white cheeks, and a brown body, with blackish wing-tips and lighter underparts. The bill and feet are black.

GRAZING

Canada geese are diurnal feeders, foraging on dry land as well as on water for plants. Now a common sight in parks outside their native range, flocks leave the water at dawn to feed on grasses, seeds, and grains.

black feet

white below blackish tail

Cereopsis novaehollandiae
Cape Barren goose

Length 75–91 cm
(30–36 in)

Weight 4–6 kg
(8 3/4–13 lb)

Plumage Sexes alike

Location S. Australia (including offshore islands and Tasmania)

Migration Non-migrant

Status Lower risk†

With reduced webbing on its toes, this dove-grey goose has adapted for life mainly on land. Its black, stubby bill is mostly hidden by a pale green cere – a fleshy pad at the base of the upper part of the bill. Aggressive and territorial, it performs several threat displays, and may bite or strike with its knobbly wings. Mating occurs on land and is followed by a "triumph" ceremony by pairs after aggressive encounters with rival males.

Alopochen aegyptiaca
Egyptian goose

Length 63–73 cm
(25–29 in)

Weight 2.5 kg
(5 1/2 lb)

Plumage Sexes alike

Location Africa (south of Sahara)

Migration Non-migrant

Status Locally common

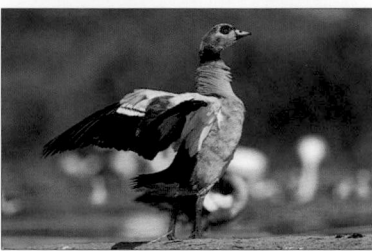

A close relative of the common shelduck (see below, left), this long-legged bird lives in lakes, rivers, and subtropical wetlands, but also spends much of its time on land. Its "bullying" behaviour towards smaller species is notable, as is its aggressive behaviour within its own species. The female chooses the most aggressive male as a mate, which means there is much fighting during the pairing season. Males have husky calls, while females make loud, nattering sounds.

Tadorna tadorna
Common shelduck

Length 58–67 cm
(23–26 in)

Weight 1–1.5 kg
(2 1/4–3 1/4 lb)

Plumage Sexes differ

Location Europe, Asia, N. Africa, E. North Atlantic, Mediterranean, W. Pacific

Migration Partial migrant

Status Locally common

An inhabitant chiefly of marine or saline waters, this duck has iridescent, greenish black, white, and chestnut plumage. The male has a knob at the top of its red bill, absent in the female (shown below). Like all shelducks, it is intermediate in behaviour between a goose and a typical duck. It forages at the tide-line, usually standing in shallow water, probing for molluscs and other marine invertebrates. Nesting often takes place in holes, such as the abandoned burrows of other animals.

pink feet

Anas platyrhynchos
Mallard

Length 50–65 cm
(20–26 in)

Weight 1–1.5 kg
(2 1/4–3 1/4 lb)

Plumage Sexes differ

Location North America, S. Greenland, Europe, Asia

Migration Partial migrant

Status Common

The mallard has a wide distribution in the northern hemisphere, its success being due to its great flexibility – it can adapt to almost all types of aquatic habitats for breeding, even urban environments, and is able to feed in a number of ways (see panel, right). The male has a low-pitched quack and a sharp whistle, while the female is more vocal and has many distinctive quacking calls.

neck ring

BREEDING MALE
The breeding male has a green head with a white neck ring, a rufous breast, grey flanks, black rump, and a white tail with curled central feathers. The bill is yellow.

UP-ENDING

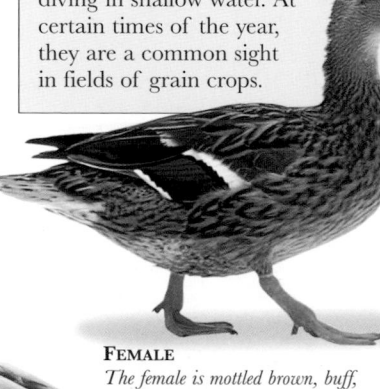

Mallards often feed by up-ending to reach submerged plants and invertebrates. They also eat by dabbling, grazing or, rarely, diving in shallow water. At certain times of the year, they are a common sight in fields of grain crops.

FEMALE
The female is mottled brown, buff, or white; the non-breeding male has a similar plumage, called eclipse plumage.

curled central feathers of tail

Anas crecca
Green-winged teal

Length 34–38 cm
(13 1/2–15 in)

Weight 350 g
(13 oz)

Plumage Sexes differ

Location North America, Europe (including Iceland), Asia, N. to C. Africa

Migration Partial migrant

Status Common

The smallest dabbling duck of North America and Europe, the teal breeds in tundra, grassland, forest, and forest edges, and in relatively small bodies of water. Its tiny bill helps it to feed on small seeds of aquatic plants; molluscs, crustaceans, and insects also form part of its diet. The male has bright breeding plumage, and utters a cricket-like call that is a double-note whistle, while females produce high-pitched quacks. Both sexes have a green wing patch called "speculum".

Aix galericulata
Mandarin

Length 41–49 cm (16–19 1/2 in)	
Weight 625 g (22 oz)	
Plumage Sexes differ	
Location N.W. Europe, E. Asia	**Migration** Partial migrant
	Status Locally common†

The male mandarin in its breeding plumage is among the most ornate and beautiful of all birds. It has a prominent crest on its head, golden hackles, and a pair of usually bright yellow, sail-shaped feathers on each inner wing. These can be erected above the flanks – their function is purely ornamental. Females and non-breeding males are mostly olive-brown. This is one of the most arboreal of all ducks, often roosting or perching on branches and nesting in tree cavities; ducklings have to jump from the nest-hole to the ground at only one day old. Like other perching ducks, its claws are sharp, to help it cling to branches, and its tail is long

and broad, acting as a brake to slow the duck down before it lands in the trees. It also has very large eyes, which help it see at night. The mandarin feeds on land, in the trees, and in water, where it dabbles, up-ends, and only rarely dives. Its diet consists of seeds and nuts, as well as some invertebrates such as insects and land snails. Courtship is highly social, with several males competing for a female's attention. Originally from East Asia, introduced birds have established themselves in western Europe.

prominent crest   *large eyes*

sharp claws

Cairina moschata
Muscovy duck

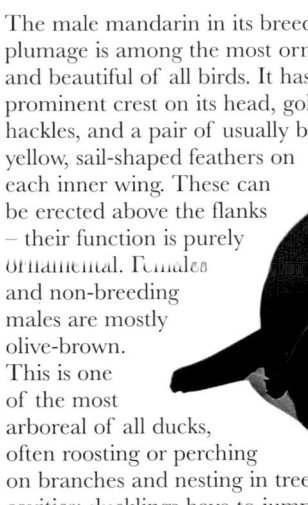

Length 66–84 cm (26–33 in)	
Weight 2–4 kg (4 1/2–8 3/4 lb)	
Plumage Sexes alike	
Location Central America to C. South America	**Migration** Non-migrant
	Status Common

This large, heavy-bodied duck is mostly blackish as an adult (brownish black when young), and has a long tail, broad wings, and sharply clawed

toes – all adaptations for perching in trees. The adult male also has a slight crest on the top of the head, a knob above the nostrils, and bare, warty skin around the eyes. Domesticated birds (below) show many colour variants. Unusually for a duck, both sexes are nearly mute.

Polysticta stelleri
Steller's eider

Length 43–47 cm (17–18 1/2 in)	
Weight 650–900 g (23–32 oz)	
Plumage Sexes differ	
Location N.W. North America, N. Europe, N.E. Asia	**Migration** Migrant
	Status Vulnerable

This is the smallest of the 4 eider species, and the only one with an orange-rust breast and underparts (breeding males only). Its grey bill, which has soft, flap-like margins on

each side, is probably adapted for scraping invertebrates off rocks, or may serve as a tactile device when foraging in deep, dark waters. It lives entirely in the open sea except during the breeding season, when it may be observed in coastal waters, just before building its nest in low, grassy tundra.

Hymenolaimus malacorhynchos
Blue duck

Length 53 cm (21 in)	
Weight 775–900 g (28–32 oz)	
Plumage Sexes alike	
Location New Zealand	**Migration** Non-migrant
	Status Vulnerable

Unique among waterfowl for its slate-blue plumage, this duck is highly adapted for life in cold, clear, rapidly flowing streams, where few other birds can survive. Its main food is the

aquatic larvae of insects such as caddisflies, mayflies, stoneflies, and midges, which it obtains by diving. It also feeds by dabbling at the surface, up-ending, and foraging around the river-bed rocks and boulders, where it uses its unique bill to scrape wet algae from the rocks.

Merganetta armata
Torrent duck

Length 43–46 cm (17–18 in)	
Weight 450 g (16 oz)	
Plumage Sexes differ	
Location W. South America	**Migration** Non-migrant
	Status Lower risk†

This highly streamlined, Andean duck never leaves rapid mountain streams, a habitat that few other ducks are able to exploit owing to the swiftness of the currents – the blue duck of New

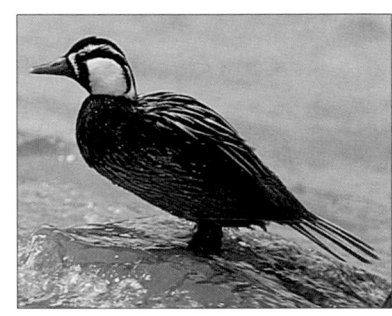

Zealand (see above) is its closest ecological counterpart. It dives and swims upstream, remaining very close to the bottom, probing under rocks and among stones for insect larvae and pupae. Females are rufous, while males are black and white.

Melanitta fusca
Velvet scoter

Length 51–58 cm (20–23 in)	
Weight 1.5–2 kg (3 1/4–4 1/2 lb)	
Plumage Sexes differ	
Location N.W., W., and E. North America, Europe, W., N., and E. Asia	**Migration** Migrant
	Status Common

The velvet scoter, also called the white-winged scoter, spends much of the year at sea, although it breeds on fresh water inland, and sometimes winters on large lakes. It forages by diving to depths of up to 7m (23ft), often remaining underwater for nearly a minute, or longer in deeper waters. It is a stout duck, with a heavy bill adapted for eating molluscs. Its large nostrils may be related to well-developed glands that excrete salts from any sea water that it swallows.

Mergus merganser
Goosander

Length 58–66 cm (23–26 in)	
Weight 1.5–2 kg (3 1/4–4 1/2 lb)	
Plumage Sexes differ	
Location North America, Europe (including Iceland), Asia	**Migration** Partial migrant
	Status Common

One of the swiftest of all waterfowl, and the largest of the 7 species of fish-eating mergansers, this duck has a streamlined body that is adapted for rapid underwater swimming and flight – its air speed may approach 100kph (60mph). It has a long, narrow, serrated bill to catch and hold slippery fish, which form the main part of its diet. Females are mostly grey above, while males have black backs.

rufous head in female

BIRDS OF PREY

PHYLUM	Chordata
CLASS	Aves
ORDER	Falconiformes
FAMILIES	5
SPECIES	307

OF ALL BIRDS, these are the chief carnivores and the most accomplished predators. Although many other birds feed on living animals, the birds of prey are set apart by their remarkably acute eyesight, muscular legs, and sharp bill and talons – as well as by their aerial ability and sophisticated hunting techniques. This large group (sometimes also referred to as raptors) includes eagles, hawks, vultures, buzzards, ospreys, and falcons, among others. Most birds of prey hunt by day, which distinguishes them from owls. As a group, raptors feed on a large assortment of living animals, from worms and snails to fish, reptiles, amphibians, mammals, and other birds. They also eat carrion. Birds of prey are found almost worldwide but are most common in open country in warm parts of the world.

HUNTING

Catching live prey takes much practice and often has a high failure rate. Adult birds are usually better hunters than young ones, but even adults miss more often than they strike. Some species vary their technique according to their prey – buzzards, for example, hover over small mammals but search for earthworms on the ground. Others, such as the osprey, are more specialized (see below). Most birds of prey hunt alone, but in a few species (including some eagles) pairs work together, one bird flushing out the prey and the other striking.

FISHING
The osprey feeds almost entirely on fish and has a unique catching technique. It approaches in a low-angled dive and throws its long legs forward to grab a fish with its talons. To help it grip, the osprey may reverse one of its 3 forward-pointing toes.

ANATOMY

Birds of prey range in size from falconets, no larger than a sparrow, to condors, which have a wingspan of up to 3.2m (10ft). Some are broad-winged and heavy, while others are slight and streamlined. Most species have a large head and short neck, although vultures have a long, bare neck that allows them to reach inside carcasses. One of the most distinctive features of the group is the bill: in almost all species, it is powerful and hooked, with sharp edges for tearing flesh. The exact form of the bill varies, reflecting differences in diet (see below, right). A bird of prey's other main tool is its feet, which are strong and muscular with long, sharp claws known as talons (see below, left). The plumage of most species is in subdued colours (such as brown, grey, black, or blue), often combined with white.

thick, scaly skin

sharp, curved talon

long, backward-pointing talon

TALONS
Most birds of prey kill with their talons rather than their bill. The talons pierce vital organs and break slender bones. Three of the 4 toes point forwards, and one points backwards. In many species (including this white-tailed sea eagle), the backward-pointing claw is the longest and sharpest. It gives the foot a powerful pincer-like action for holding and crushing prey that can be difficult for the bird to release.

hooked bill

BALD EAGLE

tooth-like ridge

HAWK

long, hooked bill

SNAIL KITE

BILLS
The distinct hook and sharp edges of a bald eagle's bill are typical of many birds of prey. Variations on this shape can be seen in other species. Hawks and falcons have a tooth-like ridge in the upper bill for breaking the spine of their prey, while the snail kite uses its long, hooked bill to pull snails from their shells.

SENSES

Birds of prey hunt mainly by sight, and they are renowned for their acute vision. It is estimated that they can see at least 4 times as much detail as a human, an ability that helps them to pick out prey from a great distance. This arises from having a high concentration of cone cells in the retina, and also from having relatively large eyes. Some species have excellent hearing, notably the harriers, which hunt over thick vegetation. A few vultures also have an exceptional sense of smell, by which they can locate hidden carrion.

deep eye cavity

forward-facing eyes

EYE PROTECTION
Like many birds of prey, the goshawk has a ridge over each eye, which provides shade and protects the eye from struggling prey. In many species, the eye is also protected by a transparent third eyelid (nictitating membrane).

BIRDS

FEEDING

Most birds of prey feed on live animals; the larger the species, the larger the prey that they are able to carry. The bald eagle, for example, is capable of taking a deer fawn if it is not too heavy. Many birds of prey are specialized feeders, including the honey buzzard (which eats wasps and their grubs), and the secretary bird (which eats snakes). As well as live prey, many species also eat carrion from time to time. A few, such as vultures and kites, feed almost entirely on dead animals. One carrion-feeder, the lammergeier, eats bones, breaking the larger ones by dropping them onto rocks. The palm-nut vulture, named after its favourite food, is unusual in that it feeds mainly on plants.

CARRION-FEEDERS
Up to 6 species of vulture may gather around a carcass at one time, each feeding on different parts of the body. For example, some may feed on soft body parts while others eat the skin and hide. The lappet-faced vulture (left) is one of the largest species and is often dominant over others.

FEEDING PERCHES
Rather than eating in flight, many birds of prey carry their kill to a favoured perch, where they carefully remove all unpalatable parts before eating it. This European sparrowhawk (above) is plucking a blackbird.

FLIGHT

All birds of prey are able fliers. They use different patterns of flight to suit the way they hunt. Using their large, broad wings to ride thermals or updrafts next to cliffs, vultures and condors can stay aloft for hours, barely flapping their wings and using their high vantage point to search for carrion. In contrast, the slim-bodied falcons attack other birds in mid-air, manoeuvring on long, thin, pointed wings and sometimes diving (or stooping) from a great height (see below). Hawks plan their strikes and then use a burst of speed to surprise or ambush prey. Harriers fly slowly forwards over the ground and then drop onto unsuspecting prey. Kestrels and buzzards can hover in one place, watching for movement below and then lunging downwards, talons first.

AERIAL KILLER
One of the most distinctive features of birds of prey is the way they kill with their feet, seizing prey from the ground, water, or air. Steller's sea eagle (seen here) uses its feet to catch Pacific salmon. It also hunts geese, hares, and young seals.

SOARING
Condors and vultures have the largest wings of all birds of prey (the lammergeier, above, has a wingspan of up to 3m/9¾ft). Both are common in mountainous areas, where rising currents of air help to keep them aloft.

CONSERVATION

Many bird of prey species are threatened. All falcons are listed by CITES, and the California condor is one of the world's most endangered animals, with a population of just 170. Birds of prey, such as the peregrine falcon (left), are vulnerable because they are at the top of the food chain, where they are sensitive to all the factors affecting their prey. Some species are routinely killed by humans because of the perceived threat that they pose to livestock.

STOOPING
Most large falcons live in open country with little cover to conceal them from their prey. To catch birds that could evade them in flapping flight, they often climb to a high altitude and then dive or stoop on their target. When diving, the peregrine falcon (right) can reach a speed of 200kph (125mph).

BIRDS

Cathartes aura
Turkey vulture

Length 64–81cm (25–32in)	
Weight 0.85–2kg (1 3/4–4 1/2lb)	
Plumage Sexes alike	
Location S. Canada to S. South America	**Migration** Partial migrant
	Status Common

Often called a "buzzard" in the USA, the turkey vulture is found in an exceptionally wide range of habitats from southern Canada through to Tierra del Fuego. This vulture and its close relatives are the only birds of prey with a good sense of smell, enabling them to locate food even in thick jungle. It soars and glides as soon as the day has warmed up, and often feeds on road kills, possibly explaining its increase in populated areas. Although little

is known about its courtship behaviour, this bird has been seen performing a ritualized dance on the ground. The female lays 2 eggs and both parents bring food to the young, which are fed by regurgitation from the parents' crops. Once past their first year, the young birds are likely to live for 12–17 years.

long, broad wings for soaring

bald head and neck

strong, bare legs

Gymnogyps californianus
California condor

Length 1.2–1.3m (4–4 1/4ft)	
Weight 8–14kg (18–31lb)	
Plumage Sexes alike	
Location W. USA (California, Arizona)	**Migration** Non-migrant
	Status Critically endangered

The California condor is the largest flying bird in North America, and also one of the most endangered. Once found from California to Florida, by 1987 the species no longer existed in the wild, but a captive breeding programme (see panel, below) is slowly helping it to recover. California condors are expert at soaring and spend much of their time circling high in the air. At one time, they fed mainly on the remains of bison and pronghorns, but with the arrival of

neck ruff

BLACK PLUMAGE
The California condor has mainly black plumage with white tips to the smaller wing feathers and a black neck ruff. The head and neck lack feathers.

Europeans, this abundant food source rapidly declined. Today, captive-bred condors that have been released feed mainly on dead cattle and deer.

CONSERVATION

Through the California Condor Recovery Program, a captively bred population of California condors – numbering just 27 in the 1980s – has grown to about 170 individuals, 40 of which live in the wild. Released birds have had mixed fortunes, but the aim of the programme is to re-establish the California condor as a wild breeding bird, a goal that is still some years away.

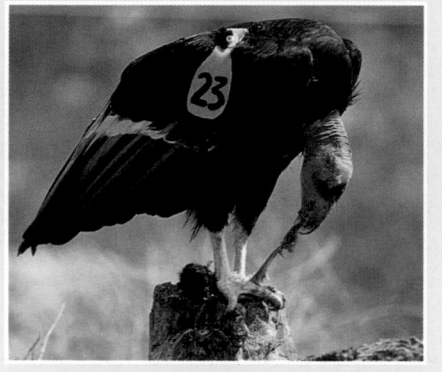

Vultur gryphus
Andean condor

Length 1–1.3m (3 1/4–4 1/4ft)	
Weight 11–15kg (24–33lb)	
Plumage Sexes differ	
Location W. South America	**Migration** Non-migrant
	Status Lower risk

The Andean condor has the largest wing area of any bird. Because of its immense size, it relies on updraughts from mountains and coastal cliffs to remain airborne, and can travel huge distances with only the occasional flap of its wings. From the ground, its silhouette is highly distinctive, with its large flight feathers spreading out like the fingers of a hand. The males are larger than the females – unusually for birds of prey – and while both sexes have bald heads, the males also have a characteristic white ruff at the base of their

necks. Andean condors feed mainly on carrion, and soar at altitudes of up to 5,500m (18,000ft) to search for food. Their diet is wide ranging, and includes not only mountain animals, but also the remains of stranded marine animals, such as seals and whales. In some places, they also feed at seabird colonies, plundering large numbers of eggs. They breed on inland cliffs and reproduce slowly, typically laying a single egg every 2 years.

outspread flight feathers

white neck ruff on male

Pandion haliaetus
Osprey

Length 1.5–1.7m (5–5 1/2ft)	
Weight 1.5–2kg (3 1/4–4 1/2lb)	
Plumage Sexes differ	
Location Worldwide (except Antarctica)	**Migration** Migrant
	Status Common

One of the most widespread birds of prey, the osprey is found on every continent except Antarctica. It has white underparts and chocolate-brown upperparts, with a dark stripe through

the eye, and strongly curved talons and bill. It plucks live fish from the water during a spectacular, feet-first dive, sometimes totally submerging itself. Its feet have evolved to hold onto slippery prey, with spiny, dry scales on the undersides and a reversible outer toe for carrying fish head-on through the air.

Milvus milvus
Red kite

Length 61–66cm (24–26in)	
Weight 750–1,000g (27–36lb)	
Plumage Sexes alike	
Location Europe, W. Asia, N. Africa, Canary and Cape Verde islands	**Migration** Partial migrant
	Status Vulnerable†

Kites are agile fliers, typically with long legs and forked tails. The red kite is the largest species, feeding mainly on small mammals, carrion, and young birds. It flies with its wings partly bent, and uses its tail like a rudder, constantly tilting it as it searches for food. Two centuries ago, the red kite was a common scavenger in European

towns and cities, but improved refuse disposal has made it much less common. It nests in trees, often using old nests of crows, laying 2–3 eggs each year.

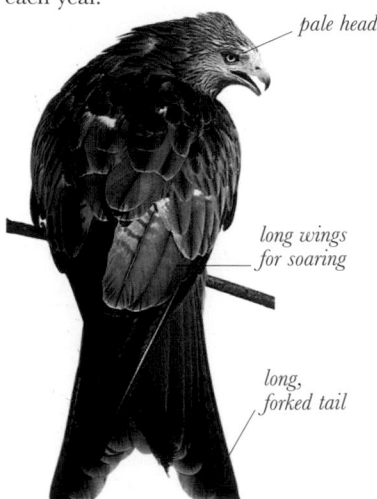

pale head

long wings for soaring

long, forked tail

Pernis apivorus
Western honey buzzard

Length 52–60 cm (20 1/2 –23 1/2 in)	
Weight 450–1,000 g (16–36 oz)	
Plumage Sexes alike	
Location Europe, W. and C. Asia, Africa	**Migration** Migrant
	Status Common

Compared to similarly sized birds of prey, the western honey buzzard has small feet, relatively undeveloped talons, and a shallow, curved bill. This reflects its unusual lifestyle as a predator of wasps and other insects. It catches its food in mid-air, and also follows insects back to their nests, to feed on their developing grubs. Because it depends on insects, it has to migrate southwards in winter.

shallow bill

relatively long tail

Rostrhamus sociabilis
Snail kite

Length 40–45 cm (16–18 in)	
Weight 350–400 g (13–14 oz)	
Plumage Sexes differ	
Location S.E. USA (Florida), Cuba, Central America, South America	**Migration** Partial migrant
	Status Locally common

Formerly called the Everglades kite, this bird is a specialist feeder on aquatic snails. It flies slowly over shallow marshland, snatching up surface-feeding snails. It inserts its fine, heavily curved, hooked bill into the snail and cuts the muscle that attaches it to its shell, allowing it to remove the animal's body. Adult males are dark grey with black primaries; females are brown with streaked buff underparts. Rare in its south Florida range, it can be abundant elsewhere, as on the Argentine pampas.

Haliaeetus vocifer
African fish eagle

Length 63–73 cm (25–29 in)	
Weight 2–3.5 kg (4 1/2 –7 3/4 lb)	
Plumage Sexes alike	
Location Africa (south of Sahara)	**Migration** Non-migrant
	Status Common

Known as the "voice of Africa", this eagle throws its head back and gives a loud, yelping cry. With its distinctive white head, it is easy to spot perching on high branches beside rivers. When it locates a fish, it swoops down to catch it with a backward swing of the feet. It will also eat small mammals, birds, and carrion, as well as steal fish from other eagles and fishermen.

black wings

white tail

Gypaetus barbatus
Lammergeier

Length 1–1.2 m (3 1/4 –4 ft)	
Weight 4.5–7 kg (10–15 lb)	
Plumage Sexes alike	
Location Europe, Asia, N., E., and southern Africa	**Migration** Non-migrant
	Status Locally common

The lammergeier is one of the largest Old World vultures, with huge wings, a conspicuous wedge-shaped tail, and distinctive beard-like feathers at the base of its bill. It eats carrion, but it also specializes at feeding on bones, carrying large ones high into the air and then dropping them onto rocks below. Once the bones have shattered, the lammergeier lands to feed on the marrow inside.

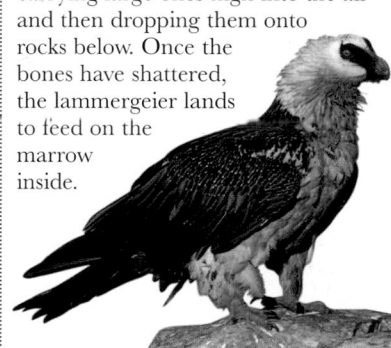

Neophron percnopterus
Egyptian vulture

Length 58–70 cm (23–28 in)	
Weight 1.5–2 kg (3 1/4 –4 1/2 lb)	
Plumage Sexes alike	
Location Europe, Africa, Asia	**Migration** Partial migrant
	Status Common

This is the smallest Old World vulture, with a much wider distribution than its name suggests. Adults have yellow faces and off-white plumage, apart from their black flight feathers; juveniles are speckled brown (shown here). Egyptian vultures scavenge all kinds of food, but are renowned for using stones to break open the eggs of ostriches and other birds. Because of their small size, they have difficulty competing with other vultures at carcasses, and are often the last to feed.

long, thin bill

off-white plumage

black flight feathers

Gyps africanus
African white-backed vulture

Length 94 cm (37 in)	
Weight 4–7 kg (8 3/4 –15 lb)	
Plumage Sexes alike	
Location Africa (south of Sahara)	**Migration** Non-migrant
	Status Common

This widespread African scavenger is a griffon vulture – one of a group of 7 species which have long necks that appear to be bald, but that are actually covered in fine down. This absence of large neck feathers allows griffon vultures to reach deep into carcasses without becoming soiled. The African white-backed vulture gets its name from a collar of white feathers at the top of its back, which contrasts with its grey neck. This species is one of Africa's commonest vultures, and is often seen in large numbers where food can be found. It makes a variety of hissing and cackling sounds as it jostles for a chance to eat.

FEEDING

White-backed vultures are legendary for their ability to find food. They have a poor sense of smell, but extremely good eyesight, enabling them to spot dead remains from high in the air. Vultures also keep an eye on each other. If one bird sees food and makes a sudden descent, others quickly follow suit.

white primary feathers

collar of feathers

downy head and neck

LARGE WINGS
The large, broad wings of the African white-backed vulture enable it to soar and circle on thermals for hours, looking for carrion.

AMERICAN ICON
The bald eagle, whose name derives from the conspicuous appearance of its white-feathered head, has been the national bird of the USA since 1782. The only eagle solely native to North America, the bald eagle has been a protected species since 1940.

Haliaeetus leucocephalus

Bald eagle

Length	71–96 cm (28–38 in)
Weight	3–6.5 kg (6½–14 lb)
Plumage	Sexes similar

Location North America

Migration Partial migrant

Status Vulnerable†

The majestic bald eagle is a large, powerfully-built bird, with a wingspan of up to 2.5m (8¼ft). Although often found well away from water when wintering, bald eagles are commonly seen close to lakes, rivers, and coastal areas, where they have ample access to fish. Bald eagles pair for life, and while sedentary birds stay together throughout the year, those that migrate separate then come together again at breeding sites. They reinforce their pair bond through often spectacular flight displays involving undulating flight, swooping at one another, and cart-wheeling through the air with clasped feet. Together they build a very large nest (in a tree or sometimes on the ground), reaching up to 4m (13ft) tall. Bald eagles start to breed at about 5 years of age, and usually 2 or 3 eggs are laid. Although the young are cared for by both parents for many weeks, a high percentage do not survive their first year.

white head and neck

large yellow bill

dark brown wings

strong talons

white tail feathers

DISTINCTIVE FEATURES
The bald eagle is easily spotted by its pure white head and tail, its broad, brown-black wings, and its large yellow bill. Juveniles take up to 5 years to gain full adult plumage.

FEEDING HABITS

The bald eagle takes food from a range of sources, live and dead, including small birds, carrion (especially during the winter months), and fish. When hunting for fish, it does not usually enter the water like the osprey (see p.288), but instead searches for dead or dying fish or those that live near the surface.

FISHING TECHNIQUE
The bald eagle uses its strong feet equipped with sharp claws to snatch fish from the water's surface.

FOOD FIGHT
As well as piratically stealing food from other predatory birds such as ospreys, bald eagles often fight between themselves over prey items.

WINTER DIET
In winter, bald eagles sometimes collect in large groups close to where salmon come to spawn.

BIRDS

Black-breasted snake eagle

Circaetus pectoralis

Length 65 cm (26 in)	
Weight 1–2.5 kg (2 1/4–5 1/2 lb)	
Plumage Sexes alike	

Location E. to southern Africa

Migration Non-migrant

Status Common

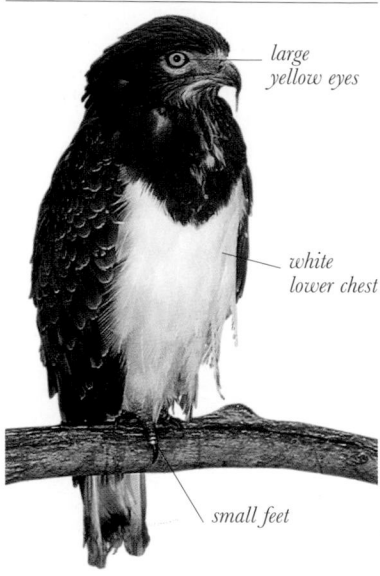

large yellow eyes

white lower chest

small feet

A specialist hunter of reptiles, the black-breasted snake eagle has tightly meshed scales on its legs and toes to protect it from biting snakes. It has a large head with large yellow eyes. The bird has strong legs, bare of feathers, and small feet. It is often found soaring over open hill slopes as it searches for prey, and will occasionally hover. It feeds primarily on snakes – usually avoiding poisonous ones – but will also take lizards, birds, bats, and even fish. The female lays one egg and incubates it for 48 days. Both parents care for the young; once fledged, the young is dependent for up to 6 months.

Bateleur eagle

Terathopius ecaudatus

Length 60 cm (23 1/2 in)	
Weight 2–3 kg (4 1/2–6 1/2 lb)	
Plumage Sexes differ	

Location Africa (south of Sahara)

Migration Non-migrant

Status Common

The bateleur is a colourful eagle with a chestnut-red mantle on its back, a black body and head, and a bare, red or orange face. Females have grey shoulders and black-edged, white secondary flight feathers; males have black secondaries. The long, pointed, almost falcon-like wings and the short tail make this bird unmistakable in flight – it tilts from side to side like a tightrope walker trying to balance. The bateleur is mainly a scavenger, but will also take small mammals, birds, reptiles, fish, eggs, and insects. Groups of these eagles may gather at termite mounds when the insects fly from their nests to breed. During courtship, the bird displays a rocking and rolling flight, almost stopping in mid-air with the wings held open. It builds its large stick nest in an open-branched tree.

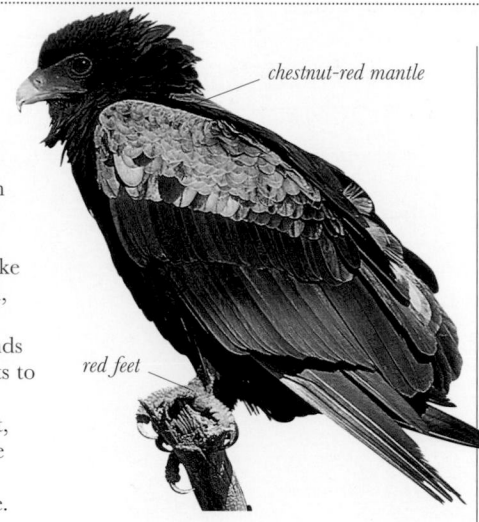

chestnut-red mantle

red feet

Hen harrier

Circus cyaneus

Length 43–52 cm (17–20 1/2 in)	
Weight 350–525 g (13–19 oz)	
Plumage Sexes differ	

Location North America to N. Central America, Europe, Asia

Migration Migrant

Status Common

Instead of soaring, the hen harrier flies close to the ground with its wings held in a shallow "V". It has keen eyesight and good hearing, which it uses to locate small animals hidden in vegetation. The female (shown above) is larger than the male, and is brown, with a white rump, whereas the male is grey. Unusually for a bird of prey, the hen harrier nests on the ground, making a nest of sticks and grass.

Northern goshawk

Accipiter gentilis

Length 48–70 cm (19–28 in)	
Weight 1–1.5 kg (2 1/4–3 1/4 lb)	
Plumage Sexes alike	

Location Canada to Mexico, Europe, Asia

Migration Partial migrant

Status Common

A high-speed hunter of woods and forests, the northern goshawk varies considerably across its very wide range. Asian birds are usually pale, while North American ones often have dark heads. Females are always larger than males – sometimes weighing up to half as much again – and young birds are brown, eventually turning grey. Northern goshawks feed on birds up to the size of crows and pigeons, and mammals up to the size of small hares. They often hunt from a perched position at the forest edge, and are shaped for pursuit at close quarters, with short, rounded wings, and a long tail that is used for steering and braking. Apart from these sudden sorties after their prey, they are secretive, and often go unseen. When courting, prospective partners call loudly as they soar into the air, and once their nest is built, they are extremely territorial, driving any intruding birds away. In cold winters, when food is scarce, they are sometimes found well to the south of their normal range.

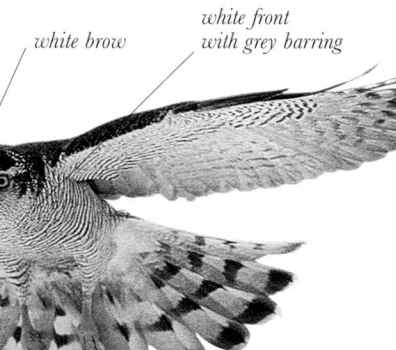

white brow

white front with grey barring

rounded wings

long tail

Common buzzard

Buteo buteo

Length 50–57 cm (20–22 1/2 in)	
Weight 525–1,000 g (19–36 oz)	
Plumage Sexes alike	

Location Europe, Asia, N. and E. to southern Africa

Migration Partial migrant

Status Common

This medium-sized raptor, also known as the Eurasian buzzard, has large, broad wings and a shortish tail, and is built for soaring on thermals. It is very variable in colour. Although large, the common buzzard catches relatively small prey such as voles, mice, and insects. It is often seen on the ground, especially in winter, feeding on insects and earthworms. During courtship, the buzzard displays a high, soaring flight with spectacular climbs and stoops, and the male passes nesting material to the female in mid-air.

large eyes

large, broad wings

Galapagos hawk

Buteo galapagoensis

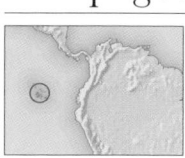

Length 55 cm (22 in)	
Weight 650–850 g (23–30 oz)	
Plumage Sexes alike	

Location Galapagos Islands

Migration Non-migrant

Status Vulnerable

This is the only diurnal bird of prey on the Galapagos Islands. It is also one of the few bird species that have cooperative breeding habits – females that have extra males helping them are more successful in producing young. The Galapagos hawk is sooty brown all over, with a grey-barred tail. The female is considerably larger than the male. This hawk hunts mainly from the air, gliding in to take prey, but can also hover. It feeds on small mammals, birds, reptiles, and insects. Both parents care for the young – usually one – and will allow it to remain close for up to 4 months before driving it away.

Aquila chrysaetos
Golden eagle

Length 75–90 cm (30–35 in)	
Weight 3–6.5 kg (6½–14 lb)	
Plumage Sexes alike	

Location North America, Europe, Asia, N. Africa

Migration Partial migrant

Status Locally common

With a wingspan of up to 2.3m (7½ft), this striking bird is one of the largest land eagles in the northern hemisphere. Its plumage is generally dark brown, but it gets its name from the tawny or gold feathers on its nape and crown. The golden eagle is skilled at soaring, but it generally catches its prey by cruising low down across suitable terrain. It breeds on cliff ledges and in tall trees, making platform nests up to 2m (6½ft) across. Widely persecuted in the past – through the mistaken belief that it attacks livestock – the golden eagle is now protected in many countries.

FEEDING THE YOUNG

This eagle has a wide-ranging diet, including carrion and living prey, such as rabbits, squirrels, and grouse, and even tortoises in the southern parts of its range. Parents feed their chicks, by tearing food up into strips, for several months after they have fledged. Young birds take 4–5 years to mature, but mortality is high, especially in the first 12 months.

BROAD WINGS
This eagle has broad wings for soaring, and "slotted" flight feathers, spread like the fingers of a hand, for increased lift. It soars at a considerable height, flapping its wings as little as possible, surveying the land below for food.

large talons

Pithecophaga jefferyi
Philippine eagle

Length 86–100 cm (34–39 in)	
Weight 4.5–8 kg (10–18 lb)	
Plumage Sexes alike	

Location Philippines

Migration Non-migrant

Status Critically endangered

Threatened by forest clearance and by hunting, this gigantic eagle is one of the world's rarest birds of prey. It feeds on a wide variety of animals, including monkeys, snatching them in a low-level attack. Attempts are being made to breed the eagle in captivity, but with a rapidly shrinking habitat, its future looks uncertain.

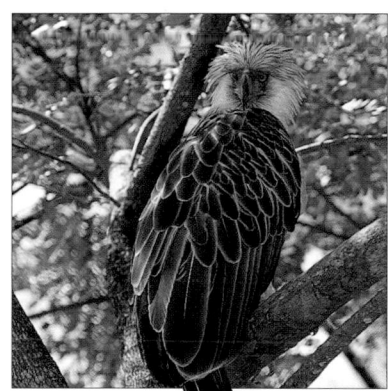

Aquila audax
Wedge-tailed eagle

Length 81–100 cm (32–39 in)	
Weight 2–5.5 kg (4½–12 lb)	
Plumage Sexes alike	

Location S. New Guinea, Australia (including Tasmania)

Migration Non-migrant

Status Common

This eagle is Australia's largest bird of prey, with dark brown plumage, and a long, graduated tail that gives it a distinctive silhouette in flight. It feeds on a wide range of animals, including other birds and rabbits, and it also eats carrion, taking the place of vultures in a continent that has no vultures of its own. Wedge-tailed eagles usually nest in trees, lining their nests with leaves.

Harpia harpyja
Harpy eagle

Length 89–100 cm (35–39 in)	
Weight 4–9 kg (8¾–20 lb)	
Plumage Sexes alike	

Location S. Mexico to C. South America

Migration Non-migrant

Status Lower risk

In habitat and lifestyle, this immense bird closely matches the Philippine eagle (see above), although it lives in a quite different part of the world. Despite its great size, it is unusually agile, which enables it to steer through the treetops to catch its prey. Sloths make up about a third of its food by weight, but it also eats a wide range of other animals, including snakes, lizards, and macaws. Harpy eagles make stick nests high up in old trees, and they often perch on exposed branches, watching the forest canopy below for signs of prey.

POWERFUL PREDATOR
The harpy eagle has grey, black, and white plumage and a long, barred tail. It has a double crest, a strong bill, and very thick, powerful legs and feet.

Polemaetus bellicosus
Martial eagle

Length 78–86 cm (31–34 in)	
Weight 3–6 kg (6½–13 lb)	
Plumage Sexes alike	

Location Africa (south of Sahara)

Migration Non-migrant

Status Vulnerable†

This is the largest of the African eagles, and one of the biggest eagles in the world. It can be found in a wide variety of open habitats, from semi-desert and steppes to foothills, and also in moderately forested areas. Its prominent brow gives it a menacing expression. The abdomen is white with grey to black speckling. The martial eagle soars for extended periods in good weather, searching for prey such as gamebirds, hares, hyraxes, small antelopes, monitor lizards, and other medium-sized vertebrates.

CONSERVATION

The population of the harpy eagle is declining as a result of habitat destruction. It needs a very large area to hunt, which makes it vulnerable when a continuous stretch of forest is divided into isolated tracts. Young birds are now being radio-tracked via satellite, to determine their range and the space they need to survive.

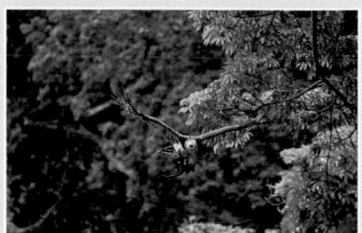

BIRDS

Sagittarius serpentarius
Secretary bird

Length 1.3–1.5 m (4 1/4 –5 ft)	
Weight 2.5–4.5 kg (5 1/2 –10 lb)	
	Plumage Sexes alike
Location Africa (south of Sahara)	**Migration** Non-migrant
	Status Common

With its stork-like legs, long, wedge-shaped tail, and crest of black feathers on the back of its head, the secretary bird is unlike any other raptor. The crest resembles a number of quill pens, as used in the past by secretaries. The flight feathers are black, and the rest of the body plumage is grey. The powerful legs are used for striking prey and running after faster quarry. The secretary bird walks up to 24 km (15 miles) a day through grassland searching for prey. Food includes grasshoppers and other large insects, small mammals, frogs, snakes, lizards, and tortoises. The bird flushes out the prey by stamping on tufts of grass. It then runs after and catches moving items, dealing them repeated blows with its strong feet. The wings act as shields when the bird attacks snakes.

crest

orange face

long, powerful legs

long tail feathers

Caracara plancus
Crested caracara

Length 49–59 cm (19 1/2 –23 in)	
Weight 0.85–1.5 kg (1 3/4 –3 1/4 lb)	
	Plumage Sexes alike
Location S. USA, Caribbean, South America	**Migration** Non-migrant
	Status Locally common

Also called the common caracara, this striking bird of prey is dark brown with a cream head, breast, and upper back, all finely barred, and a bare, orange face. It spends much of its time walking on the ground. The crested caracara is an opportunistic scavenger, and will often dig for food or chase other birds, including raptors and vultures, to steal it. It eats almost anything, from carrion, eggs, chicks, and frogs to road kills, rotting vegetables, dead and dying fish, worms, and insects.

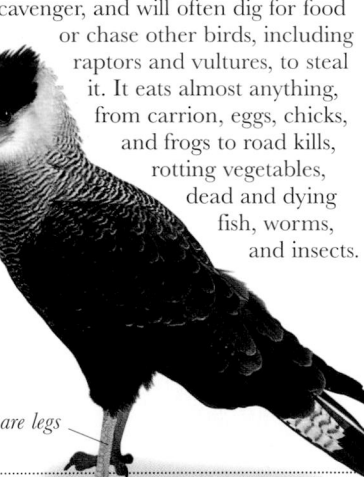

bare legs

Micrastur ruficollis
Barred forest falcon

Length 33–38 cm (13–15 in)	
Weight 150–225 g (5–8 oz)	
	Plumage Sexes alike
Location S. Mexico to N. South America	**Migration** Non-migrant
	Status Locally common

Compared to other falcons, forest falcons have relatively short, rounded wings – a shape that helps them to manoeuvre as they hunt among trees.

Falco tinnunculus
Common kestrel

Length 32–39 cm (12 1/2 –15 1/2 in)	
Weight 125–325 g (4–12 oz)	
	Plumage Sexes differ
Location Europe, Asia, Africa	**Migration** Partial migrant
	Status Common

The common kestrel is a small, chestnut-brown falcon with a black tip to its tail as well as black bars and spots. It is one of the few

This species has long and slender legs, with small feet and sharp talons. It hunts from trees, with short dashes out to catch lizards, and also catches army ants on the ground.

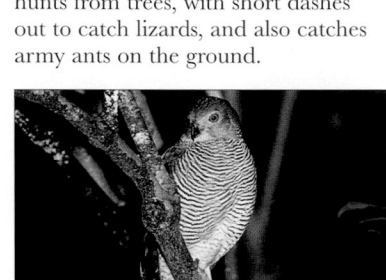

medium-sized birds that can hover for extended periods. This ability allows it to live and hunt in a variety of habitats, especially areas not often frequented by other species, such as the edges of major roads. Its diet consists mainly of small mammals (including voles and mice), insects, and amphibians.

Falco punctatus
Mauritius kestrel

Length 20–26 cm (8–10 in)	
Weight 175–225 g (6–8 oz)	
	Plumage Sexes alike
Location Mauritius	**Migration** Non-migrant
	Status Vulnerable

This chestnut-brown, cream-fronted bird nearly became extinct because of destruction of its natural habitat of evergreen primary forest, but is now being raised in zoos and encouraged to inhabit secondary forest and scrub. In the 1970s, there were thought to be only 4–8 birds left, but, due to a successful breeding programme, numbers now stand at over 500.

Falco eleonorae
Eleonora's falcon

Length 36–42 cm (14–16 1/2 in)	
Weight 350–400 g (13–14 oz)	
	Plumage Sexes alike
Location S. Europe, N. and E. Africa, Madagascar	**Migration** Migrant
	Status Vulnerable†

This fairly small falcon has the typical streamlined falcon shape, but its wings are longer and more swift-like. It has 2 colour forms – normal and melanistic: the normal plumage is dark above and cream with flecks below; the melanistic form is dark all over. This bird breeds in autumn, usually on cliffs of remote islands, feeding mainly on small birds.

Falco peregrinus
Peregrine falcon

Length 34–50 cm (13 1/2 –20 in)	
Weight 0.55–1.5 kg (1 1/4 –3 1/4 lb)	
	Plumage Sexes alike
Location Worldwide (except Antarctica)	**Migration** Partial migrant
	Status Locally common

The peregrine falcon is one of the world's fastest birds. It also has the widest distribution of any day-flying land bird, being found on every continent, except Antarctica, and on many oceanic islands. Females can be up to 30 per cent larger than males. The bird's wings have sharply pointed tips, making them both fast and highly manoeuvrable; it flies largely by flapping rather than soaring. This falcon usually attacks its prey in a steep, powerful dive or "stoop", during which it may reach a speed of 230 kph (145 mph). Courtship involves aerial displays accompanied by noisy calling. The peregrine falcon has been used in falconry for centuries. In the 1950s and 1960s, it was badly affected by DDT pollution, particularly in Europe and the USA, but is now staging a gradual recovery.

AERIAL PURSUIT

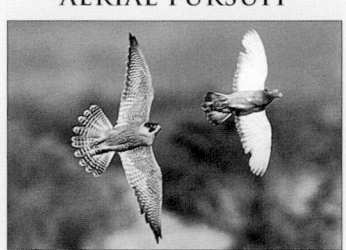

The peregrine falcon may chase its prey – especially doves and pigeons – in order to exhaust them quickly. It usually strikes the prey with its talons, and then follows it to the ground.

PLUMAGE
This falcon has white to cream to rufous underparts, and black, grey, or blue upperparts. The young are brown with buff edging on contour feathers, and vertical stripes on the breast.

pointed wings

GAMEBIRDS

PHYLUM	Chordata
CLASS	Aves
ORDER	Galliformes
FAMILIES	6
SPECIES	281

THIS GROUP OF MAINLY ground-dwelling species includes some of the birds most useful to humans. In their domestic forms (including the chicken), they provide an important food source, and many other species (such as pheasants, partridges, and grouse) are hunted for sport or food. Also in this group are the spectacular curassows and the less conspicuous guineafowl and megapodes. Gamebirds are found almost worldwide (even within the Arctic circle) in a wide range of habitats, including dense forest and high mountains.

DISPLAY
This is the display of the Palawan peacock-pheasant. The eyes on the fan are thought to attract the female.

ANATOMY

Most gamebirds are plump with a small head and short, rounded wings. Their powerful flight muscles are ideal for rapid escape, but are usually unable to support their heavy body for long distances. The short bill is slightly curved, and the feet are stout and strong for scratching and digging for food. Many species have bare areas of coloured skin or long and spectacular tails or crests.

TAKE-OFF
Like most gamebirds, Reeves's pheasant launches itself quickly into the air with a flurry of rapid wingbeats to escape danger.

CAMOUFLAGE
Ground-dwellers, like this female black grouse, need cryptic plumage to avoid detection. Another grouse species, the ptarmigan, changes colour with the seasons.

REPRODUCTION

Most gamebirds nest in a shallow depression in the ground. The megapodes are unusual in that, instead of incubating their eggs, they store them in mounds or burrows, allowing the sun's heat, microbial activity, or geothermal energy to keep them warm. Young gamebirds fly very soon after hatching – often within a week or, in the case of megapodes, within a few hours.

CLUTCH SIZE
Compared with other birds, many gamebirds lay large clutches. Some produce up to 20 eggs (a pheasant's eggs are shown here).

BIRDS

Leipoa ocellata

Mallee fowl

Length 61 cm (24 in)
Weight 2 kg (4 1/2 lb)
Plumage Sexes alike

Location W. and S. Australia	Migration Non-migrant
	Status Vulnerable

The mallee fowl and its close relatives – known as megapodes – are the only birds that do not directly incubate their eggs. Instead, eggs are laid

small head with short bill

heavily spotted flanks

in a giant heap of leaves, sticks, and bark, up to 1.5m (5ft) high and 4.5m (15ft) across, and the heat given off as this decomposes incubates the eggs. During the lengthy incubation period, lasting up to 11 weeks, the parents stay close to the mound and monitor its temperature with their bills. If it gets too hot, they take some of the vegetation away; if too cool, they add more to it. When the young hatch, they dig their way out – fully feathered and able to look after themselves. Male and female mallee fowl form pairs that last for years; they live fairly separately much of the time, but come closer together during the summer breeding season. The mallee fowl lives on a largely vegetarian diet consisting of fruits, buds, and seeds; however, it will also eat invertebrates such as ants, beetles, spiders, and cockroaches.

Ortalis motmot

Little chachalaca

Length 38 cm† (15 in)
Weight 600 g (21 oz)
Plumage Sexes alike

Location N. South America	Migration Non-migrant
	Status Locally common

This species is the smallest member of the Cracidae – a family of gamebirds from the Americas that also includes the curassows. Compared to these, chachalacas are slim-bodied and plain in colour, and they do not have crests. There are about 12 species and they are all noisy birds – their name comes from their call. They live mostly in trees, and feed on berries and other fruits.

Crax daubentoni

Yellow-knobbed curassow

Length 90 cm (35 in)
Weight Not recorded
Plumage Sexes alike

Location N. South America	Migration Non-migrant
	Status Lower risk

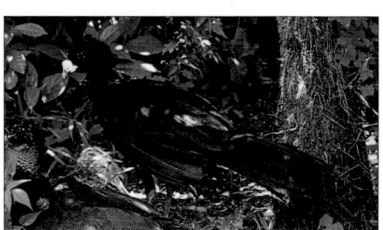

Like other curassows, this large forest bird feeds mainly on the ground, but it flies up into trees if threatened. Its most striking features are its crest, made of feathers that curl forwards, and the fleshy yellow knob at the base of its bill. It eats fruits, leaves, seeds, and small animals. Unusually for gamebirds, curassows nest off the ground, with both sexes helping in the construction. The female lays just 2 eggs – a tiny clutch compared to those of many ground-nesting gamebirds.

Meleagris gallopavo
Common turkey

	Length 1.2 m (4 ft)
	Weight 10 kg (22 lb)
	Plumage Sexes differ
Location North America	Migration Non-migrant
	Status Common

The common turkey is a large game-bird with bronze, iridescent plumage and a naked head with conspicuous blue and red fleshy ornaments. The male has a "beard" of hair-like feathers on the upper breast. For much of the year, this turkey is seen in groups of around 20, but in the breeding season the males set up individual territories. Each male mates with a number of females, courting them with tail fanned, wings spread low, and head held high, giving the characteristic gobbling call. The common turkey is omnivorous: chicks take up to 4,000 insects per day; adults eat seeds, herbs, roots, buds, and flowers, as well as insects. To defend itself, it pecks with its bill, scratches with its claws, and buffets with its wings.

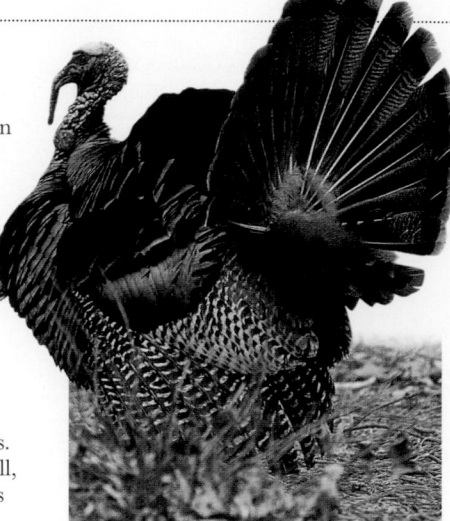

Callipepla californica
California quail

	Length 25 cm (10 in)
	Weight 175 g (6 oz)
	Plumage Sexes differ
Location S.W. Canada, W. USA, N.W. Mexico	Migration Non-migrant
	Status Common

A long, black, teardrop-shaped crest, black and white facial feathers, and scaly plumage on the belly are distinctive features of this elegant gamebird. The female is smaller than the male, has a smaller crest, and is generally drabber in appearance. Shy and elusive, the California quail is more often heard than seen. It lives in small flocks of usually 25–30, and eats a variety of seeds and bulbs, as well as leaves and buds.

scaly plumage

Tetrao urogallus
Western capercaillie

	Length 80–115 cm (32–45 in)
	Weight 4–4.5 kg (8¾–10 lb)
	Plumage Sexes differ
Location N., W., and S. Europe, W. to C. Asia	Migration Non-migrant
	Status Lower risk†

A heavy-bodied grouse that lives in and on the edges of forests, the western capercaillie relies almost entirely on pine needles for winter food. In summer, it feeds on leaves, buds, and berries. As well as feeding on the ground, it flies up into trees for food. Males have black plumage with blue-green iridescence; females are much smaller, with mottled plumage in black, grey, and buff. Up to three-quarters of the year's nestlings die during their first winter, probably because they lack experience in coping with cold and finding food.

COLOURS AND MARKINGS
The male has a glossy green breast, and dark brown wings with variable amounts of white blotching. There is bare red skin above the eye.

COURTING GROUND

Like many other grouse, male capercaillie congregate at traditional display grounds (leks) to compete for mates. Each male struts around in a circle, pointing his head upwards, and fanning his wings and tail feathers. They have an unusual vocal-ization during displays, including a sound like a bottle being uncorked.

Lagopus lagopus
Willow grouse

	Length 38 cm (15 in)
	Weight 550–700 g (20–25 oz)
	Plumage Sexes differ
Location N. North America, N. Europe, N. Asia	Migration Non-migrant
	Status Lower risk†

This exceptionally hardy gamebird – of which there are 20 subspecies – is well adapted to life in the harsh conditions of the northern winter. Like other grouse, its legs and nostrils have feathers for insulation. It tunnels in snow to keep warm, and – except in the British subspecies – the normally reddish brown plumage turns white in winter, giving the bird excellent camouflage.

Tetraogallus caspius
Caspian snowcock

	Length 60 cm (23½ in)
	Weight Not recorded
	Plumage Sexes differ
Location S.E. Europe, W. Asia	Migration Non-migrant
	Status Lower risk†

Typical of the 5 species of snowcocks, this bird has mottled brown, grey, and white plumage that provides good camouflage against bare rock. It feeds in small flocks, and migrates vertically with the seasons, often descending below the treeline in winter.

Perdix perdix
Grey partridge

	Length 31 cm (12 in)
	Weight 300–450 g (11–16 oz)
	Plumage Sexes differ
Location Europe, W. and C. Asia	Migration Non-migrant
	Status Locally common

The grey partridge is a farmland bird, feeding and nesting among crops as well as in pasture. It has a tawny head and greyish breast, and the male has a conspicuous chestnut horseshoe mark on the abdomen. Coveys of 15–20 birds live together in winter, but late in the season these begin to break up as the males become more aggressive, even fighting with one another. Although changes of mate are common early on, stable pairs soon form, usually of males and females from different coveys.

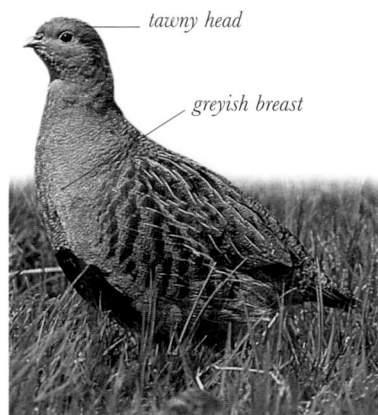

tawny head

greyish breast

Coturnix coturnix
Common quail

	Length 18 cm (7 in)
	Weight 70–150 g (2½–5 oz)
	Plumage Sexes alike
Location Europe, Asia, Africa, Madagascar	Migration Migrant
	Status Locally common

A small and secretive gamebird, this quail is heard much more than it is seen. The first sign of its presence is most often the male's repeated "whit wit-wit" call. During the breeding season, the male and female establish contact by calling, which means they can locate a mate while remaining safely hidden. The common quail roosts on the ground at night in tight groups. Its diet is remarkably varied, and includes seeds, flower buds, leaves, small fruits, and insects and other invertebrates. This is one of the few gamebirds that migrates long distances; those that breed in Europe arrive in spring from Africa.

black and buff streaks on flanks

dull grey belly

Tragopan temminckii
Temminck's tragopan

Length 63 cm (25 in)
Weight Not recorded
Plumage Sexes differ

Location S.E. Asia
Migration Non-migrant
Status Lower risk†

Male gamebirds often have flamboyant plumage, and this species is no exception. However, with Temminck's tragopan, the male's most conspicuous feature is a blue and red throat wattle, which looks like a multicoloured bib. Like the other 4 species of tragopan (all from central and southern Asia), Temminck's is a forest bird, often nesting in bushes or low down in trees, where it makes a simple platform out of sticks. It lives at altitudes of up to 4,500m (14,850ft), and feeds mainly on plants – including young shoots and berries – but also eats insects scratched up from the forest floor.

INFLATED DISPLAY

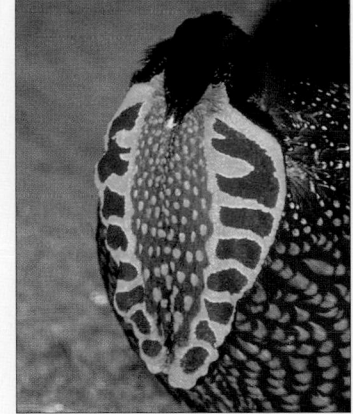

During his courtship display, the male inflates his colourful throat wattle until it covers his breast, and then shakes it to attract the female's attention. If the female is sufficiently impressed, she allows the male to mate.

FORAGER
Like many gamebirds, Temminck's tragopan uses its large feet to clear away leaves and scratch at the soil, exposing small insects that it eats.

colourful throat wattle on male

speckled plumage

strong, stout legs with large feet

Pavo cristatus
Common peafowl

Length 1.8–2.3 m (6–7 1/2 ft)
Weight 4–6 kg (8 3/4 –13 lb)
Plumage Sexes differ

Location S. Asia
Migration Non migrant
Status Locally common

The male peafowl, or the peacock, is one of the world's most spectacular gamebirds, with an iridescent blue body and a long train that is spread out like a fan when courting females. The train is not the true tail, but consists of elongated tail coverts, each ending in a colourful "eye". The female – the peahen – is relatively drab, with a shorter train lacking the "eyes". The female chooses a mate on the basis of his appearance, and he usually turns to face her, shaking the train, erect and fanned, to reinforce its effect. He mates with many females, and gives a loud, "kee-ow" call to advertise his presence. The male plays no part in building nests or raising the young. Like almost all gamebirds, the peafowl feeds on the ground, but it roosts in tall trees at night, safe from most predators.

Gallus gallus
Red jungle-fowl

Length 80 cm (32 in)
Weight 0.5–1.5 kg (1–3 1/4 lb)
Plumage Sexes differ

Location S. and S.E. Asia
Migration Non-migrant
Status Locally common

This bird is the original ancestor of the domesticated chicken, first raised in captivity at least 5,000 years ago. The male is brightly coloured, with fleshy red wattles and comb. The female is smaller and drabber. Hens and chicks use calls to keep in contact and signal danger; the male's "cock-a-doodle-doo" is used to attract females and advertise his presence to rival males.

Argusianus argus
Great argus pheasant

Length 1.9 m (6 1/4 ft)
Weight Not recorded
Plumage Sexes differ

Location S.E. Asia
Migration Non-migrant
Status Lower risk

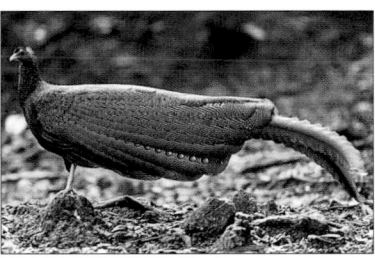

One of the world's largest pheasants, the male of this species has very large secondary flight feathers, decorated with egg-shaped "eyes", and a tail much longer than that of the female. To attract a mate, he calls loudly, raises his tail, and fans out his wings. The females with which he mates raise the young on their own.

Phasianus colchicus
Common pheasant

Length 89 cm (35 in)
Weight 0.75–2 kg (1 3/4 – 4 1/2 lb)
Plumage Sexes differ

Location North America, Europe, Asia
Migration Non-migrant
Status Common

Male common pheasants have a dark head with a purple and green gloss and red facial wattles; many also have a white neck ring and maroon breast. Compared to the drab brown female, the male is very colourful and larger, with a long tail. With more than 30 subspecies, this pheasant has been widely introduced, chiefly so that it can be hunted for food.

long tail of male

Numida meleagris
Helmeted guineafowl

Length 55 cm (22 in)
Weight 1–1.5 kg (2 1/4 – 3 1/4 lb)
Plumage Sexes alike

Location Africa (south of Sahara)
Migration Non-migrant
Status Locally common

Domesticated long ago as a source of food, the helmeted guineafowl is a flock-forming gamebird whose natural habitat is the open grassland of tropical Africa. Its most conspicuous features are its spotted plumage and bony, horn-like helmet on top of its naked head. Males have a characteristic lateral, hump-backed posture during encounters with other males and while courting. Compared to most gamebirds, it is well able to defend itself and its young – by pecking and scratching, and buffeting with its wings. A very noisy bird, when alarmed it gives a staccato "kek-kek-kekkek-kekkekkek" call.

dark plumage, dotted with white

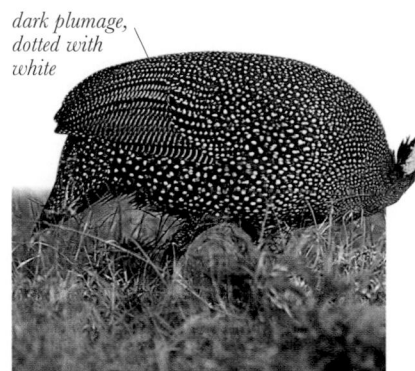

CRANES AND RELATIVES

PHYLUM	Chordata
CLASS	Aves
ORDER	Gruiformes
FAMILIES	11
SPECIES	203

THE LONG-LEGGED cranes and their relatives are an assortment of birds that look outwardly different to one another but are united by aspects of their internal anatomy – such as the lack of a crop in their digestive system. In addition to cranes themselves, this group includes rails, bustards, trumpeters, and sun-bitterns, among others. By far the largest family is formed by the rails (which include the familiar coots and moorhens). Rails are found worldwide but many of the other families are less widely distributed.

ANATOMY

Most members of this group have long legs, a slender bill, rounded wings, and modest, cryptic plumage. However, there is otherwise great variation in appearance between species, depending on their habitat and lifestyle. Those that wade in wetlands (cranes), or walk on floating vegetation (the limpkin), have long, slender toes to distribute their weight. The strictly ground-dwelling bustards have shorter toes and strong legs for running over arid land. Aquatic members of the group (coots and finfoots) have lobed feet for swimming. Cranes fly great distances on long, broad wings, while others (mesites, buttonquails, and some rails) have small, rounded wings that are poor or useless for flying.

MIGRATION

Many cranes are migratory, travelling vast distances between their breeding and wintering areas. Along the way, large flocks stop at long-established staging posts, where they may stay for a few days before moving on. Cranes breed in wetlands, typically seeking remote places such as Arctic tundra, steppes, highland plateaux, and forested swamps. Some crane species and most other birds in the group do not migrate at all.

NESTING CORNCRAKE
The corncrake is one of the few long-distant migrants among the rails, breeding in Europe in summer and wintering in sub-Saharan Africa.

MOVING NORTH
Some sandhill cranes make long-distance annual migrations between their winter feeding grounds in Texas and their summer breeding areas in Alaska, which are almost 6,500km (4,000 miles) apart.

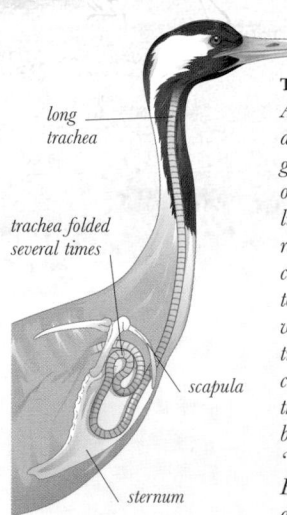

TRACHEA
An unusual feature of cranes and the limpkin is their greatly elongated windpipe, or trachea, which is coiled like a hose-pipe around the region of the sternum in the chest. This long tube helps to amplify the bird's calls, which are very loud. The trumpeting of cranes can carry several kilometres, and the calls of the limpkin have been described variously as "strangled" or "shrieking". Each species makes its own distinct, recognizable sounds.

long trachea

trachea folded several times

scapula

sternum

COURTSHIP

Several members of this group have elaborate courtship rituals. Some of the most elegant displays are performed by cranes, which mate for life. Males and females form pairs, and later maintain the bond, with carefully orchestrated dances (see below). Birds that perform similar courtship displays include rails, some bustards (see right), and the sun-bittern, in which the birds hold out their wings, revealing large eye-spots, and fan their tail, often while jumping and running in a circle. In addition to their courtship displays, rails also attract mates by making loud and distinctive calls.

BREEDING PLUMAGE
Breeding male bustards inflate their throat pouch, puff out some of their feathers to make themselves look larger, and call loudly. Many species, such as the great bustard (above), have a ruff around the neck.

CONSERVATION

Cranes and their relatives have a higher proportion of endangered species among their number than any other group of birds, and several have recently become extinct. Threats include habitat destruction, introduced ground predators, and hunting. Among the true cranes, 7 of the 15 species are endangered, including the Siberian crane and whooping crane, with more species likely to join them. Protection of cranes is made particularly difficult by their long migrations, requiring the co-operation of many countries.

COURTSHIP DANCE

1 JUMPING
Two grey crowned cranes raise their wings and jump into the air.

2 RITUAL FLIGHT
One of the pair flaps its wings and leaps towards the other.

3 WING-FLAPPING
As the leaping bird approaches, the other raises its wings.

4 STEPPING BACK
The leaping bird retreats slightly and begins to bow.

5 BOWING
The 2 birds now lower their wings and bow to one another.

BIRDS

Mesitornis variegata
White-breasted mesite

Length 31 cm
(12 in)

Weight 100 g
(3 5/8 oz)

Plumage Sexes alike

Location Madagascar

Migration Non-migrant

Status Vulnerable

This small forest bird is one of 3 species in the mesite family, all unique to Madagascar, and threatened by the destruction of their natural habitat. Its short, rounded wings and stout legs are typical of a ground-living bird. It feeds mainly on insects and spiders, flicking through fallen leaves or lifting up larger ones in order to find prey. A tangled mass of vegetation piled into a tree or bush, 1–3m (3¼–9¾ft) above the ground, serves as a nest.

Grus grus
Common crane

Length 1.2 m
(4 ft)

Weight 5–6 kg
(11–13 lb)

Plumage Sexes alike

Location Europe, Asia, N. Africa

Migration Migrant

Status Common

Like other members of its family, the common crane is a large, elegant bird, with a long bill and long legs. Grey, with a black head and neck, it has a white stripe down the nape, and a red spot on the crown. It gives a loud, trumpeting call, the sound being amplified by an enlarged windpipe which is fused with the breastbone.

long bill

long legs for foraging

Turnix varia
Painted button-quail

Length 17–23 cm
(6½–9 in)

Weight 55–95 g
(2–3⅜ oz)

Plumage Sexes differ

Location S.W. and E. Australia, New Caledonia

Migration Non-migrant

Status Locally common

With their compact bodies and camouflage plumage, button-quails look very much like true quails but they are not closely related. This species is relatively large, reddish in colour and heavily spotted. When foraging for seeds and insects, it scratches with one foot while pivoting on the other, leaving circular marks on the ground.

Grus japonensis
Japanese crane

Length 1.5 m
(5 ft)

Weight 7–12 kg
(15–26 lb)

Plumage Sexes alike

Location E. Asia

Migration Migrant

Status Endangered

With its intricate courtship dances and lifelong partnerships, this elegant bird has long been a symbol of happiness and good luck.

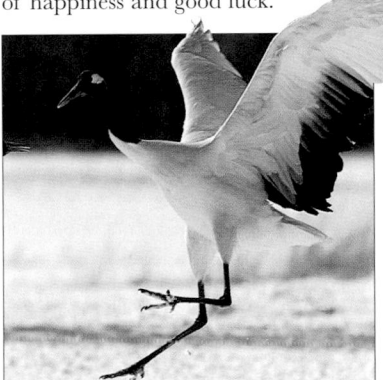

Grus canadensis
Sandhill crane

Length 1.2 m
(4 ft)

Weight 4–5.5 kg
(8¾–12 lb)

Plumage Sexes alike

Location North America, N.E. Asia

Migration Migrant

Status Locally common

The smaller of the 2 species of crane native to North America, this is also by far the most common. A grey bird with a red forehead, its relatively short

Pedionomus torquatus
Plains wanderer

Length 15–19 cm
(6–7½ in)

Weight 40–80 g
(1 7/16–2 7/8 oz)

Plumage Sexes differ

Location S.E. Australia

Migration Non-migrant

Status Endangered

Similar to a button-quail but lankier, the plains wanderer looks deceptively like a gamebird, with a compact body, small head, and well-developed legs. It will sometimes stand stretched up on tiptoe in order to see over the grasses in which it lives. The female of this species is larger than the male and more brightly coloured, with a white-speckled black collar and a

CONSERVATION

This crane probably bred on all 4 of Japan's main islands but, due to hunting (birds are shot down as they migrate) and habitat loss, it has lived only on Hokkaido since the 1890s. A protection scheme has seen its population rise from 20 in the 1920s to about 600.

The heaviest member of the crane family, it is mainly white, with black flight feathers and a black face and neck. Large flocks congregate on feeding grounds in winter, but in the breeding season, pairs of birds establish territories, defending them vigorously against other cranes.

COURTING DANCES
During courtship, the Japanese crane performs elaborate dances involving head bobbing, bowing, pirouettes, jumping, and tossing material in the air.

bill enables it to deal with a wide range of foods – plants, insects, and small animals such as mice. Northern populations move south in winter and as many as 40,000 cranes can gather at a single stop-over site, their loud calls carrying across great distances.

pale eyes

yellow bill

angular head

speckled plumage

chestnut breast patch. This bird is seriously threatened by the spread of agriculture, which has destroyed much of its grassland habitat.

Balearica regulorum
Grey crowned crane

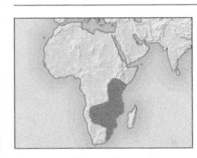

Length 1–1.1 m
(3¼–3½ ft)

Weight 3–4 kg
(6½–8¾ lb)

Plumage Sexes alike

Location E. and southern Africa

Migration Non-migrant

Status Locally common

Crowned cranes are easily identified by their golden crests or "crowns" and by their ability to perch in trees – something that other cranes cannot do. Found in both marshland and cultivated fields, this bird has a relatively short, versatile bill, unlike more aquatic cranes. It stamps its feet while foraging to flush out potential prey, and travels with large mammals to feed on the insects frightened by their movement. The black crowned crane is a very similar species found in tropical Africa.

red throat wattle

grey plumage

well-developed hind toe for perching

BIRDS

Aramus guarauna
Limpkin

Length	56–71 cm (22–28 in)
Weight	1–1.5 kg (2¼–3¼ lb)
Plumage	Sexes alike

Location S.E. USA (Florida), Central and South America	Migration Non-migrant
	Status Locally common

An ibis-like bird with a striking brown and white spangled neck, the limpkin has a long, slender bill that is sharpened and slightly twisted at

slender, sharp bill

its tip – features that enable it to extract snails from their shells with ease. Breeding pairs often stay together year after year, and the male defends the nest by charging intruders, fighting with its feet, and uttering an unearthly, mournful call.

Psophia crepitans
Common trumpeter

Length	50 cm (20 in)
Weight	1–1.5 kg (2¼–3¼ lb)
Plumage	Sexes alike

Location N. South America	Migration Non-migrant
	Status Lower risk†

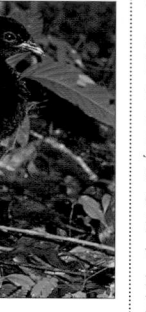

This black, humpbacked bird is one of 3 species in the trumpeter family, all found in South America. Small groups of 6–8 live on the forest floor, bathing and roosting together, and foraging for the fallen fruits that are their main food. They use a variety of calls, some loud and trumpeting, to warn of danger, beg for food, mark territory, and threaten intruders. Unusually for birds, several males mate with one female.

Porzana porzana
Spotted crake

Length	22–24 cm (9–9½ in)
Weight	55–150 g (2–5 oz)
Plumage	Sexes alike

Location Europe, W. to C., S., and S.E. Asia, Africa	Migration Migrant
	Status Locally common

The spotted crake is a small and secretive wetland bird, with a narrow body that helps it to move about among dense reeds and other water-side plants. It walks with a distinctive jerking gait, often bobbing its head and flicking its tail. Most active at dawn and dusk, it feeds mainly on insects, foraging both on land and in water. This extremely wary bird will run for cover or fly off if disturbed. It is generally quiet, but the male calls to attract a female.

Crex crex
Corncrake

Length	27–30 cm (10½–12 in)
Weight	125–200 g (4–7 oz)
Plumage	Sexes alike

Location Europe, W. to C. Asia, S.E. Africa	Migration Migrant
	Status Vulnerable

Heard far more easily than seen, this cryptically coloured bird has a loud and rasping, 2-syllable call, which the male makes both to advertise its territory and to attract females. The call becomes a growling, pig-like squeal before mating, or during encounters with other males. The corncrake is declining in many parts of its range as a direct result of changes in farming practices and because the damp pastures it inhabits are increasingly drained and ploughed.

Gallinula chloropus
Common moorhen

Length	30–38 cm (12–15 in)
Weight	175–500 g (6–18 oz)
Plumage	Sexes alike

Location Worldwide except Australia	Migration Partial migrant
	Status Common

white flank markings

The moorhen is one of the world's most widespread freshwater birds. A member of the rail family, it has distinctive white markings along the flanks and a conspicuous red "shield" and yellow-tipped red bill. Although not as shy as other rails, and often seen moving in the open, if threatened, the common moorhen will flee into dense cover.

Fulica atra
Common coot

Length	36–39 cm (14–15½ in)
Weight	300–1,200 g (11–43 oz)
Plumage	Sexes alike

Location Europe, Asia, Australia, New Zealand, N. and W. Africa	Migration Partial migrant
	Status Common

Compared to some of its relatives, the common coot is an assertive and even aggressive bird. In conflicts with rivals, it displays its white frontal "shield",

which contrasts starkly with its heavy black body. If threatened, it will either gather in tight flocks and splash water, or turn on its back and kick out with its feet. It has lobed toes that aid swimming and diving, and it presses air from its plumage before diving to reduce buoyancy.

Porphyrio porphyrio
Purple swamphen

Length	38–50 cm (15–20 in)
Weight	500–1,300 g (18–46 oz)
Plumage	Sexes alike

Location S. Europe, W., S., and S.E. Asia, Australia, Africa	Migration Non-migrant
	Status Common

One of the biggest members of the rail family – almost as big as a chicken – the purple swamphen is a powerful, robustly built bird with

purple and black plumage, a reddish orange bill, and long legs and toes. It feeds on all manner of vegetation, pulling it up with its bill, as well as on aquatic and terrestrial invertebrates. It will also eat the eggs and young of other waterbirds, climbing if necessary to reach the nest. The purple swamphen has a very large range, with several local races differing in plumage and in behaviour: some stay in one place, while others move long distances in order to find food. In many areas, males and females pair off for breeding, but in New Zealand complex breeding groups form, consisting of one or 2 females and

2–7 males. The eggs are laid in one nest and all members of the group help to incubate them and rear the young.

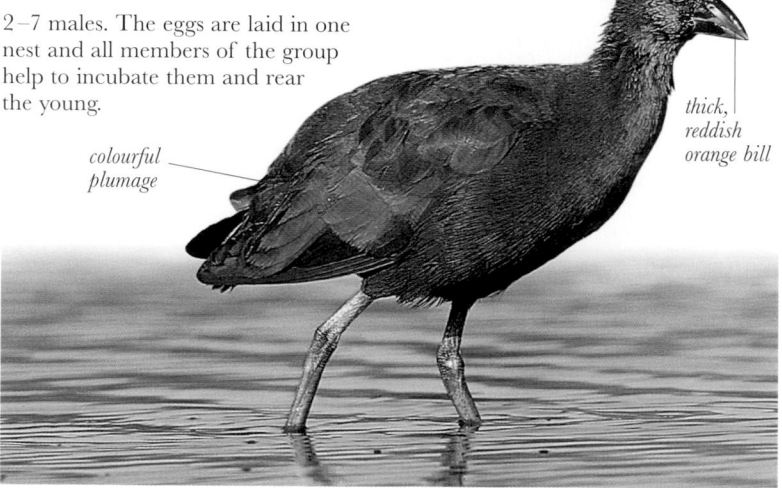

colourful plumage

thick, reddish orange bill

Heliornis fulica

Sungrebe

	Length 26–33 cm (10–13 in)
	Weight 125–150 g (4–5 oz)
	Plumage Sexes differ
Location Central and South America	Migration Non-migrant
	Status Vulnerable†

This grebe-like bird belongs to the finfoot family, 3 species named for the lobes on their feet which enable them to swim well and move easily on land. The sungrebe is unique in that the male has skin pouches in the "armpit" of each wing in which he carries the young, even in flight. This bird has a slender body and a narrow tail.

Rhynochetos jubatus

Kagu

	Length 55 cm (22 in)
	Weight 900 g (32 oz)
	Plumage Sexes alike
Location New Caledonia	Migration Non-migrant
	Status Endangered

With its stocky body, weak wings, and large, erectile crest, the kagu is a typical island species that has evolved a distinctive appearance and way of life. It lives on the ground, and often stands motionless on one leg, watching and listening for prey. If it strikes and misses, it will use its bill to dig out the prey; unique flaps of

skin over the nostrils keep out debris while the kagu forages in soil. When threatened, it raises its shaggy crest and spreads its wings to expose the shield-like pattern on its flight feathers.

bluish grey plumage

BINOCULAR VISION
The kagu has forward-facing eyes that help it to spot insects and other small animals among fallen leaves.

CONSERVATION

As a ground-dweller, the kagu is particularly vulnerable to predators. A 1991–1992 survey found fewer than 500 adult birds, and in one month in 1993 dogs killed 14 from a single population. Other animals also pose a threat to the kagu: pigs eat the eggs and compete with the birds for food.

Eurypyga helias

Sunbittern

	Length 43–48 cm (17–19 in)
	Weight 200 g (7 oz)
	Plumage Sexes alike
Location Central America, N. South America	Migration Non-migrant
	Status Vulnerable†

An inhabitant of shady rain-forest streamsides, the sunbittern is a wary bird that stalks its prey slowly and deliberately, then quickly stabs it with its slender bill. If threatened when in the nest, it moves its neck backwards and forwards, and hisses like a snake. At other times it will turn to face its predator, and fan its tail and spread its wings (like the kagu, above) to expose large, eye-like patches. The

large "eyes" and the spread of the wings and tail make the sunbittern look very big and imposing. Its mottled, cryptic plumage is soft, enabling it to fly silently. A solitary bird, even paired adults are seldom seen together.

red eyes

long bill

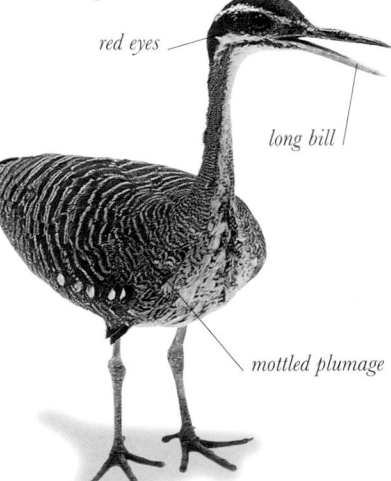

mottled plumage

Cariama cristatus

Red-legged seriema

	Length 75–90 cm (30–35 in)
	Weight 1.5 kg (3¼ lb)
	Plumage Sexes alike
Location E. South America	Migration Non-migrant
	Status Locally common

In both shape and behaviour, this species resembles the secretary bird (see p.294): both have long legs and crests, and both hunt by striding across the ground. The red-legged seriema often captures large prey in its bill, beating it on the ground to break it into pieces. It usually runs from predators, but may lie down and rely on its cryptic colouring to escape detection. Out of the breeding season it is highly vocal, and its call is one of

the most characteristic sounds of South American grasslands. Although generally solitary, this species may also be found in groups at times.

grey-brown plumage

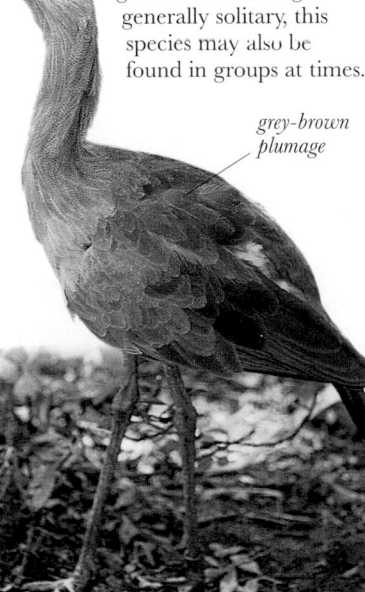

Chlamydotis undulata

Houbara bustard

	Length 65–75 cm (26–30 in)
	Weight 1.5–3 kg (3¼–6½ lb)
	Plumage Sexes differ
Location Canary Islands, N. Africa, W., C., and E. Asia	Migration Non-migrant
	Status Lower risk

This stout, long-legged bird is a typical member of the bustard family – a group of about 22 species that are found only in the eastern hemisphere. Like other bustards, it has cryptically coloured plumage, and although it can fly well, it spends most of its time on the ground. The houbara bustard searches for food as it walks through its desert homeland, eating seeds and shoots, insects, and small reptiles such as lizards. Of all bustards, it is the one best adapted to the desert, and seldom drinks, obtaining most of the water it needs from food. The male uses a traditional courtship arena for mating, adopting a curious method of

attracting the females' attention: it trots blindly about its display ground, with its ruff and crest feathers raised over its head. Although not globally threatened, the houbara bustard is declining fast across its range, and in many parts of West Asia, it is the focus of intensive conservation measures. This species is a highly valued game-bird in the region and is hunted with falcons.

tan plumage spotted with brown

Ardeotis kori

Kori bustard

	Length 1.2 m (4 ft)
	Weight 11–19 kg (24–42 lb)
	Plumage Sexes alike
Location E. and southern Africa	Migration Non-migrant
	Status Lower risk†

Weighing up to 19 kg (42 lb), the kori bustard is one of the world's heaviest flying birds. Like its relatives, however, it lives on the ground and is reluctant to fly unless in serious danger. It often associates with

large, herding animals, feeding on insects frightened by the herd as it moves. It also inhabits areas that have recently been burnt, eating the new shoots of grass and insects exposed by the lack of vegetation.

WADERS, GULLS, AND AUKS

PHYLUM	Chordata
CLASS	Aves
ORDER	Charadriiformes
FAMILIES	18
SPECIES	343

IN MANY PARTS of the world, these birds are a common sight at sea, along shorelines, and in wetlands. Most are strong fliers that feed on other animals in or near water. Waders or shorebirds (which include sandpipers, plovers, avocets, stilts, snipes, curlews, and jacanas) are long-legged birds that feed by the water's edge. Gulls (which embrace terns, skuas, and skimmers) use their flying skills to catch prey. Auks (including puffins and guillemots) dive underwater for food. Auks look strikingly similar to penguins but are able to fly and are confined to the northern hemisphere. Waders and gulls occur worldwide, but many species are threatened by habitat destruction, oil pollution, or hunting.

ANATOMY

Most of the birds in this group have subdued black, white, brown, or grey plumage, but some have colourful bare parts such as the bill, eyes, legs, and mouth linings. Many go through radical plumage changes, both between seasons and in maturing to adulthood. The 3 groups exhibit major differences, especially in the bill and legs, which are long in the waders but short and web-footed in the rotund, upright auks. Most birds in the group have salt glands above the eyes that enable them to extract the fluids they require from sea water and expel the excess salts through the nostrils.

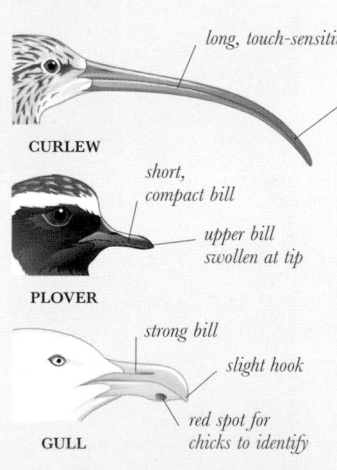

long, touch-sensitive bill

bill curves down at tip

CURLEW

short, compact bill

upper bill swollen at tip

PLOVER

strong bill

slight hook

red spot for chicks to identify

GULL

BILL SHAPES
These birds exhibit a wide variety of bill shapes, each adapted for a different feeding method – the curlew has a long, downcurved, touch-sensitive bill that is ideal for probing deep into mud, while plovers have a short, pigeon-like bill used for picking up food that is detected by sight, not touch. Gulls have a robust, multi-purpose bill, slightly hooked for tearing food.

NESTING TERNS
Most terns (including these elegant terns) breed in colonies, seeking an isolated place such as an island or reef for protection. The colonies are usually found on flat, open ground and are often densely populated.

CATCHING FISH

① FINDING PREY
A western gull detects a fish beneath the surface of the water and swoops down towards it.

② CAPTURE
With rapid wingbeats, the gull plunges its stout bill into the water and plucks out the fish.

③ TAKING OFF
Having successfully captured its prey, the gull takes off with the fish in its bill.

④ STEADY FLIGHT
The gull flies low over the water, heading towards a suitable feeding site.

BIRDS

REPRODUCTION

Most birds in this group nest on the ground, laying between one and 6 camouflaged eggs. Other species nest on cliffs, and a few, such as some terns and auks, prefer trees. Many nest colonially, in great numbers (over a million pairs in some auks), and some at extraordinary densities: incubating guillemots virtually rub shoulders with their neighbours. Numerous waders have unusual mating systems, in which either females or males have several mates. In jacanas, painted snipes, and phalaropes, the usual roles of male and female are reversed, the more colourful female initiating display, the male alone incubating the eggs and tending the young.

CLIFFSIDE COLONY
Auks often nest in extremely dense colonies. Guillemots (left) crowd together on cliff ledges at up to 20 pairs per square metre, their bodies virtually defining the extent of their territory. These birds lay a single egg, each with its own pattern – a way of detecting it among the many similar-looking nest sites.

FLIGHT

The auks, with their short, pointed wings, can fly rapidly in air, but only for limited distances. They use their wings to propel themselves underwater as they swim. The waders have long, pointed wings to help them to fly fast and efficiently, but although they are adept at rapid changes of pace and direction, they cannot soar. This skill can be seen in the gulls and their relatives, which spend hours on end on the wing, often travelling extremely long distances over water. Every year the Arctic tern, which breeds north of the Arctic Circle and winters in Antarctica, travels a distance of about 16,000km (10,000 miles) each way.

LONG-DISTANCE MIGRANTS
The Arctic tern spends much of its life in flight, feeding on the wing by catching insects in mid-air and flying low over water to take fish. It breeds and migrates in large colonies. All terns perform spectacular aerial courtship displays.

CARRYING PREY
The Atlantic puffin has an extremely large, colourful bill. As with all puffins, the upper bill and tongue are ridged with spikes that enable the bird to hold a remarkable number of fish at a time – up to 62 has been recorded.

FEEDING

Apart from seedsnipes, which eat plant matter, all members of this group feed on other animals. The waders probe for invertebrates on beaches, estuarine mudflats, and leaf litter. Gulls are opportunists, picking up fish, eggs, small birds, and mammals where they can, even scavenging at refuse tips. Gulls and skuas sometimes feed by piracy, intimidating smaller birds into giving up their food. Terns plunge-dive for fish, while skimmers fly low over the water, the lower bill just touching the surface and snapping shut when prey is touched. Auks pursue their prey underwater.

WALKING ON WATER
The jacanas (here, a comb-crested jacana) walk on floating vegetation, their long toes distributing their weight and preventing them from sinking. They eat insects and seeds that they find under the leaves.

FEEDING WADERS
These American avocets are catching their prey by swishing their bill from side to side through water or very soft mud, a technique known as scything. As the bill moves, it comes into contact with edible particles in suspension.

HOSTILE RIVALS
~enly, 2 more birds appear, harrassing the ~y squawking and flapping their wings.

6 CHANGING COURSE
Turning on its side, the gull heads off in a different direction to escape the pursuers.

7 MID-AIR SNATCH
One of the chasing birds has caught up with the gull and is pulling the fish out of its bill.

8 ESCAPE
Having successfully grabbed the fish, the "pirate" gull flees, pursued by the other 2 birds.

Jacana jacana
Wattled jacana

Length 17–25 cm
(6 1/2–10 in)

Weight 90–125 g
(3 1/4–4 oz)

Plumage Sexes alike

Location S. Central
America, South America

Migration Non-migrant

Status Locally common

The 8 species of jacanas are notable for their long legs and very large feet. They also have a distinctive spur on the leading edge of each wing. The wattled jacana feeds mostly on insects and other aquatic invertebrates, but sometimes eats seeds from rice plants.

The female, much larger than the male and weighing about 150g (5oz), mates with up to 3 males. Each male builds a nest, incubates a clutch of eggs, and rears the young. The female defends the territory from intruders.

red wattles

COLOURS
This small, black and dark chestnut bird has big red wattles around a yellow bill. When the wings are open, its yellow flight feathers make a striking contrast with the darker plumage.

extremely long toes

WALKING ON WATER

As with the other species of jacanas, the wattled jacana has extraordinarily long toes that spread its weight as it walks, enabling it to move easily over the leaves of floating waterplants. Because of this behaviour, jacanas are also called lily-trotters.

Dromas ardeola
Crab plover

Length 38–41cm
(15–16 in)

Weight 225–325 g
(8–12 oz)

Plumage Sexes alike

Location E. Africa, S.W.,
S., and S.E. Asia

Migration Partial migrant

Status Locally common

This thickset, black and white wader has an exceptionally large, powerful black bill, which it uses to crush the shells of crabs before swallowing them whole. The crab plover is also unusual among waders in that it digs a long burrow, in which it lays a single egg.

Rostratula benghalensis
Painted snipe

Length 23–28 cm
(9–11 in)

Weight 90–200 g
(3 1/4–7 oz)

Plumage Sexes differ

Location Africa, S., E., and
S.E. Asia, Australia

Migration Non-migrant

Status Vulnerable†

A medium-sized wader of freshwater marshes, the painted snipe has a distinctive broad white patch around each eye. It spends much of its time in deep cover, emerging – very

cautiously and mainly at night – to probe the mud for food. The painted snipe has a rather weak, fluttering flight, with legs dangling. The male usually incubates the eggs and rears the young, as the female goes off to find another mate.

Haematopus ostralegus
Eurasian oystercatcher

Length 40–48 cm
(16–19 in)

Weight 400–800 g
(14–29 oz)

Plumage Sexes alike

Location Europe, N.W.,
N., and E. Africa, S.W.,
C., E., and S. Asia

Migration Migrant

Status Common

A widespread, coastal and freshwater wader, the Eurasian oystercatcher is recognizable by its noisy calls and

bright orange bill. It feeds mainly on mussels, limpets, and cockles – either by cutting the muscle that holds the 2 halves of the shell together and stabbing the prey inside, or by hammering the shell open on the rocks or hard sand.

Ibidorhyncha struthersii
Ibisbill

Length 38–41cm
(15–16 in)

Weight 275–325 g
(10–12 oz)

Plumage Sexes alike

Location C. and E. Asia

Migration Non-migrant

Status Lower risk

A unique wader with a long, thin, downcurved red bill, the ibisbill is grey-brown above, and has a bluish grey breast, neck, and head, with a black face, and red legs. In spite of its striking coloration, the bird blends remarkably well with stony river banks. It feeds in mountain rivers, raking through the surface and probing under stones to find prey.

Himantopus himantopus
Black-winged stilt

Length 35–40 cm
(14–16 in)

Weight 150–200 g
(5–7 oz)

Plumage Sexes differ

Location Europe, Asia,
Africa

Migration Migrant

Status Common

In relation to its body size, this wader has the longest legs of any bird, which enables it to feed in deeper water than most other waders. The black-winged stilt hunts by sight or touch, scything its fine, straight bill through the water to detect prey. This bird also chases insects on and above water, often twisting and leaping to catch them.

Recurvirostra avosetta
Pied avocet

Length 42–45 cm
(16 1/2–18 in)

Weight 225–400 g
(8–14 oz)

Plumage Sexes alike

Location Europe, Asia,
Africa

Migration Migrant

Status Common

slender, upcurved bill

Breeding in a band from western Europe across Asia to China, this bird is the most widespread of the world's 4 species of avocet. In winter, it can be found in large flocks, mostly on the coast, in estuaries, and in nearby wetlands. It frequently roosts floating or swimming on water, and groups of birds can form large "rafts", looking from a distance rather like gulls. During the breeding season, it tends to move inland, to slightly brackish or saline marshes. The avocet defends its territory very aggressively, calling loudly if it senses danger and chasing off intruders.

FEEDING IN WATER

When feeding, the avocet sweeps its upcurved bill from side to side so that only the tip is immersed. However, if it sees something edible, it will lunge quickly forwards, sometimes submerging its head as it pursues the prey.

black wing markings

UPTURNED BILL
The pied avocet is whiter than most other waders, with black markings on the head and wings, but its most distinctive feature is the long, slender, upcurved bill, the male's being longer and straighter than the female's.

Cursorius rufus
Burchell's courser

Length	20–23 cm (8–9 in)
Weight	70–80 g (2 1/2–2 7/8 oz)
Plumage	Sexes alike
Location Southern Africa	**Migration** Non-migrant
	Status Lower risk

Unlike most waders, the 8 species of coursers live in dry habitats, and get their name from their rapid run. Burchell's courser is sandy brown,

with a slight rusty tone (as in the juvenile pictured), which provides excellent camouflage in the barren country in which it lives. When threatened, the bird stands still and upright to avoid detection, and then runs away very quickly. If forced to, it will fly far and fast. It feeds in a jerky, stop-start manner, taking insects from the ground.

Stiltia isabella
Australian pratincole

Length	21–24 cm (8 1/2–9 1/2 in)
Weight	60–70 g (2 1/8–2 1/2 oz)
Plumage	Sexes alike
Location S.E. Asia, Australia	**Migration** Partial migrant
	Status Lower risk

This Australian bird is an exception among the 8 species of pratincoles in having long legs and unusually long wings. It is pale cinnamon-brown

above and pale tan underneath except for a large, dark chestnut patch on the belly. Its downcurved bill has a red base and black tip. The Australian pratincole runs very fast to catch insects on the ground, but will also feed in the air, sometimes forming aerial flocks that hunt at considerable heights.

long legs

Charadrius hiaticula
Ringed plover

Length	18–20 cm (7–8 in)
Weight	50–70 g (1 3/4–2 1/2 oz)
Plumage	Sexes alike
Location N. North America, Greenland, Europe, Asia, Africa, Madagascar	**Migration** Migrant
	Status Locally common

A short bill, a white neck collar, and black bands on the head and chest are distinctive features of this compact plover. It breeds over a wide range of habitats, but in winter is found mostly on sandy coasts and inland wetlands. It feeds on a variety of invertebrates, foraging for them at night, especially when the moon is full. Both male and female rear the young and will perform a striking "broken wing" display to lure predators from the nest.

white neck collar

orange-yellow legs

Anarhynchus frontalis
Wrybill

Length	20–21 cm (8–8 1/2 in)
Weight	40–70 g (1 7/16–2 1/2 oz)
Plumage	Sexes differ
Location New Zealand	**Migration** Migrant
	Status Vulnerable

bill curved sideways

The wrybill is the only bird whose bill has a sideways curve. From the side, its bill looks relatively normal – if slightly long for a plover – but when seen from above or head-on, its unusual shape is very evident. The sideways curve is useful when feeding on shingle, allowing the plover to flick stones aside as it hunts for insects and worms. The plumage is soft grey above, and white below, with a narrow black breastband, more strongly defined in males than in females.

Vanellus armatus
Blacksmith plover

Length	28–31 cm (11–12 in)
Weight	125–225 g (4–8 oz)
Plumage	Sexes alike
Location Southern and E. Africa	**Migration** Non-migrant
	Status Locally common

Widespread in southern and East Africa, this plover is a medium to large wader with very striking black, white, and grey plumage and long black legs; its eyes are an intense red. Its diet consists of a wide range of invertebrates, frequently taken at night, especially when the moon is bright. Both the male and female are very protective of their young, and will chase or challenge intruders. The blacksmith plover gets its name from its sharp, metallic call.

Numenius arquata
Eurasian curlew

Length	50–60 cm (20–23 1/2 in)
Weight	450–1,350 g (16–48 oz)
Plumage	Sexes alike
Location Europe, Asia, Africa	**Migration** Migrant
	Status Locally common

Famous for its beautiful rising call, the Eurasian curlew is a large, brown-streaked wader with an extremely long, downward-curving bill. Up to 19 cm (7 1/2 in) long, the bill is ideal for extracting animals buried deep in mud or sand. Inland, the bird eats mostly insects and earthworms; on the coast it feeds on a wide range of worms, shellfish, and crustaceans, especially shrimps and crabs (which the bird usually swallows whole, after removing the legs). It breeds in a variety of habitats, from sand dunes, bogs, and fens to upland heaths and grassland, and becomes strongly territorial at this time. Both male and female defend their territory with a distinctive undulating flight, planing on wings held in a shallow "V" while giving a whistled, bubbling call. Some males will even come to blows with their wings. The Eurasian curlew is one of 8 related species that are found

Pluvialis dominica
American golden plover

Length	24–28 cm (9 1/2–11 in)
Weight	125–200 g (4–7 oz)
Plumage	Sexes differ
Location N. North America, S. South America	**Migration** Migrant
	Status Common

This medium-sized plover has a remarkable migration pattern, flying northwards over land, but southwards largely over the sea. It travels very long distances, wintering on coastal marshes and inland grassland in South America and breeding in drier areas of the Arctic tundra. Although mottled brown in winter, it has beautiful breeding plumage – a gold-spangled black back, and a black face bordered by white above the eyes and on the sides of the neck.

in different parts of the world. They gather in very large flocks for roosting, but feed in smaller groups. Most are strongly migratory – this particular species breeds as far north as the Arctic Circle, but overwinters in sandy and muddy estuaries from western Europe to East Asia. It is a long-lived bird that can survive for up to 37 years.

brown-streaked plumage

very long, downcurved bill

BIRDS

Tringa nebularia
Greenshank

Length	30–35 cm (12–14 in)
Weight	125–300 g (4–11 oz)
Plumage	Sexes alike
Location Europe, Asia, Africa (mainly south of Sahara)	Migration Migrant
	Status Common

Like its 9 other relatives in the genus *Tringa*, this long-billed wader is rarely found far from damp ground. It has dark plumage which is paler below, and long, grey-green legs. When feeding, the greenshank uses the "walk-and-peck" technique used by other members of its genus. It often runs at its prey or scythes through the water with its bill until it finds food, which includes fish, crustaceans, insects, and invertebrates. Its call, a ringing "tu-tu-tu", is sharper when the bird is disturbed.

Actitis macularia
Spotted sandpiper

Length	18–20 cm (7–8 in)
Weight	35–70 g (1¼–2½ oz)
Plumage	Sexes alike
Location North, Central, and South America	Migration Migrant
	Status Common

This small sandpiper has a horizontal posture, short legs, and a stout, 2-toned bill. Greenish brown above, it has pale underparts that are boldly spotted with brown in summer; the female has larger, blacker spots. On the ground, it bobs or teeters in its search for food. As in the grey phalarope (see right), the female courts the male, with fanned tail and quivering wings. Some mate with up to 4 males in a season, and show little maternal care.

Phalaropus fulicarius
Grey phalarope

Length	20–22 cm (8–9 in)
Weight	40–80 g (1 7/16–2 7/8 oz)
Plumage	Sexes differ
Location Circumpolar around Arctic, W. South America, W. Africa	Migration Migrant
	Status Common

Unlike most waders, which live at or near the water's edge, the grey phalarope is an excellent swimmer and spends much of its time either at sea or on muddy pools. It feeds on small animals at the surface and frequently spins around in tight circles to stir up its prey. The grey phalarope breeds in marshy ground in the high Arctic tundra, and is remarkable for its breeding behaviour. The female takes the lead in courtship, while the male incubates the eggs and rears the young.

short grey tail

ROLE REVERSAL

The grey phalarope's breeding system shows an almost complete reversal of the male and female's normal roles. The female – shown above right – is more assertive than the more drably coloured male. She establishes a territory and displays to attract a mate, and he takes care of the young.

STOUT WADER
The grey phalarope is a small, rather pot-bellied wader with very short legs and a fairly stout bill. The fleshy lobes between the toes help the bird to swim.

white face patch

black-tipped, yellow bill

Gallinago media
Great snipe

Length	27–29 cm (10½–11½ in)
Weight	150–225 g (5–8 oz)
Plumage	Sexes alike
Location N. Europe, N.W. Asia, Africa	Migration Migrant
	Status Lower risk

The 16 species of snipe are long-billed and beautifully camouflaged waders which feed mostly under cover of darkness. Like most snipe, the great

Calidris canutus
Red knot

Length	23–25 cm (9–10 in)
Weight	100–225 g (3 5/8–8 oz)
Plumage	Sexes alike
Location Worldwide, except Antarctica	Migration Migrant
	Status Common

One of the largest of all sandpipers of the genus *Calidris*, the red knot has grey plumage in winter, but during the breeding season its face and underparts

snipe has a pointed, 2-toned bill. There is extensive dark barring on the underparts, and the tail corners have more white than most other snipe. In the breeding season, male birds gather at a display ground (lek) and compete with each other, using a variety of sounds. After mating, the female nests alone.

turn brick-red, while the upperparts are dark and spotted with pale chestnut. Red knots travel up to 12,000km (7,500 miles) to breed, and form large flocks when feeding and roosting.

Chionis alba
Pale-faced sheathbill

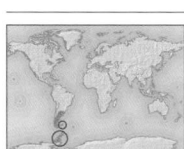

Length	34–41 cm (13½–16 in)
Weight	450–775 g (16–28 oz)
Plumage	Sexes alike
Location S.E. South America, Falkland Islands, Antarctic Peninsula	Migration Partial migrant
	Status Vulnerable†

The 2 species of sheathbill constitute the only bird family whose breeding range is entirely confined to the Antarctic and sub-Antarctic region. This species scavenges for food at penguin and seal colonies, feeding on eggs, chicks, and seals' afterbirth, and harassing penguins into regurgitating the food they bring for their young. It makes a cup-shaped nest from grass, bones, pebbles, shells, and feathers.

Catharacta antarctica
Brown skua

Length	52–64 cm (20½–25 in)
Weight	1–2 kg (2¼–4½ lb)
Plumage	Sexes alike
Location Circumpolar around Antarctica	Migration Partial migrant
	Status Common

Skuas breed in high latitudes, at times close to polar ice. The brown skua is one of 4–6 species in its genus and the most widespread species in the far south. This gull-like scavenger preys on penguins, shearwaters, and other seabirds of the southern oceans, particularly on their eggs and young, and scavenges around fishing boats and ships.

brown plumage

white wing patches

Stercorarius longicaudus
Long-tailed skua

Length 48–53 cm (19–21 in)	
Weight 225–350 g (8–13 oz)	
Plumage Sexes alike	
Location Circumpolar around Arctic and Antarctica	Migration Migrant
	Status Common

This grey and black skua differs from other skuas in having very long central tail feathers, which are clearly visible when it flies. These feathers

black cap

white underside

long tail feathers

may be half as long as the rest of the body. The long-tailed skua undertakes one of the longest of all bird migrations – from the Arctic to the southern oceans. Lemmings constitute its diet on land, whereas it feeds on fish and robs other birds while at sea.

Pagophila eburnea
Ivory gull

Length 44–48 cm (17 1/2–19 in)	
Weight 525–700 g (19–25 oz)	
Plumage Sexes alike	
Location Almost circumpolar around Arctic	Migration Partial migrant
	Status Vulnerable†

The gull family contains about 50 species, about 35 of which are found in the northern hemisphere. Many are familiar birds – both on coasts and

inland – but the ivory gull is much less well known. Its entire life is spent in the high Arctic, mostly on the edge of pack ice. This gull feeds chiefly on fish and invertebrates, but also follows polar bears to feed on the scraps they leave behind.

stout bill

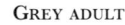

Larus argentatus
Herring gull

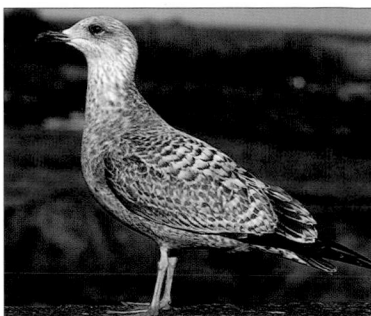

Length 55–67 cm (22–26 in)	
Weight 0.73–1.5 kg (1 1/2–3 1/4 lb)	
Plumage Sexes alike	
Location North and Central America, Europe, N.E. and E. Asia	Migration Partial migrant
	Status Common

JUVENILE
Young herring gulls have brown-streaked feathers and take 4 years to acquire adult plumage.

The herring gull is the most abundant coastal bird in much of North America and Europe, and its numbers have grown exponentially in the northern hemisphere in the past century. Adaptability is the key to its success, for it will eat almost anything, scavenging from garbage dumps, landfill sites, and sewage outflows. It also steals the eggs and young of other gulls, terns, and birds that nest in open country. Very vocal, especially in breeding colonies, adults tend to return to the same site to breed every year.

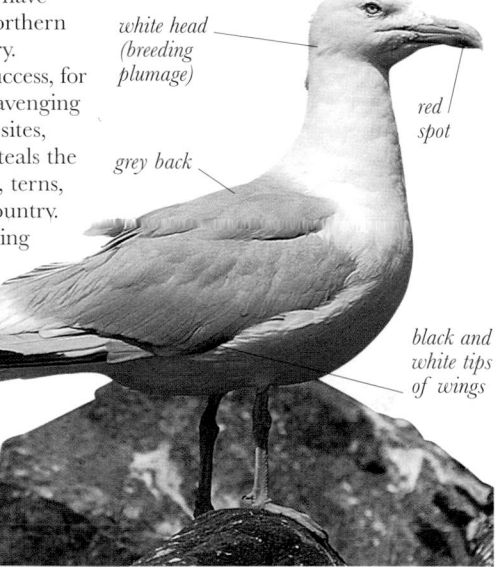

white head (breeding plumage)

red spot

grey back

black and white tips of wings

GREY ADULT
One of the larger gulls, this pink-legged bird has a grey back and wings, with black and white tips, a white head and underside, and a yellow bill with a red spot on the lower half.

Larus dominicanus
Kelp gull

Length 54–65 cm (21 1/2–26 in)	
Weight 0.9–1.5 kg (2–3 1/4 lb)	
Plumage Sexes alike	
Location South America, Antarctica, southern Africa, S. Australia, New Zealand	Migration Partial migrant
	Status Common

This large coastal bird is one of the most widespread gulls south of the equator. Its back and wings are black while the head, tail, and underside are white. Its large yellow bill has a red spot on the lower half. This opportunistic feeder catches invertebrates stirred up by whales, steals food from terns, and kills birds as large as geese. It also feeds on termite swarms, and scavenges at fish factories and slaughterhouses.

Chlidonias niger
Black tern

Length 23–28 cm (9–11 in)	
Weight 60–75 g (2 1/8–2 5/8 oz)	
Plumage Sexes alike	
Location North and Central America, N. South America, Europe to C. Asia, Africa	Migration Migrant
	Status Locally common

Terns are graceful birds, with slender bodies and forked tails. Most of them are white with a black cap, but the black tern has dark plumage. Unlike most terns, it does not breed on the coast, but on inland lakes, marshes, and bogs. Here it feeds mainly on insects plucked from plants, off the water surface, or caught on the wing. Out of the breeding season, when back in coastal waters, small marine fish make up the bulk of its diet.

black cap

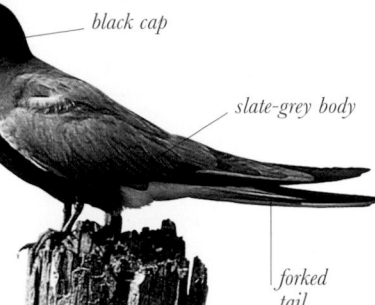

slate-grey body

forked tail

Hydroprogne caspia
Caspian tern

Length 48–56 cm (19–22 in)	
Weight 575–775 g (21–28 oz)	
Plumage Sexes alike	
Location North to Central America, Europe, Africa, Asia, Australia, New Zealand	Migration Partial migrant
	Status Locally common

With a wingspan of up to 1.4m (4 1/2 ft), this is the largest species of tern, approaching the size of larger gulls. It spends winter on lakes, coasts,

and estuaries, and in the breeding season is likely to be found in freshwater habitats. It catches fish by plunge-diving, swallowing it head-first, while in flight.

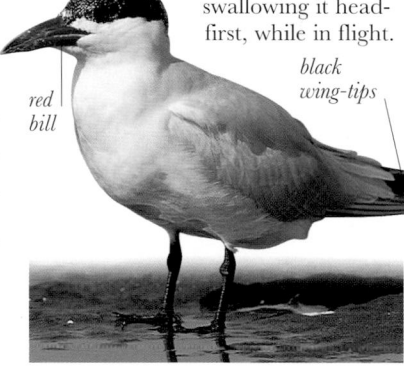

red bill

black wing-tips

Sterna paradisaea
Arctic tern

Length 33–35 cm (13–14 in)	
Weight 95–125 g (3 3/8–4 oz)	
Plumage Sexes alike	
Location Arctic, N. North America, Antarctica	Migration Migrant
	Status Common

The Arctic tern makes one of the longest migration of any bird – a journey of at least 16,000km (10,000 miles), twice a year. Each

autumn it flies from its nesting place in the north to Antarctica for the southern summer, making the most of the daylight in both hemispheres. It probably spends more time in daylight than any other creature. It feeds mainly on fish, hovering then plunge-diving, or dipping into the water.

grey wings

black crown

white cheeks

Sterna fuscata
Sooty tern

	Length 35–45 cm (14–18 in)
	Weight 150–250 g (5–9 oz)
	Plumage Sexes alike
Location Worldwide, in tropical seas	Migration Partial migrant
	Status Common

This wide-ranging, tropical tern has white undersides, but is almost completely brownish black above. A bird of the open ocean, it is often seen in large flocks, hundreds of kilometres from land. Unlike many of its relatives, it does not plunge-dive for its food. Instead, it swoops down close to the water, snapping up fish and other small animals near the surface, rarely getting wet. Sooty terns nest in remote islands throughout the tropics, often in enormous, noisy colonies.

Larosterna inca
Inca tern

	Length 39–42 cm (15½–16½ in)
	Weight 175–200 g (6–7 oz)
	Plumage Sexes alike
Location W. South America	Migration Non-migrant
	Status Locally common

white "moustache" with yellow patch

Some terns can be difficult to tell apart, but this South American species is instantly recognizable by its slate-grey colour and its remarkable white "moustache" plumes, which trail from its cheeks for 5cm (2in) or more. This bird occurs in the Humboldt Current region, feeding on the abundant small anchovies found in the nutrient-rich water. Large numbers of Inca terns congregate over feeding sea lions and humpback whales, diving for fish scraps left by these animals.

Rynchops flavirostris
African skimmer

	Length 36–42 cm (14–16½ in)
	Weight 100–200 g (3⅝–7 oz)
	Plumage Sexes alike
Location Africa	Migration Non-migrant
	Status Locally common

Skimmers are similar to terns in shape and general coloration, but their bill structure is unique, with the lower half much longer than the upper

bill with longer lower half

half, and flattened sideways rather like scissor blades. There are 3 species, and they all feed by flying low over water, ploughing the surface with the lower "blade" and snapping the bill shut when it touches a fish. The birds feed mostly at dawn and dusk. This species, like the other 2 skimmers, is black with a white underside, and its feet, legs, and bill are bright orange-red.

Alle alle
Little auk

	Length 19–23 cm (7½–9 in)
	Weight 150–175 g (5–6 oz)
	Plumage Sexes alike
Location Circumpolar around Arctic	Migration Migrant
	Status Common

The smallest auk in the Atlantic, this black and white bird has a short bill and an upright posture. Like other members of the auk family, it is a

pursuit diver, using its stubby wings like flippers to speed after its prey. It feeds mainly on plankton, its bill being specially adapted so that it can catch small prey. The little auk breeds in colonies on steep cliffs along the Arctic coast, but usually winters out at sea.

white underparts

Aethia pusilla
Least auklet

	Length 15 cm (6 in)
	Weight 85 g (3 oz)
	Plumage Sexes alike
Location North Pacific, Arctic Ocean	Migration Partial migrant
	Status Common

The smallest of all auk species, the least auklet is the same length as a house sparrow, although its body is more thickset. Breeding colonies may contain as many as one million birds, which set off in insect-like swarms to feed. Plankton is its only food – mainly copepods but also young crustaceans and fish larvae; the bird dives from the surface and swims through the water to find the drifting shoals. It lays just one egg a year, in a crack or crevice on bare rock.

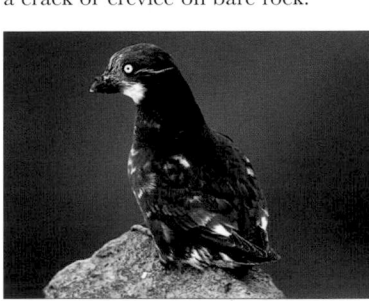

Fratercula arctica
Atlantic puffin

	Length 28–30 cm (11–12 in)
	Weight 400 g (14 oz)
	Plumage Sexes alike
Location North Atlantic, Arctic Ocean	Migration Migrant
	Status Common

The puffin is the most colourful auk in the North Atlantic, with a multicoloured bill and bright orange legs and feet. It sometimes dives to 60m (200ft) to find the shoaling fish that are its main food. These include sandeel,

relatively large head

black upperparts

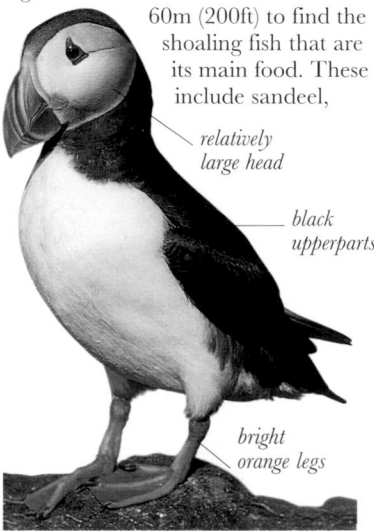

bright orange legs

PUFFIN SOCIETY

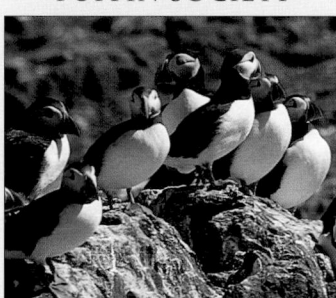

Like most members of the auk family, the Atlantic puffin is a highly social species, nesting in colonies on rocky coasts and offshore islands. On land, puffins stand in groups, but when feeding, they gather in "rafts" out to sea. Feeding areas are usually within 10km (6 miles) of a colony.

capelin, herring, and sprat, which are supplemented by animal plankton in winter. When nesting, the puffin digs a burrow 1m (3¼ft) or more deep in the ground; the burrow is often lined with feathers and plant matter. Both parents incubate the solitary egg and rear the young, carrying many small fish, packed crosswise into their bills, back to the burrow for it to eat.

BREEDING COLOURS
During the breeding season, the Atlantic puffin's triangular bill is red, yellow, and blue. It fades in late summer when its outer scales are shed.

Uria aalge
Common guillemot

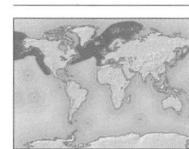

	Length 39–42 cm (15½–16½ in)
	Weight 850–1,100 g (30–39 oz)
	Plumage Sexes alike
Location North Atlantic, North Pacific	Migration Migrant
	Status Common

A relatively large, blackish brown and white auk, the common guillemot is gregarious, and often forms dense breeding colonies. Each pair produces a solitary egg, laid directly on the rock. The egg is sharply pointed so that it rolls around in a circle if disturbed, rather than falling off the nesting ledge. The chick leaves the ledge when about 3 weeks old and completes its development at sea.

upright posture

black feet

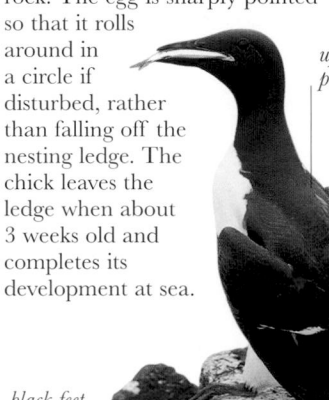

PIGEONS

PHYLUM	Chordata
CLASS	Aves
ORDER	Columbiformes
FAMILIES	1 (Columbidae)
SPECIES	309

THE FAMILIAR PIGEONS SEEN IN CITIES and on arable farms throughout the world are only a small part of a diverse group. Tropical forests are home to an immense variety of species, many of them brilliantly coloured, living on the ground or in trees. The larger kinds are generally known as pigeons and the smaller ones as doves. Adult pigeons and doves produce nutritious "crop-milk", which they feed to their young.

ANATOMY

Pigeons and doves are plump, full-breasted birds with a small head and bill; the head bobs as the bird walks, to keep it in a constant position relative to the body. These birds are strong fliers, their usually broad wings driven by powerful breast muscles that enable them to travel long distances at considerable speed. The plumage is thick and soft, although most species have a patch of bare skin around each eye.

IRIDESCENT FEATHERS

BRIGHT COLORATION

PLUMAGE
Many tropical pigeons and doves, such as the magnificent fruit pigeon (left) have bright and varied plumage. Although other species (such as the stock dove, far left) are less strikingly coloured, they have small iridescent patches.

FEEDING

Pigeons and doves eat mainly plant material. They can be broadly divided into 2 groups: seed-eaters and fruit-eaters. All have a specially adapted gut, with a well-developed crop and a strong, muscular gizzard; the latter is used for grinding food, often aided by ingested grit or stones. Fruit-eaters, taking more easily digestible food, have shorter guts than seed-eaters.

ROOSTING
Pigeons and doves often roost communally, either in spaced out groups (like these crested pigeons), or huddled together. When disturbed, several roosting birds will take off with sudden wing-beats, a shock tactic designed to alarm a predator.

FEEDING YOUNG
These young turtle doves are being fed on "milk" secreted by their mother's crop. Crop-milk is produced by both sexes and is rich in proteins and fats.

BIRDS

Columba livia
Rock dove

Length 31–34 cm (12–13 1/2 in)

Weight 200–300 g (7–11 oz)

Plumage Sexes alike

Location North, Central, and South America, Africa, Europe, Asia, Australia

Migration Non-migrant

Status Common

Originally from southern Europe, Asia, and North Africa, the rock dove is the wild ancestor of the town pigeon – one of the world's most widespread feral

birds. Whereas the wild morph is generally grey, with iridescent highlights on the neck and upper breast, town pigeons are highly variable. Paradoxically, although some feral strains have remarkable homing abilities, wild populations are largely sedentary. The wild rock dove nests on sea cliffs and among rocks; for town pigeons, bridges and window ledges provide perfect nest sites.

black bars on wing

Columba palumbus
Wood-pigeon

Length 41–45 cm (16–18 in)

Weight 275–700 g (10–25 oz)

Plumage Sexes alike.

Location Europe, N.W. Africa, W. and C. Asia

Migration Partial migrant

Status Common

The largest pigeon found in Europe, this species has prospered as a result of farming and has become an agricultural pest. It often feeds in flocks

on the ground, and is also an agile feeder in trees, clambering to branch tips to feast on fruits and seeds. Sexually active males perform a display flight, accompanied by wing claps, to attract a mate. The female usually produces only a single brood of one or 2 eggs per season. The young are cared for by both parents in the nest.

iridescent neck patch

white neck patch

Columba mayeri
Pink pigeon

Length 30–40 cm (12–16 in)

Weight 300–325 g (11–12 oz)

Plumage Sexes alike

Location Mauritius

Migration Non-migrant

Status Endangered

One of the world's rarest birds, the pink pigeon is one of many island species that have been harmed by introduced predators and habitat loss. The wild population had diminished to fewer than 20 birds in the 1980s; however, successful captive-breeding and release programmes have led to an increase in its numbers. This large pigeon has a soft pink body, a white face and forehead, usually brown wings, and a moderately long, strong bill with a hooked tip.

Streptopelia decaocto
Eurasian collared dove

Length 31–34 cm (12–13 1/2 in)

Weight 150–200 g (5–7 oz)

Plumage Sexes alike

Location Europe, Asia, N.E. Africa

Migration Partial migrant

Status Common

This slim, pinkish buff dove, with its distinctive black collar, underwent a dramatic and still unexplained expansion during the 20th century, and has become a common sight through much of Europe. It feeds mainly on the seeds and fruits of grasses (including cereals) and herbs, and occasionally on invertebrates and the green parts of plants, typically pecking food from the ground. Highly vocal in the breeding season, its specific name "decaocto" indicates the rhythm of the male's courtship call.

Zenaida macroura
Mourning dove

Length 23–34 cm (9–13 1/2 in)	
Weight 100–175 g (3 5/8–6 oz)	
Plumage Sexes alike	

Location North and Central America, Caribbean

Migration Partial migrant

Status Common

This widespread North American dove gets its name from its mournful, 4-syllable call. Small and slender, it has long, narrow wings and a long,

Goura victoria
Victoria crowned pigeon

Length 66–74 cm (26–29 in)	
Weight 2.5 kg (5 1/2 lb)	
Plumage Sexes alike	

Location N. New Guinea

Migration Non-migrant

Status Vulnerable

One of the world's largest pigeons, this species has a distinctive lacy, fan-shaped crest on top of its head. Generally

pointed tail. The female is slightly paler than the male. The latter's courtship display consists of gliding, spiralling, and flying above the female, with his wing-tips held below the body. A very rapid breeder, this dove may even reproduce in the season of its birth.

bronze sheen on neck

brown plumage

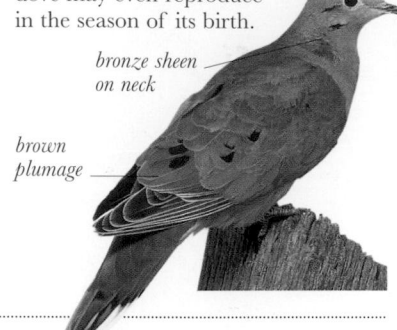

blue-grey, it has a purple-red breast and pale grey wing patches, edged with dark purple. It lives on the ground during the day, feeding on fallen fruits, and roosts in forest trees at night. This species is now vulnerable as a result of loss of its forest habitat and humans hunting it for food and its beautiful plumes. Its rarity has also made it a target for collectors.

Ptilinopus magnificus
Magnificent fruit pigeon

Length 29–55 cm (11 1/2–22 in)	
Weight 250–500 g (9–18 oz)	
Plumage Sexes alike	

Location New Guinea, N.E. and E. Australia

Migration Partial migrant

Status Vulnerable

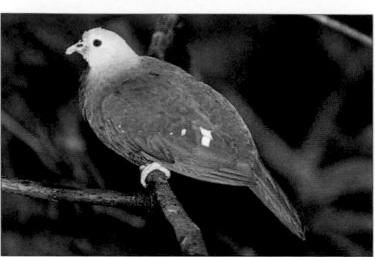

Also named the wompoo fruit dove after its "wompoo" call, this large, heavy, long-tailed fruit pigeon is threatened by loss of habitat. Despite its richly coloured plumage – mostly yellow and green, with a deep purple breast and upper belly, and a grey or greenish grey head and neck – this bird is inconspicuous and surprisingly well camouflaged in the forest canopy in which it lives. It feeds on fruits (usually figs), taken from trees, and very rarely comes to the ground.

Hemiphaga novaeseelandiae
New Zealand pigeon

Length 46–50 cm (18–20 in)	
Weight 600–800 g (21–29 oz)	
Plumage Sexes alike	

Location New Zealand

Migration Non-migrant

Status Lower risk

The largest pigeon in New Zealand, this is the only species that is native to the islands, rather than introduced. It is dark, with iridescent, bronze and green highlights on the upperparts and breast, and a white belly and lower breast. This bird feeds on a wide range of plants and fruits, which it is able to eat whole owing to its extendable gape. Several native New Zealand trees are almost totally dependent on this pigeon for their seed dispersal.

SANDGROUSE

PHYLUM	**Chordata**
CLASS	**Aves**
ORDER	**Pteroclidiformes**
FAMILIES	**1 (Pteroclididae)**
SPECIES	**16**

THESE INTRICATELY PATTERNED birds live in arid areas of Africa and Asia. They spend most of their time on the ground, where they are well camouflaged by their brown or grey, spotted or barred plumage. Sandgrouse resemble both grouse and pigeons. Like grouse, they have a small head, a squat body, and feathers on their legs, yet when they take to the air they fly strongly on long, pointed wings, with fast, steady wingbeats like those of pigeons. Their thick neck and short legs are also pigeon-like.

RETAINING WATER
Many sandgrouse (including these Burchell's sandgrouse) need to drink once or twice a day. Their belly feathers hold water, which is used to supply chicks in the nest, which are normally far from water sources.

Pterocles coronatus
Crowned sandgrouse

Length 27–30 cm (10 1/2–12 in)	
Weight 250–300 g (9–11 oz)	
Plumage Sexes differ	

Location N. Africa, W. to S. Asia

Migration Non-migrant

Status Common

This well-camouflaged bird lives in some of the most extreme desert areas, helped by its low energy and water requirements, its ability to tolerate air temperatures of over 50°C (122°F) for several hours, and its tolerance of water with a high salt content. The male has a sandy-orange crown and a black mask at the base of his bill, while the female is greyer and more barred.

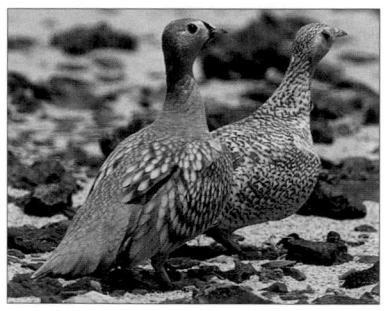

Pterocles namaqua
Namaqua sandgrouse

Length 28 cm (11 in)	
Weight 175–200 g (6–7 oz)	
Plumage Sexes alike	

Location Southern Africa

Migration Non-migrant

Status Common

The Namaqua sandgrouse shares many desert-specific adaptations with the crowned sandgrouse (see left): it needs little food or water, and its thick plumage insulates it from high and low temperatures. A pale brown head, white breastband, and mottled brown wings provide excellent camouflage against the stony ground and sand. This bird travels to areas that have received good rainfall in search of seeds, which is its sole source of food.

PARROTS

PHYLUM	**Chordata**
CLASS	**Aves**
ORDER	**Psittaciformes**
FAMILIES	**2**
SPECIES	**353**

THESE CONSPICUOUS, BRIGHTLY coloured birds have populated most of the world's warmer areas, and are particularly abundant in tropical forests. In addition to true parrots, this group includes the familiar macaws, parakeets, cockatoos, cockatiels, lorikeets, and budgerigars, among others. Noisy and sociable in the wild, parrots have long been popular as pets, prized for their beauty, intelligence, and impressive learning skills, evident in their remarkable ability to mimic human sounds. Although they may range some distance to forage, very few parrots are truly migratory.

ANATOMY

Parrots are easily recognized by their large head, short neck, and strongly hooked bill. They have hard, distinctively glossy plumage, usually predominantly green for camouflage in the forest foliage, with patches of other bright colours. Their feet, consisting of 2 toes pointing forwards and 2 backwards, are used for climbing trees, and the bill is frequently used as a third limb for climbing or holding. The wings are typically narrow and pointed, enabling parrots to fly with great speed and manoeuvrability.

upper bill in open position *upper hinge*

base of bill used for cracking food

lower hinge

hook used for pulling food

lower bill in open position

PARROT'S BILL
A parrot's hooked bill is remarkably flexible; both the upper and lower mandibles are hinged against the skull for independent movement.

NECTAR FEEDERS
Lories and lorikeets (here, a rainbow lorikeet) feed on nectar. They have a brush-like tip to their tongue to help them sweep the liquid into their mouth.

FEEDING

Virtually all parrots feed exclusively on plant material, including seeds, nuts, fruits, nectar, and flowers. Feeding is often an exercise in coordination between bill and foot, the latter grasping a food item and holding it up for the bill to work over. Although most parrots feed in the trees, many species also readily descend to the ground to forage.

SOCIAL GROUPS
Parrots (here, green-winged macaws and scarlet macaws) often form large, noisy flocks that communicate by making various calls, from harsh squawks to piercing shrieks. Most species form long-term pair-bonds.

CONSERVATION

At least a quarter of all parrot species are under threat in the wild, and a few have actually become extinct. One of the greatest threats is from the cage-bird trade, which accounts for hundreds of thousands of captures and deaths every year. Other threats to their numbers include deforestation.

Trichoglossus haematodus

Rainbow lorikeet

Length	30 cm (12 in)
Weight	150 g (5 oz)
Plumage	Sexes alike

Location New Guinea, S.E. Asia, S.W. Pacific, Australia (including Tasmania)

Migration Non-migrant

Status Common

This brightly coloured lorikeet has a stocky body and a pointed tail. It is highly variable in appearance, with 22 subspecies that differ in size or coloration, or both. Juveniles have dark brown bills, while in adults they are orange or red. Like most other lorikeets and lories, this lorikeet has a brush-tipped tongue adapted for gathering pollen and nectar from flowers. In flight, the rainbow lorikeet screeches continuously while displaying its brightly coloured underwings. However, when feeding, its subdued chatter and its cryptic coloration makes it difficult to detect among foliage. This species causes damage in orchards and vineyards, especially in Australia.

purple-blue streaked head

green upperparts

BRIGHTLY COLOURED
This startlingly bright bird varies greatly in colour. Most subspecies have green upperparts, a purple-blue streaked head, and an orange to red breast with dark scalloping.

FEEDING IN A GROUP

Flocks of rainbow lorikeets congregate to feed in flowering trees and are often found in the company of other nectar- or fruit-eating birds. This bird uses its bill to crush the flesh of fruits to extract the juice and the seeds. The rainbow lorikeet also feeds on insects and their larvae.

Eos reticulata

Blue-streaked lorikeet

Length	31 cm (12 in)
Weight	175 g (6 oz)
Plumage	Sexes alike

Location Indonesia (Tanimbar Islands, Kai Islands, Damar Islands)

Migration Non-migrant

Status Lower risk

This brightly coloured lorikeet is mainly red, with violet-blue streaking on its hindneck and back, a narrow orange-red bill, and grey legs. A noisy, conspicuous bird, it is most commonly seen in swift overhead flight, emitting shrill call-notes, but also spends much time resting or feeding in trees. Like other lorikeets, it has a brush-tipped tongue to gather pollen and nectar.

Probosciger aterrimus
Palm cockatoo

	Length 60 cm
	(23 ½ in)
	Weight 1 kg
	(2 ¼ lb)
	Plumage Sexes alike

| Location New Guinea, N.E. Australia | Migration Non-migrant |
| | Status Locally common† |

The largest of about 18 species of cockatoos, the palm cockatoo is also distinguished by its entirely black plumage, huge, hooked bill with which it can crack open palm nuts and other hard-shelled seeds, and a piercing, whistle-like call. When excited or alarmed, the palm cockatoo raises its striking crest of backward-curving, narrow feathers, and the prominent, bare crimson cheek patches deepen in colour, causing it to "blush".

elongated crest feathers

crimson cheek patches

Cacatua galerita
Sulphur-crested cockatoo

	Length 50 cm
	(20 in)
	Weight 950 g
	(34 oz)
	Plumage Sexes alike

| Location New Guinea, Australia (including Tasmania) | Migration Non-migrant |
| | Status Common |

This large white cockatoo is yellow on the undersides of its wings and tail, and has a yellow crest of narrow, forward-curving feathers. Noisy and active in the morning and late afternoon, it forages in flocks that vary in size

short, rounded tail

Eolophus roseicapillus
Galah

	Length 35 cm
	(14 in)
	Weight 325 g
	(12 oz)
	Plumage Sexes alike

| Location Australia (including Tasmania) | Migration Non-migrant |
| | Status Common |

short pink crest

grey wings

stocky body

The most widespread and numerous species of cockatoo, the galah is a familiar sight throughout Australia. Flocks of these noisy and gregarious birds are found in arid scrubland as well as in city parks. Farming has increased this parrot's food supply, leading to a steady rise in population. The galah has a relatively small crest, and its grey back and pink underparts create an alternating effect as it veers through the air.

from a few dozen birds to several hundred. At night, flocks occupy regularly used roosts, often in trees bordering water courses. This species is very popular as a pet, and both adults and chicks are regularly captured for the live-bird trade, a fact that poses a threat to the survival of some populations.

yellow crest

black bill

white upperparts

short legs

Nymphicus hollandicus
Cockatiel

	Length 32 cm
	(12 ½ in)
	Weight 90 g
	(3 ¼ oz)
	Plumage Sexes differ

| Location Australia | Migration Nomadic |
| | Status Common |

The cockatiel is the smallest species of cockatoo and the only one to have a long, pointed tail. The tapering crest is lowered when the bird is resting, or sometimes while feeding. In flight, the

Nestor notabilis
Kea

	Length 48 cm
	(19 in)
	Weight 825 g
	(29 oz)
	Plumage Sexes alike

| Location New Zealand (South Island) | Migration Non-migrant |
| | Status Vulnerable |

sharp, elongated bill

dark-edged feathers

well-developed feet for foraging on ground

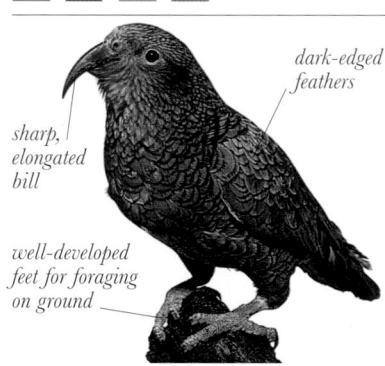

This large, stocky mountain bird is remarkable among parrots for its insatiable curiosity and wide-ranging diet. A natural opportunist, it will examine anything that appears to be edible, and often feeds on carrion, tearing it up with its unusually long bill, which resembles that of a bird of prey. Adults are largely olive-green, with orange underwings.

Platycercus elegans
Crimson rosella

	Length 36 cm
	(14 in)
	Weight 150 g
	(5 oz)
	Plumage Sexes alike

| Location E. and S.E. Australia | Migration Non-migrant |
| | Status Common |

This is one of 8 species of rosella, all of which have similar streamlined bodies and long tails, and are restricted to Australia and its outlying islands.

yellow crest of male

grey crest of female

orange patch on ear coverts

cockatiel emits a distinctive warbled chirrup, but is quieter while feeding on the ground or drinking at water holes. This species is very popular as a pet and numerous colour variants are found in captive-bred birds.

Eclectus roratus
Eclectus parrot

	Length 35 cm
	(14 in)
	Weight 500 g
	(18 oz)
	Plumage Sexes differ

| Location S.E. Asia, New Guinea, N.E. Australia, S.W. Pacific islands | Migration Non-migrant |
| | Status Locally common |

The male and female eclectus parrots look so different from each other that early naturalists identified them as separate species. Both sexes are large and stocky, with squarish tails and robust bills, but while the male is brilliant green with red flanks and underwings and a yellow bill, the female is red, occasionally with a blue underside and underwings, and has an entirely black bill.

blue hindneck of female

brilliant green body of male

violet-blue cheek patches

As its name suggests, the crimson rosella's plumage is red. It has violet-blue cheek patches and outer wing coverts, and a dark blue tail; the back and wings are mottled red and black. The bird tends to become quite tame around farms or in gardens. It has an undulating flight; it drops down towards the ground, glides upwards, and then lands.

dark blue tail

Lathamus discolor
Swift parrot

Length 25 cm (10 in)	
Weight 65 g (2 3/8 oz)	
Plumage Sexes alike	
Location E. and S.E. Australia	Migration Non-migrant
	Status Endangered

green upper-parts

red face

yellowish underparts

Long, pointed wings and a streamlined shape contribute to making this small parrot a fast flier, hence its name. Its crown is dark blue, contrasting with a red face and underwings. Found in a wide variety of wooded habitats, including suburban parks or gardens, where favoured food trees are present, it breeds only in Tasmania and overwinters in eastern Australia.

Melopsittacus undulatus
Budgerigar

Length 18 cm (7 in)	
Weight 25 g (7/8 oz)	
Plumage Sexes alike	
Location Australia	Migration Nomadic
	Status Common

The budgerigar is one of the most numerous of all parrots and is the best known because of its popularity as a pet. Unlike its domesticated counterparts, which may be of different colours, the wild budgerigar is always green, with a yellow face and blue tail. The upperparts of its body are barred black and yellow, providing excellent camouflage while it feeds in vegetation. Often nomadic, this bird is usually seen in large, noisy flocks.

Deroptyus accipitrinus
Red-fan parrot

Length 35 cm (14 in)	
Weight 225 g (8 oz)	
Plumage Sexes alike	
Location N. South America	Migration Non-migrant
	Status Vulnerable

This parrot is distinguished by dark red feathers that extend from the nape to the hindneck. When excited or alarmed, it raises its neck feathers to form a spectacular ruff or fan. Its long, rounded tail is also a distinctive

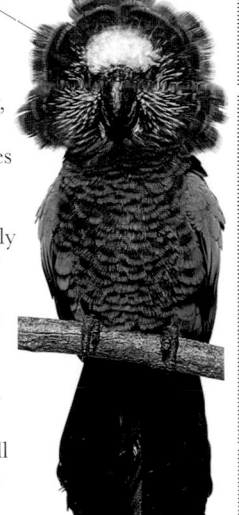

blue-edged red neck feathers

characteristic. While perching, the red-fan parrot resembles a bird of prey, with its hawk-like eyes. Usually seen in pairs or small groups (rarely of up to 20 birds), this forest-dwelling parrot is inconspicuous but is often heard well before it comes into view.

Psittacus erithacus
Grey parrot

Length 33 cm (13 in)	
Weight 400 g (14 oz)	
Plumage Sexes alike	
Location W. to C. Africa	Migration Non-migrant
	Status Locally common

This stocky, short-tailed parrot is an extremely popular cage bird because of its ability to imitate human speech and perform tricks. Despite being illegally trapped for the pet trade, it is still common in parts of its range.

yellow eyes

It is found mainly in lowland rainforest, and also in montane rainforest, at forest edges, on plantations, and on farmland or in gardens. It has unusual coloration, its grey plumage contrasting strongly with the bright red or deep maroon of the tail. In flight, members of the flock continually call to each other and also communicate visually by displaying their red tails.

Strigops habroptilus
Kakapo

Length 64 cm (25 in)	
Weight 2 kg (4 1/2 lb)	
Plumage Sexes alike	
Location New Zealand (3 offshore islands)	Migration Non-migrant
	Status Extinct in wild

One of the world's most endangered parrots, the flightless kakapo has highly unusual feeding habits. It is active at dusk and often walks long distances to feeding areas where it chews plants for their juices, leaving balls of fibrous material hanging from the plants. It also digs up or crushes rhizomes with its bill. The males gather at mating grounds (leks), where they dig hollows, and make loud, booming noises to attract females. The adults, chicks, and eggs of the kakapo are highly vulnerable to introduced predators.

CONSERVATION

Fewer than 100 kakapo survive today and a scheme is underway to save the species from extinction. The birds have been moved to 3 offshore islands where they are free from predators, and where – as with this female – their progress can be carefully monitored.

yellowish bars

short wings

FLIGHTLESS
The heaviest species of parrot, the kakapo has rounded wings that cannot sustain it in the air. Its strongly ridged bill is adapted to extracting juices from plants.

Agapornis personatus
Masked lovebird

Length 14.5 cm (5 3/4 in)	
Weight 50 g (1 3/4 oz)	
Plumage Sexes alike	
Location E. Africa (Tanzania)	Migration Non-migrant
	Status Locally common

Lovebirds get their name from their strong pair-bonds, with the male and female spending much of their time close together, frequently preening each other's feathers. All 9 species are small, short-tailed, and solidly built. The masked lovebird has a dark head, with a conspicuous white eye-ring and bright red bill. It feeds on seeds, fruits, and buds, and, unusually for a parrot, builds a nest instead of laying its eggs in an unlined hole.

white eye-ring

short wings

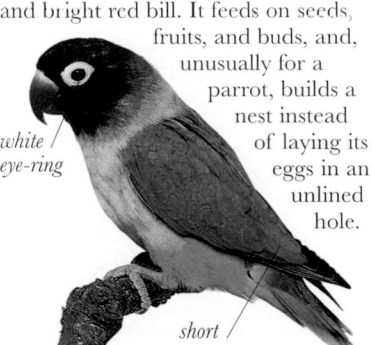

Amazona aestiva
Blue-fronted parrot

Length 37 cm (14 1/2 in)	
Weight 400 g (14 oz)	
Plumage Sexes alike	
Location C. South America	Migration Non-migrant
	Status Common

A distinctive blue forehead on a yellow face differentiates this large and stocky parrot from other members of the genus *Amazona*, of which there are about 30 species. Although conspicuous and noisy in flight, the blue-fronted parrot is unobtrusive and quiet while feeding or resting in treetops, where falling pieces of fruits may be the only indication of its presence. Birds gather in flocks at regular night-time roosts, but mated pairs always stay close together.

Psittacula krameri
Rose-ringed parakeet

Length	40 cm
	(16 in)
Weight	125 g
	(4 oz)
Plumage	Sexes alike

Location W. to E. Africa, S. and S.E. Asia

Migration Non-migrant

Status Common

This slim-bodied green bird has a rose-pink collar around its hindneck. It has the widest natural distribution of any parrot, stretching from West Africa to Southeast Asia. As a feral bird, it is also found in parts of Europe and North America. The central tail feathers are long and narrow, and the backward-sweeping wings produce a characteristically streamlined flight silhouette.

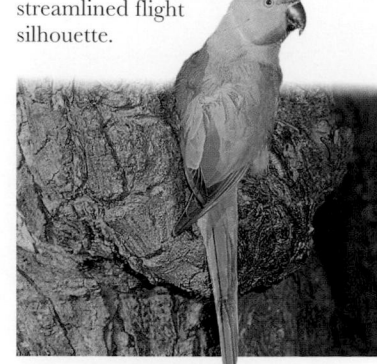

Myiopsitta monachus
Monk parakeet

Length	29 cm
	(11 1/2 in)
Weight	125 g
	(4 oz)
Plumage	Sexes alike

Location C. and S. South America

Migration Non-migrant

Status Common

The nesting habits of this green parakeet are unique among parrots. With several other pairs, it roosts in a communal nest, which forms the centre for the birds' daily activities.

green upperparts

whitish throat

The colony expands as nests are built alongside or on top of existing ones. This parakeet has a long green tail, an olive-green abdomen, and a grey-white face and throat. In some parts of South America, it is a serious agricultural pest.

Aratinga jandaya
Yellow-headed conure

Length	30 cm
	(12 in)
Weight	125 g
	(4 oz)
Plumage	Sexes alike

Location E. South America

Migration Non-migrant

Status Locally common

One of 30 species of conure, all found in Central or South America, this is a small parrot with a narrow, pointed tail. Its head and neck are yellow, gradually becoming orange-red on the breast and abdomen, and its back and wings are largely green. Usually encountered in noisy flocks of up to 15, this very active bird spends much of its time clambering among branches of trees or shrubs to find fruits. When disturbed, it swiftly takes flight while screeching loudly.

green back and wings

Anodorhynchus hyacinthinus
Hyacinth macaw

Length	1 m
	(3 1/4 ft)
Weight	1.5 kg
	(3 1/4 lb)
Plumage	Sexes alike

Location C. South America

Migration Non-migrant

Status Endangered

yellow chin patch

The largest and probably the most spectacular of all parrots, the hyacinth macaw is distinguished by its rich cobalt-blue plumage, contrasting bright yellow chin patch and ring around the eye, and long, pointed tail. The massive bill is used for crushing hard palm nuts. It feeds mainly in the trees, but comes to the ground to pick up fallen fruits and nuts. Its high value as a cage bird has caused a major decline in its numbers.

long, narrow tail

Cyanopsitta spixii
Spix's macaw

Length	55 cm
	(22 in)
Weight	Not recorded
Plumage	Sexes alike

Location E. South America

Migration Non-migrant

Status Extinct in wild†

The smallest of all blue macaws, Spix's macaw is a creature of habit, regularly sitting on favoured perches atop tall trees and daily following the same flight paths. This parrot feeds on seeds, nuts, and fruits while clambering among branches.

BLUE PARROT
Spix's macaw has a long, pointed tail, and a silvery blue head, with dark grey skin surrounding the eyes. The back and wings are darker blue while the bill is black.

CONSERVATION

While Spix's macaw has seemingly never been a common bird, settlers moving into its range and the destruction of the caraiba woodland in which it nests eventually restricted its distribution to a small area of northeastern Brazil, and pushed this macaw to the brink of total extinction. The illegal bird trade is also responsible for this bird's decline. Today, however, as a result of conservation measures begun in the late 1980s, more than 50 of these parrots have been successfully bred in captivity. With the last known wild Spix's macaw now believed to have disappeared, the survival of the species depends entirely on the success of the breeding and release programme.

Ara chloroptera
Green-winged macaw

Length	90 cm
	(35 in)
Weight	1 kg
	(2 1/4 lb)
Plumage	Sexes alike

Location N. and C. South America

Migration Non-migrant

Status Locally common

The green-winged macaw attracts attention by its spectacular, vivid coloration and strident call. A large bird, like some other macaws it has a partly red plumage and a light blue back and rump. The long tail is tipped with blue. Its wings are blue with dark green upper wing coverts, which give this species its name. The juveniles resemble adults but have shorter tails. This macaw prefers humid, lowland forest, but in the southern part of its range, where the more common scarlet macaw (*Ara macao*) is absent, it frequents open habitats, including deciduous forest, and is often seen perching on the topmost projecting branches of tall trees. Usually encountered in pairs or small groups, possibly family groups, it sometimes associates with other, more common macaws, especially at earth banks, where large numbers gather to consume exposed mineral sands. In flight, green-winged macaws call loudly to each other. This bird feeds on seeds, fruits, and nuts that are often larger or harder than those eaten by other macaws. Adult birds are killed for food or for feathers, while the chicks are taken from the nest for the live-bird trade.

white facial skin

red head and shoulders

dark red plumage

dark green wing coverts

CUCKOOS AND TURACOS

PHYLUM	**Chordata**
CLASS	**Aves**
ORDER	**Cuculiformes**
FAMILIES	**3**
SPECIES	**160**

ALTHOUGH THEY HAVE LOUD VOICES, cuckoos and turacos are secretive birds. Cuckoos, which occur throughout the world, are generally grey or brown, although some have striking patches or streaks. Turacos, which are exclusive to Africa, are sometimes brilliantly coloured, with unique red and green pigments. Both cuckoos and turacos have short wings, a long tail, and 2 pairs of toes on each foot, one pointing forwards and the other back, enabling them to climb as well as perch on trees. Besides cuckoos and turacos, this group includes the hoatzin, placed in a separate family.

BROOD PARASITES
Some cuckoos are brood parasites, laying their eggs in other birds' nests. The eggs often closely resemble those of the host. The cuckoo chick (here, a common cuckoo) may eject the host's eggs or young, and take food brought by its new "parent" (here, a reed warbler).

Clamator glandarius
Great spotted cuckoo

Length 35–39 cm (14–15½ in)	
Weight 125 g (4 oz)	
Plumage Sexes alike	
Location S. Europe, W. Asia, Africa	**Migration** Migrant
	Status Locally common†

The largest cuckoo found in Europe, this handsome bird lays its eggs in the nests of crows, magpies, and starlings. Unlike

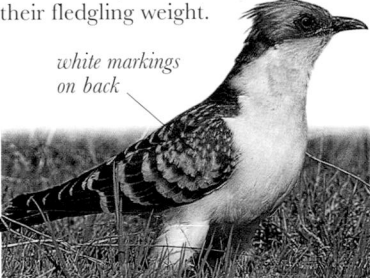
white markings on back

common cuckoos, the young do not evict their nest-mates. However, since they grow faster than the other chicks, they get a larger share of the food brought by the foster parents, with the result that within the first 8 days of their life they reach half their fledgling weight.

Tauraco erythrolophus
Red-crested turaco

Length 40–43 cm (16–17 in)	
Weight 200–325 g (7–12 oz)	
Plumage Sexes alike	
Location S.W. Africa	**Migration** Non-migrant
	Status Vulnerable†

Like other turacos, this species derives its bright colours from a copper-based pigment, unique to the turaco family. A tree-dwelling fruit-eater, it spends

most of its time hopping from branch to branch, and has a heavy, laboured flight. It usually lives in family groups, foraging in the rainforest canopy and vigorously defending both feeding and nesting sites from other turacos as well as other fruit-eating birds.

broad, dark blue tail

Opisthocomus hoazin
Hoatzin

Length 62–70 cm (24–28 in)	
Weight 700–900 g (25–32 oz)	
Plumage Sexes alike	
Location N. South America	**Migration** Non-migrant
	Status Locally common

This primitive-looking, tree-dwelling bird feeds almost entirely on leaves, a feat few other species can accomplish, because its large stomach enables it to

process plant material in a similar way to grazing mammals. The young leave the nest before they can fly and use their tiny wing claws to climb through vegetation; they dive into water if threatened. As many as 8 birds may live together, defending their common territory and helping to rear the young.

Cuculus canorus
Common cuckoo

Length 32–33 cm (12½–13 in)	
Weight 125 g (4 oz)	
Plumage Sexes alike	
Location Europe, Asia, N.W. and southern Africa	**Migration** Migrant
	Status Common

long body

black-barred underside

Usually grey with a black-barred, white underside, the common cuckoo breeds mainly in forests throughout Europe and Asia, but travels long distances to spend the winter in Africa and southern Asia. This familiar cuckoo is a classic example of a brood parasite, laying its eggs in the nests of a number of other species (see above).

Geococcyx californianus
Greater roadrunner

Length 56 cm (22 in)	
Weight 325 g (12 oz)	
Plumage Sexes alike	
Location S. North America	**Migration** Non-migrant
	Status Common

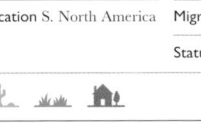

Although it can fly, this long-legged member of the cuckoo family spends most of its time on the ground. Over short distances it can reach speeds of more than 30kph (18mph). It walks and runs through the desert, trying

---

Crotophaga ani
Smooth-billed ani

Length 35 cm (11 in)	
Weight 125 g (4 oz)	
Plumage Sexes alike	
Location S.E. North America, Central America, South America, Caribbean	**Migration** Non-migrant
	Status Locally common

This odd-looking bird has a deep, arched bill and feeds mainly by following cattle, catching the insects they disturb as they move. It lives in groups of up to several pairs, which share the work of incubating the eggs and raising the chicks. These often stay in the nest once they are fledged, helping to raise the next brood.

to flush out prey – lizards, snakes, birds, and small mammals – and kills whatever it catches with a blow from its strong, pointed bill. This roadrunner also runs to avoid predators, swinging its long tail from side to side like a rudder to make turns while speeding.

heavily streaked body

OWLS

PHYLUM	Chordata
CLASS	Aves
ORDER	Strigiformes
FAMILIES	2
SPECIES	205

OFTEN HEARD YET seldom seen, owls are hunters that operate mainly at night. They resemble the day-flying predators, the hawks and falcons, in having sharp talons and hooked bills for catching and subduing prey, but in addition owls have several adaptations that help them hunt in the dark. Their eyes are very large, to gather all available light, and forward-facing, to help them judge distance. They have exceptionally acute hearing, and their soft plumage enables them to fly silently. The 2 families of owls – typical owls and barn owls – are found worldwide in most habitats, from dense forest to tundra.

REPRODUCTION

Owls do not build their own nest. Instead, they rely on the previous years' efforts of other birds or simply select a suitable cavity in the ground, a tree, a rocky crevice, or a building. The almost spherical eggs are laid in the nest or straight onto the surface. Most owls lay 2–7 eggs, which usually hatch at 2-day intervals, leading to large discrepancies in the age of the chicks in a brood. If food is scarce, the older chicks take the larger part of the food offered, while the younger chicks may starve.

YOUNG
The youngest owlet in a brood (here, snowy owls) can be 2 weeks younger than the oldest.

ANATOMY

Owls are highly distinctive, with an upright posture, a large, rounded head, and a short tail. The outer toe is reversible, allowing it to point forwards or backwards, improving the ability to perch or grab prey. Owls have excellent eyesight, which works as well in daylight as it does at night. They also have exceptional hearing, easily picking up the faint rustle made by a small mammal, even under snow. Some species can hunt in total darkness, their asymmetrical ear openings giving them a three-dimensional perception of sound. All owls have soft, dense plumage, with soft fringes on their flight feathers that muffle the sound of air turbulence.

lens
iris
large pupil
eye fixed in socket
tubular eyeball

VISION
An owl's eyeballs are tubular and cannot be swivelled. As a result, if the bird needs to look to the side, it must move its whole head. To compensate, an owl can rotate its head and neck about an arc of more than 270 degrees.

NOCTURNAL HUNTER
Under cover of darkness, a little owl closes silently upon its prey. Equipped with sharp talons, acute eyesight, and exceptional hearing, an owl is a formidable night-time predator.

FEEDING

Owls take a wide variety of living prey, the size of which depends on the owl species. Most feed on insects, birds, or small mammals, and large owls quite commonly catch and eat smaller owls. Owls that live in woods and forests tend to drop from a stationary position onto their prey, but those that occur in open country must hunt on the wing, in slow, low-level, quartering flight.

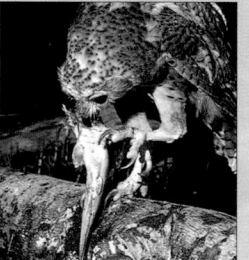
FISHING
Several species of owl, including Pel's fishing owl (right), feed on fish. They swoop down to pluck fish from the water surface, striking them with their talons and then taking hold with their bristly foot pads.

intact bones
fur
PELLETS
Owls generally swallow their food whole, taking in fur, feathers, bones, and insect chitin. Later, they regurgitate these indigestible parts (see below) in the form of compact pellets, which collect beneath their nest or roosting site.

REGURGITATION

① SWALLOWING PREY
A barn owl swallows its prey whole, including much indigestible matter, such as fur and bones.

② PREPARING TO CAST
About 6 or 7 hours after eating, the owl is about to regurgitate (or cast) the indigestible parts of its meal.

③ ROUNDED PELLETS
To ease its passage, the undigested food is rolled into a pellet, with hard parts wrapped in fur and feathers.

④ CAUGHT IN THE THROAT
Despite being rounded, the pellet causes the owl to strain as it moves up the oesophagus.

⑤ RELEASING THE PELLET
Finally, a shiny black pellet is cast from the mouth. The owl may cast a second one within 24 hours.

Tyto alba

Barn owl

Length 29–44 cm (11 1/2–17 1/2 in)	
Weight 300–650 g (11–23 oz)	
Plumage Sexes alike	

Location North, Central, and South America, Europe, Asia, Africa, Australia

Migration Non-migrant

Status Common

Found on all continents except Antarctica, the barn owl is the most widespread of all species of owl, and one of the most widely distributed of all land birds. It has a pale, heart-shaped face, long legs covered in white feathers, and a very short tail. The female lays eggs in a hollow tree or an abandoned building. She feeds the young with food brought by the male; she also broods them for up to 3 weeks from hatching, until they have acquired the down they need to keep themselves warm. Changing agricultural practices have reduced the barn owl's food supply; in some areas it is now rare.

dark eyes

serrated edges of outer flight feathers

HEART-SHAPED FACE
The barn owl may be readily identified by its pale, heart-shaped face and dark eyes.

short tail

ON TARGET
The barn owl is an expert hunter, its excellent vision and hearing enabling it to pinpoint prey in total darkness. It flies low, slowly and silently, then swoops swiftly to the ground. At the last moment, it swings its legs forwards, spreading its sharp-clawed toes to grasp and kill its prey.

long, white-feathered legs

A MEAL IN ONE

Small rodents such as mice, rats, and voles make up the bulk of the barn owl's diet. Once it has killed its prey, the owl swallows it whole, later regurgitating any parts that are undigestible in the form of pellets. If it has young to feed, the owl takes the prey back to the nest, carrying it in its bill.

Otus scops

Eurasian scops owl

Length 16–20 cm (6 1/2–8 in)	
Weight 60–125 g (2 1/8–4 oz)	
Plumage Sexes alike	

Location Europe to C. Asia, Africa

Migration Migrant

Status Locally common

Small and superbly camouflaged, this owl is heard more often than it is seen, its call being a low whistle, repeated every few seconds. The fine black flecks in its grey or reddish brown plumage make it almost invisible against the bark of a tree. When alarmed, it will stretch itself, and even sway, to imitate a branch and so remain hidden. This owl feeds mainly on insects, swooping down on them from a perch, but also eats spiders, earthworms, reptiles, bats, and small birds.

Otus lempiji

Sunda scops owl

Length 20 cm (8 in)	
Weight 100–125 g (3 5/8–4 oz)	
Plumage Sexes alike	

Location S.E. Asia

Migration Non-migrant

Status Common

The Sunda scops owl is a small, eared owl, with brown eyes. It occurs in both a brownish grey and a reddish morph. Inhabiting forests, forest edges, planta-tions, and areas with scattered trees, such as parks and villages, it spends much of the day sitting camouflaged in hiding places, coming out at dusk to hunt mainly for insects. Males and females often call together, giving out a short, mellow hooting call.

prominent ear tufts

Pulsatrix perspicillata

Spectacled owl

Length 43–52 cm (17–20 1/2 in)	
Weight 600–1,000 g (21–36 oz)	
Plumage Sexes alike	

Location S. Mexico to C. South America

Migration Non-migrant

Status Common

This owl, common in the American tropics, gets its name from the ring of white feathers, or "spectacles", around its eyes. Most often found in dense rainforest, it also lives in woodland edges and coffee plantations. Instead of patrolling on the wing like many owls, it hunts from a perch, snatching prey from the ground or foliage. It usually preys on small forest mammals and insects, but also hunts near water, taking crawfish and crabs.

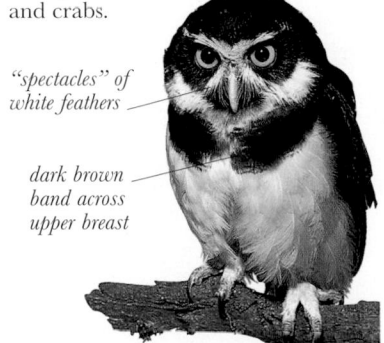

"spectacles" of white feathers

dark brown band across upper breast

Scotopelia peli

Pel's fishing owl

Length 55–63 cm (22–25 in)	
Weight 2–2.5 kg (4 1/2–5 1/2 lb)	
Plumage Sexes alike	

Location Africa

Migration Non-migrant

Status Locally common

This owl is one of 3 African species that specialize in feeding on fish, frogs, and other freshwater animals. Like its relatives, Pel's fishing owl has long, bare legs and curved talons that help to grasp its slippery prey. The plumage of Pel's fishing owl is generally light chestnut, with dark spots and bars, which help camouflage it while it roosts. This large owl lives on the wooded edges of lakes, rivers, swamps, and marshes. It hunts after dark, launching itself from a perch and swooping low over the water. Having seized its prey, it returns to a perch to feed. It makes its nest in a tree-hole, usually near water. Females lay one or 2 eggs, but normally only one young owl is reared. Both parents feed and rear the young, which may remain in the nest area for up to 8 months after fledging. A separate group of fishing owls – belonging to the genus *Keputa* – is found in Asia.

very small facial disc

dark spots and bars

curved talons

broad flight feathers

Bubo virginianus
Great horned owl

Length	50–60 cm (20–23½ in)
Weight	675–2,500 g (1½–5½ lb)
Plumage	Sexes similar
Location North, Central, and South America	**Migration** Non-migrant
	Status Common

The great horned owl, with its distinctive ear tufts or "horns", is the largest of the American owls and found throughout the continent. It occurs in a broad range of habitats, from forest to desert, and is known to nest at high altitudes, usually in old nests of other large birds, but also in tree cavities and on cliff edges. It is generally a sedentary bird and, especially during the breeding season, very territorial in its nesting and hunting ranges. Usually active between dusk and dawn, its night vision and hearing are extremely acute, making it an effective hunter. Its main prey are small mammals, but insects, reptiles, amphibians, and birds, including other owl species, are also taken. Great horned owls are very vocal, especially during courtship; their loud hoot represents the classic owl call.

large, powerful feet with sharp talons

large, horn-like ear tufts

pale yellow eyes

sharp, hooked bill

POWERFUL BUILD
This very large owl, with its sharp bill, large wings, and powerful talons, is built for hunting. Its piercing yellow eyes and prominent ear tufts add to its intimidating appearance.

BIRDS

SILENT HUNTER
As well as being the largest American owl, the great horned owl is probably also the fiercest. It mainly still-hunts from a series of favoured vantage points throughout its territory. It is very swift, and having spotted a target it glides down, in total silence, to capture its prey in its immensely powerful talons.

PARENTS AND JUVENILES

Great horned owls are very attentive parents, with both males and females tending and feeding their young for at least 6 weeks after fledging. They are also very defensive birds and have been known to drive away intruding humans from their nests.

ON THE NEST
The great horned owl lays between one and 5 eggs, with a greater number when food is abundant.

JUVENILE OWL
At about 2 months old the horned owl is almost fully feathered and capable of short flights. At this stage it can puff itself up and turn its wings forwards to look even larger when defending itself.

Nyctea scandiaca

Snowy owl

	Length 55–70 cm (22–28 in)
	Weight 1–2.5 kg (2¼–5½ lb)
	Plumage Sexes differ
Location Circumpolar around Arctic	Migration Partial migrant
	Status Common

With its all-white plumage, the male snowy owl is one of the most distinctive of owls. Long, dense feathers extend right to the toes, and even the bill is largely covered, giving the bird superb insulation against the cold. This owl is most active at dusk and dawn, but becomes diurnal in summer when daylight is continuous. It spends much of the time on the ground or perching on low rocks, and uses its extraordinary eyesight and hearing to locate distant or snow-covered prey, then ambushes silently. The snowy owl feeds on lemmings, rabbits, hares, and waterfowl. In fact, its breeding cycle is directly related to the abundance of lemmings: the population of the latter tends to rise and fall on a 3- to 4-year cycle, and the snowy owl follows suit.

MOTTLED FEMALE
The female snowy owl is mottled with black – a colour scheme that hides her in rocky outcrops when she nests after most of the snow has melted. She is substantially larger than the male.

mottled plumage

ON THE GROUND

The snowy owl nests on the tundra, forming only a slight hollow in the ground in which the eggs are laid. The female tends the young, feeding them with food brought by the male.

Strix aluco

Tawny owl

	Length 37–39 cm (14½–15½ in)
	Weight 450–550 g (16–20 oz)
	Plumage Sexes alike
Location Europe, Asia, N.W. Africa	Migration Non-migrant
	Status Common

The tawny owl is found in a broad range of habitats, wherever there is sufficient tree cover to provide a daytime roost. Its usually chestnut-brown plumage is heavily streaked and mottled, providing excellent camouflage among branches and leaves. It has a wide variety of calls, among which the best known is the "twit twoo" call produced during the breeding season. This owl hunts from a perch, and can locate its prey – mainly small mammals, birds, reptiles, and insects – by sound alone.

Athene cunicularia

Burrowing owl

	Length 19–25 cm (7½–10 in)
	Weight 125–250 g (4–9 oz)
	Plumage Sexes differ
Location North, Central, and South America, Caribbean	Migration Partial migrant
	Status Locally common

Most owls roost and nest in trees, but the burrowing owl makes its home underground – an adaptation that allows it to live in all kinds of open,

Glaucidium perlatum

Pearl-spotted owlet

	Length 17–20 cm (6½–8 in)
	Weight 50–150 g (1¾–5 oz)
	Plumage Sexes alike
Location Africa (south of Sahara)	Migration Non-migrant
	Status Common

This compact bird is Africa's most diurnal owl, hunting at almost any time of day or night. Its powerful feet enable it to catch prey even larger than itself. On the back of its head are 2 black patches ringed with white. These "false eyes" serve either to deter predators that might attack from behind, or confuse prey so that they do not know which way to flee.

treeless areas, from South American pampas to airports and golf courses. When on guard outside the burrow, it adopts a very upright stance and gives a harsh, rattling call, thought to imitate a rattlesnake. It spends much of its time hunting on the ground.

Ninox novaeseelandiae

Boobook owl

	Length 30–35 cm (12–14 in)
	Weight 150–175 g (5–6 oz)
	Plumage Sexes alike
Location Australia (including Tasmania), S. New Guinea, S.E. Asia	Migration Non-migrant
	Status Locally common

There are several races of this small, stocky owl found in Australia and Asia. One form, found in Queensland, is dark brown, while the form found in Tasmania and New Zealand, and sometimes considered a separate species, is lighter, with contrasting spots. All have a high-pitched, 2-syllable, "boobook" call. Wherever it lives, the boobook owl (also called the morepork owl) roosts in trees during the day, emerging to feed at dusk. It specializes in hunting insects and birds in mid-air.

Asio flammeus

Short-eared owl

	Length 37 cm (14½ in)
	Weight 200–500 g (7–18 oz)
	Plumage Sexes differ
Location North America, W. and S. South America, Europe, Asia, Africa	Migration Partial migrant
	Status Locally common

With its long wings and low, flapping flight, this heavily mottled owl can easily be mistaken for a hawk. Usually a solitary bird, it normally roosts on the ground, but in winter, particularly when it snows, it will perch in groups in trees. Groups can also be seen when food is scarce, as large numbers move to richer areas, or when there is an abundance of food in one particular place. This owl finds its prey – mostly small mammals such as mice and voles, and at times birds – mostly by flying, and then pouncing from mid-air. It lays its eggs in heather, grass, or crops, digging a hollow to hold them safely; unusually for an owl, the female may build a nest from sticks that are lying nearby. It is a fairly quiet owl, with a low, gentle hooting call.

IN BROAD DAYLIGHT

In many parts of its vast range, the short-eared owl – one of the most diurnal of owls – is often seen hunting by day, flying just a few metres above the ground. Its flight is butterfly-like, its large wings allowing it to fly slowly without stalling.

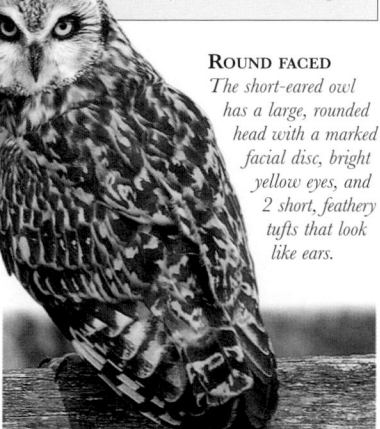

ROUND FACED
The short-eared owl has a large, rounded head with a marked facial disc, bright yellow eyes, and 2 short, feathery tufts that look like ears.

NIGHTJARS AND FROGMOUTHS

PHYLUM	Chordata
CLASS	Aves
ORDER	Caprimulgiformes
FAMILIES	5
SPECIES	118

THESE LONG-WINGED BIRDS, together with their relatives the nighthawks, potoos, owlet-nightjars, and oilbirds, are adapted for a life spent mainly in the air; indeed, most cannot walk or hop. They hunt at dusk, dawn, or at night – usually capturing insects on the wing – and roost motionless in trees or on the ground during the day. Many species are known for their loud, distinctive call. Nightjars occur almost throughout the world, in forests and open habitats; their relatives are restricted to the Americas, Asia, or Australasia.

ANATOMY

Nightjars and their relatives are round-bodied birds, with a large head and short neck. They have a remarkably large mouth, with an extremely wide gape for trapping insects. Most species have long tails and elongated wings that are ideal for flying swiftly with rapid changes of direction in pursuit of food. Except for owlet nightjars, these birds have short legs and weak, tiny feet that are unsuitable for walking. Birds in this group usually have brown or grey, cryptically patterned plumage.

WIDE GAPE
The bill of nightjars and their relatives (here, a red-necked nightjar) is almost as broad as their head and can be opened very wide. In nightjars, this makes an effective trap for flying insects, which are caught in mid-air, one at a time.

FEEDING

Most of the birds in this group feed primarily on insects, which they catch in mid-air; alternatively, they may swoop down to take them from the ground. Many species fly continually, while others (such as the potoo) make brief but regular sorties from a perch. Frogmouths feed on other birds, mammals, and amphibians, as well as insects. Oilbirds eat only fruits.

BRISTLES
Nightjars (here, the European nightjar) and some of their relatives are equipped with a set of bristles around the edge of their bills. These bristles may be sensitive to touch, and in some species help to filter insects into the mouth.

AVOIDING DETECTION
Perched on an upright branch, the common potoo flattens its wings against its body and points its head upwards with the bill slightly open. This stance, combined with the camouflage colouring, makes the bird resemble a broken branch, providing protection against predators.

BIRDS

Oilbird

Length	40–49 cm (16–19 1/2 in)
Weight	350–475 g (13–17 oz)
Plumage	Sexes alike

Location N. and C. South America	**Migration** Non-migrant
	Status Locally common†

The oilbird is a unique species – the only nocturnal, fruit-eating bird in the world. During the day, it remains deep inside caves, sometimes 1km (2/3 mile) underground. It leaves the cave at night and may travel over 75km (45 miles) in search of food. Oilbirds use sight and possibly smell to locate the fruits of trees such as laurels and palms, swooping down and plucking them from branches.

LARGE-MOUTHED BIRD
The reddish brown oilbird has white spots on the head, throat, and wings. Its large mouth enables it to carry plenty of food back to the young in the nest – a mound of mud, faeces, rotten fruits, and seeds.

ECHOLOCATION

The oilbird nests and roosts in large colonies in caves. It uses echolocation to find its way in the dark; the signals it produces are relatively low-pitched, and sound like "clicks" to the human ear. Birds also call to one another to stay in touch, filling the cave with their loud, harsh cries.

Tawny frogmouth

Length	34–53 cm (13 1/2–21 in)
Weight	175–675 g (6–24 oz)
Plumage	Sexes differ

Location S. New Guinea, Australia (including Tasmania)	**Migration** Non-migrant
	Status Common

Frogmouths look similar to nightjars, but instead of hunting nocturnal insects on the wing, they pounce on small animals and larger insects on the ground. The tawny frogmouth's large eyes help it to see well in the dark, and it flies through its territory at night, stopping at suitable foraging perches and watching for prey to move nearby. When necessary, it will beat larger animals to death before swallowing them. This species lives in pairs or family groups, and the birds call frequently to maintain contact over the large territory in which they feed and breed. The nest is a simple platform of sticks, placed in the fork of a tree.

DEFENCE POSTURE

If threatened, the tawny frogmouth stands erect, looking like a broken branch on a tree. This posture may sometimes be accompanied by a gaping display in which the bird opens its large, wide mouth.

— *large head*

— *stocky body*

CAMOUFLAGE
The tawny frogmouth's red- or brown-mottled grey plumage blends in with the colour of branches of trees and bushes.

— *short legs*

Nyctibius griseus
Common potoo

Length 33–38 cm (13–15 in)	
Weight 150–200 g (5–7 oz)	
Plumage Sexes alike	
Location S. Central America, South America	Migration Non-migrant
	Status Common

One of about 5 species of potoo found in Central and South America, the common potoo is a solitary bird that spends the day in trees, perching upright where it looks like a broken branch. It keeps its conspicuous yellow eyes shut to avoid detection by predators, but creases in the eyelids allow it to continue to watch for danger. This bird comes to life at night, darting from its perch to catch insects in the air. Birds form pairs for breeding, laying a single egg in a knot-hole or depression in a branch.

Aegotheles cristatus
Australian owlet-nightjar

Length 21–25 cm (8½–10 in)	
Weight 35–65 g (1¼–2⅜ oz)	
Plumage Sexes differ	
Location S. New Guinea, Australia (including Tasmania)	Migration Non-migrant
	Status Locally common

This tree-dwelling bird is a small, nocturnal insect-eater that looks very much like a miniature owl. Compared to other nightjars, it has well-developed feet and legs, and a long, slender tail. There are distinctive dark markings on its face and head. The barred plumage is grey in the male and brown in the female. The Australian owlet-nightjar is an acrobatic flier, with short, round wings permitting it to make brief, foraging flights from its perch to capture insects, either on the ground or in the air. It is preyed upon by mammals and monitor lizards.

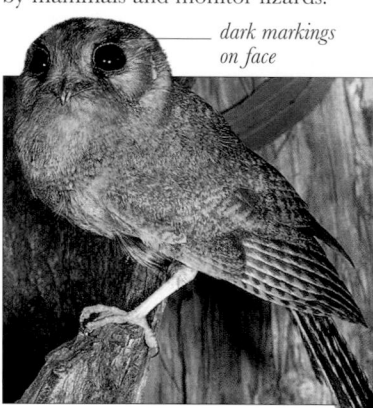

dark markings on face

Chordeiles minor
Common nighthawk

Length 22–25 cm (9–10 in)	
Weight 45–100 g (1⅝–3⅝ oz)	
Plumage Sexes differ	
Location North, Central, and South America	Migration Migrant
	Status Common

A familiar summer visitor through much of North America, this nighthawk is more active by day than most other species in the nightjar family. It is commonly seen on the wing at dawn and dusk, hawking for insects, and is recognized by its distinctive nasal call. The male has a highly acrobatic aerial display when courting – it swoops down low, making a hollow, booming sound with its wings, and almost colliding with its prospective partner. This bird nests and roosts in open places of all kinds. In some parts of the USA, it has taken to nesting on flat, gravel-covered roofs – a habitat that helps it to avoid some of its predators. The nest is often simply a shallow depression made in the ground, and the female alone cares for the eggs and chicks. When disturbed, the bird often flies up from the nest and settles nearby; occasionally it will lunge at an intruder. After breeding, large flocks move south for the winter, travelling as far as Argentina.

large eyes

mottled, brown, black, and white plumage

Phalaenoptilus nuttallii
Common poorwill

Length 18–21 cm (7–8½ in)	
Weight 30–60 g (1¹⁄₁₆–2⅛ oz)	
Plumage Sexes differ	
Location S. Canada, W. and C. USA, N. and C. Mexico	Migration Partial migrant
	Status Locally common

The smallest North American nightjar, the common poorwill gets its name from the male's repetitive 2-note courtship call. It rests on the ground during the day, the intricately patterned plumage providing effective camouflage in its often arid habitat. A night forager, it swoops on flying insects from a perch. Unusually, this bird conserves energy in winter by going into torpor – a hibernation-like state that is rarely found in birds.

Caprimulgus europaeus
European nightjar

Length 25–28 cm (10–11 in)	
Weight 50–100 g (1¾–3⅝ oz)	
Plumage Sexes differ	
Location Europe, W. to E. Asia, N.W., W., and S.E. Africa	Migration Migrant
	Status Common

This bird is the only nightjar found in northern Europe. It takes to the air after dark, catching flying insects on the wing. Males patrol their territories regularly, driving off intruders. During the breeding season, the male has a purring airborne song, which sounds like a piece of machinery. It also claps its wings, to display the white patches on the outer flight feathers. This bird does not build a nest; the eggs are laid directly on the ground, usually close to a tree or shrubs. The female rears the chicks initially, but after about 2 weeks she may leave them with the male and then lay a second clutch.

CAMOUFLAGE

Like other nightjars, this species spends the day resting on the ground or on low branches. Its dull, mottled plumage renders it almost invisible on the ground amongst vegetation.

BROWN AND MOTTLED
The European nightjar has brown plumage, mottled with black and grey, and long, pointed wings.

Macrodipteryx longipennis
Standard-winged nightjar

Length 21–25 cm (8½–10 in)	
Weight 60–90 g (2⅛–3¼ oz)	
Plumage Sexes differ	
Location W., C., and E. Africa	Migration Non-migrant
	Status Common

Male standard-winged nightjars are instantly recognizable in flight, because their breeding plumage includes 2 elongated wing feathers that can be up to 78 cm (31 in) long. They perform their aerial courtship display at canopy level, often chased by females, but will also select a perch and show off their pennants there. More active during the day than some nightjars, the birds may forage in loose flocks or in pairs, when the female follows the male; they fly very high off the ground, catching insects on the wing.

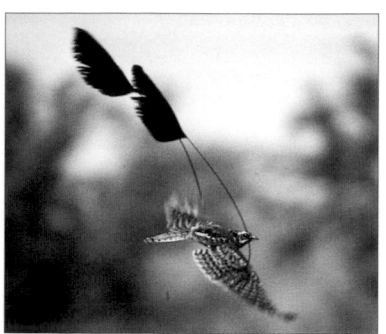

HUMMINGBIRDS AND SWIFTS

PHYLUM	Chordata
CLASS	Aves
ORDER	Apodiformes
FAMILIES	3
SPECIES	424

SWIFTS AND HUMMINGBIRDS share a unique wing structure that makes them acrobatic fliers capable of highly intricate aerial manoeuvres. However, their appearance and lifestyles differ greatly. The soberly coloured swifts rarely land, spending their lives in mid-air in search of flying invertebrates; they can sleep, and even mate on the wing. The multicoloured hummingbirds hover around flowers to feed, and perch readily. Swifts occur worldwide, but hummingbirds are restricted to the Americas.

FEEDING ON NECTAR
When feeding, a hummingbird (here, a green violetear from Costa Rica) hovers in front of a flower, manoeuvring its bill up the tube to draw up the nectar with its long tongue.

ANATOMY

Both hummingbirds and swifts have a compact, muscular body, and relatively small feet. Although swifts are dull in colour, hummingbirds are remarkable for their dazzling colours and patterns. Hummingbirds have a specialized bill designed for removing nectar from flowers. The bill length and shape are variable, and often match the shape of the flowers at which the birds feed. Swifts have a small bill with a wide gape for trapping tiny insects in flight. The hummingbird family contains many of the world's smallest birds.

long "hand"   "wrist"   shoulder
"elbow" close to body   shoulder girdle

WINGS
In hummingbirds and swifts, the joint between the upper and lower arm, the "elbow", is very close to the body, giving the wings great flexibility and leverage. This feature enables hummingbirds to hover.

FLIGHT

Hummingbirds beat their wings in a figure-of-eight pattern, which allows them great manoeuvrability; they are the only birds that can fly backwards, and even upside down. Smaller species may beat their wings 80 times a second. Swifts do not hover, but can vary the speed of their wingbeats to turn sharply.

SWIFT IN FLIGHT
Swifts resemble swallows, but are not closely related to them. They spend more time in the air, and are quicker, more erratic fliers. Swallows perch more readily than swifts.

BIRDS

Cypseloides niger
Black swift

Length 18–20 cm (7–8 in)
Weight 45 g (1 5/8 oz)
Plumage Sexes alike

Location W. North America, Central America, Caribbean

Migration Migrant
Status Locally common†

A typical, sooty black swift, with long, pointed wings and a slightly forked tail, this fast flier often soars on outstretched wings. Like other swifts, it spends most of its time in the air, and flocks are often seen wheeling through the skies, foraging for insects. Only when breeding does it come down to the ground. Choosing a site next to a waterfall or running water, it builds a half-cup nest formed of living plants – mosses and liverworts – held together with mud.

sooty black coloration

forked tail

Collocalia esculenta
White-bellied swiftlet

Length Up to 11.5 cm (4 3/4 in)
Weight Not recorded
Plumage Sexes alike

Location C., E., and S.E. Asia, E. Australia

Migration Partial migrant
Status Locally common

Also known as the glossy swiftlet, this highly acrobatic flier is one of a small group of species that nest in caves, often in colonies many thousands strong. It makes its nest entirely from dried saliva, produced by its extra-large salivary glands. Like other swifts, its diet consists of airborne insects, and it is most often seen over forests and gorges, swooping after its prey. In some parts of Southeast Asia, swiftlet nests are harvested, and are used to make birds' nest soup.

Hirundapus caudacutus
White-throated needletail

Length Up to 21 cm (8 1/2 in)
Weight Not recorded
Plumage Sexes alike

Location C., E., and S.E. Asia, E. Australia

Migration Migrant
Status Lower risk†

Needletails get their name from their tail feathers, which have spiny tips projecting beyond the end of the feathers. There are 4 species, and all of them are extremely fast fliers, reaching speeds of 125kph (80mph) during their spectacular courtship displays. The white-throated needletail lives in small flocks or larger groups, and feeds in all kinds of habitats, including farmland and towns. This bird often feeds near the ground, but it also soars high up, catching insects that are sucked high into the sky by columns of rising warm air. The white-throated needletail breeds in Asia and the Himalayas, and builds a shallow, cup-shaped nest in rocky crevices or hollow trees. In the autumn, this bird migrates south across the equator, reaching as far as Tasmania.

Chaetura pelagica
Chimney swift

Length 12–15 cm (4 3/4–6 in)
Weight 19–25 g (11/16–7/8 oz)
Plumage Sexes alike

Location E. North America, N.W. South America

Migration Migrant
Status Common

This small, smoky brown swift is the only one that breeds in eastern North America. It builds a small, half-cup nest of twigs glued together with saliva, mostly in man-made structures such as chimneys and old barns. When not breeding, chimney swifts roost in large flocks; just before and during migration, several thousand birds might roost together. Large chimneys are a favourite roosting site, and birds pour into them at dusk.

Cypsiurus parvus

African palm swift

Length	14–16 cm (5 1/2–6 1/2 in)
Weight	10–18 g (3/8–5/8 oz)
Plumage	Sexes alike

Location Africa, Madagascar	Migration Partial migrant
	Status Common

Fan palms are the preferred nesting site of this medium-sized, blackish brown swift, although it will use other palms (as well as artificial sites such as bridges). The nest (seen below) is a pad of feathers and plant fibres, glued with saliva to the vertical side of a drooping palm frond, and the eggs themselves are glued in place with saliva. Nestlings have unusually long claws, helping them to cling on as the nest blows about in the wind.

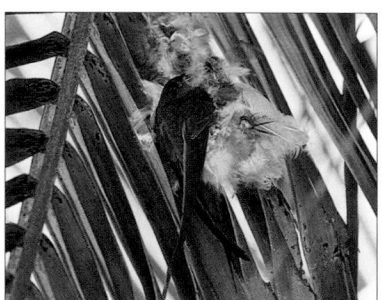

Apus apus

Eurasian swift

Length	16–17 cm (6 1/2–7 in)
Weight	35–50 g (1 1/4–1 3/4 oz)
Plumage	Sexes alike

Location N. and southern Africa, Europe, W. to C. Asia	Migration Migrant
	Status Common

One of the most aerial of birds, the Eurasian swift feeds, mates, and even sleeps on the wing. Its nest is a cup-shaped platform formed from plants and feathers that are caught in mid-air and glued with saliva. It originally nested on rock crevices but now often uses man-made structures. In cold weather, parents often fly hundreds of kilometres away from the nest. The young survive by becoming torpid; they revive when it warms up, and the parents return with food. After leaving the nest, young swifts remain airborne for up to 3 years, before landing to breed themselves. They spend a relatively short time at their summer breeding grounds.

NARROW WINGS
The Eurasian swift has distinctive narrow wings and a forked tail. It has a short bill, small feet, and blackish brown plumage.

ON THE WING

Small, noisy groups of these gregarious birds are often seen feeding and moving over towns. Like other swifts, they are fast and agile, turning in flight by beating each wing at a different rate.

long, narrow wings

forked tail

Hemiprocne longipennis

Grey-rumped tree-swift

Length	23 cm (9 in)
Weight	Not recorded
Plumage	Sexes alike

Location S.E. Asia	Migration Non-migrant
	Status Common

There are 3 species of tree-swifts, all of them found in southern Asia. Like a true swift, the grey-rumped tree-swift catches insects on the wing, but it spends less of its time in the air. Between bouts of feeding, it perches on branches – something that true swifts are unable to. This species has long, scythe-like wings and a deeply forked tail, but its most distinctive feature is a short, upright crest, attached not to its crown, but to the base of the upper part of its bill. The grey-rumped tree-swift makes a minute nest glued to a branch, and lays a single egg. The parents take turns to incubate the egg, covering the nest completely while resting their weight on the branch beneath. Often found in small flocks, nesting pairs of this bird are known to defend their territories aggressively.

Eutoxeres aquila

White-tipped sicklebill

Length	12–14 cm (4 3/4–5 1/2 in)
Weight	10–13 g (3/8–7/16 oz)
Plumage	Sexes alike

Location S. Central America to N.W. South America	Migration Non-migrant
	Status Lower risk†

This large hummingbird is green above with streaked underparts, and has white-tipped tail feathers and an unmistakable downcurved bill. The shape of the bill allows it to drink nectar from flowers that have curved throats, especially heliconias. Unlike most hummingbirds, it perches on the flower while feeding from it. It also gleans spiders from their webs and catches insects on the wing. The cup-shaped nest, loosely woven from palm fibres, is attached by cobwebs to the tip of a hanging leaf.

Lophornis ornatus

Tufted coquette

Length	7 cm (2 3/4 in)
Weight	2 g (1/16 oz)
Plumage	Sexes differ

Location Trinidad, N. South America	Migration Non-migrant
	Status Lower risk†

The tufted coquette is a tiny hummingbird with a short reddish bill and a whitish band on its bright green rump. The male has a long red crest and long, fan-like cheek tufts. When courting, it hovers in a semi-circle around the female, with cheek tufts spread out and crest feathers erect. The tufted coquette feeds primarily on nectar and small insects. To find food, the bird follows a set route from flower to flower, hovering in front of each one and collecting nectar by piercing the flower tube with its bill. The female builds a cup-shaped nest of fine plant fibres and cobwebs, attaching it to a forked twig. Only the female takes care of the young, and will do so for up to 1½ months after fledging.

Ocreatus underwoodii

Booted racquet-tail

Length	11–15 cm (4 1/4–6 in)
Weight	3 g (1/8 oz)
Plumage	Sexes differ

Location N.W. and W. South America (Andes)	Migration Non-migrant
	Status Vulnerable†

Elongated tail feathers with bare shafts ending in blue-black "racquets" distinguish the male from the female of this tiny species. When courting, the male puts on a spectacular flight display, holding up the fluffy white or brown "puffs" or leg feathers, and flicking his tail feathers up and down to produce a whip-like sound. The male and female fly a regular route when foraging for food, calling to one another to stay in contact. The bird hovers in front of a flower, its rapidly beating wings making a distinct humming noise, and pierces the flower tube with its bill to reach the nectar. Like most hummingbirds, it often becomes torpid at night, its body temperature dropping almost to that of the air in order to save energy. Hummingbirds are so tiny that they would risk starving to death if they maintained their normal body heat throughout the night.

iridescent green plumage

short black bill

brown leg puffs

shorter inner tail feathers

bare tail shafts

long tail feathers end in "racquets" in male

Crimson topaz

Topaza pella

Length 21–23 cm
(8¹/₂–9 in)

Weight 10–15 g
(³/₈–⁹/₁₆ oz)

Plumage Sexes differ

Location N. South America

Migration Non-migrant

Status Lower risk†

A relatively short, slightly curved bill allows this large hummingbird to feed from a variety of rainforest flowers. Rarely seen on the ground, it lives in the middle and upper levels of the forest, and will

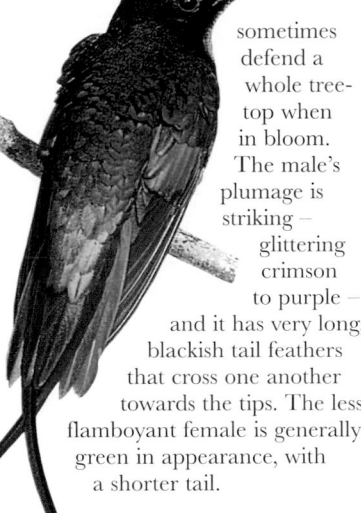

sometimes defend a whole tree-top when in bloom. The male's plumage is striking – glittering crimson to purple – and it has very long, blackish tail feathers that cross one another towards the tips. The less flamboyant female is generally green in appearance, with a shorter tail.

— *long tail feathers*

Giant hummingbird

Patagona gigas

Length 20–22 cm
(8–9 in)

Weight 18–20 g
(⁵/₈–¹¹/₁₆ oz)

Plumage Sexes alike

Location W. South America (Andes)

Migration Non-migrant

Status Locally common

This species is the largest of all hummingbirds – even so, it weighs only 19 g (¹¹/₁₆ oz). Its colouring is rather dull for a hummingbird: brownish overall with pale rump feathers. It often flies more like a swift, sometimes gliding, over dry Andean valleys and arid, steppe-like mountain slopes. Commonly found around stands of prickly pears and puyas, it hovers while feeding on the flowers, occasionally clinging to large blossoms.

Ruby-throated hummingbird

Archilochus colubris

Length 9 cm
(3¹/₂ in)

Weight 3 g
(¹/₈ oz)

Plumage Sexes differ

Location S. Canada, C. and E. USA, Mexico to S. Central America

Migration Migrant

Status Common

This is one of the few hummingbirds that migrate as far north as Canada, on an annual journey from Central America that may total over 3,000 km (1,900 miles). For some birds, migration involves a non-stop flight of about 850 km (530 miles) across the Gulf of

red throat

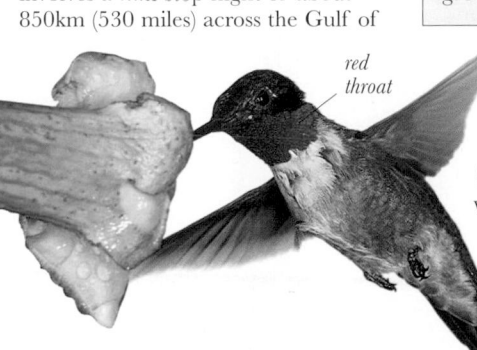

FEEDING THE YOUNG

Like other hummingbirds, this bird feeds its young largely on insects, because although nectar is a good source of energy it does not contain enough protein for growth and development.

Mexico – a huge distance for such a tiny animal. This hummingbird feeds on nectar, and also uses the holes drilled in tree trunks by sapsucker woodpeckers to feed on tree sap and on the insects this attracts.

MALE AND FEMALE
A glittering red patch on the throat helps to identify this small, iridescent green hummingbird; the female lacks the male's vivid colouring.

forked tail ——

Sword-billed hummingbird

Ensifera ensifera

Length 17–23 cm
(6¹/₂–9 in)

Weight 12–15 g
(⁷/₁₆–⁹/₁₆ oz)

Plumage Sexes alike

Location N.W. and W. South America (Andes)

Migration Non-migrant

Status Lower risk†

Dark greenish in colour, with a deeply forked, blackish tail, this bird has a slender, sword-shaped bill up to 11 cm (4¹/₄ in) long. It is the only bird whose bill is longer than the rest of its body. When resting on a perch it holds the bill almost vertically to reduce the strain on its neck. Apart from flying insects, which it catches by hawking with a wide-open bill, it feeds almost exclusively from flowers with very long tubes, such as passion-flowers and daturas. These blossoms usually hang downwards and the bird hovers immediately beneath, pushing the bill up into the flower to extract the nectar. This species follows a feeding strategy known as traplining. While some hummingbirds defend clumps of flowers from other birds and even butterflies, trapliners make regular visits to a number of scattered flowers, remembering where each one is and following a set route between them. The intervals between visits mean that the plants have a chance to produce more nectar.

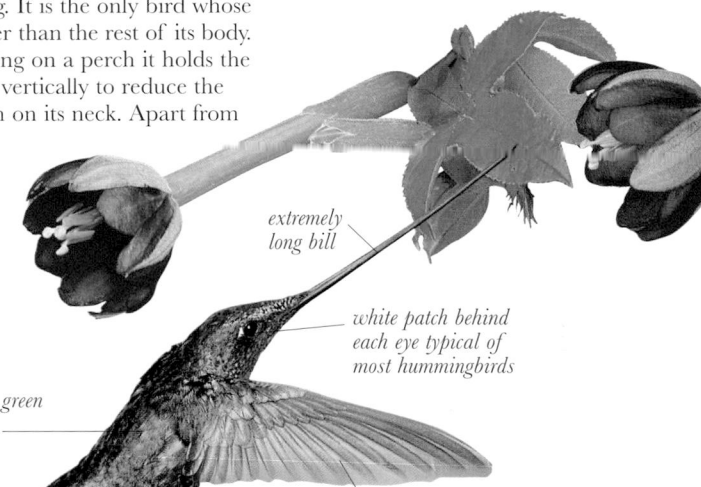

extremely long bill

white patch behind each eye typical of most hummingbirds

dark green body ——

long, pointed wings

Bee hummingbird

Mellisuga helenae

Length 5–6 cm
(2–2¹/₄ in)

Weight Up to 2 g
(¹/₁₆ oz)

Plumage Sexes differ

Location Cuba, Isle of Pines

Migration Non-migrant

Status Lower risk†

The smallest of all birds, the male bee hummingbird weighs a mere 1.6 g (less than ¹/₁₆ oz). It is green with a greyish white underside, and has an iridescent, fiery red head and bright collar. The female is slightly larger, weighing 2 g (¹/₁₆ oz), and has no iridescence on the head or neck. This bird feeds mainly on nectar, hovering with its body held horizontally and pushing its short, straight bill into each flower. The eggs are as little as 6 mm (¹/₄ in) long – smaller than a single pea.

Andean hillstar

Oreotrochilus estella

Length 13–15 cm
(5–6 in)

Weight 8–9 g
(⁵/₁₆–¹¹/₃₂ oz)

Plumage Sexes differ

Location W. South America (Andes)

Migration Non-migrant

Status Locally common

Common in rocky slopes (with grass and scrub) at altitudes of up to 5,000 m (16,500 ft), this mountain bird does not hover in front of the flowers from which it feeds, but perches on them. Nocturnal torpor is very important for it, as nights in the high Andes can be very cold, and the bird would not survive if it tried to maintain its body temperature at daytime level. The male's iridescent emerald-green collar with a black border marks it from the female.

MOUSEBIRDS

PHYLUM	Chordata
CLASS	Aves
ORDER	Coliiformes
FAMILIES	1 (Coliidae)
SPECIES	6

NAMED FOR THEIR ability to run like rodents, mousebirds scramble about branches in search of buds, leaves, or fruits. Their feet are unique – the 2 outermost toes are reversible, so can point either forwards (with the other 2 toes, for hanging), or backwards (for grasping). Mousebirds live in flocks and are found only in Africa.

PERCHING
Mousebirds, such as these white-headed mousebirds, have an unusual perching posture. Using their flexible toes, they hold on to branches with their feet held level with their shoulders. They can even perch upside down.

Colius striatus
Speckled mousebird

	Length 30–40 cm (12–16 in)
	Weight 45–75 g (1 5/8–2 5/8 oz)
	Plumage Sexes alike
Location C., E., and southern Africa	Migration Non-migrant
	Status Vulnerable

This plump bird has a long tail of narrow, stiffened feathers, a stout, softly crested head, and a stubby, downcurved bill. Primarily brown and grey, it has faint barring and speckling on its wings, neck, and throat, and red legs and feet; the bill is dark grey to blackish above and lighter grey below. Like all mousebirds, it often forms groups of 4 to 20 – usually consisting of pairs and some immature individuals – that sleep together in a tight cluster hanging from branches. They may preen and even offer food to each other. At times regarded as agricultural pests, mousebirds are often exterminated in farm areas and gardens.

crested head

TROGONS

PHYLUM	Chordata
CLASS	Aves
ORDER	Trogoniformes
FAMILIES	1 (Trogonidae)
SPECIES	35

THESE BRILLIANTLY COLOURED birds live in tropical forests in the Americas, Southeast Asia, and sub-Saharan Africa. They have short, rounded wings, a long tail, soft, often iridescent plumage, and bright patches of bare skin around the eyes. They grip branches with their small feet, which have 2 toes pointing forward and 2 behind; uniquely, it is the first and second toes that point backwards. The short bill has a wide gape for catching invertebrates in flight.

NESTING
Trogons nest in cavities, either in existing holes in tree trunks, or by digging their own holes in rotten wood or in wasps' or termites' nests. This slaty-tailed trogon is making a nest-hole in a termite nest.

Pharomachrus mocinno
Resplendent quetzal

	Length 35–40 cm (14–16 in)
	Weight 200–225 g (7–8 oz)
	Plumage Sexes differ
Location Central America	Migration Non-migrant
	Status Vulnerable

The male resplendent quetzal is widely regarded as one of the world's most beautiful birds. It has brilliantly coloured plumage – mainly iridescent emerald-green – and a rich crimson-coloured breast. Other characteristics, unique to the male, are the exceptionally long tail coverts, which extend beyond the tail to form an elegant train, and the short, bristle-like crest. Both sexes are plump-bodied and have stout heads. This bird's habit of perching motionless for long periods makes it difficult to spot. It flies among trees looking for fruits or insects. Both parents brood the eggs – the tail feathers of the male sticking out of the nest-hole as he sits on the eggs – and share in feeding the young.

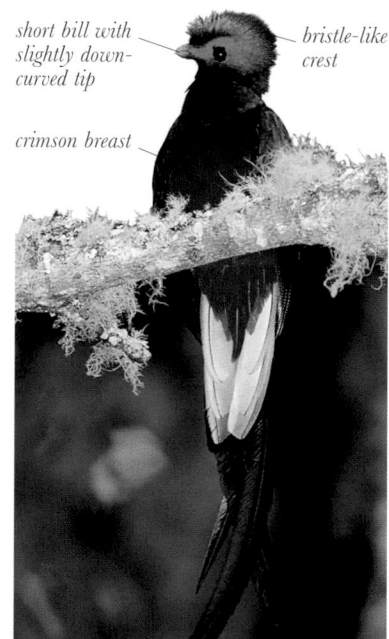

short bill with slightly down-curved tip

bristle-like crest

crimson breast

Trogon violaceus
Violaceous trogon

	Length 23–26 cm (9–10 in)
	Weight 45–65 g (1 5/8–2 3/8 oz)
	Plumage Sexes differ
Location S.E. Mexico to C. South America	Migration Non-migrant
	Status Locally common

Generally solitary, this bird is mainly found in tropical rainforest and woodland (occasionally in dry brushland and farmland), in Central and South America. Males have a black and iridescent violet-blue head, a green back and upper tail, and a yellow-orange breast, while females are primarily grey. The nest consists of an excavated hole, usually in a tree but sometimes built in old termite or wasp nests. This bird feeds on fruits, insects, and other invertebrates.

black iridescent head

barred tail feathers

KINGFISHERS AND RELATIVES

PHYLUM	Chordata
CLASS	Aves
ORDER	Coraciiformes
FAMILIES	10
SPECIES	191

FAMOUS FOR THEIR SPECTACULAR dives into water, the kingfishers are arguably the most familiar birds in this group. However, there are 10 very different families in all (including bee-eaters, todies, motmots, rollers, and hornbills), most of which are not waterside birds. They are found worldwide, mainly in woodland, and all nest in holes. These birds range in size from tiny todies, 10cm (4in) long, to huge hornbills, up to 1.5m (5ft) long.

HUNTING FOR FISH
A kingfisher perches or hovers over water, watching below for signs of movement. When it sees a fish, it plunge-dives head-first and grabs – never spears – its prey.

ANATOMY

Most members of this group have a relatively large head and bill, and a compact body. Their legs are frequently short, and their feet tend to be weak. Two of the front toes are usually partially fused together near the base. The wings of most species are broad, but in the elegant and highly aerial bee-eaters they are relatively long and pointed. Many relatives of the kingfisher, such as the motmots, ground-rollers, bee-eaters, and hornbills, have a long tail, and a large number of species have vividly coloured plumage.

long, curved bill
BEE-EATER

straight, dagger-shaped bill
KINGFISHER

casque *wide, curved bill*
HORNBILL

BILL SHAPES
Most members of this group have a strong bill, useful for dealing with animal prey. However, there is great variation in bill shape and size: it is long and curved in bee-eaters and hoopoes; dagger-like in kingfishers and todies; and vastly expanded, often with a horny casque, in hornbills.

CATCHING INSECTS
Rollers, such as this lilac-breasted roller (left), drop from an elevated perch to catch their prey on the ground. A few also catch flying insects in mid-air.

EATING FRUITS
Like other large hornbills, the great hornbill (right) is mainly a fruit-eater. It uses its long bill to reach fruits on trees and then tosses them back into its gullet.

FEEDING

Kingfishers adopt a hunting strategy that is common to many birds in this group. The bird sits still, watching for movement, and then takes off in pursuit – either into the water, or down to the ground, or in an aerial chase – before returning to its perch. Since fish are slippery and difficult to control, a kingfisher often stuns its prey by beating it on a hard surface before swallowing it. Most members of this group are meat-eaters, feeding on all kinds of animals, including other birds, fish, mammals, and insects. Hornbills supplement their diet with fruits.

Ceryle alcyon
Belted kingfisher

Length	29 cm (11 1/2 in)
Weight	150 g (5 oz)
Plumage	Sexes differ

Location North America to N. South America, Caribbean	Migration Partial migrant
	Status Common

This is one of the few species of kingfisher to be found in North America. It has a short, squarish tail and a craggy, erectile crest. The female belted kingfisher has a prominent reddish brown band across the lower breast. Conspicuous when fishing or sitting on an exposed perch, the belted kingfisher flies off over the water when disturbed, with a loud, rattling call. It feeds predominantly on fish but also eats crustaceans,

amphibians, and reptiles. During courtship, these birds circle high overhead and chase each other while uttering shrill cries. The male offers fish to the perching female as part of the courtship ritual. This kingfisher builds its nest in a chamber at the end of a burrow excavated in earth, usually in a river bank devoid of vegetation, but also in man-made earthworks. Birds in the north of the range migrate south when lakes and rivers freeze over.

Ceryle rudis
Pied kingfisher

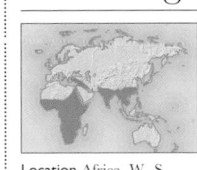

Length	25 cm (10 in)
Weight	90 g (3 1/4 oz)
Plumage	Sexes differ

Location Africa, W., S., and S.E. Asia	Migration Non-migrant
	Status Common

This large and boldly marked kingfisher has a wide distribution, stretching from western and southern Africa as far east as China. It has a shaggy crest, and a conspicuous breastband, which is double in the male, and only partly complete in the female. Unlike most other kingfishers, this species is equally at home over fresh and salt water, and sometimes ventures far out over estuaries and shallow coasts. It flies rapidly, but

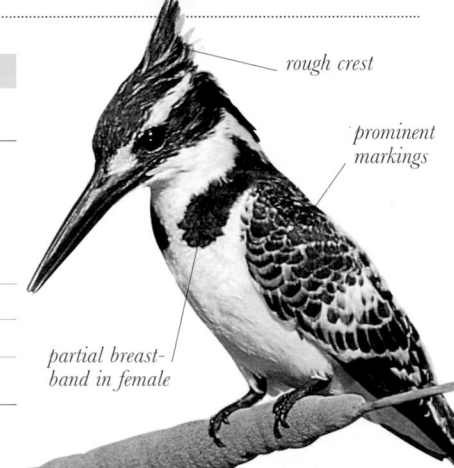
rough crest
prominent markings
partial breast-band in female

often hovers over the water when locating prey, before diving down to make a catch. The pied kingfisher nests in sandy banks, and often reproduces cooperatively, with up to 4 young non-breeding adults acting as helpers for a breeding pair. Assertive and vocal, it has a high-pitched call, which it often makes on the wing.

Alcedo atthis

Eurasian kingfisher

Length	16 cm (6¹/₂ in)
Weight	35 g (1¹/₄ oz)
Plumage	Sexes alike

Location Europe, Asia, N. Africa	**Migration** Partial migrant
	Status Common

The Eurasian kingfisher is the only kingfisher encountered in most of Europe, where it is also known as the common kingfisher. It is a small, swift, and active bird with vivid, distinctive plumage. Size and coloration vary throughout its range, which stretches from western Europe and northern Africa through to East and Southeast Asia, and 7 subspecies are recognized. In the western sector of its range it inhabits most aquatic habitats, but prefers lowland, freshwater streams or rivers. In eastern areas it is more prevalent in coastal or subcoastal habitats, notably estuaries, mangroves, and intertidal pools. At high latitudes, freezing conditions force winter migration, although in warmer areas birds move locally or are partially resident. Although it supplements its diet with crustaceans, amphibians, and insects, the Eurasian kingfisher primarily eats small fish, which it takes by plunge-diving. Each breeding pair occupies a territory of up to a kilometre (²/₃ mile) along river banks. Pair-bonds are maintained through the breeding season, and both sexes incubate their eggs then care for the chicks for up to 4 weeks, until they are ready to leave the nest.

white ear tufts

long, sharp bill

cobalt-blue feathers

cinnamon underside

DAZZLING PLUMAGE
The Eurasian kingfisher is a dazzling bird with a deep cinnamon underside, greenish blue crown, back, and wings, and dazzling cobalt-blue rump and tail. Its sharp bill is well adapted for striking and grasping fish.

COURTING COUPLE
Courtship behaviour is elaborate and includes erratic twisting and turning flights, as well as courtship feeding in which the kingfishers offer each other fish just before copulation.

PLUNGE-DIVER

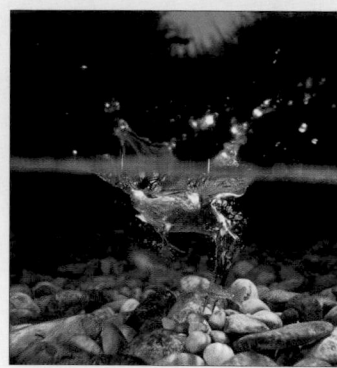

After a steep dive from a favoured vantage perch or, less often, from hovering flight, the kingfisher catches its fish usually no deeper than 25cm (10in) below the water. Its natural buoyancy and a swift downstroke of its wings send it clear of the water's surface.

LIGHTNING STRIKE
Having captured its prey by a dramatic plunge-dive, the Eurasian kingfisher then takes it back to its perch, on which it strikes the fish repeatedly before swallowing it head-first. Any undigested remains are usually regurgitated as pellets.

Dacelo novaeguineae
Laughing kookaburra

Length 42 cm (16 1/2 in)	
Weight 350 g (13 oz)	
Plumage Sexes alike	

Location E., S.E., and S.W. Australia (including Tasmania), New Zealand	**Migration** Non-migrant
	Status Common

The largest of all kingfishers, the laughing kookaburra is known throughout Australia for its raucous call, which it delivers with partly opened bill pointing skywards and its tail cocked. In urban areas, it becomes quite tame and may even be fed by hand. It is found in strong family groups, which during the day keep within sight or sound of each other, and at night, roost together. This bird has been introduced in southwest Australia, Tasmania, and New Zealand.

Clytoceyx rex
Shovel-billed kingfisher

Length 34 cm (13 1/2 in)	
Weight 300 g (11 oz)	
Plumage Sexes differ	

Location New Guinea	**Migration** Non-migrant
	Status Common

The shovel-billed kingfisher is an inhabitant of wet gullies, deep ravines, and heavily shaded, stream-side areas. A large, heavy-bodied

Todus todus
Jamaican tody

Length 11 cm (4 1/4 in)	
Weight 6 g (7/32 oz)	
Plumage Sexes alike	

Location Jamaica	**Migration** Non-migrant
	Status Common

The Jamaican tody is one of 5 species of small, vivid green, kingfisher-like birds with a large head, a long, flat, bicoloured bill, and a prominent crimson bib. It perches on small branches with its bill upturned, looking

---

FEEDING

Unlike river kingfishers, this species captures prey on the ground. It sits on an exposed perch waiting for likely victims – such as insects, snails, frogs, small birds, fish, and reptiles – and then swoops down on them. It crushes small prey in its bill; larger prey, like the snake the bird pictured above has caught, are beaten to death against a branch.

EYE-STRIPE
The kookaburra's back and wings are dark brown, while its underparts are a dusky white. Its white head has a distinctive black eye-stripe.

kingfisher, it has a dark brown head, mantle, upper back, and wings, while its lower back and rump are a brilliant pale blue. The tail is dull blue in the male and reddish in the female. Juveniles have fine dusky margins to the feathers of their underparts and a collar encircling the hindneck. The peculiarly stubby, broad bill of this bird is adapted to digging for earthworms in the forest floor. The bill is thrust into the soft soil at a slight angle and moved from side to side until it grasps a worm. Little is known of this bird's habits; shy and wary, single birds or pairs are usually encountered only when flushed from the ground to fly up into the forest canopy.

for insects on the undersides of over-hanging leaves and twigs. It usually darts out rapidly to snatch prey, but will sometimes hover. Extremely active from dawn to dusk, it has one of the highest feeding rates known in birds. Its call consists of a loud, nasal "beep".

Momotus momota
Blue-crowned motmot

Length 47 cm (18 1/2 in)	
Weight 150 g (5 oz)	
Plumage Sexes alike	

Location Central America to C. South America, Trinidad and Tobago	**Migration** Non-migrant
	Status Common

The blue-crowned motmot is a robust bird with elongated central tail feathers that are bare except for the tips. Its plumage is generally green and it has a black eye-stripe. For most of the day, this quiet, unobtrusive bird sits on a low branch, intermittently swinging its racquet-tipped tail like a pendulum. However, it is very active in the early morning and evening, emitting a far-carrying "hoot-hoot" call. The blue-crowned motmot feeds mainly on insects, either capturing prey on the ground or probing into leaf litter; it also snatches prey from tree trunks.

Merops apiaster
European bee-eater

Length 30 cm (12 in)	
Weight 70 g (2 1/2 oz)	
Plumage Sexes alike	

Location Africa, Europe, W., C., and S. Asia	**Migration** Migrant
	Status Common

One of the most aerial of all bee-eaters, this medium-sized bird has long wings and a sharply pointed bill. During the day, it perches on telegraph wires, fences, or branches. It preys on stinging insects, which it devenoms by rubbing the insect's tail-end rapidly against the perch and squeezing it in its bill to expel the venom and sting. It breeds in colonies and nests in burrows.

Merops bullockoides
White-fronted bee-eater

Length 23 cm (9 in)	
Weight 35 g (1 1/4 oz)	
Plumage Sexes alike	

Location Southern Africa	**Migration** Partial migrant
	Status Common

The white-fronted bee-eater lives in a social unit, known as a clan, of up to 16 individuals. These clans associate in nesting colonies of 500 or more birds. Although each clan vigorously defends its own foraging territory, members of different clans greet each other, and regularly visit other burrows at the nesting colony.

---

dark eye-stripe

yellow throat

bluish green tail

scarlet throat

Coracias garrulus
European roller

Length 32 cm (12 1/2 in)	
Weight 150 g (5 oz)	
Plumage Sexes alike	
Location Africa, Europe, W. and C. Asia	**Migration** Migrant
	Status Common

The European roller is a heavy-bodied bird with conspicuous, blue and tobacco-coloured plumage and black wing-tips. It spends long periods sitting on a prominent perch, such as a bare branch or a power line, looking for prey on the ground. It feeds mainly on insects, but also eats small reptiles, mammals, birds, and fruits. The nest is built in a hollow limb or hole in a tree.

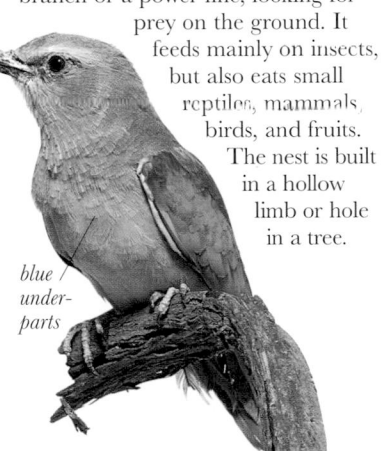

blue under-parts

Upupa epops
Hoopoe

Length 28 cm (11 in)	
Weight 75 g (2 5/8 oz)	
Plumage Sexes alike	
Location Europe, Asia, Africa, Madagascar	**Migration** Partial migrant
	Status Common

With its bold colours, long, curved bill, and fan-shaped crest, the hoopoe is unmistakable. It spends much of the day on the ground, probing the soil with its bill in search of insects. Its diet also includes earthworms, snails, slugs, and spiders. This bird has an undulating flight with strong wingbeats interspersed by pauses. Hoopoes are notoriously foul-smelling and unhygienic; the fetid state of their nests, arising from the accumulation of excreta, food remains, and the nestlings' body secretions, is thought to deter predators.

barred wings

Buceros bicornis
Great Indian hornbill

Length 1.5 m (5 ft)	
Weight 3 kg (6 1/2 lb)	
Plumage Sexes alike	
Location S. and S.E. Asia	**Migration** Non-migrant
	Status Lower risk

The most spectacular of the hornbills, this species is distinguished by its large size, resounding call, and the loud "swooshing" of its wingbeats. It has a massive bill with a prominent casque, and long, rounded wings. The bill, casque, white head, and neck are often stained yellow with oil from the preen gland. Mostly confined to the forest canopy, the hornbill moves easily through it with a series of sideways hops. It follows a regular routine, visiting certain trees at the same time each day. It feeds mainly on fruits, especially figs, but also preys on reptiles, amphibians, small mammals, and birds. Fruits and small prey are grasped with the tip of the bill and moved down the gullet by throwing the head up and back. Larger prey is crushed in the bill and battered against a branch before being swallowed. Like other hornbills, this species builds its nest in the natural hollow of a tree trunk, which it seals with mud, leaving only a narrow opening through which the male passes food to the nesting female. The female breaks out of the nest after almost 3 months, but the entrance is resealed by the chick, which is fed for a further month in the nest.

massive casque

huge bill

black face

yellow stain from preen gland

broad white tips on black flight feathers

Phoeniculus purpureus
Green wood hoopoe

Length 38 cm (15 in)	
Weight 75 g (2 5/8 oz)	
Plumage Sexes alike	
Location Africa (south of Sahara)	**Migration** Non-migrant
	Status Common

This slim bird has a long tail and a slender, down-curved bill. The plumage is black with an iridescent sheen of green and dark violet that changes colour and intensity with differing light sources. The bill is adapted to probing for insects in crevices or under bark as the bird climbs through trees. Green wood hoopoes live in close-knit groups, which include up to 16 birds headed by a dominant pair. When neighbouring groups meet, a peculiar "flag waving" display often ensues, with one or 2 birds waving pieces of bark in the air. The other members clump together, rocking their bodies and calling loudly.

Tockus flavirostris
Southern yellow-billed hornbill

Length 50 – 60 cm (20 – 23 1/2 in)	
Weight 250 g (9 oz)	
Plumage Sexes alike	
Location Southern Africa	**Migration** Non-migrant
	Status Common

This small hornbill has a curved, bright yellow bill and black wings with white spots. It feeds on insects and fruits, especially figs. This species sometimes forms cooperative foraging parties with dwarf mongooses, which flush out locusts on which the birds feed. The hornbill, in turn, warns them of approaching danger.

white-spotted wings

strongly curved bill

long black tail

Bucorvus leadbeateri
Southern ground hornbill

Length 1.3 m (4 1/4 ft)	
Weight Not recorded	
Plumage Sexes alike	
Location Southern Africa	**Migration** Not migrant
	Status Common

Together with the Abyssinian ground hornbill (*Bucorvus abyssinicus*), this turkey-sized bird is the largest hornbill, and one of the few that feeds mainly on the ground. Its plumage is almost entirely black, apart from white wing patches, and its most conspicuous feature is a patch of bare, brightly coloured skin on its face and throat. The male has a red wattle, whereas the female's is blue. Southern ground hornbills live in groups of up to 8 birds, and feed mainly on small animals. They nest in tree-holes, but in each group only the dominant pair breeds – the junior birds help to collect food and defend the nest.

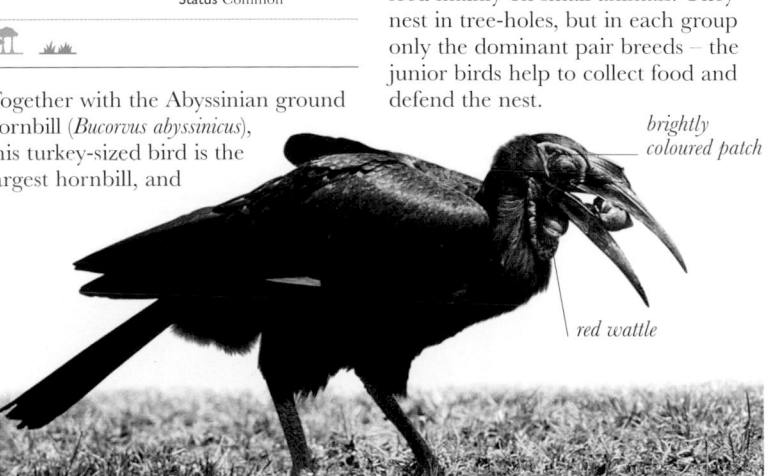

brightly coloured patch

red wattle

WOODPECKERS AND TOUCANS

PHYLUM	Chordata
CLASS	Aves
ORDER	Piciformes
FAMILIES	6
SPECIES	380

THIS GROUP OF arboreal birds includes woodpeckers, toucans, barbets, jacamars, honeyguides, and puffbirds. They all have a similar type of foot, with 2 toes pointing forwards and 2 backwards, which helps them to climb easily. All the birds in this group also nest in holes. Woodpeckers and barbets carve out their own nest-holes, the woodpeckers using their stout bill as a chisel. These are mainly tropical birds; only the woodpeckers are also widespread in temperate areas. Curiously, not a single species occurs in Australia. Toucans and barbets help to disperse the seeds of some trees by eating their seeds and then passing them through their body, while woodpeckers limit the spread of some tree diseases by feeding on the insects that carry them.

REPRODUCTION

Although all birds in this group nest in holes, not all use living trees. The smaller woodpeckers prefer to bore into the softer bark of dead and decaying trees. Jacamars, puffbirds, and barbets may dig out burrows, either in the ground or in termite mounds. Unlike the other families, the honeyguides are brood parasites, laying their eggs in the nests of other birds, including woodpeckers. The heels of young toucans are fitted with pads of spike-like projections that protect the ankles from damage against the unlined nest-floor.

NESTING
Like most barbets, the black-collared barbet nests in dead trees, sometimes communally. A few species burrow into banks or termite mounds.

ANATOMY

The main climbers in this group, the woodpeckers and barbets, have strong leg muscles and use their tail for support when holding themselves vertically upright on tree trunks; woodpeckers have specially stiffened tail feathers for this purpose. Chiselling involves delivering powerful blows to the wood surface, so the skull of wood-peckers is unusually thick to absorb the shock; the nostrils are also slit-like to prevent wood-chips from entering the airways. Woodpeckers also have a highly distensible, slightly barbed tongue that is coated with a sticky substance for catching insects. Honeyguides have thick skin to protect them from bee stings and specialized bacteria in their gut that enable them to digest beeswax.

chisel-shaped bill

WOODPECKER

conical bill

BARBET

serrated edges

TOUCAN

sharp, slim bill

PUFFBIRD

BILL SHAPE
There is great variation in bill shape in this group. A woodpecker's long, sharp bill is ideal for chiselling out wood, while the stout bill of barbets is used to handle animal prey. The large, serrated bill of toucans is adapted for grasping and tearing fruit. Puff-birds and jacamars have a sharp-tipped bill for catching flies.

FEET
The feet of woodpeckers and all their relatives are described as zygodactylous. Two toes point forwards and 2 back, an arrangement that helps them to climb and perch on tree trunks.

FEEDING

Insects form the greater part of the diet of most birds in this group. Wood-peckers take insects from both the tree surface and deep holes in bark, finding and extracting them with their long tongue. Other species pounce on insects or grab them in mid-air; jacamars, which take insects in flight, are the only birds to eat the giant morpho butterflies of the American tropics, removing the wings first. Toucans and most barbets feed mainly on fruit; honeyguides, which can break down cellulose, are the only birds that eat beeswax; they also feed on insects and fruit.

GUIDING
Honeyguides, such as the lesser honeyguide (right), take their name from their habit of leading large mammals to bees' nests.

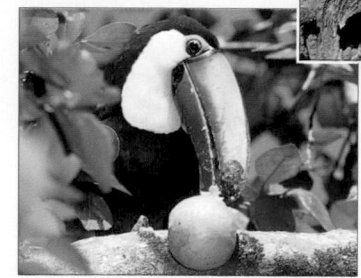

EATING FRUITS
Toucans, like the toco toucan (left), eat mainly fruits. The long bill helps them to take fruits from otherwise out-of-reach, slender branches.

NESTING AND FEEDING YOUNG
Woodpeckers use their sturdy bill to bore their own nest-holes in tree trunks and defend them vigorously against other hole-nesters. Here, a green woodpecker is feeding its young with ants that it has collected from the ground.

BIRDS

Galbula ruficauda
Rufous-tailed jacamar

Length 25 cm (10 in)	
Weight 25 g (7/8 oz)	
Plumage Sexes alike	
Location S. Central America, N.W. and C. South America	**Migration** Non-migrant
	Status Common

Found only in Central America and South America, jacamars are a family of 17 species of brilliantly coloured, iridescent birds that feed on large insects, catching most of their prey in mid-air. Their diet includes butterflies, bees, winged ants, and dragonflies.

The rufous-tailed jacamar has a metallic golden green head and upper breast, while the lower breast, belly, and tail are rufous. The dagger-like bill is long, thin, and straight. Males have a white chin, the females buff. The chicks hatch with down, unlike most piciform birds. The rufous-tailed jacamar usually forms pairs or family groups and perches upright on exposed twigs to wait for prey. Both the male and the female incubate the eggs and feed the young. This bird has an eerie, mournful call.

Bucco capensis
Collared puffbird

Length 21 cm (8 1/2 in)	
Weight 55 g (2 oz)	
Plumage Sexes alike	
Location N. South America	**Migration** Non-migrant
	Status Common

Puffbirds comprise a family of about 32 species of dull-coloured, small to medium-sized, sit-and-wait hunters. Most puffbirds hunt in treetops, and the larger species may catch small reptiles with their strong, hooked bills. The collared puffbird has an orange bill, brown upperparts, and whitish underparts with a large breastband. Its plumage is fluffy, and it has tiny feet. Inhabiting humid, lowland forests, it sits quietly most of the time; only the head moves occasionally as it scans the surroundings for prey. This behaviour has earned it names like "sleeper" and "lazy bird".

Chelidoptera tenebrosa
Swallow-wing puffbird

Length 18 cm (7 in)	
Weight 35 g (1 1/4 oz)	
Plumage Sexes alike	
Location N. South America	**Migration** Non-migrant
	Status Common

The disproportionately long wings of the swallow-wing puffbird make it look like a bat or a large butterfly in flight. It is an excellent, almost acrobatic flier and soarer. Black to dark grey, it sits more in the sun, and is easier to spot, than other puffbirds. It sits and waits on treetops for prey, which it takes on the wing in long sallies; it may also hover. Vocal utterances of this species include repeated whistles.

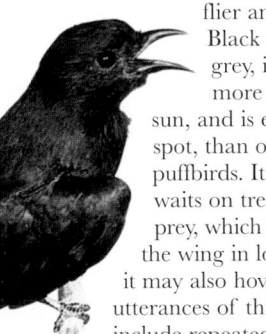

Lybius dubius
Bearded barbet

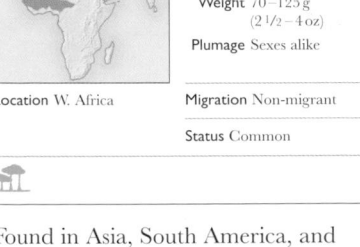

Length 25 cm (10 in)	
Weight 70–125 g (2 1/2 – 4 oz)	
Plumage Sexes alike	
Location W. Africa	**Migration** Non-migrant
	Status Common

Found in Asia, South America, and as in this species, Africa, barbets are usually solidly built birds with gaudy plumage. Most have conspicuous bristles at the base of their bill – a feature that gives them their name.

The bearded barbet has black upperparts, and red underparts with a black breastband and white flanks. Its bill is large and powerful, and it feeds mainly on fruits, showing a particular liking for wild figs. It also gleans insects from bark.

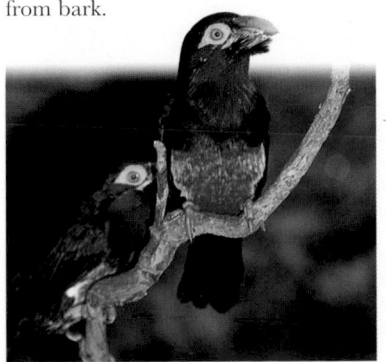

Megalaima virens
Great barbet

Length 32 cm (12 1/2 in)	
Weight 200–300 g (7–11 oz)	
Plumage Sexes alike	
Location C., S.E., and E. Asia	**Migration** Non-migrant
	Status Common

One of the largest barbets, this clumsy looking bird has a large, pale bill with numerous long bristles. Like other barbets, it often engages in duetting, the male and female singing alternate notes. While calling – a loud, musical sound that is repeated monotonously for several minutes – the bill remains closed and the throat inflates and deflates. Although usually solitary, it may at times be found in feeding flocks in treetops. The bird excavates its nest in a tree-hole; both parents feed the young.

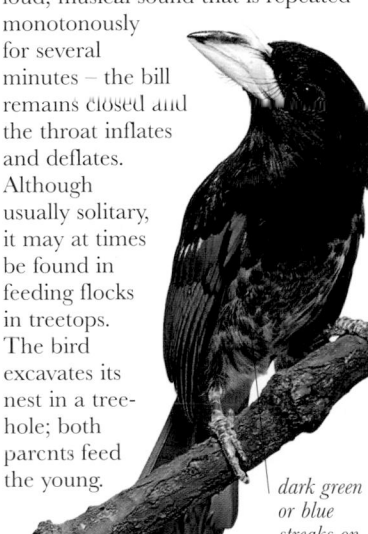

dark green or blue streaks on belly

Trachyphonus darnaudii
D'Arnaud's barbet

Length 20 cm (8 in)	
Weight 30 g (1 1/16 oz)	
Plumage Sexes alike	
Location E. Africa	**Migration** Non-migrant
	Status Common

Most barbets live in trees, but a small number – particularly in tropical Africa – are equally at home on the ground. D'Arnaud's barbet is one of these species, and is often seen hopping over bare earth or through grass as it searches for food. It may also forage in low bushes and trees for insects, fruits, and seeds. It inhabits flat terrain: grassland with scattered trees, bushland, and open woodland. D'Arnaud's barbet is not shy in the presence of humans, and is best known for the duetting of mating pairs, during which the crown feathers are erected and the tail is cocked every now and then. The bird has various other calls,

Pogoniulus bilineatus
Yellow-rumped tinkerbird

Length 10 cm (4 in)	
Weight 13 g (7/16 oz)	
Plumage Sexes alike	
Location W. to S.E. Africa	**Migration** Non-migrant
	Status Common

white facial stripes *black bill*

A wide-ranging African bird with many regional forms differing in pattern and colour, the yellow-rumped tinkerbird has a metallic call consisting of a series of popping sounds. While calling, the bird sits upright, puffing up its throat and making its rump feathers erect. This tiny barbet moves rapidly through canopy, bushes, and vines, picking up fruits and insects. It occasionally hunts insects on the wing. Rather aggressive towards other small barbets, it defends its territory vigorously. Courtship appears to involve flutter flights.

including alarm calls. While displaying, this yellow-orange, black, and red barbet spreads and swings its tail, bobs its head, and wipes its bill. It is a social bird, living in pairs or small family groups consisting of the breeding pair and subordinate individuals who may assist in raising the young. The mating pair defend their territory aggressively, marking the boundaries vocally by their duets. Both the male and female D'Arnaud's barbet dig the nest vertically into the ground, and line the nest chamber with grass.

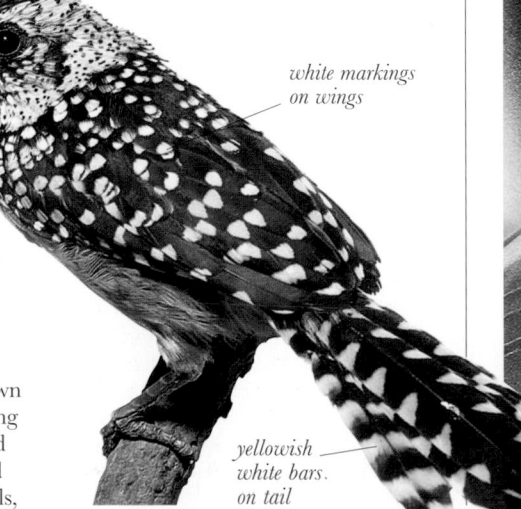

white markings on wings

yellowish white bars on tail

Indicator indicator

Black-throated honeyguide

Length	20 cm (8 in)
Weight	50 g (1 ¾ oz)
Plumage	Sexes alike

Location Africa (south of Sahara)	Migration Non-migrant
	Status Common

The black-throated honeyguide is famous for guiding humans and animals – such as honey badgers – to wild bees' nests. Once the nest has been broken open, the honeyguide flies down to it to feed. Apart from beeswax and bees, this bird also feeds on ants, termites, other insect larvae, and even the eggs of other birds. The black-throated honeyguide is a brood parasite and does not build its own nest. Nestlings have hooked bills with which they kill the nestlings of the host.

Melichneutes robustus

Lyre-tailed honeyguide

Length	9.5 cm (3 ¾ in)
Weight	55 g (2 oz)
Plumage	Sexes alike

Location C. Africa	Migration Non-migrant
	Status Common

The lyre-tailed honeyguide has a lyre-shaped tail and brownish plumage with white markings on its outer feathers. It inhabits lowland and low mountain

Aulacorhynchus prasinus

Emerald toucanet

Length	30 cm (12 in)
Weight	150 g (5 oz)
Plumage	Sexes alike

Location Mexico, S. Central America, W. South America	Migration Non-migrant
	Status Locally common

Toucanets are medium-sized members of the toucan family with relatively modest-sized bills. This species is an upland bird typically found in mountain forests. It lives in pairs and

forests and is also found in coffee plantations. Unlike the black-throated honeyguide (see left), this species is not known to lead other animals to bees' nests. It is best known for its spectacular flight displays that take place all year round. During these displays, performed by both sexes, the bird flies high above the forest, then descends swiftly in a zigzag or spiral flight. The sound heard during the flights is probably produced by the air passing through the tail feathers. This unusual behaviour is associated with feeding sites (bee hives). This honeyguide feeds mainly on beeswax, but also hunts insects in the air. Like the black-throated honeyguide, it does not build a nest, and probably parasitizes barbets.

small groups, nesting and roosting in holes. Noisy groups may form in fruiting trees. The call is loud and varied, and includes croaking sounds as well as imitations of many local species of birds. The long, black and yellow bill, white at the base, helps this toucanet to reach for fruits and animal prey.

Pteroglossus castanotis

Chestnut-eared aracari

Length	33–40 cm (13–16 in)
Weight	300 g (11 oz)
Plumage	Sexes alike

Location N. to C. South America	Migration Non-migrant
	Status Common

yellow belly

Compared to most other toucans, aracaris are smaller and more lightly built, with a slender bill. The long, brightly coloured bill of the chestnut-eared aracari has yellow, teeth-like markings. Living in the forest and forest edge, this bird may be seen restlessly hopping through the tree canopy. It usually occurs in small groups that roost huddled in tree-holes and hunt for insects and other small animals together, large flocks sometimes gathering at fruiting trees. Paired birds feed and preen each other. Like other toucans, the chestnut-eared aracari has an undulating flight.

Ramphastos toco

Toco toucan

Length	53–60 cm (21–23 ½ in)
Weight	550 g (20 oz)
Plumage	Sexes alike

Location N.E. to C. South America	Migration Non-migrant
	Status Common

black oval spot on bill

The largest of all toucan species, the toco toucan has a spectacular orange bill, up to 19cm (7½in) long. The huge bill, white rump, and red undertail coverts are especially conspicuous during the bird's undulating flight, with alternating flaps and glides. The toco toucan is found in woodland, gallery forest, wooded savanna, groves, and even open country. It usually reaches for fruits and prey from a perch, but may descend to the ground for fallen fruits. The large bill helps the bird to reach

food on the end of thin twigs that cannot support its weight. Once collected, food has to be moved from the tip of the bill into the bird's throat – a task that is achieved with a quick backward toss of the head. Less sociable than other toucans, this bird sometimes participates in periodic invasions of new areas, which occur in certain years. It also migrates locally during some winters. The toco toucan makes its nest in existing holes in dead or living trees and in earth banks; it may also use the nests of terrestrial termites. Its call, a deep, low snore, is often made high up from a branch.

bare, yellow to orange skin around eyes

glossy black plumage

Jynx torquilla

Northern wryneck

Length	16 cm (6 ½ in)
Weight	10 g (⅜ oz)
Plumage	Sexes alike

Location Europe to Asia, N. Africa	Migration Migrant
	Status Common

The northern wryneck derives its name from its habit of twisting and writhing its neck, in snake-like fashion, as a form of defensive display. The speckled, brown, grey, buff, and white plumage camouflages it well against bark and the ground. It feeds on the larvae and pupae of ants, opening anthills with its bill and using its sticky tongue to pick up its food.

Picumnus cirratus

White-barred piculet

Length	10 cm (4 in)
Weight	10 g (⅜ oz)
Plumage	Sexes alike

Location N.E. and C. South America	Migration Non-migrant
	Status Common

Piculets (about 30 species) are the smallest members of the woodpecker family. Unlike typical woodpeckers, they do not use their tails to brace themselves against trees. The white-barred piculet has a short, soft tail and brownish grey plumage, the male having a distinctive red crown. Found in a wide variety of habitats, including open woodland, savanna, brush, and forest edges, it forages at low heights on thin twigs, vines, and bamboo. This tiny bird feeds on small insects getting most of its food by vigorous hammering. It nests and roosts in holes in dead branches.

Melanerpes formicivorus
Acorn woodpecker

Length 23 cm (9 in)	
Weight 65–90 g (2 ³/₈ – 3 ¹/₄ oz)	
Plumage Sexes alike	
Location W. North America to N. South America	**Migration** Non-migrant
	Status Locally common

Dry acorns may comprise half of the daily food intake of the acorn woodpecker, which has a unique habit of storing acorns (see panel, right). It is also known to feed on other seeds, sap, and fruits. Insects form important food for the nestlings. Both sexes have red crowns, but the forecrown is black in females. The tail and bill are moderately long. Restricted to oak and pine-oak woodland, this sociable woodpecker is often found in a group or breeding unit of 3–12 individuals, who share work on the communal acorn store and on building the nest, and defend the common territory that contains foraging sites, sap trees, the group larder, roosts, and the nest site. The acorn woodpecker is characterized by a complex breeding pattern which may be described as "polygynandrous". Pairs may breed alone, but are typically assisted by males and females that usually belong to previous broods. Helping males participate in reproduction only after their mother has been displaced by another female.

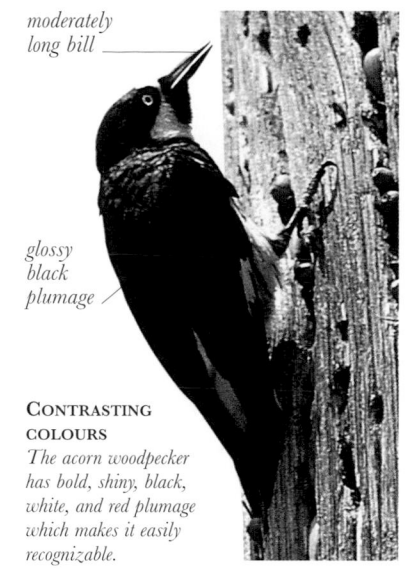

moderately long bill

glossy black plumage

CONTRASTING COLOURS
The acorn woodpecker has bold, shiny, black, white, and red plumage which makes it easily recognizable.

STORING ACORNS

The acorn woodpecker drills small, neat holes in tree bark, and jams an acorn firmly into each one. Working together, birds belonging to a single breeding unit can store up to 50,000 acorns in one "granary tree", creating a store of food to last them through the winter.

Sphyrapicus varius
Yellow-bellied sapsucker

Length 22 cm (9 in)	
Weight 60–80 g (2 ¹/₈ – 2 ⁷/₈ oz)	
Plumage Sexes differ	
Location North and Central America, Caribbean Islands	**Migration** Migrant
	Status Common

The yellow-bellied sapsucker drills holes into trees to extract sugary sap. A series of holes are chiselled into the trunks of broadleaved trees like maple, fruit trees, birch, and poplar. After a short while, the bird returns to feed on the sap oozing out, and on the insects attracted by the sap. This bird has a mewing call-note.

Colaptes auratus
Northern flicker

Length 30–35 cm (12–14 in)	
Weight 100–175 g (3 ⁵/₈ – 6 oz)	
Plumage Sexes alike	
Location North and Central America	**Migration** Partial migrant
	Status Common

The northern flicker has a long, pointed, and slightly curved bill. During confrontations, the bill points forwards, and the head and body engage in

---

Picus viridis
Green woodpecker

Length 32 cm (12 ¹/₂ in)	
Weight 200 g (7 oz)	
Plumage Sexes differ	
Location Europe, W. to C. Asia, N. Africa	**Migration** Non-migrant
	Status Locally common

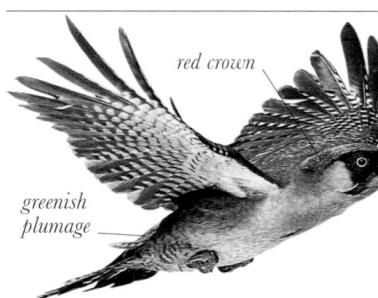

red crown

greenish plumage

Like the northern flicker (see below), the green woodpecker feeds mainly on the ground, although it uses trees as a refuge, and as places to breed. It has an exceptionally long tongue – even by woodpecker standards – and it uses it to probe into ants' nests, extracting them with its sticky tip. The green woodpecker has a loud, laugh-like call and steeply undulating flight.

swinging movements, while the tail and wings are flicked and spread. Often found on the ground, this bird feeds mainly on ants, digging into the earth and pillaging ants' nests, with the help of its long tongue. When feeding arboreally, it usually concentrates on the dead parts of a tree. The female, shown here, lacks the black moustachial stripe of the male.

long tail

spotted whitish underparts

Dendrocopos major
Great spotted woodpecker

Length 20–24 cm (8–9 ¹/₂ in)	
Weight 65–100 g (2 ³/₈ – 3 ⁵/₈ oz)	
Plumage Sexes differ	
Location Europe, Asia, N. Africa	**Migration** Non-migrant
	Status Common

This small, black and white woodpecker has a red lower belly and undertail coverts. The male has a red band across the nape. As is typical in woodpeckers, the strong feet, stiff tail, bill, and tongue of this species are adapted for climbing trees, drilling holes, and probing into cavities. The great spotted woodpecker feeds on fruits, berries, sap from trees, insects and their larvae, and even nestlings of other birds. Early in the breeding season, the bird can be easily detected by the loud drumming sound it makes.

Dryocopus martius
Black woodpecker

Length 45 cm (18 in)	
Weight 350 g (13 oz)	
Plumage Sexes differ	
Location Europe to Asia	**Migration** Non-migrant
	Status Common

This crow-sized bird is Eurasia's largest woodpecker, easily recognized both by its size and by its jet-black plumage and crimson crown. Females have a smaller crown patch than males, placed further back, but apart from this the 2 sexes are the same. The black woodpecker lives in mature forests and woodland, where it actively chisels out prey that is hidden deep within living and dead wood. It feeds mainly on tunnelling insects, such as sawfly grubs and ants, and, when probing for food, it can hack out vertical gashes up to 50 cm (20 in) long. This woodpecker's excavations produce wood chippings the size of clothes' pegs, which build up in piles on the forest floor. It also pecks at cones, often wedging them in the stumps of fallen trees. This bird lays up to 6 eggs, in holes that it carves out every year. Old holes are very useful to other birds, which take them over when their original owner moves on.

BIRDS

PASSERINES

PHYLUM	Chordata
CLASS	Aves
ORDER	Passeriformes
FAMILIES	About 80
SPECIES	5,200–5,500

MOST OF THE world's bird species are passerines. They are sometimes known as the perching birds because they have a unique type of foot that enables them to grip even the most slender branches. Passerines are also distinguished by the complex sounds, or songs, made by many species. These are produced using a vocal organ known as the syrinx (which is also found in other birds). Passerines are considered by many to be the most highly evolved of all birds, and many exhibit unusual intelligence. Most live in bushes and trees, but some are adapted to living on the ground; a few (such as swallows) lead an almost entirely aerial existence. Passerines are found throughout the world in all terrestrial habitats, from arid deserts to tropical rainforest. Many species are a familiar sight around buildings and in gardens.

ANATOMY

The passerines' specialized perching foot (see Perching) and well-developed voicebox (or syrinx, see Singing) are among their defining characteristics. These features apart, the members of this group are enormously varied. While a great number of them have subdued colouring, there are also many with spectacularly vivid and bizarre plumage – for example, the remarkable birds of paradise and the multi-coloured tanagers and finches. Males are often more brightly coloured and patterned than females. Most passerines are small birds, but they range from the large, bulky crows and ravens, up to 65cm (25in) long, to the short-tailed pygmy tyrant, only 7cm (3in) long. Another highly variable feature of passerines is the shape of the bill, which often indicates a specific food preference (see below).

GREENFINCH   **WOOD WARBLER**   **GREAT GREY SHRIKE**   **SUNBIRD**

BILL SHAPE
The shape of a passerine's bill offers important clues to its diet. Species that feed on seeds (such as the greenfinch) often have a short, conical bill, while those that eat invertebrates (such as the wood warbler) tend to have a thin bill. Like many other predatory passerines, the great grey shrike has a hooked and notched bill for subduing relatively large prey. In sunbirds, which feed on nectar, the bill is long and thin, often with a downward curve, to help them reach inside flowers.

SINGING

Passerines owe their singing ability to their complex syrinx, which is more highly developed than it is in other birds. Each species has its own distinctive song (or songs), a series of sounds uttered within a defined rhythm and structure. Many passerines have beautiful and complicated songs, among them the larks, wrens, thrushes, nightingales, and lyrebirds. In some species, the chicks learn to sing only by listening to adults of the same species; in others, the songs are innate and do not need to be learnt. It is mainly the males that sing, either to claim and keep territory or to attract females.

PERCHING

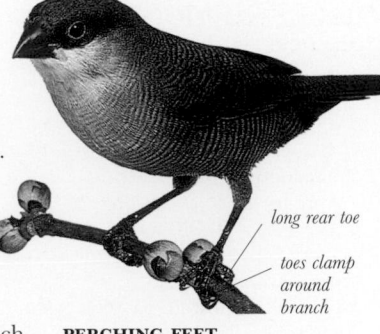

Passerines can perch securely on twigs, reeds, and even grass stems. Three of their 4 toes point forwards and the other backwards. The toes can move independently of each other, and the front toes can oppose the back one, which is especially strong. All 4 toes are level with each other, unlike in many non-passerines, which have a raised hind toe. When a passerine lands on a perch, its weight causes tendons in the leg to tighten, clamping the toes tightly shut. This grip functions even when the bird is asleep. Despite these advantages, some passerines perch much less often than others. Larks, for example, are essentially terrestrial, and have relatively flat feet to give them balance when running.

long rear toe

toes clamp around branch

PERCHING FEET
Like many passerines, the common waxbill has long, thin toes that are easily wrapped around a small stem. The backward-facing toe is strong and, in contrast to several non-passerine groups, cannot be reversed. Most passerines have moderately curved, sharp claws that can grip many surfaces.

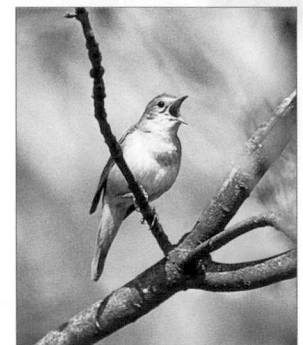

SINGING MALES
The male nightingale produces some of the most varied of all birdsongs, often delivering them with a characteristic crescendo. It sings by both day and night, using its daytime songs mainly to mark its territory and deter rivals and its nocturnal songs to attract females.

trachea

muscle

membrane

cartilage rings

bronchus

SYRINX
The syrinx is located in the trachea. Membranes in the syrinx vibrate and produce sound as air passes over them. The bird uses muscles, which are attached to rings of cartilage, to modify the sound.

REPRODUCTION

Birds of paradise and manakins perform some of the most elaborate courtship displays of all birds, the males attracting females by jumping, hanging upside down, and making extraordinary noises. However, most passerines have more subtle displays. Many are skilled nest-builders, typically constructing a cup nest from twigs, leaves, and soft material above ground, but they build many more elaborate structures, too. The chicks hatch naked, blind, and helpless, and rely entirely on their parents to feed them until they are ready to leave the nest at 10–15 days old.

ATTRACTING A MATE
Male bowerbirds (such as the spotted bowerbird, left) create elaborate "stages" on which to perform courtship displays. They select a patch of ground, or even build a special structure (the bower), and decorate it with a wide variety of items, such as seeds, grass, and moss, or even human artefacts.

HANGING NEST
Weavers, including the Baya weaver (right), build a highly elaborate domed nest that hangs from a branch, often above water. The nest has an entrance chamber, a nest chamber, and a long tube hanging below. Weavers often nest colonially.

entrance tube

COURTSHIP DISPLAY
Most passerines are monogamous. During the breeding season, a male and female generally form a pair, establish a territory, and build a nest. Among the exceptions to this pattern of behaviour are the birds of paradise (such as these Raggiana birds). Males display to females at a breeding arena. After mating, the female builds a nest and cares for her eggs and young alone.

FEEDING

Most passerines are small, and therefore require high-energy foods. This means that most species feed either on invertebrates or on seeds, or a combination of the two. Some also take nectar, especially the sunbirds. A few rainforest-dwelling families, including birds of paradise, manakins, and cotingas, have an almost exclusive diet of fruits. Among the great variety of passerines are some with more unusual feeding habits. Shrikes are meat-eaters that store their captured prey in a "larder", impaling the bodies of large insects or small vertebrates on thorns or barbed wire before eating. Crossbills feed only on the seeds of coniferous trees, using their crossed mandibles as tweezers to part the scales of a cone in order to remove the seeds.

USING TOOLS
The woodpecker finch of the Galapagos islands is one of the few animals that uses tools to catch prey. It picks up a small stick or cactus spine in its bill and uses it to prize insects from crevices in the bark of trees.

UNDERWATER FEEDING
Dippers are among the few passerines to venture into water. When feeding, they dive into shallow, fast-flowing streams, propelling themselves downwards with their wings, to catch prey as it swims. They also walk along the river bed, picking up grubs in their bill.

ANTING

A few passerines, including starlings and blackbirds in Europe, indulge in an unusual behaviour known as anting, in which a bird collects ants in its bill and rubs them over its feathers. Occasionally, the bird will crouch down and let ants run over its plumage. Birds select only ants that exude formic acid, so it is thought that the habit may help to cleanse the feathers of parasites. Also unique to passerines is ant-following, in which birds follow army-ant columns through tropical forest. The birds don't eat the ants themselves, but catch the small insects fleeing the ants' path.

COLLECTING ANTS
A Eurasian jay lies down among a swarm of ants, allowing the insects to crawl through its plumage. The reason for anting is unknown, but it is thought that acid produced by some ants may act as an insect repellent or insecticide, or as a plumage lubricant.

Order PASSERIFORMES

Family EURYLAIMIDAE

Broadbills

Length 13–28 cm (5–11 in)

Species 14

Broadbills are found across the humid tropics, from West Africa to the Philippines, in habitats including scrub and forest-edge thickets, mangrove swamps, inland forest, and mountain moss-forest. These plump birds have large heads and wide, flattened, hooked bills. Males are usually coloured a striking green, red, pink, or blue; females are duller but can be larger than males. Broadbills feed mostly on fruits and insects; one species eats lizards, and another catches crabs and even fish. Most feed in the mid-level of the forest, filter-sucking insects from leaves and branches. Some broadbills feed alone or in pairs, others gather in foraging flocks of 20–30 when not breeding. These birds build large, pear-shaped nests with a "porched" entrance in the lower half. Woven from rootlets, leaves, and twigs, and often decorated with cobwebs and lichens, these are hung from inaccessible branches. Females lay clutches of 1–8 eggs.

black wing bars

wide bill covered by feathers

short, rounded tail

LESSER GREEN BROADBILL
The Southeast Asian species
Calyptomena viridis *is 20cm
(8in) long, has long feathers around
the bill, nearly concealing it, and 3
black wing bars. Small groups feed in
the lower branches on ripening fruits
and buds. The lesser green
broadbill is also known to
feed on insects such as
flying termites.*

Family DENDROCOLAPTIDAE

Woodcreepers

Length 14–37 cm (5½–14½ in)

Species 52

Woodcreepers are an American family mostly found in South America. They behave in a similar fashion to European and American treecreepers (see p.351). These birds have a long, distinctive tail with stiffened feather shafts that curve inwards. Like treecreepers and woodpeckers, they use the tail to support themselves as they climb trees. They also have strong feet and long, sharp claws to help them climb. Generally olive-brown, with rufous wings and tail, many have a sturdy, relatively long, straight or slightly downcurved bill. Some have a thinner bill with a more defined downward curve, and the extreme is reached in the scythebills which have a long, sickle-shaped bill, up to one third of their total length. When hunting for insects and invertebrates, wood-creepers work up a tree, either straight up or in a spiral, examining crevices, loose bark, moss and other plants, flying to the base of the next tree as they finish with one. Larger species also eat lizards and small frogs. Some species follow army ant swarms, feeding on the insects disturbed by the columns of ants as they move. Wood-creepers nest in old woodpecker holes and crevices.

stout bill

olive-brown coloration

WEDGE-BILLED WOODCREEPER
*Found in humid lowland
forests, mainly in the
Amazon Basin,*
Glyphorhynchus
spirurus *is 14cm (5½ in)
long. Although rarely seen,
its sneezing "cheeyf" call
often draws attention.*

Family FURNARIIDAE

Ovenbirds

Length 15–25 cm (6–10 in)

Species About 221

A large group with many subfamilies, ovenbirds are mostly small brown birds with paler underparts, found in Central and South America. They live in a wide range of habitats, and may be foliage-gleaners, living in the canopy, or leaf-scrapers, which flick leaves in the air and skulk in the densest undergrowth. Many species, such as the true ovenbirds, or horneros, prefer open country, while the cinclodes live along water courses; a few species inhabit marshes. A number of species, such as the Patagonian earthcreeper and brown cachalote, tend to run rather than fly. Ovenbirds are mostly insect-eaters, but some also feed on seeds. Their nesting habits vary widely: while the true ovenbirds build substantial, domed mud-ovens on tree branches, many species nest in natural holes, animal tunnels, or holes they dig themselves. The white-throated cachalote builds a huge stick nest that lasts for many years after the birds have used them. Almost all ovenbirds lay white eggs, usually in clutches of 3–5; some species lay up to 9 eggs. These are mostly incubated over 15–20 days, with the young leaving the nest in 13–18 days. Old ovenbird nests are often used by other birds.

DARK-BELLIED CINCLODES
Found in southwest South America, Cinclodes
patagonicus *is always near fresh water or the sea.
It moves along rocky streams and rivers, feeding on
small aquatic animals. This bird is 21cm
(8½ in) long, and generally nests in a rock
burrow, often in a stream bank. Its
call is a sharp "tjit".*

ground-probing bill

RUFOUS HORNERO
Argentina's national bird,
Furnarius rufus *has a
misleading common name as it
is the least rufous of horneros.
Measuring 18–20cm
(7–8in) in length, it clambers
about on the ground or on
branches. Its mud-oven nest,
which it uses only once, adorns
the top of fence posts and
telephone poles, lasting for years.*

rufous tail

BUFF-FRONTED FOLIAGE-GLEANER
Philydor rufus, *19cm (7½ in) long, is locally
common in humid mountain forests from
Costa Rica to Argentina. It
searches the canopy for insect
prey, hanging upside
down athletically.*

dull plumage

whitish underparts

COMMON THORNBIRD
*Occurring in widely separated areas of
semi-arid habitat,* Phacellodomus
rufifrons *is 16cm (6½ in) long. It builds a
large, hanging stick nest, which has several
chambers and is often used by other birds.
This South American bird spends much of
its time in trees, in pairs or groups.*

stripe behind eyes

rufous wings

long tail for balance

Family FORMICARIIDAE

Antbirds

Length 8–36 cm (3¼–14 in)

Species About 236

Many of the species in this family follow army ants but rarely eat them, preying instead on the insects and small animals they flush out. When following swarms of ants, some antbirds cling to saplings and dart down for prey, while others follow on foot at the outskirts of the swarm. Found in Central and South America, antbirds range from the tiny, warbler-like antwrens to the large, hook-billed antshrikes, and include antcreepers, antpittas, ant-thrushes, and antvireos. They are often strikingly patterned. These birds live in lowland and montane tropical forests, as well as savanna and scrubland, and up to

VARIEGATED ANTPITTA
Grallaria varia, 20cm (8in) in length, has the long legs and short tail typical of the antpittas. It is found in Guyana, Surinam, French Guiana, and Brazil. This bird flicks leaves aside on the forest floor to probe for earthworms and insects.

50 species can be found within small areas of the Amazon. Most antbirds pair for life and work together to build a simple nest cup. The female lays 2, rarely 3, eggs. Both sexes incubate the eggs and feed the hatchlings. Antbird songs are simple – a series of whistles or chattering notes.

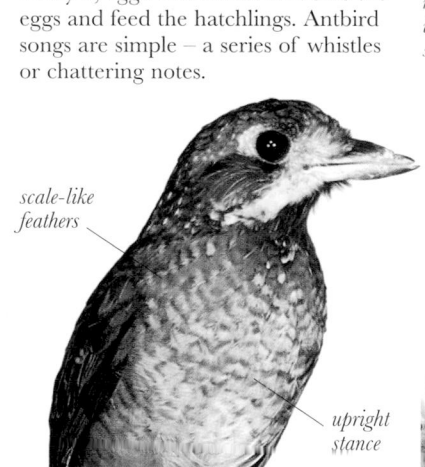

scale-like feathers

upright stance

long legs

BARE-CROWNED ANTBIRD
An inhabitant of Central America and north Colombia, Gymnocichla nudiceps is found in humid lowland forest. About 16cm (6½ in) long, it spends most of its time on the ground and pounds its tail downwards when feeding. While singing, the male's tail quivers.

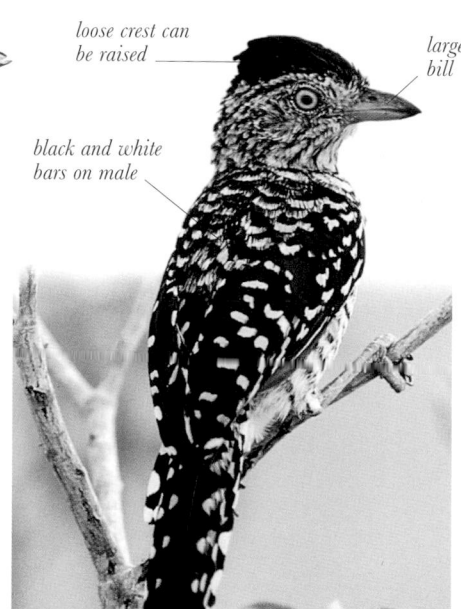

loose crest can be raised

black and white bars on male

bright blue, featherless crown

stocky body

large bill

short legs

broad tail

BARRED ANTSHRIKE
Found in the dense undergrowth and scrub of Central and South America, Thamnophilus doliatus is 16cm (6½ in) long. The male has a black crown, and is barred black and white all over. The female is rufous, with black and white barring only on the sides of the head and neck. These birds are usually found in pairs, hopping about in vegetation or peering into foliage. Both sexes sing, the female often echoing the male.

Family RHINOCRYPTIDAE

Tapaculos

Length 11–25 cm (4¼–10 in)

Species 30

Found principally in southern South America, tapaculos are mainly ground-dwelling birds. With long legs, large, powerful feet, and long tails, these birds tend to run rather than hop. Their short, rounded wings allow them to fly only briefly and weakly. A distinctive, movable flap of skin covers the nostril. Secretive and rarely seen, tapaculos can be identified by their loud calls; however, the calls can be difficult to pinpoint as they have a ventriloquial quality. These birds feed on insects and other invertebrates. They generally build their nest in a hole or crevice at ground level, or in a burrow that they dig themselves, and lay 2–4 white eggs.

CHUCAO TAPACULO
Scelorchilus rubecula superficially resembles a European robin. It has a striped black and white belly, and is 19cm (7½ in) long. It is found in south Chilean temperate forest, including bamboo thickets.

Family COTINGIDAE

Cotingas

Length 9–45 cm (3½–18 in)

Species 65

Cotingas vary widely in appearance. They include the 2 bright orange cock-of-the rock species, brilliant blue cotingas, umbrellabirds with their umbrella-like crests, and the bellbirds with their wattles. Most are found in lowland forest, but some occur in mountain forest and shrubbery in the Andes. All cotingas are fruit-eaters, and some also feed on

large crest conceals bill

black wings and tail on male

insects. Males of some species display at a lek to attract females. Some species build a tiny platform nest and lay only one egg, which takes 23–28 days to hatch.

3 wattles on male

THREE-WATTLED BELLBIRD
The male of the species Procnias tricarunculata has one of the loudest bird calls, a deafening "bock" that rings over the forest canopy. This Central American bird is 25–30cm (10–12in) long. Its bill opens wide enough to swallow large fruits whole.

LOVELY COTINGA
Cotinga amabilis, 20cm (8in) long, is found in high forest canopies of Central America. The male's outer wing feathers make a tinkling noise during courtship. The female may destroy the nest after her young have flown, to prevent predators detecting the site, which is often re-used.

ANDEAN COCK-OF-THE-ROCK
During courtship, the males of this species, Rupicola peruviana, confront each other aggressively at the lek, the display becoming frantic when females appear. This bird is 30cm (12in) long, and as its name suggests, is found in the Andes.

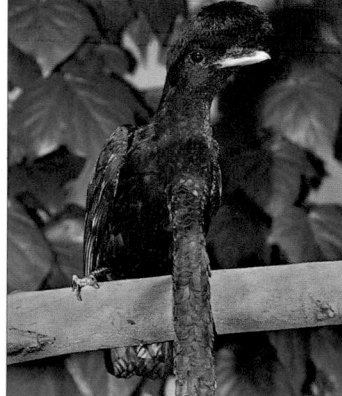

AMAZON UMBRELLABIRD
Found along the Amazon and Orinoco rivers, and in east Andean forest, Cephalopterus ornatus is 41–49cm (16–19½ in) long. While courting, males expand their crest and their enlarged wattle as they utter a deep, moaning "boom". Females are smaller.

purplish blue on throat

black bill

black markings on wings

cinnamon-rufous plumage

RUFOUS PIHA
Lipaugus unirufus, 23cm (9in) long, is an inhabitant of Central and northwest South America. Males do not display at a lek but may utter loud, whistling calls within earshot of each other.

Order PASSERIFORMES *continued*

Family PIPRIDAE

Manakins

Length 9–15 cm (3½–6 in)

Species About 51

Small birds, with short tails and bills, and bead-like eyes, manakins are found mainly in the lowlands of Central and South America. Males are often brightly coloured, while females are usually olive. Males of certain species have modified feathers that can produce a mechanical noise during courtship. The male wire-tailed manakin uses his long, curved, wire-like tail feathers to tickle the throat of a visiting female. Males of some species gather in leks to display to females, either on the ground or on a branch. Successful males mate with several females. The female alone builds the nest – a woven cup in a horizontal tree fork – incubates the eggs, and feeds the young. She generally lays 2 eggs which hatch after 12–15 days. The nestlings feed on insects, but adults feed mainly on fruits, plucking these on the wing. All manakins that have been studied are sedentary and do not migrate.

black crown

white throat feathers

orange legs

WHITE-BEARDED MANAKIN
At a lek, the male Manacus manacus, *using a sapling as the display perch, performs stylized jumps and produces noises with his wing feathers. This South American bird is 10cm (4in) long.*

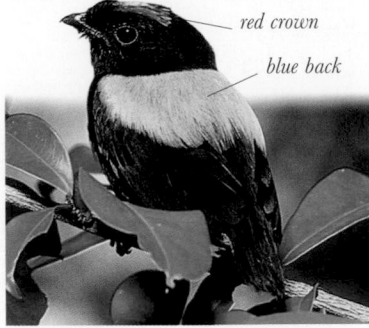

red crown

blue back

BLUE-BACKED MANAKIN
Chiroxiphia pareola, *12cm (4¾in) long, is found in South America. Two males often display by jumping over each other on branches; the winner makes short, circular courtship flights.*

Family TYRANNIDAE

Tyrant flycatchers

Length 5–38 cm (2–15 in)

Species 375–397

This large family is very diverse and some experts split it into even more species. It contains the world's smallest passerine, the short-tailed pygmy-tyrant, which is just 5cm (2in) long with a wingspan of 3.5cm (1½in). However, within this large and varied family, body sizes vary widely and include many medium-sized birds. This family is found from the taiga of Alaska south through North America and the West Indies to the southern tip of South America, and in offshore islands including the Galapagos. However, the vast majority occur in Central and South America, with only 30 species breeding in North America; all the North American species migrate south in the autumn. Tyrant flycatchers have drab plumage overall and the sexes look similar. There are exceptions, however, with some species having colourful crown patches. Most members of the family are insect-eaters, but some catch frogs and lizards, and tropical species regularly eat fruits. Most tyrant flycatchers form monogamous pairs and many have courtship displays that involve wing-stretching or flicking. These birds are not colonial nesters, and their nests vary considerably: many build a simple cup in a fork, but some create large, hanging nests suspended from a branch; quite a few nest in cavities. Female tyrant flycatchers in the neotropics lay 2 or 3 eggs, but the clutches can be larger in North America. Females incubate their eggs for 14–20 days and the nestlings fledge in 14–23 days. Usually, both sexes build the nest and feed the young.

EASTERN KINGBIRD
Tyrannus tyrannus, *20cm (8in) long, breeds in North America and winters in Central and South America. It eats insects in summer and fruits in winter. Renowned for its vigorous defence of territory, it often attacks other birds, even landing on their backs in flight to inflict blows.*

long tail

grey back

COMMON TODY-FLYCATCHER
The most common and widespread tody-flycatcher, Todirostrum cinereum *moves quickly among leaves as it catches insects. It can look quite comical as it hops and flutters about, perpetually in motion. It measures 9cm (3½in) in length, and occurs in Central America and parts of South America.*

GREAT KISKADEE
Common, widespread, and noisy, Pitangus sulphuratus *has a variety of calls that include "kis-ka-dee". It eats insects and fruits but will also feed on fish and tadpoles, diving for them from a perch. The great kiskadee occurs from Texas, USA, south to Argentina. It is 21–24cm (8½–9½in) in length.*

black and white stripes on head

yellow belly

large wings

LEAST FLYCATCHER
Empidonax minimus *breeds in eastern Canada and the USA, and winters from Mexico to Panama. This bird, 12.5–14.5cm (5–5¾in) long, is the first species of the genus* Empidonax *to migrate north and the last to fly south. It flies out from perches to catch insects and builds a frail-looking cup nest in a tree fork.*

white eye-ring

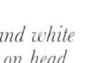

SCISSORTAIL FLYCATCHER
In its courtship sky dance, Tyrannus forficata *executes vertical and zigzag dives and tumbles, its long tail feathers streaming like ribbons. These aerobatics are accompanied by rolling cackles that sound like applause. Found in North and Central America, this species is 30–38cm (12–15in) long; the female has a shorter tail.*

long, white and black outer tail feathers

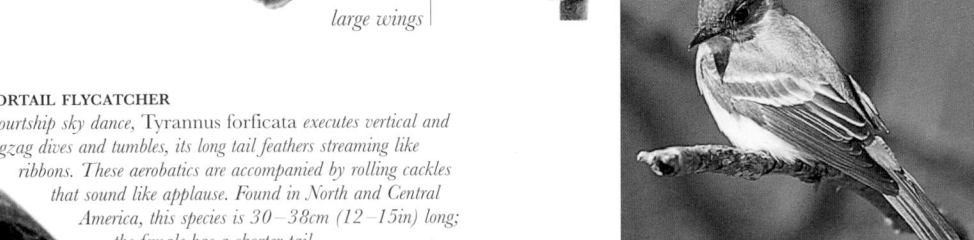

EASTERN PHOEBE
Sayornis phoebe, *which measures 18cm (7in) in length, gets its name from its hoarse "fee-bee" song. This bird is found in the USA, Canada, and Mexico. It hunts over farmland and wooded roadsides, sallying after insects and waggling its tail when it alights. It may hover briefly over water when chasing flies.*

male's blue-tipped, scarlet crest

blue spots

ROYAL FLYCATCHER
Onychorynchus coronatus *is 16cm (6½in) long and has a spectacular, brightly coloured, fan-like crest: the male's crest is scarlet with blue tips while the female's is yellow with blue tips. The crest is normally folded away and sometimes protrudes behind the head, giving the bird a "hammerhead" look. The royal flycatcher is found in Central and South America.*

Family OXYRUNCIDAE

Sharpbill

Length 17 cm (6 1/2 in)

Species 1

This family consists of only one species, *Oxyruncus cristatus*, although it should be noted that some authorities now place the species with the cotingas (see p.339), while others include it with the tyrant flycatchers (see p.340). The sharpbill occurs scattered from Costa Rica southwards to southern Brazil, in humid forest up to 1,600m (5,300ft).

Found alone or in mixed flocks, it snatches insects and invertebrates with great agility, hanging upside down like a tit (see p.351). A small, stocky bird with a sharply pointed bill, it draws attention through the male's far-carrying song: an exceptionally sharp and prolonged whistle.

scaly face

very pointed bill

Family PHYTOTOMIDAE

Plantcutters

Length 18–19.5 cm (7–7 1/2 in)

Species 3

All 3 bulky, finch-like plantcutters – the Peruvian, the rufous-tailed, and the white-tipped – have distinct ranges in different parts of South America. Found in agricultural areas, they are often regarded as pests by farmers. Included by some authorities with cotingas (see p.339), these birds derive their name from their rounded bill which has a serrated cutting edge that enables them to pluck leaves,

buds, shoots, and fruits. The nest is an untidy platform of root fibres; the female lays 2–4 greenish blue eggs with dark markings.

WHITE-TIPPED PLANTCUTTER
Phytotoma rutila, 19cm (7½ in) in length, is found in the dry scrub and acacia groves of Bolivia, Paraguay, Uruguay, and Argentina. This short-crested bird has a mechanical-sounding call.

Family PITTIDAE

Pittas

Length 15–28 cm (6–11 in)

Species 29

Stumpy-looking birds with an acute sense of smell, pittas live on the forest floor in Asia, Australia, and Africa, and feed on insects, spiders, snails, and other invertebrates. They are secretive birds and seldom seen, often blending in with the forest floor despite their highly colourful plumage. The female lays 2–7 eggs and both sexes raise the young.

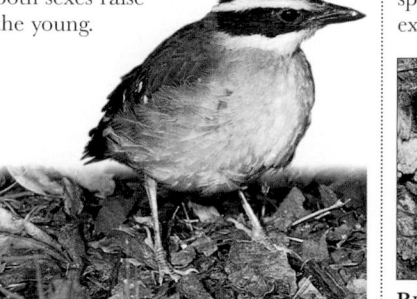

AFRICAN PITTA
Pitta angolensis, the only African species, is a migrant found in thick, evergreen forest. This usually silent bird is 20cm (8in) long.

Family ACANTHISITTIDAE

New Zealand wrens

Length 8–10 cm (3 1/4 – 4 in)

Species 3

Found only in New Zealand, this family of small and stocky birds includes the rifleman, rock wren, and bush wren. Both sexes of the rifleman and rock wren incubate eggs and feed the young. A fourth species, Stephen's Island wren, was a very weak flier and was wiped out soon after its discovery in 1894. Stead's bush wren, a subspecies of the bush wren, became extinct in 1965.

RIFLEMAN
Abundant and widespread, Acanthisitta chloris occurs on North and South Islands. It is 8cm (3¼ in) long; unusually among passerines, the female is larger. She is also more streaked.

Family PHILEPITTIDAE

Asities

Length 10–15 cm (4–6 in)

Species 4

Asities are short, stocky birds found only in dense, evergreen forest mainly in the humid east of Madagascar. Schlegel's asity is also found in the northwest of the island. The velvet asity and Schlegel's asity feed mainly on fruits plucked from bushes, while the 2 sunbird-asities are insect-eaters and also probe flowers, probably for nectar. Only the nest of the velvet asity – a pear-shaped, hanging

structure with 3 white eggs – has been found. The forest habitat of these birds is fast disappearing with the increase of cultivated land.

COMMON SUNBIRD-ASITY
Neodrepanis coruscans occurs at all levels in the forest, from leaf litter to high canopy. It is 10cm (4in) in length and has a downcurved bill.

Family MENURIDAE

Lyrebirds

Length 90–100 cm (35–39 in)

Species 2

Lyrebirds occur in rainforests and scrub in east and southeast Australia. Of the 2 species, Albert's lyrebird is the smaller and has more chestnut plumage, while the superb lyrebird is greyer. Males of both species are known for their extravagant tail, which in the superb lyrebird can be up to 60cm (23½in) long, and has 2 distinctive lyre-shaped feathers. Male lyrebirds display from midwinter; while Albert's lyrebird

treads down a display platform of vines, the superb lyrebird rakes up numerous low display mounds around his territory. The male's powerful song includes impressive mimicry. While calling, he inverts his tail over his back, shimmering it and turning around slowly.

SUPERB LYREBIRD
Menura novaehollandiae, 86–100cm (34–39in) long, is found in southeast Australia and has been introduced in Tasmania. The male of this species does not acquire his full tail until he is 6–8 years old.

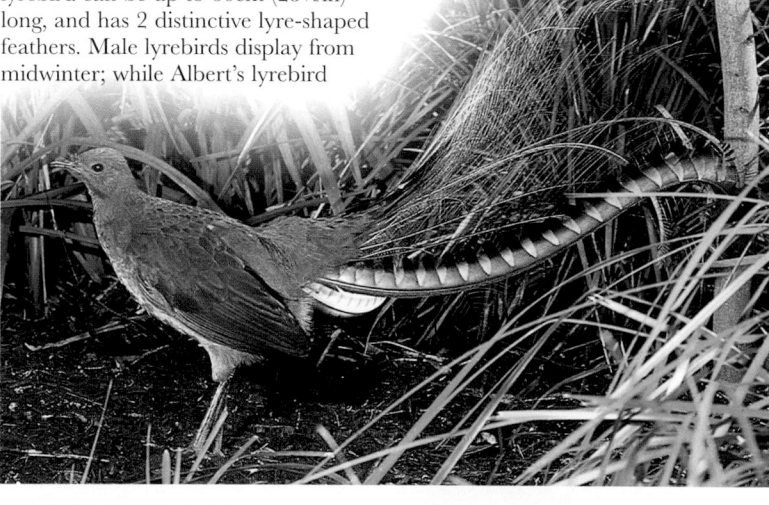

Family ATRICHORNITHIDAE

Scrub-birds

Length 16–23 cm (6 1/2 – 9 in)

Species 2

The distribution and population of these brown Australian birds have decreased since the mid-1800s. Secretive, they spend much of their time on the ground in thick cover, in wet forest. They feed on insects, lizards, and frogs. Both species of scrub-birds build domed nests.

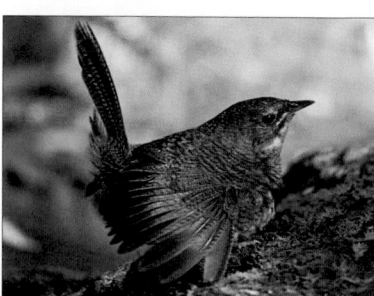

RUFOUS SCRUB-BIRD
Found in highland rainforest, in a small area of east Australia, Atrichornis rufescens has a penetrating call but is extremely hard to locate. It is 16.5–18cm (6½–7in) long, and is a weak flier that scuttles in leaf litter like a mouse.

Order PASSERIFORMES *continued*

Family ALAUDIDAE

Larks

Length 13 – 20 cm (5 – 8 in)

Species About 81

Larks are found in virtually all parts of the world – North and South America, Africa, Europe, Asia, and Australia. The skylark, famous for its far-reaching song during flight, is the most well-known member of this family. Larks are ground-dwelling birds, mostly with streaked brown, cryptic plumage which provides good camouflage on the fields and deserts they inhabit; some species, however, have black and white markings. Males and females usually appear similar. Some species, such as the skylark and crested lark, have a crest of feathers that they raise in territorial and courtship displays and when singing. Larks usually have a strong and undulating flight, although they tend to flutter over

short distances. The skylark's song-flight is a vertical ascent, high into the sky. Many other species have a climbing or circling song-flight. Larks feed on invertebrates, plant matter, and seeds. Many species flock in autumn and winter and only a few species migrate. The females lay 2 – 6 eggs, which are generally speckled, in nests made on the ground; the young chicks often leave the nest before they can fly.

rufous wings

RUFOUS-NAPED LARK
A stocky, short-tailed lark, Mirafra africana inhabits the open and scrubby grassland of southern Africa. Measuring 15 – 18 cm (6 – 7 in) in length, this bird walks with an upright stance, and has a mournful whistle.

SKYLARK
Alauda arvensis, 18 – 19 cm (7 – 7½ in) long, is a streaked, brown lark of farmland and open spaces known for its continuous song when in hovering flight. It is common in North Africa, Europe, and parts of Asia, and has been introduced in Australia and New Zealand. Northern populations migrate south in winter.

HOOPOE LARK
Measuring 18 – 20 cm (7 – 8 in) in length, Alaemon alaudipes inhabits desert and semi-desert areas of North Africa (including the Sahara) and West Asia. It often runs away instead of flying. Its striking black and white wing pattern is partly revealed in flight.

pale body

distinctive yellow and black face markings

HORNED LARK
Also known as the shore lark, Eremophila alpestris inhabits tundra and steppe in Europe, and prairie, farmland, and desert in North America. It is also found in North Africa and Asia. About 14 – 17 cm (5½ – 6½ in) long, it has little "horns" of feathers.

Family HIRUNDINIDAE

Swallows and martins

Length 12 – 23 cm (4¾ – 9 in)

Species About 81

Found all around the world, except in the frozen Arctic and Antarctic regions, swallows and martins are often dark blue or green above and paler below. They generally have a forked tail, sometimes with long tail

streamers. The bill is short but it has a wide gape that is an effective trap for catching insects on the wing. Their dependence on aerial insects makes many species migrants – they breed in temperate areas but fly to the tropics and south temperate areas when the weather gets colder and insect prey disappears. However, many species live in the tropics throughout the year. Swallows and martins have 3 types of nest: natural holes in trees, cliffs, or buildings; tunnels excavated in river banks or sand quarries; and mud cups attached to cliffs or buildings. Some species raise 2 or 3 broods, the clutch size ranging from 1 to 8 eggs.

CLIFF SWALLOW
Hirundo pyrrhonota breeds from Alaska to Mexico and winters in South America. About 13 – 15 cm (5 – 6 in) long, it nests colonially, building gourd-shaped mud nests under the eaves of a building or on a cliff. The female lays 4 – 6 white eggs.

distinctive white rump

blue-black male

PURPLE MARTIN
Progne subis nests in man-made martin houses and in old woodpecker holes in trees. Colonies can number up to 200 pairs. About 18 cm (7 in) long, it breeds in North America and winters in the Amazon Basin. The male is blue-black all over, but the female is duller above and pale grey below.

forked tail

HOUSE MARTIN
A summer visitor to North Africa, Europe, and northern Asia, Delichon urbica winters in Africa (south of the Sahara) and Southeast Asia. It has a long association with humans, building mud half-cup nests on walls under eaves of buildings, where it nests in groups. This bird is 12.5 cm (5 in) in length.

white throat

short, broad wings ideal for gliding

BROWN-THROATED SAND MARTIN
Resident in Morocco and widespread south of the Sahara in Africa, usually near inland wetlands, Riparia paludicola is also found in southern Asia east to the Philippines. About 13 cm (5 in) long, it usually has a white belly but is sometimes entirely brown. It forms very large flocks.

2 long tail streamers

BARN SWALLOW
A summer visitor to North America, North Africa, Europe, and Asia, Hirundo rustica winters in Africa south of the Sahara, and in southern Asia, Australia, and South America. It is 18 cm (7 in) long. A lively singer, its nest is a mud half-cup, often on a beam in farm outbuildings. It lays 4 – 6 red-spotted, white eggs.

Family MOTACILLIDAE

Wagtails and pipits

Length 14–19 cm (5 1/2–7 1/2 in)

Species 54–58

Wagtails and pipits are ground-dwelling birds, wagtails occurring in Europe, Asia, and Africa, while pipits are found worldwide. They have long legs and long claws, except for the hind claw, which is the shortest in species that perch on trees. Most pipits have streaked brown plumage that helps in camouflage, while wagtails have contrastingly coloured plumage that is white, black, yellow, or blue-grey.

Pipits usually nest on the ground, while most wagtails nest on ledges. Wagtails are generally found near water; some, such as the grey wagtail, are associated with fast-running upland streams, but can also be found near weirs on lowland rivers. Sexes are similar in pipits but there are more marked differences among wagtails. Female pipits and wagtails lay 2–7 eggs, often white, brown or grey, incubating them for 2–3 weeks; the young leave the nest after 12–18 days. In fact, young pipits leave the nest before they can fly. Many species in this family are migratory.

PIED WAGTAIL
The subspecies of the European white wagtail that is found in the British Isles, Motacilla alba is 18cm (7in) long, and is common near water and on mown turf. When feeding on the ground, it pauses with tail wagging, then walks rapidly forwards, picking up insects in its bill. It nests on buildings and bridges near water, laying 5–6 pale grey eggs.

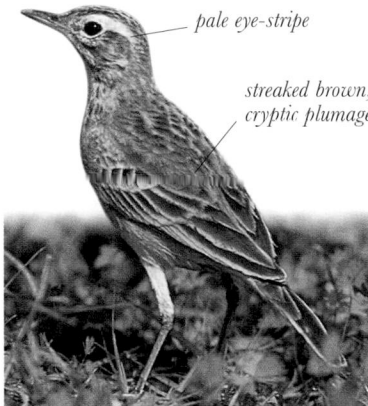

pale eye-stripe

streaked brown, cryptic plumage

RICHARD'S PIPIT
Found in Africa (south of the Sahara), Asia, and Australia, Anthus novaeseelandiae is also an occasional autumn visitor to Europe. Its song has repeated chirping and trilling phrases and it has a loud flight call that goes "prrreep". This bird nests on the ground, laying 4–6 spotted eggs. It is about 18cm (7in) long.

Family CAMPEPHAGIDAE

Cuckoo-shrikes

Length 14–40 cm (5 1/2–16 in)

Species About 70

Cuckoo-shrikes have no connection with either cuckoos or shrikes. This is mainly a tropical family consisting of 2 groups of birds ranging from Africa across southern and Southeast Asia to Australia and the islands in the western Pacific; the minivets occur only in Asia. Except for the ground cuckoo-shrike, found on the treeless Australian plains, these are usually tree-living

birds. Minivets are colourful birds, the size of wagtails; the male tends to be red and black, whereas the female is yellow or orange and black or grey. The remaining cuckoo-shrikes resemble cuckoos in flight and are often dull grey, ranging from the size of sparrows to that of pigeons. The females of the latter group are often paler, and many species have bristles growing from the base of the bill. The nest is built high in a tree and the female lays 2–5 eggs, usually incubating them alone.

dull, greyish coloration

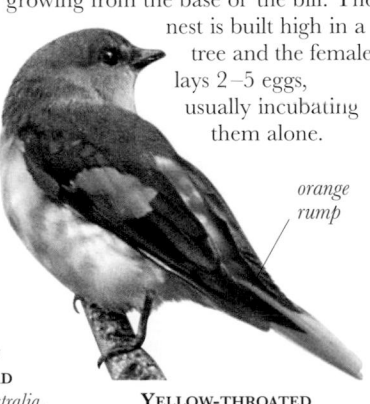

COMMON CICADABIRD
Found in Australia, Papua New Guinea, and eastern Indonesia, Coracina tenuirostris lives in the canopy of rainforest and other woodland. The female is browner with barred underparts. The name is derived from the male's harsh, descending, cicada-like buzz, which it repeats often. This bird is 24–26cm (9½–10in) long.

orange rump

YELLOW-THROATED MINIVET
Found from the Himalayas through southern China and Southeast Asia to Borneo, Pericrocotus solaris lives in open forest. It measures 19cm (7½in) in length. The male has a grey throat and an orange belly, but the female is duller with yellow rather than orange underparts.

Family PYCNONOTIDAE

Bulbuls

Length 15–27 cm (6–10 1/2 in)

Species About 124

Of the 124 species in the bulbul family, about 51 are found only in Africa or Madagascar, 2 species occur in both Africa and Asia, and the remainder are found mainly in Asia. These birds primarily inhabit forests, although some have adapted to more open, man-made habitats. The garden species are more gregarious and noisy than forest species. Often located and identified by their calls, few bulbuls have musical songs. Small to medium-sized birds, many species have dull olive, brown, or grey plumage,

sometimes with yellow underparts. Some species have a crest and most have bristles around the base of the bill, which may curve downwards towards the tip. Many bulbuls feed on fruits and berries or buds and nectar, some eat only insects, whereas others are omnivorous. The female usually lays 2 eggs per clutch, in a nest that tends to be open in structure to allow the tropical rain to drain through.

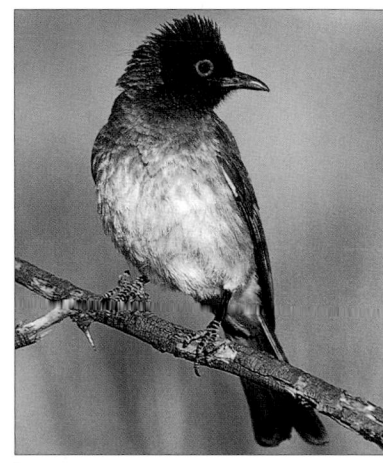

RED-EYED BULBUL
Found in Namibia, Botswana, and western South Africa, in forests near rivers and in dry bush near water, Pycnonotus nigricans is a noisy bird. About 20cm (8in) long, it nests from November to March, the female laying 3 pinkish eggs with dark markings.

RED-VENTED BULBUL
Found from southwest China and Burma to Sri Lanka, Pycnonotus cafer is a familiar garden bird in India. About 20cm (8in) long, it is noisy and active and has a chirpy call. Flocks feed on fruits and, in the evenings, on insects in flight. This bird has been introduced to Fiji, Singapore, Australia, and New Zealand.

red vent

coral red bill

RED-WHISKERED BULBUL
The red patch behind the eye in both sexes gives Pycnonotus jocosus its common name. Its other trademark feature is a black, forward-pointing crest. A common garden bird, about 20cm (8in) in length, it occurs from India to Hong Kong.

red patch behind eye

white throat

BLACK BULBUL
Found in Madagascar, the Comoros, and South and Southeast Asia, Hypsipetes madagascariensis is 23cm (9in) long. One subspecies from China and northern Vietnam has a white head and throat. This bird feeds on berries and insects in noisy flocks of up to 100. Its call resembles the cry of a kitten.

white-tipped tail

YELLOW STREAKED GREENBUL
Found in isolated locations, in mountain rainforests in central and southeast Africa, Phyllastrephus flavostriatus forages for insects, mainly in trees. Working its way round tree trunks like a woodpecker, it repeatedly flicks one wing at a time, revealing its yellow underwing. This bird is 15–27cm (6–10½in) long.

Order PASSERIFORMES *continued*

Family IRENIDAE

Fairy bluebirds

Length 25–30cm (10–12in)

Species 2

These tree-living birds are found in the evergreen or semi-deciduous forest of Asia: the blue-backed fairy bluebird in wetter parts of India, the Himalayas, southwest China, and Southeast Asia up to Borneo and Palawan in the Philippines, and the black-mantled fairy bluebird in the rest of the Philippines. They are noted for their very long upper and lower tail coverts, extending almost to the tip of the tail. Moving by day through the canopy in search of fruits, fairy bluebirds may also probe tree flowers for nectar. They have a range of loud calls, including a sharp flight call. The blue-backed fairy bluebird nests in the forks of small trees, where the female builds a platform of twigs and incubates her 2 or 3 eggs. The male helps to feed the young.

red eyes

BLUE-BACKED FAIRY BLUEBIRD
Despite the metallic blue plumage, the male Irena puella *can be inconspicuous in the canopy – but his repeated, sharp, double whistle often reveals his presence. Females are a duller blue. This bird is 27cm (10½ in) long, and gathers to feed on fig trees, often with other species.*

Family LANIIDAE

Shrikes

Length 15–35cm (6–14in)

Species About 69

These small- to medium-sized birds are the most predatory of all passerines. All have a hooked bill with a tooth-like point in the upper part, and strong legs and sharp claws for holding prey. Shrikes feed on insects, but some are known to augment their diet with lizards, small birds, and rodents. Several of them create

white to pale grey crest

yellow eye wattles

black and white plumage

WHITE-CRESTED HELMET-SHRIKE
The most common and widespread of all the species of helmet-shrikes, Prionops plumata, *which is 15–25cm (6–10in) long, is found in bushes in sub-Saharan Africa. It also occurs in areas of cultivation and open woodland. This noisy and gregarious bird gathers in flocks of 5–12, chattering loudly as it forages for food.*

larders, impaling their prey on thorns or barbed wire, which has given rise to the name "butcher birds" for these shrikes. They are birds of open scrub, and some species are very territorial, even in their winter quarters. They frequently perch on top of a bush, scanning the countryside for prey, and flicking their tail up and down and from side to side. The bush shrikes and helmet-shrikes are confined to Africa, while the true shrikes are widespread throughout Africa, Europe, and Asia, with 2 species occurring in North America. Northern species migrate south in autumn. Shrikes nest in trees or bushes, and the females lay about 2–7 eggs.

BOKMAKIERIE
Found in southern Africa, Telephorus zeylonus *occurs in the bush, open savanna, plantations, and gardens. It is 22cm (9in) long and, like a thrush, spends much of its time on the ground. This bird performs a beautiful courtship dance in spring. Its name is a transcription of one of its loud, ringing calls.*

black crown

white "V" on back

FISCAL SHRIKE
Widespread in Africa, south of the Sahara down to the Cape, in open woodland, parks, and gardens, Lanius collaris *raids other birds' nests and eats the nestlings. About 22cm (9in) long, it perches on poles and wires, scanning the ground for insects, mice, and small birds. It will attack birds that enter its territory.*

long tail

rufous wings

BLACK-CROWNED TCHAGRA
Measuring 22cm (9in) in length, Tchagra senegala *is found in northwest Africa, south of the Sahara, and in southwest Asia. A bird of open savanna woodland and the edges of cultivation, it forages on the ground for insects and their larvae, rather like a thrush (see p.346).*

black throat

yellow underparts

Family VANGIDAE

Vanga shrikes

Length 12–30cm (4¾–12in)

Species 14

Of the 14 species of vanga shrikes, 13 are found only in Madagascar; the blue vanga also occurs in Moheli in the Comoro Islands. All are tree-living birds, some being confined to the evergreen forest in the humid east, while others extend into wooded areas in the western savanna, and some even occur in semi-desert scrub. Many of them are boldly patterned in black and white, and have a heavy bill, often hooked at the tip. Although they usually eat insects, some may also feed on tree-

HOOK-BILLED VANGA
A solitary bird, Vanga curvirostris *is 25–29cm (10–11½ in) long, and has a heavy, hooked bill. It lives in evergreen forest, brush, plantations, and mangroves across Madagascar.*

frogs and small reptiles. They often form loose feeding flocks, sometimes with other species. Some species build a cup of twigs in a tree and lay 3–4 eggs; both parents raise the young.

hooked bill

black and white head

white underparts

Family BOMBYCILLIDAE

Waxwings

Length 15–23cm (6–9in)

Species 8

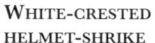

This group consists of 3 waxwings and 5 related species – the 4 silky flycatchers and the hypocolius. The 3 waxwings breed across northern Europe, northern Asia, and much of North America, wintering in the south. They are plump, fawn-brown birds with distinctive crests and silky feathers. Two species have the waxy tips to the secondary wing feathers that give the family its name. Sociable birds, waxwings nest in loose colonies, building their home from twigs and grasses in branches. They feed mainly on berries, flocks turning up even in areas where they are not normally seen and rapidly stripping berry bushes. Like waxwings, the silky

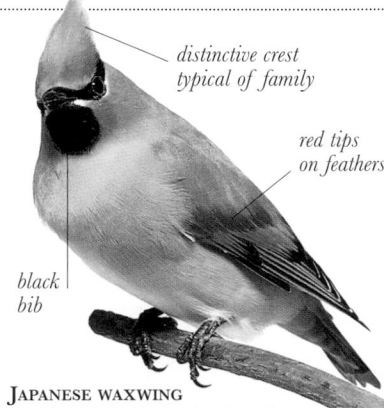

distinctive crest typical of family

red tips on feathers

black bib

JAPANESE WAXWING
Bombycilla japonica *breeds in the taiga forest of eastern Siberia and winters further south in Japan and Korea. Unlike the other 2 waxwings, its secondary feathers lack the waxy drops at their tip. This bird is 16cm (6½ in) long and is a strong flier.*

flycatchers of southwestern USA and Central America also have silky feathers and prominent crests; however, they catch insects in flight like flycatchers. The hypocolius is a fruit-eating bird with a crest, found only in West Asia.

Family DULIDAE

Palmchat

Length 18 cm (7 in)

Species 1

The single species in this family, *Dulus dominicus*, is found only in Haiti (including the islet of Gonave) and the Dominican Republic. Common

boldly streaked underparts

dark olive tail

and conspicuous in these areas, palmchats are often aggressive and very noisy, chattering loudly in chorus. Like the waxwings (see p.344), with which it is sometimes grouped, the palmchat is a fruit-eater, often seen in flocks. It never comes to the ground, spending all its time in trees, feeding on berries and flowers with its strong, heavy bill. The palmchat differs from waxwings in its nesting habits: it is known to weave a large communal nest around the trunk and lower fronds of palm trees. Woven from twigs and lined with soft bark and grass, this nest sometimes has compartments for 2 to as many as 30 pairs of birds, each pair having a private entrance from the outside. The female lays 2–4 white eggs, which are heavily spotted with grey. Outside the breeding season, palmchats use the communal nest as a night-time roost.

Family CINCLIDAE

Dippers

Length 15–21 cm (6–8½ in)

Species 5

These plump, short-tailed birds, found in Europe, Asia, and North and South America, are rather like aquatic versions of wrens (see below, left). Their tail is often cocked in the same way, and they build a similar domed

white throat

plumage repels water

nest of moss, usually lined with grass and leaves, with an entrance hole at the side. The nest is hidden in a rock crevice, among tree roots, under a bridge, or behind a waterfall. The female incubates the 3–6 white eggs for about 16 days, and both parents feed the young. As soon as they leave the nest, the young are able to feed underwater. All 5 species of dippers live near fast-flowing hill or mountain streams, catching mayflies and other insects on the bank and shore, and wading into the water to pick insect larvae, snails, and fish eggs from rocks and stream-bed pebbles. The preen gland of the dippers is 10 times bigger than that of any other passerine; this enables the birds to make their feathers waterproof.

WHITE-THROATED DIPPER
One of the few birds able to "walk" underwater, the chubby, pot-bellied Cinclus cinclus is 18–21cm (7–8½ in) long and has a rasping "strits" call. Found in Europe, North Africa, and northern Asia, it flies low over the water, its wings making a whirring sound as it flies, and nests in crevices and on ledges under bridges.

Family TROGLODYTIDAE

Wrens

Length 8–22 cm (3¼–9 in)

Species About 69

Widespread across North, Central, and South America, Europe, Africa, and Asia, wrens are renowned for their calls which are disproportionately loud for such a small bird. They carry their short tail cocked upright, and have slim, sharply pointed, sometimes

CACTUS WREN
One of the larger wrens, Campylorhynchus brunneicapillus is 20cm (8in) long, and is found in North America from Nevada to Mexico.

downcurved bills. Wrens fly straight, but are not long-distance fliers. They are usually found close to the ground, often in dense undergrowth, where they seek the insects and other invertebrates they eat. The male is the chief nest-builder; he may build half a dozen domed "cock nests" within his territory before building a more finished version that the female will probably select and line, before laying 2–10 speckled white eggs. Most females incubate alone, but both parents feed the young.

short, cocked tail

WINTER WREN
Troglodytes troglodytes is the only wren species that is found outside the Americas. It is 8cm (3¼ in) long, and is known for its loud, rattling song. In winter, northern birds migrate south.

Family MIMIDAE

Mockingbirds

Length 20–33 cm (8–13 in)

Species 30

Found in North, Central, and South America, the West Indies, and the Galapagos Islands, mockingbirds mostly live on the ground or among brush and thickets. Almost all species are known for their singing and, as the name suggests, some are very good mimics. Many sing at night during the breeding season. Generally long-tailed birds, most mockingbirds are white, grey, or brown; however, 2 species are blue and one is black. They are closely related to thrushes (see p.346), and build similar cup-shaped nests in bushes or trees, where females lay and incubate 2–5 eggs. Mockingbirds are known for their vigorous defence of the nest, and will attack and strike predators if threatened.

GALAPAGOS MOCKINGBIRD
Found in the Galapagos Islands, Nesomimus parvulus often runs instead of flying. It is 25–27cm (10–10½ in) long, and feeds in competitive groups on eggs, young birds, carrion, and fruits.

blackish tail

NORTHERN MOCKINGBIRD
Particularly associated with southern USA, Mimus polyglottos is the state bird of 5 states in the USA. It is 23–28cm (9–11in) long, famous for singing on moonlit nights and incorporating other birds' voices in its song.

long tail

white patches on wings

Family PRUNELLIDAE

Accentors

Length 14–18 cm (5½–7 in)

Species 12

Found in Europe, North Africa, and Asia, accentors are known for their inconspicuous behaviour, often shuffling forwards or hopping slowly near bushes or rocks or other cover as

they scour the ground for food – mainly insects in summer and seeds in winter. They are predominantly grey and brown, sometimes with contrasting orange or black and white markings, and their upperparts are often streaked. The adult males and females and the juveniles all look similar. Accentors have a straight, low flight, usually covering short distances. Some species have a song-flight, while others sing from a perch. In addition to simple pairs, some species of accentors – notably the dunnock –

exhibit polygyny (a male with 2 or 3 females) and polyandry (a female with 2 or 3 males). Females lay 3–6 bluish eggs which are incubated for up to 2 weeks. The young fly in another 2 weeks.

DUNNOCK
Native to Europe, Prunella modularis measures 14cm (5½ in) in length, and has a whistling call. It is often found under shrubs, hunting for insects. When a dominant male mates with several females there is a subordinate male in attendance; if the subordinate succeeds in mating, he helps feed the chicks.

streaked back

dull orange legs

Order **PASSERIFORMES** *continued*

Family **TURDIDAE**

Thrushes

Length 12.5–30 cm (5–12 in)

Species About 63

This family, which includes such familiar birds as the song thrush and the blackbird, is widely distributed in Europe, Africa, Asia, Australia, and North America. In fact, thrushes are native to almost all regions, excluding Antarctica and New Zealand. Many, such as the blackbird and song thrush, are fine singers, the most notable of them all being the nightingale. Thrushes may form a solid pair-bond, singing to proclaim ownership of territory. Species like the mistle thrush, which nests very early in the year, may be heard singing in the dead of night. Thrushes build cup-like nests of grass and moss, usually in the fork of a bush or tree.

Their courtship displays often emphasize their distinctive physical features, for instance, the robin's red breast or the song thrush's speckled breast. During courtship, males of some species, such as the robin, feed the female. However, the female is often solely responsible for nest-building, incubation, and brooding the nestlings. Some species of thrush, such as the redwing and fieldfare of Scandinavia, are migratory and fly south in winter. The migratory species often form large flocks but the non-migratory members of the family may remain in their pairs throughout the year.

brownish buff upperparts

NIGHTINGALE
Luscinia megarhynchos *is renowned for its long and varied song, heard most clearly at night. The song is delivered from within cover; in fact, the nightingale is notoriously difficult to spot. Found in Europe, North Africa, and Asia, it migrates south in autumn. Females lay 4 or 5 olive-brown eggs in a nest that is built in thick cover, near the ground. The nightingale is 16.5cm (6½ in) long.*

rusty-red tail

blue head and back

reddish chest

EASTERN BLUEBIRD
Found in North America, Sialia sialis *is viewed as the harbinger of spring, arriving in northern areas in late February or March. Males often arrive first, launching into extravagant song-flights from treetops and rising 30m (100ft) or more into the air, singing all the way. Nesting in hollow branches or nestboxes, the female lays 3–7 light blue eggs. This bird is 14–19cm (5½–7½ in) long.*

blue tail

AMERICAN ROBIN
Turdus migratorius, *called robin in North America, has adapted well to living close to humans. About 25cm (10in) long, it feeds on the ground, cocking its head to watch for the movements of worms and other invertebrates. Many populations are migratory, moving north in February; the males leave first. Both sexes build the cup nest and raise 2 or 3 broods.*

white star over eye

bright yellow body

WHITE-STARRED ROBIN
Pogonocichla stellata *has a yellow body and tail, a blue-grey head, and white patches or "stars" over both eyes. It is 15cm (6in) long, and reveals a white spot on the throat when it sings its creaking notes. It is found in highland forest in eastern Africa, from the Cape to Malawi.*

bright orange bill

orange eye-ring on male

EUROPEAN ROBIN
Perky and bold, with a distinctive red breast, Erithacus rubecula *is found in Europe, North Africa, and northwest Asia. Juveniles have a brown speckled breast and head. About 14cm (5½ in) long, the European robin has a rippling song and a sharp "tic" alarm call. It feeds on insects and berries, and when on the ground, usually hops. The female lays 4–6 red-spotted, white eggs.*

uniform brown upperparts

grey edge to red breast

all-black plumage of male

BLACKBIRD
Native to Europe, North Africa, and Asia, and introduced in Australia, New Zealand, and South America, Turdus merula *is 24–25cm (9½–10in) long. The male sings melodiously from a prominent perch, particularly in the evenings. Blackbirds are noted for their "chak-chak" calls as they go to roost. They feed on berries, fruits, worms, and insects.*

long tail

pale eye-ring

slender bill

BLUETHROAT
There are 2 European subspecies of Luscinia svecica: *the Scandinavian form (pictured below) with a red throat patch, breeding in willow and birch forest, and the continental form with a white throat patch, which nests on swampy lakesides and scrubby ditches. There are other subspecies such as* Luscinia svecica magna *in Turkey, which has an all-blue throat. All forms are 14cm (5½ in) long.*

Scandinavian form with red throat patch

clear speckles on underparts

SONG THRUSH
Turdus philomelos *is a fluent singer – often at dusk – and repeats its phrases 3 or 4 times. Native to Europe, North Africa, and northwest Asia, and introduced in Australia and New Zealand, it is a bird of gardens and woodland. About 23cm (9in) long, it feeds on berries, insects, and worms, and is noted for its use of an anvil stone to crack snail shells.*

NORTHERN WHEATEAR
Oenanthe oenanthe *is the wheatear of northern Europe, arriving for the summer from its winter quarters in Africa. Its conspicuous white rump and black "T" at the tail tip catch the eye in flight. It is 15cm (6in) long and eats insects. A bird of open uplands, it nests in crevices, old rabbit burrows, or under rocks; the female lays 5 or 6 pale blue eggs.*

rusty-red tail base

Family TIMALIIDAE

Babblers

Length 10–35 cm (4–14 in)

Species About 256

Found in North America (only one species), Africa, Asia, and Australia, most babblers are heavy-bodied birds with a stout bill. They are highly gregarious and maintain contact mainly through quiet calls. They tend to be sedentary; only a few species migrate. Many babblers are insectivores but some also eat fruits and many are omnivores. The ground feeders rummage in leaf litter, while the more arboreal species find invertebrates on leaves and bark. Nesting habits vary: a number of species build domed nests low down or at ground level; however, the song-babblers build an open nest in a bush or tree. Some species even have a cooperative breeding system whereby birds that are not breeding help the new parents with incubating the eggs and feeding the chicks. Both the incubation and the nestling period take 13–16 days. Young birds may remain with a group of adults for a year or more.

rufous crown and nape

RED-TAILED LAUGHING THRUSH
There are about 140 species of laughing thrush, of which Garrulax milnei is found in south China and Southeast Asia, in scrub, grass, and second-growth forest above 1,000m (3,300ft). It is 26cm (10in) long and has red wings and a red upperside to its tail; the underside is black.

reddish wings

red tail with black underside

rufous crown

puffed-out throat feathers

dark brown streaks on breast

SPOTTED BABBLER
A bird more often heard than seen, the whistled chattering and loud calls of Pellorneum ruficeps are familiar jungle sounds. This bird is found from India to Southeast Asia and lives in brushwood and bamboo, from sea-level to 1,300m (4,350ft). The spotted babbler, 16cm (6½in) in length, is known to travel in small flocks.

orange breast

long, downward-curving bill

LONG-BILLED SCIMITAR BABBLER
Widespread in evergreen forest and bamboo at heights of up to 2,300m (7,600ft), Pomatorhinus hypoleucos is found on the forest floor, often digging into the soil with its long bill. It is 28cm (11in) long, has a high-pitched, mellow hoot, and is found from Pakistan to south China and Southeast Asia.

COMMON BABBLER
Turdoides caudatus is a cooperative breeder – a small flock of birds helps one pair to incubate their eggs and feed the young. This bird, 23cm (9in) long, is found from Iraq eastwards to Bangladesh, and moves through the undergrowth in forests and gardens foraging for insects, seeds, and berries.

long tail

red bill

PEKIN ROBIN
A very popular cage bird, thanks to its melodious, warbling song and attractive plumage, Leiothrix lutea is found from the Himalayas to south China and Southeast Asia. The female's throat and breast are paler than the male's and she lacks his red coloration in the wings. This bird is 15cm (6in) long and forages in bamboo and scrub with other babblers.

blackish bill

black cap

rufous back

rufous underparts

RUFOUS SIBIA
Also called the black-capped sibia, Heterophasia capistrata is a Himalayan bird, found from northern Pakistan eastwards to southern China. It is 21cm (8½ in) long and favours deciduous forest, especially oak woodland, but is also seen in coniferous forest. It drinks sweet sap oozing from holes in tree bark, and is also known to feed on insects.

long tail with bands

Family ORTHONYCHIDAE

Logrunners

Length 18–30 cm (7–12 in)

Species 19

This group includes species that may or may not be closely related to each other. There are 2 logrunners, 6 quail-thrushes, 3 whipbirds and wedgebills, 4 rail-babblers, and up to 4 other species. Apart from 2 species that are found in Indonesia, they are all distributed in Australia or Papua New Guinea. By and large, logrunners are terrestrial insect-eaters, and will toss leaf litter aside to look for food items. Most species breed on or near the ground in variously designed nests. The size of logrunner clutches varies between 1 and 3. Many birds of this family can be seen moving around in small groups.

SPINE-TAILED LOGRUNNER
Seen in pairs or family parties on the floor of the rainforest in southeastern Australia, Orthonyx temminckii is 18–20cm (7–8in) long.

Family PANURIDAE

Parrotbills

Length 12–15 cm (4¾–6 in)

Species 19

Most species of parrotbill are found in the Himalayas and China, although the bearded tit is found from the UK to East China. Most parrotbills have an unusual, sinuous-edged bill and the sexes are alike. They live in flocks, foraging together, and fly feebly, but are agile when feeding. Many of them build complex nests, woven from grass and other vegetation, where females lay 2–4 eggs in a range of colours.

black, drooping moustache on male

BEARDED TIT
Found in Europe and Asia, Panurus biarmicus lacks the unusual bill of its family. A bird of reedbeds, unlike other parrotbills which prefer thick vegetation, the bearded tit flies through or above the reeds, eating insects, spiders, and reed seeds. It is 16.5cm (6½in) long and raises 3 or 4 broods a year; both sexes incubate the eggs and feed the young.

Order PASSERIFORMES *continued*

Family PICATHARTIDAE
Rockfowl

Length 39–50cm (15½–20in)

Species 2

Rockfowl are unusual in having bare skin on the head. They are found in damp, upland rainforest in West Africa up to 2,100m (6,950ft). They nest in a rock cave in their territory, sticking a mud nest to a rock face, at a height of 2–4m (6½–13ft). Here, the female lays 2 blotched, brown and grey eggs.

black and yellow skin on head

WHITE-NECKED ROCKFOWL
A sociable bird, Picathartes gymnocephalus is found from Guinea and Sierra Leone eastwards to Togo. It is 40cm (16in) long, and has a low, croaking call. This bird forages on the ground for insects, frogs, and snails, moving in springy hops.

Family POLIOPTILIDAE
Gnatcatchers

Length 10–13cm (4–5in)

Species About 13

These birds, found in North, Central, and South America, are sometimes placed with the Old World warblers (see below). They feed on insects in foliage, picking them off leaves and twigs, and fluttering after those that take flight. They build a delicate nest of petals and plant down, bound with spiders' webs, moss, and lichens, on a horizontal branch. Both sexes incubate the 4–5 eggs.

BLUE-GREY GNATCATCHER
A grey bird with a long black and white tail, Polioptila caerulea is 11–13cm (4¼–5in) long, and has a thin, nasal call. It is found from southern Canada to Guatemala and Cuba.

Family SYLVIIDAE
Old World warblers

Length 9–20cm (3½–8in)

Species 376

Old World warblers are found in Europe, Asia, Africa, Australia, and the Pacific islands. The smallest of all are the kinglets, which include some of Europe's smallest birds. Many, like the blackcap, are well known for their song. Warblers are generally birds of woodland and scrub, and sometimes grassland. The reed warbler group feeds and nests in reedbeds and has colonized islands in the Pacific. One species, *Cettia major*, can be found at 4,000m (13,200ft) in bamboo thickets in the Himalayas. The scrubwarblers nest in bushes and woodland throughout Europe.

The tiny leaf warblers feed in the tree canopy, picking insects off leaves. Most species are migratory, flying south to warmer climes in autumn after breeding in northern regions. Siberian willow warblers travel 12,000km (7,500 miles) each way. Old World warblers nest in low cover or on the ground, where females lay 3–7 eggs. Both sexes feed the young.

yellow crown in female

GOLDEN-CROWNED KINGLET
The smallest North American warbler at 8–11cm (3¼–4¼in) long, Regulus satrapa can withstand the harsh northern winter in the pine woods of New England. It breeds in spruce trees, in a dainty, globe-like nest of moss and feathers under a branch. The female lays 8–10 tiny whitish eggs.

WHITETHROAT
Found in Europe and western Asia during the summer, Sylvia communis winters south of the Sahara. It has a sprightly, blustering song delivered from the top of a bush or in a rising song-flight in which it descends back to the bush. This bird is 14cm (5½in) long, and nests in scrub or field edges, laying 4–5 pale buff eggs with dark spots.

white eyebrows

TAWNY-FLANKED PRINIA
Found in Asia and southern and eastern Africa, Prinia subflava has a monotonous "chip, chip" song and a harsh, scolding "sheee" call. A restless, sprightly bird, it is 12cm (4¾in) long, and forages in pairs or small flocks in scrub or vegetation.

FAN-TAILED WARBLER
Found in Europe, Africa, Asia, and Australia, Cisticola juncidis is only 10cm (4in) long, but is disproportionately active and noisy. It has a bouncy, circular song-flight: it beats its wings rapidly to rise, utters a "tzet" call, then drops before rising again. Its fan-shaped tail has black and white markings on the underside.

long tail

heavily streaked back

GRASSHOPPER WARBLER
Increasingly scarce, Locustella naevia is found in Europe and western Asia and winters in Africa and India. It is 12.5cm (5in) long, and favours meadows, ditches, and marshland edges, creeping among weeds and grass, and climbing up stems. Its song is a very long-lasting, insect-like reeling, with only short pauses.

red eyes

WILLOW WARBLER
A summer visitor to northern Europe and an inhabitant of Africa during winter, Phylloscopus trochilus has no particular association with willows. A small warbler, about 10–11cm (4–4¼in) long, it can only be separated from the very similar chiffchaff, Phylloscopus collybita, by its song: a melancholy verse which ends abruptly, as though cut off. It forages for insects among leaves, often hanging upside down.

pale eyebrows

long bill

fawn eyebrows

inconspicuous buff coloration

DARTFORD WARBLER
Found in western Europe and northwest Africa, Sylvia undata is one of the non-migratory warblers in this region. Winter is a critical time for this species, particularly in Britain, which lies at the northern limit of its range. It is 13cm (5in) long, with a long tail, which is often cocked, and short wings.

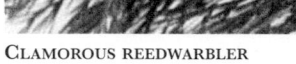

CLAMOROUS REEDWARBLER
A marshland warbler found in Asia and Australia, Acrocephalus stentoreus nests in reeds and other wetland plants, usually over the water. About 18–20cm (7–8in) long, it feeds on butterflies, dragonflies, flies, other insects, and small frogs. The female lays 3–6 dark-speckled white eggs. When singing, the male repeats phrases 3 or 4 times.

Family MUSCICAPIDAE

Old World flycatchers

Length 10–21 cm (4–8½ in)

Species About 147

Distributed across Europe, Africa, Asia, and Australia, Old World flycatchers sally out from a perch to catch flying insects in their bill, hence their name. They usually have a broad-based bill which is surrounded by bristles that help them catch their prey. Species that breed in the northern regions migrate south in autumn when insects begin to get scarce. The plumage of these agile fliers varies: some are brown, others are brightly coloured, and the sexes frequently have different plumage. The male is often larger. Some species have a crest, and a few have bright facial wattles. Although some of these birds sing well, most have harsh call notes, and the group as a whole is no match for the warblers and thrushes. Most build an open cup in the branches of a tree or bush, where females lay 1–11 mottled eggs.

black in front of eyes

INDIAN VERDITER FLYCATCHER
Muscicapa thalassina, *measuring 17cm (6½ in) long, is found from India to south China and south to Borneo. It often perches prominently and delivers a loud, sweet warble. The male is greenish blue, with darker wings and tail; the female is duller and greyer than the male.*

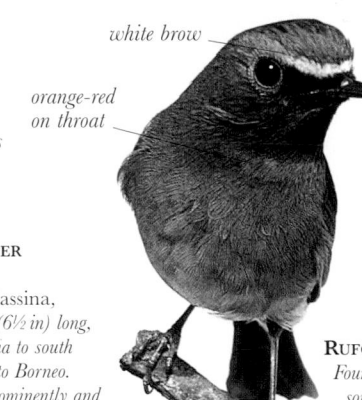

white brow

orange-red on throat

RUFOUS-BELLIED NILTAVA
Niltava sundara *occurs from the Himalayas to southwest China and is resident in Burma above 1,000m (3,300ft). About 18cm (7in) long, it hunts for insects in low scrub and forest undergrowth. The male has a black throat, the female having a white throat patch.*

RUFOUS-GORGETTED FLYCATCHER
Found from Kashmir in north India through the Himalayas to southwest China, and in parts of Southeast Asia, Ficedula strophiata occurs in forest and forest edges above 1,000m (3,300ft). This bird is 14cm (5½ in) long, the female having a smaller and paler gorget or throat patch. It has 2 calls: a low croak and a high-pitched "pink".

Family PLATYSTEIRIDAE

Wattle-eyes and batises

Length 8–16 cm (3¼–6½ in)

Species 31

All of these birds are found in Africa, mostly in forest or forest edges. Wattle-eyes have colourful wattles around their eyes, and share several features with the batises (also called puff-back flycatchers): long, fluffy rump feathers that can be raised in a puff above the back, wing-flicking sounds, and semi-musical songs. These are insectivorous birds, and the batises are skilled flycatchers. They build cup-shaped nests and lay 2–5 eggs. The males are grey, black, and white. Various species of wattle-eyes and batises look very similar and are easily confused. However, they can usually be separated on the basis of geographical distribution.

CHIN-SPOT BATIS
Found in Sudan, Kenya, southwest Africa, and Mozambique, Batis molitor frequents open woodland, darting out after insects like a flycatcher; it also hovers to search leaves. It is 10cm (4in) long, and has a grey crown.

Family MALURIDAE

Fairy-wrens

Length 12–19 cm (4¾–7½ in)

Species 24

Found in Australia, Papua New Guinea, and eastern Indonesia, fairy-wrens are brightly coloured birds that hold their tails upright or over their back in the same way as wrens (see p.345). The metallic colours of their plumage include blues, purples, reds, blacks, and whites. Sexes generally differ. Many fairy-wrens are good singers, some being good mimics as well. They hunt for insects and their larvae through the underbrush, and are often found in flocks. Their nest is usually domed, and is made of grass, spiders' webs, and a lining of feathers and plant down. Fairy-wrens lay 2–4 whitish, speckled eggs which are incubated for 12–15 days; the nestlings fledge in 10–12 days.

SOUTHERN EMU-WREN
Notable for its very long tail, formed of 6 delicate feathers, Stipiturus malachurus is 15–19cm (6–7½ in) long. It occurs along the coastal strip of southern Australia and in Tasmania.

bright blue cap

BLUE WREN
Also known as the superb fairy-wren, Malurus cyaneus is found in southeast Australia and Tasmania. It is 13–14cm (5–5½ in) long, and has a tinny song.

orange-scarlet back   *cocked tail*

RED-BACKED WREN
At 10–13cm (4–5in), Malurus melanocephalus is the smallest fairy-wren. Found in northern and eastern Australia, it occupies habitats varying from tall grass to plantations and gardens, and breeds from November to March.

Family ACANTHIZIDAE

Australasian warblers

Length 9–13 cm (3½–5 in)

Species About 65

Confined to Australia and adjacent islands, with some species occurring in New Zealand, these are very localized birds; most occur in just one part of Australasia and sometimes in quite a limited area. This fact, however, helps with identification as similar species can occur in geographically distinct areas. Usually olive, brown, or yellow, the sexes among Australian warblers are similar. Many species are insectivores, while some feed on seeds. They are often seen in pairs, in family parties, and sometimes in mixed species flocks. Most species do not migrate. The nest made by these birds is a dome of plant fibres, often with a "porch". Females lay 2–4 white, sometimes spotted, eggs. Incubation and fledging each take 15–20 days.

BROWN THORNBILL
Found in eastern and southeastern Australia and Tasmania, Acanthiza pusilla is 10cm (4in) long, and has narrow streaks on its grey throat. It forages among the twigs of lower branches and has a variety of calls, among them a scolding note. The nest is built low down in the undergrowth.

BROWN-BREASTED FLYEATER
About 9cm (3½ in) in size, Gerygone ruficollis is found in primary forest in the mountains of New Guinea, at heights of 1,100–3,300m (3,650–10,900ft). Its song is a distinctive and lengthy series of high-pitched whistles, and its nest a pendant, globular structure with a side entrance.

Order PASSERIFORMES continued

Family MONARCHIDAE

Monarch flycatchers and fantails

Length 12–30cm (4³/4–12 in)

Species About 164

Found in Africa, Asia, and Australia, these are mostly non-migratory birds of rainforest, open woodland, and shrub savanna. They feed on insects – dragonflies, grasshoppers, moths, and bees – and spiders, flushing them from vegetation by flicking their wings or fanning their tail, or catching them by making sallies into the air. Most species also eat some fruits, one eats shellfish and crabs, and another catches worms. Their songs

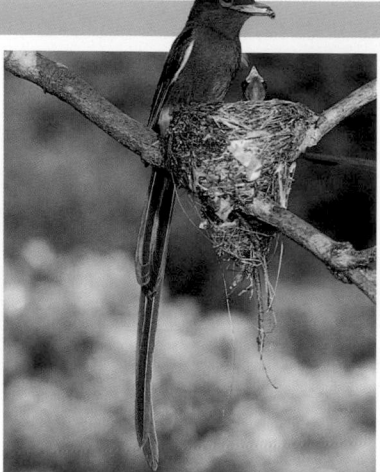

AFRICAN PARADISE FLYCATCHER
Terpsiphone viridis is 40cm (16in) long, with the male's tail up to 20cm (8in) long. The tails of both sexes trail over the side of the small nest cup. This flycatcher is found in forests and riverside woodland, in sub-Saharan Africa.

are typically weak but some species have a loud, forceful song. Sexes have different plumage in some monarch flycatchers, but are usually the same in fantails. These birds build their nests on a branch or in a fork; both parents incubate the eggs for 14–17 days.

NEW ZEALAND FANTAIL
Found from Papua New Guinea to Australia and New Zealand, in a variety of habitats, Rhipidura fuliginosa measures 14–17cm (5½–6½ in) long. It is a migrant in some parts of its range.

Family EOPSALTRIDAE

Australasian robins

Length 11–18cm (4¹/4–7 in)

Species 39

One group of these birds has red breasts that may have reminded settlers of the European robin, hence their common name. Like the robin, these small birds have an upright stance, and occasionally flick their drooping tail. The 2 sexes are often different, the female being drabber. This tree-living family generally feeds on the ground on insects and their

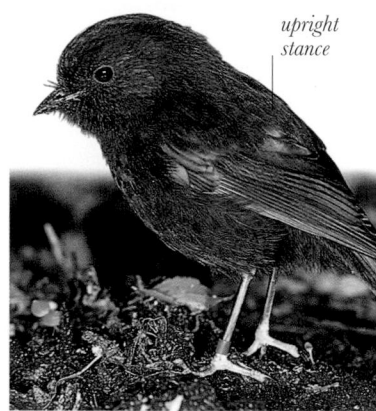

upright stance

larvae, or sallies forth after flying insects. The nest, a tiny cup of moss bound by spiders' webs, is placed on a branch. The female lays 2–4 pale blue or green eggs with red-brown to violet spots, and is known to incubate them herself for 12–14 days. Many species of Australasian robins are partial migrants.

CHATHAM ISLAND BLACK ROBIN
Petroica traversi has come back from the brink of extinction on the Chatham Islands, east of New Zealand. Once common on 4 islands, it was nearly wiped out by introduced cats. By 1976 there were only 7 birds left. Their population was boosted by captive breeding methods.

scarlet breast

SCARLET ROBIN
Found in southeast and southwest Australia, Tasmania, and Pacific islands including Fiji and Samoa, Petroica multicolor occurs in eucalyptus woods in summer but moves to a more open habitat in autumn. It is 12–14cm (4¾–5½ in) long, and darts from a stump or low branch to seize prey on the ground.

Family PACHYCEPHALIDAE

Whistlers

Length 12–28cm (4³/4–11 in)

Species About 52

Whistlers are mainly concentrated in Australia and Papua New Guinea, but some species are found from southeast India through Malaysia into the Indonesian islands, and others occur to the east of Fiji and other Pacific islands. Most species are resident. Whistlers are strong singers, with loud, whistling voices; the crested bellbird is noted for its ventriloqual calls. Many whistlers have a thick bill

CRESTED SHRIKE-TIT
Falcunculus frontatus is found in various parts of Australia, in habitats ranging from rainforests to gardens. It uses its strong bill to tear off loose bark in its search for insects. About 15–19cm (6–7½ in) long, it has a black crest. The female's throat is olive, the male's is black.

GOLDEN WHISTLER
Found in Indonesia, southern and eastern Australia, Tasmania, and Fiji, Pachycephala pectoralis shows wide geographical variation: about 73 subspecies have been described. It is 16–18cm (6½–7in) long; the grey-brown female lacks the male's black and white and much of the yellow. This bird may be found in mixed-species flocks.

with a hook at the tip, which has led to some species being called shrike-thrushes and others shrike-tits; their resemblance to shrikes (see p.344) ends here. Mostly insect-eaters, these birds pick their prey from leaves, branches, and trunks of trees. One species catches small crabs in mangroves, while another sometimes eats nestlings and birds' eggs. Several species are fruit-eaters. Information on breeding is limited, but both sexes usually incubate the eggs. Clutch sizes vary: pitohuis lay one or 2 eggs, bellbirds 3 or 4, and shrike-thrushes lay up to 4 or 5 eggs.

Family AEGITHALIDAE

Long-tailed tits

Length 11–14cm (4¹/4–5¹/2 in)

Species 7

Not true tits, this family of small birds with short, conical bills may be most closely related to the babblers (see p.347). Like the babblers, they are non-migratory, live in small groups, often breed cooperatively, and defend common territories. They are found in North and Central America, Europe, and Asia. A Himalayan subspecies occurs up to 3,400m (11,200ft), but descends lower in winter. Groups of long-tailed tits roost together, huddled along a branch. Breeding pairs sometimes have a "helper", a bird that does not mate but helps feed the young. The domed nest is lined with feathers and can take up to 20 days to build. Clutch sizes range from 4 to 12.

long black and white tail

pinkish brown feathers

LONG-TAILED TIT
About 14cm (5½ in) long, the tiny Aegithalos caudatus is found in Europe and Asia, often travelling along hedges in large flocks.

Family REMIZIDAE

Penduline tits

Length 8–14cm (3¹/4–5¹/2 in)

Species 10

These birds, found in North America, Europe, Asia, and Africa, are even more acrobatic than true tits (see p.351), and can climb along the under side of branches. They pick invertebrates off branches and foliage, and also eat seeds and fruits. Small birds with fine, pointed bills, penduline tits forage in groups and nest colonially. Their nest is an unusual pouch- or purse-like structure, with a spout-like entrance tunnel, woven from grass and roots, and hung at the end of a branch. Females incubate 5–10 white eggs for 12 days, and the young fly in 16–18 days.

PENDULINE TIT
Remiz pendulinus, 11cm (4¼ in) long, uses the down from bulrushes that grow in its marshy habitat to build its nest. Its song is unobtrusive.

Family PARIDAE

True tits

Length 11–22cm (4¹/₄–9in)

Species About 50

Many species in this family have adapted to life with humans – they come readily to bird feeders and often use nestboxes for breeding. Found across North and Central America, Europe, Africa, and Asia, these small, acrobatic birds can pluck insects from foliage, or peck at peanuts inside hanging feeders. In autumn and winter, a number of species form mixed foraging flocks and move along hedgerows, uttering frequent calls, as they search for insects. True tits generally nest in tree-holes or nest-boxes, lining these with moss before laying a large clutch. The eggs are incubated for 13–14 days and the nestlings fledge in 17–20 days.

BLACK-CAPPED CHICKADEE
The most widespread North American tit, Parus atricapillus is one of 6 species named after their "chick-a-dee-dee-dee" call. This bird is 13cm (5in) long. In spring, it finds a mate and digs a nest-hole in a dead stump, where the female lays 6–8 eggs in a cup of grass, moss, and feathers.

GREAT TIT
Parus major, 14cm (5½ in) long, is a dominant tit at feeding stations in Europe. It flocks with other tits in winter, searching for insects, and also eats seeds and fruits. It has a large variety of calls and its song is a sequence of "tee-tee-tu" whistles. The female lays 5–11 white eggs that have reddish spots.

BLUE TIT
Highly acrobatic, Parus caeruleus scours vegetation for insects and hangs upside down to feed on seed hoppers. It nests in holes and nestboxes, laying 6–12 reddish-speckled white eggs. It is 11.5cm (4½ in) long.

ASHY TIT
Found in southwest Africa, Parus cinerascens is a bird of dry thorn savanna and Kalahari scrub. It often hangs upside down when feeding. It is a restless bird, hopping from one branch to the next, and can tear acacia pods in search of insects.

brown eyes

Family CERTHIIDAE

Treecreepers

Length 10–16.5cm (4–6¹/₂ in)

Species 7

Treecreepers are arboreal birds of North and Central America, Europe, Africa, and Asia, with thin, down-curved bills, brown plumage, and pale underparts. They use their stiffened tails for balance as they climb trees in search of insects in the bark, almost always climbing upwards unlike nuthatches (see above right). Their calls and song are high-pitched, and their nest, made of twigs, bark, and moss, is shaped like a loose hammock. They lay 2–9 eggs, which are incubated mostly by the female over 14–15 days. The young, which are fed by both parents, fledge in 14–16 days.

AMERICAN TREECREEPER
Once thought to be a subspecies of the treecreeper of Europe, Certhia americana breeds from Alaska to Newfoundland and south to Nicaragua, wintering in the south of its range. It is 13–14cm (5–5½ in) long and climbs up trees in a spiral. It flies from the top of a tree to the base of another, searching for insects.

stiff tail feathers for support

Family SITTIDAE

Nuthatches

Length 10–20cm (4–8in)

Species 25

Birds in this family, found in North and Central America, Europe, North Africa, and Asia, spend their whole life in trees, inching up or down trunks and branches with equal facility. However, the 2 species of rock nuthatches are an exception – they live on rock faces or buildings. Nuthatches do not use their short, square-ended tail for

black eye-stripe

EURASIAN NUTHATCH
Bright slate-blue and orange, Sitta europaea is 11–13cm (4¹/₄–5in) long and has a carrying "peeu, peeu" song. It plasters its nest-hole with mud to make the opening smaller. This bird feeds on insects, seeds, and nuts which it wedges in bark and cracks open with its bill.

Family RHABDORNITHIDAE

Philippine creepers

Length 15cm (6in)

Species 2

This little-known family is found only on some islands in the Philippines. The more common stripe-headed creeper is found in the lowlands, while the plain-headed creeper usually lives above 900m (2,950ft). With their brown plumage and pale underparts, they resemble treecreepers (see left) but have a stouter and straighter bill, and a square-ended tail that is neither stiff nor used for climbing. They forage mainly in the canopy, but will also feed on trunks. It is thought that they nest in tree-holes.

STRIPE-HEADED CREEPER
Rhabdornis mystacalis feeds on insects picked from twigs and leaves, and on fruits. It also probes among flowers, perhaps collecting pollen with its brush-tipped tongue. About 15cm (6in) long, it forages in small flocks, sometimes with other species.

RED-BREASTED NUTHATCH
Sitta canadensis, 11–12cm (4¹/₄–4³/₄ in) long, lives in North American coniferous forests as far north as Alaska and Newfoundland. In winter, it feeds on conifer seeds and, when the harvest fails, moves south in large numbers. It smears pine resin around the nest-hole, possibly to keep ants out.

white eyebrow mark

black eye-stripe

pale rusty underparts

balance, as woodpeckers (see p.332) and treecreepers (see below) do. They have long, sturdy toes and claws, and climb by placing one foot higher than the other, balancing on the lower one. They search the bark for insects and invertebrates but also eat seeds and nuts, cracking these with their bill. They nest in holes in trees or rocks, and a few excavate their own holes in rotten wood. Females lay 4–10 red-spotted, white eggs. Many nuthatch species have blue-grey upperparts, although the 3 South Asian species are blue-green. All species, except for some red-breasted nuthatches, are non-migratory.

Family CLIMACTERIDAE

Australasian treecreepers

Length 10–17.5cm (4–7in)

Species 8

Seven members of this family are confined to Australia while the other, the Papuan treecreeper, occurs in New Guinea. They mainly eat ants, and nest in tree-holes; several species breed in communities. They lay 2–3 eggs, incubated in 16–23 days.

WHITE-THROATED TREECREEPER
Cormobates leucophaeus, 16–17.5cm (6½–7in) long, occurs in eastern and southeastern Australia in a wide range of habitats.

Order PASSERIFORMES *continued*

Family DICAEIDAE
Flowerpeckers

Length 7–19cm (2³/₄–7¹/₂ in)

Species About 48

Flowerpeckers are found in Southeast Asia and New Guinea, with 5 species – 4 species called pardalotes and one, the mistletoebird – in Australia. Some experts consider the pardalotes to be members of another family. All flowerpeckers have short bills and stumpy tails. Their habitat varies from rainforests, mountain and bamboo forests, to gardens. Although they are by and large not migratory, the mistletoebird is nomadic. Most flowerpeckers lay a clutch of 1–4 eggs. Their nests are domed and suspended from a twig, like those of sunbirds (see below). However, pardalotes build their nests in tree hollows or holes in the ground. The feeding habits of these birds vary: while some flowerpeckers eat both berries and insects, those in New Guinea eat only fruits, and pardalotes eat only insects. Some feed almost exclusively on the fruits of mistletoes and are important distributors of their seeds.

ORANGE-BELLIED FLOWERPECKER
Found in Asia, Dicaeum trigonostigma is 9cm (3¹/₂ in) long and lives in evergreen forest and mangroves. The male has a grey back, head, and upper breast, and an orange-yellow belly and rump, while the female has a grey tinge on its breast and a yellow rump.

short tail

STRIATED PARDALOTE
Endemic to Australia, Pardalotus striatus is common in eucalyptus forest and other types of woodland. It is about 10cm (4in) long. There are at least 5 geographically variable populations, with either black or striped crowns.

Family NECTARIINIDAE
Sunbirds and spiderhunters

Length 8–22cm (3¹/₄–9 in)

Species About 130

Found throughout Africa, across Asia, and into Australia, almost all male sunbirds have highly iridescent plumage, while females are usually a dull green. After the breeding season, the males moult into a dull, female-type plumage. Sunbirds have long bills for collecting nectar and insects, and some have a long tail that is almost half their body length. The 10 species of spiderhunters, although similar to sunbirds, have much longer bills and stouter bodies, and are found only in Southeast Asia. Sunbirds and spiderhunters resemble hummingbirds (see p.323) and feed in a similar fashion but with slower wingbeats. Their nests, made of fine moss and cobwebs, are oval in shape and are usually suspended from twigs and branches. These aggressive and territorial birds are not migratory, although the pygmy long-tailed sunbird moves annually from Sudan to Congo, and at least 2 other sunbirds move between seasons.

long, curved bill

SCARLET-CHESTED SUNBIRD
A dark, sturdy bird with a long, curved bill, Nectarinia senegalensis is about 15cm (6in) long. It is widespread across open woodland and gardens in West, East, and southern Africa. The male has a brilliant scarlet upper breast.

PALESTINE SUNBIRD
A scarce bird, Nectarinia osea (female shown above), is 10–11cm (4–4¹/₄ in) long. It is found in West Asia and Central Africa.

long bill

VARIABLE SUNBIRD
Widespread across open savanna woodland and gardens of West, East, and southern Africa, Nectarinia venusta is also known as the yellow-bellied sunbird. It is 11cm (4¹/₄ in) long. The male is iridescent blue above with a blue-violet upper breast, while the female is olive-brown above with a buff-white breast.

black tail with trace of blue

COLLARED SUNBIRD
Found in forests and woodland in sub-Saharan Africa, Anthreptes collaris inhabits coastal and riverine woodland. It measures 10cm (4in) in length. The male has brighter plumage than the female.

blue-violet upper body in male

STREAKED SPIDERHUNTER
Arachnothera magna is found in dense forest and overgrown clearings from India to China, and south to Malaysia. It is about 17cm (6¹/₂ in) long, and has a long, downward-curving bill.

yellow or white belly

yellow breast

Family ZOSTEROPIDAE
White-eyes

Length 10–14cm (4–5¹/₂ in)

Species About 94

White-eyes are generally found around forest edges and canopy, from sea level up to 3,000m (9,900ft), in Africa and its islands, West Asia, South and Southeast Asia up to Japan, the Indian and Pacific Oceans, and Australasia. They have slightly downward-curving, sharply pointed bills, brush-tipped tongues for feeding on nectar, and small, rounded wings. Their most distinctive feature is a white ring around the eye which varies in size and shape across the group and is the source of their name. Almost the entire family consists of pale green birds that resemble each other. They have a similar "peeuu" call, and the sexes are alike. Highly mobile birds, most white-eyes migrate away from cold weather. However, some non-migratory populations are found on islands in the Indo-Pacific, where they are vulnerable to freak weather conditions. These birds form distinct subspecies, and they show a lack of genetic variation typical of animals that live in remote places.

JAPANESE WHITE-EYE
Widespread throughout South and Southeast Asia and introduced to Hawaii, Zosterops japonica – about 10.5cm (4¹/₄ in) long – has a yellow throat and chin, and olive-green cap, back, and tail.

CAPE WHITE-EYE
Common from forests to gardens in Botswana, Lesotho, Mozambique, Namibia, South Africa, and Swaziland, Zosterops pallidus is about 11cm (4¹/₄ in) long.

Family MELIPHAGIDAE

Honeyeaters

Length 9.5–32 cm (3 ³/₄–12 ¹/₂ in)

Species About 174

Found mainly in Papua New Guinea, Australia, Asia, and the Pacific islands, honeyeaters are mostly nectar- and fruit-eating birds with a characteristic brush tongue adapted to nectar feeding. This tongue can be extended into nectar about 10 times per second. They are generally dull-coloured birds, and the sexes are similar. However, there are exceptions: the regent honeyeater is brighter, with its yellow and black plumage, and the sexes among scarlet honeyeaters differ –

BLUE-FACED HONEYEATER
Distributed across the northern and eastern sides of Australia and Papua New Guinea, Entomyzon cyanotis is 31cm (12in) long and is found in open forest and farmland. Both sexes are alike and have a black head, with a white patch on the nape, and distinctive, 2-toned, blue facial skin surrounding a yellow eye.

males are red and females are brown. Members of this family vary greatly in size. Some resemble warblers or thrushes, others look like hummingbirds, and the larger honeyeaters could almost be mistaken for magpies. Some of these birds do not have feathers on their face and develop wattles. All are found in wooded habitats, with only a few descending to the ground to feed. Outside the breeding season, these birds often migrate in search of flowering trees.

blue facial skin

olive-green back and tail

white breast and belly

NOISY MINER
Found only in eastern Australia, including Tasmania, the noisy, gregarious Manorina melanocephala is locally called the soldierbird. It is 26cm (10in) long and both the sexes look alike. It breeds in colonies from June to December.

TUI
Endemic to New Zealand, where it is found in native forest and scrub, Prosthemadura novaeseelandiae has glossy black plumage with a purple-green iridescent colour and 2 white throat tufts. An energetic bird with a rich, melodic song, it feeds on nectar, fruits, and insects. The male – 30cm (12in) long – is larger than the female.

LITTLE FRIARBIRD
Philemon citreogularis is found in Australia, Papua New Guinea, and the Lesser Sundas in Indonesia. At 27cm (10½ in) long, it is the smallest of the friarbirds – a group of 16 species within the honeyeater family.

Family EMBERIZIDAE

Buntings, sparrows, and relatives

Length 10–20 cm (4–8 in)

Species 321

This large family is found the world over: buntings in Europe, Africa, and Asia, and New World sparrows and seedeaters in North and South America. They are small birds, with medium-sized legs but large feet that are equipped for scratching the ground to locate food. Their plumage is extremely varied, although no bird in this family has particularly brilliant feathers. The tail is fairly long and sometimes forked, and the wings are long and mostly pointed. Buntings

YELLOWHAMMER
Widespread across Europe (although its numbers are declining in western regions), mainland Asia, and cultivated parts of west Asia, Emberiza citrinella was also introduced to New Zealand in the 1860s. It measures 16cm (6½ in) in length and is common in farmland and open country.

and sparrows have short bills which are conical and designed for peeling seeds. Most of the species live in open countryside but their habitat preferences vary greatly. They can be found from bleak Tierra del Fuego at the southern tip of South America to the northern tip of Greenland where snow buntings breed, and from the seashore to the high, barren plateaux of the Andes, enduring extremes of climate from very humid to very arid and very hot to very cold. Their nests are cup-shaped and often domed, and

breast spot

SONG SPARROW
Melospiza melodia is found across most of North America, except the tundra in the north and some southern states, and is common to abundant in a wide range of non-forest habitats. It is 17cm (6½ in) long, has a relatively long, rounded tail, a brown crown with a grey central stripe, and grey cheeks with brown stripes. It also has a greyish breast streaked with black, and a rufous back and tail. This bird breeds from April to July.

long, rounded tail

COMMON CACTUS FINCH
Found on most of the Galapagos Islands, Geospiza scandens is 14cm (5½ in) long. It specializes in feeding on seeds of Opuntia cacti, and has a longer bill and a more pronounced split tongue than other finches, adapted to this purpose.

are generally built low in a bush or tree or on the ground. The females lay 2–7 eggs, which usually have red, brown, or black markings on a pale background. Most sparrows and buntings are migrants. They can, occasionally, occur well away from their expected wintering grounds, but birdwatchers have to be cautious with such records, as this family is particularly popular with aviculturists, and some of the birds observed may have escaped from aviaries.

white head

white wing bar

SNOW BUNTING
Widespread across the tundra of Europe into northern Asia up to eastern Siberia, and similarly across North America, Plectrophenax nivalis breeds in the Arctic but migrates to lower latitudes in winter. It breeds from May to August. It is 17cm (6½ in) long, and the male, with his white head and wing bar, is brighter than the female.

DARK-EYED JUNCO
A highly variable species divided into 5 main subspecies, Junco hyemalis is distributed across North America except the central states. About 16cm (6½ in) long, it breeds in forest areas, and winters in many habitats in the USA and Mexico.

grey back

long tail

rufous sides

EASTERN TOWHEE
This species, Pipilo erythrophthalmus, and its western counterpart, the spotted towhee, were originally believed to be one North American species, the rufous-sided towhee. This bird (female shown) is 19cm (7½ in) long and is found in eastern America.

Order PASSERIFORMES *continued*

Family CATAMBLYRHYNCHIDAE

Plush-capped finch

Length 14cm (5½in)

Species 1

The exact placing of the species *Catamblyrhynchus diadema* in the systematic order has been debated by ornithologists for years. Historically, it has been either grouped with finches or tangers, or in a family of its own. It is found in Bolivia, Colombia, Ecuador, Peru, and Venezuela, in clearings in cloud forests between 1,800 and 2,900m (5,950–9,550ft). This bird has a bright yellow forecrown and a black nape, dark grey back, wings, and tail, and chestnut underparts.

yellow crown

dark grey wings

chestnut underparts

Family CARDINALIDAE

Saltators, cardinal-grosbeaks, & relatives

Length 12.5–22cm (5–9in)

Species About 43

Saltators and cardinal-grosbeaks are thick-billed "seed-crunchers" and differ in structure from buntings and sparrows (see p.353) which are "seed-peelers"; the dickcissel is intermediate in this grouping in that it both crunches and peels seeds. These birds are only found in the Americas, particularly within the tropics. Fourteen species of saltator, 17 species of cardinal-grosbeak, and 7 New World buntings are included within this group. The saltators are limited to Central and South America and are mostly colourful birds with large bills. Male cardinal-grosbeaks are usually brightly coloured, as are the New World buntings – particularly the male painted bunting.

red bill

long crimson tail

COMMON CARDINAL
Found along woodland edges and in gardens in southern and eastern North America and into Mexico, Belize, Guatemala, and Bermuda, Cardinalis cardinalis *measures 22cm (9in) in length. The male is bright crimson with a black throat and red bill, and the female is buff-olive. Both sexes deliver a loud, liquid, whistling song.*

brownish tail

DICKCISSEL
Spiza americana *breeds from March to July in central and southern North America and winters in South America. It is 16cm (6½in) long, and has a yellow breast, a grey head with a pale eye-stripe, yellowish eyebrow, and greyish back. The male is distinguished by a black bib.*

purple-blue hood

green back

red breast

PAINTED BUNTING
Limited in its breeding distribution to the extreme southern states of the USA, Passerina ciris *is one of the most colourful birds in North America. The male has a purple-blue hood and gaudy plumage; the female is bright green above and paler yellow-green below. This bunting is 14cm (5½in) long.*

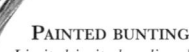

Family THRAUPIDAE

Tanagers and relatives

Length 10–28cm (4–11in)

Species 256

Confined to the western hemisphere, mostly to the tropics, this family contains some of the most colourful American birds. Relatively few tangers live in the densest parts of tropical forests; they mostly wander in mixed flocks among the canopy, particularly outside the breeding season. Most birds in this family build open, cup-shaped nests, rarely on the ground. Some feed among low bushes, but few are ground-dwelling birds. They feed on fruits and insects. The euphonias, included within this family, feed on mistletoe berries, while the honey-creepers have bills and tongues designed for sucking nectar from flowers. A few species follow columns of army ants, catching insects and spiders flushed by the ants. Many remain mated throughout the year and therefore hardly sing at all, but a few species, such as the scarlet tanager, do have attractive songs.

MAGPIE TANAGER
Widespread across South America, Cissopis leveriana *inhabits scrub in cloud- and rainforests and in cultivated and suburban areas. It is 28cm (11in) long, with a blue-black head, mantle, throat, and breast. Its long black tail is edged with white, the wings are black and white, while the belly is white.*

greyish white belly

COMMON BUSH TANAGER
Chlorospingus ophthalmicus *is 14cm (5½in) long, and has a pale throat flecked with tiny black spots. The crown and the sides of its head are dusky brown and the eye has a white ring around it. Found south from Mexico and Belize, over much of South America, including Argentina, Bolivia, and Colombia, this bird inhabits cloud- and rainforests between 900 and 3,000m (2,950–9,900ft).*

SCARLET TANAGER
Found in deciduous forests, Piranga olivacea *breeds from May to July in eastern North America, and winters in South America, from Colombia to Bolivia; it is a rare migrant in the Caribbean islands. This bird measures 17cm (6½in) in length, and the male has a bright scarlet head and body.*

scarlet head

black wings

blue hood and throat

black breast

blue-black wings

BLUE-NECKED TANAGER
Resident in Bolivia, Brazil, Colombia, Ecuador, Peru, and Venezuela, Tangara cyanicollis *is found between 300 and 2,400m (1,000–7,900ft) in open areas with isolated trees. Both sexes are alike, with a blue hood and throat, and measure 12cm (4¾in) in length. The wings are blue and black, or green and black, depending upon the bird's subspecies.*

yellow bill

bluish green body

GREEN HONEYCREEPER
Widespread and common in most countries of northern South America, Chlorophanes spiza *is also found in Trinidad and Tobago. About 14cm (5½in) long, it inhabits forest canopies below 1,500m (4,900ft). The male is bright bluish green all over, with a black mask. Both sexes have a yellow bill and red eyes.*

BIRDS

Family TERSINIDAE

Swallow tanager

Length 15 cm (6 in)

Species 1

Sometimes placed among the tanager family, *Tersina viridis* is distributed throughout the middle and north-western parts of South America as far south as Paraguay and Argentina. The male is mostly turquoise blue, with a white lower belly and black mask, while the female (pictured) is green; both

have red eyes. Found in hilly areas with humid forest, up to 1,500m (4,950ft), this bird lives in groups of over 30.

Family COEREBIDAE

Bananaquit

Length 11 cm (4 1/4 in)

Species 1

A familiar garden bird, *Coereba flaveola* is sometimes placed among New World warblers (see below). At least 41 subspecies of this widely varied species have been identified. It is widespread in northern and eastern South America and the Caribbean islands, and is common in the northern

part of its range, but relatively scarce deeper in the forest. It feeds mainly on nectar but also eats fruits. Both sexes have a black crown and mask, a white eye-stripe, and a vivid yellow breast and belly.

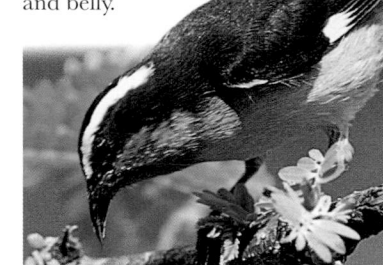

Family PARULIDAE

New World warblers

Length 10–16 cm (4–6 1/2 in)

Species About 116

New World warblers, known in USA as wood warblers, include many brightly coloured birds, with frequent splashes of orange and yellow in their plumage. New World warblers inhabit a wide range of woodland and scrub

habitat; many head from North America to Central and South America to spend the winter in tropical latitudes. The pressure of human habitation and agriculture have shrunk these areas, leading to a substantial decline in numbers of breeding birds. Although they are largely insectivorous, these birds also feed on fruits during migration. The songs of different New World warblers are generally well developed but are similar enough to confuse birdwatchers.

white tail patches

black hood and cowl in male

black and white stripes

HOODED WARBLER

Wilsonia citrina is an expert fly catcher that inhabits damp forest and wooded swamp, always on or a few metres above the ground. It breeds in eastern and central USA, and winters in Central America and the West Indies. This bird is 14cm (5½ in) long, and both sexes flaunt white tail patches as they move about.

black throat in male

BLACK AND WHITE WARBLER

Breeding in Canada and eastern USA, Mniotilta varia winters in the south, as far as northern South America. It forages along branches, probing for insects with its long bill. Its call is a high-pitched whistle. This bird is 11.5–14cm (4½–5½ in) long.

KIRTLAND'S WARBLER

Dendroica kirtlandii breeds only in jack pine woodland in Michigan, USA, and winters in the Bahamas. Habitat management has improved its nesting conditions. It is 15cm (6in) long, and both sexes (female pictured) have blue-grey upperparts and white eye-rings.

black face mask

yellow breast

COMMON YELLOWTHROAT

Distributed across North America, Geothlypis trichas – which is 13cm (5in) long – winters in southern USA and South America, and prefers low vegetation in fields and swamps. The male has a black face mask bordered with grey, and yellow throat, breast, and undertail coverts. The female is similar but lacks the face mask.

white throat in eastern race

yellow rump

grey plumage

YELLOW-RUMPED WARBLER

Dendroica coronata is one of the most familiar birds in North America, found across Canada and Alaska, and through-out the mid-west in the breeding season. It is 14cm (5½ in) long. There are 2 subspecies, both with grey plumage and a yellow rump.

Family DREPANIDIDAE

Hawaiian honeycreepers

Length 10–20 cm (4–8 in)

Species 23

Hawaiian honeycreepers are an important family from the point of view of conservation, since almost half are endangered, possibly as a result of avian malaria caused by an introduced mosquito. Although there are currently 23 species of these birds, the family was originally much bigger, and at least 15 further species have been identified through fossil

remains. Hawaiian honeycreepers have some remarkably shaped bills, adapted to different feeding habits. These range from powerful seed-cracking bills to very thin, tapering, downward-curving ones. In one species, the bill takes up one-third of the bird's entire length, and in the akiapolaau, the upper mandible is

olive-green back

yellow head

nearly twice the length of the lower mandible – the upper one being used to remove grubs and the lower to break bark. Most of these birds feed on nectar and fruits, but some eat insects. Endemic to Hawaii, they are now restricted to wet montane forests on the sides of volcanoes, up to 3,000m (9,900ft), as their original habitats at lower elevations have been destroyed. They are not migratory, but undertake feeding sorties in order to find flowering trees. Generally, they breed from February to June.

AKIAPOLAAU

Hemignathus munroi is a critically endangered species restricted to the koa forest above 1,000m (3,300ft) on the slopes of volcanoes of the Big Island of Hawaii. About 12.5cm (5in) long, the male is olive-green above, and yellow below, with a yellow head. The female is smaller and duller.

IIWI

A common and noisy species found in native forests above 700m (2,300ft) on Hawaii's Big Island, Kauai, and Maui, Vestiaria coccinea is rare on other Hawaiian islands. The adult, about 13cm (5in) long, is bright vermilion, with a long, pink, downcurved bill and a yellow ring around the eye.

Order PASSERIFORMES *continued*

Family VIREONIDAE

Vireos and relatives

Length 17–20 cm (6½–8 in)

Species About 52

The unusual name "vireo" is derived from a Latin word that translates as "to be green". Birds in this family, which includes greenlets, shrike-vireos, and pepper shrikes, weigh from 9g (¹¹⁄₃₂oz) to 25g (⅞oz). Their characteristically notched bills are larger than those of the New World warblers (see p.355), the eye-stripes bolder, and the green plumage duller. These birds inhabit broadleaved or mixed forests, and species coexist by foraging in separate areas. Although mainly insectivorous, they also eat fruits, depending on the season. They are migratory birds that breed in the USA and up to northern Canada, but move south in winter to Central or South America. Some vireos are also found in the Caribbean during the breeding season and in winter.

RED-EYED VIREO
About 15cm (6in) in length, Vireo olivaceus lives in deciduous forest in North America, and migrates to South America in winter.

TAWNY-CROWNED GREENLET
Found in the forests of northern South America, Hylophilus ochraceiceps is 11.5cm (4½ in) long and has a rusty-orange crown.

pale brown underparts

Family ICTERIDAE

New World blackbirds

Length 15–53 cm (6–21 in)

Species About 97

New World blackbirds, or troupials, range from the highly colourful orioles, blackbirds, and meadowlarks to the sombre-plumaged grackles. It is perhaps surprising that this family also includes the oropendolas, which in many ways appear to be closer to crows (see p.360). All these species are found exclusively in North and South America and the Caribbean. Many live close to human settlements, some benefitting from agriculture and, like the red-winged blackbird, often becoming crop pests. Although highly adaptable, some of the memebrs of this family are extremely rare and local. In some species, males and females differ in size and plumage. Almost all of these birds have conical, unnotched, pointed bills, but in the oropendolas the top ridge is expanded to form a frontal shield or casque. This is a noisy family (although not especially musical), and the sounds made by the oropendolas are very impressive. Cowbirds are unusual in that they parasitize the nests of other species to rear their own young. Generally, species in this family are not migratory, although the bobolink undertakes a major migration from North America to Argentina every year.

CRESTED OROPENDOLA
Psarocolius decumanus is widely distributed across northern and central South America, and Trinidad and Tobago. About 47cm (18½ in) long, it is conspicuous for its loud, rasping, gurgling call, ivory bill, blue eyes, and glossy black plumage. This bird nests in colonies.

pale eyebrows

spotted belly

EASTERN MEADOWLARK
A resident of open fields and plains in south and eastern USA, Sturnella magna is about 24cm (9½ in) in length and has a distinctive, black V-shaped band on its bright yellow breast. Its back is a mottled brown.

buff neck

BOBOLINK
Dolichonyx oryzivorous, 18cm (7in) long, is found across northern USA and southern Canada, and winters in Argentina. The males are mainly black while the females are yellowish brown and streaked.

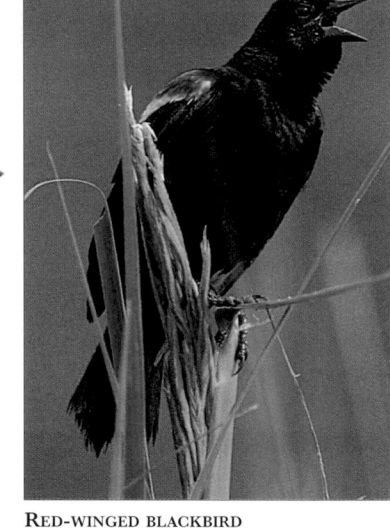

RED-WINGED BLACKBIRD
Agelaius phoeniceus inhabits marshland in much of North America, and gathers in immense flocks in winter. It is 22cm (9in) long. The male, glossy black all over, has bright red shoulder patches which he puffs out when displaying from his song post.

COMMON GRACKLE
Found abundantly in farmland, parks, and urban areas of North America, except the western states, Quiscalus quiscula, 32cm (12½ in) long, is one of the commonest birds in its area. The male has a glossy purple head and neck. Both sexes have an unusual twisted tail, which is obvious in flight.

large white wing patches

orange cheeks

NORTHERN ORIOLE
Icterus bullocki, 21cm (8½ in) long, inhabits deciduous woodland in southwest Canada, west USA, and north Mexico, but winters in Central America. In the Great Plains of the USA, it hybridizes with its eastern relative, the Baltimore oriole.

brown hood on male

glossy black plumage

BROWN-HEADED COWBIRD
Molothrus ater is about 19cm (7½ in) long, with the males distinguished by a brown hood. This bird inhabits coastal and southern USA, and is a summer visitor to the central states and Canada.

Family FRINGILLIDAE

Finches and relatives

Length 11–19 cm (4¼–7½ in)

Species About 134

Found in North America, Africa, Europe, and Asia, finches have bills that have evolved for feeding on seeds; some species have particularly strong skulls and large jaw muscles to crack very hard seeds. When feeding, they wedge the seeds in a special groove at the side of their palate, then crush them by raising the lower jaw. The husk is peeled off by the tongue and the kernel swallowed. Some finches are migrants, but will travel only a short distance away from cold weather to find food. Although most are territorial, some finches breed cooperatively.

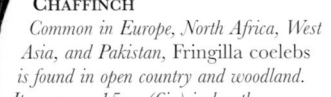

CHAFFINCH
Common in Europe, North Africa, West Asia, and Pakistan, Fringilla coelebs is found in open country and woodland. It measures 15 cm (6 in) in length.

EVENING GROSBEAK
This bulky bird, Coccothraustes vespertinus, is 20 cm (8 in) long, and inhabits coniferous forest across northern North America, moving southwards in winter. It is noisy and gregarious.

brown wings

EURASIAN GOLDFINCH
The dainty Carduelis carduelis, 14 cm (5½ in) long, with its distinctive red face and beige body, is common in the countryside of Europe and Asia.

RED CROSSBILL
A resident of North America, Europe, and Asia, Loxia curvirostra is remarkable for its bill. The upper and lower mandibles cross over each other, a feature that allows the bird to extract seeds from ripe pine cones. This bird (female shown) is 17 cm (6½ in) long.

yellow wing bars

WILD CANARY
The wild ancestor of the common cage bird, Serinus canaria is 13 cm (5 in) long and has greyish yellow plumage. It is endemic to the Canary Islands, the Azores, and Madeira, off the west coast of Africa.

grey back

pink breast in male

white undertail coverts

BULLFINCH
The attractive bullfinch, Pyrrhula pyrrhula, 16 cm (6½ in) long, can quickly strip a pear tree of its buds, making it an unpopular bird with fruit growers. The male's pink breast combined with its grey back and black and white wings make it easy to identify. This bird is found across Europe, and east into China.

Family ESTRILDIDAE

Estrildid finches

Length 9–13 cm (3½–5 in)

Species About 133

Estrildid finches, a subject of much debate with regard to taxonomy, include waxbills, grass finches, parrot finches, and mannikins, and are found in Africa, Asia, open grassland of Australia, and New Guinea. Largely sedentary, these birds move only to find food or water; they feed mainly on grass seeds. Many of these birds pair for life. Most build domed grass nests while some nest in holes. They lay 4–8 white eggs, which are incubated for 10–21 days. After the breeding season, they generally become sociable, and flock together for safety.

orange cheek patches

ZEBRA FINCH
Taeniopygia guttata is 10 cm (4 in) long, and has a black, teardrop-like line below its eye. Found across mainland Australia, it gathers in flocks near water.

red bill

JAVA SPARROW
Although mainly a bird of towns and gardens, large flocks of Java sparrows, Padda oryzivora, congregate on agricultural land and can be pests in rice and maize fields. This species, 14 cm (5½ in) long, is endemic to Indonesia from Java to Sulawesi, and has been introduced in Asia, Australia, and Tanzania.

GOULDIAN FINCH
Restricted to savanna woodland, stony hills, and dense grass in northern Australia, Chloebia gouldiae is one of the country's most endangered birds, its populations depleted in recent years by infections caused by a mite. It is 14 cm (5½ in) long. Most birds are black-headed, but rarer red- or golden-headed ones also exist.

Family PLOCEIDAE

Weavers

Length 13–26 cm (5–10 in)

Species About 114

This is a large, mainly African group of finch-like birds, most of which are known as weavers. They have short, conical bills, short, rounded wings, and plumage that is usually yellow or brown, sometimes both. Some species look confusingly similar. Most weavers build roofed nests, some with long entrance tubes, by weaving strips torn from the leaves or stems of palm trees. The sight of a male weaver hanging upside down under a recently completed nest to impress a female is very common in Africa. Most weavers are seed-eating, migratory, savanna birds, but some are insectivorous, non-migratory forest dwellers. A few species are brood parasites: they lay their eggs in nests built by other birds.

HOUSE SPARROW
One of the world's most familiar urban species, Passer domesticus, 15 cm (6 in) long, lives easily with humans. It originates from Asia, but is now seen in many countries, although population declines have been noted.

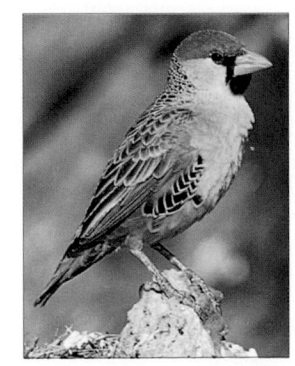

SOCIABLE WEAVER
Restricted to Namibia, Botswana, and South Africa, Philetairus socius nests in large colonies of up to 300 birds, the nests adjoining each other in a huge mass on trees or poles. This bird is 14 cm (5½ in) long.

purple sheen

black mask

RED-BILLED QUELEA
Considered by many to be the commonest bird in the world, Quelea quelea, 12 cm (4¾ in) long, is found in 37 African nations. It feeds on rice and corn, and is a crop pest.

VARIABLE INDIGOBIRD
Vidua funerea mimics the song of the African firefinch and lays its eggs in the firefinch's nest. Found from Kenya to South Africa, and in parts of West Africa, it is 12 cm (4¾ in) long.

PARADISE WHYDAH
The male of this species (shown above) is 33 cm (13 in) long and the female 12 cm (4¾ in). In the breeding season, the magnificent tail of the male adds some 21 cm (8½ in) to his length. Vidua paradisaea does not build nests, laying its eggs in the nests of the green-winged pytilia. It is a gregarious bird, common throughout the eastern and central areas of Africa.

extremely long tail on male

Order PASSERIFORMES *continued*

Family STURNIDAE

Starlings

Length 16–45 cm (6½–18 in)

Species About 114

An ability to imitate the sounds they hear is a notable feature of this group of birds, which includes the irrepressible mimics, mynahs. Starlings are found in Africa, Europe, and Asia, but the common starling has been introduced around the world and is now found in 119 nations. These birds have strong legs and bills, and mostly black plumage, often with an iridescent sheen that is particularly noticeable in the African species. Some, such as

the long-tailed glossy starling, have long tails, which can be as much as 35 cm (14 in) long. Some starlings have areas of bare skin on the face, particularly around the eyes, and the wattled starling has loose folds of skin, known as wattles, on its head. Most of these birds are very much at home on the ground, and several species live among human habitation. Most starlings are residents but some are partial migrants, perhaps moving to find food in winter; the group is by and large omnivorous. Starlings lay 2–6 pale blue-green, often brown-spotted, eggs.

yellow wattle behind eye

HILL MYNAH
Distributed across much of southern and Southeast Asia, Gracula religiosa is found in moist forest, where its diverse range of piercing calls are a distinctive feature. It is 28–30 cm (11–12 in) long, and is the typical mynah found in pet shops and aviaries.

glossy black and purple plumage

violet throat and breast

bright, blue-green iridescent plumage

SPLENDID STARLING
Distributed across much of Africa, in 28 countries, Lamprotornis splendidus is a woodland species. Measuring 27 cm (10½ in), this strikingly coloured bird is quite shy, and has declined in many places due to large-scale deforestation. Outside of the breeding season, it is known to form flocks.

COMMON STARLING
Occurring in 119 countries, Sturnus vulgaris is one of the world's most familiar species, having adapted to living alongside humans. This bird – which is 21 cm (8½ in) long – feeds and roosts in large flocks, often perching on building ledges overnight.

iridescent sheen of green, blue, or violet

RED-BILLED OXPECKER
This slender African species, Buphagus erythrorhynchus, is 20 cm (8 in) long. It feeds on blood that oozes from the wounds of large mammals such as rhinos and buffaloes. It uses its comb-like bill to scour the mammal's skin for food, and helps its host by removing ticks and leeches from its body.

METALLIC STARLING
Australia's only native starling, Aplonis metallica arrives annually from New Guinea and nearby islands to breed along the Queensland coast. Hundreds of birds may be found breeding in one tree. This bird is 22 cm (9 in) long, and has glossy green and purple plumage and bright red eyes. It feeds on native and cultivated fruits.

Family ORIOLIDAE

Old World orioles

Length 20–30 cm (8–12 in)

Species 29

The robust Old World orioles have downcurved bills and long, pointed wings. They inhabit woodland throughout Africa, Asia, New Guinea, and Australia; one species occurs in Eurasia. Most have predominantly yellow plumage, often with black either on the head or the wings; the females are generally dull and more streaked than the males. All of them feed on insects and fruits, and one species, the black-headed oriole, also takes nectar. The distinctive, fluty songs of orioles can be heard over considerable distances. Most of them build a deep cup nest, slung in a horizontal fork on a branch. A number of Old World orioles are migratory.

BLACK-HEADED ORIOLE
Found in central, eastern, and southern Africa, Oriolus larvatus is a noisy and conspicuous species (although it often sits motionless in the canopy), with a loud, fluty call. This oriole, which is 22 cm (9 in) in length, feeds on seeds, crops, fruits, and caterpillars.

Family DICRURIDAE

Drongos

Length 18–72 cm (7–28 in)

Species About 23

The drongos, which occur in Africa, Asia, and Australia, have long, forked tails, and in all but 2 cases, are glossy black. Their stout bills are arched and slightly hooked, with a small notch, and the nostrils are often concealed by dense feathers. Drongos are mainly woodland birds, but are also found in open country with scattered trees, up to a height of 3,300 m (10,900 ft). These insectivorous birds also feed on lizards and small birds. They feed like flycatchers, flying out from a branch to snatch prey and returning to devour it, and are well known for chasing birds of prey, which may be several times their own size.

CRESTED DRONGO
Confined to the Comoro Islands and Madagascar, Dicrurus forficatus is the only drongo in these countries. It is 26 cm (10 in) long, and has a striking crest which sticks up from the base of the bird's bill.

Family CALLAEIDAE

Wattlebirds

Length 25–48 cm (10–19 in)

Species 2

Wattlebirds are found in New Zealand, the best known being the kokako, *Callaeas cinerea*. Surveys in the 1970s revealed that only tiny colonies of kokako remained in the North Island. Conservationists worked on plans to ensure its survival, which involved preservation of podocarp forests and control over the predation of its eggs and chicks. There are now around 30 separate populations of kokako, but it is still not out of danger. The other extant wattlebird is the saddleback (shown right). The huia, *Heterolocha acutirostris*, last seen in 1907, was unique in that the sexes had different bill shapes: the male had a straight, pointed bill while the female had a thin, downcurved one. The huia became extinct because it was hunted for its feathers and due to habitat destruction.

SADDLEBACK
A weak flier, Creadion carunculatus – which is 25 cm (10 in) long – runs and hops through trees, probing the bark for insects. It nests on remote islands and is seldom seen. A small population has recently been introduced to an island in Auckland Harbour where visitor numbers are regulated.

Family GRALLINIDAE

Magpie-larks

Length 20–26 cm (8–10 in)

Species 2

Also known as mudnest builders or mudlarks, these birds, despite their names, are related neither to magpies nor to larks. There are only 2 species: the magpie-lark of Australia and New Guinea, and the torrent-lark of New Guinea. Magpie-larks are reliant on mud for nest-building, and are seldom found far from water.

male has black throat

long legs

MAGPIE-LARK
Common in Australia near surface water, Grallina cyanoleuca is 29cm (11½in) long with a loud, piping, 5-note call. It has adapted to urban spaces and can be seen feeding at roadsides.

Family ARTAMIDAE

Woodswallows

Length 12–20 cm (4¾–8 in)

Species 11

Woodswallows occur in parts of Asia, the Pacific islands, New Guinea, and Australia. Despite the family name, these birds are not closely related to the swallow family, although their flight silhouette is not unlike that of martins. They are 12–20cm (4¾–8in) long, and have strong bills and short legs and feet, while the wings are quite short and pointed. A characteristic family trait is the birds' habit of huddling together on branches, particularly when roosting. These insectivorous birds catch a large percentage of their food while on the wing. All members of the family are colonial, building nests mainly at the tops of stumps or large branches, or in holes. Females lay 2–4 white or cream eggs, with red-brown spots.

black mask

BLACK-FACED WOODSWALLOW
Distributed across Australia, New Guinea, the Lesser Sundas, and Timor, Artamus cinereus – measuring 19cm (7½in) in length – is the commonest bird in the family. Unlike other woodswallows, it is not very nomadic.

Family CRACTICIDAE

Bell-magpies

Length 18–53 cm (7–21 in)

Species About 12

Bell-magpies resemble crows with their strong bodies, sturdy bills, and the scales on their legs and feet. However, they are less carnivorous than crows and eat a wide range of food including fruits. They are found in Asia, Australia, and New Guinea, the commonest bird being the Australian magpie. Also included in this family are 6 butcher birds and 3 currawongs. The lowland and mountain peltops of New Guinea (considered members of the woodswallow family by some) have black plumage with a white face patch and red undertail coverts.

AUSTRALIAN MAGPIE
Found in Australia, New Zealand, and New Guinea, Gymnorhina tibicen is an opportunist feeder that scavenges on garbage and also kills smaller birds and rodents. It is 42cm (16½in) long and is mostly black-backed, although there are many regional variations.

Family PTILORHYNCHIDAE

Bowerbirds

Length 21–38 cm (8½–15 in)

Species 20

Male bowerbirds are well known for the elaborate bowers they create to attract females – these are complex structures decorated with brightly coloured objects. They also decorate their display grounds with feathers, flowers, leaves, pebbles, and even clothes pegs or coloured bits of plastic and paper. They cohabit with one or more females and the latter rear the young. These birds are spread across New Guinea and Australia, where most inhabit damp forest, although some prefer drier areas. Bowerbirds are mainly fruit eaters, but will also feed on leaves, flowers, seeds, and invertebrates, feeding at all levels from ground to canopy.

SATIN BOWERBIRD
Restricted to northeast and southeast Australia, Ptilonorhynchus violaceus is 31cm (12in) long. The glossy, blue-black male – the most photographed bowerbird – builds an "avenue" bower of sticks, decorating it with bright blue or yellow objects. This bird has a pale yellow bill.

Family PARADISAEIDAE

Birds of paradise

Length 12–100 cm (4¾–39 in)

Species About 44

Birds of paradise are renowned for the spectacularly beautiful and ornate plumage of the males, which is used in display, although in a few species the sexes are alike. Birds in this family differ in size, and have extremely variable tail lengths. Some species have short, straight bills, whereas others have long, curved ones. Most, however, have rounded wings and strong legs and feet, and use loud calls to attract mates. The majority of the birds of paradise are found in New Guinea, mostly in wet montane forests. The members of this family feed mainly on fruit. Their nests are typically bulky, made of leaves, ferns, and twigs, and are generally placed in tree forks. Their voices vary from soft, drawn-out calls to loud, explosive sounds like a gun being fired. A few species exhibit mimicry in captivity.

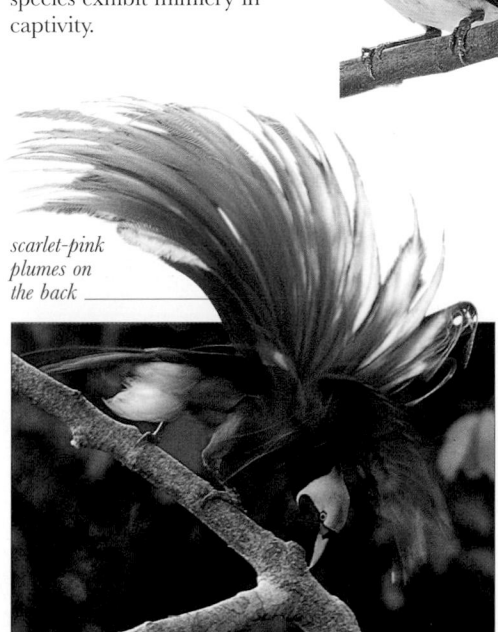

scarlet-pink plumes on the back

2 long tail feathers with racquets

flank feathers

tail streamer

KING BIRD OF PARADISE
In a spectacular courtship display, the male bird expands his feathers and appears almost spherical, waving his green racquets above his head. Cicinnurus regius is 16–17cm (6½in) long, and its 2 thin tail feathers add another 14cm (5½in).

RAGGIANA BIRD OF PARADISE
Found only in New Guinea, Paradisaea raggiana, 35cm (14in) long, is the best known bird of paradise. Adult males display communally in traditional lek sites, where up to 20 birds will congregate.

spiral tail-wires

WILSON'S BIRD OF PARADISE
Cicinnurus respublica, *16cm (6½in) long, is found in Irian Jaya but not Papua New Guinea. Both sexes have areas of bare blue skin on the hind crown, but the male is distinguished by red wings and spiral tail-wires.*

GREATER BIRD OF PARADISE
Found on the Aru Islands and in southern New Guinea, Paradisaea apoda *is 42–45cm (16½–18in) long. During courtship, 12–20 males gather in a tree to dance, spreading their wings, while the dominant male stays at the centre.*

BIRDS

Order PASSERIFORMES *continued*

Family CORVIDAE
Crows and relatives

Length 20–66 cm (8–26 in)

Species About 117

Habitat All terrestrial

Crows and their relatives are distributed across almost every part of the world, except the Arctic and Antarctic. They are the most highly developed of all birds – intelligent, sociable, and very adaptable, with most of them feeding on a wide range of food. Many members of this family have

mainly black plumage, although some of the jays and magpies are highly coloured. Found in open countryside and woodland, in both lowland and upland areas, crows are not migratory birds although they may move locally between seasons. They usually nest as separate pairs but a notable exception in this regard is the rook, which nests in colonies. A typical crow nest holds 2–8 eggs and is built of twigs, in trees or bushes, or sometimes in holes.

iridescent plumage

BLACK-BILLED MAGPIE
Pica pica, with its 12 very similar but identifiable subspecies, is found in central and western North America, Europe, North Africa, and Asia. This is a common bird, 47cm (18½ in) in length, which lives well alongside humans. Gregarious outside of the breeding season, over 100 birds may gather at a winter roost. Its nest is a domed pile of sticks, high up in a tree.

distinctive tail, about 25cm (10in) long

iridescent green-blue plumage

downcurved red bill

EURASIAN JAY
The many subspecies of Garrulus glandarius have slightly different head markings; all are 33cm (13in) long. Distributed across Europe and Asia, this bird is a fairly common inhabitant of woodland, but is also found in many other habitats, such as rainforests and taiga. Across much of its range the Eurasian jay collects and stores acorns as winter food.

blue wing feathers

RED-BILLED CHOUGH
Pyrrhocorax pyrrhocorax is a gregarious bird, particularly outside the breeding season. It forages for insects, especially ants, and also eats berries. During courtship, it performs spectacular aerobatics. The nest is built either on a cliff ledge or in a cave. Measuring 40cm (16in) in length, this bird is found in mountain and rocky coastal areas in Europe, North Africa, and South Asia.

red legs

CLARK'S NUTCRACKER
Found only in the mountains of western North America, Nucifraga columbiana inhabits juniper and pine woodland. Agile when swooping through canyons, it does not fly long distances and is usually found on treetops. It is 31cm (12in) long.

red bill

black cap

grey nape

black head

BLUE JAY
Common in woodland and parkland across central and eastern North America, Cyanocitta cristata is about 30cm (12in) long. This bird is found in pairs or small groups, and is always noisy, with a distinctive "peeah peeah" call. It uses mud to build its nest.

thin black collar

JACKDAW
Widespread across Europe, West Asia, and parts of North Africa, this is a familiar crow which often flies and feeds with rooks. A farmland bird in much of its range, Corvus monedula also inhabits gorges and sea cliffs, nesting there as well as in buildings and holes in trees. It roosts in flocks of several thousands outside the breeding season. It is about 34cm (13½ in) long.

RED-BILLED BLUE MAGPIE
This striking species, 68cm (27in) long, has a tail that itself measures 47cm (18½ in). Found up to a height of 1,500m (4,900ft) in the deciduous forests of South and Southeast Asia, Urocissa erythroryhncha hunts in flocks at lower levels. Its nest is a rough, flimsy cup in a tree.

stout black bill

COMMON RAVEN
At 65cm (26in), Corvus corax is the largest crow in the northern hemisphere, and is found in North and Central America, Europe, Asia, and North Africa. It lives in open habitats as well as in urban areas in some parts of its range, and up to 6,350m (20,900ft) on Mount Everest, often nesting when snow is still on the ground.

large bill

CARRION CROW
Regarded as the archetypal crow by many people in Europe, the all-black Corvus corone lives in many habitats including woodland, moorland, farmland, and towns. About 50cm (20in) long, it is also found (but with grey in its plumage) through West Asia and in eastern Asia including Vietnam and Korea. It is a solitary nester, usually making its home in a tree, although in mountainous areas cliff ledges are often used.

all-black plumage

wedge-shaped tail

BIRDS IN DANGER

The list on this page includes all those birds currently classed as critically endangered by the IUCN (see p.31). With non-passerines, the major threats include hunting, habitat change, and also collection for the cage-bird trade – a problem that affects parrots in particular. Many birds of prey are also in a precarious position, because they are at the top of their food chains, and therefore highly vulnerable to any changes that affect their food supply. With passerine birds, habitat change is the chief threat. For island birds, such as the native thrushes and creepers of the Hawaiian islands, this is aggravated by other problems, including competition from introduced species, and the effects of avian malaria – a disease imported from outside, to which native birds have little resistance. Hundreds of other birds (not listed here) fall into lower categories of threat.

NON-PASSERINES

Aceros waldeni (Visayan wrinkled hornbill)
Aegotheles savesi (New Caledonian owlet-nightjar)
Amaurornis olivieri (Olivier's crake)
Amazilia castaneiventris (chestnut-bellied hummingbird); *A. luciae* (Honduran emerald)
Amazona vittata (Puerto Rican parrot)
Anas nesiotis (Campbell Island teal)
Anodorhynchus glaucus (glaucous macaw); *A. leari* (indigo macaw)
Anthracoceros montani (Sulu hornbill)
Ara glaucogularis (blue-throated macaw)
Athene blewitti (forest little owl)
Aythya innotata (Madagascar pochard)
Bostrychia bocagei (dwarf olive ibis)
Buteo ridgwayi (Ridgway's hawk)
Cacatua haematuropygia (red-vented cockatoo); *C. sulphurea* (yellow-crested cockatoo)
Calyptura cristata (kinglet cotinga)
Campephilus imperialis (imperial woodpecker); *C. principalis* (ivory-billed woodpecker)
Caprimulgus noctitherus (Puerto Rican nightjar)
Carpococcyx viridis (Sumatran ground-cuckoo)
Centropus steerii (black-hooded coucal)
Charmosyna diadema (New Caledonian lorikeet); *C. toxopei* (blue-fronted lorikeet)
Chondrohierax wilsonii (Cuban kite)
Columba argentina (grey wood-pigeon)
Columbina cyanopis (blue-eyed ground-dove)
Corvus hawaiiensis (Hawaiian crow)
Crax alberti (blue-knobbed curassow)
Crypturellus saltuarius (Magdalena tinamou)
Diomedea amsterdamensis (Amsterdam albatross)
Ducula galeata (Marquesan imperial-pigeon); *D. whartoni* (Christmas imperial-pigeon)
Eriocnemis godini (turquoise-throated puffleg); *E. mirabilis* (colourful puffleg); *E. nigrivestis* (black-breasted puffleg)
Eutriorchis astur (Madagascar serpent-eagle)
Francolinus ochropectus (Djibouti francolin)
Fregata andrewsi (Andrews' frigatebird)
Gallicolumba erythroptera (white-collared ground-dove); *Gallicolumba keayi* (Negros bleeding-heart); *G. menagei* (Sulu bleeding-heart); *G. platenae* (Mindoro bleeding-heart); *G. salamonis* (thick-billed ground-dove)
Gallinula pacifica (Samoan moorhen); *G. silvestris* (San Cristobal moorhen)
Gallirallus lafresnayanus (New Caledonian rail)
Geopsittacus occidentalis (night parrot)
Geronticus eremita (northern bald ibis)
Grus leucogeranus (Siberian crane)
Gymnogyps californianus (California condor)
Gyps bengalensis (Asian white-backed vulture); *G. indicus* (long-billed vulture)
Haliaeetus vociferoides (Madagascar fish-eagle)
Hapalopsittaca fuertesi (indigo-winged parrot)
Heliangelus zusii (Bogotá sunangel)
Himantopus novaezelandiae (black stilt)
Lanius newtoni (Newton's fiscal)
Lepidopyga lilliae (sapphire-bellied hummingbird)
Leptodon forbesi (white-collared kite)
Leptotila wellsi (Grenada dove)
Lophornis brachylopha (short-crested coquette)
Megapodius pritchardii (Nevafou megapode)
Mergus octosetaceus (Brazilian merganser)
Neophema chrysogaster (orange-bellied parrot)

Ninox natalis (Christmas Island hawk-owl)
Nothoprocta kalinowskii (Kalinowski's tinamou)
Numenius borealis (eskimo curlew); *N. tenuirostris* (long-billed curlew)
Oceanodroma macrodactyla (Guadalupe storm-petrel)
Odontophorus strophium (gorgeted wood-quail)
Ognorhynchus icterotis (yellow-eared parrot)
Ophrysia superciliosa (Himalayan quail)
Otus capnodes (Anjouan scops owl); *O. insularis* (Seychelles scops owl); *O. moheliensis* (Moheli scops owl); *O. pauliani* (Comoro scops owl); *O. siaoensis* (Siau scops owl)
Papasula abbotti (Abbott's booby)
Penelope albipennis (white-winged guan)
Phapitreron cinereiceps (dark-eared brown-dove)
Pipile pipile (Trinidad white-headed curassow)
Pithecophaga jefferyi (Philippine eagle)
Podiceps taczanowskii (puna grebe)
Procellaria conspicillata (spectacled petrel)
Pseudibis davisoni (white-shouldered ibis)
Pseudobulweria aterrima (Mascarene petrel); *P. becki* (Beck's petrel); *P. macgillivrayi* (Fiji petrel)
Psittacula eques (Mauritius parakeet)
Pterodroma axillaris (Chatham Islands petrel); *P. caribbaea* (Jamaica petrel); *P. madeira* (Madeira petrel); *P. magentae* (Chatham Island taiko); *P. phaeopygia* (dark-rumped petrel)
Ptilinopus arcanus (negros fruit-dove)
Puffinus auricularis (Townsend's shearwater)
Rhinoptilus bitorquatus (Jerdon's courser)
Rhodonessa caryophyllacea (pink-headed duck)
Sapheopipo noguchii (Noguchi's woodpecker)
Sephanoides fernandensis (Juan Fernández firecrown)
Siphonorhis americanus (Jamaican paraque)
Sterna bernsteini (Chinese crested tern)
Strigops habroptilus (kakapo)
Tachybaptus rufolavatus (Delacour's little grebe)
Tadorna cristata (crested shelduck)
Thalassarche eremita (Chatham albatross)
Thaumatibis gigantea (giant ibis)
Vanellus macropterus (Sunda plover)

PASSERINES

Acrocephalus familiaris (millerbird)
Alauda razae (Raza Island lark)
Antilophia bokermanni (Araripe manakin)
Apalis fuscigularis (Taita apalis)
Aplonis pelzelni (Pohnpei mountain starling)
Atlapetes pallidiceps (pale-headed brush-finch)
Camarhynchus heliobates (mangrove finch)
Cinclodes aricomae (royal cinclodes)
Clytoctantes atrogularis (Rondônia bushbird)
Colluricincla sanghirensis (Sangihe shrike-thrush)
Conothraupis mesoleuca (cone-billed tanager)
Copsychus sechellarum (Seychelles magpie-robin)
Curaeus forbesi (Forbes's blackbird)
Cyornis ruckii (Rueck's blue-flycatcher)
Dicaeum quadricolor (Cebu flowerpecker)
Eurychelidon sirintarae (white-eyed river-martin)
Eutrichomyias rowleyi (Caerulean paradise-flycatcher)
Foudia rubra (Mauritius fody)
Geothlypis beldingi (Belding's yellowthroat)

Hemignathus lucidus (nukupu'u)
Heteromirafra ruddi (Rudd's lark)
Icterus oberi (Montserrat oriole)
Junco insularis (Guadalupe junco)
Laniarius liberatus (Bulo Burti bush-shrike)
Leucopeza semperi (Semper's warbler)
Leucopsar rothschildi (Bali starling)
Macroagelaius subalaris (mountain grackle)
Melamprosops phaeosoma (black-faced honeycreeper)
Merulaxis stresemanni (Stresemann's bristlefront)
Mimodes graysoni (Socorro mockingbird)
Monarcha boanensis (black-chinned monarch)
Myadestes lanaiensis (Molokai thrush); *M. myadestinus* (large Kauai thrush); *M. palmeri* (small Kauai thrush)
Myrmotherula fluminensis (Rio de Janeiro antwren); *M. snowi* (Alagoas antwren)
Nemosia rourei (cherry-throated tanager)
Neospiza concolor (Sao Tome canary)
Oreomystis bairdi (Kauai creeper)
Orthotomus moreaui (long-billed apalis)
Paroreomyza maculata (Oahu creeper)
Philydor novaesi (Alagoas foliage-gleaner)
Phyllastrephus leucolepis (spot-winged greenbul)
Phylloscartes ceciliae (long-tailed tyrannulet); *P. roquettei* (Minas Gerais tyrannulet)
Pitta gurneyi (Gurney's pitta)
Pomarea nigra (Tahiti flycatcher); *P. whitneyi* (Fatu Hiva flycatcher)
Psittirostra psittacea (ou)
Pyriglena atra (fringe-backed fire-eye)
Rukia ruki (Faichuk white-eye)

DALMATIAN PELICAN
One of Europe's largest waterbirds, the Dalmatian pelican lives in shallow lakes and river deltas, in southeast Europe and Russia. It faces several threats, including drainage, pollution, and persecution by local fishermen. The species is currently listed as lower risk and dependent on conservation measures.

Scytalopus psychopompus (Bahia tapaculo)
Sporophila insulata (Tumaco seedeater); *S. melanops* (hooded seedeater); *S. zelichi* (Entre Rios seedeater)
Synallaxis infuscata (pernambuco)
Telespiza ultima (Nihoa finch)
Terpsiphone corvina (Seychelles paradise-flycatcher)
Thryothorus nicefori (Nicefero's wren)
Toxostoma guttatum (Cozumel thrasher)
Turdus helleri (Taita thrush); *T. ludoviciae* (Somali thrush)
Vermivora bachmanii (Bachman's warbler)
Vireo caribaeus (Saint Andrew vireo)
Zoothera major (Amami thrush)
Zosterops albogularis (white-breasted silvereye); *Z. luteirostris* (Ganonga white-eye); *Z. modestus* (Seychelles grey white-eye); *Z. natalis* (Christmas Island white-eye); *Z. nehrkorni* (Sangihe white-eye); *Z. rotensis* (Rota bridled white-eye)

REPTILES

REPTILES

PHYLUM	Chordata
CLASS	Reptilia
ORDERS	4
FAMILIES	62
SPECIES	7,984

REPTILES ARE EGG-LAYING vertebrates that have a tough skin with a covering of scales. These ectothermic (cold-blooded) animals cannot generate internal heat. There are 4 orders of reptiles: snakes, amphisbaenians, and lizards (collectively known as squamates), crocodiles, alligators, and caimans (crocodilians), tortoises and turtles, and tuataras. Most reptiles, including those that live mainly in water, lay eggs on land. The young emerge fully formed without a larval stage.

SCALES
The Gila monster belongs to the largest group of reptiles, the lizards. Unlike amphibians, from which they are descended, all reptiles have scales rather than smooth, moist skin.

EVOLUTION

Reptiles first appeared about 340 million years ago, having evolved from early amphibians. The first reptiles differed from their amphibian ancestors in 2 significant respects: they developed a hard, scaly outer skin that protected them from abrasion and moisture loss. More importantly, they also evolved a shelled, amniotic egg, in which the embryo developed within a sac of water, protecting it from the environment. Together, these features allowed reptiles to move away from the margins of water bodies, to which the amphibians were restricted, and to colonize land.

By the Mesozoic Era (230–70 million years ago), reptiles had diversified into an enormous variety of types, and were the dominant land animals. The turtles and tortoises branched off at a relatively early stage, and by 200 million years ago, species similar to those we would recognize today were already present. Some time after this, reptiles began to diversify explosively. Among the orders that appeared at this time were the pterosaurs (flying reptiles) and the dinosaurs, along with 2 orders that survive today, the crocodilians and the tuataras. Later, another evolutionary line led to the appearance of the squamates – the lizards, amphisbaenians, and eventually the snakes. Of the 20 or so orders of reptiles known to have existed in the Mesozoic Era, only 4 survive today.

TURTLE FOSSIL
These are the fossilized remains of a turtle, dating back about 200 million years. Turtles and tortoises have changed little since that time, and are the oldest group of living reptiles.

ANATOMY

The external anatomy of reptiles shows immense variation, from the long, slender, limbless forms of snakes to the short, stout, shell-covered bodies of turtles. However, all reptiles are characterized by the presence of scales (see panel, below), which form a barrier that protects them from abrasion, attacks from predators and parasites, and dehydration. Scales differ considerably between reptiles, and in some species scales of various shapes and sizes are found on different areas of the body. Pigment below the scales gives the animal its colouring and markings, which may be for camouflage or for display. In some reptiles, especially

LIZARD SKELETON

Lizards are a highly diverse group of animals. Most have a long tail and 4 well-developed legs with long digits, as seen in this skeleton of a monitor lizard. However, some species closely resemble snakes, and do not have any limbs at all.

shoulder girdle

rib attached to spinal column

limb attached to side of body

pelvic girdle

long digits

male lizards, groups of scales have evolved into dramatic crests, horns, and other features used for display.

The internal skeleton of a reptile is made up of bony elements and provides a stiff supporting system that is considerably sturdier than that of an amphibian, making reptiles more suited to life on land. Unlike those of mammals and birds, a reptile's limbs support the body from the side, which gives them a sprawling gait when they move. Snakes, together with most amphisbaenians and some lizards, are without functional legs.

Differences in the skull reflect the evolutionary origins of the various reptile groups and are a useful feature for classification. Turtles and tortoises have no openings in their temples, while crocodilians and tuataras have 2 on each side. The squamates also have 2 per side, but they are sometimes joined to

SKIN STRUCTURE

A reptile's skin consists of 2 main layers: the epidermis (outer layer) and the dermis (lower layer). The scales, which are present only on the epidermis, are made of a horny substance, keratin, which is similar in composition to human hair and fingernails. The dermis contains nerves, blood vessels, and cells that support and nourish the epidermis. Unlike a fish's scales, those of a reptile cannot be removed individually. All reptiles replace their scales by shedding their outer skin. This allows room for growth, and also replaces skin that is worn out. Whereas snakes slough their skin in one piece, lizards, crocodilians, turtles and tortoises, and tuataras shed it in chunks or flakes. After a snake has shed its outer skin, it often appears much more brightly coloured.

| SNAKE SCALES | LIZARD SCALES | CROCODILE SCALES |

SCALE TYPES
Reptiles' scales differ greatly in size, shape, and texture. They may be smooth or rough, and may overlap, like roof tiles, butt up against each other, or have layers of stretchy skin in between. In crocodilians, the scales on the back are strengthened by bony plates.

epidermis   horny layer   scale

dermis   pigment cells

CROSS-SECTION OF SNAKE SKIN
The scales of a reptile are made from the thick, horny outer layer of skin. Each scale is joined to its neighbours by flexible, hinged areas, so that the body can move and bend. Pigment cells between the epidermis and the dermis determine the animal's coloration.

VESTIGIAL LIMBS
In some snake families, such as dwarf boas and some boas and pythons, the vestigial back limbs are still evident, indicating an evolutionary link between snakes and lizards.

form a single, large opening. In many reptiles, bone growth does not end when sexual maturity is reached, which means that some long-lived adults can grow extremely large.

SENSES

Reptiles' senses are better developed than those of amphibians, and some have sense organs that are not found anywhere else in the animal kingdom. The eyes are often large and well developed, although many snakes have poor sight, and in some burrowing squamates the eyes are reduced or absent. Turtles and tortoises, crocodilians, and most lizards have mobile eyelids, while in snakes and some lizards they are immovable. Lizards and tuataras have a light-sensitive area on top of their skull, known as the third eye, which is thought to control diurnal

EYE COVERINGS

Crocodilians, turtles, tortoises, and most lizards have 2 movable eyelids (an upper and a lower) as well as a nictitating membrane. This membrane consists of a transparent fold of skin that is drawn over the eye from the side, providing protection while also allowing the animal to see.

In snakes and some lizards, the lower lid, which is transparent, is fused with the upper one, forming a fixed, transparent protective covering over the eye known as a spectacle or brille.

SHEDDING SPECTACLES
A snake (here, a cross-marked sand snake) sheds the outer layer of its spectacles, or brilles, at the same time as it sloughs the rest of its skin.

ACTIVITY PATTERNS

Maintaining an optimum body temperature is the key to a reptile's survival and lifestyle, as this graph depicting a diurnal lizard's activity patterns shows. During the night and early morning, the animal shelters from the cold in its burrow. Later, as the temperature rises, it has to bask in the sun in order to obtain the energy required to forage. The lizard must seek shelter around noon to prevent overheating, but re-emerges later, as the air cools, to forage once more.

AIR TEMPERATURE
LIZARD'S BODY TEMPERATURE
SHELTERING TO AVOID COLD
BASKING
NORMAL ACTIVITY
SHELTERING TO AVOID HEAT

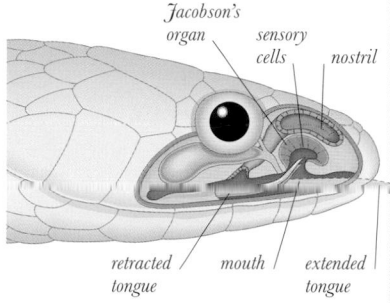

Jacobson's organ   sensory cells   nostril

retracted tongue   mouth   extended tongue

JACOBSON'S ORGAN
Snakes and most lizards extend and flicker their tongue to pick up scent molecules around them. On retracting their tongue, they transfer the molecules to the Jacobson's organ, inside the mouth, where the scents are analyzed.

and seasonal patterns of activity by measuring day length. Reptiles tend to have poor hearing. Some have no external ear opening or middle ear structure at all, and transmit sounds through the skull bones. Taste is not important to reptiles, but smell is highly developed. Some snakes have heat-detecting organs in their faces, and can detect minute temperature changes to help them to locate prey.

TEMPERATURE CONTROL

Since reptiles cannot generate heat internally, they depend on external factors to keep their temperature

within critical limits. At temperatures below their preferred range (in most species, 30–40°C/86–104°F), they will slow down and may act to raise their temperature, perhaps by basking. This involves flattening or angling their body towards the sun, or pressing their underside against a warm rock. At very low temperatures their bodily functions operate at a much reduced rate, although normally this occurs only after the animal has sought shelter, in a burrow or under rocks. In temperate regions, reptiles may shelter for prolonged periods, or hibernate, over winter. Similarly, species from hot, arid climates may shelter for the hottest part of the year, an activity known as aestivation. Reptiles that live in tropical regions rarely need to bask.

REPRODUCTION

Little is known about the courtship behaviour of reptiles, although chemical communication probably plays a large role. Several reptiles vocalize during the breeding season, and males of many lizards and some other groups indulge in visual displays using bright colours, crests, and flaps of skin (dewlaps). These displays serve to establish territories

GIVING BIRTH
The female viviparous lizard retains her eggs inside her body until they are ready to hatch. The 2–12 young break out of their eggs as soon as they are outside their mother's body and can survive without parental care.

as well as to attract females. In most cases, a female is fertilized by a male, although parthenogenesis, in which a female reproduces without the need for fertilization, occurs in several species of lizards and one species of snake.

Most reptiles lay eggs, usually on land, although a significant number of lizards and snakes give birth to live young. Reptile eggs may have a hard shell, like those of birds, but most have a leathery shell that allows water and oxygen to pass through to the developing embryo. Reptiles hide their eggs in burrows, decaying vegetation, or other similar locations. Incubation periods can last from a few days to several months, with the young of some species overwintering in their nest and emerging nearly one year later. Live-bearing species retain the eggs inside their body and in some cases nourish them through a form of placenta. Hatchling and newborn reptiles are very similar to their parents, although their colours and markings may differ. Parental care is rare except in crocodilians, where it may last for 2 years or more. Some lizards also care for their young after birth.

EMERGING FROM HIBERNATION
A group of red-sided garter snakes emerge from their communal burrow, where they have been in hibernation since the onset of cold weather in winter.

TORTOISES AND TURTLES

PHYLUM	Chordata
CLASS	Reptilia
ORDER	Chelonia
FAMILY	11
SPECIES	About 294

TORTOISES AND TURTLES are among the oldest of all living reptiles. They first appeared about 200 million years ago but have evolved little in the intervening time, so that the living species are remarkably similar to those that lived side-by-side with such animals as dinosaurs. Their most distinctive feature is the hard shell that encloses the soft parts of the body, providing protection and camouflage from predators and the elements. Tortoises and turtles have no teeth and instead use sharp jaws to cut their food. They live on land as well as in freshwater and marine habitats (although all species lay their eggs on land). The terrestrial species are commonly referred to as tortoises, while those that live in fresh water are often called terrapins. The term turtle was traditionally reserved for marine species, but most zoologists now use it to refer to all members of the order. Although they are most common in tropical regions, tortoises and turtles are also found in temperate parts of the world. Some marine species undertake long-distance migrations, either in search of food or to reach their nest sites.

FINDING THE SEA
On hatching from their eggs, young sea turtles, such as these green turtles, have to find their way to water. It is thought that they either instinctively travel down the slope of the beach or sense small differences in light levels over the land compared with the sea (the sea being brighter by both day and night). On their way, they must contend with predators such as gulls and crabs.

ANATOMY

All tortoises and turtles have a shell, 4 limbs, and a horny, toothless beak in their jaw. The shell consists of upper and lower parts (known as the carapace and plastron respectively), joined between the front and back legs on each side by a bridge. All parts of the shell have 2 layers: an underlying bony layer and an outer epidermal layer. The outer layer is made of thin, horny plates (scutes), which contain the pigment that gives each species its distinctive coloration. Some species lack scutes and have soft, leathery shells. The shape of the limbs differs between terrestrial and aquatic species: most terrestrial species have short, club-shaped legs, while in aquatic species they are either webbed or shaped like flippers. Since their ribs are fused to the shell (see below), tortoises and turtles cannot move their ribs to draw air into and out of their lungs. Instead, they use muscles at the tops of their legs to provide the necessary pumping action.

DOMED SHELL

STREAMLINED SHELL

SKELETON
Tortoises and turtles have an unusual skeleton. The ribs and some of the vertebrae are fused to the inner surface of the carapace, and the pelvic and shoulder girdles are in an unusual position inside the ribcage. The skull is heavily built, with no openings behind the eye sockets, as there are in other reptiles. The length of the neck varies greatly between species and determines how the head is withdrawn into the shell (see right).

inner bony layer
outer horny scutes
rib
pelvic girdle
carapace
vertebra
plastron
shoulder girdle

SHELL SHAPES
The shape of a tortoise's or turtle's shell reflects its way of life. Terrestrial species, like the Indian starred tortoise (above top), have a high, domed shell that is difficult for predators to bite or crush. Aquatic species, like the red-eared turtle (above), have a low, streamlined carapace that helps them slip easily through water. A few species, such as the pancake tortoise, have a flexible carapace. Others, such as box turtles, have a hinged plastron so that they can close up completely.

SIDE- AND STRAIGHT-NECKED TORTOISES AND TURTLES
Two major subdivisions of tortoises and turtles are recognized, based on the way they withdraw their head into their shell. Side-necked turtles bend their neck sideways, so that the head rests under the edge of the shell. All side-necked turtles are aquatic or semi-aquatic, freshwater species. Straight-necked species (which include all terrestrial and some aquatic forms) have a shorter neck, which they bend into a vertical S-shape inside the shell, so that their head appears to go straight back when it is withdrawn.

head under edge of shell

head drawn inside shell

SIDE-NECKED

STRAIGHT-NECKED

MIGRATION

Some sea turtles migrate long distances from their feeding grounds to the beaches where they lay their eggs. For example, some green turtles feed off the coast of Brazil and nest on Ascension Island, in the eastern Atlantic, which involves a return journey of at least 4,500km (2,800 miles). All the breeding adults arrive at the nesting beaches within a few weeks of each other. Exactly how they navigate is still poorly understood, but they probably use a combination of the earth's magnetic field, the directions of ocean currents, water chemistry, and memory. The greatest distances are travelled by the leatherback turtle, which ranges from tropical seas almost to Arctic waters in pursuit of jellyfish, its preferred food.

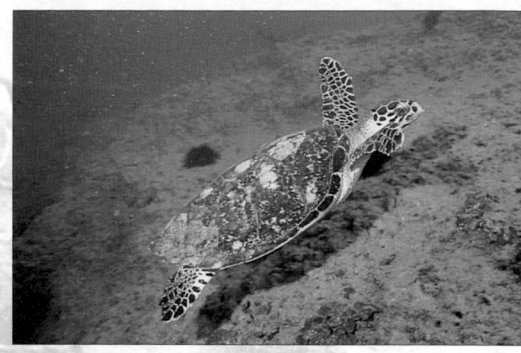

SWIMMING
When swimming, sea turtles (such as this hawksbill turtle) use their front limbs for propulsion and their back limbs for steering. The speed at which they swim varies between species, from 3 to just under 30kph (1½ to 18mph).

REPRODUCTION

All tortoises and turtles lay their eggs on land. Tropical species may lay several clutches throughout the year whereas temperate species lay only once or twice during a breeding season; females can store sperm so that they can continue to lay fertile eggs long after mating. The size of the clutch depends on the size of the species: small turtles lay 1–4 eggs, while the largest species can lay over 100. The green turtle is the most prolific species, laying up to 6 or 7 clutches of more than 100 eggs every 2 weeks. Turtles do not care for their young, although at least one species (the Burmese brown tortoise) stays at its nest for a few days after laying to protect the eggs from predators.

EGGS
The eggs of large turtles and tortoises are almost spherical, but those of small species tend to be elongated. The shells may be either hard and brittle or flexible. This leopard tortoise hatchling is breaking out of its shell using an egg tooth that will be shed shortly afterwards.

NESTING
Some female tortoises and turtles lay their eggs under rotting vegetation or in the burrows of other animals. However, most species (such as this green turtle) use their back limbs to dig a special flask-shaped chamber. Once the eggs have been laid, the sand or soil is carefully replaced and smoothed over to conceal them from predators.

FRESHWATER CARNIVORES
Snapping turtles live in shallow lakes, rivers, and swamps. This common snapping turtle (right) will prey on almost anything that comes within range of its sharp jaws and cavernous mouth, and is small enough to be swallowed.

FEEDING

Turtles and tortoises are too slow-moving to pursue active prey, although a few aquatic species – mostly those that live in murky, clouded water – hunt by ambush, remaining motionless in the hope that a fish or crustacean will inadvertently swim past. Tortoises are primarily herbivorous, grazing or browsing on leaves and fruits. Many species also eat animals, such as caterpillars, that appear incidentally in their food, and most will eat carrion given the chance. Freshwater turtles often start life as insectivores, finding enough small aquatic larvae and other small prey to survive but, as they grow, they switch to a diet consisting mainly of aquatic vegetation. Some marine turtles eat only seaweed as adults, while others feed on invertebrates, including jellyfish, sea urchins, and molluscs.

VEGETARIANS
Tortoises such as this Galapagos tortoise (left) graze on grass and other low plants, or browse on bushes and shrubs. They often eat almost continuously when active and are quick to take advantage of more nutritious fallen fruits and even animal carcasses if they are available.

Carettochelys insculpta
Pig-nosed river turtle

Length 70–75 cm (28–30 in)	
Breeding Oviparous	
Habit Aquatic	

Location S. New Guinea, N. Australia   **Status** Vulnerable

This unusual Australasian turtle lives in fresh water, but has a number of similarities with turtles that live at sea. Its limbs are broad and flipper-like, with relatively few claws, and its grey-green or greyish brown carapace is covered by a layer of soft skin. The shell has a pitted or sculpted surface. Pig-nosed turtles are active foragers, feeding on snails, small fish, and fruits. The pig-like snout gives them their name. They use the snout to breathe while submerged. Females nest in shallow holes on river banks, laying up to 22 thin-shelled eggs.

Chelonia mydas
Green turtle

Length 1–1.2 m (3 1/4 – 4 ft)	
Breeding Oviparous	
Habit Aquatic	

Location Tropical, subtropical, and temperate waters worldwide   **Status** Endangered

This graceful, streamlined swimmer is one of the world's most widespread marine turtles, found in tropical and subtropical waters across the globe. Despite its name, it is not invariably green, but it always has distinctive, light edging to the scales on its head and limbs, and on scutes around the edge of its shield-shaped carapace. The adults graze on sea grass, mangrove roots, and leaves, but young green turtles are more carnivorous, also eating jellyfish, molluscs, and sponges. In order to breed, green turtles make their way to isolated, sandy beaches, sometimes travelling across more than 1,000km (620 miles) of open sea. After mating in shallow water, the females crawl ashore after dark, laying 100–150 eggs in deep pits above the high-tide mark. The

Chelodina longicollis
Common snake-necked turtle

Length 20–25 cm (8–10 in)	
Breeding Oviparous	
Habit Semi-aquatic	

Location E. and S. Australia   **Status** Common

With its small head and extremely long neck, this turtle is one of Australia's most distinctive freshwater reptiles. Together, the head

Chelus fimbriatus
Matamata

Length 30–45 cm (12–18 in)	
Breeding Oviparous	
Habit Aquatic	

Location N. South America   **Status** Locally common

Most turtles actively seek out their prey, but the knobbly shelled matamata lies in wait in shallow, muddy water, using its remarkable camouflage to avoid being seen. Its eyes – set in a triangular head – are

and neck are often longer than the shell, allowing the turtle to lunge at passing prey, such as fish, tadpoles, and crustaceans. Its long neck also allows it to "snorkel" while resting on the beds of slow-moving rivers or streams, or in swamps and lagoons. In summer, it may make extensive overland migrations to find water. If disturbed, this turtle can emit a foul-smelling fluid from its musk glands. The female lays 6–24 brittle eggs in a nest dug in grassy or sandy areas at night or after rains.

small, but it has large external ear-drums and sensory skin flaps that help to detect moving prey. It also has a snorkel-like nose, which allows it to breathe without rising to the surface. Female matamatas lay up to 28 eggs in a single nesting.

frills of skin on neck

knobbly carapace

small head

young hatch in about 6–8 weeks, but before they can reach the water, many are eaten by crabs, gulls, and other seabirds.

LARGE AND STREAMLINED
This large turtle – weighing 65–300kg (145–660lb) – is a superb swimmer, with a smooth shell and powerful flippers. Adults are greenish brown to black, sometimes streaked with dark brown, red-brown, and yellow.

large white plastron

limbs modified into flippers

CONSERVATION

The green turtle has been the main quarry of commercial turtle fishing, which has led to its endangered status. With the disturbance of many breeding beaches, suitable nesting sites are also decreasing. Conservation measures for the green turtle include legal protection from hunting and egg harvesting, patrols of nesting beaches, and artificial incubation of eggs so that hatchlings can be given a head start, being released into the wild once past their most vulnerable stage.

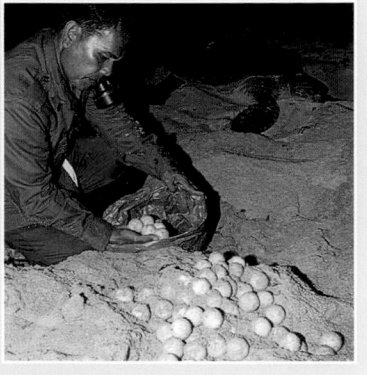

Platemys platycephala
Twist-necked turtle

Length 14–17 cm (5 1/2 – 6 1/2 in)	
Breeding Oviparous	
Habit Semi-aquatic	

Location N. South America   **Status** Common

If threatened, this turtle withdraws into its shell by twisting its head onto its side. Also called the flat-headed turtle, this little-known species has an extremely flat shell that may help it to hide under rocks and debris. The carapace is dark or chestnut-brown, or yellow with brown patches, with 2 ridges along its length. A poor swimmer, the turtle remains in shallow pools and puddles, or moves about on the forest floor. It forages for aquatic insects, worms, snails, and tadpoles. Females lay a single egg in a shallow scrape or directly on the surface, and cover it with rotting leaves.

dark brown head with yellow top

Eretmochelys imbricata
Hawksbill turtle

Length 60–80 cm (23 1/2 – 32 in)	
Breeding Oviparous	
Habit Aquatic	

Location Warmer parts of Atlantic, tropical Indo-Pacific   **Status** Critically endangered

One of the smallest marine turtles, the hawksbill is easily recognized by its carapace, which has a central keel and serrated edges. Seen closely, its scutes are beautifully marked – the reason why this species has been so widely hunted. Hawksbills are long-lived animals, and they are less migratory than other marine turtles. They use their narrow beaks to forage for sponges, molluscs, and other animals on the sea bed and among coral reefs.

smooth, non-overlapping, scutes on carapace

strong flippers

dark carapace with light markings

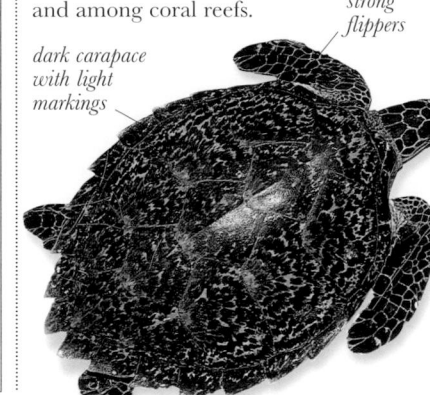

Caretta caretta

Loggerhead turtle

Length 70–100 cm
(28–39 in)
Breeding Oviparous
Habit Aquatic

Location Tropical,
subtropical, and temperate
waters worldwide

Status Endangered

Compared to most other marine
turtles, the loggerhead has a large
head and powerful jaws, capable of
crushing crabs, lobsters, and other
hard-bodied prey. In open water, it
usually floats near the surface, but it
stays near the bottom in estuaries and
bays, surfacing only to breathe.
Loggerheads breed only every 2 years
– or sometimes longer – and lay up to
5 clutches of about 100 eggs.

Dermatemys mawii

Central American
river turtle

Length 50–65 cm
(20–26 in)
Breeding Oviparous
Habit Aquatic

Location S. Mexico to
Central America

Status Endangered

The webbed feet and streamlined
shell of this olive-grey turtle are
adaptations for swimming; it is almost
helpless on land. Females have olive
tops to their heads (yellowish to
reddish brown in males), and very
short tails. They lay 6–20 eggs in
muddy river banks in summer and
autumn, burying them or covering
them with decaying vegetation. While
adults are herbivorous, the juveniles
also eat molluscs, crustaceans and,
probably, fish. This turtle is hunted
by otters and humans.

*oval
carapace*

Lepidochelys olivacea

Olive ridley turtle

Length 50–75 cm
(20–30 in)
Breeding Oviparous
Habit Aquatic

Location Tropical parts of
Atlantic, Indian, and
Pacific oceans

Status Endangered

The smallest sea turtle, the olive ridley
turtle has an unmarked, dark or light
olive carapace. An active forager, it
takes crustaceans, fish, and squid.
It migrates in large numbers to its
breeding grounds. Once, huge flotillas
of turtles would arrive at sandy
beaches to nest; human depredation
has put an end to this spectacle.

Macroclemys temminckii

Alligator snapping
turtle

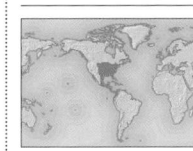

Length 40–80 cm
(16–32 in)
Breeding Oviparous
Habit Aquatic

Location S.E. USA

Status Vulnerable

This formidable animal is the
world's largest freshwater
turtle, with a maximum
weight reputed to be over
100 kg (220 lb). Like the
matamata (see p.368), it
hunts mainly by sitting
and waiting, but it is
equipped with a lure that entices fish
towards its scissor-sharp jaws. At
night, it is an active forager. Males
spend their entire lives at the bottom
of lakes or rivers, but the females
leave the water in spring to lay
clutches of 10–50 spherical eggs,
buried in mud or sand.

ROUGH AND KEELED
*The alligator snapping turtle has a rough shell,
with 3 jagged keels, and an extra row of scutes
on each side.*

*rough
carapace*

*extra row
of scutes*

*massive
head*

Chrysemys picta

Painted turtle

Length 15–25 cm
(6–10 in)
Breeding Oviparous
Habit Aquatic

Location North America

Status Common

This freshwater turtle is one of the
most widespread in North America,
found in lakes, ponds,
and slow-moving
streams and rivers
from the eastern
seaboard to the far
Midwest. There are
4 distinct subspecies:
all have a flattened,
smooth carapace, but
some have red margins
to their shells, a red stripe
on the back, or yellow or red
stripes on the neck. This turtle

*yellow stripes
on neck*

*very smooth
carapace*

Chelydra serpentina

Common snapping
turtle

Length 25–47 cm
(10–18½ in)
Breeding Oviparous
Habit Aquatic

Location E. and C. North
America, Central America,
N.W. South America

Status Common

This large, dark turtle is quick-
tempered, biting ferociously
when threatened. It has a
massive head, a small
plastron, and a rough
carapace, sometimes
covered with algae.

Inhabiting fresh and brackish water,
preferably with plenty of plant life, it
often lies buried in mud with only the
eyes and nostrils exposed. It sits and
waits for prey by day but at night is an
active forager, lunging with its mouth
open to engulf small mammals, birds,
fish, invertebrates, and plants. Mating
may occur without preliminaries, but
rituals in which the male and female
face each other with necks extended
have also been seen. Females lay one
clutch of 20–30 eggs in a season, in
a flask-shaped nest, often in a muskrat
lodge; they can retain sperm from one
breeding season to the next.

*3 raised keels
on carapace*

FISH LURE

During the day, this turtle lies
with its jaws open and lures
fish by wriggling a small, pink,
worm-like structure on the floor
of its mouth. The hooked
upper and lower beaks can
deliver a powerful bite.

basks for long periods, especially
in the morning, several individuals
piling up on one another to bask on a
favourite log. It is omnivorous and
forages actively by day, sleeping on
the bottom at night. Hunted by birds,
fish, and racoons, it defends itself by
hiding its head inside its shell, or by
burying itself in mud. The painted
turtle hibernates for variable lengths
of time, depending on its location.
Most females lay 2–20 eggs in 3
clutches in chambers dug in sandy
soil; not all lay eggs every season.

Clemmys guttata

Spotted turtle

Length 10–12.5 cm
(4–5 in)
Breeding Oviparous
Habit Semi-aquatic

Location S.E. Canada,
E. USA

Status Vulnerable

This small turtle has yellow spots on
its black carapace, head, neck, and
limbs. An omnivorous active forager,
it is itself hunted by birds and small
mammals. In summer, the semi-
aquatic spotted turtle aestivates in the
muddy beds of lakes and rivers or in
muskrat burrows; it may hibernate in
similar sites in winter.

Geochelone nigra

Galapagos tortoise

Length	Up to 1.2 m (4 ft)
Breeding	Egg-laying
Habit	Terrestrial

Location Galapagos Islands	**Status** Vulnerable

The Galapagos tortoise, the largest living tortoise, has a huge carapace, massive limbs, and a long neck. There is a great variation in shell shape and overall size depending on which of the Galapagos Islands the tortoise originates. They spend most of their time grazing in small herds and basking in pools or mud wallows. During the breeding season, males become territorial and start looking for mates, while females create nests in chambers in the ground to lay their eggs. Despite some individuals living to well over 100 years of age, the Galapagos tortoise remains vulnerable. The main threat is depredation at a young age by introduced species such as black rats and cats, as well as competition for vegetation from goats and cattle.

GENERALIZED HERBIVORE
The Galapagos tortoise has strong, toothless jaws well adapted to feeding on whatever type of vegetation it can find, including tough cacti.

saddle-backed shell allows greater neck movement

long neck can stretch up to shrubs and bushes

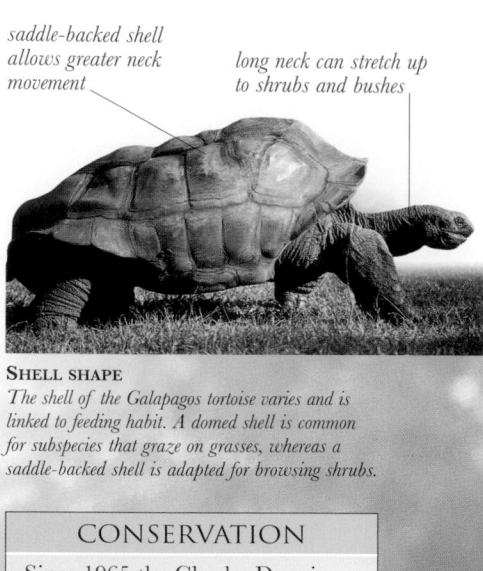

SHELL SHAPE
The shell of the Galapagos tortoise varies and is linked to feeding habit. A domed shell is common for subspecies that graze on grasses, whereas a saddle-backed shell is adapted for browsing shrubs.

CONSERVATION

Since 1965 the Charles Darwin Research Station has been running a breeding and repatriation programme to boost the dwindling population of Galapagos tortoises. By March 2000 the programme had released 1,000 tortoises.

UNMISTAKABLE GIANT
The impressive size of the Galapagos tortoise is probably due to its adaptation to living in a difficult environment with an unreliable food supply — the larger the tortoise, the more nourishment it can store.

MATING

The courtship technique of the male Galapagos tortoise is uncompromising. Having located a suitable female, he rams her into submission, nipping her legs to further immobilize her. He then climbs onto her back to mate.

Dermochelys coriacea
Leatherback turtle

	Length 1.3–1.8m (4¼–6ft)
	Breeding Oviparous
	Habit Aquatic

Location Tropical, subtropical, and temperate waters worldwide	Status Critically endangered

Weighing up to 800kg (1,770lb), the leatherback is by far the largest of the marine turtles, and one of the animal world's greatest oceanic travellers. Tagged individuals have been known to cross the Atlantic. Leatherbacks also range into cold waters at high latitudes, thanks to a limited ability to generate body heat. Physically, they differ from other marine turtles not only in size, but also in having a narrow, leathery shell, and flippers without claws. Leatherbacks feed primarily on jellyfish, and their throats have backward-pointing spines to prevent their slippery prey escaping. They usually feed near the surface, but they can dive down to 400m (1,300ft) – perhaps as much as 1,000m (3,300ft) – holding their breath for up to half an hour. Although normally found in the open ocean, leatherbacks gather inshore during the breeding season, when males scramble for females arriving to lay their eggs. As with other marine turtles, the females have a strong homing instinct that guides them to a particular stretch of beach, and they lay eggs on dark, moonless nights to avoid being seen. The hatchlings' shells are covered with small, pearly scales, which soon disappear.

CONSERVATION

Leatherback turtles dig their nests on sandy, tropical beaches. In many places, these beaches are subject to development and disturbance by tourists, and the turtle nests are raided by humans and feral dogs as well as by natural predators. Measures to protect leatherbacks include collecting the eggs, so that they may be incubated in safety and the young released into the ocean.

5 ridges along leathery carapace

EXCURSION ASHORE
Emerging from the sea to lay eggs, a female leatherback laboriously hauls herself up a beach. On average, she may spend less than 2 hours a year on land.

large head on short, thick neck

Clemmys insculpta
Wood turtle

	Length 14–19cm (5½–7½ in), max 23cm (9in)
	Breeding Oviparous
	Habit Semi-aquatic

Location S.E. Canada, E. USA	Status Vulnerable

Besides being an unusually good climber, the wood turtle is a natural wanderer, and is often found crossing fields or roads, particularly after rain. The scutes on its carapace have concentric ridges, and their sombre colours provide it with effective camouflage. Wood turtles are omnivores, eating worms, slugs, insects, and tadpoles, as well as leaves, berries, and other plant food. They mate in water, and the females lay 7–8 eggs a year in flask-shaped nests. At one time, the adult turtles were caught in large numbers for food.

pyramid-shaped scutes

reddish skin on neck and legs

Cuora flavomarginata
Yellow-marginated box turtle

	Length 10–12cm (4–4¾ in)
	Breeding Oviparous
	Habit Semi-aquatic

Location E. Asia (including Taiwan and Ryukyu Islands)	Status Endangered

yellow stripe

The yellow-marginated box turtle gets its name from a hinged plastron that allows it to be completely boxed in when withdrawn into its shell, and yellow stripes that run down its back and from each eye to the neck. It spends much of its time in rice paddies, ponds, and streams, but also basks and wanders on land. This turtle feeds on fish, crustaceans, worms, and fruits. Nesting in soil or sand, females may lay several clutches of one or, occasionally, 2 eggs a year.

Cyclemys dentata
Asian leaf turtle

	Length 15–24cm (6–9½ in)
	Breeding Oviparous
	Habit Semi-aquatic

Location S.E. Asia	Status Lower risk

The oval carapace of this turtle has a serrated edge near the tail, and is coloured light to dark brown (sometimes mahogany) with faint dark markings. The plastron, which is hinged, is more boldly marked with lines radiating from the centre of each scute. As with most turtles, males are smaller than females and have longer, thicker tails. The Asian leaf turtle is an omnivore, feeding on invertebrates and tadpoles, as well as plants. Very active both on land and in water, it prefers shallow streams in either mountains or lowlands. When danger threatens, it may withdraw into its shell or it may dive to the bottom of the water and hide in the mud. The female digs a chamber in which she lays up to 5 clutches of 2–4 relatively large eggs each year; unusually, in this turtle the plastron becomes flexible to allow eggs to be laid. Hatchlings are more aquatic than adults and have spines or points – which disappear with age – around the edge of the carapace, perhaps to deter predators.

raised keel down centre

dark brown carapace

light to reddish brown legs

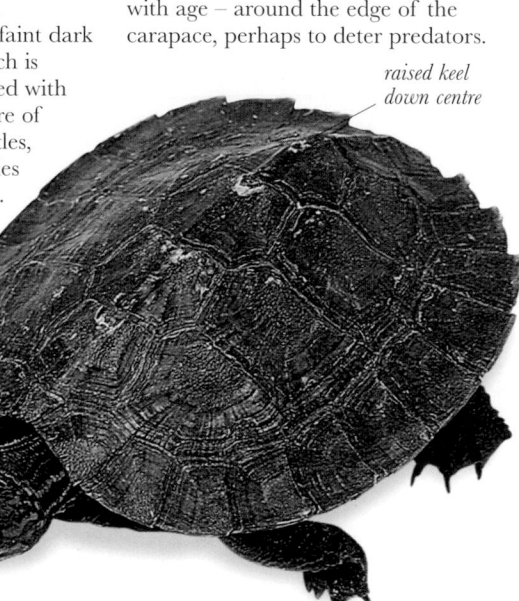

reddish brown head

Emys orbicularis
European pond turtle

Length 15–20 cm
(6–8 in),
max 30 cm (12 in)
Breeding Oviparous
Habit Semi-aquatic

Location N. Africa, Europe, W. Asia
Status Lower risk

spotted shell

This olive, brown, or black turtle is one of just 2 freshwater species that live in Europe. Its carapace is smooth, and a hinge towards the front of the plastron allows it to be raised when the turtle withdraws its head, although adults cannot close the shell completely. These turtles spend much of their time basking on stones or logs, but they dive at the first sign of danger. They eat frogs, fish, and other small animals, in water or on land.

Terrapene carolina
Carolina box turtle

Length 10–21 cm
(4–8 1/2 in)
Breeding Oviparous
Habit Semi-terrestrial

Location E. USA
Status Lower risk

Known to have lived for as long as 100 years in captivity, in the wild the Carolina box turtle prefers damp habitats, such as meadows, floodplains, or moist, forested areas. Very active in the mornings, particularly after rain, this omnivore forages for slugs, worms, mushrooms, or berries. To escape the heat, it retreats into cover or mud at midday and in midsummer, and survives the cold winter months by hibernating. Females can lay fertile eggs for several years following a single mating. The adults have few natural predators, but the young turtles fall prey to mammals such as racoons, and birds of prey.

BOXED IN

The front section of the Carolina box turtle's plastron is hinged, so that when the turtle withdraws its head into the shell, the front of the plastron can be raised, effectively shutting the animal into a protective "box".

TAN TURTLE
This individual has a tan shell, but some Carolina box turtles are mainly orange or yellow, and some may be quite dark, with radiating yellow lines.

domed shell

Trachemys scripta elegans
Red-eared turtle

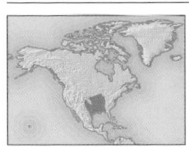

Length 20–30 cm
(8–12 in)
Breeding Oviparous
Habit Aquatic

Location S. Central USA
Status Lower risk

yellow-marked green shell   *red stripe behind eyes*

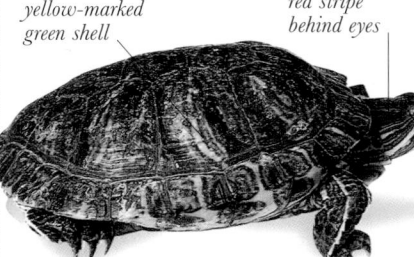

Identified by a prominent red stripe behind its eyes, this freshwater turtle is one of over 2 dozen species that live in lakes, rivers, and creeks in North America. Like many of its close relatives, it is fond of basking in sunshine, sometimes forming stacks where basking sites are in short supply, but it is quick to dive into the water if disturbed. As an adult, it lives almost entirely on plant food, but its young eat tadpoles and aquatic invertebrates.

Kinosternon flavescens
Yellow mud turtle

Length 12–16 cm
(4 3/4–6 1/2 in)
Breeding Oviparous
Habit Mainly aquatic

Location S. Central USA, N. Mexico
Status Common

In addition to withdrawing into its shell for defence, the yellow mud turtle can also release foul-smelling musk from 2 pairs of glands under its carapace, and may occasionally

bite. It has especially powerful jaws for crushing its prey, which includes snails, worms, insects, and tadpoles. Primarily diurnal, it spends most of its time in shallow, slow-moving bodies of water in grassland. In midsummer it shelters from heat in muskrat dens or by burrowing under leaf litter. It hibernates in the cooler parts of its range.

smooth, rounded shell

dark border to scutes

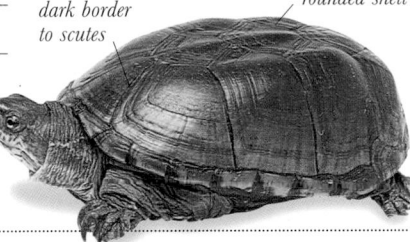

Sternotherus odoratum
Stinkpot

Length 8–13 cm
(3 1/4–5 in)
Breeding Oviparous
Habit Mainly aquatic

Location S. and E. USA
Status Common

As a form of defence, this turtle expels a foul odour from its musk glands, hence the name "stinkpot". It can also bite viciously. Rarely leaving quiet, shallow, muddy-bottomed waters, it is often

covered in algae. It is active by day, but may also hunt at night, and a pair of sensory barbels on the chin may help it to find food – mainly insects, molluscs, plant material, or carrion. It is eaten by bald eagles and red-shouldered hawks, alligators, and fish such as the largemouth bass. Females nest under tree stumps or in the walls of muskrat lodges, laying one to 5 eggs at a time.

3 slight keels on carapace
barbels on chin

Pelomedusa subrufa
African helmeted turtle

Length 20–32 cm
(8–12 1/2 in)
Breeding Oviparous
Habit Semi-aquatic

Location Africa (south of Sahara)
Status Common

flattened carapace

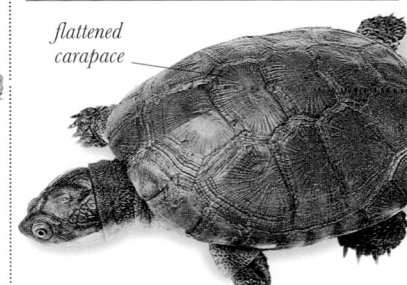

The sluggish African helmeted turtle lives in rain pools and watering holes in Africa's open country. During the rainy season, it wanders from pool to pool foraging for frogs, tadpoles, molluscs, invertebrates, and carrion. It may aestivate in dry conditions by burying itself in mud. If threatened by crocodiles or other carnivores, it withdraws into its shell, helped by a hinged plastron, and may also discharge a strong-smelling musk and the contents of its cloaca.

Pelodiscus sinensis
Chinese soft-shelled turtle

Length 15–30 cm
(6–12 in)
Breeding Oviparous
Habit Aquatic

Location E. Asia
Status Vulnerable

With its long snout and tube-like nostrils, the Chinese soft-shelled turtle can "snorkel" in shallow water. It also has webbed feet for swimming. When resting, it lies at the bottom, buried in sand or mud, lifting its head to breathe or snatch at prey. It forages at night, taking crustaceans, molluscs, insects, fish, and amphibians. If threatened, it cannot withdraw completely into its shell, but it can give a vicious bite.

pancake-like carapace with no bony scutes

REPTILES

Geochelone carbonaria
Red-footed tortoise

Length 40–50 cm
(16–20 in)
Breeding Oviparous
Habit Terrestrial

Location N. to C. South America — Status Locally common

The genus *Geochelone* includes the world's biggest tortoises, as well as many smaller species, such as this one from South America. Its legs are marked with

reddish edge to forelimb

splashes of red or yellowish orange, and the shell – in adult males – is unusually elongated, with a constriction in the middle. Like most of its relatives, the red-footed tortoise lives mainly on plant food, such as leaves and fallen fruits, but it also scavenges animal remains. Females lay clutches of 2–15 nearly spherical eggs, several times a year.

growth rings on scutes

reddish or yellow blotch on scutes

Geochelone elegans
Indian starred tortoise

Length Up to 28 cm
(11 in)
Breeding Oviparous
Habit Semi-aquatic

Location S. Asia — Status Vulnerable†

With its star-shaped markings and knobbly carapace, this Asian species is one of the world's most distinctive tortoises. Each of its scutes rises to a rounded point, and the

carapace as a whole is strongly domed. The Indian starred tortoise needs plenty of water, and it is most active during the monsoon season. In drier weather, it stirs only in the morning and late afternoon. Each year, the female lays several clutches of up to 10 eggs in a flask-shaped chamber 10–15cm (4–6in) deep.

scutes rising to rounded points

yellowish brown lines radiating to edge of scutes

Geochelone pardalis
Leopard tortoise

Length 30–70 cm
(12–28 in)
Breeding Oviparous
Habit Terrestrial

Location E. to southern Africa — Status Unconfirmed

The carapace of the leopard tortoise is yellowish, with scattered dark markings that are bolder, although fewer, in hatchlings. In some individuals, it can grow to 70cm (28in) long. This tortoise lives in a variety of habitats,

from sandy, coastal scrub to grassland and semi-desert, although it is scarce in very dry deserts. It may bury itself to hibernate in cool parts of its range, and to aestivate in the hotter parts. The leopard tortoise is herbivorous, grazing on grasses, fallen fruits, fungi, and the leaves of succulents. In the breeding season the female digs a flask-shaped pit with her hindfeet, urinating to soften hard soil, and lays 3–6 clutches of 5–30 eggs at monthly intervals. Hatchlings may have to wait in the nest chamber for several weeks for rain to soften the ground before they can burrow to the surface. The young become prey to monitor lizards, storks, crows, and small mammals.

domed carapace with vertical sides

growth rings on scutes

scattered dark markings

Gopherus agassizii
Desert tortoise

Length 25–36 cm
(10–14 in)
Breeding Oviparous
Habit Terrestrial

Location S.W. USA, N.W. Mexico — Status Vulnerable

This tortoise's thick, shovel-shaped forefeet are specially adapted for digging burrows in which to escape the extreme heat of the desert during the day and the cold at night. It shelters individually in burrows in very hot or drought conditions but may gather in large, communal dens dug into gravel banks to escape the cold. It lives for at least 50 years, mainly eating cacti and grasses, but also feeds on insects. Males have longer projections at the front of their plastrons and may attack each other on sight in the breeding season.

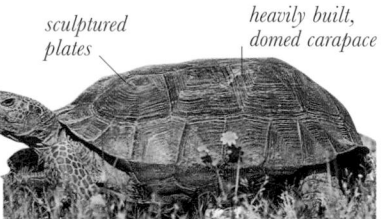

sculptured plates

heavily built, domed carapace

Kinixys erosa
Serrated hinge-back tortoise

Length 25–32 cm
(10–12 ½ in)
Breeding Oviparous
Habit Semi-terrestrial

Location W. to C. Africa — Status Unconfirmed

A hinge towards the back of the carapace enables this tortoise to shut itself in its shell completely. A fair swimmer, it seeks out marshes and

river banks in its forest habitat, but when on land, it spends much of the time buried beneath roots and logs. It is an omnivorous species, feeding on plants, fruits, and invertebrates, as well as carrion. In a breeding season, females lay several clutches of up to 4 eggs on the ground, covering them with leaves.

hinged carapace

flared marginal scutes

Malacochersus tornieri
Pancake tortoise

Length 14–17 cm
(5 ½ – 6 ½ in)
Breeding Oviparous
Habit Terrestrial

Location E. Africa — Status Vulnerable

The carapace of this tortoise is not only extremely flat but also very flexible, owing to openings in the underlying bones. This allows the tortoise to squeeze into narrow crevices in order to escape predatory

Homopus signatus
Speckled padloper

Length 6–8 cm
(2 ¼ – 3 ¼ in)
Breeding Oviparous
Habit Terrestrial

Location W. South Africa — Status Lower risk

This African species is the smallest tortoise in the world, with males being particularly tiny. Their small size makes speckled padlopers vulnerable to carnivorous mammals and birds, but it also enables them to squeeze under rocks to escape attack, and to shelter from the sun. The flattened shell is brown, orange-red, or salmon-pink, with fine black markings. Speckled padlopers live in dry habitats, and forage for small, succulent plants.

mammals or birds; it can also wedge itself in, for safety, by digging in its foreclaws and rotating the forelimbs, making extraction difficult. It forages, mainly in the mornings, for grasses, leaves, and fruits, but never moves far from the rocky outcrop in which it spends the night. In summer, this tortoise crawls under a flat stone and aestivates to avoid the heat. Females lay several clutches of a single egg throughout the breeding season.

elongated, yellow to tan carapace

Testudo hermanni
Hermann's tortoise

Length 15–20 cm (6–8 in)	
Breeding Oviparous	
Habit Terrestrial	

Location S.E. Europe, Mediterranean islands **Status** Lower risk

Easily confused with the spur-thighed tortoise (see right), this tortoise is the smallest of 3 species that live in southern Europe. It has a domed and slightly lumpy carapace, coloured yellow, olive, or brown, with a scattering of irregular dark markings. As with most tortoises, the males are slightly smaller than the females, and have longer tails and concave undersides. Hermann's tortoise is largely a vegetarian, feeding on fruits, flowers, and leaves, although it

also eats slugs, snails, and animal remains. It lives in places with dense plant cover, forcing its way through the vegetation on its short but powerful legs. It is not active during the hottest part of the day in summer, and in the extreme south it becomes dormant. Where winters are cold, it also hibernates for several months each year. In common with most tortoises, this species hisses and grunts, particularly during the breeding season. Having mated, females dig flask-shaped nesting chambers, where they lay up to 12 eggs. Like the spur-thighed tortoise, this species was once collected in large numbers for the pet trade, but is now fully protected.

rounded, domed carapace

scaly forelimbs

irregular dark markings

Testudo graeca
Spur-thighed tortoise

Length 20–25 cm (8–10 in)	
Breeding Oviparous	
Habit Terrestrial	

Location S. Europe, S.W. and W. Asia, N.W. Africa, Mediterranean islands **Status** Vulnerable

Conspicuous spurs on the hindlegs give this tortoise its name, and help to distinguish it from Hermann's tortoise (see left), which overlaps its range. Its shell is similar to the Hermann's but smoother. It lives in grassy places and sand dunes, and feeds on leaves, fruits, and occasionally carrion and mammal dung. The females lay clutches of up to 12 eggs.

Chersina angulata
Angulate tortoise

Length 15–20 cm (6–8 in)	
Breeding Oviparous	
Habit Terrestrial	

Location Southern Africa **Status** Locally common

Also known as the bowsprit tortoise, this species has a high, domed carapace. It eats grasses and succulents – hibernating during the winter – and, in spite of its armour, provides food for monitor lizards, carnivorous mammals, and even birds of prey. Unusually, the males are slightly larger than the females, and fight with rivals at the onset of the breeding season. Females lay one, or rarely, 2 eggs.

black triangles

TUATARAS

PHYLUM	**Chordata**
CLASS	**Reptilia**
ORDER	**Rhynchocephalia**
FAMILIES	**1 (Sphenodontidae)**
SPECIES	**2**

TUATARAS ARE the only surviving representatives of a group of reptiles that flourished over 200 million years ago. Found on 2 groups of small islands off the coast of New Zealand, they look similar to lizards. Tuataras live in burrows and are largely nocturnal. Compared to other reptiles, they are remarkably tolerant of cold, remaining active in temperatures as low as 10°C (50°F). They grow slowly, breed infrequently, and live to a great age (possibly over 100 years).

ANATOMY
Tuataras have a large head, a long tail, and well-developed limbs. They have a different skull structure to lizards (see below), and unlike lizards they have no eardrums, middle ear, or external male sexual organ. A "third eye" is situated on the top of the head, but there is no evidence that this is functional.

bony arch

SKULL SHAPE
There are 2 openings towards the back of the tuatara's skull; most lizards have only one such opening. The teeth are not separate structures but serrations along the edges of both the upper and lower jaws.

openings in back of head

FEEDING
Tuataras are almost entirely insectivorous and feed on the large numbers of scavenging beetles and crickets that are attracted to colonies of seabirds. They also eat other invertebrates, small lizards and, occasionally, the eggs and chicks of birds whose nest holes they share.

BASKING
Although usually nocturnal, tuataras may occasionally be seen basking on rocks on sunny days. Due to a low metabolic rate, they draw breath infrequently, and when at rest may take only one breath an hour.

BURROWING
Tuataras either dig their own burrows or inhabit the nest holes excavated by breeding seabirds, such as small petrels. They usually hunt for food just outside the entrance of the burrow.

Sphenodon punctatus
Tuatara

Length 50–60 cm (20–23 1/2 in)	
Breeding Oviparous	
Habit Burrowing	

Location New Zealand (coastal islands) **Status** Locally common

With its spiny crest and loose, scaly skin, the tuatara looks deceptively similar to some iguanas (see p.406–7). It was once widespread in New Zealand, but now survives almost entirely on small, offshore islands, where it is protected from introduced mammals. Adult tuataras feed after dark, with spiders, insects, and worms making up the bulk of their diet. They grow very slowly, and females do not breed until they are at least 20 years old. The eggs spend a year developing inside the female's body, and after being laid, they take at least a year to hatch. Until recently, this species was thought to be the only member of the tuatara family. However, genetic analysis has shown that a group of tuataras on North Brother Island form a distinct species, now known as *Sphenodon guentheri*.

reduced crest on female

SNAKES

PHYLUM	Chordata
CLASS	Reptilia
ORDER	Squamata
SUBORDER	Serpentes
FAMILIES	18
SPECIES	About 2,900

CLASSIFICATION NOTE

Snakes are closely related to lizards and amphisbaenians. There is much debate over the relationships between different types of snakes, and there is no universally accepted system of classification. Most controversy concerns the number of families. While most authorities recognize 18, others consider some of these to be subfamilies and arrive at a smaller number, combining boas and pythons in a single family, for example.

SNAKES ARE FORMIDABLE and highly evolved predators. Although they have no limbs, no eyelids, and no external ears, these versatile animals move with ease and find their prey using sophisticated senses. All snakes eat other animals, ranging from ants to antelopes, some of which they subdue by constriction or by delivering a venomous bite from specialized fangs. Although snakes cannot chew their food, the bones in their skull are lightly constructed and loosely connected, so the jaws can be opened wide and the prey swallowed whole. Snakes have established themselves on all the world's major landmasses (except Antarctica) as well as on many oceanic islands. Only about one in 10 are venomous and of these only a small proportion represent a threat to humans.

ANATOMY

The shape of a snake's body usually reflects where it lives: climbing species tend to be long and thin; burrowers are often short and stout with short tails and blunt snouts; and sea snakes have flattened, paddle-shaped tails. Unlike other reptiles, snakes have a single row of scales on their underside, the ventral scales, which are usually wide, and smaller scales on the upper surfaces. Some species have large, regular plates on the head, while others have small, fragmented scales. Each eye is covered with a transparent scale, the brille or spectacle, which is replaced when the snake sheds its skin. All scales can be smooth, ridged (keeled), or granular. The internal organs are modified to fit the elongated body. Paired organs are staggered within the body cavity, and only one lung may be functional, with the other reduced in size. Sea snakes have an enlarged lung, part of which forms a buoyancy chamber.

PRIMITIVE SNAKE *short, heavy jaw*

REAR-FANGED SNAKE *fang at rear, below eye*

FRONT-FANGED SNAKE *fixed hollow fang*

ribs along body

skull

vertebrae

no ribs in tail

SKELETON
A snake has up to 400 vertebrae that articulate on each other to give a highly flexible skeleton. Each vertebra has 2 wing-like processes to keep the spine from twisting. Ribs are attached to the vertebrae in the body but are absent in the tail. Some species have back-limb girdles and primitive snakes have vestigial back limbs in the form of small spurs, but no snakes have front-limb girdles.

SKULLS AND TEETH

Primitive snakes have a heavy skull with few teeth. Most other snakes have a lighter skull with loosely connected jawbones that can move apart. Teeth are fixed to the upper or lower jaw or to the roof of the mouth. Venom-injecting fangs may be at the front or rear of the mouth. In some front-fanged snakes, they are hinged to the upper jaw.

SENSES

Snakes have poor eyesight and hearing, and rely instead on other senses. A well-developed sense of smell is supplemented by the Jacobson's organ (see p.365), and active snakes flick their tongue constantly to sample their surroundings. Pit vipers and some boas and pythons can detect small changes in air temperature using organs on their face known as heat pits.

heat pit between eye and nostril

HEAT PITS

tongue extends through slot in closed upper jaw

FORKED TONGUE

SWALLOWING A LIZARD

1 STEALTHY APPROACH
A vine snake slides slowly and carefully along a branch towards an unsuspecting gecko.

2 THE ATTACK
In one swift movement, the snake arches over the gecko and, using its rear fangs, injects its venom.

3 HOLDING ON
Although the gecko falls from its perch, the snake's tail is wrapped around a branch and it holds on.

4 SHIFTING GRIP
Once the gecko is dead, the snake begins to manoeuvre it into position for swallowing.

MOVEMENT

Snakes have evolved several ways of moving around (see right) to compensate for not having legs. The method used involves different uses of their ribs and the muscles attached to them, but also depends on the snake's weight, the speed at which it needs to move, and the type of surface. Most snakes can use different types of locomotion as the need arises. As well as the methods shown here, some vipers use a technique called side-winding when crossing loose sand (see p.398).

LINEAR PROGRESSION

Heavy snakes and snakes that are moving slowly or stalking prey sometimes progress in a straight line, hooking the trailing edges of the ventral scales over surface irregularities. This occurs in a wave-like sequence, as the snakes pulls itself forward.

LATERAL UNDULATION

Using each point of its body in turn to push against irregularities in the surface, the snake wriggles from side to side. This method is also used for swimming and sometimes climbing.

CONCERTINA MOVEMENT

When moving through their tunnels, burrowing snakes expand one end of the body to jam it against the tunnel walls while the other end is thrust forwards.

HUNTING AND FEEDING

Some snakes are specialized feeders, while others eat a wide range of prey, including other snakes. Small animals or ones that cannot fight back are simply grasped and then swallowed; animals that are larger or better able to defend themselves must be subdued with venom or by constriction before being eaten. A large meal may take a long time to digest, starting in the mouth with powerful salivary juices. Venom itself is a cocktail of modified digestive juices that usually acts quickly on the nervous system or blood tissue of the prey.

SWALLOWING

Flexible jaws and highly elastic skin allow snakes, like this common egg-eater, to eat prey that is larger than their own head. A large meal can take several hours to swallow.

CONSTRICTION

A constrictor kills its prey by suffocation. Each time its victim breathes out, this rock python tightens its coils until breathing stops.

REPRODUCTION

Most snakes lay eggs but a significant number give birth to live young. Species from temperate regions usually mate in spring, soon after emerging from hibernation, and produce eggs or young in summer. Tropical species may breed in response to rainfall and sometimes have a long breeding season, laying several clutches of eggs each year. Males and females find each other using chemical scent trails, and courtship is rarely elaborate. Snakes show little parental care, although pythons and a few other species coil around their incubating eggs.

EGGS AND LIVE YOUNG

Young snakes, like this monocled cobra, break out of their shell using a sharp, temporary egg tooth. The snake is often coiled tightly inside its shell, and may be up to 7 times longer than the egg.

SWALLOWING HEAD-FIRST

gecko is swallowed head-first because the s follow more easily in this direction.

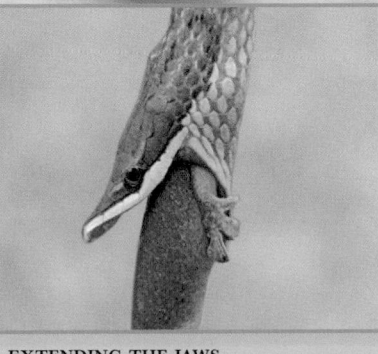

6

EXTENDING THE JAWS

The snake opens its highly flexible jaws and uses its rear fangs to pull the gecko into its mouth.

7

MUSCULAR ACTION

When most of the prey is in the throat, wave-like muscle contractions force it into the stomach.

8

DIGESTION

Having completely swallowed its prey, the snake searches for a quiet place to digest its food.

BOAS, PYTHONS, AND RELATIVES

PHYLUM	Chordata
CLASS	Reptilia
ORDER	Squamata
SUBORDER	Serpentes
SUPERFAMILY	Henophidia
FAMILIES	11
SPECIES	149

BOAS AND PYTHONS are powerful constricting snakes. They include the world's largest snakes, among them the anaconda (a boa) and the reticulated, African, and Indian pythons. Boas and pythons do not reproduce in the same way. They also have different distributions: pythons are found in Africa, Asia, and Australasia, while boas occur mainly in the Americas with a smaller number of species in Africa, Asia, and on some Pacific islands. The relatives of boas and pythons, most of which are also constrictors, include the iridescent sunbeam snakes and the aquatic file snakes.

descended. These include a back-limb (pelvic) girdle and, in most species, the remains of back limbs in the form of small claws or spurs. All species have 2 functioning lungs. Several species of boas and pythons have heat-sensitive pits in the scales bordering their mouths, which they use to locate prey in the dark.

Some of the relatives of the boas and pythons are smaller but otherwise outwardly similar. These include the wood snakes and the Round Island boas. A few species are brightly coloured and others, such as the sunbeam snakes, have scales that are highly iridescent. The file (or wart) snakes are specialized for an aquatic way of life and are effectively helpless on land. Their scales are granular and rough to the touch, an adaptation for gripping and constricting the fish on which they feed.

CONSERVATION

Several species of boas and pythons, especially the largest ones, have been adversely affected by human activities; these large predators require extensive areas of undisturbed land over which to hunt. In addition, many are relatively conspicuous and sluggish and therefore easy targets for hunters, who kill them for their flesh, their skins, or simply out of prejudice. However, some species perform a useful service to humans by controlling the populations of rodents that are attracted to towns and villages.

ANATOMY

The snakes in this group are often regarded as being relatively primitive. Their skulls are heavier and their jaws more rigid than those of the more advanced snakes (the colubrids, elapids, and vipers). They have also retained several anatomical features from the limbed animals from which they are

CONSTRICTION

Although boas, pythons, and their relatives are not the only snakes that kill by constriction (some colubrids also use this method), most of the constrictors are in this group. When the snake has selected its prey, it throws one or more coils around the animal's body. Each time the victim breathes out, the snake tightens its grip. The victim eventually dies by suffocation, either because it cannot draw breath or because its heart cannot pump blood, rather than by being crushed.

Once the prey is dead, the snake loosens its hold and

searches for the head. It swallows this end first, gradually releasing the rest of the body from its coils. Constriction is most effective for killing birds and mammals because, being warm-blooded, they have to breathe relatively frequently.

REPRODUCTION

One of the most significant differences between boas and pythons lies in the way they reproduce. Boas (with the possible exception of one species) bear live young – as do wood snakes, pipe snakes, file snakes, and shield-tailed snakes. Pythons, sunbeam snakes, and Round Island boas lay eggs. Pythons are among the small number of snakes that show parental care. The female coils around the clutch of eggs – which can number up to 100 in large species – throughout the 2–3 month period of incubation, to protect them from predators. One or 2 species of python are unique among snakes in that they can regulate the temperature of their eggs by producing metabolic heat.

FEEDING
A substantial meal can sustain a snake for several weeks or even months. Large constrictors, such as this green anaconda, are capable of killing a variety of animals, including wading birds, deer, young jaguars, and even caimans (as shown here). Despite their great size and weight, anacondas move easily in water and usually ambush their prey in the shallows.

Anilius scytale
South American pipe snake

Length 70–90 cm
(28–35 in)
Breeding Ovoviviparous
Habit Burrowing

Location N. South America

Status Locally common

The sole member of the family Aniliidae, this little-known snake leads a secretive, underground life, its cylindrical body shape adapted to its burrowing habit. The striking red and black coloration mimics that of venomous coral snakes from the same region, whose defensive behaviour the South American pipe snake also imitates. It hunts in burrows, and its diet is thought to consist of small vertebrates, including snakes. It may be preyed upon by other snakes.

Cylindrophis ruffus
Red-tailed pipe snake

Length 70–100 cm
(28–39 in)
Breeding Viviparous
Habit Burrowing

Location S.E. Asia

Status Locally common

The underside of this slender snake's tail is red, hence its name. The upper body is black, often with a red mark behind the head. Living mainly underground, this pipe snake occasionally takes to water and may be found in rice paddies and swamps. It actively forages for smaller snakes and eels. Hunted by larger snakes, birds, and mammals, its defence is to raise its tail and expose the red underside, possibly to mimic venomous species such as the krait, or to deflect attack away from its head.

red underside of tail *smooth, shiny scales*

Pseudotyphlops philippinus
Large shield-tailed snake

Length 45–50 cm
(18–20 in)
Breeding Viviparous
Habit Burrowing

Location Sri Lanka, below 1,000 m (3,300 ft)

Status Locally common

This short, stout, mainly brown snake, with yellow edges to its scales, has a tail that ends with a single shield or caudal plate and looks as though it has been chopped off at an oblique angle. The shield, covered with tiny spines that are more concentrated around the edge, is believed to plug the burrow down which the snake moves, keeping it safe from predators. Rarely coming to the surface, the snake has the cylindrical body, pointed head, and smooth scales of a burrowing species, and feeds on the earthworms it finds underground. As with other members of the family Uropeltidae, little is known about this species' social, defensive, or reproductive behaviour.

Loxocemus bicolor
Mexican burrowing snake

Length 1–1.3 m
(3 1/4–4 1/4 ft)
Breeding Oviparous
Habit Burrowing

Location S. Mexico to Central America

Status Endangered†

This constrictor has the narrow head and cylindrical, muscular body of a burrowing species. Its body is grey, with irregular patches of white scales that develop with age. This snake eats rodents, lizards, and reptile eggs, and is hunted by birds and mammals. Females lay 3–6 large, thick-shelled eggs in chambers dug in soil or leaf litter.

small, shiny scales

muscular body

Xenopeltis unicolor
Sunbeam snake

Length 1–1.3 m
(3 1/4–4 1/4 ft)
Breeding Oviparous
Habit Terrestrial

Location S.E. Asia

Status Common

cylindrical body

wedge-shaped head

shiny scales

uniform, dark grey colour

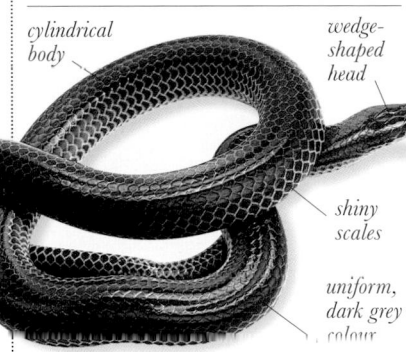

Named after its highly polished, iridescent scales, this snake has the typical body adaptations of a burrowing species. It comes above ground only at night, and eats frogs, lizards, snakes, and small mammals. Females lay 6–9 eggs in undergound nests in soil or leaf litter. The young have a distinct white collar that disappears after 2–3 moults.

Boa dumerili
Dumeril's boa

Length 1.5–2 m
(5–6 1/2 ft)
Breeding Ovoviviparous
Habit Terrestrial

Location S. and W. Madagascar

Status Vulnerable

Dumeril's boa is intricately patterned in shades of brown, with distinctive, glossy black markings around the mouth. A fairly inactive snake, it prefers to stay

Calabaria reinhardtii
Calabar ground boa

Length 0.9–1.1 m
(3–3 1/2 ft)
Breeding Oviparous
Habit Burrowing

Location W. and C. Africa

Status Common

This boa has a blunt head and a tail that are hard to tell apart. When threatened, it coils into a ball and raises its tail as a "false head" to protect its real head. The scales are black or brown, scattered with red or orange. The Calabar ground boa

Casarea dussumieri
Round Island boa

Length 1–1.5 m
(3 1/4–5 ft)
Breeding Oviparous
Habit Terrestrial

Location Round Island, in Indian Ocean

Status Endangered

Poorly known and seldom seen, the Round Island boa is one of the world's rarest snakes, with perhaps fewer than 500 left in the wild. It feeds on lizards, and its narrow head and slender body allow it to squeeze between rocks, so that it can creep up on its prey. Unlike true boas, which bear live young, this species lays eggs, in crevices where leaf litter has accumulated. A breeding programme is currently underway to save this snake, but its closest relative, *Bolyeria multicarinata* – also from Round Island – has not been seen since 1975, and is probably extinct.

dark grey above

small, keeled scales

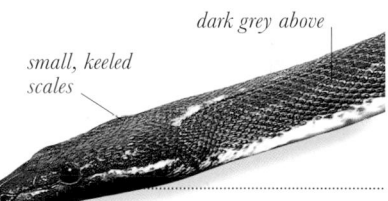

hidden in leaf litter on the floor of dry forest, its coloration allowing it to blend in almost imperceptibly. This boa ambushes and constricts its prey, which includes mammals and birds – although it lacks the heat-sensitive facial pits that some boas use to detect warm-blooded prey. The young, typically about 6 or 7, are relatively large at birth.

intricate brown markings

hunts small mammals in their tunnels and nests, and rarely comes above ground. Formerly classified as a python, it is now considered to be the only egg-laying species of boa.

patches of white scales under tail

tail raised to imitate head

HUNTING BY STEALTH
The common boa will sit and wait, watching its prey from a suitable vantage point. Timing its strike to perfection, it lunges forward and seizes its victim in its jaws before wrapping itself around it. The snake then asphyxiates its prey by tightly squeezing it – a process that can be so quick that small animals may be killed in seconds.

Boa constrictor

Common boa

Length	1–4 m (3½–13 ft)
Breeding	Ovoviviparous
Habit	Terrestrial/ Arboreal

Location Central America, South America, some Caribbean islands

Status Common

A large snake, with a narrow head and a pointed snout, the common boa varies in colour (see right) but has dark, saddle markings along the back, sometimes becoming dark red towards the tail. There may be as many as 10 subspecies (see right). The common boa is very adaptable, and uses a huge range of habitats, from tropical forest to dry savanna. It may also be found in urban areas. It climbs well but may equally hunt on the ground, especially in some of the drier habitats in which it occurs (here, it may aestivate in order to survive the worst heat of the summer months). It is also an able swimmer and often takes to the water voluntarily.

Secretive, and fairly sluggish in behaviour, it may be active by day or by night, depending on the climate. It preys on a wide variety of mammals and birds – suffocating them with its muscular coils before swallowing them whole.

COMMON BOA SUBSPECIES

The many subspecies of the common boa vary considerably in size and colour. Dwarf island races, such as the Hog Island boa, may be only 1m (3¼ft) long, while mainland boas are typically 3m (9¾ft) – although in exceptional cases some reach 4m (13ft). The colour ranges from mainly black (Argentinian boa) or olive-green (Central American boa) to the much lighter, pinkish or silver-grey specimens (Hog Island boa), often with a markedly contrasting tail. All, however, have distinctive saddle markings.

COMMON BOA FROM CENTRAL AMERICA

ARGENTINIAN BOA (JUVENILE)

HOG ISLAND BOA

CONCEALING COLOURS
The colours and patterns seen in the common boa serve a valuable purpose – helping to break up the snake's outline and enabling it to disappear into its surroundings.

characteristic saddle markings

dark stripe behind each eye

GIVING BIRTH
Depending on the subspecies (here a red-tail boa), common boas may produce 6–50 young in a litter, each 35–60cm (14–23½in) long. The newborn snakes fend for themselves immediately after breaking through the soft membrane that surrounds them at birth.

REPTILES

Candoia carinata
Pacific ground boa

Length 70–100 cm
(28–39 in)
Breeding Ovoviviparous
Habit Terrestrial/
Arboreal

Location S.E. Asia,
New Guinea, Solomon
Islands

Status Common

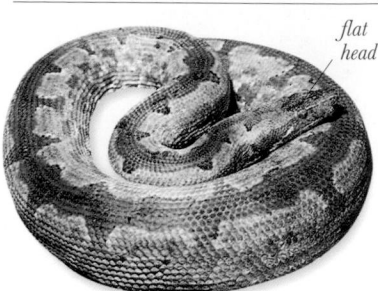

flat
head

Also known as the Pacific boa, this variable snake has 2 distinct sub-species. Shown above is subspecies *paulsoni*, which is thick and short-tailed, ranges in colour from red or greyish brown to off-white, and is completely terrestrial. Subspecies *carinata* is more slender with a longer tail, is brown or grey with an off-white patch above the cloaca, and is known to climb. Both subspecies feed on lizards, frogs, and small mammals.

Charina bottae
Rubber boa

Length 35–80 cm
(14–32 in)
Breeding Ovoviviparous
Habit Burrowing

Location S.W. Canada,
W. USA

Status Vulnerable†

The common name of this boa derives from its rubbery feel. The head and tail are both blunt, and may be hard to tell apart, as with the Calabar ground boa (see p.379). Its defensive technique is also similar: when a threatening

mammal or bird of prey approaches, the snake coils into a ball and raises its tail as a "false head", with which it may pretend to strike in order to deflect attack from its real head. However, the rubber boa is neither venomous nor aggressive. It has the typical body features of a burrowing species, and lives a secretive life below ground or under debris, hunting for birds, reptiles, and small mammals in burrows and tree holes. When winter comes, it hibernates for long periods. While mating, the male stimulates the female with his pelvic spurs and coils his tail around hers. Females give birth to 2–8 young at a time; these measure about 15cm (6in) long, and are pinkish or tan.

blunt
tail

small, smooth
scales

blunt
head

stout body

Charina trivirgata
Rosy boa

Length 60–110 cm
(23 1/2–43 in)
Breeding Ovoviviparous
Habit Burrowing

Location S.W. USA,
N.W. Mexico

Status Locally common

This snake usually has a pattern of 3 wide, black, brown, reddish brown, or orange stripes down its cream, buff, or grey body; coloration and size vary between its several subspecies. A burrowing snake, it mostly remains hidden beneath rocks and in crevices, foraging there for birds, reptiles, and small mammals. In the colder parts of its range, it hibernates in winter. While mating, the male stimulates the female with his pelvic spurs as he crawls along her back; 3–5, exceptionally up to 12, young are born.

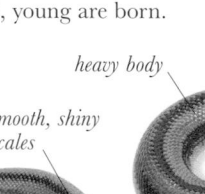

heavy body

smooth, shiny
scales

Corallus caninus
Emerald tree boa

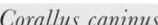

Length 1.5–2 m
(5–6 1/2 ft)
Breeding Ovoviviparous
Habit Arboreal

Location N. South
America

Status Locally common

white
markings
along back

JUVENILE COLORATION
The emerald tree boa gives birth to 3–15 young each season. They are brick-red, orange, bright red, or yellow, changing to green after about a year.

Like many members of the boa family – particularly the green tree python (see p.384) – this striking South American snake is fully adapted for life off the ground. Its intense green colour blends in with the foliage of the rainforests it inhabits, concealing it from predatory birds, while its strong, prehensile tail anchors it firmly to a branch, allowing the boa to lunge out or down to reach a bird or passing mammal. Its eyes have vertical pupils, which help it to sense movement; and it has deep pits in the scales around its mouth, for detecting heat given off by its prey. While mating, the male crawls over the female, and their tails become entwined.

READY TO STRIKE
The emerald boa typically drapes itself around a branch in a series of concentric coils, waiting, with its head hanging down, ready to strike. This boa feeds on small mammals and birds, and has long teeth which help it to grasp its prey firmly.

Corallus hortulanus
Amazon tree boa

Length 1.5–2 m
(5–6 1/2 ft)
Breeding Ovoviviparous
Habit Arboreal

Location Central and South
America, S. Caribbean

Status Common

A flattened body gives this slender, almost entirely arboreal snake rigidity when reaching out across branches; it often hangs down to snatch lizards and birds from lower branches. The Amazon tree boa may come down to

Epicrates cenchria
Rainbow boa

Length 1–2 m
(3 1/4–6 1/2 ft)
Breeding Ovoviviparous
Habit Partly arboreal

Location Central and
South America

Status Common

All subspecies of this boa have smooth, shiny, often highly iridescent scales, with colours varying from dark brown to dark orange. The back has black circles, while the flanks have

the ground at night. Coloration varies from brown or greyish with bars or blotches along the back, to plain orange, yellow, or olive.

deep,
wide
head

light-centred spots. This snake rests by day, actively foraging for birds, lizards, and small mammals at night. Only the southern subspecies hibernate. The size and number of young varies greatly between subspecies.

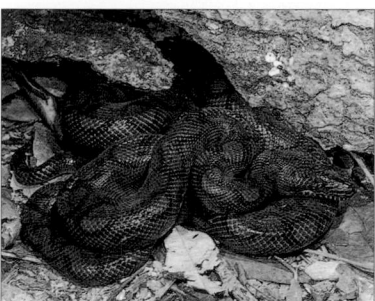

Eunectes murinus
Anaconda

Length 6–10 m
(20–33 ft)

Breeding Ovoviviparous

Habit Semi-aquatic

Location N. South
America, Trinidad

Status Common

With a maximum weight approaching 250kg (550lb), the green and black anaconda is the world's heaviest snake. It spends most of its life partially submerged in shallow water,

ADAPTED TO WATER

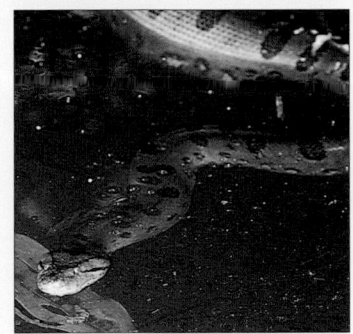

The anaconda's eyes and nostrils are positioned towards the top of its head, enabling the snake to see and breathe while partially submerged in water.

and favours areas with thick waterside vegetation, where it can move about unseen. Anacondas hunt mainly after dark. They lie in wait for mammals such as capybaras and deer, and can even kill fully grown caimans; there are also records of them attacking people, sometimes with fatal results. Males are smaller than females, but have large pelvic spurs which they use during courtship. Females produce litters of 4–80 young, which take about 6 years to reach maturity. Anacondas may have a lifespan of 25 years or more.

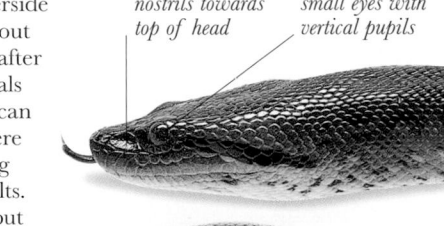
nostrils towards top of head
small eyes with vertical pupils

relatively small head

smooth scales

GIANT CONSTRICTOR
The anaconda's powerful body is strong enough to asphyxiate animals up to the size of a horse. It often ambushes its victims as they arrive to drink.

small, light-centred markings on flanks

oval black markings on olive-green back

Aspidites melanocephalus
Black-headed python

Length 1.5–2.5 m
(5–8 1/4 ft)

Breeding Oviparous

Habit Terrestrial

Location N. Australia

Status Locally common

The black-headed python is slender-bodied and has a jet black head and neck, with irregular brown and cream bars across its body. Not an aggressive snake, it retreats into a burrow or crevice if a predator threatens – usually a mammal or bird of prey. Its diet consists of small mammals, birds, and reptiles, including venomous snakes. Females lay up to 18 eggs at a time, under logs or roots, and in chambers underground. Like most pythons, they coil around the eggs to protect them and regulate their temperature until they hatch.

Antaresia childreni
Children's python

Length 75 cm (30 in),
max 100 cm (39 in)

Breeding Oviparous

Habit Terrestrial

Location N. Australia

Status Common

Named after the Victorian zoologist J.G. Children, this small, Australian python spends much of its time hidden away in caves and rock crevices. It specializes in eating lizards, but also ambushes birds and small mammals, including bats. If attacked, it defends itself by striking and biting but, like all pythons, it is non-venomous. Male Children's pythons have longer and thicker tails than the females. Juveniles have darker markings than adults.

faint dark blotches

PARENTAL CARE

Seen here coiled protectively round her eggs to regulate their temperature until they hatch, the female Children's python lays a clutch of 8–10 eggs each year, or every other year, in a hollow tree or an underground chamber.

SMALL SNAKE
Very small for a python, this snake is brown or reddish brown, with darker but often faint blotches.

Morelia spilota
Carpet python

Length 2 m (6 1/2 ft),
max 4 m (13 ft)

Breeding Oviparous

Habit Mostly terrestrial

Location S. New Guinea,
Australia

Status Common

The carpet python is one of Australia's most widespread snakes, found in an unusually wide range of habitats. It is also one of the most variable, with several subspecies, each with a different common name. One feature shared by all subspecies is a bold pattern of irregular markings – which can be brown, grey, reddish brown, or black – laid over variously coloured backgrounds. Carpet pythons are active both by night and by day, hunting lizards, birds, and small mammals. Like other pythons, they are non-venomous, but they can inflict a painful bite. Female carpet pythons lay clutches of up to 50 eggs in decaying vegetation or hollow tree trunks, coiling protectively around them until they hatch.

DIAMOND PYTHON

The diamond python, *Morelia spilota spilota*, is the most distinctive subspecies, with an intricate pattern of diamond-shaped blotches from head to tail.

JUNGLE CARPET PYTHON
The jungle carpet python, Morelia spilota cheynei, lives in Queensland's rainforests, and spends most of its time in trees.

intricate markings

large, heat-sensitive pits

REPTILES

Morelia viridis
Green tree python

	Length 1.8–2.4 m (6–7¾ ft)
	Breeding Oviparous
	Habit Arboreal

Location New Guinea and surrounding islands, extreme N. Australia | Status Locally common

bright green colour

This bright green python has the same adaptations for an arboreal lifestyle – slender shape, prehensile tail, and green coloration – as the emerald tree boa (see p.382). It also rests and hunts in the same way, coiled around a branch with its head hanging down ready to strike at prey. All the scales around the mouth have prominent heat pits. Females lay 6–30 eggs in a tree hole or among epiphytic plants. The young – which are bright yellow or occasionally red – are hunted by birds of prey.

Python molurus
Burmese python

	Length 5–7 m (16–23 ft)
	Breeding Oviparous
	Habit Terrestrial

Location S. and S.E. Asia | Status Lower risk

Also known as the Indian python, this very large snake is afforded excellent camouflage by the pattern of interlocking, dark brown blotches on its buff or grey skin. The coloration varies from area to area, but the "arrowhead" marking on the top of the head is constant. As with all pythons, there are heat pits along the jaws. Once adult, this snake has no predators, but it will hiss, strike, or bite if threatened; its bite, however, is neither venomous nor dangerous to humans. A powerful constrictor, the Burmese python feeds on mammals and birds, ambushing and constricting prey as large as a deer. It may aestivate in the drier parts of its range. Females, which are larger than males, usually lay 18–55 eggs at a time, in tree hollows or scrapes in the ground, and regulate their temperature until they hatch. By trembling its muscles, the Burmese python can generate enough heat to raise its own body temperature while it incubates its eggs. Population numbers have been seriously affected as a result of habitat destruction, and the Burmese python is now a legally protected species.

cryptic colouring

arrowhead mark

Python regius
Royal python

	Length 0.8–1.2 m (2½–4 ft)
	Breeding Oviparous
	Habit Terrestrial

Location W. to C. Africa | Status Common

The royal python is short and stocky, with intricate, dark brown markings over a tan or yellowish background, and conspicuous heat pits around its mouth. It defends itself against predators such as mammals and birds by curling up into a ball and hiding its head in the centre. In the dry season, this snake may aestivate underground. Females lay and incubate 3–8 eggs among rocks and in underground chambers.

Python reticulatus
Reticulated python

	Length 6–10 m (20–33 ft)
	Breeding Oviparous
	Habit Terrestrial

Location S.E. Asia | Status Common

This python is the world's longest snake and may weigh up to 200kg (440lb). It has a yellow or tan body with black markings, and prominent heat pits along the jaws. It can swim well but spends more time on land than in water, seldom straying far from its den. The reticulated python eats birds, mammals and, very rarely, humans. The female lays and broods 30–50 eggs in hollow trees and underground chambers. This python may live up to 30 years. It is widely hunted for its skin and larger individuals are becoming increasingly rare.

interlocking black marks

smooth scales

Tropidophis melanurus
Cuban wood snake

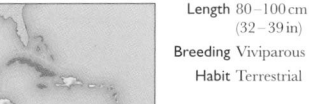

	Length 80–100 cm (32–39 in)
	Breeding Viviparous
	Habit Terrestrial

Location Caribbean | Status Locally common

This stocky snake is commonly grey, brown, or buff, with darker markings and a black-tipped tail; it may also be orange, with a yellow-tipped tail. Living mainly on the ground, occasionally climbing into bushes or onto rocks, it defends itself by rolling into a ball or releasing a foul-smelling slime from its cloaca. It feeds on frogs, lizards, smaller snakes, and rodents. Females give birth to around 8 young at a time.

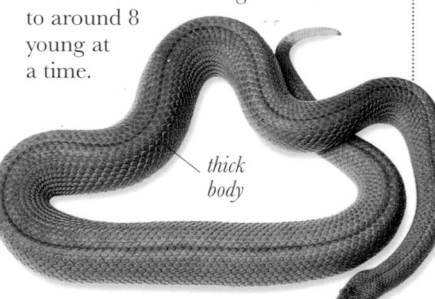

thick body

Acrochordus arafurae
Arafura file snake

	Length 1.5 m (5 ft), max 2.5 m (8¼ ft)
	Breeding Ovoviviparous
	Habit Aquatic

Location S. New Guinea, N. Australia | Status Locally common

Also called the Arafura wart snake, this snake belongs to a family of only 3 poorly known species. It is totally aquatic and is helpless out of water. It can absorb oxygen through the skin, and has enlarged tracheal lungs that enable it to remain underwater for long periods. The file snake prefers fresh or brackish water, such as rivers, lakes, swamps, and large ponds, where it rests on the bottom or anchors itself to aquatic plants with its prehensile tail. It may also be found along coasts, especially in mangrove forests. Sluggish in its movements,

loose skin

TRUNK-LIKE
The file snake resembles an elephant's trunk, with its loose-fitting skin, which is grey or dark brown with irregular dark bands, especially on the flanks.

SCALES THAT GRIP

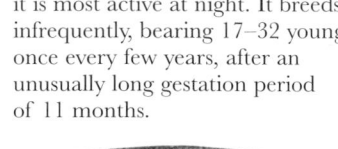

The name "file snake" is derived from the covering of tiny, rough, granular scales on the head, body, and tail of this snake. These help it to grip fish in its coils.

it is most active at night. It breeds infrequently, bearing 17–32 young once every few years, after an unusually long gestation period of 11 months.

COLUBRIDS

PHYLUM	Chordata
CLASS	Reptilia
ORDER	Squamata
SUBORDER	Serpentes
SUPERFAMILY	Caenophidia
FAMILY	Colubridae
SPECIES	1,858

SOMETIMES REFERRED TO as typical snakes, colubrids form by far the largest family of snakes, accounting for almost two-thirds of all species. They include the familiar garter snakes, grass snakes, and whipsnakes. Colubrids occur in all parts of the world except Antarctica. Measured in both numbers of species and individuals, they are the most numerous snakes on all continents except Australia. Colubrids live in habitats ranging from freshwater lakes to coastal marshes and estuarine swamps, rainforest, and arid desert. Such variation in mode of life and habitat means that they have a wide range of sizes, shapes, and colorations. All the world's rear fanged venomous snakes are colubrids.

ANATOMY

Colubrids have several anatomical features in common. All species are without a functioning left lung and a back-limb (pelvic) girdle. They also lack a small bone in the lower jaw known as the coronoid bone. These 3 characteristics separate colubrids from the more primitive blind snakes, thread snakes, and boas and pythons. Colubrids also differ from these groups in having a more flexible skull, with jaws that can be spread apart to accommodate large items of food. This is possible because the 2 halves of the lower jaw are not connected to one another. With a small number of exceptions, colubrids have large, plate-like scales covering their head. This differentiates them from boas, pythons, and most vipers. The arrangement of the head scales is usually consistent within a species and can therefore be a useful feature for identification.

VENOM

Some colubrids, notably rat snakes, kill prey by constriction. Although most colubrids lack venom, about a third of all species have a type of venom-producing apparatus known as Duvernoy's gland, which delivers venom to the base of enlarged fangs towards the back of the mouth. Most venomous colubrids have a single pair of fangs, although others have 2 or 3 pairs. The fangs create a wound, into which the venom runs by capillary action. Unlike some front-fanged snakes, colubrids have solid fangs, although the venom may travel along a groove that runs the length of the fang. Most rear-fanged snakes are of little danger to humans because their fangs are too far back to be brought into play unless they chew for a sustained period and because their venom is, in any case, weak. A few species, however, notably the boomslang and twig snakes, are dangerous and have caused human fatalities.

REPRODUCTION

Colubrids lay eggs or give birth to live young. In those species that lay eggs, clutch sizes vary from one to over 100 eggs. The young may be similar to the parents but some species have distinctive juvenile coloration and markings.

BURROWING ASPS

The burrowing asps are a group of snakes that have been placed in various families, including the Colubridae, but are now generally regarded as forming a separate group, the Atractaspidae, which consists of about 62 species. All except one species are found in Africa. These tunnelling snakes feed mainly on amphisbaenians. Some have unique fangs that hinge sideways so that the snake can expose them without opening its mouth, injecting its venom with a sideways strike of the head.

HUNTING

Compared with many front-fanged snakes, the venomous colubrids have relatively weak venom and inefficient fangs with which to deliver it. Possibly because of this, some colubrids, such as the rat snakes, kill their prey by constriction, causing suffocation. This Everglades rat snake has caught a barking treefrog. Once the victim is dead, the rat snake will manoeuvre its body until the frog's head is in position for it to be swallowed.

REPTILES

REPTILES

Atractaspis bibroni

Bibron's burrowing asp

	Length 50–75 cm (20–30 in)
	Breeding Oviparous
	Habit Burrowing

| Location E. and southern Africa | Status Locally common |

Often found in old termite mounds, this burrowing snake has a short, sharp spine at the end of its tail that helps it to push its way along its burrow, or gain purchase when travelling through smooth-sided tunnels or above ground. Active at night, Bibron's burrowing asp feeds on other snakes, small mammals, lizards, and frogs. It has large fangs, which it forces into its prey with a sideways stab, without having to open its mouth fully – an adaptation that allows this snake to bite its victims within the confines of a narrow burrow. Females lay 3–7 eggs underground in summer.

cylindrical black body

smooth, shiny scales

Bogertophis subocularis

Trans-Pecos ratsnake

	Length 1–1.3 m (3¼–4¼ ft)
	Breeding Oviparous
	Habit Terrestrial

| Location S. USA, N.E. Mexico | Status Locally common |

North America is home to about 6 species of ratsnakes, all of them slender-bodied predators that hunt after dark. The Trans-Pecos ratsnake is typical in having variable patterning: one form has large, H-shaped markings, while another – the "blond" form – has paler, circular blotches. Where their distributions overlap, these 2 forms interbreed. Large, prominent eyes indicate that this species is entirely nocturnal. This burrowing snake shelters in cracks and under slabs during the day to escape from excessively high temperatures, and hibernates in winter. When threatened by predatory birds and mammals, it emits a foul-smelling musk and may bite. The female lays 4–8 eggs, which she buries in rotting vegetation, or under rocks.

twin stripes start at neck

H-shaped markings further down back

Dasypeltis scabra

Common egg-eating snake

	Length 70–100 cm (28–39 in)
	Breeding Oviparous
	Habit Terrestrial

| Location Africa (mainly south of Sahara) | Status Common |

Many snakes eat soft-shelled reptile eggs, but this nocturnal, African snake specializes in eating the hard-shelled eggs of birds. It feeds heavily during the bird breeding season – swallowing its food whole – and then fasts for the rest of the year. This species and its close relatives are the only snakes that do not have teeth.

MIMICKING THE VIPER

This snake mimics the saw-scaled vipers (see p.395) by forming a horseshoe-shaped coil and rubbing the strongly keeled scales on its flanks together, making a rasping sound.

A DIET OF EGGS

The very flexible skull, mouth, and throat enable this snake to eat eggs. When an egg reaches the throat, projections of the snake's backbone saw through the shell, which is regurgitated.

slender, grey or reddish brown body

dark, angular markings

rounded snout

Boiga dendrophila

Mangrove snake

	Length 2–2.5 m (6½–8¼ ft)
	Breeding Oviparous
	Habit Arboreal

| Location S.E. Asia | Status Common |

The mangrove snake is mainly glossy black, with vibrant yellow markings on its lips and flanks – these vivid colours may act as a warning to predators. Its body is slightly flattened from side to side, and a distinct ridge runs down the centre of its back. If threatened, the mangrove snake draws back its flattened head in readiness to strike, and flares its yellow lip scales. A stealthy nocturnal hunter, this snake preys on lizards, frogs, birds, and small mammals. During the day, it rests in the crook of a tree, or remains coiled among foliage. Females lay 4–15 eggs in leaf litter, rotting stumps, or holes in trees.

ridge down back

Chrysopelea ornata

Golden tree snake

	Length 1–1.3 m (3¼–4¼ ft)
	Breeding Oviparous
	Habit Arboreal

| Location S. and S.E. Asia | Status Common |

Also known as the flying snake, the golden tree snake can glide among high trees in tropical forests, parks, and gardens, by spreading its ribs to form a concave underside. When threatened – usually by birds of prey or carnivorous mammals – it may launch itself from a high perch in order to escape; if cornered, it will bite. It has a slender body, with green, black-tipped scales, and a narrow head with large eyes. The male has a longer, thicker tail than the female. The golden tree snake frequently rests in trees, among foliage, but always remains alert for possible prey. An active hunter, it grasps and holds its victims – generally small mammals, birds, lizards, and frogs – until the venom takes effect. Although it has a powerful bite, this rear-fanged snake is not particularly harmful to humans. Females lay 6–14 eggs in soil, leaf litter, or rotting wood. The young are 15–20 cm (6–8 in) long when they hatch.

slender body grips prey tightly

powerful, venomous bite immobilizes prey

Coluber viridiflavus

European whipsnake

	Length 1.5 m (5 ft), rarely 2 m (6½ ft)
	Breeding Oviparous
	Habit Terrestrial

| Location S. Europe | Status Common |

Also called the western whipsnake, this slender, large-eyed snake is usually greenish yellow, but may be totally black. The young are more brightly coloured than the adults. A fast-moving and agile hunter, it relies largely on sight while hunting lizards, rodents, and smaller snakes. When attacked, it is quick to escape into holes and crevices, and may bite viciously. Several individuals may share a den, resulting in a dense population. Females lay 6–14 eggs in soil, leaf litter, or rotting wood.

greenish yellow body

broad black crossbars towards head

Dipsas indica
Snail-eating snake

Length 60–80 cm
(23 1/2 – 32 in)

Breeding Oviparous

Habit Arboreal

Location N. South America

Status Common

A specialist feeder, this snake has jaws that are adapted for extracting snails from their shells – it forces its lower jaw into the shell, hooks its teeth into the flesh, and pulls it out. It hunts at night, resting in tree hollows during the day. The snail-eating snake has a slender body, with a distinctive ridge running along its back, and a rounded head with a blunt snout and large eyes. Males tend to be smaller and have proportionately longer tails.

Dispholidus typus
Boomslang

Length 1–1.7 m
(3 1/4 – 5 1/2 ft)

Breeding Oviparous

Habit Arboreal

Location Africa (south of Sahara)

Status Common

Deriving its name from the Dutch word for tree ("boom"), this is one of Africa's most venomous arboreal snakes. A highly agile climber, it uses its binocular vision to judge distances as it slips from one branch to another, and when it prepares to strike. Boomslangs feed on lizards, especially chameleons, and also on birds. Their fangs are at the back of their mouths, and they inject their venom by chewing their prey. Adult boomslangs are highly variable in colour: they may be green, brown, or almost black, with or without contrasting markings. Males are more colourful than females. When threatened, usually by birds of prey and carnivorous mammals, they inflate their necks to appear larger, and will bite readily. Females lay clutches of up to 14 eggs in tree hollows and dead vegetation.

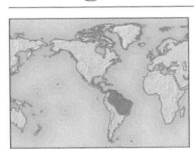

wide ventral scales

very slender body

greenish black coloration

Drymarchon corais
Indigo snake

Length 2.1–2.9 m
(7–9 1/2 ft)

Breeding Oviparous

Habit Terrestrial

Location S. USA, Central and South America

Status Locally common

Also known as the cribo, this highly variable snake is black, brown, tan, or yellowish brown, depending on where it lives (the black Florida form is shown below). It feeds on most animal groups – fish, amphibians, lizards, other snakes, birds and their eggs, and small mammals. An opportunistic feeder, it can lunge very quickly at its prey, and sometimes presses it against the side of a burrow or a solid object. Females lay 4–12 eggs in tree stumps, vegetation, or burrows.

glossy body

Elaphe guttata
Corn snake

Length 1–1.8 m
(3 1/4 – 6 ft)

Breeding Oviparous

Habit Terrestrial

Location C. and S.E. USA

Status Common

The corn snake is one of the most eye-catching of North America's ratsnakes – a group that includes the common ratsnake (see right) and the Trans-Pecos ratsnake (see p.386). Like most of its relatives, the corn snake is extremely variable, with 4 different subspecies. The subspecies shown below, *Elaphe guttata guttata*, is the most colourful, and is restricted to southeastern states; the Midwestern subspecies, *Elaphe guttata emoryi*, usually known as Emory's ratsnake, is grey with darker blotches of grey-brown. Corn snakes can be found on the ground, on trees or buildings, and under logs, rocks, and debris. As with other ratsnakes, their flanks meet their undersides at an acute angle, forming ridges that help them to grip bark or walls. Adult corn snakes feed mainly on small rodents, which makes them useful animals on farms. Although they are nocturnal, they are often active during the day at cool times of the year. When threatened, they rapidly vibrate their tails and may rear up, ready to strike. If caught, they excrete a foul-smelling musk. Female corn snakes lay clutches of up to 25 eggs in decaying vegetation.

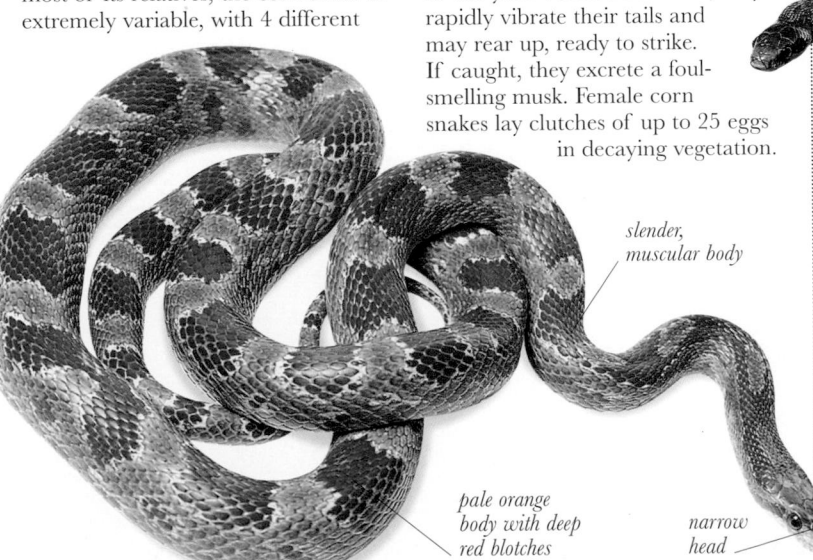

pale orange body with deep red blotches

narrow head

Elaphe longissima
Aesculapian snake

Length 1–2.2 m
(3 1/4 – 7 1/4 ft)

Breeding Oviparous

Habit Terrestrial

Location S. Europe to W. Asia

Status Common

The name of this snake derives from Aesculapius, the Greek god of medicine. It is thought to be the snake that, coiled round a staff, forms his motif, the caduceus, still used as a symbol of the medical profession today. The Aesculapian snake is olive or brown, occasionally dark grey, with a lighter patch on each side just behind the head. The scales are smooth or slightly keeled, giving it a shiny appearance. An agile swimmer and climber, this secretive snake inhabits scrubland, forest edges, and fields, and hibernates during the cooler months.

slender, muscular body

Elaphe obsoleta
Common ratsnake

Length 1.2–1.8 m
(4–6 ft)

Breeding Oviparous

Habit Terrestrial

Location S. Canada, C. and E. USA

Status Common

The common ratsnake varies greatly in coloration, ranging from bright yellow-orange to much more subdued colours, as in the grey ratsnake (*Elaphe obsoleta spiloides*), the subspecies shown below. Common ratsnakes favour rocky hillsides with open woodland, often near water, and are excellent swimmers. They feed on rodents and birds, which they kill by constriction. Females lay 5–20 eggs, occasionally up to 40, sometimes in communal sites. A ridge along the flanks helps the snake to wedge itself into the bark when climbing trees.

long head with rounded snout

strongly keeled ventral scales

REPTILES

Elaphe situla
Leopard snake

Length 70–100 cm
(28–39 in)
Breeding Oviparous
Habit Terrestrial

Location S. and E. Europe, W. Asia | Status Locally common

The leopard snake is cream or grey, with black-edged reddish blotches, or 2 wide, longitudinal reddish stripes, edged with black, along its upper side. It feeds mainly on rodents, which it catches among vegetation or in their burrows, then kills by constriction. If threatened, it may flatten its head and bite. During mating, the male holds onto the female by biting her head or neck and coiling around her. Females lay up to 8 eggs and remain with them for several days.

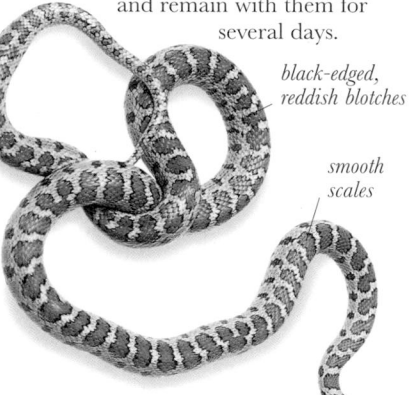

black-edged, reddish blotches

smooth scales

Erpeton tentaculatus
Tentacled snake

Length 70–100 cm
(28–39 in)
Breeding Viviparous
Habit Aquatic

Location Central S. Asia | Status Common

keeled scales form ridge

This unusual snake belongs to a subfamily of colubrids – numbering about 30 species in all – that live in water. It is easily distinguished from its relatives by the unique pair of soft, fleshy tentacles on its snout, which probably have a sensory function. Tentacled snakes are sluggish and nocturnal, and hunt by lying in weed-choked water, waiting for prey to swim by. Females give birth to 5–13 live young underwater.

Gonyosoma oxycephala
Red-tailed racer

Length 1.6–2.4 m
(5¼–7¾ ft)
Breeding Oviparous
Habit Arboreal

Location S.E. Asia | Status Common

The tail of this long, slender snake may be brown, orange, or grey, but despite its name it is never red. An active hunter, the red-tailed racer forages for birds,

bats, and small mammals at night, its bright green colour providing excellent camouflage in the trees. When threatened, usually by birds of prey or mammals, it inflates its throat vertically and adopts an S-shaped posture, in readiness to strike. Mating occurs in tree branches.

coffin-shaped head

Heterodon nasicus
Western hognosed snake

Length 40–80 cm
(16–32 in)
Breeding Oviparous
Habit Terrestrial

Location C. USA to Mexico | Status Common

With its blotched patterning and thick body, this North American snake looks more like a viper (see p.394) than a member of the colubrid family. Although its venom is toxic, it is not

considered dangerous to humans. There are 3 species of hognosed snakes in North America, all of which search through leaf litter for their prey and use their distinctive, upturned snouts for digging and rooting out burrowing toads and other amphibians. If threatened, all hognosed snakes hiss loudly, inflate their bodies, and may make mock strikes; if this fails, they often play dead.

dark blotches on cream body

Lampropeltis getula
Common kingsnake

Length 1–2 m
(3¼–6½ ft)
Breeding Oviparous
Habit Terrestrial

Location W. and S. USA, N. Mexico | Status Common

The common kingsnake is a powerful constrictor that actively hunts for prey, frequently entering rodent burrows. It also feeds on birds, lizards, frogs, and snakes, including venomous ones, to whose bite it is immune. Although mainly terrestrial, this snake may climb low vegetation and is also a fine swimmer. If threatened by a predator – mainly birds of prey and mammals – the common kingsnake may bite vigorously and, if grasped, may try to smear its captor with faeces. It may be active at night or during the day, depending on the time of year and its distribution, and it hibernates for a longer time in the north of its range than in the south. Mating

involves the male crawling along the female's back and biting her neck to hold her steady. The female lays 12 (exceptionally up to 25) eggs in the warmth of decomposing vegetation, rotting wood, or underground chambers. The young measure about 30–35 cm (12–14 in) in length. The common kingsnake can live for over 25 years in captivity.

LAMPROPELTIS GETULA CALIFORNIAE
This Californian subspecies of the common kingsnake may have a dark grey body with cream bands (shown below) or a brown body with a white stripe (shown in panel, right).

cream bands

A BEWILDERING DIVERSITY

The common kingsnake shows a remarkable variety of colours and patterns. Many herpetologists divide the species into 7 different subspecies, and some believe the total should be as many as 10. To complicate matters further, some subspecies – such as the California kingsnake – have more than one colour form.

LAMPROPELTIS GETULA NIGRITA
The Mexican black kingsnake is, in its purest form, solid jet black in colour, but it may show faint traces of a banded pattern when young. This is one of the most handsome subspecies and a popular pet.

glossy scales

yellow spots on black body

LAMPROPELTIS GETULA SPLENDIDA
This desert snake from Arizona is only active for a short period of the year, when it is neither too cold nor too hot.

LAMPROPELTIS GETULA CALIFORNIAE
The striped form of the Californian kingsnake occurs in some parts of the subspecies' range alongside the banded form (left). Both forms can hatch from the same clutch of eggs.

Lampropeltis triangulum
Milksnake

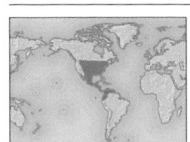

Length 0.4–2 m
(1¼–6½ ft)
Breeding Oviparous
Habit Terrestrial

Location North America,
Central America, N. South
America

Status Common

The milksnake is one of the world's most widely distributed terrestrial snakes, and also one of the most variable. About 30 subspecies have been identified, with 8 in the United States alone. Most are brightly coloured, with red, black, and yellow bands; in some subspecies the patterning is very similar to that of coral snakes (see p.392) – a feature that probably evolved as a form of defensive mimicry. Milksnakes are secretive, mainly nocturnal, and feed on invertebrates, amphibians, and small rodents, as well as other snakes. Although non-venomous, they defend themselves by biting, and by discharging a foul-smelling fluid from the cloaca. Females build nests in burrows, under rocks, in tree stumps, or in rotting vegetation, and lay up to 17 eggs.

STARTLING DEFENCE
The milksnake is not venomous. It does, however, mimic the coloration, markings, and even behaviour of venomous coral snakes and is easily confused with them in parts of its range. If uncovered from its hiding place, it may thrash around in an attempt to startle its prey with its bright colours.

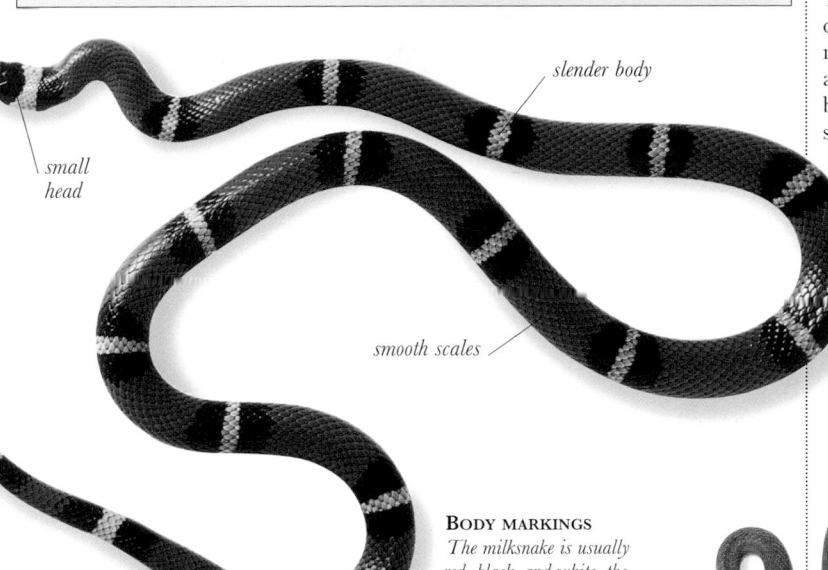

slender body

small head

smooth scales

BODY MARKINGS
The milksnake is usually red, black, and white, the markings arranged as rings around the body or as saddles over the back.

Lamprophis fuliginosus
Brown house snake

Length 0.9–1.5 m
(3–5 ft)
Breeding Oviparous
Habit Terrestrial

Location Africa (mainly
south of Sahara)

Status Common

The brown house snake is a powerful constrictor that actively forages at night for its prey – mainly rodents, but also birds and lizards. It may be tan, brown, orange, or black, with a cream stripe on either side of its head. The male is smaller than the female but has a proportionately longer tail. Females may lay 2 or more clutches of 6–16 eggs per breeding season. This snake may bite if provoked, usually by birds of prey or mammals, but it is not venomous.

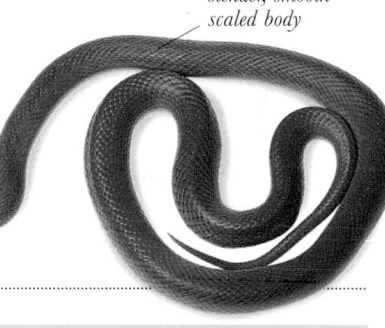

slender, smooth-scaled body

Langaha madagascariensis
Madagascan leaf-nosed snake

Length 70–90 cm
(28–35 in)
Breeding Oviparous
Habit Arboreal

Location Madagascar

Status Locally common

This snake is instantly recognizable because of its extraordinary snout. The male has a long, pointed projection on the snout tip; in the female, the appendage is more elaborate and leaf-shaped. Males are light brown with a yellow underside, while females are grey-brown with small dark markings. Both sexes have a long, slender, vine-like body, which provides excellent camouflage in the vines and branches of their forest habitat. Active during the day and at night, they ambush lizards and frogs.

Natrix natrix
Grass snake

Length 1.2–2 m
(4–6½ ft)
Breeding Oviparous
Habit Semi-aquatic

Location Europe to
C. Asia, N.W. Africa

Status Common

The grass snake is a semi-aquatic predator, spending much of its time in damp places or in still water. A good swimmer, it can sometimes be spotted rippling its way across the surface of ponds, in search of frogs and fish. It is one of the most widespread snakes in Europe. Adults are olive-brown, greenish, or grey, usually with a contrasting yellow or white collar just behind the head. Female grass snakes are larger than males, and have longer, thicker tails. After mating, they sometimes lay their eggs at shared nest sites, where the total egg count can reach 200 or more. In places where summers are cool, females often lay their

EGG-LAYING
Eggs are laid in decaying vegetation, manure or compost heaps, or other warm locations. Hatchlings measure 14–21 cm (5½–8½ in) long. Males reach maturity in 3 years, females in 4.

PLAYING DEAD
The grass snake ejects a foul-smelling fluid if handled, and sometimes reacts to extreme danger by playing dead. To make the performance as convincing as possible, it turns partly upside down, with its mouth open and tongue exposed.

body flipped onto back, feigning death

gaping mouth

eggs in compost heaps, the warmth of decaying vegetation helping incubation. The grass snake has a lifespan of approximately 15 years in captivity.

yellow collar

Nerodia fasciata
Southern water snake

Length 0.5–1.5 m
(1½–5 ft)
Breeding Ovoviviparous
Habit Aquatic

Location C. and S.E. USA

Status Common

Also known as the banded water snake, this species has a thickset body with keeled scales and highly variable coloration, sometimes with dark bands or blotches. Its eyes are situated towards the top of its head, helping it to see while partially submerged. This snake hunts for frogs and fish during the day or night, depending on the temperature. If attacked, it may smear its captor with the foul-smelling contents of its anal gland. Many males may successfully mate with the same female. The female gives birth to 2–57 live young.

dark bands across body

REPTILES

Spilotes pullatus

Tiger ratsnake

Length 1.5 – 2 m
(5 – 6 1/2 ft)
Breeding Oviparous
Habit Arboreal

Location Central America,
N. and C. South America

Status Common

The tiger ratsnake may be yellow
with black bands or spots, black
with yellow bars, or black with
yellow-centred scales. Its
triangular, deep body, which
is flattened from side to side,
allows the snake to remain

long, slender,
triangular body

distinct ridge
along back

Opheodrys aestivus

Rough green snake

Length 0.8 – 1.6 m
(2 1/2 – 5 1/4 ft)
Breeding Oviparous
Habit Terrestrial

Location S.E. USA

Status Common

Often found among low-growing
vegetation, especially near water,
this snake is bright green with
a white or yellowish green
underside. The scales are
keeled, hence the rough
texture of its skin. Although

green coloration
aids camouflage

long, thin
body

rigid while it bridges gaps between
branches. Active mainly at night,
the tiger ratsnake stalks lizards, frogs,
and small mammals, then suddenly
strikes out, catching its prey by
surprise. When threatened, this snake
flattens the front part of its body and
forms an S-shaped coil; it may also
deliver a strong bite.

it typically remains motionless,
or moves slowly through vegetation
hunting for insects, it is capable of
fast movement. The
female lays 3–13
elongated eggs.

Pituophis melanoleucus

Pine snake

Length 1 – 2.5 m
(3 1/4 – 8 1/4 ft)
Breeding Oviparous
Habit Burrowing

Location S.E. USA

Status Common

This thickset, powerful constrictor has
strongly keeled scales and a slightly
pointed snout that is adapted for
burrowing. Its coloration is
variable, ranging from uniformly
black or white to cream or
yellowish, with irregular blotches
of dark brown or black down the
back and flanks. An active forager,
it seeks out small mammals, especially
burrowing species. The female lays
3–27 eggs in burrows that she often
digs herself, using a loop in the front
part of her body to remove sand or
soil. Several subspecies are recognized,
one of which (the Louisiana
pine snake) is
endangered.

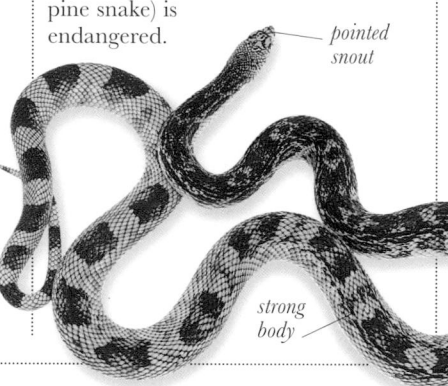

pointed
snout

strong
body

Rhinocheilus lecontei

Long-nosed snake

Length 50 – 100 cm
(20 – 39 in)
Breeding Oviparous
Habit Burrowing

Location S. USA,
N. Mexico

Status Common

black bars
with red
in between

pointed snout

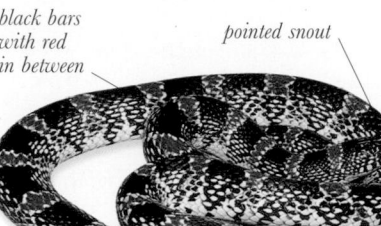

The long-nosed snake lives under
rocks and logs, and in rodent holes,
and uses its pointed snout for
burrowing. It has a markedly
underslung jaw – another adaptation
that suits it to a semi-subterranean
way of life. Mainly nocturnal, it feeds
on small mammals and birds, as well
as lizards and small snakes and their
eggs. When threatened, the long-
nosed snake hides its head in its
coils, vibrates its tail, and
discharges a foul-smelling
liquid from its anal gland.
Females lay one clutch,
or sometimes 2, of up
to 9 eggs.

Telescopus semiannulatus

African tiger snake

Length 1 – 1.2 m
(3 1/4 – 4 ft)
Breeding Oviparous
Habit Terrestrial

Location Southern Africa

Status Common

A nocturnal hunter, the African tiger
snake has large eyes that help it to
see better in the dark. Its slender,
smooth-scaled body is yellowish
brown, orange, or pinkish brown,
with black blotches that extend
across the back as far as the flanks.
This slow-mover lives mostly on
the ground, but also climbs into
shrubs, dead trees, and thatched roofs.

Thamnophis proximus

Western ribbon
snake

Length 50 – 100 cm
(20 – 39 in)
Breeding Viviparous
Habit Semi-aquatic

Location C. USA to
Central America

Status Common

Like its close relative the common
garter snake (see right), this North
American snake is fast-moving
and agile, as well as being a good
swimmer. It feeds mainly on frogs,
toads, tadpoles, and small fish, often
pursuing its prey underwater. Western
ribbon snakes give birth to up to 24
live young a year, and can reach high
densities in suitable habitats.

slender body
with keeled scales

light stripes
along body

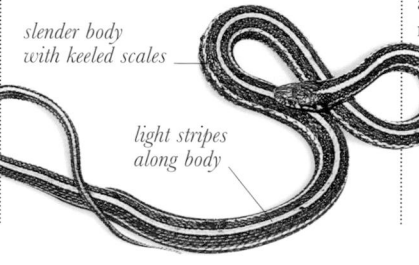

Thamnophis sirtalis

Common garter
snake

Length 65 – 130 cm
(26 – 51 in)
Breeding Ovoviviparous
Habit Semi-aquatic

Location North America

Status Common

The common garter snake invariably
lives close to water, but it has the
distinction of breeding further north
than any other snake in the Americas.
On the fringes of the Arctic, it
hibernates en masse, creating an
extraordinary spectacle when large
numbers of snakes emerge in spring.
Common garter snakes feed mainly
on earthworms, fish, and amphibians,
and are themselves attacked by
mammals and birds of prey.

PHYSICAL APPEARANCE
*The 11 subspecies of common garter
snake vary greatly in coloration. Shown
here is the vividly striped subspecies*
infernalis, *the San Francisco
garter snake.*

MATING PATTERNS

Mating systems vary according
to latitude. In the north of the
range, males scramble to compete
for females, which they often
outnumber greatly, as they
emerge from hibernation dens.
Towards the south, competition
is less frenetic.

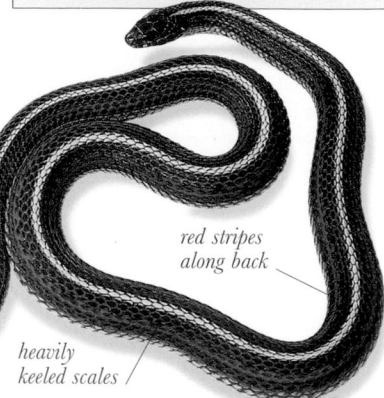

red stripes
along back

heavily
keeled scales

ELAPIDS

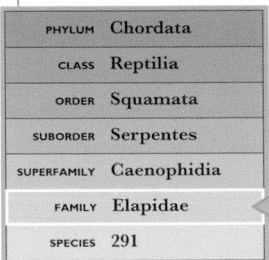

PHYLUM	Chordata
CLASS	Reptilia
ORDER	Squamata
SUBORDER	Serpentes
SUPERFAMILY	Caenophidia
FAMILY	Elapidae
SPECIES	291

ALTHOUGH ELAPIDS ACCOUNT for less than one in 10 of the world's snakes, all of them are venomous – many dangerously so. They include cobras, coral snakes, mambas, kraits, and sea snakes, and are found mainly in the tropics and in the southern hemisphere. There are ground-dwelling and semi-burrowing species of elapids, as well as some that live and hunt in trees. Some are fast-moving, diurnal hunters, while others are more secretive, hunting under the cover of darkness. Sea snakes and sea kraits are wholly marine. Elapids are related to colubrids (see p.385) and look superficially similar, but they differ from most of them in having fangs located towards the front of the upper jaw. This allows them to inject venom in a sudden and deadly strike.

ANATOMY

Elapids typically have a slender, roughly cylindrical body, with smooth, shiny scales. Sea snakes – sometimes classified in a family of their own – are noticeably different, with a flattened tail for swimming. Most terrestrial elapids have camouflage colours and markings, but coral snakes have bold markings that warn would-be predators to leave them alone. Some cobras intimidate their enemies by raising the front part of their bodies off the ground, and expanding their ribs to form a hood.

VENOM

The venom of elapids varies in its potency and the organs and body systems that it affects. Most elapids, including the cobras, release venom that acts on the nervous system, paralysing muscles used in respiration. In some elapids, the venom runs down grooves in the fangs, while in others the fangs contain an internal venom canal. The elapids include some of the world's most dangerous snakes, such as the taipan. Others, however, are inoffensive and too small to endanger humans.

REPRODUCTION

Most elapids lay eggs but a few are bearers of live young. The sea snakes, for instance, give birth in the water, whereas the sea kraits come ashore to lay eggs. A number of Asiatic cobras build nests out of dead leaves and other forest debris. Females guard the eggs fiercely, protecting them from potential predators.

SPITTING COBRA
Some cobras (including the Mozambique spitting cobra shown here), can spray their victims with venom by forcing it out at high speed through apertures in the fangs. Once sprayed into the victim's eyes, it causes great pain and temporary, or sometimes even permanent, blindness.

Acanthophis praelongus

Northern death adder

Length 30–100 cm
(12–39 in)

Breeding Viviparous

Habit Terrestrial

Location New Guinea, N. Australia

Status Common

Despite its name, this snake is not a true adder (see p.394), but an unusually stout-bodied elapid. It is one of Australia's most venomous snakes, hunting by a sit-and-wait strategy, and striking at the least provocation. It is normally active at night; during the day it relies on its camouflage markings to avoid being seen. Adults vary in colour from light grey to grey-brown or almost black, with paler rings or bands. Northern death adders feed mainly on small mammals, lizards, and birds, and they use their thin, worm-like tails to lure animals within striking range. They give birth to live young, producing litters of up to 8 each time they breed. A very similar species, the southern death adder (*Acanthophis antarcticus*), is widespread in eastern and southern Australia, and is sometimes found close to urban areas. Like its northern counterpart, it has a bite that is potentially fatal to humans.

wedge-shaped head

thick body

ADDER LOOK-ALIKE
A thick body, wedge-shaped head, and narrow neck are all features normally seen in true adders and vipers, rather than in elapids. The northern death adder also has raised "horns" over its eyes.

LIVING LURE

The tip of the northern death adder's tail is very slender, and is considerably paler than the rest of the body. When hunting, the snake lies still, and vibrates the tail tip to lure prey close enough to strike. Rapid-acting venom kills the victim, which the snake can then swallow at leisure.

Austrelaps superbus

Australian copperhead

Length 1.3–1.7 m
(4¼–5½ ft)

Breeding Viviparous

Habit Terrestrial

Location S.E. Australia, N. Tasmania

Status Common

This inhabitant of marshes and swamps is active both during the day and at night, and feeds mainly on frogs. It is grey or reddish brown to black above, and cream or yellow on the flanks. Although it hibernates in winter, it can tolerate low temperatures and is active longer than other reptiles.

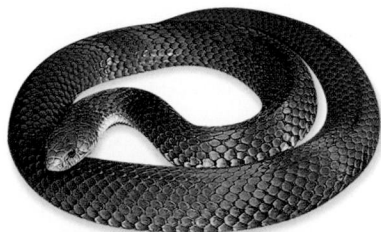

white rings around body

Dendroaspis angusticeps
East African green mamba

Length	1.5–2.5 m
	(5–8¼ ft)
Breeding	Oviparous
Habit	Arboreal

Location E. and S.E. Africa   Status Common

The bright green colour, slender shape, and long tail of the East African green mamba are all features that are ideally suited to an animal living in trees. Alert and fast-moving, this snake chases its prey through the canopy and, if disturbed, it usually moves swiftly to higher branches or passes from the branches of one tree to another; if cornered, it will turn and strike. Although highly venomous, this mamba rarely comes into conflict with humans.

smooth scales

narrow head

Dendroaspis polylepis
Black mamba

Length	2.5–3.5 m
	(8¼–11 ft)
Breeding	Oviparous
Habit	Terrestrial

Location E. and southern Africa   Status Common

As well as being one of the most poisonous snakes, the black mamba is probably the fastest, making it one of the world's most dangerous predators. In short bursts, it can reach 20kph (12½mph), which is fast enough to overtake someone moving at a brisk run. The black mamba is not actually black, but grey or brown, with a streamlined body covered with large, smooth scales. Highly territorial, it makes its home den in a rock crevice or a hollow tree. It is active during the day, and feeds on birds and small mammals. Females lay 12–17 eggs in chambers underground. The venom of the black mamba is very fast-acting, and bites are often fatal if not treated quickly.

Occasional climber
This primarily ground-living snake, mostly found in open woodland, may also occasionally climb. Despite its size, it is highly agile, effortlessly sliding through thorn bushes and trees.

FIGHTING FOR A MATE
During the breeding season, male black mambas fight by raising and intertwining the front half of their bodies, with each snake trying to force its opponent to the ground. However, the fight is ritualized and does not involve any serious harm. Following this test of strength, the winner mates with any females in his territory.

Laticauda colubrina
Sea krait

Length	1–2 m
	(3¼–6½ ft)
Breeding	Oviparous
Habit	Aquatic

Location S. and S.E. Asia   Status Common

The bluish grey sea krait is one of about 40 species of snakes that have flattened, paddle-shaped tails and live and feed at sea. It is found in coastal waters, mangrove swamps, and coral reefs, where it hunts fish – particularly eels. Like all sea snakes, it is highly poisonous, but poses no threat to humans since it does not bite. Most sea snakes spend their entire lives at sea, and give birth to live young, but the sea krait and its relatives come ashore to lay their eggs under leaf litter.

broad black bands

Micrurus lemniscatus
South American coral snake

Length	60–90 cm
	(23½–35 in)
Breeding	Oviparous
Habit	Semi-burrowing

Location N. and C. South America   Status Common

This brilliantly coloured reptile is one of about 40 species of coral snakes, all found in warm parts of the Americas. Each species has a characteristic pattern of contrasting bands, warning predators that they are poisonous. However, because coral snakes have a semi-burrowing lifestyle, generally emerging only at night, their colours are often concealed. Coral snakes feed on small lizards and snakes. They have highly toxic venom, but human fatalities are rare.

smooth scales

sequenced colour bands

small head

Naja naja
Indian cobra

Length	1.2–1.7 m
	(4–5½ ft)
Breeding	Oviparous
Habit	Terrestrial

Location S. Asia   Status Common

Renowned for its use in snake charming, this animal is one of the most dangerous snakes in India. It is responsible for a relatively small but nonetheless significant proportion of snake-bite mortalities, which total 10,000 per year in India alone. This is partly due to its preference for rice paddies and roadside banks, which are often close to villages and so bring it into contact with humans. Indian cobras are very variable in colour, ranging from brown to black, but most individuals have distinctive, pale "spectacle" markings on the back of the hood. As with other cobras, the hood is spread out when the snake feels threatened, but is folded away at other times. Although Indian cobras are sometimes seen basking in the sun, they are most active at night, and are good climbers and swimmers. They feed on small mammals, birds, lizards, and other snakes, killing them within seconds with their potent neurotoxic venom. The principal predators of Indian cobras, apart from humans, are carnivorous mammals – notably mongooses – and birds of prey.

expanded hood

broad ventral scales

Naja pallida
Red spitting cobra

Length 70–120 cm
(28–47 in)
Breeding Oviparous
Habit Terrestrial

Location N. and E. Africa | Status Common

This African snake is one of several cobras that have evolved a unique way of defending themselves. If threatened, they can squirt venom out of small apertures in their fangs, spraying a cloud of droplets 2 m (6½ ft) into the air. Sprayed venom does not kill, but it can cause permanent blindness. The red spitting cobra varies in colour from red to grey, and is active at night and early in the morning. Females lay up to 15 eggs in a burrow or in rotting vegetation.

narrow hood

black band across throat

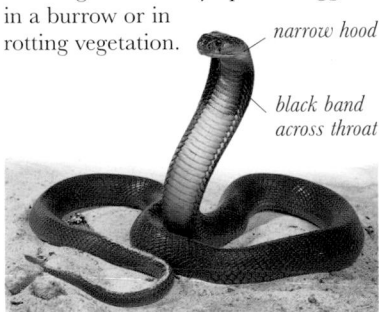

Naja haje
Egyptian cobra

Length 1–2.4 m
(3¼–7¾ ft)
max 2.5 m (8¼ ft)
Breeding Oviparous
Habit Terrestrial

Location N., W., and E. Africa | Status Common

Usually brown or greyish, sometimes with black bands, the Egyptian cobra may grow up to 2.5 m (8¼ ft) long. An inhabitant of desert, grassland, and urban areas, it is also found in fields near oases and wadis, and tends to avoid dense forest and the most arid deserts. This cobra is primarily nocturnal, but can sometimes be seen basking in the sun early in the morning. Quick to rear and spread its broad, rounded hood to intimidate an opponent, this highly venomous cobra will hiss and advance on its aggressor if the latter is undeterred, and may eventually strike, delivering a bite that is fatal for humans if an antivenom is not administered promptly. It actively pursues other snakes, small

mammals, toads, and birds, including domestic poultry, killing prey with its fast-acting venom; it also feeds on eggs. Individuals are territorial and frequently fight one another. Females lay 8–20 eggs, often in termite mounds, and the eggs hatch after an incubation period of about 60 days.

mouth open wide to intimidate

large eyes

large head

hood spreads in defence

thickset body

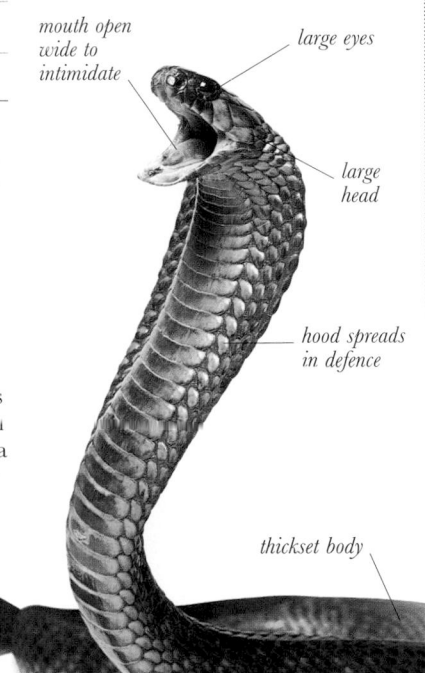

Maticora bivirgata
Blue coral snake

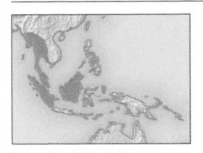

Length 1.2–1.4 m
(4–4½ ft)
Breeding Oviparous
Habit Terrestrial

Location S.E. Asia | Status Locally common

The striking, bright orange markings on the underside, head, and tail of the blue coral snake act as a warning device to keep potential predators at bay. A highly venomous species, it has poison glands along almost one-third of its entire body length. It is also known as the "100-pace snake", referring to the distance a human is thought capable of travelling before succumbing to the venom. However, this snake is generally inoffensive, and few cases of fatal bites on humans have been recorded.

Notechis scutatus
Australian tiger snake

Length 1–2.1 m
(3¼–7 ft)
Breeding Viviparous
Habit Terrestrial

Location S.E. Australia | Status Common

large scales

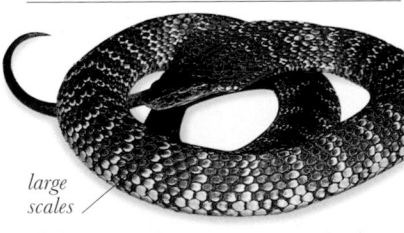

This extremely venomous species is responsible for many potentially fatal snake-bites in Australia. It may be grey, brown, dark brown, or olive-brown, often with narrow, light yellow bands. Although active mainly during the day, the Australian tiger snake may also be active early in the morning and in the evening, and on warm nights; it remains hidden in spells of cold weather. Its prey consists mainly of frogs, which it chases then subdues with its venom. In defence, it flattens its neck and rears its body slightly off the ground. Females give birth to about 30 live young.

Ophiophagus hannah
King cobra

Length 3–5 m
(9¾–16 ft)
Breeding Oviparous
Habit Terrestrial

Location S. and S.E. Asia | Status Lower risk†

The king cobra, or hamadryad, is the longest venomous snake and a specialist hunter of other snakes.
It sometimes reaches a length of more than 5 m (16 ft), which allows it to overpower and kill other snakes of a considerable size. When threatened, it raises the front third of its body so that it stands 1.5 m (5 ft) tall, erects a narrow hood, and may strike downwards, although

it rarely attempts to bite. In fact, few cases of king cobra bites on humans have been recorded, because this retiring snake shuns human contact, living mainly in deep forest. Slender and smooth-scaled, the king cobra is a good swimmer and is often found near water. Adults are plain brown, while juveniles are darker and marked with pale chevrons down their backs. Unusually for a snake, this species is at least temporarily monogamous, pairs apparently remaining together during the breeding season. Females lay 21–40 eggs, mostly in piles of dead vegetation, which are guarded by both parents until they hatch. The king cobra may live for more than 20 years in captivity.

narrow hood

long body

smooth scales

Oxyuranus scutellatus
Taipan

Length 2–3.6 m
(6½–12 ft)
Breeding Oviparous
Habit Terrestrial

Location S. New Guinea, N. Australia | Status Vulnerable†

Although nondescript in appearance, the plain tan or dark brown taipan is Australia's most venomous snake. It is shy and seldom seen, but when encountered it is capable of striking with deadly speed. It feeds mainly on mammals, but also on birds and lizards, sometimes entering burrows to trap its prey underground. Females lay 3–20 eggs each time they breed. Taipan bites can be fatal to humans, but an effective antivenom means that fatalities are now relatively rare.

keeled scales

VIPERS

PHYLUM	Chordata
CLASS	Reptilia
ORDER	Squamata
SUBORDER	Serpentes
SUPERFAMILY	Caenophidia
FAMILY	Viperidae
SPECIES	228

VIPERS ARE THE MOST HIGHLY evolved of all snakes. They have long, hinged fangs, and some (the pit vipers) also have a pair of heat-sensitive pits between their eyes and nostrils. Vipers are found in a wide range of climates, and compared to other snakes are better equipped to deal with cold conditions. Many live at high elevations or in desert areas with cold winters. Some are even found north of the Arctic Circle. Vipers live on the ground or in trees. Some species use rodent burrows as temporary shelters.

ANATOMY

Most vipers are short and stocky with wide, triangular heads and rough, keeled scales. The head has long fangs, which are folded against the roof of the mouth when not in use, and large venom glands. Vipers are generally sluggish snakes and rely on

camouflage to go undetected: they are often coloured to blend in with the surface on which they live and may have intricate geometric markings to break up their outline. Heat detection is more advanced in pit vipers than in any other snakes. Using their heat-sensitive pits, they can not only detect the presence of prey but also gauge its distance and the direction in which it lies. Rattlesnakes, which are a type of pit viper, have a unique warning device in the form of a rattle at the end of the tail.

VENOM

Most vipers ambush their prey, some using the brightly coloured tip of their tail as a lure. They lunge open-mouthed and stab the victim using their long fangs, penetrating fur and feathers, to reach the vital organs. The venom varies from species to species, but often contains proteins that break down blood cells and cause internal haemorrhaging. The venom being relatively slow-acting, vipers deliver large amounts of it, unlike cobras. For the prey, death may not be instantaneous.

REPRODUCTION

Although some vipers lay eggs, most give birth to live young. Females retain the developing eggs in their body and bask during the day to accelerate the embryos' development. Many species breed only every 2, or even 3 years, using the non-breeding years to feed and recover their body weight.

STRIKING POSITION
When preparing to strike, the desert horned viper (shown here) opens its mouth and moves its fangs down from their usual position flat against the roof of the mouth. The jaws are opened wide so that the fangs point forwards towards the prey.

Atheris hispidis

Hairy bush viper

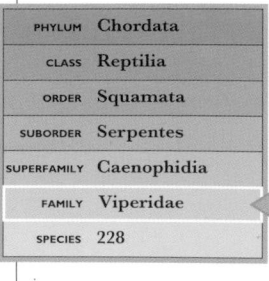

Length 50–73 cm (20–29 in)
Breeding Viviparous
Habit Arboreal

Location C. Africa (Zaire/Uganda border)
Status Lower risk†

This little-known snake is unusually slender for a viper. It is covered with very strongly keeled scales; on the head and neck, these are raised to form small spines. The male is yellowish green, while the female is olive-brown. Mainly nocturnal, this species feeds on frogs, snails, and small mammals. When threatened, it draws its head back, forms an S-shaped coil with its neck, and strikes. All existing antivenoms are ineffective against its bite. The female gives birth to 5–12 live young.

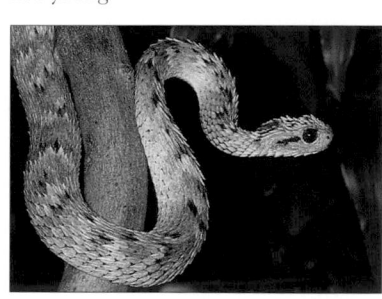

Bitis arietans

Puff adder

Length 1 m (3¼ ft), max 1.9 m (6¼ ft)
Breeding Ovoviviparous
Habit Terrestrial

Location Africa (south of Sahara)
Status Common

The puff adder is a thick-bodied snake with a wide, flattened head and a rounded snout. The male is smaller and often more brightly coloured than the female. Sluggish and slow-moving, larger individuals crawl in a straight line, like a caterpillar. The puff adder

dark-edged, white chevrons

BODY MARKINGS
The puff adder has a yellowish brown or grey body with white chevrons along its back. Some individuals may also have yellow markings.

DEFENSIVE BEHAVIOUR

Highly venomous, the puff adder is ferocious when disturbed, and defends itself by inflating its body and making a long, low hissing sound. If further provoked, it may strike, delivering a fatal bite.

hunts in the evening and at night for small mammals and, occasionally, birds and reptiles. It ambushes prey and strikes quickly, injecting large amounts of venom. Males combat and follow the trails of females in the spring. This adder's litter – of 20–40, and in exceptional cases up to 154 – is the largest of any snake. The puff adder is highly dangerous to humans, and is responsible for most of the lethal snake-bites in Africa.

broad head

Bitis caudalis

Horned adder

Length 30–50 cm (12–20 in)
Breeding Ovoviviparous
Habit Terrestrial

Location Southern Africa
Status Common

The horned adder is grey, brown, reddish, or orange, with a single horn over each eye. Its flattened body and rough scales help it to shuffle down into loose sand in order to escape the heat and to hide. When crossing loose sand, it may move in a sideways looping motion, known as sidewinding. It hunts mainly in the evening – prey may be attracted to its twitching tail.

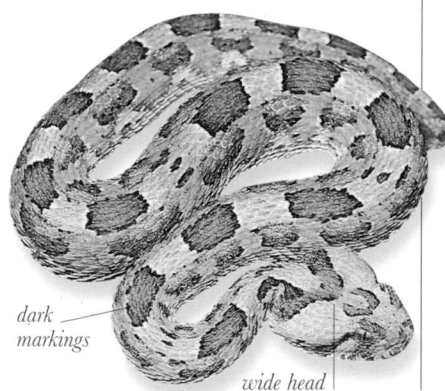

dark markings

wide head

Bitis gabonicus
Gaboon viper

Length 1.2 m (4 ft),
max 2 m (6 1/2 ft)

Breeding Ovoviviparous

Habit Terrestrial

Location W. and C. Africa   Status Locally common

The immense, thickset body, massive, triangular head with small eyes, and patterning of interlocking, geometric shapes along the back, make the gaboon viper instantly recognizable. When seen away from its natural habitat, it appears to be colourful and conspicuous, with its distinct markings in pale purple, tan, cream, and brown. However, in the wild its coloration and patterning provide excellent camouflage,

making it extremely difficult to detect. A sluggish, usually placid snake, it generally lies motionless on the forest floor, moving only when prey – namely rodents and larger mammals, birds, and frogs – comes within striking range. Large, heavy-bodied individuals normally crawl in a straight line, like a caterpillar. When disturbed, this viper gives a loud, low hiss and may bite, but only as a last resort; the bite is often fatal to humans if not treated at once. Males combat during the breeding season, and females give birth to 16–60 young every 2–3 years.

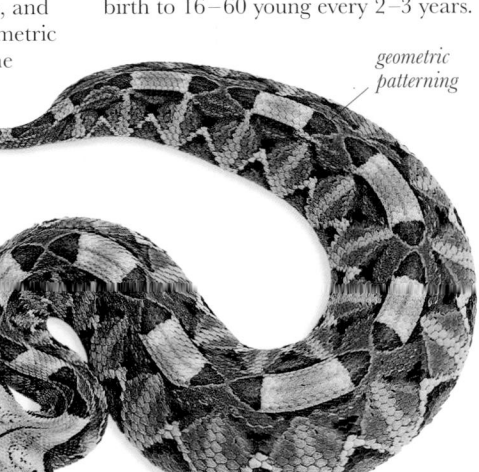

geometric patterning

venom glands behind eyes

Echis pyramidum
Northeast African saw-scaled viper

Length 30–60 cm
(12–23 1/2 in)

Breeding Oviparous

Habit Terrestrial

Location N. and N.E. Africa   Status Common

Despite its small size, this quick-moving snake is highly dangerous, with venom potent enough to kill humans. It belongs to a small genus of vipers (known as saw-scaled or carpet vipers) that have strongly keeled scales that produce a loud rasping sound

when they are rubbed together. When threatened, saw-scaled vipers use this sound as a warning signal, in much the same way as rattlesnakes (see p.396) use their tails. Like its relatives, the northeast African saw-scaled viper hunts small mammals, amphibians, reptiles, and invertebrates at night.

Females lay clutches of 6–20 eggs.

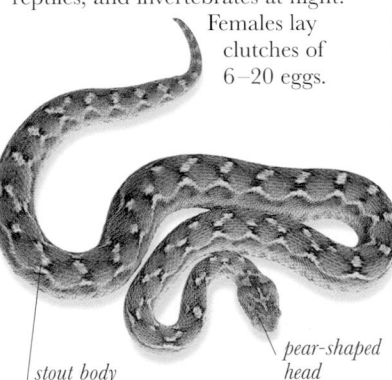

stout body

pear-shaped head

Vipera ammodytes
Nose-horned viper

Length 65–90 cm
(26–35 in)

Breeding Viviparous

Habit Terrestrial

Location S.E. Europe, W. Asia   Status Common

The light brown, reddish brown, or grey nose-horned viper is Europe's most dangerous snake, although its bite is rarely fatal if treated. It has a short, fleshy horn on its snout, the function of which is as yet unknown. It is usually

horn on snout

slow-moving, unless provoked, and often coils quietly among rocks on hillsides. Mainly diurnal, it may become nocturnal in midsummer in certain parts of its range.

zigzag line down back

Cerastes cerastes
Desert horned viper

Length 30–60 cm
(12–23 1/2 in)

Breeding Oviparous

Habit Terrestrial

Location N. Africa   Status Common

Also known as the Sahara horned viper, this short, squat snake usually has a thorn-like horn over each eye. The strongly keeled scales on its body help the snake to shuffle down into the sand, where it remains hidden

and protected from the heat. Active at night, it lies in ambush for rodents, lizards, and birds, sometimes partially buried, then launches a rapid strike. When threatened, it forms a coil, rubbing together the scales on its flanks to create a rasping sound.

dusty colour provides camouflage

Daboia russelli
Russell's viper

Length 1 m (3 1/4 ft)
max 1.5 m (5 ft)

Breeding Viviparous

Habit Terrestrial

Location S. and S.E. Asia   Status Common

Although this is a sluggish animal that usually remains coiled and well hidden during the day, Russell's viper is one of the most dangerous snakes to humans in southern Asia. It is distinguished by its light brown body

with 3 rows of oval, dark-edged brown blotches down its back. A sit-and-wait predator, it relies on its camouflage to escape detection by unwary prey, then strikes. In defence, it forms a tight coil, hisses, and strikes hard, often lifting its body off the ground in the process.

stout body with distinctive markings

Vipera berus
Adder

Length 65–90 cm
(26–35 in)

Breeding Ovoviviparous

Habit Semi-terrestrial

Location Europe, C. to E. Asia   Status Common

The adder – or common viper – is the only venomous snake in northwest Europe, and is the most widely distributed snake in the viper family. It lives in a wide range of habitats, including rocky slopes 3,000m (9,900ft) high in the European Alps, and open tundra north of the Arctic Circle. Its colour varies from grey, brown, or rust to completely black, but most individuals have a distinctive, zigzag line running down their back. Individuals may occasionally reach as much as 90cm (35in) long. Adders feed primarily on small rodents and lizards, either ambushing or actively pursuing them, but they also spend much of their time basking in the sun. In the north of their range, where winters are colder – for example in Scandinavia – they hibernate for up to 8 months each year. In spring, the snakes

emerge to mate, with males engaging in protracted combat for the chance to pair up with females. Females breed only about once every 2 or 3 years, producing litters of up to 20 young that may have been fathered by several different males. Adders are only mildly venomous. In humans, their bite can cause swelling and pain, but is rarely fatal.

flat head

thick body

zigzag line down back

REPTILES

Crotalus atrox

Western diamondback rattlesnake

Length	Up to 2 m (6¹/₂ ft)
Breeding method	Ovoviviparous
Habit	Terrestrial

Location S. USA, N. Mexico

Status Lower risk†

This large, heavy-bodied snake is North America's most dangerous rattlesnake, with the highest rate of human fatalities. A highly venomous and effective hunter, the western diamondback rattlesnake feeds mainly on small mammals, birds, and lizards, but can be deadly to larger animals or humans, if provoked. The rattle, its most distinctive feature, is a horny section at the end of the tail composed of loosely connected segments, which are added to after each moulting. Used to warn and deter intruders, the rattle consists of up to 10 segments in adults. Newborns, measuring about 30cm (12in), do not possess a rattle, but instead have a small "button" which develops with moulting. However, they are already equipped with needle-sharp fangs and extremely toxic venom. It takes between 3 and 4 years for them to reach sexual maturity.

FATAL STRIKE
The western diamondback rattlesnake usually hunts by ambushing its victim, then striking with its venomous fangs before devouring its stunned prey. It has a high yield of highly poisonous venom containing a mixture of toxins that can kill victims in seconds.

forked tongue

broad diamond shapes

segmented rattle

HEAVY-BODIED SNAKE
This bulky snake varies in colour according to its location and habitat, and may be grey, brown, olive-green, or reddish brown. The tail is marked with black and white bands just in front of the rattle.

ATTACK AND DEFENCE

The western diamondback rattlesnake is a well-armed and deadly predator that has developed effective methods for hunting prey. However, the solitary snake will often withdraw from confrontation if disturbed, using its rattle to deter intruders.

heat pits

fanged mouth

buzzing rattle

DEATH RATTLE
The rattlesnake sounds its unmistakable rattle as a means of defence and an ominous warning not to approach.

STALKING PREY
The western diamondback rattlesnake uses the pair of heat-sensitive pits between its eyes and nostrils to detect and locate warm-blooded prey.

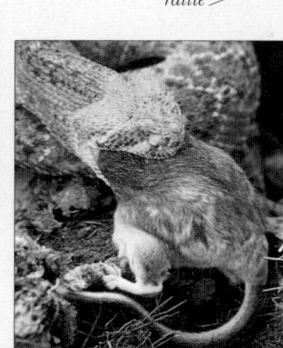

STUNNED VICTIM
Having stunned a rodent victim with its venomous bite, the rattlesnake waits for it to die before swallowing it whole.

Agkistrodon contortrix
American copperhead

Length 60–130 cm
(23 1/2–51 in)
Breeding Viviparous
Habit Terrestrial

Location C. and S.E. USA   Status Common

The American copperhead is boldly marked with chestnut-brown or orange bands on a grey, tan, or pinkish background. In spring and autumn it basks during the day, but it becomes nocturnal in summer.

This viper lies practically hidden among dead leaves or rocks waiting to ambush rodents, frogs, lizards, and invertebrates, using the heat-sensitive pits between its eyes and nostrils to locate prey in total darkness. Its bite is venomous, but rarely serious for humans. The female gives birth to 1–15 young every other year.

light brown bands

triangular head

Bothriechis schlegelii
Eyelash pit viper

Length 45–75 cm
(18–30 in)
Breeding Viviparous
Habit Arboreal

Location Central America, N. South America   Status Locally common

This tropical forest snake is one of a number of pit vipers that live in trees. Like most arboreal snakes, it has a prehensile tail and good binocular vision, but it also has distinctive "eyelashes", formed by raised scales above its eyes. Active at night, it preys on lizards, frogs, and small rodents, and lies in wait near forest flowers to catch hummingbirds as they come to feed. Eyelash pit vipers vary greatly in colour, ranging from an intense orange-yellow to greyish green.

Bothrops atrox
Common lancehead

Length 1 m (3 1/4 ft),
max 1.5 m (5 ft)
Breeding Viviparous
Habit Terrestrial

Location N. South America   Status Common

This is one of the most dangerous snakes in South America because of its rapid strike, potent venom, and its tendency to live near human habitations. Heat-sensitive pits between the eyes and nostrils enable it to detect prey in darkness. It is sometimes mistakenly known as the "fer-de-lance" – a West Indian species with a similarly pointed head. Males have longer tails than females, and may fight during the breeding season. Females may give birth to up to 100 young. Juveniles have brightly coloured tail tips, which they use to lure prey.

irregular dark blotches

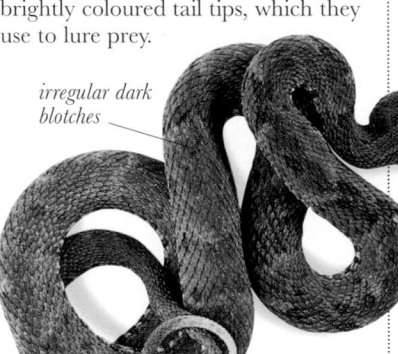

Calloselasma rhodostoma
Malaysian pit viper

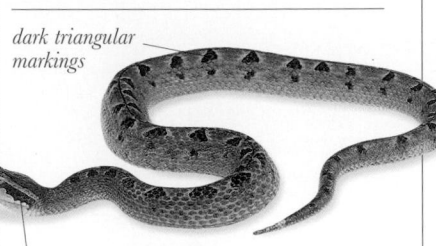

Length 70–100 cm
(28–39 in)
Breeding Oviparous
Habit Terrestrial

Location S.E. Asia   Status Common

dark triangular markings

wedge-shaped marking behind each eye

This highly venomous snake, usually found at the edges of forests and in clearings, is quick to strike, and is the cause of many snake-bite fatalities in Southeast Asia. It is purplish brown, with light-edged, darker triangles on its flanks, and has a pointed snout. A nocturnal hunter, the Malaysian pit viper uses the heat pits between its eyes and nostrils to help it to locate prey in the dark. Unlike most vipers, this species lays eggs – the female may lay 2 clutches of 13–25 eggs each in a year – and remains coiled around them throughout their incubation.

Crotalus durissus
Tropical rattlesnake

Length 1 m (3 1/4 ft),
max 1.5 m (5 ft)
Breeding Viviparous
Habit Terrestrial

Location Central America, South America   Status Locally common

strongly keeled scales

The bite of the tropical rattlesnake is highly venomous – about 75 per cent of untreated bites result in fatalities. There are 14 subspecies recognized, with highly variable colouring, but this is the only rattlesnake over much of its range. Heavily built, it has a broad head and obvious heat-sensitive pits, to detect prey in total darkness. This nocturnal species spends the day coiled, partly hidden by vegetation. The female gives birth to 6–12 young.

Crotalus cerastes
Sidewinder

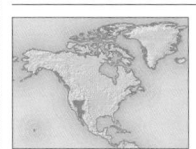

Length 45–80 cm
(18–32 in)
Breeding Ovoviviparous
Habit Terrestrial

Location S.W. USA, N.W. Mexico   Status Common

The sidewinder is the stoutest of all rattlesnakes and is the only one in which the female is smaller than the male. It has a highly distinctive way of moving – an adaptation to life on windblown desert sand. It feeds mainly on small lizards and rodents, often ambushing its prey from the cover of isolated shrubs. Although sidewinders can be aggressive, by rattlesnake standards their bite is not especially dangerous. Their young are born in late summer, in clutches of 6–12.

BODY COLORATION
The sidewinder varies from cream, grey, tan, or brown to pinkish, depending on the soil on which it lives; markings are poorly defined.

MOVING ON SAND

Sidewinding is a form of movement used by several snake species that live on loose sand. Instead of sliding forwards, with its body in contact with the sand, a sidewinding snake throws itself over the surface, following a diagonal path. This kind of movement creates characteristic parallel tracks.

flattened body

Lachesis muta
Bushmaster

Length 2.5 m (8 1/4 ft),
max 3.6 m (12 ft)
Breeding Oviparous
Habit Terrestrial

Location S. Central America, N. South America   Status Vulnerable

The bushmaster is the longest viper in the world. It is also one of the most venomous, with a 20 per cent mortality rate, even if treated; however, it generally avoids contact with humans. A secretive species, it ambushes prey at night – coiled and ready to strike – along a mammal trail. It has a slender, yellow or tan body with black markings. Unusually for a viper, females lay eggs (about 8–15), guarding them until they hatch.

Trimeresurus albolabris
White-lipped pit viper

Length 60–100 cm
(23 1/2 – 39 in)
Breeding Viviparous
Habit Arboreal

Location S.E. Asia   Status Common

This slender-bodied snake – easily confused with several other green pit vipers from the same part of the world – has a wide, rounded head with prominent heat pits. It lives in trees, hanging head-down at night in a favoured hunting place waiting for prey to come within range, then drawing back its head when ready to strike. Its bite is painful, but the venom is mild and rarely fatal to humans. Females give birth to 10–11 young.

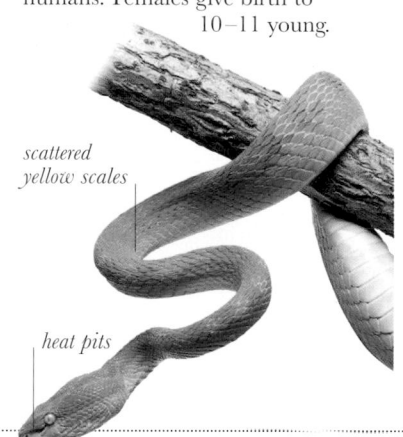

scattered yellow scales

heat pits

Tropidolaemus wagleri
Wagler's viper

Length 0.8–1.3 m
(2 1/2 – 4 1/4 ft)
Breeding Viviparous
Habit Arboreal

Location S.E. Asia   Status Locally common

Renowned for living in semi-captivity in Malaysia's Penang Snake Temple, this highly variable snake is sluggish during the day, when it rests in trees. At night, it ambushes small mammals, birds, lizards, and frogs by hanging head-down from the branches. It is a thickset viper, with a chunky head, and has various colour combinations: usually black, with green or yellow crossbars and scattered yellow spots. Juveniles are green, with yellow and/or red spots arranged into vague cross-bands; some individuals retain their juvenile markings throughout their lives. Although its bite is painful, it is seldom fatal to humans.

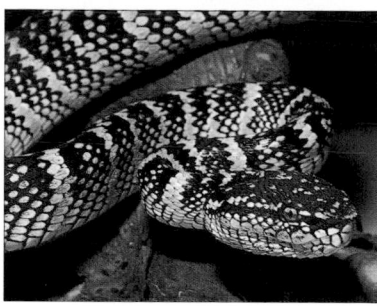

BLIND AND THREAD SNAKES

PHYLUM	Chordata
CLASS	Reptilia
ORDER	Squamata
SUBORDER	Serpentes
SUPERFAMILY	Scolecophidia
FAMILIES	3
SPECIES	319

THESE SMALL SNAKES LIVE underground, emerging above the surface only when driven out by lack of food or by flooding. They have a slender, cylindrical body and are covered with smooth, shiny scales that allow them to move easily through sand and soil. Their skull is more heavily built than in other snakes, and they have jaws that they cannot open widely, with few teeth. Their eyes are covered with scales and are barely functional. Blind and thread snakes are restricted to tropical and subtropical regions. They are often found within ant and termite mounds, and the bulk of their diet is composed of these insects and their larvae.

BURROWING
A thread snake above ground is an uncommon sight. This Peters thread snake from southern Africa has emerged from its burrow in sandy soil.

REPTILES

Leptotyphlops dulcis
Texas thread snake

Length 15–27 cm
(6–10 1/2 in)
Breeding Oviparous
Habit Burrowing

Location S. USA,   Status Common
N.E. Mexico

The slender-bodied, pinkish brown Texas thread snake lives almost exclusively below the ground, coming to the surface only at night and after rain. The smooth scales, short tail, and vestigial, scale-covered eyes are all features associated with a burrowing lifestyle. This snake feeds opportunistically on small insects and spiders, and also enters ant and termite nests in search of larvae. When the snake enters a nest, it releases pheromones that imitate that of the insects. The soldiers, deceived into thinking that the snake is part of the nest, allow it to enter and feed in safety. The chief predators of the Texas thread snake are other snakes and nocturnal birds and mammals. The female lays 2–7 eggs and coils round them until they hatch.

silvery sheen

Ramphotyphlops braminus
Brahminy blind snake

Length 15–18 cm
(6–7 in)
Breeding Oviparous
Habit Burrowing

Location W. Asia to N.   Status Common
Australia, Pacific islands,
North America, South Africa

This is the only snake known to be parthenogenetic. Males have never been found, and females lay their eggs without having to mate. Each egg is about the size of a peanut. The snake's habit of stowing away in plant pots has resulted in the spread of this species throughout warm parts of the world. Its scale-covered eyes, blunt snout, and smooth scales are all features adapted to burrowing. If in danger, this snake may give off an unpleasant, pungent fluid.

Rhinotyphlops schlegelii
Schlegel's beaked blind snake

Length 60–95 cm
(23 1/2 – 37 in)
Breeding Oviparous
Habit Burrowing

Location C. to southern   Status Common
Africa

This is the largest of approximately 180 species of blind snakes scattered throughout warm regions of the world. Like other blind snakes, it has smooth scales and reduced, scale-covered eyes that make it well adapted to life underground. Its tail ends in a short spine, and its overall colour varies from light grey, just after shedding its skin, to brown with spots and stripes. An inhabitant of dry scrubland and coastal plains, this snake is largely nocturnal, feeding on termites and other invertebrates. Females lay their eggs in underground chambers, and produce up to 60 eggs in each clutch.

LIZARDS

PHYLUM	Chordata
CLASS	Reptilia
ORDER	Squamata
SUBORDER	Lacertilia
FAMILIES	19
SPECIES	About 4,500

CLASSIFICATION NOTE

The classification of lizards, like that of snakes, is the subject of much debate, especially the way they are grouped into families. However, their division into the following superfamilies is widely accepted:

Iguanas and relatives
see pp.402–8
Geckos and snake-lizards
see pp.409–11
Skinks and relatives
see pp.412–17
Anguimorph lizards
see pp.418–22

LIZARDS ARE THE most successful group of reptiles. They form a large and varied group that is not easy to define concisely: although a typical lizard has a distinct head, 4 well-developed limbs, and a long tail, there are exceptions to all of these generalizations. Some lizards also have an unusual ability to shed their tail when attacked by a predator and then regenerate it. Most species reproduce by laying eggs, although some give birth to live young; a small number actively care for their offspring. Lizards have adapted to survive in habitats all over the world (although they are not found in Antarctica). Most live on the ground, on low-lying rocks, or in trees. There is also a small number of burrowing species and even one that feeds in sea water. A few lizards, notably the geckos, have adapted to live around humans, and in some parts of the world they are a common sight on the walls and ceilings of buildings.

DESERT LIZARDS
Deserts are an important habitat for lizards, such as these web-footed geckos. Like all reptiles, they do not require large amounts of water because they use only small amounts to expel waste. Most of the water they need is obtained from dew and fog. Webbed toes allow them to walk on soft sand.

REGENERATION

Some lizards are able to voluntarily break off part of their tail, a phenomenon known as autotomy. This is a defence mechanism that helps the lizard to escape when grabbed by the tail. To add to the effectiveness of the strategy, the tail may be brightly coloured, often blue, to draw the attacker's attention away from the head. The tail may also continue to move for several minutes after being severed, apparently to distract the predator. The tail breaks at one of several predetermined weak points (see below), and it does not have to be pulled in order to cause a break. Most lizards can regrow a broken tail, but the replacement will not have the same scale arrangement or coloration as the original. A lizard may shed its tail several times during its lifetime.

NEW GROWTH
This viviparous lizard has shed its tail and grown a replacement. The point at which the old tail was broken can be identified by the change in scale colour.

join between vertebrae

fracture plane within vertebra

FRACTURE PLANES
A lizard's tail breaks at one of several weak points, called fracture planes, across (rather than between) certain vertebrae. The muscles are arranged in such a way that they, too, will come apart neatly.

ANATOMY

As well as a distinct head, 4 legs, and a tail, most lizards have external ear openings, and eyelids that can be opened and closed. However, some lizards have short legs or none at all. There are also forms with no eyes or no ears, while others have eyes covered by a transparent scale rather than movable lids. The tongue may be short or long and is usually notched at the tip; forked tongues are found in some species. A lizard's scales may be large, smooth, and overlapping or small, rough, and stud-like, in almost any combination. Many have irregular scales or a few large scales scattered among many smaller ones, but none has the single row of wide ventral scales characteristic of snakes (see p.376). Most lizards have sharp teeth along the edges of their jaws.

SENSES
A lizard's senses are adapted to its lifestyle. Many species that live underground cannot see, while those living above ground may have good vision, but only a few are able to see colours. Lizards with external eardrums can hear airborne sounds.

nostril

eyes on side of head

external ear opening

FEET AND CLAWS
Most lizards have 5 digits on each foot. Many species (particularly those that climb, such as this monitor lizard) have sharp claws. Most lizards cannot oppose their digits, although chameleons have their digits fused into 2 opposing groups, enabling them to grip branches.

toe with sharp claw

LEGLESS LIZARDS
Over time, some lizards (such as this Burton's snake-lizard) have lost their limbs. This is an adaptation that helps them to move easily through soil burrows, sand, and leaf litter.

MOVEMENT

Most lizards move by walking, climbing, or burrowing. When walking, they usually use all 4 limbs (see right), although some species lift their front legs off the ground while running at high speed; basilisk lizards can even run for short distances over the surface of water in this way, helped by broad feet that spread their weight. Some lizards are agile climbers. To improve their grip, they may have long, sharp claws, while some have a prehensile tail. Among the most impressive climbers are geckos, which can cling to smooth, vertical or even overhanging surfaces using adhesive pads at the tips of their digits (see p.409). Burrowing lizards may use powerful front limbs for digging, although others have reduced limbs or no limbs at all and simply force their way through loose soil and sand with rapid "swimming" movements. A few lizards can glide from high trees (see below).

head moves slightly

tail swings widely from side to side

WALKING
A lizard's limbs are held at right angles to the body, rather than supporting it from underneath. This means that as the lizard walks, the weight of the body is thrown from side to side, resulting in a characteristic wriggling motion and causing its tail to swing and provide a counterbalance.

flap of skin supported by long ribs

long, whip-like tail

FLYING
The flying lizard (above) has flaps of skin, which act like wings, stretched between elongated ribs. Some other species have similar flaps between their toes. In all cases, the lizards move from tree to tree by gliding rather than powered flight.

CLIMBING TREES
Some large lizards, such as this perentie (right), are surprisingly agile climbers. Perenties usually climb only to escape predators, but other lizards, especially the chameleons, spend the greater part of their lives in trees.

CAMOUFLAGE
Chameleons (such as the Cape dwarf chameleon, left) have a remarkable ability to alter the appearance of their skin. Not only does the colour change, but markings, such as lines and bars, may appear and disappear.

DEFENCE

Lizards use various active and passive strategies to avoid the many other animals that prey upon them. Apart from running away, passive strategies include camouflage and mimicry of inanimate objects, such as sticks and leaves. Some species can alter their coloration, while others make themselves inconspicuous by pressing their bodies close to rocks or tree trunks to avoid casting a shadow. Active strategies include the voluntary loss of the tail (see Regeneration, facing page), and threat displays, which often involve warning coloration (as in blue-tongued skinks) or body enlargement (as in the frilled lizard). Large species such as monitors, as well as some smaller ones, defend themselves by biting, scratching, or lashing their tail. Two species of lizard, the Gila monster and the Mexican beaded lizard, are venomous.

FIGHTING DRAGONS
The Komodo dragon (right) is the largest of all lizards. When attacking its prey, it usually relies on its jagged teeth and sharp claws. However, when defending itself from another member of its own species, it will often turn its back and thrash its powerful tail.

IGUANAS AND RELATIVES

PHYLUM	**Chordata**
CLASS	**Reptilia**
ORDER	**Squamata**
SUBORDER	**Lacertilia**
SUPERFAMILY	**Iguania**
FAMILIES	**3**
SPECIES	**1,412**

THE IGUANAS AND their relatives include some of the most colourful and familiar of all lizards. Along with the geckos and snake-lizards, they are also the most primitive. They are divided into 3 groups: the iguanids, agamids, and chameleons. Many species, notably the chameleons, are capable of changing their skin colour. Some are highly territorial or have elaborate social rituals. Iguanids and agamids look similar to each other but are found in different parts of the world: iguanids occur throughout the Americas, in Madagascar, and on a few Pacific islands, while agamids are found in Africa, Asia, and Australasia. Chameleons occur in Africa, Madagascar, Asia, and Europe. Iguanas and their relatives occupy various habitats, including desert, grassland, and rainforest. Most species lay eggs but a small number give birth to live young.

ANATOMY

Iguanids and agamids look alike and occupy similar ecological niches. The main difference between them is in the teeth: unlike iguanids, the agamids have teeth that are fixed to the jawbone and not replaced if lost. Most species have small scales and a loose fold of skin (or dewlap) under the head. The males are often brightly coloured, and many species have ornamentation on their head and a crest down their back. They may also have a brightly coloured throat fan, which they use for communication.

Chameleons are adapted to life in the trees. Their narrow body and leaf-like profile help to make them inconspicuous among foliage, while a prehensile tail and toes that are fused in 2 opposing groups enable them to hold tightly to branches. Chameleons also have a well-known (but often exaggerated) ability to change colour (see below). Male chameleons usually have a crest or horns on their head.

COLOUR CHANGE

Two types of colour change can be seen in the lizards in this group. Chameleons change colour by expanding or contracting pigment cells (chromatophores) scattered in their skin. When the cells are expanded, the skin darkens due to the dispersal of melanin pigments in the cells; as the cells are contracted, the skin becomes lighter. A great number of different colours and patterns can be produced in this way. Chameleons change colour mainly to control their temperature (making themselves darker when they need to warm up) or to communicate with other members of their species, rather than for camouflage. At night, while they are asleep, chameleons always become paler. Small leaf chameleons and pygmy chameleons, which are brown, have limited ability to change colour.

Agamids and iguanids also change colour, but this takes place more slowly than in chameleons. Mature males become brighter in the breeding season. Females may change colour according to whether or not they are receptive to males: once they have mated, they often develop orange spots (known as gravid coloration) to deter unwanted attention from males.

COMMUNICATION

In many species of iguanas and their relatives, the males use their bright coloration for display, both to females and to other males. Often occupying a prominent vantage point, they will nod their head vigorously or bob up and down, sometimes exposing brightly coloured scales on their throat and chest. Males are highly territorial, and will chase off other males and mate with any receptive female that lives in or strays into their territory.

HUNTING

Chameleons, such as this veiled chameleon, stalk their prey slowly, locating it by sight. Each eye can swivel independently in any direction and is supported on a turret. The muscular tongue is extremely long with a thick, adhesive tip. When prey is within range, the chameleon flicks its tongue out at great speed, catching the prey on the tip and pulling it back into its mouth.

Acanthosaura crucigera
Mountain horn-headed lizard

Length 26–30 cm
(10–12 in)
Breeding Oviparous
Habit Arboreal

Location S.E. Asia | Status Common

Relying on its greenish yellow, cryptic colouring, this lizard often escapes detection from predators by remaining motionless, although

it may open its mouth wide and bite if cornered. It has long, thin limbs, a long tail, a crest of spines on its back, and larger spines on the nape of the neck and over the eyes. The male is larger, more colourful, and has a more pronounced crest than the female. This lizard lies in wait for insects, then suddenly darts out to catch them. Females lay 10–12 eggs at a time in leaf litter or debris.

black "mask"
around eyes

black collar
around throat

Chlamydosaurus kingii
Frilled lizard

Length 60–70 cm
(23½–28 in)
Breeding Oviparous
Habit Terrestrial

Location S. New Guinea, N. Australia | Status Common

Australia is home to a wide range of lizards, but this species is the most spectacular. Its name derives from the large, leathery frill around its neck, which is normally folded back over the animal's shoulders. However, if the lizard is threatened, it opens the frill like an umbrella, and at the same time rocks its body and hisses through its gaping mouth. This display is often enough to deter would-be predators, allowing the frilled lizard to make its escape.

orange patches
at base of frill

DEFENSIVE BLUFF
This lizard's frill consists of a flap of skin strengthened by rods of cartilage. It is often brightly coloured, contrasting with the animal's grey or brown body.

CLIMBING TO SAFETY

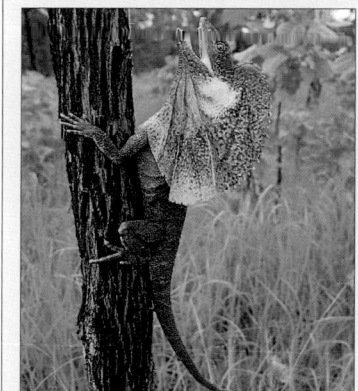

Frilled lizards forage mainly on the ground. They are good climbers, and after carrying out their defensive display they usually run up a tree trunk, heading for the highest branches.

Agama agama
Rainbow lizard

Length 30–40 cm
(12–16 in)
Breeding Oviparous
Habit Terrestrial

Location W., C., and E. Africa | Status Common

Male rainbow lizards are among Africa's most conspicuous reptiles. At night, they are drab and grey, but when they warm up in the sun they develop a bright orange head, and a blue or turquoise body. Females and juveniles are quite different, with a grey coloration that makes good camouflage. Rainbow lizards live in open habitats, and are often seen around buildings. Males have a head-bobbing courtship display, and females lay 5–7 eggs in holes in the ground.

brightly
coloured
male

long,
thin tail

Draco spilonotus
Flying lizard

Length 15–20 cm
(6–8 in)
Breeding Oviparous
Habit Highly arboreal

Location S.E. Asia | Status Locally common

The flying lizard has a pair of "wings", consisting of skin stretched between elongated ribs, that help it glide from one tree trunk to another to avoid predators such as snakes and birds. To take off, the lizard moves its

ribs outwards and opens the flaps of skin, closing them again on landing. It is generally tan or pale brown, with a throat flap for display that is large and bright yellow in the male, but small and pale yellow in the female. Highly agile, this lizard lives on tree trunks and actively hunts for ants. Females only leave the safety of trees to lay 8–12 eggs, which they bury in sand or soil.

"wings" aid
gliding between
tree trunks

Calotes versicolor
Garden lizard

Length 30–35 cm
(12–14 in)
Breeding Oviparous
Habit Terrestrial

Location W. Asia to E. and S.E. Asia | Status Common

strongly
keeled body
scales

This long-tailed lizard is one of the commonest species in southern Asia, thriving in farmland and gardens as well as in natural habitats. Its colour is highly variable, but breeding males always have a red area around the mouth and throat – a feature that explains why this lizard is sometimes known as the bloodsucker. Garden lizards are good climbers, and they hunt insects and other small animals by lying in wait, then lunging at their prey. There are many similar species throughout Asia.

Hypsilurus spinipes
Southern angle-headed lizard

Length 20–33 cm
(8–13 in)
Breeding Oviparous
Habit Arboreal

Location Australia (S.E. Queensland and N.E. New South Wales) | Status Locally common

This lizard is characterized by a large, angular head, a raised crest (more prominent in males) on the nape and back, and a long

tail. It can "freeze" to avoid detection – a useful tactic against predators such as birds, mammals, and monitor lizards – and its exceptionally long toes and claws enable it to climb trees rapidly to escape from danger on the ground. An active forager, the southern angle-headed lizard lunges quickly and suddenly at insects, its sole food source. The female lays eggs in depressions in soil or in leaf litter.

Hydrosaurus pustulatus
Sailfin lizard

Length 80–100 cm
(32–39 in)
Breeding Oviparous
Habit Arboreal

Location Philippines | Status Vulnerable†

As an adult, this large, grey or greenish grey lizard has a well-developed crest of tooth-like scales from the nape of the neck down the back. The tail also

has a crest, up to 8 cm (3¼ in) high, which helps the lizard to swim. When in danger from predators, mainly birds of prey, this lizard dashes away at great speed, often diving into water to escape. At other times it may bask on branches that overhang or are near streams. It forages for leaves and fruits, and feeds on insects opportunistically.

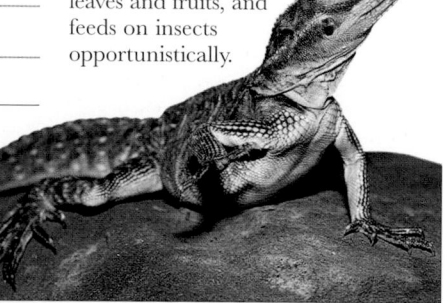

Moloch horridus

Thorny devil

Length	15–18 cm (6–7 in)
Breeding	Oviparous
Habit	Terrestrial

Location W. to C. Australia **Status** Vulnerable†

Although it is not often seen – partly because it is well camouflaged – the buff, tan, or grey thorny devil is instantly recognizable by its squat shape, its slow way of moving about, rocking backwards and forwards as it walks, and its armoury of sharp spines. These cover the whole of its body, with the largest spines on its head and back, and much smaller ones on its legs and feet. Such protection is vital for the thorny devil because, unlike most lizards, it spends a lot of time foraging in one place on the ground, leaving it vulnerable to attack while feeding. Like the desert horned lizard (see p.408), which shares a similar way of life, the thorny devil feeds during the day, when ants are on the move, and can consume up to 2,500 insects during a single meal, catching them with its projecting tongue. Females lay 3–10 eggs in underground burrows in summer.

short head

black-edged, dark brown markings

Physignathus cocincinus

Thai water dragon

Length	80–100 cm (2 1/2 – 3 1/4 ft)
Breeding	Oviparous
Habit	Terrestrial

Location Thailand, Cambodia, Vietnam **Status** Common

The Thai water dragon is a large lizard with a crest of prominent, tooth-like scales from neck to tail and a raised hump on its neck. It has a green body, but the chin and throat are white, pale yellow, or pinkish. Males become brighter in colour during the breeding season. This lizard typically basks near water, into which it may flee for safety, especially when hunted by snakes and birds of prey. It hunts primarily by ambushing its prey, which comprises invertebrates and small vertebrates such as lizards; however, it also forages actively for food, and may occasionally graze upon vegetation. Females lay multiple clutches of 10–15 eggs throughout an extended breeding season, in burrows or shallow scrapes, and cover the eggs with soil and leaves.

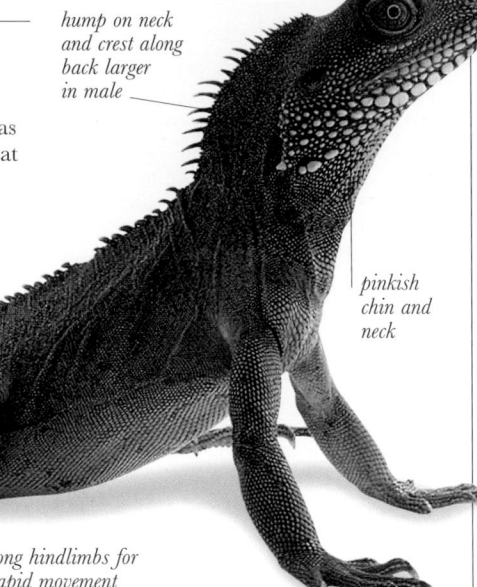

hump on neck and crest along back larger in male

pinkish chin and neck

long hindlimbs for rapid movement

Pogona vitticeps

Inland bearded dragon

Length	30–45 cm (12–18 in)
Breeding	Oviparous
Habit	Semi-arboreal

Location Australia **Status** Common

This bulky lizard inhabits dry forest, woodland, and scrubland, resting on stumps or raised humps on the ground, and foraging for small insects and vegetation near its regular perch. If confronted by a predator – usually birds of prey, monitor lizards, or snakes – it erects its black throat frill, or "beard", and presents a wide, open mouth. The males, and even the females, are territorial. The female lays several clutches of 15–30, and occasionally more, eggs throughout the breeding season, burying them in sandy soil. The hatchlings' body markings are more distinctive than those of adults, but they do not possess a beard.

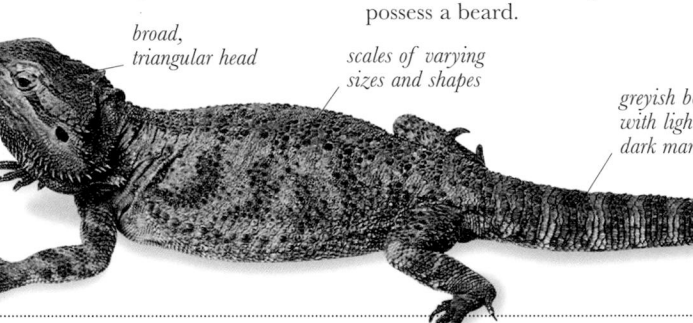

broad, triangular head

scales of varying sizes and shapes

greyish body with light and dark markings

Bradypodion thamnobates

Natal Midlands dwarf chameleon

Length	15–19 cm (6–7 1/2 in)
Breeding	Viviparous
Habit	Arboreal

Location Southern Africa (Kwa-Zulu Natal) **Status** Lower risk

Dwarf chameleons are largely restricted to southern Africa, where they live in forest and scrub. They differ from most other chameleons in giving birth to live young. The Natal Midlands dwarf chameleon is a typical species, with a high protuberance or casque covering its neck. Its coloration is highly variable, but the long-tailed males, which are often green, are more colourful than the females and juveniles, which are brown or grey. Like other chameleons, it is specially adapted for life in trees, and feeds on insects, which it captures with its long, sticky tongue.

conical scales form crest on back

Uromastyx acanthinura

Spiny dab lizard

Length	30–40 cm (12–16 in)
Breeding	Oviparous
Habit	Terrestrial/ Burrowing

Location N. Africa **Status** Locally common

The spiny dab lizard has a short, thick, armoured tail that it uses to block the entrance to its burrow to deter predatory birds or mammals. Inside the burrow, it inflates its body to wedge itself in. It may also use its tail as a club and may bite if cornered.

Males, and occasionally females, are aggressively territorial. The coloration of this lizard varies according to temperature: grey or yellowish brown in cold weather, when it hibernates, changing to orange, red, yellow, or green when active. It emerges only on warm days, basking and orientating its body to the sun, until its body temperature is high enough for foraging. Its diet consists of insects and plants, and it may walk over 1km (2/3 mile), zigzagging across rocky outcrops in the desert, to find favoured plant species. Females lay at least 2 clutches of 20–30 eggs a year, then bury them in side chambers off the main burrow, smoothing over the nest to disguise the site.

tail covered with whorls of spiny scales

very small scales

Rhampholeon spectrum
Western pygmy chameleon

Length 7–10 cm
(2 3/4 – 4 in)
Breeding Oviparous
Habit Partly arboreal

Location W. to C. Africa **Status** Locally common

This tiny chameleon – one of the world's smallest – lives in rainforests, where it is camouflaged by its leaf-like colour and shape. Unlike most chameleons, its ability to change colour is limited. It hunts in leaf litter and on low branches, hiding in shrubs at night.

tail thicker at base in male

horn on snout

Furcifer pardalis
Panther chameleon

Length 40–52 cm
(16–20 1/2 in)
Breeding Oviparous
Habit Arboreal

Location Reunion Island, **Status** Common
E. and N. Madagascar

Madagascar and its neighbouring islands have over 2 dozen species of chameleon, which is more than in any similarly sized region of the world. The panther chameleon, particularly the adult male, is one of the most brightly coloured, with a bewildering array of colour schemes, including green, brick-red, turquoise, or any combination of these hues. It can change its colour rapidly, but it usually retains its bright lateral stripe. Panther chameleons live in lightly wooded, scrubby habitats, and catch insects and other small animals by shooting out their sticky tongues, which can be longer than the rest of their bodies. They are usually slow-moving, but at the onset of the breeding season, males fight for territories. Each year females lay several clutches of 12–50 eggs, burying them in moist soil.

CHANGING COLOUR

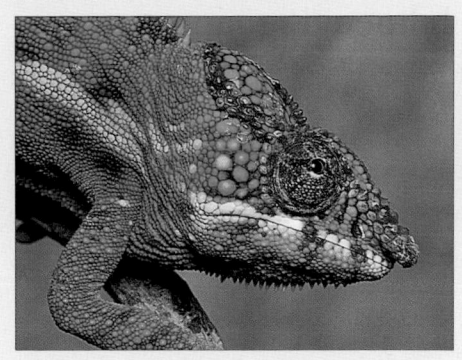

Chameleons are famous for being able to change colour for camouflage, but in most species colour is equally important as a means of communication. Male panther chameleons attract females by producing sudden flushes of colour, while females change colour when pregnant to signal that they are not receptive to mating.

bright lateral stripe usually present

prehensile tail

highly variable coloration

FLATTENED SNOUT
One of the larger chameleons, the panther chameleon has a bony, flattened region on top of its snout, fringed with enlarged scales. The male has a small nasal appendage while the female (shown left) has none.

Furcifer oustaleti
Oustalet's chameleon

Length 50–68 cm
(20–27 in)
Breeding Oviparous
Habit Arboreal

Location Madagascar **Status** Locally common

larger casque on male

fused toes for gripping, common to all chameleons

The largest of all the chameleons, this species is rather dull in colour – brown with some yellow, red, or greenish spots – and has no horns. However, it has a large, flattened casque over its head and neck. Unusually for a chameleon, the male is less colourful than the female, but he becomes more flushed with colour during courtship. Females lay approximately 50 eggs, in soil or leaf litter. Slow and deliberate in its movements, Oustalet's chameleon carefully stalks prey such as insects and small vertebrates before capturing them with its extendible, sticky tongue. It defends itself against predatory snakes and birds by inflating its body, presenting its side to the enemy, and opening its mouth wide; it may also hiss and bite.

Chamaeleo jacksoni
Jackson's chameleon

Length 20–30 cm
(8–12 in)
Breeding Ovoviviparous
Habit Arboreal

Location E. Africa, **Status** Locally common
introduced in Hawaii

With their 3 forward-pointing horns, males of this species are among the most distinctive of all chameleons, and at one time were often caught and sold as pets. Both sexes have a flattened body and serrated crest along the back, with a bony casque over the neck. Their normal colour is green, but specimens that live in East Africa's mountain forests are usually brighter and larger than ones that live lower down. Jackson's chameleon relies primarily on its camouflage to avoid being attacked. However, if cornered, it inflates its body and gives a menacing hiss. Females are ovoviviparous, producing 2 litters of up to 50 young each year. Newborns are light brown, but turn green at about 4 months. In males, horns start to form about 2 months later. In the wild, the lifespan of Jackson's chameleon is probably 2–3 years.

serrated crest along the back

HORNLESS FEMALE

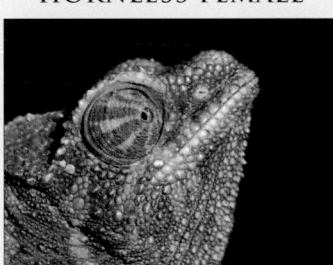

Male and female Jackson's chameleons are easily told apart: the female (shown above) is usually stockier than the male, and the horns are tiny or absent.

HORNS ON MALE
The male Jackson's chameleon has 3 long, pointed, bony horns on the front of his head, used in fights with other males over territory.

Calumma parsonii
Parson's chameleon

Length 50–60 cm
(20–23 1/2 in)
Breeding Oviparous
Habit Arboreal

Location N. and E. **Status** Locally common
Madagascar

This very large, deep-bodied chameleon is usually greenish, but may also be blue. The male is larger and more colourful than the female, with a larger casque on the top of the head and a warty, flattened horn, which is absent in females, on the snout. Parson's chameleon lives mainly in the trees and is slow-moving, relying on camouflage to escape detection. Defensive behaviour includes inflating its body, hissing, or biting, and it may become more brightly coloured to deter predators. Males of this species are highly territorial, and mate with receptive females without preliminaries. The female lays 16–38 eggs, burying them in a hole dug in moist soil.

leaf-shaped body adapted to arboreal existence

Corytophanes cristatus
Helmeted iguana

	Length 30–40 cm (12–16 in)
	Breeding Oviparous
	Habit Arboreal
Location S. Mexico to N. South America	**Status** Locally common

The most distinctive feature of the helmeted iguana is the high, narrow casque that extends from the head to the nape of the neck. Other characteristics include long, spindly legs, a narrow body, and a long, thin tail. A crest of small, tooth-like scales runs along the back and top of the tail. This iguana relies heavily on camouflage for defence, and runs away when threatened, usually by

KEY FEATURES

The high, narrow casque on the head distinguishes this iguana from all other lizards in tropical America; the precise function of this casque, or "helmet", is unknown. When this iguana is defending its territory, it extends the pouch under its head to ward off other animals.

extended pouch

casque extends from top of head to neck

birds of prey or snakes. It feeds on insects, remaining still until they come within range, then darting out to catch them. Females lay eggs in leaf litter or soil.

head small in relation to casque

FOREST CAMOUFLAGE
The helmeted iguana can change colour voluntarily, like a chameleon, from green to tan, brown, or black, helping it to escape detection in its tropical forest habitat.

long, spindly legs

Crotaphytus collaris
Collared lizard

	Length 20–35 cm (8–14 in)
	Breeding Oviparous
	Habit Terrestrial
Location C. and S. USA, N. Mexico	**Status** Common

The brown or greenish brown collared lizard is distinguished by the black and white markings around its neck. It has long hindlegs that enable it to flee rapidly from danger, often to retreat to a burrow or beneath a boulder; the front limbs may leave the ground entirely when it runs fast. Active only on hot days, this lizard perches on rocks looking for insects and smaller lizards. Its strong jaws can crush even large prey, and may also be used to bite in defence. Females develop orange patches when carrying their eggs, which they lay in burrows or under rocks.

Amblyrhynchus cristatus
Marine iguana

	Length 50–100 cm (20–39 in)
	Breeding Oviparous
	Habit Semi-aquatic
Location Galapagos Islands	**Status** Vulnerable

The marine iguana, with its large, scaly head and back, is the only lizard that forages for food in the sea. It dives for algae in the cold waters off the Galapagos Islands, and its body is specially adapted to cope both with cold and with excess salt. Large adults may dive down to 12m (39ft), and can stay underwater for over an hour, but under normal conditions they feed during shallow dives lasting less than 10 minutes. When not feeding, marine iguanas warm up by basking on rocks, and often several thousand can be seen together on the same

stretch of shore. During the breeding season, males fight aggressively for the opportunity to mate. Suitable nesting sites are scarce, so thousands of females often nest together, laying 1–6 eggs in

sandy burrows. After an incubation period of 2–3 months, the young emerge to feed in the intertidal zone, and hide in rock crevices to escape attack from gulls and other seabirds.

larger crest on male

VARYING NEIGHBOURS
Marine iguanas often have a "whitewashed" look, which comes from salt expelled by their nasal glands. There is great variation in adult weight between islands: on Fernandina Island, males can be over 11kg (24lb), while on Genovesa they rarely reach 1kg (2¼ lb).

salt encrusted on skin

GRAZING ALGAE

While sizable marine iguanas forage in the sea, small individuals do not enter the water because their body temperature would fall too quickly. Instead, they feed on algae that grow on rocks exposed at high tide. The different species of algae available on different islands may be the cause of the colour variation of marine iguanas between islands – from dark grey to black or red.

Anolis carolinensis
Green anolis

	Length 12–20 cm (4¾–8 in)
	Breeding Oviparous
	Habit Arboreal
Location S.E. USA	**Status** Common

Anolis lizards are found only in the Americas. There are over 250 species, and most – including this one – are slender animals with long tails and legs, and distinctive display fans beneath their throats. Like many of its relatives, the green anolis is a tree-dweller, feeding mainly on insects. It sleeps in bushes at night, but during the day often basks on vertical surfaces such as palm fronds, tree

trunks, fence posts, and walls, staying in place with the help of expanded pads on its toes. Although normally bright green in colour, it turns brown when it rests in shade – which explains why it is sometimes erroneously called a chameleon. Both sexes possess a pink, extendible throat fan. Males use theirs more often, signalling to

rivals and potential mates. Females lay one egg at a time, in leaf litter and moist plant debris, at frequent intervals throughout the breeding season.

extended throat fan

Anolis sagrei

Brown anolis

Length 15–20 cm
(6–8 in)
Breeding Oviparous
Habit Terrestrial/
Arboreal

Location Caribbean   Status Common

In shape, this lizard (also known as the Cuban anolis) closely resembles its relative the green anolis (see p.406). However, although it can change body colour, its overall range is limited to brown or grey. Both sexes have an orange throat fan with a white border, which the males use during courtship displays. The brown anolis is common on walls, climbing with the help of adhesive pads on its toes. It forages for flies and other insects, although it is less active in winter, and is hunted by larger lizards, snakes, and birds. Females lay one egg at a time at frequent intervals throughout spring and summer. This is a short-lived lizard, with a lifespan of only 2–3 years. However, it is prolific and adaptable, and quick to colonize new areas, often ousting the native anolis species in the process.

Basiliscus plumifrons

Plumed basilisk

Length 60–75 cm
(23 1/2–30 in)
Breeding Oviparous
Habit Arboreal

Location Central America   Status Common

The bright green or bluish green plumed basilisk rests on branches and in shrubs, waiting for insects and small vertebrates before darting out to catch them. Males are very territorial; a single male may hold territory containing a harem of females with whom he mates. Females lay about 20 eggs in a season. Plumed basilisks can live up to 10 years in captivity.

CHARACTERISTIC CRESTS
The plumed basilisk has 3 crests, located on its head, back, and tail. The back and tail crests aid swimming, helping this lizard to escape predators.

crest
on tail

crest
on back

yellow eyes

bright green body

WALKING ON WATER

The plumed basilisk lives in trees bordering ponds, streams, and rivers. When threatened, it drops from its perch and dives into the water to escape. This basilisk has the unique ability to run across the surface of standing water on its hindfeet, thus earning itself another common name – the Jesus Christ lizard.

Ctenosaura similis

Black spiny-tailed iguana

Length 70–100 cm
(28–39 in)
Breeding Oviparous
Habit Terrestrial

Location S. Mexico,   Status Common
Central America

The juvenile black spiny-tailed iguana is bright green, lives in trees, and is hunted by birds of prey. The adult, which has few if any predators, is large and grey, with a spikier tail than the juvenile. Adults bask on rocks or stumps, looking out for invertebrates and insects, and also forage for leaves, fruits, carrion, and scraps. Capable of great speed when running away, this iguana may also bite in defence. Females lay 20–30 eggs in burrows or among tree roots.

pointed head

spiky tail used
as weapon

Cyclura cornuta

Rhinoceros iguana

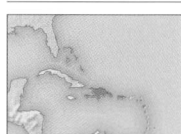

Length 1–1.2 m
(3 1/4–4 ft)
Breeding Oviparous
Habit Terrestrial

Location Caribbean   Status Vulnerable

The massive grey rhinoceros iguana is characterized by several enlarged, raised scales that resemble horns on the snout of the adult. It walks ponderously, with its head held up, foraging for leaves and fruits, but can run very fast when fleeing from danger. When threatened, it may thrash its tail and bite in defence. Males are larger than females, with larger horns and crests, and are highly territorial. Females lay 2–20 eggs in burrows and guard them aggressively. All 8 species of *Cyclura* lizards are endangered due to habitat destruction and predation from introduced mammals such as pigs, dogs, rats, cats, and mongooses.

Iguana iguana

Green iguana

Length 1 m (3 1/4 ft),
max 2 m (6 1/2 ft)
Breeding Oviparous
Habit Arboreal

Location Central America,   Status Common
N. South America

The green iguana is one of the largest lizards in the Americas. Generally greyish or green – although some local populations are orange – it has stout legs, a long tail, and a crest of tooth-like scales down its back. Adults also have a fleshy dewlap beneath the throat, which is large in males. Juveniles eat a wider range of food than adults, including insects as well as fruits, flowers, and leaves. Green iguanas defend themselves by lashing out with their tails and claws. Females lay up to 40 eggs in an underground chamber, producing several clutches through an extended breeding season. Despite being hunted for food and collected for the pet trade, the green iguana is still abundant in parts of its range.

ARBOREAL ADAPTATION

The greyish or green coloration of this large iguana provides excellent camouflage in the tall forest trees in which it lives. It has long legs, and its long claws are specially adapted to aid climbing. Since this lizard usually lives beside rivers, it may drop into the water from overhanging branches to escape from predatory birds or mammals. Shown here are a male (on the left) – easily identified by the large, fleshy dewlap beneath his throat, used for display – and a female green iguana. Adult males are highly territorial.

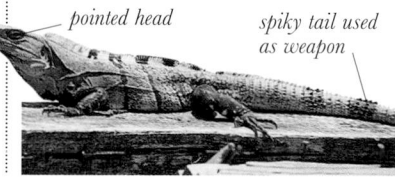

HEAVY WEIGHT
The average green iguana weighs up to 5 kg (11 lb), but an exceptional specimen may weigh up to 10 kg (22 lb). Despite their predatory appearance, adults are almost entirely herbivorous, often feeding high in trees.

green coloration
for camouflage

long tail
used as whip
in defence

long legs for
climbing

dewlap

long claws for
climbing and defence

REPTILES

Phrynosoma platyrhinos
Desert horned lizard

Length 7.5–13.5 cm
(3–5¼ in)
Breeding Oviparous
Habit Terrestrial

Location S.W. USA   Status Common

The desert horned lizard is one of 14 closely related species that are scattered across dry habitats in North and Central America. Often known as horned toads, because of their spiky scales and round, squat shape, these lizards specialize in feeding on ants, and rely primarily on camouflage for protection. The desert horned lizard

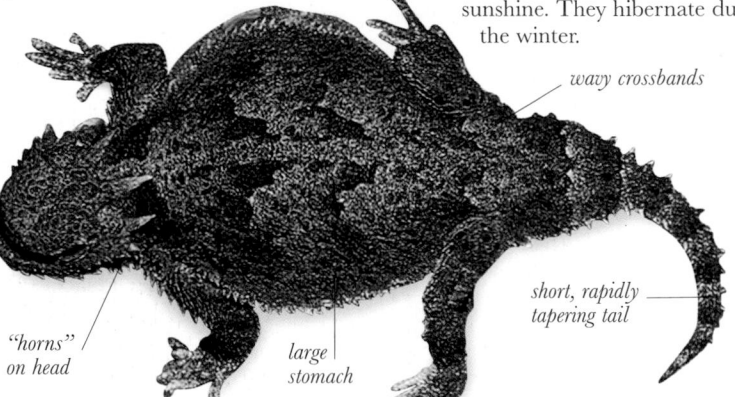

"horns" on head

large stomach

wavy crossbands

short, rapidly tapering tail

is relatively smooth-skinned compared to some of its relatives, but has large "horns" on its head, and smaller spikes on its back and tail. Its coloration – grey, reddish, or tan – helps to conceal it on sandy or gravelly ground, as it moves about slowly searching for food. When it locates a column of ants, it feeds voraciously, picking them up with its extendible tongue. If it is threatened with attack, by birds or carnivorous mammals for example, its first reaction is to "freeze" (it is capable of remaining motionless for long periods). If picked up, it inflates its body with air. Females nest in burrows in sand or loose earth, laying 2 clutches of up to 16 eggs each breeding season. Because these lizards have a round, flat shape, they warm up quickly in the morning sunshine. They hibernate during the winter.

Sauromalus obesus
Chuckwalla

Length 28–42 cm
(11–16½ in)
Breeding Oviparous
Habit Terrestrial

Location S.W. USA,
N.W. Mexico   Status Common

This lizard employs an unusual form of defence: it retreats into rocky crevices and jams itself in by inflating the folds of loose skin on its neck, throat, and flanks, making it hard for predatory birds and mammals to pull it out. It requires very hot conditions before it becomes active, and remains in the crevices in cold weather. In hot weather, it emerges to bask and forage for fruits, leaves, and flowers of succulent plants. Females lay 5–16 eggs, often only in alternate years.

tail thick at base

head black in male

Sceloporus occidentalis
Western fence lizard

Length 15–23 cm
(6–9 in)
Breeding Oviparous
Habit Terrestrial

Location S.W. USA,
N.W. Mexico   Status Common

raised, pointed scales

This common lizard belongs to a group of over 60 species that are commonly known as "blue bellies", because of the blue patches on their undersides. In this species – and more others – the blue colour is most pronounced in males. Western fence lizards typically sit on rocks, fences, and other prominent places, bobbing up and down to display their markings to rivals or to potential mates. Highly active, they forage for insects, and in between feeding bouts they bask in the sun. Females usually lay a single clutch of 3–14 eggs.

Oplurus cuvieri
Madagascan collared iguana

Length 30–37 cm
(12–14½ in)
Breeding Oviparous
Habit Arboreal

Location Madagascar,
Grand Comoro Island   Status Common

protective, spiny tail

This is a thickset lizard with a chunky head and spiny tail. Its body is greyish brown, with a distinctive black collar, and speckled brown markings. Males are more intensely coloured than females. The Madagascan collared iguana remains motionless for much of the time on a tree trunk, waiting for an insect to come into range before darting out to catch it. When threatened by birds of prey or snakes, it retreats into cracks and crevices in the trunk, using its tail to form a barrier between itself and the predator. Females lay 4–6 eggs in a season.

Liolaemus tenuis
Chilean swift

Length 19–31 cm
(7½–12 in)
Breeding Oviparous
Habit Mostly arboreal

Location S.W. South
America   Status Common

Also known as the thin lizard, the Chilean swift has a tail that is nearly twice as long as its body. It stalks its insect prey – mainly flies – with the tail held stiffly over its back. The large green scales on its body overlap like roof tiles. Chilean swifts live in small groups on trees, with one male and several females occupying a territory. Females lay their eggs in a communal site, often under bark; over 400 eggs have been found in a single site. This lizard hibernates during winter.

Tropidurus hispidus
Guianan lava lizard

Length 13–18 cm
(5–7 in)
Breeding Oviparous
Habit Terrestrial

Location N. South
America   Status Common

The Guianan lava lizard is dark brown to black above and white below, with a black collar around the neck. Its flat shape allows it to hide in narrow cracks and crevices in rocky

Laemanctus longipes
Cone-head lizard

Length 40–70 cm
(16–28 in)
Breeding Oviparous
Habit Arboreal

Location S. Mexico to
Nicaragua   Status Unconfirmed

This extremely slender lizard is green with a yellow face – coloration that provides excellent camouflage among foliage.

outcrops, an effective form of protection against predators birds, snakes, and larger lizards. The dark colour of this lizard probably helps it to warm up quickly as it basks in the sun, usually on vertical surfaces such as walls. It forages actively for insects and spiders. Females lay 3 or more clutches of 4–6 eggs in a year.

flat head

Its long legs and claws are adapted for climbing, enabling it to escape rapidly up tree trunks when threatened by predatory birds and snakes. It sits and waits for insects to come within range, then darts out to catch them. The female lays 3–5 eggs at a time in shallow scrapes in soil or leaf litter.

casque on head points backwards

green body

long tail

spindly legs

darker green crossbands

GECKOS AND SNAKE-LIZARDS

PHYLUM	Chordata
CLASS	Reptilia
ORDER	Squamata
SUBORDER	Lacertilia
SUPERFAMILY	Gekkota
FAMILIES	4
SPECIES	1,054

GECKOS ARE SMALL, VOCAL, USUALLY nocturnal lizards. Some are agile climbers, capable of finding purchase on smooth, vertical, or even overhanging surfaces, including tree trunks, rock faces, and walls and ceilings. There are also many desert species, most of which live in burrows or rock crevices, emerging at night to forage on the ground. Geckos are cosmopolitan lizards, found in most tropical and subtropical countries and on many oceanic islands. This group also includes 2 types of legless lizards that are found only in Australasia. Snake-lizards, which have scaly flaps in place of legs, live on the ground, while blind lizards have eyes that are covered with scales and live in underground burrows.

demonstrate temperature-dependent sex determination: males are born at higher temperatures, above 31°C (88°F), and females at lower ones.

VOCALIZATIONS

Geckos are unique among lizards in producing an array of sounds which are used to attract mates or to defend territory. Vocalizations are an effective way for nocturnal animals like geckos to communicate. In fact, the term "gecko" is derived from the call of a large Asian species; in other places, they are known as "chit-chat" lizards for the same reason.

ANATOMY

Most geckos are small, slender lizards with a relatively large, flat head. They are typically nocturnal and have large eyes with vertical pupils, but a small number, usually called day geckos, are diurnal. Day geckos are often brightly coloured whereas typical geckos are dull grey or brown to match the surroundings in which they live. Some have extraordinary shapes with, for example, leaf-like tails or odd frills of skin around their jaws to help them in camouflage.

Climbing geckos often have long claws and sometimes flattened pads at the end of each toe. These toe pads, usually absent in ground-dwelling species, are made of millions of microscopic structures that provide adhesion and enable geckos to climb smooth, vertical surfaces, and even to hang upside down. Some geckos (those belonging to the family Gekkonidae) do not have functional eyelids but their eyes are covered with a transparent scale, as in snakes, and they often use their tongue to lick the surface of the eye clean. Others (members of the family Eublepharidae), however, do have eyelids and can blink – they are mostly ground-dwellers from deserts, but some live in caves and one is arboreal.

Snake-lizards are long and slender and may have pointed snouts. Like geckos, they lack functional eyelids,

but can be distinguished from snakes by their external ear openings and the presence of several rows of scales along their undersides.

REPRODUCTION

Geckos' successful spread around the world may be partly due to their resilient, hard-shelled, and sticky eggs, which can become attached to uprooted trees and drifting debris. Most species lay eggs in pairs but some small geckos lay a single egg at a time. Females often lay them under loose bark, in a rock crevice or, in the case of ground-dwelling species, in burrows in the sand or soil. Several females may use a communal egg-laying site, which can contain several dozen incubating eggs as well

as the remnants of hatched ones. A few gecko species, mainly from New Zealand, give birth to pairs of live young. Eublepharid geckos and snake-lizards also lay one or 2 eggs but there are soft-shelled, like those of most other lizards. Many species in this group

CLIMBING
Geckos can climb almost any surface, including smooth rocks and glass. The structures on their toe pads are so small (only a few thousandths of a millimetre in diameter) that they can mesh with even the tiniest irregularities in the surface they are climbing and provide the gecko with purchase. This makes it possible for them to hunt for food in places where few other land-dwelling animals can reach.

Coleonyx variegatus

Banded gecko

Length	12–15 cm (4³/4–6 in)
Breeding	Oviparous
Habit	Terrestrial

Location S.W. USA, N.W. Mexico | Status Common

Unlike typical geckos, the banded gecko has prominent eyes with movable lids. It appears delicate and translucent, its tiny scales giving a silky texture to its skin.

A secretive, nocturnal lizard, it hides during the day in rock crevices, and actively forages at night for small insects and spiders. When captured, it squeaks and may discard its tail. Other defence tactics involve raising its body and curling its tail over its back to look bigger and, possibly, to mimic a scorpion. Females lay 2 soft-shelled eggs at a time.

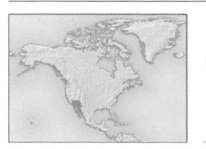
prominent eyes

dark bands broken into patches

Eublepharis macularius

Leopard gecko

Length	20–25 cm (8–10 in)
Breeding	Oviparous
Habit	Burrowing

Location S. Asia | Status Common

The leopard gecko has a thick tail that is used to store food. In times of plenty the tail increases in size, but it shrinks during droughts, when the surplus food is metabolized. It may be discarded in defence, but not as

readily as in some other lizards. This gecko forages at night. It shelters from extreme heat and cold in burrows or under rocks, hibernates in winter, and may aestivate in summer. It has been known to live up to 25 years in captivity.

warts on back and tail

black spots or bars on yellow body

broad head of male

movable eyelids

Hemitheconyx caudicinctus

African fat-tailed gecko

Length 15 cm (6 in),
max 25 cm (10 in)

Breeding Oviparous

Habit Terrestrial

Location W. Africa

Status Locally common

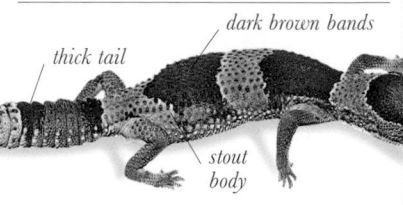

thick tail

dark brown bands

stout body

The tail of the African fat-tailed gecko is used as a food-storage organ; when the lizard is well fed, its tail becomes thicker. It has a tan body with dark brown bands across the back and a broad, dark brown head. Males have a bulge at the base of the tail. A shy animal with a lifespan of about 25 years, it lives in dry savanna, rocky hillsides, or river banks, retreating during the day into burrows and actively foraging for insects at night. It aestivates in the dry season. This gecko sometimes bites, or discards its tail in defence.

Gekko gecko

Tokay

Length 18–36 cm
(7–14 in)

Breeding Oviparous

Habit Mostly arboreal

Location S.E. Asia

Status Common

This gecko is named after the male's loud, explosive call that sounds like "to-kay". It is often found in dwellings, hiding during the day and emerging in the evening to hunt. When threatened by snakes and small, nocturnal mammals it delivers a hard bite and, if grasped, it discards its tail. The male tokay, which has a larger head than the female, is highly

NIGHT SIGHT
The tokay is a large, blue or greyish blue, nocturnal species with large yellow eyes that enable it to see well in the dark. The vertical pupils close to a slit with 3 "pinholes" in daylight.

territorial and cannibalistic, eating other tokays as well as insects and other small vertebrates. The male mates with several females. The females then attach a pair of spherical, hard-shelled eggs to a vertical surface, often behind bark, or in a crevice or a cavity in a wall.

orange spots on body

TOE PADS

Like most climbing geckos, the tokay has expanded, adhesive toe pads that enable it to gain a firm foothold and move quickly and easily over vertical surfaces such as walls and tree trunks.

Gonatodes vittatus

Striped day gecko

Length 7–7.5 cm
(2¾–3 in)

Breeding Oviparous

Habit Terrestrial

Location S. Caribbean, Trinidad

Status Locally common

The male striped day gecko is tan, with a black-edged, white stripe from the snout to the tail, while the female is grey-brown, with an indistinct white stripe (both are shown below). In both sexes, the eyes lie towards the top of the head. This diurnal gecko lives on the ground among leaf litter, but is also a good climber. It is quick to snap up small insects or spiders. Females often nest together in a communal site, where they lay 2 (rarely single) eggs, each approximately the size of a dried pea.

Hemidactylus frenatus

Common house gecko

Length 12–15 cm
(4¾–6 in)

Breeding Oviparous

Habit Mostly arboreal

Location Tropical regions worldwide

Status Common

This widespread, adaptable, grey or brown gecko mostly inhabits buildings, but may also be found on and under rocks, and in tree trunks, often in large numbers. It can climb smooth,

vertical surfaces and cling upside-down to ceilings. Rarely seen during the day, the common house gecko hunts for insects, spiders, or smaller geckos that gather near electric lights after dark; it defends its hunting area. When captured, it squeaks and discards its tail readily to escape.

Palmatogecko rangei

Web-footed gecko

Length 12–14 cm
(4¾–5½ in)

Breeding Oviparous

Habit Terrestrial

Location W. southern Africa

Status Locally common

This pink gecko is noted for its webbed toes, which enable it to run across desert dunes without sinking into the loose, windblown sand.

It spends the day in long tunnels that it digs in the sand, emerging to forage at night. Owing to lack of water in the desert, it drinks the water that condenses on its skin on foggy nights. When threatened, it raises itself on stiffened legs to appear larger.

irregular dark markings

pink, frail body

large eyes with fine red reticulations

Phelsuma madagascariensis

Madagascan day gecko

Length 22–30 cm
(9–12 in)

Breeding Oviparous

Habit Arboreal

Location N. Madagascar

Status Common

The largest of the green day geckos, this thickset lizard inhabits tree trunks and walls, perching (often head-down) on vertical surfaces with the aid of its toe pads. If it successfully evades attack from birds, it can live 10 years or more. Small groups occupy a single tree, and several females may lay eggs in the same site, under bark or in crevices. Males have a swelling at the base of the tail, and reproductive females have calcium deposits that are visible through the skin of their throat.

red spots on back

well-developed toe pads aid climbing

Ptenopus garrulus

Common barking gecko

Length 6–10 cm
(2¼–4 in)

Breeding Oviparous

Habit Terrestrial

Location Southern Africa

Status Locally common

A conspicuous feature of many deserts in southern Africa is the sound of the male common barking gecko calling from the mouth of its burrow after dusk. Both sexes have rounded heads, with a blunt snout, and are yellow or greyish yellow, with darker brown or reddish brown spots; their toes are long and strongly fringed with scales. When in danger, this gecko may "freeze", relying on camouflage to avoid detection by predators. Alternatively, it may retreat into its complex burrow system.

Ptychozoon kuhli
Kuhl's flying gecko

Length 18–20 cm
(7–8 in)

Breeding Oviparous

Habit Arboreal

Location S.E. Asia   Status Common

This tropical forest gecko is one of several lizards that escape danger by jumping from trees and gliding through the air. When not airborne, it often rests head-down on tree trunks – a posture that allows rapid take-off. It is exceptionally well camouflaged – even by gecko standards – and feeds mainly on insects, ambushing or stalking its prey. Females lay 2 eggs each time they breed.

COLOURED FOR CAMOUFLAGE
Mainly brown or greyish brown, with indistinct, dark, mottled markings, Kuhl's flying gecko is a nocturnal lizard that remains motionless during the day, relying on camouflage to escape detection.

FLYING FEET

Unlike the flying lizard (see p.403), Kuhl's flying gecko glides mainly with its strongly webbed feet. Its fall through the air is also controlled by flaps of skin along its flanks, and its flattened, frilly-edged tail.

tail with scalloped fringe

flaps of skin

webbed toes

Stenodactylus sthenodactylus
Elegant sand gecko

Length 9–10.5 cm
(3 1/2 – 4 1/4 in)

Breeding Oviparous

Habit Terrestrial

Location N. and E. Africa,   Status Locally common
W. Asia

This dainty-looking lizard has long limbs, a thin tail, and a slender, beige or sand-coloured body, with darker bands and small, pale spots on its back. Its eyes are large, and it has a slightly upturned snout. A skilful digger, the elegant sand gecko lives in holes in the ground or under stones, and buries its eggs in the sand. It becomes active at dusk, when it hunts for termites, ants, small moths, and small beetles and their larvae, walking with a stiff-legged gait to gain a better view of the surroundings. If threatened, usually by birds, this gecko may arch its back to appear larger.

Tarentola mauretanica
Moorish gecko

Length 10–16 cm
(4–6 1/2 in)

Breeding Oviparous

Habit Terrestrial

Location S. Europe,   Status Common
W. Asia, N. Africa

The stocky, greyish or brown Moorish gecko often occurs in high densities in rocky scrub, dry-stone walls, and buildings; it occupies vertical surfaces, and seldom moves far from a crack into which it can retreat. Active mainly at night, in cooler weather it may also be active during the day, and it hibernates in winter. This gecko frequently waits in likely hunting areas, often near lights, and pounces on insects, such as moths, that come near. Females may nest communally, laying their eggs behind flaking rock or plaster, or in crevices.

rough scales on tail

large, wart-like scales on back

flattened body

flat head

Rhacodactylus leachianus
New Caledonian gecko

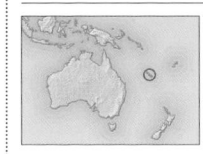

Length 30–35 cm
(12–14 in)

Breeding Oviparous

Habit Arboreal

Location New Caledonia   Status Vulnerable†

This is an extremely large gecko that can reach as much as 40cm (16in) in length, although it is usually smaller. Mottled brown, with a large head and eyes and a small, thin tail, it lives on tall forest trees, rarely venturing down to the

ground. As with most geckos, it eats insects and small vertebrates, but it also feeds on fruits. It uses its powerful jaws to crush food, as well as to deliver a painful bite in defence. New Caledonian geckos appear to live in small family groups that comprise a single male, one or more females, and their young. Females lay eggs in tree holes, behind bark.

fine scales on velvet-like skin

thick, clawed toes with adhesive pads

Sphaerodactylus elegans
Ashy gecko

Length 7–7.5 cm
(2 3/4 – 3 in)

Breeding Oviparous

Habit Terrestrial

Location Caribbean   Status Common

This tiny gecko is brown, with very fine, speckled white markings and a pointed snout. A lively but short-lived animal, it forages constantly among leaf litter and debris in woods and

near houses, catching small insects with a quick dart. Although mainly diurnal, it may also be active at night, often near street lights. Females lay one egg at a time in communal nests – behind bark, among dead leaves, or in crevices.

Phyllurus cornutus
Northern leaf-tailed gecko

Length 15–21 cm
(6–8 1/2 in)

Breeding Oviparous

Habit Arboreal

Location E. Australia   Status Locally common

Geckos are generally noted for their camouflage, but this eastern Australian species excels at this form of defence. It has cryptic, grey mottled coloration, a flattened body that is shaped to eliminate shadows, and a remarkably

leaf-like tail, which can be shed if the animal is attacked (males have a paired bulge at the base of their tail). This gecko is practically invisible when resting on tree trunks and among foliage in its tropical forest habitat, and only gives itself away when it moves. Males are territorial, and females lay 2 eggs at a time in cracks behind bark. Northern leaf-tailed geckos feed mainly at night, actively foraging for insects, which make up most of their prey.

triangular head

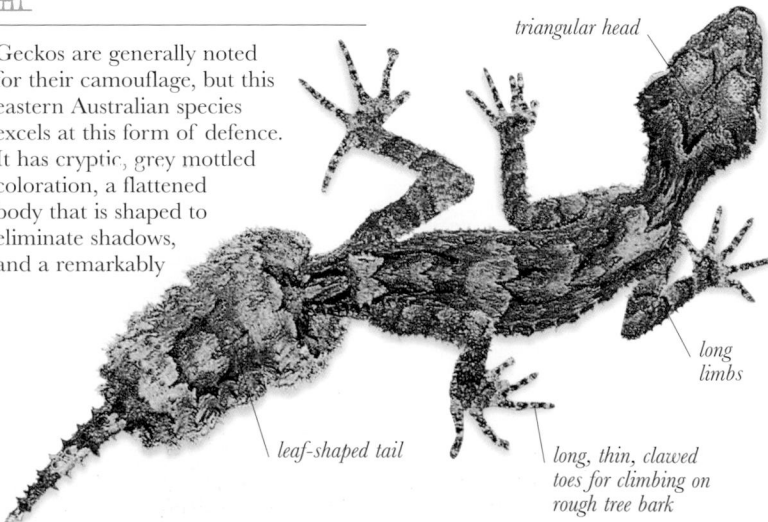

leaf-shaped tail

long limbs

long, thin, clawed toes for climbing on rough tree bark

SKINKS AND RELATIVES

PHYLUM	Chordata
CLASS	Reptilia
ORDER	Squamata
SUBORDER	Lacertilia
SUPERFAMILY	Scincomorpha
FAMILIES	7
SPECIES	1,890

SKINKS AND THEIR relatives form the largest group of lizards. The skinks are mostly long and slender, and some species are without legs. Some have a method of reproduction that is unique among lizards, the females nourishing their developing young via a placenta. Skinks occur throughout the world but are most numerous in tropical and subtropical areas. Most live on the ground, often in leaf litter, but there are also arboreal and burrowing species. The other families in this group are the girdled and plated lizards, whiptails and tegus, and wall and night lizards. Like the skinks, most of these are ground-dwelling lizards.

ANATOMY

Skinks are typically elongated, with a slightly flattened body, a long tail, and a small, wedge-shaped head with small eyes. Burrowing species may have permanently fused, transparent eyelids, which allow them to see while burrowing, and no ear openings. Some skinks have small legs or no legs at all, an adaptation to burrowing or living in dense vegetation, such as turf. When attacked, skinks readily discard their tail, which is sometimes brightly coloured to deflect the attention of predators from their head and body.

The girdled and plated lizards have large, rectangular scales, each of which has a raised keel. This keel may be drawn out into a point in girdled lizards, especially on the scales behind the head and on the tail, which is used to form a barricade when the lizard is pursued into a burrow or crevice, or even as a weapon. When threatened by a predator, the armadillo lizard grasps its tail in its mouth, presenting the attacker with a spiky hoop that it cannot easily attack.

Whiptails and tegus have small scales, a slender body, a pointed snout, and long limbs, although this family also includes some legless forms. Tegus are generally larger than whiptails.

The wall and sand lizards are outwardly similar to the whiptails, although they are found in parts of Europe, Asia, and Africa, whereas the whiptails are restricted to the Americas. Many members of both families are brightly coloured, and they are all prepared to sacrifice their tail if necessary.

REPRODUCTION

Skinks reproduce in various ways. Some species lay eggs, whereas others give birth to live young. In a few species, the developing embryo is nourished by the female via a primitive placenta. In others, the female lays eggs and guards them during incubation by coiling herself around them. She may also move the eggs if danger threatens. When the young hatch, the female licks them clean, and they may remain near her for the first few days.

Some of the relatives of skinks, notably a group of whiptails and another of wall lizards, are parthenogenetic – that is, they can reproduce without males. In these species, each young is an exact replica (clone) of the mother, and is therefore also female, producing eggs of its own as soon as it reaches maturity. Some species are entirely parthenogenetic; in others, there are some populations that include males and reproduce normally.

FEEDING
Most ground-living skinks are secretive animals that spend most of their time under the cover of leaf litter. They generally forage during daylight and take shelter under logs or stones at night. Small skinks feed on invertebrates. The larger species (including the major skink, shown here) also eat plant matter, and some may even eat rodents and birds.

REPTILES

Cordylus cataphractus

Armadillo lizard

Length 16–21 cm
(6¹⁄₂–8¹⁄₂ in)

Breeding Viviparous

Habit Terrestrial

Location Southern Africa | Status Vulnerable

The armadillo lizard is covered with stout, defensive spines on its neck and tail and squarish scales on its back. When threatened, it grasps its tail in its mouth and forms a ring with its body, presenting the spines to deter an approaching predator and making it awkward to attack. Armadillo lizards occur in scrub and rocky outcrops, hiding in large cracks, and

hibernating in winter. They live in family groups – the female gives birth to one or 2 large young, and may even feed them – a characteristic that is unusual among lizards, which rarely demonstrate parental care. Insects and spiders make up most of the diet of the armadillo lizard and it, in turn, is hunted by a variety of predators, including birds of prey. It can live for up to 25 years in captivity, slightly more in some exceptional cases.

protective spines on neck

yellowish brown colour

body formed into protective ring

tail caught in mouth

Cordylus giganteus

Sungazer

Length 28–39 cm
(11–15¹⁄₂ in)

Breeding Viviparous

Habit Terrestrial

Location Southern Africa | Status Vulnerable

Also known as the giant girdled lizard, the yellowish to dark brown sungazer can often be seen basking at the entrance to its burrow, apparently staring at the sun. Burrows usually face north or northwest, providing maximum sunshine throughout the day. This lizard possesses an armoured tail, ringed with pointed scales and resembling a fir cone, which it can use to block the entrance to its burrow. If grasped, it wedges the long spines on its head and neck into the burrow roof, making its body difficult to extract. Its diet consists of insects and small vertebrates. Females give birth to one or 2 young, which remain near the adults. The sungazer hibernates during the winter months.

Cordylus rhodesianus

Zimbabwe girdled lizard

Length 12–17 cm
(4³⁄₄–6¹⁄₂ in)

Breeding Viviparous

Habit Terrestrial

Location Southern Africa | Status Locally common

This olive-brown lizard is often found in kopjes – the piles of eroded rock that are scattered across southern Africa's grasslands. Its flattened body enables it to hide in narrow cracks or under rocks when threatened, and it uses its tail, which is covered in rings of pointed scales, to block the cracks and deter predators such as snakes, birds, and mammals. The Zimbabwe girdled lizard lives in small family groups – with the young remaining near the parents – and hibernates in winter.

keeled, slightly spiny, scales

Gerrhosaurus major

Rough-scaled plated lizard

Length 40–48 cm
(16–19 in)

Breeding Oviparous

Habit Terrestrial

Location C., E., and southern Africa | Status Common

The body of the rough-scaled plated lizard is covered with rectangular plates, arranged in rows, that help the reptile wedge itself into cracks

in rocky outcrops for protection against birds of prey, snakes, and small mammals. This lizard is light to medium brown, but the male develops a pinkish throat in the breeding season. An active forager, its diet comprises fruits, flowers,

invertebrates, and small vertebrates, including smaller lizards. The male is highly territorial, particularly during the breeding season, when he fights with other males. The female lays 2–4 relatively large, oval eggs in rock crevices, in moist soil, and under logs.

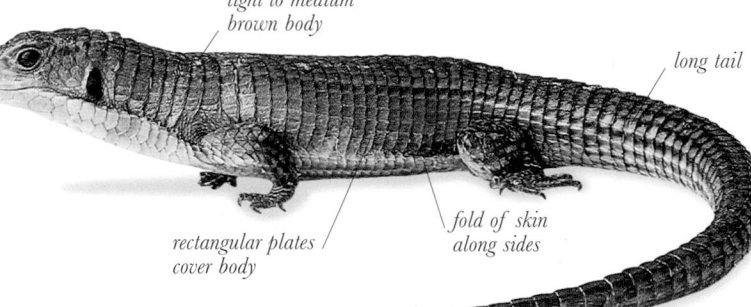

light to medium brown body

long tail

rectangular plates cover body

fold of skin along sides

Platysaurus broadleyi

Broadley's flat lizard

Length 15–20 cm
(6–8 in)

Breeding Oviparous

Habit Terrestrial

Location Southern Africa (Augrabies Falls) | Status Locally common

The extremely flattened head, body, and tail of this lizard enable it to squeeze into very narrow rock crevices to escape from predatory birds of prey, particularly the rock kestrel. Once wedged in, it is almost

impossible to extract. This lizard spends most of the day hunting for insects, especially the small black flies that swarm near waterfalls, darting rapidly to catch the flies when they land, or leaping upwards to catch them in flight. It also eats ripe berries. The male has a bluish head, green back, yellow or orange front limbs, and an orange-tan tail. The female and juveniles have dark brown backs with 3 cream stripes and a pale yellow tail. The female lays 2 clutches of 2 eggs each summer, in deep cracks between large boulders and under

exfoliating flakes of rock, especially where leaves or debris have accumulated.

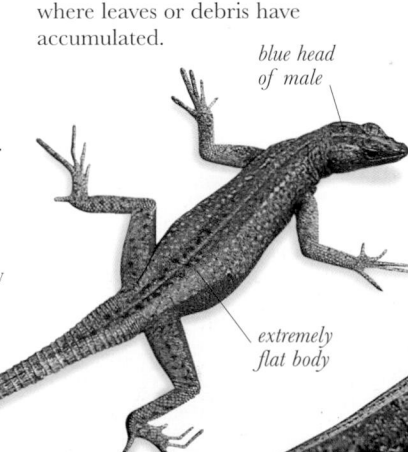

blue head of male

extremely flat body

Lacerta agilis

Sand lizard

Length 18–22 cm
(7–9 in)

Breeding Oviparous

Habit Terrestrial

Location Europe, C. Asia | Status Locally common

This stocky lizard belongs to a group of species that are common in Europe and Asia, sometimes extending north of the Arctic Circle. In Great Britain it is found in heaths and sand dunes, but further south it lives in a much wider range of habitats, including gardens. It has a blunt head and relatively short legs, and its colour is very variable, although males are at their brightest during the breeding season. Sand lizards feed on small insects and spiders, and rapidly shed their tails if they are attacked, usually by snakes, birds, or mammals, especially domestic cats. Females lay clutches of 3–14 eggs in the spring and summer, and adults hibernate in winter.

REPTILES

Schreiber's green lizard

Lacerta schreiberi

Length	36 cm (14 in)
Breeding	Oviparous
Habit	Mostly terrestrial
Location S.W. Europe	**Status** Lower risk

Schreiber's green lizard is typical of the numerous localized species of *Lacerta* that are found in different parts of Europe. The male is green, with scattered, small black spots, and a blue throat or head (sometimes both) in the breeding season. The female is brown or bright green, with large black markings and a brown head. The young are green with white or yellow, black-edged bars on their flanks. This lizard can be found basking in the open in fields, moorland, or hillsides, often near streams, occupying a position with a good, all-round view. It mostly sits and waits for insects to come within range, but also forages in cracks. When disturbed, it retreats into a burrow or dives into water, hiding under stones on the stream bed. If attacked, it can easily discard its tail. This lizard has a lifespan of over 10 years in captivity. The female lays 6–12 eggs at a time.

Viviparous lizard

Lacerta vivipara

Length	10–12 cm (4–4¾ in)
Breeding	Viviparous
Habit	Terrestrial
Location Europe, C. to E. Asia (including Japan)	**Status** Common

Also known as the common lizard, this species has one of the largest continuous ranges of any terrestrial reptile, reaching well into the Arctic. It is the only member of its genus that bears live young – an adaptation that helps it to survive in regions that are too cold for eggs to develop. In the far north of its range, it can hibernate for up to 8 months at a time and may breed only every second or third summer; however, it makes up for this by having a relatively long lifespan. Viviparous lizards live on the ground, usually in dense vegetation, and feed mainly on insects and spiders. They often bask on a sunny bank, flattening their body and spreading out their legs.

BREAKING OUT

Young viviparous lizards are born in an egg membrane from which they escape almost immediately. They are about 4cm (1½in) long and black. In the southern parts of their range – in the Pyrenees – viviparous lizards sometimes lay eggs.

thick neck
pregnant female

COLOURS AND PATTERNING
Although the viviparous lizard is usually brown or olive, some specimens are entirely black or light brown. Males tend to be spotted, while the female (shown) is usually striped.

Eyed lizard

Lacerta lepida

Length	40–80 cm (16–32 in)
Breeding	Oviparous
Habit	Terrestrial
Location S.W. Europe	**Status** Locally common

Also known as the ocellated or jewelled lizard, this thickset green animal, with blue eye-spots running along its flanks, is Europe's largest lizard. It inhabits hillsides, fields, and clearings, foraging for large insects, birds' eggs, nestlings, other lizards, and small mammals. Although it hibernates in winter, it may emerge on warm days. If grasped, it bites, discards its tail, and retreats into its burrow. Females lay a single clutch of 6–16 eggs a year.

black markings on body

Lilford's lizard

Podarcis lilfordi

Length	18–22 cm (7–9 in)
Breeding	Oviparous
Habit	Terrestrial
Location Balearic Islands	**Status** Vulnerable

This tough, adaptable lizard can survive in remarkably hostile environments. Although it has few predators (there are no snakes on many of the islands it inhabits), vegetation is sparse and food is scarce, posing the greatest threat to this lizard's survival. It is an active forager, scrambling over rocks and shrubs in a relentless search for food – insects, fish dropped by seabirds, and food discarded by tourists, as well as flowers and leaves. Some populations are entirely black, while others are brown or green. The female lays about 4 eggs in crevices filled with soil or leaves, or under shrubs.

turnip-shaped tail stores fat

Ibiza wall lizard

Podarcis pityusensis

Length	15–21 cm (6–8½ in)
Breeding	Oviparous
Habit	Terrestrial
Location Balearic Islands	**Status** Vulnerable

head with pointed snout
tail twice as long as body

With its slender body and long toes, the Ibiza wall lizard is an agile climber. It is most often seen basking on walls, rocky slopes, and fallen trees – sometimes in large numbers – but it invariably scuttles away rapidly if disturbed. Males are generally blue with black markings, while females often have dark stripes, but the colours and patterns of both sexes vary across the species' range. Wall lizards feed on insects, and they lay their eggs in burrows or under stones. The Ibiza wall lizard can live for up to 6 years.

Eyed skink

Chalcides ocellatus

Length	30 cm (12 in)
Breeding	Oviparous
Habit	Burrowing
Location S. Europe, N. and N.E. Africa, W. Asia	**Status** Unconfirmed

Also known as the ocellated skink, the eyed skink has a stocky, light brown or yellowish brown body with small white eye-spots surrounded by black. Its smooth, glossy scales and small limbs are adaptations to a burrowing lifestyle. Although this lizard is mainly active during the day, it may also be active at dusk in very hot weather. It is often found basking near old ruins or dry-stone walls near olive groves, never straying far from cover. If threatened by birds, snakes, and cats, it darts into its burrow or into crevices between rocks; if attacked, it may shed its tail. An active forager, the eyed skink searches under stones for small insects and spiders. It lives in loose colonies, and the female lays 3–10 eggs.

small limbs aid burrowing
head narrower than body
no distinct neck

Corucia zebrata
Solomon Islands tree skink

Length Up to 75 cm (30 in)
Breeding Ovoviviparous
Habit Arboreal

Location Solomon Islands
Status Vulnerable†

This olive-green or greyish green skink is a highly unusual lizard in many ways. It is very large, with a massive, wedge-shaped head and a long, prehensile tail. Unlike other skinks, the Solomon Islands tree skink is entirely herbivorous, feeding exclusively on leaves and fruits in its forest habitat. Active at night, it shelters in tree holes during the day. If threatened or provoked, it will hiss while raising itself up, and may even bite. This

wedge-shaped head

well-developed limbs

clawed toes for climbing

long, rounded tail

skink is a long-lived species. Although little is known about its reproductive and social behaviour, it is likely to be monogamous and appears to live in small family groups or loose colonies. Males and, to some extent, females are highly territorial. A female gives birth to one, occasionally 2, huge young – about one-third of the size of the adult – which stays with its very protective mother for several weeks or even months. The low reproductive potential of this skink makes it vulnerable, although it is not a protected species at present.

Eumeces schneiderii
Berber skink

Length 40–45 cm (16–18 in)
Breeding Oviparous
Habit Terrestrial/ Burrowing

Location N. Africa, W. Asia
Status Locally common

The Berber skink is the largest member of its genus, which is found in North Africa, the Middle East, Asia, and the Americas. Bulky and with a squarish cross-section, it has distinctive orange scales on its blue-grey back. It is

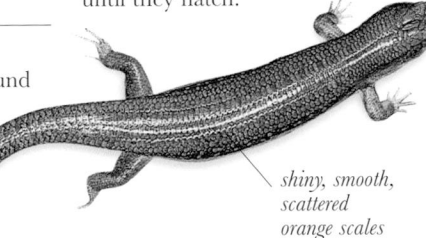

shiny, smooth, scattered orange scales

often found in cultivated land and semi-desert, where it digs burrows at the base of shrubs. When attacked, it retreats into its burrow, or may jump into water. Active and quick-moving, the Berber skink forages for insects, spiders, snails, and smaller lizards. It shelters from the sun in the middle of the day and may not emerge on cold winter days. Females lay 3–20 eggs, coiling around them until they hatch.

Eumeces lagunensis
Pink-tailed skink

Length 16–20 cm (6½–8 in)
Breeding Oviparous
Habit Terrestrial

Location Mexico (Baja California)
Status Locally common

Also known as the San Lucas skink, this species is characterized by a long pink tail that is bright in juveniles but faded in adults. The reason for this coloration is unclear, but it may help to divert predators towards the tail – which can be discarded – instead of towards more vulnerable parts of the body. The pink-tailed skink is a secretive animal, scuttling about

among leaf litter or vegetation – where it searches for insects and spiders – or hiding under stones. It is active year-round, although less so during Baja California's brief winters. The female lays 2–6 eggs in tunnels in loose soil. She stays with them until they hatch, and may even move them if they are threatened by unfavourable conditions, such as floods.

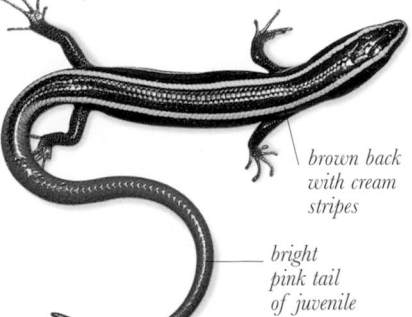

brown back with cream stripes

bright pink tail of juvenile

Lamprolepis smaragdina
Emerald tree skink

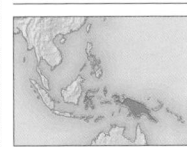

Length 18–25 cm (7–10 in)
Breeding Oviparous
Habit Arboreal

Location S.E. Asia, New Guinea, Pacific islands
Status Common

The bright green emerald tree skink has a stocky body with a long, tapering tail, a pointed snout, and smooth, shiny scales. It lives on the trunks and among the foliage

of large forest trees, and actively forages for insects, which it catches in a rapid, darting movement. If attacked by birds of prey or snakes – its main predators – this lizard will shed its tail in order to escape from its aggressor. Males are territorial and females lay 9–14 eggs.

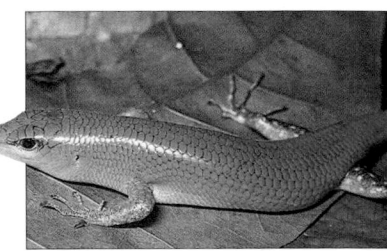

Egernia frerei
Major skink

Length 60–70 cm (23½–28 in)
Breeding Viviparous
Habit Terrestrial

Location S. New Guinea, N. and E. Australia
Status Locally common

The major skink is one of 20 closely related species that are found in Australia and New Guinea. Like most of its relatives, it has a chunky appearance, with a body that is squarish in cross-section, a thick neck, short head, and muscular limbs.

Its coloration is dark brown, becoming darker on the flanks, with a white, yellow, or orange underside. The body scales are smooth and glossy, and there are fine lines running down its back. The major skink can be seen basking in the sun or foraging for insects and small vertebrates during the day around the bases of large, buttressed trees, logs, or near rocky outcrops, never too far from cover. Females give birth to up to 6 young, which have scattered white dots on their flanks – these spots disappear as the lizard matures. Within Australia, the major skink is commonly found in forests in northeast New South Wales, islands of the Torres Straits, and Arnhem Land (Northern Territory).

Mabuya longicauda
Long-tailed skink

Length 30–35 cm (12–14 in)
Breeding Viviparous
Habit Terrestrial

Location S.E. Asia
Status Common

The most distinctive feature of this slender skink is its extremely long tail, which can be up to twice the length of the body and head combined. The body scales are shiny and brown, and the flanks are darker brown or black. This lizard lives in clearings and gardens, often near water, and also

in coastal scrub. An active forager, it frequently pokes its snout into leaf litter and debris where insects and spiders may be found. Little is known about its reproductive behaviour, but most related species give birth to 5–10 young. The genus *Mabuya*, to which it belongs, contains egg-laying as well as live-bearing species.

REPTILES

Mabuya striata
African striped skink

Length 18–25 cm
(7–10 in)
Breeding Viviparous
Habit Mostly terrestrial

Location E. and southern Africa

Status Common

The streamlined body of the African striped skink is highly variable in colour, but is often brown, with paler stripes running along the back. It has a pointed head, smooth, shiny scales, and transparent "windows" on its lower eyelids. Males of some populations have an orange head and a yellowish orange throat. This skink lives in a wide variety of habitats, from mangrove swamps to arid areas, and is sometimes found in high densities. Very active during the day, it climbs into shrubs and trees, rocky outcrops, and buildings in search of prey such as small insects and spiders, stopping frequently to bask. When attacked, it can shed its tail. Females give birth to 3–9 young, each about 6–8cm (2¼–3¼in) long, which take 15–18 months to mature.

Lygosoma fernandi
African fire skink

Length 22–37 cm
(9–14½in)
Breeding Oviparous
Habit Terrestrial

Location W. and C. Africa

Status Unconfirmed

Also called the fire-sided skink, and formerly known as *Riopa fernandi*, the African fire skink has a short head and stocky body with a brown back, red and black bands on the flanks and face, and blue and black bands on its tail. It lives among leaf litter, and is active at twilight, when it forages for insects and spiders. Like many other lizards, this skink can shed its tail when attacked by a snake or other predator. The female lays up to 8 eggs in leaf litter or rotting wood. Since the African fire skink is rarely seen in the wild, most information comes from observations of captive animals.

smooth, shiny scales

blue and black bands on tail

Tiliqua rugosa
Stump-tailed skink

Length 30–35 cm
(12–14 in)
Breeding Viviparous
Habit Terrestrial

Location Australia

Status Locally common

Also known as the shingleback, this skink has a short, stumpy tail that looks like a fir cone, and enlarged, raised dorsal scales. The thick tail acts as a food-

storage organ, while the heavily armoured body provides protection. Coloration is variable, ranging from dark brown to cream, sometimes with light or dark spots. This slow-moving, omnivorous lizard rests under logs or in leaf litter, and is an active forager. It is temporarily monogamous: pairs remain together for about 8 weeks in the breeding season, then separate, often reuniting the following year. Females give birth to 1–3 relatively large young.

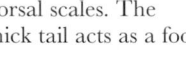

Tiliqua gigas
Giant blue-tongued skink

Length 50–62 cm
(20–24 in)
Breeding Viviparous
Habit Terrestrial

Location S.E. Asia, New Guinea and satellite islands

Status Common

A close relative of the blue-tongued skink (see below), this large, Indonesian lizard uses the same

defensive technique, displaying its startlingly blue tongue to ward off attack. If this fails, it may bite. This skink constantly pokes its snout about in leaf litter, actively foraging and scavenging for insects, molluscs, fruits, and leaves. It lives in colonies believed to consist of a dominant male and several females and juveniles. Females produce 8–13 young, which often remain close to their mother for the first few days, possibly weeks.

short, thick tail

stocky olive-green body with thin black crossbands

Tiliqua scincoides
Blue-tongued skink

Length 45–50 cm
(18–20 in)
Breeding Viviparous
Habit Terrestrial

Location N., E., and S.E. Australia

Status Common

This large, slow-moving animal is one of Australia's most familiar reptiles. It lives in a wide variety of habitats, and is often seen on roads, where its stubby shape and camouflage colouring make it look like a piece of fallen wood. The blue-tongued skink has a wide-ranging diet, and actively forages for snails, insects, carrion, flowers, fruits, and berries; it will even scavenge for leftovers at picnic sites. During the breeding season, males are territorial; mating takes place after a brief chase, at the end of which the male bites the female on the back of the head. The female gives birth to up to 25 young per litter, following a gestation period of about 150 days. The young take about 3 years to reach maturity. This

DEFENSIVE DISPLAY

The blue-tongued skink gets its name from its large, bright blue tongue, which it rolls out of its mouth when threatened. Combined with loud hissing, this visual display is often enough to ward off an attack. This skink is generally harmless, although it has a powerful – if toothless – bite.

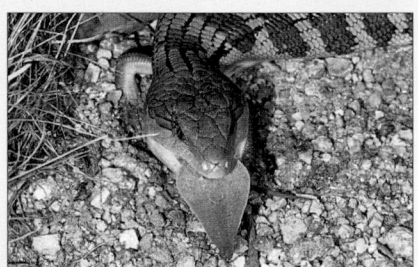

wide, dark bar through eyes

oblique tan bars on flank

PHYSICAL APPEARANCE
The blue-tongued skink is a bulky reptile, with a wide head, thickset body, and relatively short limbs. Light grey or brown above, it has irregular, dark brown crossbars on the body and tail.

short limbs

skink can live for up to 25 years in captivity. Several similar species of skink live in Australia, one of them also occurring in New Guinea and Sumatra.

Tribolonotus gracilis
Crocodile skink

Length 15–20 cm
(6–8 in)
Breeding Oviparous
Habit Terrestrial

Location New Guinea

Status Vulnerable

The crocodile skink has a bony, triangular head, 4 rows of spines on the back and tail, and orange, spectacle-like rings around its eyes. The glands present on the soles of its feet and on the abdomen are believed to be associated with scent secretion and communication. It is slow-moving and secretive, hiding in leaf litter and debris. If disturbed, it gives out a loud screech. The male usually lives with one or more females, which lay one very large egg in leaf litter or other rotting plant matter.

Tropidophorus grayi
Gray's keeled water skink

Length 20–25 cm (8–10 in)

Breeding Viviparous

Habit Semi-aquatic

Location Philippines | Status Locally common

This slender skink has strongly keeled scales on its back which end in a spine, producing a spiky appearance. Its tail is about the same length as the head and body. It lives in cool conditions, in and around mountain streams, often under rotting logs. If in danger, it may jump into the water and hide under stones on the stream bed. Although it is a relatively common skink, little is known about its social and reproductive behaviour, except for the fact that the female gives birth to 1–6 young.

strongly keeled scales

dark brown body with darker crossbands

Ameiva ameiva
Common ameiva

Length 45–50 cm (18–20 in)

Breeding Oviparous

Habit Terrestrial

Location Central America, N. South America | Status Common

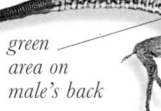

green area on male's back

The common ameiva has a streamlined body with a pointed head and a very long tail that may be twice the length of the body. The male, which is more slender and brightly coloured than the female, has a bright green area down the centre of his back. The female lays several clutches of 2–6 eggs throughout an extended breeding season (March–December). Common ameivas live in loose colonies in forest openings, on the edges of roads and paths, and in cultivated land, usually near human habitation. They bask in patches of sunlight, and shelter under leaf litter or logs on cool days.

Cnemidophorus uniparens
Desert grassland whiptail lizard

Length 15–23 cm (6–9 in)

Breeding Oviparous

Habit Terrestrial

Location S. USA, N. Mexico | Status Locally common

Found in deserts, grassland, and scrub, this is one of over a dozen very similar lizards that live in North America. It has a long, slender, striped body and whip-like tail, but its most remarkable feature – shared by several other whiptails – is that it is an entirely female species: males

long tail

STRIPED BODY
The desert grassland whiptail lizard is brown with 6 or 7 lighter stripes running along its back. Its pointed snout and slender body taper to a very long tail, which gives this lizard its name.

ALL-FEMALE LIZARDS

This lizard reproduces by parthenogenesis, laying fertile eggs without having to mate. Adults and young are always female. Research suggests that this species, and others like it, evolved from single individuals. Each species is a clone – a group of genetically identical animals.

are unknown. The desert grassland whiptail lizard is a diurnal hunter that feeds on insects. It also basks in the sunshine for short periods of time throughout the day. Adults bury 1–4 eggs in soil or beneath rocks or logs, and the newly hatched young have bright blue tails.

light stripes on brown background

pointed snout

Tupinambis teguixin
Black tegu

Length 80–110 cm (32–43 in)

Breeding Oviparous

Habit Terrestrial

Location N. to C. South America | Status Common

One of South America's largest terrestrial reptiles, the black tegu shows some striking similarities with monitors (see p.422) – a group of lizards that are found only in Africa, Southeast Asia, and Australasia. It has powerful limbs and long claws for digging, and it uses its long, forked tongue to "taste" the air. Like most monitors, it has a wide-ranging diet, including insects and invertebrates, birds, small mammals, other lizards, and carrion. Black tegus live in forest

clearings and on river banks, and are good swimmers. When confronted by predators – usually large, carnivorous mammals such as cats, or lizards and snakes when juvenile – they bite and scratch with their claws, and also use their tails as clubs. They walk on their hindlegs, and communicate with each other by making loud, snoring noises, which carry a long way in still air. Mating involves the male grasping the female by her neck, and twisting his body under hers. After mating, the female lays 7–12 eggs, sometimes in burrows, but often at the base of a termite mound. The temperature in the mound is kept constant by the termites, which soon repair the mound, sealing in the eggs until they hatch 3 months later. The black tegu is occasionally eaten by humans, and its skin is used in the fashion trade.

glossy body

thick tail

powerful limbs

Dracaena guianensis
Caiman lizard

Length 0.9–1.1 m (3–3 1/2 ft)

Breeding Oviparous

Habit Semi-aquatic/ Arboreal

Location N. South America | Status Vulnerable

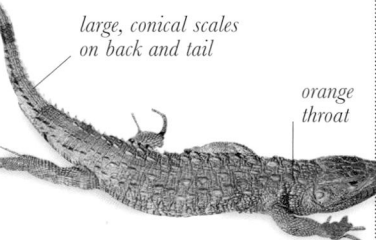

large, conical scales on back and tail

orange throat

This relatively large lizard resembles a short-headed crocodilian. It is greenish or brown, with yellow or tan blotches on its flanks. The male has an orange and black throat, while the female's is grey. The caiman lizard dives to the bottom of shallow rivers, lakes, and streams to find water snails, which it brings to the surface before crushing the shells with its large, flattened teeth, and swallowing the soft contents. Little is known about this rare reptile, but it is thought to lay eggs.

Xantusia henshawi
Granite night lizard

Length 5–7 cm (2–2 3/4 in)

Breeding Viviparous

Habit Terrestrial

Location S.W. USA (S. California), Mexico (Baja California) | Status Locally common

This nocturnal lizard has a flattened body that enables it to squeeze under exfoliating rock flakes. It is covered with tiny scales that are yellowish in colour, with many large black spots.

This lizard actively forages over rock surfaces at night, retreating into narrow crevices when threatened by predators. Although it is not particularly rare, this species is protected because collectors destroy its habitat by removing rock flakes to expose the lizards.

ANGUIMORPH LIZARDS

PHYLUM	Chordata
CLASS	Reptilia
ORDER	Squamata
SUBORDER	Lacertilia
SUPERFAMILY	Anguimorpha
FAMILIES	6 or 7
SPECIES	173

THE ANGUIMORPHS INCLUDE the largest of all lizards, the monitors, and the only venomous lizards, the Mexican beaded lizard and the Gila monster. This diverse group also includes the slow worm and the glass lizards, among others. Anguimorphs are thought to be more evolved than other lizards, and it is likely that they were the ancestors of the snakes – many species have fang-like teeth and a long, forked tongue. Anguimorph lizards are distributed almost worldwide. Opinion is divided over whether to classify them in 6 or 7 families.

ANATOMY

The largest group of anguimorph lizards are the anguids. These slender lizards have smooth scales, and many have small legs or no legs at all (as, for example, in the slow worm).

Monitor lizards are easily recognized by their long neck, narrow head with a pointed snout, powerful limbs, and muscular tail. They have a long, forked tongue, which they use to test their surroundings. This family contains some very large species, including the Komodo dragon.

The only venomous lizards are the beaded lizards of North America. These heavily built lizards have a wide head with a blunt snout, short legs, and a swollen tail. Their venom enters wounds made by sharp fangs in the lower jaw. (Snakes' fangs are in the upper jaw.)

HUNTING AND FEEDING

Anguimorph lizards use various methods to find and overpower their prey. Like most lizards, the smaller species feed mainly on insects but some of the larger monitors can subdue large mammals, including pigs and deer. Monitors also feed on carrion and have an acute sense of smell, which they use to locate decaying flesh. The beaded lizards, too, hunt by smell, and will track prey over long distances, using their tongue to follow scent trails left by small mammals and birds. They also feed on the eggs of ground-nesting birds.

REPRODUCTION

Anguimorph lizards may lay eggs or give birth to live young. Several species of monitors incubate their eggs in termite nests, using their strong claws to break into the mounds. When the termites repair the nest, the eggs are entombed in a controlled environment, protected from predators. They often hatch at the start of the rainy season, when the soil softens. Females of some species are thought to return to the mound to help the young break out.

HUNTING
For such large reptiles, monitor lizards can be remarkably agile. The mangrove monitor (shown here) spends a large part of its time in water but also hunts on land. It uses its strong claws to climb trees in search of prey such as this tree snake.

Anguis fragilis
Slow worm

	Length 30–40 cm (12–16 in), max 50 cm (20 in)
	Breeding Ovoviviparous
	Habit Terrestrial/ Burrowing
Location Europe, W. Asia, N.W. Africa	**Status** Common

With its smooth, scaly body and flickering tongue, this widespread legless lizard looks much more like a snake than a worm. At close quarters, it is easily distinguished from snakes by eyelids that can be closed, and – when threatened – by its ability to shed the end of its tail in order to escape the grip of a predator. Once shed, the tail is very slow to regenerate, leaving many adults with a truncated appearance. Young slow worms are often brightly coloured, with a metallic sheen and central stripe. Females tend to keep the stripe as adults, but males are usually a plain coppery brown or grey, although some may have blue spots. The slow worm is a secretive animal, living in habitats that offer plenty of cover, and hiding under logs, flat stones, or piles of rubbish. It may occasionally bask during the day but is active mainly at dusk, when it emerges to feed on slugs and other invertebrates – a diet that makes it a useful visitor to gardens. Males become fiercely territorial during the breeding season and, after mating, the females give birth to 6–12 live young. Slow worms have a long lifespan, but in the north of their range, up to half of it is spent in hibernation.

smooth scales facilitate burrowing

Elgaria kingii
Arizona alligator lizard

	Length 19–31 cm (7½–12 in)
	Breeding Oviparous
	Habit Terrestrial
Location S.W. USA, N.W. Mexico	**Status** Locally common

The Arizona alligator lizard has a light brown body, with darker crossbars, and a long tail. Its scales are shiny, and it has a fold of skin running along each flank. The lizard's long, slender body and short limbs are adaptations for moving quickly through grass and other dense vegetation. A secretive species, it is active throughout the day, and sometimes ventures out at dusk. It prefers to remain in moist areas in otherwise dry woodland, and in mountains and upland grassland, often near streams, where it forages among dead leaves and under debris for insects and spiders. It hibernates when food becomes scarce in winter. Preyed upon by snakes, birds, and small mammals, it discards its tail if grasped and may smear its enemy with faeces. Alligator lizards live in loose colonies, the males becoming territorial during the breeding season. Females lay 9–12 eggs, buried in moist sand or soil; the hatchlings are boldly banded.

folds of skin along flanks

shiny scales

dark crossbars

short limbs

Ophisaurus apodus
European glass lizard

Length 1–1.2 m
(3¼–4 ft)
Breeding Oviparous
Habit Terrestrial/
Burrowing

Location S.E. Europe,
W. Asia

Status Common

long, slender body with groove along each side

tail 1.5 times as long as body

With a length exceeding that of many snakes, this legless lizard is the largest member of the family Anguidae – a group of reptiles more common in the Americas than in Europe. As an adult, it is uniformly brown on the upperside, but its young are grey with dark bars. It usually lives in dry habitats, and it feeds on slugs, snails, and invertebrates. If caught, it may discard its tail or smear its captor with faeces. Females lay 8–10 eggs, and bury them in damp sand or soil.

Ophisaurus ventralis
Eastern glass lizard

Length 45–108 cm
(18–43 in)
Breeding Oviparous
Habit Terrestrial/
Burrowing

Location S.E. USA

Status Common

This legless lizard is one of 4 closely related species found in North America. Brown with black flanks, it lives mainly in damp meadows – usually on the surface, unlike some of its relatives, which spend most of

their time underground. Its tail is over twice as long as the rest of its body, and is easily shed. Active early in the morning and after rain, the eastern glass lizard feeds on slugs, snails, insects, and small vertebrates. The female lays 8–17 eggs in damp soil, and guards them until they hatch.

Anniella geronimensis
Baja California legless lizard

Length 10–15 cm
(4–6 in)
Breeding Viviparous
Habit Burrowing

Location Mexico (Baja California)

Status Locally common

The Baja California legless lizard is a small, slender species with a silvery to light brown body. It does not create permanent burrows, but "swims" through loose sand just below the surface, especially around the base of low-growing shrubs. It may come above ground at night or at twilight, leaving behind trails in the sand. This little-studied lizard eats spiders and small insects. If attacked, it can shed its tail in order to escape.

thin, dark line down back

legless body aids burrowing

Heloderma suspectum
Gila monster

Length 35–50 cm
(14–20 in)
Breeding Oviparous
Habit Terrestrial/
Burrowing

Location S.W. USA,
N. Mexico

Status Vulnerable

The Gila monster is one of North America's most distinctive reptiles, and one of only 2 poisonous lizards in the world. Its bulky body is vividly marked with orange, pink, or yellow, contrasting with bands of black – a colour scheme that warns potential aggressors that it has a toxic bite. Gila monsters live in semi-desert or scrub with rocky outcrops, in places where there is ready access to moisture. Compared to most other lizards they are slow-moving, hunting by day in the spring but at dusk or after dark during the summer heat. The tail stores fat in the same way as a camel's hump; it shrinks when food is hard to find, especially in cold weather when the gila monster hibernates. Females lay their eggs in late summer, burying clutches of up to 8 in burrows scooped out of damp sand.

FEEDING HABITS

The Gila monster feeds on small mammals and the eggs of quails, doves, and reptiles. It hunts by smell and also "tastes" its surroundings with its tongue. Venom is produced in glands in the lower jaw, not the upper jaw as in snakes. The lizard bites fiercely with its sharp teeth and holds onto its victim while the poison enters the wound. The bite, although painful, is not fatal for adult humans.

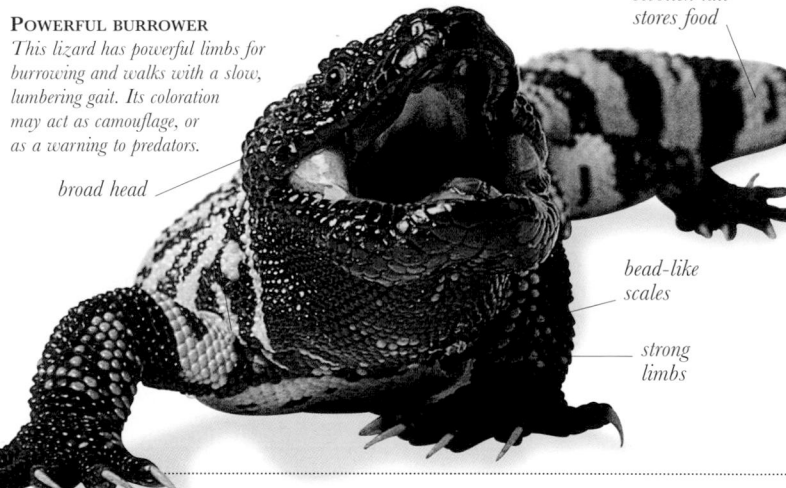

POWERFUL BURROWER
This lizard has powerful limbs for burrowing and walks with a slow, lumbering gait. Its coloration may act as camouflage, or as a warning to predators.

swollen tail stores food

broad head

bead-like scales

strong limbs

Heloderma horridum
Mexican beaded lizard

Length 70–100 cm
(28–39 in)
Breeding Oviparous
Habit Terrestrial/
Burrowing

Location W. Mexico

Status Vulnerable

A close relative of the Gila monster (see above), the Mexican beaded lizard, too, is poisonous, but its colours are more subdued. It is often dark brown with paler blotches, but in the south of its range it is entirely black. This lizard hides in burrows, emerging to forage for birds, especially fledglings, as well as eggs, and small mammals. It bites and injects venom into its prey with the sharp, grooved teeth in its lower jaw. Although painful, its bite is not often fatal to humans. Females lay 4–10 large, elongated eggs in burrows.

thick, powerful limbs for digging

Lanthanotus borneensis
Borneo earless lizard

Length 40–45 cm
(16–18 in)
Breeding Oviparous
Habit Burrowing/
Semi-aquatic

Location S.E. Asia

Status Vulnerable

This efficient swimmer inhabits stream-side burrows, which may become flooded. Its short limbs and blunt snout are adaptations for burrowing, while the transparent eyelids and nostrils that can be closed by valves are adapted for living in water. Being nocturnal and secretive, it is hardly ever seen. Only about 100 specimens have been collected since it was first described in 1878 and it remains poorly known. Once believed to be related to the heloderms (the Gila monster, above, and the Mexican beaded lizard, left) and, later, the monitors (see pp.420–2), it is now thought not to have any living close relatives.

REPTILES

Varanus komodoensis

Komodo dragon

	Length 2–3m
	(6¹/₂–9³/₄ft)
	Breeding Egg-laying
	method
	Habit Terrestrial

Location Indonesia
(Komodo, Rinca, Padar,
W. Flores)

Status Vulnerable†

With an average weight of about 70kg (155lb), and sometimes double this in captivity, the Komodo dragon is the world's heaviest lizard. It has a long body, well-developed legs, and a deeply forked tongue which it flicks out as it searches for food. Juveniles are boldly marked, with grey or cream bands, but lose these markings as they mature, becoming uniformly greyish brown, with thickly folded, scaly skin.

Komodo dragons live on scrubby hillsides and in open woodland and dry river beds, where they feed entirely on live animals or carrion. They have a keen sense of smell, and can detect the scent of decaying remains from up to 5km (3 miles). However, they hunt mainly by ambushing prey. Young Komodo dragons attack snakes, lizards, and rodents, but adults target much larger prey, including wild pigs, water buffalo, and deer. They are also cannibalistic – one reason why juveniles spend much of their time in trees.

Adults are mainly solitary, although groups may gather at the site of a kill. During the breeding season, males compete for the chance to mate, wrestling in an upright position with their tails acting as props. After mating, females dig nests in sandy ground, laying clutches of up to 25 eggs. When the eggs hatch – about 9 months later – the young are left to fend for themselves. Komodo dragons take about 5 years to become sexually mature, and in the wild they have a maximum lifespan of about 40 years.

ON THE SCENT OF FOOD

Komodo dragons have good eyesight, but they find most of their food by smell. Like snakes, they "taste" the air with their tongues, which collect scent molecules from the air. They have sharp, serrated teeth, but because they cannot chew, they tear off pieces of food and then throw them backwards into their mouths. Komodo dragon saliva is rich in toxic bacteria that thrive on traces of flesh, and when a dragon bites its prey, these bacteria contaminate the wound. Even if the victim does not succumb to the initial attack, it often dies from the resulting infection.

CONSERVATION

Komodo dragons are the world's most localized large predators, restricted to a few Indonesian islands. Estimated at 2,500–5,000, the population is a fraction of its size 50 years ago. Its decline is due to hunting, loss of prey, and habitat change. However, Komodo dragons have become increasingly important as a tourist attraction, a practical incentive for their conservation.

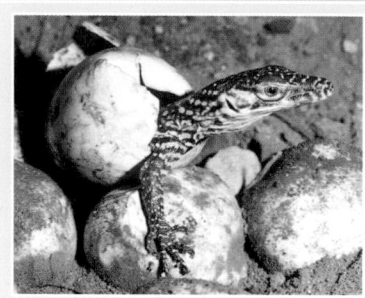

*relatively small
head with
wide jaws*

*skin folds
on neck*

*body held off
ground when
walking*

A GIANT PREDATOR

Komodo dragons look ungainly, but they are capable of running at up to 18kph (11mph) in short bursts. They are also good swimmers. They have a huge capacity for food, eating up to half their own weight in a single meal.

*small scales are
greyish brown
in adults*

*sharp claws used for digging
burrows and unearthing food*

*long, muscular tail can be used
as a weapon, and as a prop
when standing on hindlegs*

KOMODO DRAGONS AT A KILL

①

FIRST ARRIVAL

Guided by scent, a Komodo dragon homes in on a deer ambushed several hours earlier. The deer escaped, but has since died from the dragon's bite.

②

FEEDING BEGINS

The dragon begins to feed on the deer's body. It has unusually flexible joints between its jaw and skull, enabling it to swallow large chunks of food.

③

QUICK WORK

The dragon feeds rapidly, swallowing flesh, skin, and even bones. Meanwhile, other dragons begin to head for the carcass as its scent drifts downwind.

④

FIGHTING FOR SHARES

As more dragons arrive at the scene, the larger animals try to intimidate the smaller ones into backing away from the kill. If food is scarce, fighting often breaks out.

Varanus albigularis
Savanna monitor

Length 1 m (3 1/4 ft),
max 1.8 m (6 ft)
Breeding Oviparous
Habit Terrestrial/
Burrowing

Location E., C., and
southern Africa

Status Common

elongated head with   *bead-like scales*
domed snout

crossbands

Also known as the white-throated
monitor, this lizard has powerful limbs
for digging burrows, in which it lives;
it also lives in hollow tree trunks.
Territorial, with a range of up to 18
square km (7 square miles), it forages
for birds, insects, snails, and invertebrates. Its chief predators are martial
eagles and ratels, and it is sometimes
eaten by humans, despite legal protection. In defence, the savanna monitor
puffs up its throat and body, lashes out
with its tail, and bites. In a year, a
female may lay 2 clutches of as
many as 50 eggs.

Varanus varius
Lace monitor

Length 1.5–2 m
(5–6 1/2 ft)
Breeding Oviparous
Habit Arboreal

Location E. Australia   **Status** Common

Dark grey or bluish, with white or
cream markings, the lace monitor
has a long neck and tail, and long
claws used for scaling trees. It
bites and lashes out with its tail
if threatened. It forages on the
ground and in trees for insects,
reptiles, birds and their eggs,
and small mammals. Females
lay 6–12 eggs in holes in termite
mounds, which the termites seal;
reportedly, females sometimes return
to help the young dig their way out.

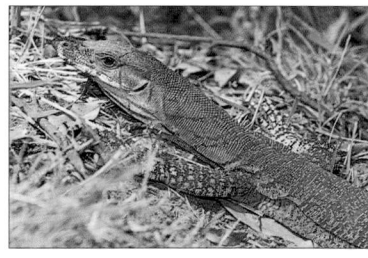

Varanus dumerilii
Dumeril's monitor

Length 1–1.3 m
(3 1/4 – 4 1/4 ft)
Breeding Oviparous
Habit Mostly
terrestrial

Location S.E. Asia   **Status** Common

indistinct light
markings

narrow
head

Although this
slender grey lizard
is mostly terrestrial, it
    is also a good swimmer, and can
    close its nostrils under water
when in pursuit of crabs to eat. It also
forages actively for insects, birds, and
bird and turtle eggs. It may
occasionally enter the sea to escape
from predators, mainly mammals and
snakes, which threaten young lizards
in particular, or may escape by
climbing into trees and shrubs. Males
are territorial during the breeding
season; after mating, females lay their
eggs in soil or leaf litter. This monitor
prefers mangrove forests but also
inhabits forests away from the coast.

Varanus niloticus
Nile monitor

Length 1.4–2 m
(4 1/2 – 6 1/2 ft)
Breeding Oviparous
Habit Semi-aquatic

Location Africa (mainly
south of Sahara)

Status Common

whip-like tail used for
swimming and defence

The greyish brown Nile monitor is
never found far from water. It basks
on nearby rocks or tree stumps, and it
feeds on crabs, molluscs, fish, and
frogs, as well as birds, eggs, and
carrion. Hunted by crocodiles and
pythons, it uses its tail, claws, and
teeth to defend itself. In the cooler
part of its range, it hibernates in
communal "dens". Females lay 20–60
eggs in termite mounds, and the
hatchlings dig their way out after rain
has softened the earth.

bead-like scales

Varanus prasinus
Green tree monitor

Length 75–100 cm
(30–39 in)
Breeding Oviparous
Habit Arboreal

Location New Guinea   **Status** Locally common

Well equipped for its arboreal way
of life, this species has a light, slender
body that enables it to support itself
on narrow branches, and a coloration
that helps to camouflage it against the

trees. It feeds on insects, small
vertebrates, crabs, and birds' eggs.
When threatened, it flees through
vegetation and may bite if cornered.
One of the few monitors that are
social by nature, it appears to live in
small groups made up of a dominant
male, several females, and subordinate
males and juveniles. Females lay
clutches of up to 6 eggs in a hole in
arboreal termite nests or leaf litter.

long claws grip bark

Varanus flavirufus
Sand monitor

Length 75–100 cm
(30–39 in)
Breeding Oviparous
Habit Terrestrial/
Burrowing

Location S. New Guinea,
Australia

Status Common

Also known as Gould's monitor or
Gould's goanna (a contraction of the
word "iguana"), this impressive reptile
is one of the largest of about 20
monitors that live in Australia.

Its slender but powerful body varies
in colour from pale yellow to dark
grey, but it almost always has a dark
line extending from the back of the
eye to the neck. The monitor's long
and muscular tail works as a prop
when it stands on its back legs to
inspect its surroundings. When
threatened, it uses its tail as a whip or
club, and also strikes out with its teeth
and claws. A voracious feeder, the
sand monitor eats mammals, birds,
and other reptiles, as well as smaller
animals and carrion. It shelters in
burrows and in hollow trees, and
females often lay their eggs in termite
mounds, clawing them open to place
the eggs inside.

Shinisaurus crocodilurus
Chinese crocodile lizard

Length 40–46 cm
(16–18 in)
Breeding Viviparous
Habit Semi-aquatic

Location E. Asia (Kwangsi
Province)

Status Vulnerable

This rare, little-known lizard is olive
in colour, with rows of enlarged, bony
scales down its back and tail, like
those of a crocodile. The male is
more colourful than the female but

perhaps only in the breeding season.
To escape detection, the lizard may
"freeze" for several hours at a time,
even in mid-stride – a tactic that,
along with its coloration, helps it to
merge with its surroundings. It adapts
in unusual ways to cool temperatures:
it may, for instance, be able to "shut
down" its system for hours to conserve
energy during very cold nights. The
female gives birth to 2–10 young,
sometimes over several days.

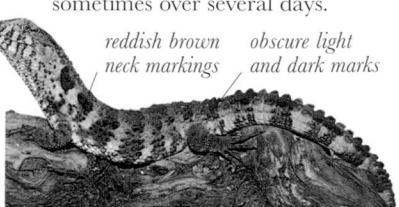

reddish brown   *obscure light*
neck markings   *and dark marks*

AMPHISBAENIANS

ALTHOUGH THEY ARE SOMETIMES called "worm lizards", amphisbaenians are neither worms nor lizards. They are, however, closely related to lizards and belong to the same order as both lizards and snakes. Adapted to an underground life, amphisbaenians are seen above ground only after heavy rain has flooded their tunnels. Their ability to regulate body temperature is limited, which confines them to tropical and subtropical regions.

PHYLUM	Chordata
CLASS	Reptilia
ORDER	Squamata
SUBORDER	Amphisbaenia
FAMILIES	3
SPECIES	158

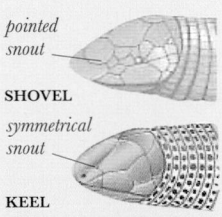

SHOVEL

KEEL

pointed snout

symmetrical snout

HEAD SHAPE
Amphisbaenians use their thick, heavy skulls to dig tunnels. The shape of the snout varies according to the method of tunnelling.

TUNNELLING

Amphisbaenians construct their own burrows, down which they move in search of worms, insects, and larvae. They make fresh tunnels by using their skulls as rams. Their nostrils point backwards, and are thus prevented from being clogged up with soil while burrowing. The lower jaw is recessed, so that, as the head is forced through the soil, the mouth is held tightly closed.

ANATOMY

Amphisbaenians are worm-like in appearance with cylindrical bodies and rings of scales that look like body segments. They have rudimentary eyes and smooth heads that look wedge-shaped in side view. They are all without legs, except for 3 species that have front limbs, positioned close to the head and ending in long claws for burrowing. Many amphisbaenians lack pigmentation and are pinkish brown; a few are more colourful.

SCALES
Unlike the scales of snakes, which overlap each other, those of amphisbaenians are arranged in concentric rings.

EYES AND EARS
An amphisbaenian's eyes are covered with translucent skin, and like many burrowing animals they have poor vision. They lack external ear openings.

LIMBS
Amphisbaenians spend almost all their lives underground, burrowing in soil, sand, or leaf litter. Most have lost all traces of limbs. However, the members of one family, the Mexican worm lizards (shown here), have retained their forelimbs.

REPTILES

Amphisbaena fuliginosa
Black and white amphisbaenian

Length 30–45 cm (12–18 in)
Breeding Oviparous
Habit Burrowing

Location N. South America, Trinidad
Status Common

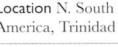

This South American species is one of the most easily recognized amphisbaenians, because of its conspicuous markings.

It lives almost exclusively in burrows (although it may come to the surface at night), using its head to ram a tunnel through the ground and its body rings to help it concertina through the tunnel. It feeds on virtually any small vertebrates and insects, attacking them with its strong jaws. If threatened, it may shed part of its tail, which does not regenerate. Like other amphisbaenians, this species is difficult to study since it spends most of its life underground.

white head
short tail
black and white, cylindrical body

Blanus cinereus
European worm lizard

Length 10–20 cm (4–8 in) max 30 cm (12 in)
Breeding Oviparous
Habit Burrowing

Location S.W. Europe
Status Common

The European worm lizard has a small, pinkish, or purplish brown body, with a short head, pointed snout, rudimentary eyes, and ring-like

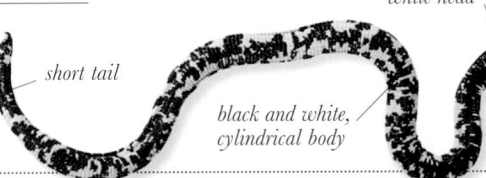

grooves, all of which help it to burrow. Mostly found under rocks and logs, it may appear on the surface at night, especially after heavy rain. It hunts along underground tunnels for earthworms and small insects. *Blanus tingitanus*, found in Morocco and Algeria, is sometimes considered to be the same species.

forelimbs
short tail
pink body ringed with furrows

Bipes biporus
Ajolote

Length 17–24 cm (6½–9½ in)
Breeding Oviparous
Habit Burrowing

Location Mexico (Baja California)
Status Common

The ajolote, or mole lizard, is one of the world's most remarkable reptiles. It has the typical amphisbaenian adaptations for a life spent burrowing – a blunt head, cylindrical body, and short tail – but it also has a pair of very short yet powerful front legs. Equipped with clawed feet, they help the ajolote to excavate its way through the ground. The ajolote rarely comes to the surface – and then only after heavy rain – and it feeds by ambushing lizards and other small animals, usually dragging them underground. There are 2 similar species, also found only in Mexico.

Rhineura floridana
Florida worm lizard

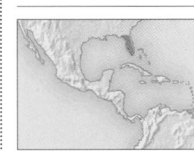

Length 25–35 cm (10–14 in)
Breeding Oviparous
Habit Burrowing

Location S.E. USA (Florida)
Status Locally common

pinkish body resembles earthworm
scales arranged in rings around body

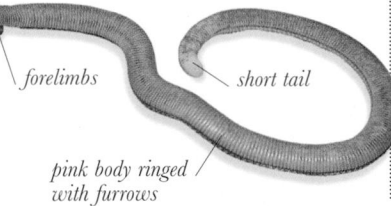

The Florida worm lizard looks like an earthworm, with a pinkish body that has no limbs, no external eyes, and no ear openings. Exceptional specimens can be up to 40cm (16in) long. The scales are ill-defined and arranged in rings around the cylindrical body. This worm lizard leads its life almost entirely underground but may be forced to the surface because of heavy rain or ploughing. An opportunistic species, it feeds on invertebrates encountered while burrowing. It is preyed upon by birds – mockingbirds occasionally unearth it when hunting for earthworms.

CROCODILES AND ALLIGATORS

PHYLUM	Chordata
CLASS	Reptilia
ORDER	Crocodilia
FAMILIES	3
SPECIES	23

THESE LARGE reptiles are among the few survivors from the time of the dinosaurs, having changed little in the last 65 million years. Collectively known as crocodilians, they include crocodiles, alligators, and caimans, and one species of gharial. Crocodilians are formidable, semi-aquatic predators. Most of them live in freshwater rivers, lakes, and lagoons, while a few species inhabit tidal reaches and may venture out to sea. Alligators and caimans (with the exception of one species) are found only in North, Central, and South America. Crocodiles occur mainly in Asia, Africa, and Australia, with some species in Central and South America.

REPRODUCTION

Male crocodilians are territorial and mate with several females. Each female lays her eggs near water, either in a mound of vegetation and mud or in an underground chamber. The females guard the eggs and help the hatchlings out of the nest when they hear them calling. Females stay with their young for several months (a year or more in American alligators) to protect them from predators.

PARENTAL CARE
In most species, the female uses her mouth to carry her hatchlings to a shallow pool or backwater.

HUNTING LARGE PREY
Sometimes hunting cooperatively, large crocodilians such as the Nile crocodile (shown here) are capable of overcoming animals as large as wildebeest or buffalo. After the kill, the carcass may be stored under an overhanging bank or log below the waterline to be eaten later.

ANATOMY

All species of crocodilians have a wide, slightly flattened body, a long, vertically flattened, muscular tail, powerful jaws, and eyes and nostrils on top of the head so that they can see and breathe while partly submerged. A transparent third eyelid can be closed to protect their eyes when underwater. The large, bony scales have raised ridges, and are shed singly or in pieces, unlike those of other reptiles. Most are dull olive, grey, brown, or black, although juveniles may have lighter markings that disappear with age. The 3 groups of crocodilians differ in the shape of their snout and the arrangement of their teeth (see below).

relatively short snout

ALLIGATOR

fourth lower tooth

CROCODILE

long, narrow snout

GHARIAL

SNOUT SHAPES
Alligators and caimans have shorter, broader snouts than crocodiles. The fourth lower tooth fits into a pit in the upper jaw and is not seen when the mouth is closed. Crocodiles, however, have a notch in their upper jaw, into which the fourth lower tooth fits so that the tooth remains visible when the mouth is shut. The gharial has a long, narrow snout with enlarged front teeth followed by smaller teeth of equal size.

flap of skin covering windpipe

ADAPTATIONS FOR DIVING
Crocodilians have 2 adaptations to stop water entering their lungs when they are underwater. By closing a flap of skin at the back of the throat, they can open their mouth to catch prey without water passing down the windpipe. Similar flaps can also be closed over the external openings of the nostrils and ears.

HUNTING AND FEEDING

Crocodilians are carnivores that eat a mixture of live prey and carrion. They use various techniques when hunting but the most common is to lie in wait at the edges of rivers and lakes for mammals to come within range to drink or try to cross. Another method is to drift towards prey, such as waterfowl, hoping to catch it off guard. Fish-eating species, such as the gharial, take prey with a rapid sideways swipe of their narrow jaws, which offer little resistance to the water. Birds, small mammals, and fish are swallowed whole, but as crocodilians cannot chew, they must dismember larger prey by spinning violently about their own axis while underwater, holding part of the carcass in their jaws.

HUNTING BY STEALTH
By drifting with only its eyes and nostrils above the water, a crocodilian can move to within striking range without being seen.

STORING FOOD
Having drowned this antelope, the crocodile may eat some of it before securing the remainder in an underwater "larder" to rot, making it easier to dismember.

Alligator mississippiensis
American alligator

Length 2.8–5 m
(9 1/4–16 ft)

Breeding Oviparous

Habit Aquatic

Location S.E. USA | Status Common

Once widely hunted for its skin, the American alligator became seriously endangered in the 1950s, but under legal protection has staged a strong recovery. A large, powerful reptile, it is black with a broad head and rounded snout, and has a heavily plated back. It also has a characteristic large fourth tooth that fits into a socket in the upper jaw. This alligator feeds in or near fresh water, taking animals of all kinds, including birds snatched from low branches. During the breeding season, the males roar to attract females, and mating takes place underwater. Each female lays 25–60 eggs in an enormous nest made of vegetation and mud, digging it open when she hears the calls of the hatching young; she may stay with her brood for as long as 3 years.

ALLIGATOR HOLES

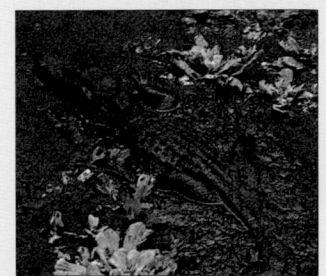

The American alligator prefers to float, partially submerged, in lakes, swamps, and marshes. In midsummer, as the water table falls, it retreats into "alligator holes" – keeping a small body of water open by digging out the sand or mud at the bottom. It feeds on fish that are trapped in the diminishing pool and animals that visit it.

THE BLACK ALLIGATOR
Both adult and juvenile American alligators are black, but the juveniles have bold yellow crossbands.

keeled scales on tail

webbed feet for swimming

Caiman crocodilus
Spectacled caiman

Length 2–2.5 m
(6 1/2–8 ft)

Breeding Oviparous

Habit Aquatic

Location Central America, N. South America | Status Common

Also called the common caiman, this dull olive crocodile has a bony ridge in front of its eyes. It is found in most freshwater habitats, rarely leaving the water unless driven out by drought, when it will burrow into the mud. It floats on the surface during the day, becoming more active at night. Adults feed on other reptiles, fish, amphibians, and waterbirds.

Males are territorial and establish a dominance hierarchy. Females lay 14–40 eggs in mounds of decaying plants and soil at the edge of water, or on rafts of floating vegetation. Several females may share the same nest and guard it against predators. Although heavily hunted for its skin, this crocodilian has benefited from artificial bodies of water, such as reservoirs, resulting in increases in local populations.

bony ridge

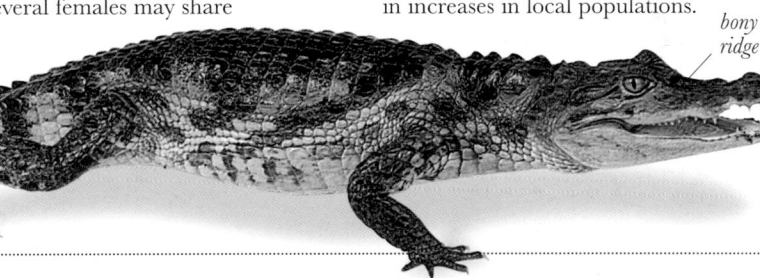

Crocodylus niloticus
Nile crocodile

Length 3.5 m (11 ft),
max 6 m (20 ft)

Breeding Oviparous

Habit Aquatic

Location Africa, W. Madagascar | Status Common

The Nile crocodile has a dark olive to grey body with darker crossbands. It favours larger rivers, lakes, and swamps, and is also found in estuaries and river mouths; it may come ashore to bask on hot days. It feeds on fish, antelope, zebras, and even buffaloes, leaping up to snatch nesting birds or pull drinking animals into the water. Although solitary by nature, several crocodiles may congregate for feeding, working together to herd fish into shallow water. Males are territorial and establish a dominance hierarchy. The female lays 16–80 eggs in a hole dug on a bank well above the water level and uses the same nest site for the whole of her life. She guards the eggs throughout the incubation period. The young chirp when they are about to hatch and the female digs them out, gently manoeuvring the hatchlings into her mouth, and carrying them into the water in batches. They remain together for 6–8 weeks, then gradually disperse. For the first 4 or 5 years they live in burrows up to 3 m (9¾ ft) long.

JAWS AND TEETH

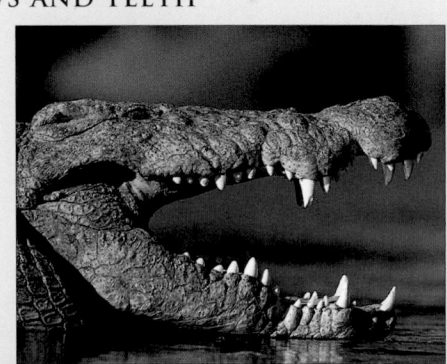

The Nile crocodile has long jaws, with teeth that are exposed even when its mouth is closed. Like other crocodilians, it can bite but it cannot chew, which presents problems when dealing with large prey such as zebras and buffaloes. Nile crocodiles deal with these by dragging them underwater, and then spinning around to tear off chunks of flesh.

pair of raised keels down tail

eyes on top of head

dark crossbands on body

powerful tail for swimming

bony, plate-like scales

AQUATIC ADAPTATIONS
Like most crocodiles, the Nile crocodile has eyes and nostrils on the top of its head, helping it to see and breathe in water; it also has a flat, powerful tail and webbed hindfeet that help it to swim.

Crocodylus porosus
Saltwater crocodile

Length 5–7 m
(16–23 ft)
Breeding Oviparous
Habit Aquatic

Location S.E. Asia,
N. Australia

Status Common

Also known as the estuarine or Indo-Pacific crocodile, this massive grey, brown, or black animal is the world's largest crocodilian. The biggest specimens on record weighed over 1 tonne (2,200lb), although decades of hunting

mean that such giants are now very rare. Equally at home in fresh water and the sea, these nocturnal reptiles feed on a wide variety of animals, including mammals, birds, and fish. They present a real threat to humans, and have been responsible for many fatalities. They mate in water, and the females lay about 60 eggs in mound nests on the river bank, high above the watermark. Each female guards her clutch, and stays with her hatchlings for the first few weeks of life.

WALKING ON LAND
The saltwater crocodile has strong enough legs to raise its body off the ground and so can walk on land.

broad snout pitted with deep pores

irregular mottling

large, plate-like scales

2 keels on tail

BASKING

The saltwater crocodile may haul itself onto a bank on hot days and bask with its mouth agape. Having its mouth open stops it overheating, and it allows small birds to peck at parasites and bits of food lodged between its teeth.

Melanosuchus niger
Black caiman

Length 4–6 m
(13–20 ft)
Breeding Oviparous
Habit Aquatic

Location N. South America

Status Lower risk

Hunted for its skin, the black caiman has become extinct in many parts of its range, and its numbers are thought to have been reduced by 99 per cent in the past 100 years. It has an almost parallel snout tapering sharply at the tip and grey bands on the lower jaw. It floats on the surface by day but may come ashore to feed at night. Fish, waterbirds, capybara, the occasional domestic animal, and even humans may fall prey to it. The female lays and guards 30–65 eggs, and stays with the young after hatching.

Osteolaemus tetraspis
Dwarf crocodile

Length 1.7 m (5 ½ ft),
max 1.9 m (6 ¼ ft)
Breeding Oviparous
Habit Aquatic

Location W. and C. Africa

Status Vulnerable

blunt, often upturned snout

bony plates cover body

This relatively short black crocodile lives in permanent pools and swamps in the African rainforest. It hides in burrows and among submerged tree roots during the day, and forages at the water's edge at night, taking its prey with a sideways snap of the jaws. In areas with seasonal flooding, fish are the main food in the wet season, and crustaceans for the rest of the year. Frogs and toads are also eaten. The female lays usually 10 eggs in a mound of soil and vegetation, guards them, and protects her young in the water. This crocodile is probably common in some parts of its range, but in others its numbers are depleted because of habitat destruction.

Gavialis gangeticus
Gharial

Length 4–7 m
(13–23 ft)
Breeding Oviparous
Habit Aquatic

Location N. Indian subcontinent

Status Endangered

Instantly recognizable by its unique narrow snout, the gharial – or gavial – is a large but slender crocodilian that spends most of its life in water. Compared to other crocodilians, its legs are relatively weak, and its feet are broadly webbed. Gharials feed primarily on fish, although they also eat waterbirds. During the breeding season, the male becomes strongly territorial, and assembles a harem of females. The female excavates a nest well away from the water's edge, and lays up to 50 eggs, which at about 150g (5oz) each are unusually large. She protects the hatchlings, but does not carry them to water – probably because of the unsuitable shape of her jaws. The gharial almost became extinct in the 1970s as a result of habitat loss, hunting, and fishing, but a programme of captive breeding is slowly helping it to recover.

A MULTI-FUNCTIONAL SNOUT

The gharial's long, slender snout may look fragile, but it is armed with small, extremely sharp teeth. It catches fish sideways on, and then flicks them within its jaws so that it can swallow them head-first. The bulbous growth at the end of the nose is present only in males, and is used to produce sounds and bubbles during courtship.

LIFE IN WATER
The highly aquatic gharial inhabits slow-moving backwaters of large rivers, leaving the water only to bask on mud or sand banks, and to lay eggs. It submerges itself in deep water for safety. Mating, too, takes place in water. The gharial has poorly developed legs and is unable to walk on land, barely managing to push itself over the mud.

grey or olive skin

bulbous growth

REPTILES IN DANGER

The list on this page includes all those reptiles that are currently classed as critically endangered by the IUCN (see p.31). Like most other vertebrates, reptiles have suffered through habitat change, but for this group of animals – perhaps more than any other – hunting has played a critical part in bringing many species to the edge of extinction. Of greatest concern in this respect are all the world's marine turtles. Their need to breathe makes them easy to catch at sea, and their nesting behaviour renders both the females and their eggs highly vulnerable to hunters and collectors onshore. Some freshwater turtles are endangered for similar reasons, but with snakes and lizards, environmental problems are sometimes a greater threat. Here, many critically endangered species are inhabitants of oceanic islands, where habitat change can have disastrous effects. Crocodiles and alligators are hunted primarily for their skins; although only four of the world's 23 species are listed as critically endangered, many others are either endangered or vulnerable.

TORTOISES AND TURTLES

Apalone ater (Cuatro Cienegas softshell)
Aspideretes nigricans (black soft-shell turtle)
Batagur baska (four-toed terrapin)
Callagur borneoensis (painted terrapin)
Chelodina mccordi (Roti Island snake-necked turtle)
Chelonia mydas (Green turtle – Mediterranean subpopulation)
Chitra chitra (striped narrow-headed softshell turtle)
Cuora aurocapitata (yellow-headed box turtle); *C. galbinifrons* (Indochinese box turtle); *C. mccordi* (McCord's box turtle); *C. pani* (Pan's box turtle); *C. trifasciata* (Chinese three-striped box turtle); *C. zhoui* (Zhou's box turtle)
Dermochelys coriacea (leatherback turtle)
Eretmochelys imbricata (hawksbill turtle)
Geochelone nigra (1 subsp: Hood Island tortoise); *G. platynota* (flatback tortoise)
Heosemys depressa (Arakan forest turtle); *H. leytensis* (Philippine pond turtle); *H. yuwonoi* (Sulawesi forest turtle)
Kachuga kachuga (Bengal roof turtle)
Lepidochelys kempi (Atlantic ridley)
Mauremys annamensis (Annam leaf turtle)
Phrynops dahli (Dahl's toad-headed turtle)
Pseudemydura umbrina (western short-necked turtle)
Rafetus swinhoei (Shanghai softshell turtle)
Testudo graeca (1 subsp: Mediterranean tortoise)
Trionyx triunguis (African softshell turtle – Mediterranean subpopulation)

SNAKES

Alsophis antiguae (Antiguan racer); *A. ater* (black racer)
Bothrops insularis (golden lancehead), *B.* species (lancehead species)
Chironius vincenti (St Vincent blacksnake)
Coluber gyarosensis (racer species)
Crotalus unicolor (Aruba Island rattlesnake)
Dipsas albifrons (1 subsp: Sauvage's snake eater)
Liophis cursor (Lacepede's ground snake)
Macrovipera schweizeri (Cyclades blunt-nosed viper)
Natrix natrix (2 subsp: Sardinian grass snake, European grass snake subspecies)

Opisthotropis kikuzatoi (Kikuzato's brook snake)
Vipera bulgardaghica (Mount Bulgar viper); *V. darevskii* (Darevsky's meadow viper); *V. pontica* (black sea viper); *V. ursinii* (1 subsp: Orsini's viper)

LIZARDS

Abronia montecristoi (Monte Cristo arboreal alligator lizard)
Ameiva polops (Saint Croix ground lizard)
Anolis roosevelti (Culebra Island giant anole)
Brachylophus vitiensis (Fiji crested iguana)
Bradypodion pumilum (1 subsp: Cape dwarf chameleon); *B. taeniabronchum* (Smith's dwarf chameleon)
Celestus anelpistus (Altagracia giant galliwasp)
Cyclura carinata (Bahamas rock iguana); *C. collei* (Jamaica ground iguana); *C. nubila* (2 subsp: Grand Cayman iguana, Lesser Cayman iguana); *C. pinguis* (Anegada ground iguana); *C. ricordi* (Hispaniolan ground iguana); *C. rileyi* (2 subsp: White Cay iguana, San Salvador iguana)
Diploglossus montisserrati (Montserrat galliwasp)
Eumeces longirostris (Bermuda rock skink)
Gallotia simonyi (Hierro giant lizard)
Lepidoblepharis montecanoensis (Paraguanan ground gecko)
Lerista allanae (Allan's lerista)

CROCODILES & ALLIGATORS

Alligator sinensis (Chinese alligator)
Crocodylus intermedius (Orinoco crocodile); *C. mindorensis* (Philippines crocodile); *C. siamensis* (Siamese crocodile)

REPTILES

HAWKSBILL TURTLE
Prized for its meat and its shell, the hawksbill turtle is critically endangered. It feeds in shallow water around coral reefs, which makes it an easy target for spear-fishers. The hawksbill is the traditional source of tortoiseshell – an animal product that is now banned by CITES (see p.33).

AMPHIBIANS

AMPHIBIANS

PHYLUM	Chordata
CLASS	Amphibia
ORDERS	3
FAMILIES	44
SPECIES	About 5,000

AMPHIBIANS ARE INTIMATELY associated with water, most of them spending part of their life in water and part on land. These ectothermic (cold-blooded) vertebrates are divided into 3 groups: newts and salamanders, frogs and toads, and the worm-like caecilians. The life cycle of most amphibians involves a transformation from aquatic larvae (which breathe using gills) to terrestrial adults (which take in oxygen through lungs). Most live close to water in tropical and temperate regions, although some have adapted to survive cold and drought.

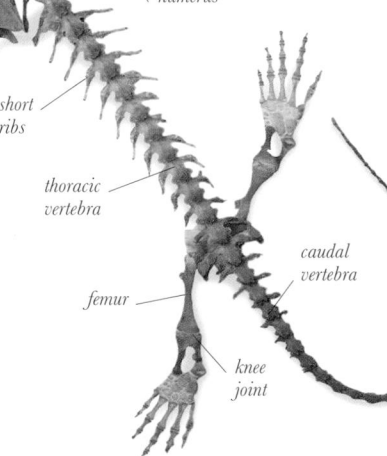

TREEFROG
This giant treefrog, from the rainforests of New Guinea and northern Australia, has moist skin that is well suited to its wet and humid habitat.

EVOLUTION

The first amphibians, which appeared about 370 million years ago during the Devonian Period, were descendants of fleshy-finned fish. The first vertebrates to colonize the land, early amphibians represent an intermediate stage between aquatic fish and fully terrestrial reptiles. Like the fish from which they descended, they

AMPHIBIAN ANCESTOR
Diplocaulus lived about 280 million years ago and probably resembled a modern newt, but for an unusually shaped head.

were equipped with lungs and breathed air, but they also shared a number of fish-like features, such as scaly skin and a finned tail.

Amphibians were the first animals to evolve features vital to terrestrial life. Most importantly, they were the first quadrupeds, evolving 2 pairs of jointed limbs from the fins of their fish ancestors. These enabled them to lift their body off the ground, providing them with a more efficient means of locomotion. However, they were heavily built animals, and their movements on land were slow and clumsy compared to the agile movements of living amphibians. Amphibians were also the first animals to evolve eyelids (which keep the eyes moist in

air), a good sense of smell, and a tongue (which enabled them to moisten food before swallowing).

By about 280 million years ago, amphibians had become the dominant land animals. However, over the next 70 million years, amphibians were gradually superseded by terrestrial reptiles (which had evolved the ability to lay shell-covered eggs on land) and most species became extinct.

SKELETON
This skeleton of a juvenile Japanese giant salamander highlights an evolutionary trend toward simplifying the amphibian skeleton. It contains fewer bones than in other vertebrates, and far fewer than in its fish ancestors.

ANATOMY

Compared to other vertebrates, amphibians have a fairly simple body structure, well adapted for life in water and on land. Most adult amphibians have 4 limbs, with 4 digits on the front limbs and 5 on the back. Newts and salamanders have a relatively small head, a long, slender body, with short limbs of equal length, and a long tail. However, some salamanders have very small limbs with

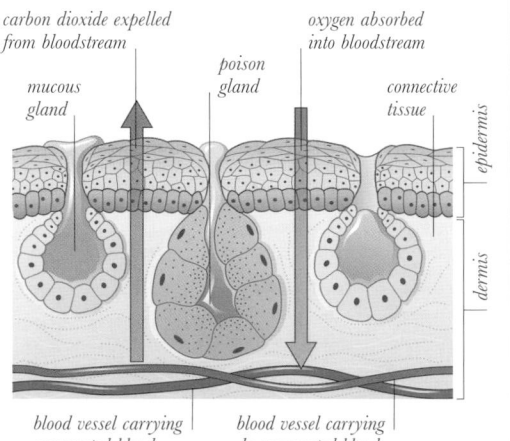

reduced numbers of digits, or lack rear limbs altogether. Most frogs and toads have a distinctive compact body shape, with a large head, short back, small front limbs, large, muscular back limbs, and no tail. Caecilians have a worm-like, limbless body, with a pointed head and a short tail. Apart from caecilians, which use smell, most amphibians hunt using sight and have large eyes that help them to locate prey at night. They also have an extremely

SKIN

The skin of an amphibian is naked and smooth, with no scales or hair to protect it. It is permeable to water, and, although numerous mucous glands help to keep the skin moist, most amphibians quickly dry out if they are not in a damp place. All amphibians have poison glands in their skin, and many produce secretions that are distasteful or highly toxic to potential predators. The skin usually contains numerous pigment cells, and many amphibians are brightly coloured. This is especially common in poisonous species, in which striking skin patterns serve as a warning. Many amphibians use their skin for breathing, which means that it has to be kept moist at all times. This may also place a limit on how large amphibians can be, since they need to maintain a high ratio of surface area to volume, to absorb an adequate amount of oxygen.

IN SUNLIGHT

IN SHADE

CHANGING SKIN COLOUR
The base colour of some amphibians' skin changes in response to variations in temperature, light level, or mood. When in sunlight, the skin of the White's treefrog (left) is light green but on entering shade gradually turns brown.

BREATHING THROUGH THE SKIN
Many amphibians can breathe through their skin, absorbing oxygen directly into their bloodstream and expelling carbon dioxide. As well as being packed with blood vessels – some only a few cells' width from the surface – the skin is very thin, enabling the easy transfer of gases. Mucous glands keep the surface damp, helping the gases to pass through.

WORM-LIKE AMPHIBIANS
With their long bodies, body rings, and lack of limbs, caecilians are often mistaken for large earthworms. Most are found in loose soil or deep leaf litter in tropical forest.

wide mouth, which enables them to consume relatively large prey.

Although most adult amphibians breathe using lungs, they do not have a rib cage and a diaphragm to inflate them. Many amphibians can absorb oxygen into their blood through other organs and tissues, including gills, the skin, and the lining of the mouth. The relative importance of each route varies among different life stages and adult forms. Young amphibian larvae breathe using external gills, as do those salamanders that retain the larval form into the adult stage. Most terrestrial frogs and toads, and some salamanders, have lungs, but also absorb a certain amount of oxygen through the skin and the mouth (see panel, facing page). One group of salamanders, the plethodontids, have no lungs and rely entirely on the skin and the mouth to take in oxygen.

LIFE CYCLES

An amphibian's life cycle involves 3 stages – egg, larva, and adult, with the transition from larva to adult involving a radical transformation known as metamorphosis.

Most amphibians deposit their eggs in water. However, some species lay their eggs on land, and others show a variety of ways of retaining them inside their body. In nearly all frogs and toads, fertilization is external, with sperm entering the eggs outside the female's body. During mating, the male clasps the female, usually from above, and sheds sperm into the water at the same moment as

KEEPING GUARD
This marbled salamander guards its clutch of eggs at the edge of a pond. The female curls her body round her eggs to protect them from predators and fungal infection.

the female expels her eggs. In most newts and salamanders, fertilization is internal, the sperm being transferred to the female in small, gelatinous capsules called spermatophores. This requires precise coordination between the movements of male and female, and many newts and salamanders have elaborate courtship behaviour to achieve this. In caecilians, the male has a penis-like organ, which he uses to insert his sperm directly into the female.

The eggs of amphibians have no shell, but are enclosed in a gelatinous envelope that quickly shrivels up in dry conditions. Eggs may be laid singly, in strings, or in clumps; they may be carefully attached to plants or other solid objects, or just left to drift in the water. Several amphibians show parental care, guarding their eggs against predators and infection with fungal diseases. The number of eggs laid varies enormously: some amphibians lay only one or 2 eggs

PARENTAL CARE
European newts lay their eggs individually, wrapping them in leaves to make them less conspicuous to predators.

TAKING TO LAND
Swarms of newly metamorphosed European common frogs gather by the side of a pond before leaving the water for the first time.

at a time; others lay up to 50,000. The eggs hatch into minute larvae, which retain sufficient yolk to sustain them for a few days.

The larvae of amphibians (called tadpoles in frogs and toads) are very different in appearance, diet, and lifestyle from adults. Metamorphosis involves a major reconstruction of all parts of the body (see panel, below). The larvae of newts and salamanders are carnivorous, feeding on tiny invertebrates. In contrast, the tadpoles of frogs and toads are typically herbivores, either filter-feeding on microscopic plants that live suspended in water or grazing on algae that cover plants, rocks, and other submerged objects.

The larval stage is an important growth phase for amphibians, but different species vary considerably in the amount of total lifetime growth that is completed in the larval rather than the adult stage. Nearly all

amphibians actually become smaller during metamorphosis.

Some amphibians show direct development, in which the larva is not free-swimming but completes its development while still inside the egg and emerges as a tiny version of the adult form. Direct development occurs in a number of species that lay their eggs on land, and also in some that retain their eggs inside the body. Such species, by having no free-swimming larval stage, are not dependent on standing water.

BREEDING MIGRATIONS

Most amphibians live on land as adults but return to water each year to breed. Breeding is usually stimulated by a combination of environmental factors, including temperature, rainfall, and day length. In some species, referred to as explosive breeders, there is a mass migration of adults, all moving towards a breeding pond. Evidence shows that various species locate the same breeding site year after year by using their sense of smell, an ability to navigate by the sun, or by detecting tiny variations in the earth's magnetic field.

GROWTH AND DEVELOPMENT

The 3 stages of amphibian development are clearly illustrated by the European common frog. Embryos start to develop inside the eggs, which are laid in clumps. About 6 days after fertilization, the eggs hatch into tiny tadpoles with a spherical body, well-developed tail, and external gills. Within 4 weeks, these gills have been absorbed and replaced by internal gills. Between 6 and 9 weeks, the back limbs start to develop, the head becomes more distinct, and the body takes on a more streamlined shape. At this stage the legs are fully functional and the internal gills have been replaced by lungs. By 9 weeks, when the front legs have emerged, the tadpole has the compact body shape of an adult. The tail is then gradually absorbed into the body, until, by about 16 weeks, it has disappeared altogether.

FROG SPAWN

external gills

RECENTLY HATCHED

internal gills

4 WEEKS OLD

EGG TO ADULT
The European common frog's development, from egg to adult, typically takes about 16 weeks, but varies according to temperature levels and food supply.

near-adult body shape

shrinking tail

12 WEEKS OLD

back limbs appear

6–9 WEEKS OLD

long tail remains

front limbs appear

9 WEEKS OLD

FROGS MIGRATING
Some species of amphibian, such as these European common frogs, make annual mass migrations to breeding sites, often from great distances and across rough terrain.

NEWTS AND SALAMANDERS

PHYLUM	Chordata
CLASS	Amphibia
ORDER	Caudata
FAMILIES	10
SPECIES	470

NEWTS AND SALAMANDERS have a slender body with a long tail and, typically, 4 legs of similar size. Of the 3 main groups of living amphibians, they most closely resemble the animals from which all amphibians are descended. They also have some of the most complex and varied life histories. Some salamanders live their entire lives in water, while others are wholly terrestrial. Newts spend most of their lives on land but return to water to breed – apart from this, there is no scientific distinction between salamanders and newts. In all but the most primitive species, fertilization is internal (rather than external, as in most frogs and toads). The male does not have a penis but instead passes capsules of sperm to the female during mating. The larvae of newts and salamanders are carnivorous, and have a long, slender body, a deep, fin-like tail, and large, feathery external gills. Newts and salamanders are found in damp places and are largely confined to the northern hemisphere.

ANATOMY

Newts and salamanders differ from frogs and toads in that they retain their tail into adulthood. They also have a relatively small head and small eyes, smell being the most important sense for finding food and social interaction. Most newts and salamanders have 4 legs (with 4 digits on the front limbs and 5 on the back), although some aquatic salamanders have no back legs and only small front legs. Many salamanders have distinct moisture-maintaining "costal" grooves that run around the body. Newts and salamanders vary in the way that they take in oxygen. Some are able to breathe through their smooth, moist skin, while many of those species that spend all their life in water retain the large, feathery external gills characteristic of the larval stage.

LUNGLESS SALAMANDERS
The plethodontids are a group of typically wholly terrestrial salamanders. They have no lungs but instead breathe through their skin and their mouth and throat, which have a rich blood supply.

SIRENS
The sirens, of which there are 4 species, spend their entire lives in water. As well as having lungs, like some other aquatic salamanders, they take in oxygen through large, feathery external gills on the side of their head. In periods of drought, they can survive for several weeks or months, encased deep inside a cocoon of dried mud. During this time, they feed on their large fat reserves.

SALAMANDER HUNTING
Salamanders have wider mouths than newts and can catch larger prey like this earthworm. Many have a long tongue that flicks out then opens up to engulf the prey.

FEEDING

All newts and salamanders are carnivorous and feed on living prey. They find their prey using a combination of smell and vision; in addition, some aquatic species are very sensitive to water currents made by moving animals. Since they are rarely very active, newts and salamanders do not have a high energy requirement and do not need to feed very often, especially if they have eaten a large amount. When food is abundant, they put down fat deposits, which enable them to survive periods when it is too dry or too cold for them to feed. The larvae are also carnivorous, eating a wide variety of aquatic invertebrates. In some species, faster-growing larvae become cannibalistic, eating smaller larvae of their own species.

COURTSHIP
The courtship of newts and salamanders is varied and often elaborate. Courtship precedes and accompanies the transfer of sperm from male to female in capsules known as spermatophores. The male stimulates the female by administering compounds known as pheromones. These are usually delivered via the nose but in some species they are passed directly through the skin. European newts – such as these Alpine newts (here the male is on the left) – use visual displays rather than contact to stimulate the female, but in many salamanders the male holds the female in a mating embrace known as amplexus while he stimulates her.

LIFE CYCLES

Newts and salamanders have complex life cycles, which typically involve 3 distinct stages: the egg, larva, and adult. The female lays eggs that contain yolk to support the growth of the developing embryo. Some species lay many small eggs, others a few large eggs, each containing a lot of yolk. The eggs hatch into larvae that have feathery external gills and feed on small animals. The larvae grow until they reach the stage at which they undergo metamorphosis into the adult form; this involves the loss of the gills and a switch to breathing air via the lungs and skin. The juvenile stage, which in some is known as the eft stage, is spent on land and may last for several years. In the adult stage, individuals become sexually mature and start to breed; this may involve a brief return to water. There are many variations on this basic pattern, which usually coincide with whether the species is amphibious, terrestrial, or aquatic (see right).

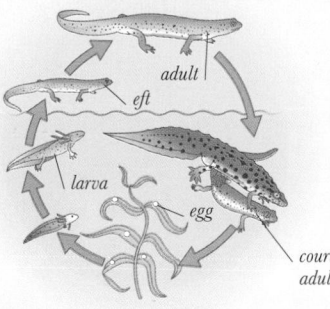

LAND

adult

eft

larva

egg

courting adults

WATER

LAND

adult

eggs

juvenile

larva developing inside egg

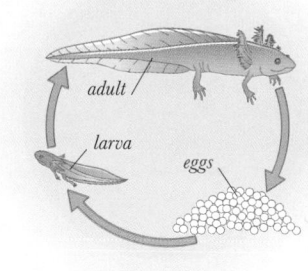

adult

larva

eggs

WATER

AMPHIBIOUS
This life cycle is typical of American, European, and Asian newts. Adults spend most of their life on land but return to water each spring to breed. This involves a partial reversal of meta-morphosis, as individuals develop skin that can absorb oxygen from water and a tail like that of the larva, enabling aquatic newts to swim powerfully.

TERRESTRIAL
The eggs of terrestrial salamanders are typically laid on land. There are usually relatively few, large eggs, each containing a lot of yolk, which may be protected by the mother. The larval stage is completed within the egg, which hatches to produce a miniature terrestrial adult. In some species, the eggs may be retained inside the female's body. She gives birth, either to aquatic larvae or to terrestrial juveniles.

AQUATIC
In the aquatic life cycle, all the life stages are completed in water. The female typically produces a large number of small eggs and does not care for them. The adults become sexually mature while retaining many of the anatomical and physiological features of the larval stage. In many species, there are individual populations that have an aquatic life history, while others have an amphibious one.

fin extends from back along length of tail

external gills

NEOTENIC NEWTS AND SALAMANDERS
In neotenic salamanders and newts, individuals become sexually mature while still having the larval form. They are fully aquatic and have large, feathery external gills. The best known example is the axolotl (shown here), which can be artificially made to metamorphose into a terrestrial adult form by injecting it with the hormone thyroxin.

DEFENCE

For most newts and salamanders, the risk of being attacked by a predator is greatly reduced by being active only at night. Many species, however, also produce distasteful or toxic secretions from glands in their skin. These glands may be concentrated on the head or the tail. The skin secretions of Californian newts are particularly lethal. Such species are often brightly coloured to warn predators, although the effectiveness of this coloration depends on predators learning to associate it with unpleasant consequences. A few species do not produce noxious secretions but mimic the colour patterns of those that do. Several species are able to shed their tails when attacked. The detached tail, which twitches after it is dropped, distracts the attention of a predator away from the animal's body.

UNKEN POSITION
Some species of newts and salamanders adopt bizarre defensive positions. This Italian spectacled salamander has taken up the contorted "unken" position to alarm predators and expose its bright warning coloration.

COLORATION
Many newts and salamanders have skin patterns and colours that provide camouflage. Others are very brightly coloured, usually indicating their capacity to produce noxious or toxic skin secretions. However, in some species (such as this alpine newt), males become more brightly coloured to attract females during the breeding season.

Siren lacertina
Greater siren

Length 50–90 cm (20–35 in)	
Habit Wholly aquatic	
Breeding Early spring season	
Location E. and S.E. USA	**Status** Unconfirmed

Sirens are among the largest amphibians in North America. There are 2 species, both with long, eel-like bodies, and only one pair of legs, behind their feathery external gills. The greater siren is usually grey or olive-brown, and this colour provides it with good camouflage in the muddy water of ponds, lakes, and slow-flowing rivers. This siren spends the day resting on the bottom, but at night forages for food, either crawling over mud or sediment, or swimming with a sinuous movement of its tail. Like most amphibians, greater sirens are almost entirely carnivorous – snails, insect larvae, and small fish make up most of their food. However, during droughts, they encase themselves in mud cocoons, and can survive without food for up to 2 years. Greater sirens lay their eggs in early spring. When their larvae hatch, they are little bigger than most other amphibian tadpoles.

feathery external gills

Pseudobranchus striatus
Northern dwarf siren

Length 10–22 cm (4–9 in)	
Habit Wholly aquatic	
Breeding Not known season	
Location S.E. USA	**Status** Locally common

This small, eel-like siren is brown or black above, with one or more yellow or tan stripes along its body. Its forelimbs are small, each with 3 toes, and it has no hindlimbs. The feathery external gills help it to breathe in water with a low oxygen content. This nocturnal amphibian inhabits swamps, ditches, and marshes, and aestivates in mud during the dry season. It feeds on aquatic invertebrates, and is preyed upon by wading birds, water snakes, alligators, and turtles. The female attaches her eggs to water plants.

Batrachuperus pinchonii
Western Chinese mountain salamander

Length 13–15 cm (5–6 in)	
Habit Largely aquatic	
Breeding Spring and season summer	
Location E. Tibet, China (Sichuan Province)	**Status** Locally common

This robust amphibian has a pale brown or olive-green body, usually with darker spots, and about 12 costal grooves along its flanks. It has short legs, with 4 toes on each limb, and a short tail that is flattened from side to side. This nocturnal animal lives in cold, clear, fast-flowing streams at high altitudes. Since there are few fish in these cold water courses, the eggs and larvae of this salamander are relatively safe from predators. The female lays 7–12 eggs in hollows under rocks in the streams.

stocky body

glossy, brown or green coloration

Cryptobranchus alleganiensis
Hellbender

Length 30–74 cm (12–29 in)	
Habit Wholly aquatic	
Breeding Autumn season	
Location E. USA	**Status** Unconfirmed

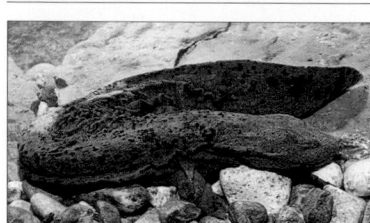

The hellbender is the only North American member of the giant salamander family – a group of 3 species notable for their wrinkled skin and exceptional size. Green, brown, or grey, often with darker spots, it has a head flattened like a spade, small eyes, and a long, paddle-like tail. It is nocturnal, and uses its head to dig out crayfish and other prey. The hellbender has a strong bite, and its skin secretes a noxious slime. Females lay their eggs in autumn, and the males guard them until they hatch.

Necturus maculosus
Mudpuppy

Length 29–49 cm (11½–19½ in)	
Habit Wholly aquatic	
Breeding Autumn in north, season winter in south	
Location S. Canada. C. and E. USA	**Status** Common

The mudpuppy has a long, cylindrical body, a flattened tail, feathery external gills, and 2 pairs of limbs, each with 4-toed feet. It is brown, grey, or black, usually with black spots and blotches. An inhabitant of various freshwater habitats, it feeds on small animals, mainly at night. Females lay their eggs in early summer, digging a nest under rocks and logs. They guard them until

BODY SHAPE
The flattened, eel-like body shape of the mudpuppy enables it to burrow under rocks and logs.

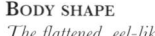

Andrias japonicus
Japanese giant salamander

Length Up to 1.4 m (4½ ft)	
Habit Wholly aquatic	
Breeding Late summer season	
Location Japan	**Status** Vulnerable

Together with the Chinese giant salamander (*Andrias davidianus*), the Japanese giant salamander is the largest amphibian in the world. It is grey or brown, with a flattened body, small eyes, and sprawling legs. Its head and throat are warty, and the skin along its sides is deeply folded – a characteristic that helps it to absorb oxygen from the surrounding water. It is active mainly at night, feeding on fish, worms, and crustaceans. During the breeding season, the male digs a spawning pit, in which the female deposits strings of about 500 eggs. The male then guards the eggs, aggressively fending off predators. Japanese giant salamanders can live for over 50 years, but they are a traditional delicacy, and have become increasingly rare.

ADAPTABLE GILLS

small front legs

red external gills

The gills of the mudpuppy adapt to the oxygen content of the animal's habitat. In cool, oxygen-rich water they are usually small, but in stagnant, warm water they tend to be considerably larger.

they hatch, about 2 months later. The mudpuppy has disappeared from many polluted or silted rivers.

dark blotches

slimy skin

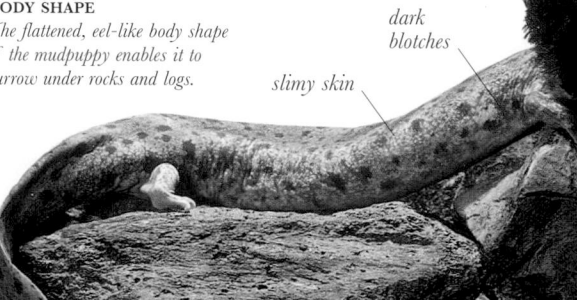

Proteus anguinus

Olm

Length 20–30 cm (8–12 in)	
Habit Wholly aquatic	
Breeding season Not known	
Location S. Europe	**Status** Vulnerable

The olm is one of the few amphibians that has become adapted to life in caves. It has a long and very slender body, a pinkish skin almost devoid of pigment, rudimentary eyes, and 2 pairs of tiny limbs – the front pair with 3 toes, and the rear pair with only 2. It also has red feathery external gills, which it retains throughout its life. Olms inhabit streams in caves, sometimes several kilometres from a cave opening, and feed on small invertebrates.

Dicamptodon tenebrosus

Pacific giant salamander

Length 17–34 cm (6½–13½ in)	
Habit Mostly terrestrial	
Breeding season Spring and autumn	
Location S.W. Canada, N.W. USA	**Status** Unconfirmed

This robust, dark brown to black salamander, with light brown mottling or spots, lives in wooded habitats in and around streams and is active at night. It progresses through an aquatic larval stage that typically lasts for several years. When individuals reach maturity, they may become terrestrial. However, some never leave the water, becoming aquatic, gilled adults. Those that do leave the water are the world's largest terrestrial salamanders. In defence, this animal secretes noxious mucus from its tail. Populations have been adversely affected by logging and the silting of streams.

12 or 13 costal grooves

Salamandra salamandra

Fire salamander

Length 18–28 cm (7–11 in)	
Habit Wholly terrestrial	
Breeding season Spring, summer, after heavy rain	
Location N.W. Africa, Europe, W. Asia	**Status** Locally common

Similar in appearance to the tiger salamander (see p.436), this striking, robust amphibian has an eye-catching coloration that warns predators that it is poisonous. It produces a toxic secretion from glands behind its eyes, and this is distasteful to anything attempting to eat it. The fire salamander inhabits forests and woodland on hills and mountains; it spends the winter underground. Breeding involves the male grasping his mate from below and depositing a spermatophore; he then flips to one side so that she falls on it. The eggs develop inside the female until the larval stage, and are then released into ponds or streams. During their development in the oviduct, larvae are occasionally cannibalistic, eating their smaller siblings as eggs or larvae. In some high-altitude populations, the young develop to the adult form inside the mother's body.

FEEDING HABITS

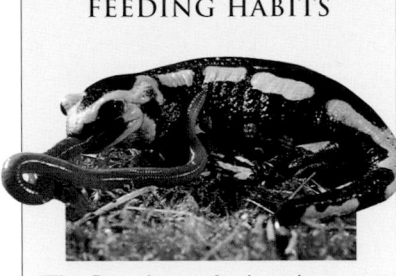

The fire salamander is active mainly at night, especially after rain, when it emerges from beneath logs and stones to feed on worms, slugs, insects, and insect larvae.

short tail

poison glands on head

BODY COLOUR
The body colour is variable, depending on geographical location. It can be black and yellow, or yellow with black spots or stripes. In some areas, the yellow may be replaced by orange or red.

prominent eyes

Amphiuma tridactylum

Three-toed amphiuma

Length 46–110 cm (18–43 in)	
Habit Wholly aquatic	
Breeding season Winter and spring	
Location C. and S. USA	**Status** Locally common

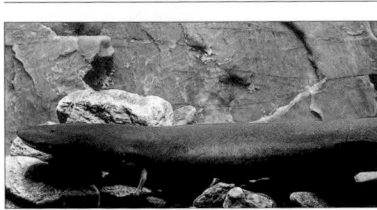

This large, slimy salamander has a slender, eel-like body that is black, grey, or brown above, and pale grey below. The tail is flattened from side to side, and each of the tiny limbs has 3 toes. It inhabits ditches, swamps, streams, and ponds and, in periods of drought, it can survive buried in mud. Active at night, it feeds on worms and crayfish. This animal can deliver a painful bite.

Triturus cristatus

Great crested newt

Length 10–14 cm (4–5½ in)	
Habit Mostly terrestrial	
Breeding season Spring and summer	
Location Europe, C. Asia	**Status** Lower risk

Also known as the warty newt, this secretive animal lives mostly on land but spends about 3–5 months of the year in ponds, lakes, and ditches. It is dark brown above, with a characteristically warty skin, and bright orange with black spots below. During the breeding season, the male develops a large, jagged crest on its back, and a white or blue streak along the tail. Like many other newts, the male courts the female with a complex underwater dance, and concludes the display by transferring a spermatophore to the female. The female lays her eggs singly and wraps them in leaves, probably for protection. These newts are widespread throughout Europe, but in many areas their numbers have decreased because their breeding habitat has been drained.

crest on male

Triturus alpestris

Alpine newt

Length 6–12 cm (2¼–4¾ in)	
Habit Mostly terrestrial	
Breeding season Spring	
Location Europe	**Status** Locally common

The alpine newt has a small body, blue above and bright orange below, and a short tail. In the breeding season, the male develops a low, black and white crest along his back, black and white spots on his sides and tail, and a swollen cloaca. As with most newts, the male attracts the female by fanning his tail and wafting a pheromone towards her snout. In the Balkans, some of these nocturnal animals are permanently aquatic, retaining larval features, such as external gills, into adulthood.

Taricha torosa
California newt

	Length 12.5–20 cm (5–8 in)
	Habit Mostly terrestrial
	Breeding Winter and **season** spring
Location W. USA (California)	**Status** Common

California is particularly rich in salamanders and newts, and this species is one of over 20 that live in the state. Its rough, warty body is brown

or brick-red above, and yellow to orange below. During the breeding season, when these newts carry out mass migrations to ponds and streams, the males develop a smooth, slimy skin, black patches on their feet, and a swollen vent (cloaca). The California newt is protected by a toxic secretion that makes both its skin and flesh extremely poisonous. It is nocturnal and feeds on invertebrates of all kinds.

warty skin

long tail

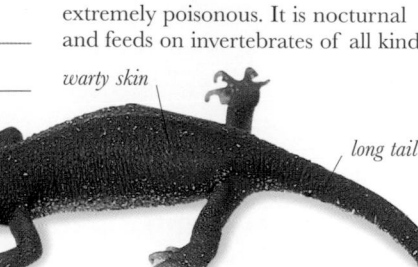

Pleurodeles waltl
Sharp-ribbed newt

	Length 15–30 cm (6–12 in)
	Habit Wholly aquatic
	Breeding Any time, except **season** cold weather
Location S.W. Europe, N.W. Africa	**Status** Unconfirmed

This mottled grey-green newt is one of Europe's largest tailed amphibians. Like most salamanders, it is protected by toxic skin secretions, but it also has sharply

pointed ribs that project through its skin when it is held. Together with its chemical defences, these help to deter most predators. Sharp-ribbed newts are mainly nocturnal and spend their lives in water, except in times of drought, when they survive by hiding in mud until wetter conditions return. During the breeding season, females lay 200–300 eggs singly or in clumps, fastening them to water plants.

plump body *warts on body* *long, flattened tail*

Tylototriton verrucosus
Crocodile newt

	Length 14–18 cm (5½–7 in)
	Habit Mostly terrestrial
	Breeding Monsoon **season**
Location S. and S.E. Asia	**Status** Endangered†

Also known as the mandarin salamander, this robust newt has a large head and thickset body, with rough black skin. The prominent orange warts, ridges, and glands indicate that the crocodile newt will exude distasteful skin secretions if attacked. This newt lives most of its life on land – spending winter and dry periods underground – but during the monsoon it migrates to breeding ponds, where it attaches its eggs to water plants. Active at night, the crocodile newt feeds on invertebrates.

prominent ridge

rough, warty skin

Chioglossa lusitanica
Golden-striped salamander

	Length 12–14 cm (4¾–5½ in)
	Habit Mostly terrestrial
	Breeding Early spring **season**
Location S.W. Europe	**Status** Vulnerable

The slender body and long tail enable the golden-striped salamander to run very fast, like a lizard. Its coloration is dark brown, with 2 yellow, copper, or golden stripes on its back that merge to form a single stripe on the tail. This salamander has large eyes and a long, sticky tongue for catching insects. If attacked, it may shed its tail.

Notophthalmus viridescens
Eastern newt

	Length 6.5–11.5 cm (2½–4½ in)
	Habit Largely terrestrial
	Breeding Variable **season**
Location E. Canada, E. USA	**Status** Common

This common North American newt has a complex life cycle. It breeds in a wide variety of freshwater habitats, but instead of developing directly into adults, its larvae turn into specialized juvenile forms, called red efts. Red efts live on land – usually in grassland, mountain forests, and other damp habitats – and then return to water after 1–4 years, where they reach maturity. Outside the breeding season, adults live on land. During the adult and eft stages, these newts are protected from predators by highly toxic skin secretions.

JUVENILE COLORATION
The young eastern newt, or the red eft, is bright red, orange, or brown all over, with black-edged red spots along its back, and rough skin. This terrestrial juvenile stage lasts for 1–4 years, after which the red eft returns to water to become adult.

ADULT STAGE
This newt completes its development in water. Its upperside turns greenish yellow, with black-edged red spots, and its underside becomes yellow. The breeding male has a very deep tail, a swollen cloaca, and rough pads on the inside of its thighs.

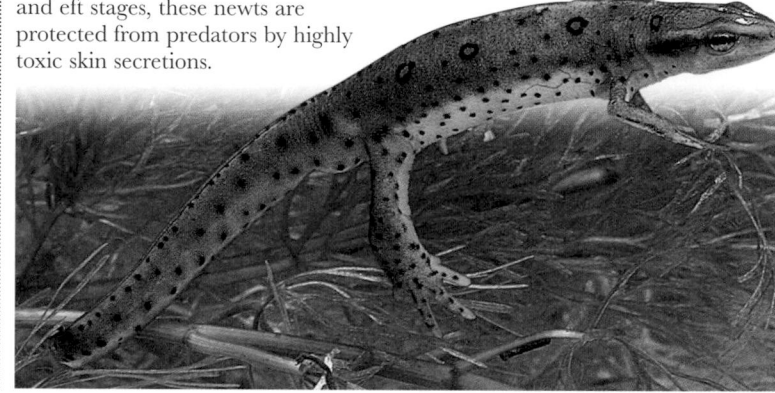

Ambystoma tigrinum
Tiger salamander

	Length 18–35 cm (7–14 in)
	Habit Mostly terrestrial
	Breeding Variable, **season** after heavy rain
Location North America	**Status** Locally common

The stout-bodied tiger salamander is one of North America's most colourful amphibians, and is also one of the most locally varied. At least 6 subspecies are known: the eastern subspecies has a dappled pattern of black and yellow markings, but those further west – including the one shown here – have larger yellow blotches or vertical bars. Adult tiger salamanders generally live on land, and hibernate in burrows dug by other animals. In early spring, once any ice has melted, they migrate to ponds, lakes, pools, and reservoirs to breed, sometimes in large numbers. The females lay up to 7,000 eggs in

a single breeding season, and their larvae feed mainly on invertebrates, although some individuals may become cannibalistic, growing larger mouths with extra teeth. In most areas, the young leave the water once they have developed adult bodies, but in some regions – particularly in the west – they may retain their larval form permanently, reproducing without taking up life on land. The chief predators of the tiger salamander are fish, which eat the larvae; however, in recent years the species has suffered through pollution and habitat loss.

small eyes *rounded snout*

large head

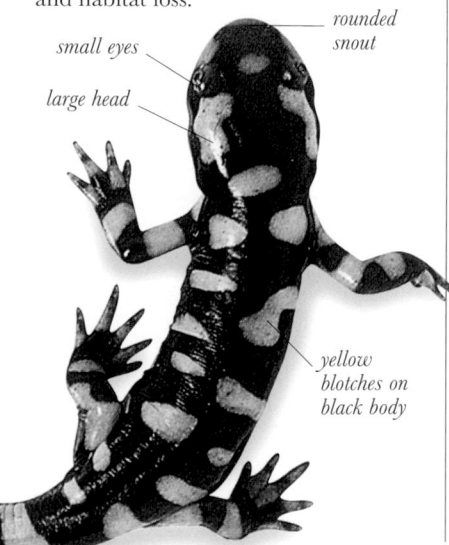

yellow blotches on black body

cylindrical tail

Ambystoma mexicanum
Axolotl

Length 10–20 cm
(4–8 in)

Habit Wholly
aquatic

Breeding Summer
season

Location Mexico (Lake
Xochimilco)

Status Vulnerable

This plump amphibian is a classic example of neoteny, or the ability to breed without ever developing a fully adult body. It has a flattened tail and large external gills – features that most salamanders lose when they become mature and take up life on land. Largely nocturnal, the axolotl feeds mainly on invertebrates, and is preyed upon in turn by waterbirds. Axolotls are vulnerable because they live in a single lake, in an area that has seen rapid urbanization in recent years. Once esteemed as a delicacy, they are now legally protected.

ALBINO
One colour variant of the axolotl is the albino. This has a smooth white body and red gills.

red external gills

rounded snout

smooth white body

CAPTIVE FORMS

Typically black in the wild, many colour variants of the axolotl have been bred in captivity. They may be albinos (white with red gills), grey, or mottled black and white. This species is more commonly seen in captivity than in the wild. If the captive animals are injected with a thyroid hormone, they lose their gills and become terrestrial.

Ambystoma macrodactylum
Long-toed salamander

Length 10–17 cm
(4–6 1/2 in)

Habit Mostly
terrestrial

Breeding Autumn and
season spring

Location S.W. Canada,
N.W. USA

Status Locally common

The long-toed salamander is found in a wide variety of terrestrial habitats, where it spends much of its time underground. However, in spring large numbers of this nocturnal species migrate to ponds to breed. It has a slender grey or black body, with a yellow, tan, or green dorsal stripe or blotches on the body. When attacked, it coils its body and lifts its tail in the air. Numbers have reduced because of habitat loss and the introduction of fish, which eat the larvae, to its breeding areas.

white speckles on flanks

Plethodon jordani
Jordan's salamander

Length 8.5–18.5 cm
(3 1/4–7 1/2 in)

Habit Wholly
terrestrial

Breeding Summer
season and autumn

Location E. USA
(Southern Appalachian
Mountains)

Status Locally common

Also known as the red-cheeked salamander, this nocturnal species has a long, slender body and tail, a large head with prominent eyes and, typically, red patches on its legs or cheeks, although this characteristic varies considerably between races. Both males and females are territorial. If threatened this salamander produces a distasteful slimy secretion from its tail if threatened, and may shed its tail if attacked. It mainly feeds on millipedes, beetles, and insect larvae. The male breeds every year, while the female breeds only in alternate years.

Plethodon glutinosus
Slimy salamander

Length 11.5–20 cm
(4 1/2–8 in)

Habit Wholly
terrestrial

Breeding Variable
season

Location E. USA

Status Locally common

This large, slender-bodied salamander gets its name from the glue-like slime that it exudes when handled. The slime is extremely sticky – taking several days to wash off – and it gives

the salamander highly effective protection against most predators. Adults are dark blue or black, with many small silver or gold spots. In the breeding season, males develop a swollen vent (cloaca), and a large gland under the chin. The slimy salamander is nocturnal, living in wooded hillsides and ravines.

costal grooves

small silver spots on tail

long tail

Plethodon cinereus
Red-backed salamander

Length 6.5–12.5 cm
(2 1/2–5 in)

Habit Wholly
terrestrial

Breeding Autumn to
season early spring

Location S.E. Canada,
C. and E. USA

Status Locally common

This small salamander is dark grey or brown, usually with a broad red or red-brown stripe along its body and tail. Males use faeces to mark their territory and to attract females. In the

breeding season, females inspect the droppings, preferring to mate with males that have eaten a lot of termites. Males can reproduce every year, but females are fertile only in alternate years. The eggs are attached to the roofs of underground chambers.

Desmognathus ochrophaeus
Mountain dusky salamander

Length 7–11 cm
(2 3/4–4 1/4 in)

Habit Wholly
terrestrial

Breeding Spring, summer,
season and autumn

Location E. USA

Status Common

The mountain dusky salamander belongs to a group of about a dozen North American species that are so similar in their appearance, that it takes expert knowledge to tell them apart. Like most of its relatives, it has a long, slender body, protruding eyes, and a long, cylindrical tail. It is

usually mottled dark grey, with legs that vary from brown to red (pictured below is a rare, red-legged variant that may be a mimic of the red-legged Jordan's salamander). The mountain dusky salamander lives on land once it has matured, and spends daylight hours underground. At night, it emerges to feed on insects and other small animals. It is protected by a toxic skin secretion and, if threatened, it sometimes sheds its tail. During the breeding season, adults mate on land. The male stimulates his prospective partner by rubbing a pheromone onto her skin, and abrading the skin's surface with specialized teeth. The female deposits a cluster of 12–20 eggs in a cavity beneath a tree trunk or underground, guarding them until they hatch. The larvae develop in streams. Adults are particularly active after rain, but they spend the winter underground.

brown to red legs

long, slender tail

AMPHIBIANS

Batrachoseps attenuatus
California slender salamander

Length 7.5–14 cm
(3–5 1/2 in)
Habit Wholly terrestrial
Breeding Not known season

Location W. USA
(California, Oregon)
Status Locally common

This salamander lives mainly underground and under logs and rocks, emerging at night after rain. Its extremely long, slender body enables it to enter very small crevices. It is dark brown or black, with a brown, yellow, or red stripe along the back, and tiny, 4-toed limbs. When in danger, the California slender salamander coils and uncoils its tail rapidly, and may shed it if attacked.

Pseudotriton ruber
Red salamander

Length 10–18 cm
(4–7 in)
Habit Mostly terrestrial
Breeding Any time of year, season usually summer

Location E. and S.E. USA
Status Locally common

Brilliantly coloured when young, but becoming progressively darker with age, the red salamander is found in damp places, springs, and cool streams in many eastern states of the USA. It has a thick body, a relatively short tail and legs, and a dense scattering of black spots over its red or orange-red skin. Largely nocturnal, it feeds mainly on invertebrates and other salamanders. To ward off predators, the red salamander raises its tail and waves it, but its colour may also act as a defence. Although it is harmless, its red markings closely resemble those of the toxic young eastern newts (see p.436) – a similarity that may be a form of protective mimicry. The female lays about 70 eggs, attached to the underside of logs or rocks, and sometimes in water. Adult red salamanders usually spend the winter underground.

Aneides lugubris
Arboreal salamander

Length 11–18 cm
(4 1/4–7 in)
Habit Terrestrial/ Arboreal
Breeding Not known season

Location W. USA
(California)
Status Locally common

This long, slender, brown and white salamander has flattened toe tips and a prehensile tail that enable it climb trees with ease. The relatively large, triangular head contains prominent jaw muscles that are used in fights between individuals, or in defence – for instance against predatory snakes.
In winter and dry weather, the arboreal salamander lives below ground. The female lays 12–18 eggs in an underground chamber.

Bolitoglossa mexicana
Mexican mushroom-tongue salamander

Length 12–15 cm
(4 3/4–6 in)
Habit Arboreal/ Terrestrial
Breeding Any time, except season dry/cold periods

Location S. Mexico to Central America
Status Unconfirmed

With its prehensile tail, feet with adhesive pads on the base, and small size, this nocturnal animal is specially adapted for climbing trees and moving around on the smallest of branches. It has a mushroom-shaped tongue, which it flicks out to catch insects. The bronze or pink colouring, with black marbled patterning, acts as a warning that this salamander will emit noxious secretions from its skin if attacked.

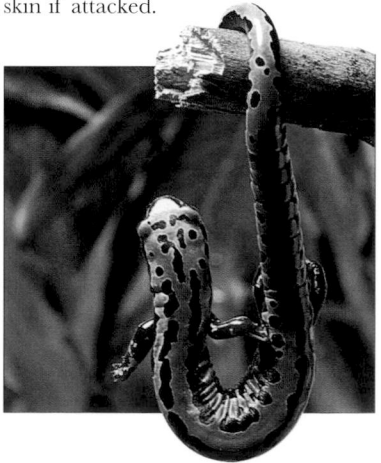

Ensatina eschscholtzii
Ensatina salamander

Length 6–8 cm
(2 1/4–3 1/4 in)
Habit Wholly terrestrial
Breeding Autumn to spring season

Location W. USA
Status Locally common

This nocturnal animal lives in damp leaf litter and under logs, and is especially active after rain. It has a short body, long legs, and prominent eyes. Its colour depends on the area in which it lives: it may be uniformly brown, or black with white, yellow, orange, or pink blotches. The round tail, which narrows abruptly at the base, is used in defence. The salamander lifts it up and waves it around, emitting a distasteful secretion to deter predators; it may also shed its tail if attacked. During the breeding season, the male has a longer tail and develops a swollen cloaca. The female lays 8–20 eggs in clusters under logs.

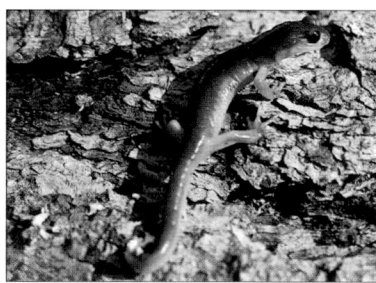

Hydromantes platycephalus
Mount Lyell salamander

Length 7–11 cm
(2 3/4–4 1/4 in)
Habit Wholly terrestrial
Breeding Not known season

Location W. USA
(California)
Status Unconfirmed

Restricted to high ground in California's Sierra Nevada, this salamander belongs to a small group of web-footed species that are unusually good climbers. They all live in rocky places, and have no difficulty running over broken boulders, or even up vertical cliffs. This species has a broad head and cylindrical body, with relatively large feet. The male's head is wider than the female's, and in the breeding season he develops a gland underneath the chin, which produces a hormonal secretion that encourages his partner to mate. The Mount Lyell salamander feeds at night on insects, spiders, and other invertebrates. It has a mushroom-shaped tongue, which extends over half its body length to enable it to capture prey. If threatened, it defends itself with toxic skin secretions, and by rolling up into a ball. In winter and dry periods the Mount Lyell salamander lives underground or in crevices.

Eurycea guttolineata
Three-lined salamander

Length 10–18 cm
(4–7 in)
Habit Wholly terrestrial
Breeding Autumn and season early winter

Location E. USA
Status Locally common

A long and slender body with 13 or 14 costal grooves, prominent eyes, and a very long tail characterize this salamander. It is yellow, orange, or brown, with dark brown or black stripes down the middle of its back and along the sides. It lives under logs and rocks in damp areas near springs and streams; winters and dry periods are spent underground. Spiders, flies, beetles, and ants are the main diet of this predator, which feeds at night, especially after it has rained. During the breeding season, the male develops 2 fleshy projections on the upper lip, 2 protruding teeth, and a swollen cloaca. Eggs are laid in damp cavities.

dark stripe along back

CAECILIANS

PHYLUM	Chordata
CLASS	Amphibia
ORDER	Gymnophiona
FAMILIES	5
SPECIES	About 170

THESE LIMBLESS, WORM-LIKE animals form the smallest of the 3 major groups of amphibians. They are rarely seen by humans, because they live either in soil burrows or underwater, and are found only in humid, tropical areas. Caecilians have varied life cycles: some species lay eggs, but in others the eggs are retained in the female's body and nourished by secretions produced by the oviduct until the eggs hatch and the young are released by live birth.

ANATOMY

Caecilians have long, thin, black or brown to pink bodies and no limbs. They have rudimentary eyes and rely on a good sense of smell to find food and mating partners. Some resemble very large earthworms and burrow into soft soil or mud, using their pointed, bony head as a shovel. The aquatic caecilians resemble eels; they have a fin on the tail that enables them to swim powerfully.

MOVEMENT
A caecilian moves by using its body to generate a curving motion. As the sides of the curves encounter resistance from surface irregularities, the caecilian pushes its body forwards.

SENSORY TENTACLES
A unique feature of caecilians is the small tentacle located just below each eye. This collects chemical information, which is used to locate prey. Caecilians feed mainly on earthworms, which they catch with their sharp, curved teeth.

tentacle below eye

Ichthyophis glutinosus
Ceylon caecilian

Length Up to 45 cm (18 in)	
Habit Terrestrial/ Burrowing	
Breeding Not known season	
Location Sri Lanka	**Status** Locally common

With its long, thin body covered with well-defined rings, and a short tail, the Ceylon caecilian resembles a large earthworm. It is brown with a bluish sheen, and has a pair of small, retractable tentacles between the nostrils and the eyes that enable it to pick up odours. This caecilian lives underground, mostly in muddy soil and swamps, where it feeds on worms and other invertebrates. The female coils around her eggs to protect them.

Caecilia tentaculata
Linnaeus' caecilian

Length 45–63 cm (18–25 in)	
Habit Terrestrial/ Burrowing	
Breeding Not known season	
Location S. Central America to N. South America	**Status** Unconfirmed

This large caecilian has a thick, flattened body, grey or black above and light brown below, and a wedge-shaped snout. It has no tail and, unlike other caecilians, its rear end is covered by a hard shield. Tiny scales are present in the skin in the 100–300 grooves that cover the body, and the tentacles, which are lower than in other caecilians, are not visible from above. Very little is known about its behaviour, since it lives buried in loose soil, but it is known to be a solitary animal that lays eggs.

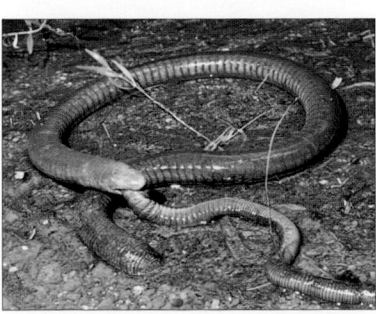

Dermophis mexicanus
Mexican caecilian

Length 10–60 cm (4–23½ in)	
Habit Terrestrial/ Burrowing	
Breeding Not known season	
Location Mexico to N. South America	**Status** Locally common

This caecilian lives in the same region of the world as several amphisbaenians (see p.423), and is superficially similar to them, although it lacks legs. Grey, brown, or olive-green above, it has well-defined rings around its body, and a pointed, burrowing snout. It feeds chiefly on invertebrates, but also on lizards, and is completely sub-terranean, living in a variety of habitats where there is loose soil. The egg and larval stages are completed inside the female's body. After an extended gestation period, the mother gives birth to live young.

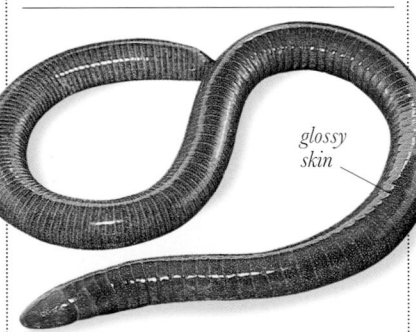

glossy skin

Siphonops annulatus
Ringed caecilian

Length 20–40 cm (8–16 in)	
Habit Wholly terrestrial	
Breeding Not known season	
Location N. South America	**Status** Locally common

Also known as the South American caecilian, this forest-dwelling amphibian lives in soil and resembles a large, thick earthworm. It has a relatively short, stocky, dark blue body, encircled by white stripes and

Typhlonectes compressicauda
Cayenne caecilian

Length 30–60 cm (12–23½ in)	
Habit Wholly aquatic	
Breeding Not known season	
Location N. South America	**Status** Locally common

An inhabitant of rivers, lakes, and streams, the Cayenne caecilian resembles an eel in shape and the way in which it swims. It has a long, slender,

numerous rings. Like other caecilians, it uses its tentacles to pick up the odours of nearby prey – usually worms and other invertebrates. When attacked, it produces a distasteful secretion. The ringed caecilian lays its eggs in soil, where they hatch into miniature adults.

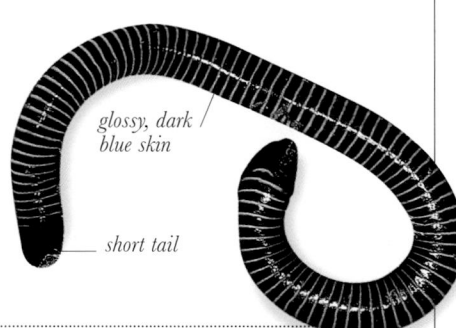

glossy, dark blue skin

short tail

glossy black body, dark grey below, encircled by 80–95 rings. Its tail is flattened from side to side, with a slight fin along the top. The Cayenne caecilian produces toxic secretions when it is attacked, usually by fish. The eggs develop into larvae inside the body of the female, who gives birth to miniature adults.

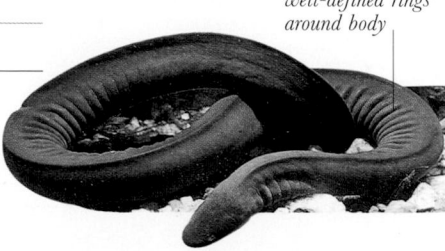

well-defined rings around body

AMPHIBIANS

FROGS AND TOADS

PHYLUM	**Chordata**
CLASS	**Amphibia**
ORDER	**Anura**
FAMILIES	**29**
SPECIES	**4,380**

FROGS AND TOADS form much the largest and best known of the 3 groups of amphibians. As adults, they have no tail – unlike other amphibians, their tail is absorbed during metamorphosis from the larval to the adult stage. There is no clear distinction between frogs and toads: the term "toad" is often used to refer only to members of the genus *Bufo*, but it is also used more widely to describe any slow-moving, terrestrial species with a squat body and rough, warty skin. The larvae of frogs and toads, known as tadpoles, mainly feed on algae or plant material and characteristically have a spherical body that contains the long, coiled gut required by such a diet. In many adults, the back limbs are modified for jumping; they are much longer and more muscular than the front limbs. Most frogs and toads live in damp habitats, near the pools and streams in which they breed, but there are several species that live in arid areas. The greatest diversity of species is found in the tropics, particularly in rainforests.

ANATOMY

Frogs and toads have a short, rigid body, a large, wide head, and, in most species, front limbs that are much smaller than the back limbs. The back limbs contain an extra section, just above the foot, that is not present in other amphibians. Most species have large, protruding eyes and a conspicuous eardrum (tympanum) on each side of the head, reflecting the importance of the senses of vision and hearing. The mouth is large, and many species have an adhesive tongue that they can flick out at high speed to catch their prey. The female is usually larger than the male, but males of many species have thicker and more muscular front limbs. These enable the male to grip the female securely during mating.

webbing extends to ends of digits

WEBBING

circular adhesive pad

widely spread digits for gripping large area

DISCS

large tubercle, used for digging

CLAWS

SKIN
The skin of frogs and toads is highly variable. At one extreme, that of some toads is so tough that it is used as a substitute for leather; at the other, there are frogs with skin so thin that their internal organs can be seen through it. Toads typically have rough, dry skin that is covered with warts and, in some species, small spines. Frogs generally have smooth, moist skin.

FEET
Frogs and toads have 4 digits on the front limbs and 5 on the back. In many species, especially those that spend much of their life in water, the digits on the limbs have webs of skin between them. Many frogs that live in trees have disc-shaped, adhesive pads on the ends of their digits, which enable them to gain a secure foothold on smooth, vertical surfaces. Burrowing species have horny protuberances, called tubercles, on their back feet, which they use to excavate soil.

FROG SKIN

TOAD SKIN

FEEDING

All frogs and toads are carnivores that feed on live prey rather than carrion. Unable to chew or break up food in their mouth, they have to swallow their prey whole. However, many species can eat large animals (including mice, birds, and snakes), and a single meal can meet their energy needs for a long time so that they do not need to feed often. Few species are active hunters, the most common feeding method being to wait until prey comes within range, either of their long, sticky tongue or of a forward, open-mouthed lunge.

CATCHING

SWALLOWING

SWALLOWING FOOD
Frogs and toads typically shut their eyes when swallowing their food. As their eyelids close, the eyeballs are rolled and pushed downwards to increase pressure in the mouth.

SPAWNING SITES
Because bodies of water suitable for breeding are often scarce, many frogs and toads, such as these European common frogs, form large breeding groups, with frogs often converging from a wide area. The collective mass of eggs is also able to retain more heat, enabling the tadpoles to hatch sooner. Ponds and streams that dry up for part of the year are often used, as they do not support other animals that eat eggs and tadpoles.

FINDING A MATE
Many male frogs and toads call to attract females to the spawning site. The calls made by males enable a female to find a mate. More importantly, each species has its own characteristic call, which helps a female to find a mate of the same species in breeding pools used by several species at once. In many species (such as the aquatic frog, right), the effectiveness of the call is enhanced by one or more vocal sacs. The male fills himself with air, which is moved back and forth between the lungs and vocal sacs; sound is produced as it passes over the vocal cords in the larynx.

REPRODUCTION

Once males and females have located one another (see facing page), mating can occur. The male adopts a position known as amplexus, in which he clasps the female from above (see right). Fertilization takes place while the pair are in amplexus, which may last for a period ranging from a few minutes to several days depending on the species. In all but a small number of species, fertilization is external, the male shedding sperm onto the eggs as they emerge from the female. Each species has a characteristic pattern of egg production and uses a specific type of site for egg deposition. Eggs may be produced singly, in clumps, or in strings. The female usually places them in water, because their development can occur only in moist conditions. Eggs may be dispersed into the water, where they may float or sink, they may be wrapped around vegetation, or they may be glued to plants, logs, or rocks.

AMPLEXUS

Amplexus may take place on land or in water – these treefrogs (right) mate in the branches of trees. Depending on the species, the male may clasp the female just behind her front limbs or just in front of her back limbs. In many species, males outnumber females in breeding groups, and males in amplexus are often attacked by rivals seeking to displace them.

EGGS

The number of eggs produced by a female varies greatly among species, from less than 20 to many thousand. The eggs of the European common toad (left), which are suspended in long strings of jelly, are wrapped around the underwater stems of aquatic plants.

DEFENCE

Frogs and toads typically lack weapons, such as large teeth or claws, that they can use to defend themselves. Most rely on being able to escape by jumping, usually towards open water. Many produce secretions from glands in their skin that make them distasteful or poisonous, sometimes lethally so. Others rely on being well camouflaged. In general, tadpoles lack defences and are eaten by various predators, notably fish and the larvae of insects, such as dragonflies. Being produced in large numbers means that at least a few survive.

WARNING COLORATION

Poison-dart frogs (left) have had to evolve highly poisonous skin secretions because many of their predators (including snakes and spiders) have evolved a resistance to milder toxins. Many species that are distasteful or toxic are vividly coloured, which acts as a warning to predators.

CAMOUFLAGE

Most frogs and toads are active only at night and remain entirely motion-less by day. Species such as this leaf frog (right) have skin that is coloured to match their background and are patterned in such a way that their outline is broken up.

MOVEMENT

Frogs and toads move in a wide variety of ways. Most toads are ground-dwellers and walk, run, or hop over the ground; few are able to leap large distances. Leaping is rarely used for moving from place to place but is primarily a means of escape from predators. Not all frogs and toads jump: many have back limbs adapted for other types of movement, such as swimming, climbing, and (in a few species) gliding. Burrowing is important for many species, enabling them to hide during the day, or for many weeks or months when it is too cold, hot, or dry for them to be active. They may simply push their way under rocks or logs, or into piles of plant debris, or they may be able to dig deep into the ground. Some desert species can survive being buried for up to 2 or 3 years.

legs straightened in mid-air to form streamlined shape

eyes narrowed or closed for protection

CLIMBING

Some frogs (such as the European treefrog, left) spend their entire life in tree canopies or among other types of vegetation. Their long, slender limbs enable them to jump or straddle gaps between branches, while the adhesive discs on their digits allow them to climb vertical surfaces.

LEAPING

Frogs and toads launch themselves into the air with an explosive burst of energy from their long back legs. On landing, the front legs are pushed forwards to cushion the impact.

Ascaphus truei
Tailed frog

Length 2.5–5 cm
(1–2 in)
Habit Mostly terrestrial
Breeding Spring to season autumn
Location S.W. Canada, N.W. USA
Status Locally common

This small green, brown, grey, or reddish amphibian is the only aquatic frog that breeds by internal fertilization. The unique "tail" – which is found only in males – is an extension of the cloaca, and is used to insert sperm into the female's cloaca. Females lay strings of eggs in cold mountain streams, attaching them to the underside of rocks. The tadpoles take 2 years to mature, and have sucker-like mouths that enable them to cling to rocks in fast-flowing water.

rough skin

dark stripe through eyes

Xenopus laevis
African clawed toad

Length 6–13 cm
(2¼–5 in)
Habit Mostly aquatic
Breeding Rainy season season
Location Southern Africa
Status Common

With its flattened body and large, webbed feet, the African clawed toad (also known as the platanna) is ideally suited to a bottom-dwelling lifestyle in lakes and ponds. Like its close relative the Surinam toad (see p.443), it has eyes and nostrils positioned on the top of its head, and camouflaged skin to protect it from herons and other predators. During the breeding season, females lay numerous very small eggs, attaching them to various underwater objects. This species is often raised in laboratories for teaching and research.

FORELIMBS WITH FINGERS
The small forelimbs possess 3 fingers, which are used to shovel invertebrates and small fish into the mouth of this voracious feeder.

3 fingers on each forelimb

clawed toes

muscular hindlimbs

Leiopelma hochstetteri
Hochstetter's frog

Length 3.5–5 cm
(1½–2 in)
Habit Mostly terrestrial
Breeding Summer season
Location New Zealand (North Island)
Status Lower risk†

This squat brown frog – an inhabitant of damp ground – is one of just 3 species of frog that are native to New Zealand. Its eggs, which are enclosed in large, water-filled capsules, hatch into non-feeding tadpoles that stay within the capsule until they mature. When they are fully developed, they emerge as froglets. Hochstetter's frog is still abundant in some localized areas, but the other 2 New Zealand species are now very rare, and are confined to offshore islands.

SENSE ORGANS

The African clawed toad has a line of white "stitch marks" along its sides. These contain special sense organs that detect vibrations, enabling the toad to locate food and predators in murky water.

Bombina orientalis
Oriental fire-bellied toad

Length 3–5 cm
(1¼–2 in)
Habit Mostly aquatic
Breeding Spring and season summer
Location E. and S.E. Asia
Status Locally common

The most distinctive feature of this squat, slightly flattened toad is its bright red underside with black mottling. When threatened, it arches its back, flattens its body, and lifts its limbs over its head to display this vivid coloration. It also produces a distasteful secretion through its skin. The Oriental fire-bellied toad lives in mountain streams in coastal areas, and hides under rocks and logs in winter and during dry spells. The male calls to attract a mate. The female lays small clutches of large eggs under rocks in streams.

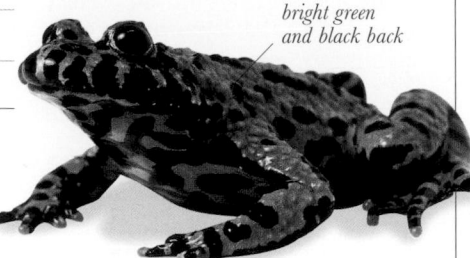

bright green and black back

Alytes obstetricans
Midwife toad

Length 3–5 cm
(1¼–2 in)
Habit Mostly terrestrial
Breeding Spring and season summer
Location W. and C. Europe
Status Locally common

The midwife toad is notable for its highly unusual method of breeding. The female lays large, yolk-filled eggs in strings, which are transferred to the male during mating. The male then wraps these strings of eggs around his legs, sometimes carrying 2 strings laid by different females. Once the eggs are ready to hatch, the male deposits them in ponds and pools. This toad is small and plump, with short legs and sandy, pale brown, or grey skin, with darker spots. It inhabits woodland, gardens, drystone walls, quarries, sand dunes, and rock slides, hiding in crevices in winter and during dry periods. Active at night, it feeds on invertebrates. Males attract females by calling from a crevice or burrow, making a distinctive, high-pitched "poo-poo" sound.

rough skin

prominent eyes with vertical pupils

powerful forelimbs for digging

eggs carried by male

Alytes muletensis
Majorcan midwife toad

Length 3.5–4.5 cm
(1½–1¾ in)
Habit Mostly terrestrial
Breeding Spring and season summer
Location Balearic Islands (Majorca)
Status Critically endangered

One of the world's rarest toads, this tiny species (also known as the ferreret) has been the subject of a very successful conservation programme, in which animals bred in captivity are released into suitable habitats on Majorca. It has a yellowish or pale brown body, with dark mottling. Both the males and females call to locate mates. The female transfers a batch of eggs to the male, who carries them until they are ready to hatch. Their tadpoles grow very large, and little growth occurs after they become adults.

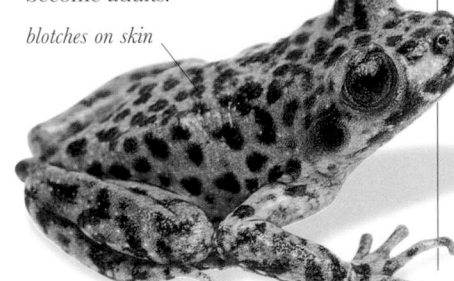

blotches on skin

Pipa pipa
Surinam toad

Length 5–20 cm (2–8 in)	
Habit Wholly aquatic	
Breeding Rainy season season	
Location N. South America	**Status** Locally common

The Surinam toad is grey above and paler below, with a flattened body and flat, triangular head. Like other members of the family Pipidae, it lacks a tongue. An inhabitant of turbid, muddy, and slow-moving water, the Surinam toad is highly adapted to an entirely aquatic existence, with powerful hindlimbs to aid swimming, sense organs along its sides that detect vibrations in the muddy water, tentacle-like projections on its fingers for feeling prey, and upward-pointing eyes for seeing above the surface. This toad exhibits very unusual mating and breeding patterns: the male clasps the female from above, and the pair turn upside-down repeatedly. The eggs are released, fertilized, and trapped in the space between the male's belly and the female's back. They are then absorbed into the skin on the female's back, where they develop in capsules and emerge as miniature frogs.

Pelobates fuscus
Common spadefoot toad

Length 4–8 cm (1½–3¼ in)	
Habit Mostly terrestrial	
Breeding Spring season	
Location C. and E. Europe, W. Asia	**Status** Vulnerable†

This plump, variable toad has a pale "spade" on each hindfoot, enabling it to dig backwards into the soil. It lives in sand dunes, on heathland, and in cultivated areas, preying on insects and other invertebrates. When threatened, it squeals, inflates its body, and raises itself up on all fours to appear larger. It often emits a garlic-like odour, possibly to deter predators.

Megophrys montana
Asian horned frog

Length 7–14 cm (2¾–5½ in)	
Habit Mostly terrestrial	
Breeding Rainy season season	
Location S.E. Asia	**Status** Locally common

This tropical frog is one of the world's most effectively camouflaged amphibians. It lives on the forest floor, mimicking dead leaves. Its brown colour and its shape are remarkably leaf-like, with a sharp snout, pointed "horns" over its eyes, and folds of skin that look like leaf edges. Female Asian horned frogs lay their eggs under rocks in streams. After hatching, the tadpoles hang vertically from the water's surface, and feed on very small organisms using their large, umbrella-shaped mouths.

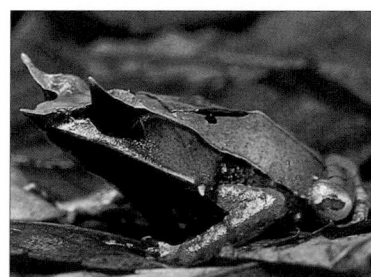

Rhinophrynus dorsalis
Mexican burrowing frog

Length 6–8 cm (2¼–3¼ in)	
Habit Terrestrial/ Burrowing	
Breeding Rainy season season	
Location S. USA to Central America	**Status** Locally common

With its bloated body, pointed snout, and small legs, this unusual-looking frog is adapted to a life spent largely underground. It lives in low-lying regions where soft soil makes it easy to burrow, and it emerges only after heavy rain. Adults breed in ponds, and the males have internal vocal sacs that they use to call to females. After the eggs hatch, the tadpoles feed by filtering food from the water, using barbels around their mouths.

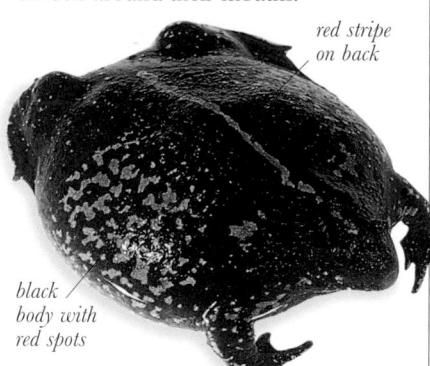

red stripe on back

black body with red spots

Scaphiopus couchii
Couch's spadefoot toad

Length 5.5–9 cm (2¼–3½ in)	
Habit Mostly terrestrial	
Breeding Spring and season summer	
Location S. USA, Mexico	**Status** Locally common

One of 6 similar species, all from North America, this yellow or yellowish green toad, with darker mottling, is a classic example of an amphibian that survives drought by burrowing underground. Like its relatives, it has spade-like black ridges on its hindfeet, which it uses to dig its way 1m (3¼ft) or more into loose, sandy soil. Once below ground, it encases its body in a watertight cocoon made by shedding several layers of skin, and then remains dormant, waiting for rain. When rain does fall – sometimes many months later – the cocoon breaks down, and this toad makes its way to the surface to breed.

large eyes with vertical pupils

Pelodytes punctatus
Parsley frog

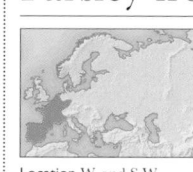

Length 3–5 cm (1¼–2 in)	
Habit Mostly terrestrial	
Breeding Spring season	
Location W. and S.W. Europe	**Status** Locally common

The small, agile parsley frog lives in damp, vegetated habitats near ponds. It has an unusual method of climbing, which involves clinging onto vertical surfaces such as trees, rocks, and walls, using its belly as a kind of suction pad. Its name derives from the green flecks on its pale skin that resemble chopped parsley. Like most frogs, it is active at night, spends winter and dry spells buried underground, and feeds on insects and other invertebrates. During the breeding season, males call out from underwater to attract females, who call back. Females lay their eggs in broad strips in ponds. The tadpoles sometimes grow to be larger than the adult frogs.

prominent eyes with vertical pupils

long limbs

Notaden bennetti

Holy cross toad

Length 4–7 cm (1 1/2 – 2 3/4 in)

Habit Mostly terrestrial

Breeding After heavy season rain

Location E. Australia

Status Locally common

Named after the cross-like pattern of warts on its back, this small, globular toad belongs to a family of over 100 species that are found only in Australia. Like many of its relatives, it spends dry periods underground, and emerges to feed and breed when it rains. This toad feeds on ants and termites, and produces a sticky, defensive secretion when handled. Its tadpoles develop in temporary pools.

short, stumpy legs

Limnodynastes peronii

Brown-striped frog

Length 3–6.5 cm (1 1/4 – 2 1/2 in)

Habit Mostly terrestrial

Breeding Spring and season summer

Location E. Australia

Status Locally common

Also known as the Australian grass frog, this long-limbed amphibian, with dark longitudinal stripes along its body, lives in and around swamps, feeding at night on insects and other invertebrates. During winter and dry spells, it buries itself underground. The male makes a click-like call to attract the female. During mating, the pair make a floating foam nest in which 700–1,000 eggs are deposited. The eggs hatch quickly, and the tadpoles develop very rapidly.

powerful hindlegs

Rheobatrachus silus

Southern gastric brooding frog

Length 3.5–5.5 cm (1 1/2 – 2 1/4 in)

Habit Wholly aquatic

Breeding Summer season

Location Australia (S.E. Queensland)

Status Critically endangered

Also known as the platypus frog, this extremely rare amphibian has a remarkable method of reproduction. The female lays 18–25 eggs and broods them in her stomach, apparently swallowing them once they have been fertilized. Secretions from the eggs suppress the digestive activity of the mother's stomach, ensuring that the eggs develop normally. The tadpole stage occurs in the egg and, once hatched, the tiny frogs emerge from the mother's mouth. Southern gastric brooding frogs are aquatic, but little else is known about their life in the wild. They were discovered in 1973, but have not been seen for some years, and may now be extinct.

Pseudophryne corroboree

Corroboree frog

Length 2.5–3 cm (1–1 1/4 in)

Habit Mostly terrestrial

Breeding Summer season

Location Australia (S.E. New South Wales)

Status Endangered

This rare animal, also known as the corroboree toadlet, is one of Australia's most colourful amphibians. It lives in moss-covered bogs in the Australian Alps, and lays its eggs in the moss rather than in water. The reason for its vivid colours – yellow or yellow-green, with black stripes – is unclear: unlike most brightly coloured amphibians, it does not have toxic skin.

short, unwebbed digits

Crinia insignifera

Sign-bearing toadlet

Length 1.5–3 cm (1/2 – 1 1/4 in)

Habit Mostly terrestrial

Breeding Winter season

Location S.W. Australia

Status Locally common

The sign-bearing toadlet is a small Australian frog with a slender body and limbs. It is grey to brown, and often has a dark triangular patch between its eyes; its skin may be smooth or warty. This frog inhabits temporary swamps that form after winter rain, preying on insects and other invertebrates, and spends dry periods underground. During the breeding season, the male calls to attract a mate. Females lay about 60–250 eggs at a time in water, either singly or in clumps; the eggs fall to the bottom of the pond.

darker spots or stripes

long toes

Uperoleia lithomoda

Stonemason toadlet

Length 1.5–3 cm (1/2 – 1 1/4 in)

Habit Mostly terrestrial

Breeding Winter season

Location N. Australia

Status Locally common

The click-like courtship call of this small, squat frog sounds like 2 stones being hit together, hence its name. It is dull brown or grey, with a gold line or line of patches along each side of its body, short limbs, and a horny tubercle on its hindfeet that enables it to dig into the soil. It spends dry periods buried underground. The female lays eggs in clumps that fall to the pond floor.

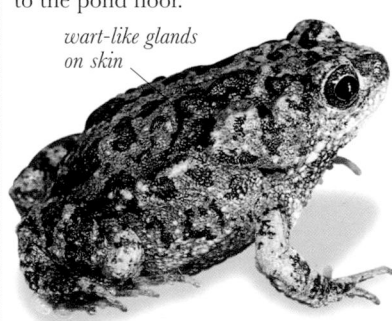

wart-like glands on skin

Heleophryne purcelli

Cape ghost frog

Length 3–6 cm (1 1/4 – 2 1/4 in)

Habit Mostly aquatic

Breeding Summer season

Location Southern Africa (Cape region)

Status Locally common

The Cape ghost frog is ideally suited to its aquatic lifestyle. Its long legs and webbed hindfeet enable it to swim in fast-flowing streams, and its flattened toe tips give it a strong grip on

Sooglossus gardineri

Gardiner's Seychelles frog

Length 1–1.5 cm (3/8 – 1/2 in)

Habit Wholly terrestrial

Breeding Rainy season season

Location Seychelles

Status Vulnerable

This tiny, creamy white to yellowish green frog is notable for its unusual breeding pattern. The eggs, which are laid in clumps on the ground, are guarded by the male and hatch into froglets instead of tadpoles. The froglets crawl onto their father's back and remain there, glued on by mucus, until they have used up their yolk and their legs are fully developed.

slippery rocks and boulders. During the breeding season, males develop tiny spines on their bodies. The tadpoles have large, sucker-like mouths to attach themselves to rocks.

green, yellow, or brown colouring with mottling

flattened body

smooth skin

Ceratophrys cornuta
Surinam horned frog

Length 10–20cm
(4–8in)
Habit Mostly
terrestrial
Breeding Rainy season
season

Location N.W. South
America

Status Locally common

This bulky and intricately patterned frog is a formidable hunter, with an exceptionally large mouth. It ambushes its prey and, if threatened, defends itself ferociously, using its large, sharp teeth. It has a well-camouflaged body, relatively short legs, and a "horn" above each eye. Males are slightly smaller than females and, in the breeding season, they produce a mating call that sounds like cattle lowing. Females lay up to 1,000 eggs, wrapped around aquatic plants. Unlike most tadpoles, Surinam horned frog tadpoles are predatory from the moment they hatch, and attack the tadpoles of other frogs and even each other. They have large, horny beaks and muscular jaws.

HUNTING BY AMBUSH

The Surinam horned frog catches other frogs, lizards, and mice using a sit-and-wait technique. It partly buries itself in soft ground, sometimes with only its eyes projecting above the surface, and then waits for animals to pass by. Like other amphibians, it is triggered into action by movement. Once its prey is within range, it bursts out of hiding, immobilizes its victim with its fangs, and then swallows it whole.

horn-like
projections

wide mouth

STOCKY BUILD
The Surinam horned frog is a massive, plump-bodied frog, with a huge head and a very wide mouth. It has conspicuous horn-like projections over its eyes.

Physalaemus pustulosus
Tungara frog

Length 3–4cm
(1¼–1½in)
Habit Mostly
terrestrial
Breeding Rainy season
season

Location Central America

Status Common

With its dark grey or brown warty skin, this Central American frog resembles a toad. The male calls loudly to attract females, repeatedly inflating his large vocal sacs to produce a sound that consists of a basic "whine" followed by one or more "chucks". The more chucks he makes, the more likely he is to attract a female; however, his call can also attract predatory bats. The female lays her eggs in floating foam nests in temporary pools.

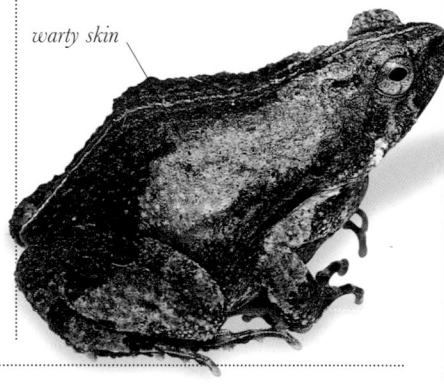

warty skin

Physalaemus nattereri
Four-eyed frog

Length 3–4cm
(1¼–1½in)
Habit Mostly
terrestrial
Breeding After heavy rain
season

Location E. South America

Status Unconfirmed

The four-eyed frog is characterized by the 2 black and white eye-spots on its rear. When threatened, it inflates its body and lifts up its rear end, displaying these false eyes to startle the predator. It can also secrete a distasteful substance from glands near its groin to further deter its attacker. This frog has a plump, toad-like body, with short limbs. It is mid-brown, with marbled patterning in darker and lighter shades of brown.

eye-spots

Leptodactylus pentadactylus
South American bullfrog

Length 8–22cm
(3¼–9in)
Habit Mostly
terrestrial
Breeding Rainy season
season

Location Central America,
N. South America

Status Locally common

This powerfully built frog is one of the largest members of its genus – a group of about 50 species that are mainly confined to the American tropics. It is yellow or pale brown, with dark markings, long legs, large eyes, and conspicuous external eardrums. The male has muscular forearms, with a sharp black spine on his thumbs, which he uses to fight off other males. The female lays eggs near water, depositing them in a foam nest that the male has created by whipping up mucus with his back legs. South American bullfrogs are active at night, feeding on insects and other invertebrates as well as other animals, and spend dry periods hidden under logs or underground. If picked up, they emit a loud scream to startle the predator into loosening its grip. The hindlimbs of this frog are sometimes eaten by humans.

smooth, pale
skin

Eleutherodactylus planirostris
Greenhouse frog

Length 2.5–4cm
(1–1½in)
Habit Wholly
terrestrial
Breeding Summer
season

Location S.E. USA,
Caribbean

Status Locally common

This tiny frog – one of the world's smallest – belongs to a family of about 400 species that are found throughout warm parts of the Americas. It is brown or tan, with darker mottling or stripes on its back, and well-developed adhesive discs on its fingers and toes. The greenhouse frog usually lives in wooded habitats, but also enters gardens. Like its many relatives, it lays eggs in moist places on the ground, and the eggs hatch directly into fully formed frogs.

Telmatobius culeus
Lake Titicaca frog

Length 8–12 cm
(3 1/4 – 4 3/4 in)

Habit Wholly
aquatic

Breeding Summer
season

Location South America
(Lake Titicaca)

Status Locally common

This frog has several adaptations that
suit it to a life in the cold, oxygen-
impoverished waters of Lake Titicaca.
It has long hindlegs with fully webbed
hindfeet for swimming, and very small
lungs that enable it to dive. Owing to
the small size of its lungs, it relies
mainly on oxygen taken in through
the skin. This uptake is increased by
the extensive folds in its skin, which
increase the animal's surface area,
and the presence of numerous blood
vessels and red blood cells.

*folds
on skin*

Bufo bufo
European common toad

Length 8–20 cm
(3 1/4 – 8 in)

Habit Mostly
terrestrial

Breeding Spring
season

Location N.W. Africa,
Europe to C. Asia

Status Common

One of Europe's most widespread
amphibians, the European common
toad is a robust animal with a warty
or spine-covered skin. Its colour ranges
from brown or green to brick-red, and
females are generally substantially
larger than males, particularly in early
spring, when they are laden with eggs.
At this time of year, males and
females congregate in ponds and,
after pairing up, the female lays
strings of eggs around underwater
plants. During
the rest of the
year, this toad

APPEARANCE
*This common toad
has large, copper- or
gold-coloured eyes. During
the breeding season, the
male develops pads on his
front feet for gripping
the female.*

specimen has been recorded. As the
species has only ever been seen in a
small region of high-altitude cloud
forest, just 10 square km (4 square
miles) in extent, it is possible that it is
now extinct, probably as a result of
climatic changes. Golden toads are
remarkable for the striking colour
difference between males and females.
Females are greenish yellow to black,
with red spots edged with yellow,
while males are bright orange.
During the breeding season, males
and females converge on small,
temporary pools, and it is here that
naturalists are focusing their attention,
in the hope that the species will be
seen once more.

It spends winter and dry periods
under logs, rocks, or buried in rotting
vegetation. Active at night, it feeds
mainly on insects. During courtship,
the male makes a high-pitched call.
A female green toad may lay between
10,000 and 12,000 eggs at a time.

warty skin

IN SELF-DEFENCE

The European common toad has
a large parotoid gland behind each
eye, which exudes a distasteful
secretion to ward off predators. If
threatened, it often swells up by
gulping air, and stands on tiptoe
to make itself look larger. Despite
its menacing appearance, it rarely
bites and is harmless to humans.

spends much of its time away from
water, feeding on insects, slugs, and
other small animals. It is mainly
nocturnal, hiding by day under logs
and in other damp places.

*warty or
spiny skin*

Bufo americanus
American toad

Length 5–9 cm
(2–3 1/2 in)

Habit Mostly
terrestrial

Breeding Spring
season

Location E. Canada,
E. USA

Status Locally common

*robustly built
body*   *parotoid gland*

Like the European common toad (see
left), the American toad has warty
skin, prominent eyes, and a large
parotoid gland on each side of its
head. It shows great variation in
colour, from brown to brightly
coloured, and has a spotted chest.
An adaptable animal, it is found in
all habitats, including gardens and
mountains, and is active at night.
During courtship, males produce
a musical trill lasting 3 to 60
seconds to attract mates. Females
lay eggs in long strings in ponds,
wrapped around vegetation.

Bufo periglenes
Golden toad

Length 4–5.5 cm
(1 1/2 – 2 1/4 in)

Habit Mostly
terrestrial

Breeding Rainy season
season

Location Central America
(C. Costa Rica)

Status Critically endangered

The golden toad provides a striking
example of a catastrophic population
decline: 1,500 toads were counted in
1987, one in 1989, and one the
following year. Since then, not a single

Bufo viridis
Green toad

Length 6–9 cm
(2 1/4 – 3 1/2 in)

Habit Mostly
terrestrial

Breeding Spring and
season summer

Location E. Europe, Asia

Status Locally common

This robust toad has pale skin with
green marbling and red spots. It
prefers sandy habitats in lowlands
and dry places, but is also found in
mountain regions and in urban areas.

*green
marbling*

warty skin

Bufo gutturalis
Guttural toad

Length 5–10 cm
(2–4 in)

Habit Mostly
terrestrial

Breeding Spring
season

Location Southern Africa

Status Locally common

prominent eyes

The guttural toad is found in a variety
of sites that provide cover, close to
ponds, dams, and streams. In urban
areas, it can often be seen under street
lights at night, feeding on insects.
However, it also preys on other
invertebrates and on flying termites
as they leave their nest. The female
is larger than the male, but the
latter has longer forelimbs.
Both sexes are pale brown
with dark brown patches. The
male makes a deep guttural
"snore" to attract a mate, and
the female lays eggs in water.

Bufo rangeri
Raucous toad

Length 5–11 cm
(2–4 1/4 in)

Habit Mostly
terrestrial

Breeding Spring
season

Location Southern Africa

Status Locally common

This yellow or pale brown, spotted
toad inhabits grassland and woodland
close to rivers, streams, and ponds. In
cool and dry spells it hides under logs
or rocks, or in rotting vegetation. When
cornered by a predator, it produces a
sticky, distasteful secretion through its
parotoid glands. During the breeding
season, males let out loud, harsh
"quacks" to attract females. Mating
activity, such as calling and fighting, is
extremely energetic, and males lose
weight at this time of year.

large parotoid gland

*warty
skin*

Atelopus zeteki
Golden frog

	Length 3.5–6 cm (1 1/2–2 1/4 in)
	Habit Mostly terrestrial
	Breeding Rainy season season
Location S. Central America (Panama)	Status Endangered†

This vivid yellow or orange toad, sometimes with black markings, provides a good example of warning coloration. Its skin produces poison when it is attacked and predators learn to avoid it, having experienced the pain once, associating bright colour with a noxious stimulus. Like many frogs and toads living in Central America, its numbers have declined drastically in recent years.

vivid warning coloration

black markings

long limbs

Brachycephalus ephippium
Gold frog

	Length 1–2 cm (3/8–3/4 in)
	Habit Mostly terrestrial
	Breeding Not recorded season
Location E. South America	Status Locally common

Also known as the saddleback toad, this species is an inhabitant of mountain rainforest, where it lives mainly among leaf litter, preying on insects and other small invertebrates. During dry spells, it hides in crevices. The gold frog is a ground-dweller – it cannot jump, nor can it climb well, since its stubby toes lack adhesive discs. It has a saddle-like, bony plate embedded in the smooth skin of its back, and is bright orange above, with a yellow underside. The female lays 5 or 6 large eggs on land; there is no tadpole stage, the eggs hatching directly into small frogs.

Nimbaphrynoides occidentalis
West African live-bearing toad

	Length 1.5–2.5 cm (1/2–1 in)
	Habit Wholly terrestrial
	Breeding Spring and season summer
Location W. Africa (Mount Nimba)	Status Endangered

This toad has a highly unusual mode of reproduction: the eggs are fertilized and the young develop inside the mother. After a 9-month gestation period, she gives birth by inflating herself with air and pushing out 2–16 tiny toadlets. The West African live-bearing toad hides in rocky crevices in the dry season.

Rhinoderma darwinii
Darwin's frog

	Length 2.5–3 cm (1–1 1/4 in)
	Habit Mostly terrestrial
	Breeding Year round season
Location S. South America	Status Locally common

An inhabitant of mountain forest in the southern Andes, Darwin's frog is famous for brooding its eggs in its vocal sacs – a form of parental care seen in few other amphibians. Brown or green above with a black underside, it is easily recognized by its small size and sharply pointed

fleshy proboscis

thin legs

Pedostibes hosii
Boulenger's Asian tree toad

	Length 5–10 cm (2–4 in)
	Habit Mostly terrestrial
	Breeding Rainy season season
Location S.E. Asia	Status Locally common

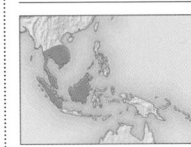

Unusually for a toad, Boulenger's Asian tree toad is a good climber, with adhesive discs on all its toes. It has warts and small spines on its skin, and varies in colour from greenish brown to black, with yellow spots; some females also have a purple tinge. It is usually found along rivers and streams in

forest, and is active at night, feeding on ants. The female lays numerous small eggs in strings in flowing water. The tadpoles have sucker-like mouths that enable them to cling to rocks.

prominent eyes

warty skin

adhesive discs on toes

snout. Adults live in damp, shady valleys close to flowing water, and are largely nocturnal, feeding on insects and other small animals. During the breeding season, the male attracts the female with a soft, bell-like call. Once the eggs have been laid, the male looks after them until they hatch.

PHYSICAL FEATURES
This small frog has a characteristic fleshy proboscis on the end of its nose. Other features include thin legs, long fingers, webbing on its toes, small eardrums, and eyes with horizontal pupils.

PARENTAL CARE

After a female Darwin's frog has deposited a clump of up to 40 large eggs on the ground, the male guards them for 2–3 weeks. When the embryos begin to move, he swallows up to 15 eggs and retains them in his vocal sacs, where they hatch into tadpoles. The tadpoles require only their own egg yolk in order to develop into froglets. These then emerge from the father's mouth.

Pseudis paradoxa
Paradoxical frog

	Length 5–7 cm (2–2 3/4 in)
	Habit Wholly aquatic
	Breeding Rainy season season
Location E. South America, Trinidad	Status Locally common

In most frogs, adults are considerably larger than mature tadpoles, but in this species the situation is reversed. Paradoxical frog tadpoles have an unusually long life, and they can grow up to 25cm (10in) long – about 4 times longer than the adult frog. During the final stage of metamorphosis, the animal shrinks, largely by absorbing its very long tail. Most of the features of the adult frog are adaptations for a

highly aquatic lifestyle: it has upward-pointing eyes and nostrils that enable it to float just beneath the water's surface, fully webbed hindfeet for swimming, and slimy skin. It lives in

ponds, lakes, and swamps, and feeds on aquatic invertebrates, stirring up mud with its feet to disturb prey before catching it. The female lays eggs in a floating foam nest.

powerful, muscular hindlimbs

slimy skin

long fingers

webbed toes help in swimming

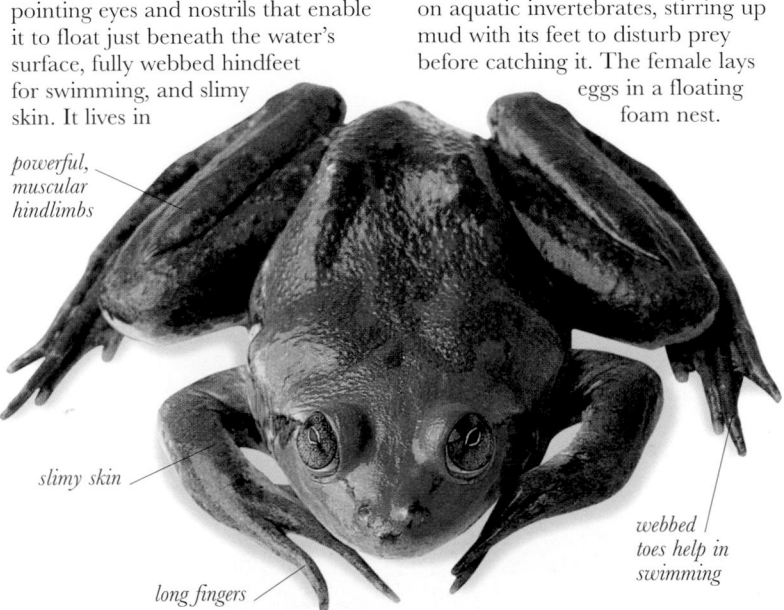

AMPHIBIANS

Bufo marinus

Marine toad

	Length 5–23 cm (2–9 in)
	Habit Mostly terrestrial
	Breeding After rain season
Location Central America, South America. Introduced to Australia and elsewhere	**Status** Locally common

The marine toad is the world's largest toad. A heavily built creature, it has very tough, warty brown skin that feels leathery to the touch. It has a massive head, with bony ridges above the eyes, and prominent parotid glands that secrete a milky poison whenever the toad is in danger. A hungry feeder, it eats mainly ants, termites, and beetles, but will readily consume a wide variety of other insects and invertebrates as well as other frogs. The marine toad is usually solitary, except in the breeding season. Then males call to females, drawing their attention with a purring trill that sounds like a running motor. The female, which is larger than the male, lays her eggs in long, gelatinous strings, draping them over plants or debris in water. Each clutch may consist of up to 20,000 eggs. In ideal conditions, toadlets will reach adult size within a year of leaving the water. This species has been deliberately introduced into some countries in the hope of controlling insects devastating sugarcane crops – and here it is usually known as the cane toad.

CONSERVATION

This toad's numbers have grown so rapidly in Australia since it was introduced in 1935 that it is now considered a serious pest. It has had a serious impact on native frogs and toads, with which it may compete for shelter and breeding sites, as well as on other animals. Efforts are being made to control the spread of the toad – and predators appear to be learning to eat this highly toxic intruder safely, or to avoid it completely.

VORACIOUS FEEDER
Emerging at night to feed, the marine toad will eat virtually anything it can fit into its mouth. In urban areas, these toads often gather under street lamps, eating insects attracted by the light.

FORMIDABLE SURVIVOR
The marine toad is highly adaptable, in terms of both survival and reproduction. It is able to thrive in almost any habitat, and can even tolerate brackish water. It feeds on any animal it can catch, and can survive without water for long periods. Breeding, which produces large quantities of eggs, occurs throughout the year.

bony ridge

distinct visor over each eye

very large parotid glands

olive to reddish brown skin

brown-speckled, white or pale yellow underside

no webbing on front toes

SHORT AND SQUAT
With its heavy, squat body and short limbs, the marine toad is best suited to life on the ground. It usually crawls along or makes short hops.

VENOMOUS DEFENCE

The marine toad is extremely toxic to other animals. If threatened or squeezed, such as when in the mouth of a predator, it oozes venom from the large glands on its shoulders. Smaller glands over the body also produce venom. Anything swallowing this poison may die very quickly – in as little as 15 minutes in some cases.

Litoria caerulea
White's treefrog

Length 5–10 cm (2–4 in)

Habit Terrestrial

Breeding Spring and **season** summer

Location S. New Guinea, N. and E. Australia

Status Common

Also known as the green treefrog, this widespread Australasian species is well known for its tame behaviour, and its habit of living in or near buildings. Its colour varies from green to greenish blue, sometimes with white spots. Like most treefrogs, it has adhesive discs on its toes. In wet conditions, its loose, deeply folded skin allows the frog to

take in a large amount of water – an ability that makes it unusually tolerant of drought. White's treefrog is a nocturnal hunter, preying on insects and other invertebrates, and occasionally larger animals such as mice. Males are smaller than females, and give out a harsh, bark-like courtship call. After mating, the female lays 2,000–3,000 eggs in water. These frogs are of considerable interest in medicine: their skin produces several useful anti-bacterial and anti-viral compounds, and a substance that has been used to treat high blood pressure in humans.

plump body

large head

adhesive discs

Litoria infrafrenata
Giant treefrog

Length 10–14 cm (4–5 1/2 in)

Habit Mostly terrestrial

Breeding Spring and **season** summer

Location New Guinea, N.E. Australia

Status Locally common

Plain green above, with a distinctive stripe along its lower lip, this tropical species is the largest frog in Australia, and one of the biggest treefrogs in the world. Found in forests and gardens, it feeds after dark on insects and other invertebrates. It has large adhesive discs on its toes, and spends most of its life in trees. During the breeding season, adults descend and females lay up to 400 eggs in water.

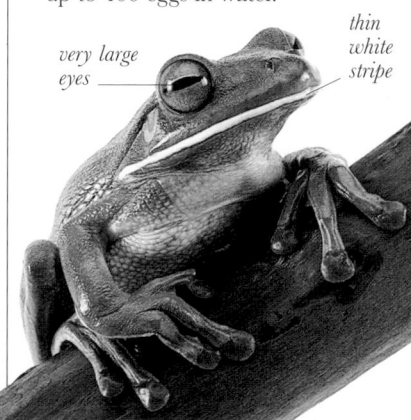

very large eyes

thin white stripe

Acris crepitans
Northern cricket frog

Length 1.5–4 cm (1/2–1 1/2 in)

Habit Mostly terrestrial

Breeding Spring **season**

Location E. and S. USA

Status Common

This frog is able to leap particularly large distances owing to its short, powerful hindlimbs. Its other characteristics include rough skin, and a distinctive, dark, triangular mark between the eyes. It is found among foliage near swamps, ponds, streams, and lakes, and often basks in the sun. In early spring, large numbers converge in ponds, where the males call loudly to attract females, with a series of metallic clicks reminiscent of the sound of crickets. In some areas, tadpoles have black tips to their tails. These distract predatory dragonfly larvae, and so divert attack from more vulnerable parts of the body.

Pseudacris crucifer
Spring peeper

Length 2–3.5 cm (3/4–1 1/2 in)

Habit Mostly terrestrial

Breeding Spring **season**

Location S.E. Canada, E. USA

Status Locally common

Across much of eastern North America, the distinctive call of this small but highly agile frog is a sign that spring is underway. It belongs to a group of species known as chorus frogs, which are adept climbers, with reduced webbing, and adhesive discs on their toes. The spring peeper lives in woodland with temporary or permanent ponds, spending winter and dry spells under logs or rotting vegetation.

INFLATED VOCAL SAC

Male spring peepers gather in shrubs and trees next to water in early spring to form choruses, giving out high, piping whistles to attract females. The male's vocal sac amplifies his call as the air is passed back and forth between the sac and lungs over the vocal cords.

BODY COLORATION
The spring peeper is brown, grey, or olive-green, with darker patches that usually form an X-shaped pattern on its back.

Hyla cinerea
Green treefrog

Length 3–6 cm (1 1/4–2 1/4 in)

Habit Mostly terrestrial

Breeding Spring **season**

Location S.E. USA

Status Common

This bright green frog can change colour quite rapidly, from yellow when calling out for mates, to grey when the weather is cold. In the breeding season, males form choruses around pond edges; some call out to attract mates,

while others, called "satellite males", are silent and attempt to intercept females attracted to the calling males. This species is also known as the "rain frog", because it calls just before and during wet weather. It often appears at night at windows, where it feeds on insects attracted to the light.

golden eyes

pale stripe along sides

Hyla arborea
European treefrog

Length 3–5 cm (1 1/4–2 in)

Habit Terrestrial

Breeding Spring and **season** summer

Location Europe, W. Asia

Status Lower risk

Europe is home to just 2 species of treefrogs – this one is the most widespread. Usually bright green, although sometimes yellow or brown, it has a

conspicuous dark horizontal stripe on its head and body. It lives in overgrown areas near water, and hibernates in winter. During the breeding season, males make a loud quacking sound, and this can quickly build up into a noisy chorus.

adhesive discs for climbing

Hyla chrysoscelis
Grey treefrog

Length 3–6 cm
(1 1/4 – 2 1/4 in)
Habit Mostly
terrestrial
Breeding Spring
season

Location S. Canada,
C. and E. USA

Status Locally common

The blotchy grey colouring of this frog enables it to blend perfectly with a background of lichen-covered tree trunks. It can change colour in different climatic or lighting conditions: turning darker in cold, dark conditions, and paler in bright light. Its blood contains glycerol, an "anti-freeze" that enables it to survive sub-zero temperatures. The grey treefrog is almost identical to another species, *H. versicolor*, which has a deeper-voiced call and twice the number of chromosomes.

Agalychnis callidryas
Red-eyed treefrog

Length 4–7 cm
(1 1/2 – 2 3/4 in)
Habit Terrestrial
Breeding Summer
season

Location Central America

Status Locally common

The jewel-like coloration and red eyes make this one of the most striking frogs. In the breeding season, males congregate on branches overhanging a pond, and call out to females with a series of clicks. The male clambers onto the female's back to mate, and she descends to the pond, carrying the male with her, to take up water, then climbs back up into the tree to lay a batch of about 50 eggs on a leaf over the water. The process is usually repeated several times, the female drawing up water between each new batch of eggs. The eggs hatch after about 5 days, and the tadpoles fall into the pond.

COLORATION

This multi-hued frog has a green body, with blue and white stripes on its flanks; the insides of its legs are red and yellow. By day, it rests on leaves, with its legs folded along its sides in order to hide all the bright parts of its body, its green back providing camouflage.

prominent red eyes

GOOD CLIMBER
Like all treefrogs, the red-eyed treefrog has well-developed pads on all its toes to aid climbing.

Phyllomedusa hypochondrialis
Orange-sided leaf frog

Length 2.5–4.5 cm
(1–1 3/4 in)
Habit Wholly
terrestrial
Breeding Rainy season
season

Location N. to C. South America

Status Locally common

An inhabitant of arid regions, this long, slender frog secretes a waxy substance from special glands, and rubs it all over its body to reduce water loss. Females surround their fertilized eggs with fluid-filled capsules in order to provide extra water for the developing eggs. The orange-sided leaf frog produces an unpleasant odour to ward off potential predators, and may also feign death.

green back

orange underside

Gastrotheca monticola
Mountain marsupial frog

Length 4–6 cm
(1 1/2 – 2 1/4 in)
Habit Wholly
terrestrial
Breeding Spring and
season summer

Location N.W. South America

Status Locally common

The female mountain marsupial frog is characterized by a brood pouch on her back. During mating, the male places the eggs in the

pouch, where they mature. The developing young are joined to the mother's blood system by a link similar to the placenta of mammals. After heavy rain, the mother releases young tadpoles into water (in some closely related species, the young are released as tiny adults). This frog is active at night, and hides under logs during winter and dry periods.

broad head

Cyclorana australis
Giant frog

Length 7–10.5 cm
(2 3/4 – 4 1/4 in)
Habit Terrestrial/
Burrowing
Breeding Rainy season
season

Location N. Australia

Status Locally common

This large frog is able to survive underground for very long periods by burrowing into the soil and shedding numerous layers of skin, which create a cocoon around its

body. It can retain large amounts of water in its bladder, body cavity, and beneath its skin. A related species, the water-holding frog (*C. platycephala*), is used as a "living well" by aborigines, who dig up the frog and extract drinking water by squeezing it. The giant frog is pale grey, brown, or bright green in colour, with a dark brown or black stripe through its eyes. It has a broad head, short, muscular limbs, horny tubercles on its hindfeet, and a pair of skin folds on its back. During the breeding season, the male calls to the female from the edge of a pool. She lays up to 7,000 eggs in clumps in water, and the eggs sink to the bottom of the pond.

Scinax ruber
Red-snouted treefrog

Length 2.5–4 cm
(1–1 1/2 in)
Habit Mostly
terrestrial
Breeding Rainy season
season

Location S. Central America to N. South America, Caribbean

Status Locally common

The red-snouted treefrog has a slender yellow, silver, or grey body. During the breeding season, males call very loudly to attract mates. The females, which are slightly larger than the males, select their mates on the basis of size, preferring those that are about 20 per cent smaller than they are. This ratio allows closest contact with the male cloaca, so that the maximum number of eggs can be fertilized. The female scatters the eggs around the pond to reduce the impact of predation.

pointed snout

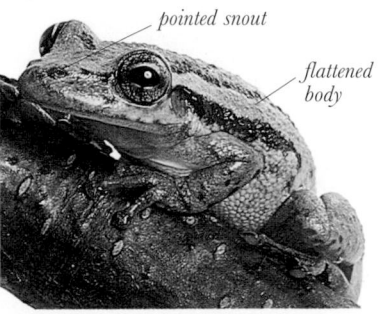

flattened body

Triprion petasatus
Duck-billed treefrog

Length 5.5–7.5 cm
(2 1/4 – 3 in)
Habit Mostly
terrestrial
Breeding Rainy season
season

Location Mexico (Yucatan region), N. Central America

Status Locally common

duck-like snout

darker spots on grey body

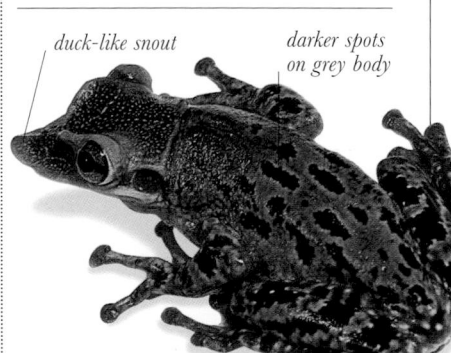

Characterized by the unique shape of its head, the duck-billed treefrog has a flattened, duck-like snout, and saddle-like flaps behind the eyes, the function of which is unknown. It is grey, brown, or green, with darker spots, and has webbed toes and well-developed adhesive pads on its fingers and toes. This treefrog lives in shrubs and trees that occur in patches in grassland. Active only at night, by day it commonly hides in tree holes, which also provide a refuge during cold and dry weather.

Hyalinobatrachium valerioi
White-spotted glass frog

Length 2–3.5 cm
(3/4–1 1/2 in)

Habit Mostly terrestrial

Breeding Rainy season season

Location S. Central America to N.W. South America

Status Locally common

This small, delicate frog has translucent skin that gives it the appearance of being made from glass. It is an excellent climber, with large adhesive discs on its fingers and toes. Females attach their eggs to foliage overhanging streams. When the pink or red tadpoles hatch, they drop into the water and burrow into the mud or sand on the stream bed.

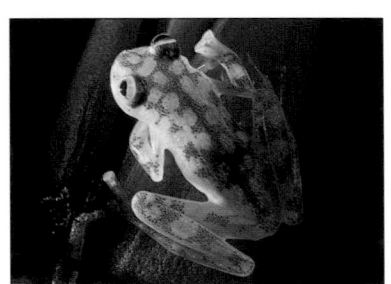

Dendrobates auratus
Green poison-dart frog

Length 2.5–6 cm
(1–2 1/4 in)

Habit Wholly terrestrial

Breeding Rainy season season

Location S. Central America to N.W. South America

Status Locally common

This brilliantly coloured, tropical American frog belongs to a family of about 120 species that includes the most poisonous amphibians in the world. Like its relatives, it lives on the ground, and uses the poison as a defence. Its bright colours show

PATERNAL CARE

The female green poison-dart frog lays clutches of 5–13 eggs in leaf litter, but it is the male that guards the eggs, sometimes caring for more than one clutch at a time. When the eggs hatch, he carries the tadpoles on his back, to small pools in tree holes.

predators that it is toxic, allowing it to hop about unmolested on the forest floor. Courtship among poison-dart frogs is conducted in a highly unusual way. Females take the lead, enticing the males to mate by beating their partners on the back with their hindfeet.

WARNING COLORATION
The green poison-dart frog is small with a pointed head. A brightly coloured species, it is black with green patches, and may sometimes have a golden sheen. This colouring functions as a warning to predators.

Dendrobates pumilio
Strawberry poison-dart frog

Length 2–2.5 cm
(3/4–1 in)

Habit Wholly terrestrial

Breeding Rainy season season

Location S. Central America

Status Locally common

Although not the most toxic poison-dart frog, this Central American species is still highly poisonous. Like other poison-dart frogs, it relies on its chemical defence system to keep it out of trouble. Its colours are very variable: in some places it is bright red, but in others it can be brown, blue, or green. Females lay eggs in batches of 4–6, and deposit them singly in water-filled tree holes.

short limbs

Colostethus inguinalis
Common rocket frog

Length 2–3 cm
(3/4–1 1/4 in)

Habit Mostly terrestrial

Breeding Rainy season season

Location S. Central America to N.W. South America

Status Locally common

This small frog – like all *Colostethus* species – is related to the poison-dart frogs, and occurs in many of the same locations. However, unlike poison-dart frogs it has brown and black cryptic coloration and is non-toxic.

Rana temporaria
European common frog

Length 5–10 cm
(2–4 in)

Habit Mostly terrestrial

Breeding Spring season

Location Europe, N.W. Asia

Status Locally common

The common frog is one of the most widespread amphibians in northern Europe, with a range that extends well beyond the Arctic Circle. An

inhabitant of damp places, it spends most of its adult life on land, returning to water only to escape attack and to breed. In early spring, large numbers of these frogs make their way to shallow ponds. The males call the

females with a deep, growl-like croak, and after mating the females lay floating clumps of several thousand eggs. In warm ponds, the tadpoles reach maturity in 6 weeks, but in cool water they can take 4 months.

long, powerful hindlegs

black spots on body

black bands on hindlegs

Rana pipiens
Northern leopard frog

Length 5–9 cm
(2–3 1/2 in)

Habit Terrestrial/ Aquatic

Breeding Spring season

Location S. Canada, North USA

Status Locally common

This frog is green or brown with a pointed snout, and 2 or 3 rows of large, irregular, dark, spots that have pale borders.

Found in meadows near ponds, lakes, and marshes, it feeds on insects and other invertebrates, and spends winters and dry spells hidden under rocks and logs. The male's call is a loud, long, deep, rattling snore, interspersed with grunts. This frog breeds any time from March to May, depending on the latitude.

pale stripe along upper jaw

long, powerful hindlegs

Rana albolabris
White-lipped river frog

Length 6–10 cm
(2 1/4–4 in)

Habit Mostly terrestrial

Breeding Summer season

Location W. and C. Africa

Status Locally common

Frogs belonging to the family Ranidae, often referred to as the "true" frogs, number more than 600 species, and are found in all the world's major

continents. African members of this family are generally poorly known and include 11 species, living south of the Sahara, previously known as Hylarana. Characterized by a white stripe on its upper lip, this African frog has a pointed snout, prominent folds of skin down each side of its back, and large adhesive discs on its toes. The male has a larger eardrum than the female and, in the breeding season, develops large glands on the upper forelimbs, which enable him to maintain a firm grip on the female while mating. The white-lipped river frog lives in rainforest and woodland areas. Its tadpoles have a fringe of skin along their lower lip, the function of which is not known.

Rana catesbeiana
North American bullfrog

Length 9–20 cm (3 1/2–8 in)

Habit Mostly aquatic

Breeding Spring and **season** summer

Location S.E. Canada, W., C., and E. USA

Status Locally common

Renowned for its voracious appetite, the North American bullfrog is the largest frog in North America. It lives in lakes, ponds, and slow-flowing streams, and spends most of its life either in water, or very close to it. Its coloration is variable: in some parts of its range it is plain green above, while in others it is boldly patterned. During the breeding season – in spring and summer – males defend their territories, and produce a deep, sonorous call from a vocal sac under their throats. After mating, females lay several thousand eggs. Their tadpoles can take up to 4 years to become adult.

brown markings

AMPHIBIAN INVADER

The North American bullfrog has a varied diet, including mammals, reptiles, and other frogs. Originally from eastern North America, it has been introduced into several regions further west, where it has had a harmful effect on local freshwater wildlife.

PHYSICAL APPEARANCE
The most prominent features of this large frog are its long, powerful hindlimbs and prominent eardrums. It is green above, with brown markings, and white below.

large eardrums

Rana dalmatina
Agile frog

Length 5–9 cm (2–3 1/2 in)

Habit Mostly terrestrial

Breeding Spring **season**

Location N., C., and S. Europe

Status Locally common

With its exceptionally long hindlegs, the agile frog is a remarkable jumper, capable of leaping several metres. It is pale brown, with brown-striped yellow flanks, and has dark brown spots on its back that are usually arranged in an inverted "V" pattern. Males can be distinguished from females by their thicker forelimbs. This frog

inhabits open woodland and swampy meadows, spending winter and dry periods under logs and rocks, or buried under vegetation, although males often overwinter under ice in ponds and lakes. They breed as soon as the ice begins to melt in spring, and the eggs are laid in clumps in water.

very long hindlimbs

pointed snout

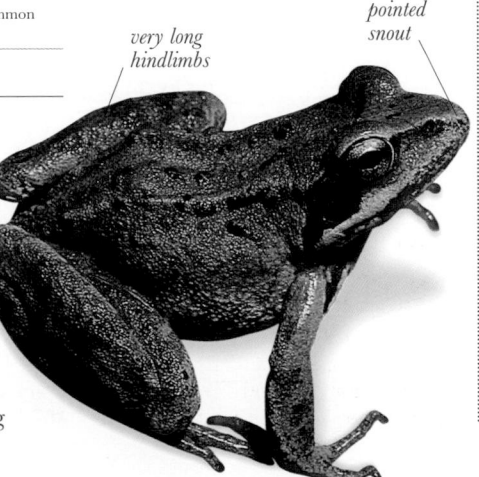

Rana sylvatica
Wood frog

Length 3.5–8 cm (1 1/2–3 1/4 in)

Habit Mostly terrestrial

Breeding Spring **season**

Location Canada, E. USA

Status Locally common

Although the wood frog is usually found in temperate woodland in the southern part of its range, it can survive well into the Arctic Circle, further north than any other frog in North America. Its blood contains chemicals that act as an "anti-freeze", enabling it to live in sub-zero

temperatures. It often breeds while the snow is still on the ground and the breeding ponds are partially frozen. The wood frog varies in colour from pink through various shades of brown to black, with darker markings. It has stripes on its legs and a characteristic dark patch, called the "robber's mask", behind its eyes.

Rana ridibunda
Lake frog

Length 9–15 cm (3 1/2–6 in)

Habit Mostly aquatic

Breeding Spring **season**

Location Europe, W. and S.W. Asia

Status Locally common

Europe's largest frog, this brown or green species has typically long, powerful hindlegs, longitudinal folds of skin on its back, and close-set eyes. An inhabitant of lakes, ponds, ditches,

and streams, the lake frog feeds on fish, lizards, snakes, mice, and other frogs, as well as on insects and other invertebrates. Females lay up to 12,000 eggs in water.

black spots on back

black bands on legs

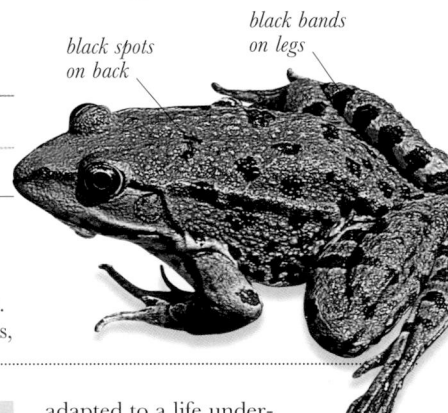

Conraua goliath
Goliath bullfrog

Length 10–40 cm (4–16 in)

Habit Mostly aquatic

Breeding Rainy season **season**

Location Cameroon, Equatorial Guinea

Status Vulnerable

Found exclusively in West Africa, this is the world's largest frog. With its powerful hindlimbs, long, webbed toes, and smooth, slippery skin, it is well

adapted to a life under-water, and is an excellent swimmer and diver. The tips of its toes are dilated, but the function of this feature is not known. The goliath bullfrog lives along streams in jungles, and feeds on other frogs, small reptiles, and mammals. It rarely emerges on land, but when it does it is quick to dive back into water whenever it is threatened or disturbed. Unusually among frogs, the male of this species is larger than the female and does not call to attract mates. Little is known about the breeding habits of the goliath bullfrog.

AMPHIBIANS

Ceratobatrachus guentheri

Solomon Islands horned frog

Length 5–8 cm
(2–3 1/4 in)

Habit Mostly terrestrial

Breeding Rainy season season

Location Solomon Islands   Status Locally common

This forest-floor animal strongly resembles the Asian horned frog (see p.443), although it is not closely related. It also uses the same kind of camouflage to avoid being seen: its projections, or "horns", help to break up the outline of the frog's head, making it difficult to detect against a background of leaves.

Unusually for a frog, the lower jaw of the Solomon Islands horned frog contains tooth-like projections, made of bone, which it uses to hold its prey. These "tusks" are larger in males than in females. The female lays clumps of large eggs, which hatch directly into froglets, in damp soil. The froglets have folds in their skin, which help the absorption of yolk from the eggs.

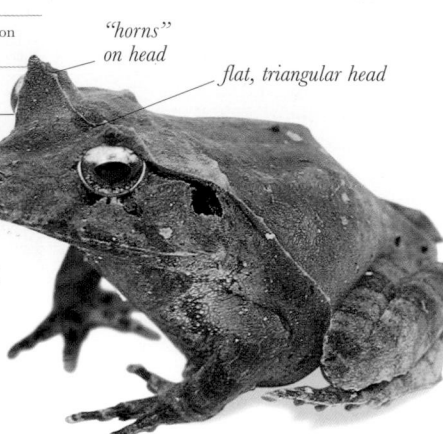

"horns" on head

flat, triangular head

Ptychadena oxyrhynchus

Sharp-nosed grass frog

Length 4–7 cm
(1 1/2–2 3/4 in)

Habit Mostly terrestrial

Breeding Spring season

Location W. and C. to S.E. Africa   Status Locally common

This highly athletic frog can move very fast through grass and in water, its powerful hindlegs enabling it to take prodigious leaps. It is yellowish brown above, with brown or black spots and patches, and a white underside. A pale triangular patch is present on the top of its very pointed snout. This

frog inhabits open savanna and thorn veld in areas where temporary ponds form after heavy rain, and feeds on insects, especially crickets, and other invertebrates. During dry and cold weather, it remains hidden. The male makes a high-pitched trill to attract the female, and mating takes place on the surface of water after heavy rain. The eggs are fertilized in the air and ejected onto the surface of the water in small batches.

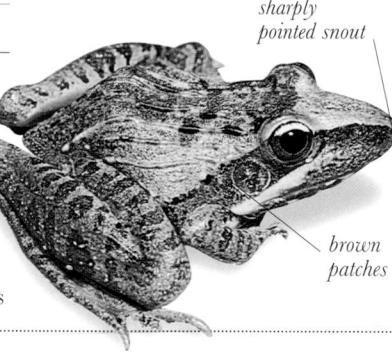

sharply pointed snout

brown patches

Mantella viridis

Green mantella

Length 2–3 cm
(3/4–1 1/4 in)

Habit Wholly terrestrial

Breeding Rainy season season

Location N. and E. Madagascar   Status Vulnerable†

The mantellas are 9 species of brightly coloured frogs, exclusive to Madagascar, which are all threatened with extinc-tion due mainly to the destruction of their forest habitat. Their numbers are further depleted by the worldwide pet trade. The green mantella, which lives near streams in tropical forest, is

yellow or pale green on its back and head, black along its sides, and has a white stripe running along its upper lip. Active during the day, it feeds on insects and other invertebrates. The male calls out with a series of clicks to attract the female. After mating, which takes place on land during the rainy season, the female lays eggs close to streams.

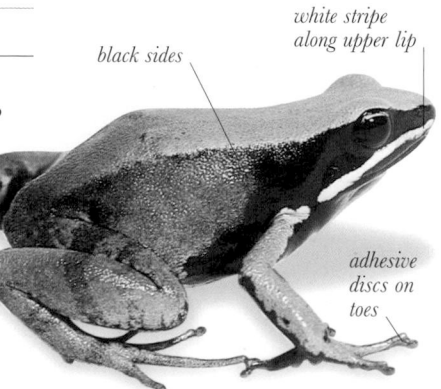

white stripe along upper lip

black sides

adhesive discs on toes

Pyxicephalus adspersus

African bullfrog

Length 8–23 cm
(3 1/4–9 in)

Habit Mostly terrestrial

Breeding Rainy season season

Location C. to southern Africa   Status Locally common

The African bullfrog inhabits wet and dry savanna, where it may remain buried underground for very long periods, even up to several years if there is no heavy rain. When

underground, it encases itself in a cocoon that reduces water loss. It feeds on other frogs as well as insects and other invertebrates. This bullfrog is very large, with a massive head and horny tubercles on its hindfeet, for digging. During the breeding season, the male, which is larger than the female, develops a pair of tooth-like tusks on the lower jaw, which he uses in fights over territory with other males. He produces a loud, booming call to attract the female. After mating in flooded pools, the male guards both the eggs and the tadpoles. As the tadpoles develop, the male digs channels to enable them to swim away to larger expanses of water.

green or brown coloration

warts and ridges on skin

tubercles on hindfeet

very wide mouth

Mantella aurantiaca

Golden mantella

Length 2–3 cm
(3/4–1 1/4 in)

Habit Wholly terrestrial

Breeding Rainy season season

Location W. central Madagascar   Status Vulnerable

Strikingly similar to poison-dart frogs (see p.452), which live in the American tropics, this Madagascan frog has evolved bright colours for the same reason – to warn potential predators that it is toxic. It is active during the

day, and searches for small animals on the forest floor. During the breeding season, adults mate on land, and females lay their eggs in moist leaf litter. Once the tadpoles have hatched, they are swept into pools by rain. Like other mantellas, this species is threatened by deforestation.

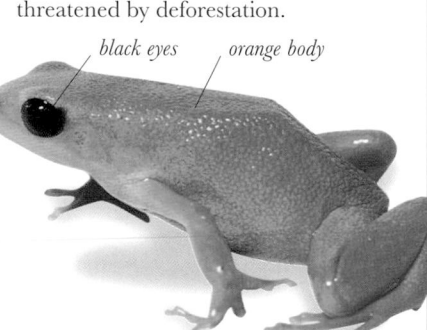

black eyes   orange body

Platymantis vitiensis

Levuka wrinkled ground frog

Length 3.5–5 cm
(1 1/2–2 in)

Habit Mostly terrestrial

Breeding Spring and season summer

Location Fiji   Status Endangered

The Levuka wrinkled ground frog is a slender-bodied amphibian that is encountered only in the tropical forests of Fiji. It has a flat head with large eardrums, wrinkled skin, and adhesive discs on its unwebbed fingers

and toes. The female lays a small number of large eggs on the ground, which hatch directly into small froglets, the tadpole phase being completed within the egg. Little is known about the behaviour and life cycle of this frog.

Hemisus marmoratus
Mottled shovel-nosed frog

Length 3–4 cm
(1 1/4–1 1/2 in)
Habit Terrestrial
Breeding Rainy season season

Location W. and C. to S.E. Africa | Status Locally common

Unusually for a burrowing frog, this squat species digs into soil head-first, using its sharp, tough, spade-like snout, rather than digging backwards with its hindfeet like most other burrowing frogs. It preys on ants and termites. Females lay up to 2,000 eggs in underground chambers close to pools. They guard the eggs until they hatch, then dig a tunnel to release the tadpoles into the water.

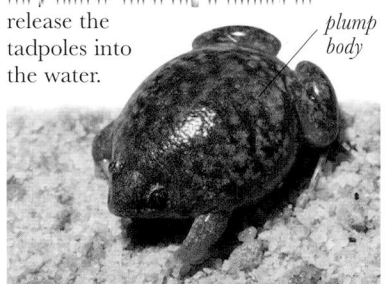

plump body

Hyperolius marmoratus
Painted reed frog

Length 2.5–3.5 cm
(1–1 1/2 in)
Habit Terrestrial
Breeding Spring and season summer

Location E. to S.E. Africa | Status Locally common

The painted reed frog is a small, graceful animal with highly variable coloration – it may be striped, spotted, or plain brown. A good climber, its fingers and toes are equipped with adhesive discs as well as webbing. It inhabits areas close to permanent ponds, lakes, and swamps; like most frogs, it spends dry spells hidden under logs and rocks. Active at night, it feeds on insects and other invertebrates. During its long breeding season, which extends through spring and summer, males form choruses close to ponds, and call very loudly to attract mates. Females may come to the chorus more than once in a season, laying a clutch of eggs in small clumps in water at each visit. Males vary in the number of nights they spend in a chorus; those that call on the greatest number of nights get the most matings.

Afrixalus fornasinii
Greater leaf-folding frog

Length 3–4 cm
(1 1/4–1 1/2 in)
Habit Wholly terrestrial
Breeding Spring and season summer

Location S.E. Africa | Status Locally common

Many treefrogs lay their eggs above the ground, but this species has a particularly skilful technique. The female lays her clutch on a leaf overhanging water, and then folds the leaf, fastening it with a glue that she secretes. When the eggs hatch, the tadpoles drop into the water below – a system that gives them a headstart in the struggle to avoid being eaten. Greater leaf-folding frogs are brown with dark, longitudinal stripes, and have slender limbs with adhesive discs as well as webbing on their fingers and toes. They live in dense vegetation around pools and swamps, and are active at night. At the onset of the breeding season, large males call from an elevated position in vegetation, making rapid clicks to attract mates. However, smaller males often stay silent, waiting by the callers. When females approach, the non-calling males sometimes manage to intercept them and mate.

long limbs

pale stripes along back

Leptopelis modestus
African treefrog

Length 2.5–4.5 cm
(1–1 3/4 in)
Habit Terrestrial
Breeding Spring and season summer

Location W. Africa, central E. Africa | Status Locally common

Also known as the modest treefrog, this species has a broad head with a wide mouth, large eyes, and conspicuous eardrums. Grey or pale brown, it has a darker hourglass-shaped patch on its back. Males call to females from trees, producing a deep "clacking" sound while displaying their bright blue or green throat. They will fight to defend their calling positions.

forward-pointing eyes

long limbs

Kassina senegalensis
Bubbling kassina

Length 3–5 cm
(1 1/4–2 in)
Habit Mostly terrestrial
Breeding Rainy season season

Location W. and C. to S.E. Africa | Status Locally common

Adapted for walking rather than hopping, the bubbling kassina has slender limbs and slight webbing on the toes of its hindfeet. Its plump, elongated body varies in colour from beige or yellow to grey, with bold, dark brown or black stripes and blotches. An inhabitant of low-lying savanna and sand dunes, it lives close to ponds and pools. In the breeding season, males move towards the water in the afternoon, calling from hidden places on the ground; after dark, they call from trees. Mating occurs in water, and the eggs are attached individually to aquatic plants, often in clusters.

Rhacophorus nigropalmatus
Abah River flying frog

Length 7–10 cm
(2 3/4–4 in)
Habit Terrestrial
Breeding Rainy season season

Location S.E. Asia | Status Locally common

There are over 50 species of *Rhacophorus* frogs, several of which – including the Abah River flying frog – use their feet to glide from tree to tree. The partial webbing between their toes acts as a parachute, enabling them to "fly" across wide gaps to escape predators, steering themselves in mid-air. The Abah River flying frog is bright green with white spots, and has a flat head with large, protruding eyes and visible eardrums, and large adhesive discs on its fingers and toes for climbing. The male is smaller than the female, and has pads on his thumbs so that he can grasp her during mating. Active at night, the Abah River flying frog is found mostly in brush and forest, and is also frequently found close to human habitation. During the breeding season, the female lays up to 800 eggs in a foam nest attached to foliage overhanging water. When the eggs hatch, the tadpoles emerge and fall into the water below.

prominent eyes with horizontal pupils

flat head

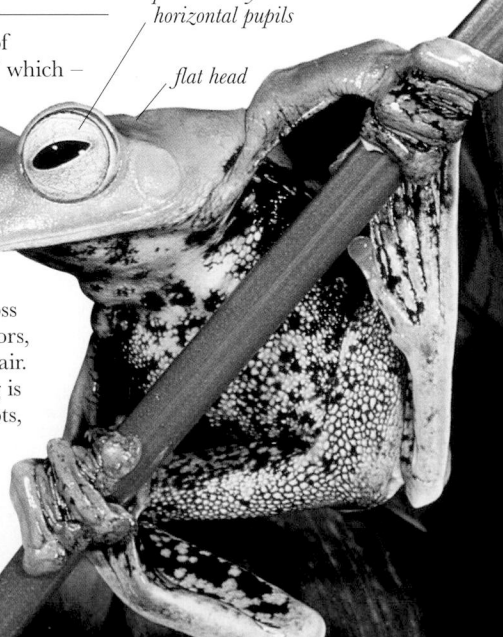

Chiromantis xerampelina
Grey foam-nest frog

Length 5–9cm
(2–3¹/₂ in)

Habit Mostly
arboreal

Breeding Rainy season
season

Location Southern Africa   Status Locally common

This treefrog and its close relatives are remarkable for their breeding method and their ability to survive in savanna that has a long dry season. Its skin is adapted to reduce water loss, and it excretes dry droppings rather than watery waste. Usually grey or pale brown, this frog can change colour for camouflage – at night it is usually dark-coloured, but it can become almost white at midday.

LIVING IN TREES
This foam-nest frog has a compact body and prominent eyes with horizontal pupils. It possesses the typical adaptations for an arboreal lifestyle: long, slender limbs, and adhesive discs on all its toes. Its fingers and toes are also webbed.

adhesive
discs

BUILDING A FOAM NEST

During the breeding season, the male grey foam-nest frog – which can be distinguished from the female by the pads he develops on the first and second toes of his fore-limbs – calls out to females from branches overhanging a pool. He may be joined in this by one or more other males. When a female arrives, one of the males clasps her, they mate, and she produces a secretion that one or several males whip into a foam with their legs. The female then ejects her eggs – up to 1,200 at a time – into the foam, and one or more of the males shed sperm onto them. The exterior of the nest hardens to protect the eggs, but the inside remains moist. After the eggs hatch, the tadpoles burrow out of the nest and fall into the water below.

Microhyla ornata
Ornate narrow-mouthed toad

Length 2–3cm
(³/₄–1¹/₄ in)

Habit Mostly
terrestrial

Breeding Rainy season
season

Location S. and S.E. Asia   Status Locally common
(including Japan)

This small, Asian toad has a plump, smooth-skinned body with short forelimbs and long hindlimbs, and a small head. It is yellow or ochre, with a darker marbled pattern and dark stripes along its flanks. The ornate narrow-mouthed toad inhabits rain-forest and cultivated areas, notably rice paddies, and is also found in grass or leaf litter, preying on insects and other invertebrates. It remains buried underground during dry and cold weather. Breeding takes place in ponds, rice paddies, temporary pools, and slow-moving streams. The male – which is smaller than the female – uses its large vocal sacs to call very loudly from water to encourage females to mate. The female ornate narrow-mouthed toad lays approx-imately 270–1,200 eggs at a time in water. The eggs, which form a film on the surface, hatch quickly, and the tadpoles develop very rapidly, before the breeding pool dries up.

Breviceps adspersus
Bushveld rain frog

Length 3–6cm
(1¹/₄–2¹/₄ in)

Habit Terrestrial

Breeding Rainy season
season

Location S.E. Africa   Status Locally common

The bushveld rain frog lives mostly underground, emerging to feed and mate only after rain. It has a stout body – either light or dark brown, with rows of lighter, yellowish or orange patches with dark borders – and the short, stout limbs typical of most burrowing species. When cornered, this frog inflates its body and lodges itself firmly in its burrow. The females are much larger than the males. Since the male is too small to clasp the female during mating, the female emits a secretion from her back that acts as a glue, keeping the mating pair together.

globular body

flattened
face

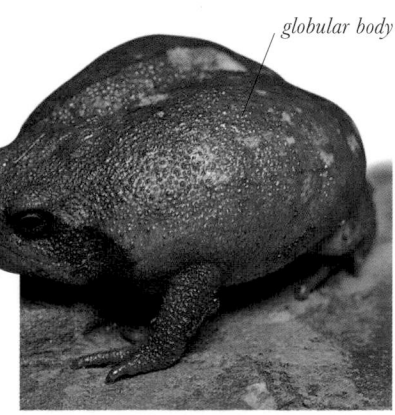

Phrynomantis bifasciatus
Banded rubber frog

Length 4–6cm
(1¹/₂–2¹/₄ in)

Habit Mostly
terrestrial

Breeding Rainy season
season

Location S.E. Africa   Status Locally common

Unusually for a frog, the short-legged banded rubber frog walks and runs rather than hops or jumps. It is black and glossy, with mainly pink or red spots and stripes down its back; these change colour throughout the day, turning paler in bright light. Its head is pointed, and it has small eyes with round pupils. This frog feeds on insects, mainly termites and ants. If threatened, it raises itself on outstretched limbs and inflates its body. It also produces a highly toxic substance from its skin, which may be lethal to humans if ingested. Males produce a trilling courtship call from the pond edge, and females lay clumps of up to 600 eggs, which they attach to aquatic plants.

Gastrophryne olivacea
Great Plains narrow-mouthed toad

Length 2–4cm
(³/₄–1¹/₂ in)

Habit Mostly
terrestrial

Breeding Spring and
season summer

Location C. and S. USA,   Status Locally common
N. Mexico

Also known as the Western narrow-mouthed toad, this species has a narrow, pointed head with small eyes, a stout body, and short legs that allow it to burrow and hide in small crevices. Brown or grey in colour, it is sometimes distinguished by the dark, leaf-shaped pattern on the surface of its back. In spring, after heavy rain, Great Plains narrow-mouthed toads are known to gather in large numbers around temporary pools. The males call out to the females, making a protracted buzzing sound. After mating, the females lay their eggs in the pool. The tadpoles develop rapidly, before the water dries up.

AMPHIBIANS IN DANGER

The list on this page includes all those amphibians currently classed as critically endangered by the IUCN (see p.31). Although the list is relatively short, many other species (not listed here) are endangered or vulnerable, and the declining populations of frogs and toads in particular are a major source of concern. Exactly why so many amphibians should be in simultaneous decline is not entirely clear. Some are undoubtedly suffering from habitat change, but the worldwide scope of the problem suggests that other factors such as climate change may also be at work. Amphibians are also threatened by introduced fish, and, in a limited number of cases, by other amphibians. This has happened where large, aggressive species, such as the marine toad or American bullfrog, have been accidentally or deliberately introduced into new habitats.

NEWTS & SALAMANDERS

Ambystoma lermaense (Lake Lerma salamander)
Batrachoseps aridus (desert slender salamander)
Eleutherodactylus karlschmidti (Karl's robber frog)
Euproctus platycephalus (Sardinian brook salamander)
Salamandra atra (1 subsp: golden salamander)

FROGS & TOADS

Alytes muletensis (Majorcan midwife toad)
Bufo periglenes (golden toad)
Holoaden bradei (Itatiala highland frog)

Litoria flavipunctata (yellow-spotted tree frog);
L. lorica (amoured frog); *L. nyakalensis* (Nyakala frog); *L. piperata* (peppered tree frog)
Paratelmatobius lutzii (Lutz's coastal frog)
Platymantis hazelae (Hazel's forest frog);
P. insulatu (island forest frog); *P. isarog* (Isarog forest frog); *P. levigatus* (smooth-skinned forest frog); *P. negrosensis* (Negros forest frog); *P. polillensis* (Polillo forest frog); *P. spelaeus* (Negros cave frog)
Rheobatrachus silus (southern gastric brooding frog)
Taudactylus acutirostris (sharp-nosed torrent frog); *T. diurnus* (Mt Glorius torrent frog); *T. rheophilus* (tinkling frog)
Thoropa lutzi (Lutz's river frog);
T. petropolitana (Petropolis river frog)

GOLDEN FROG
This inhabitant of Panama's tropical rainforests is typical of many amphibian species currently in decline. Some scientists believe that amphibians' thin skin may make them particularly susceptible to increased levels of ultraviolet radiation due to ozone decline, and perhaps to other physical or chemical changes in their environment.

AMPHIBIANS

FISH

FISH

PHYLUM	Chordata
CLASSES	Myxini, Cephalaspidomorphi, Chondrichthyes, Osteichthyes
ORDERS	62
FAMILIES	491
SPECIES	About 24,500

FISH WERE THE first backboned animals to appear on earth, and they form by far the largest group of vertebrates. Unlike the other major divisions of vertebrates, however, fish are not a natural group; instead, they are an informal collection of 4 classes that are only distantly related to one another. A typical fish breathes using gills, has a body covered with scales, manoeuvres using fins, and is ectothermic (cold-blooded). Most species live in either fresh water or the sea, but a few species move between both environments.

UNDERWATER LIFE
Most fish (such as these butterfly fish) spend their entire life in water, although a small number of species can survive short periods out of water.

EVOLUTION

Living fish are often divided into 3 groups: jawless fish (of which there are 2 classes) and cartilaginous and bony fish (each forming a single class). Each of these groups had a different ancestor and evolved separately from the others.

The first fish appeared more than 500 million years ago, having probably evolved from soft-bodied, filter-feeding invertebrates. Like living jawless fish, these early fish had a round, fleshy mouth without jaws (although they did have teeth).

The first fish (indeed, the first animals) with movable jaws arose about 440 million years ago. Known as acanthodians (or spiny sharks),

their jaws evolved from their front gill arches. Linked to the development of jaws was the appearance of teeth. Having jaws and teeth made it possible for acanthodians to feed on a wide range of food. They also had several pairs of spines along the lower sides of their body, from which the paired fins of fish evolved. Unlike the early jawless fish, their body was covered with scales.

Fish with a cartilage skeleton first appeared about 370 million years ago. These were the ancestors of living sharks, skates, and rays.

Some 50 million years before the appearance of cartilaginous fish, a group of fish with a bony internal skeleton had appeared. There are 2 main groups of bony fish: fleshy-finned fish have fins that are supported by lobes of flesh; in ray-finned fish, the fins are supported by lengths of bone. The fleshy-finned fish almost certainly gave rise to the first 4-limbed land vertebrates. The ray-finned fish appeared at about the same time but have since become a far larger and more successful group. They gradually developed features that

made them more efficient swimmers, including thinner scales, highly flexible fins, and a symmetrical tail. Also important were the evolution of more mobile jaws and gill chambers that could be more easily ventilated.

ANATOMY

Fish have several physical adaptations for life in water. The body of a fish is usually stream-lined, covered with smooth scales, and has fins to

provide power, steering, and stability. All fish have gills for extracting oxygen from water.

All fish have an internal skeleton, although this takes different forms in each of the 3 main groups of fish.

FIN TYPES
A fish's dorsal, anal, and tail fins occur singly along the midline of the body, while the pectoral and pelvic fins are paired. The common carp (shown here) has only one dorsal fin; however, some other fish have 2 or 3 dorsal fins.

CEPHALAPSIS

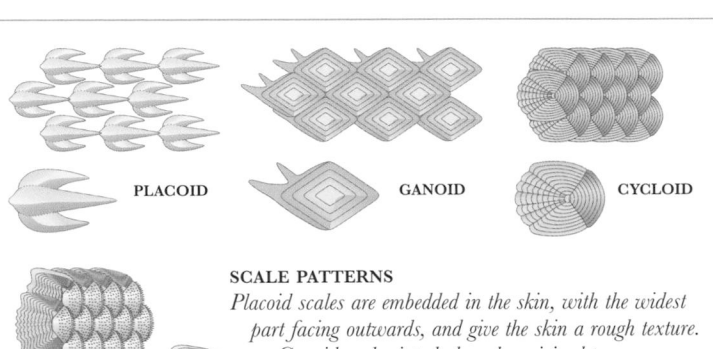

STICHOCENTRUS
FISH FOSSILS
Among the first fish to appear was a form called Cephalapsis (top). It had a jawless mouth, and its head and gills were protected by a large, bony shield. By contrast, a fossil of the bony fish Stichocentrus (above) reveals jaws, scales, and a much lighter and more flexible internal skeleton.

gill cover · dorsal fin · fin rays · left pectoral fin · left pelvic fin · anal (ventral) fin · tail (caudal) fin

SCALES

Fish have several types of scales, made of various materials. Cartilaginous fish have tooth-like placoid scales (dermal denticles). Some primitive bony fish have thick, relatively inflexible scales. They include diamond-shaped ganoid scales, found in gars, and layered cosmoid scales, seen in coelacanths. Like placoid scales, these are made of dentine and a substance similar to enamel. Bony fish have thin scales made of bone, with one end embedded in the skin and the other exposed: cycloid scales have a smooth exposed surface, while ctenoid scales are rough or spiny.

PLACOID · **GANOID** · **CYCLOID**

CTENOID

SCALE PATTERNS
Placoid scales are embedded in the skin, with the widest part facing outwards, and give the skin a rough texture. Ganoid scales interlock and are joined to one another by fibres. Cycloid and ctenoid scales are arranged so that the exposed surfaces overlap with one another, forming a smooth, flexible covering.

GILLS

A fish's gills are located in internal chambers on either side of the body, just behind the mouth. In most species, each gill consists of a framework of bone or cartilage, supporting tissues that contain densely packed capillaries. As water enters the gill chambers, it is first sieved clean by structures called gill rakers. Behind the rakers are supports called gill arches, which provide attachment points for gill filaments. To increase the efficiency of respiration, the filaments have folds on their surface, known as lamellae. Gases are exchanged across the surface of the lamellae, before being distributed to the rest of the fish's body. In some fish, the area of the lamellae is 10 times greater than that of the body's outer surface.

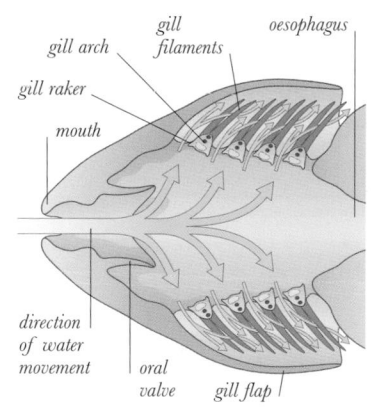

gill
filaments
gill arch   *oesophagus*
gill raker
mouth
direction of water movement
oral valve   *gill flap*

GAS EXCHANGE

Water enters a fish's gills through its mouth and leaves through the gill openings. As it passes through the gills, dissolved oxygen is transferred across the thin outer membrane of the capillaries and enters the bloodstream. At the same time, carbon dioxide is expelled from the capillaries.

eyes can be rolled back in sockets to keep them moist

muscular fin

BREATHING OUT OF WATER

By keeping their gill chambers moist, mudskippers can survive out of water for long periods. They live in coastal swamps, and at low tide can be found using their well-developed pectoral fins to move around on exposed mudflats.

In jawless fish, the body is supported by a simple rod called the notochord; some species also have rudimentary vertebrae. Sharks, skates, and rays have a skeleton of cartilage hardened by calcium carbonate, as well as fully developed vertebrae. Most fish have a skeleton made of bone. In addition to a bony skull and vertebral column, they also have a separate bony fin skeleton.

Although some fish have smooth skin, most are protected by scales, plates of bone, or spines. Scales, which are the most common covering, provide protection and promote the efficient movement of water over the body, while still allowing the fish to move freely (see panel, facing page). Glands in the skin secrete mucus, which protects the fish from bacteria and, in some species, helps to reduce drag.

Almost all fish have fins. There are 2 main types: median (or unpaired) and paired fins. Median fins are found singly, on the dorsal or ventral midline of the body. They include the dorsal, anal (ventral), and tail (caudal) fins. The paired fins – the pelvic and pectoral fins – are arranged in twos, one on each side of the body. Fins are used mainly for locomotion (see p.462) but they also have other uses: colours and patterns on their surface can be used as signals to warn predators, attract mates, defend territories, or lure prey; the fins of some fish have spines, which are sometimes poisonous, to protect them from predators.

All fish have gills, which they use to take oxygen into their bloodstream (see panel, above).

Some fish also have other ways of taking in oxygen. Lungfish, for example, breathe air using primitive, lung-like organs. Other fish can absorb oxygen and release carbon dioxide through their skin.

SENSES

Fish collect information about their environment using some sensory systems that are also found in other types of animals – these include vision, hearing, touch, taste, and smell. However, fish also have some unique sense organs.

Although most fish have eyes, their ability to perceive light, colour, shape, and distance varies greatly between species. This variation can often be related to habitat. Fish that live in clear water tend to have good vision, while many fish that live

TASTE

Catfish are named after the whisker-like barbels found in many species. These are covered by taste buds and are used to find food. This is especially useful for species such as the squeaker (shown here), which lives in dark waters at the bottom of deep lakes.

in dark conditions (such as muddy water, caves, or the deep sea) either have poor vision or have lost their eyes altogether. However, some deep-sea fish have large eyes so that they can collect as much light as possible.

Water is an effective medium for transmitting sound, and most fish are able to respond to sound waves. Sound vibrations are usually transmitted through the bones and tissues of the head to the inner ear, although in some species they may be amplified by the swim bladder.

In aquatic animals, the distinction between taste and smell is blurred because the same sense organs often respond to chemicals that are dissolved in water and contained in food. Fish smell using sensory pits (called nares), which are lined with tissues that contain olfactory receptor cells. Most fish have a good sense of smell, and some can detect chemicals in minute concentration. Taste receptors are usually found in or near the mouth, although some fish have them on their fins, skin, or whisker-like barbels, allowing them to taste food on or near the bottom.

Almost all fish have what is called a lateral-line system, which they use to detect vibrations (including those made by their predators and prey) and changes in water pressure and currents. These changes are detected by organs known as neuromasts, which are usually housed in canals in the bones of the head and in a canal that runs along the side of the body beneath the scales and is connected to the exterior via pores that run through the scales.

Many fish are also able to detect electrical waves and impulses. In cartilaginous fish, signals are

TUBULAR EYES

LARGE EYES

EYES

Fish that live in habitats with little light often have unusual eyes. Spookfish (top) are deep-water fish with tubular eyes that point upwards. The pinecone soldierfish (above) lives in shallow water but avoids direct sunlight. It feeds at night, using its large eyes to search for food.

received by structures called the ampullae of Lorenzini. These specialized nerve cells, located in small pores at the surface of the skin, contain a conductive gel and can detect the weak electrical currents produced by other animals. Some bony fish can also detect electrical signals. Certain species have organs that produce electrical fields, which they use to detect objects or other fish in low visibility, and to communicate.

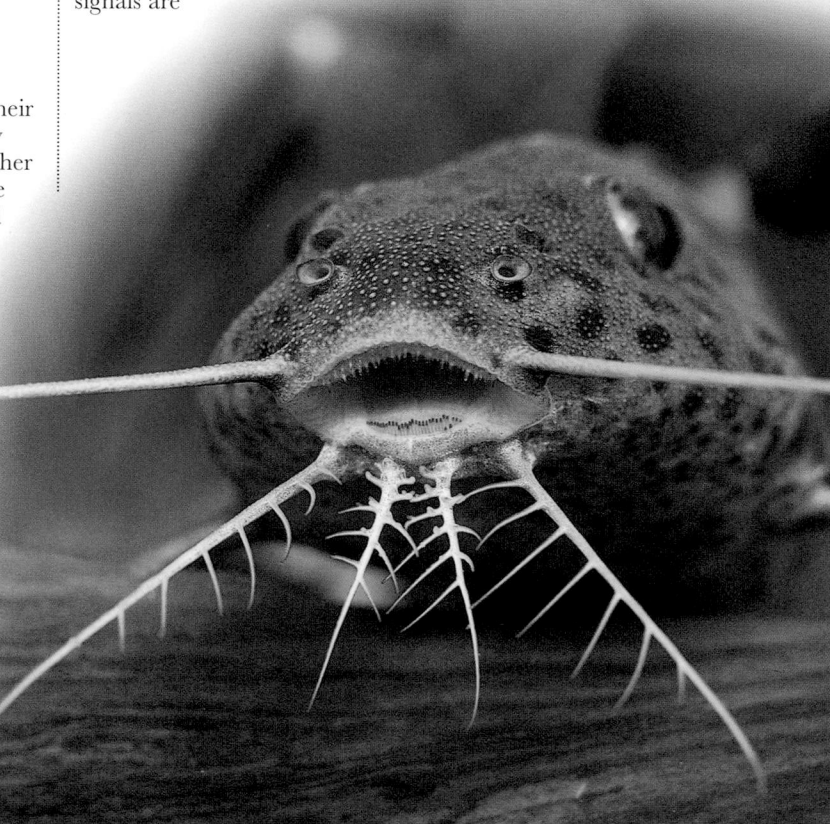

FISH

FISH

RAY SWIMMING
Most fish propel themselves with sideways movements of their body and tail. However, rays (including the thornback ray, shown here) use vertical movements of their large pectoral fins to generate forward movement.

SWIMMING AND BUOYANCY

A fish propels itself through water using the action of its muscles. Along either side of the backbone are bundles of muscle fibres (known as myotomes). By contracting these muscle bundles in sequence, one a fraction of a second after the other, the fish creates a wave motion that travels along its body from front to back and ultimately causes the tail to move from side to side. It is this wave motion that propels the fish through the water. The wave usually begins near the rear third or half of the body so that only the back end moves from side to side. In some fish (usually fast-moving ones), only the very end of the tail moves. The movement is more pronounced in fish that have a long, slender body. Some fish, including eels, can reverse the direction of the wave movement and so swim backwards.

Fins play an important part in propulsion and movement. The dorsal and anal fins act in a similar way to the keel of a boat, stabilizing the fish's body. The paired fins have various uses. Most fish use them as control surfaces, adjusting the angle of the fins to move up and down in the water. In sharks and some bony fish, the paired fins also provide lift as the fish moves forwards, while some fish use their paired fins to "walk" on the bottom and, on rare occasions, on land. The tail fin is used mainly for propulsion but it also has a role in steering.

In general, fish move relatively slowly, rarely exceeding 5 kph (3 mph). However, some fish are capable of moving at much greater speed – for example, the wahoo (a close relative of tunas) can reach 75 kph (50 mph) for short periods.

A fish's body shape usually reflects the way it swims. Fish that need to swim at high speed in open water for long periods typically have a torpedo-shaped (fusiform) body. A body that is tall and flattened from side to side is less efficient but is common among reef fish, which need to make sudden turns around dense vegetation or rocky surfaces.

Fish with a slender, cylindrical body, such as eels, can wriggle with ease into crevices to find food or escape from predators. Bottom-dwelling fish typically have a body that is compressed from top to bottom, helping them to remain inconspicuous on the sea floor.

Many fish are neutrally buoyant. That is, their density is similar to that of the surrounding water, so they can remain suspended at a constant depth in the water column. Most other fish are slightly negatively buoyant – their fins give them lift as they move, but they will sink if motionless. For bottom-dwelling fish, such as skates and rays, being negatively buoyant is an advantage. As they

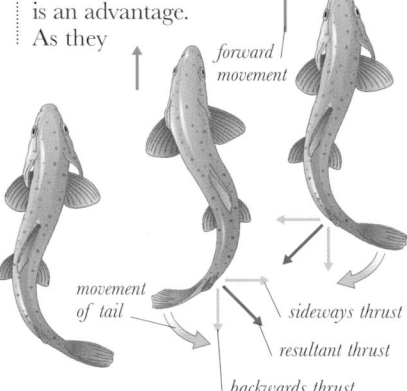

forward movement

movement of tail

sideways thrust

resultant thrust

backwards thrust

PROPULSION
As a fish's tail moves, it exerts a sideways thrust and a backwards thrust on the water. The resultant force acts at an angle halfway between these two. Movement of the tail from side to side means that the resultant thrusts to left and right produce a net backwards thrust. Because every force in nature produces an equal and opposite reaction, this backwards thrust propels the fish forwards.

SURVIVING IN FRESH AND SALT WATER
The European eel can survive in both fresh and salt water. It spends most of its adult life in rivers but returns to sea to breed. The opposite is true of many salmon and trout, which move from the sea to fresh water to breed.

move up and down in the water column, most fish can adjust their buoyancy using a gas-filled organ known as the swim bladder. To make themselves more or less buoyant, gas is added to or extracted from this bladder (see p.479). Cartilaginous fish do not have a swim bladder but they do have a large, oil-rich liver, which is less dense than water and hence increases their buoyancy.

MARINE AND FRESHWATER FISH

Fish need mechanisms to keep the concentration of water and salts in their body in a stable state. Those fish that move between salt and fresh water, including salmon, lampreys, and sharks and rays, have the same concentration of water and salts in their body as the water around them. However, most fish have a different internal salt balance to the surrounding water. This difference presents a potential problem, because osmosis, which occurs across tissues such as those in the mouth and gill chambers, could result in a harmful, possibly fatal, change in their internal chemistry.

In marine fish, the concentration of salts is lower inside the body than it is in sea water, so the fish face a tendency to lose water, raising their internal salt concentration. To counteract this, most marine fish drink large quantities of sea water and excrete most of the salt but only a little of the water. Some bony fish have cells called chloride cells in their gills that remove excess salt from the body. By contrast, freshwater fish are at risk of losing salt to the environment. They drink only a small amount and obtain extra salt from their food. They also have cells in their mouth and gills that move salts against the normal direction of diffusion. Those fish that have a salt balance the same as the water around them do not need to drink at all.

Fish that move between salt and fresh water can transfer salts both in and out of their body. When moving from one environment to the other, they often spend a short period of adjustment in an area of brackish water.

TEMPERATURE CONTROL

Fish are ectothermic (cold-blooded) animals, which means that their internal temperature is the same as that of the surrounding water. Some species modify their behaviour to make themselves warmer or cooler than the ambient water. They may bask in the sun, seek shade, or rise and fall between warmer and colder depths. Other species regulate their temperature by altering their colour, using pigments that change between dark, heat-absorbing colours and light, heat-reflecting shades.

A few fish – including tuna and the great white shark and its relatives – have some similarities with warm-blooded animals. Some of the heat generated by their large swimming muscles is conserved, so that their internal temperature is higher than that of the water around them. This makes it possible for these fish to remain active in cold water.

SURVIVING COLD
Sea water freezes at a lower temperature than fish blood. However, the blood of some fish, such as this icefish, remains liquid even in extremely cold water because it contains proteins that act like anti-freeze.

LAYING EGGS
The perch, like many fish, releases large numbers of eggs to ensure that at least some of them will eventually produce adult fish. This is known as broadcast spawning.

LARVA

ADULT
METAMORPHOSIS
When a fish emerges from its egg, its yolk sac is still attached to its body and provides nourishment during the first weeks of its life. The larva of a brown trout (top) takes 3–4 years to mature to the adult form (above).

REPRODUCTION

The reproductive behaviour of fish is varied. Although in most cases fertilization takes place outside the female's body and the young emerge from the eggs as larvae, there is also a significant number of species in which fertilization is internal and females give birth to live young.

Some fish breed regularly, often once a year. However, others reproduce only once in their lifetime, and may die shortly afterwards. The timing of reproduction is controlled by external factors (including changes in temperature, light levels, or day length) or by internal cycles (such as changes in hormone levels).

Some fish have preferred spawning grounds, laying eggs or releasing their young in places where they are most likely to survive. In order to reach these areas, they may have to undertake migrations of thousands of kilometres.

In some species, large numbers of males and females gather in shoals to breed, without any kind of courtship behaviour. However, some fish conduct complex courtship rituals to increase their chances of attracting a suitable mate. As a prelude to breeding, the male may change colour. For example, the male dragonet develops more intense colours, which are displayed to potential mates during elaborate circling movements.

In most fish (as in many other aquatic animals), fertilization occurs externally. The male releases sperm (milt) over the eggs (roe) as they are released from the female's body. Milt has a thick consistency, which prevents the sperm cells from dispersing too quickly. The eggs of most marine fish are made buoyant by droplets of oil and float freely as part of the plankton. The eggs of freshwater fish are typically heavier and have a sticky surface so that they adhere to objects in the water.

Most freshwater fish make a nest for their eggs, and in some species they guard the eggs (see panel, right). In almost all cases, the eggs hatch to release larvae, which are incompletely formed and gradually develop their skeleton, fins, and some of their organ systems.

External fertilization is a viable way for fish (and other aquatic animals) to reproduce partly because the greater density of water makes it a much more suitable medium than air for transferring eggs and sperm. Water also provides the developing eggs with nutrients and dissolved oxygen. However, the chances of an embryo surviving to reach adulthood are relatively low, particularly for marine fish. To compensate for this, females often produce large numbers of eggs (as many as 5 million in some species).

In a small number of species, reproduction begins with internal fertilization. In sharks, for example, the male's pelvic fins are modified into a copulatory organ, which is inserted into the female's cloaca. Sea water flushes the male's sperm along a groove in this organ and into the female's body.

Where fertilization is internal, the young may be alive when they are released from the female's body. In some species that give birth to live young, the eggs simply hatch inside the mother's body. In others, including some species of sharks and rays, there is a connection between the developing embryo and the female, along which nutrients are passed to the embryo.

Live-bearing females invest a relatively large amount of energy in carrying and nourishing their developing young. However, compared with fish that reproduce by external fertilization, their young are more developed at birth and have a greater chance of survival. This means that live-bearing fish can produce and fertilize fewer eggs and yet maintain a stable population.

Compared with other groups of vertebrates, fish include a relatively large number of hermaphrodites, all of which are bony fish (see p.479). Some fish, including certain species of carp and loach, reproduce parthenogenetically, the eggs being fertilized without a male. Other species reproduce by a process called gynogenesis. These include the Amazon molly, which is a hybrid of 2 other species and has a population that consists only of females. Reproduction occurs when a female is fertilized by a male of either of the 2 parent species.

eye of embryo

EGGS
Inside its egg, an embryonic fish is nourished by fluids contained in a yolk sac. The egg case is transparent (allowing the eyes of these salmon embryos to be seen from the outside). The time taken for the eggs to hatch often depends on the water temperature.

PARENTAL CARE

Having released their eggs and sperm into water, many fish have no further contact with their developing young. However, other species actively care for their eggs or young. Besides protecting their eggs by placing them in a nest, adult fish may care for them by chasing away predators, cleaning them to prevent infection, or fanning them to provide oxygenated water. When the eggs hatch, the parents may use various ways to protect the vulnerable young (see below). A small number of species also feed their young, including the discus fish, which secretes a nutritious fluid through its skin.

adult male

juvenile

SEAHORSES
Male seahorses (right) play an unusual role in caring for their young. The female places her eggs in a pouch in the front of the male's abdomen, where they are fertilized. The young are released when they hatch, 2–6 weeks later.

MOUTHBROODING
Several hundred species of cichlids brood their eggs in cavities in their mouth and throat. When the eggs hatch, the parents also defend the young from predators by allowing them to shelter inside their mouth.

JAWLESS FISH

PHYLUM	**Chordata**
CLASSES	**Myxini, Cephalaspidomorphi**
ORDERS	**2**
FAMILIES	**4**
SPECIES	**About 90**

JAWLESS FISH FIRST APPEARED more than 500 million years ago, before any other group of living fish. Although they were once diverse, they are now represented by 2 relatively small groups: the lampreys and hagfish. Both groups have an elongated body, smooth, scaleless skin, and a jawless mouth, which they use to hold onto and rasp away at their food. Lampreys live in temperate areas, in coastal marine or fresh water as adults, and breed in fresh water. Hagfish are entirely marine, inhabiting deep waters in the tropics, and shallower seas in cool temperate areas.

PARASITISM
Adult sea lampreys use their sucker-like mouthparts to attach themselves to other fish. Once attached, they are difficult to dislodge, and the host often dies as a result of loss of blood or damage to its tissues.

ANATOMY

Jawless fish differ from all other groups of living fish in lacking true, movable jaws; instead, they have an open oral area armed with sharp, tooth-like projections used in feeding. While they lack a bony internal skeleton, their eel-like bodies are reinforced by a tube of cartilage (notochord). Their gill openings consist of one or more pore-like openings near the head. Like other fish, their tail ends in a flattened, fin-like paddle, and they also have median fins for stabilization, but lack paired fins. The scaleless skin is coated with mucus. Hagfish can secrete copious amounts of slime to deter predators. Lampreys have moderately good vision, although hagfish are essentially blind; however, both groups have acute senses of smell and taste.

BODY SECTION
The bodies of lampreys (shown here) and hagfish are supported by a notochord. Both have some rudimentary vertebrae, although these are better developed in lampreys.

dorsal fin   spinal cord

notochord   gill openings   round, fleshy mouth

REPRODUCTION

Most lampreys spend their adult lives in marine waters, migrating to fresh waters to breed, but some species spend their entire adult life in fresh water. Spawning pairs excavate shallow nests in gravel beds, lay their eggs, and then die. Young lampreys go through several larval stages before becoming adults and heading out to sea. Hagfish live and breed only in marine waters. Having no larval stages, young hagfish emerge from egg capsules as miniature versions of the adults.

HAGFISH EGGS
Hagfish eggs are protected by a tough case and are attached to one another or anchored to the sea floor by small filaments on the outer surface of the case. When the larvae are ready to hatch, the egg case splits along a line of weakness.

Myxine glutinosa
Hagfish

Length	Up to 40 cm (16 in)
Weight	Up to 750 g (27 oz)
Sex	Male/Female

Location North Atlantic, Mediterranean

Status Locally common

Also known as the slime eel, this marine, bottom-dwelling fish produces large quantities of glutinous slime from pores running along both flanks. It feeds mainly on dead and dying fish, and uses its slime chiefly for

fin near tail

eel-like body

defence. This fish can knot its body to free itself from slime, to escape capture, and to feed. When feeding, the knot that is formed near the tail works its way up towards the head and forces the fish's mouth away from the prey so that some flesh is torn away with it. Like all jawless fish, the hagfish lacks a bony skeleton, but has a spine formed by nodules of cartilage.

Lampetra fluviatilis
Brook lamprey

Length	18–49 cm (7–19 1/2 in)
Weight	30–150 g (1 1/16 – 5 oz)
Sex	Male/Female

Location North Atlantic, N.W. Mediterranean, Europe

Status Lower risk

The brook lamprey, or European river lamprey, spends around a third of its life, between 4 and 7 years, at sea.

Unlike other lampreys, which are blood-sucking parasites, adult brook lampreys bite their prey, which are fish – especially herrings and sprat – and even dead remains. The adult's teeth are sharp, except at sexual maturation, when it ceases to feed. During the larval stage – lasting up to 6 years – this fish is toothless and blind and feeds on algae and particles of organic matter which are filtered from the water sucked in through the mouth. Females are longer and heavier than males, but become smaller while spawning. Adults breed just once, in the mid-upper reaches of rivers, and then die.

Petromyzon marinus
Sea lamprey

Length	Up to 1.2 m (4 ft)
Weight	Up to 2.5 kg (5 1/2 lb)
Sex	Male/Female

Location North Atlantic, Mediterranean, North America, Europe

Status Lower risk†

The adult sea lamprey is a parasite that attacks a wide range of salt- and freshwater fish, including salmon, trout, herring, mackerel, and some sharks. Clinging to prey by its teeth, it sucks blood using its rough-surfaced tongue. It migrates from the sea to breed in fresh water, but is landlocked in the North American Great Lakes. The opening of the Welland Ship Canal in 1929 saw the proliferation of lampreys throughout the Great Lakes,

and these had a devastating effect on populations of lake trout and other fish. Since then the sea lamprey population has been controlled.

EEL-LIKE BODY
Both the adult and the larvae of the sea lamprey have eel-like bodies, which are rounded near the head but flattened at the tail.

SUCKERS

Sea lampreys have oral discs called "suckers" in place of their mouths. The sucker, which is wider than the fish's body, has a fringed edge and contains many small teeth arranged in concentric rows, with larger teeth surrounding the mouth opening.

CARTILAGINOUS FISH

PHYLUM	Chordata
CLASS	Chondrichthyes
ORDERS	14
FAMILIES	50
ORDERS	About 810

ALTHOUGH OFTEN DESCRIBED as primitive, cartilaginous fish include some of the largest and most success-ful of all marine predators. They are divided into 3 groups: sharks, skates and rays, and a group of deep-sea fish called chimaeras. All possess a skeleton made of cartilage (rather than bone), specialized teeth that are replaced throughout their lifetime, and skin covered with small, dense, tooth-like scales. Most cartilaginous fish are marine, but some sharks and rays enter fresh water and certain tropical species live exclusively in fresh water.

FEEDING

All cartilaginous fish are carnivorous; however, plankton feeders also take in microscopic plants. The greater part of the diet of most species consists of live prey, including other fish, invertebrates, and occasionally marine mammals. Most species also feed on the remains of dead animals, although only a few species rely solely on carrion. Several of the largest species of sharks and rays are filter feeders. The teeth of these species have been reduced in size, and stiff projections in the gill arches have developed into sieve-like organs for straining small animals out of water.

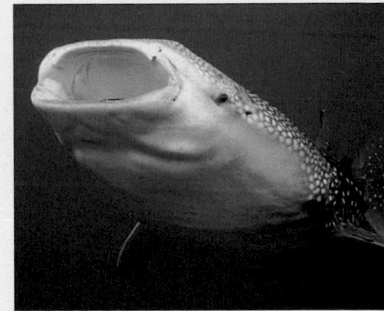

FILTER FEEDING
Whale sharks drift slowly through water, catching plankton, small fish, and squid. They live in warm tropical waters, where the food supply is large enough to sustain them.

ANATOMY

All the fish in this group have an internal skeleton made of tough, flexible cartilage. This may be strengthened by mineral deposits, and some species have hard, bone-like dorsal spines. However, the skeleton of a cartilaginous fish is much more flexible than that of a bony fish (see p.478), which has a much higher mineral content. On land, a cartilage skeleton would not be rigid enough to support the weight of a large animal, but in water – which has a much higher density than air – cartilage provides an effective skeleton for animals that are up to 10m (33ft) long. Most cartilaginous fish have skin consisting of thousands of interlocking scales, called placoid scales or dermal denticles (see p.460). These have a similar composition to teeth and give the skin a sandpaper-like texture. Some rays have large, thorn-like scales but other rays and all chimaeras lack these entirely. Unlike bony fish, cartilaginous fish do not have a gas-filled swim bladder.

CUTTING
serrated edge

GRIPPING
sharp point
accessory cusp
teeth arranged in rows

GRINDING
jawbone
flat, interlocking teeth

TEETH
Cartilaginous fish have teeth that are shaped to suit their diet. Serrated teeth are used for cutting, while pointed teeth are used to hold prey. Many rays have flat teeth for grinding food. As teeth are lost, they are replaced by new ones growing behind them.

GILL OPENINGS
Sharks and rays have 5–7 pairs of gill openings, while chimaeras have only one. When water enters the mouth, the gill slits are closed. As the water passes out of the open gill slits, the mouth is held closed.

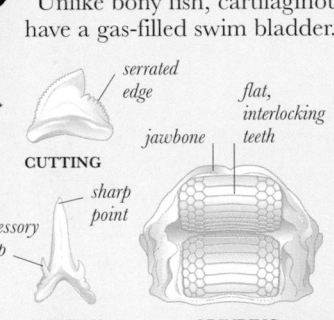

nostril
mouth
paired gills covered by flaps of skin

SWIMMING AND BUOYANCY
Cartilaginous fish, such as the hammer-head shark, have an oil-rich liver that adds to their buoyancy. However, most are still negatively buoyant and must keep swimming to avoid sinking.

SENSES

Cartilaginous fish have acute senses that can be used to locate prey, even if it is a long way off or buried by sediment. All species have a system of pores, called the ampullae of Lorenzini, which they use to detect the weak electrical signals emitted by other animals. Most also have an effective lateral-line system that responds to very small vibrations, as well as acute vision and a sense of smell that can detect odours in even the weakest solutions.

ELECTRICAL SIGNALS
The ampullae of Lorenzini (seen here on the snout of an oceanic whitetip shark) are deep pores connected to electro-receptive nerve endings.

REPRODUCTION

In all cartilaginous fish, fertilization occurs internally. The male passes sperm into the female's cloaca via a modified pelvic fin. The young are produced by one of 3 processes. The females of some species release leathery egg cases. In other species, the young hatch from the egg inside the female's body and are then released alive. In the remaining species, the young develop inside a placenta-like structure, from which they have a direct nutritional connection with the female. In all cases, there is no larval stage; instead, the young are born as miniature versions of the adults.

LIVE BIRTH
Young lemon sharks hatch inside the female's body before being released tail-first.

EGG CASE
This egg case contains the embryo of a small-spotted catshark. It is anchored to seaweed by tendrils on the outside.

FISH

SHARKS

PHYLUM	Chordata
CLASS	Chondrichthyes
SUBCLASS	Elasmobranchii (part)
ORDERS	9
FAMILIES	33
SPECIES	About 330

SHARKS ARE FORMIDABLE hunters with few natural predators when fully grown. Most have a streamlined body, powerful jaws with several rows of sharp teeth, and acute senses, especially smell. They range in size from the whale shark (the largest of all fish) to squaloid sharks less than 30cm (12in) long. Sharks are most abundant in warm temperate and tropical waters, their habitats ranging from coastal areas to open oceans and deep basins. Some even venture into tropical rivers.

ANATOMY

Most sharks have a torpedo-shaped body that allows them to cut easily through water, although several bottom-dwelling species have a flattened body, similar to that of rays. Unlike bony fish, sharks have rigid fins that cannot be flexed or folded, yet their fins are nonetheless efficient for steering, propelling, and stabilizing the body during swimming. A shark's teeth, which are arranged in rows, are replaced throughout its lifetime. The shape of the teeth varies according to diet (see p.465), although some species have 2 or more types of tooth.

Sharks do not have a swim bladder, but their large, oil-filled liver adds to their buoyancy and helps to prevent them sinking. Many open-ocean sharks need to swim constantly to force water over their gills, but other species can pump water across their gills while motionless. Sharks usually have 5 gill slits on either side of their head; a few species have 6 or 7 pairs.

FOOD AND FEEDING

Sharks use various tactics to catch prey. Predatory sharks rely on short, high-speed pursuits through open water, or on ambush and surprise. Other species simply search an area for sedentary or slow-moving prey, or excavate burrowing animals. Some sharks are scavengers, using an acute sense of smell to locate carcasses (they can detect a scent in a solution as weak as one part per million). The 3 largest species of sharks are filter feeders, and at least one species, the cookie-cutter shark, is an ectoparasite on other sharks, large fish, and marine mammals.

REPRODUCTION

Coastal bays, reefs, and atolls are the favoured areas for most female sharks to lay eggs or give birth (pupping). These areas provide a calm, food-rich environment for young sharks to grow and develop. Baby sharks are

CONSERVATION

Over the last few decades, rising human populations and falling stocks of many food fish have led to an increase in the popularity of sharks as a food source for humans. Compounding this pressure are the growth in the popularity of shark fishing as a sport and a rise in demand for shark fins in Asian cuisine. About 100 million sharks are caught annually, of which 6.2–6.5 million are blue sharks, killed only for their fins. Conservation measures include regulation of catches, protection of threatened species, and protection of pupping grounds.

born or hatch as miniature adults, even armed with fully functional teeth, and are immediately ready to hunt. Sharks grow slowly to adulthood, often taking a decade or longer to reach maturity. During this time, many young sharks fall prey to larger predators.

HUNTING
Sharks of the open oceans (such as the blue shark, shown here) typically have a streamlined body and a powerful tail fin that propels them quickly through the water. The mainstays of their diet are squid and fish. When hunting, they often circle their prey several times before attacking from below.

Chlamydoselachus anguineus
Frill shark

Length Up to 2 m (6 1/2 ft)	
Weight Not recorded	
Breeding Ovoviviparous	
Location E. Atlantic, S.W. Indian Ocean, W. and E. Pacific	**Status** Unconfirmed

This unusual, deep-water species looks more like an eel than a shark, with frilled gill slits, a single small dorsal fin, and a much larger anal fin. It feeds on soft-bodied animals such as squid as well as fish. Similar to some long-extinct species, it has no close living relatives.

needle-like teeth

Echinorhinus cookei
Bramble shark

Length Up to 4 m (13 ft)	
Weight Over 220 kg (490 lb)	
Breeding Ovoviviparous	
Location W., C., and E. Pacific	**Status** Unconfirmed

The most interesting feature of this shark is the thorny denticles that cover its body. These have star-shaped bases and sharp tips, resembling bramble thorns. Large and slow-moving, the bramble shark spends most of its life foraging on or near the sea bed. It feeds on other sharks, ratfish, bony fish, cephalopods, and the egg cases of sharks, skates, and rays. It may use its large pharynx and small mouth to create a vacuum effect to suck in prey.

dorsal fins near tail

Oxynotus centrina
Angular roughshark

Length Up to 1.5 m (5 ft)	
Weight Not recorded	
Breeding Ovoviviparous	
Location E. Atlantic, Mediterranean	**Status** Unconfirmed

has a compressed, triangular trunk and furrowed "lips" with lance-like teeth in the upper jaw. There is a large spiracle behind the eye. The expanded body cavity and large, oily liver enable it to maintain buoyancy, allowing it to hover above the ocean floor in search of polychaete worms and other invertebrates. It is one of 5 species in the roughshark family.

This shark has unusually high, sail-like dorsal fins with spiny tips. It also

sail-like dorsal fins

Squaliosus laticaudus
Pygmy shark

Length Up to 25 cm (10 in)	
Weight Not recorded	
Breeding Ovoviviparous	
Location North Atlantic, W. South Atlantic, W. Indian Ocean, W. Pacific	**Status** Unconfirmed

One of the world's smallest sharks, this species is unique in having a single dorsal spine. It has well-developed light-producing organs (photophores) on its belly and sides, but hardly any on the back, giving this shark a luminescent underside. The arrangement of these photophores is thought to eliminate shadows, making it harder for predators (such as swordfish) to see it from below. The first known pygmy sharks were caught in 1908 off Japan, but no further specimens were seen until the 1960s.

light fin margins

Hexanchus griseus
Sixgill shark

Length 4.8 m (16 ft) or more	
Weight 600 kg (1,320 lb) or more	
Breeding Viviparous	
Location Tropical and temperate waters worldwide	**Status** Vulnerable†

This reclusive, primarily deep-water shark has a broad head and cylindrical body. It has 6 gill slits – unlike most other sharks, which have 5 – and sharp, comb-like teeth. The sixgill feeds on rays, squid, bony fish, and even seals. It also forages opportunistically for slow-moving animals on the sea bed by positioning its body, head down, at an angle of 45–60 degrees to the bottom. With its open mouth directly above the prey, it probably sucks in its food.

single dorsal fin near tail

6 gill slits

powerful tail

Squalus acanthias
Spiny dogfish

Length 1 m (3 1/4 ft), max 1.6 m (5 1/4 ft)	
Weight Over 9 kg (20 lb)	
Breeding Ovoviviparous	
Location Worldwide outside the tropics	**Status** Lower risk

Also known as the spurdog, this small, bottom-dwelling shark lives in coastal waters, usually where the temperature is at or below 15°C (59°F). A slow-moving fish, it gets its name from the spines on the front of its dorsal fins, which can inflict painful wounds. It eats crustaceans, sea anemones, and other fish, and often forms large, single-sex shoals. Spiny dogfish give birth after a gestation period of nearly 2 years – a long time for any shark.

Etmopterus spinax
Lanternshark

Length 45 cm (18 in), max 60 cm (23 1/2 in)	
Weight Not recorded	
Breeding Ovoviviparous	
Location E. Atlantic	**Status** Locally common

Sometimes called the velvet belly, this small, shoal-forming shark is one of about 18 similar species. It feeds on invertebrates and small fish. Tiny luminescent organs on its skin produce light, which help to attract potential prey or other lanternsharks, frighten predators, and disrupt shadows, making it less conspicuous.

FIRST MEAL

Like many sharks, young lanternsharks are born with an attached yolk sac. This provides them with food during their first few days at sea.

SPINED FINS
The lanternshark has a sharp spine at the front of each dorsal fin; its second dorsal fin and fin spine are larger than the first.

spine

Pristiophorus japonicus
Sawshark

	Length Up to 1.5 m (5 ft)
	Weight Not recorded
	Breeding Ovoviviparous
Location W. North Pacific	Status Vulnerable†

The sawshark's body is flattened from top to bottom and has 2 dorsal fins. Unlike rays, it has gills on the side of its head. It has a pair of barbels on a distinctive, saw-shaped snout. Taste sensors on the barbels and snout are used to probe the sea bed for small fish or invertebrates. Its exceedingly sharp teeth are presumably used for destroying prey as well as for defence. Young sawsharks are born with large teeth, but these remain folded while the young are inside the mother to avoid injuring her. The sawshark is not to be confused with sawfish (see p.476).

Heterodontus portusjacksoni
Port Jackson shark

	Length Up to 1.7 m (5 1/2 ft)
	Weight Not recorded
	Breeding Oviparous
Location Australia (including Tasmania), New Zealand	Status Locally common

Heterodontus francisci
Horn shark

	Length 97 cm (38 in), max 1.2 m (4 ft)
	Weight Over 10 kg (22 lb)
	Breeding Oviparous
Location E. North Pacific	Status Locally common

This slow-moving, generally solitary, nocturnal shark is found near rocky or sandy bottoms and kelp beds. It can "walk" across the sea bed using its muscular, flexible, paired fins. During the day it lies motionless among rocks and in caves, often resting with its head in rock crevices. This shark has a pig-like snout and a small mouth. Its enlarged, flattened teeth, located at the back of the mouth, are used to crush

Like the horn shark (see below), this shark belongs to a family of 8 species that have bulky heads, powerful jaws, and crushing teeth. Its mouth points downwards, enabling it to feed on starfish, sea urchins, and other sea-bed animals. The Port Jackson shark lives in inshore waters, and lays large eggs with spiral cases.

TAPERED BODY
This shark's body becomes progressively narrower from the head to the tail. It has spines on both dorsal fins, and the dark stripes on its back and sides form a harness-like pattern.

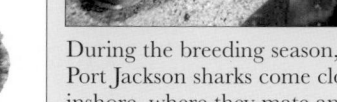

tapering body

prey such as sea urchins or crabs and, sometimes, bony fish. Horn sharks mate in December or January. The female lays about 30 eggs which she places under rocks or in crevices. The egg cases have a highly sculptured, spiral shape, which helps to keep them firmly lodged.

ridges above eyes   *black spots on body*   *rough skin*

INSHORE MATING

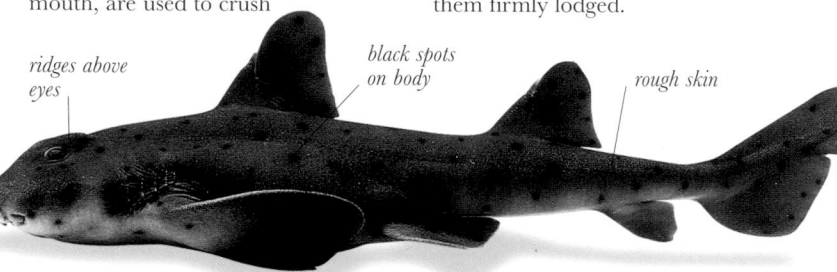

During the breeding season, Port Jackson sharks come close inshore, where they mate and lay eggs. Females sometimes wedge the eggs in rocky crevices.

Orectolobus maculatus
Spotted wobbegong

	Length 1.8 m (6 ft), max 3.2 m (10 ft)
	Weight Not recorded
	Breeding Viviparous
Location Japan, South China Sea, Australia	Status Locally common

A large, slow-moving shark with a flattened body, the spotted wobbegong lives in shallow water close to the shore. Instead of actively hunting for food, it rests on the bottom, and ambushes animals that come within striking range. When the tide falls, it may haul itself from one rock pool to another, partly exposing itself to the air. Although generally unaggressive, spotted wobbegongs can inflict serious injuries if accidentally stepped on.

dark saddle

PERFECT DISGUISE

The spotted wobbegong's mottled colouring and flattened, textured body provide perfect camouflage on the sea bed. Lured by the weed-like flaps of skin around its snout, lobsters, crabs, and octopuses may swim right up to this shark's waiting mouth.

SPOTTED AND PATTERNED
The spotted wobbegong is elaborately patterned, with light, O-shaped spots and dark saddles on an olive background.

Chiloscyllium plagiosum
Whitespotted bambooshark

	Length Up to 95 cm (37 in)
	Weight Not recorded
	Breeding Oviparous
Location Indo-Pacific	Status Locally common

Its small size and attractive appearance have made this shark particularly popular with marine aquarists. It has a slender body, with white spots over a dark background and darker transverse bands. Short barbels on its mouth help it to forage for food; the mouth is located well in front of the eyes. The whitespotted bambooshark uses its thickened pectoral and pelvic fins to clamber onto rocks.

thickened fins   *white spots*

Hemiscyllium ocellatum
Epaulette catshark

	Length Up to 1 m (3 1/4 ft)
	Weight Not recorded
	Breeding Oviparous
Location New Guinea, N. Australia	Status Locally common

The epaulette catshark, also known as the blind shark, is easily recognized by the large black spot behind each pectoral fin; there are smaller black spots all over the body. Found near coral reefs in shallow water or tidepools, it uses its thick, muscular, pelvic and pectoral fins to crawl across the ocean floor. It forages on the sea bed, trapping prey such as crabs, shrimps, worms, and fish in rocky crevices, and relies on stealth and camouflage for defence. Epaulette catsharks mate between July and September, and the female produces eggs in pairs.

black spot

Stegostoma fasciatum
Zebra shark

Length 2.8 m (9 1/4 ft), max 3.5 m (11 ft)

Weight Over 30 kg (66 lb)

Breeding Oviparous

Location Indo-Pacific   Status Locally common

Beige with brown spots, the zebra shark has an unusually flexible body that allows it to hunt out shrimps, crabs, and small bony fish from tight crevices in the coral reefs where it lives. Its mouth, located just behind the snout, points downwards, enabling it to feed on molluscs which live on the sea bed. During inactive periods, this shark rests on the sea floor, propped up on its erect pectoral fins, facing the prevailing current. It is the only species in its family.

Ginglymostoma cirratum
Nurse shark

Length 3 m (9 3/4 ft), max 4.3 m (14 ft)

Weight Over 150 kg (330 lb)

Breeding Viviparous

Location E. Pacific, W. and E. Atlantic   Status Locally common

The nurse shark has very tough skin, and a pair of barbels below the mouth for sensing the invertebrates on which it feeds. The small mouth and large pharynx form a powerful suction mechanism, which enables this shark to draw in prey at high

speed. During the day, the nurse shark spends its time resting on the sea bed, in rocky crevices, or in caves, where it may be found in groups of several dozen individuals, sometimes lying on top of one another. An inhabitant of shallow, inshore areas, it can "walk" along the sea bed using its pectoral fins as limbs. The nurse shark is not generally a threat to people but, if provoked, will hold onto its victim with a bulldog-like tenacity.

Rhincodon typus
Whale shark

Length 12 m (39 ft), max 14 m (46 ft)

Weight Over 12 tonnes (11 3/4 tons)

Breeding Viviparous

Location Tropical and temperate waters worldwide   Status Vulnerable

The whale shark is by far the largest fish in the world. However, despite its fearsome appearance, it is quite harmless to humans, and lives almost entirely on plankton, filtering its food with its gills. Its mouth is at the end of its snout – an unusual position for a shark – and although its jaws can be over 1m (3 1/4 ft) across, they are armed with the tiniest of teeth. Whale sharks usually feed by cruising slowly near the surface, where they show little alarm at being approached. They have also been reported to feed in a vertical position, using their mouths

like giant buckets to trap food. They sometimes gather in large numbers where plankton is abundant, but they otherwise lead solitary lives. Little is known about their breeding behaviour, or about their movements across the world's tropical seas.

GENTLE GIANT
The largest of all fish, the whale shark is a slow-moving plankton-feeder. Its blue-green skin is mottled with pale patches, and it has conspicuous ridges along its body.

SUCTION FEEDING

In order to feed, the whale shark creates a suction effect by lifting its head above the surface, then sinking quickly to draw water into its open mouth. The water that pours in is filtered for plankton, fish, and squid by screens on the shark's internal gill slits. This method allows for a large intake of prey.

FISH

Mitsukurina owstoni
Goblin shark

Length 3.3 m (11 ft)

Weight Over 160 kg (350 lb)

Breeding Ovoviviparous

Location North Atlantic, E. South Atlantic, S. Indian Ocean, W. and E. Pacific   Status Unconfirmed

Formally described in 1898, this deep-water shark is white or greyish, with a pointed snout. It has small eyes, and probably detects its prey by sensing electrical fields. It has no close living relatives, and is often described as a living fossil, having changed relatively little for millions of years.

Carcharias taurus
Sandtiger shark

Length Up to 4.3 m (14 ft)

Weight Over 150 kg (330 lb)

Breeding Viviparous

Location W. and E. Atlantic, S. Indian Ocean, W. Pacific   Status Vulnerable

Light brown or beige with darker brown blotches distributed randomly over its body and fins, the sandtiger shark has a distinctive, "snaggle-toothed" appearance. It is a large, cumbersome-looking fish, with a large mouth, and dorsal fins situated further back than most sharks. This shark

swallows air at the surface and retains it in the stomach to regulate buoyancy. A slow, strong swimmer, it feeds upon a variety of bony fish, as well as on squid, crabs, and lobsters, and shoals of sandtigers have been observed working cooperatively to surround shoals of prey. The embryos of the sandtiger shark are

known to be cannibalistic. After eating other eggs inside their egg case, surviving embryos hatch, develop teeth, and eat other embryos within the uterus. Only one embryo survives from each of the 2 uteri.

large mouth

FISH

Carcharodon carcharias
White shark

Length	6 m (20 ft) max 8 m (26 ft)
Weight	2 tonnes (2 tons) or more
Breeding	Viviparous

Location Temperate and tropical waters worldwide; occasionally in cold waters

Status Vulnerable

Also known as the great white or white pointer, the white shark is blamed for more attacks on humans than any other shark – but research shows that most human victims are quickly released. A powerful swimmer, the white shark cruises through the water, either at the surface or just off the bottom, covering long distances very quickly. It also excels at short, fast chases, and may leap spectacularly out of the water. Like other members of its family, it can maintain its muscles at an unusually high temperature, helping it to swim more efficiently. Although usually solitary, it may sometimes be seen in pairs or small groups feeding at a carcass, with larger individuals eating first. The sharks may also swim in a variety of patterns in order to establish their dominance hierarchy. The primary prey is seals, sea lions, dolphins, and large fish, including other sharks – but this species will feed on any large creature that it can catch. It attacks with great ferocity initially, then retreats while the injured prey weakens, when the shark returns to eat in safety. Females, which are larger than males, give birth to 4–14 live young, 1.2m (4ft) or more long. While in the uterus, the young are nourished by eating unfertilized eggs or other embryos. Although this species is protected in some parts of the world, it is a popular sport fish, and its numbers are dropping rapidly.

triangular dorsal fin
grey to black upper body
conical snout
crescent-shaped tail fin
sickle-shaped pectoral fins
pale underside

SWIMMING POWER
The white shark powers through the water with sideways beats of its tail. Its fixed pectoral fins stop it nose-diving, and its large dorsal fin aids stability.

AWESOME WEAPONS

Armed with its large, triangular, coarsely serrated teeth, the white shark is superbly equipped for ripping into the flesh of its unfortunate prey. In some situations, this shark has been known to swim along with its teeth bared, which may serve to warn off competitors for food, or rival sharks intruding on its personal space.

THE GREAT "MAN-EATER"
This very large, formidable-looking shark is greatly feared as a man-eater, partly due to the media fascination that followed the release of the film Jaws. *However, according to the International Shark Attack File, there have been only 311 human deaths attributed to white shark attacks since records began.*

FISH

Crocodile shark

Pseudocarcharias kamoharai

Length 1.1 m (3½ ft) or more	
Weight Not recorded	
Breeding Ovoviviparous	
Location Pacific, E. Atlantic, W. and N. Indian Ocean	**Status** Lower risk

This small, cylindrical shark, rarely seen, is an oceanic species. It is almost always found far from land, although it sometimes comes inshore. It has a pronounced snout, with large teeth, and is able to extend its jaws unusually far forwards to snatch at prey. It migrates vertically, following its prey towards the surface at night and then sinking again at dawn, its large eyes enabling it to see in dimly lit conditions. Like many sharks, this species is an intra-uterine cannibal, with embryos eating each other before they are born. Most sharks that develop in this way produce only one pup per uterus, the others all having been eaten by the survivor. However, in this shark cannibalism reduces the number of embryos to 2 per uterus – at this stage, the cannibalism stops, and both embryos normally survive.

Megamouth shark

Megachasma pelagios

Length Up to 5.5 m (18 ft)	
Weight Up to 790 kg (1,740 lb)	
Breeding Unconfirmed	
Location W. and E. Atlantic, S. Indian Ocean, W. and E. Pacific	**Status** Unconfirmed

The megamouth is one of the world's least known large sharks. It was discovered in 1976 when one specimen became trapped by a ship's anchor off the Hawaiian Islands. Only 13 specimens have been seen since. This shark is characterized by an extraordinarily large mouth, which facilitates filter feeding. One of only 3 plankton-eating sharks, the megamouth shark is believed to follow the movement of plankton, rising from deep water to the surface at night, and returning to the depths during the day. Its great size means it has few enemies, sperm whales (see p.176) being one of the few predators big enough to tackle this huge fish. However, the megamouth is harmless to humans.

FEEDING AIDS

Distinctive colouring in and around this shark's huge mouth may be aids to feeding. A luminescent stripe on the margin of the upper jaw and a silvery roof to the mouth – visible as the shark swims along with jaws open – may serve to attract prey.

COLOUR CONTRASTS
The megamouth's back is uniformly grey to blue-black in colour, while the underside of the body is whitish. A white stripe runs along the margin of the upper jaw.

small dorsal fins

long upper tail lobe

large, flipper-like pectoral fins

white underside

large mouth

Basking shark

Cetorhinus maximus

Length 10 m (33 ft), max 15 m (49 ft)	
Weight 6 tonnes (5⅞ tons) or more	
Breeding Ovoviviparous	
Location Cool to warm temperate waters worldwide	**Status** Vulnerable

The basking shark is the world's second largest fish, after the whale shark (see p.469), and like it, is a harmless, migratory plankton-feeder. It often basks in the sun near the surface of the water, sometimes lazily rolling completely upside down. Although primarily solitary, groups may gather to feed in areas where plankton blooms are dense.

PLANKTON TRAPPER

This filter feeder moves through the ocean with its mouth agape, enabling it to take in large volumes of water and plankton. As the water is expelled through the enlarged gill slits, mucus-laden gill rakers trap the plankton – which the shark swiftly swallows.

WATER OUTLETS
This species has very large gill slits that nearly encircle its head. It sheds its gill rakers in winter.

Shortfin mako

Isurus oxyrinchus

Length Up to 4 m (13 ft)	
Weight 570 kg (1,260 lb) or more	
Breeding Ovoviviparous	
Location Temperate and tropical waters worldwide	**Status** Lower risk

Also called the blue pointer, mackerel shark, or snapper shark, this is possibly the world's fastest shark. The tail is adapted for speed, with well-defined keels on either side of the base – a common feature in fast-swimming fish. This shark has large, unserrated, dagger-like teeth that enable it to stab and grip slippery, fast-moving prey such as mackerel, tuna, bonito, and squid. It can jump several times its length out of the water when pursuing prey, or when hooked.

pointed snout

white underside

keeled tail

Thresher shark

Alopias vulpinus

Length Up to 5.5 m (18 ft)	
Weight 450 kg (990 lb) or more	
Breeding Ovoviviparous	
Location Temperate and tropical waters worldwide	**Status** Vulnerable†

The thresher shark is uniquely shaped, the upper lobe of its tail fin being almost as long as the rest of the body. The shark uses this extended lobe to herd fish together into tight shoals, and then stuns them with powerful slaps of the tail. A number of threshers may cooperatively encircle shoals of fish such as herring and small tuna, corralling them with their tails before the final attack. The thresher has also been known to use the tail as a means of defence against humans. Large individuals boated by fishermen may slap furiously with their tails, one such tail swipe reportedly resulting in decapitation. It is likely that this shark uses its tail to defend itself against marine predators as well. The thresher shark is one of 3 similar species, and is brown with a white underside. It is generally confined to deeper water, but the young and some adults may be found in shallower regions around coasts. It may at times congregate in shoals made up entirely of either males or females.

long upper lobe of tail

abrupt, pointed snout

pointed pectoral fins

distinctive black markings

Mustelus mustelus
Smooth-hound

Length Up to 1.6 m (5¼ ft)	
Weight 13 kg (29 lb) or more	
Breeding Viviparous	
Location E. Atlantic, Mediterranean	**Status** Common

The relatively smooth skin of the smooth-hound is grey or grey-brown above

slender body *short head*

and lighter below, with few or no markings. This shark's flat, slab-shaped teeth are designed for crushing molluscs, crabs, lobsters, and other invertebrates, as well as bony fish. It often feeds in shoals. Active after dark, the smooth-hound generally swims near the sea bed in the shallow, intertidal zone. One of 20 similar species, this shark is commercially fished around Europe, the Mediterranean, and West Africa.

Triakis semifasciata
Leopard shark

Length Up to 2.1 m (7 ft)	
Weight Up to 32 kg (71 lb)	
Breeding Viviparous	
Location E. North Pacific	**Status** Lower risk

The leopard shark is distinguished by a pattern of conspicuous black markings on its sides and fins, over a light brown background. Its

powerful jaws and small, sharp teeth help it to capture a wide variety of prey, including burrowing invertebrates, which it pulls out by quickly grasping any exposed parts; the shark is thought to feed both by day and by night. Harmless to humans, this is one of about 6 similar species.

Galeocerdo cuvier
Tiger shark

Length 5.5–7.5 m (18–25 ft)	
Weight 900 kg (1,990 lb) or more	
Breeding Ovoviviparous	
Location Tropical and warm temperate waters worldwide	**Status** Lower risk

A known man-eater and one of the world's most dangerous sharks, the tiger shark has distinctive teeth shaped like a rooster's comb, and a head that is disproportionately large for its slender, streamlined body. This shark generally prefers coastal waters, often moving inshore to feed at night, and sometimes exploring the very shallow waters of bays and estuaries.

Recent studies, however, indicate

that it may also cross large expanses of open ocean. It makes forays from the tropics to higher latitudes during the warmer months. The tiger shark is the only ovoviviparous member of the family of requiem sharks; females may secrete uterine "milk" to provide additional nutrition for the embryos. Although not highly valued for its flesh, it is often harvested in commercial shark fisheries.

tiger-like stripes *large upper tail lobe for speed*

large head *white underside*

STRIPED YOUNG
The juveniles and the young adults of this species are conspicuously marked with vertical, tiger-like stripes. These may fade or be altogether absent in adults.

EATING HABITS

A primarily nocturnal hunter, the tiger shark has the most indiscriminate diet among sharks: it employs quick bursts of speed to catch live prey, such as fish and a variety of marine reptiles, invertebrates, and mammals. It also scavenges on carrion, and may even feed on rubbish.

Cephaloscyllium ventriosum
Swell shark

Length Up to 1 m (3¼ ft)	
Weight Not recorded	
Breeding Oviparous	
Location E. North and South Pacific	**Status** Locally common

When threatened, the swell shark may expand itself by taking in water or air. In addition to making it appear more imposing to potential predators, this defensive technique allows the shark to wedge itself tightly into crevices for safety. This large member of the catshark family has a variable pattern of small spots and dark brown markings. Highly nocturnal, it usually lies motionless in crevices by day, and swims slowly through algae beds or near the sea bed at night, presumably preying on diurnal fish that are less alert at night. The swell shark is one of about 8 similar species.

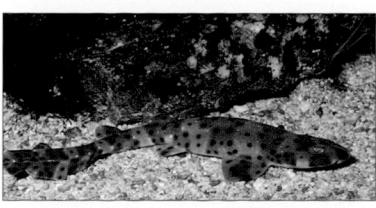

Scyliorhinus canicula
Small-spotted catshark

Length Up to 1 m (3¼ ft)	
Weight 3 kg (6½ lb) or more	
Breeding Oviparous	
Location E. North Atlantic, Mediterranean	**Status** Common

Also called the lesser spotted dogfish, this is the commonest European shark. It is small and slender, with

relatively smooth skin

gill slits *numerous dark spots aid camouflage* *slender body*

a smooth skin covered with many dark spots and fewer light or white ones. The colours and patterns vary greatly, and help conceal this shark from predatory fish. Normally solitary, when it does form shoals, males and females are rarely found together. This shark is active by both day and night, hunting by scent and electrical sense for worms, molluscs,

crustaceans, and small bony fish, close to the sea bed. Females deposit egg capsules on beds of algae; these are often washed up on the shore and are commonly known as mermaid's purses. They are rectangular, with tendrils at each corner that anchor them to weeds. In Europe, the small-spotted catshark is commercially fished for oil as well as food.

Carcharhinus leucas
Bull shark

Length Up to 3.4 m (11 ft)	
Weight 230 kg (510 lb) or more	
Breeding Viviparous	
Location Tropical and subtropical waters worldwide	**Status** Lower risk

Like all members of the requiem shark family, the bull shark, also known as the river whaler, freshwater

small, rounded second dorsal fin *large first dorsal fin* *stubby, rounded snout*

large, angular pectoral fins *large gill slits*

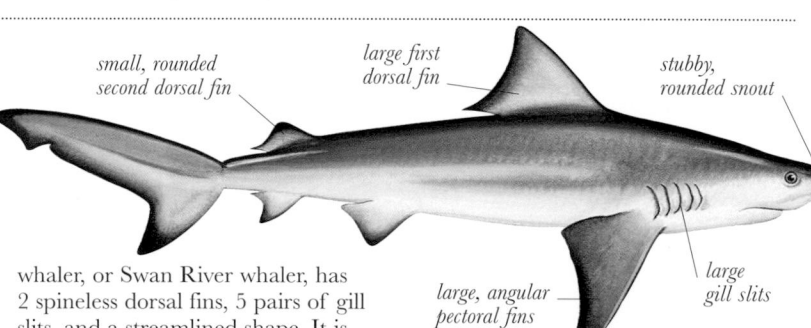

whaler, or Swan River whaler, has 2 spineless dorsal fins, 5 pairs of gill slits, and a streamlined shape. It is one of the few sharks that often swims up rivers – it has been found over 3,000 km (1,870 miles) from the sea in the Amazon. Its diet is unusually wide-ranging and includes bony fish, invertebrates, mammals, and other

sharks, including young bull sharks. One of the world's most dangerous sharks, the bull shark has been involved in numerous provoked and unprovoked attacks on humans.

FISH

Carcharhinus melanopterus
Blacktip reef shark

	Length Up to 2 m (6 1/2 ft)
	Weight Over 45 kg (99 lb)
	Breeding Viviparous
Location Tropical Indo-Pacific	**Status** Lower risk

The blacktip reef shark is the most commonly encountered species of shark in the tropical Indo-Pacific region.

Negaprion brevirostris
Lemon shark

	Length Up to 3.4 m (11 ft)
	Weight Over 185 kg (401 lb)
	Breeding Viviparous
Location E. Pacific, W. and E. Atlantic	**Status** Locally common

The lemon, or sharptooth, shark adapts well to captivity and hence is one of the most-studied members of the requiem shark family. It is capable of a powerful bite and may attack

This shark usually occurs in shallow water, and the dorsal and upper tail fin may project above the surface of the water. It is responsible for numerous attacks on humans, both provoked and unprovoked, and is often curious when it spots divers, although it can be driven away. The blacktip reef shark has a streamlined body, which makes it an active and powerful swimmer. As its name suggests, the tips of the fins are black.

black-tipped fins

humans if provoked. Yellowish to light brown in colour, this fish has large fins and an abrupt snout. The lemon shark is adapted to survive in brackish waters and areas of low oxygen. It feeds mainly on bony fish, guitarfish, and stingrays, but may also eat crustaceans, molluscs, and even seabirds. Adults are most active at night; the juveniles by day.

Triaenodon obesus
Whitetip reef shark

	Length 1.6 m (5 1/4 ft), max 2.1 m (7 ft)
	Weight Over 18 kg (40 lb)
	Breeding Viviparous
Location Tropical Indo-Pacific	**Status** Lower risk

While most sharks must swim continuously, mouth agape, to force oxygenated water across their gills, the whitetip reef shark, or blunthead, can pump water across its gills and can therefore rest on the ocean floor. It relies heavily on sound and smell for catching prey – often associating the sound of boat engines with the opportunity to steal fish, it doggedly pursues speared fish and may bite fishermen. Nevertheless, it is generally non-aggressive.

white-tipped dorsal fin

CONSERVATION

Despite their reputation for being dangerous, sharks have more to fear from humans than humans do from sharks. Many of the world's inshore species – including the whitetip reef shark – have been heavily overfished in recent years, and are rapidly declining as a result. Sharks are often fished for their fins alone, the rest of the animal being thrown away.

WHITE MARKERS
This slender shark has distinctive white tips on its dorsal fin and upper tail lobe.

Prionace glauca
Blue shark

	Length 3.8 m (12 ft) or more
	Weight Over 200 kg (440 lb)
	Breeding Viviparous
Location Tropical and temperate waters worldwide	**Status** Lower risk

The blue shark is believed to be the widest ranging of all vertebrates in the world.

It migrates seasonally, moving from cooler to warmer waters. Blue sharks are highly dangerous, and are known to attack humans. They sometimes circle potential prey before attacking, and large numbers may gather around a whale or porpoise carcass in a feeding frenzy, biting at chunks of floating debris. They may also be seen following trawlers, and feeding on the trapped fish.

cobalt-blue body

wing-like pectoral fins

Squatina dumeril
Angelshark

	Length Up to 1.5 m (5 ft)
	Weight Over 27 kg (60 lb)
	Breeding Ovoviviparous
Location W. North Atlantic	**Status** Locally common

This flattened fish is often found partially buried in sand on the sea floor, its nondescript coloration serving as camouflage. Mainly a sit-and-wait predator, it uses its protrusible

mouth to strike with lightning speed at unwary prey, consisting chiefly of crustaceans and other fish. One of about 14 similar species, the angelshark (or sand devil) is commercially fished for its flesh, skin, and oil.

wing-like pectoral fins

flattened body

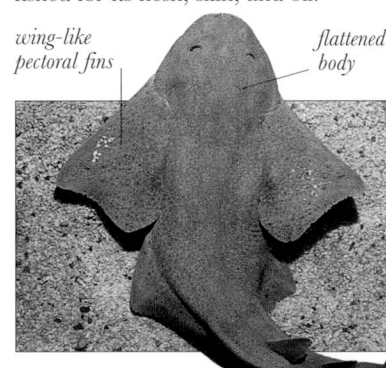

Sphyrna zygaena
Smooth hammerhead

	Length Up to 4 m (13 ft)
	Weight Up to 400 kg (880 lb)
	Breeding Viviparous
Location Tropical, subtropical, and temperate waters worldwide	**Status** Lower risk

The smooth hammerhead is one of 8 species of hammerhead sharks, instantly identifiable by their bizarre heads, which have eyes at the end of wing-like flaps. It is common in

inshore waters and is often seen at the surface, with its large dorsal fin slicing through the water. It has an unpredictable temperament, and is known to attack humans. Like its relatives, the smooth hammerhead feeds on a wide variety of fish, including other sharks. A powerful swimmer, it

moves north to cooler waters during the summer. Unlike other sharks, hammerheads often form shoals, probably for protection from predators; the young are sometimes observed migrating in huge numbers.

IDENTIFYING FEATURES
Species of hammerhead shark can be difficult to tell apart. In the smooth hammerhead, the blades of the "hammer" are narrow.

narrow head flaps

long upper tail lobe

UNIQUE ANATOMY

The reason for the hammerheads' strange anatomy is uncertain. It provides these sharks with extra lift as they swim, but it may also help them to track prey, by improving the acuity of their sense of smell, and by giving them a wide visual field.

SKATES AND RAYS

PHYLUM	Chordata
CLASS	Chondrichthyes
SUBCLASS	Elasmobranchi (part)
ORDERS	4
FAMILIES	14
SPECIES	About 450

BROAD BODIES AND wing-like fins distinguish skates and rays from other cartilaginous fish. Their flattened shape is generally an adaptation for life on the sea bed, although several species, including the mantas and eagle rays, swim in open water. Skates and rays are found throughout the world. Rays are more diverse in the tropics, while skates are more widespread in temperate waters. Most skates and rays live along coasts, but some range far into deep-sea basins. Although some rays range between fresh- and saltwater estuaries, there are others that live exclusively in fresh water.

ANATOMY

Skates and rays have flat bodies with large pectoral fins that are used for propulsion. Larger species flap these fins like wings. They have an oil-filled liver like that of sharks, which helps to maintain buoyancy, but it is smaller and allows them to remain on the sea bed. Most species are coloured to match the sea bed but they may also partially bury them-selves in the sediment. When buried, they breathe through openings called spiracles located behind the eyes. Water enters the spiracles and leaves through the gill openings, bypassing the mouth, which is on the underside. Many skates and rays either lack scales or have large, thorn-like scales for protection. Rays have a barbed spine on their tails that is coated with toxic compounds.

FEEDING

All species of skates and rays are carnivorous, and most are predators on small, bottom-dwelling fish and invertebrates, although prey varies considerably between species. These fish have teeth that are designed to grasp, rasp, or crush food, and interlock in a pattern known as pavement dentition. A few species, such as the mantas, are filter feeders that strain small fish, invertebrates, and plankton with their sieve-like gill rakers.

REPRODUCTION

Skates and rays have different modes of reproduction. Most skates breed by releasing large, leathery egg capsules containing one or more developing offspring. Some rays produce eggs that hatch inside the female's body. The young live off the yolk inside the female before being released. In other species, the embryos are further nourished by a fluid produced by the mother and delivered by the membranes of the uterus.

CHIMAERAS

Ratfish, rabbitfish, and their relatives – collectively known as chimaeras – make up a small group of cartilaginous fish that are distinct from sharks, skates, and rays. They form the sub-class Holocephali, containing 30 species. Their soft, scaleless bodies are usually long and narrow, and they have a bulky head with large, iridescent eyes. Found in the waters of the Arctic and Antarctic, they live at depths up to 8,000m (26,400 ft).

SHOALING
Most rays lead sedentary lives or move slowly along sea or river beds. However, some, such as these bat rays, inhabit open waters, propelling themselves with their large pectoral fins. They form shoals that may contain several hundred individuals.

Raja undulata

Undulate ray

Length	Up to 1m (3 1/4 ft)
Weight	3–4kg (6 1/2 – 8 3/4 lb)
Breeding	Oviparous

Location E. Atlantic, Mediterranean

Status Common

mottled body surface

scattered whitish spots

Found on mud or sand at depths of 200m (650ft), the undulate ray relies on its intricately patterned, reddish or brown skin to hide from predators. During the breeding season, females lay up to 15 eggs, each in a tough capsule up to 9cm (3½in) long.

Raja clavata

Thornback ray

Length	Up to 90cm (35in)
Weight	2–4kg (4 1/2 – 8 3/4 lb)
Breeding	Oviparous

Location E. Atlantic, Mediterranean

Status Common

The thornback ray occurs in varying shades of brownish grey with darker marbling and spotting; its underside is of a lighter colour. Both sexes have big spines on their back – hence the common name of this species. As with most rays, it lies camouflaged on the ocean floor.

dark marbling

Raja batis

Common skate

Length	Up to 2.5m (8 1/4 ft)
Weight	50–100kg (110 – 220lb)
Breeding	Oviparous

Location E. Atlantic, Mediterranean

Status Common

On the upper surface, the common skate is brownish grey with darker marbling and lighter spotting, sometimes with a large, oval spot on the surface of each pectoral fin. Its underside is often light grey. As with the thornback ray (see left), the male tends to be smoother, the female having small prickles on the underside. She lays 10 or more egg cases per year, and these are about 20cm (8in) long, with a dark, thick, flexible shell and a pair of tendril-like thorns on either side. Unlike most other skates, the common skate is active by day and night. It feeds on sea-bed animals, especially fish, worms, and crustaceans.

single row of spines along body and tail

Rhinobatos productus
Shovelnose guitarfish

Length	Up to 1.5 m (5 ft)
Weight	15–18 kg (33–40 lb)
Breeding	Viviparous
Location E. Pacific	**Status** Common

Named for its unusual shape, the shovelnose guitarfish has a broad head with a clear, cartilaginous area on either side of the snout. Its pectoral fins are wide, like those of a typical ray, but the rest of its body is cylindrical, like that of a shark. Commonly found in the warmer areas of the eastern Pacific Ocean, it tends to move inshore in the summer months. The shovelnose guitarfish is generally solitary but sometimes gathers in large numbers, possibly for breeding. It is often found in shallow water around beaches, bays, and estuaries, where it lies partially buried in sand or mud. It feeds on small, bottom-dwelling fish, and on crustaceans and worms, which it uncovers from the sea floor. The female produces 5–25 young, which are about 15cm (6in) long at birth.

cylindrical, shark-like body

triangular tail fin

rounded pectoral fins

Pristis pectinata
Smalltooth sawfish

Length	6 m (20 ft) or more
Weight	250–300 kg† (550–660 lb)
Breeding	Viviparous
Location W. and E. Atlantic, Indo-Pacific	**Status** Endangered

This long-bodied fish has a remarkable saw-like snout, which has 24–32 pairs of pointed teeth on either side. Even though it is equipped with this large saw, the smalltooth sawfish is not generally aggressive and is dangerous only if mishandled when caught or when harassed by divers. It has a flattened head and small eyes on the sides. The mouth and gill slits are located on the underside of the head. While most common in the shallow, sandy, and muddy water of beaches and bays, the smalltooth sawfish may occasionally be found at the mouth of rivers and freshwater streams. It feeds by patrolling the sea bed and sucking up small organisms, sometimes using the saw to probe the bottom or excavate buried prey. It also slashes into shoals of fish with its sawlike snout, feeding on dead or maimed fish that fall to the bottom. The female gives birth to 15–20 young, which are born with a protective membrane over their saws, an adaptation that protects the female while giving birth. One of 6 similar species, the small-tooth sawfish is subject to overfishing in many areas as its flesh is edible and the saw is sold as a souvenir.

snout 30 per cent of body length

triangular dorsal fins

stout pectoral fins

Torpedo marmorata
Marbled electric ray

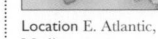

Length	Up to 60 cm (23 ½ in)
Weight	10–13 kg (22–29 lb)
Breeding	Viviparous
Location E. Atlantic, Mediterranean	**Status** Common

brownish grey colour acts as camouflage

Many cartilaginous fish use weak electrical signals to detect prey, but this ray can generate shocks, powerful enough to kill other fish, from specially modified muscles at the base of its pectoral fins. Although such shocks are not known to be fatal to humans, contact with this ray can be dangerous. Active during the day and in the evening, the marbled electric ray generally lies on the ocean bed, well camouflaged against the sediment. Females give birth to 5–35 young every year; larger females produce more young.

Manta birostris
Manta ray

Length	4–7 m (13–23 ft)
Weight	Up to 1.8 tonnes (1¾ tons)
Breeding	Viviparous
Location Tropical and sometimes warm temperate waters worldwide	**Status** Vulnerable

Measuring up to 7m (23ft) across, the manta – also known as the devil ray – is the world's largest ray, swimming through the open ocean like a large bird on slowly flapping wings. However, despite its sinister look, it is a harmless plankton-eater, filtering food with its gills. Its mouth, unlike those of other rays, is positioned in front of its body, allowing it to feed continuously as it moves along. Normally a leisurely swimmer, it can suddenly accelerate if threatened, and can even leap out of the water to avoid large predators, such as sharks and killer whales. Mantas are usually solitary, although they sometimes swim in small, loosely organized shoals. They prefer warm waters, and may enter inshore regions in summer. During the breeding season, the male chases the female, swimming beneath her so that their undersides face each other, and inserts his claspers. The female usually gives birth to 1 or 2 young per year, each about 1.2m (4ft) wide.

FUNNELLING FOOD
Manta rays have large, flap-like lobes on either side of their heads, and they use these to funnel prey into their mouths. The water passes out through the ray's gills, but the prey is trapped by the gill rakers and then swallowed. Mantas eat small, shoaling fish, as well as smaller planktonic animals.

triangular pectoral fins

"wingspan" over 6.5m (21ft)

flap-like lobes

UNDERWATER FLIGHT
The manta has large, triangular pectoral fins with pointed tips, and a short tail without a fin or sting. In normal circumstances, its fins beat up and down about once every 4 or 5 seconds, but the manta can accelerate rapidly, somersaulting out of the water and crashing back onto the surface with a massive impact.

Myliobatis aquila
Eagle ray

Length	2.5 m (8 1/4 ft) or more
Weight	20–30 kg (44–66 lb)
Breeding	Viviparous

Location E. Atlantic, Mediterranean, S.W. Indian Ocean **Status** Common

The eagle ray has wide but narrow pectoral fins, a blunt snout, and an exceptionally long tail armed with a venomous spine. The body of this fish varies in colour from greyish brown to bronze or blackish, and its mouth, positioned on its underside, is equipped with 7 rows of flat, crushing teeth. Like its numerous relatives, which are found in oceans all over the world, it feeds on the sea bed, but is equally at home in open water. The eagle ray is believed to prey on a wide variety of small animals, and it excavates these from the sediment by flapping its fins, or by blowing jets of water to clear away mud and sand. A good swimmer, the eagle ray can leap clear of the water's surface to escape attack from predators. Female eagle rays give birth to 3–7 live young each year.

bill-like snout

flattened body

stinger

pointed pectoral fins

Taeniura lymma
Blue-spotted stingray

Length	Up to 2 m (6 1/2 ft)
Weight	Up to 30 kg (66 lb)
Breeding	Viviparous

Location Indo-Pacific **Status** Lower risk

One of the handsomest members of the stingray family, this species has blue spots scattered over a contrasting greenish or yellowish body. In addition, it has blue stripes along the side of its tail and a lighter underside. Like all stingrays, this fish has a toxic spine located at the base of its tail and is capable of inflicting a nasty sting if stepped on or mishandled. Although mainly diurnal, it is sometimes active at night. It moves into shallow areas with the advancing tide to feed on small fish and invertebrates, especially crustaceans and worms. It is often found in sandy areas next to reefs, and rests in reef caves and crevices when not feeding. The female gives birth to 3–7 young and, like the freshwater stingray (see right), she nourishes her young by secreting uterine "milk". The colourful appearance of the blue-spotted stingray has made it a popular fish in the aquarium trade and a favourite subject of underwater photographers.

LYING IN WAIT

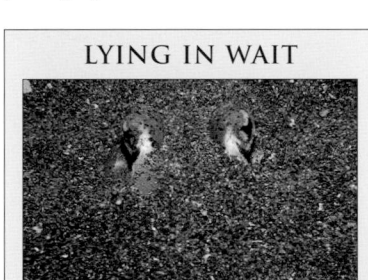

Blue-spotted stingrays normally actively forage for food, but they rest with their bodies largely hidden by mud or sand, with just their eyes projecting above the surface. In shallow water, hidden rays are easily stepped on.

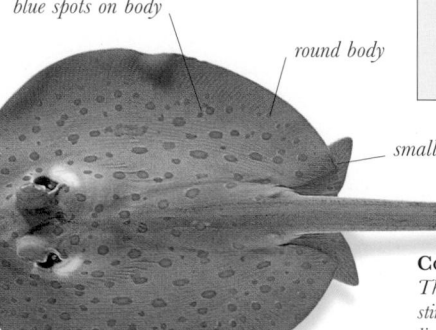

blue spots on body

round body

small pelvic fins

stripes on tail

COLOUR AND FORM
The small eyes of the colourful blue-spotted stingray are located on top of its head. Its mouth lies on the underside of the round, extremely flattened body. The tail is relatively long.

Urolophus halleri
Round stingray

Length	50–60 cm (20–23 1/2 in)
Weight	2–4 kg (4 1/2–8 3/4 lb)
Breeding	Viviparous

Location E. North Pacific **Status** Common

Unlike typical stingrays, the round stingray and its relatives, sometimes known as stingarees, have an almost circular body. The body is very flattened, and the tail is relatively short, with a rounded fin at its tip. During warm seasons, large numbers sometimes move into shallow beaches where they may sting bathers who are unfortunate enough to step on them. Although not fatal, the sting of this fish can be intensely painful.

sting partway along tail

tail fin

Potamotrygon motoro
Freshwater stingray

Length	Up to 1 m (3 1/4 ft)
Weight	3–5 kg (6 1/2–11 lb)
Breeding	Viviparous

Location South America **Status** Common

This round-bodied ray is the most common and widespread freshwater stingray in South America. It is a member of an entire family of stingrays that evolved in fresh water from a marine ancestor. The short tail of the freshwater stingray has no fin, while its small pelvic fins are tucked under its pectoral fins. The freshwater stingray has few predators, except some larger fish and caiman, and feeds on small fish and invertebrates.

brownish colour with dark spots aids camouflage

finless tail

rounded pectoral fins

Hydrolagus colliei
Spotted ratfish

Length	Up to 95 cm (37 in)
Weight	2–6 kg (4 1/2–13 lb)
Breeding	Oviparous

Location E. North Pacific **Status** Common

Ratfish are found in deep, inshore waters in most parts of the world, but the spotted ratfish is one of only 2 species that live on the Pacific coast of North America. Like other ratfish, it has a disproportionately large head, and a downward-facing mouth that is equipped with large, crushing teeth, and scaleless, slippery skin. Males have a retractable, club-like appendage between their eyes, which is probably used during courtship. Adults are usually dark brown or greyish, with green, yellow, or bluish hues, and they have silvery white spots along their sides. Spotted ratfish are slow swimmers, and rely on a venomous dorsal spine as their main defence against attack. They patrol the ocean floor in search of prey, feeding on invertebrates and sea-bed fish. Females usually lay 2–5 eggs a year; each one is contained in a long, spindle-shaped case. Ratfish liver contains an oil that was once used as a machine and gun lubricant, but is today of little commercial value.

BONY FISH

PHYLUM	Chordata
CLASS	Osteichthyes
ORDERS	46
FAMILIES	437
SPECIES	About 23,500

CLASSIFICATION NOTE

Bony fish account for more species than any other class of vertebrates. They are split into 2 subclasses: fleshy-finned fish (Sarcopterygii) have fins that look like flippers and are joined to the body by fleshy lobes; ray-finned fish (Actinopterygii) have fins supported by bony spines. The ray-finned fish are by far the larger group and are divided into 9 superorders. Classification of bony fish is in a state of constant change as new species are discovered and zoologists learn more about the relationships between them.

BONY FISH FORM overwhelmingly the largest and most varied group of fish, accounting for more than 9 out of 10 species. Of the 4 classes of fish, they evolved most recently and are usually regarded as the most advanced. Although most are small, they vary greatly in size and shape. All bony fish have a light but strong internal skeleton, made wholly or partly of bone, which supports flexible fins that enable the fish to control its movements with precision. Most also have a gas-filled swim bladder that allows them to adjust their buoyancy within narrow limits. Bony fish are found in almost all aquatic habitats, including marshes, lakes, rivers, coasts, reefs, and deep oceans. Many have adapted to extreme conditions, and species of bony fish occur in high-altitude lakes, polar coasts, hot springs, high-salinity ponds, acidic streams, and low-oxygen swamps.

ANATOMY

The skeleton of a bony fish consists of 3 main units: the skull, backbone, and fin skeleton (see below). The gills of bony fish are paired and located behind the head and below the cranium. Unlike in other groups of fish, the gill openings are covered by a bony flap, called the operculum, and the lower gill chamber contains bony supports, called branchiostegal rays. These 2 structures enable the fish to take a gulp of water into its mouth and then pump it over its gills: the branchiostegal rays help to open the mouth and regulate intake, while the operculum acts as a seal over the gills to control the outflow. In this way, a bony fish can respire while stationary. The same 2 structures are also used in seizing and gulping food. Most bony fish have light, flexible, cycloid or ctenoid scales (see p.460), usually covered by a thin layer of skin that secretes mucus. The mucus repels parasites and disease-causing organisms; in some species, it also prevents moisture loss. Other bony fish have large protective scales or no scales at all (although their skin still produces mucus).

SKELETON
A bony fish's skull encases the brain and supports the jaws and gill arches. Teeth may be found in the jaws, in the throat, on the roof of the mouth, or on the tongue. The backbone, consisting of articulated vertebrae that are linked to the ribs in the abdomen, provides support for the body. Each fin consists of a bony base embedded in the body, from which rod-like structures extend to form the outer fin.

neural spine of vertebra

streamlined skull

symmetrical tail

fin ray

bony fin base

vertebral column

rib

operculum

SENSES

Most bony fish have keen senses of vision and hearing. Both of these senses are used in communication and social interaction. They are also thought to contribute to bony fish's well-developed ability to shoal. The eyes are generally set on the sides of the head, giving the fish a wide field of view. Rods and cones in the retina give them good colour vision, and many species have bright coloration, which helps them to attract mates and defend territory. Sound is an effective means of underwater communication. Some species of bony fish can produce sounds, either using their swim bladder or by rubbing parts of their body together.

lateral-line pore

LATERAL LINE
The lateral-line system, which is used to detect movement, is highly developed in bony fish. This is thought to be another reason why they can move effectively in large shoals.

SWIMMING
Bony fish, such as these French grunts, can manoeuvre precisely. Each fin is controlled by a separate set of muscles so that it can move independently of the others. The symmetrical tail typical of bony fish is an efficient tool for propelling the fish through water.

FISH

SHOALING

1 ATTRACTING PREDATORS
A shoal of fish is an easy target for predators. These yellowtail horse mackerel have attracted a marlin.

2 STAYING TOGETHER
As the marlin swims towards the shoal, the mackerel are able to change direction and stay together.

3 CLOSING RANKS
While the marlin circles close by, the mackerel pack themselves even more closely together.

4 USING THE LIGHT
The marlin stays below the shoal, where it is easier to pick out the fish against the light coming from above.

5 ON THE ATTACK
The marlin attacks again. If the fish are forced to split into 2 groups, they will try to rejoin each other quickly.

BUOYANCY

Most bony fish control their buoyancy using an outgrowth of the intestine known as the swim bladder. This bladder is filled with gas, making it less dense than water. The fish can adjust its buoyancy by regulating the volume and pressure of the gas in the bladder. The swim bladder of freshwater fish has a greater capacity than that of marine fish, because fresh water is less dense than sea water and does not give as much support to the fish's body. Having a swim bladder means that bony fish (unlike cartilaginous fish) do not need to use their fins for buoyancy control and, instead, can use them entirely for manoeuvring.

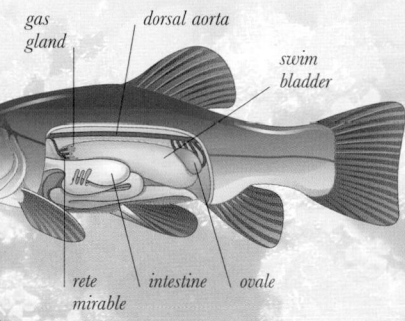

gas gland · dorsal aorta · swim bladder · rete mirable · intestine · ovale

SWIM BLADDER
Apart from some primitive species, most bony fish adjust their buoyancy by moving gas between the bloodstream and the swim bladder. Gas enters the bladder through the gas gland, which is supplied with blood by a network of capillaries called the rete mirable. Using another structure, known as the ovale, gas is removed from the bladder and re-absorbed into the blood.

REPRODUCTION

In most bony fish, fertilization occurs outside the body but in a few species it is internal. Marine fish that live in open water usually produce eggs that float as part of the plankton, often in large numbers because only a small number will actually hatch. Coastal marine and freshwater fish tend to produce fewer eggs, which are laid in the sediment, in nests, or attached to plants. A few species care for their eggs or even their young. This usually takes the form of protecting them from predators, but some species provide their offspring with food. Bony fish include some hermaphroditic species. Some have both male and female reproductive organs. Others change sex during the course of their adult life: males that become female may do so to become larger and accommodate more eggs (which are larger than sperm); by becoming males, females may assume dominant positions in social groups.

PARENTAL CARE
The male yellowhead jawfish incubates its eggs by storing them inside its mouth. Some other bony fish (including some cichlids) use their mouth to protect their young from predators.

LARVAE
Most bony fish hatch from eggs and begin life as larvae. These larvae of the European eel are in the process of metamorphosis, during which they are referred to as glass eels.

SHOALING

Many bony fish swim in large groups that often contain tens of thousands of individuals. When the fish are grouped together but behave independently, the group is referred to as an aggregate. In a shoal (or school), however, the movement of the fish is much more harmonized, to the extent that the shoal resembles a single organism. Only bony fish can swim in this coordinated way. Each fish usually follows a course parallel to its neighbours, maintaining its position using its vision, hearing, and lateral-line system. Fish in a shoal are safer than solitary ones, because there are more individuals to look out for danger and it is difficult for a predator to pick out a single fish from a large mass (see left). It is also easier for fish in a group to find mates and locate food.

STAYING TOGETHER
Shoaling fish pack tightly together, each one typically keeping a distance to its nearest neighbour of no more than one body length. A silvery coloration (seen in these sweepers) helps the fish to see each other's movements, as a small change in direction produces a large change in the amount of light reflected.

FLESHY-FINNED FISH

PHYLUM	Chordata
CLASS	Osteichthyes
SUBCLASS	Sarcopterygii
ORDERS	3
FAMILIES	4
SPECIES	8

FLESHY-FINNED FISH HAVE AN ENLARGED fin base that consists of muscles connected to an internal skeleton. These fins are used for swimming, or to "walk" on the sea bed. In some of the fish related to the ancestors of this group, the fins developed into limbs for movement on land, and the living fleshy-finned fish are perhaps the closest relatives of the early terrestrial 4-legged animals. Fleshy-finned fish are divided into 2 groups. The eel-like lungfish have both gills and one or 2 primitive lungs, and can breathe air directly. They can live in oxygen-poor swamps and even survive out of water. One species, the West African lungfish, can reside in a mud cocoon at the bottom of a dry lake, awaiting the next wet season to replenish the water. The other group, the coelacanths, were thought to be extinct since the time of the dinosaurs until discovered in 1938; they live in deep areas of the Indian Ocean.

AIR-BREATHING FISH
The adult South American lungfish, which has small gills and 2 lungs, can only breathe air. In the breeding season, the male provides oxygen to its young by releasing it into the water through blood-rich filaments in its thread-like fins.

Latimeria chalumnae
Coelacanth

Length 1.5–1.8m (5–6ft)
Weight 65–98kg (145–220lb)
Sex Male/Female

Location W. Indian Ocean | Status Endangered

First seen alive by scientists in 1938, the coelacanth is a classic example of a "living fossil", as it belongs to a group that was thought to have died out over

65 million years ago. It is found at depths of 150–700m (500–2,300ft), along rocky slopes with submarine caverns, swept by strong oceanic currents. It has large, thick, heavy scaling with iridescent white flecks, muscular paired fins, and an unusual tail with an additional central lobe. The pectoral fins of the coelacanth are highly mobile, and it uses these to manoeuvre into crevices to reach fish and other prey. Populations are poorly documented and estimates are difficult, but susceptibility to capture by humans and a restricted habitat range have made the coelacanth an endangered species.

Latimeria menadoensis
Indonesian coelacanth

Length Up to 1.6m (5¼ft)
Weight 65–98kg (145–220lb)
Sex Male/Female

Location Pacific (Celebes Sea) | Status Endangered

The Indonesian coelacanth was discovered in the late 1990s, and as yet, relatively little is known about its behaviour or ecology. However, since it is physically very similar to the

coelacanth of southern Africa (see above), it almost certainly has a comparable way of life. Molecular analysis has shown that the 2 species probably diverged from each other between 4.7 and 6.3 million years ago. Since then, they have been kept apart by the geology of the sea bed, and by currents that confine them to their own particular parts of the world. Like its African counterpart, the Indonesian coelacanth appears to be a solitary fish in open water, but it has also been found in groups in caves. So far, population estimates are not available, but it is likely to be endangered. The chief threat to this coelacanth's survival is fishing, which can have a severe effect on a species that is geographically highly isolated.

Neoceratodus forsteri
Australian lungfish

Length Up to 1.8m (6ft)
Weight Up to 45kg (99lb)
Sex Male/Female

Location E. Australia | Status Lower risk†

Unlike other lungfish, which often live in pools that sometimes dry up, the Australian lungfish is found in permanent bodies of water associated with dense vegetation, such as large, deep pools, reservoirs, and slow-flowing rivers. A large, freshwater fish with a heavy body, it has a specialized swim bladder which functions as a single lung. Although it normally

breathes through its gills, when oxygen levels in the water fall, it gulps air at the surface, breathing through its lung. Unlike African and South American lungfish, this species has paddle-shaped, paired fins. Its dorsal and anal fins are continuous with its tail fin. The Australian lungfish becomes less active during dry conditions, and can survive for months if kept moist by a covering of damp leaves or mud. It feeds on frogs, crabs, insect larvae, molluscs, and small fish, crushing its prey between its strong toothplates.

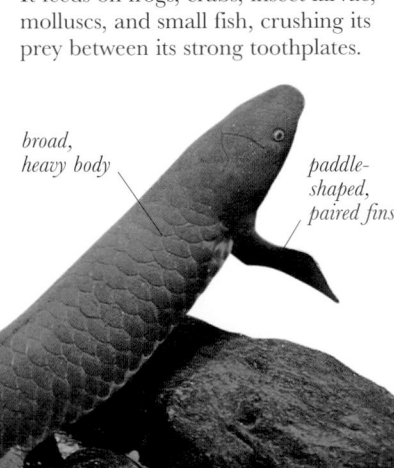

broad, heavy body

paddle-shaped, paired fins

long, tapering tail

Protopterus annectens
West African lungfish

Length 1.8–2m (6–6½ft)
Weight Up to 17kg (37lb)
Sex Male/Female

Location W. to C. Africa | Status Common

The West African lungfish has long, thread-like, paired fins, and is the largest of the 4 lungfish found in Africa. Like the other 3 species, it breathes through a pair of lungs. At

the onset of the dry season, this fish burrows into mud, forming a mucus-filled cocoon. It lives in a variety of freshwater habitats, and is carnivorous, capturing its prey by stealth, rather than by rapid pursuit.

PRIMITIVE RAY-FINNED FISH

PHYLUM	Chordata
CLASS	Osteichthyes
SUBCLASS	Actinopterygii
ORDERS	Polypteriformes
	Acipenseriformes
	Lepisosteiformes
	Amiiformes
FAMILIES	5
SPECIES	43

AMONG THE LEAST advanced of all bony fish, primitive ray-finned fish have an unusual skeleton composed partly of bone and partly of cartilage. They are divided into 4 groups: the sturgeons and paddlefish, gars, bichirs, and the bowfin (a single species that is placed in a separate order). Primitive ray-finned fish are among the largest of all freshwater species, with sturgeons reaching lengths of 8m (26ft). Unique to the northern hemisphere, most species live in fresh water, although some are able to live alternately in fresh and salt water.

rows of large, thickened scales (bony scutes) protect an otherwise scaleless body. Paddlefish have a long, paddle-shaped rostrum, which may act as a sensory organ or improve the flow of water to the mouth during feeding. These filter-feeding fish use their gaping mouth and large gill chambers to strain small aquatic organisms from the water.

Gars are cylindrical and highly predatory fish that catch prey through ambush or bursts of speed. The elongated rostrum and jaws are

studded with sharp teeth, resembling the snout of a crocodile. Gars are covered by smooth, thickened ganoid scales (see p.460).

Bichirs have a long, slender body protected by thick scales. Most species (sometimes referred to as the true bichirs) have a row of small finlets on their back and a pectoral fin divided into fleshy lobes.

The bowfin is much smaller than other primitive ray-finned fish. A predatory fish, it has a blunt head and small teeth.

ANATOMY

In most primitive ray-finned fish, the skull and some fin supports are made of bone, while the body and tail are supported by a cartilage notochord with some rudimentary vertebrae. The tail is asymmetrical, with the end of the notochord extending into an upper lobe that is longer than the lower one. Gars, some bichirs, and the bowfin have a swim bladder lined with blood vessels that can be used as a lung when levels of oxygen in the water are low.

Sturgeons are large, bottom-feeding fish with a flattened snout, a protrusible mouth, and sensitive barbels for locating food. Several

PADDLEFISH
There are 2 species of paddlefish, both found in large river systems with an ample supply of plankton. The American paddlefish (shown here) lives in the Mississippi river system, where it feeds at night and spends the daylight hours resting at the bottom of deep pools.

FISH

Acipenser sturio
European sturgeon

	Length Up to 3.5 m (11 ft)
	Weight Up to 315 kg (690 lb)
	Sex Male/Female

Location North Atlantic, Mediterranean, Europe | **Status** Locally common

flattened head

Heavily overfished for its flesh and eggs – caviar is made of the unshed eggs of females – the European sturgeon is now a rare species. One of the largest European fish that swim up rivers to breed, it is greenish brown and, like other sturgeons, has a downward-facing mouth. It often migrates more than 1,000km (620 miles) from the sea, laying eggs that stick to the river bed.

5 rows of bony scutes along body

Polyodon spathula
American paddlefish

	Length 1.2–1.8 m (4–6 ft)
	Habit Up to 24 kg (53 lb)
	Sex Male/Female

Location C. to S.E. North America | **Status** Common

One of the few freshwater fish that feed by straining plankton from water, the American paddlefish is often described as a living plankton net. It sweeps through the water with its lower jaw dropped and the sides of the head inflated to form a funnel-like opening, filtering large amounts of water and extracting plankton with its gills. Its distinctive paddle-shaped snout gives it stability while swimming. This fish has been heavily exploited for its flesh and eggs, which can be sold as caviar.

bluish grey, scaleless skin

Polypterus ornatipinnus
Ornate bichir

	Length 40 cm (16 in) or more
	Weight Up to 500 g (18 oz)
	Sex Male/Female

Location W. to C. Africa | **Status** Common

Bichirs are cigar-shaped fish with triangular finlets arranged in a row down the rear half of their back. They have armour-like scales and tubular nostrils, and can breathe atmospheric air since their swim bladder functions much like a primitive lung. The ornate bichir – one of 11 in the bichir family – is beige, covered by black, net-like markings. It is a slow swimmer that stalks its prey, such as small fish, amphibians, and crustaceans, quickly sucking them in when in range.

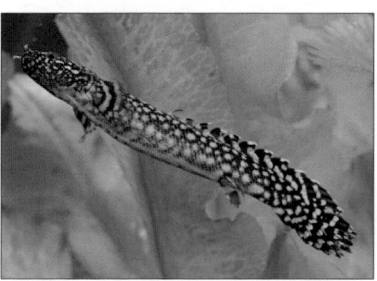

Lepisosteus osseus
Longnose gar

	Length 1.2–1.8 m (4–6 ft)
	Weight Up to 15 kg (33 lb)
	Sex Male/Female

Location C. to E. North America | **Status** Common

diamond-shaped scales

The longnose gar is one of a group of primitive, North American fish that have long jaws armed with sharp teeth. Its body is long, and its propulsive fins are set far back – much like those of pikes (see p.494). It is primarily a freshwater fish, but in the southern part of its range, adults are frequently found in brackish water. The longnose gar hangs motionless in the water, hidden by vegetation, and waits for prey to come within striking distance. Then, with a sudden thrust, it takes the prey cross-wise in its mouth, often holding it for several minutes. Its predatory habits make it unpopular among fishermen since it can become entangled in nets and cause a good deal of damage.

BONY-TONGUED FISH

PHYLUM	Chordata
CLASS	Osteichthyes
SUBCLASS	Actinopterygii
SUPERORDER	Osteoglossiformes
FAMILIES	5
SPECIES	215

THE BONY-TONGUED FISH are named after the tooth-studded tongue and palate found in most species. These relatively large fish – up to 3m (9¾ft) long – live mainly in fresh water in the tropics. This group contains such varied forms as the large Amazonian arapaima, the peculiar elephantnose fish (which has an extended, downward-curving mouth), the eel-like knifefish, and the delicate and ornate freshwater butterflyfish. Most are carnivores, feeding chiefly on invertebrates, but a few species eat detritus and plant material.

ANATOMY

Although some species also have teeth in their jaws, all bony-tongued fish have many small, sharp teeth on both their tongue and the roof of their mouth, which are used to exert the main biting motion when seizing prey. Most species in this group have a relatively elongated body with prominent eyes and large, hard scales, often with fine ornamentation, although some have reduced scales. The dorsal and anal fins are usually placed far back on the body and are often very long. Some fish in this group, particularly the elephantnose fish and the knifefish, can produce and detect electrical signals, which they use for navigation and communication. In some species, such as the arapaima and the freshwater butterflyfish, the swim bladder has a lung-like lining so that it can be used as a temporary lung in oxygen-poor water.

ELECTRIC ELEPHANTNOSE
Elephantnose fish navigate using electrical signals. An electrical field is created around the fish's body by modified muscles, and disturbances in this field are detected by receptor cells.

Notopterus chitala
Clown knifefish

Length	Up to 87 cm (34 in)
Weight	Not recorded
Sex	Male/Female

Location S. and S.E. Asia   **Status** Common

tiny, feather-like dorsal fin

With a body flattened from side to side, adult clown knifefish are conspicuously hump-backed, whereas the young are relatively slender. The unusually small dorsal fin is feather-like, giving rise to this fish's other common name, the featherback. This fish swims equally well forwards or backwards, with an undulating motion of its large anal fin – the main propulsive organ. It has a swim bladder modified to form an accessory breathing organ, and frequently rises to the surface to obtain air. It is often seen at the surface, splashing as it rolls over, exposing its silvery flanks.

Pantodon bucholzi
Freshwater butterflyfish

Length	10–13 cm (4–5 in)
Weight	Up to 500 g† (18 oz)
Sex	Male/Female

Location C. Africa   **Status** Common

This fish is a surface-feeding insect-eater, able to leap a metre (3¼ft) or more out of water and glide for short distances using its enlarged, wing-like pectoral fins. In addition to hanging motionless at the surface, it is often seen standing for long periods in shallow water, using its pelvic fin rays as stilts. It has a large, lung-like swim bladder, believed to function as an additional air-breathing organ.

dark bands on all fins

pelvic fins with elongated rays

Arapaima gigas
Arapaima

Length	2.5 m (8¼ ft), max 4.5 m (15 ft)
Weight	Up to 200 kg (440 lb)
Sex	Male/Female

Location N. South America   **Status** Common

One of the largest purely freshwater fish in the world, the arapaima has a streamlined, grey to dark greenish yellow body with long dorsal and anal fins set back close to its tail, a common feature among fish that make sudden lunges to catch their prey. Apparently unable to fulfil its respiratory requirements with its gills alone, it takes in air from the surface of the water.

light grey-green juvenile coloration

powerful, rounded tail fin

Gnathonemus petersi
Elephantnose fish

Length	Up to 23 cm (9 in)
Weight	Up to 1 kg† (2¼ lb)
Sex	Male/Female

Location W. to C. Africa   **Status** Common

Commonly found in still, murky, muddy waters of rivers, lakes, and swamps, this fish possesses a weak electrical system that functions as a kind of radar to detect obstacles, food, and mates. The highly mobile, finger-like appendage on its chin is used for probing the muddy bottom in search of food. The elephantnose has an unusually large brain that, in relation to body mass, is equalled only by that of humans. Perhaps for this reason, it has a remarkable ability to learn, and its playful behaviour makes it a popular aquarium fish.

constricted tail base

elongated, compressed body

TARPONS AND EELS

PHYLUM	Chordata
CLASS	Osteichthyes
SUBCLASS	Actinopterygii
SUPERORDER	Elopomorpha
ORDERS	4
FAMILIES	24
SPECIES	730

THIS GROUP CONSISTS of 3 main lines: the eels, the tarpons, and the halosaurs. Although they look and behave differently as adults, they all start life as a transparent larva (known as a leptocephalus larva). These long, thin larvae may drift in ocean currents for several years before reaching adulthood. Most eels are marine, but a few live mainly in fresh water and return to sea to breed. Tarpons are largely confined to coastal areas of the tropics, while halosaurs occur in oceans worldwide, down to depths of 1,800m (6,000ft).

ANATOMY

Eels are elongated and snake-like. The spinal column consists of more than 100 vertebrae, which makes the body very flexible. Long dorsal and anal fins may extend most of the length of the body, and the pelvic, pectoral, and tail fins are often very small or lost entirely. Most eels are scaleless, but some species have fine scales embedded in their skin. Tarpons look more like other fish than like eels. Their bodies are deeper and less elongated, with large, reflective scales, and well-developed and differentiated fins. The halosaurs, however, are more eel-like in shape, with a large head that tapers into a tubular body and elongated tail. They are covered in small scales and have reduced fins, some of which have sharp spines.

BREEDING AND MIGRATION

The breeding cycles of some eels involve remarkable migrations. Most marine eels spawn near where they live, but freshwater eels and some marine species have separate spawning grounds, usually far out to sea. For example, North American and European eels, both of which live in freshwater lakes and rivers, travel over 6,400km (4,000 miles) to different parts of the Sargasso Sea in the western Atlantic. It is thought that during this journey, which takes 4–7 months, they do not eat. Instead, they survive by consuming stored fat and muscle tissue. They may also absorb nutrients through the lining of their mouth. Spawning occurs in deep water, after which the adults die. The leptocephalus larvae are carried passively towards coastal waters by ocean currents. After about 3 years, they develop fins, scales, and pigmentation before beginning their journey upriver to the areas where they grow as adults.

HUNTING AND FEEDING

Tarpons, halosaurs, and most species of eel are predators. The fast-swimming tarpons use speed to catch other fish. Eels are more likely to ambush their prey, using rocks or reef crevices for cover. Many eels have long, sharp teeth for seizing prey; others have thickened teeth for crushing shells. A few species filter feed, and some search the bottom for buried prey or scavenge on dead animals. Deep-sea gulper eels have gigantic mouths and distensible stomachs, and can eat prey much larger than themselves. Halosaurs feed on invertebrates.

HUNTING
Most eels (such as the moray eel shown here) ambush prey rather than chase it as tarpons do. Moray eels are found in shallow coastal waters of tropical seas. They often hide in crevices, from which they launch surprise attacks when a suitable prey is within reach. The wide mouth is armed with sharp teeth, making it difficult for prey to escape once it has been caught.

Elops saurus
Ladyfish

Length	Up to 1 m (3¼ ft)
Weight	14 kg (31 lb)
Sex	Male/Female

Location W. Atlantic, Gulf of Mexico, Central America, Caribbean **Status** Common

Also known as the ten-pounder, this sleek, blue-grey fish is covered with fine scales. It attacks small fish in shoals, and is well known for its habit of skipping along the surface and for leaping when hooked. Adult ladyfish move out from the coast into open water to spawn in shoals; exact locations are not known, but they are believed to spawn up to 160km (100 miles) offshore. The transparent, eel-like larvae eventually drift or migrate back to inshore waters.

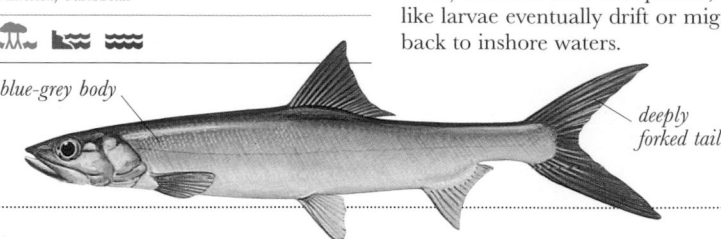

blue-grey body

deeply forked tail

Megalops atlanticus
Tarpon

Length	1.3–2.5 m (4¼–8¼ ft)
Weight	160 kg (350 lb)
Sex	Male/Female

Location W. and E. Atlantic **Status** Common

A large and powerful fish, with a deep body and an upturned mouth that extends far beyond the eyes, the tarpon looks like an unusually large herring. It is found in coastal waters and estuaries, and sometimes also in fresh water, and feeds almost exclusively on shoaling fish such as sardines, anchovies, and mullets. Its lung-like swim bladder enables it to overcome the occasional shortages of oxygen that occur in estuaries and similar kinds of habitat, by breathing air at the surface. It is a popular sport fish, and makes spectacular leaps when hooked. Between late April and August, adult tarpons move into open water to spawn. The transparent, eel-like larvae ultimately end up in estuarine nursery grounds, but also frequently occur in pools or lakes disconnected from the sea.

long ray at rear end of dorsal fin

unusually large eyes

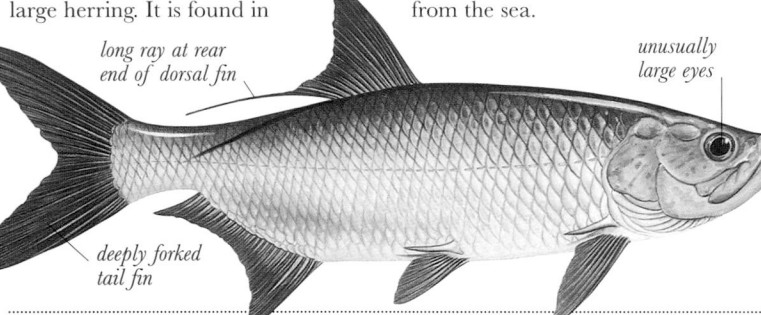

deeply forked tail fin

Avocettina infans
Snipe eel

Length	Up to 75 cm (30 in)
Weight	0.5 kg† (1 lb)
Sex	Male/Female

Location Tropical, subtropical, and temperate waters worldwide **Status** Lower risk

A deep-sea eel with a remarkably slender, brown or black body ending in a whip-like tail, the snipe eel belongs to a family of about 9 species. Its long, thin jaws cannot be fully closed and have numerous teeth. This eel probably feeds exclusively on small crustaceans. Immature male and female snipe eels look very similar, but as the males approach sexual maturity, they undergo a marked transformation. Their jaws become drastically shortened, and all their teeth are lost. At this stage of the life cycle, males and females look so different that at one time they were classed in separate families. Very little is known about the snipe eel and its relatives. Individuals are usually only captured by specially designed, deep-sea fishing gear during scientific research expeditions.

whip-like tail

large eyes

open jaws

slender, fragile, tapering body

Anguilla anguilla
European eel

Length	Up to 1 m (3¼ ft)
Weight	4.5 kg (10 lb), max 13 kg (29 lb)
Sex	Male/Female

Location E. North Atlantic, Mediterranean, Europe **Status** Common

This common European fish has a remarkable life history. Adults are believed to spawn in the area of the Sargasso Sea, the larvae being transported over a period of 2½ years to European coastal waters by ocean currents. There they metamorphose into cylindrical, unpigmented "glass eels" before entering rivers in their millions, as silvery, pigmented juveniles called "elvers". The freshwater stage is a feeding and growing phase; as the eel matures sexually, it descends towards the sea. Mature "silver eels" undertake their spawning migration usually from late summer to winter and mainly on dark, moonless, stormy nights. At this stage, the eyes enlarge, the snout becomes more narrow and pointed, and the pectoral fins more lance-shaped – changes that are adaptations to the ocean's depths. As with all true eels, the European eel's skin appears scaleless, and a single long fin runs down its back, around the tail, and along its underside.

ADULT COLORATION
In its freshwater stage, this fish has a yellowish or even golden underside. As it approaches sexual maturity, it becomes silvery below and almost black above (as shown here).

cylindrical body

single fin

OUT OF WATER

European eels can survive out of water for several hours in damp, cool conditions. Adults usually travel overland on rainy nights, when the darkness gives them some protection from attack.

Gymnomuraena zebra
Zebra moray

Length	Up to 89 cm (35 in)
Weight	10 kg (22 lb)
Sex	Male/Female

Location Pacific, Gulf of California, Indian Ocean **Status** Locally common

Morays are thick-bodied and often powerful eels that typically lurk in clefts in rocks or coral with only their heads showing. These fish have no pectoral or pelvic fins. There are over 100 species, and many, including the zebra moray, are brightly coloured. This snake-like fish has a long, cylindrical, highly muscular body, and a large mouth with blunt and pebble-like teeth, adapted for crushing prey. It usually adopts a lie-and-wait feeding technique, and typically lurks in rock or coral crevices, lunging outwards to catch its prey. However, it may occasionally leave its hole at night in search of prey. The zebra moray feeds on large and heavily armoured prey, particularly crabs and other crustaceans, found around reefs. There is evidence that it can eat larger crabs than can other morays of comparable size. An aggressive fish, it is known to defend its territory strongly against intruders. Some morays, although not necessarily this species, are noted for the presence of toxins on their skin.

blunt snout

scaleless skin

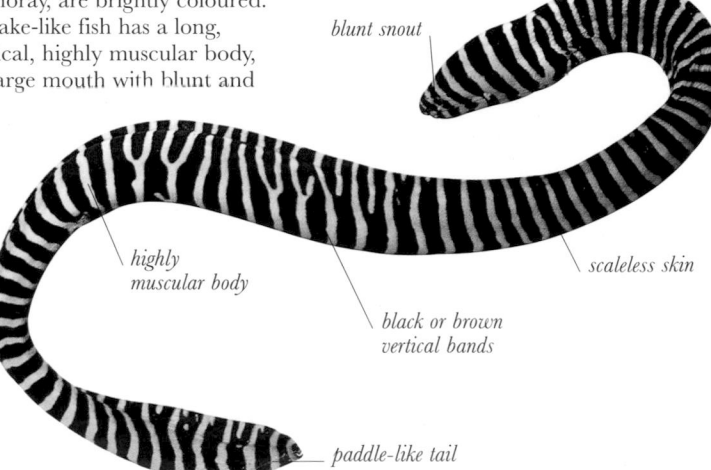

highly muscular body

black or brown vertical bands

paddle-like tail

Eurypharynx pelecanoides
Gulper eel

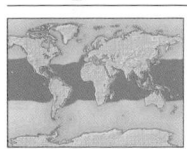

Length 60–100 cm
(23 1/2 – 39 in)
Weight 1 kg†
(2 1/4 lb)
Sex Male/Female

Location Tropical and subtropical waters worldwide

Status Lower risk†

This deep-sea fish is one of the most unusual of eels, with a thin, tapering body, but enormous jaws equipped with tiny teeth. Its huge gape and expandable stomach enable it to swallow and digest fish almost as large as itself. Due to its odd shape and fragile build, the gulper eel is almost certainly a poor swimmer and may not pursue other fish. Instead, it probably relies heavily on the luminescent organ at the tip of the tail to attract prey.

large jaws

body tapers to thin filament

Aldrovandia affinis
Halosaur

Length Up to 55 cm
(22 in)
Weight 6 kg†
(13 lb)
Sex Male/Female

Location Tropical, subtropical, and temperate waters worldwide

Status Lower risk†

This relatively rare fish is found hovering in mid-water only between 700 and 2,000m (2,300–6,600ft). This species is usually only captured by specially designed, deep-sea fishing gear during scientific research expeditions. Little is known about the halosaur but its well-developed nasal organs, particularly among the males, seems to indicate that sense of smell is a means of communication between the sexes.

well-developed nose

large, smooth scales

anal fin extends to tip of tail

Saccopharynx ampullaceus
Swallower

Length Up to 1.6 m
(5 1/4 ft)
Weight 1 kg†
(2 1/4 lb)
Sex Male/Female

Location North Atlantic

Status Lower risk†

distensible stomach

The swallower is a relative of the gulper eel (see above), with a similar small head and large jaws (although smaller than the gulper's). It also has a luminescent organ at the tip of its tail which, if dangled in front of its mouth, may be used to entice prey. It is thought that this fish swims slowly, with its enormous jaws swung open so that its prey literally swims into its mouth. A little-studied fish, the swallower belongs to a rare family, known from fewer than 100 specimens.

Heteroconger hassi
Spotted garden eel

Length Up to 36 cm
(14 in)
Weight 2 kg†
(4 1/2 lb)
Sex Male/Female

Location Red Sea, Indian Ocean, W. Pacific

Status Locally common

This eel lives in large colonies, with the lower half of its body buried in the sandy sea bed. The upper half of its body projects into the water, creating "gardens" that sway gracefully with the current.
When a colony is approached by a diver, the eels nearest the diver slowly sink, each into its individual burrow, to hide from the intruder, in perfect gradation from one end of the "garden" to the other. The spotted garden eel feeds by picking up small organisms, such as tiny planktonic invertebrates or larval fish, one by one, from the water flowing by.

tail used as anchor in burrow

dark spots behind head

Albula vulpes
Bonefish

Length Up to 1 m
(3 1/4 ft)
Weight 9 kg
(20 lb)
Sex Male/Female

Location Atlantic

Status Common

With its slender, streamlined body, long head, and high dorsal fin, the bonefish is very much a "standard" fish. Found in shallow, coastal marine habitats, typically over mud flats and mangrove lagoons, it is more important as a sport fish than as food. It feeds on a rising tide, working its way to the muddy bottom with its head down and tail up, blowing jets of water to uncover its prey – usually clams, crabs, and shrimps. Its large size protects it from marine predators.

faint vertical stripes on adults

conical, scaleless head

Conger conger
Conger eel

Length 2.7 m
(8 3/4 ft)
Weight 65 kg
(145 lb)
Sex Male/Female

Location E. North Atlantic, Mediterranean

Status Common

This dark grey, thick-bodied fish is one of more than 100 species in the conger family. As with most eels, it has no scales or pelvic fins. With its large size and well-equipped jaws providing protection from most predators, it hunts at night for fish, crustaceans, and cephalopods. The conger eel is considered an excellent food fish and is caught by anglers in large numbers. It can be found on rocky or sandy bottoms, at depths of up to 100m (330ft), but in summer, adult fish migrate into deeper, offshore waters to spawn. The female conger eel lays 3–8 million eggs, which hatch to produce transparent, eel-like larvae.

IN HIDING

Like most of its relatives, the conger eel hides during the day in crevices, emerging only at night to ambush prey. Small conger eels are common in tide pools on rocky shores, as well as in rocky offshore habitats while large ones often inhabit wrecks of sunken ships.

These drift inshore for 1–2 years before growing into juvenile eels; they reach sexual maturity at 5–15 years.

SNAKE-LIKE FISH
The conger eel, with its snake-like body, has a long snout, and its gill openings are restricted to a small, crescent-like slit on each side of its body.

Myrichthys colubrinus
Ringed snake eel

Length Up to 88 cm
(35 in)
Weight 3 kg†
(6 1/2 lb)
Sex Male/Female

Location Indian Ocean, W. Pacific

Status Locally common

There are some 175 species of snake eels worldwide and they occur in a wide range of colours and shapes. They are mostly less than 1m (3 1/4 ft) long, and burrow in shallow, sandy and muddy bottoms of lagoons, sandy flats, and reefs. This species has a hardened, sharply pointed tail, modified for burrowing tail-first into the sand. Primarily active at night, it feeds on sand-dwelling fish and crustaceans, and seems to detect its prey by smell.

hard, pointed tail

25–32 black rings on body

HERRINGS AND RELATIVES

PHYLUM	**Chordata**
CLASS	**Osteichthyes**
SUBCLASS	**Actinopterygii**
SUPERORDER	**Clupeomorpha**
ORDERS	**1 (Clupeiformes)**
FAMILIES	**4**
SPECIES	**363**

THESE SMALL, STREAMLINED fish are widespread and abundant. Often forming vast shoals, they are an important part of marine food chains and vital to many coastal economies. The herring family includes sprats, shads, and pilchards; it also includes sardines, although this term is often used to refer to the young or adults of other species. Herrings and anchovies occur from the tropics to temperate zones, in shallow, coastal marine waters and freshwater streams and lakes. Some species live in marine waters but spawn in brackish or fresh waters.

ANATOMY

Members of the herring family are similar to one another in overall appearance. They have streamlined bodies that are flattened from side to side, with large scales and well-developed fins without any supporting spines. Most are silver with green or blue hues on the back. Anchovies are generally longer and thinner than herrings, and have a large mouth with a prominent snout. In both anchovies and herrings, the swim bladder is linked to the ear apparatus, which is thought to enhance their hearing. At one time, herrings and their relatives were considered to be related to tarpons and eels because of similarities between their larvae, but now the consensus is that they are sufficiently different to form a group of their own.

FEEDING

Almost all herrings and their relatives feed on planktonic organisms (mainly crustaceans and their larvae), which they filter out of the water by opening their large mouths to expose their gill rakers. Following the daily vertical movements made by plankton, they usually feed near the surface at night and move to lower depths during the day. Their feeding habits often vary seasonally; herrings and some other species cease to feed during the breeding season. All members of this group typically feed in dense groups that often contain tens of thousands of fish, thus attracting predators, including other fish (such as salmon and tuna), seabirds, and marine mammals. By contributing to the diet of so many other species, herrings and anchovies form a key element in many marine ecosystems.

CONSERVATION

Herrings and anchovies have long been valued by humans, and there is a long history of commercial exploitation. Easy to catch and preserve, they are an important source of food and oil. During the days of sailboat fishing, catches were relatively stable, but with great advances in fishing technology, some areas were overfished to the point of economic collapse. Herring catches are now widely regulated, with many fishing areas closed at some or all times of the year.

SHOALING
Herring shoals are at their largest during the breeding season, which usually falls in the warmer months of the year. Each female may release up to 40,000 eggs, which are attached to the sea bottom.

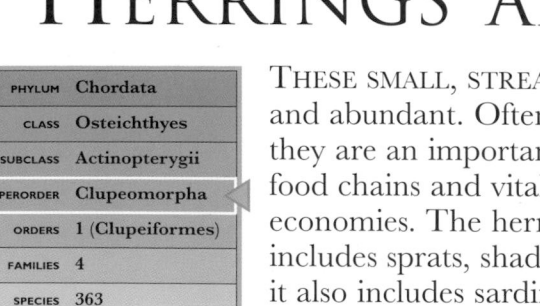

Clupea harengus

Atlantic herring

Length Up to 40 cm (16 in)

Weight Up to 700 g (25 oz)

Sex Male/Female

Location North Atlantic, North Sea, Baltic Sea

Status Common

For centuries, the sleek, silver-scaled herring has been one of the most important commercially fished species in the North Atlantic. Its numbers declined precipitously during the 20th century, due to improved fishing techniques, but are now recovering slowly because of active management. Like most of its relatives, the Atlantic herring is a plankton feeder, and lives in large, highly mobile shoals. It comes to the surface at night, but spends the day in deeper water. Across its wide range, the species is divided into many local races, which differ from each other in habits and in size. Each race uses a number of traditional spawning grounds. Young herrings, which initially resemble tiny eels, hatch on the sea bed, but soon swim upwards to feed close to the surface. They are eaten in vast numbers by other fish, and only a tiny proportion of them survive to become adults.

dorsal fin in centre of body

strongly forked tail

heavy lower jaw

Pristigaster cayana
Hatchet herring

Length 14.5 cm
(5 3/4 in)
Weight Not recorded
Sex Male/Female

Location N. South
America

Status Unconfirmed

This narrow-bodied, freshwater fish has a distinctive profile, with a deeply curving belly. Like many members of the herring family, its underside has a line of modified scales – or scutes – which have sharp, backward-facing points. Hatchet herrings tend to remain in shoals for safety and are omnivorous, filtering plankton from the water through their gill rakers. They inhabit the Amazon river basin as far west as Peru and Colombia. They are also believed to enter the slightly saline waters of the river mouth, but this is not well documented.

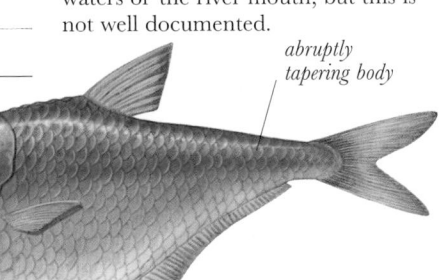

abruptly tapering body

Chirocentrus dorab
Wolf herring

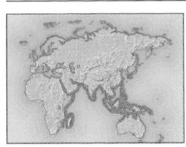

Length 1 m
(3 1/4 ft)
Weight 400 g†
(14 oz)
Sex Male/Female

Location Indo-Pacific

Status Common

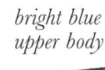

bright blue upper body

This is the largest member of the herring family. It has an elongated, bright blue body and numerous teeth, with 2 fang-like canines protruding outwards from its upper jaw. Unlike most herrings, it actively hunts other fish. Little is known about the behaviour of this species, but the fact that it is a large carnivore strongly suggests that it hunts alone or in small groups, rather than in large shoals. Along with the closely related whitefin wolf herring, it is extensively fished in some parts of the world.

Denticeps clupeoides
Denticle herring

Length 8 cm
(3 1/4 in)
Weight Not recorded
Sex Male/Female

Location W. Africa

Status Unconfirmed

This small silver-coloured fish is the most primitive member of the herring group, being virtually indistinguishable from fossil forms. Modified scales along the underside form backward-pointing serrations, and tooth-like denticles cover the head. Unlike most herrings, it lives in fast-flowing, fresh water. It is also the only herring to have a sensory line running along the side of its body. This fish is omnivorous.

Sardinops caeruleus
California pilchard

Length 25–36 cm
(10–14 in)
Weight 475 g
(17 oz)
Sex Male/Female

Location E. North Pacific

Status Common

This medium-sized silvery fish can be recognized by its distinctive black spots, arranged in rows along its body. In California, it is canned and marketed as the sardine, and is also extensively used to produce oil and fish meal. At one time shoals consisting of an estimated 10 million individuals were found. During the 1930s, over half a million tonnes a year were caught, but stocks dwindled during the 1950s, and collapsed completely around 1967. Today, the numbers of California pilchard are recovering, but stocks are still far from their original levels.

Alosa sapidissima
American shad

Length 50–60 cm
(20–23 1/2 in)
Weight 2 kg
(4 1/2 lb)
Sex Male/Female

Location W. North
Atlantic, E. North
America, E. and W. Pacific

Status Common

With its streamlined silvery body, the American shad looks like many other fish in the herring order but it is distinctive in that it enters rivers to spawn. Beginning at the age of 4 or 5, American shad migrate to their natal river every year to spawn. Spawning occurs near the shore, after sunset, usually in late summer. Males arrive first, followed by the females, and pairs swim erratically near the surface with their dorsal fins showing.

Adults return to the ocean after spawning, and the young shad hatched out in the river migrate to the oceans in autumn. The numbers of American shad have suffered because of dam construction and water pollution. The Pacific populations of this fish have been introduced.

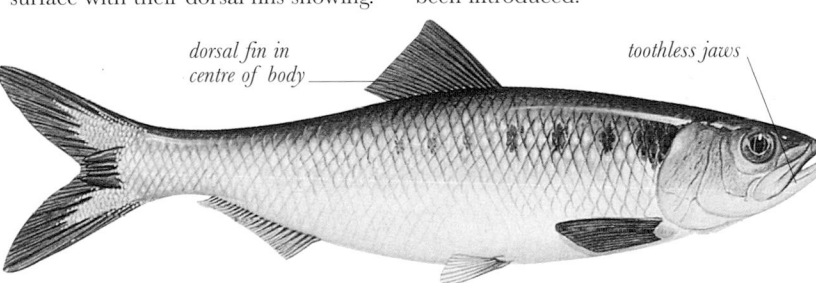

dorsal fin in centre of body

toothless jaws

Engraulis ringens
Peruvian anchoveta

Length Up to 20 cm
(8 in)
Weight 25 g
(7/8 oz)
Sex Male/Female

Location E. South Pacific

Status Common

Like its close relative the northern anchovy (see right), this small silvery fish feeds on plankton, and forms enormous shoals where upwelling currents create a good supply of food. Following the plankton as they migrate, they descend to depths of about 50m (165ft) during the day and rise to the surface at night. Peruvian anchovetas also shoal in huge numbers for safety. This fish spawns largely from winter to early spring.

SLIM FISH
This slender fish, with a pointed snout and deeply forked tail, has a marked resemblance to the northern anchovy. It swims with an open mouth, filter feeding with the help of its fine gill rakers.

CONSERVATION

Peruvian anchoveta numbers are susceptible to sudden crashes, triggered both by El Niño – a periodic rise in the temperature of the Pacific – and by over-fishing. In the 1960s, the total catch of this fish often exceeded 10 million tonnes a year, a huge figure for a fish that individually weighs about 25g (7/8oz). Recently, the stocks of Peruvian anchovetas collapsed completely, with severe repercussions for local people and the wildlife in its food chain.

Engraulis mordax
Northern anchovy

Length Up to 25 cm
(10 in)
Weight 50–60 g
(1 3/4–2 1/8 oz)
Sex Male/Female

Location E. North Pacific

Status Common

This small but abundant fish is important commercially and a vital link in the marine chain, providing food for other fish and fish-eating birds. Slender, with a distinctly pointed snout, it swims with its mouth open, filtering plankton from the water with its gill rakers. The northern anchovy is one of about 150 species in the anchovy family, many of which are poorly known and difficult to tell apart. Anchovies are found in all the oceans of the world.

deeply forked tail

CATFISH AND RELATIVES

PHYLUM	**Chordata**
CLASS	**Osteichthyes**
SUBCLASS	**Actinopterygii**
SUPERORDER	**Ostariophysi**
ORDERS	4
FAMILIES	62
SPECIES	About 6,000

CATFISH AND THEIR relatives account for three-quarters of all species of freshwater fish. Although they are a varied group, almost all species have a bony internal structure, known as the Weberian apparatus, that is used in sound reception and contributes to their acute hearing. They are divided into 4 groups: milkfish and relatives, cyprinids (the most widespread of which are carp and minnows), characins (including tetras, tigerfish, and piranhas), and catfish themselves. Although mostly confined to fresh water, some species are found in brackish water or the sea.

ANATOMY

Catfish and their relatives vary greatly in size and shape. Most are small fish, less than 10cm (4in) long, although some species reach 5m (16ft). The Weberian apparatus that unites most members of this group consists of modified vertebrae that act as a set of levers to transmit sound waves received by the swim bladder to the inner ear. This feature gives catfish and their relatives an acute sense of hearing, and many species use sound to communicate with each other.

Catfish typically have a flattened head with a wide mouth that is usually surrounded by whisker-like sensory barbels. These are used for touch and taste, and they compensate for poor eyesight. Most species have smooth, scaleless skin but in several groups the scales have developed into a dense armour.

Nearly all species have sharp or serrated pectoral and dorsal spines, some of which can release venom.

Characins and cyprinids look similar to one another, although most characins are small and often brightly coloured, while cyprinids are generally dull cloured and can reach lengths of up to 2m (6½ft). Another distinguishing factor is the arrangement of the teeth: characins have teeth in their jaws, often with a replacement row directly behind the functional teeth, whereas cyprinids have toothless jaws but usually have teeth in the pharynx. Characins, unlike cyprinids, also have a small fin (called an adipose fin) between the dorsal fin and tail, which is not supported by rays and stores fat.

Milkfish and their relatives (which include the beaked salmon) are the only members of this group that do not have a Weberian apparatus. Instead, vibrations are transmitted from the swim bladder to the inner ear by a modified set of ribs.

FEEDING

This group includes voracious predators, scavengers, herbivores, and filter feeders. Catfish are generally predators or scavengers, living close to the bottom in murky water, although a few species are herbivores. Most characins are shoaling or free-swimming fish that eat plants or prey on invertebrates or other fish. Most species of cyprinids have downward-turned, often sucker-like mouths, with one or more pairs of sensory barbels on either side. They usually feed on detritus or on small invertebrates and plants in the sediment; few are predatory. Milkfish are filter feeders but some of their relatives feed on small animals.

NIGHT HUNTER
Most catfish are active at night and feed alone. The barbels around the mouth are covered with taste buds, which help them to find food. This long-whiskered (or pimelodid) catfish is found in tropical rivers in South America. The larger members of this group feed on fish and other animals, including monkeys that fall into the water.

Chanos chanos
Milkfish

Length Up to 1.8 m (6 ft)	
Weight Up to 14 kg (31 lb)	
Sex Male/Female	

Location E. Pacific, North and South America, Indo-Pacific, Asia, Australia

Status Common

A large, fast-moving, filter feeder, the milkfish has a streamlined silvery body and an unusually large, deeply forked tail. It can tolerate a wide range of salinity but spends most of its life in fresh water, swimming out to the sea to spawn. The larvae and juveniles develop in warm, coastal wetlands, in brackish and freshwater environments. Milkfish larvae are collected at sea and cultured in ponds for food in Southeast Asia and the west Pacific; adults are also prized as food fish.

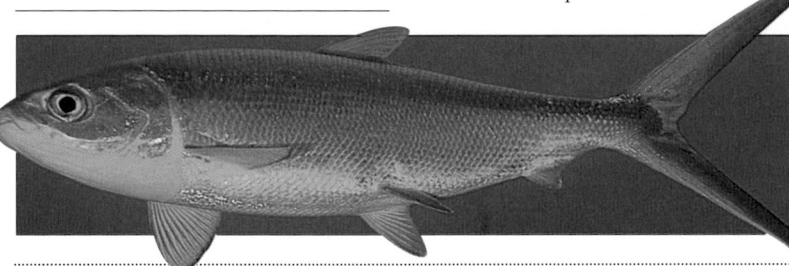

Gonorhynchus gonorhynchus
Beaked salmon

Length Up to 60 cm (23 1/2 in)	
Weight Not recorded	
Sex Male/Female	

Location Mozambique Channel to Cape of Good Hope

Status Locally common

Also known as the beaked sandfish – a more accurate name since it is not a true salmon – this fish is adapted for life on the sea bed, where it spends the daytime hiding in sand or mud. Its upperside is well camouflaged, and its eyes are protected from abrasion by a thin covering of skin. It has silvery flanks, and the tips of its anal and tail fins are tinged with orange or pink. Its dorsal fin is set far back on the body. The beaked salmon is found on sandy bottoms along the southern shores of Africa, at depths of up to 200m (660ft). It feeds on invertebrates that live on the sea bed, and its small, downward-pointing mouth is surrounded by small sensory flaps, or papillae, that help it to find food.

Cyprinus carpio
Common carp

Length Up to 1.2 m (4 ft)	
Weight Up to 37 kg (82 lb)	
Sex Male/Female	

Location W. Europe to S.E. Asia

Status Common

Originally from Europe and Asia, this deep-bodied, freshwater fish has been raised in semi-captivity for centuries, and introduced to rivers, lakes, and ponds in many parts of the world. It uses its protrusible mouth to grub through the bottom sediment for plant and animal food.

protrusible mouth

ORNAMENTAL CARP

Selective breeding has created different varieties of common carp over the centuries. Mirror carp have large scales, while leather carp have no scales at all. Leather carp are raised for food; the multi-coloured variety shown here is kept as an ornamental fish.

TOOTHLESS JAWS
The jaws of the common carp have no teeth. It grinds food with the pharyngeal teeth at the back of the throat.

Rhodeus amarus
Bitterling

Length Up to 11 cm (4 1/4 in)	
Weight Not recorded	
Sex Male/Female	

Location C. and E. Europe, E. Asia

Status Locally common

This small, freshwater fish lives in overgrown waterways, where oxygen levels are often too low for most fish. The females are silvery, but during the breeding season the males become an iridescent silvery-blue. Bitterlings have an unusual way of reproducing: using a long egg-tube, or ovipositor, the female lays her eggs in the mantle cavity of a freshwater mussel, over which the male sheds his sperm. The young grow in this "living nursery", without harming their host.

fins with rays

deep body

small head

Myxocyprinus asiaticus
Chinese sucker

Length Up to 60 cm (23 1/2 in)	
Weight Not recorded	
Sex Male/Female	

Location S.E. Asia

Status Locally common

This slow-moving, hump-backed fish is one of the few Asian species in the sucker family. It has protrusible, fleshy lips, which it uses to collect small animals from river-bed mud. Its body is triangular in cross-section, with a pointed back, or hump, and a flat underside. It faces upstream to feed, so that the current pushes it against the river bed – an adaptation that prevents it from being swept away.

dark bands

silver to orange coloration

Botia macracanthus
Clown loach

Length Up to 30 cm (12 in)	
Weight Not recorded	
Sex Male/Female	

Location S.E. Asia

Status Common

A popular aquarium fish, the clown loach is a bottom feeder, feeling for aquatic invertebrates with the 4 pairs of barbels that hang down like a moustache from its mouth. Many loaches – including this one – can defend themselves by flicking out a sharp spine in front of each eye. They use these spines against potential predators, and sometimes against each other. The clown loach spawns in fast-flowing rivers, at the onset of the rainy season.

moustache-like barbels

3 broad, wedge-shaped black bands

Brachydanio rerio
Zebrafish

Length Up to 6 cm (2 1/4 in)	
Weight Not recorded	
Sex Male/Female	

Location S. Asia

Status Common

elongated body

4 dark, lateral stripes

Also known as the zebra danio, this highly active, freshwater fish has a slender, silvery-yellow body, with a characteristic pattern of 4 horizontal stripes running from its head to the tip of its tail. This tiny fish feeds on freshwater invertebrates, including worms, insect larvae, and crustaceans. The female is slightly larger than the male, and breeds by scattering her eggs in open water, or on the bottom of ponds and pools. Zebrafish are widely used as a model organism in studies of genetics. Several varieties of this popular aquarium fish have been created by breeders.

FISH

Hydrocynus vittatus
Tigerfish

	Length 1 m (3 1/4 ft)
	Weight Up to 18 kg (40 lb)
	Sex Male/Female
Location Africa	Status Common

This African fish is a fierce, freshwater predator, and a well-known gamefish, prized for its habit of fighting back when hooked. It has a sleek silvery body, marked with dark horizontal stripes, a deeply forked tail edged with black, and large, fang-like teeth. Tigerfish move in shoals while feeding, and swallow their prey whole, head first. They can eat fish up to half their length. Apart from humans, the African fish eagle is their only significant enemy. The giant tigerfish

(*Hydrocynus goliath*), a relative of this species and also from Africa, is one of the largest characins, growing up to 1.8m (6ft) long.

dark spot on top of second dorsal fin

large eyes   *pelvic fin directly below dorsal*

Gyrinocheilus aymonieri
Siamese algae-eater

	Length 28 cm (11 in)
	Weight Not recorded
	Sex Male/Female
Location S.E. Asia	Status Common

Found on solid surfaces at the bottom of flowing water, this slender fish has a modified mouth and lips that it uses to attach itself to rocks and plants, scraping plant material from them. It is golden green above and silvery below, and has a large dorsal fin with 9 branched fin rays. The Siamese algae-eater is commercially exploited for food and the aquarium trade.

dark spots or stripe along sides

Anostomus anostomus
Striped headstander

	Length 16 cm (6 1/2 in)
	Weight Not recorded
	Sex Male/Female
Location N. South America	Status Locally common

This slender, torpedo-shaped fish is golden yellow, with 3 broad, black horizontal bands running along the body, from the snout to

the tail fin. The slender head has an upturned snout, and all fins have reddish bases. This fish gets its name from its habit of swimming head-down, at an oblique angle to the beds of rivers and streams. It swims in small shoals (containing up to 40 individuals), often remaining motionless among vegetation. Its upturned mouth enables it to feed from leaves and stems.

3 dark stripes across body   *reddish base of fins*

Copella arnoldi
Splash tetra

	Length 8 cm (3 1/4 in)
	Weight Not recorded
	Sex Male/Female
Location N. South America	Status Locally common

smaller, lighter fins of female   *dark blotch on female*

Also called the jumping characin, this finger-sized fish has a slender, creamy yellow body, with a black stripe on the head and through the eye. It belongs to a group of freshwater species, the lebiasinids, often kept in aquariums. This particular species gets its name from its distinctive reproductive behaviour. The female lays her eggs on the underside of overhanging leaves, so that when the water level drops, they are safe from most predators. The male then splashes them with water to keep them alive and the young fish drop back into the water after they have hatched. Splash tetras feed on worms, insect larvae, and crustaceans.

Astyanax mexicanus
Mexican tetra

	Length 12 cm (4 3/4 in)
	Weight Not recorded
	Sex Male/Female
Location S. Central USA to Central America	Status Locally common

Found in springs, creeks, and small rivers, the Mexican tetra typically has a silvery body with red and yellow fins. This fish is of interest to biologists because a variant of the species is found in caves. The subterranean form, called blind cave characin (pictured below), looks very different, with its reduced pigmentation and non-functioning eyes covered by skin. Like other cave-dwelling fish, it finds food by using its pressure-sensing lateral line and its keen sense of smell. Despite the differences in colour and eye development, the 2 forms of Mexican tetra are similar at a genetic level, and can interbreed.

reduced pigmentation

Pygocentrus nattereri
Red piranha

	Length 33 cm (13 in)
	Weight 1 kg (2 1/4 lb)
	Sex Male/Female
Location N., C., and E. South America	Status Common

Notorious for its predatory, pack-feeding behaviour, the red piranha is extremely variable in appearance, displaying a wide diversity of colour and shape in different parts of its range. Red piranhas typically feed at dawn and dusk, lurking and then dashing at their prey. They generally eat animals smaller than themselves, such as

insects, aquatic invertebrates, or other fish, and will also feed on seeds and fruits. However, given an opportunity, these extremely aggressive, voracious predators can kill much larger prey by hunting as a group, feeding in a frenzied rush reminiscent of the behaviour of some sharks. These mass attacks can kill animals as large as capybaras and horses, and

RED BELLY
The red piranha is usually silvery, with silvery red-flecked eyes, and a head that is dark grey above and orange-red below. In large individuals, the underside is often intensely red – giving this fish its name.

have also resulted in some human deaths. During the breeding season, female red piranhas lay up to 1,000 relatively large eggs, and attach them to the roots of trees trailing in the water. Both parents, particularly the male, guard the eggs until they hatch, which takes about 9–10 days.

thick fins   *numerous body spots*

SLICING TEETH

Red piranhas have small but sharp, triangular teeth set in powerful jaws. The teeth interlock when the fish's mouth is closed, enabling it to slice off chunks of flesh. The blunt snout and underslung lower jaw also help it to attack, allowing it to bite with unusual force.

Thoracocharax stellatus
Spotfin hatchetfish

Length	6.5 cm (2 1/2 in)
Weight	Not recorded
Sex	Male/Female

Location N. to C. South America — Status Common

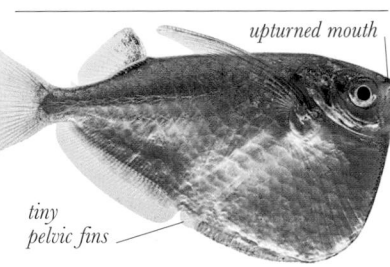

upturned mouth

tiny pelvic fins

The spotfin hatchetfish gets its name from its narrow shape, and deep, blade-like keel. The most unusual feature of this silver-grey fish is its highly enlarged pectoral fins. When threatened, or when chasing prey, the spotfin hatchetfish uses these fins to gain speed in the water and even to take off into the air. Marine hatchetfish (see p.499) belong to a different group of fish, and have quite a different lifestyle.

Synodontis contractus
Upside-down catfish

Length	9.5 cm (3 3/4 in)
Weight	Not recorded
Sex	Male/Female

Location C. Africa — Status Locally common

This African catfish gets its name from its habit of swimming upside down near the surface of rivers and streams, particularly when taking food from the surface or feeding from the undersides of broad leaves and rocks. Its diet includes plant material, aquatic invertebrates, and insect larvae. The upside-down catfish is entirely brown or brown-violet, with a series of small spots scattered across its body. Most fish are dark above and light beneath – a pattern called countershading that serves as a means of camouflage. In this fish, the arrangement is often reversed, with the underside being darker than the back. The upside-down catfish makes sounds, which are possibly used to communicate with others of its species.

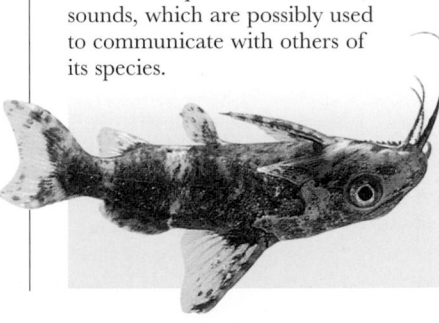

Hoplias macrophthalmus
Giant trahira

Length	Up to 1 m (3 1/4 ft)
Weight	Up to 2 kg (4 1/2 lb)
Sex	Male/Female

Location Trinidad, Central America to C. South America — Status Common

The giant trahira has a thick, stocky body with a blunt snout, very large eyes, and a somewhat rounded stomach. It has 2 or 3 canines in its upper jaw, in addition to a series of conical teeth. This fish can tolerate low oxygen conditions and has been known to move across land at night between bodies of water. An important food fish, it is raised in aquaculture.

Silurus glanis
Wels

Length	Up to 5 m (16 ft)
Weight	Up to 300 kg (660 lb)
Sex	Male/Female

Location C. Europe to C. Asia — Status Lower risk

This huge, bottom-dwelling catfish is one of the largest freshwater fish in the world. The biggest specimen on record, caught in the 19th century in the River Dnieper in southern Russia, was over 4.5m (15ft) long, and weighed over 300kg (660lb). However, it is unlikely that any wels of a similar size exist today since they have been heavily fished in most parts of their range. The wels is a solitary fish, lurking at the bottom of rivers and large lakes. It moves about mainly at night, camouflaged by its greenish grey markings which conceal it against river-bed mud. Like most catfish, it is a voracious predator, eating waterfowl and aquatic mammals, besides crustaceans and smaller fish. Females lay their eggs when the water temperature rises above 20°C (68°F). The male initially guards both the eggs and the young.

large, flat head

long anal fin

Kryptopterus bicirrhis
Glass catfish

Length	15 cm (6 in)
Weight	Not recorded
Sex	Male/Female

Location S.E. Asia — Status Common

Many catfish are camouflaged by their dark markings, but this slender species is protected by being almost transparent. The only clearly visible parts of its body are the backbone, the eyes, and the region behind the

Ameiurus nebulosus
Brown bullhead

Length	Up to 52 cm (20 1/2 in)
Weight	2.5–3.5 kg (5 1/2–7 3/4 lb)
Sex	Male/Female

Location E. North America — Status Common

fleshy barbels — *mottled body*

This common catfish belongs to a large family of freshwater fish found in North America. Like other members of the family, it is valued as a food fish. It has a scaleless body, and a large mouth surrounded by 4 pairs of fleshy barbels, which it uses to find food. Its jagged, dagger-like pectoral spines with attached venom glands are a formidable weapon against predators. Both parents guard the eggs in the nest, fanning to increase the air flow around them, and herd the young in tight shoals until they reach a length of about 5cm (2in).

FOOD SENSORS

The wels has 2 very long barbels attached to its upper jaw, and 4 smaller ones below its mouth, which help it to find food.

HIDDEN GIANT
Renowned for its extraordinary size, this well-camouflaged fish has a large, flat head and a long fin that extends over half the length of its underside.

head that contains its digestive and reproductive organs. Its transparency is the result of thin skin, and of body tissues that are suffused by oils, which make its flesh translucent. The glass catfish's body is flattened from side to side, and its dorsal fin is reduced to one or 2 rays, or is missing altogether. As in many other catfish, it has a very long anal fin, which stretches from the region behind its head and merges

with the base of its tail. This fish inhabits lowland flood plains and large rivers with turbid waters. Individuals gather in small groups at an oblique angle to the water surface. Several other groups of fish contain species that have transparent or nearly transparent bodies. The Indian glassfish, which actually comes from Southeast Asia, is often kept in aquariums.

deeply forked tail — *transparent body*

extremely long anal fin

Ancistrus dolichopterus
Bushymouth catfish

	Length Up to 13 cm (5 in)
	Weight Not recorded
	Sex Male/Female
Location N. South America	**Status** Common

The bushymouth catfish and its close relatives get their name from the distinctive fleshy tentacles on the upperside of their snouts. The males have more tentacles than the females, giving rise to an as yet unproven theory that they are linked to their fitness to breed. When threatened, the bushymouth catfish erects its pectoral and dorsal fin spines, as well as the spines on the sides of its head.

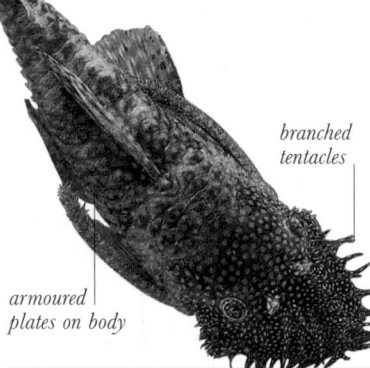

branched tentacles

armoured plates on body

Vandellia cirrhosa
Candiru

	Length Up to 2.5 cm (1 in)
	Weight Not recorded
	Sex Male/Female
Location N. South America	**Status** Common

A tiny catfish with a slender, almost transparent body, the candiru has an unusual way of life. Instead of feeding in open water, it swims into the gill cavities of larger fish – particularly larger catfish and characins – using the large, hooked spines of its own gill cover to lodge itself in the gill tissue of its host. It then bites off pieces of gill with its fine, comb-like teeth, and feeds on the released blood. Candiru are notorious for swimming up the urethras of mammals, even humans, urinating in rivers. It is thought that, in these instances, the candiru mistakes the flow of urine for the stream of water expelled from the gills of larger fish. This behaviour has extremely painful results.

Farlowella acus
Twig catfish

	Length Up to 15 cm (6 in)
	Weight Not recorded
	Sex Male/Female
Location N. to C. South America	**Status** Unconfirmed

long snout

bony scutes cover body

Although they are almost all bottom dwellers, catfish exhibit a remarkable diversity of body form. The twig catfish is unusually long and slender, with a pointed snout, and is camouflaged to resemble a piece of sunken wood. It lives in slow-flowing rivers, and feeds mainly on river-bed algae after dark. During the day, it relies on its cryptic colour and shape to protect it from attack. Females lay their eggs on the river bed, and the males guard them until they hatch.

Bagre marinus
Gafftopsail catfish

	Length Up to 1m (3¼ft)
	Weight Up to 4 kg (8¾lb)
	Sex Male/Female
Location W. Central Atlantic, Gulf of Mexico, Caribbean	**Status** Common

Commercially exploited as a food source, this Atlantic catfish is mainly marine, but sometimes swims into estuaries that have relatively high salinity levels. Its head has a hard, bony shield, but its most conspicuous features are the very long, flattened barbels attached to the sides of its mouth, and the long spines that arch outwards from its dorsal and pectoral fins. These serrated spines are also venomous and can inflict painful wounds.

Electrophorus electricus
Electric eel

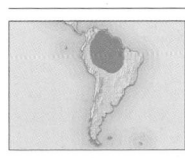

	Length Up to 2.5 m (8¼ft)
	Weight Up to 20 kg (44 lb)
	Sex Male/Female
Location N. South America	**Status** Locally common

Despite its name, this fish is not a true eel, but a giant member of the knifefish order with an eel-like body. When fully grown, it can be as thick as a human thigh, and is one of the largest freshwater fish in South America. It has a continuous fin that runs along its underside, but no fins on its back; its tail tapers to a sharp, finless point.

Clarius batrachus
Walking catfish

	Length Up to 40 cm (16 in)
	Weight Not recorded
	Sex Male/Female
Location S. and S.E. Asia	**Status** Unconfirmed

This long-bodied catfish gets its name from its ability to crawl overland, using its pectoral fins to gain purchase. While on land, it breathes through its specially modified gills, which have strengthened filaments, preventing them from collapsing when exposed to the air. The walking catfish lives in slow-moving water, walking overland to reach suitable habitats. It is a popular aquarium species; the piebald variety shown here is one of several kinds that have been raised.

long dorsal fin

barbels

Gymnotus carapo
Knifefish

	Length 60 cm (23½ in)
	Weight Not recorded
	Sex Male/Female
Location Central America to C. South America	**Status** Common

A close relative of the electric eel (see below), this South American, fresh-water fish has a similar cylindrical body tapering towards its finless, rod-shaped tail. It produces weak electric currents with which it senses its environment, although these are much milder than those of the electric eel. The dark and pale stripes on its body help to camouflage it in murky water, and give it its other common name, the banded knifefish. It moves by rippling the long fin on its underside.

The gafftopsail catfish erects its dorsal and pectoral spines when threatened. This fish is a bottom feeder, eating crustaceans and other fish. During the breeding season (between May and August), the male gafftopsail catfish fertilizes the female's eggs and broods them in his mouth until they hatch.

deeply forked tail

venomous, serrated spines

tapering, finless tail

Electric eels have poor eyesight, and find their way by using weak pulses of electricity. However, using modified muscles, or electric organs, that run the entire length of their bodies, they can also produce sudden jolts of up to 600 volts – powerful enough to kill other fish and potentially lethal to humans.

continuous fin on underside

thin, tapering body

SALMON AND RELATIVES

PHYLUM	Chordata
CLASS	Osteichthyes
SUBCLASS	Actinopterygii
SUPERORDER	Protacanthopterygii
ORDERS	3
FAMILIES	14
SPECIES	316

DURING THEIR LIFETIME, many species of salmon and trout make impressive journeys from the oceans (where they live as adults) to freshwater lakes and rivers (where they spawn). Their close relatives include a group of freshwater fish known as galaxiids. Among their other relatives are the freshwater pikes and a group of marine fish called argentoids (which include tubeshoulders and spookfish, among others). Salmon, trout, and pikes are native to North America, Europe, and Asia, although they have been introduced elsewhere. Galaxiids occur only in the southern hemisphere, while argentoids are found from moderate depths to deep water in all the world's oceans.

ANATOMY

Salmon and their relatives have a slim, tapering body, usually with well-developed swimming muscles. The fins are relatively small, except for the tail fin, which is usually large and powerful, enabling the fish to swim and manoeuvre quickly. Most species have a small, fleshy fin, known as an adipose fin, near the base of the tail, and pelvic fins that are located far back on the body. Scales in these fish are either small or entirely absent. As carnivores that feed on a wide variety of prey, most species have a large mouth studded with many sharp teeth. The highly predatory pikes have long teeth for seizing prey and an especially large mouth that allows them to swallow prey almost half their own size. Salmon, trout, and pikes often ambush prey or attack with a burst of speed over a short distance.

Most galaxiids are small fish, less than 25cm (10in) long, with a tubular, scaleless body. They have a square tail and, unlike most other fish in this group, no adipose fin.

Some of the marine fish in this group have unusual adaptations for life in deep water. The argentoids, for example, include some species with large, tubular eyes that point forwards or upwards (giving good binocular vision). Other argentoids have light-emitting organs.

LIFE CYCLE

Some species in this group complete their entire life in either the sea or in fresh water. However, other species, including many salmon and trout, spend most of their life in the ocean but move to fresh water to breed, a life cycle described as anadromous. Males and females spawn in freshwater rivers. When the young hatch, they grow for a short time in fresh water and then move out to sea. After a period that varies from a few months to several years, the fish mature into adults and return to fresh water to spawn, usually to the stream where they hatched. To complete the journey, which may be several thousand kilometres long, they often have to negotiate fast currents or rough terrain, requiring them to leap clear of the water to progress upstream. Breeding adults usually cease feeding before beginning their migration. Having used all their energy on the journey, they usually die soon after spawning. How they find their way to their hatching site is not entirely understood, but the sense of smell and perhaps celestial navigation are thought to be important. Not all salmon and trout follow this type of life cycle: those belonging to landlocked populations spend all their life in fresh water.

Some species of galaxiids are also anadromous. The tiny larvae hatch in fresh water and are swept out to sea. After a few months, they return as juveniles to fresh water, where they mature into adults.

BREEDING MALES
During the migration to their breeding grounds, the males of many species of salmon undergo drastic physical changes. These include the appearance of new colours and markings. Some species also develop a hump on their back and an enlarged lower jaw that bends upwards. These pink salmon from the Pacific Ocean are swimming up a stream in Alaska, USA.

Esox lucius
Northern pike

Length Up to 1.3m (4 1/4 ft)	
Weight Up to 34 kg (75 lb)	
Sex Male/Female	

Location North America, Europe, Asia   **Status** Common

This powerfully built predator is one of the few freshwater fish found across the whole of the northern hemisphere, except in the Arctic coastal plains. The northern pike is superbly camouflaged with light markings on its body. Like its close relative the muskellunge (see below), it has a long head with a shovel-like snout, a large, slightly forked tail, and a single dorsal fin. Pike generally spawn during the day, in early spring after the ice has melted. They initially spawn as a pair, but either the male or the female may take on a new partner later. While courting, the male prods the female on her head and pectoral fins with his snout. Female northern pikes grow faster and larger than the males.

STEALTHY HUNTER
The single dorsal fin of the northern pike, set well back near its tail, allows it to approach prey on the surface of lakes without making any warning ripples.

SILENT AMBUSH

The well-camouflaged northern pike hunts by lying in wait among waterweeds, bursting out of cover the instant its prey comes within range. It feeds mainly on other fish, but also eats insects, frogs, and crayfish, as well as animals swimming on the surface, such as water voles and young waterfowl.

head 25 per cent of body length

shovel-like snout

dorsal fin

underslung lower jaw

light markings on dark body

Esox masquinongy
Muskellunge

Length Up to 1.8m (6 ft)	
Weight Up to 45 kg (99 lb)	
Sex Male/Female	

Location E. North America   **Status** Locally common

The muskellunge or "musky" is the largest member of the pike family, and is notorious among anglers for putting up a ferocious fight when hooked. In common with other pikes, it has a very large mouth, with a sloping forehead and beak-like jaws, and a torpedo-shaped body which can accelerate very rapidly with a sudden flick of the tail. A lie-in-wait predator, it ambushes and feeds on animals as large as muskrats, catching its prey sideways in its mouth, before turning them around and swallowing them head-first. Muskellunges spawn over vegetation in spring, after the ice melts; a female may pair with more than one male. This fish is found in heavily vegetated lakes and rivers, generally in warmer waters than those inhabited by northern pikes (see above), although the 2 fish may be found in the same habitat and occasionally even interbreed. The resulting hybrids are a robust breed with strongly barred markings. The male hybrid is sterile but females are often fertile.

Plecoglossus altivelis
Ayu

Length Up to 70 cm (28 in)	
Weight Not recorded	
Sex Male/Female	

Location W. North Pacific, E. Asia   **Status** Common

The ayu is flattened from side to side, and has a dorsal fin set high on its back, and a forked tail. A vegetarian, it feeds on river algae with its modified teeth and digests food with the help of an extra-long gut. The ayu's specialized kidneys enable it to move between sea and fresh water without being affected by the changes in salinity. In spring, young ayu form shoals in the river mouth and then make their way upstream to feed.

dark markings on light body

long head

large anal fin

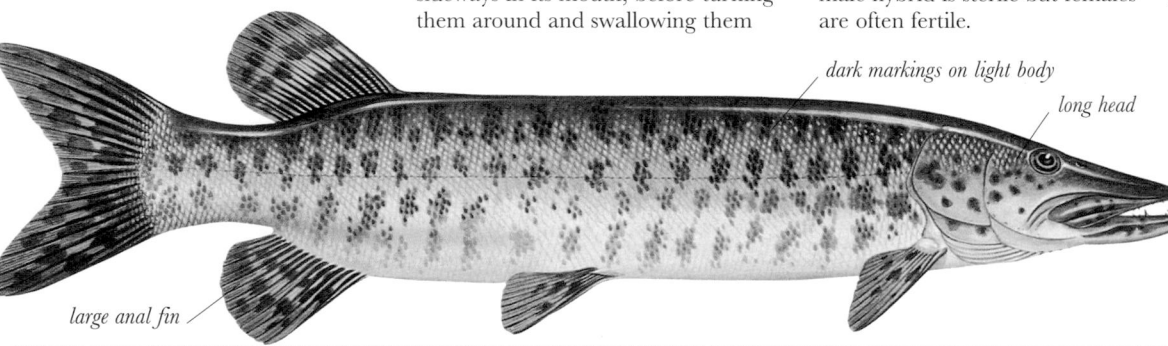

Opisthoproctus soleatus
Barrel-eye

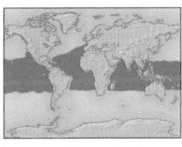

Length Up to 10.5 cm (4 1/4 in)	
Weight Not recorded	
Sex Male/Female	

Location Tropical and subtropical waters worldwide   **Status** Locally common

Barrel-eyes are relatively small fish named for their tubular, upward-facing eyes, an adaptation that probably helps them to hunt other fish from below. They belong to the same family as the spookfish (see right) and, like them, have most of their fins – apart from the pectoral pair – set close to the tail. This fish is dark above and silvery below, and has a luminescent organ on its rectum that contains light-producing bacteria. The light shines through a lens, and is reflected by the fish's flattened underside or "sole", which spreads along the length of its body. This diffused light matches the dim light from the surface, making the barrel-eye practically invisible from below. Barrel-eyes are found mainly in tropical waters, down to a depth of about 800m (2,650ft).

barrel-shaped eyes

large dorsal fin

Dolichopteryx binocularis
Spookfish

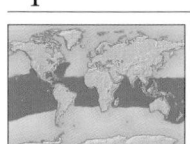

Length At least 8.5 cm (3 1/4 in)	
Weight Not recorded	
Sex Male/Female	

Location Tropical and subtropical waters worldwide, W. North Atlantic   **Status** Unconfirmed

This bizarrely shaped, slow-growing fish is one of about 15 species in the spookfish family. It is distinguished from its relatives by its long, filament-like pectoral fins, which are over half as long as the rest of its body, but its most striking feature is the highly unusual anatomy of its eyes. Tubular in shape, they point obliquely upwards, like a pair of binoculars. This makes light directed at the side of the eyes hard to detect and so, located on the side of each tubular eye is a second eye, complete with retina and lens, making this the only truly 4-eyed fish. Like many deep-sea fish, the spookfish is very fragile, and prone to damage if recovered from great depths.

filament-like pectoral fins

transparent white body

5 large dark spots

Legless searsid

Platytroctes apus

Length Up to 18 cm
(7 in)

Weight Not recorded

Sex Male/Female

Location North Pacific,
North Atlantic, N. Indian
Ocean

Status Unconfirmed

The legless searsid belongs to a family
of fish called the tubeshoulders,
which get their name from a unique
"shoulder" gland, located over the
pectoral fin, that emits a luminescent
liquid. The liquid forms a glowing
greenish cloud in the water, probably
helping the fish to distract and escape
from predators. The legless searsid
feeds on plankton, possibly using its
long gill rakers to sieve them from the
surrounding water.

leaf-like shape to minimize
shadows and escape
detection

sideways-flattened body

tiny
pectoral fins

Capelin

Mallotus villosus

Length Up to 20 cm
(8 in)

Weight Up to 40 g
(1 7/16 oz)

Sex Male/Female

Location North Pacific,
North Atlantic, Arctic
Ocean

Status Locally common

The capelin lives and breeds at sea,
forming large shoals that are an
important source of food for seabirds,
seals, dolphins, and whales; in years
when capelin numbers are low,
seabird rookeries may entirely fail to
reproduce. Slender and with a
protruding lower jaw, it is olive-green
on the back, merging with silver on
its sides and underside, and has a large
adipose fin – one of the characteristic
features of salmon and their relatives.
Although the capelin feeds mainly on
zooplankton, using its comb-like gill
rakers to filter them from the water, it
is also known to feed on marine worms
and other fish. During the breeding
season, males can be identified by the
swollen base of their anal fin and by
a velvety band of modified scales
(villi) along the sides of their body.
When courting, the male stimulates
the female by caressing her with his
villi, which induces her to spawn.
Capelin migrate inshore to spawn on
beaches, where the female lays up to
60,000 eggs at high tide. These are
buried in the sand and subsequently
exposed by the action of the waves.
This fish is a major food source for
the indigenous populations of Alaska
and Canada.

adipose fin

olive-green back

protruding lower jaw

Inconnu

Stenodus leucichthys

Length At least 1.5 m
(5 ft)

Weight 28–40 kg
(62–88 lb)

Sex Male/Female

Location North Pacific,
N.W. North America,
E. Europe to N.E. Asia

Status Endangered

This large member of the salmon
family has a long, broad, and shallow
head and a large mouth with very fine
teeth. As an adult, it feeds mainly on
other fish, but its young eat insect
larvae and planktonic animals.
Inconnus live in 2 different ways:
some – for example in Canada's
Great Slave Lake – are permanent
freshwater fish which remain in their
home range to breed, but most
overwinter in brackish water or
estuaries, and move upriver in
spring. They remain there until
autumn, when they spawn. The
female is accompanied by one or 2
males during spawning bouts, which
take place between dusk and night-
time, and probably occur only every
third or fourth year. She lays between
130,000 and 400,000 eggs. Although
it is valued as a sport fish, the inconnu
is more important in commercial
fisheries in Canada, Alaska, and
Siberia, where it is commonly found.

long,
shallow head

large
mouth

high, pointed
dorsal fin

anal fin

Cisco

Coregonus artedi

Length Up to 57 cm
(22 1/2 in)

Weight Up to 3.5 kg
(7 3/4 lb)

Sex Male/Female

Location N. North
America

Status Locally common

Essentially a freshwater fish, the cisco
may also be found in brackish water
and salt water. It is usually found in
large shoals in the mid-waters, the
depth varying with the seasons.
Basically a plankton feeder, it will also
consume a wide variety of other foods.
Spawning occurs in autumn and is
dependent on the temperature of
water. Males are usually smaller than
the females and, during the breeding
season, they develop small bumps
(pearl organs), which are used to
stimulate the female to lay her eggs.

steel-blue back

Eulachon

Thaleichthys pacificus

Length Up to 23 cm
(9 in)

Weight Up to 60 g†
(2 1/8 oz)

Sex Male/Female

Location E. North Pacific,
W. North America

Status Common

The eulachon is a member of the
smelt family – a group of slender,
silvery fish that also includes the
capelin (see left). Like the capelin, it
has a long body and underslung lower
jaw, but is blue to blue-brown above.
Adult eulachons live at sea, where
they feed on plankton, filtering their
food from the water. However, they
travel upstream in medium- to large-
sized rivers to spawn. Few survive
spawning, which limits the average
lifespan of this fish to between 2 and
4 years. Eulachons are both a sport
and a subsistence resource, and when
caught in the mouth of rivers are so
oily that they will actually burn;
hence their common name
"candlefish". Their oil used to be
traded by native North Americans, the
trade routes often called "grease trails".

Grayling

Thymallus thymallus

Length Up to 60 cm
(23 1/2 in)

Weight Up to 6.5 kg
(14 lb)

Sex Male/Female

Location W., C., and N.
Europe

Status Lower risk

This freshwater member of the
salmon family lives in cool lakes and
fast-flowing rivers and streams, where
it feeds on insects, small worms, and
crustaceans. It has an unusually long
and tall dorsal fin, which the male
curls over the female's back during
egg-laying. The grayling is a useful
indicator of water quality because it is
one of the first fish to disappear when
rivers and streams become polluted.

FISH

Oncorhynchus nerka

Sockeye salmon

Length	Up to 84 cm (33 in)
Weight	Up to 7 kg (15 lb)
Sex	Male/Female

Location N.E. Asia, N. Pacific, N.W. and W. North America

Status Locally common

Also known as the "blueback" and the "red" salmon, the sockeye salmon undergoes a remarkable physical transformation at the time of spawning. Pre-spawning sockeye salmon of both sexes have steel-blue heads and backs and silvery sides. However, between June and September, when they leave the ocean and enter their spawning streams, their heads turn bottle-green and their bodies bright red. Of all species of Pacific salmon, the sockeye is unique in requiring a lake to rear in, which makes it dependent on rivers that have a lake in their watershed. The lake-locked form of sockeye salmon – the kokanee – is typically one-sixth the weight of its sea-going kin.

JUMPING THE RAPIDS
The sockeye salmon's determination to reach its native spawning ground is demonstrated by the dramatic way it launches itself over rapids.

spotless dorsal fin

humped back

DISTINGUISHING FEATURES
The sockeye salmon has a streamlined body and a blunt, conical snout. It is distinguished from other salmon species by its fins, which usually lack definite spots, and its dramatic colour change at spawning.

bright crimson spawning coloration

CRIMSON TIDE
The spectacular annual run of the sockeye salmon – sometimes comprising millions of fish – has become a popular attraction, with viewing platforms being built across spawning rivers.

SPAWNING

Usually about 4 years after hatching, all adult sockeye salmon eventually return to their native streams to spawn and then die – a final journey that can involve travelling up to 1,500km (930 miles).

SMALL FRY
When sockeye salmon fry emerge from the gravel, they embark on their first journey to a nearby lake. They will stay there for at least a year before heading out to sea.

NEW LIFE
Having reached its spawning ground, the female sockeye salmon digs a nest in the gravel river bed and lays up to 4,300 bright pink eggs, which are fertilized by the male.

RIVER GRAVEYARD
After spawning, all sockeye salmon die, their bodies littering the same stretch of river in which they themselves were spawned.

RETURNING SOCKEYE SALMON
The sockeye salmon is anadromous – living in the sea but entering freshwater systems to spawn. As the salmon return to their spawning rivers, the bodies of both males and females turn red and their heads turn green; the male also develops a humped back and a hooked nose.

FISH

Salmo salar
Atlantic salmon

Length	Up to 1.5 m (5 ft)
Weight	Up to 45 kg (99 lb)
Sex	Male/Female

Location N.E. North America, W. and N. Europe, North Atlantic

Status Locally common

The Atlantic salmon is one of the world's most highly prized sport fish, as well as being widely farmed as a source of food. Wild fish start their lives as tiny fry – known as alevins – that hatch out among stream-bed gravel. The fry feed on a diet of insect larvae and other small animals, and after a freshwater life of between 1 and 4 years, they begin their downstream migration to the sea. On entering salt water, they lose their dark colour and develop a silvery sheen, and for up to 4 years they roam widely throughout the North Atlantic, preying on other fish. At the end of this period, they begin to become sexually mature, and make their way back to the rivers and streams where they originally hatched. Like other salmon, they are powerful swimmers, able to leap over almost all obstacles in their way. They rarely feed on their upstream journey, and by the time they arrive at their spawning grounds, the females are plump with eggs, and the males have developed distinctively hooked jaws. The females are known to use their tails to excavate hollows in gravel, and the males lie alongside while the females lay eggs. Adult Atlantic salmon may die after spawning, but unlike other salmon, a large number of the adults often survive. Emaciated and exhausted, these fish make their way back to the sea, where they feed and recover. They return to breed again at intervals of one or 2 years.

EARLY DAYS
At the age of a few months, young Atlantic salmon are known as parr. At this stage, they are about 15cm (6in) long and have distinctive dark markings or "fingerprints" along their backs.

black spots on the side

blue-black back

silvery sheen on body

READY TO BREED
When they re-enter rivers, Atlantic salmon are silvery, with blue-black backs and black spots along their sides. Their powerful tails enable them to leap up waterfalls and weirs on their way to their breeding grounds.

Oncorhynchus gorbuscha
Pink salmon

Length	Up to 76 cm (30 in)
Weight	Up to 6.5 kg (14 lb)
Sex	Male/Female

Location N.E. Asia, North Pacific, W. North America

Status Common

Of all salmon, this species spends the least time in fresh water; it migrates soon after hatching, travelling over 1,000km (620 miles) at sea. Normally metallic blue, with the approach of the breeding season it becomes olive-green or yellow, with a distinct red or pink tinge along the sides. The male also develops a high humped back and a hook at the end of the upper jaw. The pink salmon has a lifespan of 2 years, which means that fish born in even years and those in odd years hardly ever interbreed.

large, oval black spots on back

tail fin with 13–19 rays

Salvelinus alpinus
Arctic char

Length	Up to 96 cm (38 in)
Weight	Up to 12 kg (26 lb)
Sex	Male/Female

Location N. North America, N. Europe, N. Asia, Arctic Ocean

Status Locally common

The Arctic char is the most northerly of all freshwater fish. It is found in lakes in the Arctic tundra, where it spawns during autumn, as well as in mountains further south, where spawning takes place in winter. Its colour varies according to its breeding habits: adults that migrate towards the sea become brightly coloured during the breeding season, but landlocked specimens, usually found in lakes, often have bright undersides all year round. The Arctic char is a popular sport fish among anglers in Canada and Scandinavia.

light spots on dark background

Oncorhynchus mykiss
Rainbow trout

Length	Up to 1.2 m (4 ft)
Weight	Up to 24 kg (53 lb)
Sex	Male/Female

Location N.E. Asia, North Pacific, North America

Status Common

This black-spotted member of the salmon family is one of the world's most widely introduced freshwater fish. Originally from rivers and lakes west of the Rocky Mountains, it has been spread eastwards in North America, and to countries as far apart as the British Isles and New Zealand, wherever well-oxygenated, cool water can be found. Like all its relatives, the rainbow trout is a predator, feeding on insects, snails, and crustaceans, and – when adult – on other fish. It takes most of its food from the bottom, but also rises to the surface for flying insects – a habit exploited by anglers using artificial flies. Introduced rainbow trout often spend their entire lives in fresh water, but in their natural habitat some adults spend part of their time at sea, returning to rivers to spawn. Compared to freshwater rainbow trout, these migratory adults – called "steelheads" in North America – grow faster, live longer, and are more productive.

TROUT FARMING

In many parts of the temperate world, rainbow trout are raised in captivity. Some trout "farms" supply fish for food but an increasing number raise this trout – a popular sport fish – for angling, using them to stock lakes where they can be caught by line. Farmed trout are fed on protein-rich pelleted food, and grow much more rapidly than they would in the wild.

RAINBOW COLOURS
Freshwater forms of rainbow trout range from bluish green to brown on the back and sides, and are white or yellowish on the belly. Spawning males develop a vivid red or pink band on the sides.

tail fin slightly indented

hooked snout on male

red band on spawning male

Dallia pectoralis
Alaska blackfish

Length Up to 33 cm (13 in)	
Weight Up to 375 g (13 oz)	
Sex Male/Female	
Location N.E. Asia, N.W. North America	**Status** Locally common

The Alaska blackfish belongs to the mudminnow family – a group of fish that can survive in very cold and stagnant water by gulping mouthfuls of air. This species has a slender body with rounded fins, a blunt head, and a large mouth with a protruding lower jaw. It is dark green or mottled brown, and the males develop a red outline to their fins during the breeding season. This slow-moving fish feeds on insect larvae and snails, and smaller fish which it ambushes and hunts with a darting movement. Found largely in the areas around the Bering Sea, Alaska blackfish aggregate in large numbers under the ice in winter, making them easy to trap. They can even survive being partly frozen in the ice for periods of several weeks. In spring, they migrate to shallower water. Once used as dog food, Alaska blackfish are still a local food resource.

Argyropelecus affinis
Marine hatchetfish

Length Up to 7 cm (2 3/4 in)	
Weight Not recorded	
Sex Male/Female	
Location Tropical, subtropical, and temperate waters worldwide	**Status** Common

Like freshwater hatchetfish, this fish gets its name from its shape, which resembles the blade of a tiny axe. Its body is deepest just behind its head, and is extremely narrow when seen from above. It has a complex collection of tubular, downward-pointing, light-producing organs, or photophores, which may disguise its silhouette when seen from below. The photophores blink on and off, probably helping the fish to confuse predators, and also to communicate with its own kind. The marine hatchetfish feeds largely on invertebrates, rather than on other fish. Its mouth opens up to suck in its food and it has a very large gape – a useful adaptation in environments that are relatively sparsely inhabited. Found at greater depths during the day, it swims to shallower waters at night-time to feed. There are about 40 similar species of hatchetfish, largely distinguished by details of their shape and the distribution pattern of their photophores.

silvery sides

body extremely deep behind head

large, tubular eyes

continuous lines of photophores

Argentina silus
Greater argentine

Length Up to 60 cm (23 1/2 in)	
Weight Over 450 g (16 oz)	
Sex Male/Female	
Location North Altantic, Arctic Ocean	**Status** Common

Argentines are open-water or bottom-dwelling fish with slender bodies and large eyes. Their scales are large and have tiny spines. The greater argentine is brown, with an olive-coloured back, white belly, and silvery sides that have a gold lustre. It feeds on crustaceans, squid, comb jellies, and small fish, using its keen eyesight to spot and ambush prey in the dim conditions that prevail several hundred metres below the ocean surface. The greater argentine moves in shoals and can swim away very rapidly to escape bigger fish that prey on it. The eggs laid by the female float in mid-water currents in the upper regions of the open sea and, after hatching, the young make their way to shallower water, where they are left to survive on their own. The greater argentine is fished for food in Europe, Japan, and Russia.

sharply pointed head

olive back

forked tail fin with pointed tips

white belly

Alepocephalus rostratus
Risso's smooth-head

Length Up to 47 cm (18 1/2 in)	
Weight Up to 825 g (29 oz)	
Sex Male/Female	
Location E. North Atlantic, Mediterranean	**Status** Unconfirmed

Smooth-heads – also known as slickheads – are medium-sized, deep-sea fish. There are about 30 species, found in temperate and tropical waters throughout the world. Risso's smooth-head is typical of the family, with its long and compressed body. its dorsal and anal fins are set well back on the body and accentuates its elongated shape. Dark grey or brown in colour, this fish has a slimy and scaleless head, while the rest of the body is covered with smooth scales of moderate size. It is an ambush predator, snapping up its prey, such as invertebrates and crustaceans, from the sea bed or from open water.

large, scaleless head

dorsal fin near tail

forked tail fin

Lepidogalaxias salamandroides
Salamanderfish

Length Up to 6 cm (2 1/4 in)	
Weight Up to 25 g (7/8 oz)	
Sex Male/Female	
Location S.W. Australia	**Status** Vulnerable

Discovered in 1961, this tiny silver-brown fish is a remarkable example of adaptation to a highly localized habitat. It is found only in the southwest corner of Western Australia, where it lives in small pools of water that lie over peaty sand. Although the water is fresh, it is black and acidic, and dries up entirely in the summer months. Salamanderfish survive during summer by burrowing up to 60cm (23½in) deep. To facilitate this, they have unusually flexible backbones which allow them to wriggle through the damp sand. They remain dormant for up to 5 months, breathing through their skin. Salamanderfish cannot move their eyes, because they lack external eye muscles, but they can bend their "neck" to look around. At the onset of the breeding season, males mate with females, and the eggs are fertilized inside the female's body. Each female lays up to 100 eggs, and the complete lifespan of the species is often just one year. Salamanderfish feed mainly on the larvae of aquatic insects.

Galaxias maculatus
Common jollytail

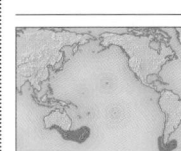

Length Up to 20 cm (8 in)	
Weight Not recorded	
Sex Male/Female	
Location S. Australia, New Zealand, South Pacific, S. Atlantic, S. South America	**Status** Unconfirmed

Also known as whitebait, this wide-spread fish of the southern hemisphere has a small, slender, almost cylindrical body, clear olive-grey to amber above, with a blunt head, and dorsal and anal fins set back close to its tail. It has a high tolerance for sudden salinity changes, linked to its life cycle which is divided between fresh water and the sea. The eggs are laid on estuarine plants and hatch when the next high tide occurs. The larvae spend up to 7 months at sea, before returning to mature in fresh water.

DRAGONFISH AND RELATIVES

PHYLUM	Chordata
CLASS	Osteichthyes
SUBCLASS	Actinopterygii
SUPERORDER	Stenopterygii
ORDERS	1 (Stomiiformes)
FAMILIES	4
SPECIES	About 250

DRAGONFISH AND THEIR RELATIVES are deep-sea predators with unusual adaptations for catching prey, including large, flexible jaws, sharp teeth, and light-emitting lures. They also have luminous organs on their body known as photophores. As well as dragonfish, this group includes bristlemouths, viperfish, slackjaws, and snaggletooths. Although their names reflect their ferocity, these are relatively small fish whose prey consists of other deep-sea fish and invertebrates. These fish are found in all the world's oceans, but their distributions are generally discontinuous.

ANATOMY

Most of the fish in this group have an elongated body and a relatively large head. The head usually contains modified jaws with a very wide gape. For example, slackjaws have no floor to their mouth. Viperfish have an extra joint in the head that increases their gape; as the jaw opens, organs such as the heart and gills are pushed backwards and downwards, out of the way of incoming food. The jaws are usually armed with numerous needle-like teeth that are used to seize prey or to prevent food from escaping. Apart from viperfish, the fish in this group have a generally scaleless body that is coloured black or brown to match the darkened waters. However, some species are silvery or even transparent.

Many species have light-producing organs called photophores. They share this feature with the lanternfish (see p.501), although in dragonfish the photophores are generally larger and more numerous. They are usually located along the sides and underparts of the body or near the eyes. Each is stud-like in shape and linked to a nerve. Light is reflected through a silvery backing and focused through a lens-like thickening of the overlying scale. They either shine constantly, or flash on and off. Species that live in deeper water – at depths to about 300m (1,000ft) – generally have fewer and weaker photophores, and the light output seems to vary according to the amount of light received from above. Thus, photophores may hide the fish from predators below them by matching the background illumination from above. Almost all species in this group have a light organ at the end of a chin barbel or a specialized fin ray. This is used as a lure to attract prey within striking distance.

FEEDING
Dragonfish, such as this black dragonfish, are impressive predators. Many species feed near the surface at night and move to deeper water to rest during the day. The body is illuminated by photophores.

Chauliodus sloani
Sloane's viperfish

Length Up to 35 cm (14 in)
Weight Up to 30 g (1 1/16 oz)
Sex Male/Female

Location Tropical, subtropical, and temperate waters worldwide

Status Common

This deep-sea fish has a bluish black, elongated body, a large mouth, and exceptionally long, transparent fangs. The largest fangs are slightly barbed, and are too big to fit into the mouth – instead they protrude outside it when the jaws are closed. The dorsal fin has a long, arching ray that acts as a lure, and the fish also has photophores along its body and around its mouth, which help to entice prey towards its jaws. This viperfish spawns all year round. It is one of about 6 species of viperfish, found at great depths in oceans all over the world.

arching ray on dorsal fin

head deeper than body

barbed fangs

photophores line both sides

Idiacanthus antrostomus
Black dragonfish

Length Female up to 38 cm (15 in)
Weight Up to 55 g (2 oz)
Sex Male/Female

Location E. Pacific

Status Unconfirmed

This bizarre, deep-water marine fish has a black, snake-like body and extremely large teeth that must be rotated to allow the mouth to open and close. Females are about 4 times larger than males. To attract prey, the black dragonfish uses a movable light organ located at the end of a barbel that protrudes from the front of the lower jaw. Other light organs (photophores) run along the belly. Dragonfish feed by night, moving nearer the surface to prey on small fish and crustaceans. The larvae are remarkable in having eyes set on long stalks and intestines that extend beyond the tail.

movable light organ

Malacosteus niger
Slackjaw

Length Up to 24 cm (9 1/2 in)
Weight Not recorded
Sex Male/Female

Location Tropical, subtropical, and temperate waters worldwide

Status Unconfirmed

The slackjaw is unique in having no membranes joining its jaws and tongue – a feature that may help it to make a rapid strike at prey. The head can pivot and the jaws are protractile, which allows it to draw in prey – fish and crustaceans – much larger than itself. A deep-sea fish, it migrates to the surface at night to feed. While many species in this family have light-producing organs (photophores) arranged in rows along the body, the slackjaw has them only around its mouth.

short, blunt snout

scaleless skin

LANTERNFISH AND RELATIVES

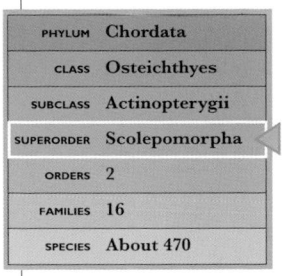

PHYLUM	Chordata
CLASS	Osteichthyes
SUBCLASS	Actinopterygii
SUPERORDER	Scolepomorpha
ORDERS	2
FAMILIES	16
SPECIES	About 470

LANTERNFISH SHARE WITH dragonfish (see p.500) the ability to emit light from organs called photophores on their head and body. They are sometimes regarded as the deep-water equivalent of herrings because they are small, swim in large, dense shoals, and form a vital element of deep-water ecosystems by providing food for many other fish, seabirds, and marine mammals. Lanternfish are found in great numbers in the middle layers of deep waters off the continental shelves. Their relatives include the slender lancetfish and tripod- and lizardfish, which use their fins to rest on or crawl across the sea floor. These fish are found in both coastal waters and deep oceans.

ANATOMY

All the fish in this group have thin scales – some silvery and others darkly pigmented – that often fall off when they are caught. Lanternfish have bioluminescent photophores on their flanks, undersides, and heads, which emit light in shades of green, yellow, or blue. Each species has a unique pattern of photophores,

which helps groups of a particular species to maintain cohesion in darkly lit waters. Lanternfish often make vertical migrations from deep water to surface waters at night to feed on plankton, returning to deep water as the sun rises. They have large eyes and good vision that help them detect changes in light intensity and recognize photophore patterns in dark water. The other fish in this group are quite different from lanternfish in anatomy and behaviour. Most species have no photophores. Lancetfish are slender deep-water fish. They have a large mouth containing sharp, pointed teeth and feed on other deep-water organisms. These include lanternfish, and the lancetfish make similar daily migrations as they follow their prey between deep water and the surface. Lizardfish are bottom-living predators that hunt by ambush in tropical and temperate coastal areas. They have many sharp teeth and are camouflaged to match the colour and texture of the sea floor. Tripodfish are so called because they stand on the tips of 3 elongated fins. They are found on the bottoms of deep ocean basins.

LIGHT PRODUCTION
A lanternfish's photopores resemble small, bright studs (a white-spotted lanternfish is shown here). The males and females of some species have photopores arranged in different patterns, which helps them to recognize one another in dark water.

FISH

Bathypterois grallator
Tripodfish

Length Up to 36 cm (14 in)
Weight Not recorded
Sex Hermaphrodite

Location Atlantic, Mediterranean, Indian Ocean, W. Pacific
Status Unconfirmed

This fish perches over the ocean floor by forming a tripod with elongated rays from its pelvic and tail fins.

Suspended above the sea bed and facing into the current, it feeds on small crustaceans. Its eyes are very small, and its mouth has a large gape, the tip of the upper jaw extending back beyond the orbit. The bat-like pectoral fins have an elaborate nerve supply, suggesting that they have a sensory function.

Alepisaurus ferox
Longnose lancetfish

Length Up to 2.8 m (9 1/4 ft)
Weight 4.5 kg (10 lb)
Sex Hermaphrodite

Location Pacific, North Atlantic, Mediterranean
Status Locally common

This large and distinctive deep-sea fish has a spindle-shaped, iridescent body, a deeply forked tail, and a long dorsal

elongated dorsal rays

sail-like dorsal fin

fin that stretches for about two-thirds of its length, and is held upright like a sail. The function of this fin is unknown. The large mouth has big, pointed teeth for snagging slippery prey. This fish is a useful source of deep-sea specimens for scientists, because the fish in its stomach are usually intact.

Myctophum asperum
Prickly lanternfish

Length Up to 7 cm (2 3/4 in)
Weight Not recorded
Sex Male/Female

Location North and South Pacific, W. and E. Atlantic, Indian Ocean
Status Common

light organs on body

One of the most numerous of the deep-sea fish and an important link in ocean food chains, lanternfish eat mainly planktonic crustaceans, and are in turn eaten by seabirds, seals, and larger fish such as tuna. They derive their name from the light organs on the head and body. The arrangement of these light organs is an important feature for the identification of the 250 or so species. Differing arrangements of larger light organs near the tail indicate whether the fish is male or female.

Synodus variegatus
Variegated lizardfish

Length 25–35 cm (10–14 in)
Weight Not recorded
Sex Male/Female

Location Pacific, Indian Ocean
Status Common

With its pointed snout and large mouth, this fish's head looks very much like a lizard's, when viewed from the side. It often sits on the sea floor, propped on its pelvic fins, with its head raised in a very lizard-like fashion. It may also bury itself in the sand with only its snout and eyes visible, and lunge at small fish passing by. The variegated lizardfish has a slender, tubular body and its brown, orange, or reddish coloration effectively camouflages it against the sea floor.

COD AND ANGLERFISH

PHYLUM	Chordata
CLASS	Osteichthyes
SUBCLASS	Actinopterygii
SUPERORDER	Paracanthopterygii
ORDERS	8
FAMILIES	46
SPECIES	About 1,260

COD FORM A crucial link in marine ecosystems, since they are both predators and prey to many other species. Their close relatives include hake, pollock, and the widespread grenadiers. Anglerfish are bottom-dwelling or deep-sea predators. They owe their name to a dorsal spine, at the end of which is a lure, which is used to attract prey. As well as cod and anglerfish, this group contains several other fish, including cusk eels (which include pearlfish and ventfish) and trout-perches. Apart from the last group, they are mostly marine, with a small number found in brackish water. Most species live on the sea floor but some (such as cod, haddock, and hake) form vast shoals.

ANATOMY

Cod and their relatives typically have a long, narrow body. The fins, which are supported by relatively weak spiny rays, have a distinctive arrangement: the ventral fins are positioned well forward, and the long dorsal fin is often divided into smaller units. True cod have small, elongated scales, a mouth that opens vertically for feeding on bottom-dwelling animals, and a single sensory barbel on the lower jaw. Hake look similar to cod but have scales on their body and a large mouth with numerous sharp teeth. They are predatory fish that move in large shoals in pursuit of smaller fish and invertebrates. Grenadiers are covered in rough scales and have a large, bony head that tapers into a thin body and a filament-like tail (giving rise to the alternative name rat-tails). These deep-water fish scavenge and hunt for small prey on the ocean floor.

The bottom-living anglerfish are usually coloured and patterned to blend into the sea floor. They are relatively inactive fish, with a stout, unstreamlined body and a large head with an enormous mouth. One of the fin rays is modified into a whip-like rod with a lure at the end. The lure is often shaped to look like a small marine organism, while in some deep-water species it has a light-emitting organ at the end.

A typical cusk eel has a small head and a tapered body. The dorsal, anal, and tail fins are often joined to form a single fin. Cusk eels and their close relatives have adapted to live in some unusual habitats, including caves and areas around volcanic vents on the ocean floor. Some species even live inside the bodies of invertebrates, including molluscs, sea cucumbers, and sea squirts.

True trout-perches are so called because they have a dorsal fin with spines at the front, similar to that of perches, and an adipose fin, like that found in trout. Their close relatives include several species that are found only in limestone caves in the eastern USA.

LIFE CYCLE

In winter and spring each year, cod migrate to established breeding grounds to spawn. They reproduce in great numbers (a single female can produce several million eggs each year), which explains why they have been such an important food for other fish, marine mammals, sea-birds, and humans. In recent years, however, overfishing has greatly decreased the population of many commercial species, and may have even eliminated cod from some of their spawning areas.

Some species of deep-sea angler-fish overcome the difficulties of finding mates in the darkness of the deep oceans by reproducing in a unique way. Soon after developing from the larval state, the male bites into the body of a much larger female. A permanent attachment is formed, the tissues and circulatory systems of the 2 fish becoming fused. The male lives on nutrients from the female's circulatory system and releases sperm when she needs to fertilize her eggs. A single female may have one or several males attached to her.

FEEDING

Anglerfish have a sophisticated feeding method that consumes little energy. The well-camouflaged fish (here, an angler or monkfish) uses its lure to draw prey towards its mouth. Once it is close enough, the anglerfish opens its mouth, creating a strong suction current. The mouth and stomach can both be expanded, so even large prey can be swallowed whole.

FISH

Amblyopsis spelaea
Northern cavefish

Length Up to 10.5 cm (4 1/4 in)
Weight Not recorded
Sex Male/Female

Location USA (Kentucky, Indiana)
Status Vulnerable

Like most cave-dwelling fish, the northern cavefish has pale skin, is completely blind – in this case only vestigial eye tissues remaining under the skin – and has a very localized distribution. It is restricted to the limestone caves straddling the state line between Kentucky and Indiana, an area of less than 100 square km (40 square miles). It feeds on tiny, cave-dwelling invertebrates, detecting them with the sensory papillae on its head and body.

Carapus acus
Pearlfish

Length Up to 20 cm (8 in)
Weight Not recorded
Sex Male/Female

Location Mediterranean
Status Locally common

Like most of the other species in its family, this slim, scaleless, silvery white fish spends its adult life inside sea cucumbers. Attracted to them by their shape, and the chemicals that they release, the pearlfish enters its host, tail first, through the animal's anus, and takes up residence in the body cavity. After dark, it goes out to feed on other animals, but it may also feed parasitically on its host's internal organs.

Lota lota
Burbot

Length Up to 1.2 m (4 ft)
Weight Up to 34 kg (75 lb)
Sex Male/Female

Location N. North America, Europe, N. Asia
Status Locally common

This long, slender fish is the only cod to spend the whole of its life in fresh water. It is usually a dull, dirty green to brown above, with dark mottling, and yellowish to creamy below. The fish hides under tree roots or in crevices by day, and is most active at dawn and dusk, when it hunts for insect larvae, crustaceans, and other fish. It spawns in winter, and does so – unusually for a fish – in the middle of the night, and often under the ice.

Gadus morhua
Atlantic cod

Length Up to 1.4 m (4 1/2 ft)
Weight Up to 25 kg (55 lb)
Sex Male/Female

Location North Atlantic, Arctic Ocean
Status Vulnerable

This large, shoal-forming fish is one of the world's most commercially important species. It lives in water over the continental shelf, and usually feeds at 30–80m (100–250ft) above areas of flat mud or sand. Other fish, especially herrings, sprats, and capelin, form its main prey. It usually breeds in early spring, with females releasing millions of eggs into the water. Adult cod migrate extensively and keep together, making them relatively easy to catch.

Bythites hollisi
Ventfish

Length Up to 30 cm (12 in)
Weight Not recorded
Sex Male/Female

Location Galapagos Islands
Status Unconfirmed

First identified in 1990, this rare fish is found only around hydrothermal vents on the deep-sea bed. It has a short, stout body and a sizable mouth, but small, poorly developed eyes. Like other vent animals, it depends for its survival on the bacteria that thrive in the hot, mineral-laden water that gushes out of vents. It is not known whether it feeds on the bacteria directly, or on other animals that use them for food. The ventfish belongs to a family of about 85 species, known as live-bearing brotulas, which generally live in shallow water, and sometimes in caves. All give birth to live young.

Merluccius productus
Pacific hake

Length Up to 90 cm (35 in)
Weight Up to 5 kg (11 lb)
Sex Male/Female

Location North Pacific
Status Common

Closely related to the Atlantic cod (see below), the Pacific hake has a similar shape, with an elongated body, large head, and projecting lower jaw. It is bluish grey on its back, and silvery on its sides. Much like the Atlantic cod, it forms large shoals and spawns in early spring, when the females lay millions of eggs. These drift on the surface and are widely distributed by ocean currents. Eaten by marine animals such as porpoises, seals, and swordfish, the Pacific hake has also been heavily harvested by humans of late.

pointed snout

deep notch in fin

Ophidion scrippsae
Basketweave cusk-eel

Length Over 28 cm (11 in)
Weight Not recorded
Sex Male/Female

Location E. North Pacific
Status Common

Cusk-eels are not true eels, but close relatives of pearlfish (see left) and brotulas (see ventfish, left). The basketweave cusk-eel lives at depths of up to 100m (330ft), but some specimens have been found at over 8,000m (26,400ft) – the greatest depth for any fish. Like other cusk-eels, this North American species has a blunt head, large mouth, and an elongated body that tapers almost to a point. The scales on its body are arranged in a criss-cross pattern, which give the species its name. Basketweave cusk-eels are active only at night; during the day they bury themselves in sand with only the tip of their snout showing. At dusk, they partly emerge from the sand, waiting for small fish or invertebrates to pass by. These fish may also swim along the ocean floor in search of their food.

CONSERVATION

The Atlantic cod has been heavily harvested for many years. In the 19th century, specimens of up to 90kg (200lb) were sometimes caught, but with today's sophisticated fishing techniques, stocks have shrunk dramatically; fish weighing over 15kg (33lb) are rare.

FULL-BODIED
A long-bodied fish, the Atlantic cod has a bluntly pointed snout, with one barbel hanging from its lower jaw. Coloration ranges from greenish to sandy brown.

FISH

Coryphaenoides acrolepis
Pacific grenadier

Length	Up to 87 cm (34 in)
Weight	Up to 10 kg† (22 lb)
Sex	Male/Female

Location North Pacific **Status** Common

Also known as rat-tails or whip-tails, grenadiers are the dominant fish on the sloping edges of the world's continental shelves. Like its 300 or so relatives, the Pacific grenadier has a large, bulbous head, a sharp snout, and a body that tapers rapidly into a narrow tail. Its eyes are large, enabling it to see prey on the sea bed. The males make surprisingly loud sounds with the muscles in their swim bladder.

very large eyes

long, filamentous, scaly tail

Antennarius maculatus
Warty frogfish

Length	Up to 11.5 cm (4 1/2 in)
Weight	Not recorded
Sex	Male/Female

Location W. Pacific, Indian Ocean **Status** Locally common

This fish belongs to the anglerfish order – a group noted for their often bizarre appearance. Most anglerfish live in the deep sea, but frogfish live in shallow water – at depths of only a few metres – usually in coral reefs. The warty frogfish's dorsal fin has a long spine that ends in a flap of coloured skin, which the fish wriggles to mimic a small fish and so entice prey towards its jaws.

various spots and patches on body

Histrio histrio
Sargassum fish

Length	Up to 18.5 cm (7 1/2 in)
Weight	Up to 400 g† (14 oz)
Sex	Male/Female

Location Tropical and subtropical seas worldwide **Status** Common

Superbly camouflaged, with highly variable colouring and patterns, this member of the frogfish family lives among drifting weed and is hence very widely distributed. It hunts prey by using its dorsal fin spine as a lure. Sargassum fish exhibit unusually elaborate courtship behaviour, with the males violently nipping and chasing the females. Except in the spawning season, adults are highly cannibalistic.

leg-like pectoral fin

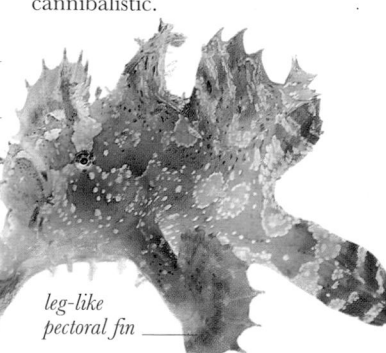

Melanocetus johnsoni
Blackdevil

Length	Female up to 18 cm (7 in)
Weight	Up to 600 g† (21 oz)
Sex	Male/Female

Location Pacific, Atlantic, Indian Ocean **Status** Lower risk†

This deep-sea anglerfish hunts in open water rather than on the sea bed. The female is rounded, with an enormous head and jaws, and a lure on her dorsal fin. The male has no lure, and is less than one-fifth as long as his mate. He attaches himself to the female with his jaws, but, unlike other deep-sea anglers, swims away after she has spawned.

smooth, scaleless skin

Lophius piscatorius
Angler

Length	Up to 2 m (6 1/2 ft)
Weight	Up to 40 kg (88 lb)
Sex	Male/Female

Location E. North Atlantic, Mediterranean, Black Sea **Status** Locally common

Also known as the monkfish, the angler is dull greenish brown, with a broad and flattened body that tapers rapidly towards the tail. A highly specialized sit-and-wait predator, it is very well camouflaged (see panel right), and has a fleshy lure on its dorsal fin that can be suspended above its mouth. If anything swims close by, the fish opens its jaws and sucks it in.

DEEP COVER

The superbly camouflaged angler is almost invisible against the sea bed. Its darkly marbled skin blends into the sediment and the outline of the body is difficult to discern because of the flaps of skin that fringe it.

"ELBOWS"
The well-developed pectoral fins of the angler are set on arm-like bases and have sharp "elbows" that enable it to shuffle over the sea bed.

Ceratias holboelli
Kroyer's deep-sea anglerfish

Length	Female up to 1 m (3 1/4 ft)
Weight	Up to 50 kg† (110 lb)
Sex	Male/Female

Location Pacific, Atlantic, Indian Ocean **Status** Lower risk†

This fish shows an extraordinary difference between the sexes. The female (pictured) has a large head and mouth, but the male is tiny. When the adult male finds a mate, he bites into her flesh, and then lives parasitically on her. The pair's blood systems merge so that nutrients flow into the male's body – a bizarre lifestyle found only in deep-sea anglers.

Linophryne arborifera
Bearded angler

Length	Female up to 10 cm (4 in)
Weight	Up to 300 g† (11 oz)
Sex	Male/Female

Location Atlantic **Status** Lower risk†

Female bearded anglers have a luminous lure attached to the tip of their snouts, and also a complex branching barbel, resembling a piece of seaweed, hanging beneath their chins. Like the lure, the barbel gives off light that attracts prey.
As in Kroyer's deep-sea anglerfish (see left), the male is much smaller than the female, and lives parasitically on his partner.

barbel emits light

SPINY-RAYED FISH

PHYLUM	Chordata
CLASS	Osteichthyes
SUBCLASS	Actinopterygii
SUPERORDER	Acanthopterygii
ORDERS	15
FAMILIES	259
SPECIES	About 13,500

SPINY-RAYED FISH FORM the largest and most recently evolved group of fish, accounting for about half of all species. With such a large number of species they have a remarkable range of shapes, colours, behaviour, and specialized adaptations. They vary in size, from tiny gobies 1cm (⅜in) long, to huge oarfish reaching lengths of over 8m (26ft). Spiny-rayed fish are found in almost all aquatic habitats. They dominate inshore marine waters, and are widespread in lakes, ponds, and slow-moving streams, but are also found in the open ocean, including deep waters.

ANATOMY

Although spiny-rayed fish form a large and varied group, they do share many of the same anatomical features. They typically have stiff, bony spines in or near their first dorsal fin. This differentiates them from other ray-finned fish, which have fins supported by jointed bone segments. However, in some species, the spines are small or missing altogether; in others, such as lion-fish and stonefish, the spines are armed with venom glands. Most species have ctenoid scales (see p.460), with minute spines on the surface of each scale, giving them a rough texture. In other species, the scales have evolved into bony plates or stout spines, and many have lost their scales completely. Most spiny-rayed fish have a mobile mouth that can be extended forwards, but mouth shape and types of tooth are extremely variable.

Spiny-rayed fish include a great variety of specialized predators, scavengers, and herbivores, which have evolved a remarkable range of body shapes for varying lifestyles. Tunas have a torpedo-shaped body for maximum swimming efficiency. Flatfish, such as plaice and sole, have an asymmetrical body, with both eyes on one side, and swim with their underside flat against the sea bed. Seahorses have a long snout, a head positioned at right angles to the body, and a prehensile tail to grasp marine plants. By far the largest group of spiny-rayed fish is the perch-like fish, which form the largest order of vertebrates (with over 9,000 species). As well as perch, this group contains many familiar fish such as bass, snappers, mackerel, and butterfly fish.

FIN ADAPTATION
These shoaling mullet snappers belong to the largest group of spiny-rayed fish, the perch-like fish. These fish share the fin structure of other spiny-rayed fish, but in many the pelvic fin is positioned level with the pectoral fins, increasing mobility and improving their ability to catch prey.

Leuresthes tenuis

California grunion

Length Up to 19cm (7½ in)
Weight Up to 100g (3⅝ oz)
Sex Male/Female

Location E. North Pacific
Status Common

The California grunion is one of over 160 species in the silverside family, and like most of its relatives, lives close to the shore in large shoals. It displays remarkable breeding behaviour: during spring and early summer, shoals of grunion strand themselves en masse on the moist sand of beaches to mate – an extraordinary sight that attracts many spectators. Spawning takes place at night for 2–6 days after a full moon, at high tide (see panel right). The hunting of spawning grunion is a popular sport in southern California. However, only licensed fishermen are permitted to catch the stranded fish, using only their hands. The California grunion is thought to feed on plankton and other micro-organisms, possibly capturing these tiny organisms by quickly projecting its mouth.

SINGLE STRIPE
A small, slender fish, the California grunion has a silvery blue stripe running the length of its green body, and is silver-white below. It has no teeth, and its mouth is tube-like when projected.

MATING

California grunion beach themselves between waves of surf and mate on the moist sand. The female creates a shallow depression for the eggs, and the male wraps himself around the female, fertilizing the eggs as they are laid.

Melanotaenia boesemani
Boeseman's rainbowfish

Length Up to 15 cm (6 in)	
Weight Not recorded	
Sex Male/Female	
Location N.W. New Guinea	**Status** Locally common

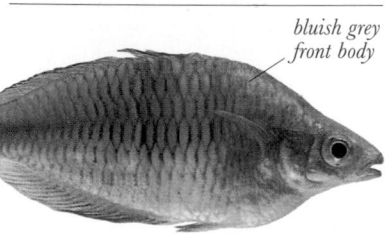

bluish grey front body

Boeseman's rainbowfish is found in the Ajamaru Lakes region in the centre of the Vogelkop Peninsula, New Guinea. The colour pattern of the males of this species is completely different from most other rainbowfish. The head and front portion are a brilliant bluish grey, while the fins and the posterior are bright orange-red. There are alternating light and dark vertical bars in between these 2 areas. Females are less brightly coloured and have a shorter dorsal fin ray than the males. They lay 100–200 eggs.

Chelon labrosus
Thicklip mullet

Length Up to 75 cm (30 in)	
Weight Not recorded	
Sex Male/Female	
Location E. North Atlantic, Mediterranean, W. Black Sea	**Status** Locally common

The common name of this fish derives from its prominent, thick upper lip. It feeds on organic and algal material found on surface sediments, using its gill rakers to filter the more indigestible material. The muscular stomach and long intestines make digestion more efficient. Found in dense shoals in relatively shallow sea water and estuaries, the thicklip mullet is popular both as a sport fish and a commercial species. It belongs to a family of about 80 species and was once considered to be a relative of the barracuda because of similarities in the arrangement of the fins.

Tylosurus acus
Needlefish

Length 90 cm (3 ft), max 1.5 m (5 ft)	
Weight Up to 3.5 kg (7 3/4 lb)	
Sex Male/Female	
Location Tropical and warm temperate waters worldwide	**Status** Locally common

The largest member of the needlefish family, this species gets its name from its beak-like jaws and slender body. It is dark blue or green above and silvery below, and its jaws contain an impressive array of sharp teeth. To escape predators, this agile fish may jump clear of the water. While leaping, it may inadvertently spear boats or even boaters, causing severe injury due to its size and strength. The young may defend themselves by floating motionlessly on the surface in an attempt to mimic seagrass and escape detection.

Cypselurus heterurus
Atlantic flyingfish

Length Up to 40 cm (16 in)	
Weight Not recorded	
Sex Male/Female	
Location North Atlantic	**Status** Locally common

Also known as the four-wing flyingfish, this coastal species lives in the surface waters and feeds on plankton and other small fish. Unlike other flyingfish, both the pectoral and pelvic fins are modified for "flight". By opening out these fins, the four-wing flyingfish can glide over long distances – up to 200m (655ft) – to escape predators. In order to propel itself clear of the water surface, it may reach speeds of 60kph (37mph). The eggs have long, adhesive filaments by which they attach themselves to seaweed to avoid sinking.

Exocoetus volitans
Tropical two-wing flyingfish

Length Up to 18 cm (7 in)	
Weight Not recorded	
Sex Male/Female	
Location Tropical and subtropical waters worldwide	**Status** Locally common

Despite its name, this fish does not actually fly; instead, it glides over the water on specially enlarged and stiffened fins (see panel right). Some species of flyingfish have 2 pairs of "wings", consisting of enlarged pectoral and pelvic fins (see Atlantic flyingfish above right) but this species glides on its 2 pectoral

wing-like pectoral fins

protruding lower jaw

large, evenly curved scales

FISH OUT OF WATER

To achieve lift-off, flyingfish beat their tails rapidly beneath the water for the initial thrust, then extend their pectoral fins to lift them clear off the surface. The lower tail lobe continues to oscillate. At full speed, the fish can glide for up to 12 seconds.

fins alone. Flyingfish use this technique to escape predators such as dolphins, billfish, and tuna, and may achieve speeds of up to 65kph (40mph) in the air.

DISTINCTIVE FINS
The tropical two-wing flyingfish is dark blue above and silver below. Its distinctive, enlarged pectoral fins and asymmetrical tail are evident only in adults.

Aphyosemion amieti
Amiet's killifish

Length Up to 7 cm (2 3/4 in)	
Weight Not recorded	
Sex Male/Female	
Location W. Cameroon (lower Sanaga system)	**Status** Unconfirmed

Amiet's killifish, or Amiet's lyretail, remained undescribed until 1976, perhaps because of its very limited range. It is found in the stoneless, slow-moving, swampy parts of rivers in the rainforests of Cameroon. The female lays her eggs on the muddy river floor towards the end of the rainy season. Adult fish die after only one year, leaving the eggs dormant until the next rainy season. The onset of rains stimulates the eggs to develop and hatch. The fish mature in only one month and so quickly repopulate the river.

Jordanella floridae
American-flag fish

Length Up to 7.5 cm (3 in)	
Weight Not recorded	
Sex Male/Female	
Location S.E. USA (Florida)	**Status** Unconfirmed

This fish belongs to the pupfish family – a group of small freshwater fish that are particularly common in the Americas. Like most pupfish, the sexes differ in colour, and the male's distinctive pattern and coloration gives the fish its common name. Normally a peaceful eater of algae, the male becomes aggressive during courtship and when guarding eggs. He attracts a female by flashing his red, unpaired fins. After the female has laid her eggs, he drives her away and guards the eggs until they are ready to hatch after about a week.

green and orange mottled body

Anableps anableps
Foureyed fish

Length Up to 32 cm
(12 1/2 in)
Weight Not recorded
Sex Male/Female

Location Central to
N. South America

Status Locally common†

This long-bodied fish has protruding, frog-like eyes that are each divided horizontally, allowing it to see above the water and below it at the same time. This enables the fish to hunt equally effectively for aquatic insects underwater or flying insects that have fallen onto the water's surface. It swims rapidly to escape predators like king-fishers and wading birds, and may jump clear of the water to escape predatory fish. Female foureyed fish can store sperm and so can lay multiple clutches of eggs.

Cyprinodon diabolis
Devil's hole pupfish

Length Up to 2.5 cm
(1 in)
Weight Not recorded
Sex Male/Female

Location USA (Nevada)

Status Vulnerable

Found only in a single desert spring in Nevada, this tiny fish has the most restricted range of any vertebrate. The entire population numbers fewer than 500 individuals and occupies an area of about 20 square m (215 square ft). Water pollution and fluctuations in the water table threaten its survival. The smallest of all pupfish, it lacks pelvic fins and the vertical barring seen in most of its relatives. It breeds throughout the year, but rarely lives for more than 12 months.

Poecilia reticulata
Guppy

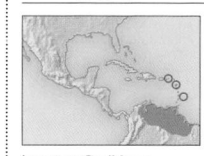

Length 6–7 cm
(2 1/4 – 2 3/4 in)
Weight Under 5g
(3/16 oz)
Sex Male/Female

Location Caribbean,
N. South America

Status Common

A popular aquarium fish, the guppy belongs to a family of about 140 species that are notable for bearing live young. In the wild, the male guppy is very variable in colour, often having blotches of blue, red, orange, yellow, and green pigments, as well as occasional spots, bands, or stripes of black. Such variation is greater in the aquarium fish. Very fertile, females may have over 100 young at a time, although the typical brood is 20–40. Adept at feeding on mosquito larvae, the guppy has been widely introduced as an agent of mosquito control.

subdued colour
of female

Poecilia latipinna
Sailfin molly

Length 1.5–5 cm
(1/2 – 2 in)
Weight Not recorded
Sex Male/Female

Location E. North America

Status Locally common†

The sailfin molly derives its name from the large dorsal fin that stands up along almost the entire length of the body. A close relative of the guppy (see above), this fish is also a popular aquarium species. Although it is generally olive-green, many different colour forms have been developed. The back, sides, and dorsal fin have rows of spots, which may merge to look like stripes.

sail-like
dorsal fin

Lampris guttatus
Opah

Length Up to 1.8 m
(6 ft)
Weight 50 kg
(110 lb)
Sex Male/Female

Location Tropical,
subtropical, and temperate
waters worldwide

Status Lower risk†

Also known as the moonfish, this large fish is an oceanic predator and has large eyes, an oval, silvery blue body, much deeper than it is wide, with strongly contrasting, deep red fins. Unlike most fish, which swim using their tail, it swims with its pectoral fins, beating them up and down like a pair of wings. The fins are long and stiff, with a narrow profile that minimizes drag as they slice through the water. Despite its toothless mouth and large size – the maximum recorded weight of a specimen is 73kg (160lb) – the opah is well known as a very efficient predator, energetically pursuing small fish and squid. It is found throughout the world but is most common in the North Atlantic and North Pacific Oceans. An inhabitant of mid-water levels, the opah is taken on long lines and with gill nets, and is a valuable food fish in Hawaii and on the west coast of the USA.

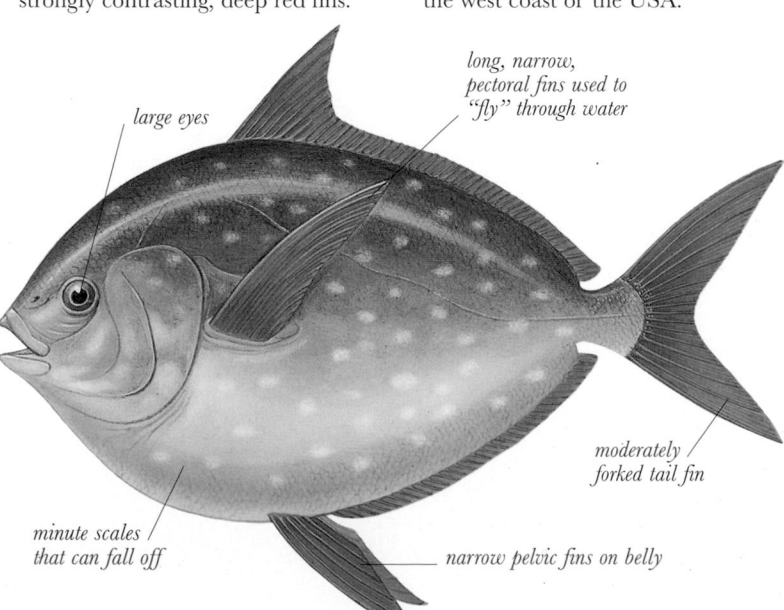

large eyes

long, narrow,
pectoral fins used to
"fly" through water

minute scales
that can fall off

moderately
forked tail fin

narrow pelvic fins on belly

Regalecus glesne
Oarfish

Length Up to 8 m
(26 ft)
Weight Not recorded
Sex Male/Female

Location Tropical,
subtropical, and temperate
waters worldwide

Status Unconfirmed

Although it lives worldwide in tropical and temperate seas, this huge, long, ribbon-like, deep-sea fish is rarely captured, or even seen alive. Hence little is known about its behaviour. Its extraordinary length protects it against most predators and has made it the subject of stories involving sea serpents and marine "monsters". This fish has a silvery body with oblique dusky bars, a short bluish head, and deep red fins. Its dorsal fin runs along almost the full length of its body, and has long rays forming a crest over the fish's head. The pelvic fins are attached just behind its gill cover, and each consists of a long, single ray, with an expanded tip that looks like the blade of an oar, which is how the oarfish gets its name.

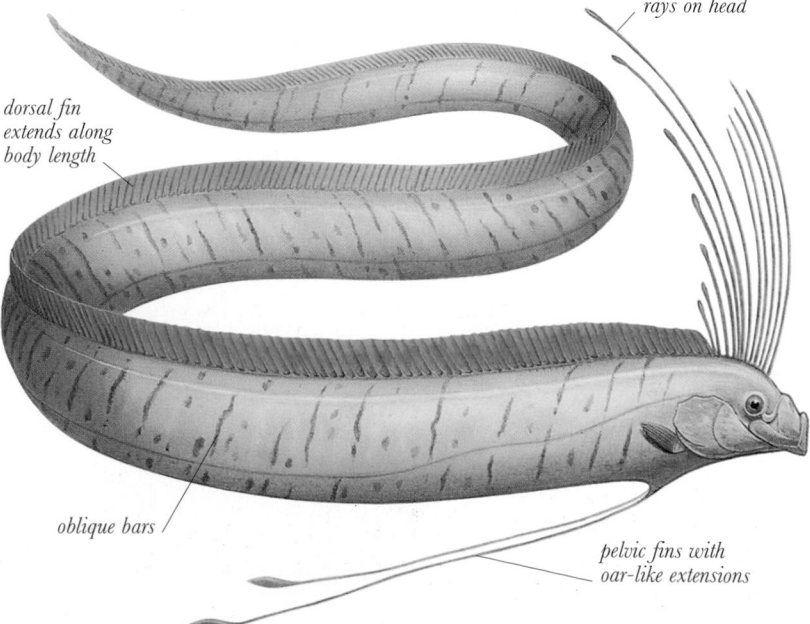

crest of long
rays on head

dorsal fin
extends along
body length

oblique bars

pelvic fins with
oar-like extensions

Trachipterus trachypterus
Ribbon fish

	Length Up to 1.6 m (5 1/4 ft)
	Weight Not recorded
	Sex Male/Female
Location Mediterranean, E. Atlantic, Indian Ocean, C. and W. Pacific	Status Lower risk†

A close relative of the oarfish (see p.507), this widespread species differs from it not only in being smaller, but also in having a body that tapers rapidly towards its tail,

ending in a tiny, upturned, fan-like fin. It usually has a pattern of 1–4 widely spaced dark blotches on the upper body and one or 2 blotches on the underside; it has no scales. As in the oarfish, its dorsal fin is deep red in colour. This carnivore feeds on mid-water fish and squid, and depends on its large size for protection against predators. Its eggs and larvae are often found in Mediterranean waters but there is little information on its reproductive behaviour. The ribbon fish family contains about 16 species; all are rare, and their biology is poorly known.

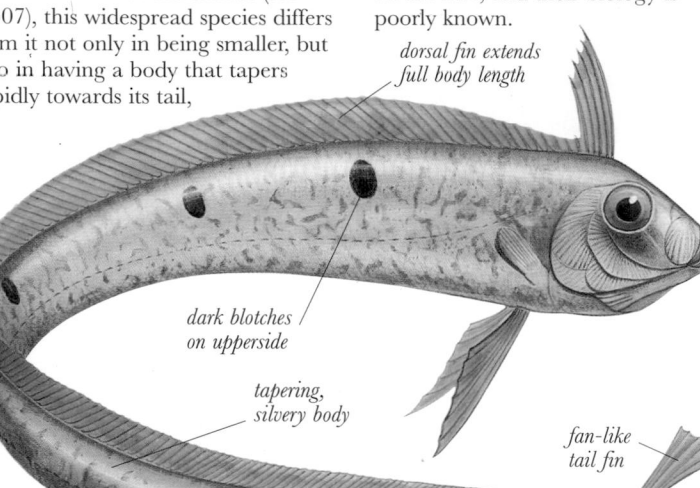

dorsal fin extends full body length

dark blotches on upperside

tapering, silvery body

fan-like tail fin

Barbourisia rufa
Red whalefish

	Length Up to 39 cm (15 1/2 in)
	Weight Not recorded
	Sex Male/Female
Location Tropical, subtropical, and temperate waters worldwide	Status Common†

This rarely sighted, deep-sea fish has a large mouth, the jaws having many small, depressible teeth. It is reddish orange, appearing black in deep waters, and its elongated body has a flabby feel. The skin of the red

whalefish is also slightly prickly – a characteristic produced by its scales, each of which has a tiny, central spine. Its fins are placed far back along its body, suggesting that the fish is an ambush predator that attacks with a rapid, headlong rush. The fact that only individual specimens have been collected suggests a solitary lifestyle. This species has not been well studied, as few specimens have been collected, but their size suggests that, like many deep-sea fish, the red whalefish is found at different depths at varying stages of its life. Small young fish have been found in middle depths, whereas older fish have been brought up from the sea bed, indicating that this is where they spend their adult lives.

Anoplogaster cornuta
Fangtooth

	Length 18 cm (7 in)
	Weight Not recorded
	Sex Male/Female
Location Tropical, subtropical, and temperate waters worldwide	Status Locally common†

Like many deep-sea fish, the fangtooth has a small body and a disproportionately large head – an adaptation for eating prey that can be

almost as big as itself. This cruising predator uses its large teeth to impale its prey, which it traps with rapid, lunging movements. Young fangtooths look very different from their parents, and at one time were mistakenly classified as a separate species. Adults have a brown to black body, head, and fins, while juveniles are silvery. Adult fangtooths inhabit greater depths than the young.

lateral line on body

Myripristis murdjan
Pinecone soldierfish

	Length Up to 30 cm (12 in)
	Weight Not recorded
	Sex Male/Female
Location Pacific, Indian Ocean	Status Locally common

The pinecone soldierfish lives in shallow lagoons close to reefs, hiding in crevices during the day and coming out at night to feed on plankton, such as crustacean larvae. It swims with quick, jerky movements, propelled by

a deeply forked tail. Like many nocturnal fish, it has large eyes. The body scales are large, sharp, and very rough, and the spines on the fins and gill covers help to deter predators. Its red fins are edged with white. The pinecone soldierfish is capable of producing a chattering noise, but why it does this is not known.

Cleidopus gloriamaris
Pineapple fish

	Length 28 cm (11 in)
	Weight 500 g (18 oz)
	Sex Male/Female
Location Indo-Pacific	Status Locally common

The pineapple fish gets its name from its shape, and from its large, black-edged scales, which look like the segments of pineapple skin. It belongs to a family of fish that are protected by bony armour, and by sharp spines on their dorsal and pelvic fins. This

particular species has a bioluminescent organ on the lower jaw – orange during the day, glowing blue-green at night – covered by the upper jaw when the mouth is closed. As this mainly solitary fish ventures out at night from the depths of dark caves, curious small fish, crustaceans, and other invertebrates are attracted by the glowing organ on its lower jaw, and are caught and eaten. There is only one other species in this family – the pinecone fish – which has its bioluminescent organ on the upper instead of the lower jaw. The bizarre appearance of the pineapple fish makes it extremely popular in the aquarium trade. However, only a skilled aquarist is capable of keeping it alive and well in captivity.

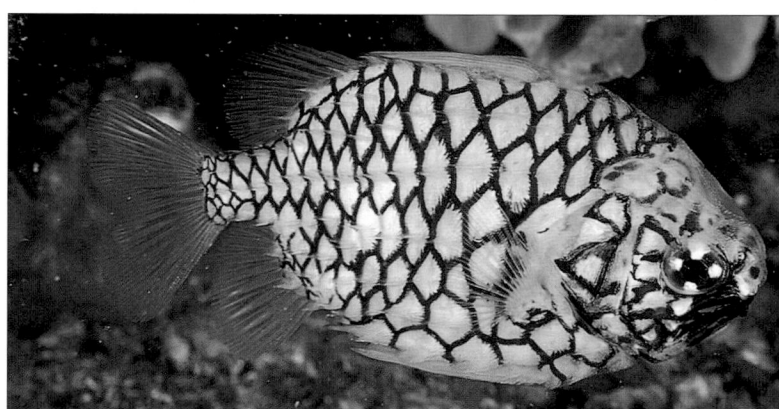

Hoplostethus atlanticus
Orange roughy

	Length 50–60 cm (20–23 1/2 in)
	Weight Not recorded
	Sex Male/Female
Location Atlantic, Indian Ocean, W. Pacific	Status Locally common†

Roughies live mainly at the edges of continental shelves, or on the floor of the open ocean. The orange roughy

has a large, blunt head with an upturned mouth, and small teeth arranged in bands. Its body is orange-red, although it appears dark at great depths, which affords it protection from predators. Like its close relative, *Hoplostethus mediterraneus*, the orange roughy is often caught as a food fish, and is very vulnerable to overfishing.

Photoblepharon palpebratus
Flashlightfish

Length Up to 12 cm (4 3/4 in)	
Weight Not recorded	
Sex Male/Female	

Location W. Pacific | **Status** Locally common

This small, blunt-nosed fish is one of the best examples of bioluminescence – or light production – in the animal world (see panel, right). One of only 7 known species of flashlightfish, it remains relatively inactive in caves and deeper waters during the day, and moves to shallower areas over reefs at night, where it feeds on small planktonic animals. Individuals defend small territories on the reef.

LIGHT CONTROL

Symbiotic bacteria, contained in an organ under the eye, produce this fish's lime-green light. The light helps it to find prey, startle predators, and communicate with other individuals. Using a black "eyelid" to cover and expose the eye, the fish can synchronize shoaling, indicate the presence of another fish, or defend territory.

MARKINGS
This blackish fish has reflective areas on the shoulder girdle and along the midline and fin edges.

Gasterosteus aculeatus
Three-spined stickleback

Length 5 cm (2 in), max 10 cm (4 in)	
Weight Not recorded	
Sex Male/Female	

Location Pacific, Atlantic, Mediterranean, North America, Europe, E. Asia | **Status** Locally common

This widespread, freshwater and coastal fish is easily identified by the 3 movable spines on its back. Its breeding behaviour is remarkably complex, with the males building underwater nests for the females to enter. When properly stimulated, the females lay a few eggs in the nests for the males to fertilize, and swim out again. The males aerate the eggs and guard the young.

brownish green back

Hippocampus guttulatus
Seahorse

Length Up to 16 cm (6 1/2 in)	
Weight Not recorded	
Sex Male/Female	

Location E. North Atlantic, Mediterranean, Black Sea | **Status** Vulnerable

Seahorses are remarkable fish, with pipette-shaped mouths, horse-like heads, and bodies covered with bony plates. They feed on planktonic animals, and swim upright, propelled by tiny fins. Most species live among eel grass and seaweed, and anchor themselves with their prehensile tails. Like their close relatives, the pipefish, seahorses have unusual breeding behaviour. In the spotted seahorse, the female lays the eggs in the male's brood pouch, at the base of his tail. He carries these until the young hatch.

Phyllopteryx taeniolatus
Weedy seadragon

Length Up to 46 cm (18 in)	
Weight Not recorded	
Sex Male/Female	

Location S. Australia | **Status** Locally common

The weedy or common seadragon is one of the largest seahorses, and one of the most bizarrely shaped. Like other seahorses, its body is covered with toughened plates, but it also has leaf-like flaps, which help to provide camouflage. Usually found just above the sea bed, the weedy seadragon hides among seaweed and plants growing on rocky reefs. Its bright coloration – brown to orange-red with yellow spotting – does, however, make

LEAFY PROTRUSIONS
The weedy seadragon derives its name from the leaf-like appendages on its body. Unlike other seahorses, its tail cannot be coiled around objects.

tough plates on body

leaf-like appendages

long tail

DRAGON'S EGGS

After mating takes place, the fertilized eggs are attached to the wrinkled undersurface of the male's tail and carried around by him. As the eggs hatch, the baby seadragons swim away.

it relatively easy to spot, and it is occasionally sighted in the open, away from reefs. The leafy seadragon, *Phycodorus eques*, has more elaborate camouflage than this species, and it is less commonly seen. The weedy seadragon searches for small invertebrates, including shrimps, among weeds and sponges near the ocean bed. It also feeds on planktonic animals in mid-water. Numbers of this species are decreasing due to indiscriminate collection.

Aulostomus chinensis
Trumpetfish

Length Up to 80 cm (32 in)	
Weight Not recorded	
Sex Male/Female	

Location Pacific, Indian Ocean | **Status** Locally common

Aeoliscus strigatus
Shrimpfish

Length Up to 15 cm (6 in)	
Weight Not recorded	
Sex Male/Female	

Location W. Pacific | **Status** Locally common

This fish spends much of its time swimming with its head vertically downwards. Shoals of shrimpfish move in synchrony

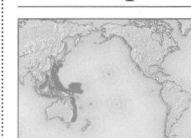

The trumpetfish is a common reef predator that often conceals itself by floating with its head vertically down among the branches of sea fans and other corals. It has a long, slender, flattened body, and a small mouth at the end of a very long snout. It is generally brown, but some are completely yellow. The trumpetfish hunts by ambush, lying motionless in water like a piece of drifting wood; it is galvanized into sudden action by passing shoals of fish.

long body

while feeding on minute crustaceans that live among the corals and between the spines of sea urchins. The long, narrow body and elongated snout are adaptations to this lifestyle. A brown stripe runs along its body, which also has scattered spots – the patterning provides the shrimpfish with effective camouflage. The body is almost entirely covered by a transparent armour; only the underside of the tail region is free of this armour to allow for fin movement. The belly has a sharp edge, and this gives the shrimpfish its other common name, the razorfish.

Chelidonichthys lucerna
Tub gurnard

	Length Up to 75 cm (30 in)
	Weight Not recorded
	Sex Male/Female
Location E. North Atlantic, Mediterranean, Black Sea	**Status** Locally common

A reddish coloured fish with a bony plated head, the tub gurnard has adapted to life on the sea floor. It has 3 isolated, long, and highly mobile rays on each of its pectoral fins that act as legs to support its body on the ocean floor. These fin rays are also highly senstitive and are used to probe the sea bed for molluscs and crustaceans, which, along with fish, form its main diet. The tub gurnard is capable of producing grunting sounds by contracting the muscles associated with its swim bladder. These sounds may help individuals stay together.

wing-like pectoral fins

Synanceia horrida
Estuarine stonefish

	Length Up to 60 cm (23 1/2 in)
	Weight Not recorded
	Sex Male/Female
Location Indo-Pacific	**Status** Endangered

This fish is a stout, slow-moving predator, with superb camouflage and deadly venom produced by glands at the base of its 13 dorsal spines. The spines are equipped with sharp tips, and these can easily penetrate human feet, sometimes with fatal results. The stonefish hunts by lying in wait for its prey, sometimes half burying itself in the sea bed. When partially buried, its upturned head and protruding eyes allow it to locate prey close by.

vertical mouth

SCALELESS FISH
The stonefish has an oddly coloured, scaleless body which camouflages it very effectively against the rocky or gravelly sea bed.

HIDDEN DANGER

Making the most of its cryptic colouring, the stonefish makes itself even less conspicuous by scooping out a shallow depression on the sea floor with its pectoral fins. It then piles up sand or mud around its body, and waits for prey.

dorsal spines

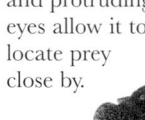

tail fin

Scorpaena porcus
Scorpionfish

	Length Up to 25 cm (10 in)
	Weight Not recorded
	Sex Male/Female
Location E. North Atlantic, Mediterranean, Black Sea	**Status** Locally common†

The scorpionfish, or rascasse, has a large, bulbous head with large eyes and mouth, a brown body with mottled markings, and several moss-like appendages between its mouth and eyes. It is a sit-and-wait predator that lurks among weed-covered rocks, using its highly effective camouflage to avoid being seen. All of its fins have rounded margins, and the spines of its dorsal, pelvic, and anal fins have poison glands, which, when erect, can inflict a harmful sting on its predators.

poison glands on spines

Cottus gobio
Bullhead

	Length Up to 18 cm (7 in)
	Weight Not recorded
	Sex Male/Female
Location W. Europe	**Status** Lower risk

The bullhead is a solitary freshwater species that lives under stones or in dense aquatic vegetation where its dark brown, marbled skin provides excellent camouflage. It feeds on the eggs and larvae of fish and small invertebrates. Spawning occurs between March and May, depending on the locality. The male excavates a small depression under a stone for the female to lay eggs in, and guards the eggs until they hatch.

fan-shaped pectoral fins

wedge-shaped head

Anoplopoma fimbria
Sablefish

	Length Up to 1.1 m (3 1/2 ft)
	Weight Up to 57 kg (125 lb)
	Sex Male/Female
Location North Pacific	**Status** Locally common†

Although it belongs to the scorpion-fish family, this widespread, North Pacific fish looks superficially like a cod (see pp.502–3), with its long, streamlined body and widely spaced fins. Adults are blackish to greenish grey above, often with blotches or stripes, and have a white underside; juveniles are blue-black above. Young fish live near the surface, usually near coasts, but adults live close to the sea bed. The sablefish makes extensive migrations along coasts, moving to deeper water in winter. It is an ambush predator which cruises in groups, chasing prey. It feeds on crustaceans, worms, and small fish such as lanternfish. An important food source, caught in trawls and traps, and on long lines, the sablefish is commercially exploited and widely marketed, mostly in Asia.

Cyclopterus lumpus
Lumpfish

	Length Up to 60 cm (23 1/2 in)
	Weight Up to 9.5 kg (21 lb)
	Sex Male/Female
Location North Atlantic	**Status** Locally common†

Also called the lumpsucker, this stout fish has an adhesive disc, formed from modified pelvic fins, on its underside. This acts as a sucker and anchors the slow-moving lumpfish to rocks and seaweed, allowing it to live in places where the current is strong. Found in water up to 300m (1,000ft) deep, it swims into much shallower waters to breed. In the breeding season, the male turns reddish, whereas the female is blue-green. Its eggs are an important source of inexpensive caviar.

PARENTAL CARE

Female lumpfish lay large masses of pink eggs, attaching them to rocks in shallow water. The male guards the eggs constantly, and fans water over them to keep them oxygenated. The eggs hatch after 6–8 weeks.

ARMOUR PLATING
Lumpfish get their characteristic knobbly shape from bony plates scattered over the body. Arranged in rows, these run along the fish's back and sides.

rounded body

"lumps"

Agonus acipenserinus
Sea poacher

	Length Up to 31 cm (12 in)
	Weight Not recorded
	Sex Male/Female

| **Location** North Pacific | **Status** Locally common† |

The sea poacher is one of about 50 species of poacher, and derives its other name, the sturgeon poacher, from its sturgeon-like shape. Like all poachers, it has a large head and elongated snout, and uses its pectoral fins like oars and its tail like a rudder to move along the sea floor. All poachers also have a distinctive cluster of finely divided barbels, called cirri, that hang from their downward-pointing mouths. The taste buds on the cirri are used to probe the sea floor for crustaceans and small animals. Slow-moving, poachers are easy fish to catch.

cirri

irregular blotches on body

rudder-like tail

Cromileptes altivelis
Pantherfish

	Length Up to 70 cm (28 in)
	Weight Up to 3.5 kg (7 3/4 lb)
	Sex Sequential hermaphrodite

| **Location** Indo-Pacific | **Status** Locally common |

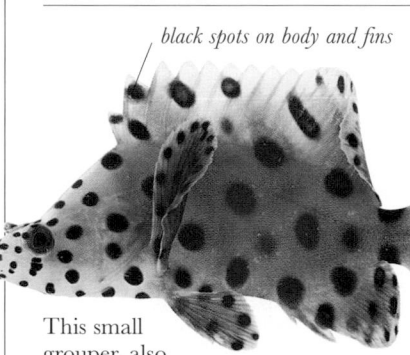

black spots on body and fins

This small grouper, also known as the Barramundi cod or humpback grouper, has very distinctive coloration, with a whitish head, body, and fins all covered with small, widely spaced black spots. Its back rises steeply behind its head in a "hump". Groupers typically change from female to male as they get older. Apart from the fact that it lays eggs, little is known about the breeding biology of this species.

Heteropriacanthus cruentatus
Glasseye

	Length Up to 50 cm (20 in)
	Weight Up to 2.5 kg (5 1/2 lb)
	Sex Male/Female

| **Location** Tropical waters worldwide | **Status** Common |

This solitary fish gets its name from the brilliant reflective layer – called the tapetum lucidum – in its large, conspicuous eyes. It spends the day under, or near, ridges of rock formations or reefs, emerging only at night to feed on a variety of prey including small fish, shrimp, octopus, crabs, and marine worms. Young glasseyes, unlike the adults, spend their time in the open sea.

Lophiolatilus chamaeleonticeps
Tilefish

	Length Up to 1.2 m (4 ft)
	Weight Up to 30 kg (66 lb)
	Sex Male/Female

| **Location** W. Central Atlantic | **Status** Locally common† |

This large, blunt-headed fish belongs to a family of about 30 species, all notable for building mounds or tunnelling into the sea bed. The tilefish creates funnel-shaped tunnels that can be up to 5m (16ft) across and 3m (9¾ft) deep. It uses these as a refuge, and also to attract prey, such as crustaceans and small fish. Each tilefish makes an individual burrow, which it enters head-first, and from which it emerges tail-first.

yellow spots on fins

blue-olive back

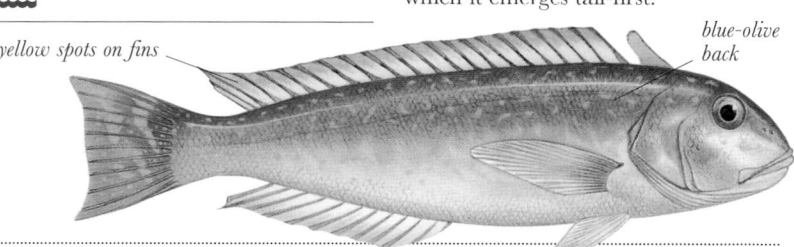

Anabas testudineus
Climbing perch

	Length Up to 25 cm (10 in)
	Weight Up to 150 g (5 oz)
	Sex Male/Female

| **Location** S. and S.E. Asia | **Status** Locally common |

The climbing perch is a small freshwater fish, adapted to life in stagnant lakes and ponds where the oxygen supply is often poor. It gulps atmospheric oxygen at the surface of the water or climbs out to do so (see panel, right). In some areas, it buries itself in mud in the dry season. The climbing perch is a popular aquarium species, and is also sold as a food fish.

CLIMBING

The climbing perch often uses the stiff spines near the gills and the pectoral fins to climb out and "walk" from one water body to the next on humid nights. It can survive out of water for several days at a time.

SPINY FINS
This small, stocky fish is uniformly grey, olive, or brown. Its fins are short, the dorsal and anal fins having prominent spines.

spiny dorsal fin

Caranx hippos
Crevalle jack

	Length Up to 1.2 m (4 ft)
	Weight Up to 32 kg (71 lb)
	Sex Male/Female

| **Location** E. and W. Atlantic | **Status** Locally common |

This fish can be recognized by its large, blunt head, the black spot near the edge of the gill cover, and the narrow-based, deeply forked tail. The largest of all jack fish, it can tolerate a broad range of salinities. Juveniles, which congregate in large shoals, tend to favour brackish water while older fish are found in deeper water, often cruising in pairs. It is an active predator of small fish: when it attacks a shoal of baitfish at the surface, the commotion caused can be seen from a distance.

narrow tail base

Chiasmodon niger
Black swallower

	Length Up to 25 cm (10 in)
	Weight Not recorded
	Sex Male/Female

| **Location** Tropical and subtropical waters worldwide | **Status** Locally common† |

The black swallower belongs to a small group of marine deep-water fish that are able to catch and eat fish larger than themselves. They can do this because their mouths and stomachs can be greatly distended to accommodate prey. Once the contents of the stomach are digested, the fish regain their normal shape. The black swallower lives in the deep ocean but may move to slightly shallower waters to feed. It has 2 dorsal fins, a long anal fin with a single spine, and a forked tail.

forked tail

distended stomach

large mouth

PATIENT HUNTER
The giant grouper usually lurks close to the sea bottom, often lying among coral or rocks, until an unwary crustacean or slow-moving fish swims close. Capture follows a quick lunge and opening of the cavernous mouth to suck in the prey. Numerous, depressible, needle-like teeth secure the quarry in the mouth.

Epinephelus lanceolatus

Giant grouper

Length	Up to 2.5 m (8¼ ft)
Weight	Up to 400 kg (880 lb)
Sex	Sequential hermaphrodite
Status	Vulnerable

Location Indian Ocean, W. and C. Pacific

A dark behemoth of a fish – one of the largest of the groupers – the giant grouper lives in warm, shallow, coastal waters; it may also be seen in estuaries, able to survive in the brackish conditions. Usually solitary, it stays close to coral and rocky reefs, remaining largely in the same area, usually around caves and wrecks. The giant grouper feeds mainly on crustaceans, with a favourite food being the spiny lobster, but it is also known to eat crabs, a variety of fish, including small sharks and rays, and young sea turtles. As juveniles, they are preyed upon by other fish, but as mature adults they are typically at risk only from humans. Their often massive size and fairly sluggish behaviour has made them an easy target of spearfishing. Since their great size requires a large food source and takes many years to develop, their density on reefs is extremely limited. They are easily fished out of entire regions, with little chance of their reappearance unless they are strictly protected. The giant grouper's close relation, the jewfish or Goliath grouper (*Epinephelus itajara*), is now so scarce that, in US waters at least, it is protected, with all harvesting of it prohibited.

elongated dorsal fin

small eyes　*heavy body*　*rounded tail fin*

ROBUST BUILD
Also known as the Queensland grouper, this fish has a characteristically heavy body and large mouth. Its typical coloration is brownish, with light mottling.

NOT-SO-GENTLE GIANT
The giant grouper is usually easily approached by divers – but some larger specimens are believed to have attacked and killed humans.

Opisthognathus aurifrons
Yellowhead jawfish

Length 10 cm (4 in)	
Weight Not recorded	
Sex Male/Female	
Location Caribbean to N. South America	**Status** Locally common†

yellow head

translucent body

This finger-sized, shallow-water fish lives on the sea bed, usually in coral sand and rubble. Found in clear, coastal waters, largely in the tropics, it has a bright yellow head, but its most notable characteristic – shared by other jawfish – is its habit of building burrows. This fish usually hovers outside the burrow, looking for prey; it mates inside. The male incubates the eggs in his mouth, spitting them out into the burrow before feeding.

Perca fluviatilis
Perch

Length Up to 51 cm (20 in)	
Weight Up to 4.5 kg (10 lb)	
Sex Male/Female	
Location Europe, Asia	**Status** Lower risk

The perch is a predominantly freshwater fish that inhabits slow-flowing rivers, deep lakes, and ponds. Juveniles feed on small planktonic creatures and invertebrates; adults eat invertebrates and small fish such as sticklebacks, minnows, and even other perch. Smaller fish tend to group together, while larger individuals are more solitary, preferring to lie close to a large rock or among aquatic vegetation.

STRIPED
A deep-bodied fish, the perch has 2 dorsal fins and characteristic dark vertical stripes, usually 6 in number.

dark stripes

EGGS STRINGS

Female perch lay their eggs in lace-like strings, wrapping them around river-bed stones or underwater plants. A single clutch can contain up to 300,000 eggs, connected by their sticky mucus. The eggs hatch in about 3 weeks.

Important both commercially and as a sport fish, it has been introduced to non-European countries such as Australia. In some places, this has caused significant ecological damage.

Lepomis cyanellus
Green sunfish

Length Up to 31 cm (12 in)	
Weight 975 g (35 oz)	
Sex Male/Female	
Location E. USA	**Status** Common

large head

fins edged in white or yellow

At home in a wide variety of habitats, from sluggish streams to lakes and ponds, the green sunfish is one of the most common large freshwater fish in North America. It forms relatively large breeding colonies, and is found in groups around aquatic vegetation and shelter. During the breeding season, between late spring and early summer, the male uses his tail to scrape out a shallow nest in which the female lays her eggs. The eggs stick to the gravel bed and the males guard them and the newly hatched young for a few days.

Sphaeramia nematopterus
Pajama cardinalfish

Length Up to 8 cm (3 1/4 in)	
Weight Not recorded	
Sex Male/Female	
Location Indo-Pacific	**Status** Common†

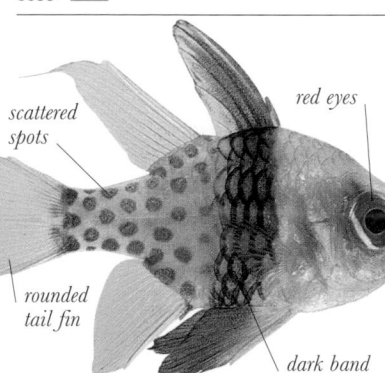

scattered spots

red eyes

rounded tail fin

dark band

This strikingly marked tropical fish belongs to a family of about 200 species, almost all of which inhabit coral reefs and lagoons. Both sexes have a yellow head, a dark "waistband", and scattered spots towards the tail, but the female is more brightly coloured. Males of this species incubate the eggs in their mouths, protecting them until they hatch.

Pomatomus saltatrix
Bluefish

Length Up to 1.2 m (4 ft)	
Weight Up to 14 kg (31 lb)	
Sex Male/Female	
Location Atlantic, Mediterranean, Black Sea, Indo-Pacific	**Status** Common

The sole species in its family, this primitive, perch-like fish has an exceptionally wide distribution in coastal and estuarine waters across much of the world. Known as the tailor in Australia, it has strong jaws and rows of razor-sharp teeth, making it a powerful killing machine. Operating in packs, bluefish attack smaller baitfish from behind, often herding their prey into tight shoals. This voracious feeder may even eat its own young. The bluefish is not territorial, and forms large shoals which migrate to warmer waters to mate and spawn. The larvae and young are then carried back to cooler waters by ocean currents.

powerful tail

Remora remora
Sharksucker

Length Up to 86 cm (34 in)	
Weight Up to 6 kg (13 lb)	
Sex Male/Female	
Location Tropical, subtropical, and temperate waters worldwide	**Status** Locally common

sucking disc

spineless dorsal fin

Nomeus gronovii
Man-of-war fish

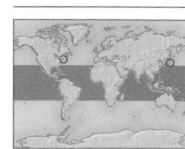

Length Up to 25 cm (10 in)	
Weight Not recorded	
Sex Male/Female	
Location Tropical waters worldwide	**Status** Lower risk†

During its early life, this blue-spotted fish lives among the stinging tentacles of the Portuguese man-of-war (see p.532) – a habit that keeps it beyond the reach of most predators. Remarkably, it possesses only

Sharksuckers, or remoras, are easily identified by the sucking disc above their head, which they use to fasten themselves onto whales, dolphins, and large fish. This species is most often found attached to sharks. The disc enables it to save energy by using the swimming power of its host for locomotion and gill ventilation. The host protects it from predators, and provides food scraps; in return, the sharksucker gleans parasitic crustaceans from the host's skin and gill cavity.

limited immunity to its host's toxins, and relies on its own swimming agility to avoid being stung. It is known to feed on the tentacles, and so is arguably a parasite of its host. Adults are believed to abandon the relative safety of their hosts for deep waters.

Coryphaena hippurus

Dolphinfish

Length Up to 2.1 m (7 ft)	
Weight Up to 40 kg (88 lb)	
Sex Male/Female	

Location Tropical, subtropical, and temperate waters worldwide | **Status** Common

Also known as the dorado, this blunt-headed fish has a single dorsal fin stretching along almost the whole length of its elongated body. It also possesses a large, deeply forked tail – a characteristic feature of an oceanic species that relies on speed to catch food and avoid being eaten. It has unusual coloration: a back that is a brilliant metallic blue or green, silvery sides with a golden sheen, and irregular rows of dark or golden spots. An exceptionally rapid swimmer, it

CAMOUFLAGE COLORATION
The countershading on its body helps this fish blend both into the dark depths and the well-lit open sea above, making it invisible to predators approaching from above or below.

graded colouring

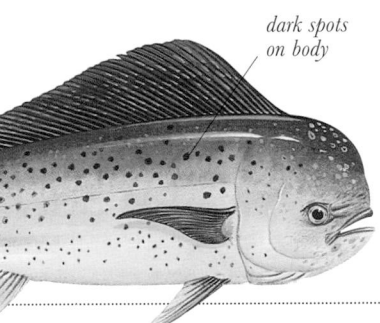

dark spots on body

FEEDING

The dolphinfish eats mainly fish, but also crustaceans and squid associated with floating objects. It follows ships and may form small groups below floating sargassum and other seaweed.

can notch up speeds of 60kph (37mph). This prized food fish is generally restricted to waters warmer than 20°C (68°F) and breeds in the open ocean. Spawning occurs round the year, in the evening or at night.

Mullus surmuletus

Red mullet

Length Up to 40 cm (16 in)	
Weight Up to 1kg (2¼ lb)	
Sex Male/Female	

Location E. North Atlantic, Mediterranean, Black Sea | **Status** Common

Highly prized as a food fish, the red mullet is reddish in colour, and has a darker red stripe along its midline, with several

yellowish stripes on its lower flanks. It has a downward-pointing mouth and a pair of barbels under the chin, equipped with sensory organs that help it to find food. These barbels, which give the fish its other name – the goatfish – are independently movable. They are thrust into sand and mud to detect small, sea-bed animals – crustaceans, worms, molluscs, and fish.

separated dorsal fins

forked tail fin

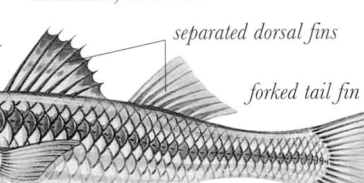

Toxotes chatareus

Archerfish

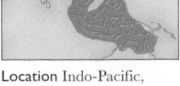

Length Up to 40 cm (16 in)	
Weight Up to 1kg (2¼ lb)	
Sex Male/Female	

Location Indo-Pacific, S.E. Asia, N. Australia | **Status** Locally common†

Found in brackish water as well as in freshwater streams and lakes, this deep-bellied, silver and black fish is famous for its ability to "shoot" insect prey off

overhanging vegetation, using a precisely aimed jet of water of up to 1.5m (5ft). It does this by pressing its tongue against the palate, then rapidly constricting its gill chamber. Little is known about its breeding biology; its eggs float on the surface.

spots and bands on body

tinted fins

Naucrates ductor

Pilotfish

Length Up to 70 cm (28 in)	
Weight Not recorded	
Sex Male/Female	

Location Tropical, subtropical, and temperate waters worldwide | **Status** Locally common

The pilotfish is a small or medium-sized fish with a silvery body marked with 6 or 7 black vertical bands. It derives its name from its habit of

black bands

swimming just in front of much larger fish, such as sharks and rays. This arrangement seems to benefit both the pilotfish and its host. The pilotfish feeds off the parasites on sharks and, in turn, benefits from the protection afforded by the larger fish it "pilots"; it also becomes less visible to the smaller fish on which it preys.

Sciaena umbra

Croaker

Length 70 cm (28 in)	
Weight Not recorded	
Sex Male/Female	

Location E. North Atlantic, Mediterranean, Black Sea | **Status** Common

A deep-bodied fish with a strongly arched back but a nearly straight underside, the croaker lives in inshore waters. It belongs to a large family of around 270 species. The

overhanging snout

croaker eats small fish and sea-bed invertebrates, usually after dark. Its most notable feature is its ability to produce loud sounds, used for communication. This fish also has a sharp sense of hearing. The male croaker is known to use the muscles on the highly evolved swim bladder to make a drumming sound as a mating cue. This fish seems to have remarkable buoyancy control and appears to swim almost effortlessly.

arched back

Pomacanthus imperator

Emperor angelfish

Length Up to 40 cm (16 in)	
Weight Not recorded	
Sex Sequential hermaphrodite	

Location Indian Ocean, Pacific | **Status** Locally common

Angelfish are among the most flamboyant inhabitants of coral reefs, with intense colours and bold patterns that often differ markedly in juveniles and adults. The emperor angelfish is a typical species, with a deep body flattened from side to side, a small mouth with comb-like teeth, and a backward-pointing spine on each cheek. Like other angelfish, it has the ability, rare among fish, to digest the tough flesh of sponges, which are plentiful on coral reefs. This fish sometimes

STRIPES AND MARKINGS
Adults have alternating, yellow and purple oblique stripes on the body. On the head and around the pectoral fin are light blue, dark blue, and black markings.

JUVENILES
Juveniles have a strikingly different colour pattern; they are dark blue to black with many concentric, curved to semi-curved white lines on their sides.

produces a loud thumping sound when alarmed. If the male dies, one female of his harem of 2–5 will change sex and assume dominance.

single, continuous dorsal fin

Chelmon rostratus

Copperband butterfly fish

Length Up to 20 cm (8 in)	
Weight Not recorded	
Sex Male/Female	

Location Indo-Pacific **Status** Common

eye-spot on dorsal fin

elongated snout

With their sideways-flattened bodies and long, pincer-like snouts, butterfly fish are ideally shaped for picking small animals from cracks and crevices in coral reefs. This species has 3 orange bands on its sides, and a narrow black band at the base of its tail. These striking markings, and the prominent black eye-spot on the dorsal fin, help to confuse predators. At night, the copperband butterfly fish takes on a blotchy coloration to blend in with the background of the reef.

Amphiprion frenatus

Tomato clownfish

Length 7.5–14 cm (3–5 1/2 in)	
Weight Not recorded	
Sex Hermaphrodite	

Location W. Pacific **Status** Common

This small fish, like other anemonefish, lives with sea anemones in a symbiotic relationship. It protects itself from the anemone's stings by the mucus covering its body. If the tomato clownfish leaves the anemone, it must, on its return, re-establish its immunity through a series of brief contacts with the anemone, after which it can again immerse itself among the stinging tentacles without harm. It receives shelter from the anemone, and feeds on plankton floating by and algae growing around the anemone. A family unit will usually occupy a single sea anemone. The largest 2 fish on an anemone are the male and female that do all the spawning; the rest are males, one of which, when the female dies, changes sex and takes her place. Anemonefish are members of the damselfish family, one of the most colourful groups of fish on coral reefs.

Oxycirrhites typus

Long-nosed hawkfish

Length Up to 13 cm (5 in)	
Weight Not recorded	
Sex Sequential hermaphrodite	

Location Indian Ocean, Pacific **Status** Lower risk†

The 35 species of hawkfish inhabit tropical waters and are found on coral or rocky reefs. Resting on the sea bed as lie-and-wait predators, they ambush unwary invertebrates and small fish. If threatened, they use their thickened, lower pectoral fin rays to wedge themselves into crevices, making them difficult to extract. The slender long-nosed hawkfish has whitish and red stripes and spots that form a cross-hatch pattern, and an elongated snout used to pick up prey. This fish is mostly found on its own, but the male is territorial and may maintain a harem within his territory. If a male hawkfish dies, one of the females may change sex and take over as the male of the harem.

long spines on single dorsal fin

red stripes and spots

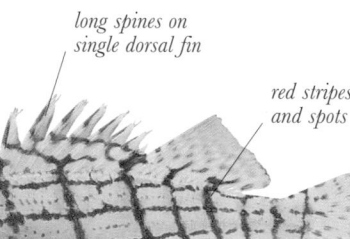

Labroides dimidiatus

Cleaner wrasse

Length Up to 11.5 cm (4 1/2 in)	
Weight Not recorded	
Sex Sequential hermaphrodite	

Location Indo-Pacific **Status** Common

The cleaner wrasse, an important member of coral-reef communities, feeds on parasites infesting other fish, as well as on their mucus and scales. "Client" fish queue up at special cleaning stations, which are usually occupied by a pair of wrasses or a group of females with an adult male. Once the dominant male dies, he is replaced not by another male, but by the dominant female, which changes sex.

dark stripe hides eye

Pseudotropheus zebra

Malawi cichlid

Length Up to 12 cm (4 3/4 in)	
Weight Not recorded	
Sex Male/Female	

Location East Africa (Lake Malawi) **Status** Common†

This small, metallic blue fish is one of hundreds of cichlid species that have evolved in the relative isolation of Africa's large lakes. It has a small mouth with thickened lips, and scrapes algae from rocks. During breeding, the female picks up her eggs in her mouth, and the male then fertilizes them. After the eggs hatch, the female protects the young for about a week.

pale blue-white fins

dark, vertical bars on body

Cetoscarus bicolor

Bicolour parrotfish

Length Up to 80 cm (32 in)	
Weight Not recorded	
Sex Sequential hermaphrodite	

Location Indian Ocean, Pacific **Status** Common

Parrotfish get their name from their bright colours and bill-like teeth. As these herbivorous fish scrape encrusting algae from coral rocks, they also remove and ingest part of the rock. To process this, parrotfish are also equipped with upper and lower rows of strong, molar-like teeth in the throat. These help in grinding down the rock, transforming it into sand, which is then defecated. Common on coral reefs, parrotfish are major producers of sand sediments on reefs.

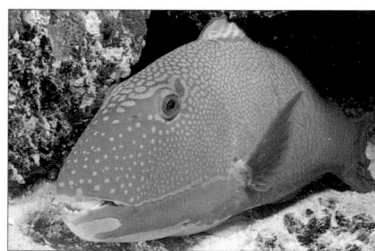

Astronotus ocellatus

Velvet cichlid

Length Up to 40 cm (16 in)	
Weight Up to 1.5 kg (3 1/4 lb)	
Sex Male/Female	

Location South America (Rio Ucayali drainage and upper Amazon river) **Status** Common

The cichlid family is a huge group – of perhaps more than 2,000 species – of deep-bodied freshwater fish restricted to South America, Africa, and Sri Lanka. The velvet cichlid, found in South America, has eye-spots – large, orange-ringed black spots at the tail base – that serve to confuse predators. A single spawning contains up to 2,000 eggs. Like most cichlids, this species tends and guards its young.

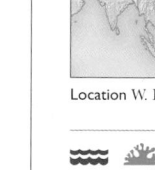

Anarhichas lupus
Wolf fish

Length Up to 1.5 m
(5 ft)
Weight Up to 24 kg
(53 lb)
Sex Male/Female

Location North Atlantic,
Arctic
Status Common

The wolf fish is a solitary, shallow-water predator that has no pelvic fins and only a rudimentary pelvic girdle. In contrast, the teeth are very well developed: they are conical at the front of the jaws, and molar-like at the sides. This fish uses its teeth and a bony plate in the roof of the mouth to crush mollusc shells and the carapaces of crabs, lobsters, sea urchins, and other echinoderms on which it feeds. Spawning takes place in deeper water during winter, and the eggs are guarded by the male until they hatch. Although originally caught only as a byproduct of trawler fishing, the delicately flavoured flesh of the wolf fish has led to an increase in its commercial importance in recent years.

large, broad head

small tail fin

Chaenocephalus aceratus
Blackfin icefish

Length Up to 72 cm
(28 in)
Weight Up to 3.5 kg
(7 ³/4 lb)
Sex Male/Female

Location E. South Pacific,
W. South Atlantic
Status Common

The blackfin icefish produces a natural antifreeze, enabling it to survive in the subzero waters of the Antarctic. It lacks red blood cells and hence looks rather pale, but has excellent blood circulation, and a strong heart which weighs as much as that of a small mammal. Its large, toothy mouth led to it being called the crocodile fish by 19th-century whalers.

Dissostichus mawsoni
Antarctic toothfish

Length Up to 2.2 m
(7 ¹/4 ft)
Weight Up to 120 kg
(260 lb)
Sex Male/Female

Location Antarctic Circle
Status Common

This bottom-dwelling fish can survive in cold Antarctic waters because it produces special proteins in its tissues and blood that act as an antifreeze. Like many creatures from this region, it grows very slowly, reaching sexual maturity at 8–10 years of age. It may sometimes swim up from the sea bed to feed on a range of small fish, squid, crabs, and prawns. Due to its large size, this fish can be eaten only by a few marine animals such as sperm whales, killer whales, and elephant seals. The Antarctic toothfish does not have a swim bladder, and achieves neutral buoyancy by having light bones and a high content of body fat.

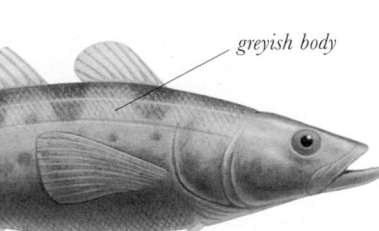

greyish body

Echiichthys vipera
Lesser weeverfish

Length Up to 15 cm
(6 in)
Weight Not recorded
Sex Male/Female

Location E. North Atlantic,
Mediterranean
Status Common

venomous dorsal fin

This venomous fish, frequently found near European beaches, is known to inflict painful wounds if stepped upon. Spines on the first dorsal fin and gill cover have poison glands at their bases; if disturbed, this fish erects these as a warning, or uses them in defence against larger fish. This species spends much of its time buried in the sea bed with only its eyes and the dorsal fin-tip exposed.

Periophthalmus barbarus
Atlantic mudskipper

Length Up to 25 cm
(10 in)
Weight Not recorded
Sex Male/Female

Location E. Atlantic
Status Common

Mudskippers have tapering bodies, large heads, and protruding eyes. Like other mudskippers, this species feeds on small animals on the mud surface, and spends most of its time out of the water. It breathes through the skin, which it has to keep moist. It can crawl on its pectoral fins, and also skip and jump across the mud by quick, flexing movements of its body.

Naso lituratus
Striped-face unicornfish

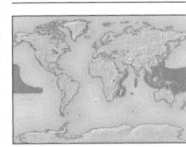

Length 45 cm
(18 in)
Weight Not recorded
Sex Male/Female

Location Pacific, Indian
Ocean
Status Common

The striped-face unicornfish belongs to a family of about 72 species known as surgeonfish because of the knife-like spine or spines on the side of their tail. This species uses its bright orange spines both in defence against predators and as a weapon in fights for dominance, inflicting deep wounds with rapid sideways swipes of the tail. The striped-face unicornfish has a moderately high, compressed body, but no horn-like projection on the forehead, unlike most other unicornfish. It grazes on leafy algae attached to the sea bed, nipping with its incisor teeth. A valued food source, it is caught off coral reefs.

tail spines

Nemateleotris magnifica
Firefish

Length Up to 8 cm
(3 ¹/4 in)
Weight Not recorded
Sex Male/Female

Location Pacific, Indian
Ocean
Status Common

This finger-sized tropical fish is one of about 60 related species that construct burrows in sand or coral rubble on reefs. Its coloration is distinctive, with a pale head and a body that grades smoothly into a deep orange-red near the tail. Adults hover nervously, alone or in pairs, over their burrows, facing the current while feeding on zooplankton. They quickly dart into the burrows when threatened.

deep orange-red posterior

dorsal fin held upright

FISH

FISH

Siganus vulpinus
Rabbitfish

	Length Up to 24 cm (9 1/2 in)
	Weight Not recorded
	Sex Male/Female
Location Indo-Pacific	Status Common

deep, compressed body

Living close to the bottom in the shallow water above coral reefs, this brightly coloured fish feeds primarily on algae that grow at the dead bases of corals. A single row of flat, spade-like teeth enables it to snip off bits of its food. Its single dorsal fin has 13 sharp spines. The conspicuous black, white, and yellow coloration may serve to warn off predators, while venom in its fin spines also forms part of its defensive armour. Males and females apparently pair for life.

Sphyraena barracuda
Great barracuda

	Length Up to 2 m (6 1/2 ft)
	Weight Up to 50 kg (110 lb)
	Sex Male/Female
Location Tropical and subtropical waters worldwide	Status Common

The great barracuda is a formidable, fast-moving predator, with few natural enemies apart from large sharks. Its large head and powerful jaws conceal knife-like teeth. This fish remains relatively motionless, then lunges forward quickly to grab its prey. It has been known to attack swimmers, often causing serious injury. However, these mishaps have usually occurred in murky waters where a human limb or shiny jewellery can be mistaken for other fish. Attacks have also been reported in clear waters when the fish has been provoked with a spear. It is curious and will swim close up to divers. The great barracuda is a popular game fish and is sometimes eaten, but large individuals can cause ciguatera poisoning – a potentially fatal condition caused by toxins that barracudas ingest with their food.

SOCIAL JUVENILES

Adult great barracudas are, by and large, solitary cruisers, found typically at or near the surface of open tropical and subtropical seas, in warm and shallow waters. They have been caught occasionally at depths of 100m (330ft). The young, however, sometimes swim in small groups inshore, in shallow, sheltered reefs and mangrove swamps. This allows them to hunt in groups and also provides them with some protection from predators.

projecting lower jaw

LONG BODY
The fish has an elongated body with a long snout. Its body length and associated musculature are especially adapted to produce short, rapid bursts of speed.

first dorsal fin set far from second one

dark bars on upper sides

inky blotches on lower body

Scomber scombrus
Atlantic mackerel

	Length Up to 60 cm (23 1/2 in)
	Weight Up to 3.5 kg (7 3/4 lb)
	Sex Male/Female
Location North Atlantic, Mediterranean, Black Sea	Status Common

Atlantic mackerel are streamlined, predatory fish that inhabit cold waters and continental shelf areas. In spring, they migrate closer to the shore. Mackerel group in large shoals according to size, feeding on fish larvae and small crustaceans. Spawning occurs from March to June when females lay as many as 450,000 eggs in a season. This commercially important species is extensively fished throughout its range.

Thunnus thynnus
Bluefin tuna

	Length Up to 4.3 m (14 ft)
	Weight Up to 910 kg (2,010 lb)
	Sex Male/Female
Location Tropical, subtropical, and temperate waters worldwide	Status Vulnerable†

The bluefin tuna is one of the world's largest and fastest bony fish, reaching speeds of at least 70kph (43mph). It has a powerful and streamlined body, with an acutely pointed snout, sickle-shaped pectoral fins, and a tall, crescent-shaped tail. Its pectoral and pelvic fins and its first dorsal fin can be recessed into grooves to minimize drag. Swimming in shoals, the tuna migrates as far as 10,000km (6,250 miles) across ocean basins, feeding on open-water fish.

finlets

crescent-shaped tail

Xiphias gladius
Swordfish

	Length Up to 4.5 m (15 ft)
	Weight Up to 590 kg (1,300 lb)
	Sex Male/Female
Location Tropical, subtropical, and temperate waters worldwide	Status Vulnerable†

A popular food and game fish, this fast-swimming predator gets its name from its long, sword-like snout. Unlike billfish, such as the blue marlin (see p.519), it certainly uses its snout as a weapon, stunning or impaling smaller fish and squid. It then slashes the crippled victim into pieces or swallows it whole. The swordfish occasionally attacks small boats – including those with copper-sheathed hulls – with devastating results. Its chief predators are the mako and blue shark, several species of tuna and marlin, and sometimes sea lampreys. Swordfish spawn in tropical waters throughout the year, peaking between April and September, depending on the area.

SOLITARY HUNTER

Adult swordfish usually hunt alone. Like tuna and marlin, they have 2 distinct kinds of swimming muscles with separate functions: dark red muscles are used for steady cruising, while faster-acting white muscles are used for short bursts of speed.

broad, flat "sword"

large eyes close to mouth

HIGH-FINNED
Grey-blue above and almost white below, the body of the swordfish tapers to a narrow tail base. The dorsal and anal fins are high, like sails, enabling efficient cruising.

tiny second anal fin

Makaira nigricans
Blue marlin

Length Up to 4.3 m (14 ft)	
Weight Over 900 kg (1,985 lb)	
Sex Male/Female	

Location Tropical, subtropical, and temperate waters worldwide

Status Vulnerable†

The blue marlin is renowned among anglers for the way it fights when caught. Hooked blue marlin breach the water surface with extraordinary leaps. The largest member of the billfish family, some individuals can weigh over 900 kg (1,985 lb). The upper jaw forms a large bill which is apparently used to stun or kill prey. The lateral keels at the base of its tail and the pelvic fins set into grooves make this fish a powerful and fast swimmer, capable of explosive bursts of speed and long-distance travel. Although blue marlin may stray to greater depths,

EGGS AND LARVAE
A single spawning produces millions of eggs, each 1mm (1⁄32 in) in diameter. These possibly hatch within a week. Eggs and larvae are planktonic.

they prefer the warmth of surface waters, where they feed on other fish such as mackerel, tuna, dolphinfish, and squid.

CONSERVATION

Game fishing can locally affect the numbers of this fish, but there is a much more serious threat in commercial fishing. A number of developing countries, particularly in East Asia, whose offshore waters abound with this species, exploit the blue marlin as a popular food fish. Due to their size, adult blue marlin are generally safe from predators other than humans, while juveniles may fall prey to other sea fish such as tuna, mackerel, and sharks.

high first dorsal fin

bill-like upper jaw

streamlined body

PATTERNED
This fish is dark blue above with silvery white sides crossed by bars made up of blue spots.

lateral keels

Hippoglossus hippoglossus
Atlantic halibut

Length Up to 2.5 m (8 1⁄4 ft)	
Weight Up to 315 kg (690 lb)	
Sex Male/Female	

Location North Atlantic, Arctic

Status Common

Highly sought after as a food fish, the Atlantic halibut is one of the largest flatfish – a group of over 500 species that have a highly distinctive anatomy, suiting them to life on the sea bed. Instead of being flattened from top to bottom, like many sea-bed fish, flatfish are flattened sideways, but habitually lie on one side. The Atlantic halibut lies on its left side, and both its eyes are positioned facing upwards, on its right. Compared to most other flatfish, Atlantic halibuts are unusually active swimmers, and often catch other fish in mid-water.

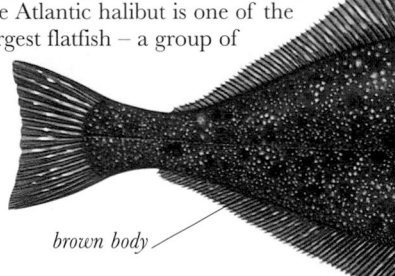

brown body

camouflaged upper surface

Pleuronectes platesa
Plaice

Length Up to 1 m (3 1⁄4 ft)	
Weight Up to 7 kg (15 lb)	
Sex Male/Female	

Location North Atlantic, Arctic, Mediterranean, Black Sea

Status Vulnerable†

As a newly hatched fish larva, the plaice lives for about 6 weeks near the surface of the sea. As in all flatfish, a remarkable transformation then occurs that turns it into a bottom dweller. The left eye migrates across the head to the right side, so that the blind left side can lie against the sea floor. There the plaice feeds on a diet of thin-shelled molluscs and marine worms. The upper side, capable of rapid and varied colour changes, camouflages this fish from potential predators. This is the most important commercial flatfish in Europe.

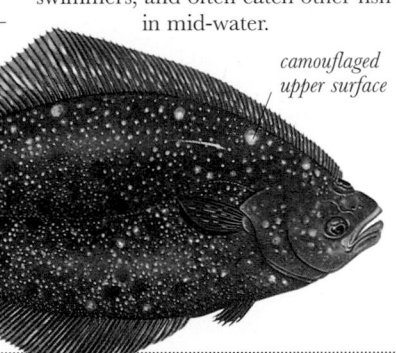

Scophthalmus maximus
Turbot

Length Up to 1 m (3 1⁄4 ft)	
Weight Up to 25 kg (55 lb)	
Sex Male/Female	

Location E. North Atlantic, Mediterranean, Black Sea

Status Common

The almost round body of this flatfish has a number of bony tubercles scattered over the surface of the upper side instead of scales. Both the eyes are on the left side, and the mouth is asymmetrical, with teeth only on the blind side. The jaws on the eyed side act like a siphon, drawing in water that is free of sand, thereby making respiration more efficient. Females, which are less numerous than males, produce 10–15 million eggs.

body width equal to length

Paralichthys dentatus
Summer flounder

Length Up to 94 cm (3 ft)	
Weight Up to 12 kg (26 lb)	
Sex Male/Female	

Location W. North Atlantic, E. North America

Status Common

Most flatfish are marine. The summer flounder is unusual in that it is also found in estuaries and in fresh water. It is a bottom-dwelling predator that relies on camouflage, speed, and sharp teeth to catch the small fish and other sea animals that make up its diet. The eggs of the summer flounder are covered in oil droplets; when the fish spawns, the eggs float up towards the surface, where they hatch into ordinary fish larvae. When a larva reaches about 1.5cm (1⁄2 in) in length, its body becomes flattened and the right eye migrates to the left side. The larva, which lacks a swim bladder, then settles on the bottom.

flattened body

FISH

Solea solea
Sole

Length Up to 70 cm (28 in)	
Weight Up to 30 kg (66 lb)	
Sex Male/Female	
Location E. North Atlantic, Mediterranean, Black Sea	**Status** Common

A commercially important flatfish, the sole lives in marine and estuarine environments, where it burrows into sandy and muddy floors. Like other flatfish, it is a highly asymmetrical animal, with both eyes on its camouflaged right-hand side, which always faces upwards as it moves over the sea bed. The mouth is twisted onto the blind side and has tiny teeth. The sole is carnivorous, feeding on molluscs and worms, which it locates by smell. The blind side of its head has special filamentous tubercles instead of scales. These are thought to enhance the sole's sense of taste.

Diodon holocanthus
Spiny puffer

Length Up to 30 cm (12 in)	
Weight Not recorded	
Sex Male/Female	
Location Tropical and subtropical waters worldwide	**Status** Common

If threatened, this spine-covered fish swallows large amounts of water, inflating itself until it becomes almost spherical, and making its spines, which are usually folded back, stand on end. But the spiny armour may hinder its ability to swim, and it is often found entangled in nets, much to the annoyance of fishermen. Brown above and yellow below, spiny puffers are reef dwellers. Their powerful jaws, armed with fused teeth, enable them to crack the shells and skeletons of many marine invertebrates.

pectoral fins

prominent eyes

erected spines

Mola mola
Oceanic sunfish

Length Up to 4 m (13 ft)	
Weight Up to 2 tonnes (2 tons)	
Sex Male/Female	
Location Tropical, subtropical, and temperate waters worldwide	**Status** Locally common†

The world's heaviest bony fish, the oceanic sunfish has a highly distinctive shape. Its scientific name comes from the Latin for millstone as it looks

DISC SHAPE

Seen in profile, this fish resembles a huge disc. Its mouth has a padded upper lip, and it has small eyes on either side. The tail is little more than a narrow frill attached directly to the body.

ERECT FINS
The triangular dorsal and anal fins are placed far back, and are held erect like a pair of giant blades. This fish moves by sculling with these fins.

almost circular when seen from the side. The young are elliptic in shape. The oceanic sunfish feeds primarily on jellyfish, using its teeth to nip pieces off prey. Although generally slow-moving, it may jump clear of the water if alarmed. Sunfish are occasionally seen basking horizontally at the surface; they probably do this only when they are sick or dying. Hugely prolific, they breed in the open sea, laying up to 300 million eggs, which drift away with the current. Most sightings of sunfish are in tropical waters, but specimens are occasionally seen at much higher latitudes, where currents keep the temperature relatively warm.

Balistoides conspicillum
Clown triggerfish

Length Up to 50 cm (20 in)	
Weight Up to 2 kg (4 1/2 lb)	
Sex Male/Female	
Location Pacific, Indian Ocean	**Status** Common

This exceptionally colourful, coral-reef species belongs to a family of fish that can lock their dorsal fins upright, enabling them to wedge themselves into holes or coral crevices. Its head is covered with tough, protective scales, and its teeth are strong and adapted to crushing the shells of prey such as molluscs and crustaceans. Like other triggerfish, it can be very territorial, especially during the mating season.

white-spotted lower body

Takifugu niphobles
Fugu

Length Up to 15 cm (6 in)	
Weight Not recorded	
Sex Male/Female	
Location W. North Pacific	**Status** Locally common

Despite the fact that the fugu contains a poison more potent than cyanide, it is considered a great delicacy in Japan, where it is prepared with great care before being eaten. To deter predators, this fish may secrete its powerful neurotoxin into the water, or it may inflate its body – by pumping water or air into an inflatable sac in the stomach – and erect the spines on its sides. The fugu breeds between May and July, 1–5 days after the full and new moons. Thousands of these fish arrive to spawn on pebbly beaches at high tide.

Lactoria cornuta
Boxfish

Length Up to 46 cm (18 in)	
Weight Not recorded	
Sex Male/Female	
Location Pacific, Indian Ocean, Red Sea	**Status** Common

The boxfish derives its name from its rigid outer skeleton, composed of fused bony, thick scales, which make it a clumsy swimmer. Small, horn-like projections above each eye account for its other name, the long-horned cowfish. This fish hunts by blowing jets of water into the sand on the ocean floor to expose the bottom-living invertebrates on which it feeds. Boxfish are caught and dried to sell as ornaments.

bony projections

dorsal fins towards back

FISH IN DANGER

The list on this page includes all those fish currently classed as critically endangered by the IUCN (see p.31). At present, relatively few marine species appear on the list, but others may well join this category if the rapid growth in sea fishing is not curtailed. Other marine species under threat include ones that live in coral reefs. Again, few of these are currently listed as critically endangered, but the total is certain to rise as reefs are affected by climate change. By contrast, freshwater species make up a large part of the list, and include fish as different as sturgeons and small tropical aquarium favourites. The preponderance of freshwater species is partly because they are much better known than marine fish, and partly because they face a wider range of threats. These include pollution, fishing, collection, competition from introduced species, and also habitat change.

CARTILAGINOUS FISH

Glyphis gangeticus (Ganges shark)
Himantura chaophraya (giant freshwater whipray – Thailand subpopulation)
Pristis microdon (freshwater sawfish – S.E. Asian subpopulation); *P. pectinata* (smalltooth sawfish – N. and S.W. Atlantic subpopulation); *P. perotteti* (largetooth sawfish)
Rhinobatos horkeli (Brazilian guitarfish)

BONY FISH

Acipenser dabryanus (Dabry's sturgeon); *A. nudiventris* (ship sturgeon); *A. sturio* (European sturgeon)
Adrianichthys kruyti (duck-billed buntingi)
Allochromis welcommei (cichlid species)
Allotoca maculata (opal goodeid)
Aphanius splendens (pupfish species); *A. sureyanus* (pupfish species); *A. transgrediens* (pupfish species)
Astatotilapia "dwarf bigeye scraper" (cichlid species); *A. latifasciata* (cichlid species)
Austroglanis barnardi (Barnard's rock-catfish)
Barbus erubescens (Twee River redfin); *B. euboicus* (petropsaro); *B. trevelyani* (border barb)

Betta miniopinna (betta species); *B. persephone* (betta species); *B. spilotogena* (betta species)
Botia sidthimunki (ladderback loach)
Brachionichthys hirsutus (spotted handfish)
Callionymus sanctaehelenae (St Helena dragonet)
Cephalakompsus pachycheilus (cyprinid species)
Chasmistes cujus (cui-ui)
Chela caeruleostigmata (leaping barb)
Chilatherina sentaniensis (Sentani rainbowfish)
Chiloglanis bifurcus (Incomati rock catlet)
Chlamydogobius micropterus (Elizabeth Springs goby); *C. squamigenus* (Edgbaston goby)
Chondrostoma scodrensis (cyprinid species)
Clarias cavernicola (cave catfish); *C. maclareni* (catfish species)
Coregonus reighardi (shortnose cisco)
Cottus petiti (sculpin species); *C. pygmaeus* (pygmy sculpin)
Cyprinella alvarezdelvillari (cyprinid species); *C. bocagrande* (sardinita bocagrande)
Cyprinodon bovinus (Leon Springs pupfish); *C. meeki* (cachorrito de

mezquital); *C. pachycephalus* (cachorrito cabezon); *C. pecosensis* (Pecos pupfish); *C. verecundus* (cachorrito de dorsal); *C. veronicae* (cachorrito de charco azul)
Cyprinus carpio (common carp)
Danio pathirana (barred danio)
Dionda mandibularis (sardinita quijarrona)
Encheloclarias curtisoma (catfish species); *E. kelioides* (catfish species)
Enterochromis paropius (catfish species)
Epinephelus drummondhayi (speckled hind); *E. itajara* (jewfish); *E. nigritus* (Warsaw grouper)
Galaxias fontanus (swan galaxias); *G. fuscus* (barred galaxias); *G. johnstoni* (Clarence galaxias); *G. pedderensis* (Pedder galaxias)
Gambusia eurystoma (Guayacon bocon)
Gila modesta (charalito saltillo); *G. nigrescens* (charalito chihuahua)
Girardinichthys viviparus (mexclapique)
Glossogobius ankaranensis (goby species)
Glossolepis wanamensis (Lake Wanam rainbowfish)
Hampala lopezi (cyprinid species)
Haplochromis "ruby" (cichlid species); *H. annectidens* (cichlid species)
Harpagochromis guiarti (cichlid species); *H. plagiostoma* (cichlid species), *H. worthingtoni* (cichlid species)
Hubbsina turneri (splitfin species)
Huso huso (beluga – Sea of Azov stock)
Ilydon whitei (splitfin species)
Kiunga ballochi (glass blue-eye)
Konia dikume (dikume); *K. eisentrauti* (konye)
Labeo lankae (cyprinid species); *L. seeberi* (Clanwilliam sandfish)
Latimeria chalumnae (coelacanth)
Lepidomeda albivallis (White River spinedace)
Leuciscus ukliva (cyprinid species)
Lipochromis "backflash cryptodon" (cichlid species); *L.* "black cryptodon" (cichlid species); *L.* "parvidens-like" (cichlid species); *L.* "small obesoid" (cichlid species); *L. maxillaris* (cichlid species); *L. melanopterus* (cichlid species)
Lucania interioris (sardinilla cuatro cienegas)
Maccullochella peelii (Mary River cod)
Macropleurodus bicolor (cichlid species)
Mandibularca resinus (bagangan)
Moapa coriacea (Moapa dace)
Myaka myaka (Myaka myaka)
Notropis cahabae (Cahaba shiner); *N. mekistocholas* (Cape Fear shiner); *N. moralesi* (sardinita de tepelmene)
Noturus baileyi (smoky madtom); *N. trautmani* (Scioto madtom)

Oncorhynchus apache (apache trout); *O. formosanus* (Formosan salmon)
Oreoleuciscus handlirschi (çiçek)
Ospatulus truncatus (bitungu)
Pandaka pygmaea (dwarf pygmy goby)
Paralabidochromis beadlei (cichlid species); *P. victoriae* (cichlid species)
Paretroplus maculatus (Damba mipentina); *P. petiti* (kotso)
Physiculus helenaensis (skulpin)
Poblana alchichica (charal de alchichica)
Poecilia latipunctata (molly del Tamesi); *P. sulphuraria* (molly del Teapa)
Prognathochromis mento (cichlid species); *P. worthingtoni* (cichlid species)
Psephurus gladius (Chinese paddlefish)
Pseudobarbus burgi (Berg River redfin); *P. quathlambae* (Maluti minnow)
Pseudophoxinus egridiri (Yag baligi)
Pseudoscaphirhynchus fedtschenkoi (Syr-Dar shovelnose sturgeon); *P. hermanni* (small Amu-Dar shovelnose sturgeon)
Ptychochromoides betsileanus (Trondo mainty)
Ptyochromis "rainbow sheller" (cichlid species); *P.* "Rusinga oral sheller" (cichlid species)
Pungitius hellenicus (ellinopygóstcos)
Pungu maclareni (pungu)
Puntius amarus (pait); *P. bandula* (bandula barb); *P. baoulan* (baolan); *P. clemensi* (bagangan); *P. disa* (disa); *P. flavifuscus* (katapa-tapa); *P. herrei* (cyprinid species); *P. katolo* (katolo); *P. lanaoensis* (kandar); *P. manalak* (manalak); *P. tras* (tras)
Rheocles wrightae (zona)
Rhodeus ocellatus (Nippon baratanago)
Romanichthys valsanicola (asprete)
Salmo platycephalus (ala balik); *S. trutta* (Aral Sea trout)
Sarotherodon caroli (fissi); *S. linnellii* (unga); *S. lohbergeri* (Leka keppe); *S. steinbachi* (kululu)
Scaphirhynchus suttkusi (Alabama sturgeon)
Scaturiginichthys vermeilipinnis (red-finned blue-eye)
Sebastes paucispinus (Boccacio rockfish)
Speoplatyrhinus poulsoni (Alabama cavefish)
Spratellicypris palata (palata)
Stereolepis gigas (giant sea bass)
Stomatepia mariae (nsess); *S. mongo* (mongo); *S. pindu* (pindu)
Syngnathus watermeyeri (river pipefish)
Thunnus alalunga (Albacore tuna – South Atlantic stock); *T. maccoyii* (southern bluefin tuna); *T. thynnus* (bluefin tuna)
Tilapia guinasana (Otjikoto tilapia)
Totoaba macdonaldi (totoaba)
Weberogobius amadi (Poso bungu)
Xenopoecilus poptae (Popta's buntingi)
Xiphophorus couchianus (Monterrey platfish)
Xystichromis "Kyoga flameback"; *X. phytophagus* (cichlid species)
Zingel asper (asper)

SAND TIGER SHARK
Currently listed as vulnerable, the sand tiger shark is one of several species that have suffered through being hunted for their fins. As a way of producing food, this form of hunting is extremely wasteful, as the rest of the shark is thrown away once its fins have been removed.

INVERTEBRATES

INVERTEBRATES

INVERTEBRATES ARE ANIMALS that do not have a backbone. Although many of them are small – and therefore easily overlooked – they are immensely varied and wide-spread, accounting for about 97 per cent of all known animal species. While vertebrates form a single phylum, invertebrates are an informal collection of more than 30, and the members of just one phylum – the arthropods – probably outnumber all other animals on earth. Invertebrates are found in every conceivable type of habitat, but they are most plentiful in the oceans, which is where animal life first arose.

EVOLUTION

Invertebrates were the first animals to evolve, although exactly how this happened is not yet known. There is little doubt that their ancestors were single-celled, food-eating microorganisms, similar to living protozoans, and many biologists think that at some point, groups of these began to form permanent

large eyes

symbiotic partnerships. When this occurred, animal life began.

The earliest animals had soft bodies and left few visible remains, so the first signs of animal life are indirect ones – features such as fossilized tracks and burrows, which date back about 1 billion years. The earliest fossils of invertebrates themselves come from late Precambrian times, about 600 million years ago. At the start of the Cambrian Period, about 545 million years ago, there was an acceleration of evolutionary change, known as the Cambrian Explosion. Soft-bodied animals, such as jellyfish, were joined by ones with hard body cases and shells. Hard body parts fossilize well, so from this point onwards the development of invertebrate life is much easier to

PAST SUCCESS

Trilobites were among the most successful invertebrates that have ever lived. They existed for about 300 million years, leaving huge numbers of fossils, but died out in a mass extinction about 245 million years ago.

trace. Although it happened long ago, the Cambrian Explosion brought into being all the invertebrate phyla that exist today. Studies of living species show that these can be divided into 2 groups: the mollusc–annelid–arthropod line and the echinoderm–chordate line. The second of these groups gave rise to animals with backbones.

ANATOMY

In anatomical terms, invertebrates are united principally by things that they lack: as well as not having a backbone, they do not have a bony skeleton, and they do not have true jaws. As adults, some invertebrates look like plants, and spend their lives in one place; others are instantly recognizable as animals, because they are always on the move.

Despite this bewildering variety, invertebrates do follow some under-lying patterns, particularly on the

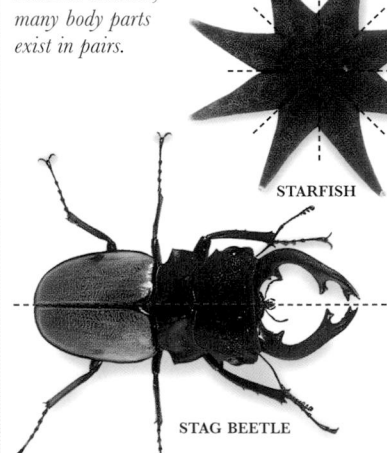

LIGHTWEIGHT LIFE
Butterflies are typical invertebrates, with a small body, a high reproductive rate, and a relatively short lifespan. Unlike some vertebrates, invertebrates are ectothermic (cold-blooded). They cannot remain active when temperatures drop to low levels.

SYMMETRY
A starfish has radial symmetry, which means that equal body parts radiate from a central point. A stag beetle has bilateral symmetry, with a body divided in two about a single axis. In bilateral animals, many body parts exist in pairs.

STARFISH

STAG BEETLE

cellular level. In the simplest kinds, such as sponges, individual cells are specialized but operate as separate units, rather than in groups. Cnidarians are more complex, with similar cells arranged in tissues. But in the third and highest grade of organization – seen in most inverte-brates – tissues themselves are arranged in organs, and organs into body systems, a pattern that is also followed in the vertebrate world.

This cellular segregation begins as soon as a fertilized egg cell starts to divide. In one species of nematode worm, the adult body consists of exactly 959 cells. Each one is the

SKELETONS

Many invertebrates have a skeleton, which may be internal or external – examples of both types are found in animals that live in water and on land. Some skeletons have evolved chiefly for protection and can withstand powerful impacts, but others are more flexible and exist mainly to keep the body in shape. Skeletons are made of various materials: hard structures often consist of crystalline minerals, while the cuticle of insects is made largely of a plastic-like material called chitin. Once formed, many external skeletons cannot grow and have to be periodically shed and replaced.

SHELLS
A bivalve shell is a two-part external skeleton. It is relatively heavy but exceptionally hard and grows in step with its owner.

SPICULE SKELETON
Most sponges have an internal skeleton of mineral slivers (spicules), set in a lattice of protein-based fibres. In this glass sponge, they are arranged with almost mathematical precision.

HYDROSTATIC SKELETON
The fluid-filled body cavity of an earthworm is bounded by a flexible wall. The pressurized fluid acts like the air inside a tyre.

fluid-filled cavity

JOINTED SKELETON
Arthropods, such as this pill millipede, have an external skeleton made of hard plates. Joints between the plates make the skeleton flexible. This kind of skeleton cannot grow.

BACK TO FRONT
Octopuses have elements of both radial and bilateral symmetry. Like bilaterally symmetrical animals, they have a definite head, with a pair of eyes. However, they also have arms that are arranged radially around the mouth. When they swim, the rear part of the body leads.

product of a precise sequence of cell divisions and ends up in a location encoded in the animal's genes.

On a larger scale, one of the fundamental differences between different invertebrates lies in the symmetry of their bodies. Animals in some groups – such as cnidarians – have radial symmetry. Their body parts are arranged in the same way as the spokes or rim of a wheel, usually with the mouth at the centre. These animals are often fixed in place, but if they do move, any part of the "wheel" can lead. By contrast, animals with bilateral symmetry usually have a distinct head, which leads the way when they move. These animals have an imaginary central dividing line, which splits them into 2 more or less equal parts.

During the course of evolution, some invertebrates have developed a combination of these body plans. Starfish, for example, show almost perfect radial symmetry, except for a small off-centre opening (or madreporite) that connects their water vascular system to the outside.

Internally, invertebrates show a wide range of body plans. Sponges – uniquely in the animal kingdom – are full of perforations, or pores. By drawing water through these pores, they filter out tiny particles of food. Some simple invertebrates have a bag-like anatomy, with a mouth that leads to a central digestive cavity. This arrangement, seen for example in cnidarians and flatworms, has one major drawback: indigestible waste has to be ejected through the mouth. Most invertebrates avoid this problem by having a digestive tract that runs through the body, in effect forming a tube. This plan – which is also found in vertebrates – allows the digestive system to work like a production line, with different parts being specialized for separate tasks.

Although invertebrates do not have bones, many have a skeleton, which protects and supports them. The skeleton may be built of hard

parts, but it may also work by fluid pressure (see panel, facing page). Known as a hydrostatic skeleton, this way of supporting a soft body is found only in the invertebrate world.

SENSES

For those invertebrates that are permanently fixed in one place the sensory world is relatively simple. These animals reach out towards anything that gives off the scent of food, and retreat from anything that might be a threat. They are sensitive chiefly to dissolved chemicals, direct contact, or changes in pressure. These animals have specialized nerve cells, or receptors, scattered throughout the body, and a simple nervous system without a brain.

This system works well enough for organisms that rarely move, but mobile invertebrates need more sensory information. They have more elaborate nervous systems, with sense organs to gather information and a brain to process it and trigger the most suitable response. In some of

TUNING IN
A male mosquito's bushy antennae are used for hearing. This is an example of hearing of a precise kind: the antennae respond to the exact frequency of the female's whining flight. The frequency used varies from one species to another.

MULTIPLE ATTRACTION
Attracted by colour and airborne odours, flies congregate on ripe fruit. They taste this sugary food with their mouthparts, and with chemical receptors on their feet.

these animals, such as flatworms, the brain is tiny, but in octopuses and other cephalopods it is highly developed and can be bigger than that of some vertebrates.

Vision is often one of the most important senses for invertebrates that move. Some of them, such as flatworms, have rudimentary eyes without lenses, which simply detect changes in the overall level of light. But many invertebrates have much more complex eyes, equipped with lenses that produce an image. One form, known as a compound eye, is widespread among arthropods, including insects. A compound eye consists of many self-contained units, each with its own receptors and lens system. Each unit responds to part of the field of view, and the total number of units determines how much detail the animal sees. A smaller number of invertebrates – principally cephalopods – have eyes that resemble our own. In these, a single lens focuses light onto a sensitive surface, the retina, which converts the image into signals that travel to the brain.

For invertebrates that live on land, hearing is also a key sense, partly because it is often used for locating

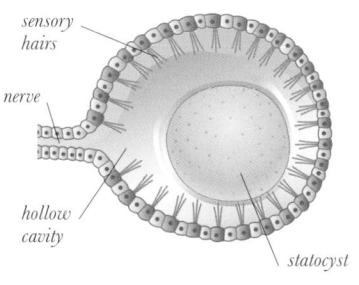

sensory hairs

nerve

hollow cavity

statocyst

DETECTING GRAVITY
Bivalves have microscopic sensors that respond to the direction of gravitational pull. The sensor consists of a chamber lined with nerve cells that detect the movement of a mineral weight (known as a statocyst).

MIMICRY

Mimicry occurs in the vertebrate world, but invertebrates are the true experts at this form of deception. Invertebrates mimic various objects, from pebbles and twigs to bird droppings, largely to avoid being eaten. They also mimic each other, either as a defence – by resembling species that are dangerous or poisonous – or, more rarely, as a form of camouflage for launching an attack. Mimicry is not only visual: some insects mimic the behaviour and even smell of others, so that they can enter their colonies and feed on their young without being attacked.

UNLIKELY MEAL
A postman butterfly caterpillar's first line of defence is its resemblance to bird droppings. If another animal approaches, it waves its menacing-looking "horns", which give off a disagreeable smell.

a mate. Unlike vertebrates, invertebrate hearing apparatus is not necessarily located on the head. Grasshoppers, for example, have eardrums in their abdomen, while crickets have them on their legs. The same is true of taste organs: flies and butterflies have chemical sensors on their feet, which means they can decide whether something is edible simply by standing on it.

Mobile invertebrates also need to know about their orientation and motion. Most have gravity sensors that tell them which way up they are, and tension sensors that are triggered when they move body parts such as legs and wings.

BUDDING PARENT
An adult hydra can produce young asexually by developing small buds, which grow into new individuals. This parent has grown a single bud, which is now several days old. Eventually, the young animal will break away and become independent.

REPRODUCTION

Invertebrates have a wide range of reproductive techniques, and include animals that reproduce both asexually and sexually. Asexual reproduction usually involves just a single parent. In sexual reproduction, 2 separate individuals are normally needed, although sometimes a single animal can simultaneously take on both male and female roles.

Few invertebrates rely solely on asexual reproduction, because it can lead to problems by creating young that are genetically identical to one another and their parents (see p.28). However, for many species – particularly sap-sucking insects, such as aphids – asexual reproduction is a valuable way of boosting numbers when food is in plentiful supply. Internal parasites, such as flukes, also use asexual reproduction at certain stages of their life cycle. For them, this way of multiplying is particularly important, because their complex life histories mean that each individual offspring has a very small chance of survival. However, with almost all these animals, sexual reproduction intervenes at some point in the life cycle, creating the

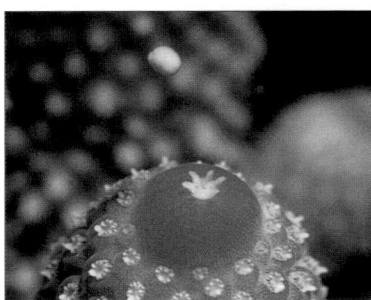

NEW LIFE
Released by a coral polyp, a parcel of eggs and sperm floats upwards. At the sea surface, the parcel will separate, the eggs and sperm will mingle with those of other individuals, and fertilization will occur.

genetic variations that allow a species to adapt to changing conditions around them.

Although asexual reproduction is widespread among invertebrates, most species use sexual reproduction alone. With vertebrates, sexual reproduction almost always involves a male and female animal pairing up, but in the invertebrate world things are often quite different. Animals with male and female reproductive organs – known as hermaphrodites – are common, and these may either pair up, as in the case of earthworms, or they may fertilize themselves.

Even where individual animals are either male or female, they may reproduce without ever having to meet. This apparently paradoxical situation is common among marine invertebrates, such as sea urchins, and it happens because these animals reproduce by external fertilization. They release their sex cells into the water, and the sperm and egg cells mingle, producing large numbers of fertilized eggs. For this form of reproduction to succeed, the eggs and sperm must be released with precise timing, which is often achieved by co-ordinating with the changing phases of the moon.

On land, most invertebrates use internal fertilization, with the male and female pairing up, and the male inserting his sperm into the female's body. This can involve complex courtship displays, and sometimes – for example, in spiders – careful behaviour on the part of the male, who runs the risk of being eaten by his mate. In the insect world, some male bugs adopt a less ceremonious approach: they use "traumatic insemination", piercing the female's abdomen with an organ that works like a dart. However, internal fertilization does not necessarily involve the direct transfer of sperm. In some scorpions, for example, the male produces a packet of sperm, known as a spermatophore, which it fixes to the ground. The male then presses the female onto the spermatophore, forcing it into her internal genital chamber, where fertilization takes place.

PARENTAL CARE

Once a female's eggs have been fertilized, most invertebrates leave their young to develop on their own. As a result, their young face an uphill struggle during their first few weeks of life, and a large proportion fail to survive. These losses are offset by the large number of eggs that are produced. A female starfish, for example, can release 2.5 million eggs in 2 hours.

METAMORPHOSIS

Most invertebrates change shape as they grow up. This change, known as metamorphosis, allows the young and adults to live in different ways, which can improve their chances of finding food. It also helps species to disperse, because one stage – either the larva (juvenile) or the adult – can have adaptations that enable it to travel. Metamorphosis may be gradual, or rapid and abrupt. In the first type, the young often resemble their parents, but in the second, they look completely different. Metamorphosis is controlled by hormones.

pupal case

wings showing through

COMPLETE CHANGE
During the pupa stage, a caterpillar's body is broken down and an adult butterfly formed in its place. The change occurs inside a chrysalis, or pupal case.

EARLY STAGE   **LATE STAGE**
DRIFTING YOUNG
Sea-urchin larvae have slender spikes that help them to drift in the plankton. As a larva matures, the spikes are slowly absorbed.

SAFE PASSAGE
With her family crowded on its back, a female scorpion guards her brood. At this stage, the young have pale, soft exoskeletons – these darken and harden as they age.

At the other extreme, some invertebrates are attentive parents. Female earwigs, for example, remain with their eggs, periodically cleaning them to ensure that they are not attacked by parasites or by mould. Female octopuses fix their eggs to rocky recesses on the sea bed, and

remain nearby, squirting jets of water over them to keep them clean and well oxygenated. Invertebrates may also take care of their young after they have hatched. Many arachnids carry their young on their back, while some freshwater crayfish carry theirs on their legs.

COLONIES

Even if they are looked after, most young invertebrates soon disperse to take up life on their own. But with colonial species, animals from the same species remain together, forming permanent groups or colonies. In a colony, the members are usually closely related to one another, and they often divide up the work needed for survival.

Colonial life is a recurring theme in invertebrate evolution, and it can take a variety of different forms. In

COLONY AFLOAT
A Portuguese man-of-war is an animal colony, formed by a collection of specialized polyps that live and work together.

some colonies – particularly in the sea – the members are physically joined, and may be so highly integrated that they look and behave like a single animal. This kind of organization is seen in some hydrozoans, including the Portuguese man-of-war. This animal looks much like a jellyfish but is actually a colony of polyps, one of which forms a bag-like float (see facing page). However, floating colonies are comparatively rare: many more invertebrates – including corals, bryozoans, and

CENTRE OF ATTENTION
Surrounded by workers, a queen honey bee controls all the activity in her hive. She does this by giving off volatile hormones – called pheromones – that evaporate and spread throughout the nest.

sea squirts – form fixed colonies, attached to all kinds of objects, from shells to the sea bed. These animals are often connected to one another by a shared network of nerves, allowing sensory signals to be passed from one individual to its neighbours.

Invertebrate colonies are less common on land, and the members of terrestrial colonies are separate, rather than joined. They reach their greatest development in social insects, which include termites, as well as ants, bees, and wasps. Despite their apparent independence, the members of these colonies rely on each other for survival, just as much as if they were fastened together. In each colony, one individual – the queen – normally lays all the eggs, while the role of the other members is to ensure that the colony is fed, housed, and defended from attack.

EATING AT LEISURE
Gripping the 2 halves of a mussel's shell, a starfish slowly prises them apart. It can evert its stomach through a paper-thin aperture to feed on the mussel's body.

FEEDING

Collectively, the world's invertebrates eat almost everything that is – or once was – alive. They include herbivores and carnivores, as well as a host of species that feed on dead remains. Their appetites embrace everything that we consume as food, as well as some highly specialized foodstuffs, such as rotting seaweed, feathers, fur, and even animal tears.

In water, a common feeding method among invertebrates involves filtering out edible matter, using body parts that act like a sieve. This way of life supports vast numbers of static invertebrates that live on the sea bed or near the shore. In these animals, evolution has crafted filtering apparatus out of a wide range of body parts: barnacles, for example, use their legs, while bivalves use their gills. Many of these filter feeders actively pump water through their filters, to increase the size of their catch. For these animals, pumping often has the added advantage that it supplies oxygen as well as food. Filter feeding is also used by many planktonic animals, including krill, which in turn provide food for baleen whales – the largest filter feeders of all. Underwater "grazing", in which vegetation is scraped from a solid surface, is also common. It is a technique used by limpets, which have rows of microscopic teeth, and also by sea urchins, which have a set of 5 chalky jaws.

On land, there is no equivalent of filter feeding, although some web-spinning spiders come close to mirroring this way of life, using their webs to trap airborne animals. Instead, most terrestrial invertebrates have to actively forage for food. The smallest herbivores – including many wood-eating species – often burrow through their food, a lifestyle that helps to protect them from attack, but most predatory invertebrates move about in the open. In herbivores

and carnivores alike, mouthparts are often shaped to deal with particular sources of food. Caterpillars bite off and swallow their food in substantial chunks, but many other insects have a liquid diet. Moths and butterflies, for example, are equipped with a coiled proboscis or "tongue" up

GATHERING FOOD
Dwarfed by its load, a leafcutter ant carries food back to its nest. In some parts of the world, ants eat as much food as all the plant-eating vertebrates in their habitat.

Dozens of dung beetles clamber over a pile of animal droppings. The beetles feed on the dung, and they also roll it away and bury it to provide food for their young.

to 30cm (12in) long, which can be unrolled to drink nectar from flowers or juices from fruits, while horseflies have a pair of blade-like jaws that can slice through skin, allowing them to lap up oozing blood. Despite being carnivorous, spiders are also liquid feeders. They have a tiny mouth, and feed by pouring digestive juices onto their prey or injecting it into its body, and then sucking up the slurry that results.

Given their immense numbers, many invertebrates have an important ecological impact as they feed. Earthworms play a key part in maintaining the fertility of soil, while insects have a major role in pollinating flowers and breaking down dead remains. In many habitats, invertebrates have a much greater effect than larger animals, but unless they are pests, their activity often goes unnoticed by humans.

PARASITES

Invertebrates make up the vast majority of the world's parasitic animals. Some – including ticks and leeches – live on the outside of their hosts. Many others, such as tapeworms and roundworms, live internally, feeding on their host's tissues or on the food that it consumes. External parasites can often survive with a single host, but many internal parasites have complex life cycles, involving several different hosts in sequence. These multiple hosts often include a vertebrate. Most parasites do not kill their hosts, but parasitoids are normally lethal. These animals, which include many insects, lay their eggs in or near the bodies of other animals. When the young hatch, they devour the host from inside.

LIFE CYCLE OF A BLOOD FLUKE
The human blood fluke is a parasitic flatworm that uses water snails as intermediate hosts. Its eggs reach water via untreated sewage, and produce free-swimming larvae. These multiply asexually inside snails, producing a further swimming stage that attacks humans.

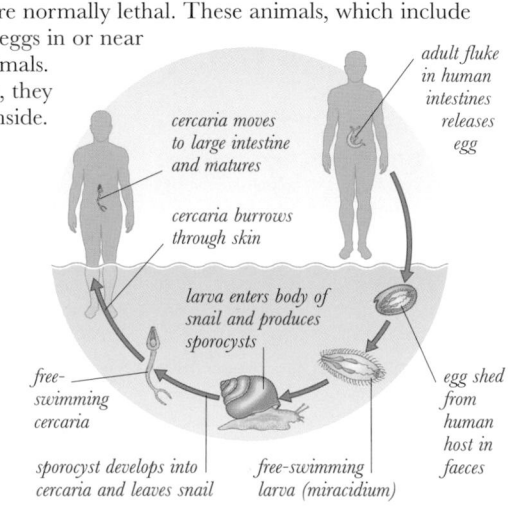

adult fluke in human intestines releases egg

cercaria moves to large intestine and matures

cercaria burrows through skin

larva enters body of snail and produces sporocysts

free-swimming cercaria

egg shed from human host in faeces

sporocyst develops into cercaria and leaves snail

free-swimming larva (miracidium)

SPONGES

PHYLUM	Porifera
CLASSES	4
ORDERS	18
FAMILIES	80
SPECIES	About 10,000

SPONGES, WHICH AT ONE TIME were thought to be plants, are the simplest of all living animals. Unlike most animals, which move through their habitat, sponges are immobile, living attached to solid surfaces and feeding by setting up currents of water from which they filter out particles of food. In other words, their aquatic environment is made to move through their body rather than the other way around. Most sponges are marine animals, although some species have adapted to life in fresh water.

ANATOMY

The body of a sponge lacks many features typical of other animals. Although the cells carry out specialized functions, they do not form true tissues or organs, and there are no nerve and few muscle cells. However, sponges do have a skeleton – a network of organic collagen fibres or inorganic structures (called spicules) made of silicon dioxide or calcium carbonate. Sponges have no system of symmetry, nor do they have clearly distinct body parts. The simplest sponge is essentially a tube, closed at one end, although sponges can take on other forms, including spheres and branching and thread-like structures.

SIMPLE SPONGE

osculum
central cavity
collar cell
flagellum
spicule
pore
collar

COLLAR CELL

BODY SECTION
Water enters a simple sponge through minute pores in the outer surface. It then passes into a central cavity before being expelled through a larger aperture (the osculum).

FEEDING

Sponges are remarkably effective filter feeders. The cells lining the body's inner surface are known as collar cells, each consisting of a ring (or collar) of tentacle-like structures surrounding a whip-like feature called the flagellum. Movement of the flagella creates the current of water that moves through the body. As the water passes through, plankton and particles of organic matter are trapped in the collars.

MARINE SPONGES
Most sponges are found in environments with stable temperature regimes. Like the leaf sponges shown here, they are often found on hard, rocky surfaces, where they may dominate large areas of the sea floor. They are especially common in caves.

CARNIVOROUS SPONGES
Found in Mediterranean caves, this is an unusual carnivorous sponge. It catches live prey on hook-like spicules at the end of long filaments and digests it externally.

Class CALCAREA

Calcareous sponges

Occurrence c.100 spp. worldwide; in shallow water (in crevices, under overhangs, on seaweed, in caves)

These are the only sponges that have skeletal spicules made of calcium carbonate, the mineral that molluscs use to make their shells. Compared to other sponges, they are often small – less than 10cm (4in) high – and drab, and are easy to overlook.

LEUCETTUSA LANCIFER
This sponge, less than 5cm (2in) high, has the classic compound vase-shape of calcareous species. Members of the genus are widespread in shallow waters from the tropics to the sub-Antarctic.

Class HEXACTINELLIDA

Glass sponges

Occurrence 500 spp. worldwide in cold water; mostly at depths of 200–2,000m (650–6,600ft) in seas

Glass sponges can be up to 1m (3¼ft) across and are less than 1m high. Their spicules are usually fused together to form rigid, often beautiful, glass-like masses. These structures are fragile, however, and the skeletons of the species living in deep water tend to collapse when they are dredged up and brought to the surface.

skeleton made of glass-like spicules

VENUS' FLOWER BASKET
Glass sponges are often objects of remarkable beauty. Species in the genus Euplectella, such as the one shown here, are found in tropical oceans below a depth of about 150m (500ft), and are commonly collected and dried for decorative purposes. They grow up to 30cm (12in) long.

Class DEMOSPONGIAE

Demosponges

Occurrence 9,500 spp. worldwide; in shallow and deep water; on solid surfaces of all kinds

These sponges make up over 90 per cent of all sponge species, and include the familiar bath sponge, whose populations are now badly affected by over-collection. Demosponges may be more than 1m (3¼ft) in height or breadth, and are often brilliantly coloured, creating underwater "gardens" that rival corals for their vivid hues. Many are mound-shaped or encrusting, but some of the smaller species bore their way into mollusc shells, leaving them riddled with pin-sized holes.

TUBE SPONGE
This large, tubular sponge from the Caribbean can reach a length of over 50cm (20in). In sponges with this kind of shape, water is drawn in through pores in the sides of the tube, and expelled through an opening at the top.

NIPHATES DIGITALIS
Many demosponges form irregular masses, but many others, like the Caribbean Niphates digitalis (above), have a definite shape. It is 30cm (12in) tall, and has a skeleton of siliceous spicules embedded in organic fibres.

tube open at end

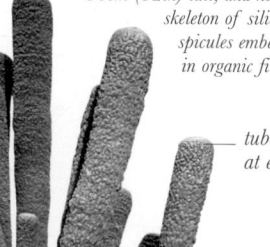

wall punctured by pores

CNIDARIANS

PHYLUM	Cnidaria
CLASSES	4
ORDERS	27
FAMILIES	236
SPECIES	8,000–9,000

THIS GROUP OF simple aquatic animals includes sea anemones, corals, jellyfish, and hydroids. The radially symmetrical body is essentially a tube that is closed at one end and open at the other. This tube is either flattened into a bell shape (a medusa) or elongated with the closed end attached to a hard surface (a polyp). In general, medusae swim freely, while polyps live on the sea bed. All cnidarians have tentacles around their mouth that contain stinging cells, known as cnidocytes. Most species are marine, although a small number are found in fresh water.

LOCOMOTION

Cnidarians can change shape by co-ordinating the contraction of some of their muscle cells. In polyps, the water inside the body cavity can act in a similar way to a skeleton, providing a resistant structure against which the cells are contracted. A polyp can lengthen, shorten, or bend its body – for example to reach towards potential prey or to retreat from predators. By alternately relaxing and contracting groups of muscle cells, free-living medusae such as the jellyfish (shown right) can move using a weak form of jet propulsion.

bell fully relaxed, ready to push forwards

bell begins to relax to take in water

elastic bell contracts to expel water

JELLYFISH MOVING
Jellyfish propel themselves through water by contracting their bell, obtaining lift from the water being forced out of the body cavity. The bell then relaxes, water re-enters, and the cycle is repeated.

ANATOMY

Some cnidarians (such as sea anemones and corals) exist only as polyps. Others (the hydroids and some jellyfish) exist as both polyps and medusae at different stages of their life cycle. In both forms, the wall of the body tube consists of just 2 layers of cells: an outer epidermis and an inner gastrodermis, separated by a jelly-like matrix (the mesoglea). The central cavity of the tube acts as a gut, and the single opening is used both to take in food and to expel waste. Around the mouth, the body wall is drawn out into one or more whorls of tentacles. These contain the cnidocytes, which are unique to cnidarians and give rise to their name. Cnidocytes are used to inject venom and to draw prey into the mouth (see below).

tentacle
cnidocytes
mouth
gut
mesoglea
epidermis
gastrodermis
basal disc

POLYP
The mouth of a polyp faces outwards. At the opposite end of the body is a basal disc, which is often attached to a solid surface. Many polyps, such as this hydra, do not have a well-developed mesoglea. Polyps may occur either singly or in colonies.

gastrodermis
velum (shelf of muscle)
epidermis
mesoglea
gut
mouth
tentacle

MEDUSA
The body of a medusa looks like a bell, with the mouth at the centre of the lower surface and the tentacles at the edges. The mesoglea is thicker than it is in most polyps. Some medusae have a shelf of tissue that is rich in muscle cells and used in locomotion.

nematocyst
coiled thread
epidermal cell

BEFORE DISCHARGE

uncoiled hollow thread

barbs

AFTER DISCHARGE

CNIDOCYTES
Each cnidocyte contains a bulb-like structure, called a nematocyst, which houses a coiled, barbed thread. When triggered by touch, the thread explodes outwards and pierces the victim's skin. The tentacles are then used to pull the prey in.

COLONIES

Most cnidarians can reproduce asexually (without mating) by producing buds from their own body. In some cases, the new polyp splits completely; in others, the budding is incomplete so that several individuals remain joined together to form a colony. Colonies can reach great size, as for example in reef-forming corals. Some planktonic colonies are highly complex. For example, colonies of siphonophores consist of polyps and medusae integrated into what appears to be an individual organism, with specialized polyps forming the equivalent of organ systems.

polyp projecting above hard skeleton

connecting sheet between 2 polyps

cup

exoskeleton

CORAL POLYP
The polyps in a coral colony secrete an exoskeleton of hard calcium carbonate, which may accumulate over time to form a reef. The exoskeleton includes structures called cups, into which the living polyps can withdraw.

REEF-BUILDING
Many corals are host to algae that contribute to their metabolism and nutrition. This symbiotic relationship allows the corals to grow quickly enough to accumulate the material to form reefs.

CONSERVATION

About a dozen cnidarian species (mainly corals) are classified as threatened. Many species that form and live on coral reefs are also endangered. Threats include rising sea levels and temperatures, discharge of silt in coastal areas, and collection for souvenirs. If the sea temperature rises above a certain level, the corals' symbiotic algae are expelled, and the corals become bleached (see below).

INVERTEBRATES

Class ANTHOZOA

Anthozoans

Occurrence 6,500 spp. worldwide; on the sea bed, in soft sediments

The word "anthozoa" means "flower animal", and with their bright colours, waving tentacles, and fleshy trunks, these animals often look remarkably like plants. However, they are carnivorous – as are all cnidarians – although some species supplement their diet with food produced by microscopic algae harboured in their bodies. Anthozoans vary greatly in size, ranging from minute species less than 1mm (¹/₃₂in) across, to those over 10m (33ft) across. When adult, they form polyps – simple, tubular animals with the tube closed at one end and open at the other. The open end is drawn out into one or more whorls of tentacles containing stinging cells, used for feeding and for defence against predators as varied as gastropods, polychaetes, sea spiders, and starfish. Anthozoans multiply asexually and sexually. Many species are solitary, but this group also

includes colony-forming corals – animals that are responsible for reefs, the biggest structures in the living world. These are formed by the animals' internal or external skeleton, which may be horny, rubbery, or calcareous. In addition to corals, this group of animals includes sea anemones, sea pens, and sea fans.

DEAD MAN'S FINGERS
Soft corals – members of the order Alcyonacea – occur in all oceans. The North Atlantic species Alcyonium digitatum (above) grows up to 20cm (8in) high, and lives in shallow water. Its polyps protrude from a white to pink or orange, finger-like mass – hence its ghoulish common name. The lobes contract if touched.

REEF SOFT CORALS
Like dead man's fingers, Sarcophyton species belong to the order Alcyonacea, and have the same fleshy texture. Members of this genus are most common in warm parts of the Indian and Pacific Oceans, and form colonies up to 1m (3¼ft) across.

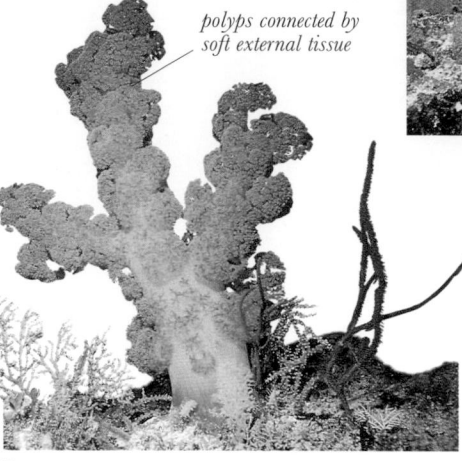

polyps connected by soft external tissue

BUSH-LIKE SOFT CORALS
The fleshy masses of most soft corals contain calcareous structures (sclerites) that support them. In the bush-like species (above), the sclerites may protrude when the colony is expanded. A member of the genus Dendronephthya, this Indo-Pacific coral grows to 1m (3¼ft) high. Soft corals like these are either attached to hard surfaces or anchored in soft sediments.

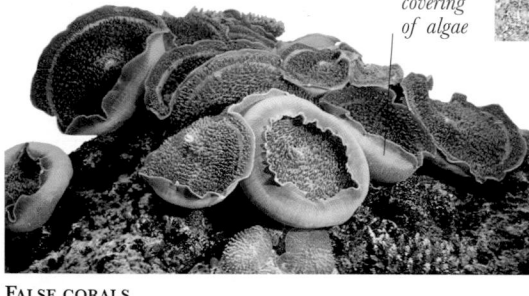

covering of algae

FALSE CORALS
Living in tropical waters worldwide, false corals are intermediate in form between true corals and sea anemones. Their polyps closely resemble those of corals, but lack an external hard framework. Like reef corals, they have symbiotic algae that supply them with nutrients. Amplexidiscus fenestrafer (above) attains a diameter of at least 30cm (12in).

HORMATHIID SEA ANEMONES
These widespread sea anemones commonly form symbiotic relationships with other marine invertebrates, living on mollusc shells or the carapaces and even claws of crabs. The anemone gains food scraps, while its partner is protected by the anemone's stinging tentacles. The species above, Stylobates aenus, grows up to 20cm (8in) tall.

SEA FANS
Growing 3m (9¾ft) or more high, sea fans such as Annella mollis (left) occur widely, but are especially common in tropical seas. They attach themselves to hard surfaces by creeping rootlets, and have a tough but flexible stem, with several branches that bear polyps.

fan-like mass

flexible stem

quill-like shape

SEA PENS
With a central axial polyp bearing many lateral polyps, often on regularly arranged side branches, sea pen colonies – found throughout the world – look like old-fashioned feather quills. The lower end of the axial polyp anchors the colony to the sea bed. The Pennatula species on the right grows to 50cm (20in) high, but some are double this size. Some of the secondary polyps of sea pens – such as the Pteroides species on the left, which grows to 30cm (12in) – pump water through the colony.

lateral polyp

axial polyp

base of axial polyp used for anchoring

stinging tentacles

central mouth

SEA ANEMONES
Unlike most anthozoans, sea anemones live as solitary polyps; many of these, however, reproduce by dividing and forming clones. The Atlantic snakelocks anemone, Anemonia viridis (shown right and below), has tentacles up to 10cm (4in) long. It needs well-lit conditions because a large part of its food is supplied by symbiotic photosynthetic algae.

long, snake-like tentacles

sucker-like base

PORITES SPP.
These massive corals, found in tropical waters, are second only to the staghorns as reef builders. They grow in the form of mounds, sometimes with large, blunt, finger-like projections. Some Porites domes on the Great Barrier Reef are almost 10m (33ft) in diameter, and may be 200 years old. The individual polyps, however, are only about 3mm (⅛in) across.

MUSHROOM CORALS
While most corals are colonial, mushroom corals consist of a single polyp usually less than 50cm (20in) across. They live in tropical waters and, when young, are attached to the sea bed by a stalk. When adult, as in the Fungia species above, they break away from the stalk and rest on the bottom.

PAVONA SPP.
This species in the genus Pavona has a diameter of 25cm (10in) and is a common and conspicuous member of reefs from the Red Sea to the eastern Pacific. It has fine, star-like markings on the surface, and grows in spreading colonies that form columns, plates, or fronds.

STAGHORN CORALS
The staghorns, belonging to the genus Acropora, are among the fastest growing corals on reefs. They would quickly swamp most other species if not for the fact that they are fragile and easily damaged in storms. Found in tropical seas, many staghorn colonies resemble antlers or bushes – like the branching species on the left, which may be 2m (6½ ft) tall. Some are more spreading and resemble the flat-topped acacia trees on the East African savanna. The plate-like growth of the species below – up to 3m (9¾ ft) across – intercepts light for its symbiotic algae and shades out potential competitors beneath.

large polyps

BRAIN CORALS
The individual polyps of this Lobophyllia colony are relatively large – 3cm (1¼ in) in diameter or more. They are arranged in meandering rows, giving the coral a brain-like form. Mature colonies of brain coral may weigh over a tonne. Many species inhabit the Pacific Ocean.

antler-like growth

plate-like shape

Class **SCYPHOZOA**

Jellyfish

Occurrence 215 spp. worldwide; in marine plankton, a few on seaweed or the sea bed

With bodies that can be over 2m (6½ft) across and tentacles that may extend many metres, jellyfish are among the largest animals found in plankton. A few species, however, attach themselves to seaweeds, and some spend much of their time on the sea bed. A jellyfish's body or "bell" is shaped like a saucer or hollow disc, and has a thick layer of "jelly" between the 2 cell layers surrounding a central hollow, which serves as the gut. The mouth is in the centre of the underside, and tentacles dangle from the lower margin. These simple animals have complex life cycles. They usually begin life as small polyps on the sea bed. Dozens of tiny medusae – free-swimming forms – bud off from each polyp, and drift away to start life on their own, growing into sexually reproducing adults. Although jellyfish can swim by contracting their muscles, they make little headway against even a gentle current, and are often washed up on beaches after storms. Most are carnivorous, catching prey with their tentacles. This class includes species such as the moon jelly and lion's mane jellyfish.

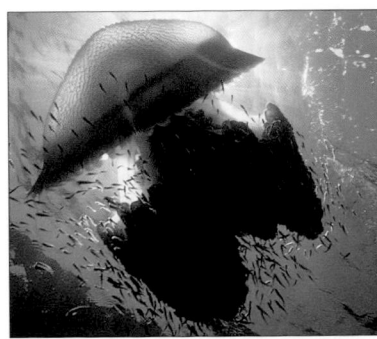

MASTIGIAS SPP.
These broad-bodied jellyfish, some up to 19cm (7½ in) across, are most common in the tropics. They have many small mouth openings in the highly folded lobes that hang from the bell.

LION'S MANE JELLYFISH
With a bell that can be over 2m (6½ ft) across, the lion's mane jellyfish, Cyanea capillata, which lives in northern waters, is one of the biggest in the world. Its numerous long tentacles pose a great danger to swimmers, and even when stranded on a beach, this jellyfish retains its ability to sting for a long time.

SEA WASP
Chironex fleckeri – also known as the Australian box jellyfish – grows up to 25cm (10in) across, and is one of the most poisonous of all cnidarians. Common off northeast Australia, it is almost transparent, and has long, trailing tentacles that can inflict potentially lethal stings.

translucent, domed bell

CASSIOPEIA XAMACHANA
A roughly disc-shaped jellyfish, 30cm (12in) across, this species lives in warm, shallow waters, and usually lies quietly pulsating upside down on the sea bed. For nutrition it relies on the symbiotic, photosynthetic algae that fill the bladder-like structures around its mouth.

feeding tentacles

Class HYDROZOA

Hydrozoans

Occurrence 3,300 spp. worldwide; in plankton, on sea, river, and lake beds

This group of cnidarians contains a diverse array of animals, with various body shapes and lifestyles, and sizes ranging from as small as 1mm (¹⁄₃₂in) to as large as 1m (3¼ft) across. In the course of their lives, they usually pass through both an attached polyp stage and a free-living medusa stage. Most are colonial, containing tens to thousands of polyps. Some colonies have hard skeletons and resemble corals, while others are often mistaken for jellyfish, drifting through the surface waters. Colonial forms that remain on the sea bed mostly have polymorphic polyps, with different polyps being shaped for different

tasks. The result is a colony that looks and works like a single animal. Planktonic colonies may also have polymorphic polyps, with some catching prey and others digesting the catch. Mostly carnivorous, hydrozoans use their tentacles to capture organisms ranging from minute planktonic animals to sizable fish; some are dependent on symbiotic, photosynthetic algae. Most animals in this class – which includes the sea fir and by-the-wind-sailor, as well as the freshwater hydras – are harmless to humans, but a few, including the notorious Portuguese man-of-war, have potentially deadly stings. They are preyed on by some molluscs, gastropods, polychaetes, sea spiders, starfish, and fish. Reproduction occurs both sexually and asexually.

PORTUGUESE MAN-OF-WAR
Found almost worldwide in tropical and temperate seas, the Portuguese man-of-war, Physalia physalis (right), has a well-deserved reputation for being dangerous. It looks like a single animal, but it is actually a highly integrated colony of polyps, some of which pack a powerful sting. A large gas-filled float keeps the colony at the surface, and supports trailing tentacles up to 20m (66ft) long. These tentacles contain polyps that catch prey, ones that digest food, and others that enable the colony to reproduce.

POLYORCHIS SPP.
This delicate hydrozoan jellyfish has a transparent bell up to 4cm (1½in) across, with about 90 tentacles trailing from it. Common on the west coast of North America, it feeds on planktonic animals.

FEATHER HYDROIDS
Species in the widespread genus Aglaophenia live in shallow water, and their dead remains are often washed up by the tide. Each colony can be up to 60cm (23½in) high, and contains numerous feeding and stinging polyps, as well as larger reproductive polyps – the solid yellow structures in the picture (right) – which release floating larvae into the water.

sail
float

BY-THE-WIND-SAILOR
The coin-shaped Velella velella *– 10cm (4in) in diameter – is a highly modified hydroid polyp that, like the Portuguese man-of-war, drifts on the ocean surface. It is moved by the wind acting on the small sail attached to the float and, after storms, is often washed up on tropical shores.*

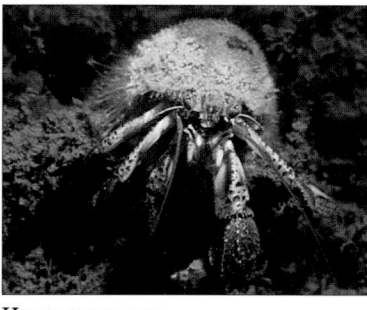

retractable tentacles

HYDRACTINIA SPP.
Members of the genus Hydractinia, *distributed worldwide, are small, colonial hydroids, typically 2cm (¾in) high, that often form encrustations on animals. The hydroids get improved access to food by being carried about, and in return, protect their host with their stings.* Hydractinia echinata *grows on shells of hermit crabs (see above).*

HYDRA SPP.
Often used as a textbook example of a hydroid, the genus Hydra *is, in fact, unusual because its members are solitary, not colonial, and are not surrounded by a calyx – a secreted tube. These tiny animals, 2.5cm (1in) high at most, live attached to stones, submerged wood, or plants, in bodies of fresh water throughout the world. A large number contain photosynthetic algae, which give them a green colour.*

DISTICHOPORA VIOLACEA
In some hydroids, the polyps can withdraw into cup-shaped extensions of the colonial "exoskeleton". The colonies are attached to the sea bed by a stalk. While many of them are quite small, some achieve a very large size, containing millions of individual polyps. The Indo-Pacific species Distichopora violacea *(left) is 5cm (2in) high.*

branching growth form

flask-shaped polyps with 2 rows of tentacles

polyp's stomach extends down stem

TUBULARIA SPP.
Found in shallow waters around the world, Tubularia *species look remarkably like plants. Their colonies grow from "roots" that cling to rocks and wrecks, and consist of a dense tangle of branching "stems", with a polyp at each tip. Most colonies are less than 15cm (6in) high.*

FLATWORMS

PHYLUM	Platyhelminthes
CLASSES	4
ORDERS	35
FAMILIES	360
SPECIES	About 13,000

FLATWORMS ARE THE simplest of all the animals that have bilateral symmetry. Their bodies are solid rather than containing an internal cavity, and they have no blood or circulatory systems, and no organs for exchanging gases with the environment. Most flatworms, such as the flukes and tapeworms, are parasites. However, many are free-living, and they are abundant in freshwater habitats and in the sea, particularly on rocky coasts and reefs.

BODY SECTION
Flatworms have a simple body that is not divided into segments. The space between the internal organs is filled with spongy connective tissue.

connective tissue
gut
longitudinal muscle
dorsoventral muscle
gut branch

FREE-LIVING FLATWORMS
Since flatworms rely on diffusion to take in oxygen, their bodies have to be thin. This requirement makes large species – such as this marine turbellarian – much more fragile than other animals of a comparable size.

Class TURBELLARIA
Turbellarians

Occurrence 4,000+ spp. worldwide; on rock and sediment surfaces in water, in damp soil, under logs

Turbellarian flatworms show a variety of different shapes. Although they have a distinct front end, the mouth is positioned some way back on the underside of the body. Most turbellarians are translucent, black, or grey, but some marine species – particularly ones that live in coral reefs – have brightly coloured markings. They may be less than 1mm (1/32in) to 50cm (20in) long. The largest species are paper-thin, and this helps them to absorb oxygen directly from their surroundings. Unlike other flatworms, which are parasitic, most turbellarians are free-living and have simple sense organs to help them negotiate their way around their habitat. Many of them move by creeping, but some of the bigger species swim by rippling their bodies. The majority of turbellarians are predators, feeding on other small invertebrates; some are parasitic, some are commensal, and a few depend on symbiotic, photosynthetic algae. Almost all are hermaphrodites.

black and yellow bands

MARINE FLATWORMS
Some marine turbellarians are large and brightly coloured. This species – Pseudobiceros zebra – is common on coral reefs in the Indian and Pacific Oceans, and grows to 5cm (2in) long. It shows a variety of colour schemes, including black and yellow traverse bands and a line of black blotches.

broad head with many eyes

TERRESTRIAL FLATWORMS
Land-dwelling turbellarians often have elongated bodies, with head flaps bearing numerous small eyes. The Bipalium species shown here is one of the largest terrestrial forms, sometimes reaching 35cm (14in). Originally from the tropics, it has been introduced throughout the world via potted plants, and is now a widespread pest.

ACOEL FLATWORMS
The acoels are the simplest types of flatworms known, lacking even a true gut. All are small, marine, and inconspicuous, many living in the surface layers of soft sediments. Several are coloured green by symbiotic algae and undertake small-scale migrations to give their algae light. The small olive-green blotches on the bubble coral, Plerogyra (right), are flatworms from the genus Waminoa, inhabitants of corals in the Indo-Pacific; they measure only 5mm (3/16in) in length.

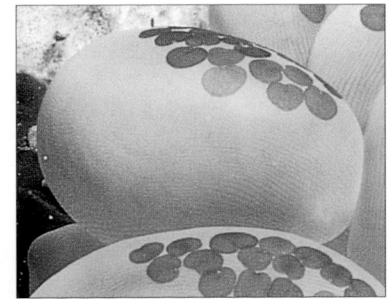

Class CESTODA
Tapeworms

Occurrence 3,500 spp. worldwide; within other animals, mostly cartilaginous fish

Highly specialized for their parasitic way of life, tapeworms are usually less than 1mm (1/32in) thick, but can be up to 30m (100ft) long. As adults, they live in the intestines of vertebrates – including humans – and they absorb food directly through their body wall. Tapeworms grow from a region just behind the head, forming a long line of segments called proglottids, each of which houses a complete reproductive system. As the proglottids age, they break away from the worm's "tail" and pass out of the host in the faeces, carrying thousands of eggs with them. If eaten by another animal, the eggs develop – and the cycle begins again.

TAENIA TAPEWORMS
The widespread Taenia species – some over 10m (33ft) long – live in humans and other mammals. An adult specimen is shown above, while the close-up on the right shows the head, or scolex. It is equipped with hooks and suckers, with which the worm fastens itself to its host's intestines.

suckers on scolex

Class TREMATODA
Flukes

Occurrence 12,000 spp. worldwide; within or on other animals

Like tapeworms, flukes are wholly parasitic, but they have a quite different shape. They are typically cylindrical or leaf-like, with hooks or suckers at one or both ends, or on the underside. Their length ranges from less than 1cm (3/8in) to, in exceptional cases, as much as 6m (20ft). Unlike tapeworms, flukes have mouths and digestive systems, and they often burrow through their hosts' tissues. While some attach themselves to internal organs, others are external parasites. Their life cycles are remarkably complex, involving a variety of larval stages, and up to 4 different hosts. The intermediate hosts are often molluscs, but the final host – which harbours the adult stage – is usually a vertebrate. Flukes can cause serious diseases, including bilharzia, which affects millions of people in the tropics.

LIVER FLUKES
Several species of fluke live in the livers and bile ducts of mammals, including humans, causing severe damage and even death. The species shown above is the widespread sheep liver fluke, Fasciola hepatica, which grows to about 3cm (1¼in) long. Its intermediate hosts are freshwater snails.

BLOOD FLUKES
Blood flukes are estimated to infect over 200 million people, besides occurring in other mammals and in birds, causing schistosomiasis or bilharzia; larvae attempting to bore through human skin also cause "swimmer's itch". Unusually, the sexes are separate, although the male lives wrapped around the female. This tropical species, Schistosoma mansoni, is 1–1.5cm (3/8 – ½in) long.

slender male

INVERTEBRATES

SEGMENTED WORMS

PHYLUM	Annelida
CLASSES	3
ORDERS	31
FAMILIES	130
SPECIES	About 12,000

THE BODIES OF SEGMENTED WORMS – or annelids – differ from other worms in being divided into a series of linked but partly independent sections. Each segment contains the same set of organs, and acts as a separate functional unit. The segments are filled with pressurized fluid, which gives the body its shape. Segmented worms have various modes of life, but most are burrowers, living either in soil or in sediments beneath fresh water or the sea.

EARTHWORM BURROWS
Feeding mainly on decaying vegetable matter, earthworms are present in soils throughout the world. They are usually found close to the surface but rarely leave their burrows, to avoid being eaten by birds and other animals.

ANATOMY

A segmented worm's body consists of a head at one end, a tail at the other, and many trunk segments in between. Each segment is divided from the others by partitions and contains (at least in primitive forms) a complete set of organs. The mouth is on the first segment, and the gut passes through all the segments to end in an anus in the tail. On the side of each segment are hooks and bristles (chaetae). Some marine worms have sensory tentacles; others, such as the earthworms, have no obvious projections.

body cavity · *longitudinal muscle* · *chaeta* · *circular muscle* · *gut*

FRESHWATER PARASITES
Leeches are parasitic worms. They use 2 suckers, one at each end of the body, to attach themselves to their hosts.

BODY SECTION
The body cavity of an earthworm is filled with fluid and contains a large gut. Short chaetae project through 2 layers of muscle to the body's outer surface.

MOVEMENT

A segmented worm moves by coordinating the action of muscles in different parts of its body. In each segment, there is one layer of longitudinal muscle and another of circular muscle. When contracted, these muscles make the segment either long and thin or short and fat. As the worm moves through its burrow, short, fat segments anchor part of the body against the wall, while long, thin segments penetrate further through the soil.

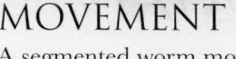

WAVE MOVEMENT
When a segmented worm moves, its segments alternately narrow and widen, forming waves of muscular contractions that pass along the body.

Class **POLYCHAETA**

Bristleworms

Occurrence 8,250 spp. worldwide; in burrows and tubes, amid seaweed and animal colonies on the sea bed

Bristleworms, also called polychaetes, form the biggest and most varied class of segmented worms, ranging from less than 1mm (¹⁄₃₂in) to 3m (9¾ft) long. Lugworms, bloodworms and ragworms, tubeworms, and bamboo worms are all members of this class. Despite being abundant and sometimes brilliantly coloured, bristleworms often escape attention because of their burrowing or tube-dwelling lifestyles and their quick reactions when disturbed. The exceptions to this rule are the active hunters, such as ragworms, which can sometimes be spotted in rock pools or under stones. Most bristleworms are long, slender, and worm-shaped, but some may be almost circular. They have parapodia – flaps bearing bristles or chaetae – or at least ridges projecting from the sides of each segment, and heads that bear appendages. Two general types occur: free-living and often carnivorous species, which have well-developed parapodia along the body; and sedentary species, which dwell in burrows or tubes, are deposit or suspension feeders, and have very specialized parapodia. A few species are parasitic. Most bristleworms are solitary animals, but some tube-worms may form large colonies or reefs. The sexes are usually separate.

feeding tentacles extending from tube

tube made by worm

tentacles fanned to capture food

FANWORMS
Some polychaetes live within a tube of their own making, which may be attached to a hard surface or, as in this fanworm, partially buried in soft sediment. Their movement is largely restricted to within the tube, but their feeding tentacles extend out into the water. The widely distributed species shown here – Sabella pavonina – is 20cm (8in) long.

paddle-like appendages

SEA MOUSE
Somewhat inappropriately named after the Greek goddess of love, Aphrodite aculeata has a felt-like mass of chaetae covering its dorsal surface and scales. This marine scaleworm is up to 20cm (8in) long, and is widespread.

eyes

coarse chaetae on sides · *felt-like dorsal chaetae*

RAGWORMS
Many marine polychaetes, like the ragworm, live in burrows in soft sediment, but can swim well and actively pursue their prey. The predatory lifestyle of the Atlantic species Nereis virens (right) – which may be up to 50cm (20in) long – is reflected in the numerous sensory structures, including eyes and large jaws, at its head end.

eyes

chaetae

PARCHMENT WORMS
Some marine polychaetes live in a U-shaped tube mostly hidden within sediment. They draw in a current of water by beating special paddle-like appendages; potential food particles are filtered from the current by a bag of mucus threads. Seen here out of its tube, the widespread parchment worm, Chaetopterus variopedatus, is large, up to 25cm (10in) long, but fragile.

Class CLITELLATA

Clitellates

Occurrence 6,000 spp. worldwide; in soil, freshwater or marine sediments, on marine fish

Habitat All except dry regions

Clitellates are characterized by the presence of the "clitellum", a largely dorsal area of glandular skin that secretes a cocoon where fertilized eggs are placed. The eggs hatch into miniature versions of the adults – an adaptation to life on land where aquatic-style larvae cannot survive. This class contains some of the world's most familiar invertebrates – the common earthworm and its relatives. Earthworms belong to the

biggest subgroup, the oligochaetes (comprising about 3,000 species), which live mainly in soil, silt, and fresh water. Leeches form a smaller group (about 500 species), and are either predatory or parasitic. Unlike polychaetes, clitellates lack parapodia and head appendages, and they are hermaphrodites. They may be up to 3m (9¾ft) long.

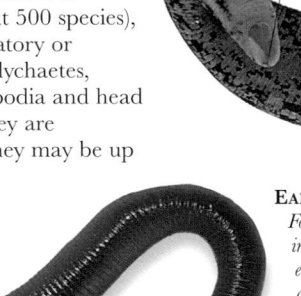

small sucker on head

narrow segments

clitellum

EARTHWORMS
Found almost worldwide, these familiar invertebrates are often seen when they emerge from their burrows in damp conditions, or are pulled from safety by birds. As Charles Darwin pointed out, their importance in aerating and enriching soil is enormous. Lumbricus terrestris (left) can attain a length of 35cm (14in).

MEDICINAL LEECH
Most leeches capable of penetrating human skin live in tropical rainforests, but Hirudo medicinalis *(left) lives in temperate fresh water, and grows to over 12cm (4¾ in) long. It was once commonly used by the medical profession as a bloodletting agent. When it bites, a leech may inject both anti-coagulants and anaesthetics.*

large sucker on tail

HORSE LEECH
Leeches that feed on mammals and fish are ectoparasites whose hosts survive the attacks. The horse leech, Haemopsis sanguisuga *(above), found throughout Europe, Asia, and North America, can measure 6cm (2¼ in) long when at rest.*

ROUNDWORMS

PHYLUM	Nematoda
CLASSES	4
ORDERS	20
FAMILIES	185
SPECIES	At least 20,000

ALSO KNOWN AS NEMATODES, these worms are among the most abundant of all animals. Roundworms are either free-living or parasitic. The parasitic forms, which include hookworms, pinworms, and threadworms, live inside almost all types of plants and animals, and some are a significant cause of disease. Free-living roundworms are also widespread in all aquatic habitats, including small films of water on land. Their bodies are unsegmented, round in cross-section, and tapered towards each end. They have a complex cuticle that is replaced 4 times by moulting, between hatching from an egg and reaching sexual maturity.

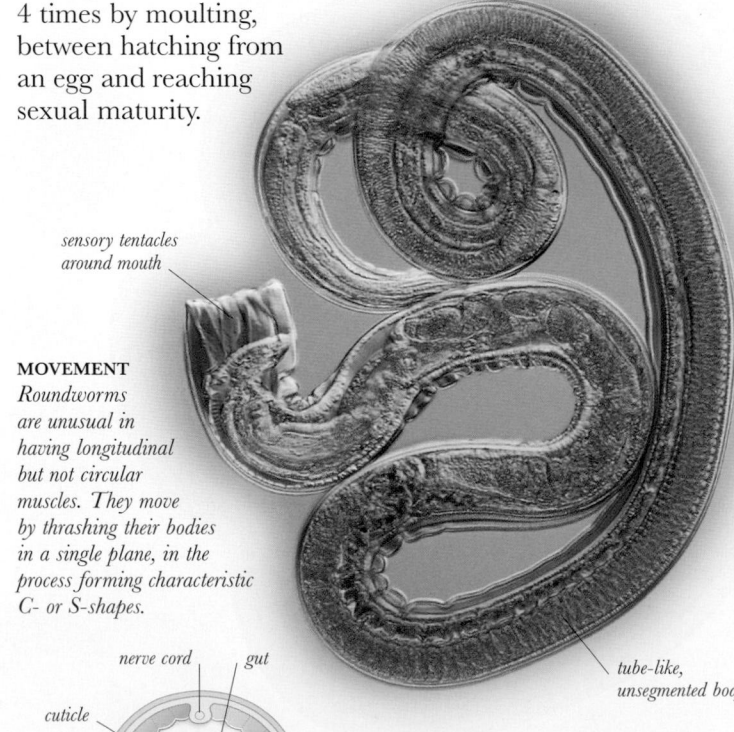

sensory tentacles around mouth

MOVEMENT
Roundworms are unusual in having longitudinal but not circular muscles. They move by thrashing their bodies in a single plane, in the process forming characteristic C- or S-shapes.

tube-like, unsegmented body

nerve cord   gut

cuticle

excretory canal

body cavity

longitudinal muscle

BODY SECTION
The large central cavity in a roundworm's body is occupied by fluid-filled cells. The high internal pressure in this cavity, together with a lattice of fibres in the cuticle, prevents the body from shortening when the muscles contract.

Class SECERNENTEA

Secernenteans

Occurrence 8,000 spp. worldwide; in water-filled areas around vegetation and in soil, on plants, in animals

The physical differences between secernenteans and other roundworms such as adenophoreans (see below) are very technical, relating, for example, to types of sense organs. Secernenteans are mostly parasites, some causing serious diseases. They vary greatly in size, from microscopic to several metres long. The largest known secernentean, up to 9m (30ft) in length, lives in the placentas of female sperm whales and is believed to be the largest member of the roundworm phylum. However, the harmful effect of these parasites is not necessarily related to their size: microscopic roundworms, such as pin-worms, filarial worms, and hookworms, can multiply rapidly in the host, causing great damage (as in elephantiasis).

WUCHERERIA BANCROFTI
This microscopic roundworm is the cause of elephantiasis in humans, a tropical disease that produces disabling enlargement of soft tissues, particularly in the legs. The worm lives in the lymphatic system, and is spread by mosquitoes.

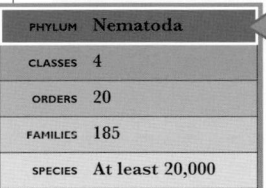

mouthparts with piercing style

hard outer cuticle

ASCARIS LUMBRICOIDES
About a sixth of the world's population suffers from ascariasis, an infection by this parasitic roundworm, which can grow up to 40cm (16in) long. The infection starts when the eggs are inadvertently taken into the gut with food. The larvae migrate to the lungs, returning to the gut as adults.

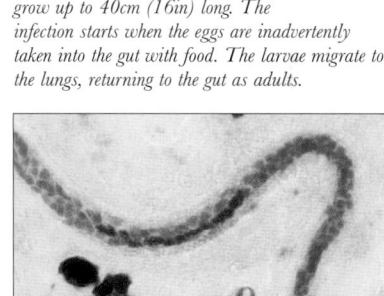

Class ADENOPHOREA

Adenophoreans

Occurrence 12,000 spp. worldwide; in water-filled areas around vegetation and in soils, on plants, animals

Habitat All except desert regions

Adenophorean roundworms are mainly free-living rather than parasitic, and are commonly found in most kinds of habitat throughout the world, particularly in marine sediments. They range in size from microscopic to as long as 1m (3¼ft) in exceptional cases. Although only 12,000 species of adenophoreans have been described in total, surveys of sea-bed mud suggest that vastly more may be awaiting discovery.

TRICHINA WORM
Trichinella spiralis *is a small roundworm, 1mm (¹⁄₃₂ in) long, which lives mainly in rats and other small mammals. Pigs may pick it up while scavenging and, by eating undercooked pork, each year about 40 million humans do the same.*

MINOR PHYLA

THE INVERTEBRATE WORLD is classified into about 30 major groups, or phyla. Some contain common and familiar animals but well over half contain organisms that go largely unnoticed – because they are rare, because they are microscopic, or because they live in inaccessible habitats, such as sea-bed mud. These animals differ from each other as much as invertebrates differ from vertebrates, and they have widely divergent ways of life. Some minor phyla, such as arrow worms, comprise just a handful of species, while a few – such as bryozoans – contain thousands of species, yet both groups are ecologically important. A selection of minor phyla is shown on these 2 pages.

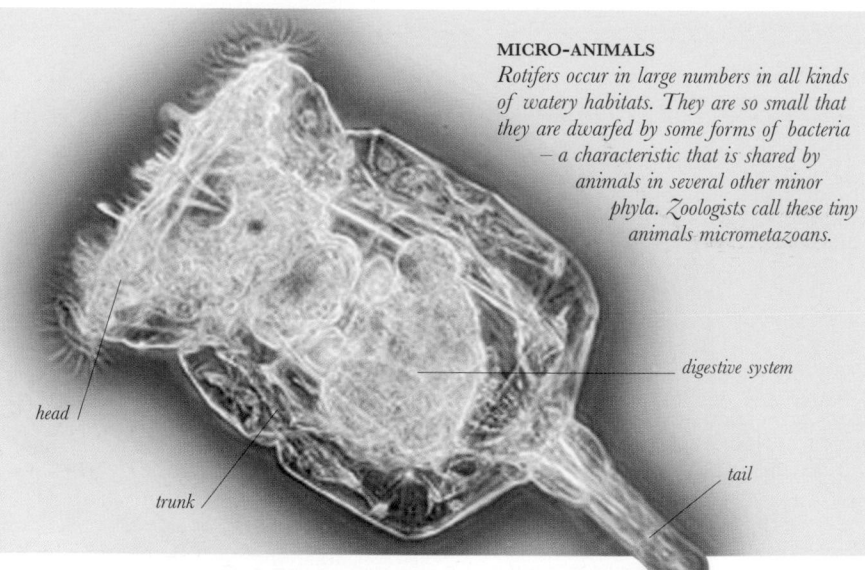

MICRO-ANIMALS
Rotifers occur in large numbers in all kinds of watery habitats. They are so small that they are dwarfed by some forms of bacteria – a characteristic that is shared by animals in several other minor phyla. Zoologists call these tiny animals micrometazoans.

head

trunk

digestive system

tail

Phylum BRACHIOPODA
Brachiopods

Occurrence 350 spp. worldwide; attached to hard surfaces or buried in soft sediments on the sea bed

Encased in bivalved shells, usually with a stalk (pedicle) that may anchor the animal to a hard surface or be used in burrowing, brachiopods look remarkably like bivalve molluscs with

small tentacles

stalks attached, but the 2 groups are not related. Also known as lamp shells, brachiopods are especially abundant in colder waters. They are less than 10cm (4in) long, and most of the space in the shell is taken up by a loop or spiral of hollow, fringed tentacles (lophophore) surrounding the mouth, with which the animal suspension feeds. Brachiopods are extremely common as fossils: some rocks are made up entirely of brachiopod shells. Once an important group with more than 25,000 species, brachiopods appear to have declined as a result of competition with bivalve molluscs.

LIOTHYRELLA UVA
Although brachiopods resemble bivalves, the 2 halves of the shell are dorsal (upper) and ventral (lower), not left and right. The species featured, Liothyrella uva, is common in Atlantic waters, and has a shell that is up to 2cm (¾ in) long.

Phylum BRYOZOA
Bryozoans

Occurrence 4,300+ spp. worldwide; attached to hard surfaces or moving across the sea bed

Also called sea mats, ectoprocts, or polyzoans, bryozoans are tiny, aquatic animals, less than 1mm (¹⁄₃₂in) long, that live inside box-like cases, which are joined together to form colonies. Some bryozoans grow

mound-like shape

as encrustations on rocks and seaweeds, but others develop a branching, plant-like shape and wave backwards and forwards in the water. Once dead, these branching colonies are often washed up on the shore, where they may be mistaken for dried-out seaweed. This phylum is ecologically important as a source of food for some marine invertebrates.

PENTAPORA FOLIACEA
Bryozoans bud continuously until thousands of interconnected "zooids" are produced. The North Atlantic Pentapora foliacea *may form colonies up to 2m (6½ ft) in diameter.*

Phylum CHAETOGNATHA
Arrow worms

Occurrence 90 spp. worldwide; in plankton, one genus on the sea bed

Growing only up to 12cm (4¾in) long, arrow worms are relatively small, highly active predators that can eat a third of their own body weight a day. They catch their prey – animal plankton including larval fish – with movable spines around their mouths and kill it by injecting tetrodotoxin, a potent poison produced by bacteria, and used by very few animals.

SAGITTID SPP.
The small, torpedo-shaped, transparent arrow worms are carnivores. Like the Sagittid *species above, they live among the plankton in all the oceans – where their large numbers can have great ecological impact, for instance on fish larvae.*

Phylum HEMICHORDATA
Hemichordates

Occurrence 85 spp. worldwide; in marine mud and sand, other intertidal and subtidal habitats

Hemichordates are unusual among invertebrates in that they have some vertebrate characteristics: a nerve cord that runs along the back, and pharyngeal perforations – structures that have the same anatomical origin as fish gills. The length of the 3-part bodies – comprising a proboscis, a collar, and a long trunk – ranges from less than 1.2cm (½in) to 2.5m (8¼ft). These animals are found in 2 forms: sedentary, polyp-like pterobranchs, whose sac-like trunk ends in a stalk

that may be prehensile or connect individuals to form colonies; and free-living, worm-like forms, in which the trunk is stalkless and bears a terminal anus. The sexes are separate, and reproduction occurs either sexually or asexually, by fragmentation or budding.

worm-like body

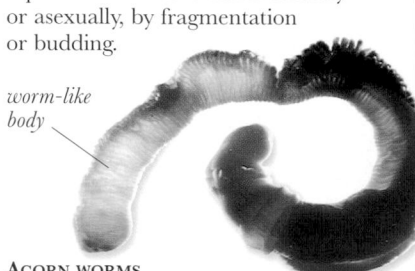

ACORN WORMS
Slow-moving and fragile, acorn worms live in U-shaped burrows in intertidal and subtidal habitats in all parts of the world. The large trunk of Balanoglossus australiensis *(above), which may be 1m (3¼ ft) long, bears numerous gill slits.*

Phylum CTENOPHORA
Comb jellies

Occurrence 100 spp. worldwide; planktonic, a few on the sea bed

With their tentacles extended, comb jellies or sea gooseberries are some of the most attractive and graceful animals in the sea. Although they can swim, by beating their comb-like plates of fine cilia (structures similar to hairs), they make little headway against the current, and are often carried along in large swarms by the tide or the wind. As the cilia rows beat, they produce beautiful, iridescent colours. Most comb jellies are small in size, but some can grow to 2m (6½ft) long.

PLEUROBRACHIA PILEUS
Comb jellies come in a variety of shapes, from spheres to ribbons and discs. The Atlantic-dwelling Pleurobrachia pileus *(right) is 2cm (¾ in) long and has a pair of tentacles 15cm (6in) long.*

Phylum SIPUNCULA
Peanut worms

Occurrence 350 spp. worldwide; in soft sediments, mollusc shells, rock crevices, holes in coral or wood

These plump, unsegmented worms, 0.2–70cm (1/16–28in) long, have a peanut- or sausage-shaped body, with walls formed of layers of circular and longitudinal muscles. Their most notable feature is their retractable trunk, known as an introvert, which is extended with the help of hydraulic pressure, and retracted by muscles. Peanut worms are suspension or deposit feeders, and collect organic particles by means of a ring of hydraulic tentacles around the mouth. They appear to be most closely related to molluscs.

stout body

GOLFINGIA VULGARIS
This species of peanut worm, found on northwest Atlantic coasts, lives in mud below the low-tide mark. It is 10cm (4in) long.

Phylum NEMERTEA
Ribbon worms

Occurrence 1,200 spp. worldwide; on bed and in surface or middle waters of seas, rivers, and lakes, in forests

A key feature of these mainly marine worms is their unique proboscis. This muscular tube, housed above the gut, is pushed out by hydraulic pressure in order to catch prey; in one large group of species, it bears piercing barbs that allow it to inject toxins. These worms range from less than 0.5mm (1/64in) in length to over 50m (165ft). All have a circulatory system and most have eyes – up to 250 in some species. Many are brightly coloured. Two groups of these worms are commensal – one within the shells of bivalve molluscs, the other on and within the shell of a variety of crabs.

UNIDENTIFIED SP.
Ribbon worms are noted for their flat and sometimes extraordinarily long bodies. The worm shown here belongs to a group that includes the world's longest animal, which reaches a length of 55m (180ft). This is twice as long as the average adult blue whale.

Phylum ONYCHOPHORA
Velvet worms

Occurrence 70 spp. in tropical and southern temperate zones; in moist forests, leaf litter, under stones

Sometimes described as "half-worm, half-arthropod" animals, these are the only survivors of a group of worms that were abundant in the Cambrian period. They have cylindrical, worm-like bodies, up to 15cm (6in) long, 14–43 pairs of short, unjointed, fleshy legs, each terminating in a pair of claws, and an antennae-bearing head with jaws similar in form to the claws on the legs. These carnivorous animals catch their prey by spraying them with a mucus-like substance from a pair of slime glands opening on either side of the mouth; they use the same technique to defend themselves from their predators.

MACROPERIPATUS TORQUATUS
Velvet worms, such as Macroperipatus torquatus (below), inhabit warm and humid regions. Their anatomy is basically that of a worm but, like the arthropods, they have pairs of legs serially repeated along the body. The legs are fleshy, unjointed, and hydraulically operated.

Phylum ROTIFERA
Rotifers

Occurrence 1,800 spp. worldwide; in vegetation in lakes, rivers, and seas, micro-freshwater habitats on land

Rotifers are among the smallest animals, with even the largest being just 3mm (1/8in) long. They live in water, and are either free-swimming, or attached to solid surfaces. The trumpet-shaped to spherical body usually has a "wheel organ" in front, formed by 2 whorls of hair-like structures (cilia), which is used for feeding and locomotion. Many rotifers are transparent, and several are parthenogenetic, males among these being unknown.

crown of tiny hairs

UNIDENTIFIED SP.
Although some rotifers are marine, most live in freshwater habitats, including temporary pools such as those in gutters and water films around leaves and soil particles. Like water bears (see right), they are specially adapted to survive drought. They can exist in a dried-out state for many years, coming back to life as soon as they are rehydrated.

Phylum ECHIURA
Spoonworms

Occurrence 150 spp. worldwide; in burrows or other cavities on the sea bed

Spoonworms or echiurans are sedentary, marine animals with bulbous, sausage-like bodies. Females are up to 50cm (20in) long, but have a forked proboscis that can be extended several times their own length to sweep up algae and other kinds of food. Males are minute, and live parasitically on their partners. Most spoonworms live in sand or mud close to the shore.

BONELLA VIRIDIS
This European spoonworm has bright green, toxic skin. Females are up to 15cm (6in) long, with a proboscis that can reach out for food over about 1m (3 1/4 ft). The proboscis rapidly contracts when touched.

Phylum PHORONA
Horseshoe worms

Occurrence 20 spp. worldwide; in soft sediments or attached to hard surfaces on the sea bed

These sedentary, marine worms are generally less than 20cm (8in) long, and usually remain encased in a secreted chitinous tube, with a horseshoe-shaped crown of as many as 15,000 feeding tentacles around the mouth. The sexes are separate in some species, while others are hermaphrodite. Horseshoe worms are most closely related to brachiopods or lamp shells (see p.536).

Phylum TARDIGRADA
Water bears

Occurrence 600 spp. worldwide; in fresh water, salt water, and damp and wet terrestrial habitats

Habitat All except dry ones

Water bears – or tardigrades – are plump-bodied, microscopic animals with 4 pairs of stubby, clawed legs. Their lumbering movements are remarkably bear-like, but their closest relatives are probably velvet worms. They live in wet and damp habitats of all kinds, and terrestrial species are especially good at surviving droughts, turning into a dormant form in which they can remain for years. Water bears reproduce by laying eggs. In some species, males are unknown and the females are parthenogenetic.

UNIDENTIFIED SP.
Water bears or tardigrades inhabit water films on land and cracks and crevices on the sea bed; many live in the clumps of moss found on roofs and in gutters. By being dormant for extended periods, they can probably live for over 50 years.

PHORONIS SPP.
Members of the widespread genus Phoronis can be 25cm (10in) long, but most are much smaller. They have an unsegmented body, usually encased in a chitinous tube buried in sand or mud. The body wall has layers of longitudinal and circular muscles, which enable the worm to move up the tube to feed, and to retreat afterwards.

MOLLUSCS

PHYLUM	Mollusca
CLASSES	8
ORDERS	35
FAMILIES	232
SPECIES	About 100,000

OF ALL invertebrates, molluscs have arguably the widest range of body forms. They include gastropods (slugs and snails), bivalves (such as oysters and clams), cephalopods (octopuses, squid, and cuttlefish, among others), chitons, and several less familiar forms. All molluscs have one or all of the following: a horny, toothed ribbon in the mouth (the radula) that is unique to molluscs; a calcium-carbonate shell or other structure covering the upper surface of the body; and a mantle or mantle cavity, typically with a characteristic type of gill. Molluscs are adapted to life on land and in water; there are also several parasitic species.

MOVEMENT

Most molluscs use their broad, flat foot to creep over hard surfaces, often on a thin layer of mucus. Some bivalves are fixed to the surfaces on which they live, and all are relatively inactive, but they often have a well-developed foot, which some species use to dig burrows in sediment. The most mobile and fastest-moving of all molluscs are the cephalopods, especially the octopuses and squid. Using undulations of fin-like extensions of their skin, they can make a slow swimming motion but they are also able to move more rapidly by squirting a jet of water from the mantle cavity.

MUCUS TRAILS
To reduce surface friction, slugs and snails secrete mucus from their foot. Slugs (such as this striped slug) have lost their shells, and generally move more quickly than snails.

ANATOMY

Most molluscs consist of 3 parts: a head, a soft "body" mass, and a muscular foot, which is formed from the lower surface of the "body" mass and used for locomotion. In some molluscs, the head is well-developed and supports sophisticated sense organs, but in others (such as the bivalves) it is effectively absent. The upper surface of the "body" mass is covered by a layer called the mantle, from which the shell is secreted. Retractor muscles can be used to pull the shell over the body for safety. Besides protection, the shell may be used as an organ for burrowing or controlling buoyancy. A fold in the mantle forms a cavity that in aquatic molluscs houses the gills.

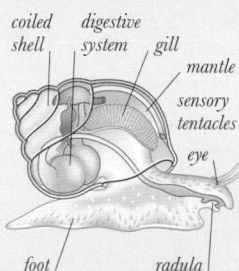

coiled shell, digestive system, gill, mantle, sensory tentacles, eye, foot, radula

GASTROPOD
Gastropod snails have a coiled shell, a large foot, and a distinct head, with eyes and sensory tentacles. Slugs have no shell.

mantle, shell, hinge ligament, digestive system, siphon, adductor muscle, gills, foot

BIVALVE
Bivalves are enclosed within a pair of shells, from which the foot and siphons can be extended. The gills are often large and folded.

arm, jaw, radula, funnel, shell, feeding arm, gills, digestive system, mantle cavity

CEPHALOPOD
In cephalopods, part of the head and foot are modified to form arms, and a funnel into the gill chamber. The shell is small or absent altogether.

SQUID SHOAL
All cephalopods are marine. Octopuses tend to be solitary but squid commonly form shoals. This shoal of luminescent squid have light-emitting bacteria living symbiotically in the light organs along their bodies.

FEEDING

Molluscs include predators, grazers, and browsers, as well as species that feed on deposited or suspended material. Many bivalves and a few gastropods are filter feeders, using their gills to strain particles down to the size of bacteria from water. Cephalopods are carnivores. They use their long arms, which are often equipped with suckers, to catch fish and crustaceans. These relatively intelligent animals have been known to stalk prey or to use their arms to lure it towards them. Besides their radula, cephalopods have a parrot-like beak, which they use to dismember their prey. Most other molluscs feed on food ranging from attached algae, plants, and animals to planktonic jellyfish and even slow-moving fish. Parasitic species obtain nourishment from their host.

HUNTING
Some octopuses are capable of subduing large and formidable prey. This one has overpowered a zebra moray eel.

RADULA
The tiny, hook-shaped teeth on the radula, which may number up to several thousand, are used like a rasp to collect small particles of food.

BIVALVE SHELL
The 2 halves of a bivalve's shell are joined along one edge by elastic material and closed by contracting the adductor muscles.

GASTROPOD SHELL
In most snails, the shell is secreted unevenly from the mantle, which causes it to coil into a clockwise spiral.

Class **BIVALVIA**

Bivalves

Occurrence 15,000 spp. worldwide; in aquatic habitats, mostly on the sea bed, but some in fresh water

Bivalves are the only molluscs that have a shell consisting of 2 parts, called valves, hinged together. In most cases, the valves can be closed tightly to protect all or a large part of the animal's body. Mussels, clams, oysters, and scallops are among the most familiar bivalves, providing a valuable source of food for humans as well as other creatures. Some species, such as the piddock and shipworm, have small, sharp-edged valves, used like drill-bits to bore through wood or soft

rock. Sizes of bivalve molluscs vary greatly, from less than 2mm (¹⁄₁₆in) to as much as 1m (3¼ft) across. The largest species (giant clam) is about 2 billion times heavier than the smallest. Bivalves are usually static, although some, for example scallops, can swim. Most are filter feeders, using modified gills to strain food from the water current. The water is often drawn in and pumped out through tubes or siphons – especially important for species that live buried in mud or sand. The siphons may be so large as to be permanently outside the shell. Bivalves have tiny heads, and they do not have a radula – the mouthpart that other molluscs use to rasp away at their food. The sexes are separate, although some bivalves may be hermaphrodite.

distinctive, fan-shaped shell

SCALLOPS
Unusually among bivalve molluscs, scallops can swim using a form of water-jet propulsion, to which they usually resort to escape predatory starfish. This North Atlantic species, Chlamys opercularis, *may attain a diameter of 9cm (3½ in).*

flat shell

THORNY OYSTERS
The tropical thorny oysters, found mostly on rocky outcrops and coral reefs, grow up to 20cm (8in) in diameter and have shells covered with elongated spines, as in the Spondylus *species pictured above. Their shells often become covered with algae, which may serve as camouflage.*

SADDLE OYSTERS
Like true oysters, saddle oysters or jingle shells live with one valve attached to a hard surface. The method of attachment is peculiar: a calcified plug passes from the upper valve through a hole in the lower one onto the surface. The Atlantic Anomia ephippium *(above) is about 6cm (2¼in) long.*

MUSSELS
Mussels attach themselves to hard surfaces by means of small threads, allowing them to hold fast against wave action and water currents. They often live in large clusters, and can cause problems for coastal power stations by blocking their cooling systems. Mussels are often reared in aquaculture systems. Mytilus edulis *(above), a species found in northern waters, may be 15cm (6in) long.*

EDIBLE OYSTERS
The edible oyster cements its left-hand valve directly onto a surface such as a rock, shell, or mangrove root, so the animal lives lying on its side. A valued food source since prehistoric times, oysters are among the most popular of edible molluscs. This North Atlantic Ostrea *species may grow up to 10cm (4in) in diameter.*

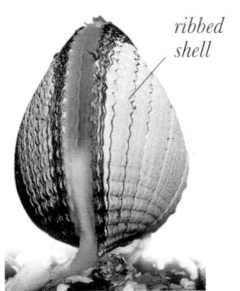
ribbed shell

COCKLES
Cockles live in sandy sediments, burrowing just below the surface so the body is half in and half out of the sand. The shell is usually strongly ribbed, stout, and semi-spherical. Several species, including the Atlantic Cerastoderma edule *(above) – which is 5cm (2in) high – are harvested commercially.*

long, slender valves

RAZOR CLAMS
Razor clams are so-called because of their resemblance to closed "cut-throat" razors. Occurring in almost all latitudes, they inhabit deep, vertical burrows in sandy sediments, using the shell valves to wedge themselves in. The species whose shell is shown here, Ensis siliqua, *may grow to 25cm (10in) long.*

SWAN MUSSELS
Most bivalves are marine, but some – such as this swan mussel – inhabit fresh water. A species of the genus Anodonta, *it is found in North America and Eurasia, and can grow to 23cm (9in) across. Their shells are thin, mainly because calcium is less readily available in fresh water. In their larval stages they are parasitic on fish.*

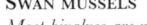

GIANT CLAMS
The tropical, reef-dwelling giant clam, Tridacna gigas *(below), is the largest bivalve; it may measure more than 1m (3¼ft) across and weigh over 200kg (440lb). Contrary to popular myth, it is harmless to humans. As with all giant clams, most of its dietary needs are met by symbiotic algae that live in its fleshy lips; using sunlight, the algae produce a supply of energy-rich food. The clam also filters microscopic particles of organic matter from the sea.*

large, fluted shell

respiratory siphon

fleshy mantle

sculpted shell

PIDDOCKS
Using their elliptical shells like drill-bits, piddocks bore into materials such as rock, wood, peat, and clay, and live in the excavated burrow. Some piddocks are phosphorescent, glowing with a ghostly blue-green light at night. This North Atlantic species, Pholas dactylus *(left), may be up to 15cm (6in) across.*

SHIPWORMS
These wood-boring bivalves occur in all seas, but are particularly common in warm waters. The relatively small shell is used to gouge out a tunnel in which to live, and the shipworm lines it with a chalky tube. Their bodies may extend up to 2m (6½ft), although the common shipworm, Teredo navalis *(pictured), rarely exceeds a tenth of this size.*

Class GASTROPODA

Gastropods

Occurrence 75,000 spp. worldwide; in aquatic habitats (mostly bottom-dwelling), in dry regions on land

Habitat All

This vast and variable group of invertebrates includes over four-fifths of the molluscs alive today. The familiar garden snail, slug, winkle, sea hare, sea slug, abalone, pond snail, and conch are all gastropods. Typical species have a spiral shell and move by creeping on a sucker-like foot, but there are many exceptions to this rule: some, for example, have no shell at all. Other characteristic features include a large head, with eyes and tentacles, and a ribbon-shaped mouthpart called a radula, which is armed with small, often microscopic teeth and is used to rasp away at food. These molluscs vary widely in size, and can grow up to 1m (3¼ft) long. The shape of the shell is highly variable – some form a tall spire, while others flare outwards. In some species, an operculum or "door" shuts the shell opening when the animal withdraws inside. Gastropods are characterized by torsion, a process in which the mantle cavity is rotated anti-clockwise up to 180 degrees, until it faces forwards and is positioned over the head. This occurs early in larval development and means that the digestive, reproductive, and excretory systems all discharge through the one opening in the shell. Gastropods feed in a wide variety of ways: they may be hunters, browsers, or grazers; some are suspension or deposit feeders; some are parasitic. Although sexes are separate in most gastropods, several groups are hermaphrodites, while in other groups, animals change from one sex to the other during their lifetime.

iridescent interior

holes to expel water and waste

ABALONES
Also known as ormers, awabis, or ear-shells, abalones are grazing animals that can be over 20cm (8in) across. Their shells have a distinctive series of holes, through which the water current drawn into the gill chamber leaves, as do waste products. The iridescent shells of species such as Haliotis rufescens (above), found off the west coast of North America, are used to make jewellery, and the animals inside are often eaten.

LIMPETS
Found on rocky shores, limpets, like abalones, graze on young algae, using their conical shells to protect themselves from the waves. Of the 2 Atlantic species shown here, the common limpet, Patella vulgata (left) – 5cm (2in) across – wanders over rocks, while the blue-rayed limpet, Helcion pellucida (below) up to 2cm (¾in) across – creeps over kelp, feeding on small epiphytic algae and on the kelp itself. Many rock-dwelling limpets are strongly territorial, making circular scars in the rock where they stay locked in position at low tide.

blue "ray"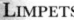

conical shape

spreading lip

HELMET SHELLS
The stout, colourful, and attractive helmet shells resemble the helmets worn by the gladiators of ancient Rome. The shells may be more than 30cm (12in) long. The animals that make them inhabit sandy sediments in the tropics, feeding on other molluscs and sea urchins. The shells of species such as Cassis cornuta (left) are much used as ornaments and for carving.

long spines

mucus raft

COWRIES
The beauty of their glossy shells ensures that the mostly tropical cowries are particularly sought after by collectors. In life, the shell is covered by a flap of skin, which is a part of its brightly coloured mantle. The shell aperture is long and narrow, which helps to protect the cowrie from attacks by predatory crabs. This species, Cypraea ocellata, can grow up to 6cm (2¼in) across.

VIOLET SEA SNAIL
The violet sea snail, Janthina janthina, is one of the few shelled gastropods that inhabit the ocean surface. It does this by secreting a raft of mucus bubbles that keeps it afloat. This warm-water snail is blind and has a paper-thin shell, 2cm (¾in) across. It feeds on by-the-wind-sailors (see p.532) and other colonial hydroids that drift through the water.

MUREXES
These gastropods are predators, often consuming other molluscs and barnacles after boring a hole through their shells. Several secrete a purple substance that has been used for centuries as a dye. Murex troscheli (right), which is found in the Indo-Pacific, has a shell that is 15cm (6in) long, with elongated, spiny extensions.

HARP SHELLS
The globular harp shells have smooth, regularly spaced ribs, which resemble the strings of a harp. Up to 10cm (4in) long, the shells are often brown, red, or pinkish. Species such as Harpa major, shown here, are mainly nocturnal, and inhabit tropical sandy sediments, into which they burrow. When disturbed, they may discard part of the large foot, in much the same way as a lizard discards its tail.

smooth ribs

patterned shell

CONE SHELLS
Cone shells are highly dangerous predators: using their harpoon-like mouthparts, they attack fish and other animals, injecting a venom so potent that it has been known to cause death in humans. The West African garter cone, Conus genuanus, shown here, is a typical example, with a highly decorated, tapering shell up to 7cm (2¾in) long.

SEA HARES
Most sea hares have a thin shell (not visible externally), a large, well-developed head with 2 pairs of tentacles, and a large foot. Sea hares inhabit beds of seaweed and sea grasses, where they feed on large, fleshy algae. The North Atlantic species shown above, Aplysia punctata, may attain a length of more than 30cm (12in).

INVERTEBRATES

SACOGLOSSANS

Sacoglossans are slug-like marine gastropods, 1cm (⅜ in) or so long, that graze on algae, sucking up the contents of their cells. This species, belonging to the widespread genus Elysia, *gets its colour from chlorophyll – the bright green pigment that algae use to harness the energy in sunlight.*

SEA SLUGS

True sea slugs lack shells and are often brightly coloured with striking patterns. Many species, like Bonisa nakasa *above – which is 8cm (3¼ in) across and lives on reefs off the southern coast of Africa – have colourful protuberances (papillae) on the back. Most sea slugs live on the sea bed, although a few are planktonic.*

colourful papillae

finger-like tentacle

heavy shell

FRESHWATER SNAILS

Both the European ramshorn snail, Planorbarius corneus *(pictured below), which can grow up to 3cm (1¼ in) across, and the great pond snail,* Lymnaea stagnalis *(above), which is up to 5cm (2in) tall, have a single lung instead of gills. Because they breathe air, they can survive in slow-flowing or stagnant water, where oxygen levels are low. The great pond snail and its many relatives are common in fresh water throughout the northern hemisphere.*

eyes near base of tentacles

flattened, spiral shell

TERRESTRIAL SLUGS

Unlike sea slugs, terrestrial slugs breathe air through their mantle cavity, which functions as a lung. In this European black slug, Arion ater, *usually 15cm (6in) long, the opening leading to the cavity can be seen just behind the animal's head.*

respiratory opening

stalked eyes

TERRESTRIAL SNAILS

Land-dwelling snails differ from sea snails in having thin, relatively light shells. Many also lack an operculum – a hard plate that seals the shell's opening. The species pictured right is the giant African land snail, Achatina fulicula, *which can reach a length of 30cm (12in) with its body extended. Originally from East Africa, it has been introduced to many parts of southern Asia, and in some regions is a serious agricultural pest.*

thin, whorled shell

eyes at tip of tentacles

large, muscular foot

WINKLES

Winkles are small marine grazers common on rocky shores, in mangroves and salt marshes, and more rarely on mudflats. Some can live at the very top of the shore, where they are wetted only by high spring tides; these species have a gill chamber partly converted into a lung. The European Littorina obtusata, *pictured above, like most other species of winkle, is usually less than 3cm (1¼ in) tall.*

coiled tip

WORM SHELS

These mainly tropical gastropods start life with tightly coiled shells, but as they grow older, the coils become wider and more irregular, making the shells look as if they have been unwound. This Caribbean species, Vermicularia spirata, *grows up to 10cm (4in) long. Like many other worm shells, it secretes a mucus filter from its foot, and uses this to trap small animals.*

irregular adult growth

red shell

siphonal canal

TOP SHELLS

Top shells are grazing marine gastropods with a conical shell up to 3cm (1¼ in) high, which has a flat, circular base. They occur on many rocky shores and reefs. Some, like the widely distributed Calliostoma *species shown above, are brightly coloured. Males have no penis, and fertilization is external.*

HETEROPODS

The species of the warm-water genus Carinaria *and its relatives are among the few gastropods that have taken up life in the open sea, where they make up part of the plankton. These unusual molluscs, up to 5cm (2in) long, have gelatinous bodies, and either very thin shells or no shell at all. They appear to live upside down because the shell, if present, is located beneath the body, and the fin – a foot modified for swimming – is uppermost.*

shell

TULIP SHELLS

The spindle-shaped tulip shells or horse conches are predators on other molluscs and worms. Although some live in shallow water and in high latitudes, most prefer warm water and the sublittoral zone (below the low watermark). The body is often red, as in the Fasciolaria *species above, and the shell, up to 25cm (10in) long, has a long, tubular projection – the siphonal canal – at the front.*

operculum

INVERTEBRATES

Class APLACOPHORA

Aplacophorans

Occurrence 250 spp. worldwide; on the sea bed or in association with other animals (especially cnidarians)

With their thin worm-like bodies, aplacophorans bear very little resemblance to any other molluscs. Typically less than 5mm (³⁄₁₆in) long, they have no trace of a shell, but are covered by a tough cuticle that contains a large number of small, mineral spicules. They have poorly developed heads and no eyes, and move by creeping or by burrowing through sediment. The creeping species feed on cnidarians, while the burrowers feed on tiny animals, or on particles of organic matter. Some species have separate sexes, but most are hermaphrodites. Some keep their eggs until they hatch, but little else is known about their breeding habits.

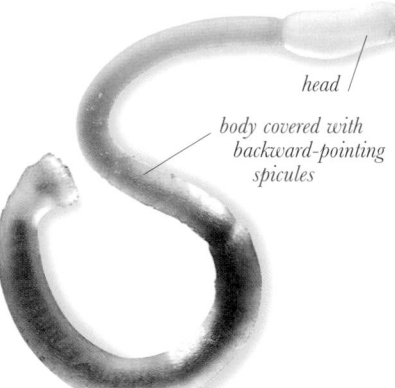

head

body covered with backward-pointing spicules

CHAETODERMA SPP.
Members of the genus Chaetoderma, *found in the northern hemisphere, are relative giants among aplacophorans, growing up to 8cm (3¼in) long. Like other aplacophorans, they lack a foot. They breathe with gills, located in a cavity at the back of the body.*

Class MONOPLACOPHORA

Monoplacophorans

Occurrence 15 spp. in Atlantic, Pacific, and Indian Oceans; at depths of 200–7,000m+ (650–23,100ft)

Monoplacophoran molluscs were thought to have become extinct more than 300 million years ago, until a handful of specimens were dredged up in the eastern Pacific Ocean in 1952. Since then, they have been found on the deep sea bed in widely scattered parts of the world. Monoplacophorans look like limpets (see p.540), but they each have up to 8 pairs of foot retractor muscles, whereas limpets and other gastropods have just one pair. A horseshoe-shaped mantle cavity containing 3–6 pairs of gills surrounds the foot. The shell is conical or cap-shaped and the head is small, with a large mouthpart (radula). There are no eyes or tentacles. Sexes are separate. These molluscs grow up to 3cm (1¼in) long.

Class POLYPLACOPHORA

Chitons

Occurrence 550 spp. worldwide; on the sea bed

Chitons resemble gastropods in having a sucker-like foot, but their shells are quite distinctive, being made of 8 plates. These are arranged in a line, and they arch over the animal's back. These fat, oval to elongated molluscs are covered by a thick "girdle", which often flanks the plates like a lip. It has a rubbery texture, and in many species is covered with bristles or calcareous spicules. Chitons can be up to 40cm (16in) long. They have a large, muscular foot, a small head with a well-developed rasping mouthpart (radula), but without eyes or tentacles, and a U-shaped mantle cavity around the foot with 6–88 pairs of gills. In both sexes, the reproductive organs (gonads) are fused into a single structure.

Chitons graze algae from rocks on the sea bed. They creep slowly over the surface – often after dark or at high tide – and usually return to a particular site where they fit snugly against the rock, helping to protect themselves from attack, and from desiccation if exposed to the air.

thick girdle

slight gloss on plates

CHITON STRIATUS
Found on the west coast of South America, this chiton is large and well-arched and surrounded by a broad girdle. It measures 9cm (3½in) in length and inhabits intertidal rocks. The central portion of each plate has crowded, straight riblets which are usually worn smooth at the highest point. Head and tail plates have coarse, radiating ribs, and the girdle is covered with polished granules.

ACANTHOPLEURA GRANULATA
This is a strongly humped chiton from the Caribbean. It is surrounded by a thick girdle which is covered with coarse spines. The plates are brown when unworn, but usually appear greyish brown with darker brown crests and sides. Up to 6cm (2¼in) in length, it inhabits intertidal rocks.

head end

MARBLED CHITON
Tonicella marmorea, a North Atlantic chiton up to 4cm (1½in) long, has a broad, leathery girdle surrounding the shell plates. It can be found from the intertidal zone to depths of 200m (650ft). It occurs on rocks, stones, and pebbles, and in some Scottish sea lochs can attain densities of 50 per square metre.

Class SCAPHOPODA

Tusk shells

Occurrence 550 spp. worldwide; on the sea bed, in soft sediments

Widely distributed in all the world's oceans, these burrowing molluscs are easily recognized by the shape of their shells, which are like tapering, hollow tusks, open at both ends, with one opening larger than the other. Tusk shells live head downwards in sediment, the head and the foot protruding out of the larger of the shell openings. The small head has numerous fine, ciliated feeding filaments, which contract to transport food particles from the sediment into the mouth. The smaller aperture projects above the sediment surface; respiratory water current is drawn in and let out through this opening.

There are no gills; instead, gases are exchanged through the mantle surface. The sexes are separate, and there is a single reproductive organ. Living tusk shells are rarely seen, but their empty shells, which can be up to 15cm (6in) in length, are often washed up on the shore.

raised, longitudinal ribs

ELEPHANT TUSK
The solid and slightly glossy shell of the species Dentalium elephantinum *has a characteristic notch at its narrower end. This mollusc inhabits offshore sands in the Indo-Pacific, and is 7.5cm (3in) long.*

ELONGATED TUSK
Antalis *species are a small group of marine molluscs from the North Atlantic that live in shells 6cm (2¼in) long. They feed on micro-organisms, using numerous club-shaped tentacles.*

smaller aperture helps in respiration

small pipe at rear end

alternating colour rings

BEAUTIFUL TUSK
This species, Pictodentalium formosum, *found on Indo-Pacific coasts, grows to 7.5cm (3in) in length. It differs from other large tusk shells in its head end being only slightly broader than its rear end; it is also straighter than most. Its longitudinal ribs and riblets are crossed by concentric growth lines. A small, notched pipe protrudes from the rear end.*

Class CEPHALOPODA

Cephalopods

Occurrence 660 spp. worldwide; most inhabit open water, some live on or near the sea bed

Cephalopods include the largest, fastest, and most intelligent of all invertebrates. Unlike other molluscs, they have sucker-bearing arms or tentacles, and they can move with speed by a form of jet propulsion, which involves squirting water out of the mantle cavity. The tentacles are used to catch fish and crustaceans and also to transfer sperm from the male to the female. Cephalopods have a rasping mouthpart (radula), a bird-like beak, a large brain, and well-developed eyes. Early cephalopods had coiled, external shells, but among the existing species, only the nautiluses show this characteristic. In all other cephalopods, such as cuttlefish, squid, and octopuses, the shell is either internal or absent altogether. In the open-water squid and cuttlefish, the body is reoriented, so that the animal effectively lies on its side, helping to give it a streamlined shape. Cephalopods vary greatly in size: spirulas are less than 7cm (2¾in) long, while the giant squid exceed 20m (65ft). Many cephalopods breed only once and then die.

SPIRULAS

coiled, chambered shell

These small relatives of cuttlefish are less than 7cm (2¾in) long. Their bodies contain a coiled and chambered shell that acts as a float, enabling them to hang head-down in the water (see right), at depths of up to 2,000m (6,600ft). Using their small arms, they catch planktonic animals that drift within reach. When they die, their shells sometimes float to the surface, and are often carried onto the shore. Shells of the common spirula, Spirula spirula, shown, are a familiar sight on tropical beaches.

NAUTILUSES

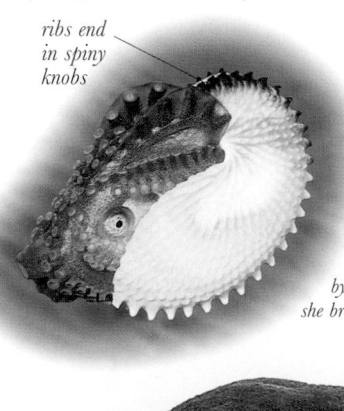

Nautiluses are the only surviving cephalopods with a true external shell, which may be up to 20cm (8in) across. It contains a series of gas-filled chambers that act as a buoyancy device, and a large final chamber that houses the animal itself. The pearly nautilus, Nautilus pompilius (above), lives in the Indo-Pacific.

GIANT SQUID

Members of the genus Architeuthis, *distributed worldwide, giant squid can be 20m (65ft) or more long. No giant squid has ever been captured alive; their cast-up bodies are sometimes found on beaches, and their undigested beaks in the guts of sperm whales.*

VAMPIRE SQUID

black, gelatinous body

Members of the genus Vampyroteuthis, *the vampire squid are less than 5cm (2in) long. Like octopuses, they have 8 arms, and 2 additional, retractile tentacles, like cuttlefish. They live in tropical waters at depths of up to 3,000m (9,900ft).*

PACIFIC SQUID

Squid are torpedo-shaped hunters, with internal shells that are reduced to a small, horny "pen". Some hunt alone, but many medium-sized species – such as the Pacific Loligo opalescens, *above, up to 20cm (8in) long – feed in large shoals. All swim by rippling their fins, or by jet propulsion. Like cuttlefish, squid have 2 long, prey-catching tentacles, but these cannot be retracted.*

PAPER NAUTILUS

ribs end in spiny knobs

A species in the genus Argonauta, *the pantropical paper nautilus (shown left) is really a type of octopus. It is 20cm (8in) long. The breeding female has a superficial, nautilus-like, thin shell – a case secreted by 2 of the arms – in which she broods the fertilized eggs.*

CUTTLEFISH

undulating fins along sides

Cuttlefish have a flat, internal shell – the familiar "cuttlebone" that is often washed up on the shore. Their bodies are short and broad, with 8 short arms and 2 much longer tentacles that can shoot out to catch prey. Cuttlefish are known for their ability to change colour according to their moods. The species shown above, Sepia officinalis, *is common around the North Atlantic and grows up to 30cm (12in) long.*

tentacles can be extended to reach out to prey

BLUE-RINGED OCTOPUS

blue rings on body

The blue-ringed octopus, Hapalochlaena lunulata, *inhabits coral reefs in the Pacific and Indian Oceans. It is small – 10cm (4in) long – and beautiful, but has a venomous bite which can kill humans in 15 minutes.*

long arms to grasp prey

eyes similar to human eyes, used to spot prey

COMMON OCTOPUS

Octopuses differ from cuttlefish and squid in having more rounded bodies, and in lacking the 2 long, prey-capturing tentacles and any trace of the ancestral shell. Most are bottom-dwelling, although some "sail" bottom currents using a web of skin between the arms. The octopus on the left is an Atlantic species, Octopus vulgaris, *1m (3¼ft) long.*

2 rows of powerful suckers to grip rock

Enteroctopus dolfeini
Giant Pacific octopus

Occurrence 1 sp. in the north Pacific Ocean, down to depths of 750m (2,500ft); mostly on the sea floor

The largest species of octopus in the world, the giant Pacific octopus usually ranges from 3 to 5m (9³⁄₄–16ft) long and weighs 10–50kg (22–110lb). It lives on the rim of the north Pacific Ocean, from Japan up to the Aleutian Islands and south to California, where it crawls about on the bottom, using its long, sucker-covered arms. It seeks out rocky dens on the sea floor, under boulders or in crevices; youngsters will often dig holes under rocks in sand or gravel. Here, the octopus can take refuge from predators – seals, sea otters, sharks, and other large fish – too big to slip through the den mouth. Foraging mainly at night, this giant octopus looks especially for crabs and lobsters, but also takes shrimps and shellfish, smaller octopuses, and fish.

Often it will return to its den to feed, depositing empty shells and other inedible fragments of prey in piles at the entrance. Like its relatives, this octopus mostly lives alone, except for a brief period (which may occur at any time of the year) when adults come together for mating. The male deposits sperm packets up to 1m (3¼ft) long in the female's mantle cavity, by means of a specially modified arm. The female lays her eggs in a sealed den, and will tend them until they hatch, which may take as long as 8 months, depending on the temperature of the water. She will not feed in all this time – and will die soon after her young emerge. Lifespans range up to 4 years.

EMERGING YOUNG
As the eggs hatch, the mother blows the young out of the den and they swim towards the surface. They will spend their first few months among the plankton before settling to the sea bed.

AGILE GIANT
The giant Pacific octopus is remarkably agile for its size, its suckered arms enabling it to move easily over the ocean floor, and its soft, flexible body allowing it to pass through the slimmest of openings.

sac-like body

purplish red
skin

pale
underside

2 rows of
suction pads
on each arm

DEFENCE AND ATTACK

When threatened, the giant Pacific octopus may change colour, or flee hidden by an inky cloud. When it catches prey, it grasps it firmly and pulls it apart, bites it open, or drills through the shell with its beak.

FLEEING FROM DANGER
When a predator gives chase, the octopus ejects an inky substance that forms a dark cloud in the water. Hidden from view, the octopus can swiftly make its escape.

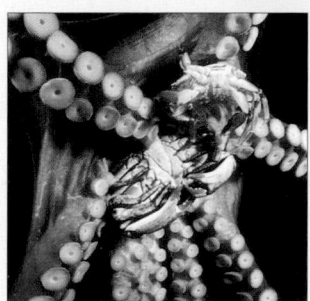

DEALING WITH A CRAB
Holding a crab firmly in its suckers, the octopus makes a hole in the shell with its beak, then injects saliva to dissolve the connective tissue. The meat can then be pulled out and eaten.

DEVOTED MOTHER
The female lays many thousands of eggs and stays with them for 5–8 months, guarding them until they hatch. She continuously washes them with a stream of water from her siphon, and grooms them with her arms to keep them free of parasites.

ARTHROPODS

PHYLUM	Arthropoda
CLASSES	17
ORDERS	99
FAMILIES	About 2,140
SPECIES	About 1.1 million

CLASSIFICATION NOTE

Arthropods are divided into 2 subphyla: the chelicerates have pincer-like appendages and no antennae, while mandibulates have chewing mouthparts and one or 2 pairs of antennae.

MANDIBULATES
Hexapods *(insects and other six-legged arthropods) see pp.548–77*
Centipedes and millipedes *see p.578*
Crustaceans *see pp.579–85*

CHELICERATES
Sea-spiders *see p.585*
Horseshoe crabs *see p.585*
Arachnids *see pp.586–93*

ARTHROPODS FORM THE largest phylum of living organisms and account for more than 3 out of 4 known species of animals. They include insects, centipedes and millipedes (myriapods), crustaceans, and arachnids, among others. All arthropods have an exoskeleton that covers a body divided into segments, and all have jointed legs. The first arthropods were marine animals, and many (especially the crustaceans) are still found in the sea. However, the insects, arachnids, and myriapods have successfully adapted to life on land. Insects, which are by far the most numerous of all arthropods, are the only invertebrates capable of powered flight, having evolved functional wings more than 100 million years before flying reptiles or birds. By providing food for many larger animals, arthropods have become essential to the functioning of most of the world's ecosystems. Krill, copepods, and other crustaceans form the foundation of marine food chains, supporting fish and marine mammals. On land, insects provide food for most other animal species.

MIGRATION

Arthropods are among the most mobile of invertebrates. Some species make use of this to migrate in response to changes in environmental conditions and food supply. Each year, many spiny lobsters make mass migrations between relatively warm and cool waters. Typically moving in single file, they maintain contact with one another by touch (using their antennae) or sight (picking out the light-coloured spots on the abdomen of the individual in front of them). The autumn migration is often triggered by violent storms.

ANATOMY

Arthropods share several common features. All species have a bilaterally symmetrical body that is divided into segments (see below). Arising from these segments are several jointed legs, which are arranged in pairs. The body is covered by a tough exoskeleton, which is produced by the epidermis and is composed of protein and a material called chitin. To allow for movement, the exoskeleton articulates at joints and hinges, where the cuticle is soft and flexible. In large marine species, the exoskeleton is strengthened by calcium carbonate, while land species have a thin layer of waterproof wax to stop them drying out. All arthropods have an open circulatory system: their organs are bathed in a fluid called haemolymph, which is moved around the body by the heart. Gas exchange is carried out by gills, organs known as book lungs, or a system of tracheae. The nervous system consists of a brain that is connected via paired nerve cords to networks of nerve cells in the thorax and abdomen.

antenna

head

first trunk segment

jointed leg

flexible cuticle between segments

SEGMENTATION

The body of an arthropod is divided into segments. Arthropods probably evolved from marine worms that had unspecialized segments. Over time, eyes, antennae, and appendages appeared, and some segments were fused into functional units, of which the most common is the head. Primitive species, such as myriapods (a centipede is shown here), have a head and trunk. In arachnids, the head and thorax are fused to form a cephalothorax. Insects, the most advanced arthropods, have a head (made of 6 fused segments), a thorax (that has 3 segments) and an abdomen (with 11 segments).

BODY SEGMENTS

Each segment (seen in cross-section, below) is essentially a box, with a top (tergum), bottom (sternum), and sides (pleura). These outer surfaces are made from tough, relatively rigid plates (sclerites), which are joined by soft cuticle to allow for movement. Muscles between the tergum and sternum allow the segment to be flattened in shape, while those linking adjacent segments allow the body to flex sideways, curl up, or telescope lengthways. The nerve cord, digestive system, and heart pass through most segments.

toughened outer part of cuticle

cuticle

dermal gland

epidermis

CUTICLE

The cuticle (seen in cross-section, left) protects the body of an arthropod from damage and repels pathogens. Although it is a composite of several layers, cuticle can be repaired: a small wound is quickly sealed; more extensive damage can be repaired by the production of a new cuticle at the next moult. Some species can also regenerate lost limbs.

heart

tergum

gut

pleuron

muscle for moving leg   *muscle connecting segments*   *nerve cord*   *sternum*   *muscle for changing shape of segment*

MOVEMENT

One of the key features that separates arthropods from most other invertebrates is their jointed limbs. Most arthropods are active animals – moving by walking, swimming, or jumping – which enables them to look for food or mates, colonize new habitats, and escape from predators or adverse local conditions. The number of legs varies from 3 pairs in insects to several hundred pairs in millipedes. Each leg usually ends in a claw of some kind that improves the arthropod's grip on the surface. Many species also have adaptations, such as bristles or adhesive pads, for walking on smooth vertical surfaces or on water. In some species, the joints between leg segments contain an elastic protein, called resilin. When compressed, this acts as a store of energy that can be released when the animal is on the move. As well as movement, arthropods also use their legs to catch prey, to mate, and to communicate.

point of attachment to body segment

coxa

basis

ischium

FORWARDS AND BACKWARDS MOVEMENT

MOVEMENT UP AND DOWN

merus

movement at each joint can occur in only one plane

carpus

dactylus

propodus

LEG MOVEMENT
An arthropod's limb, such as the first walking leg of a crab (left), consists of a set of hollow, rigid sections. These are joined by areas of flexible cuticle and moved by muscles that connect the segments. The ends of the segments are shaped so that movement at each joint occurs in a different plane to movement at the adjacent joints. The result is a limb that can be moved into almost any position.

WALKING
Most insects, including this beetle, walk with what is known as a tripod gait. As each leg is lifted off the ground, the body is supported by one leg on the same side of the body and 2 legs on the opposite side.

GROWTH

One disadvantage of having an exoskeleton is that it must periodically be replaced with a larger one as the animal grows. Arthropods grow and develop in various ways. Crustaceans, myriapods, and arachnids moult throughout their adult life. Larval crustaceans look nothing like the adults. In crabs, for example, the egg hatches into a planktonic zoea larva. This becomes a bottom-dwelling megalopa larva, which in turn develops into the adult. Young myriapods typically look like small adults and become longer by adding trunk segments at each moult, although some centipedes hatch with the adult number of segments. Young arachnids typically look like small versions of their parents. With the exception of bristletails and silverfish, insects stop moulting on reaching sexual maturity.

young adult climbs up plant stem to leave water

discarded skin of nymph

MOULTING
When moulting, an arthropod's body is soft and unprotected. An immature damselfly (known as a nymph) moults several times underwater. When ready for the final moult, it crawls out of the water. Once exposed, it must shed its skeleton and take to the air in a couple of hours if it is to avoid being eaten by a predator.

PARENTAL CARE
It is not unusual for adult insects, centipedes, and arachnids to care for their eggs and young. Female spiders wrap their eggs in silk and carry or guard them until they hatch. Scorpions brood their eggs and carry their young on their back after they hatch.

FEEDING

Arthropods eat a wide range of foods, from decaying organic matter to the tissues of other animals and plants. Many species are specialist feeders, the shape of their limbs and mouthparts reflecting the food they eat. Strong claws and jaws can tear and cut tough material, such as skin, cuticle, and plant tissue, while needle-like mouthparts are used to suck up liquids, such as sap, blood, and nectar. Many aquatic crustaceans have feather-like structures on their legs or around the mouth that act like strainers to filter minute organisms or organic particles from water. Most arthropods use saliva to lubricate their food as it passes into the digestive system.

PREDATORS
Most arachnids eat other animals. However, they can ingest only liquid food. Having subdued a cockroach, this scorpion will release digestive enzymes over its body and then consume the resulting liquid.

PLANT-EATERS
Plant-eating arthropods can eat large quantities of food. A swarm of locusts can consist of up to 50 billion individuals, theoretically capable of consuming 100,000 tonnes of food in a day.

INSECTS

PHYLUM	Arhropoda
SUBPHYLUM	Mandibulata
SUPERCLASS	Hexapoda
CLASS	Insecta
ORDERS	29
FAMILIES	949
SPECIES	At least 1 million

INSECTS ARE THE MOST successful animals on earth. They account for more species than any other class: over a million have so far been identified, but it is thought that the true number may be between 5 and 10 million. Insects belong to a group of arthropods called the hexapods, all of which have 3 pairs of legs. Many insects also have wings, which makes them the only arthropods capable of powered flight. Together with their small size and ability to survive in dry environments, flight has enabled insects to colonise a vast range of habitats. Most live on land or in the air, but they are also numerous in fresh water. Without insects, many other forms of life would not exist: the majority of flowering plants are largely dependent on them for pollination, and they form the main component of the diet of many animals. During their lifetime, many insects undergo complete metamorphosis, passing through several physical stages before reaching adulthood.

NON-INSECT HEXAPODS

As well as insects, the hexapods include 3 smaller classes: the springtails, proturans, and diplurans. They are collectively known as the non-insect hexapods, and a selection are described in this section. All non-insect hexapods are without wings, and many also lack eyes and antennae. Unlike insects, their mouthparts are enclosed in a pouch located on the underside of their head.

SPRINGTAILS
Springtails are small hexapods that occur in great numbers in soil and leaf litter all over the world, including the Arctic and Antarctic.

ANATOMY

The body of an insect is divided into 3 segments – the head, thorax, and abdomen – each of which carries out different functions. The head supports the mouthparts and much of the sensory apparatus, including the compound eyes and antennae. The thorax, which bears the legs and in many species the wings, is important in locomotion, while the abdomen contains the organs for digestion, excretion, and reproduction. All adult insects breathe air, which enters the body through openings (called spiracles) on the sides of the abdomen and thorax. Immature stages of aquatic species often have gills. The internal organs are bathed in a fluid called haemolymph, which transports nutrients and waste and is pumped around by a tube-shaped heart.

segmented antenna
tarsus
tibia
head
compound eye
prothorax
femur
coxa
mesothorax
hindwing
folded hindwing
forewing
metathorax
abdomen

BODY SEGMENTS
The body of a jewel beetle, like that of other insects, is protected by a rigid exoskeleton. There is some flexibility between segments – for example, the head can be moved independently. The thorax and abdomen are further subdivided: the thorax consists of the pro-thorax, mesothorax, and metathorax, while the abdomen has up to 11 segments.

COMPOUND EYES
Insects have 2 types of eyes: simple eyes (ocelli) and compound eyes. This hornet's ocelli are on top of its head. The forward-facing compound eyes consist of hundreds of light-sensitive units connected by nerves to the brain.

MOUTHPARTS
The mouthparts of insects have evolved into an amazing array of forms. Like most insects, ground beetles have jaws suited for cutting and chewing. Houseflies use a sponge-like structure to absorb exposed liquid. Mosquitoes use needle-like mouthparts to pierce the skin of other animals.

biting jaws
sponge-like organ
piercing tube

HOUSEFLY   **GROUND BEETLE**   **MOSQUITO**

POPULATIONS

There are several reasons why insects have become so successful. Being small, they are able to occupy microhabitats that are inaccessible to other animals. Given the right conditions, they are also able to breed very rapidly, enabling them to respond quickly to increases in the availability of food. For example, a single pair of bruchid beetles have the theoretical potential to produce enough offspring to occupy the entire volume of the earth within 432 days. In practice, this doesn't happen because there are limits to the food supply and because of competition between individuals of the same species and between different species.

SPECIES NUMBERS
High reproduction rates have enabled insects to evolve so rapidly that they make up over half of all animal species alive today. Bugs (such as these firebugs) alone account for 82,000 species. Insects are found on land, in fresh water, and even on the ocean surface.

LIFE CYCLES

Insects begin life as an egg. However, once a young insect hatches, it may follow one of several routes to adulthood (see right), varying from a simple series of moults to a complete physical transformation of its body parts. The length of each stage varies greatly between species. For example, one North American cicada takes 17 years to reach adulthood, while some fruit flies may take less than 2 weeks to mature. Once insects reach sexual maturity, they begin to mate. Courtship displays may involve the production of sexual odours, sounds, and even light displays. In almost all insects, fertilization is internal (see pp.28–9). The fertilized eggs are usually laid close to a source of food. However, in parasitic species, they may be laid either on or inside the body of a host animal.

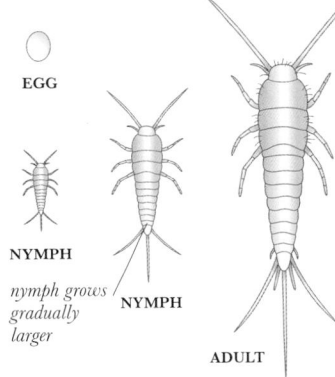

EGG

NYMPH

nymph grows gradually larger

NYMPH

NYMPH

ADULT

AMETABOLOUS DEVELOPMENT
In some wingless insects, there is little difference in form between young and adults. The nymph develops by shedding its exoskeleton, growing larger after each moult.

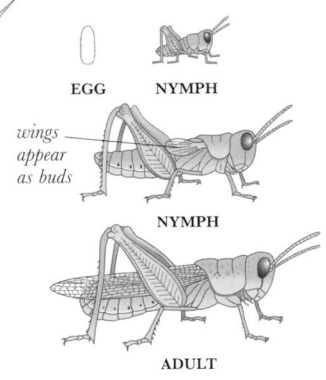

EGG   NYMPH

wings appear as buds

NYMPH

ADULT

INCOMPLETE METAMORPHOSIS
In some winged insects, such as grass-hoppers, the change from the immature stages (nymphs) to the adult is gradual. Nymphs look similar to adults but have no wings or reproductive organs.

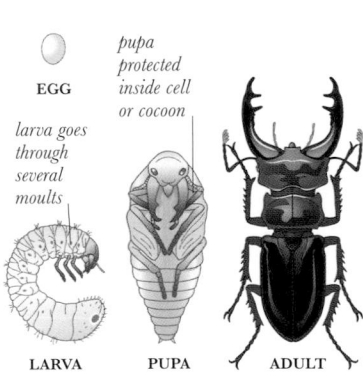

EGG

pupa protected inside cell or cocoon

larva goes through several moults

LARVA   PUPA   ADULT

COMPLETE METAMORPHOSIS
In other winged insects, the immature stages (larvae) moult several times. During the last moult, the larvae pupate and the larval tissues are transformed into adult structures.

MIGRATION
Many insects survive low temperatures by becoming dormant. However, some of the larger winged insects are capable of migrating over great distances to avoid winter cold. One of the most celebrated examples is the monarch butterfly (shown here). Each year, tens of millions of monarchs travel up to 4,000 km (2,500 miles) from Canada and the eastern United States to winter in roosting sites in California and Mexico.

FLIGHT AND WINGS

Insects were the first animals to evolve the power of flight, a development that made it easier for them to evade predators and to find food and mates efficiently. An insect's wings, which are made of cuticle, are hinged to the thorax. Most insects have 2 pairs of wings. These are usually hooked together but in some species, such as dragonflies, they beat alternately, which improves manoeuvrability. True flies are the only insects that have one pair of wings; they are also the most agile fliers, capable of flying backwards, sideways, and even upside-down. Almost all flying insects can fold their wings, which means that when they are not flying they can squeeze into small spaces, such as crevices in bark or gaps under stones.

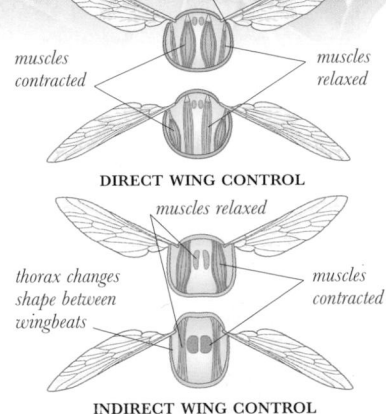

muscle attached to wing

muscles contracted

muscles relaxed

DIRECT WING CONTROL

muscles relaxed

thorax changes shape between wingbeats

muscles contracted

INDIRECT WING CONTROL

DIRECT AND INDIRECT WING CONTROL
Winged insects use various mechanisms to control their wings. Some control them directly, using muscles attached to the wing bases to produce the upstroke. Other insects move their wings up and down by changing the shape of the thorax. Insects that use this indirect control mechanism usually beat their wings more quickly (100–400 times per second) than those that use direct control (usually about 50 times per second).

TAKING OFF
Beetles protect their delicate hindwings under tough, case-like forewings (elytra). It takes a little time for a beetle to lift its elytra and unfold the hindwings before taking off. If the air is cold, many insects vibrate their flight muscles to warm them up.

NON-INSECT HEXAPODS

Class COLLEMBOLA

SPRINGTAILS

ALTHOUGH SMALL AND EASILY overlooked, springtails are present in vast numbers in grassland, leaf litter, and soil. On the underside of the abdomen, they have a ventral tube that allows them to regulate water and grip smooth surfaces, and a fork-shaped jumping organ (furcula), which can be released like a spring to flick them away from predators. There are 18 families (6,500 species) of springtails.

Family PODURIDAE

Water springtail

Occurrence 1 species in the northern hemisphere; on the surface of freshwater ditches, ponds, canals, bogs

The single species in this family – the common springtail *Podura aquatica* – is up to 2mm (1/16in) long. It varies from brown or reddish brown to dark blue or almost black. The furcula is quite flat and long, and reaches the abdominal ventral tube, enabling the springtail to cling to the water's surface and jump effectively. A scavenger, it feeds on small, floating particles of food. Its eggs are laid among vegetation in or around water bodies.

PODURA AQUATICA
This abundant springtail is well adapted to life on water. It may gather in such numbers on freshwater surfaces that the water appears dark.

Family SMINTHURIDAE

Globular springtails

Occurrence 900 spp. worldwide; in leaf litter, fungal fruiting bodies on water surfaces, in caves

Also known as garden springtails, these hexapods are 1–3mm (1/32–1/8in) in length. Ranging from pale to dark brown or green, they have almost spherical bodies. The long, elbowed male antennae in many species are designed to hold the female during mating. Eggs are laid in small batches in soil, and development to sexual maturity may take as little as one month. Several species of these herbivores, such as *Sminthurus viridis*, are pests of crop seedlings.

SMINTHURIDES AQUATICUS
This tiny hexapod – only 2mm (1/16in) long – is common on the surface of ponds, but does not flock in such large numbers as the water springtail.

Class PROTURA

PROTURANS

LIKE SPRINGTAILS, PROTURANS are sometimes found in leaf litter, but most species live in the soil. These often microscopic animals do not have eyes, tails (cerci), or antennae, and instead they use their front legs as sensory organs. Their mouthparts, which can both pierce and suck, are kept in a pouch and are pushed out during feeding. There are 4 families of proturans, containing a total of about 400 species.

Family EOSENTOMIDAE

Eosentomids

Occurrence 90 spp. worldwide; in soil, leaf litter, mosses, humus, decaying wood

These pale and soft-bodied hexapods are 0.5–2mm (1/64–1/16in) long, and have a conical head and an elongated body. They eat dead organic matter and fungi. The eggs, laid in soil or leaf litter, are spherical and have patterns or warts on them. Larvae are identical to adults, but smaller.

EOSENTOMON DELICATUM
This often abundant species is native to Europe, although the genus is found all over the world. It lives in soil, especially chalky soil.

Order DIPLURA

DIPLURANS

THESE BLIND, ELONGATED, AND SOFT-BODIED hexapods are sometimes called two-tailed bristletails due to their 2 terminal cerci. They have a large, distinct head and biting, eversible mouthparts. There are 9 families and about 800 species.

Family JAPYGIDAE

Japygids

Occurrence 200 spp. worldwide; in soil

These species are slender-bodied and pale, ranging in length from 0.6 to 3cm (1/4–11/4in). The end of the abdomen carries a pair of distinctive, tough, forceps-like cerci which are used to catch small arthropod prey. Eggs are laid in batches in soil.

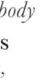

long, slender body

CATAJAPYX DIVERSIUNGUIS
Also called the slender dipluran, this common North American species has stout, terminal cerci. Its abdominal segments are pale around the edges.

cerci

INSECTS

Order ARCHAEOGNATHA

BRISTLETAILS

BRISTLETAILS ARE WINGLESS insects that look hump-backed in side view. They have simple mouthparts, large eyes that touch each other, and 3 tails (cerci), the middle being the longest. Like silverfish, they do not undergo metamorphosis. There are 350 species in 2 families.

Family MACHILIDAE

Jumping bristletails

Occurrence 250 spp. worldwide; under stones, in leaf litter, decaying vegetation

These species are elongated and covered with patterns of drab brown or dark grey scales. Up to 1.2cm (3/8in) long, many can run and jump. They feed on algae, lichen, and plant debris. Eggs are laid in small batches in crevices, and the young mature in 2 years. They may live up to 4 years.

PETROBIUS MARITIMUS
This fast-running insect lives on rocky shorelines in the northern hemisphere, often in large numbers.

Order THYSANURA

SILVERFISH

DISTINGUISHED FROM BRISTLETAILS by their widely separated eyes, and 3 equally sized tails (cerci), silverfish have elongated, flattened bodies, often covered in silvery scales. There are about 370 species in 4 families; some kinds are a familiar sight in houses.

Family LEPISMATIDAE

Lepismatids

Occurrence 190 spp. worldwide, especially in warmer regions; in tree canopies, caves, human dwellings

Lepismatids are brownish, nocturnal insects with compound eyes but no ocelli. They are 0.8–2cm (5/16–3/4in) long. Courtship is simple, and fertilization is indirect. Eggs are laid in small batches in cracks or crevices. Domestic species feed on flour, damp textiles, wallpaper paste, and book bindings. Lepismatids may live up to 4 years.

LEPISMA SACCHARINA
The common silverfish emerges only at night, prefers damp places, and can be found in kitchens, bathrooms, and basements worldwide.

Order EPHEMEROPTERA

MAYFLIES

THE MOST PRIMITIVE WINGED insects, mayflies are unique in undergoing a final moult after their wings have formed. Their underwater nymphs live for 2–3 years, but adults do not feed, and often live for just one day. There are about 2,500 species in 23 families.

Family SIPHLONURIDAE

Primitive minnow mayflies

Occurrence 175 spp. worldwide, but mainly in the northern hemisphere; mostly in and near running water

These mayflies have a wingspan of up to 5cm (2in). The forewings are quite narrow. Nymphs are agile swimmers and, when grown, pull themselves out of the water onto a stone or stem.

typically large hindwings held upright

SIPHLONURUS LACTUSTRIS
Anglers call this species the summer mayfly. It is one of the first to colonize a new water course and is tolerant of acidic water.

Family EPHEMERIDAE

Common burrowing mayflies

Occurrence 50 spp. worldwide except Australia; in and on vegetation around fresh water

These are large mayflies, 1–3.4cm (³⁄₈–1½in) long, with clear or brownish, rarely dark-spotted, triangular wings and 2 or 3 long tails at the end of the abdomen. Females lay thousands of eggs, which they

EPHEMERA DANICA
Also known as the greendrake, this is a large, widespread, European species. It breeds in rivers and lakes with sandy or silty bottoms.

drop directly into water. The nymphs burrow into the silt at the bottom of the water, using their tooth-like mandibles, and eat organic material present in the silt. Adults do not feed. Artificial flies used in trout fishing are modelled on these mayflies.

large, veined, triangular forewings

small hindwings

3 tails of equal length

Order ODONATA

DAMSELFLIES AND DRAGONFLIES

WITH THEIR LARGE EYES and superb flying ability, these insects are highly effective aerial predators. Damselflies are slender, and tend to rest with wings folded, while the stouter dragonflies rest with wings held out. There are about 5,500 species, in 30 families. All start life as aquatic nymphs.

Family AESHNIDAE

Darners

Occurrence 420 spp. worldwide; in and near still water, ponds, swamps with vegetation

Also known as hawkers, darners are some of the biggest, most powerful dragonflies, with a wingspan of mostly 6.5–9cm (2½–3½in). Their large eyes touch on top of the head. Males hold a territory and patrol it regularly, while females cut slits in plants to lay their eggs. Their nymphs are large and well camouflaged, and hunt animals as large as tadpoles and young fish.

ANAX IMPERATOR
Found in parts of Europe and Asia, the adult emperor dragonfly is a strong flier often seen hunting far from water.

hairy thorax

long, thin pterostigma

long abdomen

BRACHYTRON PRATENSE
The adult hairy dragonfly typically emerges in May, flying low, with a characteristic zigzag path.

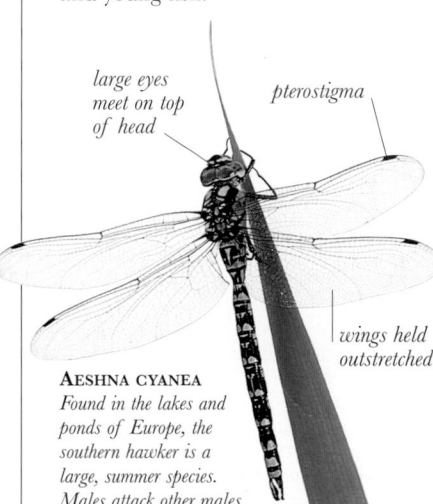

large eyes meet on top of head

pterostigma

wings held outstretched

AESHNA CYANEA
Found in the lakes and ponds of Europe, the southern hawker is a large, summer species. Males attack other males in their territory on sight.

AESHNA MIXTA
The migrant hawker, widespread in central and southern Europe, hawks along a regular beat to catch prey. It flies in summer and autumn and often migrates across Europe in large numbers.

Family COENAGRIONIDAE

Narrow-winged damselflies

Occurrence 1,000 spp. worldwide, especially in temperate regions; in ponds, bogs, streams, brackish water

These slender-bodied damselflies mostly have a wingspan of 2–4.5cm (¾–1¾in). Females make slits in aquatic plants and lay eggs in them. Nymphs hunt on underwater plants.

narrow wings folded above body

thoracic segments angled backwards

long abdomen

COENAGRION PUELLA
The azure damselfly is common in small ponds in Europe. Typical of the family, the male tends to be more brightly coloured than the grey or greenish female (shown above).

Family LESTIDAE

Stalk-winged damselflies

Occurrence 160 spp. worldwide; in swampy or boggy places, pools, lakes, slow-flowing streams

These are also called spread-winged damselflies, these insects have sturdy, bright blue or green bodies, with a metallic sheen. At rest they adopt a vertical pose with the head up and the wings, spanning 3–7.5cm (1¼–3in), held out. Males have forceps-shaped claspers at the end of the abdomen. Females lay eggs inside the aerial parts of aquatic plants.

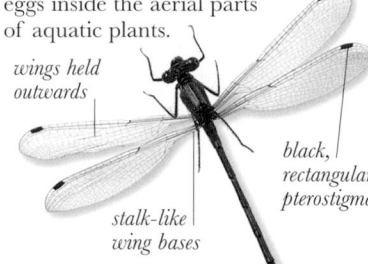

wings held outwards

black, rectangular pterostigma

stalk-like wing bases

LESTES SPONSA
The emerald damselfly is found in Europe and Asia. It has a dark green body and inhabits canals, ponds, and lakes.

Family LIBELLULIDAE

Common skimmers

Occurrence 1,300 spp. worldwide; in forests near slow-flowing streams, ponds, bogs

Also known as darters due to their unpredictable flight pattern, these stout-bodied, often colourful insects have a wingspan of mostly 4.5–7.5cm (1¾–3in). The broad, flat abdomen makes them highly manoeuvrable. Females hover over water and dip down to release their eggs. Nymphs hunt for prey in mud and debris.

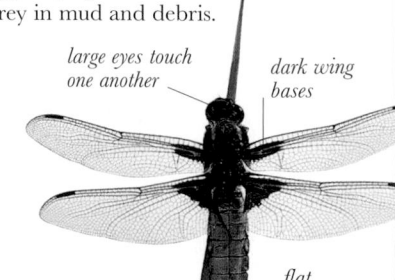

large eyes touch one another

dark wing bases

LIBELLULA DEPRESSA
The broad-bodied chaser is a European species. Adults fly over ponds and lakes, in June and July.

flat, broad abdomen

Order ORTHOPTERA

CRICKETS AND GRASSHOPPERS

WITH OVER 20,000 SPECIES in 28 families, crickets and grasshoppers are widespread in all but the coldest parts of the world. Most species have large wings – the front pair, or tegmina, being tough and leathery – but instead of flying away from danger they often jump away on their powerful hindlegs. They have chewing mouthparts, but their diets vary: crickets, katydids, and their relatives (suborder Ensifera) are predators or omnivores, while grasshoppers and locusts (suborder Caelifera) are entirely vegetarian. All species undergo incomplete metamorphosis, and most adult males stridulate or "sing" to attract mates.

Family ACRIDIDAE

Grasshoppers

Occurrence 10,000 spp. worldwide; on ground, in vegetation

Grasshoppers are 1–8cm (⅜–3¼in) long and have camouflage colouring and patterning. However, some have brightly coloured bodies, or wing bands, to warn off predators. Some also produce noxious chemicals. The antennae are always short. Males sing during the day to attract mates. Females are nearly always larger than males and do not have a conspicuous ovipositor. After mating, they lay egg pods containing 15–100 eggs in the ground and protect them by secreting a foamy substance around them. Grasshoppers are agricultural pests; species that periodically form nomadic swarms – known as locusts – destroy crops in some parts of the world.

drab coloration

CHORTHIPPUS BRUNNEUS
The common field grasshopper inhabits grassland from northern Europe to Spain and Italy. It has a very distinctive song – a series of chirps repeated every 2 seconds or so.

saddle-shaped pronotum

short antennae typical of family

STENOBOTHRUS LINEATUS
The stripe-winged grasshopper lives in dry, sunny grassland and heaths in Europe. It has a high-pitched, alternately loud and soft, song.

strong hindlegs for jumping, and steering in flight

toughened forewings

SCHISTOCERCA GREGARIA
Found from Africa to India, the desert locust is 6cm (2¼in) long. It has been a crop pest since agriculture first developed. Swarms of up to 50 billion individuals move with prevailing winds and can devour up to 100,000 tonnes of food in a day.

large, membranous hindwings

Family GRYLLIDAE

True crickets

Occurrence 4,000 spp. worldwide; on ground, under stones, in litter, on trees

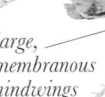

True crickets are 0.5–5cm (³⁄₁₆–2in) long, with slightly flattened bodies. Some species have loud, chirping songs, and are kept as pets. Most true crickets, except the carnivorous tree crickets, are either plant-eaters or scavengers.

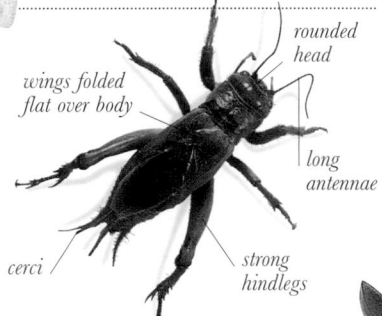

wings folded flat over body

rounded head

long antennae

cerci

strong hindlegs

BRACHYTRUPES SPP.
Sometimes called tobacco crickets, these species are distributed across the African savanna. Relatives include pests that attack tea, tobacco, and cotton.

Family GRYLLOTALPIDAE

Mole crickets

Occurrence 60 spp. nearly worldwide; burrowing in damp sand or soil near streams, ponds, lakes

Like mammalian moles, these crickets are stout and covered with short, velvety hairs. Measuring 2–4.5cm (¾–1¾in) in length, they have short, broad, toothed front legs used for digging. Females dig chambers in the soil to lay their eggs; males make special burrows to amplify their songs, which can be heard up to 1.5km (1 mile) away. Adults and nymphs eat parts of plants as well as small prey, feeding underground during the day and on the surface at night.

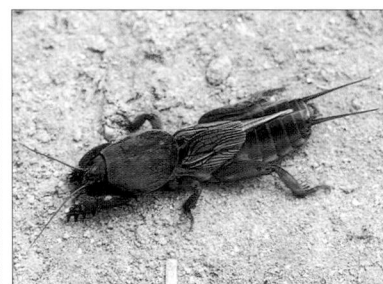

GRYLLOTALPA GRYLLOTALPA
The European mole cricket is attracted to lights after dark. It uses its front legs for digging, and hindlegs for pushing soil along burrows. It can be a crop pest.

Family TETTIGONIIDAE

Katydids

Occurrence 6,000 spp. worldwide, mainly in tropical regions; on vegetation

Also known as bush crickets, these insects get their name from a North American species whose song sounds like "kate-she did". They measure 1.5–7.5cm (½–3in) in length and are brown or green, usually with big wings that slope over their sides. Most species mimic leaves or bark. They may also flash bright hindwing colours to startle predators. Males sing using a file and scraper system at the base of the forewings. Females lay eggs in plants or soil. Most katydids feed and sing only at night.

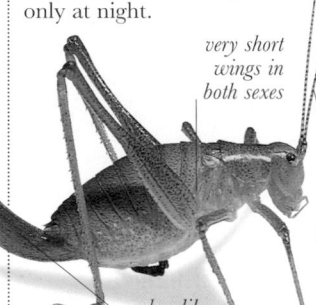

very short wings in both sexes

long, segmented antennae

sabre-like ovipositor

LEPTOPHYES PUNCTATISSIMA
The speckled bush cricket is distributed across Europe, and can be found in bushes, trees, and grassland.

Family RAPHIDOPHORIDAE

Cave crickets

Occurrence 500 spp. widespread, especially in warmer regions; in caves, humid areas, under logs and stones

Squat, hump-backed, and wingless, cave crickets are 1.3–3.8cm (½–1½in) long. They have long hindlegs and even longer antennae that help detect predators. These crickets are drab brown or grey in colour. Some cave-adapted species have reduced eyes and soft bodies, and lay their eggs in cave-floor debris. Nymphs start searching for food as soon as they hatch; some scavenge, while others catch living prey.

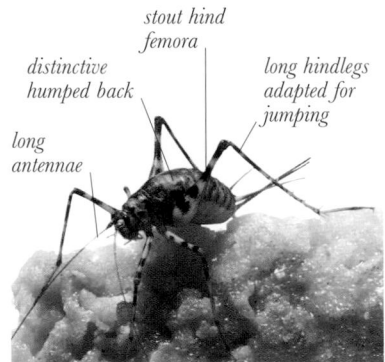

stout hind femora

distinctive humped back

long hindlegs adapted for jumping

long antennae

PHOLEOGRYLLUS GEERTSI
This species is native to North Africa and parts of southern Europe.

very long antennae

uneven wing outline to mimic leaves

brown, leaf-like colour

sickle-shaped ovipositor

OMMATOPTERA PICTOFOLIA
Found in Brazil, this bush cricket lives in trees and bushes, and is an excellent mimic of dead leaves.

distinctive ovipositor

MECONEMA THALASSINUM
The oak bush cricket can be found in Europe, mostly in oak woodland. It is mainly nocturnal and flies readily. Males signal to females by drumming their hindlegs against leaf surfaces.

Order PLECOPTERA

STONEFLIES

STONEFLIES ARE FLAT, slender-bodied insects that start life in water, developing by incomplete metamorphosis. Adults have 2 pairs of membranous wings, but are reluctant fliers. Most are short-lived, often not feeding at all. There are about 2,000 species in 15 families.

Family PERLIDAE

Common stoneflies

Occurrence 400 spp. worldwide except Australia; in lakes, rivers, vegetation near running water

As adults, common stoneflies are weak fliers, and often rest on stones close to the water's edge. They are 1–4.8cm (3/8–2in) long, and are coloured in shades of brown or yellow. The adults do not feed, but the nymphs are often carnivorous.

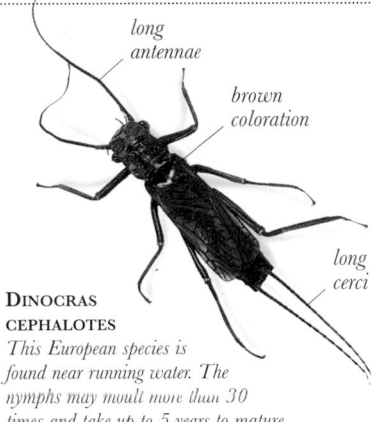

long antennae

brown coloration

long cerci

DINOCRAS CEPHALOTES
This European species is found near running water. The nymphs may moult more than 30 times and take up to 5 years to mature.

Order GRYLLOBLATTODEA

ROCK CRAWLERS

ROCK CRAWLERS WERE first discovered in the Canadian Rockies in 1913. Forming one family of 25 species, these small, wingless insects are adapted to cold conditions, and eat dead, windblown, or torpid prey, moss, and other plant matter. Metamorphosis is incomplete.

Family GRYLLOBLATTIDAE

Rock crawlers

Occurrence 25 spp. in cooler regions of the northern hemisphere; in limestone caves, decaying wood

Mainly nocturnal, rock crawlers, or ice crawlers, are 1.2–3cm (3/8–1¼in) long, have small, compound eyes, and simple, biting mouthparts. Females may lay eggs more than 2 months after mating. The nymphs may take more than 5 years to develop.

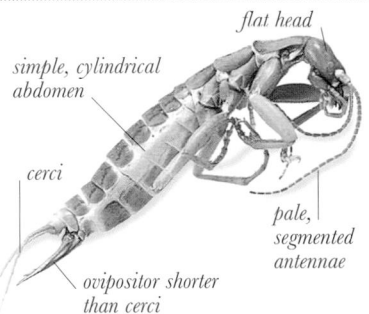

flat head

simple, cylindrical abdomen

cerci

pale, segmented antennae

ovipositor shorter than cerci

GRYLLOBLATTA CAMPODEIFORMIS
The northern rock crawler is found at high altitudes in the USA and Canada. It lives among rocks near glaciers.

Order DERMAPTERA

EARWIGS

THIS ORDER CONTAINS 10 families (1,900 species) of scavenging or plant-eating insects with characteristic abdominal "pincers". Most have short front wings, and fan-shaped hindwings that can be folded. Maternal care is well developed. Metamorphosis is incomplete.

Family FORFICULIDAE

Common earwigs

Occurrence 450 spp. worldwide; in leaf litter and soil, under bark or in crevices

Common earwigs are 1.2–2.5cm (3/8–1in) long and usually dark brown. Two forceps-like appendages are relatively straight in females and curved in males. Females lay eggs in soil, guard them, and lick them clean.

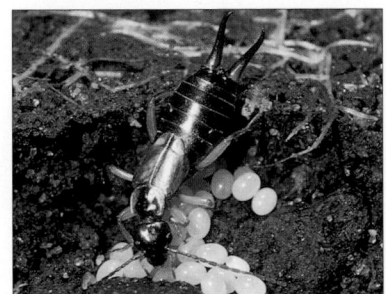

FORFICULA AURICULARIA
The European earwig, now found worldwide, is a pest of vegetables, cereals, fruits, and flowers.

Order PHASMATODEA

STICK AND LEAF INSECTS

THESE STRIKINGLY STICK-LIKE or leaf-like insects live and feed on plants, and use their remarkably effective camouflage to avoid being eaten. The males – which are rare in some species – have wings, but the females are frequently wingless. They develop by incomplete metamorphosis, and are usually nocturnal. There are about 2,500 species in 3 families.

Family PHASMATIDAE

Stick insects

Occurrence 2,450 spp. mainly in warm regions; among vegetation or foliage of trees and shrubs

Also called walking sticks, these insects can adopt a twig-like posture that camouflages them superbly; some can even sway in the breeze to blend with the background. Stick insects vary from 2.5 to 29cm (1–11½in) in length and have a spiny or warty, brown or green body, with short, tough front wings to protect the large, membranous hindwings. Some species are wingless. When threatened, some stick insects adopt a scorpion-like posture, while others flash their wing colours, or give off noxious chemicals. If attacked, they can shed their legs, which soon grow back. In many species, males occur either rarely or not at all, and the females typically scatter their seed-like eggs on the ground.

short antennae

leaf-like expansions on leg segments

EXTATOSOMA TIARATUM
Native to Australasia, Macleay's spectre is one of the many species that is often reared in captivity on bramble leaves. The female (shown right) is wingless, larger than the male, and lays a dozen eggs every day of her adult life, flicking them away with her abdomen.

spiny, leaf-like expansions at sides of abdomen

curve allows front legs to be held at side of head

square head

long metathorax

spined, ridged leg segments

slender body

distinctive ovipositor

PHARNACIA SPP.
Found in India, this genus contains extremely slender insects. The females, like the one shown here, are wingless. They can hold their legs close to their body to enhance their twig-like appearance.

Family PHYLLIIDAE

Leaf insects

Occurrence 30 spp. in Mauritius, Seychelles, Southeast Asia, Australasia; on various plants in well-vegetated areas

Flat, expanded abdomens, extended leg segments, and brown or green colouring all give these insects an uncanny resemblance to leaves. The disguise is helped further by blotches of colour and texture, and an ability to sway in the breeze. Some species specialize in looking like dead, wrinkled leaves. Leaf insects range from 3 to 11cm (1¼–4¼in) in length. At rest, their veined forewings cover the transparent hindwings. The antennae are short and smooth in females, and long and slightly hairy in males. The eggs are seed-like and are dropped on the ground. Because of their odd appearance, leaf insects are bred as pets all over the world.

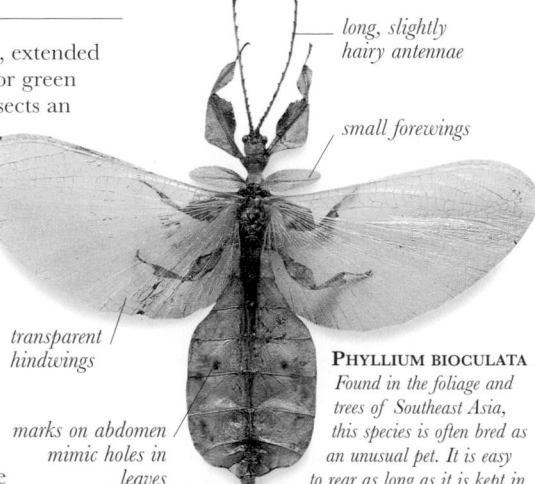

long, slightly hairy antennae

small forewings

transparent hindwings

marks on abdomen mimic holes in leaves

PHYLLIUM BIOCULATA
Found in the foliage and trees of Southeast Asia, this species is often bred as an unusual pet. It is easy to rear as long as it is kept in warm conditions.

THREAT DISPLAY
When disturbed, mantids like this flower mantid often rear up into a defensive pose to display their vivid coloration. As well as using its excellent eyesight, the mantid avoids attack by bats by using an ultrasonic ear on the thorax between the middle legs.

Order **MANTODEA**

MANTIDS

THE COMMON NAME "praying" mantid is derived from the distinctive way in which mantids hold their front legs up and together, as if in prayer. Although variable in shape, mantids are distinguished by their triangular heads and large eyes – they are the only insects able to turn their head around to look behind them – and an elongated prothorax with the front pair of legs distinctively modified for catching live prey. Most mantids are diurnal and eat a large range of arthropods and, occasionally, even vertebrates such as frogs and lizards. Mantids reproduce by laying eggs in a papery or foam-like egg case which, in some species, is guarded by the female. Among the 8 different mantid families, there are 2,000 individual species.

Family **HYMENOPODIDAE**

Flower mantids

Occurrence 185 spp. in tropical regions, all around the world except Australia; on a wide variety of vegetation

Vivid coloration including bright reds and greens (see left) allows these species to blend in perfectly with the plants on which they rest, awaiting the arrival of prey. Some parts of the body, such as the legs, often have broad extensions that resemble leaves.

The forewings of flower mantids may have coloured bands or spirals, or circular marks that look like eye-spots. In some species the coloured markings on the front wings can be asymmetrical. The wings of the female flower mantid are sometimes shorter than those of the male. Young nymphs are immediately predatory and are able to tackle prey as soon as their cuticles have hardened. As well as having an ultrasonic ear, used to detect predators, some species have a second ear that is responsive to much lower frequencies, but its function is unknown.

Family **MANTIDAE**

Common praying mantids

Occurrence 1,400 spp. in warmer regions, all around the world; on any vegetation where there is ample prey

Most praying mantids belong to this family, and many members vary very little in appearance. Most have green or brown coloration and a close resemblance to leaves and twigs. Unlike families such as the flower

mantids, the wings of common praying mantids are seldom patterned, although they may have wing-spots. Females often have smaller wings than males and some lack wings altogether. Some species of common praying mantids can grow up to 15cm (6in) in length, and the larger species are known to prey on small reptiles, salamanders, and frogs.

thread-like antennae

large, forward-facing eyes

elongated prothorax

leaf-like, veined wings

raptorial front legs

slender legs

ELEGANT PROFILE
The common praying mantid's forward-facing eyes give it true binocular vision, allowing it to calculate distances accurately. This ability, together with its extendable front legs, makes it an effective hunter.

head and part of thorax eaten

large female

DANGEROUS ENCOUNTERS
Female common praying mantids are famed for eating the male after mating (see above). However, although mating can be dangerous for smaller individuals, males are usually cautious and seldom get eaten by their partner in the wild.

STEALTHY HUNTER

1

2

3

4

LYING IN WAIT
The mantid lies patiently in wait for a potential meal. Perching on the end of a stick, forelegs together, it focuses on a likely landing site for prey.

THE POUNCE
As an unsuspecting fly comes within range, the mantid acts quickly. It launches itself at its victim, lunging forwards and reaching out its long forelegs.

THE CAPTURE
At full stretch the mantid seizes and impales its victim with its spiky legs before it is even aware of the mantid's presence.

ALL OVER IN SECONDS
Clasped in a strong grip, the helpless fly has no means of escape. Having retreated back to its stick, the mantid begins to eat its prey.

Order BLATTODEA

COCKROACHES

COCKROACHES ARE LEATHERY INSECTS, typically with oval and flattened bodies that enable them to squeeze through tight spaces in search of food or to escape predators. They are sensitive to vibration and can quickly flee from danger. The head is often covered by a shield-like pronotum and there are generally 2 pairs of wings. The order Blattodea comprises 6 families and 4,000 species, of which less than 1 per cent are pests; the rest are useful scavengers in many habitats. The pest species thrive in warm conditions and where there is poor hygiene and sanitation, and can carry many disease-causing organisms on their bodies.

Adults may produce sexual pheromones to attract mates. The females lay up to 50 eggs in 2 rows surrounded by tough, protective egg cases called oothecae. Metamorphosis is incomplete.

Family BLABERIDAE

Live-bearing cockroaches

Occurrence 1,000 spp. in tropical regions; in caves, underground

These large insects, usually measuring 2.5–6cm (1–2¼in) in length, often have well-developed, pale brown wings with dark markings. In many species, however, the females are wingless and burrow under wood and stones. The pronotum may be very wide. During courtship some species use sounds to communicate with a mate. All species are ovoviviparous and give birth to live young. The egg sacs are fully extruded from the end

of the abdomen, rotated, and then drawn back inside to be brooded within the body of the female. Adults and nymphs are mainly saprophagous or eat dung.

leathery abdomen

thoracic humps used in male-to-male combat

short, thick legs

GROMPHADORHINA PORTENTOSA
Better known as the Madagascan hissing cockroach, this insect is 6–8cm (2¼–3¼in) long, and startles predators by squeezing air out of its spiracles to produce a loud hiss.

Family BLATTELLIDAE

Blattellids

Occurrence 1,750 spp. worldwide in warmer areas; in woodland litter, debris, waste tips, buildings

Many of these fully winged species are pale brown but a few are olive-green. Their body length varies from 0.8 to as long as 10cm (⁵⁄₁₆–4in). Adults and nymphs are scavengers. Females carry their egg cases, which protrude from their bodies, until the eggs are about to hatch. Thousands of eggs may be produced in a lifetime.

long, thread-like antennae

males and females fully winged

MEGALOBLATTA LONGIPENNIS
This species from Peru, Ecuador, and Panama is the largest winged cockroach in the world. It has a wingspan of up to 20cm (8in).

Family BLATTIDAE

Common cockroaches

Occurrence 600 spp. mainly in tropical and subtropical regions; common in warehouses, sewers, rubbish dumps

Most species of common cockroaches are brown or reddish or blackish brown with markings. They are 2–4.5cm (¾–1¾in) long, very active, and can run very fast or fly. Some species emit repellent chemicals which can cause skin rashes. Females can produce up to 50 egg cases, each containing 12–14 eggs, which are stuck to concealed surfaces. Both adults and nymphs are nocturnal and saprophagous.

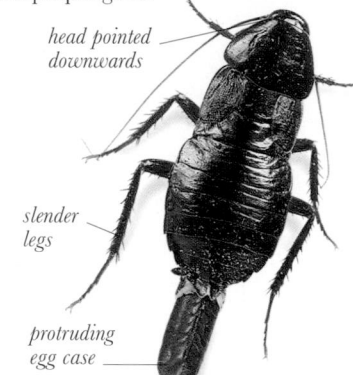

head pointed downwards

slender legs

protruding egg case

BLATTA ORIENTALIS
The oriental cockroach is a common domestic pest. Like many species, it has spread from its native region by travelling on board ships.

Order ISOPTERA

TERMITES

LIKE ANTS (SEE P.576) – WITH WHICH they are sometimes confused – termites are social insects, living in colonies that can be over a million strong. There are about 2,750 species, in 7 families, and most live in the tropics, where they build impressive nests. They are usually pale and soft, with chewing mouthparts and short antennae, but their body form varies with their role in the colony. Workers are small, while soldiers have large heads and jaws. Queens, which lay all the colony's eggs, can be as big as a human finger. Some termites feed on wood or plants, but others eat fungi that they raise in under-ground "gardens". Although important in food chains as nutrient recyclers, many are destructive pests. Metamorphosis is incomplete.

Family RHINOTERMITIDAE

Subterranean termites

Occurrence 345 spp. worldwide in warmer regions; in soil or in damp wood touching ground

As adults and nymphs, subterranean termites feed mainly on wood, digesting it with the help of microbes which they harbour in their intestines. They are voracious feeders, and in warm parts of the world are highly

damaging pests. Generally cream or light brown, they are typically 6–8mm (¼–⁵⁄₁₆in) long.

RETICULITERMES LUCIFUGUS
This termite has distinctive antennae. It makes its nest under the ground, inside damp wood, and is commonly found throughout southern Europe.

Family TERMITIDAE

Higher termites

Occurrence 1,950 spp. worldwide; in trees, soil, underground

Three-quarters of the world's termite species belong to this family. Highly variable, both in appearance and habits, they range in length from 4 to 14mm (⁵⁄₃₂–½in). Queens are much bigger than the workers and soldiers, and may produce thousands of eggs a day. Nests vary from small structures built in trees, to vast earthen towers, with underground "fungus gardens". Many species are crop pests.

MACROTERMES SPP.
Termites in this abundant African genus often form giant colonies. At certain times of the year, winged reproductives fly off to found new colonies.

broad head

large, pale wings

GLOBITERMES SULPHUREUS
This Southeast Asian pest makes thick-walled nests partly underground. The nests' soldiers can "explode" to cover intruders in a sticky fluid.

TRINERVITERMES TRINERVOIDES
The snouted harvester termite lives in pastures in Central and southern Africa. It is eaten by the aardwolf (see p.207) – an animal that has evolved immunity to its defensive secretions.

Order EMBIOPTERA

WEB-SPINNERS

THIS RELATIVELY SMALL ORDER consists of 8 families and only 300 species. Web-spinners are gregarious, and live in soil, leaf litter, and under bark. They make extensive protective tunnels, using the silk made by glands in their front tarsi. Metamorphosis is incomplete.

Family CLOTHODIDAE

Clothodids

Occurrence 14 spp. worldwide in tropical and subtropical regions; in soil, leaf litter, under bark and stones

Clothodids are typically elongated insects with short legs. They are 0.5–2cm ($^3/_{16}$–$^3/_4$ in) long, and have small eyes, segmented antennae, and simple, biting mouthparts. As adults, males do not feed but use their mandibles to hold onto the female

during mating. Females and nymphs are saprophagous. Females cover their eggs with silk and detritus and feed the nymphs with pre-chewed food.

CLOTHODA URICHI
Seen here inside its silken nest, this species is a native of the Caribbean island of Trinidad.

Order ZORAPTERA

ANGEL INSECTS

THESE SMALL, DELICATE, GREGARIOUS, termite-like insects make up a single family of 30 species. They are light straw-yellow to dark brown or blackish in colour, with short, abdominal tails (cerci) and unspecialized mouthparts. Metamorphosis is incomplete.

Family ZOROTYPIDAE

Angel insects

Occurrence 30 spp. worldwide in tropical and warm temperate regions except Australia; in rotting wood, sawdust

Angel insects are 2–3mm ($^1/_{16}$–$^1/_8$ in) long, with a distinctive triangular head and a pair of antennae. The nymphs and adults eat fungal threads and small arthropods. Sexual behaviour may be complex: males of some species give the females mating gifts

in the form of secretions from glands in their heads; in other species, the males fight for mates by kicking each other.

short, stout legs

bead-like segments

cerci with one segment

ZOROTYPUS HUBBARDI
The rare Hubbard's angel insect is native to eastern and southern North America. This slide-mounted specimen is stained red.

Order PSOCOPTERA

BARKLICE AND BOOKLICE

THE ORDER PSOCOPTERA COMPRISES 35 families and 3,500 species of soft-bodied, drably coloured insects. The head is large, with bulging eyes, a bulbous forehead, and long, thread-like antennae. Metamorphosis is incomplete, typically with 5 nymphal stages.

Family LIPOSCELIDAE

Liposcelid booklice

Occurrence 150 spp. worldwide; in dry leaf litter, under bark, in nests, inside buildings, in food stores

These insects have flattened bodies, with swollen femora on the hindlegs. Most species are wingless. The length of the body ranges from 0.5 to 1.5mm ($^1/_{64}$–$^3/_{64}$ in). Some species gnaw on or chew stored flour, cereals, or paper.

Eggs are laid in crevices in tree bark, leaf litter, and bird nests. The nymphs look like small adults.

LIPOSCELIS TERRICOLIS
Found in leaf litter and in a variety of stored, dry produce, this widespread species can become a pest.

Order PHTHIRAPTERA

PARASITIC LICE

FLATTENED AND WINGLESS, the 6,000 species (in 25 families) of parasitic lice are ectoparasites, living permanently on the bodies of birds and mammals without killing them. The mouthparts are used to bite skin, fur, or feathers, or to suck blood. Different lice are linked with specific hosts, and many are restricted to particular areas of the body. Parasitic lice develop by incomplete metamorphosis.

Family MENOPONIDAE

Bird lice

Occurrence 650 spp. worldwide; on a variety of bird hosts

Bird lice are 1–6mm ($^1/_{32}$–$^1/_4$ in) long, with a large and roughly triangular head, and biting mandibles. The abdomen is oval, and the legs are short and stout. Each leg has 2 claws. Adults and nymphs of most species feed on feather fragments but some also ingest blood and skin secretions. Eggs are glued in masses to the base of feathers. Some species, such as *Menopon gallinae*, or the shaft louse, can be serious poultry pests.

broad head

short, stout legs

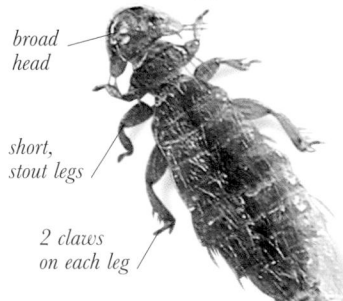

2 claws on each leg

MENACANTHUS STRAMINEUS
The chicken body louse is a widespread species that lives on poultry, bringing about feather-loss and infection.

Family PEDICULIDAE

Human lice

Occurrence 2 spp. worldwide; on humans, apes, monkeys

Human lice are small, pale, and elongated, measuring 2–6mm ($^1/_{16}$–$^1/_4$ in) in length, with short, strongly clawed legs for gripping onto their hosts. They have a narrow head and a pear-shaped, flattened body. Adults and nymphs feed on blood. The human louse has 2 subspecies: the body louse, which glues its eggs to clothing, and the head louse, which glues its eggs (nits) to hair. Outbreaks of head lice are common among young schoolchildren. The body louse carries the organism that causes typhus.

pear-shaped, flattened body

PEDICULUS HUMANUS CAPITIS
A female human head louse can lay 9–10 eggs a day, fastening each one separately to a hair. Once cemented in position, the eggs are difficult to dislodge.

Family PTHIRIDAE

Pubic lice

Occurrence 2 spp. worldwide; wherever their hosts (humans and gorillas) live

Also known as crab lice, pubic lice have a squat, flattened body, 1.5–2.5mm ($^3/_{64}$–$^3/_{32}$ in) long. The middle and hindlegs are especially stout and have very strong, enlarged claws for gripping hair shafts. The nymphs and adults suck the host's blood, leaving bluish marks on the skin. They move very slowly, often hardly at all. A female may lay 30 eggs in

her lifetime, and uses a waterproof glue to stick them singly to pubic hair. It takes 4 weeks for the eggs to mature into adults.

PTHIRUS PUBIS
This louse may be found in armpits and beards as well as the groin. Although unpleasant, it is not known to transmit disease.

eye

small head

antennae

nymph

squat body

strong, vice-like claws for holding onto hair

Order HEMIPTERA

BUGS

BUGS VARY IN SIZE FROM SPECIES a few millimetres long to giant aquatic predators large enough to catch fish and frogs. They are found all over the world, in every terrestrial habitat as well as in fresh water. The order comprises 134 families and 82,000 species. There are 4 suborders: Coleorrhyncha, consisting of a single family; Heteroptera or true bugs (including plant bugs); Auchenorrhyncha (including treehoppers, lantern bugs, and cicadas); and Sternorrhyncha (including whiteflies and scale insects). All bugs have 2 pairs of wings; in the Heteroptera, the front ones are typically larger and thickened. The mouthparts, which form a long rostrum or "beak", pierce and suck, and many sap-sucking species are serious crop pests. Metamorphosis is incomplete.

Family ALEYRODIDAE

Whiteflies

Occurrence 1,200 spp. worldwide, especially in warmer regions; on a wide range of plants, including crops

These delicate, moth-like bugs are 1–3mm (1/32–1/8in) long. They carry a pair of 7-segmented antennae, and 2 pairs of transparent, white, or mottled wings of similar size. Females lay their eggs on tiny stalks on the undersides of leaves. Many of these plant sap-suckers are serious pests.

TRIALEURODES VAPORARIORUM
The greenhouse whitefly is a common pest of cucumbers and tomatoes grown under glass. It may also attack field crops in warm conditions.

Family APHIDIDAE

Common aphids

Occurrence 2,250 spp. worldwide, especially in northern temperate regions; on most wild and cultivated plants

Their huge reproductive potential makes aphids among the most destructive of all plant-eating insects, affecting almost all crop species by feeding and transmitting diseases. Also called greenflies or blackflies, these are pear-shaped, soft-bodied insects, mostly 5mm (3/16in) long, and typically green, pink, black, or brown. The abdomen usually has a pair of cornicles, short tubes that emit a defensive secretion. Reproduction is often parthenogenetic or only seasonally sexual.

MACROSIPHUM ALBIFRONS
The American lupin aphid is a pest of garden flowers, both in North America and Europe. Here, one female (top left) is giving birth.

Family BELOSTOMATIDAE

Giant water bugs

Occurrence 150 spp. worldwide, especially in tropical and subtropical regions; in standing water, slow streams

Also known as electric light bugs due to their attraction to lights, this family contains the largest bugs, able to eat frogs and fish. They are brownish insects with a broadly oval body that is 1.5–10cm (1/2–4in) long. The front legs are modified as prey-capturing pincers. Uniquely among insects, the female glues her eggs to the male's back and the latter broods the eggs.

curved tibiae with sharp claws for seizing prey

large eyes

hairs on middle and hindlegs aid swimming

LETHOCERUS GRANDIS
This widespread giant water bug is caught and eaten by humans in certain parts of Southeast Asia.

Family CERCOPIDAE

Frog-hoppers

Occurrence 2,400 sp. worldwide, especially in warmer regions; on shrubs, trees, herbaceous plants

These squat, round-eyed, plant bugs are very good jumpers. Many species have drab coloration but some are coloured in bright red and black or yellow and black patterns. Body length ranges from 0.5 to 2cm (3/16–3/4in). Eggs, up to 30 in some species, are laid in soil or in plant tissue. Nymphs feed on plant sap. They surround themselves in foam, which hides them and stops them from drying.

pronotum wider than head

distinctive, black and red coloration

CERCOPIS VULNERATA
This frog-hopper inhabits grassy places and meadows in Europe and Asia.

stout hindlegs with spines

Family CICADELLIDAE

Leaf-hoppers

Occurrence 16,000 spp. worldwide; almost anywhere with vegetation, abundant in agricultural habitats

Leaf-hoppers are typically slender-bodied and 0.3–2cm (1/8–3/4in) long. Many species are brown or green; some may be brightly striped or spotted. Males attract females with low-amplitude courtship calls, which travel through leaves and stems, instead of being carried through the air. Once a female has mated, she lays hundreds of eggs over a period of 6–8 weeks. Leaf-hoppers are highly destructive pests, attacking many of the world's staple crops, such as rice, maize, cotton, and sugar cane.

triangular head with black markings

green front wings

CICADELLA VIRIDIS
Widespread in Europe and Asia, the common green leaf-hopper can be a pest of fruits, rice, wheat, and sugar cane.

Family CICADIDAE

Cicadas

Occurrence 2,500 spp. worldwide in warmer regions; on suitable host trees or shrubs

Cicadas include some of the world's noisiest insects, producing calls that can be heard up to 1.5km (1 mile) away. The adults are usually dark brown or green, with squat bodies 2.2–5.5cm (3/4–2 1/4in) long. They have short antennae, and prominent eyes set far apart. Only the males sing, using a pair of drum-like tymbal organs located on the sides of the abdomen. Females lay eggs by cutting slits in plants; after hatching, the nymphs feed on roots underground. Some species can live for up to 17 years.

CICADA ORNI
This species, common in southern Europe and the Mediterranean, has a greyish bloom, clear front wings with 11 black spots, and a long rostrum. The adults are eaten by birds such as jays.

short antennae

prominent eyes

membranous, yellow front wings

hindwings smaller than front wings

male's abdomen acts as resonator, to amplify songs

ANGAMIANA AETHEREA
Like all cicadas, this Indian species produces songs to signal aggression or to attract mates.

PYCNA REPANDA
This Indian species is commonly found resting and feeding among a variety of tree species, in forests and woodland.

Family CIMCIDAE

Bed bugs

Occurrence 90 spp. worldwide; on bird and mammal hosts, in nests, caves, crevices in buildings

These flat, oval, reddish brown bugs are 4–12mm (⁵⁄₃₂–³⁄₈in) long, and have remnant front wings and no hindwings. Their compound eyes have a small number of facets. All bed bugs are blood-sucking, temporary ectoparasites that live on birds, bats, and even humans. They live for 2–10 months depending on feeding and temperature. Males practise traumatic insemination, injecting sperm by penetrating the female's body with a sharp sexual appendage. Females may lay more than 150 eggs in their adult life, with 2 or 3 eggs laid per day.

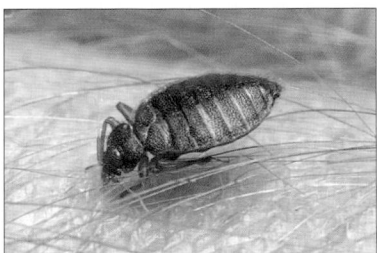

CIMEX LECTULARIUS
The common bed bug feeds at night and senses the presence of a host using temperature, odour, and carbon dioxide concentration as cues.

Superfamily COCCOIDEA

Scale insects

Occurrence 7,000 spp. worldwide, especially in tropical and subtropical regions; on all parts of many plants

Scale insects are so highly adapted to a sap-sucking life that they hardly resemble insects at all. The females are often wingless and legless, and spend their adult lives permanently attached to plants. Most are under 1cm (³⁄₈in) long, with an oval body covered by a smooth or waxy scale. The males look very different, and often have wings. Many scale insects reproduce asexually, and can be prolific crop pests.

mealy-bugs clustered under bud

PLANOCOCCUS CITRI
The common citrus mealybug is a pantropical pest that attacks coffee, soya bean, and guava crops. It attaches itself mainly to the plants' roots, but also to their stems.

Family FULGORIDAE

Fulgorids

Occurrence 800 spp. in tropical and subtropical regions; mostly on trees and woody shrubs in well-vegetated areas

Many tropical fulgorids are notable for their bizarrely shaped, bulbous heads, and are known as lantern bugs. Their eyes are located at the sides of the head, above the antennae. The thorax is often quite large, as are the wings, which have cross veins. These sap-suckers are 0.8–10cm (⁵⁄₁₆–4in) long and have eye-spots on the hindwings that can be flashed to deter predators. Adults rest and feed during the day and fly at night. Eggs are laid on host plants and covered by a frothy secretion, which hardens to protect them.

elongated head

large eye-spots on hindwings

FULGORA LATERNARIA
Also called the lanternfly or peanut bug, this species has a bulbous head that was once thought to glow. It occurs in South America and the West Indies.

FULGORA SERVILLEI
Found in the forests of South America, this species has a very enlarged head with tooth-like markings. Like most species in the family, it is a sap-sucker.

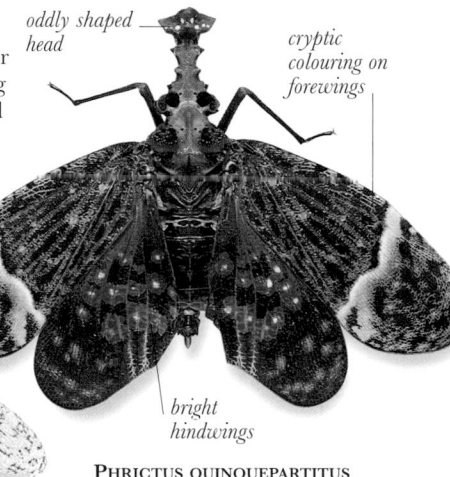

oddly shaped head

cryptic colouring on forewings

bright hindwings

PHRICTUS QUINQUEPARTITUS
The wart-headed bug has colourful wings and an unusual head. It is found in woodland and forests of Panama, Brazil, and Colombia.

Family GERRIDAE

Pond skaters

Occurrence 500 spp. worldwide; in water bodies ranging from pools, ponds, streams, rivers, lakes, to warm oceans

Also known as water-striders, these fast-moving bugs are 0.2–3.5cm (¹⁄₁₆–1½in) long, dark brown or black, and covered with short, velvety hairs. The elongated body has short, stout front legs used for grasping prey, and long middle and hindlegs that help to spread its weight over the water film. These predatory bugs detect prey using ripple-sensitive hairs. Mating

follows ripple courtship signals, and eggs are laid on floating objects or embedded in water plants.

GERRIS SPP.
These bugs are widespread and live on water surfaces. Ripples made by prey attract the bugs, which move quickly to attack.

Family HYDROMETRIDAE

Water measurers

Occurrence 120 spp. worldwide; in standing water bodies, on floating plants

Like pond skaters, water measurers, also called marsh-treaders, live and feed on the surface film, but they move in a slow and stealthy way, creeping up on their prey. They are 0.3–2.2cm (⅛–¾in) long, with an elongated head and bulging eyes. Most species have either short wings or no wings at all. Water

measurers are predators and scavengers in all life stages, using a spearing and sucking technique to feed; they prefer dead or dying prey.

eyes far from pronotum *long, thread-like legs*

HYDROMETRA STAGNORUM
Like all water measurers, this European bug is sensitive to surface vibrations and can locate small prey, which it spears with sharp mouthparts.

Family MEMBRACIDAE

Treehoppers

Occurrence 2,500 spp. worldwide, mainly in warmer regions; especially in understorey layers in tropical forests

The large, distinctive pronotum, often shaped like a thorn, gives this family the alternative name of thorn bugs. These plant sap-suckers are mostly green, brown, or brightly coloured, measuring 0.5–1.5cm (³⁄₁₆–½in) in length. Female treehoppers lay egg masses in slits made in the host plant's bark and, depending on the species, a single female may lay up to 300 eggs. The female frequently guards them until they hatch. Ants protect many species in return for honeydew.

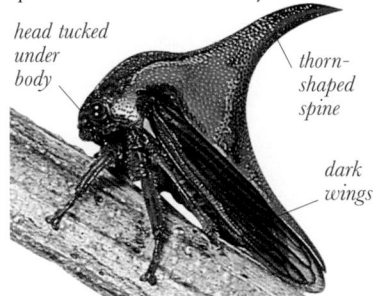

head tucked under body

thorn-shaped spine

dark wings

UMBONIA SPP.
Members of this genus are found in woodland and forests in North and South America and Southeast Asia. The pronotum is tough and can puncture skin or penetrate shoes.

Family MIRIDAE

Plant bugs

Occurrence 8,000 spp. worldwide; on all aerial parts of a wide range of plants, including crops

Also known as capsid bugs, these fragile, elongated or oval insects form the largest family of true bugs. They show a great variety of colours and markings – some resemble ants – and are generally 2–16mm (¹⁄₁₆–½in) long. Most are sap-suckers, but some scavenge or hunt live prey. Plant bugs protect themselves by having cryptic colouring, or by giving off a noxious scent when threatened. Eggs are inserted inside plant tissues; a female may lay a total of 30–100 eggs during her adult life. Many species are fruit and crop pests.

oval body

LYGOCORIS PABULINUS
The common green capsid bug is a European pest that attacks fruit and vegetable crops. Feeding by this bug leaves raised, warty spots on fruits.

INVERTEBRATES

Order HEMIPTERA continued

Family NOTONECTIDAE

Back-swimmers

Occurrence 350 spp. worldwide; in still water, mostly small pools, ponds, edges of lakes

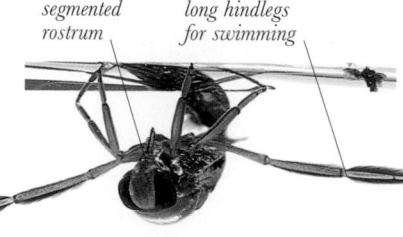

segmented rostrum

long hindlegs for swimming

These spindle-shaped bugs, 2–17mm (1/16–1/2 in) long, are good fliers, but spend much of their lives underwater, swimming upside down. They use their forelegs to catch prey and their hindlegs for swimming.

NOTONECTA GLAUCA
Like other back-swimmers, this European species stores air under its wings. It detects its prey by sight, and by sensing surface ripples.

Family PENTATOMIDAE

Stink bugs

Occurrence 5,500 spp. worldwide; on herbaceous plants, shrubs, trees

Also known as shield bugs because of their shield-like shape, these bugs are 0.5–2.5cm (3/16–1in) long. The name stink bugs comes from their ability to give off strong defensive odours from their thoracic glands. Many species are green or brown but some may be brightly coloured. The pronotum is broad and may have sharp corners.

black dorsal surface of head

red and black pronotum

large black mark on scutellum

EURYDEMA DOMINULUS
The brassica bug has bold red or orange colouring that serves to warn predators of its distastefulness. Found across Europe, it is a pest of brassica crops.

Females lay clusters of barrel-shaped eggs, up to 400 at a time, on plants. There are 5 nymphal stages. Nymphs are mostly herbivorous to begin with but some turn predatory later, or become mixed feeders. Several species are serious pests of a wide-ranging variety of crops such as pulses, potato, cotton, and rice.

3 pale spots on edge of large scutellum

shield-like shape

sculptured body surface

NEZARA VIRIDULA
The green vegetable bug is abundant in tropical, subtropical, and warm temperate regions. It is a pest of more than 100 plants, mostly vegetables and legumes.

PALOMENA PRASINA
This is the green shield bug, found commonly in woodland and on shrubs across Europe. It is a minor pest of beans and alfalfa crops.

Family PSYLLIDAE

Jumping plant lice

Occurrence 1,500 spp. worldwide; on stems, leaves, and bark of herbaceous and woody shrubs, trees, crops

Looking like small leaf-hoppers (see p.558) with long antennae, jumping plant lice are 1.5–5mm (3/64–3/16 in) long, and have 2 pairs of oval wings, held roof-like over the body. They feed on a wide range of plants, and some are pests. The females lay stalked eggs on or in their foodplants.

CACOPSYLLA PYRICOLA
Abundant in the northern hemisphere, the pear sucker is a major pest of pear trees, from the leaves of which adults and nymphs suck sap.

Family REDUVIIDAE

Assassin bugs

Occurrence 6,000 spp. worldwide, especially tropical and subtropical regions; on ground, in leaf litter, on plants

Assassin bugs get their name from their predatory habits, although some live by sucking blood. They are 0.7–4cm (7/32–1½in) long, typically with dark, slender bodies. Many prey on insect pests; some blood-sucking bugs, particularly in the tropics, attack humans and transmit Chagas' disease.

stout front legs designed to hold prey

bright markings

PLATYMERIS BIGUTTATA
This is a large, highly predatory Afrotropical species, 3–3.6cm (1¼–1½ in) long. Its toxic saliva may cause temporary blindness.

Family TINGIDAE

Lace bugs

Occurrence 2,000 spp. worldwide; on foliage of herbaceous plants, trees

Lace bugs are small greyish insects, recognizable by the complex, lace-like patterns on their wings and upper body. They are 2–5mm (1/16–3/16 in) long. The pronotum can extend, hood-like, over the head. Females of some species take care of the eggs and nymphs. Many species are pests but some are used in weed control.

TINGIS CARDUI
The spear-thistle lace bug is common in Great Britain. It eats certain types of thistles. A covering of light, powdery wax gives it a pale grey look.

Order THYSANOPTERA

THRIPS

THIS ORDER CONSISTS OF 5,000 species within 8 families. Thrips are small and slender, with 2 pairs of narrow, hair-fringed wings. They have large eyes, short antennae, and sucking mouthparts, with one small and one needle-like mandible. A sticky, inflatable structure between the claws aids grip. Metamorphosis is unusual, with 2 nymphal stages and one or more pupa-like stages.

Family THRIPIDAE

Common thrips

Occurrence 1,750 spp. worldwide; in leaf litter, on leaves, flowers, and fruits of a wide range of plants

In warm, humid weather, clouds of common thrips or thunderbugs form part of the aerial plankton – myriad tiny, winged insects that are blown far and wide by the breeze. Common thrips vary from pale yellow to brown or black, and have very narrow, feathery wings. Typical species are less than 2mm (1/16 in) long. Some common thrips reproduce asexually, and in most, the female inserts eggs into plant tissues or flowers using a tiny, saw-like ovipositor. The nymphs develop either on plants or in the soil. Adults and nymphs generally live by sucking plant juices, although a few attack other insects. Some species are serious pests of tobacco, cotton, beans, and other crops.

THRIPS FUSCIPENNIS
This dark-coloured species, with distinctive hair on its body, is abundant in the northern hemisphere, and is found on a wide range of plant species.

THRIPS SIMPLEX
The gladiolus thrips was originally from South Africa, but is now also widespread in Europe and North America, wherever gladiolus flowers grow.

Order MEGALOPTERA

DOBSONFLIES AND ALDERFLIES

DOBSONFLIES AND ALDERFLIES are the most primitive insects that exhibit complete metamorphosis. Forming 2 families with a total of 300 species, they are weak fliers and never move far from water. Adults do not feed, but the aquatic larvae are predatory.

Family CORYDALIDAE

Dobsonflies

Occurrence 200 spp. worldwide, especially in temperate regions; in running water

Measuring 2.5–7.5cm (1–3in) in length, with a wingspan of up to 15cm (6in), dobsonflies are large, soft-bodied, grey or brown insects, with 3 ocelli. Males may have large mandibles, used not for feeding but for combat or for holding the female. The latter lays masses of hundreds of thousands of eggs near water.

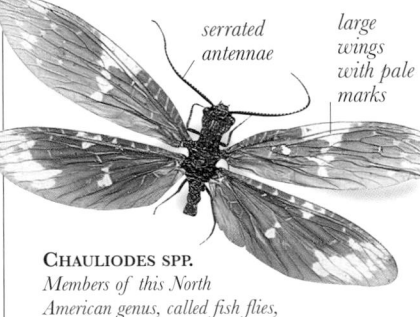

serrated antennae

large wings with pale marks

CHAULIODES SPP.
Members of this North American genus, called fish flies, have distinctly rounded corners on their heads.

Family SIALIDAE

Alderflies

Occurrence 100 spp. worldwide, especially in temperate regions; in muddy ponds, canals, slow-moving water

Alderflies are dark and smoky-winged; most species are less than 1.5cm (½in) long. They have long, thread-like antennae. Unlike dobsonflies they have no ocelli. Females lay egg masses near water, from which hatched larvae drop. These have fully formed legs for crawling, and often mature in a year.

long antennae

square head

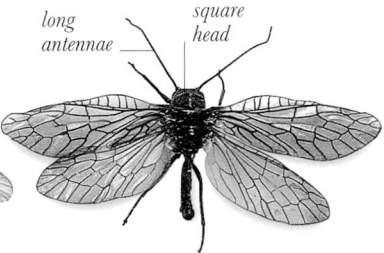

SIALIS SPP.
Common in the northen hemisphere, insects of this genus have typically smoky wings with dark veins. Fully grown larvae crawl out of water and pupate in damp soil above the water margin.

Order RAPHIDIOPTERA

SNAKEFLIES

SNAKEFLIES GET THEIR NAME from the snake-like way in which they catch their prey – raising their head, which is fixed to a neck-like pronotum, and seizing the prey with a sudden lunge. There are 2 families and 150 species of snakeflies. Metamorphosis is complete.

Family RAPHIDIIDAE

Snakeflies

Occurrence 85 spp. primarily in the northern hemisphere; among vegetation

Snakeflies are dark, with a distinct neck formed by a long pronotum. Body length varies between 0.6 and 2.8cm (¼–1¼in). The head is broad in the front and tapered at the rear. Females, slightly larger than males, use their long ovipositor to lay eggs in slits in bark. Larvae are elongate and found under loose bark or leaf litter. Like the adults, they feed on beetle larvae and soft-bodied insects.

AGULLA SPP.
These snakeflies are found in the woodland of the USA and Canada, from the Rocky Mountains to the Pacific coast.

long pronotum

small pterostigma on front edge of wings, towards the tip

net-like pattern of veins

transparent wings

long, slender ovipositor of female

XANTHOSTIGMA XANTHOSTIGMA
This species has the order's characteristic long pronotum, a distinct, broad head that tapers to the rear, and 4 wings of almost equal size. It is found in Europe and Asia.

Order NEUROPTERA

ANTLIONS, LACEWINGS, AND OTHERS

THIS ORDER COMPRISES 17 FAMILIES and about 4,000 species, characterized by large eyes, chewing mouthparts, long antennae, and 2 pairs of filmy, net-veined wings held over the body when at rest. Metamorphosis is complete. The adults are generally predatory and nocturnal, while the larvae are usually predatory or parasitic.

Family CHRYSOPIDAE

Common lacewings

Occurrence 1,600 spp. worldwide; in a range of microhabitats including vegetation, ants' nests

Common lacewings are green or brown, with iridescent wings that span 1–5cm (⅜–2in) and carry a complex pattern of veins. The wings also have ultrasound sensors to detect predators such as bats. Adults are nocturnal and prey upon aphids, thrips, and mites.

NOTHOCHRYSA CAPITATA
Like other lacewings, this European species lays eggs on long stalks. The stalks help to protect the eggs from attack by ants and other predators.

Family MANTISPIDAE

Mantispids

Occurrence 300 spp. worldwide, mainly in warm temperate to tropical regions; on plants in well-vegetated areas

The front legs of mantispids resemble those of praying mantids (see p.554), and are used in the same way. Their wingspan is 1–5.5cm (⅜–2¼in), and they lay eggs on tree bark. The larvae of most species feed on spider eggs.

CLIMACIELLA SPP.
Members of this Central and South American genus have bright, wasp-like, protective coloration.

Family MYRMELEONTIDAE

Antlions

Occurrence 1,000 spp. worldwide, in semi-arid areas in subtropical and tropical regions; in woodland, scrub

Antlions are soft, large, and slender insects which resemble damselflies. The long, narrow wings, which span 3.5–12cm (1½–4¾in), may have brown or black patterns and produce a weak, fluttering flight. The head is broader than the pronotum, with large, conspicuous eyes. The club-headed antennae are about as long as the head and thorax together. The abdomen in males has terminal sexual clasping organs that look like earwigs' forceps. Eggs are laid in soil or sand, singly or in small groups. Adults and larvae catch and eat insects and spiders; some species construct conical pits to trap prey. The larvae, sometimes called doodlebugs, live in these with only their spiny mandibles showing, and flick sand grains at prey to knock them into the pit. Other larvae live on tree trunks, in soil and debris, or under stones.

lemon-yellow coloration

club-headed antennae

black spots on head

large eyes

long, narrow wings

brown patterning on wings

TOMATARES CITRINUS
Named after its distinctive lemon-yellow wings, this species is probably distasteful to predators such as birds. It lives in Africa, like many members of its genus.

slender abdomen

Order COLEOPTERA

BEETLES

ABOUT ONE IN 3 INSECTS IN EXISTENCE today is a beetle. Collectively they make up an order containing 370,000 known species in 166 families, and have successfully colonized every sort of habitat on land and in fresh water. They range in size from insects that are just visible to the naked eye to tropical giants 18cm (7in) long. Beetles are distinguished by their toughened front wings, or elytra, which fold over their membranous hindwings like a case. Because the hindwings are protected, beetles can squeeze into confined spaces. Most species are herbivorous, but the order also includes many predators and scavengers, and some parasites. Beetles develop by complete metamorphosis.

Family ANOBIIDAE

Woodworm

Occurrence 1,500 spp. worldwide, especially in temperate regions; in wood, warehouses, stores, houses

These small, elongated to oval beetles are pale or reddish brown to black, and are mostly 2–6mm (1/16–1/4 in) long. The head is often hooded by the pronotum. Eggs are laid in suitable food, into which the larvae bore tunnels. Several species are pests of cereals, pulses, spices, and tobacco.

branched antennae in males

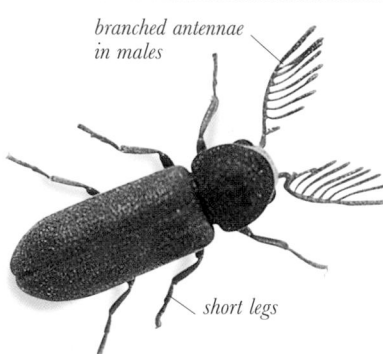

short legs

PTILINUS PECTINICORNIS
Found in Great Britain and central Europe, this species is especially associated with beech trees and may attack wooden furniture.

Family CARABIDAE

Ground beetles

Occurrence 29,000 spp. worldwide; on ground, under stones, logs, in debris, leaf litter, tree foliage

Measuring 0.2–8cm (1/16–3 1/4 in) long, these elongated and flattened beetles can be dull or shiny, brown or black, often with a metallic sheen. The head, thorax, and abdomen are clearly differentiated and the legs are adapted for quick escape from predators. A few species can deter predators using blasts of hot, caustic chemical substances expelled with an audible "pop" from the end of their abdomen. Most species are nocturnal hunters; adults and larvae are mostly predatory. Eggs are laid on the ground, on vegetation, and on decaying wood and fungi.

contrasting markings advertise distastefulness

striated elytra

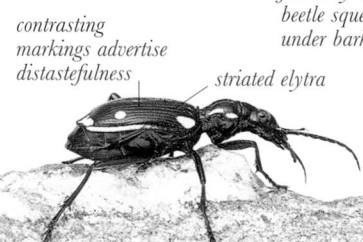

THERMOPHILUM SEXMACULATUM
Commonly called the domino beetle, this Afrotropical species hunts during the day in semi-arid areas.

long, slender legs for fast running

CICINDELLA CAMPESTRIS
Also known as the green tiger beetle, this fast-running, diurnal species can be found from April to September on sandy soils in Europe.

extremely long head

very long, thread-like antennae

flat body lets beetle squeeze under bark

body outline resembles a violin

MORMOLYCE PHYLLODES
The violin beetle is found in the tropical forests of Southeast Asia, and feeds on insect larvae and snails.

Family BUPRESTIDAE

Jewel beetles

Occurrence 15,000 spp. worldwide, primarily in tropical regions; in sunny glades, on bark or flowers

Most jewel beetles are a brilliant, metallic green, red, or blue, with spots, stripes, and bands. They are 0.2–6.5cm (1/16–2 1/2 in) in length and taper towards the rear. They lay eggs in wood – some species have heat sensors that can detect freshly burned areas where they may mate and lay eggs – and the larvae are typically wood-feeders. Many species are timber beetles. Adult beetles feed on nectar, flowers, and pollen. Their bright wing cases are used in headdresses or as ornaments.

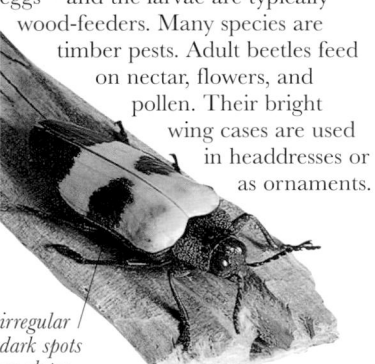

irregular dark spots on elytra

CHRYSOCHROA BUQUETI
This beetle is found in tropical forests of Vietnam and Thailand and is highly prized by collectors. In places this has led to a decline in local populations.

Family CERAMBYCIDAE

Longhorn beetles

Occurrence 30,000 spp. worldwide; on flowers, at sap flows

Also known as timber beetles, these insects are serious pests of wood. Their larvae, responsible for the damage, tunnel through living trees, and also through timber in buildings. Once adult, they feed on sap, pollen, nectar, or leaves. Adult beetles have strong, toothed mandibles, parallel-sided bodies, and are 0.3–15cm (1/8–6in) long. They are often brightly coloured, but their most striking feature is their antennae, which may be up to 4 times as long as the body. Females lay their eggs singly under bark. Like many wood-boring species, the life cycle can take several years to complete.

bright coloration

PHOSPHORUS JANSONI
This species is a native of African forests. It lays eggs in trees, and its larvae attack economically important species such as the cola tree.

strong, toothed mandibles

eyes notched where antennae arise

STERNOTOMIS BOHEMANNI
This Afrotropical species has the typical longhorn shape. The females use their mandibles to cut through bark, under which they lay eggs.

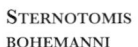

parallel-sided body

XIXUTHRUS HEROS
This large longhorn beetle is 8–10cm (3 1/4–4in) long and inhabits tropical South American forests. The males are territorial.

Family CHRYSOMELIDAE

Leaf beetles

Occurrence 35,000 spp. worldwide; widespread on most types of plant

Although some eat pollen, adults in this family are mostly leaf-eaters – hence the name. This makes many species serious pests, although a few have been used in biological control of weeds. Leaf beetles range from cylindrical and elongated, to round-backed. They measure 1–3cm (3/8–1 1/4 in) in length, with the antennae less than half the body length. Many species are brightly coloured. Eggs are laid in groups on host plants or in soil.

robust antennae

squarish pronotum

SAGRA SPP.
Found in Africa and Asia, the jewelled frog beetles are named for their strong hindlegs, used in male-to-male combat.

curved tibiae

Family COCCINELLIDAE
Ladybirds

Occurrence 5,000 spp. worldwide; almost anywhere on foliage where insect prey is present

Also called lady beetles or ladybugs, these short-legged beetles are brightly coloured, often with spots or stripes. They are 1–15mm (1/32–1/2in) long, rounded and convex, with the pronotum often hiding the head. They are either shiny and smooth, or hairy. Antennae are segmented, with a short, terminal club. Eggs are glued singly or in small groups to plants. Although some species are herbivorous, most adults and larvae are predators of soft-bodied insects. This makes them useful as controlling agents of pests.

hindwings extended only in flight

bright colours deter predators

ANATIS OCELLATA
Also known as the eyed ladybird, this unmistakable, large, European species feeds on jumping plant lice and other small prey.

Family CURCULIONIDAE
Weevils

Occurrence 48,000 spp. worldwide; associated with almost every terrestrial and some aquatic plant species

Habitat Terrestrial

Weevils, also known as snout beetles, form one of the largest families in the animal kingdom. Most have a rostrum, which is an extension of the head and carries the mouthparts. The body is 0.1–9cm (1/32–3 1/2in) long, and may be cryptically or brightly coloured, with patterns and metallic scales. The antennae are usually 11-segmented, and are elbowed or bent. Eggs are laid inside plant tissue or in soil. Mostly diurnal, adults chew and eat soft plants, while larvae mostly feed inside plant tissue. Many weevils are crop pests.

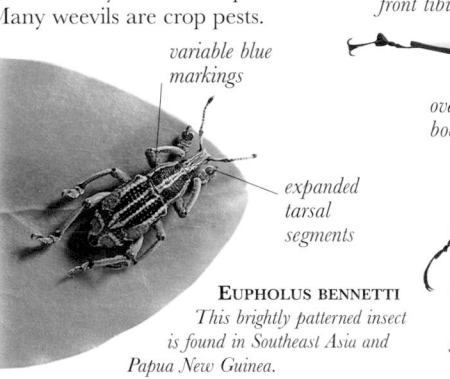

variable blue markings

expanded tarsal segments

EUPHOLUS BENNETTI
This brightly patterned insect is found in Southeast Asia and Papua New Guinea.

elbowed antennae

cryptic markings

CRATOSOMUS RODDAMI
The larvae of this South American weevil bore into the stems of their host plants.

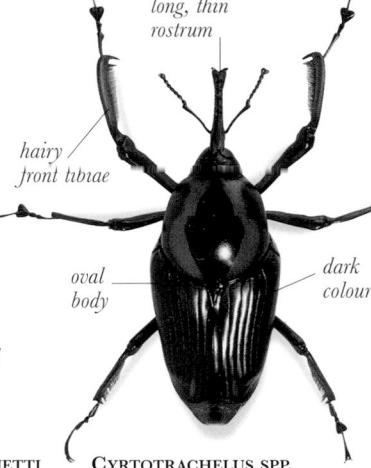

long, thin rostrum

hairy front tibiae

oval body

dark colouring

CYRTOTRACHELUS SPP.
The males of this Southeast Asian genus have long, hairy front tibiae, used in courtship displays.

Family DERMESTIDAE
Larder beetles

Occurrence 950 spp. worldwide; mostly associated with dried animal remains

Many species of larder beetles are serious pests of food or animal matter stored in buildings; larvae scavenge on dried meat or fish, and on plant material, woollens, silks, or furs. These beetles are either rounded or slightly elongated, and measure 2–12mm (1/16–3/8in) in length. Dull brown or black, they are usually covered with patterns of coloured hairs or scales. Females lay up to 150 eggs, in about 2 weeks, on foodstuffs. The hairy larvae feed at a great rate and, in hot countries, mature in a few weeks. The adults feed on pollen and nectar. These beetles are also called skin or museum beetles.

pronotum hides head

DERMESTES LARDARIUS
Found worldwide, this species inhabits bird nests, warehouses, dry corpses, and animal matter. It feeds on dried meat and cheese in houses.

Family DYTISCIDAE
Predacious diving beetles

Occurrence 3,500 spp. worldwide; in streams, shallow lakes and ponds, brackish pools, thermal springs

These beetles are 0.2–5cm (1/16–2in) long, oval, and mostly black or dark brown. They are fierce predators that attack everything from other insects to molluscs or vertebrates. Larvae are called water tigers. Well adapted to aquatic life, the adults can carry air supplies beneath their wing cases.

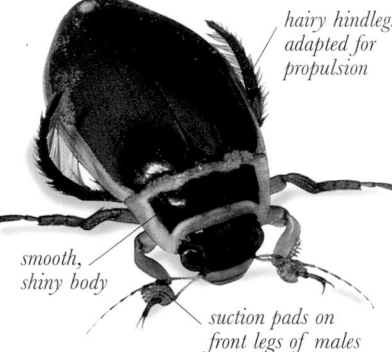

hairy hindlegs adapted for propulsion

smooth, shiny body

suction pads on front legs of males

DYTISCUS MARGINALIS
The giant diving beetle inhabits ponds and shallow lakes in Europe and Asia. The male has suction pads on its front legs for holding onto the female during mating.

Family ELATERIDAE
Click beetles

Occurrence 9,000 spp. worldwide; around plants, in leaf litter, rotten wood, soil

Click beetles, also called skip jacks, are able to propel themselves into the air, making a loud click that frightens predators. Their rapid jump is made possible by powerful thoracic muscles and a unique "peg and joint" catch mechanism on the underside of the thorax. They are dull in colour, elongated to oval, and parallel-sided, tapering at the rear. Body length

ranges from 0.2 to 7cm (1/16–2 3/4in). Eggs are laid in soil and plant matter. The larvae, called wireworms, are generally predatory, although some feed on plant roots and tubers.

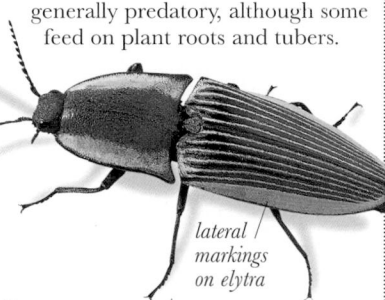

lateral markings on elytra

CHALCOLEPIDIUS LIMBATUS
Found in South America, this species eats plants as well as insects. Larvae develop in soil, litter, or rotting wood.

Family HISTERIDAE
Hister beetles

Occurrence 3,000 spp. worldwide; in dung, carrion, under bark, in tunnels of wood-boring insects, ants' nests

These tough-bodied beetles have a rounded or oval outline; some species may be flattened. Most species are black or metallic, and have striations on the elytra and other body parts. They are mostly less than 1.5cm (1/2in) in length. Eggs are laid in foodstuff. Adults and larvae are predatory and feed on fly and beetle larvae, and mites.

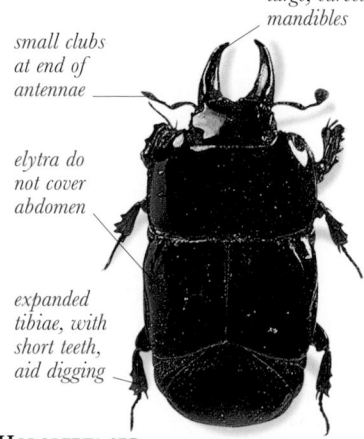

large, curved mandibles

small clubs at end of antennae

elytra do not cover abdomen

expanded tibiae, with short teeth, aid digging

HOLOLEPTA SPP.
Commonly called flat hister beetles, the species in this genus have a very flat and hard body surface. They live under the bark of fallen trees.

Family LAMPYRIDAE
Fireflies

Occurrence 2,000 spp. worldwide; on vegetation in woodland and moist grassland, in soil, under stones

Also known as lightning-bugs, many fireflies use species-specific flashes of cold green light made by luminous organs on their abdomens in order to attract mates. The females of some species even imitate the flashing of closely related species to lure males and eat them. Fireflies are 0.5–3cm (3/16–1 1/4in) long. Males are usually fully winged whereas some females are wingless and may resemble larvae. The larvae eat a small range of snails and other invertebrates. Most adults feed on nectar or dew, although some are predatory.

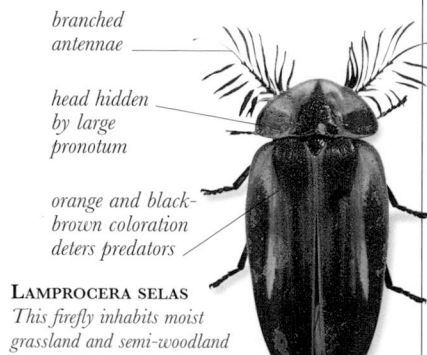

branched antennae

head hidden by large pronotum

orange and black-brown coloration deters predators

LAMPROCERA SELAS
This firefly inhabits moist grassland and semi-woodland in tropical South America.

Order COLEOPTERA *continued*

Family LUCANIDAE

Stag beetles

Occurrence 1,300 spp. worldwide, in deciduous woodland and forests; in and on trees

Stag beetles are smooth, black or reddish brown, and mostly 0.6–8.5cm (¼–3¼in) long. Males are larger than females and have enlarged mandibles used in male-to-male combat over females. Eggs are laid on decaying tree stumps or roots, and larvae pupate inside a cell of chewed wood fibres. Adults are nocturnal and either do not feed, or take fluids such as nectar.

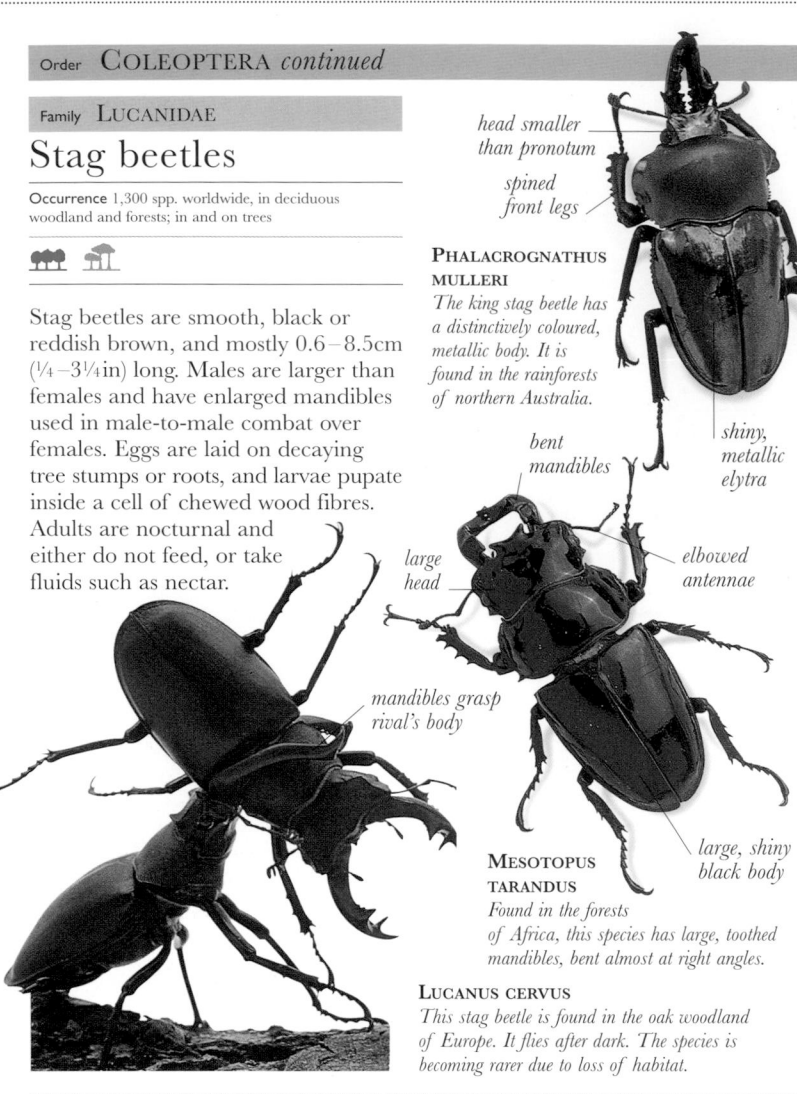

head smaller than pronotum

spined front legs

PHALACROGNATHUS MULLERI
The king stag beetle has a distinctively coloured, metallic body. It is found in the rainforests of northern Australia.

shiny, metallic elytra

bent mandibles

large head

elbowed antennae

mandibles grasp rival's body

MESOTOPUS TARANDUS
Found in the forests of Africa, this species has large, toothed mandibles, bent almost at right angles.

large, shiny black body

LUCANUS CERVUS
This stag beetle is found in the oak woodland of Europe. It flies after dark. The species is becoming rarer due to loss of habitat.

Family SILPHIDAE

Carrion beetles

Occurrence 250 spp. worldwide, mainly in northern hemisphere; on ground near carcasses, dung, and fungi

Carrion beetles are usually flat and soft-bodied, 0.4–4.5cm (⁵⁄₃₂–1¾in) long. They are black or brown, often with bright yellow or red markings. In carcass-burying species, such as *Nicrophorus* (burying beetles), the elytra are shortened, exposing a few abdominal segments. These beetles bury corpses of small animals and lay eggs on them. Adults often feed larvae on regurgitated carrion.

clubbed antennae

flat, broad body

coarse bumps on elytra

SILPHA AMERICANA
Like all members of its family, this beetle is attracted by the odour of carrion. It is found in North America and has a distinctive broad shape with fairly bright coloration.

Family STAPHYLINIDAE

Rove beetles

Occurrence 29,000 spp. worldwide; on ground, in soil, fungi, leaf litter, decaying plants, carrion, ant colonies

Most rove beetles are small – usually under 2cm (¾in) in length – smooth, and elongated. They are often brown or black, although some have bright colours and a sculptured body surface. They all have short elytra, with hindwings folded beneath them, and an exposed abdomen. The smaller species tend to be diurnal, while the larger ones are nocturnal. Eggs are laid in soil, fungi, or leaf litter.

flexible, exposed abdomen

flat head and prothorax

matt black body

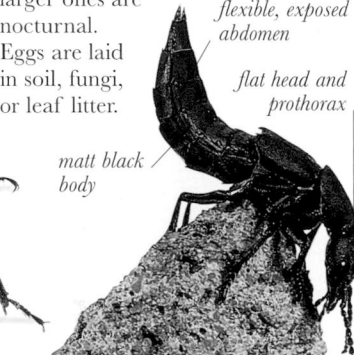

STAPHYLINUS OLENS
Known as the devil's coach horse, this is a large, black European species, 2–3.4cm (¾–1½in) long. If threatened, it assumes a scorpion-like posture by curving its abdomen upwards.

Family PASSALIDAE

Bess beetles

Occurrence 500 spp. in tropical regions, mainly woodland of South America and Asia; in dead wood

Also called patent-leather or betsy beetles, these insects are shiny black or brown, with a flat body that measures 1–8.5cm (⅜–3¼in) in length. The elytra have noticeable striations, and the head often has a short horn. Adults make high-pitched sounds in aggression or during courtship. Eggs are laid in dead wood.

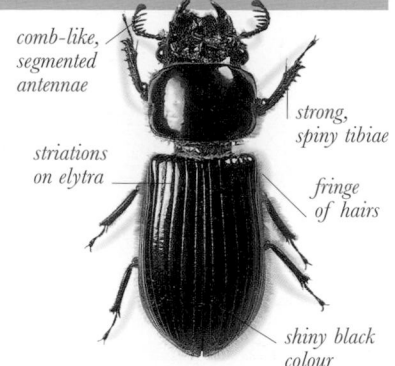

comb-like, segmented antennae

strong, spiny tibiae

striations on elytra

fringe of hairs

shiny black colour

ACERAIUS RECTIDENS
Found in the woodland of Southeast Asia, this beetle has curled, 10-segmented antennae, and a short peg on the head.

Family SCARABAEIDAE

Scarab beetles

Occurrence 16,500 spp. worldwide; on dung, carrion, fungi, vegetation, under bark, in ant or termite nests

Members of this family are extremely varied in shape, size, and colour; subfamilies include dung beetles, Hercules beetles, and leaf chafers. The body length ranges from 0.2 to 17cm (¹⁄₁₆–6½in). The ends of the antennae have a distinct club made up of 3–7 flat, movable plates. Males of many species have an enlarged head or horns to fight over mates. Eggs are laid – and larvae can be found – in soil, rotten wood, or dung (which dung beetles roll into a ball and bury).

shovel-shaped head

red metallic sheen

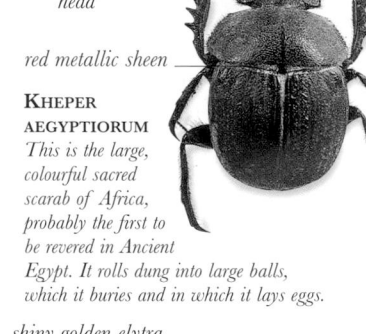

KHEPER AEGYPTIORUM
This is the large, colourful sacred scarab of Africa, probably the first to be revered in Ancient Egypt. It rolls dung into large balls, which it buries and in which it lays eggs.

shiny golden elytra

strong claws

stout hind femora

PLUSIOTUS RESPLENDENS
Sometimes called precious metal scarabs due to their gold or silver colouring, this rare scarab and its relatives are found in South America.

antennal clubs made of plates

golden hairs

folding, membranous hindwings

MELOLONTHA MELOLONTHA
The common cockchafer, sometimes called the May beetle, is found in Europe and temperate Asia. It is attracted to lights at night, and can damage buds of fruit and other trees. Larvae feed on plant roots.

Family TENEBRIONIDAE

Darkling beetles

Occurrence 17,000 spp. worldwide; on ground in all terrestrial habitats, especially in desert and arid regions

Darkling beetles vary considerably in colour (black or brown body, or white elytra), shape (parallel-sided or large and oval), and texture (smooth and shiny or dull and rough). They are 0.2–5cm (¹⁄₁₆–2in) long. Many species have reduced hindwings and do not fly. These beetles eat decaying vegetable or animal matter, although some are pests of stored products. Certain species produce a foul-smelling, defensive secretion.

pronotum rounded at sides

smooth, rounded body

strong legs

pointed elytra

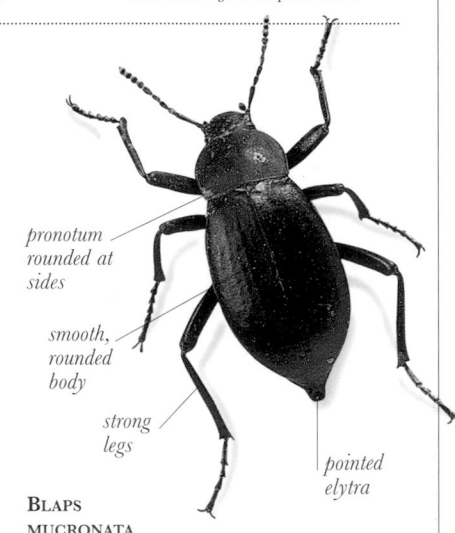

BLAPS MUCRONATA
The cellar or churchyard beetle is a non-flying, European species which favours dark places.

Order STREPSIPTERA

STREPSIPTERANS

WITH THEIR SMALL SIZE, PARASITIC lifestyle, and extreme sex differences, strepsipterans make up one of the most unusual orders of insects. Males usually have wings, the hindwings often being twisted, but females are grub-like, and spend their lives lodged in the bodies of larger insects, such as bugs, bees, or wasps. There are about 560 species, in 8 families, and all undergo complete metamorphosis.

| Family STYLOPIDAE |

Stylopids

Occurrence 260 spp. worldwide; in vegetated habitats where hosts are found, females in bees and wasps

Male stylopids are small and dark, with bulging eyes and bodies that are 0.5–4mm (1/64–9/32in) long. Like all strepsipterans, their forewings are tiny and their hindwings relatively large and fan-like. Females have neither legs nor wings, and never leave the body of the host insect. After mating, thousands of eggs hatch inside the female; some species may lay up to 70,000 eggs. The active, tiny, 6-legged larvae (triungulin) leave the female through a special passage and crawl onto flowers to wait for the next host. They may cling to the host's body or be ingested with nectar. In many cases, once inside the host's nest, they leave the body or are regurgitated and proceed to parasitize the eggs or larvae. The sexual organs of the parasitized host also degenerate.

conspicuous eyes

hindwings twisted in appearance

STYLOPS SPP.
Males of these species have a highly distinctive shape, with branched antennae and the order's characteristic berry-like eyes.

Order MECOPTERA

SCORPIONFLIES

THE MALES OF SOME SPECIES IN THIS ORDER have a scorpion-like, slender abdomen – hence the common name. Metamorphosis is complete. There are 9 families and 550 species of scorpionflies.

| Family BITTACIDAE |

Hangingflies

Occurrence 170 spp. in the southern hemisphere; in damp, wooded or well-vegetated, shady areas

Hangingflies use their front legs to hang from vegetation and hindlegs to catch prey. The adults are 1.5–4cm (1/2–1 1/2in) long. Eggs are laid in soil.

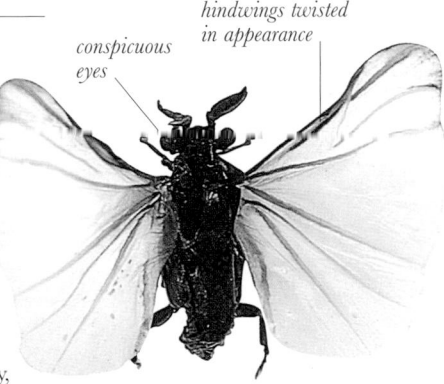

spine at end of tibiae

slender brownish wing

long abdomen

HARPOBITTACUS AUSTRALIS
This is an Australian species with brownish wings and orange-red banded legs.

hind tarsi can fold around prey

| Family PANORPIDAE |

Common scorpionflies

Occurrence 360 spp. mostly in the northern hemisphere; among vegetation in woodland and scrub

These insects are 0.9–2.5cm (11/32–1in) long, often with mottled wings. They lay eggs in soil or damp leaf litter; the larvae pupate underground. Adults eat nectar, fruits, and dead insects.

slender antennae

brown wings with white patches

PANORPA LUGUBRIS
This species is common in parts of North America. Males are larger than the females.

upturned, swollen genitalia in males

Order SIPHONAPTERA

FLEAS

FLEAS ARE BROWN, SHINY, WINGLESS INSECTS, superbly adapted for a parasitic lifestyle feeding on mammalian blood. Their bodies are flattened sideways, which helps them to slip through fur, and they are very tough, which makes them hard to kill. They leap onto suitable host animals using their enlarged hindlegs, which store energy in special pads of rubber-like protein. There are about 2,000 species (18 families) of fleas; all undergo complete metamorphosis. Larvae scavenge on debris, dried blood, and adult flea excrement.

| Family PULICIDAE |

Common fleas

Occurrence 200 spp. worldwide; on mammalian hosts, in lairs, burrows

Common fleas parasitize humans and a wide range of other mammals, including dogs, cats, and other carnivores, as well as hedgehogs, rabbits, hares, and rodents. Some of them have a broad host range, occurring on up to 30 different animal species. Common fleas are 1–8mm (1/32–5/16in) long and often have combs of backward-pointing bristles on their cheeks – features that are an important aid in species identification. Adult females lay eggs as they feed, dropping them into their hosts' nests, burrows, or bedding, where they hatch to produce pale, worm-like larvae that have tiny, biting jaws. The larvae typically live for 2–3 weeks, before pupating in a minute silk cocoon. Once they have become adult, the fleas remain inside their cocoons for weeks or months until they sense vibrations from a nearby animal. They then emerge from their cocoons and jump aboard their prospective host, attracted by its body heat. However, if they fail to find a host, or become dislodged, they can survive for a long time without food. Many species of common fleas, such as the cat flea, dog flea, and human flea, are of medical importance. Their bite causes severe itching and allergic reactions, as well as spreads diseases and parasitic worms. The bacterium that caused bubonic plague in medieval Europe was carried by various types of rat flea.

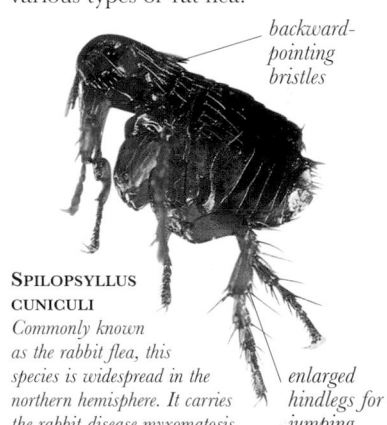

backward-pointing bristles

SPILOPSYLLUS CUNICULI
Commonly known as the rabbit flea, this species is widespread in the northern hemisphere. It carries the rabbit disease myxomatosis.

enlarged hindlegs for jumping

eyes touch cheek combs

CEDISPSYLLA SIMPLEX
This species is found on cotton-tail rabbits in North America.

spiny hindlegs

PULEX IRRITANS
This species was once the most common flea attacking humans; today, cat and dog fleas are more dominant. It also attacks pigs, goats, foxes, and badgers, and can be a carrier of bubonic plague.

CTENOCEPHALIDES CANIS
Very common in human habitations the world over, the dog flea is reddish brown, with a head that is more domed than that of the cat flea. It is a carrier of the dog tapeworm, which can also affect cats and humans.

CTENOCEPHALIDES FELIS
The cat flea is a common domestic species found in temperate regions. A cat may have only a few adult fleas feeding on it but its bed and lair support thousands of larvae. Hungry cat fleas can jump up to 34cm (13 1/2 in) and readily bite humans.

INVERTEBRATES

Order DIPTERA

FLIES

FLIES HAVE ONLY A SINGLE PAIR of wings (although some ectoparasitic species are wingless). The hindwings are reduced to small, club-shaped balancing organs called halteres. Flies are ecologically essential as pollinators, parasites, predators, and decomposers, but many damage crops, and the disease-carrying species have a huge impact on wild and domesticated animals, as well as on humans.

There are 130 families and 122,000 species of flies, divided into 2 suborders, Nematocera (26 families) and Brachycera (104 families). The former are delicate, with thread-like antennae and slender bodies, while the latter are more robust, with stout antennae.

Family AGROMYZIDAE

Leaf-mining flies

Occurrence 2,500 spp. worldwide; on leaves, stems, seeds, roots of a wide range of plants, including crops

Habitat Terrestrial

This family gets its name from the ability of larvae to chew mines (channels) in leaves. These crop pests also feed inside stems, seeds, or roots. Grey, black, or greenish yellow, they vary from 1 to 6mm ($\frac{1}{32}$–$\frac{1}{4}$in) in length. Females have a pointed ovipositor, and lay up to 50 eggs a day inside plant tissue. These flies may live for as little as 2 weeks in tropical areas.

smoky tint on wings

HEXOMYZA SPP.
These species occur throughout the northern hemisphere, and in South Africa and Australia. The larvae make galls on the twigs of poplars.

Family CALLIPHORIDAE

Blow flies

Occurrence 1,200 spp. worldwide; on flowers, vegetation, carcasses, cooked and raw food

---

Bluebottles and greenbottles are familiar members of this widepread and sometimes troublesome family of flies. Like their many relatives, they have stout, bristle-covered bodies, up to 1.5cm ($\frac{1}{2}$in) long, a metallic sheen, particularly on the abdomen, and a noisy, buzzing flight. Blow flies breed in decaying animal matter, dung, or rotten fish. Adults feed on liquids from rotting material, nectar, or fruits. The larvae of some species are predators of ants, termites, and the larvae and eggs of other insects. Several species burrow into human flesh and many carry diseases which they can transmit to livestock or humans.

metallic blue body

stout bristles

minute claws

POLLENIA RUDIS
Common in the northern hemisphere, cluster flies congregate in large numbers in attics and unheated buildings. In the spring adult females lay their eggs on earthworms or near fresh worm casts.

black, bristly thorax

sponge-like mouthparts for lapping up fluids

metallic blue abdomen

CALLIPHORA VOMITORIA
This bluebottle species is common in the countryside in the northern hemisphere. Its females may lay many hundreds of eggs in their lifetime.

CALLIPHORA VICINA
This widespread species is common in towns and cities, where the maggots develop inside the corpses of dead rats, mice, and pigeons.

Family ASILIDAE

Robber flies

Occurrence 5,000 spp. worldwide, especially in tropical and subtropical regions; in a variety of microhabitats

---

This is a family of predatory flies; both the adults and larvae are predatory although the latter eat decaying matter as well. Robber flies use their forward-pointed, stout and sharp proboscis to stab prey and inject a paralysing saliva. They then suck up the body contents of the paralysed insect. They are 0.3–5cm ($\frac{1}{8}$–2in) long, either slender or stout and bee-like. Their head is slightly hollow between the eyes, and the face has a tuft of long hairs. Eggs are laid in soil, rotting wood, or inside plants.

sharp proboscis

broad wingspan

stout, bristly legs

BLEPHAROTES SPLENDIDISSIMUS
This large Australian fly has plate-like tufts of hair at the sides of its flat abdomen.

Family CECIDOMYIIDAE

Gall midges

Occurrence 5,000 spp. worldwide; anywhere near decaying matter, fungi, a wide range of plants

---

With bodies typically less than 4mm ($\frac{5}{32}$in), gall midges are seldom seen, but their larval homes – plant galls – are much easier to spot. Looking like miniature fruits or malformed buds, the galls are growths triggered by some gall midges, for protection and as a source of food. Like gall wasps (see p.574), gall midges are species-specific, always attacking particular kinds of plants. Adult gall midges have long, slender legs, and wings with a few unbranched veins. A number are serious pests of crop plants.

thread-like antennae

CECIDOMYIA SPP.
Each species in this genus makes a characteristic gall.

very long, thin legs

Family CERATOPOGONIDAE

Biting midges

Occurrence 4,000 spp. worldwide, mainly in the northern hemisphere; near margins of water bodies

Habitat All

Also called punkies or no-see-ums, these small flies – under 5mm ($\frac{3}{16}$in) long – have a bite that causes severe irritation. They have short, strong legs, dark patterns on their wings, and piercing mouthparts shaped to suck blood and insect body fluids. They mate while flying in a swarm. Eggs are laid in groups of 30–450, and are covered with a protective, jelly-like substance. Some biting midges transmit worm parasites to humans or carry animal diseases. They are pests but are also important pollinators of tropical crops such as cocoa or rubber.

Family BOMBYLIIDAE

Bee flies

Occurrence 5,000 spp. worldwide, in open, tropical, or subtropical semi-arid regions; around flowers, on ground

---

Bee flies are stout and hairy insects that strongly resemble bumble bees (see p.573). Between 0.2 and 3cm ($\frac{1}{16}$–$1\frac{1}{4}$in) long, they are brown, red, or yellow, sometimes with bright markings. They have narrow wings, and when flying produce a high-pitched whine. Like bees, they have a long proboscis that they use to suck nectar from flowers. However, instead of landing on flowers as bees do, bee flies hover in front of them, steadying themselves against the petals with their front legs. Some bee fly larvae eat grasshopper eggs, but most are parasitic and attack beetles, moths, wasps, or other flies.

long proboscis

stout, hairy body

BOMBYLIUS MAJOR
This common species inhabits meadows, hedgerows, and open areas from Europe to China, and North Africa.

CULICOIDES IMPUNCTATUS
This notorious, European species breeds in boggy areas. The female's bite is extremely itchy and painful.

Family CHIRONOMIDAE
Non-biting midges

Occurrence 5,000 spp. worldwide; near ponds, lakes, and streams; swarms occur at dusk

These delicate, pale brown, or greenish flies resemble mosquitoes, but do not have functional mouthparts. They are 1–9mm ($\frac{1}{32}$–$\frac{11}{32}$in) long. The males have a slender body and feathery antennae, while the females are stout, with hairy antennae. Most of the 2–3-year life cycle is spent as larvae; adults live no longer than 2 weeks. They mate on the wing in a swarm, and lay their eggs in a mass of sticky

jelly on water or plants. The larvae are plentiful in fresh water; up to 100,000 larvae have been recorded in a single square metre (11 square ft).

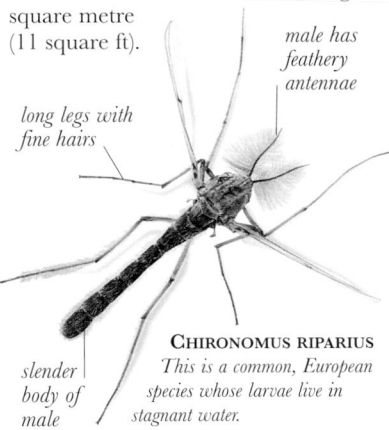

male has feathery antennae

long legs with fine hairs

slender body of male

CHIRONOMUS RIPARIUS
This is a common, European species whose larvae live in stagnant water.

Family CULICIDAE
Mosquitoes

Occurrence 3,100 spp. worldwide, especially in warmer regions; everywhere near water

Habitat All

Mosquitoes are among the world's most dangerous pests, spreading malaria, dengue, and several other potentially fatal diseases. They are 0.3–2cm ($\frac{1}{8}$–$\frac{3}{4}$in) long, with narrow bodies, a humped thorax, and slender legs. Males often feed at flowers, but females feed on blood

chiefly from mammals and birds – and have syringe-like mouthparts for piercing skin. They lay their eggs on stagnant water.

humped thorax

long proboscis

broad abdomen

CULEX SPP.
These mosquitoes rest with their bodies parallel to the surface; the malaria-carrying Anopheles *species rest in a head-down position.*

Family DROSOPHILIDAE
Pomace flies

Occurrence 2,900 spp. worldwide; near decaying vegetation, fruits, fungi, fermenting liquids

Habitat All

Also known as lesser fruit flies or vinegar flies, these small, yellow, brown to reddish brown, or black species have light or bright red eyes. The thorax and the abdomen may be striped or spotted. Pomace flies are 1–6mm ($\frac{1}{32}$–$\frac{1}{4}$in) long. Males make a buzzing sound with their wings to attract mates. Eggs are laid near a food source at the rate of 15–25 eggs per day. Larvae feed on bacteria and fungi, while the adults look for fermenting liquids and rotting fruits.

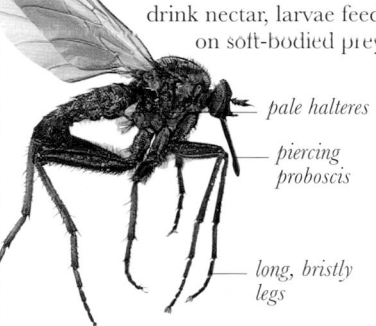

translucent wings

brownish coloration

DROSOPHILA MELANOGASTER
The best-known pomace fly, the laboratory fruit fly is often used for genetics studies as it breeds quickly and has large chromosomes in its salivary glands.

Family EMPIDAE
Dance flies

Occurrence 3,000 spp. worldwide, mainly in the northern hemisphere; adults on tree trunks, larvae in rotten wood

The common name of this family comes from the dancing motions of mating swarms. Dance flies are 1.5–15mm ($\frac{3}{64}$–$\frac{1}{2}$in) long. Most species have a stout thorax, an elongated abdomen, a rounded head with large eyes, and a downward-pointing, sharp proboscis with which they can stab prey. Courtship often involves males offering prey as food to females. Eggs are laid on soil, dung, or plant matter. While adults eat small flies and drink nectar, larvae feed on soft-bodied prey.

pale halteres

piercing proboscis

long, bristly legs

EMPIS SPP.
These flies inhabit damp vegetation. The long proboscis enables the fly to "stab" its prey and extract fluids on which it feeds.

Family HIPPOBOSCIDAE
Louse flies

Occurrence 200 spp. worldwide, mainly in tropical and subtropical regions; ectoparasites of birds and mammals

These dull brown, stout, flat, and hairy flies, mostly 4–7mm ($\frac{5}{32}$–$\frac{9}{32}$in) long, have a short proboscis and strong, clawed legs for gripping hairs or feathers. Some have fully formed wings, but in others the wings are vestigial, or absent altogether. Females lay mature larvae (prepupae) one at a time; these develop inside a "uterus", feeding on secretions from special "milk" glands. Adult louse flies are blood-suckers.

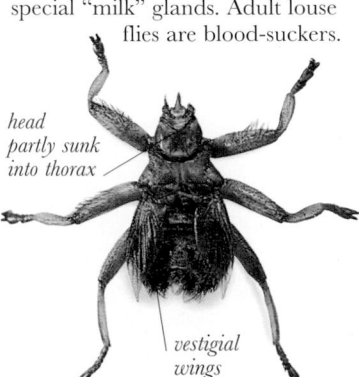

head partly sunk into thorax

vestigial wings

CRATAERINA PALLIDA
Also called the swift louse fly, this European species is parasitic on swifts in their nests. Three-quarters of all louse fly species parasitize birds.

Family MUSCIDAE
Muscids

Occurrence 4,000 spp. worldwide; on flowers, excrement, decaying matter, near mammalian hosts

Habitat Terrestrial

The muscid family includes one of the world's most successful insects – the common house fly – together with its many relatives. Muscids are usually drably coloured, with bristly bodies up to 1.2cm ($\frac{3}{8}$in) long, and mouthparts that are sponge-like for lapping fluids, or piercing for sucking blood. Female houseflies can lay 100–150 eggs per day in rotting matter, excrement, fungi, water, or plants. Maggots grow fast and, in ideal conditions, pupate in just over a week. These flies can carry infections such as typhoid and cholera.

yellowish base to wings

yellow femora

fine bristles

MYDAEA CORNI
This species, found in Europe and Asia, inhabits dung. The larvae are predators on small organisms inside the dung.

MUSCA DOMESTICA
The common house fly is found throughout the world, and spreads bacterial and viral diseases.

slender legs

orange patches on sides of abdomen

large eyes

translucent wings

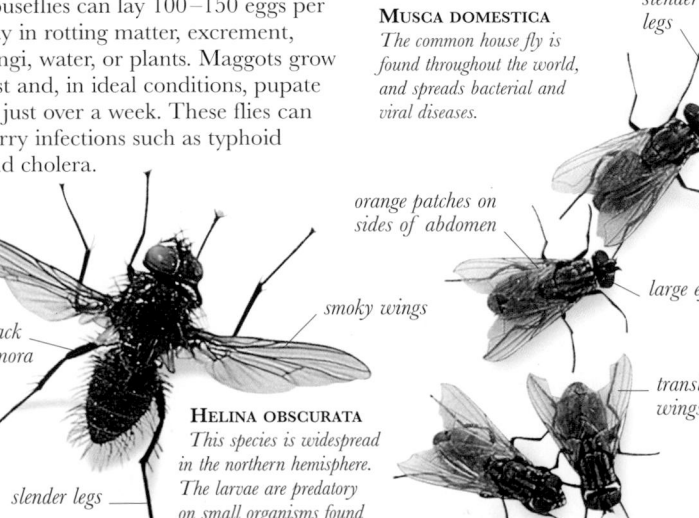

black femora

smoky wings

HELINA OBSCURATA
This species is widespread in the northern hemisphere. The larvae are predatory on small organisms found in soil and humus.

slender legs

Family MYCETOPHILIDAE
Fungus gnats

Occurrence 3,000 spp. worldwide; in swamps, bogs, moist places in forest, forest fungi

Fungus gnats are small, mosquito-like flies, usually brown, black, or yellowish in colour; some species are brightly coloured. The thorax is characteristically humped, and the legs are long and slender. The body is 2–14mm ($\frac{1}{16}$–$\frac{1}{2}$in) long. Eggs are laid in fungi, on or under bark, on cave walls, and in nests. The larvae eat fungal material and, sometimes, tiny insects and worms. Pupation occurs inside the food or in a silk cocoon. Adults may feed on fluids. Some species are pests in mushroom cultivation.

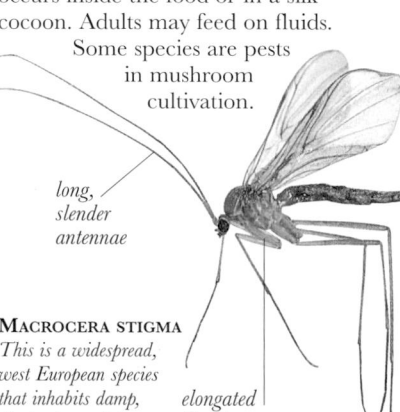

long, slender antennae

elongated first leg segments

MACROCERA STIGMA
This is a widespread, west European species that inhabits damp, shady places. It has hairs on its wings.

Order DIPTERA *continued*

Family OESTRIDAE
Bot flies and warble flies

Occurrence 70 spp. worldwide, especially in the northern hemisphere and Africa; near mammalian hosts

Many members of this heavy-bodied family look like bees. They are 0.8–2.5cm (5/16–1in) long. Because the adults are short-lived and do not feed, the mouthparts are either tiny or absent. The larvae are all intestinal parasites of mammals, especially horses, goats, and camels. The females of some species lay larvae directly in the nostrils of hosts, where they feed and grow in the sinuses. In warble flies, eggs are laid on the hair of the host, and the larvae burrow under the skin. Fully grown larvae are sneezed out by the host or chew their way out and pupate in the soil.

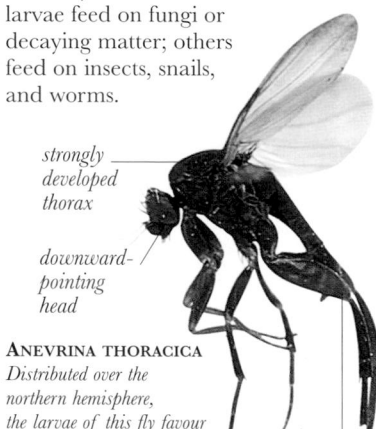

hairy body

small antennae in grooves on head

HYPODERMA BOVIS
The ox warble fly or northern cattle grub is common in the northern hemisphere, widespread near cattle. It damages hides, and reduces animal vigour and milk production.

Family PHORIDAE
Scuttle flies

Occurrence 3,000 spp. worldwide; in a wide variety of microhabitats

Habitat All

Also known as hump-backed flies, the brown, black, or yellowish scuttle flies have characteristically fast, jerky movements while running. They are 0.5–6mm (1/64–1/4in) long with a strongly down-turned head. The antennae appear to have only one segment and the hind femora are often very stout. Conspicuous bristles are often present on the body; these look feathery when highly magnified. Adults live from a few days to a week, and may take fluids. Some of the larvae feed on fungi or decaying matter; others feed on insects, snails, and worms.

strongly developed thorax

downward-pointing head

ANEVRINA THORACICA
Distributed over the northern hemisphere, the larvae of this fly favour soil, the corpses of animals, and moles' nests.

strong femora

Family PSYCHODIDAE
Sand flies and moth flies

Occurrence 1,000 spp. worldwide, especially in warmer regions; in water, rotting matter, moist, shady habitats

These flies measure 1.5–5mm (3/64–3/16in) in length. Their bodies, wings, and legs are covered with long hairs or scales. The broad wings are held tent-like or partly spread in moth flies, or together above the body in sand flies. Sand and moth flies have a weak, fluttery or hopping flight pattern. Female sand flies are blood-suckers and, particularly in the tropics, several species spread diseases.

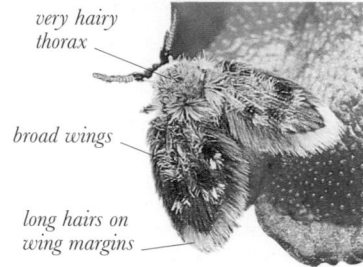

very hairy thorax

broad wings

long hairs on wing margins

PERICOMA FULIGINOSA
The larvae of this dark, widely distributed moth fly are, like sand flies, saprophagous. They develop in shallow water and in rot holes in trees.

Family SARCOPHAGIDAE
Flesh flies

Occurrence 2,300 spp. worldwide, especially in the northern hemisphere; foliage, flowers, carrion

Some of these robust-looking flies lay larvae in body cavities or wounds in vertebrates, including humans. They are mostly dull or silvery grey or black, and are 0.2–2cm (1/16–3/4in) long. The thorax typically has 3 dark longitudinal stripes on a grey background, and the abdomen appears chequered or spotted. Most females give birth to larvae which are laid or dropped in flight. The larvae are very varied, and live on a range of different foods. Some are predators, attacking aphids, sawfly larvae, and other soft-bodied insects, while others feed on rotting vegetation or animal dung. A few live as scavengers in the nests of bees and wasps, where they eat the dead larvae and pupae of many species feed on carrion, and some are parasitic on other insects, spiders, snails, and worms.

stout bristles

pale markings

dark hairs

male genitalia

SARCOPHAGA MELANURA
This species is common near coastal areas in western Europe. Its larvae are found in rotting meat and may parasitize snails or insects.

Family SIMULIIDAE
Black flies

Occurrence 1,600 spp. worldwide; larvae in flowing water, adults on plants near water

Black flies, also called buffalo gnats or turkey gnats, are 1–5mm (1/32–3/16in) long, and usually black, orange-red, or dark brown. Males drink nectar and water while females feed on vertebrate blood. They lay their eggs on rocks or vegetation near water. The aquatic larvae, attached to submerged rocks or plants by a hook-like holdfast, filter food from the water with special bristles around the mouth.

SIMULIUM SPP.
Although known as black flies, some members of this genus are not entirely black. Many species are carriers of the roundworm that causes river blindness in Africa and tropical America.

Family SYRPHIDAE
Hover flies

Occurrence 6,000 spp. worldwide; common on flowers, especially umbelliferous ones

Hover flies include some of the most skilled fliers in the insect world. They feed on pollen and nectar, and can often be seen hovering over flowers, or darting after each other in a high-speed chase. Ranging from 1 to 2cm (3/8–3/4in) in length, they have large eyes and slender wings, and are often coloured yellow or orange and black. This coloration – together with their overall shape – makes them look very much like stinging bees or wasps, and deters many predators from attacking them. Hover fly larvae

eyes in females well separated

VOLUCELLA ZONARIA
This migratory species is distributed mainly over Europe. It is stout and distinctively banded. Its larvae scavenge inside wasps' nests.

distinctive, broad yellow bands

SYRPHUS RIBESII
Common in Europe, this wasp-like species often gathers in large numbers, in flower-rich meadows. Its larvae eat greenflies.

short antennae

false margin

their hosts. The larvae of one group, known as drone flies (*Eristalis* spp.), live in stagnant water, and use a telescopic, posterior "snorkel" to breathe.

SERICOMYIA SILENTIS
This species is found on acid heathland in Europe. The larvae live in boggy pools, such as those formed after cutting peat, and feed in the decaying rhizomes of reedmace.

eyes in males usually meet on top of head

wasp-like shape and colouring

wings with false veins

broad wedges of yellow at sides of abdomen

Family TABANIDAE

Horse flies

Occurrence 4,000 spp. worldwide; near mammals; many favour damp or wet habitats

Horse flies are stout and hairless with a distinctive, flattened, hemispherical head and colourful, patterned eyes. Most are black, grey, or brown, with brightly marked abdomens, and measure 0.6–2.8cm (¼–1¼in) in length. Females have blade-like mouthparts for cutting skin; they feed on the blood of mammals and birds. Eggs are typically laid in a mass of 200–1,000 on debris overhanging water, leaves, or rocks. Larvae are predatory or saprophagous.

large eyes occupy most of head

stout, dark body

TABANUS SPP.
These dark and robust flies often mimic large bees, such as the carpenter bee, in order to deter predators.

parallel-sided abdomen

TABANUS SUDETICUS
This horse fly has a bee-like abdomen with distinctive, pale, triangular marks down the middle. Some species transmit diseases to animals as well as humans.

CHRYSOPS RELICTUS
This fly inhabits bogs and ponds in heather moorland in the northern hemisphere. Its bite is very painful.

Family TEPHRITIDAE

Fruit flies

Occurrence 4,500 spp. worldwide; on a range of plants, crops, rotting matter

Habitat Terrestrial

Most fruit flies are under 1.5cm (½in) long and recognizable by their distinctive wing patterns, which can take the form of bands, patches, and zigzag markings. Females have a pointed ovipositor which can be longer than the rest of the body. Males display to females by waving their wings. Eggs are laid singly or in groups under the surface of fruits. Adults take fluids, plant sap, nectar, or liquid from rotting matter; the larvae are herbivorous. Many species are crop pests.

patterned wings

broad abdomen

ICTERICA WESTERMANNI
Found in hedgerows, waste ground, and grassy places in Europe, this species of fruit fly feeds on the flowerheads of ragworts.

Family TIPULIDAE

Crane flies

Occurrence 15,000 spp. worldwide; usually in moist microhabitats

Habitat Terrestrial

Also known as daddy-long-legs, most of these fragile, slender, and elongated flies measure 0.6–6cm (¼–2¼in) in length. They shed their legs very easily if caught. The adults of many are short-lived and feed on nectar and other fluids. The tough-bodied larvae, called leatherjackets, live in soil, rotting wood, bird nests, and bogs, and are an important food for many animals and birds.

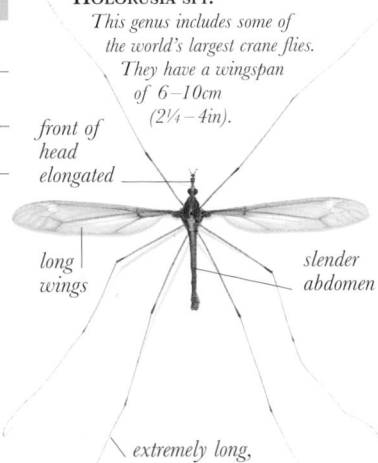

HOLORUSIA SPP.
This genus includes some of the world's largest crane flies. They have a wingspan of 6–10cm (2¼–4in).

front of head elongated

long wings

slender abdomen

extremely long, fragile legs

Family TACHINIDAE

Parasitic flies

Occurrence 8,000 spp. worldwide; where insect hosts are found

Stout and robust, parasitic flies are 0.5–2cm (³⁄₁₆–¾in) long, and are very variable in appearance. Most are dark; some are metallic or pale. The abdomen is very bristly, especially towards the rear. Many species look like bristly house flies (see p.567) and some larger species are almost bee-like. Adults feed on sugary fluids, but their larvae are endoparasites of insects. Some species locate crickets to parasitize by listening out for their mating songs. Males and females often congregate on hilltops for mating, and the eggs are laid on foliage eaten by hosts, or inside the hosts' bodies. Parasitic flies are mostly diurnal. Many species are used as biological control agents of herbivorous insect pests.

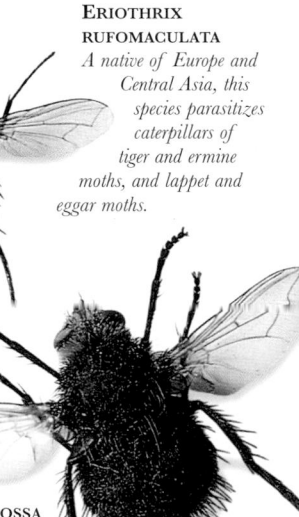

ERIOTHRIX RUFOMACULATA
A native of Europe and Central Asia, this species parasitizes caterpillars of tiger and ermine moths, and lappet and eggar moths.

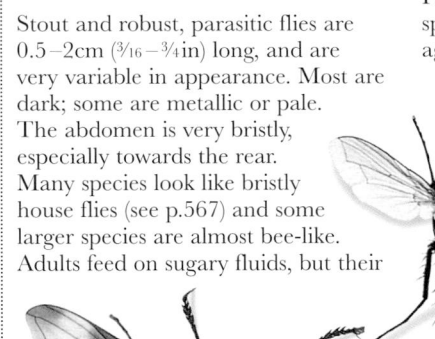

mottled, angular wings

bristles on body

PHASIA HEMIPTERA
An inhabitant of meadows and woodland in parts of Europe, this species parasitizes shield bugs.

TACHINA GROSSA
Larger than most of its relatives, this fly measures 1.9cm (¾in) in length. It is found in Europe and Asia.

bristles on end of body

Order TRICHOPTERA

CADDISFLIES

THERE ARE ABOUT 8,000 SPECIES of caddisfly, in 43 families, and they can be found almost anywhere there is fresh water. The adults are slender, dull-coloured insects, with a hairy body and wings, and long antennae. The larvae are soft-bodied, and live in water, often inside a protective, portable case, which each species builds in a characteristic way. They develop by complete metamorphosis.

Family LIMNEPHILIDAE

Northern caddisflies

Occurrence 1,500 spp. mainly in the northern hemisphere; around lakes, temporary pools, ditches

Northern caddisflies are reddish, yellowish, or dark brown, with dark wing markings, mostly under 2.4cm (1in). Larvae eat organic detritus and algae, and make cases from sand, pebbles, vegetation, or snail shells. Adults may feed on liquids.

LIMNEPHILUS LUNATUS
A widespread species, its larvae are found in a wide variety of freshwater habitats. Coloration ranges between drab shades of black and brown.

Family PHRYGANEIDAE

Large caddisflies

Occurrence 450 spp. mainly in the northern hemisphere; near ponds, lakes, bogs, slow-flowing streams, rivers

Members of this family have light brown or grey markings, and may look mottled. They are 1.2–2.6cm (³⁄₈–1in) long and have at least 2 tibial spurs on the forelegs and 4 on the mid- and hindlegs. Larvae make cases out of plant fragments and fibres.

PHRYGANEA GRANDIS
This is the largest caddisfly found in Great Britain. The male, smaller than the female (shown here), lacks the distinctive, dark stripe on the wing.

INVERTEBRATES

Order LEPIDOPTERA

MOTHS AND BUTTERFLIES

WITH OVER 165,000 SPECIES IN 127 families, moths and butterflies are among the most diverse insects in the world. They have minute, overlapping scales on their bodies and wings, and mouthparts in the form of a proboscis. There is no clear-cut difference between butterflies and moths – and no scientific basis for separating them – but as a general rule, butterflies are brightly coloured and day-flying, with club-ended antennae, while moths are nocturnal and drab. In both, courtship often involves airborne scents and displays. The larvae, called caterpillars, have cylindrical bodies and chewing jaws. They develop by complete metamorphosis, often forming a pupa (or chrysalis) inside a silk cocoon.

Family ARCTIIDAE

Tiger and ermine moths

Occurrence 2,500 spp. worldwide; in well-vegetated areas where host plants occur

Habitat Terrestrial

The hairy and heavy-bodied tiger moths are brightly coloured, while ermine moths tend to be pale with black spots or patches. They have a wingspan of 2–7cm (¾–2¾in).

Females lay eggs on or around host plants. The caterpillars are herbivorous, and many are poisonous. Adults take nectar and liquids.

pale orange margin on hindwings

ARCTIA CAJA
Also called the garden tiger, this moth is highly unpalatable, and its bright, contrasting coloration advertises this to predators.

Family HESPERIIDAE

Skippers

Occurrence 3,000 spp. worldwide except New Zealand; in open habitats such as cultivated fields and grassland

Heavy-bodied and moth-like, these butterflies get their name from their rapid, darting flight. Most are drab brown with white or orange markings, but a few can be brightly coloured. Their wings span 2–8cm (¾–3¼in). The antennae end in a long, curved, and pointed club. Females lay single eggs on host plants. Caterpillars are herbivorous and feed at night.

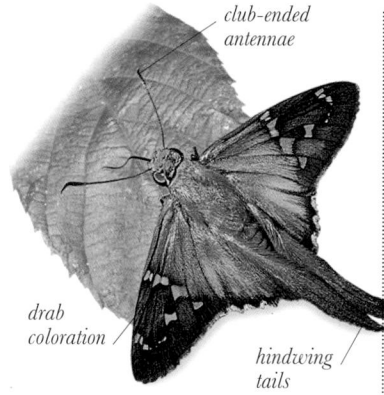

club-ended antennae

drab coloration

hindwing tails

URBANUS PROTEUS
The long-tailed skipper is common in parts of South and North America. It has characteristically long hindwing tails and an erratic flight pattern.

Family LASIOCAMPIDAE

Lasiocampids

Occurrence 2,000 spp. worldwide except New Zealand; on foliage of a wide range of deciduous trees

These insects, also known as lappet and eggar moths, are heavy-bodied and very hairy; most are yellowish brown, brown, or grey. The wingspan is mostly under 4cm (1½in); females are bigger than males. Eggs are laid in a band of 100–200 around a twig of a host tree. The stout caterpillars may live communally in silk tents or webs

which catch the sun's rays. They pupate inside tough, egg-shaped, papery cocoons.

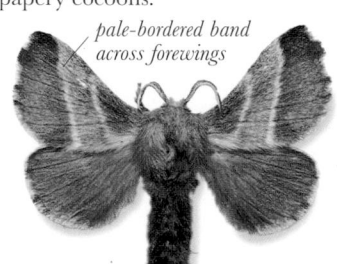

pale-bordered band across forewings

MALACOSOMA AMERICANUM
Also called the eastern tent caterpillar moth, this is an orchard pest. It inhabits forest and woodland areas in the northern USA and southern Canada.

Family BOMBYCIDAE

Silk moths

Occurrence 100 spp. in tropical areas of Southeast Asia; on mulberry trees, other plants

Long valued as a source of silk, these stout, pale cream, grey, or brown moths have wings that span 2–6cm (¾–2¼in). The smooth caterpillars have prolegs on all abdominal segments and usually a short, horn-like structure at the tail. Caterpillars of the commercial silk moth feed on mulberry; other species feed on fig and related plants. Pupation takes place inside a dense silk cocoon. Functional mouthparts are absent in adults, which do not feed.

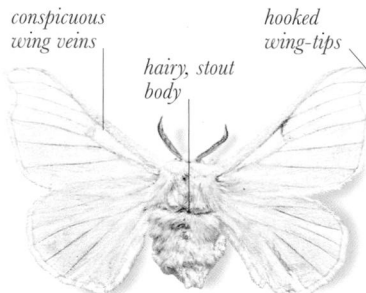

conspicuous wing veins

hooked wing-tips

hairy, stout body

BOMBYX MORI
The silk moth originated in China but is now cultivated in silk farms the world over. It is, however, extinct in the wild.

Family LYCAENIDAE

Blues, coppers, and hairstreaks

Occurrence 6,000 spp. worldwide, particularly in warmer regions; in association with ant nests or host plants

The butterflies in this family are generally small, with a wingspan of 1.5–5cm (½–2in), but they are notable for colour differences between males and females. Males often have brilliantly coloured, iridescent wing surfaces, while females are generally dull. Most species have squat caterpillars, which feed on plants or small insects. Some produce a special fluid that is eaten by ants – in return the ants guard the caterpillar, and sometimes take it into their nest.

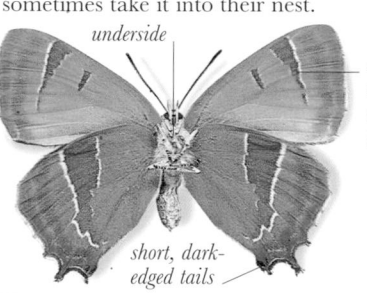

underside

orange markings on forewings of female

short, dark-edged tails

THECLA BETULAE
The brown hairstreak butterfly inhabits woodland in Europe and temperate Asia. The caterpillars eat the leaves of sloe, plum, and birch trees.

Family GEOMETRIDAE

Geometer moths

Occurrence 20,000 spp. worldwide; on deciduous and coniferous trees, woody shrubs, herbaceous plants

The wings of these slender, generally nocturnal moths are rounded, with a span of 1.4–7.4cm (½–3in), and complex patterns of fine markings. The females of some species are wingless. These moths are coloured brown or green for camouflage; coloration often varies between sexes. Females lay eggs singly or in groups, on bark, twigs, or stems of host plants. The characteristic looping motion of the caterpillars gives them the name "loopers" or "inchworms". Many species are agricultural or forestry pests and can cause severe damage.

OOSPILA VENEZUELATA
This green species is found mainly in the forests of Venezuela.

---

upper wing surfaces bright and iridescent

tails on hindwings

THECLA CORONATA
Also called the Hewitson's blue hairstreak, this species inhabits tropical forests of South America. It is one of the largest and most brilliantly coloured members of the family.

white rings on antennae

orange edge to hindwings

LYCAENA PHLAEAS
The small copper butterfly is common in grassy areas up to 2,000m (6,600ft) in the northern hemisphere. The caterpillars feed on dock leaves.

Family LYMANTRIIDAE

Tussock moths

Occurrence 2,600 spp. worldwide; oh foliage of deciduous and coniferous trees, shrubs

Tussock moths resemble noctuids (see right) and although most are dull in colour, tropical species can be colourful. The wingspan is 2–6cm (¾–2¼in); males are slightly smaller than the females, which are sometimes wingless. Females lay eggs in clumps on the bark of host trees and shrubs. Adults lack a proboscis and do not feed. The caterpillars are hairy, often brightly coloured, and feed in groups on foliage. They may induce an allergic reaction if touched.

distinctive, V-shaped mark on forewings

broad, hairy body

LYMANTRIA DISPAR
Commonly called the gypsy moth, this species is a native of Europe and Asia. Introduced to North America to produce inexpensive silk, it escaped from captivity and has become a pest.

Family NOCTUIDAE

Noctuid moths

Occurrence 22,000 spp. worldwide; on all types of vegetation, including crops

Almost every important crop plant in the world is attacked by one or more noctuid moth species. These moths have narrow forewings and broad hindwings and are mostly dull in colour, although the hindwings of some are brightly coloured and patterned. The wingspan is usually under 5cm (2in). Females lay eggs singly or in groups of 50–300, at the bases of host plants or in the soil. These moths have thoracic hearing organs to detect their main predators, bats.

black wing markings

dark, wavy line near margin of hindwings

AGROTIS IPSILON
The dark sword grass moth inhabits crops and grassy areas across the world. Its caterpillars attack potato, tobacco, lettuce, and cereal crops.

Family PAPILIONIDAE

Swallowtails

Occurrence 600 spp. worldwide in warmer regions; in flower-rich, open or shaded areas

This family includes the birdwings – the world's largest butterflies, now protected by law. True to their name, many swallowtail species have tails on their hindwings. The wings are typically dark with bands, spots, or patches of white, yellow, orange, red, green, or blue, and wingspan ranges from 4.5 to 28cm (1¾–11in). Round eggs are laid singly on host plants and pupation also occurs here. The caterpillars have a forked, fleshy scent gland that emits an odour which protects them from predators. They are herbivorous and eat a range of foliage. Adult swallowtails feed on nectar and other liquids.

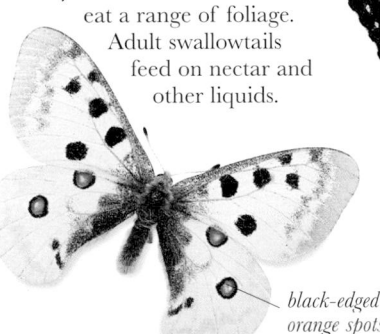

black-edged orange spots on hindwings

PARNASSIUS APOLLO
Also known as the Apollo, this species is found in mountainous areas of Europe and parts of Central Asia.

scalloped wing markings

tails on hindwings

PAPILIO GLAUCUS
Commonly called the tiger swallowtail, this North American species inhabits woodland and gardens. Its caterpillars feed among high foliage in many deciduous trees.

ORNITHOPTERA PRIAMUS
Also called the Cairns birdwing, this species inhabits tropical forests in the Moluccas, Papua New Guinea, the Solomon Islands, and northeastern Australia. Females are larger than males, and are drably coloured.

bright green patterns on dark background

black wing-spots

yellow abdomen

Family NYMPHALIDAE

Nymphalid butterflies

Occurrence 5,000 spp. worldwide; in flower-rich meadows, woodland clearings

This family includes some of the world's most eye-catching butterflies, their most characteristic feature being their very short, brush-like front legs, which have a sensory function. These are held off the ground, close to the head. The wingspan is 3–15cm (1¼–6in); the upper wing surfaces are brightly coloured but the undersides have cryptic colouring to camouflage the butterfly at rest. Males and females may be differently coloured, and there may be seasonal variation. Females lay groups of rounded eggs on the foliage of trees, shrubs, and herbaceous plants. The pupae often have warty, conspicuous bumps and hang from the host plant by a group of terminal hooks called cremasters. When young, the caterpillars may feed communally on foliage. Adults feed on nectar and liquids; some are attracted to fruits, dung, carrion, urine, and even petrol.

white patches on head

white patches around wing margins

DANAUS PLEXIPPUS
The monarch butterfly is famous for its long-distance migrations, the longest of which extends from Mexico to Canada. Its caterpillars feed on milkweed plants; a substance in these plants makes them unpalatable to predators.

MORPHO PELEIDES
Also called the common morpho, this butterfly inhabits forests in South and Central America. With a wingspan of 9.5–12cm (3¾–4¾in), it is one of the larger nymphalids. Its caterpillars can produce a noxious defensive odour.

dark bands with white marks on wings

pale spots on margin of forewings

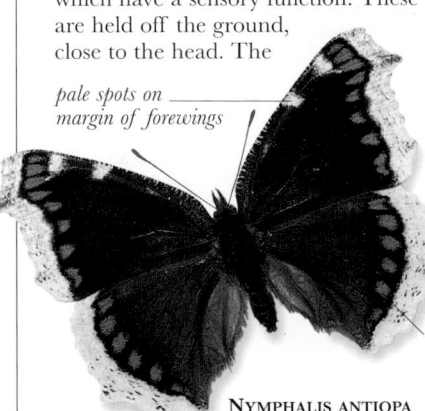

row of purple-blue spots

NYMPHALIS ANTIOPA
Called the Camberwell beauty in Great Britain, and the mourning cloak in North America, this butterfly inhabits grassland in the northern hemisphere.

distinctive white spots on forewings

broad orange bands

black-spotted hindwings

VANESSA ATALANTA
The red admiral is widespread in the northern hemisphere. Its caterpillars feed on nettles and hops.

Order LEPIDOPTERA *continued*

Family PIERIDAE

Whites and sulphurs

Occurrence 1,200 spp. worldwide; anywhere, often in groups at bird droppings, urine, or puddles in sunshine

The wings of these very common butterflies are usually white, yellow, or orange, with black and grey markings. The wingspan is 2–7cm (¾–2¾in) and the pigments in the wing scales are derived from by-products of the food eaten by caterpillars. Eggs are laid singly or in groups of 20–100 on host plants. The pupae, which have a distinctive, spiny projection on the head, are held upright on the host plant by a silk belt.

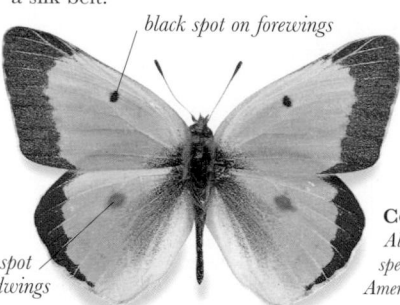

black spot on forewings

orange spot on hindwings

distinctive orange tip on male's forewings

ANTHOCHARIS CARDAMINES
The orange tip inhabits damp, grassy meadows and open woodland in Europe and temperate Asia. The undersides of the hindwings are dappled pale green.

COLIAS EURYTHEME
Also called the orange sulphur or the alfalfa sulphur, this species inhabits grassy, open areas and cultivated land in North America. Caterpillars feed on clover and can be pests of alfalfa.

PIERIS NAPI
The green-veined white inhabits grassy, open areas and cultivated land in Europe, Asia, and North America.

dark wing margins

Family PYRALIDAE

Snout moths

Occurrence 24,000 spp. worldwide, especially in warmer regions; on a range of plants, from aquatic plants to trees

Some species of snout moths have a short "snout" on the front of their head. Most are drably coloured. Their wings span 1–4.6cm (⅜–1¾in); the forewings may be broad or narrow, but the hindwings are broad and rounded. Eggs are laid on the undersides of leaves.

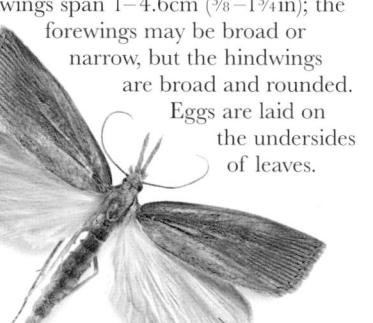

long fringe on hindwings

CHILO PHRAGMITELLA
This species is found in reedbeds, where its caterpillars feed on the stems of reeds belonging to the genus Phragmites.

Family SATURNIIDAE

Saturniid moths

Occurrence 1,200 spp. worldwide, especially in wooded tropical and subtropical regions; on foliage

Variously known as emperor, moon, royal, and atlas moths, these are large and often spectacular insects with broad, often conspicuously marked wings. The wings span 5–30cm (2–12in) and the centre of each may have an eye-spot; some species have hindwing tails. Males have broad, feathery or comb-like antennae. Fully grown caterpillars make dense cocoons, attached to the twigs of host plants, and feed on the foliage of a wide range of trees and shrubs. Adults do not feed.

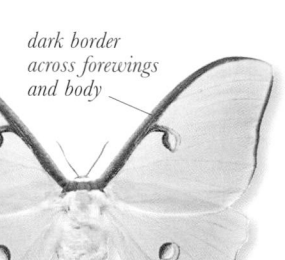

dark border across forewings and body

large, furry body

ACTIAS LUNA
The American moon moth is found from Mexico to the Canadian border. The caterpillars feed on the leaves of deciduous trees.

long tail with red border

curved wing-tip

broad, feathery antennae of male

ATTACUS ATLAS
This Southeast Asian moth, commonly called the Atlas moth, has the largest wing area of any insect. It inhabits forest and open, grassy areas. Although easy to rear, it is vulnerable and protected in some countries.

black, wavy line on wing margins

triangular, translucent patches on wings

CALLOSAMIA PROMETHEA
The promethea moth inhabits woodland and cultivated areas of southern Canada and the USA. Females are heavy-bodied and nocturnal; males are dark, day-flying, and resemble Nymphalis antiopa (see p.571).

eye-spots on forewings

Family SESIIDAE

Clear-winged moths

Occurrence 1,000 spp. worldwide, especially in northern temperate regions; around flowers or near host plants

These usually diurnal moths resemble wasps or bees; some even pretend to sting. They are black, bluish, or dark brown, with yellow or orange markings. The wings, large areas of which are transparent, span from 1.4 to 4cm (½–1½in) and produce a buzzing sound in flight. Many species are pests of fruit and other trees and shrubs.

transparent areas on wing

SESIA APIFORMIS
With its transparent wings, the hornet moth, from Europe and Asia, convincingly mimics a hornet. Its caterpillars bore inside willows and poplars.

Family SPHINGIDAE

Hawk moths

Occurrence 1,100 spp. worldwide, especially in tropical and subtropical regions; on foliage

With their streamlined bodies and long, narrow wings, hawk moths are among the fastest flying insects, reaching speeds of over 50kph (30mph). Their wings measure 3.5–15cm (1½–6in) from tip to tip. Adult hawkmoths mostly suck nectar – some hovering like hummingbirds as they do this. They have a long, curled proboscis, up to 25cm (10in) long. Although some species are diurnal, most are active after dark. Hawkmoths often migrate long distances to breed. Their caterpillars are known as hornworms, because they have a spine-like process at the end of the abdomen. Some are very serious crop pests.

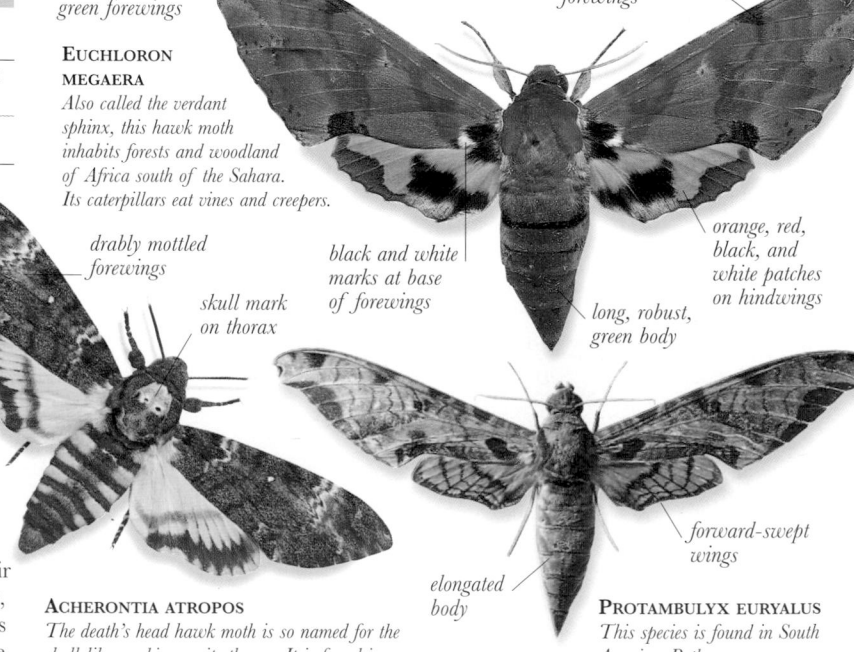

distinctive, dark green forewings

EUCHLORON MEGAERA
Also called the verdant sphinx, this hawk moth inhabits forests and woodland of Africa south of the Sahara. Its caterpillars eat vines and creepers.

drably mottled forewings

skull mark on thorax

black and white marks at base of forewings

dark triangles on forewings

orange, red, black, and white patches on hindwings

long, robust, green body

ACHERONTIA ATROPOS
The death's head hawk moth is so named for the skull-like marking on its thorax. It is found in Africa and Asia, but also migrates to Europe.

elongated body

forward-swept wings

PROTAMBULYX EURYALUS
This species is found in South America. Both sexes are very similar in appearance.

Family TINEIDAE

Clothes moths and relatives

Occurrence 2,500 spp. worldwide; in rotting wood, fungi, dried organic matter, woollens, fabrics, dry foodstuffs

Many of these small, drab moths are pests, and are variously known as grain moths, case-bearing moths, and tapestry moths. Typically dull brown in colour, some have a shiny golden appearance. Their head has a covering of raised, hair-like scales or bristles and the proboscis is short or absent. The wings, 0.8–2cm (5/16–3/4in) across, are narrow and are held at a steep angle over the body at rest. Females lay 30–80 eggs over a period of 3 weeks. Caterpillars are saprophagous. Some of them make portable cases of silk and debris, others make a silk web wherever they feed. Adults do not generally feed. Many species will not readily fly, scuttling away from danger instead.

TINEOLA BISSELLIELLA
Also known as the common clothes moth, this widespread species inhabits clothing, carpets, and furs. The female lays eggs in the folds of clothes.

Family TORTRICIDAE

Tortricid moths

Occurrence 5,000 spp. worldwide; on leaves, shoots, buds, fruits

These small moths are also known as bell moths. While most are brown, green, or grey, and patterned to blend in with bark, lichen, and leaves, some can be brightly coloured. Their head is covered with rough scales. The wingspan is 0.8–3.4cm (5/16–1½in) and the forewings, unusually rectangular, are held roof-like over the body at rest. A female can lay up to 400 eggs in a week, singly or in small groups, on the surface of fruits or leaves. The caterpillars feed on leaves, shoots, and buds of plants and often tie or roll leaves with silk. Being fruit- and stem-borers or gall-formers, many species are pests that cause damage to trees and fruit crops.

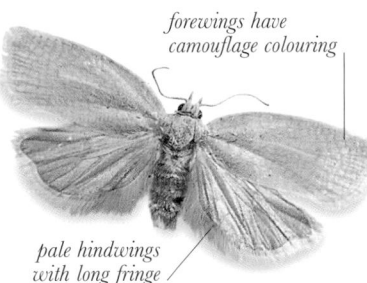

forewings have camouflage colouring

pale hindwings with long fringe

CLEPSIS RURINANA
Found in woodland and forests of Europe and Asia, the caterpillars of this species roll the leaves of deciduous trees and feed inside.

Family URANIIDAE

Uraniid moths

Occurrence 100 spp. in tropical and subtropical regions; on plants, especially euphorbias

This family includes large, long-tailed, colourful, day-flying moths with iridescent wing scales, and dull, nocturnal species without tails. The South American and Malagasy species are so colourful and large that they are often mistaken for butterflies. The wings typically span 6–10cm (2¼–4in). Adults are often migratory, sometimes forming large swarms, in response to changes in larval food quality. The caterpillars of many species eat plants of the family Euphorbiaceae. They avoid ant-predation by dropping from leaves on silk lines.

LYSSIDIA ZAMPA
This large, Southeast Asian moth has a wingspan of 10–10.6cm (4–4¼in), a little larger than most species in the family.

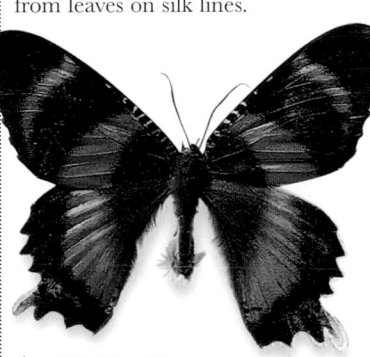

forewings have pointed tip

iridescent markings

ALCIDES ZODIACA
This day-flying moth from the rainforests of North Queensland, Australia, feeds on flowers during the day and flies back to the canopy in the evening.

hindwings have distinct spotting and 3 tails

CHRYSIRIDIA RIPHEARIA
Also known as the Madagascan sunset moth, this is a brightly coloured day-flying species that inhabits woodland and forests of Madagascar.

INVERTEBRATES

Order HYMENOPTERA

BEES, WASPS, ANTS, AND SAWFLIES

THE ORDER HYMENOPTERA is a vast group of insects that contains at least 198,000 species in 91 families. It is divided into 2 suborders: plant-eating sawflies (Symphyta), and wasps, bees, and ants (Apocrita) which – unlike sawflies – have a narrow "waist" and, in females, an ovipositor that may sting. Most species have 2 pairs of membranous wings, joined in flight by tiny hooks. Ecologically, these insects are of tremendous importance in acting as predators, parasites, pollinators, or scavengers. Social Hymenoptera, including ants and bees, are the most advanced insects on earth.

Family AGAONIDAE

Fig wasps

Occurrence 650 spp. in subtropical and tropical regions; on fig trees

Fig wasps have evolved a symbiotic relationship with fig trees. The female wasps, no more than 3mm (1/8in) long, crawl inside figs, which are lined with tiny flowers. They pollinate some flowers, and lay eggs in others. After developing, the wingless males fertilize the females, which fly away to find new trees.

BLASTOPHAGA PSENES
The common fig wasp is found in Asia, South Europe, Australia, and California, USA. It pollinates Ficus carica, the common fig. Females are 20 times more common than males.

Family APIDAE

Honey bees and relatives

Occurrence 1,000 spp. worldwide; in all well-vegetated, flower-rich areas

This family includes one of the world's most useful insects – honey bees – together with orchid bees and bumble bees. They provide honey and wax, and pollinate most of the world's plants. Typically, they are 0.3–2.7cm (1/8–1in) long. Most females have a venomous sting and a basket to carry pollen on the hindlegs. Honey bees and bumble bees are social insects, living in complex colonies consisting of a queen, males, and sterile female workers. The honey bee nest is a vertical array of wax combs divided into thousands of cells for rearing the young and storing pollen and honey. Queens may lay more than 100 eggs a day. Bumble bee nests are made of grass with wax cells, and the queens raise fewer offspring.

queen

drone

worker

BOMBUS TERRESTRIS
Like all its relatives, the European buff-tailed bumble bee has a body covered by "fur", enabling it to fly in the cool conditions of early spring.

APIS MELLIFERA
Originally from Southeast Asia, the honey bee is now raised all over the world. A honey bee colony has strict division of labour. The queen lays all the colony's eggs, while the workers gather food and tend the larvae. The queen is fertilized by a drone.

Order HYMENOPTERA *continued*

Family BRACONIDAE

Braconid wasps

Occurrence 25,000 spp. worldwide; on suitable caterpillar hosts

Like ichneumon wasps (see below), braconids are parasitoids – animals that develop inside a living host, eventually killing it. Braconids mainly attack the caterpillars of moths and butterflies, often laying over 100 eggs on or in their victims. The developing larvae feed on the caterpillar's tissues, pupate, and fly away as adults. Adult braconids are typically less than 7mm (7/32in) long, and reddish brown or black. Many species are used as biological controls against insect pests.

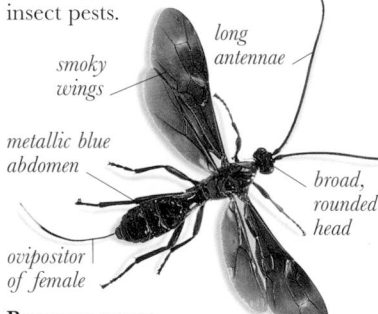

smoky wings
long antennae
metallic blue abdomen
broad, rounded head
ovipositor of female

BATHYAULAX SPP.
Larger than most of its family, these braconids measure 1.5–2cm (1/2–3/4in), and are found in Africa and Southeast Asia. They are used to control stem-boring caterpillar pests of cereal crops.

Family ICHNEUMONIDAE

Ichneumon wasps

Occurrence 60,000 spp. worldwide, especially temperate regions; wherever suitable insect hosts are found

Ichneumon wasps are of great ecological importance, because they parasitize other insects, including many pests. Using their slender ovipositors, the females lay eggs in the larvae and pupae in a wide range of hosts, sometimes drilling through several centimetres of wood to reach them. On hatching, the ichneumon larvae eat their hosts from inside. Adult ichneumons measure 0.3–4.2cm (1/8–1½in) long, and are typically slender, with a narrow waist.

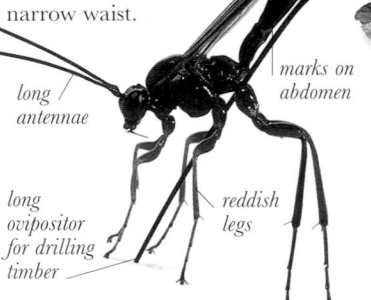

long antennae
marks on abdomen
long ovipositor for drilling timber
reddish legs

RHYSSA PERSUASORIA
This large ichneumon wasp uses its long, drill-like ovipositor to reach the larvae of horntail sawflies deep in pine timber.

Family CHALCIDIDAE

Chalcid wasps

Occurrence 1,800 spp. worldwide; on suitable insect hosts

Mostly under 8mm (5/16in) long, chalcids make up a small but important family of parasitic insects. Many species lay their eggs in beetle larvae, or in the caterpillars of butterflies and moths, but some are hyperparasites, attacking the eggs or larvae of other parasitic insects. Adult chalcids are typically dark brown, black, red, or yellow, sometimes with a metallic sheen. Their larvae – like those of other wasps – are white and grub-like, without legs. Some chalcid species are bred and released to control insect pests.

slight vein pattern on wings
curved hind tibiae
enlarged hind femora
punctured body surface

CHALCIS SISPES
This species is native to Europe and Asia. Its larvae parasitize the larvae of soldier flies.

slender antennae
black and yellow wing patterns
ovipositor

PARACOLLYRIA SPP.
These African wasps have a bright colouring that works as a warning to predators and is mimicked by other insects.

swollen antennal segments
black pterostigma on forewings
orange markings on hind femora
black rear half of abdomen

JOPPA ANTENNATA
This species is characterized by swollen antennal segments and contrasting, wasp-like coloration.

Family CHRYSIDIDAE

Jewel wasps

Occurrence 3,000 spp. worldwide; on suitable insect hosts

Jewel wasps, also known as cuckoo wasps or ruby-tailed wasps, have very hard and dimpled bodies that protect them from bee and wasp stings. They are 0.2–2cm (1/16–3/4in) long. Most species are a very bright metallic blue, green, or red, or combinations of these colours. The abdomen is hollowed underneath, allowing the wasp to curl up if attacked. Typically, the female lays an egg in the nest of a solitary bee or wasp, and the larva immediately hatches. The jewel wasp larva then feeds on the host larva and also eats the host's provisions. The adults feed on nectar and liquids.

bright metallic coloration
distinctive lobe on hindwings
end of abdomen metallic purple
hard, dimpled body surface

STILBUM SPLENDIDUM
This jewel wasp is a large, North Australian species that parasitizes solitary mud-nesting wasps.

Family CYNIPIDAE

Gall wasps

Occurrence 1,250 spp. mostly in the northern hemisphere; on trees, plants

Gall wasps are shiny, reddish brown or black, and usually fully winged. They measure 1–9mm (1/32–11/32in) in length; males are usually smaller than females. The latter have a laterally flattened abdomen and a humped thorax. Many species, by a process not yet fully understood, induce the growth of species-specific galls on oaks and related trees by laying eggs inside the plant tissue. The galls protect and nourish the developing larvae. Life cycles of gall-forming species involve sexual and asexual generations.

ANDRICUS SPP.
These gall wasps are widespread in Europe, and use various oak trees as their hosts.

shiny, punctured surface
shiny, smooth abdomen

Family MUTILLIDAE

Velvet ants

Occurrence 5,000 spp. worldwide, especially subtropical and tropical regions; females found in dry habitats

Despite their name, these are not ants. Their name velvet ant comes from the soft, velvety hairs that cover the ant-like, wingless females; males have fully developed wings. Velvet ants are 0.3–2.5cm (1/8–1in) long. They use the larvae of other wasps or bees as food for their larvae. Females bite open a suitable host cell and, if the larva in it is fully grown, lay their eggs on it and reseal the cell. The larvae then eat the host larva, and pupate inside the cell. Female velvet ants have powerful stings.

dimpled thorax
female is wingless

MUTILLA EUROPEA
The European velvet ant, widespread in the warmer regions of Europe, parasitizes species of bumble bee.

Family MYMARIDAE

Fairyflies

Occurrence 1,400 spp. worldwide; in a wide variety of habitats, wherever insect hosts are found

The smallest flying insects on earth belong to this family. Fairyflies are 0.2–5mm (1/16–3/16in) long, and are dark brown, black, or yellow. The narrow front wings have a fringe of hairs but no obvious venation. The hindwings appear stalked, and are strap shaped. The females of all species parasitize the eggs of other insects, mostly of plant-hoppers and other bug families. Several fairyfly species are used as biological control agents against insect pests.

ANAGRUS OPTABILIS
This is a specialist parasitoid of the eggs of certain plant-hoppers. Related species have been used to control plant-hoppers that attack rice crops.

Family POMPILIDAE

Spider-hunting wasps

Occurrence 4,000 spp. worldwide, especially tropical and subtropical regions; wherever spider hosts are found

True to their name, these wasps can paralyse spiders with their strong venom. Females fly or run along the ground in search of spiders, drag the immobilized spider to a nest prepared in mud or in a crevice, and lay a single egg on it before sealing the nest. Spider-hunting wasps have a painful sting. Most species are dark blue or black (although the wings may be dark yellow), and measure 0.5–8cm (³⁄₁₆–3¼in) long. Males are smaller than the females.

amber wings

POMPILUS SPP.
This genus is common in warm regions, particularly where the ground is sandy. Like other spider-hunting wasps, pompilus wasps are solitary.

yellow patches on abdomen · *visible male genitalia*

matt black body

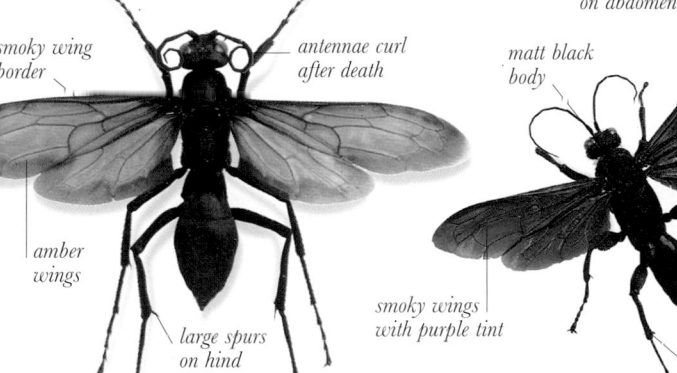

smoky wing border · *antennae curl after death*

amber wings

large spurs on hind tibiae

PEPSIS HEROS
The tarantula hawk is the largest spider-hunting wasp, up to 8cm (3¼in) long. As its name suggests, it preys on tarantulas.

smoky wings with purple tint

spurs on tibiae

MACROMERIS VIOLACEUS
The violet spider-hunting wasp, so named for its iridescent purplish wings, is found in Southeast Asia. It is 5–5.5cm (2–2¼in) long.

Family SPHECIDAE

Digger wasps

Occurrence 8,000 spp. worldwide; in a variety of microhabitats – plant stems, soil, rotten wood

Some species of this family are also known as solitary hunting wasps, sand wasps, or mud dauber wasps. They are 0.4–4.8cm (⁵⁄₃₂–1¾in) long, relatively hairless, and often brightly coloured. Digger wasps are solitary, and nest in plant stems, soil, or rotten wood.

Family SCOLIIDAE

Mammoth wasps

Occurrence 350 spp. worldwide, especially in tropical regions; wherever scarab beetle hosts are found

As their name suggests, these are large wasps, measuring 1–5.6cm (³⁄₈–2¼in) in length, with stout bodies that are densely covered with hair. They are bluish black with reddish brown markings. Males are smaller and slimmer than females, with longer antennae. After mating, females hunt for the larvae of scarab beetles. They then sting, paralyse, and lay a single

The female catches an insect or a spider, paralyses it, and buries it in a nest where she lays her egg. In this way the emerging larva has a ready supply of food.

wings held flat over body

spiny legs · *blue-green, metallic coloration*

AMPULEX SPP.
These are cockroach-hunters that may enter houses in search of their quarry. This genus is found in tropical parts of the world.

egg on each larva. The wasp larva pupates in a cocoon that it spins inside the beetle larva's remains.

wrinkles at end of wings

hairy legs

SCOLIA PROCERA
Found in Java, Borneo, and Sumatra, wherever scarab beetle larvae exist, this insect has a painful sting but is not very aggressive.

Family VESPIDAE

Social wasps

Occurrence 4,000 spp. worldwide; found in a variety of habitats, in nests and foraging widely for prey

The most familiar social wasps are the paper wasps and yellow jackets, which make nests out of chewed wood or other fibres. The queen overwinters, makes a nest in spring, and rears the first brood herself. Workers co-operate with her in caring for subsequent broods. Larvae develop in horizontal "combs", and are fed chewed-up insects by sterile female workers. The adult wasps, which measure between 0.4 and 3.6cm (⁵⁄₃₂–1½in) in length, roll or fold their wings longitudinally, rather than hold them flat over the body. Nearly all social wasps have warning coloration. They can also have a painful sting.

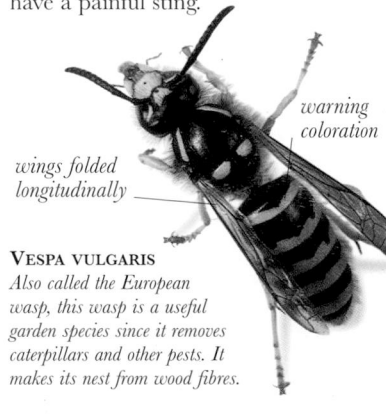

warning coloration

wings folded longitudinally

VESPA VULGARIS
Also called the European wasp, this wasp is a useful garden species since it removes caterpillars and other pests. It makes its nest from wood fibres.

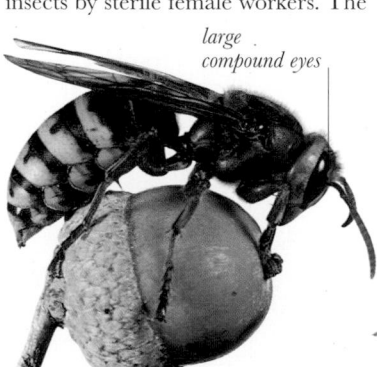

large compound eyes

VESPA CRABRO
Commonly known as the European hornet, this wasp makes its nest in hollow trees. Its colonies have only a few hundred workers.

yellow patches on thorax

black dots on abdomen

VESPULA GERMANICA
Colonies of the German wasp may be perennial, with more than one queen. This wasp lives mostly in the northern hemisphere and was introduced to North America in the late 1800s.

Family SIRICIDAE

Horntails

Occurrence 100 spp. in temperate areas of the northern hemisphere; on and near deciduous and coniferous trees

Horntails or woodwasps are the largest members of the sawfly suborder (see right). Despite their menacing appearance, they do not sting – the "horn" at the end of their abdomen is a harmless spine. They are reddish brown, black and yellow, or metallic purple, and measure 1.8–4.2cm (¾–1½in) in length. Females have a long ovipositor and use it to lay their eggs inside tree trunks, in the process infecting them with fungi. Larvae burrow into heartwood and eat fungus and wood. Pupation occurs in a cocoon of silk and chewed wood.

yellow patch behind eye

large, black and yellow body

spine

UROCERUS GIGAS
The horntail shown above is a female common horntail or giant woodwasp. The male of this harmless species is elusive.

Family TENTHREDINIDAE

Common sawflies

Occurrence 6,000 spp. worldwide, except New Zealand; on a wide range of plants in gardens, pastures, woodland

Sawflies often look like wasps, but are more primitive insects, classified in a suborder of their own. Unlike wasps, bees, and ants, they do not have a narrow waist, do not sting, and are solitary. Common sawflies are 0.3–2.2cm (⅛–¾in) long. The females use their ovipositor to lay eggs in plants, and the caterpillar-like larvae, which are sometimes brightly coloured, usually feed in the open on leaves. Common sawflies often attack fruits, vegetables, and trees, and can be highly destructive pests.

long, narrow, yellow wings

wasp-like, black and yellow abdominal stripes

yellow hindlegs

TENTHREDO SCROPHULARIAE
This is a wasp-mimicking species, found in Europe and Asia. Its larvae feed on mullein and figwort plants.

Family FORMICIDAE

Ants

Occurrence 9,000 spp. worldwide; in virtually all regions except in Antarctica and on a few oceanic islands

Habitat Most terrestrial habitats

Ants are found almost everywhere on Earth and have an immense impact on terrestrial ecosystems. In many, ants move more earth than earthworms and are vital in nutrient recycling and seed dispersal. Like termites and some bees and wasps, ants are highly social and live in colonies ranging from a handful of individuals sharing a tiny shelter to tens or hundreds of millions inhabiting structures reaching 6m (20ft) below ground. Two or more generations of ants overlap at any one time, and adults take care of the young. There are different classes (or castes) of ant within a colony, each with a specific task to perform: some are workers, who also defend the nest; others are dedicated to reproduction. Caste is determined by diet: a rich diet producing reproductives, a poor one producing sterile workers. Mating may take place on the wing or on the ground, after which the males die and the females lose their wings. All ants have glands that secrete pheromones – chemical messages used for communication between colony members as well as for trail-making and defence. Numerous species of insect and plant have evolved symbiotic and often highly complex relationships with ants. In some cases plants provide ants with food or homes; in return the ants may protect the plants by disturbing potentially damaging herbivores, or stripping leaves from encroaching vines. The ant family is divided into 10 distinct subfamilies, of which the biggest by far are the Myrmicinae and the Formicinae. Some myrmicine ants have stings while formicines defend themselves by spraying formic acid. The infamous driver and army ants of tropical regions belong to the subfamily Dorylinae. Their colonies move in massive columns, up to several millions strong, often raiding termite or ant nests.

EATING HABITS

Some ants are herbivorous seed gatherers or fungus eaters. Other species are carnivorous or omnivorous, while some rely exclusively on the honeydew produced by sap-sucking bugs.

WORKERS WITH PREY
Worker wood ants, Formica rufa, *may cooperate in dragging a large item of prey back to the nest, where it will be cut up.*

large worker

LEAF CUTTERS
Leafcutter ants, Atta *spp., chew pieces of leaves into pulp, which they use to fertilize fungus beds – their sole source of food.*

DIFFERENT CASTES

A colony will contain 3 castes of ant: workers, which are always wingless sterile females, queens, and males (which are usually winged and form mating swarms with queens at certain times of the year). The physical differences between the castes of the African driver ant *Dorylus nigricans* (shown here at relative sizes) are typical of most ants, although queens in this species are wingless.

MALE ANT
Because of its body shape, the male is called a sausagefly. Males are often attracted to lights after dark.

sausage-shaped body

broad forewings

FEMALE WORKERS
Worker ants are all female. Large workers carry items of prey, while the smaller workers tend to carry liquid.

QUEEN ANT
The queens of this species are the largest of any ant and can lay 1–2 million eggs every month.

large abdomen

elbowed antennae

constricted waist

chewing mouthparts

WINGLESS FEMALE
Sterile, wingless females do the work of the colony, and in some species may show amazing adaptations to the head and jaws for crushing seeds, blocking nest entrances, dismembering enemies, and other purposes.

ON THE DEFENSIVE
This worker wood ant, Formica rufa, *has taken up a defensive posture with its jaws wide open and abdomen curled under and forwards. In this position it will be able to spray formic acid towards an attacker.*

CENTIPEDES AND MILLIPEDES

PHYLUM	Arthropoda
SUBPHYLUM	Mandibulata
SUPERCLASS	Myriapoda
CLASSES	Chilopoda Diplopoda
ORDERS	16
FAMILIES	144
SPECIES	About 13,700

CENTIPEDES AND MILLIPEDES ARE terrestrial arthropods with many legs, each of their numerous body segments carrying one or 2 pairs. Their heads have biting mandibles and a single pair of antennae. Far less abundant and diverse than insects, they are confined to humid microhabitats because they do not have a waterproof cuticle. However, like insects, they breathe by taking gases directly into the body through openings in the cuticle.

Centipedes have a single pair of legs on each trunk segment. They can run fast, although some soil-living species move slower. They are carnivorous and have venomous claws to kill prey. Millipedes are all slow-moving, with elongated, cylindrical bodies. Most of their trunk segments are fused together in pairs, called diplosegments, each bearing 2 pairs of legs. They are mostly herbivores or scavengers, using their toothed mandibles to chew rotting organic matter, soil, plants, algae, and moss.

FOREST CARNIVORES
Most centipedes, such as this red giant centipede, are forest-dwelling carnivores. Their long legs allow them to move quickly over ground in search of other invertebrates and to escape from predators. Most species remain hidden during the day and become active at night.

Class CHILOPODA

CENTIPEDES

ELONGATED AND FLATTENED, CENTIPEDES have a trunk made up of at least 16 segments, most of which carry one pair of legs; the last pair of walking legs is always longer than the rest. Most species are yellowish or brownish and all have fine, sensory hairs on the body. There are 4 orders, 22 families, and about 3,000 species.

Class DIPLOPODA

MILLIPEDES

DESPITE THEIR NAME, MILLIPEDES never have a thousand legs. Some have as many as 750, but even these species are slow runners. To defend themselves, most millipedes rely primarily on their tough exoskeletons, which are reinforced by calcium carbonate. When touched, many also coil up or produce toxic chemicals. This class has at least 10,000 species, classified in 13 orders and 115 families.

Family LITHOBIIDAE
Lithobiids

Occurrence 1,500 spp. worldwide, especially in temperate regions of the northern hemisphere; in cracks, crevices

Lithobiids are mostly reddish brown but some can be brightly coloured. The tough and flattened body is 0.6–4.5cm (¼–1¾in) long. There are 15 pairs of legs, and thread-like antennae with 75–80 segments. The young emerge from the egg with 7 pairs of legs

and add a leg-bearing trunk segment at each moult. Lithobiids have a lifespan of 5 to 6 years.

dark brown trunk segments

broad head

venomous claws

LITHOBIUS VARIEGATUS
This common, European species inhabits leaf litter in deciduous woodland. It does climb trees in search of food but most often sits and waits on the ground for prey to pass by.

light and dark bands on rear legs

Family SCOLOPENDRIDAE
Scolopendrids

Occurrence 400 spp. worldwide, especially in subtropical and tropical regions; in soil, leaf litter, cracks, crevices

The largest centipede in the world, *Scolopendra gigantea* from South America, belongs to this family. The body length of scolopendrids ranges from 3 to 30cm (1¼–12in). Nocturnal hunters, they are robust enough to catch and kill mice and frogs with their venomous claws, the smaller species being the most deadly. Scolopendrids are typically brightly coloured and may be yellow, red, orange, or green, often with dark stripes or bands.

They have 21 or 23 pairs of legs, and thread-like antennae with fewer than 35 segments. Usually, there are 4 ocelli on either side of the head. Females lay their eggs under soil, rocks, or loose bark.

yellow legs

contrasting bands of colour

venomous claws

SCOLOPENDRA SPP.
Tiger centipedes owe their common name to the markings on their body and their predatory habits. They are found in Southeast Asia, living under stones, rotting wood, and loose bark.

Family JULIDAE
Cylinder millipedes

Occurrence 450 spp. in the northern hemisphere; in soil, leaf litter, caves, under stones and rotting wood

These millipedes have rounded bodies which can range from 0.8 to 8cm (⁵⁄₁₆–3¼in) in length. Most are dull-coloured but a few have red, cream, or brown spots. The antennae are fairly long and slender. As in all millipedes, the females lay their eggs in nests in soil. There are typically 7 nymphal stages. Cylinder millipedes are common in Europe and Asia.

TACHYPODOIULUS NIGER
This species is commonly known as the black snake millipede. It has a black trunk and white legs, and is up to 4.8cm (2in) long. It inhabits woodland in western Europe and climbs walls, trees, or shrubs to feed after dark.

Family GLOMERIDAE
Pill millipedes

Occurrence 200 spp. in warm and cool temperate regions of the northern hemisphere; in soil, caves

Measuring 0.2–2cm (¹⁄₁₆–¾in) in length, these broad-bodied millipedes are either small and drably coloured or large and brightly marked. The trunk is made up of 13 segments, and the shape of the dorsal plates allows the millipede to roll into a tight ball if threatened. Adults have 15 pairs of legs. Eggs are laid in soil, and when the young hatch, they initially have only 3 pairs of legs. Some members of this family, such as *Glomeris* species, may live up to 7 years.

GLOMERIS MARGINATA
This compact millipede can be found in soil and leaf litter in Europe, and parts of Asia and North Africa. When completely rolled up, it can be confused with the pill woodlouse (see p.584), although it has a shinier body.

CRUSTACEANS

PHYLUM	Arthropoda
SUBPHYLUM	Mandibulata
SUPERCLASS	Crustacea
CLASSES	6
ORDERS	37
FAMILIES	540
SPECIES	More than 40,000

CRUSTACEANS ARE A VERY DIVERSE GROUP, ranging from water fleas and copepods – only just visible to the naked eye – to heavy-bodied crabs and lobsters, which include the largest arthropods alive today. Although some crustaceans, such as woodlice, have successfully adapted to life on land, the vast majority of species live in fresh water or in the sea. Aquatic crustaceans include some of the most abundant animals on earth, and they play a key part in many food chains.

ANATOMY

Crustaceans differ from insects, centipedes, and millipedes in a number of ways. They have 2 pairs of antennae, compound eyes on stalks, and a cuticle that is often strengthened with calcium carbonate, especially in larger species. The head and thorax are often covered by a shield, or carapace, and the front of this often extends to form a projection called the rostrum. The thorax has a variable number of segments. The abdomen ends in a tail-like telson. In crabs, the abdomen is short, flat, and tightly curled around to fit under the broad carapace.

Crustacean appendages – what we commonly understand as "legs" – have 2 branches and are specialized for a number of functions such as movement, sensing the environment, respiration, and egg-brooding. The first pair of thoracic legs may be enlarged to form chelipeds, with strong claws that are used for defence, handling food, and even sexual signalling. The thoracic appendages, called pereopods, typically have gills. The basal part of some appendages helps in walking, while the abdominal segments usually carry paired swimming appendages called pleopods, or swimmerets.

FEEDING

Crustaceans show a wide range of feeding strategies. Most large species capture their prey and kill it by stunning, crushing, or tearing it apart. Many species are filter feeders: they use their thoracic appendages to set up currents in the water and filter small particles of food, which are amassed and pushed into the mouth. Krill feed while swimming and trap particles in hair-like structures called setae on their thoracic legs. Small crustaceans may simply graze on particles of sediment, taking micro-organisms from the surface.

LIFE CYCLE

Most species of crustaceans have separate sexes. They copulate, lay eggs, and brood the eggs, either within a brood chamber or attached to their legs. Typically, eggs hatch into tiny larvae which float away in the water, feed, and grow, developing trunk segments and legs.

RED CRAB MIGRATIONS
Christmas Island, in the Indian Ocean, has a population of about 100 million red crabs. Each year, triggered by the first rains of the wet season, they migrate from their inland habitats to mate and spawn by the coast. The males, which arrive first, dip themselves in sea water to replenish lost salt and moisture. After mating, females remain by the sea for about 2 weeks, waiting for the eggs to mature before releasing them along the coast.

INVERTEBRATES

Class **BRANCHIOPODA**

Branchiopods

Occurrence 800–1,000 spp. worldwide; mainly in freshwater bodies, some in saline or marine conditions

Branchiopods comprise 4 orders: tadpole shrimps (Notostraca), water fleas (Cladocera), clam shrimps (Conchostraca), and brine shrimps (Anostraca). There are 22 families. Although some branchiopods reach a length of 10cm (4in), most species are much smaller. The body may be covered by a shield-like carapace which, in some species, takes the form of a 2-halved (bivalve) shell. The first pair of antennae and the first pair of mouthparts (maxillae) are small. The appendages have a flat, leaf-like appearance, and carry fringes of fine bristles for filtering particles of suspended organic food from the water. Branchiopods swim either rhythmically, using their appendages, or jerkily, using the second pair of antennae. Many species possess the respiratory pigment haemoglobin and may appear pink as a result. Females may produce more than one batch of eggs, each consisting of up to several hundred eggs. In some species, these are very drought-resistant. In this class, parthenogenesis is common.

lateral, compound eyes on stalks

leaf-like appendages

BRINE SHRIMP
Found worldwide, except in the Arctic and Antarctic, brine shrimps (family Artemiidae) live in saline lakes and pools. Their life cycles are rapid, suiting them to habitats that can dry up quickly. Like all brine shrimps, members of the genus Artemia, *such as the one shown here, lay eggs that can be dormant for up to 5 years.*

large, branched antennae used for locomotion

indistinct body segmentation

WATER FLEA
Using their branched, feathery antennae like oars, water fleas (family Daphniidae) swim jerkily through lakes and ponds. In many areas, these tiny crustaceans undergo a population explosion every spring, providing a huge supply of animal food for fish. The species above belongs to the genus Daphnia.

Class OSTRACODA

Mussel shrimps

Occurrence 6,000 spp. worldwide; in marine, brackish water, freshwater bodies, a few species are terrestrial

Mussel shrimps, also known as seed shrimps, are small, grey, brown, or greenish crustaceans that can measure less than 1mm (¹/₃₂in) to just over 3cm (1¼in) in length. They make up 60 families in 6 orders. Their indistinctly segmented body is fully enclosed in a hinged carapace, which can be variously shaped: circular, oval, or rectangular-oval. Its surface may be smooth or sculptured depending upon the species. Mussel shrimps resemble bivalves (see p.539) in having a carapace with 2 halves, strengthened by calcium carbonate, that can be closed by a muscle. The head is the largest part of the body. All species

have a single simple eye, the nauplius eye, in the middle of the head; some also have a pair of compound eyes. Most species have separate sexes, which mate in order to reproduce; however, some are parthenogenetic. Some mussel shrimps emit light to attract mates, and the males of some even synchronize their flashing. Eggs are released into water or stuck to plants; they may be very tough and resistant to drying. Mussel shrimps are very important in aquatic food chains and are found worldwide, from shallow to very deep water, as crawlers, burrowers, or free-swimmers. They may scavenge, or feed on detritus or particles suspended in the water; some are predatory. There are more than 10,000 ostracod fossil species, and they are important in oil prospecting since different types of fossils indicate different rock strata.

CYPRID MUSSEL SHRIMPS
Found in lakes, pools, and swamps, cyprid mussel shrimps (family Cypridae) include the majority of freshwater species. Some swim, while others crawl on the bottom. They often reproduce parthenogenetically, and in some species males have never been found. Their eggs are resistant to drying.

CYPRIDINID MUSSEL SHRIMPS
The largest mussel shrimps, including the deep sea genus Gigantocypris, belong to this widespread, marine group (family Cypridinidae). Cypridinids are good swimmers. They propel themselves with their second antennae, which project from a small notch in the front end of their carapace valves.

Class MAXILLOPODA

Maxillopods

Occurrence 10,000 spp. worldwide; in oceans (sediments to deep sea trenches), fresh water, hot springs, on land

Copepods (Copepoda) and barnacles (Cirripedia) form the most significant subclasses of this large and diverse group of 12 orders and about 107 families. Copepods are abundant everywhere in the ocean – their combined

mass exceeds billions of tonnes, and they are a major source of food for fish. They are also found in fresh water, from mountain lakes to hot springs. Usually under 5mm (³/₁₆in) in length, some parasitic species may grow up to 10cm (4in). Generally pale, they have no carapace and lack compound eyes. The head is fused with the trunk, and the first pair of thoracic appendages is used for feeding, while the rest help in swimming. Fertilization is external; the male simply attaches a sperm packet to the female's genitals.

Barnacles are entirely marine crustaceans they may be stalkless, stalked, or parasitic. They have a shell made up of calcareous plates that surround the body, and 6 pairs of thoracic appendages. Most species are hermaphrodite but some are dioecious, with sexes separated in individuals. Barnacles may grow in such numbers on the underside of a ship, that they reduce its speed by about one third.

short antennae

long, branched first antennae

2 tails used as rudders

CYCLOPOID COPEPODS
These copepods (family Cyclopidae) grasp prey with mouthparts called maxillae, and rip off small pieces using their mandibles. Some species live for more than 9 months. A female may produce more than 10 pairs of egg sacs in a lifetime, each containing up to 50 eggs. Shown above is a member of the freshwater genus Cyclops.

reduced abdomen

enlarged head

CALIGID COPEPODS
Members of the family Caligidae are ectoparasites on the bodies of marine and freshwater fish. They have an enlarged head, with modified appendages to grip the host, and a reduced abdominal region. Some are red due to the presence of haemoglobin in their body. Caligus rapax (shown here) pierces and sucks blood from the skin or gills of the host.

some backward-pointing spines

rear-pointing proboscis

suckers

FISH LICE
Belonging to the subclass Branchiura, members of the family Argulidae are commonly called fish lice; their mouthparts are modified for sucking, enabling these parasites to feed on marine or freshwater fish. A fish infested by them may get fungal infections or even die. Fish lice have a cephalothorax (the head fused with the first thoracic segment), a 3-segmented thorax, and a 2-lobed abdomen. The species shown above belongs to Argulus, the largest genus in the family.

crab host

pink brood sac or "externa"

SACCULINIDS
The pink swelling on the underside of this crab is part of a sacculinid (family Sacculinidae) – a parasitic cirripede. Adults have highly branched bodies that spread through the tissues of their hosts, absorbing nutrients. Some, including the species of Sacculina pictured, can infest half the host population in an area.

GOOSE BARNACLES
Goose barnacles (family Lepadidae) inhabit surface waters from polar to tropical oceans. The body has 2 parts: the flexible stalk and the main body. The larvae are free-swimming but adults attach themselves to algae, floating wood, turtles, and ships. The largest barnacle, Lepas anatifera, shown here, is up to 80cm (32in) long.

roughly conical shell

4 – 6 overlapping, calcareous plates surround body

TYPICAL BARNACLES
Members of the family Balanidae can be up to 10cm (4in) in width and height, although most are much smaller. They encrust surfaces in large numbers and filter passing food particles. Semibalanus balanoides (pictured) can withstand freezing in the Arctic tidal zones, and can survive out of water for up to 9 hours in summer.

Class MALACOSTRACA

Malacostracans

Occurrence 20,000 spp. worldwide; in marine intertidal to abyssal zones, fresh water, terrestrial habitats

The world's largest arthropod, the Japanese spider crab (*Macrocheira kaempferi*) belongs to this large and diverse group of crustaceans, which comprises 13 orders and 349 families. It has a legspan of 4m (13ft) – in marked contrast to some other malacostracans, which are less than 1mm (1/32 in) long. Despite this vast difference in size, malacostracans have many features in common. They are often brightly coloured, and have tough exoskeletons strengthened with calcium carbonate. Their carapace – if present – acts as a gill chamber, and never covers the abdomen. They often have stalked eyes, and their antennae are prominent. The thorax is made up of 8 segments while the abdomen has 6 segments and a flattened tail fan (telson) which can be moved rapidly in order to propel the animal away from predators. In many species, the first 1–3 thoracic appendages have become modified into maxillipeds, which are used in feeding; the remaining are used for locomotion. The abdominal appendages help in swimming, burrowing, mating, and egg-brooding. Most species belong to the order Decapoda, which contains all the familiar shrimps, crayfish, crabs, and lobsters. They often have an enlarged first pair of appendages equipped with claws. Important in the marine food chain, many of the bigger species are of commercial value.

MANTIS SHRIMPS

Mantis shrimps (family Squillidae) use a unique spearing (sometimes smashing) technique to catch prey. The second pair of thoracic legs is large and specially designed for spearing, and is capable of striking very fast indeed. Mantis shrimps have a flattened, flexible body and stalked, compound eyes. Squilla species, such as the one pictured right, inhabit tropical and subtropical seas, and are territorial.

KRILL

Krill (family Euphausiidae) are slender, shrimp-like crustaceans that filter food particles from seawater using the long hairs on their thoracic legs. They are highly gregarious, and also luminescent. The species shown left, Antarctic krill (Euphausia superba), forms very large swarms that sometimes have a collective weight of more than 2 million tonnes. In the Southern Ocean, these swarms are an important food for baleen whales.

luminous organs

eyes

thoracic legs

soft, transparent body

well-developed, compound eyes

PRAWNS

Members of the family Palaemonidae have a carapace that extends forwards to form a rostrum, which may have toothed edges. Most species are colourless and transparent. They use their claws to pick up food. The common prawn, Palaemon serratus (shown left), has very long antennae, up to 1.5 times longer than the body, that warn it of the presence of predators. This species is caught in huge numbers since it is a common human food.

long antennae

upturned rostrum

transparent body

spine-like rostrum

first 2 pairs of walking legs have claws

SHRIMPS

Pigment cells (chromatophores) present in their body allow these bottom-dwelling omnivores (family Crangonidae) to change colour. The carapace is extended to form a short spine-like, smooth-edged rostrum. The common shrimp, Crangon crangon (pictured left), is fished commercially along European coastlines.

tail fan

long antennae

BURROWING SHRIMPS

These shrimps (family Callianassidae) make u- or y-shaped, or branched, networks of burrows in silt and mud in shallow water. They are soft-bodied crustaceans, predatory on small organisms and worms. These shrimps are collected as fish bait in some parts of the world. The picture on the left shows a species from the genus Callianassa. In this genus one of the first thoracic legs may be much larger than the other.

dark claw tips

CANCRID CRABS

When adult, cancrid crabs (family Cancridae) typically live on the sea bed, where they prey on other invertebrates and scavenge for dead remains. The species shown here is the European edible crab, Cancer pagurus, which has a distinctive "pie crust" edge to its carapace. Found off the European Atlantic, Mediterranean, and West African coasts, it is up to 30cm (12in) across. It is commercially fished in many areas and large specimens are now rare.

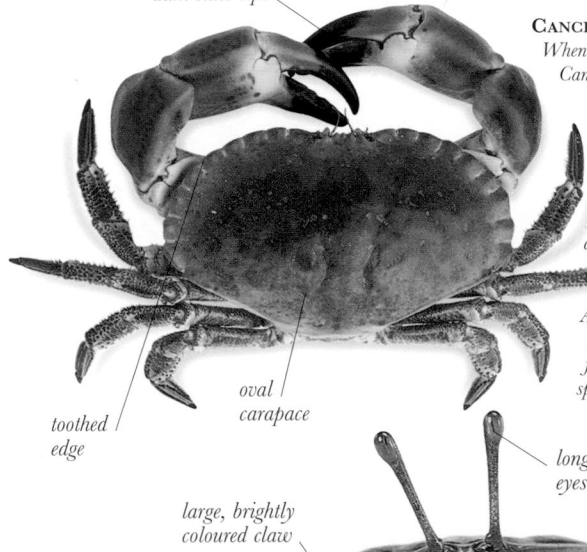

oval carapace

toothed edge

SWIMMING CRABS

Swimming crabs have paddle-like hindlegs that help them to swim efficiently. As in other crabs, the short abdomen is folded like a tail under the body. The common shore crab, Carcinus maenas (pictured), buries itself in mudflats and sandy regions. Known as the green crab in North America, this species is not as good a swimmer as other members of the family Portunidae.

wide, flattened carapace

notches on lateral border of carapace

large, brightly coloured claw

long-stalked eyes

GHOST AND FIDDLER CRABS

In the tropics, ghost and fiddler crabs (family Ocypodidae) abound on low-lying coasts. Ghost crabs live in burrows on sandy beaches, and emerge after dark, feeding on debris washed up by the tide. They have keen eyesight, and are extremely rapid runners, disappearing into their burrows at the first hint of danger. Fiddler crabs, which are smaller, burrow into the mud of mangrove swamps. Males have one large claw, which they wave to attract females and to deter rivals. The claw often makes up half the male's weight. Fiddler crabs feed on particles of organic matter present in silt. There are at least 75 species; the one shown here is the male Uca vocans, from Africa and Southeast Asia.

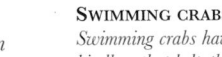

Class **MALACOSTRACA**

Family **MAJIDAE**

Spider crabs

Occurrence 950 spp. in all but polar oceans and seas; from the intertidal zone to waters 2,000 m (6,500 ft) deep.

Spider crabs are recognized by their triangular carapace, a shape formed by a narrowing of the front edge, which is often extended as an elongated rostrum. As the common name implies, many species have long, slender legs. In most cases, the pincer-bearing legs (chelipeds) are not much longer than the other legs. Spider crabs use their chelipeds to attach sponges, seaweed, hydroids, and even detritus such as wood fragments and broken worm tubes to hooked hairs on their carapace and legs. In some species the entire body can be covered, making the crab very difficult to see and protecting it from predators. If forced to move to a different location, some will change their camouflage to match their new surroundings. This behaviour has given these species the alternative names of decorating or masking crabs. Spider crabs

range from small species with a carapace length of around 8mm (^{5}/₁₆in) to the massive Japanese island crab – also known as the giant spider crab – which has a carapace up to 50cm (20in) long (see below).

GIANT AMONG ARTHROPODS
The Japanese island crab, Macrocheira kaempferi, *is the world's biggest living arthropod. Its legs may be up to 1.5m (5ft) long.*

pointed rostrum

fine yellow lines

long, slender legs

round eyes

violet-tipped claws

YELLOWLINE ARROW CRAB
Yellow lines on the carapace, running backwards from the eyes, and the arrow-shaped rostrum give this species, Stenorhynchus seticornis, *its common name.*

DIMINUTIVE FORM
The dainty yellowline arrow crab, at most 6.5cm (2½in) long, is found in the Atlantic Ocean from the Caribbean to as far south as Brazil. Males can tell how close a female is to laying eggs, and, after they have mated, they will guard the female from the attentions of other males.

Class MALACOSTRACA *continued*

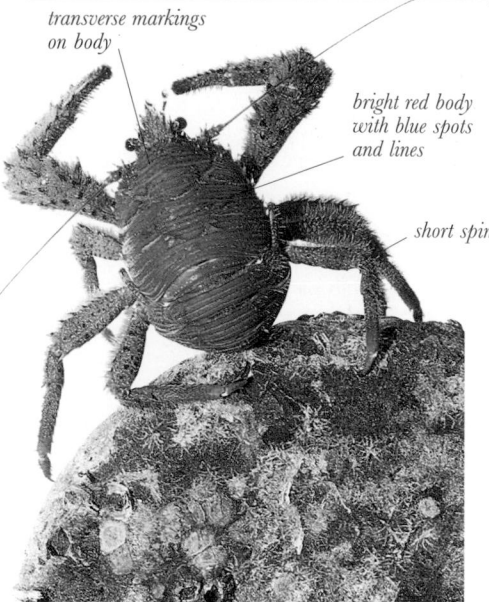

transverse markings on body

bright red body with blue spots and lines

short spines

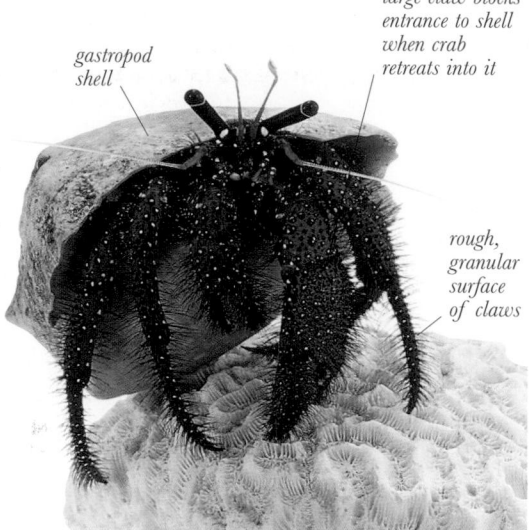

gastropod shell

large claw blocks entrance to shell when crab retreats into it

rough, granular surface of claws

SQUAT LOBSTERS
Squat lobsters (family Galatheidae) are relatively slow movers but most species use their abdomen to swim rapidly backwards to escape predators. Many species are mainly nocturnal and hide under rocks and in cracks during the day. Shown here is Galathea strigosa, which is common in shallow seas in the British Isles, Scandinavia, Spain, and the Mediterranean.

HERMIT CRABS
The most characteristic feature of hermit crabs (family Paguridae) is that they carry the shells of gastropods around as homes, changing an old shell for a bigger one as they grow. The common hermit crab, Pagurus bernhardus, shown above, inhabits the lower shores of sandy and rocky, northern and western European coasts.

LAND HERMIT CRABS
Like the hermit crabs, land hermit crabs (family Coenobitidae) also carry shells around. An exception is the coconut or robber crab, Birgus latro, pictured fighting above, which does not have a shell. It is the heaviest terrestrial crustacean, with a weight of up to 4kg (8¾ lb). It lives in the tropics, and feeds mainly on the fruit of coconut palms. A good climber, its slow speed means it is easy to hunt and is becoming rare in some places.

blue upper body

spiny rostrum

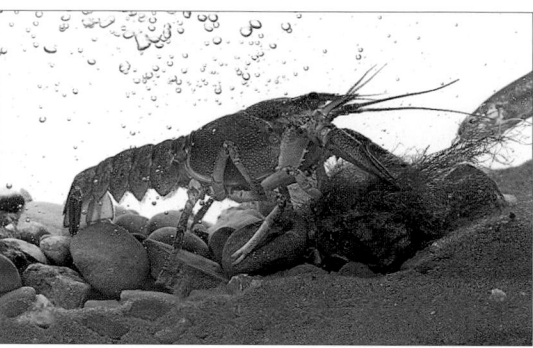

CRAYFISH
Members of the family Astacidae, freshwater crayfish resemble small lobsters. The first 3 pairs of legs have claws, the front pair being the largest. They make burrows in silt or mud where they hide during the day. Austropotamobius pallipes, shown above, inhabits well-aerated streams and limestone areas in southwest Europe.

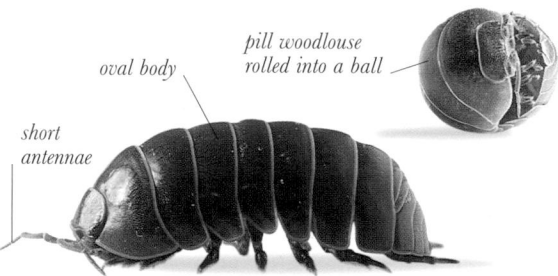

oval body

pill woodlouse rolled into a ball

short antennae

LOBSTERS
The family Nephropidae includes the largest and most commercially important crustaceans, the lobsters. Measuring up to 1m (3¼ ft) in length, lobsters have a roughly cylindrical carapace with transverse and oblique grooves. The abdomen is elongated with a broad tail fan for rapid backward escape. The common lobster, Homarus gammarus, shown above, has a blue upper and a yellowish lower surface. Using its large claws, it cracks open molluscs and feeds on them at night.

segmented brownish body

PILL WOODLICE
Pill woodlice or pillbugs (family Armadillidiidae) have the ability to roll into a tight ball for protection. They have a rounded back as well as a rounded hind margin. Usually terrestrial, they have 2 pairs of lung-like organs (pseudotracheae) on the abdominal appendages for breathing air. As in many woodlice, the newly hatched young have one pair of legs fewer than they will have as adults. Pictured here is the common pill woodlouse, Armadillidium vulgare.

ONISCID WOODLICE
Most oniscid woodlice (family Oniscidae) are terrestrial, living in damp microhabitats such as rotting wood, under stones, and in caves. They have a flattened body with clear segmentation and no carapace. Females have a brood pouch formed from plates at the bases of the first five pairs of walking legs, or pereopods. The common shiny woodlouse, Oniscus asellus, pictured left, has a shiny, mottled greyish body with 2 rows of pale patches.

long first antennae

OPOSSUM SHRIMPS
Shrimps of the family Mysidae have distinctive movement sensors at the base of the inner pair of flap-like appendages on either side of the tail fan. Most of them live worldwide in estuarine or marine waters (from the coast to the deep sea), and are free-swimmers. Elongated and soft-bodied, many are pale or translucent but deep sea species are red. Many form shoals and are important food for both fish and humans.

SKELETON SHRIMPS
Skeleton shrimps (family Caprellidae) do not swim, and instead cling to seaweeds, catching food that drifts within reach. Their bodies are extremely slender, with a long thorax and a tiny abdomen. The specimen shown here belongs to the genus Caprella which, with over 140 species, is the largest in the family. Close relatives of these animals live on whales and dolphins.

swimming appendages

BEACH-HOPPERS
Beach-hoppers and their relatives (family Gammaridae) live on the coast and in fresh water, often among rotting vegetation. Most species – like this one from Europe – have a curved abdomen that can flick the animal away from danger.

SEA-SPIDERS

PHYLUM	Arthropoda
SUB PHYLUM	Chelicerata
CLASS	Pycnogonida
ORDERS	1 (Pantopoda)
FAMILIES	9
SPECIES	About 1,000

SEA-SPIDERS ARE UNUSUAL marine arthropods with a small, cylindrical body and long, slender legs that give them a superficial resemblance to terrestrial spiders. They occur throughout the world's oceans, occupying habitats ranging from shallow, coastal areas to deep ocean waters. Most live on the sea bottom, although some can swim. The greatest numbers of species and the largest individuals – with legspans of up to 75cm (30in) – are found in deep water.

ANATOMY

Sea-spiders have a small head, or cephalon, 3 trunk segments, and a short abdomen. The head has a raised projection which supports 2 pairs of eyes. It also bears a proboscis used for feeding, 2 clawed appendages (or chelifers), and a pair of palps. Most sea-spiders have 4 pairs of walking legs arising from the trunk segments, although some have 5 or 6 pairs. The front pair is used for carrying eggs and grooming. Sea-spiders have no respiratory or digestive organs: gases and dissolved substances are absorbed and released by diffusion.

STRADDLING PREY
This sea-spider is feeding on a colony of hydroid polyps. It keeps its legs in one position while moving its body over the polyps.

FEEDING

Most sea-spiders are carnivores, feeding on soft-bodied marine animals such as sponges, anenomes, corals, hydroids, and lace corals. When feeding, they usually straddle their prey and either suck its tissues up through their proboscis or break off small pieces of its body with their chelicerae and then pass them to their mouth.

Family **NYMPHONIDAE**

Nymphonid sea-spiders

Occurrence 175 spp. in the coastal Atlantic region; in intertidal to shallow water

Slender-bodied, and with a stout proboscis, nymphonids are more active and swim around more than other sea-spiders. They measure 1–8mm ($\frac{1}{32}$–$\frac{9}{16}$in) in length. The chelifers – a pair of appendages at the front – are typically clawed. Both males and females have egg-carrying legs that have 10 segments and may carry masses of up to 1,000 eggs. There may be 4–6 pairs of normal walking legs. These sea-spiders eat soft-bodied invertebrates such as hydroids and bryozoans. Most species inhabit deep water, although some are found in shallow, coastal water.

thin carapace   *long walking legs*

NYMPHON SPP.
These smooth-bodied animals are often overlooked because they are so small. Their legs may be 3 or 4 times longer than their body.

HORSESHOE CRABS

PHYLUM	Arthropoda
SUB PHYLUM	Chelicerata
CLASS	Merostomata
ORDERS	1 (Xiphosura)
FAMILIES	1 (Limulidae)
SPECIES	4

ALSO KNOWN AS KING CRABS, these are the only living representatives of a once diverse and successful group of animals that flourished over 300 million years ago. Although they are called crabs, they are not crustaceans, but close relatives of arachnids. They live in shallow waters off the coasts of North America and Southeast Asia, and are most evident in spring, when they emerge from the sea to mate and lay their eggs.

ANATOMY

Horseshoe crabs have a fused head and thorax, jointly called the cephalothorax. This is covered by a tough, horseshoe-shaped shell, or carapace, and supports 6 pairs of appendages: one pair of chelicerae, one pair of palps, and 4 pairs of walking legs. The abdomen bears a long, spine-like tail and 5 pairs of leaf-like gills. Horseshoe crabs have a pair of simple eyes at the front and a pair of compound eyes on either side.

REPRODUCTION
Every spring, horseshoe crabs gather along the Atlantic shoreline of the USA to mate. Eggs are deposited in the sand and then fertilized by the males.

FEEDING

Horseshoe crabs are scavengers, burrowing through sediment and mud in search of food such as molluscs, worms, and other marine animals. They catch prey using their chelicerae, or their first 4 pairs of walking legs, and then transfer it to the mouth.

Family **LIMULIDAE**

Horseshoe crabs

Occurrence 4 spp., 1 in North America, 3 in Southeast Asia; on the sea bed

The 4 living species of horseshoe crabs all belong to the family Limulidae. They get their name from the horseshoe-shaped shield, or carapace, that covers all of the body except the tail. Coloured a drab brown, horseshoe crabs are up to 60cm (23½in) long. The abdomen is hinged to the cephalothorax and has leaf-like gills on the underside; the tail spine, at the rear, is used to right themselves when overturned. Components of their blue blood have been used in medical research.

LIMULUS POLYPHEMUS
The American horseshoe crab is not yet an endangered species but is thought to be declining in numbers due to exploitation, habitat loss, and disturbance of egg-laying sites. Its eggs are a major food for many seabirds.

ARACHNIDS

PHYLUM	**Arthropoda**
SUB PHYLUM	**Crustacea**
CLASS	**Arachnida**
ORDERES	**12**
FAMILIES	**450**
SPECIES	**75,500**

THIS DIVERSE GROUP includes the familiar spiders, scorpions, and harvestmen. However, by far the most common and widespread members of the group are the ticks and mites. All arachnids have 2 main body segments and 4 pairs of legs (unlike insects, which have 3 main segments and 3 pairs of legs). Most are terrestrial animals, although about 10 per cent of ticks and mites (and one species of spider) are found in fresh water.

ANATOMY

An arachnid's body is divided into 2 sections. The head and thorax are fused to form a cephalothorax, which is joined to the abdomen, in some species by a narrow stalk.

The cephalothorax bears 6 pairs of appendages. The first pair, known as the chelicerae, are pincer- or fang-like and used mainly for feeding. The second pair, called the pedipalps, are leg-like or enlarged with claws at the end. The other 4 pairs are walking legs, although in some species the first pair are long and have a mainly sensory function.

The abdomen may be divided into smaller segments, or have a tail-like extension. It houses the respiratory organs: gases are exchanged via the trachea or organs known as book lungs. In spiders, the abdomen also contains silk glands.

FEEDING

Most arachnids are predacious, but a few are scavengers and some mites are parasitic. Large arachnids usually rely on strength to subdue their prey but spiders, scorpions, and pseudoscorpions can also inject venom. Since arachnids have narrow mouths, they cannot ingest large pieces of food. They usually release digestive enzymes over the prey and then suck up the resulting liquid. In scorpions, small pieces of prey are partly digested in a small chamber in front of the mouth.

Some spiders rely on sight and stealth to catch their prey. Others use silk to build traps. For example, some spiders build spirals of sticky silk, called orb webs; those of some tropical species are large and strong enough to entrap small birds. Net-casting spiders make small webs, which they hold in their legs and drop over passing prey.

REPRODUCTION

In most arachnids, the male passes a packet of sperm (or spermatophore) to the female using his pedipalps, chelicerae, or legs. The female releases the eggs, which in some scorpions hatch immediately. In some species, the young develop or hatch inside the female's body. Many arachnids show parental care by guarding their eggs.

HUNTING
Like most arachnids, this wolf spider eats live prey. It has impaled this grasshopper with its chelicerae, which each end in a hinged fang. The adjacent pedipalps are being used to manipulate the prey. Wolf spiders have acute vision and usually hunt at night.

Order SCORPIONES

SCORPIONS

THE MOST ANCIENT GROUP OF ALL ARACHNIDS, scorpions are recognizable by their large, claw-like pedipalps, and their sting-bearing tails. They hunt after dark, often finding prey by touch. Scorpions sometimes use their stings to paralyse prey, but the sting's key function is defence. Females give birth to live young, which they carry on their backs. There are about 1,400 species in 9 families.

Family BUTHIDAE

Buthids

Occurrence 520 spp. worldwide, in warmer regions; in rock cracks, under stones, logs, bark

Despite their relatively small size, 8–12cm (3¼–4¾in) in length, buthids are the most dangerous scorpions. Their potent venom can paralyse the heart and the respiratory system.

large tail

sting

long, claw-like pedipalps

spurs on tibiae

BUTHUS OCCITANUS
An inhabitant of scrubby, open ground, this scorpion is found in parts of North African and West Asia. Antidotes to its venom are available.

Family SCORPIONIDAE

Scorpionids

Occurrence 130 spp. in Africa, Asia, and Australia; in cracks, caves, under stones, logs

This family includes the world's largest scorpions, measuring up to 21cm (8½in) long. Although they look menacing, the larger species tend to be less venomous than the smaller ones, and rely primarily on physical strength to overpower their prey. Many of these dig to locate prey, which could include spiders, lizards, and even small mammals. Scorpionids sting not only to paralyse their victim, but may also sting mates as part of their complex mating ritual. A litter of 30–35 young stay with the mother for some time after their first moult. Scorpionids have a 5-sided sternum in contrast to the buthids' triangular one.

PANDIUS IMPERATOR
This is the imperial scorpion, some specimens of which can exceed 60g (2⅛oz) in weight. It can be found among leaf litter in the tropical forests of Central and West Africa.

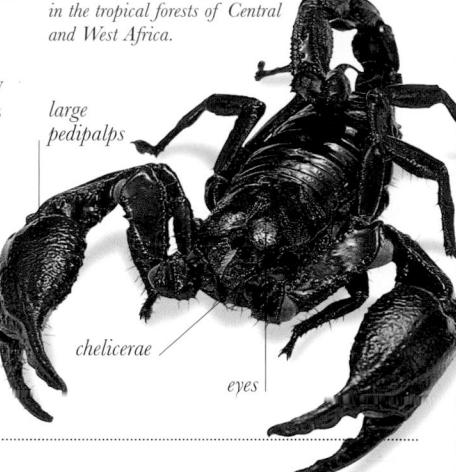

large pedipalps

chelicerae

eyes

Order PSEUDOSCORPIONES

PSEUDOSCORPIONS

ALSO CALLED FALSE SCORPIONS BECAUSE they lack a sting and tail, members of this order of 23 families (3,300 species) are otherwise similar to scorpions in shape. They have venom glands in pincer-like parts in their pedipalps and prepare silk nests in which to moult, brood their young, and hibernate. Eggs are laid in a special brood pouch. Females may cling to other animals to disperse to new habitats.

Family NEOBISIIDAE

Neobisiids

Occurrence 500 spp. worldwide, especially in the northern hemisphere; in leaf litter, soil, caves

Habitat Terrestrial

The members of this family are very small, measuring 1–5mm (¹⁄₃₂–³⁄₁₆in) in length. They usually have 4 eyes but cave-living species may have fewer eyes or none at all. In colour, neobisiids vary from olive to dark brown, with red, yellow, or cream tinges. The legs are often slightly green, with 2-segmented tarsi. The claw of each pedipalp has one venom gland located in the fixed finger; after paralysing its prey, the neobisiid shreds it with its large mouthparts.

red-brown pedipalps

NEOBISIUM MARITIMUM
Measuring 3–3.5mm (⅛–⁹⁄₆₄in) long, this species is found in cracks in rocks and under stones in coastal areas of Ireland, England, and France.

Family CHELIFERIDAE

Cheliferids

Occurrence 300 spp. worldwide, especially in warmer regions; in leaf litter, on tree bark

Unlike in neobisiids, in cheliferids both fingers of the pincers have glands that produce the venom they use to paralyse their prey. There are no teeth on the inner surfaces of the pincers. Cheliferids are 1.5–5mm (³⁄₆₄–³⁄₁₆in) long and usually have 2 eyes. Coloration varies from pale to dark brown or black, often tinged with red or olive, with dark markings. As in many arachnids, courtship can be complex; males and females dance holding each other's pedipalps.

pedipalp

brown coloration

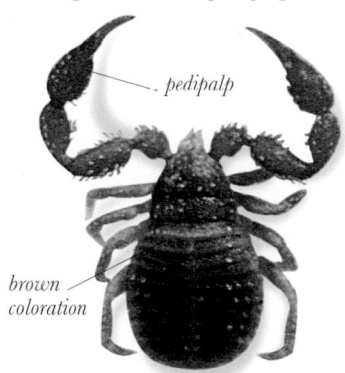

DACTYLOCHELIFER SPP.
While most of these species are distributed across the northern hemisphere in a variety of habitats, some are confined to coastal habitats.

Order SOLIFUGAE

SUN-SPIDERS OR WIND-SCORPIONS

NEITHER TRUE SPIDERS NOR SCORPIONS, the animals in this order lack poison glands. Instead, they kill and chew up their prey with their very large, pincer-like chelicerae. Their leg-like pedipalps have suction pads that hold down the prey so that it can be eaten. There are about 1,000 species, in 12 families, and most are found either in North America or in the tropics.

Family SOLPUGIDAE

Solpugids

Occurrence 200 spp. in Africa, parts of West Asia; on ground, under stones, logs, in burrows in sandy soil

Pale, brown or yellowish, sometimes with bright markings, these sun-spiders are 0.6–6cm (¼–2¼in) in length. Their 4 pairs of legs have a variable number of tarsi and all, except the first pair, have claws with a smooth surface. They have forward-facing chelicerae to kill prey; small species and nymphs eat termites. Females lay masses of 20–100 eggs.

Some solpugids are active by day; others burrow in sandy soil or hide in cracks or under stones during the day, and forage in tree canopies at night.

METASOLPULGA PICTA
This species is found in the Namib desert in Namibia. Like all solpugids, it will bite if handled carelessly, but does not have venom glands.

Family AMMOTRECHIDAE

Ammotrechids

Occurrence 72 spp. in the warmer regions of South, Central, and North America; in rotten tree stumps

This family comprises relatively slender species which measure 0.4–2cm (⁵⁄₃₂–¾in) in length. They are coloured in shades of brown. The front margin of the head is rounded, and the tarsi of the first pair of legs do not have claws. Many species are nocturnal and dig into the soil during the day. Some of the smaller ones hide in termite galleries or inside the tunnels of wood-boring insects.

These arachnids are predatory on a range of arthropods. Females may lay 2 or more masses of 20–150 eggs in their lifetime.

AMMOTRECHELLA STIMPSONI
This species is found in Florida and other areas in southern North America, where it lives in the bark of rotten, termite-infested trees.

Order UROPYGI

WHIP-SCORPIONS

THESE FLAT-BODIED ARACHNIDS have a whip-like tail (telson) at the end of their abdomen and can, in defence, squirt formic and acetic acids from a pair of abdominal glands. Their chelicerae resemble spider fangs. There are 2 families, with 99 species.

Family THELYPHONIDAE

Vinegaroons

Occurrence 75 spp. in N.E. South America, North America; in leaf litter, under stones, rotten wood

Vinegaroons hunt after dark, hiding in underground burrows during the day. They do not have a sting but can squirt acid over 60cm (23½ in), and can give a painful pinch with their pedipalps. Excluding the tail, they are up to 7.5cm (3in) long.

THELYPHONUS SPP.
Dark brown, with a tough, flattened body, these arachnids are found in Southeast Asia.

strong pedipalps for digging burrows

segmented telson

Order AMBLYPYGI

WHIP-SPIDERS

ALSO KNOWN AS TAILLESS WHIP-SCORPIONS, members of the 3 families (130 species) in this order do not sting or bite. They are nocturnal and cave-dwelling, with fang-like, segmented chelicerae, large, shiny pedipalps, and an extremely long first pair of legs.

Family PHRYNIDAE

Phrynids

Occurrence 52 spp. in tropical and subtropical regions; under stones, bark, in leaf litter and caves, among rocks

Brown, with dark markings, phrynids have a body 0.5–6cm (³⁄₁₆–2¼ in) long. They scuttle rapidly sideways if disturbed. Females carry young for months in a sac under the abdomen.

PHRYNUS SPP.
This phrynid, from Central America, has bent its long front legs backwards – a typical resting pose.

Order OPILIONES

HARVESTMEN

HARVESTMEN – OR DADDY-LONG-LEGS – are often mistaken for spiders, but they have rounded bodies, without a slender waist. Their legs are long and slender, and end in microscopic claws. Their mouthparts are small but they produce noxious secretions as a defence. Harvestmen are unusual among arachnids in that fertilization is direct. There are at least 5,000 species, in 40 families.

Family COSMETIDAE

Cosmetids

Occurrence 450 spp. mainly in tropical regions of North and South America; under stones, among debris

Like most harvestmen, cosmetids are generally dull in colour, although tropical species can be green or yellow, and a few can change colour over a matter of weeks to blend with their surroundings. They range from 5 to 11mm (³⁄₁₆–³⁄₈ in) in length. Males tend to be smaller than females but have longer appendages and larger chelicerae. The eyes are located close together on a small "turret" called an ocularium. Females use their ovipositors to lay eggs in cracks and crevices in the soil.

small pedipalps, close to chelicerae

VONONES SAYI
This harvestman is found in Panama. It uses its front legs to smear toxic secretions on attackers.

long, slender legs

Family PHALANGIIDAE

Phalangiids

Occurrence 200 spp. worldwide; under stones, among leaf litter in woods, grassy areas

Phalangiids can have a legspan of up to 10cm (4in), but their soft bodies are rarely more than 1cm (³⁄₈ in) long. Like other harvestmen, their eyes are close together, and are on a "turret" above the rest of the body. Males and females differ; the enlarged chelicerae of the males are especially distinct. Females lay 20–100 eggs at a time, burying them in soil or under bark. Most species are nocturnal and prefer humid habitats. Phalangiids are both predatory and saprophagous.

long second pair of legs

PHALANGIUM OPILIO
Found in the northern hemisphere, this white-grey to yellow, diurnal harvestman likes sunny spots, and inhabits woods, gardens, and grassland.

Order ACARI

TICKS AND MITES

WITH OVER 30,000 SPECIES AND about 300 families, ticks and mites are present in almost every habitat, particularly on land. They include many pests of crops and stored food, as well as parasites that attack mammals, birds, and reptiles. Most are less than 1mm (¹⁄₃₂ in) long – although ticks can grow much bigger after feeding on blood – and unlike spiders, their bodies have no distinct divisions.

Family IXODIDAE

Hard ticks

Occurrence 650 spp. worldwide; in association with birds, mammals, and some reptiles

Hard ticks have a tough dorsal plate, which fully covers the males, but covers only the front half in females. The abdomen is soft and flexible to allow for large blood meals. The front part of their head projects forwards and carries the mouthparts. Hard ticks are 0.2–1cm (¹⁄₁₆–³⁄₈ in) long but grow larger after feeding.

Colours vary from yellow to red- or black-brown; some species are highly marked. Some hard ticks carry viral diseases that affect humans, including encephalitis and Lyme disease, as well as diseases affecting cattle and poultry.

mouthparts
pitted surface

AMBLYOMMA AMERICANUM
The lone star tick is common in the grassland of Central America.

4 similar pairs of legs

Family TROMBIDIIDAE

Velvet mites

Occurrence 250 spp. worldwide, especially in tropical regions; in or on soil, some associated with fresh water

Named after their dense, velvety "fur", these bright red or orange mites eat insect eggs when they are adult, but start life as parasites, attacking insects, spiders, and harvestmen. The adults often emerge from the soil after rain to mate and lay eggs.

TROMBIDIUM SPP.
Trombidium velvet mites are found in many parts of the world. This species, from South Africa, grows up to 1cm (³⁄₈ in) long.

Family TETRANYCHIDAE

Spider mites

Occurrence 650 spp. worldwide; on a range of trees, plants, and shrubs

Habitat All

These soft-bodied mites are orange, red, or yellow, and 0.2–0.8mm (¹⁄₁₂₈–¹⁄₃₂ in) long. Many species produce silk. Eggs are laid on leaves, and the young are protected by silk webs. Large numbers of these sap-feeders infest host plants, which may wither and develop blotches.

TETRANYCHUS SPP.
Abundant in temperate regions, red spider mites may produce 4–5 generations per year. They are pests of deciduous fruit trees.

Order ARANEAE

SPIDERS

SPIDERS CAN BE DISTINGUISHED FROM other arachnids by their distinctive appearance and by their ability to produce silk. A typical spider has 8 eyes, and a body divided into 2 parts – the cephalothorax and the abdomen. The mouth is flanked by a pair of venom-injecting fangs, and a pair of leg-like pedipalps. These have a sensory function and, in males, are also used to transfer sperm. All spiders are predatory, injecting venom into their prey. Some use silk to catch prey, but spiders also use silk to protect their eggs, lower themselves through the air, and even to sail on the breeze. There are at least 40,000 species, in about 100 families, and they live in every terrestrial habitat, from tropical rainforests to cellars and caves.

Family ARANEIDAE

Orb web spiders

Occurrence 4,000 spp. worldwide; in grassland, meadows, woodland, forests, gardens

Habitat Terrestrial

As the name implies, many of these spiders weave circular webs made up of a central hub with radiating lines and spirals of sticky and non-sticky silk. They typically catch prey in the web, wrap it in silk, cut it out, and take it to a retreat to eat it. Some species spin opaque bands of matted silk in their webs so that birds can see them and do not accidentally destroy the webs by falling into them. However, there are a few species that do not make webs; these prey after dark, using a single thread with a bead of glue at the end to snare moths. The strength of the webs can be such that in Papua New Guinea, the huge webs of the *Nephila* species are used as fishing nets. Orb web spiders often have very large, brightly coloured, and patterned abdomens; in tropical species the abdomen may have unusual, angular shapes. The legs have 3 claws which can be very spiny. Of the 8 eyes, the 4 in the middle often form a square. Orb web spiders are 0.2–4.6cm (¹⁄₁₆–1¾in) long. Males approach females on their webs and pluck threads to attract them. Females lay eggs in silk cocoons which can be camouflaged in webs and litter, or stuck to vegetation or bark.

bright colours deter birds

stout abdominal projections

GASTERACANTHA SPP.
These tough-bodied, spiny-bellied orb weavers are found in North and South America. They are often seen hanging in the middle of their webs. They inhabit shrubs and trees, and their webs can also be seen on walls of houses.

cross-shaped abdominal markings

banded legs

ARANEUS DIADEMATUS
Variably coloured (from pale reddish to almost black), the garden orb web spider, or the garden cross spider, inhabits woods and gardens in Europe and North America. Its webs can be nearly 50cm (20in) across.

spine-like projections on abdomen

MICRATHENA GRACILIS
Called the spiny orb-weaving spider, this species inhabits woodland and forest in North America.

mottled camouflage colouring

Family AGELENIDAE

Funnel weavers

Occurrence 700 spp. worldwide; in meadows, gardens, bushes, houses, among stones, on walls

Funnel weavers are often long-legged, and have minute, feathery hairs on the body, 8 eyes, and a slender abdomen with dark bars, chevrons, or spots. The body is 0.6–2cm (¼–¾in) long. The funnel weaver's web takes the form of a flat, tangled silk sheet with a tunnel on one side. Females may feed their offspring by regurgitation and, in some species, the spiderlings may eat the mother when she dies.

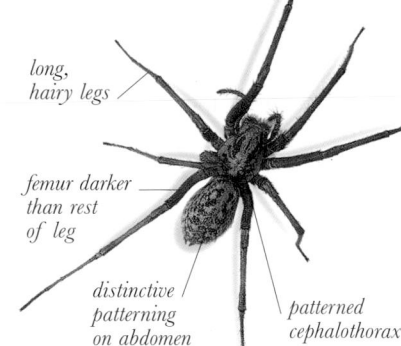

long, hairy legs

femur darker than rest of leg

distinctive patterning on abdomen

patterned cephalothorax

TEGENARIA GIGANTEA
This large spider can have a legspan of up to 8cm (3¼in), and can run very fast. Members of the genus Tegenaria *are called house spiders. The large spiders that fall into domestic baths usually belong to this genus.*

Family ARGYRONETIDAE

Water spider

Occurrence 1 sp. in Europe and parts of Asia; in slow-flowing or still water

This species is unique in living under the water surface. It ferries air, trapped by the dense hair on its abdomen, to a dome-shaped, silk "diving bell" which it anchors to underwater plants. The bell is used for feeding, courtship, and hibernation, and the eggs (wrapped in silk) are also stowed inside. The water spider is 0.7–1.5cm (⁷⁄₃₂–½in) long.

ARGYRONETA AQUATICA
This spider lives permanently in still or slow-moving water, and the air trapped by the body hairs around the abdomen gives it a silvery appearance.

Family DIPLURIDAE

Funnel-web spiders

Occurrence 250 spp. in tropical regions of the Americas, Africa, Asia, and Australia; in trees, on ground

Funnel-web spiders belong to a group known as mygalomorphs, which also includes the tarantulas (see p.593). Unlike other spiders, mygalomorphs are relatively primitive, with fangs that strike downwards, instead of closing sideways like a pair of pincers.
Funnel-webs are 1–3cm (³⁄₈–1¼in) long and, usually, dark brown. Most make a funnel-shaped retreat that leads into crevices in tree stumps or rocks. Some are highly venomous.

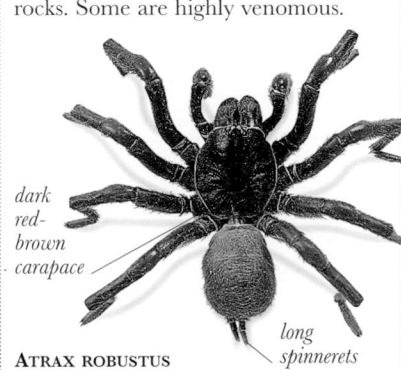

dark red-brown carapace

long spinnerets

ATRAX ROBUSTUS
The Sydney funnel-web, a native of South Australia, is an aggressive spider which has a painful, venomous bite that may prove fatal.

Family LINYPHIIDAE

Dwarf spiders

Occurrence 4,200 spp. mostly in temperate regions of the north, some in Arctic; on stones, among vegetation

Also known as money spiders, these tiny animals are rarely more than 1cm (³⁄₈in) long. They have large chelicerae with sharp teeth. The legs have strong bristles and the carapace of males may have extensions, which sometimes carry the eyes. Dwarf spiders live in leaf litter or make characteristic, horizontal, non-sticky, sheet-like webs among vegetation. Egg sacs are attached to stones, plants, and other surfaces. These spiders travel over long distances by "ballooning" on long silk threads.

bright red-orange cephalothorax

grey, oval abdomen

red-orange legs

GONATIUM SPP.
These small spiders are widespread in the northern hemisphere, inhabiting scrubland and grassland. They prefer low, shaded bushes.

INVERTEBRATES

Order ARANEAE *continued*

Family SALTICIDAE

Jumping spiders

Occurrence 5,000 spp. worldwide; in grassland, meadows, woodland, forests, gardens

The 5,000 species of jumping spider (salticids) make up one of the largest of the spider families. Most jumping spiders have stout, hairy bodies, 2–16mm (1/16–1/2in) long, and are often drab in appearance, although many tropical species are vividly coloured, with elaborate markings. Some species are very ant-like, mimicking their prey in both looks and behaviour. Although the different species do vary greatly, a distinctive feature common to all is their prominent eyes. At the front of the carapace, which is rectangular, with a rounded rear and square front, sits a row of 4 large eyes. The middle 2 eyes, which are very much larger than the others, resemble old-fashioned car headlamps. These eyes enable jumping spiders to judge distance, shape,

light and dark bands on legs

compact body

drab coloration

EUOPHRYS SPECIES
Members of the genus Euophrys are often found on the ground, among grass and low-growing plants, or under stones. As in many species, males fight for access to a female, the larger male usually winning.

JUMPING

Salticids jump not only to catch prey but also to escape from predators. The hind pair or 2 hind pairs of legs are extended rapidly by hydraulic pressure.

ATTACKING PREY
Before it leaps, a jumping spider attaches a safety line of silk in case it gets carried off course. This species is known as the zebra spider on account of its striped markings. The mouthparts (chelicerae) of males are larger than those of the females.

STRIKING APPEARANCE

Some tropical jumping spiders are very brightly coloured, with iridescent hairs arranged in patches and spots. In some, the whole of the upper surface of the abdomen may be bright red or orange. Males often have enlarged front legs and ornamented palps, which they use to attract the attention of females.

PHIDIPPUS REGIUS
Males and females of this species, native to S.E. North America, communicate using more than 20 different signals made with their legs and palps. Other spiders in the genus do the same.

CHRYSILLA SPECIES
These spiders catch and eat ants in preference to any other kind of insect. Some species jump on their prey from behind, while others tackle the ant head-first.

brightly coloured carapace

thickened front legs

and movement very accurately, and give them the keenest vision of any spider. The back of the eye capsule can be moved inside the head to keep the image of the prey centred on the retina. The smaller, outside eyes and a pair towards the rear of the carapace simply detect movement. Jumping spiders are active during the day, stalking insect prey on the ground, on walls, and in bushes, especially in warm, sunny locations. Their characteristic hunting method of jumping on their prey to seize it gives them their common name. When the weather is bad, they retreat to a small, silken nest they make in a crevice or crack. The spiders use similar shelters over winter and when they moult or lay eggs. The eggs, laid among vegetation, bark, and moss, are wrapped in silk, and the female guards them until they hatch.

ACUTE EYESIGHT
The 2 outside eyes of this jumping spider (Salticus sp.) can detect moving prey up to 25cm (10in) away. By turning its body, the spider brings the prey into the field of vision of the 2 middle eyes, which give a large, high-resolution image. The spider then moves towards the prey and jumps at it from a distance of 5–10cm (2–4in).

Order ARANEAE *continued*

Family LYCOSIDAE
Wolf spiders

Occurrence 3,000 spp. worldwide, even in Arctic regions; mostly on ground, among leaf litter

Habitat All terrestrial

Wolf spiders are 0.4–4cm (³⁄₃₂–1½in) long and have excellent eyesight which helps them in their nocturnal hunts. They are pale grey to dark brown, with markings such as bands, stripes, or spots. Females often carry their egg sacs attached to their spinnerets or, in burrowing species, keep the sacs in a silk-lined burrow. They carry spiderlings on their backs. Some wolf spiders are important predators in field ecosystems. This family includes the true tarantula, *Lycosa tarentula*, of southern Europe.

PARDOSA AMENATA
This European species is variable in appearance – the abdomen, for example, may be brown or grey.

Family PHOLCIDAE
Daddy-long-legs spiders

Occurrence 350 spp. worldwide; in caves, leaf litter, dark corners of buildings

Pale legs that are much longer than the body give these spiders a spindly appearance. Coloured grey, green, or brown, they have eyes arranged in 2 groups of 3 each, with another pair in between; a few cave-dwelling species, however, are blind. These spiders are 3–14mm (⅛–½in) long

dark "knees"

conspicuous palps in male

brown body

PHOLCUS PHALANGIOIDES
In the British Isles, this species is found in rooms and cellars of buildings. In the rest of Europe, it inhabits caves.

and have flexible tarsi, with many false joints. Also known as cellar spiders, daddy-long-legs spiders make tangled, irregular webs and quickly wrap prey in silk before biting it. When disturbed, they vibrate their webs rapidly, making themselves appear blurred and difficult to spot. Females carry 15–20 pale, pinkish grey eggs, held with silk, in their chelicerae.

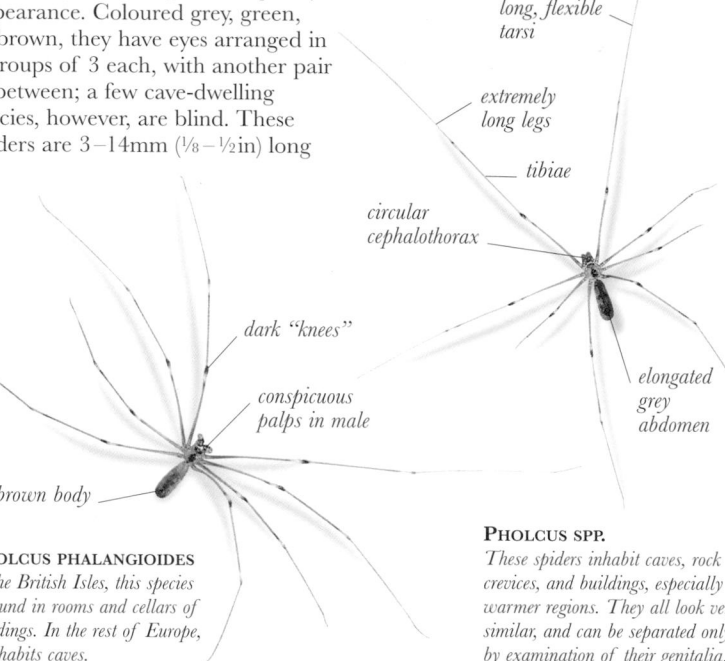

long, flexible tarsi

extremely long legs

tibiae

circular cephalothorax

elongated grey abdomen

PHOLCUS SPP.
These spiders inhabit caves, rock crevices, and buildings, especially in warmer regions. They all look very similar, and can be separated only by examination of their genitalia.

Family PISAURIDAE
Nursery-web spiders

Occurrence 550 spp. worldwide; on ground, surface of still water, aquatic plants

Habitat All

These large spiders resemble wolf spiders (see left) in appearance and hunting technique – running and catching prey on the ground rather than using webs. The carapace is oval and has longitudinal markings; the body is 1–2.6cm (⅜–1in) long. The females carry the egg sacs in their chelicerae slung below the body, placing them in tent-like nursery webs at hatching time. The sacs and spiderlings are guarded by the females until the second moult, when they leave the webs and disperse.

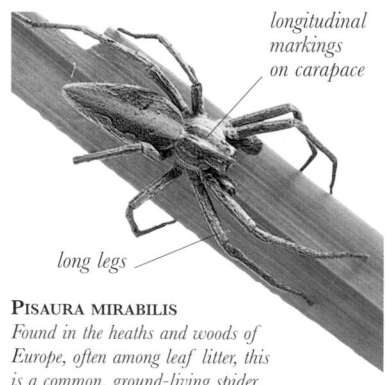

longitudinal markings on carapace

long legs

PISAURA MIRABILIS
Found in the heaths and woods of Europe, often among leaf litter, this is a common, ground-living spider.

Family SCYTODIDAE
Spitting spiders

Occurrence 180 spp. worldwide, except Australia and New Zealand; under rocks, in buildings

The common name of these 6-eyed spiders comes from their unique prey-capturing technique. Using a rapid side-to-side movement of the chelicerae, they squirt 2 zigzag streams of sticky, glue-like substance from close range to stick the prey down. Spitting spiders measure 4–12mm (⁵⁄₃₂–⅜in) long and are cream or

yellow-brown with black markings. Females carry the egg sacs under their body until the young emerge.

SCYTODES THORACICA
Often found inside buildings in North America and Europe, this spider can squirt glue over 1cm (⅜in). Males are slightly smaller than females.

Family SICARIIDAE
Six-eyed crab spiders

Occurrence 100 spp. in warm parts of N. and S. America, Europe, Africa; in shaded areas in rocks and bark

Also known as brown spiders because of their colour, these arachnids have 6 eyes, arranged in 3 pairs, and distinctly hairy body and legs. They are 0.6–1.8cm (¼–¾in) long and, in most species, have a violin-shaped mark on the carapace. They make sticky, sheet-like webs, to which some species keep adding as they grow. Females lay 30–300 eggs per sac and

keep the sacs at the rear of the web. The bite of these spiders can be very dangerous, causing tissue degeneration.

LOXOSCELES RUFESCENS
Common in Europe and introduced to Australia, this spider may bite humans and cause unpleasant lesions that are slow to heal.

Family HETEROPODIDAE
Huntsman spiders

Occurrence 1,000 spp. in tropical and subtropical regions; on ground, tree trunks

These drably coloured spiders are efficient, nocturnal hunters, able to move sideways with great agility. The larger species, also known as giant crab spiders, can easily tackle lizards. The carapace and abdomen are flattened. Legs may span up to 15cm (6in) and the body can be 1–5cm (⅜–2in) long. There are 8 eyes of equal size, 4 of which point forwards from the front edge of the carapace.

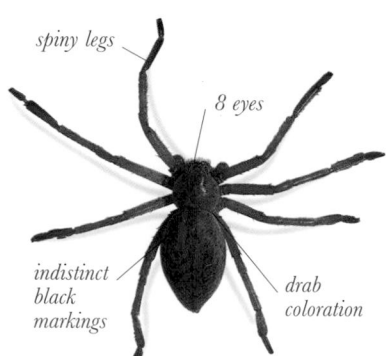

spiny legs

8 eyes

indistinct black markings

drab coloration

HETEROPODA SPP.
Huntsman spiders are pantropical species. Often found in crates of bananas, they are known to bite warehouse workers but are not dangerous.

Family THERIDIIDAE
Comb-footed spiders

Occurrence 2,200 spp. worldwide; in vegetation, under stones, in leaf litter, cracks, crevices

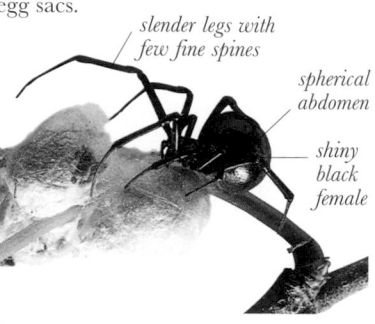

Also called cobweb spiders, these brown to black arachnids have a comb-like row of stout bristles on their hindlegs. The body is 0.2–1.5cm (¹⁄₁₆–½in) long and has a very rounded abdomen. Comb-footed spiders construct irregular webs, some of which may have sticky lines extending to the ground. Females are venomous and a few, such as the black widow, are dangerous to humans. They lay up to 1,000 eggs per year, which they often guard in round, silk egg sacs.

slender legs with few fine spines

spherical abdomen

shiny black female

LATRODECTUS MACTANS
The male of this species often dies shortly after mating and may be eaten by the venomous female; hence the common name, black widow.

hairy tarsi

prey

Family THERAPHOSIDAE
Tarantulas

Occurrence 850 spp. worldwide, especially in subtropical and tropical regions; in ground burrows, trees

The name tarantula was originally applied to a wolf spider from southern Europe (see p.592), but nowadays it is far more often used for this family. These are the giants of the spider world, with bodies up to 12cm (4¾in) long, and a legspan of up to 28cm (11in). Their bodies and legs are covered in bristly hairs, and their coloration varies from pale brown to black, with markings in pink, brown, red, or black. They are mygalomorphs, like funnel-web spiders (see p.589), with fangs that move up and down. When threatened, a tarantula will display its downward-pointing fangs. The large size of tarantulas has led to the belief that their bites are fatal but only some species are venomous; in fact the venomous species are often relatively small. Tarantulas are nocturnal hunters, relying on their large

size to subdue prey, which include small vertebrates such as frogs, lizards, and birds. They crush their prey with their large fangs, pour digestive juices over the body, and then suck up the resulting liquid. Females lay a batch of eggs in a burrow. An egg sac can be the size of a golf ball and contain about 1,000 eggs. Spiderlings stay in the burrow until their first moult, and then disperse to find food and make their own burrows. Several species of this family live for 10–30 years, and some are kept as pets.

BRACHYPELMA VAGANS
The Mexican red-rumped tarantula is a ground-living species that hunts at night. It is found in semi-desert and arid regions of Central America.

POECILOTHERIA REGALIS
Commonly called the tiger or Indian ornamental spider, this tarantula is the world's largest tree-dwelling spider, with a legspan of 17cm (6½in). It is found in India and Sri Lanka. It hides in a silk-lined retreat by day and hunts birds, lizards, and arthropods at night.

pedipalps

cheliceral fangs may be 1cm (⅜ in) long

beige and brown leg markings

legspan of 10cm (4in)

legs covered with hair

spinnerets

dark-edged, pale stripe down centre of abdomen

BRACHYPELMA EMILIA
The Mexican red-legged tarantula lives in burrows in desert and semi-desert regions of Central America. It kicks up the hairs of its abdomen as a defensive reaction.

black triangle at front of carapace

very hairy tarsi

black, hairy femora

robust, hairy body

red hairs on tibiae

GRAMMOSTOLA ROSEA
Found in South America, especially in Chile, the ground-living Chilean rose is one of the species most commonly kept as a pet because of its docile nature.

Family THOMISIDAE
Crab spiders

Occurrence 2,500 spp. worldwide; in meadows, gardens, on flowers, other parts of plants, tree bark

The common name derives from these spiders' characteristic crab-like sideways movement (although they can also move forwards and backwards) and usually squat shape; however, some species are elongated. Crab spiders range from 4 to 14mm (⁵⁄₃₂–½in) in length. The carapace is nearly circular, and the abdomen is short, often blunt-ended. The first 2 pairs of legs, used to seize prey, are larger and spinier than the other 2 pairs. Crab spiders often ambush insects landing on flowers – species that do so have very powerful venom which can easily kill a bee much larger than themselves.

Many species are coloured white, bright pink, or yellow, to match the flowers, and a few can change colour. The female keeps her eggs in a flat sac attached to plants, and guards them until she dies.

long front pairs of legs

distinctive eye pattern

squat shape

abdomen broad at rear

hindlegs can be held straight out

small, dark spots on abdomen

MISUMENA VATIA
This is one of the species of crab spider with the ability to change its colour in order to merge with the surroundings. Mostly found on white or yellow flowers, it is usually one of these 2 colours (see above and left), but can also be pale green. Common throughout Europe and North America, it can tackle large prey such as butterflies and bees. The female is much larger than the male.

TIBELLUS OBLONGUS
Sometimes called the running crab spider, this fast-moving hunter inhabits grassland and heaths in Europe. It often conceals itself on grass blades by holding its front 2 pairs and hind pair of legs out straight. The third pair grips around the blade. Some authorities place this spider in a separate family – Philodromidae.

ECHINODERMS

THIS GROUP OF MARINE invertebrates includes starfish, brittlestars, sea urchins, and sea cucumbers. Their bodies are typically spiny and usually divided into 5 equal parts arranged symmetrically around a central point. All echinoderms have a skeleton (or test) made of calcium-carbonate plates. They also have a unique internal network, called the water-vascular system, that helps them to move, feed, and take in oxygen. Echinoderms are mostly mobile, bottom-dwelling animals, found on shores, reefs, and the sea bed.

PHYLUM	Echinodermata
CLASSES	6
ORDERS	36
FAMILIES	145
SPECIES	About 6,000

ANATOMY

Echinoderms have a wide range of external shapes: their bodies may be drawn out into arms (as in starfish), branched and feathery (as in sea lilies), or spherical to cylindrical (as in sea urchins and sea cucumbers). In most, spiny or knob-like extensions of the skeleton project from the body. The water-vascular system consists of a network of canals and reservoirs, and tentacles that form external appendages called tube feet.

calcium-carbonate plate · anus · tube feet (cover whole body)
gonad · spine
intestine / mouthparts / water-vascular canal

BODY SECTION
An echinoderm's body consists of a central cavity enclosed by calcium-carbonate plates, which either fit together to form a rigid structure (as in sea urchins) or remain separate (as in starfish). Echinoderms have no head and no tail. The mouth is in the centre of one surface (the oral end), and the anus, which is sometimes missing, is in the centre of the opposite (aboral) surface.

MOVEMENT

While sea lilies lead stationary lives on the sea floor, most other echinoderms are mobile. Starfish and sea urchins move with their tube feet, spines, or the test itself. Sand dollars and sea cucumbers use their feet to burrow into the sediment.

WALKING
Sea urchins walk on the sea floor by moving water into and out of their tube feet. These extend beyond the spines and grip the surface as the animal pulls itself along.

PROTECTIVE SPINES
Many echinoderms are protected from predators by their long, sharp spines. In the crown of thorns starfish shown here, these spines are also poisonous. This starfish feeds on coral polyps and has destroyed large areas of the Great Barrier Reef in Australia.

FEEDING

Most echinoderms feed on small organisms or particles of organic matter. Some starfish, however, are carnivores, and extend their stomach over the prey, digesting it externally. Most sea urchins are grazers that use elements of their test as a series of teeth.

DEPOSIT FEEDING
Sea cucumbers use small tentacles around their mouths to take in mud and sediment, from which they extract food particles.

Sea lilies and feather stars

Occurrence 630 spp. worldwide; permanently or temporarily attached to the sea bed

Sea lilies and feather stars belong to an ancient group of echinoderms, which are abundant as fossils. They have rounded bodies and are heavily armoured with 5 arms, which often branch repeatedly. All begin life attached to a slender stalk. Sea lilies, which inhabit depths of more than 100m (350ft), remain attached to the stalk throughout their lives; in contrast, feather stars break away from the stalk when young and become free-living. Both forms resemble starfish in their basic anatomy, except that their mouths face upwards. They feed on plankton, caught by the tube feet on the arms and passed to the mouth. The arms may have a span of up to 70cm (28in), and the stalks can reach 1m (3¼ft) in length. Like some other echinoderms, they are able to regrow lost limbs. The sexes are separate.

arms spread out to collect food

OXYCOMANTHUS BENNETTI
Although stalked sea lilies dominate the fossil record, most surviving forms of crinoids are not permanently attached to the sea bed, and some can even swim. Feather stars are filter feeders that occur in all seas, particularly on reefs. This beautiful species has a crown of arms up to 40cm (16in) across.

Starfish

Occurrence 1,500 spp. worldwide; on the sea bed

Thanks to their distinctive shape and abundance along coasts, starfish, also called sea stars, are familiar marine invertebrates. Most have a circular body that grades into 5 equal arms, but some have more than 40 arms, and others have arms so short that the animal looks like a cushion. Starfish may swallow small prey whole; some can extend their stomach through the mouth and onto larger animals and digest them externally. When feeding on bivalves, they use powerful suckers on the tube feet to pull the shell valves apart. Sizes vary from less than 2cm (¾in) to 1m (3¼ft) in diameter. Starfish are famous for their ability to regenerate missing body parts such as lost arms.

long arms

tube feet on undersides of arms

LUIDIA CILIARIS
This starfish has 7 long arms instead of the standard 5. Up to 60cm (23½in) across, it lives on coarse sediments in the North Atlantic, where it feeds voraciously on other starfish, sea urchins, and brittlestars.

CULCITA NOVAEGUINEAE
This Southeast Asian species is a typical cushion star, about 25cm (10in) across, with arms that barely project from its body. Cushion stars can be found in a variety of marine habitats, but are particularly common on reefs.

Class OPHIUROIDEA

Brittlestars and basket-stars

Occurrence 2,000 spp. worldwide; on the sea bed

Although they are often mistaken for starfish, brittlestars and basket-stars form a separate and distinctive group of echinoderms. They have 5 long, narrow arms which, unlike those of starfish, are sharply demarcated from the small, central disc. The disc measures up to 10cm (4in) across, while the entire body can be up to 50cm (20in) in diameter. Brittlestars are the most mobile of all echinoderms and, like many of their relatives, can regenerate their arms if these are lost. Unlike starfish, brittlestars are either scavengers, feeding on dead animals, or suspension feeders, collecting small particles of food drifting down to the sea bed, helped by a sticky mucus that is strung between the spines on their arms. Brittlestars are extremely abundant on some parts of the ocean floor, forming writhing masses that contain millions of animals. In contrast, basket-stars feed on their own, holding their intricately divided and branched arms at right angles to the current, so that they can catch small animals drifting past.

OPHIOTHRIX SPP.
These brittlestars may occur in dense aggregations on both rocky and sandy areas of the sea bed. They have a diameter of up to 20cm (8in). Their spine-fringed arms project up into the water and filter feed.

fine spines

OPHIOTHRIX FRAGILIS
This small, northern brittlestar has a small disc and 5 long, sinuous arms with which it walks. Less than 15cm (6in) in diameter, it is spiny all over.

central disc

sinuous arms

ASTROBOA NUDA
Each arm of a basket-star branches repeatedly to give rise to a mass of armlets up to 75cm (30in) in diameter. In the Indo-Pacific species shown above, each armlet ends in tendrils. At night the basket of arms is held up into the water to catch small swimming animals.

Class HOLOTHUROIDEA

Sea cucumbers

Occurrence 1,150 spp. worldwide; almost all buried in the sea bed, or on the sea-bed surface

With their sausage-like shape and leathery texture, sea cucumbers look very different from other echinoderms, yet their internal anatomy is typical of this group, with organs present in sets of 5. They have 8–30 tube feet around the mouth, arranged in 5 rows; these are modified into finger- or bush-like feeding tentacles. Sea cucumbers vary from less than 3cm (1¼in) to 1m (3¼ft) in length. Some sea cucumbers discharge sticky, sometimes toxic, threads through their anus; these can entangle a would-be predator.

PSEUDOCOLOCHIRUS TRICOLOR
This brightly coloured, reef-dwelling sea cucumber from the Indo-Pacific has a ring of branched tentacles around its mouth. After feeding, each tentacle in turn is inserted into the mouth to be wiped clean. This animal is 10cm (4in) long.

INVERTEBRATES

Class ECHINOIDEA

Sea urchins

Occurrence 975 spp. worldwide; on or buried beneath the sea bed

Sea urchins typically have a spherical skeleton (test) of calcareous plates, covered by movable spines. Five rows of tube feet – often hidden by the spines – help these animals to creep over rocks, so that they can graze on algae and small animals. The mouth is on the underside, in the centre, and is armed with the "Aristotle's lantern" – a set of 5 large, calcareous plates that work like jaws. After feeding, many sea urchins return to refuges that they have excavated over a period of time in soft rock; in the tropics, pencil urchins use their extra-thick spines to wedge themselves into crevices in coral reefs. The burrowing heart urchins and sand dollars are more disc-shaped, and often have very short spines, in some species looking like backswept fur. Burrowing species are deposit feeders, mostly passing food to the mouth using the tube feet and spines. Sea urchins are usually 6–12cm (2¼–4¾in) in diameter, but some can grow to 35cm (14in).

spines and tube feet arranged in rows

ECHINUS ESCULENTUS
The reproductive organs (gonads) of this large sea urchin are prized as food around the shores of the North Atlantic. It occurs down to depths of 1,000m (3,300ft), grazing on animals and algae found on rocky surfaces. This species measures nearly 20cm (8in) in diameter.

PENCIL URCHINS
Phyllacanthus imperialis *(right), found in the Indo-Pacific, has only a few blunt but very stout spines, which are often covered by colonies of encrusting organisms. Up to 15cm (6in) in diameter, the spherical test is easily visible between the spines.*

blunt, stout spines

test

DIADEMA SPP.
These warm-water sea urchins, 30cm (12in) in diameter, have long, thin, brittle spines covered by glandular tissue that secretes a poisonous substance, which can cause severe irritation to human skin. Small fish and crustaceans often live among the sea urchin's spines, which offer them protection from predators.

SAND DOLLARS
Burrowing sea urchins such as sand dollars lack the spherical test and long spines of non-burrowing species. Like the tropical species pictured, Encope michelini, *they have flattened tests, up to 15cm (6in) across, with a slight dome and very short spines.*

INVERTEBRATE CHORDATES

PHYLUM	Chordata
SUB PHYLA	Urochordata
	Cephalochordata
CLASSES	4
ORDERS	9
FAMILIES	47
SPECIES	About 2,000

CHORDATES ARE ANIMALS that have an internal skeletal rod, called a notochord. Most chordates are vertebrates, in which the notochord is largely replaced by interlocking vertebrae. However, there are 2 groups of chordates that have no vertebral column, and are therefore classified as invertebrates. Both these groups, the lancelets and the tunicates, are marine animals. Lancelets have a long, narrow body and superficially resemble fish, while most tunicates have a bag-shaped body covered by a transparent tunic.

ANATOMY

Lancelets retain their notochord into adulthood, while in tunicates the notochord is present only in the larval stage. All species (apart from some tunicates) also have a hollow nerve cord above the notochord, a muscular tail (although in some species this is found only in the larvae), and a perforated throat (or pharynx). The pharynx forms an internal filtering mechanism to trap particles of food: water enters through a mouth-like opening (known as the branchial siphon) and leaves via openings that extend through to the body surface.

branchial siphon
pharynx
tunic
atrium
direction of water movement
intestines

notochord   nerve cord

pharynx   **LANCELET**   atrium   anus   muscles

TUNICATE

BODY SECTION
Most tunicates (left) are filter feeders that live attached to the sea floor. Water enters through the mouth and leaves through an opening called the atrium. The body of a lancelet (far left) is specialized for swimming: it is long, narrow, flexible, and muscular.

TUNICATE COLONIES
Colonies of tunicates often form a brightly coloured mass on the sea floor. Although the individual members of the colony have separate mouths, they sometimes share an opening for expelling water and waste products.

Sea squirts

Occurrence 2,000 spp. worldwide; attached to coastal rocks and the sea bed

Sea squirts are bag-like tunicates that live fastened to coastal rocks or to the sea bed. Some species are predators, but most feed by pumping water through a mesh-lined pharynx, which filters out particles of food. The water is drawn in through one tube, the inhalant siphon or mouth, and expelled through another, the exhalant siphon or atrium. The body is surrounded by a large, thick,

mouth
atrium

POLYCARPA AURATA
This solitary sea squirt from the Indo-Pacific is covered by a thick, leathery, sometimes brightly coloured, protective test or tunic, 2–15cm (¾–6in) high. Two siphons project from it, and these are the entry and exit sites of the water current from which the animal filters its food.

spherical to cylindrical tunic, or test, of cellulose-like material. Sea squirts are sometimes solitary, but many species form colonies, in which individual animals (zooids) are arranged in groups and the exhalant siphon is shared. The zooids range from 0.1 to 60cm (1/32 –23½in) in height, while colonies may reach several metres high or across. Sea squirts are often of particular interest because of their affinities with vertebrate chordates. As adults, the pharynx is their only visible chordate feature, but sea squirt larvae (which look like tadpoles) have a tail that contains a notochord and a hollow, dorsal nerve cord – both vertebrate characteristics.

BOTRYLLUS SCHLOSSERI
In some sea squirt colonies, such as the one above, the zooids are arranged so that each has its own site of water intake, but groups share a communal water exit. In this widespread species, the groups are arranged in star patterns, and the colour of the zooids – which are just 2mm (1/16 in) across – contrasts with the colony's jelly-like matrix.

Pelagic tunicates

Occurrence 70 spp. worldwide; planktonic in ocean

This class of tunicates contains free-living organisms, drifting among the plankton in the open sea, that have inhalant and exhalant openings at opposite ends of their test. Like sea squirts, pelagic tunicates have a large, perforated pharynx, which occupies most of the body mass. There are 2 basic kinds: pyrosomes, which are colonial, and salps and doliolids, which have alternating phases of solitary and colonial existence. Individuals in the colony (zooids) are up to 25cm (10in) long, but some complete colonies may exceed 14m (46ft) in length.

PYROSOMA ATLANTICUM
This colonial, luminescent, warm-water pyrosome, measuring more than 1m (3¼ft) in length, resembles a giant planktonic test tube. It filters food from the water, and uses its watery "exhaust" to propel itself slowly through the sea.

Lancelets

Occurrence 24+ spp. in tropical and temperate zones; in shallow sand or gravel

Lancelets are small – usually around 5cm (2in) long – semi-transparent animals with several vertebrate-like features. As well as a notochord, they have gill-like pharyngeal slits, a single dorsal fin, and – like many fish – muscle blocks arranged in V-shaped patterns down the body. Lancelets also have small tentacles around the mouth. They suck in water and food particles by beating microscopic hairs (cilia) that line their pharyngeal slits. The sexes are separate, and they reproduce by external fertilization. In some places, they are caught for food.

BRANCHIOSTOMA LANCEOLATUM
This European species grows up to 5cm (2in) long. Like all lancelets, it swims by wriggling its body from side to side. In its normal feeding position, most of its body is buried, with just its head projecting out of the sea bed.

INVERTEBRATES IN DANGER

The list below identifies all those invertebrate genera that contain species currently listed as critically endangered by the IUCN (see p.31). Because invertebrate life is relatively little known, the list almost certainly represents only a small fraction of species in imminent danger of extinction. The main threats to invertebrates are habitat change and pollution, and the introduction of competing species. Habitat change is a major problem for forest and grassland species, and also for those that spend some or all of their lives in fresh water. Here, widespread drainage of wetlands has had serious effects, particularly for damselflies and dragonflies. Pollution also has its greatest impact on aquatic species. Introduced species pose a grave threat to invertebrates living in isolated locations. Many of the 116 species listed below live in restricted habitats on Pacific islands, where browsing goats and other introduced animals have stripped away the vegetation that they need for survival. With nowhere to retreat to, their populations have rapidly collapsed. In the future, climate change will probably add many more names to this list. This is because invertebrates often have highly specialized ways of life, depending, for example, on a narrow range of temperatures, or on a single species of foodplant.

CRUSTACEANS
(57 species in 38 genera)

Amphipoda (sandhoppers *et al.*) 7 species in *Aguadulcaris, Bogidiella, Cocoharpinia, Gammarus, Idunella, Ingolfiella, Pseudoniphargus*
Anostraca (fairy and brine shrimps) 6 species in *Branchinecta, Branchinella, Dexteria, Streptocephalus*
Calanoida (calanoids) 4 species in *Antriscopia, Erebonectes, Nanocopia, Paracyclopia*
Cyclopoida (cyclopoids) 1 species in *Speleoithona*
Decapoda (crabs *et al.*) 25 species in *Afrithelphusa, Barbouria, Cambarus, Fallicambarus, Globonautes, Louisea, Orconectes, Pacifastacus, Procambarus, Procaris, Somersiella, Typhlatya*
Halocyprida (halocyprids) 1 species in *Spelaeoecia*
Isopoda (woodlice *et al.*) 7 species in *Atlantasellus, Bermudalana, Curassanthura, Thermosphaeroma*
Mictacea (mictaceans) 1 species in *Mictocaris*
Misophrioida (misophrioids) 2 species in *Speleophria*

Mysidacea (oppossum shrimps) 2 species in *Bermudamysis, Platyops*
Podocopida (podocopids) 1 species in *Kapcypriodopsis*

INSECTS
(49 species in 46 genera)

Anoplura (sucking lice) 1 species in *Haematopinus*
Coleoptera (beetles) 10 species in *Colophon, Elaphrus, Hydrotarsus, Meladema, Nicrophorus, Polposipus, Prodontria*
Dermaptera (earwigs) 1 species in *Labidura*
Diptera (flies) 1 species in *Edwardsina*
Hymenoptera (bees, wasps, ants, and sawflies) 3 species in *Adetomyrma, Aneuretus, Nothomyrmecia*
Lepidoptera (moths and butterflies) 8 species in *Atrophaneura, Chrysortis, Euproserpinus, Hellconius, Lepidochrysops, Parantica, Pieris, Polyommatus*
Odonata (damselflies and dragonflies) 17 species in *Argiocnemis, Austrocordulia, Boninagrion, Boninthemis, Brachythemis, Burmagomphus, Chlorogomphus, Coenagrion, Indolestes, Libellula, Mecistogaster, Megalagrion, Palpopleura, Platycnemis, Rhinocypha, Seychellibasis, Trithemis*
Orthoptera (crickets and grasshoppers) 8 species in *Aglaothorax, Hemisaga, Idiostatus, Ixalodectes, Nanodectes, Pachysaga, Schayera, Tettigidea*

BIVALVES
(56 species in 26 genera)

Unionoida (mussels *et al.*) 57 species in *Alasmidonta, Amblema, Arkansia, Cyprogenia, Dromus, Elliptio, Epioblasma, Fusconaia, Hemistena, Lampsilis, Lasmigona, Lemiox, Lexingtonia, Margaritifera, Medionidus, Obovaria, Pegias, Plethobasus, Pleurobema, Popenaias, Potamilus, Ptychobranchus, Quadrula, Quincuncina, Toxolasma, Villosa*

GASTROPODS
(169 species in 83 genera)

Archaeogastropoda (archaeogastropods) 5 species in *Neritina, Ogasawarana*
Basommatophora (basommatophores) 9 species in *Ancylastrum, Biomphalaria, Ceratophallus, Gyralulus, Lantzia, Neoplanorbis, Paludomus, Siphonaria, Stagnicola*

Mesogastropoda (mesogastropods) 39 species in *Angrobia, Belgrandiella, Bellamya, Boucardicus, Bythinella, Bythiospeum, Coahuilix, Elimia, Fenouilia, Hemistoma, Jardinella, Lanistes, Leptoxis, Pyrgulopsis, Somatogyrus, Tomicha*
Stylommatophora (stylommatophores) 116 species in *Achatinella, Amastra, Armsia, Caseolus, Cecilioides, Cerion, Cookeconcha, Discula, Draparnaudia, Endodonta, Erepta, Euchondrus, Falkneri, Geomitra, Gonospira, Gonyostomus, Gulella, Gulikia, Hermogenanina, Helicostyla, Helix, Helminthoglypta, Hemicycla, Hirinaba, Inflectarius, Kondoconcha, Laminella, Lampedusa, Leiostyla, Leptaxis, Leuchocharis, Mautodontha, Megalobulimus, Micrarionta, Napaeus, Natalina, Opanara, Orangia, Oxyloma, Partula, Partulina, Pene, Plactostylus, Radiocentrum, Radioconus, Rhysoconcha, Rhytida, Ruatara, Samoana, Thapsia, Trochochlamys, Trochoidea, Tropidoptera, Xerosecta, Zilchogyra, Zingis*

OTHER INVERTEBRATES
(3 species in 3 genera)

Haplotaxida (earthworms) 1 species in *Phallodrilus*
Nerillida (polychaete worms) 1 species in *Mesonerilla*
Onychophora (velvet worms) 1 species in *Speleoperiapatus*

LARGE COPPER BUTTERFLY
During its caterpillar stage, Lycaena dispar *feeds on marsh plants, which have steadily retreated following drainage for farming. Once widespread in Europe, it is now declining and listed as near threatened.*

GLOSSARY

A

ABDOMEN The hind part of the body. The abdomen lies below the ribcage in mammals, and behind the thorax in arthropods.

AESTIVATION A period of dormancy in hot or dry weather. Aestivation is common in animals that live in the subtropics, where there is a long dry season each year.

AMPLEXUS A breeding position adopted by frogs and toads, with the male using its front legs to hold the female. Fertilization usually occurs outside the female's body.

ANTENNA (pl. ANTENNAE) A sensory feeler on the head of an arthropod. Antennae are always present in pairs, and they can be sensitive to touch, sound, heat, and also taste. Their size and shape varies widely, according to the way in which they are used.

ANTLER A bony growth on the head of deer. Unlike horns, antlers often branch, and in most cases they are grown and shed every year, in a cycle linked with the breeding season.

ARBOREAL Living fully or partly in trees.

ARTICULATION A joint, for example between adjacent bones.

ASEXUAL REPRODUCTION A form of reproduction that involves just one parent. Asexual reproduction is most common in invertebrates, and is used as a way of boosting numbers quickly when conditions are favourable. See also *Sexual reproduction*.

ATRIUM A hollow entrance chamber.

B

BACULUM A bone found in the penis of some mammals.

BALEEN A fibrous substance used by some whales for filtering food from water. Baleen grows in the form of plates with frayed edges, which hang from a whale's upper jaw. The baleen plates trap food, which the whale then swallows.

BARBEL A sensory appendage attached to an animal's lips, or near its mouth. Many bottom-dwelling fish have barbels, which they use for finding food hidden on gravel or mud.

BEAK A set of narrow protruding jaws, usually without teeth. Beaks have evolved separately in many groups of vertebrates, including tortoises and turtles, and some whales.

BILATERAL SYMMETRY A form of symmetry in which the body consists of 2 equal halves, on either side of a midline. Most animals show this kind of symmetry.

BILL A bird's jaws. A bill is made of bone, with a horn-like outer covering.

BINOCULAR VISION Vision in which the 2 eyes face forwards, giving overlapping fields of view. This allows an animal to judge depth.

BIOME A characteristic grouping of living things, together with the setting in which they are found.

BIPEDAL Moving on 2 legs.

BLOWHOLE The nostrils of whales and their relatives, positioned on top of the head. Blowholes can be single or paired.

BREACHING Leaping out of the water (usually the sea), and landing back in it with a splash. Breaching is a characteristic form of behaviour shown by many large whales.

BROOD PARASITE An animal – often a bird – that tricks other species into raising its young. In many cases, a young brood parasite kills all its nest-mates, so that it has sole access to all the food that its foster parents provide.

BROWSING Feeding on the leaves of trees and shrubs, rather than on grasses. See also *Grazing*.

C

CAECUM A blind-ended part of the digestive tract, which is often used in the digestion of plant food.

CALCAREOUS Containing calcium. Calcareous structures – such as shells, exoskeletons, and bones – are formed by many animals, either for support or for protection.

CAMOUFLAGE Colours or patterns that enable an animal to merge with its background. Camouflage is very widespread in the animal kingdom – particularly among invertebrates – and is used both for protection against predators, and for concealment when approaching prey. See also *Mimicry, Cryptic coloration*.

CANINE TOOTH In mammals, a tooth with a single sharp point that is shaped for piercing and gripping. Canine teeth are towards the front of the jaws, and are highly developed in carnivores.

CARAPACE A hard shield on the back of an animal's body.

CARNASSIAL TOOTH In mammalian carnivores, a blade-like premolar tooth that has evolved for slicing through flesh.

CARNIVORE Any animal that eats meat. The word carnivore can also be used in a more restricted sense, to mean mammals in the order Carnivora.

CARRION The remains of dead animals.

CARTILAGE A rubbery substance that forms part of vertebrate skeletons. In most vertebrates, cartilage lines the joints, but in cartilaginous fish – for example sharks – it forms the whole of the skeleton.

CASQUE A bony growth on the head of an animal.

CAUDAL Relating to an animal's tail.

CELLULOSE A complex carbohydrate found in plants. Cellulose is used by plants as a building material, and it has a resilient chemical structure that animals find hard to break down. Plant-eating animals, such as ruminants, digest it with the help of microorganisms.

CEPHALOTHORAX In some arthropods, a part of the body that combines the head and the thorax. Animals that have a cephalothorax include crustaceans and arachnids.

CERCUS (pl. CERCI) A paired, hair-like appendage that is attached to the end of the abdomen in some invertebrates, particularly insects.

CHAETA (pl. CHAETAE) A short, stiff hair or bristle, found in many invertebrates. Chaetae sometimes have a sensory function, but in annelid worms they are more often used in movement.

CHEEK TOOTH see *Molar tooth, Premolar tooth*.

CHELICERA (pl. CHELICERAE) In arachnids, each one of the first pair of appendages, at the front of the body. Chelicerae often end in pincers, and in spiders they are able to inject venom. In mites, they are sharply pointed, and are used for piercing food.

CHORDATE An animal belonging to the phylum Chordata, which includes all vertebrates. A key feature of chordates is their notochord, which runs the length of their bodies; it reinforces the body, yet allows it to move by bending.

CHRYSALIS A hard and often shiny case that protects an insect pupa. Chrysalises are often found attached to plants, or buried close to the surface of the soil.

CLASPERS see *Pelvic spurs*.

CLASS A level used in classification. In the sequence of classification levels, a class forms part of a phylum, and is subdivided into one or more orders.

CLOACA An opening towards the rear of the body that is shared by several body systems. In some vertebrates – such as bony fish and amphibians – the gut, kidneys, and reproductive systems all use this single opening.

CLOVEN-HOOFED Having hooves that look as if they are split in 2. Most cloven-hoofed mammals, such as deer and antelope, actually have 2 hooves, arranged on either side of a line that divides the foot in 2.

COCOON A case made of open, woven silk. Many insects spin a cocoon before they begin pupation, and many spiders spin one to hold their eggs. See also *Pupa*.

COLD-BLOODED see *Ectothermic*.

COLONY A group of animals, belonging to the same species, that spend their lives together, often dividing up the tasks involved in survival. In some colonial species, particularly aquatic invertebrates, the colony members are permanently fastened together. In others, such as ants, bees, and wasps, members forage for food independently, but live in the same nest.

COMMENSALISM A partnership between 2 organisms of different species, in which one benefits, but the other neither gains nor loses.

COMPOUND EYE An eye that is divided into separate compartments, each with its own set of lenses. Compound eyes are a common feature of arthropods. The number of compartments they contain varies from a few dozen to thousands.

CORNICLE In aphids, one of the pair of narrow tubes at the end of the abdomen. The cornicles secrete wax that helps to protect aphids from some predators and parasites.

COSTAL Relating to an animal's ribs.

COVERT A feather that covers the base of a bird's flight feather.

CREMASTER A collection of minute hooks on the hind end of a butterfly or moth pupa. The pupa hangs by these hooks from a solid support.

CRYPTIC COLORATION Coloration and markings that makes an animal difficult to see against its background.

D

DELAYED IMPLANTATION In mammals, a delay between fertilization of an egg and the subsequent development of an embryo. This delay allows birth to occur when conditions are favourable for raising young.

DEPOSIT FEEDER An animal that feeds on small specks of organic matter that have drifted down through water, and settled on the bottom.

DETRITIVORE An animal that eats dead plant or animal matter.

DEWLAP A fold of loose skin hanging from an animal's throat.

DIASTEMA A wide gap separating rows of teeth. In plant-eating mammals, the diastema separates the biting teeth, at the front of the jaw, from the chewing teeth at the rear. In many rodents, the cheeks can be folded into the diastema, to shut off the back of the mouth while the animal is gnawing.

DIGIT A finger or toe.

DIGITIGRADE A gait in which only the fingers or toes touch the ground. See also *Plantigrade, Unguligrade*.

DIURNAL Active during the day.

DOMESTICATED Relating to an animal that lives fully or partly under human control. Some domesticated animals look identical to their counterparts in the wild, but many have been bred to produce artificial varieties, which are not found in nature. See also *Feral*.

DORSAL On or near an animal's back.

DREY The nest of a squirrel, which is built in a tree.

E

ECHOLOCATION A method of sensing nearby objects by using pulses of high-frequency sound. Echoes bounce back from obstacles and other animals, allowing the sender to build up a "picture" of its surroundings. Echolocation is used by several groups of animals, including mammals and a small number of cave-dwelling birds.

ECLIPSE PLUMAGE In some birds, particularly waterfowl, an unobtrusive plumage adopted by the males once the breeding season is over.

ECOSYSTEM A collection of species living in the same habitat, together with their physical surroundings.

ECTOPARASITE An animal that lives parasitically on the surface of another animal's body, often by sucking its blood. Some ectoparasites spend all their lives on their hosts, but many – including fleas and ticks – develop elsewhere, and climb onto the host in order to feed.

ECTOTHERMIC Having a body temperature that is dictated principally by the temperature of the surroundings. Also known as cold-blooded.

ELYTRON (pl. ELYTRA) A hardened forewing in beetles, earwigs, and some bugs. The 2 elytra usually fit together like a case, protecting the more delicate hind wings underneath.

EMBRYO A young animal or plant in a rudimentary stage of development.

ENDOPARASITE An animal that lives parasitically inside another animal's body, either feeding directly on its tissues or stealing some of its food. Endoparasites frequently have complex life cycles, involving more than one host.

ENDOSKELETON An internal skeleton, typically made of bone. Unlike an exoskeleton, this kind of skeleton can grow in step with the rest of the body.

ENDOTHERMIC Able to maintain a constant, warm body temperature, regardless of external conditions. Also known as warm-blooded.

EPIPHYTE A plant that grows off the ground, often in the branches of trees. In the tropics, epiphytes create important micro-habitats for many animals, particularly insects and amphibians.

EVERSIBLE Capable of being turned inside out.

EXOSKELETON An external skeleton that supports and protects an animal's body. The most complex exoskeletons, formed by arthropods, consist of rigid plates that meet at flexible joints. This kind of skeleton cannot grow, and has to be shed and replaced at periodic intervals. See also *Endoskeleton*.

EXTERNAL FERTILIZATION In reproduction, a form of fertilization that takes place outside the female's body, usually in water.

F

FAMILY A level used in classification. In the sequence of classification levels, a family forms part of an order, and is subdivided into one or more genera.

FEMUR (pl. FEMORA) The thigh bone in 4-limbed vertebrates. In insects, the femur is the third segment of the leg, immediately above the tibia.

FERAL Relating to an animal that comes from domesticated stock, but which has subsequently taken up life in the wild. Common examples include city pigeons, cats, and horses.

FERTILIZATION The union of an egg cell and sperm, which creates a cell capable of developing into a new animal. In external fertilization, the process occurs outside the body (usually in water), but in internal fertilization it takes place in the female's reproductive system.

FILTER FEEDER An animal that eats by sieving small food items from water. Many invertebrate filter feeders, such as bivalve molluscs and sea squirts, are sessile animals that collect food by pumping water through or across their bodies. Vertebrate filter feeders, such as baleen whales, collect their food by trapping it while they are on the move.

FLAGELLUM (pl. FLAGELLA) A long, hair-like projection from a cell. A flagellum can flick from side to side, moving the cell along. Sperm cells use flagella to swim.

FLEHMEN RESPONSE A characteristic curling of the upper lip shown by many male mammals when they detect a scent.

FLIGHT FEATHERS A bird's wing and tail feathers, used in flight.

FLIPPER In aquatic mammals, a paddle-shaped limb. See also *Fluke*.

FLUKE A rubbery tail flipper in whales and their relatives. Unlike the tail fins of fish, flukes are horizontal, and they beat up and down instead of side to side.

FOETUS A developing animal that is partly formed, and approaching the time when it will be born.

FOOD CHAIN A food pathway that links 2 or more different species in that each forms food for the next species higher up in the chain. In land-based food chains, the first link is usually a plant. In aquatic food chains, it is usually an alga or other form of single-celled life. See also *Food web*.

FOOD WEB A collection of food chains from a particular habitat. In a food web, each species is often involved in several different food chains.

FORM A depression in the ground used by hares for concealing themselves or their young.

FOSSA (pl. FOSSAE) A pit or depression, for example in an animal's skull.

FOSSORIAL Burrowing underground.

FRUGIVOROUS Fruit-eating.

FURCULA The forked, spring-like organ attached to a springtail's abdomen.

G

GALL A tumour-like growth in a plant that is induced by an animal (typically an insect) or some other organism. By triggering the formation of galls, animals provide themselves with a safe hiding place, and a convenient source of food.

GENUS (pl. GENERA) A level used in classification. In the sequence of classification levels, a genus forms part of a family, and is subdivided into one or more species.

GESTATION PERIOD In mammals, the interval between fertilization and birth, when the developing young is nourished by the mother via the placenta.

GILL An organ used for extracting oxygen from water. Unlike lungs, gills are out-

growths of the body. They are usually positioned on or near the head, or – in aquatic insects – at the end of the abdomen.

GONAD A male or female sex organ.

GRAZING Feeding on grass. See also *Browsing*.

GUARD HAIR A long hair in a mammal's coat, which projects beyond its underfur. Guard hairs protect the underfur, and also help to keep it dry.

H

HALTERES In true flies, the 2 knob-like organs that take the place of functioning hind wings. Halteres act like gyroscopes during flight, helping flies to manoeuvre with great precision.

HEAT PITS Heat-sensitive depressions that some snakes have above the front of the upper jaw. These pits detect heat from warm-blooded animals, allowing the snakes to hunt in darkness.

HERBIVORE An animal that feeds on plants or plant-like plankton.

HERMAPHRODITE An animal that has both male and female sex organs, and so can fertilize itself. See also *Sequential hermaphrodite*.

HIBERNATION A period of dormancy in winter. During hibernation, an animal's body processes drop to a low level.

HOME RANGE The area that an animal, or group of animals, uses in its daily activities, for example when foraging for food. See also *Territory*.

HORN In mammals, a pointed growth on the head. True horns are hollow, and are often curved.

HOST An animal on or in which a parasite feeds.

HYPERPARASITE A parasite that attacks another parasite.

I

INCISOR TOOTH In mammals, a flat tooth at the front of the jaw that is shaped for slicing or gnawing.

INCUBATION In birds, the period when a parent sits on the eggs and warms them, allowing them to develop. Incubation periods range from under 14 days to several months.

INSECTIVORE An animal that feeds on insects. The word can also be used in a more restricted sense, to mean animals in the mammalian order Insectivora.

INTERNAL FERTILIZATION In reproduction, a form of fertilization that takes place inside the female's body. Internal fertilization is a characteristic of many land animals, including insects and vertebrates. See also *External fertilization*.

INTRODUCED SPECIES A species that humans have accidentally or deliberately brought into an ecosystem in which it does not naturally occur.

INVERTEBRATE An animal without a backbone.

IRRUPTION A sporadic mass movement of animals outside their normal range. Irruptions are usually short-lived, and

occur in response to severe conditions, such as winter cold.

JK

JACOBSON'S ORGAN An organ in the roof of the mouth that is sensitive to airborne scents. Snakes often employ this organ to detect their prey, while some male mammals use it to find females that are ready to mate.

KEEL In birds, an enlargement of the breastbone that anchors the muscles used in flight.

KERATIN A tough structural protein found in hair, claws, and horns.

KINGDOM In classification, one of the 5 fundamental divisions of the natural world.

L

LARVA (pl. LARVAE) An immature but independent animal that looks completely different from an adult. A larva develops the adult shape by metamorphosis; in many insects, the change takes place in a resting stage that is called a pupa.

LATERAL-LINE SYSTEM A sensory system in fish and some amphibians that detects pressure waves travelling through water, allowing the animal to sense movement. The system consists of sensors arranged in a line along both sides of the body.

LEK A communal display area used by male animals – particularly birds – during courtship. The same location is often revisited for many years.

LOPHOPHORE A crest of short tentacles found in bryozoans and some other marine invertebrates.

M

MANDIBLE The paired jaws of an arthropod, or a bone that makes up all or part of the lower jaw in vertebrates.

MANTLE In molluscs, an outer fold of skin covering the mantle cavity.

MANTLE CAVITY In molluscs, a cavity between the mantle and the rest of the body. The cavity is filled with water in aquatic molluscs, and often contains the gills.

MEDUSA In cnidarians, a bell-shaped body form, which is often free-living. During the complete life cycle, this form may alternate with a form called a polyp, which is often fixed in one place.

MELON A bulbous swelling in the heads of many toothed whales. The melon is filled with a fatty fluid, and is believed to focus the sounds used in echolocation.

METABOLIC RATE The overall rate of an animal's metabolism. Metabolic rate is affected by many factors, including an animal's size, its activity levels, and whether it is cold- or warm-blooded. The higher its metabolic rate, the more food energy an animal needs in order to function.

METABOLISM The complete array of chemical processes that take place inside an animal's body. Some of these processes release energy by breaking down food, while others use energy, for example by making muscles contract. See also *Metabolic rate*.

METACARPAL In 4-limbed animals, one of a set of bones in the front leg or arm, forming a joint with a digit. In most primates, the metacarpals form the palm of the hand.

METAMORPHOSIS A change in body shape shown by many animals – particularly invertebrates – as they grow from juvenile to adult. In insects, metamorphosis can be complete or incomplete. Complete metamorphosis involves a total change in shape during a resting stage, called a pupa. Incomplete metamorphosis embraces a series of less drastic changes; these occur each time the young animal moults. See also *Larva, Nymph*.

METATARSAL In 4-limbed animals, one of a set of bones in the hind leg, forming a joint with a digit. In most primates, the metacarpals form the sole of the foot.

METATHORAX The second segment of an insect's thorax.

MIGRATION A journey to a different region, following a well-defined route. Most animals that migrate do so in step with the seasons, so they can take advantage of good breeding conditions in one place, and good wintering ones in another. See also *Partial migrant*.

MIMICRY A form of camouflage in which an animal resembles another animal or an inanimate object, such as a twig or a leaf. Mimicry is very common in insects, with many harmless species imitating ones that have dangerous bites or stings.

MOLAR TOOTH In mammals, a tooth at the rear of the jaw. Molar teeth often have a flattened or ridged surface, and deep roots. They are normally used for chewing.

MONOGAMOUS Mating with a single partner, either during the course of one breeding season or throughout life. Monogamous partnerships are common in animals that care for their young.

MORPH A physical variant of a species, sometimes known as a phase. Some species have several clearly defined morphs, which may differ in colour or patterning. See also *Polymorphic*.

MOULT Shedding fur, feathers, or skin so that it can be replaced. Mammals and birds moult to keep their fur and feathers in good condition, to adjust their insulation, or so that they can be ready to breed. Arthropods – such as insects – moult in order to grow.

MUTUALISM see *Symbiosis*.

N

NEUROTOXIN A substance that deactivates nerves or that disrupts the way that they work. Neurotoxins can be lethal in very small amounts. They are found in snake venom and in poisons produced by some other animals, such as frogs, toads, fish, and insects.

NEW WORLD The Americas.

NICHE An animal's place and role in its habitat. Although 2 species may share the same habitat, they never share the same niche.

NICTITATING MEMBRANE In birds and many reptiles – apart from snakes – a transparent or semi-opaque "third eyelid", which moves sideways across the eye. Aquatic birds often use the membrane as an aid to vision when swimming underwater.

NOCTURNAL Active at night.

NOTOCHORD A reinforcing rod that runs the length of the body. The notochord is a distinctive feature of chordates, although in some it is present only in early life. In vertebrates, the notochord becomes incorporated in the backbone during the development of the embryo.

NUCHAL CREST A crest at the nape of the neck.

NYMPH An immature insect that looks similar to its parents but which does not have functioning wings or reproductive organs. A nymph develops the adult shape by metamorphosis, changing slightly each time it moults.

O

OCELLUS (pl. OCELLI) In insects, a small, simple eye that detects overall levels of light. Many insects have 3 ocelli, positioned in a cluster on top of the head.

OLD WORLD Europe, Asia, Africa, and Australasia.

OMNIVORE An animal that eats both plant and animal food.

OOTHECA (pl. OOTHECAE) An insect egg case. In some insects – for example cockroaches – the ootheca has a thin covering, but in others it is surrounded by a jacket of foam that hardens soon after it has been produced.

OPERCULUM A cover or lid. In some gastropod molluscs, an operculum is used to seal the shell when the animal has withdrawn inside. In bony fish, an operculum on each side of the body protects the chamber containing the gills.

OPPOSABLE Able to be pressed together from opposite directions. An opposable thumb – found in many primates – can be pressed against the other fingers.

ORDER A level used in classification. In the sequence of classification levels, an order forms part of a class, and is subdivided into one or more families.

ORGAN A structure in the body, composed of several kinds of tissue, that carries out specific tasks.

OSSICLE A minute bone. The ear ossicles of mammals, which transmit sound from the eardrum to the inner ear, are the smallest bones in the body.

OVIPAROUS Reproducing by laying eggs. See also *Ovoviviparous, Viviparous*.

OVIPOSITOR In female insects, an organ at the end of the abdomen, used for laying eggs. The ovipositor often has a tubular or blade-like shape, so that it can insert the eggs in the ground, in plants, or in other animals. In some insects, the ovipositor is modified to form a sting.

OVOVIVIPAROUS Reproducing with eggs that hatch as they are laid or soon afterwards. In an ovoviviparous animal, the eggs are retained in the mother's body while they develop. The egg contains all the nutrients that the embryo needs, and there is no direct connection between it and the mother. See also *Oviparous, Viviparous*.

P

PAPILLA (pl. PAPILLAE) A small fleshy protruberance on an animal's body. Papillae often have a sensory function, for example detecting chemicals that help to pinpoint food.

PARAPODIUM (pl. PARAPODIA) A leg- or paddle-like flap found in some worms. Parapodia are used for moving, or for pumping water past the body.

PARASITE An animal that lives on or inside another animal, and that feeds either on its host animal or on food that its host has swallowed. The majority of parasites are much smaller than their hosts, and many have complex life cycles involving the production of huge numbers of young. Parasites often weaken their hosts, but generally do not kill them. See also *Ectoparasite, Endoparasite, Parasitoid*.

PARASITOID An animal that is a parasite during its early life but which becomes free-living as an adult. Parasitoids are common among insects. Unlike true parasites, parasitoids usually kill their hosts – typically by eating them.

PARATOID GLAND In amphibians, a gland behind the eyes that secretes poison onto the surface of the skin.

PARTHENOGENESIS A form of reproduction in which an egg cell develops into a young animal without having to be fertilized. Parthenogenesis is common in invertebrates, and it produces offspring that are genetically identical to the parent. In animals that have separate sexes, young produced by parthenogenesis are always female.

PARTIAL MIGRANT A species in which some populations migrate, while others are sedentary. This situation is common in birds with a wide distribution, because the different populations often experience very different climatic conditions. See also *Migration*.

PATAGIUM In bats, the flap of double-sided skin that forms the wing. This term is also used for the parachute-like skin flaps of colugos and other gliding mammals.

PECTORAL FIN One of the 2 paired fins positioned towards the front of a fish's body, often just behind its head. Pectoral fins are usually highly mobile and are normally used for manoeuvring.

PECTORAL GIRDLE In 4-limbed vertebrates, the collection of bones that anchors the front limbs to the backbone. In most mammals, the pectoral girdle consists of 2 clavicles or collar-bones, and 2 scapulae or shoulder-blades.

PEDIPALPS In arachnids, the second pair of appendages, near the front of the body. Depending on the species, they are used for walking, sperm transfer, or attacking prey.

PELVIC FINS The rear paired fins in fish, which are normally positioned close to the underside, sometimes near the head but more often towards the tail. Pelvic fins are generally used as stabilizers.

PELVIC GIRDLE In 4-limbed vertebrates, the collection of bones that anchors the front limbs to the backbone. The bones of the pelvic girdle are often fused, forming a weight-bearing ring called the pelvis.

PELVIC SPURS Spurs near the rear of the body that some male sharks and rays use during mating. Also known as claspers.

PELVIS see *Pelvic girdle*.

PHASE see *Morph*.

PHEROMONE A chemical produced by one animal that has an effect on other members of its species. Pheromones are often volatile substances that spread through the air, triggering a response from animals some distance away.

PHOTOSYNTHESIS A collection of chemical processes that enable plants to capture the energy in sunlight and convert it into chemical form.

PHYLUM (pl. PHYLA) A level used in classification. In the sequence of classification levels, a phylum forms part of a kingdom, and is subdivided into one or more classes.

PINNA (pl. PINNAE) The external ear flaps found in mammals.

PLACENTA An organ developed by an embryo animal that allows it to absorb nutrients and oxygen from its mother's bloodstream, before it is born. Young that grow in this way are known as placental mammals.

PLACENTAL MAMMAL see *Placenta*.

PLANKTON Floating organisms – many of them microscopic – that drift in open water, particularly near the surface of the sea. Planktonic organisms can often move, but most are too small to make any headway against strong currents. Planktonic animals are collectively known as zooplankton.

PLANTIGRADE A gait in which the sole of the foot is in contact with the ground. See also *Digitigrade, Unguligrade*.

PLASTRON The lower part of the shell in tortoises and turtles.

POLYGAMOUS Relating to a reproductive system in which individuals mate with more than one partner during the course of a single breeding season.

POLYGYNOUS Relating to a reproductive system in which males often mate with several females during the course of a single breeding season.

POLYMORPHIC Relating to a species that occurs in several different forms, known as morphs. The morphs may be geographically separated, so that there is only a single one in any particular place, or they may occur together, often in set proportions. Morphs persist even though parents from different morphs often interbreed.

POLYP In cnidarians, a body form that has a hollow cylindrical trunk, ending in a central mouth surrounded by a circle of tentacles. Polyps are frequently attached to solid objects by their base. See also *Medusa*.

PREDATOR An animal that catches and kills others, known as its prey. Some predators catch their prey by lying in wait, but most actively pursue and attack other animals.

PREHENSILE Able to curl around objects and grip them.

Premolar tooth In mammals, a tooth positioned midway along the jaw, between the canines and the molars. In carnivores, specialized premolars act like shears, slicing through flesh. See also *Carnassial tooth*.

Prepupa (pl. **Prepupae**) A resting stage that some insects pass through before forming a pupa.

Prey Any animal that is eaten by a predator.

Proboscis An animal's nose, or a set of mouthparts with a nose-like shape. In insects that feed on fluids, the proboscis is often long and slender, and can usually be stowed away when not in use.

Pronotum The upper surface of the front segment in an insect's thorax. In some insects, it is expanded to form a shield.

Prothorax The front segment of an insect's thorax.

Pterostigma A small, coloured spot near the leading edge on some insects' wings.

Pulmonate Having lungs. Pulmonate molluscs are air-breathing species that often live on land.

Pupa (pl. **Pupae**) In insects, a stage during which the larval body is broken down, and rebuilt as an adult. During the pupal stage, the insect does not feed and usually cannot move, although some pupae may wriggle if they are touched. The pupa is protected by a hard case, which is itself sometimes wrapped in silk. See also *Chrysalis*, *Cocoon*, *Larva*.

QR

Quadrupedal Moving on 4 legs.

Radial symmetry A form of symmetry in which the body is shaped like a wheel, often with the mouth at the centre.

Radula A mouthpart that many molluscs use for rasping away at their food. The radula is often shaped like a belt, and is armed with many microscopic teeth.

Rostrum In bugs and some other insects, a set of mouthparts that looks like a beak.

Ruminant A hoofed mammal that has a specialized digestive system, with several stomach chambers. One of these – the rumen – contains huge numbers of microorganisms, which help to break down plant food. To speed this process, a ruminant usually regurgitates its food and rechews it, a process called "chewing the cud".

Rutting season In deer, a period during the breeding season when males clash with each other for the opportunity to mate.

S

Saprophagous Feeding on dead and decaying matter.

Scape The first segment in an insect's antennae, nearest its head.

Scolex The head of a tapeworm. The scolex is usually equipped with suckers or hooks that enable the worm to fasten itself to its host.

Scute A shield-like plate or scale that forms a bony covering on some animals.

Scutellum A triangular shield behind the pronotum in many insects.

Sebaceous gland A skin gland in mammals that normally opens near the root of a hair. Sebaceous glands produce substances that keep skin and hair in good condition.

Sedentary Having a lifestyle that involves relatively little movement. This term is also used to refer to animals that do not migrate. See also *Migration*.

Sequential hermaphrodite An animal with both male and female sex organs that mature at different times.

Sessile Fastened to a solid surface, instead of being able to move from place to place. Most sessile animals are invertebrates that live in water.

Sexual dimorphism Showing physical differences between males and females. In animals that have separate sexes, males and females always differ, but in highly dimorphic species, such as elephant seals, the 2 sexes look very different and are often quite unequal in size.

Sexual reproduction A form of reproduction that involves fertilization of a female cell or egg, by a male cell or sperm. This is by far the most common form of reproduction in animals. It usually involves 2 parents – one of either sex – but in some species individual animals are hermaphrodite. See also *Asexual reproduction*.

Silk A protein-based fibrous material produced by spiders and some insects. Silk is liquid when it is squeezed out of a spinneret but turns into elastic fibres when stretched and exposed to air. It has a wide variety of uses. Some animals use silk to protect themselves or their eggs, to catch prey, to glide on air currents, or lower themselves through the air.

Simple eye An eye that has a single lens. Simple eyes are found in vertebrates and cephalopod molluscs, and, despite their name, are often highly complex. Rudimentary simple eyes are also found in insects. See also *Ocellus*.

Species A group of similar organisms that are capable of interbreeding in the wild, and of producing fertile offspring that resemble themselves. Species are the fundamental units used in biological classification. Some species have distinct populations that vary from each other. Where the differences are significant, and the populations biologically isolated, these forms are classified as separate subspecies.

Spermatophore A packet of sperm that is transferred either directly from the male to the female, or indirectly – for example, by being left on the ground. Spermatophores are produced by a range of animals, including salamanders (Urodeles) and some insects.

Spicule In sponges, a needle-like sliver of silica or calcium carbonate that forms part of the internal skeleton. Spicules have a wide variety of shapes.

Spinneret The nozzle-like organs that spiders use to produce silk.

Spiracle In rays and some other fish, an opening behind the eye that lets water flow into the gills. In insects, an opening on the thorax or abdomen that lets air into the tracheal system.

Stereoscopic vision see *Binocular vision*.

Sternum The breastbone in 4-limbed vertebrates, or the thickened underside of a body segment in arthropods. In birds, the sternum has a narrow flap.

Stridulate Producing sound by rubbing together 2 parts of the body.

Subclass A level used in classification, between a class and an order.

Subfamily A level used in classification, between a family and a genus.

Suborder A level used in classification, between an order and a family.

Subspecies see *Species*.

Suspension feeder An animal that feeds on particles of organic matter that are suspended in water. Suspension feeders include brittlestars, some cnidarians, and many annelid worms.

Swim bladder A gas-filled bladder that most bony fish use to regulate their buoyancy. By adjusting the gas pressure inside the bladder, a fish can become neutrally buoyant, meaning that it neither rises nor sinks.

Symbiosis A partnership between individuals of 2 different species, in which both gain. Animals form symbiotic partnerships with other animals, with plants, and with microorganisms. The term is also sometimes used to mean a partnership of any kind between 2 species.

T

Tarsus (pl. **Tarsi**) A part of the leg. In insects, the tarsus is the equivalent of the foot, while in vertebrates it forms the lower part of the leg or the ankle.

Tegmen (pl. **Tegmina**) The leathery forewing of some insects, such as grasshoppers, crickets, and cockroaches.

Terrestrial Living wholly or mainly on the ground.

Territory An area defended by an animal, or group of animals, against other members of the same species. Territories often include useful resources that help the male to attract a mate.

Test In echinoderms, a skeleton made of small calcareous plates.

Thorax The middle region of an arthropod's body. The thorax contains powerful muscles, and bears the legs and wings, if the animal has any. In 4-limbed vertebrates, the thorax is the chest.

Tibia (pl. **Tibiae**) The shin bone in 4-limbed vertebrates. In insects, the tibia forms the part of the leg immediately above the tarsus, or foot.

Tissue A layer of cells in an animal's body. In a tissue, the cells are of the same type, and carry out the same work. See also *Organ*.

Torpor A sleep-like state in which the body slows to a fraction of their normal rate. Animals usually become torpid to survive difficult conditions, such as cold or lack of food.

Trachea (pl. **Tracheae**) A breathing tube, known in vertebrates as the windpipe.

Tracheal lung In arthropods, a lung that connects to the tracheal system. Tracheal lungs are also found in snakes.

Tracheal system A system of minute tubes that insects and arachnids use to carry oxygen into their bodies. Air enters the tubes through openings called spiracles, and then flows through the tracheae to reach individual cells. Carbon dioxide flows in the opposite direction.

Tubercle A hard swelling on an animal's body.

Tusk In mammals, a modified tooth that often projects outside the mouth. Tusks have a variety of uses, including defence and digging up food. In some species, only the males have them – in this case their use is often for sexual display.

Tymbal organ A drum-like organ on a cicada's body that can vibrate to produce a piercing sound. The 2 tymbal organs are on either side of the abdomen.

Tympanum The external ear membrane of frogs.

U

Underfur The dense fur that makes up the innermost part of a mammal's coat. Underfur is usually soft, and is a good insulator. See also *Guard hair*.

Unguligrade A gait in which only the hooves touch the ground. See also *Digitigrade*, *Plantigrade*.

Uterus In female mammals, the part of the body that houses developing young. In placental mammals, the young are connected to the wall of the uterus via a placenta.

V

Ventral On or near the underside.

Vertebrate An animal with a backbone. Vertebrates include fish, amphibians, reptiles, birds, and mammals.

Vestigial Relating to an organ that is atrophied or non-functional.

Viviparous Reproducing by giving birth to live young. See also *Oviparous*, *Ovoviviparous*.

WX

Warm-blooded see *Endothermic*.

Warning coloration A combination of contrasting colours that warns that an animal is dangerous. Bands of black and yellow are a typical form of warning coloration, found in stinging insects.

Water vascular system In echinoderms, a system of fluid-filled tubes and chambers that connects with the tube feet. The fluid in the water vascular system is under pressure, giving the tube feet their shape.

Weaning In mammals, the period when the mother gradually ceases to provide milk for her young.

Xylophagous Wood-eating.

Z

Zooid An individual animal in a colony of invertebrates. Zooids are often directly linked to each other, and may function like a single animal.

Zooplankton see *Plankton*.

INDEX

In this index, page references in italics identify where the main illustrations for a particular animal or subject appear. Page references in bold type refer to introductions to animal groups.

ACKNOWLEDGMENTS

DORLING KINDERSLEY would like to thank Daniel Crawford, Emily Newitt, Patricia Woodburn, Richard Gilbert, and Tamara Baillie for administrative help; Steve Knowlden for original design work; Pauline Clarke, Corinne Manches, Janice English, and Emily Luff for additional design assistance; Peter Cross for location research; Clare Double for additional editorial help; Robert Campbell for DTP support; and Andy Samson, Anna Bedewell, Charlotte Oster, Jason McCloud, Martin Copeland, Richard Dabb, and Rita Selvaggio for picture reseach. Thanks also to the Berlin Zoo, Chester Zoo, Dr. David I. Harrison, Drusillas Zoo, Exmoor Zoo, Hunstanton Sea Life Centre, IUCN/SSC Canid specialist group, Kate Edmonson at Partridge Films, Marwell Zoological Park, Mike Jordan, Millport Marine Centre, NASA/Finley Holiday Films, the Natural History Museum (London), Paradise Park, Rebecca Tansley at Natural History New Zealand, Robert Oliver, Tropical Marine Centre, Twycross Zoo, Weymouth Sea Life Centre, and World of Birds (South Africa).

Picture Credits
The publisher would like to thank the following for their kind permission to reproduce their photographs:

tfl	tl	tc	tr	tfr
flac	acl	ac	acr	frac
cfl	cl	c	cr	cfr
flbc	bcl	cb	bcr	frbc
bfl	bl	bc	br	bfr

Abbreviations key:
a = above, b = below,
c = centre, f = far, l = left, r = right,
t = top

Prof R.J. van Aarde: 63tc, 306bfr, 446bcr; Kelvin Aitken: 467acl; Bryan and Cherry Alexander Photography: 34c; Allofs Photography: 228frbc; Heather Angel: 496tc; Animals Animals/Earth Scenes: 19flac, 25fr, 30frac, 30bfr, 33tfl, 157bcr, 190frbc, 247bfr, 319c, 381tr, 497c, 583tfr; A.N.T. Photo Library: 393frbc; APB Photographic: 271ac; Ardea London Ltd: 4c, 14cb, 18cfr, 25tfr, 32c, 32br, 37cr, 38ac, 40frbc, 41tfr, 43tc, 44frac, 44bl, 46tc, 48bl, 53cl, 56acr, 61tfl, 63tl, 65cfl, 66tr, 69tl, 69flbc, 71br, 76cfr, 77tr, 80cr, 80br, 87flac, 87frbc, 87frac, 88frbc, 89flbc, 96bcr, 99acr, 103frbc, 104bfl, 104bfr, 106bcl, 107frbc, 108acr, 113cl, 115acr, 121bl, 123cfl, 134flac, 137c, 139tc, 139bl, 144cfr, 147bcr, 148cl, 148bl, 150bl, 152bcl, 153tr, 154acl, 155bl, 161bcr, 161bcr, 163cl, 164cr, 167cfr, 172cl, 173tfr, 178bcr, 187cfr, 195bl, 195bfr, 199acr, 201br, 203frbc, 204bfr, 205bcr, 209bcr, 209bfl, 210bl, 212c, 213frac, 214frac, 220acr, 223tc, 223frac, 239acl, 251cfl, 254bfr, 263tfr, 265bc, 272bcr, 272bfr, 272flbc, 279c, 279ac, 283bcl, 291bcr, 291frbc, 294br, 296acl, 297tl, 302bfr, 311bcr, 319bc, 321cfl, 321frbc, 327cfl, 332tfr, 333bcl, 335tfr, 338bfr, 338cfr, 339bfr, 341frac, 346cfl, 347cl, 349cfr, 350flbc, 354cfl, 355acr, 356tfr, 359br, 361tfr, 364cb, 365bc, 390frbc, 392frac, 404cl, 412c, 415bcr, 422cr, 422bfr, 425cfrbc, 426cl, 433frbc, 435bfr, 441tfr, 441c, 441bc, 444cb, 444tfr, 455flac, 463br, 468cfl, 468tfr, 471cfr, 473frac, 477cb, 477br, 487bcr, 491cfr, 515frac, 521bc, 522c, 524cb, 524bl, 531bfl, 538bl, 539bcr, 579c, 583frac; Auscape: 91tfl, 91cr, 91bcr, 93frbc, 94frbc, 96flbc, 96acr, 119c, 185tr, 280frac, 341bfr, 511acl; Robert E. Barber: 55c, 197frbc, 198acr, 217frac, 217bl, 308bfl, 326br; David Barnes: 23c, 217bfl, 218bfr, 219c, 271frac, 273cr, 276bcr, 307cl, 528bcl, 533frbc, 537acr, 537frac; Kevin H. Barnes: 104bcl; Bat Conservation International: 108tr, 112bcr, 112bfr, 113cr, 113bcr; Fred Bavendam: 75bfr, 223br, 239frbc, 474cfr, 501bfr, 528bfr, 536cfr; BBC Natural History Unit: 15bfl, 16bcl, 16cb, 17cb, 33tfr, 44flbc, 49bc, 49cr, 52tr, 64bfr, 67bl, 71tl, 72bl, 73br, 74bfl, 77tfl, 77cfr, 80tr, 81C, 82c, 86bc, 86tfr, 90ac, 92cr, 104bcr, 112cfl, 120frbc, 121cl, 125tfr, 125cl, 129frbc, 131tfr, 133tfr, 135frac, 153flac, 153bac, 156frac, 160cl, 163tl, 163bfr, 164cfr, 165bl, 181tc, 184bcl, 185bcr, 189bcl, 189acl, 191acr, 191bfr, 198acl, 199frbc, 211frac, 219br, 219bl, 223frbc, 223tfr, 227frbc, 232bl, 233bcr, 234bc, 235c, 239bcl, 248tc, 264cl,, 266bfr, 269cfr, 275tr, 275cfr, 295frac, 302frbc, 309cfl, 313bcl, 321frac, 331c, 332c, 336c, 359bl, 367cfr, 377ccr, 403tr, 403frbc, 403cl, 417tfr, 422acr, 422flbc, 424c, 426cfr, 427bc, 432c, 435tc, 437cfr, 450tfr, 452tc, 460tfr, 463flbc, 464frbc, 465cfl, 467frbc, 472cr, 475c, 475c, 486cb, 496tl, 532ac, 544bcl, 581cfl, 582c, 588cfr, 590ac; Derek A. Belsey: 38tc, 268cl, 268br, 269c, 285cb, 342tfr, 348frac, 353bfl; Niall Benvie: 255cb, 300acr, 317cfr; William Bernard Photography: 50cb, 139cfr, 194frbc, 218ac, 240bcr; BIOS Photo: 17c, 39cl, 99cfr, 103cr, 107tfr, 107cfl, 116frbc, 132bc, 133bfr, 136bfl, 139bfl, 165cfl, 179frbc, 198bcr, 200frac, 243tfr, 428c, 445acl, 488bc, 502bc, 577c; Dr Alison Blackwell: 566frbc; Dr W.R. Branch: 379cfr, 464bfr; Edmund D. Brodie, Jr: 433bcr, 435cfl; C.K. Bryan: 184cr, 246acl; John Cancalosi: 93bcl, 100acl, 100cfl, 101acr, 101tfr,

123acr, 295bl, 308bl, 308bfr, 354c, 354frbc; Prof. Mike Claridge: 574frbc; Bruce Coleman Ltd: 15c, 18bfr, 25cl, 26bfr, 29flbc, 32bfr, 37bfl, 38cb, 39bcr, 41frac, 42cl, 43cl, 48tr, 49tr, 50cfr, 52tc, 52bc, 63c, 65cfr, 65tc, 68cfr, 70c, 77tfl, 88br, 89bfr, 89bcr, 103acl, 110acl, 114frac, 115tfr, 115cr, 119bcl, 120acl, 125tl, 130tfr, 135tfr, 136bcl, 138ac, 140cb, 140bl, 144bcr, 148tr, 150tfr, 171tfr, 176bcr, 198frac, 200flac, 200br, 205bfr, 210frac, 211acl, 211bfr, 213br, 225bcr, 229tl, 229acr, 233tc, 233frac, 241frac, 248cfl, 256frbc, 262bfr, 263flbc, 266cr, 277cfr, 281bc, 295ac, 299cb, 301frac, 301bc, 306bcr, 311cfl, 318c, 321c, 335bfr, 336tr, 339cr, 340tfr, 346frac, 353br, 367tfr, 375bcr, 377frac, 396cb, 397c, 434br, 436cr, 437br, 438flac, 440c, 441acr, 441cr, 443bcr, 443cl, 446frac, 446tc, 450bcl, 452flac, 456acr, 461bfr, 463frbc, 465tfr, 481br, 482bcr, 496tr, 496frbc, 505frbc, 534flac, 543frac, 548acr, 550acfl, 555bcr, 565bcr, 565frbc, 569frbc, 569tl, 584tfr; Bruce Coleman Inc: 18cfr, 243frac, 291bfr, 511frac, 576bfl; Wendy Conway: 241ac, 269bcr, 278bfr, 305cfr, 356c; Corbis: 178c; Cornell University of Ornithology: 27cfl; Peter Cross: 31frac, 55c, 58bc, 100bcl, 100bcr, 147flac, 157tfr, 158bcl, 201cfl, 207frbc, 217acl, 227bc, 230acr, 239frac, 240tl, 240acl, 251acr, 253cl, 253bl, 254bl, 255cfr, 256frac, 265frac, 266cfr, 279bfr, 283bfl, 284cr; Dennis Cullinane: 106bfr; Manfred Danegger: 141cr; Richard Davies: 556cfr; David M. Dennis: 369bfl, 418br, 423frbc, 434acl, 435flbc, 446cfl; Nigel Dennis: 42tc, 67tc, 120bc, 120tfr, 121cr, 121bfr, 121flbc, 131flac, 140br, 184bfl, 185frac, 204acr, 207flac, 222acr, 230tc, 233bc, 247acl, 250bc, 250flac, 250frac, 252bl, 252bfr, 253acr, 253bcr, 253frac, 254acl, 271br, 279flac, 279frac, 284frac, 289cr, 289frac, 292flac, 292cfr, 300br, 301bfr, 304bcl, 305tl, 305acr, 306cfl, 307frbc, 308tfr, 310cl, 310frbc, 317br, 320cl, 330bfr, 331ac, 334cfl, 343frac, 344cr, 351bcl, 352bfr, 357bl, 357frbc, 358acr, 358cfr, 358flbc, 410frbc, 556frbc, 588bcr; Dr Frances Dipper: 532cfr, 532flbc, 594bfl, 594bfr, 595bcr, 595bfl; DK Picture Library: Natural History Museum London 19flbc, 19flbc; NASA/Finley Holiday Films: 37br; Philip Dowell: 214 bc, 215ac, 267 frac; Ecoscene: 25bcl, 30flbc; Brock Fenton: 110frbc, 110frac, 111bfl, 111acl, 113cr, 113bfl, 113frac; David B. Fleetham: 470c; Michael & Patricia Fogden: 16cfr, 243bfr, 323ac, 376bcr, 399cr, 439bcl, 447cfr, 447flac, 456frac, 559tfr; Fotomedia: 119acl, 119frac, 185bfl, 227acl, 229bc, 239acr, 250flac, 253bfr, 314cl, 347tfr; Foto Natura: 101bbl, 116tc, 116bcr, 116frac, 144bl, 149cfr, 149bfr, 181tfl, 198cl, 224bcl, 236cfr, 576c, 576flbc; Jurgen Freund: 503ac; Tim Gallagher: 268tr; David George: 533bl; Getty Images Stone: 12c; Brian Gibbs: 199flbc; Michael P. Gillingham: 142bfl; Roy Glen: 353acl; Francois Gohier: 164cfr; Chris Gomersall Photography: 28bcl, 31frac, 62bc, 71tc, 266tl, 266bfr, 268cfr, 268frbc, 270bc, 271bfr, 271frbc, 273br, 273bc, 273ac, 276cfr, 278cfr, 283cfl, 284bfr, 285bc, 285bcr, 285cfr, 289tr, 289cfl, 292cl, 296cr, 296bfr, 299frac, 299bfl, 299flac, 304acl, 305cl, 305bcl, 306c, 306ac, 306ac, 307cl, 307bcr, 308cfl, 309bcl, 310cfr, 310acr, 315acl, 315bl, 315cr, 320frbc, 320bfr, 322tc, 324c, 324cr, 325c, 330cfr, 331br, 333ac, 334br, 334bfr, 335tr, 338bc, 338bfl, 338acl, 339bl, 340tc, 340cfl, 340bl, 341cl, 341bfl, 342cfr, 342frac, 342acl, 342bfl, 342acr, 343acl, 343cb, 343bfr, 344bl, 344tfr, 345acr, 345cfr, 346c, 348bcl, 348bcr, 348frbc, 348bfl, 349tfr, 349bfr, 350cr, 350cfl, 350bfr, 350cb, 351br, 352bl, 352c, 352cfr, 352tr, 353bcl, 353bfl, 353tr, 353frac, 353bcr, 354frac, 354cr, 355tl, 355cfr, 355cfl, 355bcl, 356tr, 357tr, 358flbc, 359c, 360bfr, 360flbc; Granada Wild 108-109b, 116-117b, 160-161b, 220-221b, 224-225b, 298b, 302b, 316b, 376-377b, 420b, 555b; Jonathan R. Green: 371tr; Derek Hall: 32frac, 36ac, 36acl, 47cl, 51br, 51cfl, 67tl, 266bcr; Howard Hall Productions: 474cl, 476br; Tim Halliday: 438frac, 438bcl; Martin Harvey Photography: 210acr, 215frac, 227tfr, 230cfr, 234tc, 247tfr; Derek Harvey: 127cl; Lawrence Heaney: 103frac; Dr C. Andrew Henley-Larus: 92bfr, 93cl, 93ac, 94acr, 97acl, 97cfr, 217bfr; Daniel Heuclin: 379c, 393bcr, 443bfr; Dr G. H. Higginbotham: 346bc; Thomas Holden: 296bl, 305bfr, 322bl, 335flbc, 340acr, 342bc, 344cl, 350bfr; Holt Studios International: 552acr, 552bcl, 556bcl, 557bcl, 560bfr, 560frbc, 569bcr; Images Colour Library: 0c, 1c, 8c, 14frbc, 71c, 78c, 128c, 182acr, 182tfr, 186b, 192bl, 216c, 249c, 258c, 290c, 302bcr, 318tc, 371c; Impact Photos: 503frbc; Innersance Visions: 160c, 161acr, 174ac, 177c, 202c, 465br, 474bfr, 496cfl; Jacana Hoa-Qui: 18frbc, 31cfl, 92bfl, 126bcr, 140cr, 140acl; F Jack Jackson: 23cb, 205flac, 531tfr, 532bfl, 596cl; Jack Jeffrey: 355bl; Mike Johnson: 506bcl, 515tl, 520flbc; Joint Nature Conservation Committee: 530frac; Mike Jordan: 103br, 105br, 148flbc, 149flbc, 151ac, 151bfl, 151bfr, 152cr, 152frbc, 154cfl, 155bcr, 155bfr, 156bl, 156bl, 156cfr, 156frbc, 256acl; Ken Kates: 98c; Hiromitsu Katsu: 142frbc; Paul Kay: 539frac, 540c, 540flac, 540frac; Kos Pictures Source: 519tfr; Mark Kostich Photography: 204bcr; FLPA - Images of Nature: 27flbc, 29tfl, 61bfl, 66c, 87br, 89br, 105tfr, 116bcr, 124flac, 124cbl, 127cfr, 131bfr, 138bfr, 138frac, 144acr, 149acl, 150tfr, 158flac, 182bcl, 190cfl, 197cfr, 198bl, 207acr, 210bcl, 210cfr,

218flbc, 219frac, 230frbc, 231frbc, 231acr, 235frbc, 252tc, 255bl, 261bfr, 269bcl, 272bfl, 294ac, 297cr, 297bl, 298bcr, 301frbc, 315tfr, 322bfr, 324cfl, 325bfl, 326acl, 327frbc, 333frac, 336bl, 339br, 339cfr, 343bfl, 348c, 349frbc, 350tl, 354bfl, 354tr, 355flbc, 357ac, 359tl, 359cr, 365tfl, 367br, 380c, 383cfr, 410bcl, 422frac, 423cl, 430bfr, 435bc, 436bfr, 453frac, 462bfr, 498tfr, 525tfl, 526ac, 526acr,, 587flbc, 588frbc; Vanessa Latford: 245flbc; Legend Photography - Andy Belcher: 514c, 515cfr; Ken Lewis: 269bc, 335bl; Look: 282cr; Mammal Images Library: 92cl; Andrew Martinez: 528frbc, 531frbc; Maslowski Photo: 278bcl; Chris Mattison Nature Photographics: 15bfr, 55cb, 58ac, 59bc, 96acl, 100bfl, 101frbc, 111flac, 142acl, 146bfr, 153bl, 197flac, 239bl, 241bfr, 251cr, 254bcl, 254frac, 256br, 278frac, 283bcr, 283fr, 288frbc, 292bfr, 293bc, 296cl, 299bcr, 300tl, 307cfr, 313tl, 321bfr, 345flbc, 345frbc, 353cfr, 359cfr, 365frbc, 368cr, 368cfr, 369bfr, 373tr, 374frac, 375cfr, 375frac, 383cb, 384cfr, 386bcr, 388cfl, 389tr, 393frac, 394flbc, 395frbc, 404bfr, 405cr, 405tfr, 408bc, 410tr, 410cfr, 411bfl, 411cfr, 411br, 413br, 413frac, 414cl, 414bcr, 414frac, 415bc, 415bfr, 416bl, 417bfr, 419cfl, 419bfr, 422cl, 423cb, 425tfr, 430tc, 436acl, 437tc, 438ac, 439frbc, 442bfr, 443bcl, 444cl, 445cfr, 445frbc, 450cr, 450frbc, 451bcl, 451cfr, 451frac, 452acl, 452frac, 455cb, 456acl, 456bcr; Photomax: 480bfr; Dr Luis A. Mazariegos: 325cr; McDonald Wildlife Photography: 53tc, 108c, 147tfr, 207bl, 222bfl; Dr Fridtjof Mehlum: 308cfr; Mr W.E. Middleton: 558cfr; Carlos Minguell: 469frac, 509bcr, 511bcl, 530tr, 530acr, 596bcr; Minden Pictures: 544ac, 544bfl, 544cfl, 590c; National Geographic Society: 26frac; Natural History Museum London: 175c, 538br, 550frac, 557cl; Natural History New Zealand: 478b; Nature Photographers: 66c, 406cb, 437cr, 444cfr, 453cfr, 550bcl; Natural Science Photos: 42c, 50bc, 114cfr, 222cfr, 272tfr, 330cl, 330cfl, 403bfr, 411bc, 430tfr, 437bc, 505tc; Natural Visions: 24frbc, 89acl, 159bl, 167bfr, 263cr, 364bcr, 366c, 375cb, 375bcl, 461tfr, 464frac, 498frbc, 526bfl; Naturefocus, Australian Museum: 95frac; N.H.P.A.: 14flbc, 17acl, 27cfr, 28frac, 29bfl, 29frac, 29frbc, 31tfr, 38c, 43bcr, 44acr, 46frbc, 48cl, 49bl, 50ac, 53bc, 56cfr, 56cfr, 60bfl, 61bc, 64flbc, 67bc, 67bcr, 68br, 69bcr, 72tr, 72c, 74frbc, 75cb, 75bcr, 80bl, 90acl, 90bcl, 92flac, 93c, 94bfr, 94bfl, 95bcr, 96bfr, 97cfl, 99tr, 100cfr, 106cr, 106bfl, 111frbc, 112bfr, 113tl, 114acl, 123fcr, 124bfr, 131bfl, 133frac, 133bl, 134cfr, 139cfl, 141bbr, 142bcl, 143bfl, 144bfr, 146cfr, 155tfr, 157cr, 159frac, 161frac, 163cfr, 181bl, 182bfr, 183bcr, 183cfl, 183acr, 184tfr, 186tr, 190bcr, 191tfr, 194acr, 195tl, 197br, 198bfr, 199acl, 200frbc, 200bfl, 203frac, 204bcl, 218bfr, 222cfl, 222frbc, 224flbc, 229frbc, 241bl, 255bfr, 257bfr, 262tfr, 262cfl, 265acfr, 266tfr, 268bcl, 273tfr, 275bcr, 277cb, 277acr, 277bc, 280br, 280flbc, 280tr, 281cl, 284bl, 284bcr, 288cl, 288bl, 289bcr, 293cfr, 301tr, 302cfr, 304tr, 307tfr, 314bl, 314bfl, 317acr, 320tfr, 321cb, 322frac, 322tfr, 323frac, 326frbc, 327cl, 330tc, 332bl, 332cl, 332cb, 334frbc, 336acr, 338frac, 339acr, 341bl, 342bfr, 348bc, 348tl, 349br, 351bfr, 362c, 365flbc, 367frbc, 368cfl, 368br, 376c, 378c, 379tl, 381frbc, 382bcl, 383frbc, 386bfr, 389flbc, 391c, 391cb, 391bfr, 392cfr, 394tc, 394cr, 398bcr, 402c, 403bl, 403bcl, 406tr, 407tr, 407frbc, 411frac, 411tl, 416cb, 418ac, 419bc, 419frac, 423frac, 424bl, 434tr, 435tfr, 439acl, 439bfl, 439bfr, 439frac, 444tl, 445bc, 444acl, 448cl, 451tr, 453br, 453cl, 454bfr, 456bl, 462cfr, 463tfr, 463frac, 463tfl, 464frac, 480cr, 481frac, 487bfr, 490frbc, 494trf, 507tfl, 509cb, 509cl, 510frbc, 514tr, 519tl, 526frac, 529c, 531cb, 534frac, 536cr, 537bl, 538frac, 538bfl, 541cfl, 543flac, 544c, 551tl, 551bcl, 557acl, 558acl, 559flac, 560frac, 565cfr, 579bcr, 584bfl, 585bc, 590tc, 592cl, 594cfr; Mark Norman: 543acl; Gary Nuechterlein: 269tfr; Mark O'Shea: 368bfr, 375acr, 382frbc, 382bfr, 382flac, 384bfr, 386acl, 387flac, 387bfr, 388frbc, 388bfr, 389br, 390frac, 390flbc, 392cfl, 392bcl, 392bfr, 394bfr, 395cfr, 395frac, 399bfr, 398bc, 398cl, 398flbc, 399frac, 399bbr, 405flac, 406acl, 407cfr, 408flac, 408frbc, 408cr, 410flac, 415frac, 416flac, 417acl, 417bl, 419bl, 423br, 426tr; Oxford Scientific Films: 15flbc, 22tfr, 26bcl, 28cfl, 29acl, 33br, 37bcr, 44c, 46tr, 55cl, 55bcr, 56cfl, 57bfl, 59cl, 62cr, 63bl, 63bc, 64ac, 65flac, 65bfr, 66br, 68flbc, 70tfr, 71c, 71c, 73cl, 75br, 87cfr, 90frac, 92frac, 93cfr, 94tl, 105flac, 105bl, 107acl, 107br, 113bl, 119cfl, 121frac, 125bcr, 131acr, 131bcl, 134tfr, 135bc, 138frac, 139bcr, 139frbc, 140cl, 141tfr, 142bcr, 143tl, 144tr, 144c, 146cl, 148cfr, 150grac, 151bcr, 153acr, 165frbc, 168tfr, 172frac, 176frbc, 179tfr, 179bc, 181frac, 191bc, 195frac, 200bcl, 201frac, 203br, 213tc, 215frbc, 219tfl, 220cl, 222tfr, 224c, 229flbc, 223bfr, 246cfr, 254bcr, 261acr, 264bl, 265cb, 269tc, 270tfr, 270cfl, 275bfr, 275frbc, 276tfr, 280bl, 285cfr, 293frbc, 298tfr, 299acr, 300cfl, 302acr, 304frbc, 306cfr, 309cfl, 309c, 310bcr, 315frbc, 321bl, 322c, 323bcl, 324cb, 325cb, 325br, 330tr, 330bl, 336frac, 336cfr, 336bfr, 347cfl, 347frbc, 347bl, 348tfr, 351tfr, 355frbc, 356flbc, 359cb, 369cfl, 369cfr, 370cfl, 370tfr, 372tfr, 372c, 377frbc, 386bcl, 393flbc, 396flbc, 406frbc, 414tfr, 423acl, 424br, 426bcr, 430cr, 430bcr, 432bl, 438bfl, 442tl, 442acl, 442bcl, 443bfl, 444bl, 447c, 447flbc, 448bfl, 453tc, 455bcr, 455bfr, 456bfr, 457bc, 464br, 465bl, 466c, 467cr, 469bcr, 472tfr, 476ac, 481bl, 482frac, 483c, 493bc, 499bfr, 506frbc, 514frbc, 514bfr, 516frbc, 517bcl, 520acl, 520c, 524bcl, 526bfl, 528bfl, 532flac, 533frac, 533cfr, 533bcl, 533bc, 535bcl, 535frbc, 536cr, 536bfl, 537cfl, 537bfr, 538c, 539bfr, 540cb, 541bfr, 542flac, 543bcl, 548bcl, 548cb, 548bc, 548br, 549frac, 549bfr,

549frbc, 549bfr, 550acl, 553c, 555frbc, 578tr, 578frbc, 580tfr, 580cfl, 580bc, 580cr, 580bcl, 580cb, 581tfr, 585frac, 585frbc, 587cfr, 589frac, 589bfr, 590bfl, 596bfr; Panda Photo: 280c; Otto Pfister: 239bcr, 255acr, 256cr, 300cb, 304bfl, 343cr 346bfr; Mark Picard: 142bfr, 184bfr, 240bc, 241cb, 241cfr, 278bc, 296tr, 327bcl, 356bcr, 356frac; Linda Pitkin: 75br, 504cfr, 530cl, 530bl, 530acl, 530bfl, 530cfl, 531tl, 531tr, 531cr, 531ac, 531flac, 533cfl, 536bl, 539acr, 570frac, 595flac, 596flbc; Planet Earth Pictures: 27bfr, 40tr, 41tfl, 41bfl, 44fbr, 47bc, 54cr, 59c, 60cfr, 73tr, 91acr, 104cfr, 107bcl, 124tfr, 126frac, 135frbc, 141tfl, 142cl, 143frbc, 143bfl, 153frbc, 153bfr, 158frbc, 170cfr, 171cfr, 171bl, 178acr, 179tc, 181flbc, 184tl, 189tfr, 201tl, 213bfr, 215cb, 220tl, 220bcl, 220bcr, 220bfr, 223bl, 225cb, 232c, 240bfl, 253tl, 270cb, 281tfr, 293tl, 295frbc, 308bcr, 309frac, 310bcl, 311bfl, 327frac, 333bl, 376frbc, 383bfr, 385c, 409cr, 424tfr, 424frbc, 432tr, 432flbc, 454cl, 464cfr, 465bfr, 468cl, 468bl, 469acl, 469cfr, 473bcr, 474frac, 476frbc, 480tfr, 480cl, 481bcl, 484tfr, 489acl, 489cl, 500cr, 501cr, 504tfr, 507frbc, 508bcr, 509tc, 510tr, 510cfr, 514flac, 515tfr, 518tfr, 520bl, 528tc, 529bfr, 533tfr, 537bcr, 538frbc, 539cfl, 541bfl, 541bc, 543tfr, 543acr, 548c, 551frac, 568tfr, 585c, 587frbc, 588cl, 592flac, 594tfr, 594ac, 596cfr; Louis W. Porras: 419bfr, 455cr; Premaphotos Wildlife: 561tfr, 561bcl, 570cl, 592bfl; John E. Randall: 506cr, 508tfr; Jean Yves Rasplus: 560bcl, 573bcl; Galen B. Rathbun: 114bl, 114frbc; Reinhard - Tierfoto: 270ac; Neil Rettig Productions Inc: 293bfr, 296c; Rex Features: 518bfr; RSPCA: 421c; Save-Bild: 99frbc; Kevin Schafer: 116frbc, 122frbc, 129tfr, 157acl, 235tfr, 326bcl; Heinz Schimpke: 458c; Science Photo Library: 14bfr, 89fl, 123bl, 260br, 364bcl, 525bcl, 526tfl, 557frbc, 558tfr, 580tr; Herb Segars: 485bcr; Hong Chalk Seng: 554c; Anup and Manoj Shah: 2c, 6bc; Wendy Shattil: 43ac, 158br, 197bl, 200cl, 214flac, 272c, 315bfr, 340c, 351tl, 374cr; C. Andrew Smith: 142tfr; Smithsonian Institution: 595bcl; South American Pictures: 281cr; Sue Scott: 532tfr, 539cr, 539flac, 540bfr, 542c, 580bcr, 584bcr, 594bc, 595ac, 595tr; Still Pictures: 62tr, 9lbfr, 95flac, 116acl, 130cr, 130frac, 134frbc, 136tc, 138cfr, 161flbc, 161frbc, 169frbc, 173cfr, 174cfl, 190tfr, 191frbc, 211cfl, 216tfr, 230bfr, 241tr, 396bl, 465cr; K. Sugawara: 352br; Harold Taylor: 79tl; Telegraph Colour Library: 24acr, 26bfl; The Roving Tortoise: 216bcr; Jamie Thom: 2c; Michael P. Turco: 203cfr; Jean Vacelet: 528cfr; R.W. Van Devender: 434bcl, 439ac, 442flac, 434bl; Andre van Huizen: 362ac, 245bfr, 266bbl, 266bfl; Colin Varndell: 147bc, 271flac, 276bfl, 285bfr, 340bcr, 340cfr, 340frbc, 341acl, 345flac, 355frac, 360frac; Vireo: 325tl, 339acl, 339bfl, 339tfr, 340frbc, 341acl, 345flac, 355frac, 360frac; Prof. Peter Vogel: 104acr, 106frbc; Judith Wakelam: 130flbc; Terry Wall: 276flac, 297bfr; Bernard Walton: 129br, 129bfr; Dave Watts: 63cb, 90br, 93br, 94bcl, 94frac, 95bfr, 96frac, 97acr, 97bcr, 97bfr, 100frac, 101bcr, 120flbc, 131acl, 218frbc, 255cfl, 256bl, 279cfr, 293bcl, 294frac, 341bcr, 349bcr, 350tfr, 358bl; Dr Jane Wheeler: 237tcl, 237tfr; Elizabeth Whiting & Associates: 79bl; Jack Williams: 507cl; Windrush Photos: 49br, 265bfr, 306bcr, 330acr; Art Wolfe: 8c, 88cfr, 88cfr, 88cfr, 108frbc, 116c, 135bcr, 158tfr, 243bl, 295tfr, 298cr, 302c, 496flbc; Woodfall Wild Images: 27fl, 30cfl, 32flbc, 57bfr, 58cr, 59bcr, 63br, 70c, 73tl, 88bfl, 88tc, 89frbc, 154cfr, 225tfr, 248flbc, 274c, 292bcr, 485cfr, 525bfr, 597bc; P. A. Woolley and D. Walsh: 92bl, 97bl, 97bfr; Norbert Wu: 464bcl, 472cl; Jerry Young 16cl, 42c, 51c, 54cb, 62bc, 67cb, 62tc, 78cb, 90bfl, 110tl, 112acr, 125bfr, 125bfl, 126flac, 127acl, 130flac, 133ac, 181ac, 183bl, 183frac, 183acl, 185bfr, 185acl, 187tfr, 189acr, 207br, 209flac, 210flac, 230bfl, 279tr, 299bfr, 320tc, 357cfr, 360bcr, 366c, 368ac, 369frac, 374acl, 374cl, 374bl, 374frbc, 403acl, 410c, 411acl, 416frac, 419frbc, 426flab, 441bfr, 442frac, 446ac, 446bl, 447bfr, 451ac, 456bfr, 473tfr, 477bfl, 490bc, 514flbc, 516bfl, 516frac, 516flac, 517bfr, 520bfr, 524frbc, 547tfr, 563flac, 584cfr, 584cfr; Gunter Ziesler: 116acr

Jacket pictures
Front jacket: Toni Angermayer
Back jacket: Robert R. Barber
Spine: Toni Angermayer
Inside flap: Animals Animals/Victoria McCormic

Additional photography
Max Alexander; Peter Anderson; Irv Beckman; Geoff Brightling; Jane Burton; Peter Chadwick; Andy Crawford; Peter Cross; Geoff Dann; Philip Dowell; Alistair Duncan; Mike Dunning; Andreas von Einsiedel; Ken Findlay; Neil Fletcher; Christopher and Sally Gable; Peter Gardner; Peter Gathercole; Steve Gorton; Frank Greenaway; Peter Hiscock; Colin Keates; Dave King; Cyril Laubscher; Bill Ling; Mike Linley; Kevin Mallet; Maslowski Photo; Chris Mattison; Jane Miller; Tracy Morgan; Gary Ombler; Nick Pope; Rob Reichenfeld; Tim Ridley; Kim Sayer; Tim Shepard; Karl Shone; Harry Taylor; Kim Taylor; M.I. Walker; Mathew Ward; Laura Wickenden; Alan Williams; Jerry Young

Additional illustrations
Martin Camm; Colin Newman

All other images © Dorling Kindersley
For further information see: www.dkimages.com